Axel Haupt

Organic and Inorganic Fluorine Chemistry

Also of Interest

Polymer Synthesis
Modern Methods and Technologies
Wang, Junjie, 2021
ISBN 978-3-11-059634-2, e-ISBN 978-3-11-059709-7

Supramolecular Chemistry
From Concepts to Applications
Kubik, 2021
ISBN 978-3-11-059560-4, e-ISBN 978-3-11-059561-1

Chemical Technologies and Processes
Staszak, Wieszczycka, Tylkowski (Eds.), 2020
ISBN 978-3-11-065627-5, e-ISBN 978-3-11-065636-7

Host–Guest Chemistry
Supramolecular Inclusion in Solution
Wagner, 2020
ISBN 978-3-11-056436-5, e-ISBN 978-3-11-056438-9

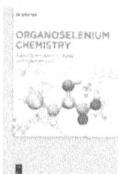

Organoselenium Chemistry
Ranu, Banerjee (Eds.), 2020
ISBN 978-3-11-062224-9, e-ISBN 978-3-11-062511-0

Axel Haupt

Organic and Inorganic Fluorine Chemistry

Methods and Applications

DE GRUYTER

Author
Dr. Axel Haupt
Jacobs University Bremen gGmbH
Department of Life Sciences and Chemistry
Campus Ring 1
28759 Bremen
Germany
a.haupt@jacobs-university.de

ISBN 978-3-11-065929-0
e-ISBN (PDF) 978-3-11-065933-7
e-ISBN (EPUB) 978-3-11-065950-4

Library of Congress Control Number: 2020951339

Bibliographic information published by the Deutsche Nationalbibliothek
The Deutsche Nationalbibliothek lists this publication in the Deutsche Nationalbibliografie;
detailed bibliographic data are available on the Internet at http://dnb.dnb.de.

© 2021 Walter de Gruyter GmbH, Berlin/Boston
Cover image: by the author
Typesetting: VTeX UAB, Lithuania
Printing and binding: CPI books GmbH, Leck

www.degruyter.com

Preface

Fluorine is a fascinating element of extremes, as it can form very reactive but also almost inert compounds, which holds for both, organic and inorganic derivatives. Since the first discovery and isolation of elemental fluorine by Henry Moissan in 1886, the chemistry of fluorine has largely evolved. Several decades ago, fluorine chemistry was still a rather exotic field, which had been worked on by a rather small number of specialists. However, the development of new, significantly milder fluorination techniques formed the basis of a beginning triumphal march, especially of fluoroorganic compounds. Nowadays, the latter play an important role in many fields of applications, such as pharmaceuticals or high-performance materials.

Of course, the increasing importance of fluorinated compounds has also been reflected in the scientific literature. To date, there are numerous excellent reviews and book publications about special topics in fluorine chemistry. However, most of the literature is limited on either inorganic or organic compounds and comprehensively presents synthetic accesses, properties as well as applications of a certain class of fluorine-containing compounds. So far, there has been no general and joint introductive review on the preparation of organic and inorganic compounds. Therefore, this manuscript rather focuses on preparative aspects and does only provide a small background on the properties and applications of fluorinated derivatives. It mainly targets students but also researchers in academia and industry and contains study questions for each chapter, which shall help to underline and repeat the most important aspects.

The present book is divided into three main parts: A, B and C, where Part A gives some general introduction to fluorine chemistry. This part for example covers the very important compounds F_2 and HF, as well as a number of general aspects, which apply for fluorine compounds. Consecutively, inorganic and organic fluorine compounds are discussed in Parts B, and C, respectively. The inorganic part is divided after the groups of the periodic table and presents the most relevant binary fluorine compounds, oxyfluorides, as well as a few other important derivatives for each element. In contrast, the organic part does not focus on special compounds, but describes general fluorination techniques of the most important substrate classes. Due to the steadily increasing amount of synthetic procedures, especially in organofluorine chemistry, it is impossible to prepare a comprehensive review. Thus, the selection of the presented techniques and compounds is certainly not an ultimate composition. Instead, the inclusion or omission of the one or other synthetic method has often been a matter of taste of the author and shall not be understood as any type of qualitative evaluation.

I am very grateful to Gerd-Volker Röschenthaler, who always supported me during the preparation of this book, but also far beyond. In addition, I owe my gratitude to Dieter Lentz for all the help and encouragement over the past years, which made me become a part of the worldwide fluorine community. Moreover, a lecture on or-

https://doi.org/10.1515/9783110659337-201

ganic and inorganic fluorine chemistry, which was jointly held by Dieter Lentz and Konrad Seppelt at Freie Universität Berlin clearly served as inspiration for this book. Most of all, I want to thank my wife, Laura, for her patience and support during the past 21 months. During this time, I have definitely spent most of my free time with the preparation of this manuscript and not with her, which she understandingly accepted without complaints.

Axel Haupt

Bremen, November 2020

Foreword

Fluorinated compounds are indispensible in our daily life. Their unique properties enable their wide-spread application in multiple areas, ranging from material science to medicinal chemistry as high performance polymers, battery components, pharmaceuticals, liquid crystal molecules, refrigerants, inhalation anesthetics, to name just a few. The study of fluorinated chemicals is nowadays internationally a major research topic in many laboratories connecting disciplines, scientists and countries. The growing need for a concise overview is met by Axel Haupt's *Organic and Inorganic Fluorine Chemistry*, the first text and study book for graduate students and scientist in both academia and industry. Axel Haupt is providing knowledgeably the fundamentals for the current directions and aspects in fluorine chemistry addressing also the problems caused by some perfluorinated compounds. I wish Axel Haupt's book broad acceptance and do believe that it will become a motivating valuable guide for promoting this field of research for a better world.

Gerd-Volker Röschenthaler

Bremen, February 2021

https://doi.org/10.1515/9783110659337-202

Contents

Part B: Inorganic fluorine chemistry

Part C: Organic fluorine chemistry

Part A: Introduction to fluorine chemistry

The existence of fluorides has been known a long time before elemental fluorine was discovered and electrochemically prepared by Henry Moissan in 1886. However, it was certainly due to the extreme reactivity of F_2, that it could not be obtained earlier in elemental form. After its discovery, pioneering work was performed in the following decades. This especially required the development of appropriate working techniques and materials for the handling of reactive fluorides and F_2 itself. In the 1940s, fluorine chemistry has been gaining a lot of interest and importance in both the inorganic and organic field. For example, natural abundant uranium was separated into its isotopes using UF_6. Apart from that, fluorinated polymers, oils and greases with a high chemical resistance have also been prepared. Later, elemental fluorine became commercially available in steel cylinders, which significantly facilitated the work with it, and thus gave another boost to fluorine chemistry. With new and more precise analytical techniques and instrumentation, it has been possible to investigate many structural details and other properties of fluoride compounds in the following period of time. In 1986, exactly 100 years after the discovery of elemental fluorine by Moissan, Karl Christe successfully prepared F_2 by a purely chemical synthesis, using starting materials that did not require F_2 for their preparation. With increasing impact of quantum chemistry, more and more valuable preparative targets have been detected and successfully tackled in the last decades. Using theoretical methods, it became also possible to shine some light on obscure chemical problems, for example, the (non)existence of a certain compound or oxidation state. A number of these findings has also been experimentally proven by matrix techniques. For sure, the one or other outstanding species, which is currently still unknown, will be prepared in future. Organic fluorine chemistry is even more flourishing nowadays. There are plenty of different, important fluorinated groups in industrial products, so that there is a large demand for the continuous development and optimisation of efficient preparative methods.

This section will give a general introduction to the field of fluorine chemistry, covering general aspects as, for example, typical properties of fluorine-containing compounds, natural occurrence of fluorides and environmental aspects. A general view over a variety of different preparative methods will be presented as well as a brief summary of the most important structural motifs, which are found in solid-state structures of inorganic fluorides. Furthermore, an introduction to ^{19}F-NMR spectroscopy will be given, the most common and important analytical tool for fluorine chemists.

https://doi.org/10.1515/9783110659337-001

1 Sources of fluorine in nature

1.1 The fluorosphere [1]

Fluorine is more widespread, than one might think in the first moment, as it is actually all around us in oceans, atmosphere and earth-crust. In the latter, it amounts to 0.065 %, thus being more than three times as abundant as chlorine (0.02 %). Among all elements, it is on position 13. The estimated quantities of fluorine in the atmosphere, hydrosphere and lithosphere amount to 7.5, about $2 \cdot 10^6$ and $1.4 \cdot 10^{10}$ million tons, respectively, which is also visualized in Figure 1.1. The abundance of fluorine is of course not influenced by anthropogenic factors; however, the distribution is strongly affected. This especially holds for fluorinated compounds in the atmosphere, which are largely represented by artificial fluoroorganic molecules, that is, chlorofluorocarbons (CFCs), hydrochlorofluorocarbons (HCFCs) and perfluorocarbons (PFCs). This has some significant effects on the environment, which will be discussed in more detail in Section 7.2. However, there are also natural sources of atmospheric fluorine compounds, the most important of them being volcanoes [2]. It is estimated that they liberate between 60.000 and 6 million tons of HF and fluorosilicates per year, the actual amount being of course strongly dependent on the world-wide volcanic activity and the number of eruptions. Further volcanic gases include the halocarbons $CFCl_3$, CF_2Cl_2 and $HCFCl_2$.

A significant influence of anthropogenic fluoride sources is also found in the hydrosphere and the lithosphere. Some of them are strongly persisting in the environment and accumulate in organisms, for example, perfluorooctyl sulphonate $C_8F_{17}SO_3^-$ (PFOS), due to which this type of compounds has been banned from industrial applications (see also Section 7.2). Apart from saturated fluorocarbons, a number of compounds with fluoroaromatic moieties are introduced to the environment, as they are used as agrochemicals or drugs. However, they are far more easily degraded and are thus not that persistent.

As depicted in Figure 1.1, the disposal of fluorinated waste in landfills also plays a role in the fluorosphere, as readily soluble fluorine-containing compounds can in principle be washed out by rain and ground water, which should be avoided, of course. The degradation of a few organic fluorine compounds may also take place to result in the formation of stable, nondegradable compounds, such as trifluoroacetic acid. More detailed descriptions of the relevant processes in the fluorosphere can be found in the original review article by Tavener and Clark [1].

1.2 Fluoride sources and elemental fluorine in minerals

Due to its reactivity, the natural occurrence of fluorine is mainly restricted to simple and complex fluorides. However, there are a few minerals, where elemental F_2 can ac-

https://doi.org/10.1515/9783110659337-002

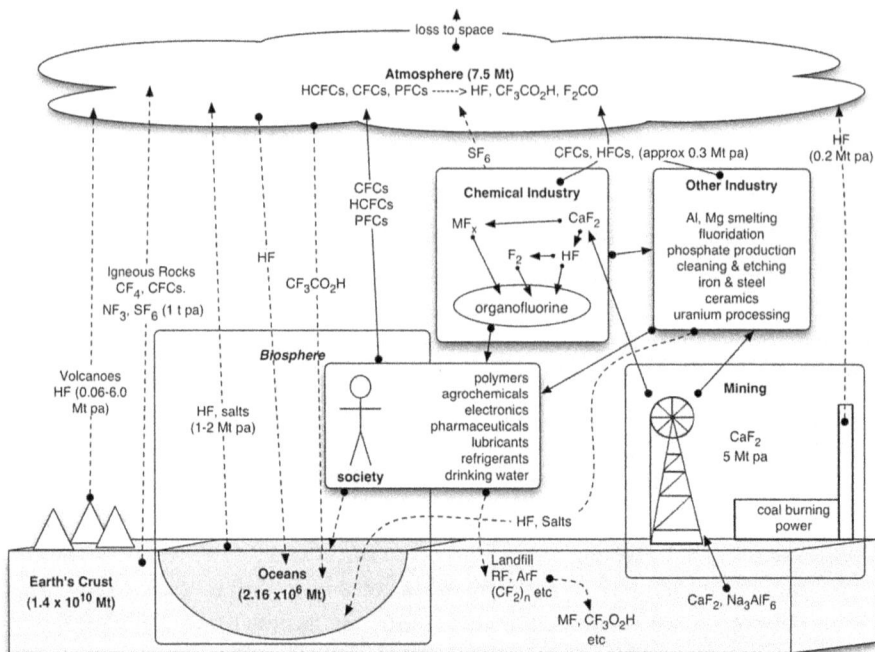

Figure 1.1: The fluorosphere: An overview of natural and anthropogenic fluorine fluxes and reservoirs on earth. (Reproduced with permission from: S. J. Tavener and J. H. Clark, Advances in Fluorine Science, Chapter 5: "Fluorine: Friend or Foe?" in A. Tressaud (ed.): Advances in Fluorine Science, Elsevier, 2006, 177–202.)

tually be found in traces, for example, in a black variety of fluorite (CaF_2, antozonite), which is found in Wölsendorf in Upper Palatinate in Bavaria, Germany. If it is shattered or grinded, one can smell elemental F_2. Its formation in this mineral is explained by the impact of the accompanying radioactive components, that is, uranium minerals, which radiolytically split CaF_2 into the elements. Due to structural defects, Ca is isolated from F_2, so that they do not directly recombine to form CaF_2 [3]. Traces of molecular F_2 have also been identified in red NaF (villiaumite) from the Kola Peninsula in Russia. However, therein the F_2 concentration is too small, so that it cannot be smelled upon grinding of the mineral [4].

The most important natural sources of fluoride are fluorspar (fluorite, CaF_2), cryolite $Na_3[AlF_6]$ and fluoroapatite [$3Ca_3(PO_4)_2 \cdot CaF_2$], the phosphate minerals being the largest reserves among them. CaF_2 is produced on a world-wide scale of about 4–5 million tons per year, more than the half of it coming from China. Other important producers are Spain, France, Mongolia, Mexico and the United Kingdom. In the United States, about 600.000 tons of CaF_2 are consumed per year, about 90 % of it are used for the production of HF, AlF_3 and Na_3AlF_6. Another important source of fluoride is obtained as a by-product of the preparation of phosphoric acid. Therein, the contained fluoride is liberated as SiF_4 and HF, forming H_2SiF_6 on a global scale of about 65.000 tons per year. Fluorine can also be found in a few silicates and aluminates as well as in traces in the soil (ca. 100 ppm) and all waters (seas, lakes, rivers and springs). Due to that, fluorine is also an essential trace element for plants, animals and humans, where it is incorporated in bones and teeth.

1.3 Natural organofluorine compounds [1, 5]

It is often believed, that organic fluorine compounds are entirely man-made. This is actually not the case, as there are a few biogenic sources of fluoroorganic molecules. In total, there are about 30 identified fluorinated natural products, including compounds like fluoroacetate, fluorocitrate and 4-fluorothreonine (Figure 1.2), whereas there are far more than a million ones, which have been prepared by chemists. It might be surprising, that the overall number of natural fluoroorganic compounds is rather small, despite the abundance of fluoride in the earth's crust.

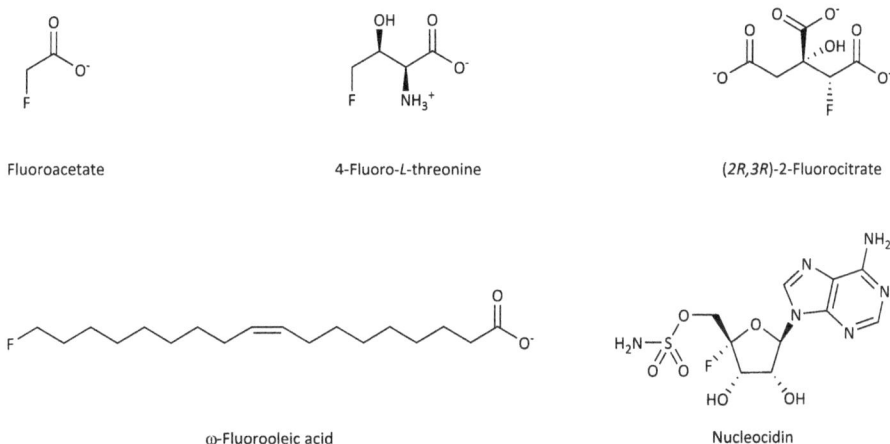

Fluoroacetate   4-Fluoro-L-threonine   (2R,3R)-2-Fluorocitrate

ω-Fluorooleic acid   Nucleocidin

Figure 1.2: Identified fluorine-containing natural products.

The prototype for a fluorinated natural product is fluoroacetate FCH_2COO^-, which is produced and accumulated in low concentration by over 40 higher plants. Typically, they grow in soils with an increased fluoride concentration and can be found in tropical and subtropical regions of Africa, South America and Australia. Since fluoroacetate is extremely toxic to animals, the plants use it as a defense mechanism. Along with the amino acid 4-fluoro-(L)-threonine, fluoroacetate is also biosynthesized by a bacterium. Fluorocitrate is sometimes found in toxic plants that contain fluoroacetate, however, in much smaller concentrations. Actually, fluorocitrate is the reason for the toxicity of fluoroacetate, as it is a metabolite, which is generated in the citrate cycle.

Upon binding of the fluorinated species, the citrate cycle is blocked. ω-Fluorooleic acid has also been identified in plants in Western Africa, where it is the toxic component of seeds (Figure 1.2). Along with the parent ω-fluorooleic acid, a number of minor ω-fatty acids have been found, which clearly are metabolites. It is assumed, that the fluorine substituent originates from fluoroacetate, which is incorporated as fluoroacetyl-CoA in the fatty-acid biosynthesis. Another major natural fluoroorganic compound is the antibiotic nucleocidin, which has also been identified in a bacterium (Figure 1.2). The biosynthesis of this compound is still not fully understood. Two more fluorinated molecules are often reported as natural abundant species, being fluoroacetone and 5-fluorouracil. However, their natural occurrence is strongly doubted, as fluoroacetone has only tentatively been claimed and 5-fluorouracil is very likely to result from human wastewater, since it is an important chemotherapeutic agent. However, it can be expected, that more fluorine-containing natural products will probably be identified in the future.

The bacterial production of fluoroacetate and 4-fluorothreonine has further been investigated, revealing that in total six enzymes contribute to the generation of the fluoroorganic compounds. The most interesting one is fluorinase, which is capable of forming the C-F bond from an inorganic fluoride. Due to this, the enzyme has been isolated and further characterized, prior to using it for *in vitro* fluorination reactions. As enzymatic reactions are generally very interesting due to their selectivity and efficiency, the use of fluorinase might become an important synthetic tool for fluoroorganic chemists.

Apart from the few natural occurring compounds, many efforts have been made to prepare artificial, fluorinated biomolecules. For example, a large number of amino-acid derivatives has been described to date [6]. Based on them, fluorinated peptides may be prepared, which in turn are used to understand or modify biosynthetic processes.

1.4 Study questions

(I) Where can one generally find fluorine in nature? Name some typical, highly abundant compounds.

(II) Most of the naturally occurring fluorine is bound as fluoride in minerals. How are fluorinated compounds entering the atmosphere? Comment on the role of anthropogenic influences, too.

(III) Is it possible to find elemental fluorine in nature? Comment on the amounts of occurring F_2 and give a short explanation for this observation.

(IV) Name the most important natural organofluorine compounds. Where can one find them and what is their role in nature?

(V) What is the major feature of the enzyme fluorinase, which makes it perspectively interesting for synthetic applications?

2 Elemental fluorine

Fluorine is a very special element, which forms numerous compounds with almost all of the other elements in the periodic table. Its name goes back to the mineral fluorspar, which has already been used in early times as additive in metallurgic processes. Fluorine is the most electronegative element; thus, its formal oxidation state in compounds is always -I per definition. However, the bonding situation in fluoride-compounds can differ from almost exclusively ionic to strongly covalent. Fluorine has the ability to stabilize elements in their highest possible oxidation states. The applications of fluorine compounds are very diverse. Certainly, the most important fluorine compound is HF, which is used as a solvent and fluoride source. For this reason, an own subchapter is dedicated to this compound (cf. Chapter 3).

2.1 Properties and general reactivity

Fluorine F_2 is a gas (mp.: $-219.62\,°C$; bp.: $-188.13\,°C$) with a characteristic pungent odour, which may be compared to a mixture of chlorine and ozone. Color descriptions are contradictory and vary from yellow to greenish or colorless. The green color has been attributed to the presence of chlorine impurities, whereas the yellow color is believed to result from O_2 impurities, which lead to the formation of O_2F. The actual color of pure, gaseous F_2 is still under investigation. In the liquid-state it is light yellow, if it is contaminated with oxygen, O_2F is formed, which causes an orange color [7]. In the solid-state, fluorine is colorless and occurs in an α-modification, which is found below $-227.6\,°C$ and a β-form, which is observed between $-227.6\,°C$ and the melting point [8]. F_2 is the most reactive element, which forms compounds with every other element except the light noble gases helium and neon. Apart from that, it will react with any compound, unless it is already a fluoride in its highest oxidation state, for example, NaF, CaF_2 or WF_6. In most of the cases, the reactions occur spontaneously at room temperature and are often accompanied by the appearance of fire or explosions. It is an extremely strong oxidizer with a potential of 3.05 V (acidic medium) or 2.87 V (alkaline medium), respectively. F_2 already reacts with H_2 at ambient temperature, even in the dark, accompanied by ignition or violent explosions. Sulphur and phosphorous vigorously react at $-185\,°C$ and highly-dispersed carbon is ignited at room temperature. Similarly, alkali and alkaline earth metals are inflamed at room temperature to result in the formation of the respective fluorides M^IF (M^I = alkali metal) and $M^{II}F$ (M^{II} = alkaline earth metal). Fluorine has a very high affinity to hydrogen, so that it readily and mostly violently abstracts it from any hydrogen-containing compound, for example, H_2S, NH_3, H_2O. The same reactivity is observed with organic materials. However, in contrast to HF, dry F_2 does not react with glass.

The extraordinary properties of F_2 in comparison to the other halogens may be explained by three facts, being (a) the low dissociation energy ($158\ kJ\ mol^{-1}$), (b) the

https://doi.org/10.1515/9783110659337-003

high bond energy of many element-fluorine bonds and (c) the relatively small atomic and ionic radii of the F-atom and -ion. The small value for the dissociation energy is mainly a result of the short nuclear distance and the repulsion of the two F-atoms in the F_2-molecule [9–12]. An overview over bonding energies and -distances of the most important element-fluorine single bonds is given in Table 2.1. In halogenation reactions, the large energies of element-fluorine bonds commonly lead to an increased reactivity, since the reaction enthalpies of fluorination reactions are often significantly higher than those for other halogenations.

Table 2.1: Molar bonding energies and bond distances of selected element-fluorine single bonds.

	E [kJ mol^{-1}]	d [pm]		E [kJ mol^{-1}]	d [pm]
B-F	646	143	N-F	278	138
Si-F	595	170	Cl-F	255	161
H-F	565	92	Br-F	238	176
P-F	496	166	O-F	214	141
C-F	489	136	F-F	158	144
S-F	368	163			

As fluorine is the element with the highest electronegativity, it exclusively appears in oxidation state -I in any compound. Thus, fluoride is mostly connected only once by a single bond; in some compounds it also forms F-bridges, so that the coordination number is increased to two or even three (cf. Section 3.5). The bonding situation in fluoride compounds differs a lot and can be largely ionic as well as strongly covalent. However, the ionic contribution in fluorides is always larger than that in other halides. For this reason, the covalent bonds are generally strengthened. Furthermore, due to the small size and the high charge-density of the fluoride anion, it is capable of deforming the electron shells of its bonding partners. Comparable to the oxide- and the hydroxide-ion, F^- itself is not really polarized. As a consequence, the structures of ionic fluorides and oxides show large similarities, crystallizing in the same type of lattice. The structural similarities of fluorides and hydroxides are also visualized by comparison of the bonding energies of many fluoro- and hydroxo complexes [13]. The most important structure types of inorganic fluorine compounds are separately summarized in Chapter 5. In addition, some structural properties of element fluorides will be discussed in more detail in the respective chapters of the elements in Part B of this manuscript.

In nature, one can only find the nuclide ^{19}F, all other isotopes with the masses 17, 18 and 20 are unstable and artificial. However, the positron emitter ^{18}F has a half-life of about 110 minutes and is commonly used in PET-imaging. Thus, the selective introduction of ^{18}F is also important in preparative fluoroorganic chemistry.

2.2 Polyfluoride anions

In the series of interhalogen compounds, a large number of fluoro-cations and -anions has been described (cf. Sections 9.7.5 and 9.7.6). Isopolyhalide cations as well as anions are also well known for the heavier halogens [14]. Generally, the stability and accessibility of these species significantly increases from chlorine to iodine. However, polyfluoride anions have also been described. By quantum-chemical methods, the linear trifluoride anion $[F_3]^-$ has been predicted to be stable in the gas phase, where it then has been identified by mass-spectrometry [15, 16]. In its ground state, the bonding situation in $[F_3]^-$ is based on a situation described as $[F^.F^-F^.]^-$, resulting in three-electron bonding. Two resonance structures can be written based on $[F_2]^-$ and $F^.$, the latter two also being the dissociation products of the trifluoride anion [17]. Vibrational spectra of $[F_3]^-$ under matrix conditions have proven the existence of an isolated anion [18]. However, the formation of ion pairs, especially $Cs[F_3]$ is also observed [19]. Upon treatment of $[F_3]^-$ with F_2 in neon matrices, the V-shaped pentafluoride anion $[F_5]^-$ is obtained [20, 21].

2.3 Preparation of F_2

Indeed, F_2 can be prepared by a purely chemical synthesis. For this purpose, the hexafluoromanganate(IV) complex $K_2[MnF_6]$ is treated with SbF_5, which leads to the formation of unstable MnF_4. The latter readily decomposes to MnF_3 and F_2 [22]. However, on industrial scale, it is far more efficient, to synthesize F_2 by electrochemical synthesis. Due to its high normal potential, its preparation is limited to the anodic oxidation of fluoride anions, which must not contain any other anion in the electrolyte. Pure HF is inconvenient for this purpose, as it is only slightly dissociated to $[H_2F]^+$ and $[HF_2]^-$. For this reason, KF is added as conducting salt. Nowadays, a melt of $KF \cdot 1.8$–2.5 HF is commonly used, where the electrolysis is performed at 70–130 °C. The electrolytically decomposed HF is replaced by fresh HF, in order to keep the temperature and composition of the electrolysis bath constant. A schematic representation of a cell for electrolytic F_2-production is depicted in Figure 2.1. The HF-electrolysis is performed in cascaded Monel- or steel cells with anodes made of graphite-free petroleum coke. The claddings of the cells serve as cathodes, where hydrogen is formed. To avoid the contact of H_2 and F_2, which would lead to violent reactions, the electrodes are spatially separated by iron plates, which are inserted into the electrolyte to a certain depth. The electrolysis is operated at a current of 4–15 kA, a potential drop of 8–12 V per cell and current densities of about 0.10–0.15 A cm^{-2}. The prepared F_2 contains up to 10 % HF, which can mostly be removed by cooling to −100 °C. Still remaining HF may be absorbed by NaF. Elemental F_2 is commercially available and is sold in 50 l-steel cylinders under a maximum pressure of 28 bar (100 % F_2) or 200 bar (10 % F_2 in N_2), respectively.

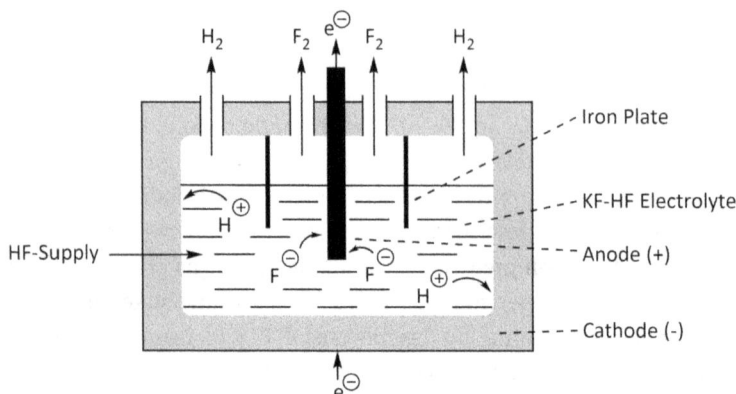

Figure 2.1: Schematic representation of an electrolysis cell for F_2 production.

F_2, which is generated by the method described above still contains impurities, being some larger amounts of HF as well as traces of O_2, OF_2, CF_4, CO_2 and others. For the separation of HF, the crude mixture is initially cooled to $-70\,°C$, afterwards cooling to $-140\,°C$ is applied in order to remove carbon-fluorine compounds. A small amount of HF is still not condensed at this temperature and can subsequently be removed at $100\,°C$, using an absorbing column filled with NaF. Commercially available F_2 typically has a purity of 99.0–99.5 % and is sold in steel cylinders under a pressure of 28 bar at maximum. The main contaminant is HF, however, for most purposes, the purity of F_2 will be sufficient. Further purification of F_2, especially for spectroscopic investigations and a few special applications is difficult. Very pure fluorine gas ($\geq 99.7\,\%$) is obtained after absorption on a 1:3-mixture of NiF_2 and KF under pressure at $250\,°C$ and subsequent thermal decomposition of the generated complex [23]. O_2 is removed by treatment with SbF_5 at $300\,°C$, which yields the salt $[O_2][Sb_2F_{11}]$. After cooling to $-183\,°C$, the mixture is distilled into a vessel cooled to $-196\,°C$. Further double distillation from $-196\,°C$ to $-210\,°C$ removes remaining impurities and yields spectroscopically pure F_2 [24].

Globally, elemental fluorine is produced on a scale of about 10,000 tons per year. The most of it is used for the preparation of inorganic and organic fluorine compounds, which are not accessible by other methods, for example, the halogen fluorides ClF, ClF_3, BrF_3, IF_5; high valent element fluorides, such as SF_6 or UF_6 as well as perfluorinated alkanes like C_3F_8. In many cases, it is not necessary to isolate F_2 prior to the fluorination of the substrate. The electrofluorination process (cf. Section 12.2) is a convenient alternative for this purpose. Therein, the reactant is dissolved in HF and fluorinated on large anodes, whereas H_2 is evolved at the cathode. The largest amount of F_2 is consumed for the production of UF_6 (nuclear fuel industry) and SF_6 (dielectric material), minor amounts are used in the surface fluorination of plastics as well as in the production of "graphite fluoride" (cf. Section 9.4.1), which is a suitable electrode material.

2.4 Working with elemental fluorine and fluorinated compounds [7]

Without any doubt, fluorine belongs to the hazardous compounds. However, using appropriate equipment and following strict safety precautions, it can be handled. The same holds for most of the fluorine compounds, especially volatile ones. Apart from that, there are regulatory requirements in order to reduce potential risks and hazards. One of the most important factors of working with fluorine and its compounds is certainly the choice of appropriate materials for reaction vessels, etc. Therefore, one always has to carefully consider the reactivity of the reactants and, of course, the reaction conditions. For tubing and valves passivated copper or Monel are typically used. Reactions with nonvolatile fluoride compounds can be performed in standard laboratory glassware, if they do not react with glass and if the reaction temperature is not too high. The use of quartz vessels is advantageous as those are more resistant; however, the formation of HF has to be strictly avoided. At low temperature and in an inert solvent, it is also possible to perform fluorination reactions with diluted F_2 in glassware. The reaction temperature can be increased, if quartz is used instead. In either case, the presence of HF and water has to be strictly excluded, as H_2O hydrolyzes the respective fluoride to yield HF, which in turn attacks SiO_2 under formation of water and SiF_4. Thus, a "cyclic" hydrolysis starts, where every step yields more hydrolyzing species. For a similar reason, the contact with silicon-based standard laboratory grease strictly has to be avoided. Instead, special fluorine-resistant poly(chlorotrifluoroethylene) grease (PCTFE, Kel-F) or poly(tetrafluoroethylene) (PTFE) sleeves can be used. Along with perfluoroalkoxy (PFOA) and fluorinated ethylene propylene (FEP), the aforementioned fluoropolymers can also be used as materials for reaction vessels. Again, the choice of the polymer depends on the reaction conditions. PTFE is relatively inert up to 200 °C, even against oxidizing fluorides; however, it is not transparent and rather difficult to handle. In contrast, Kel-F based reactors are transparent and precisely processable, but can already be pyrolyzed at temperatures of about 100 °C. Any of the aforementioned perfluorinated plastics will be burned to CF_4, if F_2 is flowing through them too fast or under elevated pressure. Eventually, these decomposition reactions may occur explosively. For high-temperature and high-pressure reactions, metal vessels have to be used. Their usage is also required, when pure F_2 is handled. Metal reactors are generally made from stainless steel, copper or nickel and its alloys, especially Monel. They are suitable in different temperature ranges; up to 100 °C, copper vessels are mostly sufficient. For reactions at higher temperatures, vessels made from nickel (up to 650 °C) and Monel (up to 550 °C) are preferentially used. The advantage of Monel over Ni is the higher resistance against corrosion, if moisture or HF are present. For diluted F_2, stainless steel or the abovementioned perfluorinated plastics can be used as reaction vessel materials. However, all the materials have to be free from any grease or other impurities and the metal vessels need to be properly passivated prior to use. In some cases, noble metal reactors are used as well (gold, platinum), along with exotic materials like sapphire or fluorspar. Still, one has to be aware of the fact, that any material used will not be indefinitely stable against F_2 at high pressures and temperatures. At red heat, even gold and platinum are heavily attacked. In addition, fluorine fires generally cannot be extinguished; the only way to stop them, is shutting down the F_2-supply.

Another important point about handling elemental fluorine is the appropriate temperature of cooling baths. As the boiling point of F_2 (−188 °C) lies above that of N_2 (−196 °C), working at liquid nitrogen temperature can become very dangerous. Liquid fluorine should definitely be avoided in any reaction vessel. An exception to this is the condensation of defined amounts of F_2 for the set-up of high-pressure reactions in appropriate metal vessels. To circumvent the liquification of F_2, in former times, liquid O_2 (bp.: −183 °C) has been used as medium for cooling baths. Overall, the safest way is the application of gaseous, diluted F_2 (mostly 10 % F_2 in N_2 or Ar) in flow reactions. However, in some cases, more drastic conditions are required.

2.5 Study questions

(I)   Elemental fluorine is extremely reactive. Comment on its general reactivity against inorganic and organic materials and give some examples.

(II)   Which are the most important facts that explain the extreme reactivity of fluorine?

(III) Which are the most important isotopes of fluorine? Do they occur in nature?

(IV) In comparison with the other halogens, only very few polyfluoride anions are known. What is their composition and structure?

(V)   How can one prepare elemental fluorine? Briefly describe the industrially relevant process.

(VI) How pure is commercially available F_2? How can a higher purity be achieved?

(VII) Which materials may be used for reaction vessels when working with fluorine and fluorinated compounds? Which limitations have to be taken into account?

3 Hydrogen fluoride

Hydrogen fluoride is certainly the most important fluorine compound and practically the source of every other fluorine-containing material. For this purpose, it may either be used directly or indirectly, when it serves for the preparation of F_2. Nevertheless, it is not only a convenient source of fluorine, but also a small molecule with some interesting and useful properties.

3.1 Preparation, general properties and applications

Under standard conditions, anhydrous hydrogen fluoride **HF** is a colorless, low-boiling liquid (mp.: −83.36 °C; bp.: 19.51 °C) with a pungent odor, which is industrially prepared by treatment of fluorspar (CaF_2) with concentrated H_2SO_4 at 200–250 °C. The purification is achieved by repeated distillation processes. Typically, HF obtained by this method has a purity of about 99.5 %, the main impurities being SiF_4, H_2SO_4, SO_2 and H_2O. In the laboratory, further purification can be achieved by careful fractional distillation. The natural abundant fluoroapatite $Ca_5(PO_4)_3F$ is only sparingly used for HF-production, due to its low fluoride content of only 2–4 %. Apart from that, about one-third of the produced HF is lost as H_2SiF_6 by the reaction with SiO_2 impurities in the apatite mineral. The direct reaction of H_2 and F_2 for the generation of HF is very uncommon. In the laboratory, anhydrous HF is conveniently liberated by heating acidic fluorides of the type MF · HF, for example, $K[HF_2]$. The specific conductivity of HF at 0 °C amounts to $1.6 \cdot 10^{-6}$ Ohm^{-1} cm^{-1}, but it is drastically increased, if impurities are present. For example, traces of water already increase it by about factor 10. HF is used for the preparation of elemental fluorine and inorganic fluorides, like NaF, BF_3, HBF_4, AlF_3, SnF_4, UF_4 or NH_4F. Especially the synthesis of AlF_3 and synthetic cryolite $Na_3[AlF_6]$ on industrial scale are important for the aluminium industry. Furthermore, HF plays a role in the synthesis of organic fluorine compounds, for example, CCl_2F_2, $CClF_3$ or $(CF_2)_n$. Generally, these chlorofluorocarbons (CFC) are used as refrigerants or solvents. Apart from the synthesis of basic fluoroorganic compounds, HF is applied as etching and polishing agent in the glass industry, for pickling of stainless steel, as a catalyst in alkylation reactions and in the production of semiconductors.

3.2 Structural properties

The bonding distance in HF amounts to 91.7 pm with a dissociation energy of about 574 kJ mol^{-1}. In solid hydrogen fluoride, polymeric $(HF)_x$ is found, where single molecules are unsymmetrically bridged via linear [F-H$\cdots$F]-groups to form planar zig-zag chains (Figure 3.1). At its boiling point, gaseous HF consists partially of hexameric $(HF)_6$ molecules and monomeric HF. $(HF)_6$ has a cyclic structure with a corru-

https://doi.org/10.1515/9783110659337-004

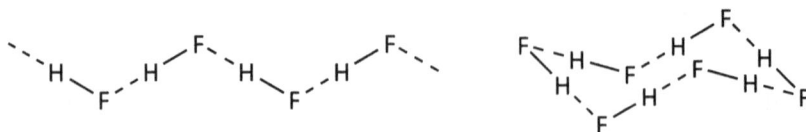

Figure 3.1: Structures of polymeric $(HF)_x$-chains (left) and hexameric $(HF)_6$-rings (right).

gated shape with asymmetric H-bridges (Figure 3.1) [25]. With increasing temperature, the amount of the hexamer decreases, so that above 90 °C monomeric HF is found exclusively. In liquid HF, chains of approximately seven HF molecules are found, which are connected by nonlinear H-bridges. The chains are typically not branched but interact with each other. The strong association of HF by H-F bridges is the reason for its relatively high melting and boiling point. Apart from that, HF has a high dielectric constant (83.5 at 0 °C), which is comparable to that of water (78.3 at 25 °C). As a consequence, hydrogen fluoride is one of the best solvents for both organic and inorganic salt-like compounds, such as metal and nonmetal fluorides, carbohydrates or proteins, even though its use requires special safety precautions.

3.3 Anhydrous and aqueous hydrogen fluoride

In liquid anhydrous HF, a self-dissociation equilibrium toward $[H_2F]^+$ and $[HF_2]^-$ is found, due to which a large number of acidic salts exists, containing the bifluoride anion $[HF_2]^-$. The formation of stable associated anions is one of the reasons for the very high acidity of anhydrous HF, which can further be increased by the addition of strong fluoride acceptors. For example, the addition of SbF_5 and SO_3 generates a super acid, as the formation of the very stable $[SbF_6]^-$ anion is favored, which in turn significantly increases the H^+-concentration. Thus, the system HSO_3F/SbF_5 is commonly used for the protonation of weakly basic, inorganic or organic compounds [26].

HF is very hygroscopic and can be mixed with water in any ratio. The resulting aqueous solution is known as "hydrofluoric acid," the azeotrope having a boiling point of 112 °C ($w(HF) = 38\,\%$; $w(H_2O) = 62\,\%$). At low temperature, the hydrates 4 HF · H_2O (mp.: -100.4 °C), 2 HF · H_2O (mp.: -75.5 °C) and HF · H_2O (mp.: -35.5 °C) can be obtained from solutions with respective concentrations. They are ionogenic and may therefore also be formulated as $[H_3O][H_3F_4]$, $[H_3O][HF_2]$ and $[H_3O]F$, respectively. Due to this, the dissolution of HF in water is not a physical, but a chemical process. HF dissolves many metals under formation of the respective fluorides and H_2. However, gold and platinum are resistant to hydrofluoric acid; lead is only partially attacked on the surface.

In diluted, aqueous systems, HF is only a weak acid ($pK_a = 3.19$). Actually, HF is completely dissociated in such media, but it does not form hydronium ions as it is observed for other acids. Instead, less acidic ion pairs $[H_3O]^+ \cdot F^-$ are formed due to the high H-F affinity. Upon concentration of diluted hydrofluoric acid, its acidity is increased, which is due to the increasing formation of stable $[HF_2]^-$ anions and comparably more acidic $[H_3O]^+$ cations. The characteristic property of hydrofluoric acid is certainly its ability to etch glass, which is due to the formation of gaseous SiF_4 out of SiO_2. As a consequence, it may only be stored and handled in containers and vessels made of lead, platinum, polyethylene, polypropylene or other comparable materials.

3.4 Salts of hydrogen fluoride

The corresponding metal salts of HF, the fluorides, are mostly water-soluble, however, a few of them only sparingly, for example, CaF_2, SrF_2, BaF_2, CuF_2 and PbF_2 [27]. Some of the fluorides can be obtained in anhydrous form from aqueous solutions, such as LiF, NaF, NH_4F, the alkaline earth fluorides $M^{II}F_2$ (M^{II} = Mg, Ca, Sr, Ba), SnF_2, PbF_2, SbF_3 as well as the gaseous compounds SiF_4 and GeF_4. In some other cases, hydrates are formed, especially for the heavier alkali metals ($KF \cdot 2\,H_2O$, $RbF \cdot 3\,H_2O$, $CsF \cdot 1.5$ H_2O) and many late transition metals ($AgF \cdot 4\,H_2O$, $CuF_2 \cdot 4\,H_2O$, $ZnF_2 \cdot 4\,H_2O$, $CdF_2 \cdot 4\,H_2O$, $HgF_2 \cdot 2\,H_2O$, $M^{II}F_2 \cdot 6\,H_2O$ (M^{II} = Fe, Co, Ni)). The trifluorides of group 13 form the trihydrates $M^{III}F_3 \cdot 3\,H_2O$ (M^{III} = Al, Ga, In).

Additionally, some fluorides also incorporate one or more molecules of HF. As a typical example, $KF \cdot HF$ (= $K[HF_2]$, potassium hydrogendifluoride, "acidic KF", mp.: about 225 °C) crystallizes from aqueous HF solutions containing KF. It exhibits one of the strongest H-bridges, where the bifluoride anion $[HF_2]^-$ has a linear and symmetric structure (d_{HF} = 113 pm; d_{FF} = 226 pm). From molten $KF \cdot HF$ in presence of different amounts of HF, the higher aggregates $KF \cdot 2\,HF$ (mp.: 72 °C), $KF \cdot 3\,HF$ (mp.: 66 °C) and $KF \cdot 4\,HF$ are accessible, too. Generally, these salts are useful sources of HF in the laboratory, as they can be thermally decomposed. Analogously to KF, a few more acidic fluorides $M^IF \cdot n\,HF$ (M^I = Li, Na, Rb, Cs, H_3O, NH_4, $[N(CH_3)_4]$, NO, pyridinium) are known. The respective salts are based on complex anions of the general composition $[H_nF_{n+1}]^-$ (n = 1–5, 7) (Figure 3.2). Therein, one to four HF molecules coordinate to F^- in a linear, bent, trigonal or tetrahedral manner. Alternatively, $[HFH]^-$ is surrounded by two or four HF-units by either cis-coordination or forming a tetragonal arrangement. The X-ray crystal structure analysis of $K[H_2F_3]$ has shown that $[H_2F_3]^-$ is unsymmetric and bent by 135° [28]. The anion $[H_3F_4]^-$ exists in three isomeric forms. A chain-like cis-configuration is found for example in $K_2[H_2F_3][H_3F_4]$ (= $KF \cdot 2.5\,HF$) and $[H_3O][H_3F_4]$. In contrast, a trigonal-planar arrangement is observed in $K[H_3F_4]$, whereas the anions in the ammonium salts $[NR_4][H_3F_4]$ and $NH_4[H_3F_4]$ show a trigonal-pyramidal geometry. In $[H_7F_8]^-$, three groups of $[F\text{-}H \cdots F\text{-}H]$ are connected with $[HFH]^-$ by H-bridges. The $F \cdots F$-distances in the $[F\text{-}H \cdots F]$-units are elongated with increasing HF-content. The limit is found to be 250 pm in polymeric HF, compared to the van der Waals F-F distance of about 270 pm. Both, neutral HF molecules as well as its polyanions $[H_nF_{n+1}]^-$ show ligand properties in metal complexes. Due to their geometry and the capability

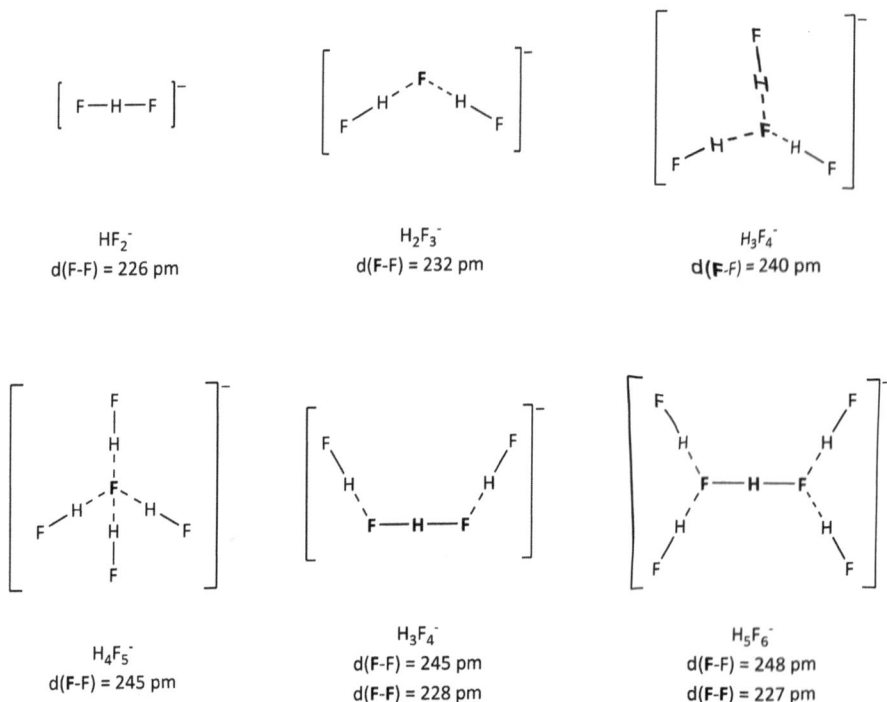

Figure 3.2: Selected structures of complex anions $[H_nF_{n+1}]^-$.

of forming strong hydrogen-fluorine bonds, the anionic species especially lead to the formation of three-dimensional networks. Of course, neutral HF ligands also exhibit H-bonding properties, resulting in the formation of room structures [29]. In the system HF/SbF_5, the fluoronium cations $[H_2F]^+$ and $[H_3F_2]^+$ have been identified and characterized by X-ray crystallography. As a result, a *trans*-orientation has been found for the $[H_3F_2]^+$ cation [30].

3.5 Naked fluoride and ligand properties

The "naked" fluoride ion is nonexisting in condensed phase and has strongly Lewis basic properties. However, in anhydrous CsF and especially $[N(CH_3)_4]F$, the fluoride ion is comparably naked, so that these compounds are strong fluoride donors. The tetramethylammonium salt is slightly soluble in acetonitrile and may be prepared from $[N(CH_3)_4]OH$ and aqueous HF as hydrate $[N(CH_3)_4]F \cdot x\,H_2O$. Subsequent drying in vacuo at 150 °C gives rise to the anhydrous compound. Instead of Cs^+ or $[N(CH_3)_4]^+$, other weakly coordinating cations may be used as well, such as $[PPh_4]^+$, $[AsPh_4]^+$, $[N(C_4H_9)]^+$ and many more. The strong fluoride donors, especially $[N(CH_3)_4]F$ are conveniently used to prepare anionic fluorocomplexes, for example, $[XeF_5]^-$, $[XF_6]^-$ (X = Cl, Br, I), $[IF_8]^-$, $[IOF_6]^-$ or $[TeF_7]^-$. Such complexes are known with many metal as well as nonmetal fluorides. However, F^- is not exclusively a monodentate (μ^1) ligand, but can also have μ^2- (less common) and μ^3-bridging functions (rare). Two metal or nonmetal cations can be either bridged in a bent or diagonal manner, whereas three cations can be connected in a trigonal-planar or trigonal-pyramidal way (Figure 3.3). Mostly, the bridges in the aforementioned systems are symmetric, but they may also be asymmetric in some cases. In salt-like compounds, four or six metal ions are often tetrahedrally or octahedrally surrounded by fluoride ions. The bridging of two metal-centers by two or three F-ligands is also known.

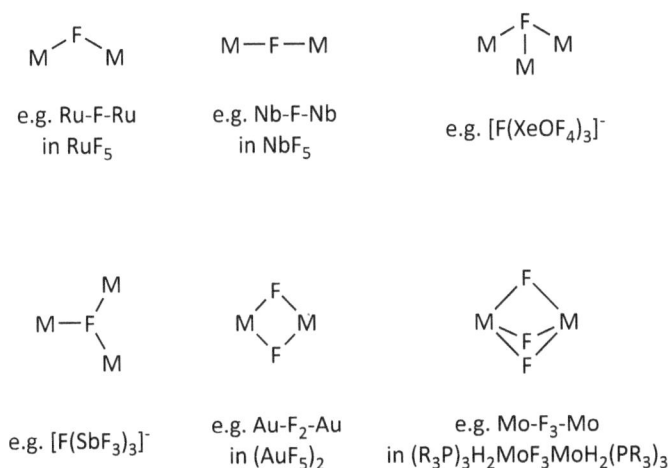

e.g. Ru-F-Ru
in RuF$_5$

e.g. Nb-F-Nb
in NbF$_5$

e.g. [F(XeOF$_4$)$_3$]$^-$

e.g. [F(SbF$_3$)$_3$]$^-$

e.g. Au-F$_2$-Au
in (AuF$_5$)$_2$

e.g. Mo-F$_3$-Mo
in (R$_3$P)$_3$H$_2$MoF$_3$MoH$_2$(PR$_3$)$_3$

Figure 3.3: Bridging modes of fluoride ligands.

3.6 Anhydrous hydrogen fluoride as reaction medium

As already mentioned above, HF is a versatile solvent [31]. The high dielectric constant of anhydrous HF is the reason for the good solubility of ionic fluorides (Table 3.1) [27]. Upon dissolving, the fluorides act as Lewis bases and donate fluoride ions. Thus, the metal fluorides M^IF (M^I = Li-Cs, Ag, Tl), $M^{II}F_2$ (M^{II} = Mg-Ba, Pb, Ag, Hg), SF$_4$ and SF$_6$ dissolve in HF under formation of M^{n+} and $n[HF_2]^-$, thus generating "basic HF." As a general trend, the solubility of the ionic fluorides decreases with increasing oxidation state, monofluorides being better soluble than difluorides, the latter being in turn better dissolved than trifluorides. For example, AgF is about 1700 times more soluble than AgF$_2$; TlF even 7.000 times better soluble than TlF$_3$. The solubility also drops with decreasing ionic radii of the metal, as CsF is about 20 times more soluble than LiF, and BaF$_2$ is about 300 times better soluble than BeF$_2$. Surprisingly, also XeF$_6$ dissolves very well in HF. Due to this behavior, it may be considered as monofluoride according to [XeF$_5$]$^+$F$^-$. The halogen trifluorides are other excellent fluoride donators.

Table 3.1: Solubility of selected ionic element fluorides in anhydrous HF at 12 °C.

Fluoride	Solubility [g per 100 g aHF]	Fluoride	Solubility [g per 100 g aHF]
TlF	580.0	CeF_4	0.10
CsF	199.0	TlF_3	0.08
AgF	83.2	AgF_2	0.05
NaF	30.1	CeF_3	0.04
SrF_2	14.8	CoF_2	0.04
LiF	10.3	NiF_2	0.04
BaF_2	5.6	MgF_2	0.03
PbF_2	2.6	ZnF_2	0.02
Hg_2F_2	0.88	BeF_2	0.02
CaF_2	0.82	BiF_3	0.01
HgF_2	0.54	ZrF_4	0.009
SbF_3	0.54	FeF_3	0.008
CoF_3	0.26	FeF_2	0.006
CdF_2	0.20	AlF_3	0.002
MnF_3	0.16		

Many of the rather covalent element fluorides are also well soluble in HF. Some of the pentafluorides, for example, PF_5, AsF_5, SbF_5, NbF_5 or TaF_5 act as Lewis acids and accept fluoride ions. This leads to the formation of "acidic HF", containing fluoronium cations $[H_2F]^+$ and a stable, six-fold coordination in the respective $[MF_6]^-$ anions. BF_3 and SiF_4 also show fluoride accepting properties, however, they are rather weak Lewis acids. The same holds for some other tetrafluorides [32].

The fluorides CrF_3, SbF_3, AlF_3 and BeF_2 show amphoteric properties in liquid HF [32]. In contrast, other element fluorides are dissolved molecularly without dissociation, as for example, XeF_2, XeF_4, VF_5, UF_6, OsF_6, SF_6 or ReF_7. The neutralization reaction of $[BrF_2]^+[HF_2]^-$ with $[H_2F]^+[SbF_6]^-$ yields the salt $[BrF_2]^+[SbF_6]^-$, along with HF, thus generally proving the ionization equilibria of ionic fluoride species. In a solvolysis reaction at 0 °C, HF liberates nondissociated H_2SO_4 and $K[HF_2]$ from K_2SO_4. At higher temperatures, the latter system generates fluorosulphonic acid. Generally, solvolysis reactions in HF are suitable methods to generate anhydrous fluorides from chlorides, bromides, oxides, hydroxides, carbonates, sulphites and other salts. In the same way, in the reaction with nitric acid or nitrates, HF forms $[NO_2]^+$, $[H_3O]^+$ and $[HF_2]^-$; similar fluorination reactions are observed for the oxoanions PO_4^{3-}, CrO_4^{2-}, MoO_4^{2-} and MnO_4^-, which lead to the formation of H_2PO_3F, HPO_2F_2, $[H_3O][PF_6]$, CrO_2F_2, MoO_2F_2 and MnO_3F, respectively.

Analogously to aqueous systems, an electrochemical series also exists for redox reactions in anhydrous HF (Table 3.2) [33]. The redox pairs above H_2/H^+ may dissolve in liquid HF under H_2 evolution, if the reactions are not hindered. A preparatively useful reaction is the electrochemical liberation of F_2 in presence of organic or inorganic substrates . Such fluorination reactions allow, for example, the preparation

Table 3.2: Electrochemical series of selected redox pairs in anhydrous HF and H_2O.

Redox Pair	ε_0 [V] in aHF	ε_0 [V] in H_2O
Cd/Cd^{2+}	−0.29	−0.40
Pb/Pb^{2+}	−0.26	−0.13
H_2/H^+	±0.00	±0.00
Cu^+/Cu^{2+}	+0.52	+0.16
Fe^{2+}/Fe^{3+}	+0.58	+0.77
Hg/Hg^{2+}	+0.80	+0.80
Ag/Ag^+	+0.88	+0.80
Ag^+/Ag^{2+}	+2.27	+1.96
F^-/F_2	+2.71	+3.05

of NF_3, HNF_2 and H_2NF from NH_4F; OF_2 from H_2O; SF_4 and SF_6 from H_2S; $N(CF_3)_3$ from $N(CH_3)_3$; CF_3SF_5 and $(CF_3)_2SF_4$ from CS_2 or CF_3CN and $C_2F_5NF_2$ from CH_3CN, respectively. Apart from that, redox reactions in HF are only scarcely explored. However, there are some general trends, which can be derived for the reaction of transition metal fluorides with nonmetal compounds. The reduction potential decreases in the series $PF_3 > AsF_3 > SbF_3 \approx SeF_4 > SF_4$, whereas the oxidation potential decreases in the following way: $VF_5 \approx CrF_5 > UF_6 > MoF_6 > ReF_6 > WF_6 > TaF_5 \approx NbF_5$.

Due to its strongly acidic properties, anhydrous HF is also an excellent solvent for the preparation and handling of high oxidation state materials, for example, Ag(III) or Ni(IV) compounds. However, the actual appearance of these species strongly depends on the acidity or basicity of the solvent, respectively. For example, the complex fluoroanions $[AgF_4]^-$ and $[NiF_6]^{2-}$ are present in basic anhydrous HF. In contrast, the polymeric, binary fluorides AgF_3 and NiF_4 are precipitated from the neutral HF medium. Acidification finally generates solvated, cationic Ag(III) and Ni(IV) species, being the strongest oxidants known. The nature of anhydrous HF also determines the reactivity of the high oxidation state materials, even if the reaction temperature remains constant. In basic HF at room temperature, $[O_2]^+$-salts readily oxidize AgF_2 to $[AgF_4]^-$. Acidification of $[AgF_4]^-$-solutions with AsF_5 precipitates neutral AgF_3, which is converted to solvated, short lived, cationic Ag(III) upon further addition of fluoroacidic AsF_5. Cationic Ag(III) is an extremely strong oxidizer, transforming molecular oxygen to the dioxygenyl cation $[O_2]^+$ [34].

3.7 Study questions

(I)   Which methods are typically used to prepare anhydrous HF (aHF)? How is it purified?

(II) Describe the structures of solid, liquid and gaseous aHF.

(III) Compare the acidities of liquid anhydrous HF (aHF) and hydrofluoric acid. Give an explanation for this behavior. How can the acidity of aHF further be increased?

(IV)   What does the term "acidic fluoride salt" mean? Give an example.

(V)   Does the "naked" fluoride ion exist? Which compounds are typically considered as "naked" fluorides and what is their main feature?

(VI)   Anhydrous HF is an excellent solvent for inorganic fluorine compounds. What do the terms "acidic anhydrous HF" and "basic anhydrous HF" mean? How does one prepare these media?

(VII) How does the fluoroacidity or fluorobasicity determine the species that may be found in a reaction medium? Give an example.

(VIII) Which general trends are observed for the solubility of element fluorides in anhydrous HF?

4 Preparation of fluorinated compounds [35]

In principle, the starting material for the preparation of fluorides is always HF. It is either used directly for fluorination reactions or it is needed for the preparation of F_2 and other fluorinating agents. For many inorganic fluorides, some special procedures have been established for their preparation, nevertheless, a few general reaction types can be derived, which are summarized in Table 4.1.

4.1 Transformations with gaseous fluorinating agents

As many fluorinating (e. g., F_2, ClF_3, BrF_5) and fluoridating agents (e. g., HF, SF_4) are gases, many different preparative methods are known for gas-phase transformations. In principle, the same starting materials, including halides, hydrated fluorides, carbonates, oxides and hydroxides as well as decomposable ternary compounds can be either treated with oxidizing agents to obtain fluorides in higher oxidation states or, alternatively, with nonoxidizing compounds, which will then lead to halogen exchange reactions under retention of the oxidation state of the starting material. However, for the oxidation reactions it is also suitable to start from pure elemental substrates. The direct treatment of the elements, their oxides, halides or carbonates with elemental fluorine typically yields binary fluorides in the highest oxidation states. For example, IF_7, SF_6, AsF_5, CoF_3 and AgF_2 are prepared on this way. Mainly, these reactions are carried out under flow-conditions, which also allow the control of the reaction by dilution of the reagents with inert gases, such as nitrogen or argon. The application of other strong fluorinating agents apart from F_2 itself, especially the halogen fluorides ClF, ClF_3, BrF_3 and IF_5, has some major drawbacks. On the one hand, F_2 is required for their preparation and on the other hand, free halogen is liberated during the fluorination reaction, which can lead to the formation of side-products. Nevertheless, the halogen fluorides are typically easier to handle in the laboratory. In some cases, more drastic conditions are required to prepare high-oxidation state fluorides. Their preparation is readily achieved by high-pressure reactions, where the fluorinating agent, mostly pure F_2, is condensed to the substrate and heated subsequently. This method can also be used for the preparation of oxyfluorides, if O_2 is added to the reaction mixture or, in some cases, if oxides are used as starting materials. However, in order to reduce the amount of impurities in the final product, it is recommended to "pre-fluorinate" the substrate under mild conditions prior to high-pressure fluorinations. A special variety of the gas flow-reactions are those, which are performed in a fluidized bed. This method allows the preparation of fluorides in larger amounts by passing a flow of F_2 through a vertical reactor, which is equipped with a fine mesh on its bottom. As starting materials, grained metals are loaded on the mesh and fluidized by the gas flow. Another preparative method are hot wire reactions, which are still the best approaches to prepare AuF_3 and especially PtF_6. For this purpose, a wire of pure metal is

https://doi.org/10.1515/9783110659337-005

Table 4.1: General techniques for the preparation of different substrate classes of fluorinated compounds.

Reaction type	Substrates	Reagents	Products	Comments
Reactions with Gaseous Fluorinating Agents				
Halogen exchange with gaseous fluorides	hydrated metal fluorides, other halides (usually chlorides), oxides, hydroxides, carbonates or cyanides; decomposable ternary compounds	nonoxidizing gaseous fluorides, e. g., HF, SF_4	oxidation state of substrate is retained	H_2 may be added to prevent oxidation by O_2 or other impurities
Fluorination in flow	elements; hydrated metal fluorides, other halides (usually chlorides), oxides, hydroxides, carbonates or cyanides; decomposable ternary compounds	strongly oxidizing gaseous fluorides e. g., F_2, ClF_3, BrF_5	substrate is oxidized to higher, mostly the highest possible oxidation state	reagents may be diluted by inert gases, e. g., N_2 or Ar
High-pressure fluorinations	elements, element oxides, lower fluorides	typically F_2, sometimes together with HF or O_2	highly oxidized materials, including complex fluorides and oxyfluorides	most drastic reaction conditions
Fluidized-bed fluorinations	grained metals	F_2	high oxidation state fluorides	convenient and fast method, also on larger scale
Hot-wire reactions	pure metal wires, e. g., Pt or Au	F_2 (100 %)	highest oxidation state fluorides, e. g., PtF_6 or AuF_3	reactions require initiation by external heat or electric current
Reduction of high-oxidation state compounds	fluorides in high (mostly highest) oxidation state	reducing agents, e. g., H_2, CO, elemental Si	fluorides in reduced oxidation state	method for generation of fluorides, which are not accessible when starting from low oxidation states
Solid-State Transformations				
Preparation of polynary fluorides	stoichiometric mixtures of (mostly binary) fluorides	not required	polynary fluorides of defined stoichiometry	also gives access to oxidation states, which are unstable in aqueous systems

Table 4.1: (continued)

Reaction type	Substrates	Reagents	Products	Comments
Decomposition reactions	complex fluorides with volatile ligands (e. g., H_2O); thermally unstable complexes	not required	anhydrous fluorides; degraded or rearranged complexes	useful method, especially if the starting materials are easily accessible from aqueous systems
High-pressure transforma-tions	low-pressure modification of a fluoride	not required	high-pressure modification of a fluoride	rarely used for preparative purposes, but structural investigations
Preparations in Flux				
Preparation of polynary fluorides in chloride flux	stoichiometric mixtures of binary fluorides dissolved in molten chloride (mixtures)	not required	polynary fluorides of defined stoichiometry	reactions may be conducted at lower temperatures than solid-state reactions
Preparation of polynary fluorides in KHF_2-flux	metal oxides	KHF_2 (reagent and flux)	fluorometallate complexes	efficient preparation of fluorocomplexes from oxides
Solution Phase Syntheses				
Precipitation from aqueous solution	metal salts, mostly chlorides, carbonates, oxides or hydroxides	aqueous HF, alkali metal fluorides, AgF	insoluble, non-hydrolysable metal fluorides, in some cases hydrated	simple and robust technique for the preparation of hydrolytically stable compounds
Precipitation from organic solvents	metal salts, ideally soluble in organic solvents	soluble fluoride salts, e. g., NH_4F, $[N(CH_3)_4]F$	insoluble metal fluorides, in some cases solvated	solubility of starting materials is usually problematic in organic solvents
Halogen exchange reactions In inert solvents (e. g., anhydrous HF, CH_3CN)	element (oxy)-halides (usually chlorides), oxides, hydroxides, carbonates or cyanides; organic halogen compounds	anhydrous HF, alkali metal fluorides, AgF, SF_4; organic fluoride salts e. g., $[N(CH_3)_4]F$	hydrolysable, binary fluorides and oxyfluorides; fluorinated organic compounds	most convenient method for the preparation of fluorinated compounds in solution for both, inorganic and organic substrates (see Part C for more details and variants)

Table 4.1: (continued)

Reaction type	Substrates	Reagents	Products	Comments
Fluorination in inert solvents (e. g., HF, CH_3CN, CCl_3F)	organic and inorganic compounds	diluted F_2, XeF_2	selectively fluorinated organic and inorganic compounds	outcome of the reaction is strongly influenced by temperature control
Sol-gel synthesis	metal alkoxides (in nonaqueous solution)	HF (gas or nonaqueous solution)	amorphous high surface fluoride materials	mixed alkoxide fluorides (and thus (hydro)oxyfluoride precursors) are formed, if HF is sub-stoichiometrically added
Solvothermal reactions	metal salts, e. g., phosphates, carbonates or oxalates	HF (aqueous or anhydrous), BrF_3, BrF_5, IF_5, SeF_4, SbF_5	unusual framework structures with bridging ligands; high-pressure phases of fluoride compounds, often in high oxidation states	high-pressure conditions give access to otherwise unavailable compounds and structures
Electrochemical fluorination	inorganic and especially organic substrates, e. g., hydrocarbons	Ni-anodes, HF, occasionally KF as conducting salt	inorganic fluorides and perfluorinated fluoroorganic compounds, e. g., perfluoro-carbons	versatile method for the preparation of fluorinated organic and inorganic compounds

placed in an atmosphere of excess fluorine and burned. This method requires thermal initiation by either passing electric current through the wire or supplying external heat energy. In a few cases, the target fluoride is not readily accessible by fluorination of the respective element or low oxidation state compounds. This is especially observed for a number of pentafluorides, when a (more) stable hexafluoride exists. In this case, it is convenient to start from the readily available high oxidation state fluoride and reduce it with reagents like H_2, CO or elemental Si. Often, a comproportionation reaction is also suitable, that is, using stoichiometric amounts of the metal M to reduce its high valent fluoride MF_n to obtain a medium oxidation state.

4.2 Solid-state reactions and fluorinations in flux

The class of solid-state reactions is rather limited to the synthesis of polynary fluorides. For example, heating of stoichiometric mixtures of different binary fluorides is
an excellent method to prepare ternary or even quaternary compounds. Of course, this
requires that none of the components are either volatile or unstable. However, thermal
decomposition reactions of unstable compounds may also give rise to target fluoride
compounds. This includes, for example, the cleavage of solvent molecules, for example, H_2O in hydrates or the separation of a polynary fluoride into its components. If
solid-state reactions are conducted under high pressure conditions, this may give rise
to structural modifications of the starting material. Staying in the field of polynary fluorides, flux-methods have to be mentioned. In principle, the reactions are similar to
the solid-state reactions of several binary fluorides with each other; however, in this
case the transformation is performed in a melt. Typically, chlorides or chloride mixtures are used for this purpose since they have a reasonably low melting point and
show improved solubility properties. Instead of chlorides, bifluoride fluxes are also
used. In the reactions of oxides with KHF_2, the bifluoride takes both functions, fluoridating agent and flux.

4.3 Fluorination in solution phase

The last big class of preparative methods is covered by solution-phase syntheses. The
simplest preparation for binary, nonhydrolysable fluorides, such as PbF_2, Hg_2F_2, AgF
or SbF_3, is the reaction of oxides, hydroxides or carbonates with aqueous HF, alkali
metal fluorides or AgF. In some cases, this also leads to the formation of hydrated
salts, which can be obtained in anhydrous form upon heating, for example, ZnF_2,
CdF_2, NiF_2, CoF_2, FeF_3, CrF_3. The (hydrated) fluorides are obtained by precipitation
from aqueous solution. Sometimes, especially for better soluble compounds, precipitation from organic solvents, such as methanol or acetonitrile is a suitable alternative.
However, the solubility of ionic fluorides in organic solvents is very poor in most cases,
so that it is convenient to prepare the respective fluorocompound in aqueous solution prior to addition of organic solvents. Nevertheless, there are of course metal salts,
which are soluble in organic solvents, for example, alkoxides; the same holds for fluoride salts, typically represented by the group of tetraalkylammonium compounds. As
soon as the fluorides can be hydrolyzed, their preparation requires halogen-exchange
reactions with anhydrous HF, alkali metal fluorides, AgF, SF_4 or organic fluorides such
as $[NR_4]F$ (R = alkyl) in HF or inert organic solvents like acetonitrile. In any case, the
fluoride source and the solvent have to be anhydrous. Using this method, binary fluorides like TiF_4, ZrF_4, SnF_4, VF_3, NbF_5 or SbF_5 can be generated. The reaction of fluoridating agents in inert solvents can be easily extended to a large variety of organic
substrates. For the organic part, the different types of reactions and substrates will be

covered in detail in Part C of this manuscript. Inert solvents are also often used as media for the fluorination of inorganic and organic substrates, where diluted F_2 or XeF_2 are typically used as oxidizing agents. Both reagents can be very mild, if the reactions are performed at low temperatures and under high-dilution conditions. Furthermore, it is possible to obtain products and sometimes also unstable intermediates, which are not accessible by any other method. A typical example for the temperature dependence is the fluorination of iodine. At about 250 °C, IF_7 is generated, whereas IF_5 is obtained at 20 °C. At –40 °C in CCl_3F, the formation of IF_3 is observed, and finally, IF can be prepared at even lower temperatures [36, 37]. Another representative example for a mild fluorination by F_2 is the reaction with CS_2 in helium at –120 °C, which exclusively yields F_3S-CF_2-SF_3, leaving the C-S-bonds intact [38]. In anhydrous HF, XeF_2 is a medium oxidizer, which can often be used for the stepwise and selective preparation of fluorinated compounds [39].

The list of solution-phase preparations of fluorinated compounds is continued with some more sophisticated techniques. The sol-gel synthesis is a very elegant way to prepare amorphous high-surface materials, which may be used as catalysts, for example. Typically, solutions of metal alkoxides are treated with HF under nonaqueous conditions. The resulting sol or gel is subsequently evaporated to yield the metal fluoride powder. If the stoichiometry of added HF is reduced, it is also possible to obtain mixed alkoxyfluoride species, which may be hydrolyzed to give oxy or hydroxyfluorides. Solvothermal reactions are carried out by putting a mixture of a metal salt and a fluoride source, usually HF (anhydrous or aqueous), BrF_3, BrF_5, IF_5, SeF_4 or SbF_5 under high pressure. This method gives access to a number of high-pressure structures and compounds, which are otherwise not available. If the transformation is carried out in a strongly oxidizing medium, it is also possible to obtain unusual species in high oxidation states. Finally, another very popular and useful method is the electrochemical fluorination. It can be used for the fluorination of both, inorganic and organic substrates and is performed in anhydrous HF, which sometimes contains KF as conducting salt. The electrolysis is run at 20 °C or lower temperatures at a voltage of less than 8 V. This prevents the evolution of F_2 at the Ni-anode. The dissolved substances are anodically fluorinated, where the nature and the yield of fluorinated products are strongly influenced by the reaction conditions.

4.4 Study questions

(I) How can one prepare element fluorides in their highest oxidation states? Which types of substrates and reagents are commonly used?
(II) Which method is conveniently used to prepare fluorides of medium oxidation states? What is the prerequisite for this approach?
(III) Which types of fluorides are in most cases prepared by solid-state reactions?
(IV) What is the major advantage of transformations in flux compared to solid-state reactions?

(V)   How does the solution-phase preparation of hydrolysable and nonhydrolysable fluorides differ? Give examples for suitable fluorinating agents and solvents.

(VI) Which solvents and reagents are typically used for fluorination reactions in solution-phase? Which requirements exist for this type of reactions?

5 Common solid-state structures of inorganic fluorides [35]

The large majority of element fluorides has been structurally characterized by X-ray crystallographic methods. As expected, the compounds can be assigned to the well-known lattice-types of inorganic compounds, in some cases they are derived thereof by slight structural adaptations. The most important structure types are summarized in Table 5.1.

Table 5.1: Common solid-state structures of binary, inorganic fluorides.

Lattice type	Coordination number and geometry of the cation(s)	Coordination number and geometry of the anion(s)	Example compounds
NaCl	6 (octahedral)	6 (octahedral)	LiF-CsF
SiO_2 (quartz)	4 (tetrahedral)	2 (bent)	BeF_2
TiO_2 (rutile)	6 (octahedral)	3 (trigonal planar)	MgF_2, VF_2, CrF_2, MnF_2, FeF_2, CoF_2, NiF_2, CuF_2, ZnF_2
CaF_2 (fluorite)	8 (cubic)	4 (tetrahedral)	SrF_2, BaF_2, β-PbF_2, CdF_2, EuF_2, SmF_2
ReO_3	6 (octahedral)	2 (linear)	NbF_3, ScF_3
VF_3	6 (octahedral)	2 (bent; about 150°)	α-AlF_3, GaF_3, TiF_3, CrF_3, FeF_3, CoF_3
RhF_3	6 (octahedral)	2 (bent; 132°)	IrF_3, PdF_3, PtF_3
LaF_3 (tysonite)	9 + 2 (five-capped trigonal prismatic)	3 (trigonal planar)	CeF_3, PrF_3, NdF_3, PmF_3, SmF_3, EuF_3
YF_3	8 + 1 (distorted three-capped trigonal prismatic)	3 (trigonal planar)	GdF_3, TbF_3, DyF_3, HoF_3, ErF_3, TmF_3, YbF_3, LuF_3, β-BiF_3, TlF_3
SnF_4	6 (octahedral)	2 (linear)	PbF_4, NbF_4
VF_4	6 (octahedral)	2 (bent)	RuF_4
PdF_4	6 (octahedral)	2 (bent)	RhF_4, ReF_4, OsF_4, IrF_4, PtF_4
β-ZrF_4	8 (square antiprismatic)	2 (bent)	CeF_4, PrF_4, TbF_4, HfF_4, ThF_4-BkF_4
α-UF_5	6 (octahedral)	2 (linear)	BiF_5
VF_5	6 (octahedral)	2 (bent; 152°)	CrF_5
NbF_5	6 (octahedral)	2 (linear)	MoF_5
RuF_5	6 (octahedral)	2 (bent; 132°)	OsF_5, PtF_5, RhF_5

5.1 Binary fluorides

The simplest structure type is represented by the NaCl lattice, where both, cations and anions, are octahedrally coordinated, as it is, for example, found for the alkali

https://doi.org/10.1515/9783110659337-006

metal fluorides M^IF (M^I = Li-Cs). AB_2-compounds may be represented by three different structure types. The quartz-type is, for example, represented by BeF_2, where $[BeF_4]$-tetrahedra are linked via common corners to form a network. More frequently, the rutile and fluorite lattice are observed for binary element fluorides. Their occurrence mainly depends on the size-ratio of the respective cation toward F^-. Relatively smaller cations will prefer the octahedral coordination in the rutile structure, resulting in a trigonal-planar environment around fluorine. In contrast, larger cations tend to have a cubic coordination geometry, leading to tetrahedrally coordinated F-atoms.

Several possible structure types also exist for trifluorides. The simplest one is the ReO_3 type, where one can find $[MF_6]$-octahedra, which are linearly bridged via all corners. If the $[MF_6]$-octahedra are twisted, the bridging F-atoms are not linearly coordinated anymore, instead the M-F-M-angle decreases. In the range of about 150°, the VF_3 structure type is observed, further twisting to an angle of 132° gives rise to the RhF_3 type. The latter is also observed, for the "trifluorides" PdF_3 and PtF_3, where M^{II}- and M^{IV}-centers are alternatingly packed. For the heavier element trifluorides, especially the lanthanides, two additional structure types are important. In the LaF_3 type, the metal center is surrounded by overall eleven fluorine atoms. Nine of them are in rather close proximity, creating the geometry of a three-capped trigonal prism, the other two are capping the two remaining faces of the trigonal prism. However, their distance to the metal center is rather large. In the YF_3 type, a nine-fold coordination in form of a distorted three-capped trigonal prism is observed for the cation.

For the tetra and pentafluorides, the occurring structure types are in most cases not too different from each other, at least with respect to the coordination numbers and geometries of fluoride anions and cations. The majority of the structures show an octahedral coordination environment for the metal centers. The differences are found in the packing of the $[MF_6]$-units. In the SnF_4 type, the octahedra are linearly bridged via four common corners to form two-dimensional layers. As a consequence, the two terminal fluorides per building unit are *trans* to each other. A slight deviation is observed in the VF_4 structure type, where the terminal fluorides are still occupying *trans*-positions, but the octahedral $[MF_6]$-units exhibit tilting against each other. In contrast, the PdF_4 structure type shows a three-dimensional network, built up from hexameric $[MF_6]$-meshes. Therein, the F-bridges are non-linear, so that the terminal fluoride atoms at each $[MF_6]$-unit are in *cis*-positions. For a few other tetrafluorides, multiple chain-structures are observed, as for example, in α-CrF_4 (double chain), TiF_4 (triple chain) or β-CrF_4 (quadruple chain). Again, larger cations prefer a coordination number higher than six, which is achieved by a square antiprismatic coordination environment in the β-ZrF_4 type. Therein, a highly connected, three-dimensional network is observed, based on eight-coordinated metal centers.

The simplest way to arrange $[MF_6]$-octahedra in solid-state structures of pentafluorides is the formation of linear chains in the α-UF_5 type. However, if the F-bridges are a little bent, the octahedra become twisted, resulting in the VF_5 type. Accordingly, a similar behavior is found for the linear, tetrameric ring-structures of the NbF_5 type and the RuF_5 type , which is based on twisted $[MF_6]$-units, and thus bent F-bridges in the tetrameric rings. Two other major pentafluoride structures include the formation of molecular dimers, as observed for $(AuF_5)_2$ or, in case of large metal centers, the adaptation of the β-UF_5-type. Therein, square antiprismatic $[UF_8]$-units are linked by common edges and corners to build up a three-dimensional network.

5.2 Ternary and quaternary fluorides

For ternary and quaternary fluorides, the structure types can be mostly derived from those in Table 5.1. However, the most important building motifs in fluorine chemistry are certainly those of the $CaTiO_3$ (perovskite) and the Na_3AlF_6 (cryolite) type. The perovskite motif is generated from the ReO_3-structure, when the cuboctahedral cavities are occupied by another cation. If all edges of the unit cell are doubled, atoms of different elements may be equally distributed in the cell, so that the stoichiometry of the atoms can be enlarged from ABX_3 to A_3BX_6, which represents the cryolite structure type. Another variation is the elpasolite type (K_2NaAlF_6), where the positions of cation A in the cryolite lattice are occupied by two different types of cations. More detailed descriptions of the relationships between various ternary and quaternary fluorides have been comprehensively summarized elsewhere and will not be covered herein [35].

5.3 Study questions

(I)   Describe the coordination environment around cations and fluoride anions in the rutile and the fluorite structure type. On which factor does the occurrence of the lattices mainly depend? Give examples.

(II) How do the ReO_3, VF_3 and RhF_3 structure types differ from each other? What is their common structural feature?

(III) Which two structure types are typically found for trifluorides with large cations? Briefly describe them regarding the coordination environment around cations and fluoride anions.

(IV) Which structural units are found in the SnF_4 and the PdF_4 type and how are they connected? Which consequences does this have regarding the arrangement of the respective units?

(V)   What is the difference between the α-UF_5 and the VF_5 type? Is there a similar relationship between other structure types of pentafluorides?

6 Properties of fluorinated compounds

Fluorine is definitely a very special element, which is underlined by the properties of fluorinated species. On the one hand, fluorine gives rise to very stable compounds. For example, the ionic NaF is the most stable salt of the AB-type with a very high melting point, whereas the molecular fluorides CF_4 and SF_6 are extremely stable against hydrolysis. Another feature in terms of stability is the strength of H-F-bridges, the strongest one found in HF itself. However, fluorine compounds may also be very reactive, which is represented by the extreme oxidizer PtF_6, the only existing Pt(VI)-compound. Fluorine generally enables very high oxidation states, not only in the aforementioned PtF_6, but also in IF_7 and ReF_7, which are the only existing, binary AB_7-compounds. In some cases, where high-valent binary element fluorides are unknown or not accessible, fluorine enables the existence of related compounds by forming oxyfluorides or fluorocomplexes. Interestingly, many of the high-valent, neutral fluoride compounds are volatile. The typical example for this feature may be UF_6, which is extremely important for the isotope enrichment in nuclear fuel industry.

Looking at rather organic systems, the list of special features may be easily continued. Fluoroacetic acid CH_2FCOOH is a weak acid, but very toxic. If incorporated, it will readily enter the citrate cycle, which is then blocked (cf. Section 7.1). With two fluorine atoms more, trifluoroacetic acid CF_3COOH is a stronger acid, but nontoxic. CF_3SO_3H, actually a derivative of sulphuric acid, is 15 orders of magnitude stronger than concentrated H_2SO_4. It is also called "magic acid." In general, organic fluorine compounds are so manifold, that their properties cannot simply be summarized in a few words. Nevertheless, an attempt is made to give a short overview over important compound classes and their properties in the following paragraph. Some more details and of course preparative approaches toward fluoroorganic molecules will then be the subject of Part C of this manuscript.

According to produced amounts, the group of volatile hydro and halofluorocarbons is certainly one of the most important compound classes in fluoroorganic chemistry. These compounds have commonly been used as blowing agents, however, for environmental reasons, this application is strongly regulated nowadays. An important property of these molecules is their enthalpy of vaporization, due to which they are still used as refrigerants. Another very big part of compounds can be categorized in the family of fluoropolymers, the most famous one being polytetrafluoroethylene (PTFE) without any doubt. They are important materials with a high chemical durability, as they can in part even withstand diluted F_2 under mild conditions. Apart from that, they are used for numerous purposes, as for example, the construction of reaction vessels in chemistry, the production of high-performance parts in engineering or just the simple nonstick coating for kitchen utensils. Due to their chemical durability, some of them are also commonly used as a sealing material. Fluorinated polymers

https://doi.org/10.1515/9783110659337-007

also play a role in rechargeable batteries, for example, as membrane materials. Leaving the class of polymeric compounds, a few more fluorine-containing additives are typically found in batteries, for example, as flame retardants. Even if it may not yet be that relevant in industrial processes, organofluorine compounds are also used as solvents, which is represented by the growing field of fluorous phase chemistry. In this case, the (in)miscibility properties of perfluoroalkanes with many other typical organic solvents as well as of course with water are made use of.

The most sophisticated fluoroorganic molecules are developed for the use in drugs and agrochemicals. The main advantages of the introduction of fluorine can be summarized regarding three major aspects. Primarily, the substitution of labile C-H functions by metabolically stable C-F- or C-R^F-groups (R^F = perfluorinated group, e. g., CF_3) increases the bioavailability of the compound by reduced metabolism and a potentially higher membrane permeability, as fluorinated groups typically render the drug more lipophilic. A second important point is the often improved activity of fluorinated drugs, as those bind more efficiently to the respective target receptors. Organic fluorine chemistry is also important for medicinal imaging techniques, which is especially represented by positron emission tomography (PET), the latter being conveniently enabled by the isotope ^{18}F. Typically, labeled fluorodeoxyglucose is used for this diagnostic technique; however, for special purposes, other ^{18}F-containing molecules are designed. Fluorine has also found his way into artificial peptides, which may also be used for several medical purposes. Again, the introduction of fluorine drastically increases the metabolic stability of the compounds and further gives rise to additional analytical techniques.

Overall, one can conclude, that fluorinated compounds can be very multifaceted. There are very reactive molecular inorganic fluorides with extreme oxidation potentials but also extremely stable ionic ones. The situation is a little different for the organic derivatives, as there are of course always covalent C-F bonds. However, the electronic effects of fluorine and fluorine-containing groups can drastically render the chemical behavior of a molecule in a way, which is not always predictable.

7 Health and environmental aspects of fluorinated compounds

7.1 Impact of fluorinated compounds on humans and animals [7]

Elemental fluorine is very dangerous for humans and animals. Direct contact with the extreme oxidizer leads to severe skin burns and thermal harm to the body. However, even in very small concentrations of about 0.001 % it can be easily identified by its smell, which is comparable to a mixture of O_3 and HF. As a consequence, poisoning with F_2 does not occur often. More care has to be taken to avoid any contact with the skin, as the severe burns, which are caused by the released heat of reaction with skin tissue and moisture, are accompanied by very painful corrosive injuries, due to HF-formation and F^--poisoning resulting from the generated fluorides. Besides elemental F_2, many of the organic and inorganic fluoride compounds are toxic, harmful or at least an irritant [40]. This especially holds for readily soluble and volatile fluorides, which may be incorporated by swallowing or inhaling, respectively. In some cases, skin contact is also hazardous. Depending on the concentration of F^-, fluoride poisoning can lead to the precipitation of CaF_2 and fluoroapatite $Ca_5(PO_4)_3F$ from the blood serum. Higher concentrations will also lead to the formation of insoluble MgF_2. In any case, the reduction of Ca^{2+}- and Mg^{2+}-cations in the blood system will have severe effects, that is, hypocalcaemia and hypomagnesaemia. None of them is directly life threatening, but the following transfer of K^+-ions from the cells into the serum leads to increased concentrations of potassium cations (hyperkalaemia). The latter is definitely life-endangering, as it can cause ventricular fibrillation and cardiopulmonary arrest. The consumption of several tubes of fluoride-containing toothpaste may cause such acute toxic effects, which can be very dangerous, especially for children. The lethal dose for adults amounts to 32–64 mg fluoride per kg bodyweight. Similarly, the skin contact with pure or aqueous HF can be as dangerous, depending on amount and concentration of HF as well as the affected skin area. A long-term poisoning with fluoride leads to bone and tooth fluorosis. However, it is only observed, if the fluoride uptake is surpassingly high over months or even years. This will certainly happen, if a tube of fluoride-containing toothpaste is consumed every day.

A typical example for an extremely toxic fluoroorganic compound is fluoroacetate FCH_2COO^-. It is produced and accumulated by plants on fluoride-rich soils and serves as defence mechanism. Fluoroacetate poisoning in humans leads to respiratory failure. Furthermore, coma-like symptoms may occur. However, if the poisoning is survived, a complete recovery is typically observed after a short period of time. Interestingly, in areas, where fluoroacetate-producing plants grow, some rodents and birds of prey have evolved a tolerance against the toxic effects of fluoroacetate [1]. In contrast to species that accumulate fluoroacetate, there are also investigations on negative effects of fluorides to the growth of plants. For example, soil pollution with fluoride leads to

https://doi.org/10.1515/9783110659337-008

the formation of necroses at the tips of needles of the lodgepole pine [41]. The most toxic organofluorine compound known so far is perfluoroisobutene $(CF_3)_2C=CF_2$, with an LC_{50} value of less than 1 ppm. The compound targets lungs and liver, where it is believed to attack the thiol group in glutathione, an intracellular antioxidant. The toxicity of perfluoroisobutene most likely arises from its extreme electrophilicity, which makes it very susceptible to nucleophilic attacks [42].

Nevertheless, fluoride is essential for the human body, as about 10 mg F^- are contained per kg body tissue, mainly in teeth and bones. The daily consumption of about 1 mg F^- per kg tissue helps to avoid tooth decay and osteoporosis, but more than the double amount will lead to osteosclerosis. The oral uptake of more than 100 mg fluoride at once is acutely toxic (see above), and causes nausea, vomiting and diarrhoea, which may be treated by injections of Ca^{2+}, typically as calcium gluconate.

7.2 Fluorinated compounds in the environment [1]

As discussed above (cf. Section 1.1), fluorine is all around us, the so-called "fluorosphere" being also strongly influenced by anthropogenic factors. Herein, the impact of fluorinated compounds on the environment will be discussed in a little more detail. In the earth's crust, the naturally abundant fluorine is typically bound in poorly soluble fluoride minerals. However, the industrial use of fluorinated compounds increases the amount of readily available fluorine in nature by several different pathways (cf. Figure 1.1). The most important source of anthropogenic fluoride has certainly been caused by the extensive use of volatile fluoroorganic compounds as refrigerants, etc. As these compounds accumulate in the atmosphere, they are not directly threatening the health of humans or animals. Nevertheless, they cause severe effects, which will be discussed below. Especially the use of fluorinated agrochemicals increased the amount of artificial fluorinated compounds in soil and water. This is problematic, because still little is known about decomposition or metabolism of most of the compounds. However, in most cases it can be excluded, that the degradation to HF plays a significant role [43]. Apart from that, it is suggested, that CF_3-containing aromatic compounds are degraded to trifluoroacetic acid CF_3COOH and its derivatives. Generally, fluoroaromatics are more readily decomposed than saturated fluorocarbons, which are persistent in many cases (see below). An important issue is the disposal of waste from fluorine-related industries. Clearly, it has to be avoided, that any fluorocompound is released to the environment. This especially holds for volatiles and water-soluble sources of fluoride ions. As discussed above in Section 7.1, increased uptake of F^-, for example, by contaminated water sources will lead to health problems. Sophisticated methods for the defluoridation of water have been developed; however, their use is rather expensive [44–46]. Another very problematic group of pollutants are compounds with long perfluoroalkyl chains, especially perfluorooctanoic acid $C_7F_{15}COOH$ (PFOA) and perfluorooctyl sulphonate $C_8F_{17}SO_3^-$ (PFOS), which are persistent in the environment. The use of PFOA and PFOS has been strongly restricted since 2020, in former times they have been widely used as water-, oil- and dirt-proof impregnation for textiles and, especially, in fire-extinguishing foams. It is proven, that those compounds accumulate in the environment as well as in the human body, where they are toxic to the liver and, furthermore, cause fertility problems. It has also been shown, that this class of compounds is not removed by wastewater treatment, so that it remains in the drinking water and can further accumulate in human bodies. Recently, charged and partly fluorinated hydrogels have been shown to be promising candidates for the uptake of perfluoroalkylated contaminants from water [47]. In the same way, ionic fluorogels are capable of removing this class of compounds from the environment [48].

As already indicated above, volatile fluorocompounds also cause environmental issues. The first group of industrially used compounds have been chlorofluorocarbons (CFCs), which played an important role as refrigerants and blowing agents. Their use has been banned by the Montreal Protocol in 1987, as they significantly contribute to the destruction of atmospheric ozone, due to the formation of chlorine radicals via photolysis. For a short time, CFCs have been replaced by hydrochlorofluorocarbons (HCFCs);, however, they still contain chlorine, and thus have an ozone-depletion potential, so that they have been replaced by hydrofluorocarbons (HFCs). HFCs do not contain chlorine, and hence do not destroy the atmospheric ozone; nevertheless, they still act as greenhouse gases and have a global warming potential (GWP, Table 7.1). For this reason, research is ongoing to reduce the application and especially the liberation of long-lived compounds with a high GWP. Nevertheless, suitable and effective replacements have to be found before currently used HFCs can be completely banned [49]. Special attention is given to SF_6, which is used as electrical insulator in switch gear and as blanket gas in magnesium smelting. By far, it has the largest GWP on this list of fluorinated compounds. However, to date, there are no adequate replacements, so that the liberation of SF_6 may be reduced and regulated, but unfortunately not completely avoided yet. A potential replacement, SF_5CF_3, is not significantly less problematic in terms of GWP. Its chemical behavior has been investigated, concluding that it is most likely not degraded by UV-photolysis in the stratosphere, thus having a long atmospheric life time [50]. However, even if their GWP may be strongly increased compared to CO_2, the overall contribution of fluorine-containing compounds to the total GWP is still rather limited. This is due to the relatively small amounts of fluorinated molecules compared to the huge amount of all other anthropogenic emissions, such as CO_2 or CH_4. Nevertheless, a responsible dealing with volatile fluorine compounds is strongly required [1]. Since the use of CFCs and HCFCs is banned or at least strongly regulated, some hundreds of thousands of tons of waste have to be dealt with. Attempts have been made to degrade them to either synthetically useful chemicals or at least nonhazardous compounds. This is generally achieved by treatment of the compounds with basic reagents like NaOH or $NaOCH_3$ in alcoholic solutions at room temperature and atmospheric pressure under UV irradiation. Doing so, some

Table 7.1: Global warming potentials and atmospheric residence times for selected compounds.

Compound	Global warming potential	Residence time in years
CO_2	1 (per definition)	
CH_4	23	10
CF_3H	12,000–13,000	243
$H_2C=CF_2$	550–710	5.6
$CHF_2\text{-}CF_3$	3,400–3,600	32.6
$CH_2F\text{-}CF_3$	1300	13.6
$CF_3\text{-}CH_2\text{-}CF_3$	6,300–8,400	226
CF_4	5,700	50,000
$CF_3\text{-}CF_3$	9,200	10,000
$CF_3\text{-}(CF_2)_4\text{-}CF_3$	9,000	3,200
CCl_3F	4,500	45
CCl_2F_2	10,600	100
NF_3	10,800	740
SF_6	23,900	3,200
SF_5CF_3	18,000	1,000

valuable compounds have been prepared, such as the difluoromethylether CH_3OCHF_2, which is considered as potential replacement for some more problematic refrigerants [51].

Many of the HCFCs and HFCs contain trifluoromethyl groups, which are not degraded in the environment. Compounds containing those CF_3-substituents are finally transformed to trifluoroacetic acid (TFA, CF_3COOH), which is rained out. In nature, it is not known to be highly toxic to animals or plants, however, it will certainly not be beneficial to any organism. TFA is not metabolized by humans and has a half-life in the body of about 16 h. Accordingly, the pyrolysis of fluoropolymers as well as the metabolism of CF_3-containing drugs will likely lead to the formation of TFA.

The probably least hazardous fluorine-containing waste is resulting from products, which are made from fluorinated polymers. Under standard conditions, these compounds are stable against degradation to toxic and/or volatile species, however, they still need to be properly disposed in landfills or incinerated. A few years ago, an "up-cycling" process has been established for fluoropolymer waste treatment in the European Union. This method allows the chemical recycling of fully fluorinated polymers by converting them back into monomers, using pyrolysis techniques. The thermal degradation products may then be purified and reused for polymerization. Overall, "up-cycling" is a sustainable approach, which helps to save resources and additionally reduces waste by converting it into valuable raw-materials [52].

7.3 Perspectives in the use of fluorinated compounds [1]

Despite the two previous chapters, which may give the strong intention that fluorinated compounds are rather a significant burden for human health and environment, there are some very beneficial aspects about fluorine chemistry. Indeed, there are several developments, which will prospectively lead to cleaner synthetic processes, and thus less environmental pollution and accompanying health issues. An important contribution is represented by fluorous-phase chemistry, which is based on the property that perfluoroalkanes do not readily mix with either aqueous or organic solvents. For example, a biphasic mixture at room temperature might contain the substrate in the organic phase and the catalyst in the fluorous phase. Upon heating and stirring, the phases mix, so that the substrate is in contact with the catalyst. After completion of the reaction, cooling leads to a phase separation again, where the catalyst is recovered in the fluorous phase, whereas the product is easily isolated from the organic phase. Since the solubility of molecular oxygen in the fluorous phase is quite high (e. g., 52.1 ml O_2 in 100 ml C_8F_{18}), these systems are especially suitable for oxidation reactions. As a consequence, the use of metal-based oxidants, typically $KMnO_4$ or $K_2Cr_2O_7$, can be excluded, which is beneficial for both, economic and ecological reasons. Furthermore, fluorous compounds enable the use of supercritical CO_2 as a clean reaction medium. In this case, the fluorinated molecules take the role of cosolvents, surfactants or ligands.

Fluorinated compounds have also advanced the production of several end-products, which is due to the very special properties of the fluorine atom, namely its small size and its outstanding electronegativity. Based on this, it has been possible to produce molecules with dielectric properties that have found application in liquid-crystal engineering. More specifically, this has given rise to improved visual display units, especially by the development of TFT (thin film transistor) flat screens. As a consequence, a miniaturization has been achieved, saving a lot of resources in manufacturing, packaging and transporting the new technology screens. Ongoing studies have also shown that recovery of the fluorinated liquid crystal molecules is possible.

Finally, the introduction of fluorine substituents and fluorinated groups is important for the optimization of the properties of pharmaceuticals and agrochemicals. Again, the beneficial effects mainly go back to the small size and high electronegativity of the single fluorine atom. In addition, a low hydrophobicity and high lipophilicity is achieved, if groups like $-CF_3$, $-CF_2H$, $-OCF_3$, $-SCF_3$ or $-SF_5$ are attached to aromatic scaffolds. At first sight, it may be contradictive that just the use of fluorinated drugs is economically and ecologically advantageous over the use of non-fluorinated molecules. However, the fluorine-containing variants are in most cases more stable against metabolism (blocking of labile C-H sites), more readily available (improved membrane permeability) and bind more effectively to the target site. Overall, this gives rise to far more efficient drugs, leading to reduced doses for the patient. Consequently, less resources and energy are required for the preparation along with a reduced amount of generated waste upon production.

Overall, one can conclude, that the use of fluorinated compounds is nothing that should be demonized. In contrast, there are many beneficial properties of fluorine-containing molecules, which can certainly be further exploited in the design of new, efficient drugs and highly functional materials. However, the release of highly persistent and nondegradable species to the environment has to be strictly avoided. This is preferentially achieved by proper recycling processes.

7.4 Study questions

(I)   Why does poisoning with F_2 hardly occur? What is the main hazard of elemental fluorine?

(II)   What is typically the lethal dose of fluoride for an adult human? Why is HF so dangerous?

(III)   Comment on the toxicity of fluoroacetate and perfluoroisobutene. Why is the latter far more dangerous?

(IV)   Give examples for persistent fluorinated compounds in the environment. What have they been applied for?

(V)   What is the Montreal Protocol?

(VI)   Despite not containing chlorine substituents, hydrofluorocarbons still have negative effects on the atmosphere. Why? Which measures are taken to reduce their impact on the atmosphere?

(VII)   What is "up-cycling"?

(VIII)   Which property is exploited in fluorous chemistry? Which major benefits has fluorous chemistry?

8 ^{19}F-NMR – an important analytical tool for fluorine chemists [53, 54]

In the following section, some aspects of ^{19}F-NMR spectroscopy will be discussed, with a strong focus on chemical shifts and coupling constants of fluorine compounds in solution. However, this does not include a general introduction to physical and experimental basics of NMR-spectroscopy. Furthermore, advanced topics, such as solid-state NMR-spectroscopy, will not be subject of this manuscript.

8.1 General aspects and reference substances

Fluorine belongs to the monoisotopic elements, the only naturally occurring nuclide being ^{19}F. This is very beneficial for analytical purposes, as this isotope has a nuclear spin of I = 1/2 and can be compared to the proton in terms of its sensitivity (0.83 (^{19}F) versus 1 (^{1}H)). The gyromagnetic ratio of ^{19}F amounts to 0.94 of that of ^{1}H. As a consequence, it is very convenient to use ^{19}F-NMR as a fast, analytical technique, for example, for controlling reaction progress, unless a fluorinated solvent is used. Of course, it is also suitable for standard analytics, including two-dimensional NMR-techniques for nearly every element-fluorine derivative. For fluorine nuclei, the dispersion is far bigger than for protons, as the chemical shifts for organic, inorganic and organometallic fluorine compounds range over 1300 ppm. Thus, the chemical shift in the ^{19}F-NMR spectra can already give quite a lot of information about the investigated compound.

Nowadays, the standard substance defining the zero-point is $CFCl_3$, or more precisely, the isotopomer $CF^{35}Cl_2{}^{37}Cl$, however, in older literature, other substances have been used for this purpose, too. Furthermore, ^{19}F-NMR spectra are commonly referenced to a few more compounds with precisely known chemical shift (Table 8.1). It has been shown, that internal referencing is the most reliable method to obtain correct data [55]. As can be seen in Table 8.1, the chemical shift is also influenced by the solvent, apart from that, the temperature plays a role.

Table 8.1: Common reference substances for ^{19}F-NMR spectroscopy in selected solvents.

Substance	Chemical shift [ppm] in solvent			
	$CDCl_3$	C_6D_6	Acetone-d_6	CD_3CN
$CFCl_3$	0.65	−0.19	−1.09	−1.14
C_6H_5-F	−112.96	−113.11	−114.72	−114.81
C_6H_5-CF_3	−62.61	−62.74	−63.22	−63.10
C_6F_6	−161.64	−163.16	−164.67	−164.38
CF_3COOH	−75.39	−75.87	−76.78	−76.72

https://doi.org/10.1515/9783110659337-009

8.2 Chemical shifts

Certainly, the chemical shift δ is the most important information that can be gained by ^{19}F-NMR-spectroscopy. It can be either positive or negative, the latter indicating higher shielding and a translation to lower frequencies. In older literature, the scale is sometimes inverted, so that higher shielded nuclei appear at positive shift values. Some general trends for the dependence of chemical shifts have also been established. For binary fluorides, the resonance frequency, and thus the chemical shift, is increasing with the electronegativity, for example, $\delta(SbF_5) < \delta(AsF_5)$ and $\delta(TeF_6) < \delta(SeF_6)$. The same observation holds for fluoroorganic compounds, as for example, $\delta((CF_3)_2S)$ $< \delta((CF_3)_2O)$ and $\delta(CF_3CH_2Br) < \delta(CF_3CH_2Cl)$. In poly and perfluoroalkanes, a behavior opposite to that of nonfluorinated alkanes is found for nuclei in primary, secondary and tertiary positions, being $\delta(CF) < \delta(CF_2) < \delta(CF_3)$. Furthermore, the structure of the molecule influences the chemical shift of the ^{19}F-nuclei, which is especially observed for fluorinated olefins. For compounds containing *trans*-CF$_3$-groups, the resonances are shifted to higher field with respect to *cis*-CF$_3$-containing moieties. In aromatic systems, fluorine nuclei are generally more shielded than those found in perfluoroethylene derivatives. However, the chemical shift again depends on the degree of substitution, as $\delta(C_6F_6) < \delta(1,2\text{-}C_6H_4F_2) < \delta(C_6H_5F)$. A variety of chemical shifts of typical inorganic as well as organic fluorinated compounds is presented in Table 8.2 and Table 8.3, respectively [53, 54, 56]. Nevertheless, these values just represent a general trend and depend, of course, more or less on the abovementioned parameters, that is, solvent and temperature. In some fluxional molecules, for example, SF$_4$, the axial and equatorial F-atoms can be differentiated, if the ^{19}F-NMR spectrum is recorded at sufficiently low temperature, whereas a single resonance is usually observed at room temperature. If available, the data for the low-temperature resonances are represented for nonrigid species in Table 8.2.

A selection of fluoroorganic compounds is listed in Table 8.3. As can be seen, the range of their ^{19}F-NMR shifts is not that large (about 300 ppm), compared to the inorganic derivatives in Table 8.2 (about 1300 ppm). Furthermore, for most of the typical organic compound classes, a rather narrow region is observed (see Table 8.4). However, fluorides of the type R$_3$CF cover almost the complete range, which is due to the strong dependence on the nature of the substituents R. If they are halides, the fluorine is only poorly shielded and the chemical shift appears in rather high-field regions of +10 to about −30 ppm. The situation is completely opposite with increasing number of protons, as $\delta(^{19}F)$ is then moved to strongly negative values, the extreme example being represented by methyl fluoride CH$_3$F.

The total number of fluoroorganic compounds is steadily increasing. A list of chemical shifts for a few basic moieties as represented in Table 8.3 might be useful, however, it is impossible and not expedient to list every known compound. More conveniently, general shift regions can be derived (Table 8.4). As already stated above, chemical shifts of fluorides of the type R$_3$CF are found in almost the complete spectrum

Table 8.2: Chemical shifts of selected inorganic fluorine compounds ($\delta(CFCl_3) = 0$ ppm).

Compound	δ [ppm]	Compound	δ [ppm]
F_2O_2	+865	TeF_6	−57
$XeOF_4$	+555	CF_4	−63
XeF_6	+550	$[PF_6]^-$	−65
XeF_4	+438	AsF_5	−66
F_2	+423	$[AsF_6]^-$	−70
ClF_5	+412; +247	PF_5	−76
ReF_7	+345	$SiFBr_3$	−77
ClO_3F	+287	$C_6H_5-PF_2$	−92
XeF_2	+258	CH_3-PF_2	−93
F_2O	+249	SiF_2Br_2	−95
ClO_4F	+226	SbF_5	−108
IF_7	+170	$[SbF_6]^-$	−109
WF_6	+166	SiF_3H	−109
NF_3	+145	SiF_3Br	−125
BrF_5	+142; −43	KF	−125
SF_5Cl	+125; +62	Si_2F_6	−126
SF_4	+118; +71	WF_5Cl	−126; −182
ClF_3	+116; −4	$[SiF_6]^{2-}$	−127
F_2CS	+108	TeF_4	−128
$ReOF_5$	+102; −6	BF_3	−131
SOF_2	+75	F_3SiCH_3	−135
$C_6H_5-SO_2F$	+66	F_2HSiCH_3	−138
SeF_4	+64	$[BF_4]^-$	−150
IF_5	+60; +20	SiF_2H_2	−151
SeF_6	+55	$[SnF_6]^{2-}$	−155
SF_6	+50	FCN	−156
$OsOF_5$	+50	$(SiF_3)_2O$	−159
HF	+40	SiF_4	−164
BrF_3	−23	GeF_4	−180
PF_3	−35	$SiFH_3$	−217
AsF_3	−40	MoF_6	−278
SbF_3	−53	ClF	−448

of fluoroorganic compounds. The situation is still somewhat comparable for R_2CF_2, where the shift is again strongly influenced by the nature of the substituents R. For other halide substituents, $\delta(^{19}F)$ will be about +10 to −10 ppm, in perfluoroalkyl chains it is rather found at −100 to −140 ppm. The most negative shift is again observed for the proton-substituted CF_2H_2. Another very important compound class of fluoroorganic compounds is represented by trifluoromethyl derivatives. Their fluorine nuclei generally appear in a rather narrow region between −30 and −90 ppm. The actual value

Table 8.3: Chemical shifts of selected organic fluorine compounds (δ(CFCl$_3$) = 0 ppm).

Compound	δ [ppm]	Compound	δ [ppm]
CFBr$_3$	+7	(E)-BrFC=CBrF	−113
CF$_2$Br$_2$	+7	para-C$_6$H$_4$F$_2$	−113
CF$_2$Cl$_2$	−8	H$_2$C=CFH	−114
CF$_3$Br	−21	C$_6$H$_5$F	−116
CF$_3$Cl	−33	2-Fluoronaphthalene	−117
N(CF$_3$)$_3$	−59	(E)-ClFC=CClF	−120
C(CF$_3$)$_4$	−63	1-Fluoronaphthalene	−123
C$_6$H$_5$-CF$_3$	−64	cyclo-C$_5$F$_{10}$	−133
CFCl$_2$CFCl$_2$	−67	cyclo-C$_6$F$_{12}$	−133
CF$_3$C(O)CH$_3$	−74	cyclo-C$_4$F$_8$	−135
CF$_3$COOH	−78	F$_2$C=CF$_2$	−135
CF$_3$H	−79	ortho-C$_6$H$_4$F$_2$	−139
H$_2$C=CF$_2$	−81	CF$_2$H$_2$	−144
CF$_3$CCl$_3$	−82	cyclo-C$_3$F$_6$	−151
CF$_3$C(O)CF$_3$	−85	C$_6$F$_6$	−163
CF$_3$C(OH)$_2$CF$_3$	−93	(Z)-HFC=CFH	−165
FC≡CF	−95	(E)-HFC=CFH	−183
(Z)-BrFC=CBrF	−95	C$_6$H$_5$-CH$_2$F	−207
meta-C$_6$H$_4$F$_2$	−104	CFH$_3$	−272
(Z)-ClFC=CClF	−105		

for δ(^{19}F) depends on the scaffold CF$_3$ is attached to. If it is directly bond to hetero atoms (Se, S, P, N, O), the chemical shift is moved to less-negative values of about −15 to −85 ppm. Aromatic CF$_3$-groups typically appear around −65 ppm, whereas those in aliphatic compounds are found between −75 and −90 ppm. In fluoroolefins, the chemical shift of vinylic fluorine nuclei is again quite dependent on the nature of the other substituents, and, of course, on the structure of the molecule. Similar to saturated compounds, the substitution with protons generally moves the chemical shift to more negative values (compare C$_2$Br$_2$F$_2$ and C$_2$Cl$_2$F$_2$ with C$_2$H$_2$F$_2$ in Table 8.3). Furthermore, (E)-isomers are found at slightly more negative values than the respective (Z)-isomers. Carbonyl- and thiocarbonyl fluorides are basically the only fluoroorganic compound classes, where the chemical shift appears in the positive region. The chemical shift of fluoroarenes depends on the number of fluorine substituents, typically an increasing amount of F leads to decreasing δ(^{19}F). Their values range from about −115 ppm for monofluorinated aryl moieties to −180 ppm for pentafluorophenyl substituents. The same overall region holds for fluorinated pyridines and polyaromatics; however, in some cases the upper limit is exceeded to about −65 ppm. Nonaromatic heterocyclic compounds are usually found between −60 and −180 ppm in the ^{19}F-NMR spectrum.

Table 8.4: Typical shift regions for selected fluoroorganic compound classes ($\delta(CFCl_3) = 0$ ppm).

Compound class	δ [ppm]
R_3CF	−270 to +10
R_2CF_2	−145 to +10
RCF_3	−90 to −30
-OCF_3	−75 to −55
-SCF_3	−70 to −40
-NCF_3	−85 to −50
-PCF_3	−80 to −45
-$SeCF_3$	−40 to −15
vinylic F in olefins	−190 to −65
-C(O)F	−40 to +50
-C(S)F	+120 to +160
derivatives of C_6H_5F	−140 to −90
derivatives of $C_6H_4F_2$	−175 to −90
derivatives of $C_6H_3F_3$	−175 to −95
derivatives of $C_6H_2F_4$	−180 to −110
R-C_6F_5	−180 to −120
pyridines	−175 to −65
polyaromatics	−170 to −100
nonaromatic O-heterocycles	−150 to −60
nonaromatic N-heterocycles	−180 to −65

8.3 Coupling constants

The chemical shift is not the only information that can be gained by ^{19}F-NMR spectroscopy. The other important detail that allows further characterization of fluorinated compounds is given by coupling constants. Again, like the chemical shift, the coupling constants are dependent on solvent and temperature. Some general rules can also be applied in terms of their ranges. In binary fluorides, it has been shown that the one-bond coupling constants $^1J(X\text{-}^{19}F)$ generally increase with the atomic number of X; for late transition metals like platinum, it can have a value of several thousands of Hz. For one-bond couplings with ^{13}C, typically a coupling constant around 300 Hz is observed, whereas those with ^{31}P are settled in a region between 500–1500 Hz. In HF, which represents one-bond coupling with a proton, $^1J(^1H\text{-}^{19}F)$ amounts to 530 Hz.

The geminal coupling constants $^2J(X\text{-}E\text{-}^{19}F)$ do not only depend on the electronegativity of E; furthermore, the X-E-F bond angle and of course the stereochemistry of the compound play a role. In most of the cases, the atom E will be carbon. The coupling constants $^2J(^1H\text{-}E\text{-}^{19}F)$ are typically found in a range between 50–80 Hz, whereas the respective values for $^2J(^{13}C\text{-}E\text{-}^{19}F)$ are observed to amount to 8–15 Hz. For geminal couplings, the coupling constants $^2J(^{19}F\text{-}E\text{-}^{19}F)$ largely depend on the hybridization of X and the electronegativity of the substituents attached to it. If E is a sp^2-hybridized carbon, $^2J(^{19}$F-C-^{19}F) typically ranges from < 5 to about 110 Hz, whereas it is significantly larger in saturated carbon centres, showing values of 100–350 Hz. In vicinal couplings, the coupling constants $^3J(X$-E-E-^{19}F) are dependent on the electronegativity of the substituents as well as on bonding and dihedral angles. In cyclic organic molecules (E = carbon), the coupling constant between protons and fluorine $^3J(^1$H,^{19}F) depend almost in the same way on the dihedral angle as the respective proton-proton systems $^3J(^1$H,^{1}H). The maximal coupling constants are approached for dihedral angles of 0 and 180°, whereas minimums are found for 90°. Apart from that, another general trend is found for olefinic compounds, where $^3J(^1$H,^{19}F)$_{trans}$ is always larger than $^3J(^1$H,^{19}F)$_{cis}$. As a representative example for some of the aforementioned types of coupling, trifluoroethylene is depicted in Figure 8.1, showing its ^2J and $^3J(^{19}$F,^{19}F) coupling constants as well as the chemical shifts of the fluorine nuclei [57]. Long range couplings are also widely observed; their coupling constants strongly depend on structure and stereochemistry of the compound. Thus, it is hardly possible to summarize general trends for this type of interaction.

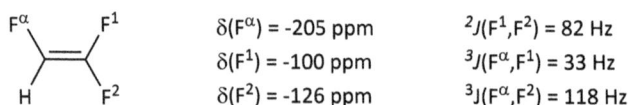

$\delta(F^\alpha) = -205$ ppm $^2J(F^1,F^2) = 82$ Hz
$\delta(F^1) = -100$ ppm $^3J(F^\alpha,F^1) = 33$ Hz
$\delta(F^2) = -126$ ppm $^3J(F^\alpha,F^2) = 118$ Hz

Figure 8.1: ^{19}F-chemical shifts and ^{19}F-^{19}F coupling-constants in trifluoroethylene.

In conclusion, ^{19}F-NMR spectroscopy is a versatile analytical method to characterize compounds, which may be small molecules, transition metal complexes, polymers or even proteins. The high natural abundance and sensitivity of ^{19}F further allow the online analysis of reaction mixtures to screen reaction progress. Often, relatively large changes in the chemical shift are observed upon chemical transformations, so that the product formation is unambiguous. Finally, ^{19}F-NMR can serve as a tool to investigate other properties, such as exchange processes in fluxional molecules like SF_4, but also in more complex organic derivatives.

8.4 Study questions

(I)   Why is ^{19}F-NMR spectroscopy an important analytical tool? What are its main benefits?
(II)   Comment on reference substances and the most important factors that chemical shifts of fluorinated compounds may depend on.
(III) How does the electronegativity influence the chemical shift of ^{19}F-nuclei?
(IV)   How does the ^{19}F-NMR spectrum of SF_4 look at room temperature? Is the same spectrum observed at low temperature? Explain this behavior.
(V)   Why are the chemical shifts of compounds R_3CF and R_2CF_2 rather disperse? Comment on the influence of substituent R.

(VI)   How does the chemical shift of fluoroarenes generally depend on the number and substitution pattern of fluorine substituents?

(VII)   How can coupling constants help to characterize fluorinated compounds? Give some examples for structural motifs that may be identified.

Part B: **Inorganic fluorine chemistry**

The following part covers the chemistry of inorganic fluorine compounds. This includes the description of synthetic approaches toward binary element-fluorine compounds, oxyfluorides as well as some further relevant derivatives, which are partly composed of multiple elements. Apart from the synthetic access, the most important properties of the compounds are highlighted.

https://doi.org/10.1515/9783110659337-010

9 Main group element fluorides

9.1 Group 1: Lithium, sodium, potassium, rubidium, cesium

9.1.1 The alkali metal fluorides

The alkali metal fluorides **LiF**, **NaF**, **KF**, **RbF** and **CsF** are formed upon treatment of the respective metal hydroxides or carbonates with aqueous hydrofluoric acid and exist exclusively as monofluorides. They are all colorless, ionic solids, crystallizing in the cubic NaCl type with high lattice-energy, however, the latter significantly drops from LiF to CsF (Table 9.1). Due to the size of the Cs^+ cation, the structure of CsF is described by an inverted NaCl-lattice.

Table 9.1: Physical properties of the alkali metal fluorides.

	LiF	NaF	KF	RbF	CsF
melting point [°C]	848	995	858	795	703
boiling point [°C]	1676	1704	1505	1410	1251
structure type	NaCl	NaCl	NaCl	NaCl	inv. NaCl
lattice energy [kJ mol^{-1}]	1025	903	800	767	715
d(MF) [pm]	191	231	267	282	301

LiF is nonhygroscopic, only poorly soluble in water (0.148 % at 20 °C) and does not form isolable hydrates. This behavior changes for the heavier homologues KF, RbF and CsF, which are strongly hygroscopic. Consequently, the sensitivity toward hydrolysis decreases from LiF to CsF. At elevated temperatures, NaF is readily hydrolyzed by humidity to form NaOH and HF. Aqueous solutions of KF also have a slightly basic character. To obtain a reasonable solubility of alkali metal fluorides in organic solvents, crown ethers or cryptands are commonly used [58].

In the gas-phase, the alkali metal fluorides are molecular monomers and dimers, in case of LiF and NaF trimeric or even tetrameric (NaF)-clusters exist as well. The dimers have also been investigated by vibrational spectroscopy in an Ar-matrix [59, 60]. A planar, rhombic structure with D_{2h}-symmetry could be assigned. When treated with fluoride ion acceptors, all the alkali metal fluorides form salts with complex fluoroanions. In the gas phase, LiF reacts with NaF to give LiNaF$_2$. In presence of HF, the alkali metal fluorides form adducts with different stoichiometry. For example, the sodium hydrogen fluorides NaF · nHF (n = 1–4) are known and structurally characterized [61].

Alkali metal fluorides are used as additives in industrial melting processes, as for example, LiF is used as additive to Al$_2$O$_3$-Na$_3$[AlF$_6$] melts in Al-production. Apart from that, LiF is also an important part of eutectic mixtures or fluxes for soldering

https://doi.org/10.1515/9783110659337-011

low-weight metals. In the same way, NaF and KF are applied as additives in melting processes. Furthermore, applications as preservatives for wood and glues have been described. The alkali metal fluorides, especially NaF and KF are commonly used in laboratories as absorbing material for HF. Additionally, alkali metal fluorides are used as fluorinating agents, especially in organic synthesis (cf. Part C). This holds mainly for the readily available sodium and potassium compounds. Even if LiF is not directly involved, fluorinated compounds also play important roles in high-performance Li-batteries. For example, they may be used as electrolyte salts, solvents, cosolvents or additives [62].

9.1.2 Study questions

(I) How are the alkali metal fluorides prepared?
(II) Describe the solid-state structures of the alkali metal fluorides.
(III) Comment on the solubility of the alkali metal fluorides. Are they stable toward hydrolysis?

9.2 Group 2: Beryllium, magnesium, calcium, strontium, barium, radium

The alkaline earth metals exclusively form difluorides. In general, the chemical properties of the heavier homologues MgF_2, CaF_2, SrF_2 and BaF_2 are comparable, as they are all high-melting compounds (Table 9.2), which do not show a rich chemistry. In contrast, BeF_2 is capable of forming many structurally interesting derivatives, especially fluoroberyllates.

Table 9.2: Physical properties of the alkaline earth metal fluorides.

	BeF_2	MgF_2	CaF_2	SrF_2	BaF_2
melting point [°C]	552	1263	1418	1477	1368
boiling point [°C]	1283	2239	>2500	2489	2137
structure type	SiO_2	TiO_2	CaF_2	CaF_2	CaF_2

9.2.1 Beryllium

Beryllium fluoride **BeF$_2$** is formed upon treatment of BeO with anhydrous HF at 220 °C, or more conveniently by thermal decomposition of $(NH_4)_2[BeF_4]$ at 900–1100 °C in a CO_2-stream [63]. The tetrafluoroberyllate salt is prepared by dissolving beryllium or beryllium hydroxide in an aqueous solution of ammonium bifluoride $NH_4[HF_2]$ and subsequent evaporation of the solvent. BeF_2 has some structural relationship to SiO_2. As a consequence, BeF_2 does not conduct electricity very well and solidifies, like SiO_2, as a transient glass with a structure similar to quartz. In solid phases, $[BeF_4]$-tetrahedra are present, however there are several temperature-dependent phase transitions in crystalline BeF_2, comparable to the situation in SiO_2 [64, 65]. In the gas phase, BeF_2 forms linear monomers, dimeric structures are also known and have been identified in matrix experiments [66]. In contrast to SiO_2, beryllium fluoride is hygroscopic and well soluble in water. In aqueous solution it forms a dihydrate $[BeF_2(H_2O)_2]$ along with the hydrated ions $[Be(H_2O)_4]^{2+}$, $[BeF(H_2O)_3]^+$ and $[BeF_3(H_2O)]^-$ as well as $[BeF_4]^{2-}$ [67–69]. With an excess of HF in aqueous solution, H_2BeF_4 is formed, too [70]. BeF_2 cannot be isolated from aqueous solutions, as hydrolysis occurs before dehydration takes place. The alkaline hydrolysis with NaOH occurs in two steps. Initially, neutral BeF_2 is converted to anionic $[BeF_4]^{2-}$ and $Be(OH)_2$. The tetrafluoroberyllate anion further reacts with NaOH to yield NaF and again $Be(OH)_2$, which is overall the single hydrolysis product. BeF_2 reacts with liquid NH_3 to form the diammine adduct $[BeF_2(NH_3)_2]$, which consists of almost C_{2v}-symmetric molecules [71].

BeF$_2$ is only sparingly soluble in HF but it readily acts as fluoride ion acceptor to form fluoroberyllates of the type $[BeF_4]^{2-}$, $[BeF_3]^-$, $[BeF_5]^{3-}$, $[Be_2F_5]^-$ and $[Be_2F_7]^{3-}$. Similarly, the formation of $[NH_4(NH_3)][BeF_3(NH_3)]$ is observed upon treatment of BeF_2 with NH_4F in liquid ammonia [72]. The tetrafluoroberyllate anion $[BeF_4]^{2-}$ is directly accessible by treatment of Be-metal with aqueous solutions of HF or NH_4F [73]. Double salts containing the $[BeF_4]^{2-}$ anion are also known, for example, 2 $Na_2[BeF_4]$ · $Na_2[SiF_6]$ or 2 $Na_2[BeF_4]$ · $Na[HF_2]$ [74, 75]. Most of the fluoroberyllate anions have some isoelectronic relatives in the groups 4 to 6 (Table 9.3). Due to this, the fluoroberyllates can be considered as model compounds for silicates. The advantages of the beryllium compounds are their lower melting points. The fluoroberyllates are partly hydrolyzed in aqueous solution, if no excess of fluoride ions is present. The solubility of $[BeF_4]^{2-}$ salts is comparable to that of the respective sulphates SO_4^{2-}, which can be explained by the similar radii of the ions. As a result, $Ba[BeF_4]$ is used for the gravimetric assay of beryllium. The tetrafluoroberyllates are isotypic to the silicates, thus both $Na_2[BeF_4]$ and $Ca_2[SiO_4]$ exist in five different structural types [76, 77]. Furthermore, $Li_2[BeF_4]$ is isotypic with hexagonal $Be_2[SiO_4]$ [78]. Generally, the $[BeF_4]$-units have a tetrahedral structure and are mostly not isolated but relatively strongly associated to the cation via, for example, H- or F-bridging [79, 80]. In the molten state, $Na_2[BeF_4]$ dissociates to NaF and $Na[BeF_3]$; in vapors over $Na_2[BeF_4]$ NaF, BeF_2 and dimeric $[NaBeF_3]_2$ are found.

In the system LiF-BeF_2, only the tri- and tetrafluoroberyllates $Li[BeF_3]$ and $Li_2[BeF_4]$ have been unambiguously identified [81, 82]. The same species are observed if the binary fluorides are reacted under matrix conditions. The isolated products also contain dimeric $[LiBeF_3]_2$ [83]. The pentafluoroberyllates $M_3^I[BeF_5]$ (M^I = K, Rb, Cs) are isotypic with Sr_3SiO_5 [84]. The respective diberyllates $M^I[Be_2F_5]$ are only stable in the form of their Rb- and Cs-salts, for the other alkali metals (M^I = Li, Na, K),

Table 9.3: Isotypy of fluoroberyllates and silicates.

	Isoelectronic compounds	isotypy between	
$[BeF_4]^{2-}$	$[SiO_4]^{4-}$, $[PO_4]^{3-}$, $[SO_4]^{2-}$	$Li_2[BeF_4]$	$Be_2[SiO_4]$
		$Na_2[BeF_4]$	$Mg_2[SiO_4]$
		$Ca[BeF_4]$	$Zr[SiO_4]$
$[Be_2F_7]^{3-}$	$[Si_2O_7]^{6-}$, $[P_2O_7]^{4-}$, $[S_2O_7]^{2-}$	$Na_2[LiBe_2F_7]$	$Ca_2[MgSi_2O_7]$
$[BeF_3]^-$	$[SiO_3]^{2-}$, $[PO_3]^-$, SO_3	$Li[BeF_3]$	$Mg[SiO_3]$
		$Na[BeF_3]$	$Ca[SiO_3]$
$[Be_2F_5]^-$	$[Si_2O_5]^{2-}$, P_2O_5		

they decompose below their melting point. The trianion $[Be_2F_7]^{3-}$ consists of two $[BeF_4]$-tetrahedra, which are connected by a common corner. It is obtained, if a mixture of NaF, LiF and BeF_2 is molten in stoichiometric quantities [85]. It has been found that the $[BeF_3]^-$ anion can act as a phosphate analogue in phosphorylated proteins [86].

9.2.2 Magnesium, calcium, strontium, barium and radium

The alkaline earth fluorides **MgF_2**, **CaF_2**, **SrF_2** and **BaF_2** are, for example, formed upon treatment of the corresponding carbonates with HF. They are resistant against hydrolysis and their solubility in water increases in the following order: $CaF_2 < SrF_2 < MgF_2 < BaF_2$. The most important of these compounds is unambiguously the poorly soluble CaF_2, which can be found in significant amounts in nature in the mineral fluorspar (cf. Part A). In pure form, fluorspar is colorless, but it may also be colored due to impurities, for example, $CaCO_3$, $BaSO_4$, SiO_2, PbS or ZnS. The solubility of alkaline earth fluorides can be increased when they are treated with HF to form acidic salts, for example, $CaF_2 \cdot 2\,HF \cdot 6\,H_2O$.

MgF_2 crystallizes in the rutile-lattice, thus having the coordination numbers 6(Mg) and 3(F). The other alkaline earth fluorides CaF_2, SrF_2 and BaF_2 are structurally described by the fluorite-lattice with 8:4-coordination. In the gas phase, they exist as linear (MgF_2) or bent (CaF_2, SrF_2, BaF_2) monomers. The chemistry of radium fluoride **RaF_2** is limited due to its radioactivity. However, it is obtained as a white solid upon treatment of $RaCO_3$ with aqueous HF ($w = 10\,\%$). Structurally, RaF_2 is isotypic with BaF_2, thus it also crystallizes in a fluorite lattice [87]. Quantum-chemical calculations predict a significantly bent structure for molecular RaF_2 [88, 89].

Apart from its source for HF production, CaF_2 has some minor applications as flux in metallurgical processes and as blurring agent in enamel industry. Further, as a result of its good dispersion and UV-permeability, CaF_2 is used as material for lenses and prisms in spectrometers.

9.2.3 Study questions

(I) How is BeF_2 prepared? Is it possible to isolate it from aqueous solution?

(II) Describe the structure of BeF_2 in the solid-state and the gas-phase. To which compound may it be compared?

(III) Comment on the fluoride accepting properties of BeF_2. Is the tetrafluoroberyllate anion easily accessible?

(IV) How are the alkaline earth metal fluorides MF_2 (M = Mg-Ra) prepared? Are they sensitive toward hydrolysis?

(V) Describe the solid-state structures of the alkaline earth metal fluorides MF_2 (M = Mg-Ra). Which general structure type(s) do they represent?

9.3 Group 13: Boron, aluminium, gallium, indium, thallium

Again, the chemistry of the first element in the group differs significantly from that of the others. Boron forms a number of binary fluorides, which are all covalent in nature, whereas the heavier homologues mainly exist as ionic compounds, which in addition tend to form fluorometallate complexes.

9.3.1 Boron

Boron forms many fluorides, which in general can be divided into the two classes of boron subfluorides and their derivatives as well as boron trifluoride and its derivatives.

Subfluorides

Boron monofluoride **BF**, which is an isoelectronic compound to carbon monoxide CO, exists only at high temperatures and low pressure in its monomeric form. Its bonding distance amounts to 126.5 pm, which is a value between a typical B-F single- and a B-F double bond. BF is formed, when BF_3 is passed over granulated, elemental boron in a graphite tube at 2000 °C in vacuo (Scheme 9.1) [90]. If it is condensed with unreacted BF_3 in liquid N_2, it forms boron subfluorides of the general formula B_nF_{n+2}. The resulting mixture can be fractionated into B_2F_4, B_3F_5 and higher boron subfluorides, which can then be transformed to B_2F_4, B_3F_5 and B_4F_6 upon heating. Along with the aforementioned volatile compounds, the yellow, nonvolatile and polymeric $(BF)_n$ is formed (Scheme 9.1). The latter also results from the thermal decomposition of B_2F_4 [91]. If BF is condensed with B_2F_4, unstable and very reactive B_3F_5 is generated as main product [90].

Diboron tetrafluoride **B_2F_4** is a colorless gas (mp.: –56 °C; bp.: –34 °C). Beside the preparation from BF as described above, it is also conveniently synthesized by fluori-

B_2X_4 (X = Cl, Br)

halogen exchange | SF_4, SbF_3, $CFBr_3$

BF_3 $\xrightarrow{\text{B} \atop 2000\,°C}$ $[BF]$ $\xrightarrow{BF_3 \atop -196\,°C}$ B_nF_{n+2} $\xrightarrow{\text{fract.}}$ B_2F_4 + B_3F_5 + higher homologues

$\downarrow B_2F_4$

B_3F_5

$\downarrow$ -30 °C

B_2F_4 + $[B_4F_6]$ $\xrightarrow{2x}$ B_8F_{12}

rt $\downarrow$

$(BF)_n$ + BF_3

Δ $\downarrow$

B_2F_4
+
B_3F_5
+
$[B_4F_6]$
+
$(BF)_n$

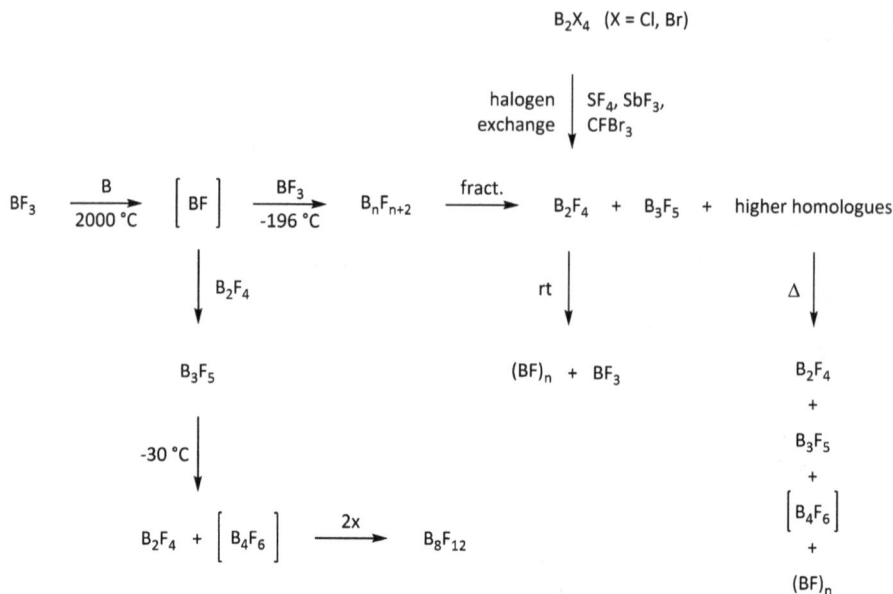

Scheme 9.1: Selected synthetic routes toward boron subfluorides.

nation of B_2Cl_4 or B_2Br_4 with SbF_3 or SF_4 (Scheme 9.1) [92–94]. Another approach is the fluorination of boron monoxide $(BO)_n$ or diboronic acid $(B(OH)_2)_2$ and its deriva-tives with SF_4 [95]. Additionally, B_2F_4 is generated in high yield by the reaction of B_2X_4 (X = Cl, Br) with $CFBr_3$ or B_2Cl_4 with $Hg(CF_3)_2$ [96]. B_2F_4 is perfectly planar in the crystalline phase (d(B-F): 132 pm; d(B-B): 167 pm), having D_{2h}-symmetry [97, 98]. As a liquid and in the gas-phase, the molecule may still have D_{2h}-symmetry, however, due to a very small rotational barrier (about 1.75 kJ mol^{-1}), B_2F_4 easily adapts a nonpla-nar D_{2d}-structure [99–102]. At room temperature, B_2F_4 slowly disproportionates to BF_3 and $(BF)_n$ [91, 103]. It is very reactive and reacts explosively with O_2 under formation of BF_3 and solid $(B_2O_3F)_n$. B_2F_4 slowly attacks glass, where SiF_4 and OF_2 are liberated [103]. It also reacts with Cl_2 to give BF_3 and BCl_3 and is hydrolyzed by water resulting in boric acid $B(OH)_3$, H_2 and HF. Halide exchange is observed, if B_2F_4 is treated with BCl_3 [104]. Moreover, B_2F_4 is a Lewis-acid and forms 1:1- and 1:2-complexes, for example, with NMe_3 [93, 105]. Furthermore, it undergoes addition reactions to olefins, resulting in the formation of saturated compounds containing BF_2-groups [106, 107]. Upon reac-tion with 1,3-butadiene, 1,4-diborylated 2-butene is obtained [108]. Some mechanistic details for these addition reactions to unsaturated compounds have been explored by quantum-chemical calculations [109]. Transition metal complexes with BF_2-ligands are also known. For example, the platinum compound cis-$[Pt(BF_2)_2(PPh_3)_2]$ is acces-sible by the reaction of $[Pt(PPh_3)_2(\eta\text{-}C_2H_4)]$ with B_2F_4 [110]. A few more Pt- and Ir-complexes with difluoroboryl ligands have also been prepared by oxidative addition of B_2F_4 [111, 112]. The treatment of B_2F_4 with the fluoride donor $[NBu_4][Ph_3SiF_2]$ gives rise to the anion $[B_2F_6]^{2-}$, which is isoelectronic to hexafluoroethane. Accordingly, the two $[BF_3]$-units of the dianion are in a staggered conformation [113].

Triboron pentafluoride **B_3F_5** is a colorless compound which already disproportionates at −30 °C, which results in the formation of B_2F_4 and B_4F_6 (Scheme 9.1) [92]. Tetraboron hexafluoride B_4F_6 does not exist in monomeric form and immediately dimerizes to give B_8F_{12}. By infrared and NMR spectra it could be shown that B_3F_5 has a chain-like structure. It reacts explosively with water or air and is reduced by H_2 at −70 °C to form HBF_2, B_2F_4 and diborane B_2H_6 [90].

Octaboron dodecafluoride **B_8F_{12}** is a yellow, volatile oil, which decomposes above −10 °C. In both, the solid state and the gas phase, its structure can be compared to diborane B_2H_6, where the hydrogen substituents have been replaced by BF_2-groups. Similarly, four of the six groups are connected in a terminal fashion to the remaining two boron atoms, whereas the other two BF_2-units adapt a bridging function. Thus, one could rewrite the formula of B_8F_{12} as $B_2(BF_2)_6$. However, the central $[B(BF_2)_2B]$-group is nonplanar, unlike the $[BH_2B]$-group in diborane (Figure 9.1) [92, 114, 115]. Like the other boron subfluorides, B_8F_{12} reacts explosively with air and water and in addition is reduced by H_2 at −80 °C to form BF_3, HBF_2 and a nonvolatile solid. Furthermore, it reacts with Lewis bases; whereas strong bases (NMe_3, $MeCN$, Et_2O) yield BF_3-complexes, soft bases (CO, PF_3, PCl_3, AsH_3, PH_3, Me_2S) symmetrically cleave B_8F_{12} and lead to complexes of the type $(BF_2)_3B$-base [92, 116]. This also reminds of borane-adducts, for example, $BH_3 \cdot Me_2S$. At −90 °C, it is also possible to generate the respective amine complex $(BF_2)_3B \cdot NMe_3$, however, it decomposes with BF_3-evolution upon warming to −23 °C [117].

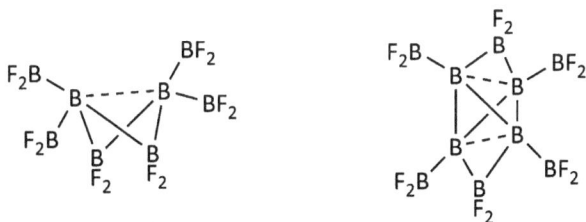

Figure 9.1: Schematic structures of the boron subfluorides B_8F_{12} (left) and $B_{10}F_{12}$ (right).

Decaboron dodecafluoride **$B_{10}F_{12}$** is obtained as colorless crystalline compound from the low-temperature condensate of BF. It consists of a distorted tetrahedral B_4-cage with a BF_2-group linked to each corner (Figure 9.1). The other two BF_2-units are bridging opposing edges of the tetrahedron [114]. The molecule has D_{2d}-symmetry. Again, one could reformulate $B_{10}F_{12}$ as $B_4(BF_2)_6$ and consider BF_2 as a replacement for H. But in this case, this reveals a big difference to the boranes, because *closo*-B_4H_6 is nonexisting.

Boron trifluoride and its derivatives

Boron trifluoride BF_3 is clearly the most important boron fluoride. It is a colorless gas with a suffocating smell (mp.: $-128.4\,°C$; bp.: $-99.9\,°C$), which is formed in many reactions of boron compounds with fluorinating agents. On technical scale, BF_3 is generated by the reaction of boron(III)-oxide B_2O_3, boric acid $B(OH)_3$ or borax $Na_2B_4O_7$ with CaF_2 and $H_2SO_4 \cdot xSO_3$. In the first step, HF is formed, which then fluorinates the boron compound taking advantage of the dehydrating effect of oleum. The formation of a complex with diethyl ether $BF_3 \cdot Et_2O$ (mp.: $-57.7\,°C$; bp.: $125\,°C$) allows a more convenient handling of the toxic and corrosive compound. The ether-adduct as well as a few other complexes are commercially available. Very pure BF_3 is generated when phenyl diazonium tetrafluoroborate $[Ph\text{-}N_2][BF_4]$ is thermolyzed. In contrast to borane BH_3, boron trifluoride exists in monomeric form and does not tend to oligomerize. Only in the solid state one can find some weak intermolecular interactions. Hence, boron trifluoride is a trigonal planar molecule with a B-F distance of only 130 pm, which is remarkably short for a B-F single bond. The reason for this is the overlap of the p_z-orbital of boron with one of the p-orbitals of fluorine, which results in p_π-p_π-interactions and hence, π-backdonation from fluorine to boron. As a consequence, one can write four resonance structures for BF_3 (Figure 9.2). It might look irritating, when fluorine, the most electronegative element, has a positive charge in the mesomeric structures. However, one should keep in mind that these are only formal charges which have nothing to do with real charges. In reality, the charge distribution is as expected and amounts -0.47 at each fluorine-atom and $+1.41$ at the boron center.

Figure 9.2: Resonance structures of BF_3.

Boron trifluoride is a strong Lewis acid and a very efficient catalyst in many organic reactions. There exist several hundreds of very stable BF_3 adducts with organic and inorganic compounds. In any adduct, the neighboring atoms of the donor atom are more positively polarized than in the free donor molecule. In the case of proton-active donors, for example, water, alcohols, ammonia or amines, the Brønsted acidity is raised upon addition to BF_3. The hydrate $BF_3 \cdot H_2O$ (mp.: $6.0\,°C$) may be isolated as a colorless, oily, unstable liquid. It is a strong acid that adds another water molecule to form the colorless and liquid $BF_3 \cdot 2\,H_2O$ (mp.: $6.2\,°C$). The 2:1-adduct is a molecular hydrate below its melting point; above, it is described as oxonium salt (Scheme 9.2) [118–121]. Apart from the 1:1-adduct $BF_3 \cdot NH_3$, a higher ammoniate with the composition $[BF_3(NH)_3] \cdot 3\,NH_3$ also exists. It forms a three-dimensional network with a large number of N-H- and H-F-bridges [122].

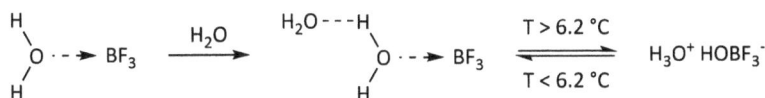

Scheme 9.2: Hydration of BF_3.

Furthermore, BF_3 can react under replacement of its fluorine substituents, as for example, the reaction with B_2O_3 lead to the formation of the cyclic trifluoroboroxine $B_3O_3F_3$, whereas Grignard reagents yield organoboranes. With other boron trihalides, BF_3 readily undergoes redistribution reactions, which is a result of its monomeric presence. However, it is not possible to isolate mixed boron halides, as they are in reversible equilibrium with each other [123]. BF_3 is sensitive toward hydrolysis, which results in the formation of boric acid $B(OH)_3$ and HF. The latter reacts with BF_3 and forms tetrafluoroboric acid HBF_4. As the bond energies of B-F and B-OH are comparable, the hydrolysis of BF_3 is an equilibrium process and can be inverted under dehydrating conditions.

BF_3 and its adducts are used on large scale as catalysts in organic synthesis. Further, it is applied as flux, smoking agent and for the preparation of other boron compounds. The isotope labeled $^{10}BF_3$ also has been applied in neutron counting devices.

Tetrafluoroboric acid **HBF_4** is a colorless, toxic and strong acid, which does not exist in pure form. It is highly corrosive but does not attack glass. From aqueous systems, it can only be isolated as oxonium salt $[H_3O][BF_4]$. In very diluted solution, it gets partly hydrolyzed to $H[BF_3(OH)]$ and to a very small extent to the further hydrolysis products $H[BF_2(OH)_2]$, $H[BF(OH)_3]$ and $H[B(OH)_4]$ [124]. The acidity of the hydrolyzed species decreases with every fluoride displacement. The hydrolysis can be reverted, when the HF-concentration is raised. HBF_4 is mainly used as a strong acid to catalyze organic reactions.

The tetrafluoroborate anion $[BF_4]^-$ is a very weak Lewis base. For this reason, $[BF_4]^-$ is often used as a weakly coordinating anion. It is possible to synthesize anhydrous tetrafluoroborate salts $M[BF_4]$ from concentrated HBF_4 solutions and metal oxides, hydroxides or carbonates. $K[BF_4]$ may also be generated by reacting KF with B_2O_3 in BrF_3 or by reaction of KCl and BF_3. In the series of the alkali metals, the stability of the $M[BF_4]$ salts increases with the size of the metal cation, for example, the sodium salt $Na[BF_4]$ decomposes at 384 °C, whereas $Cs[BF_4]$ melts at 550 °C without decomposition. Tetrafluoroborates show a comparable solubility to the corresponding perchlorates $M[ClO_4]$. Another similarity can be found in the solid-state structures. Generally, the $[BF_4]^-$ anion has a tetrahedral structure with a B-F distance of 142 pm. In the crystalline phase; this distance is often shortened, which gives a hint to partial covalent bonding character, for example, in $NH_4[BF_4]$ (140.6 pm) or $[SF_3][BF_4]$ (138.6 pm) [125]. When stoichiometric amounts of BF_3 are added to $[BF_4]^-$-solutions at very low temperature, the heptafluoro diborate anion $[B_2F_7]^-$ can be observed by NMR spectroscopy. Structurally, it consists of two $[BF_4]$-tetrahedra, which are bridged by a fluorine atom

[126, 127]. Apart from the use of $[BF_4]^-$ as a weakly-coordinating anion, salts with the tetrafluoroborate anion find application as flux in the galvanic metal deposition and as flame retardants.

Mixed fluorohydridoborate anions of the composition $[BH_xF_{4-x}]^-$ are also known. Out of this family, $[BHF_3]^-$ is available by deboronation of 1,7-dicarba-*closo*-dodecaborane with anhydrous $[NMe_4]F$. Surprisingly, the $[BHF_3]^-$ anion does not undergo dismutation to $[BH_4]^-$ and $[BF_4]^-$, even if these species are thermodynamically favored, as has been indicated by quantum-chemical calculations [128].

9.3.2 Aluminium, gallium, indium, thallium

The heavier elements of group 13 have a stronger metallic character than boron. Thus, their fluorides are more ionic. Moving from Al to Tl, the stability of the monofluorides increases, whereas the trifluorides are destabilized.

Monofluorides
Aluminium monofluoride **AlF** is formed as short-lived, monomeric molecule (d(Al-F): 165.5 pm), when aluminium trifluoride AlF_3 reacts with Al at about 1000 °C [129]. In condensed phase, it disproportionates back to AlF_3 and Al at room temperature. Gallium monofluoride **GaF** and indium monofluoride **InF** are formed upon reaction of the metals with AlF_3 or CaF_2 at 900–1200 °C [129–131]. They exist as mixtures of monomers and dimers in a noble gas matrix. The simultaneous condensation of GaF and O-atoms in an argon matrix leads to the formation of GaOF [132].

The most stable group 13 monofluoride is thallium fluoride **TlF** (mp.: 322 °C; bp.: 826 °C). It can be synthesized by treatment of aqueous solutions of Tl_2CO_3 with hydrofluoric acid. TlF is a colorless, crystalline solid, which is stable under aerobic conditions. It crystallizes in an orthorhombic lattice with a distorted octahedral coordination geometry at the Tl-centers [133]. At slightly elevated temperature as well as high pressure, TlF undergoes phase transitions [134]. Vibrational spectra of solid TlF indicate that the lattice also contains monomeric molecules [135]. In the gas phase and in an Ar-matrix one can also find $(TlF)_2$-dimers, which can be assumed to have a planar, rhombic structure based on vibrational spectra [136, 137]. The ionic radius of Tl^+ (154 pm) is comparable to K^+ (144 pm) and Rb^+ (158 pm) and not too different from Ag^+ (127 pm). Due to that the chemistry of TlF is similar to that of the alkali metal fluorides and AgF. It is well soluble in water as well as in anhydrous HF and forms several different solvates in the latter [27]. TlF reacts with halogens and the interhalogens BrCl, ICl and IBr to form mixed Tl(III) halides. There exist numerous fluorometallates, which are products of the reaction of fluoride acceptors with TlF, for example, $Tl[M^{II}F_3]$, $Tl_2[M^{II}F_4]$, $Tl[M^{III}F_4]$, $Tl_2[M^{III}F_5]$, $Tl_2[M^{IV}F_6]$ and $Tl[M^VF_6]$. Thermal treatment of equimolar mixtures of TlF and TlF_3 gives rise to the mixed valent phases

Tl_2F_3 (= $Tl_3[TlF_6]$), Tl_2F_4 (= $Tl[TlF_4]$), Tl_3F_5 (= $Tl_2[TlF_5]$) and Tl_3F_7 (= $Tl[Tl_2F_7]$), which have all been identified by X-ray crystallography [138, 139].

Trifluorides

Aluminium trifluoride **AlF_3** is generated, when HF is passed over Al or Al_2O_3 at red heat. It is insoluble in water, acids and alkaline solutions. The colorless, crystalline solid contains octahedrally coordinated aluminium atoms in a distorted ReO_3-lattice. Fluorine is almost linearly coordinated by two Al-atoms. In the vapor phase, an equilibrium between monomeric AlF_3 and dimeric $(AlF_3)_2$ exists. The AlF_3 molecule has a trigonal planar geometry as could be proven by vibrational spectroscopy [140]. Gaseous AlF_3 is stable against HF below 330 °C, but in the condensed phase it can be dissolved in aqueous HF to give the trihydrate $AlF_3 \cdot 3\ H_2O$. AlF_3 forms several hydrates, out of which the mono, tri and nonahydrate are very well known. The nonahydrate $AlF_3 \cdot 9\ H_2O$ is soluble in water, but unstable above 8 °C. The trihydrate exists in two different forms, the monohydrate is formed at elevated temperatures of about 200 °C. In aqueous AlF_3-solutions, species of the general formula $[AlF_x(H_2O)_{6-x}]^{3-x}$ are detected by NMR spectroscopy, at high fluoride concentrations $[AlF_6]^{3-}$ can be observed, too [141, 142].

AlF_3 and the mineral cryolite $Na_3[AlF_6]$ are very important products for industrial processes, especially in the fused salt synthesis of Al. As the natural cryolite sources are limited, $Na_3[AlF_6]$ is also produced synthetically via intermediary hexafluoroaluminic acid, which is obtained upon treatment of Al_2O_3 with HF. Subsequent neutralization with Na_2CO_3 yields cryolite, along with H_2O and CO_2. More recently, $Na_3[AlF_6]$ has been prepared by the reaction of NH_4F with aluminium and sodium hydroxide. The structure of cryolite is described by a fcc-packing of octahedral $[AlF_6]^{3-}$ anions, wherein the Na^+ cations occupy all tetrahedral- and octahedral voids.

$H_3[AlF_6]$ slowly decomposes under precipitation of $AlF_3 \cdot 3\ H_2O$ in concentrated solutions. The precipitate also contains minor amounts of the hydrolysis products $AlF(OH)_2 \cdot 6\ H_2O$ and $AlF_2(OH) \cdot 6\ H_2O$. With transition metal cations, for example, Ag^+, Cu^{2+}, Zn^{2+} or Cd^{2+} hydrated hexafluoroaluminates are formed [143].

Mixed aluminium fluorohalides AlF_nX_{3-n} are obtained in form of THF-adducts, when halogen-substituted alanes of the type H_nAlX_{3-n} or Alk_nAlX_{3-n} are treated with HF in THF [144]. The chlorofluorides $AlClF_2$ and $AlCl_2F$ can be identified in the mass spectra of the system $Al-AlF_3-AlCl_3$. Nanoscopic compounds in the aforementioned system are interesting tools in fluoroorganic synthesis. Both, the mixed chlorofluoride $AlCl_xF_{3-x}$ ($x = 0.05–0.3$) as well as AlF_3 are capable of activating C-F bonds, which overall leads to hydrodefluorination reactions. This method may for example be used for the transformation of hydrofluorocarbons with a high greenhouse potential to synthetically useful, fluorinated building blocks [145].

AlF_3 does not react with strong fluoride acceptors like AsF_5 or SbF_5 to result in molecular adducts or even cationic species [146]. However, numerous fluoroaluminates of the types $M^I[AlF_4]$, $M^I_2[AlF_5]$ or $M^I_3[AlF_6]$ (M^I = alkali metal) are accessible. In all of them, aluminium is octahedrally coordinated by fluorine, whereas fluorine is almost linearly coordinated by two aluminium neighbors. The difference is the linkage of the subunits in the fluoroaluminates, for example, the $[AlF_6]$-octahedra in the $[AlF_5]^{2n-}_n$ anion are linked via two common corners, resulting in a one-dimensional chain-like structure. In contrast, the octahedra of $[AlF_4]^{n-}_n$ are linked via four common corners, which then gives rise to a two-dimensional sheet structure. Interestingly, the reaction of AlF_3 with $[NMe_4]F$ in anhydrous acetonitrile yields the ammonium salts $[NMe_4][AlF_4]$ and $[NMe_4]_2[AlF_5]$, which contain isolated $[AlF_4]^-$ and $[AlF_5]^{2-}$ ions, respectively. $[AlF_4]^-$ is isoelectronic to SiF_4, and thus has a tetrahedral structure, the trigonal-bipyramidal $[AlF_5]^{2-}$ has the same electron count like $[SiF_5]^-$ or PF_5. In the vapor-phase over cryolite melts or mixtures of NaF-AlF_3 or TlF-AlF_3, the respective monomeric $M^I[AlF_4]$ and dimeric $(M^IAlF_4)_2$ are observed as molecular species [147]. Electron diffraction experiments on $Na[AlF_4]$ and $K[AlF_4]$ have shown, that the alkali metal cation is side-on oriented to the edge of the $[AlF_4]$-tetrahedron, resulting in a C_s-symmetric structure [148]. Isolated octahedral units are generally found in salts with the $[AlF_6]^{3-}$ anion [149]. There are also a few quaternary salts of the type $M^IM^{II}[AlF_6]$ (M^I = K, Rb, Cs; M^{II} = Ag, Cu) as well as compounds with different alkali metal cations $M^I_2M^{I'}[AlF_6]$ (M^I and $M^{I'}$ = alkali metal, Ag, Tl), which are prepared by heating of mixtures of the binary fluorides [150–152]. The natural occurring mineral weberite Na_2MgAlF_7 does not comprise $[AlF_7]$-polyhedra but octahedral $[AlF_6]$-units. An overview over the structural properties of fluoroaluminate anions is given in Table 9.4.

Table 9.4: Structural properties of fluoroanions of trivalent group 13 metals.

	$[MF_4]^-$	$[MF_5]^{2-}$	$[MF_6]^{3-}$
$M = Al^{III}$, Ga^{III}, In^{III} in (alkali) metal salts	sheets of $[MF_6]$-octahedra in polymeric $[MF_4]^{n-}_n$	chains of $[MF_6]$-octahedra in polymeric $[MF_5]^{2n-}_n$	isolated $[MF_6]^{3-}$-octahedra
$M = Al^{III}$ in $[NMe_4]^+$-salts	isolated $[AlF_4]^-$-tetrahedra	isolated trigonal-bipyramidal $[AlF_5]^{2-}$ anions	
$M = Tl^{III}$ in (alkali) metal salts	layers of distorted $[TlF_6]$-octahedra; $[TlF_7]$-polyhedra in $K[TlF_4]$	unknown	isolated $[TlF_6]^{3-}$-octahedra

Gallium trifluoride **GaF$_3$** and indium trifluoride **InF$_3$** are directly generated by treatment of the corresponding metal with HF at higher temperature. The structural behavior of GaF_3 and InF_3 is very similar to AlF_3. They both have the characteristic octahedral coordination geometry at the metal centers and also tend to form hexafluorometallates of the type $M_3^I[GaF_6]$ and $M_3^I[InF_6]$. Accordingly, higher aggregates are known, which also have structures similar to the aluminates, for example, the chain-like species $[GaF_5]_n^{2n-}$ and $[InF_5]_n^{2n-}$ or the sheet-structured anions $[GaF_4]_n^{n-}$ and $[InF_4]_n^{n-}$ (Table 9.4). In aqueous solution, GaF_3 and InF_3 form trihydrates. $GaF_3 \cdot 3\,H_2O$ may also be generated by the reaction of elemental gallium with an aqueous mixture of hydrofluoric acid ($w = 40\,\%$) and hydrogen peroxide ($w = 30\,\%$). However, in presence of water, the hydrates partially tend to hydrolyze, so that the hydrated ions Ga^{3+}, $[GaF]^{2+}$, $[GaF_4]^-$, $[GaF_5]^{2-}$ and, with an excess of fluoride ions, also $[GaF_6]^{3-}$ can be observed by NMR spectroscopy [153]. The same observations hold for InF_3. For both Ga(III) and In(III), the hexafluorometallate salts contain isolated, octahedral anions in $K_3[M^{III}F_6]$ (M^{III} = Ga, In) [149]. Apart from that, compounds of the compositions $M^I M^{II}[GaF_6]$ and $M^I M^{II}[InF_6]$ (M^I = K, Rb, Cs; M^{II} = Ag, Cu) are known, as well as $M_2^I M^{I'}[InF_6]$ (M^I, $M^{I'}$ = alkali metal) [150, 151, 154, 155]. At 170 °C, thermal decomposition of the ammonium complexes $(NH_4)_3[MF_6]$ (M = Ga, In) mainly yields the tetrafluorometallates $NH_4[MF_4]$ [156]. At 300 °C, $NH_4[InF_4]$ splits off NH_4F and is transformed to InF_3. GaF_3 does not show any fluoride donor properties since it does not react with AsF_5 or SbF_5. However, InF_3 forms a stable 1:3-adduct with SbF_5, which consists of infinite chains of In-centers connected by each three $[SbF_6]$-units [146]. It is also possible to substitute the fluoride-substituents of GaF_3 and InF_3 by azido-groups. This is achieved by treatment of the trifluorides with Me_3SiN_3 in SO_2- or CH_3CN-solution. Further treatment of the highly explosive $Ga(N_3)_3$ and $In(N_3)_3$ with $[PPh_4]N_3$ gives rise to the anionic azidometallate species $[Ga(N_3)_5]^{2-}$ and $[In(N_3)_6]^{3-}$, respectively [157]. Similarly, the preparation of the respective cyanometallates is achieved by treatment of GaF_3 and InF_3 with Me_3SiCN and subsequent reaction with $[PPh_4]CN$ [158].

Thallium trifluoride **TlF$_3$** is a colorless crystalline solid, which can be obtained by fluorination of Tl_2O_3 with F_2, BrF_3 or SF_4 at about 300 °C. TlF_3 forms orthorhombic crystals and is isostructural to β-BiF_3 and YF_3. Thus, thallium is nine-fold coordinated by fluorine, where six F-atoms build a distorted trigonal prism, the other three F-atoms are capping the faces of the prism. One of the distances between Tl and the F-caps is slightly elongated, so that the coordination number is more correctly noted as $8+1$ [159, 160]. TlF_3 readily hydrolyzes to $Tl(OH)_3$ and forms complexes comparable to the Ga- and In-compounds. The fluorination of $M^I TlO_2$ or an equimolar mixture of Tl_2O_3 and MCl at 450–500 °C yields the tetrafluorothallates $M^I[TlF_4]$ (M^I = K, Rb, Cs). In general, the structural behavior of the fluorothallate(III) anions differs from that of the lighter homologues of group 13 (Table 9.4). The potassium salt $K[TlF_4]$ crystallizes in a trigonal space group, wherein Tl is heptacoordinated [161, 162]. In contrast, $Rb[TlF_4]$, $Cs[TlF_4]$ and also $Tl[TlF_4]$ contain layers of distorted $[TlF_6]$-octahedra [163]. Salts containing the octahedral anion $[TlF_6]^{3-}$ are also known [149]. Furthermore, a few quaternary fluorides $M^I M^{II}[TlF_6]$ (M^I = K, Rb, Cs; M^{II} = Ag, Cu) and $M_2^I M^{I'}[TlF_6]$ (M^I, $M^{I'}$ = alkali metal) have been described, as well as the weberite-type compounds

$M_2^I M^{II}[TlF_7]$ (M^I = Na, Ag; M^{II} = Mg, Zn, Mn) [150, 151, 154, 155, 164]. With strong fluoride acceptors, TlF_3 yields the adducts $TlF_3 \cdot 3\,SbF_5$ and $TlF_3 \cdot AsF_5 \cdot 2\,HF$. Interestingly, Tl shows a different structural behaviour in these compounds. $TlF_3 \cdot 3\,SbF_5$ exhibits nine-fold Tl-coordination, resulting in the shape of a tricapped trigonal prism. In contrast, rectangular rings of alternating Tl- and F-atoms are found in $TlF_3 \cdot AsF_5 \cdot 2\,HF$. Overall, Tl is seven-fold coordinated in this compound, the remaining three axial F-substituents are contributed by two $[AsF_6]$-units and one HF-molecule [146]. Analogously to the lighter homologues, the fluoride-substituents in TlF_3 may be substituted by azide-ligands upon treatment with Me_3SiN_3 to result in highly explosive $Tl(N_3)_3$. Addition of $[PPh_4]N_3$ exclusively yields the hexaazidothallate(III) anion $[Tl(N_3)_6]^{3-}$ [157]. Accordingly, cyano-groups are introduced, if TlF_3 is treated with Me_3SiCN. Subsequent treatment with $[PPh_4]CN$ gives rise to anionic $[Tl(CN)_5]^{2-}$ [158].

9.3.3 Study questions

(I)   Which type of compounds are described by the term "boron subfluorides"? Are these species stable? How are they typically prepared?

(II) Briefly describe the structures of B_8F_{12} and $B_{10}F_{12}$. Are there comparable borane species to which these structures may be compared?

(III) How does one prepare BF_3 on technical scale? Which of its properties is exploited to obtain a compound that is less volatile, and thus better to handle?

(IV) Describe the structure of BF_3. Are their similarities with BH_3?

(V)   What is the most important property of the tetrafluoroborate anion in preparative chemistry? How are anhydrous tetrafluoroborate salts prepared?

(VI) Which are the most stable fluorides of the group 13 metals (Al, Ga, In, Tl)? How can one explain this?

(VII) How can one prepare TlF? To which fluoride compounds does it show close similarities regarding its chemical behavior?

(VIII) Describe the structure of AlF_3 in the solid-state and in the gas phase. How is it synthesized?

(IX) Which important industrial process requires cryolite? How is "synthetic cryolite" prepared?

(X)   Which products are observed, when AlF_3 is treated with fluoride donors? Comment on the structure of the obtained species.

(XI) What happens if the hexafluorometallate salts $(NH_4)_3[MF_6]$ (M = Ga, In) are heated?

(XII) How is TlF_3 prepared? Briefly describe its solid-state structure. Which structure type does it represent?

9.4 Group 14: Carbon, silicon, germanium, tin, lead

9.4.1 Carbon

There are numerous fluorinated carbon compounds, the overwhelming majority of them being described by organic derivatives. Thus, these compounds will generally be covered in Part C of this manuscript. However, there also exist a few carbon fluorides, which are considered to be inorganic compounds, namely $(CF)_x$, $(C_4F)_x$ and CF_2 as well as their derivatives. The final product of fluorination reactions of carbon or C-containing compounds is always carbon tetrafluoride **CF_4**. It is a colorless gas with a low boiling point (mp.: $-183.5\,°C$; bp.: $-128.5\,°C$), that is chemically inert. Regarding thermodynamic aspects, the hydrolysis of the C-F bonds of the tetrahedral carbon tetrafluoride is very exothermic and, therefore, favorable, but due to kinetic stabilization, this reaction is impossible. CF_4 is produced by fluorination of graphite or chlorofluoromethanes.

The simplest of the remaining carbon fluorides is "carbon monofluoride," also called "graphite fluoride" **$(CF_x)_n$**, which is obtained as grey, hydrophobic product upon fluorination of graphite at $400–600\,°C$ [165]. The stoichiometric composition of $(CF_x)_n$ varies between $x = 0.676$ to 0.99. The reaction is catalyzed by HF and the fluorine content strongly influences the color and permeability of the product (Table 9.5) [166, 167]. Pure, white products $(CF_{1.12})_n$ are obtained, when the fluorination is carried out at about $630\,°C$. Above $630\,°C$, graphite and graphite fluorides are burned by F_2 and result in CF_4 and carbon black-like products [168]. The general formula $(CF_x)_n$ is a result of the addition of a fluorine atom to each carbon atom. Defects in the layer and at the edges of the layers allow the partial addition of more than one F-atom. Thus, the maximum value for x is larger than one (1.12). The fluorinated graphite compounds have some technical relevance for the anode effect in aluminium synthesis. They are fully stable up to $600\,°C$, do not conduct electricity and have excellent lubricant properties [169]. Furthermore, they are chemically fairly inert, as they are not attacked by strong acids or bases and do not react with H_2 below $400\,°C$. The structure of $(CF)_n$ consists of layers of waved, six-membered rings, wherein every C-atom is sp^3-hybridized and covalently connected to one fluorine substituent. The distance between the layers can vary (580 to 615 pm) but is almost doubled in comparison to graphite (335 pm) [166, 167].

Table 9.5: Colors and permeabilities of $(CF_x)_n$ in dependence of the fluorine content.

Fluorine content x	Colour	Permeability
0.676–0.715	black	opaque
0.773	grey	opaque
0.780–0.805	silver-grey	opaque
0.858–0.885	silver-grey	almost transparent
0.983–0.988	silver-white	transparent

"Tetracarbon monofluoride," also called "graphite subfluoride" **$(C_xF)_n$** ($x = 3.6$ to 4), is formed as black solid, when graphite is treated with fluorine and HF at room temperature [166, 167, 170]. Chemically, it is also quite inert, but it is thermally less stable than $(CF_x)_n$, as it already decomposes at $100\,°C$. It consists of planar carbon layers with a covalently bound fluorine atom to every fourth C-atom. Its ability to conduct electricity is only about 1 % as large as that of graphite. The distance between the layers amounts to 550 pm. Graphite subfluoride $(C_xF)_n$ finds application in electrodes of button cells with high energy density.

Monomeric difluorocarbene **CF$_2$** can only exist at high temperatures or in an inert gas matrix. In absence of other reactants, it readily dimerizes or polymerizes. CF_2 is a bent molecule (angle$_{(FCF)}$: 105°), which is very reactive and undergoes numerous reactions, especially additions to unsaturated organic compounds (cf. Section 13.4.10) [171]. Due to similar sizes of the p-orbitals of carbon and fluorine, a good overlap is enabled. Thus, fluorine can donate electron-density to carbon by π-backdonation and stabilize the carbene. Consequently, the free electrons of CF_2 cannot occupy different orbitals. Hence, in contrast to CH_2 (triplet-carbene), CF_2 is a singlet-carbene in its ground-state [172]. Despite its short lifetime and high reactivity, difluorocarbene is an important intermediate in fluoroorganic chemistry (cf. Section 11.2.6).

It is also possible to fluorinate fullerene C_{60} with gaseous F_2 via various $C_{60}F_x$ intermediates to yield the final product "fullerene fluoride" **C$_{60}$F$_{48}$** [173–176]. The yield of $C_{60}F_{48}$, which is obtained by fluorination of C_{60} is improved, when transition metal fluorides, for example, MnF_2 are present or even used as matrix materials [177]. Analogously to C_{60}, other fullerenes C_n may be fluorinated, for example, C_{70}. Interestingly, the fluorination of C_{60} stops at $C_{60}F_{48}$, it is not possible to obtain the hypothetical, perfluorinated $C_{60}F_{60}$ [173, 176, 178]. Despite the presence of six double bonds, further fluorination of $C_{60}F_{48}$ results in the breakup of the carbon cage. The reason for this is the shielding of the double bonds by the 48 fluorine atoms, which cover the C_{60} surface completely. The double bonds are located in six of the twelve C_5-rings, each three of them located on opposite faces of C_{60} [174, 179–185]. The chemical fragmentation of $C_{60}F_{48}$ upon extensive high-temperature fluorination has been investigated by both experimental and computational methods. It has been shown, that the product mixtures are not randomly composed and generally consist of numerous fluorinated compounds with 10–34 carbon atoms, mainly fused, cyclic fluorocarbons [186]. $C_{60}F_{48}$ is stable in chlorinated hydrocarbons, however, it is decomposed in diethyl ether, tetrahydrofuran and acetone and forms colored charge-transfer complexes in aromatic solvents, such as benzene or toluene [175]. In linear, aliphatic hydrocarbons C_nH_{2n+2} (n = 5–11), molecular solvates of $C_{60}F_{48}$ are observed [187]. The dissolution of $C_{60}F_{48}$ in ionic liquids has also been studied and compared with that of nonfluorinated C_{60}. As a consequence, it has been found, that cations with long-chained alkyl substituents are beneficial for the solvation of $C_{60}F_{48}$, underlining the hydrophobic nature of this compound in comparison with unsubstituted C_{60} [188]. Fullerenes with a high fluorine content show oxidizing and fluorinating properties. For example, they readily liberate I_2 from NaI-solutions, oxidize isopropanol to acetone or fluorinate arenes in the presence of catalysts like BF_3 [189]. In terms of applications, it has also been shown that fluorinated fullerenes are capable of inducing conductivity on diamond surfaces [190].

9.4.2 Silicon, germanium, tin, lead

The binary fluorides of the heavier homologues of group 14 mainly exist as di- and tetrafluorides. The stability of the difluorides increases from silicon to lead. In the same direction, the tetrafluorides are getting less stable. Apart from the di- and tetrafluorides, several subfluorides and numerous derivatives exist.

Difluorides

Silicon difluoride SiF_2 is obtained by thermal comproportionation, when SiF_4 is re-acted with elemental Si at 1100–1400 °C *in vacuo* (0.1–0.2 mbar) [191]. Alternatively, it can be generated by thermal disproportionation of Si_2F_6 at 700 °C in high vacuum (Scheme 9.3) [192]. It only exists in the gas phase or at low temperatures in an inert gas matrix. Monomeric SiF_2 is a bent molecule with a singlet ground state, which is rather stable in comparison to other carbenes like CF_2 or CH_2 (SiF_2: $\tau_{1/2} = 150$ s; CF_2: $\tau_{1/2}$ ca. 1 s; CH_2: $\tau_{1/2} < 0.1$ s). Thus, SiF_2 can be stored for a short time as diluted gas at room temperature [172, 191, 193, 194]. SiF_2 is very reactive and reacts with numerous organic and inorganic compounds under insertion of SiF_2-groups to σ-bonds or addition to π-bonds [191]. It is thermolyzed above 200 °C and ignites on air.

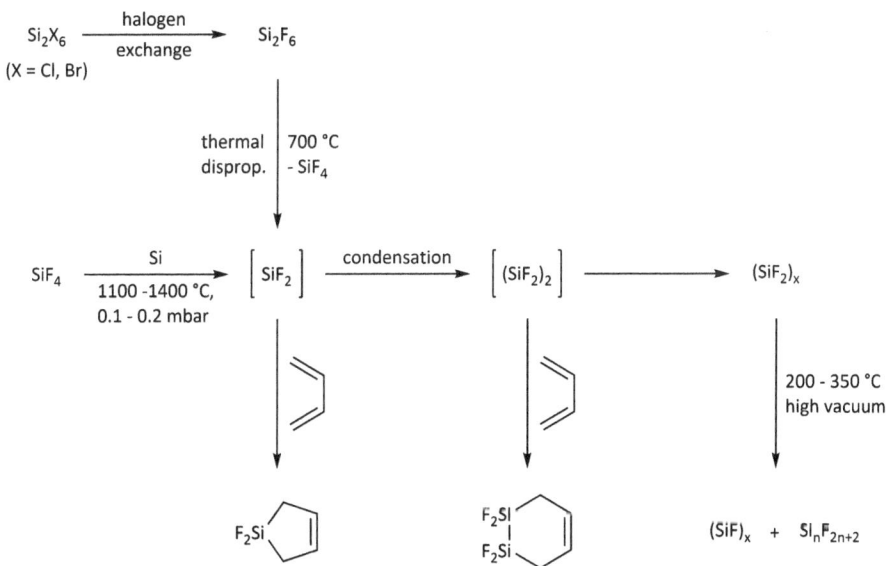

Scheme 9.3: Selected synthetic routes toward silicon subfluorides.

When it is frozen out, SiF_2 polymerizes via the biradical dimer $(SiF_2)_2$ to yield the color-less, waxy $(SiF_2)_x$ (Scheme 9.3) [191]. In low temperature reactions with, for example, BF_3 or HBr one usually observes products with at least two SiF_2-groups, for example, $BF_2(SiF_2)_nF$ ($n = 2$ to 5) or SiF_2BrSiF_2H [195, 196]. Hence, it can be assumed, that silicon difluoride di- or polymerizes in the condensed phase to form the more reactive $(SiF_2)_2$ diradical. The presence of the latter has been proven by ESR spectroscopy and a test reaction with butadiene. Whereas silacyclopentene is obtained from the reaction with gaseous SiF_2, the similar reaction yields 1,2-disilacyclohexene in the condensed phase (Scheme 9.3) [197–200].

When heated to 200–350 °C in high vacuum, $(SiF_2)_x$ melts and decomposes to colorless $(SiF)_x$ and colorless perfluorosilanes Si_nF_{2n+2} (Scheme 9.3) [191, 194]. Polymeric silicon monofluoride (**SiF**)$_x$ can also be obtained from the reaction of $SiBr_3F$ with Mg in ethereal solution as colorless solid, which decomposes explosively into Si and SiF_4 above 400 °C [201]. The perfluoropolysilanes Si_nF_{2n+2} are known with chain lengths up to $n = 14$ [194]. Hexafluoro disilane **Si$_2$F$_6$** can also be prepared by the fluorination of Si_2Cl_6 or Si_2Br_6 or substitution of Si_2Cl_6 with ZnF_2. It is a colorless gas and has a staggered conformation, the same can be assumed for the higher perfluorosilanes [202]. Upon hydrolysis of Si_2F_6, hexafluoro silicic acid H_2SiF_6, silicic acid $Si(OH)_4$ and H_2 are formed. The perfluorosilanes generally differ from their carbon homologues, as they are very reactive and sensitive toward hydrolysis and oxidation. Further, they have a limited solubility in most organic solvents. Perfluorosilanes may be used as starting materials for other silicon compounds.

The difluorides GeF_2, SnF_2 and PbF_2 are different to their carbon and silicon analogues, as they are stable at room temperature. Germanium difluoride (difluorogermylene) **GeF$_2$** is formed upon reduction of GeF_4 with elemental Ge above 120 °C or directly from Ge and anhydrous HF at 225 °C (Scheme 9.4) [203–205]. It is also accessible by the reaction of HgF_2 with elemental germanium at 160 °C under reduced pressure [206]. GeF_2 is a colorless, hygroscopic and crystalline solid (mp.: 112 °C; bp.: 160 °C) with a singlet ground-state, wherein the [GeF$_2$]-units are linked by F-atoms, so that polymeric chains are found in the solid-state structure. Therein, parallel chains are linked by F-bridges, so that every Ge-center has a ψ-trigonal bipyramidal coordination with two axial and two equatorial fluorine substituents as well as an equatorial lone pair [172, 207]. In total, it can be described as a room structure of see saw shaped [GeF$_4$]-units. In the gas phase, GeF_2 is mono- and dimeric [208]. Monomeric GeF_2 has a bent geometry, the dimer has a centro-symmetric, nonplanar structure with C_{2h}-symmetry as has been shown by vibrational spectroscopy in a N_2-matrix [209, 210].

Germanium difluoride is a strong reducing agent; it transforms SeF_4 to elemental Se, I_2 to GeI_4 and SO_3 to SO_2 [205]. Acidification of GeF_2-solutions leads to the formation of H_2; addition of concentrated solutions of M^IF (M^I = K, Cs) gives rise to the trifluorogermanates(II) $M^I[GeF_3]$ (Scheme 9.4). Salts containing the [GeF$_3$]$^-$ anion are hydrolytically more stable than the respective tin-compounds. However, they are readily oxidized by O_2. Hence, GeF_2 is converted to the germanate(IV)-salt $Cs_2[GeF_6]$ in solutions of CsF in aqueous HF ($w = 20$ %) (Scheme 9.4) [203]. Apart from the aforementioned reactivity, GeF_2 shows a strong carbene character. Thus, it readily inserts into a large variety of σ-bonds [211, 212]. It is also capable of adding to carbon-carbon

GeF$_4$

Ge,
T > 120 °C

H$_2$ + GeF$_4$

H$^+$

HgF$_2$, 160 °C,
reduced pressure

Ge ──────────── GeF$_2$ ────────── M[GeF$_3$]

or conc. KF or CsF

aHF, 225 °C

CsF, O$_2$,
aqueous HF

SeF$_4$

Se + GeF$_4$

Cs$_2$[GeF$_6$]

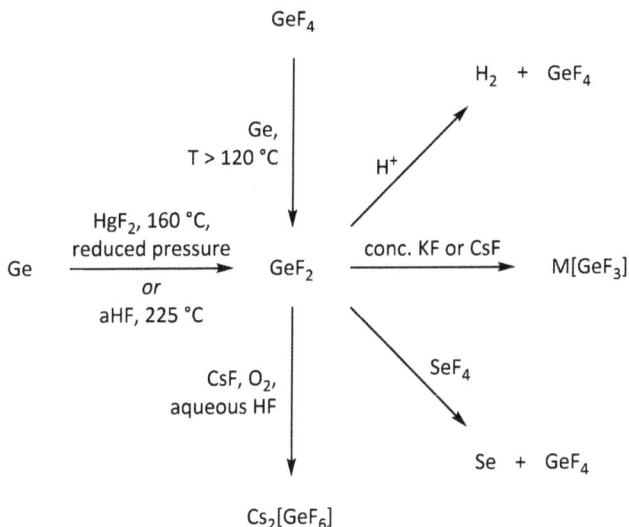

Scheme 9.4: Preparation and selected reactions of GeF$_2$.

multiple bonds under intermediary formation of cyclic species, which are then converted to polymeric compounds [213]. Cycloaddition reactions are observed with conjugated substrates like 2,3-dimethylbutadiene or 1,3,5-hexatriene [214]. Analogously, the reaction with orthoquinones gives rise to substituted 2-germa-1,3-dioxolanes [215]. Similar to the heavier homologues tin and lead, germanium also forms a mixed valent fluoride compound Ge$_3$F$_8$. It is synthesized from Ge-powder and GeF$_4$ under moderate heating and pressure. The structure of this species is described by pyramidal [Ge^{II}F$_3$]- and octahedral [Ge^{IV}F$_6$]-units, which are connected by F-bridges [216].

Tin difluoride **SnF$_2$** is obtained in form of colorless, monoclinic crystals (mp.: 213 °C; bp.: 853 °C), when SnO is dissolved in hydrofluoric acid (w = 40 %) and evaporated. Alternatively, it is directly prepared by treatment of Sn-metal with anhydrous HF at 200 °C (Scheme 9.5) [203, 204]. Tin difluoride has a singlet ground-state, since its triplet state is much higher in energy [172]. The solid-state structure comprises a room structure, which can be described as distorted rutile type. In the subunits, closed chains in form of eight-membered, waved (-Sn(F)-F-)$_4$ rings are present. Every Sn-atom has a distorted octahedral coordination geometry [217]. In the gas phase, one can find monomeric SnF$_2$, but also di- and trimers, which tend to dissociate to the monomer with increasing temperature [208]. Monomeric SnF$_2$ is a bent molecule. It is water-soluble and has Lewis-acidic properties, for example, the formation of 1:1-adducts is observed with Lewis bases like water, DMSO or pyridine (Scheme 9.5). Furthermore, SnF$_2$ is a strong fluoride acceptor and forms numerous fluorostannates of the general formulas M^I[SnF$_3$] (M^I = Na, K, Rb, Cs, NH$_4$) and M^I[Sn$_2$F$_5$] (M^I = Na, K, Rb, NH$_4$, Tl) [203, 218–222]. Trifluorostannates [SnF$_3$]$^-$ may be isolated from aqueous solutions or melts, however, upon contact with air,

SnO

aq. HF

Sn $\xrightarrow{\text{aHF, 200 °C}}$ SnF$_2$ $\xrightarrow[\text{H}_2\text{O, dmso, py}]{\text{Lewis base, e.g.}}$ SnF$_2$ · base

O$_2$ or F$_2$, HF

F$^-$

Sn$_3$F$_8$

$[\text{Sn}_2\text{F}_5]^-$ + $[\text{SnF}_3]^-$ $\xrightarrow{\text{air}}$ $[\text{SnF}_6]^{2-}$

PbCO$_3$

HF, then Δ

Pb $\xrightarrow{\text{aHF, 160 °C}}$ PbF$_2$ $\xrightarrow{\text{F}^-}$ $[\text{PbF}_3]^-$ + $[\text{PbF}_6]^{4-}$

aHF, 250-500 °C

BrF$_3$

Pb$_3$O$_4$ $\xrightarrow{\text{aHF, T < 100 °C}}$ Pb$_3$F$_8$

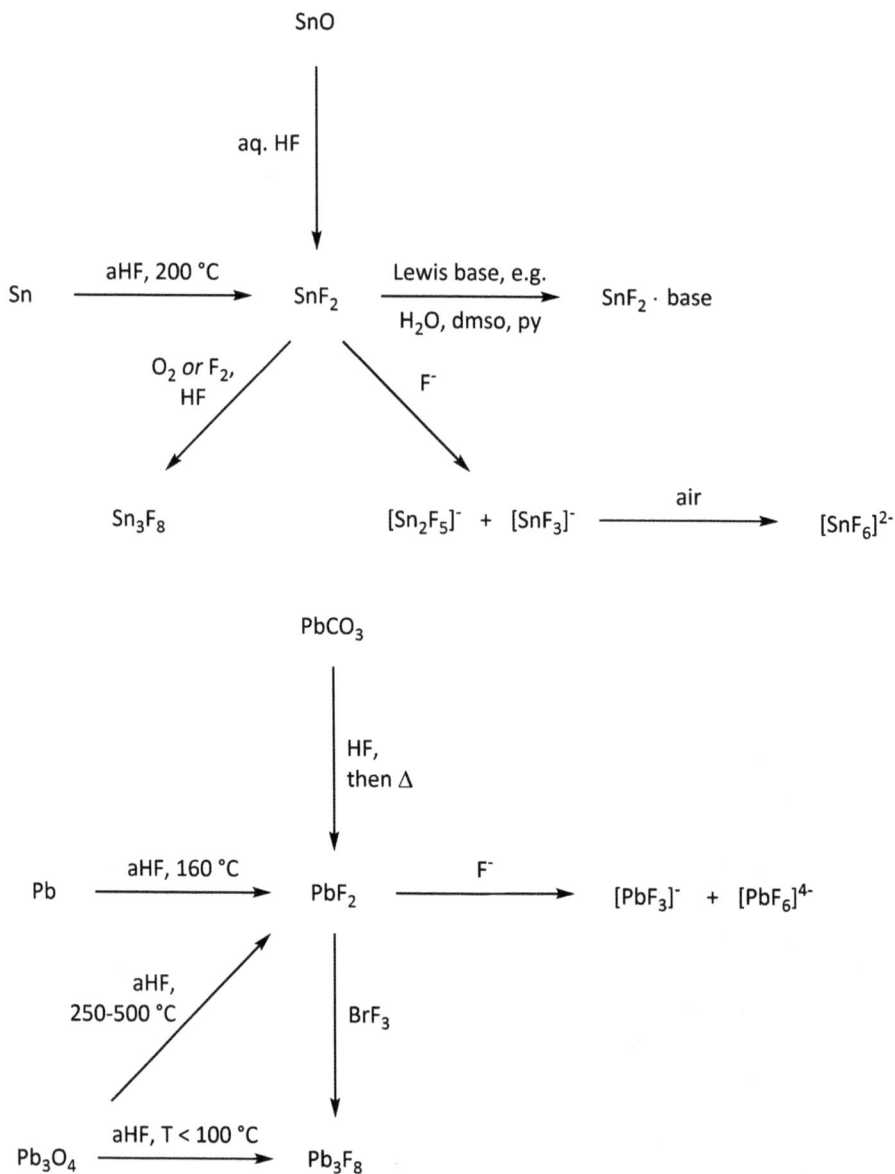

Scheme 9.5: Preparation and selected reactions of SnF$_2$ (top) and PbF$_2$ (bottom).

they are oxidized to $[\text{SnF}_6]^{2-}$ (Scheme 9.5). Additionally, solutions of $[\text{SnF}_3]^-$ deposit SnO upon standing [203]. The $[\text{SnF}_3]^-$ anion has a pyramidal structure due to its sterically active, nonbonding electron pair [218, 223, 224]. The pentafluorodistannate anion $[\text{Sn}_2\text{F}_5]^-$ is isolated from concentrated SnF$_2$-solutions or -melts, and consists of two SnF$_2$-units, which are bridged by a fluoride ion [219, 220]. The fluorostannates $Na_4[Sn_3F_{10}]$ and $M^{II}[SnF_4]$ (M^{II} = Pb, Sr, Ba) are also known [225, 226]. The decafluorotristannate anion $[Sn_3F_{10}]^{4-}$ is composed of three distorted tetragonal-pyramidal $[SnF_4]$-units, which are linked by common F-atoms [225]. The $[SnF_4]^{2-}$-salts crystallize in the $PbCl_2$ structure type [226]. With strong fluoride acceptors, for example, SbF_5, AsF_5 or BF_3, the polymeric fluorine-bridged $[SnF]_n^{n+}$ cation is generated [227]. Oxidation of SnF_2 with O_2 or F_2 in HF yields the mixed valent Sn_3F_8 (Scheme 9.5). The latter is composed of $trans$-F-bridged $[Sn^{IV}F_6]$-units, which are linked by polymeric $[Sn^{II}F]$-chains [228]. Along with Sn_2F_6, Sn_7F_{16} and $Sn_{10}F_{34}$, Sn_3F_8 is also found in melts of SnF_2 and SnF_4 under dry nitrogen atmosphere [229].

Lead difluoride **PbF₂** is a white, crystalline powder (mp.: 818 °C; bp.: 1292 °C) that is generated upon dissolving $PbCO_3$ in HF and subsequent fast melting of the intermediate bifluoride. Alternatively, it can be synthesized from elemental lead and HF at 160 °C (Scheme 9.5) [204]. Like all the other group 14 difluorides, it has a singlet ground-state [172]. In cold water, PbF_2 is almost insoluble. Gaseous PbF_2 is a bent molecule that is only present in monomeric form [208]. In the solid state, one can find a low temperature form, which crystallizes in the $PbCl_2$ lattice. Therein, Pb is coordinated by nine F-atoms in form of a distorted three-fold capped trigonal prism. The $[PbF_9]$-polyhedra are linked via common faces. Above 316 °C, a high temperature polymorph exists that can be described by the cubic fluorite structure. The fluoroplumbates(II) $K_4[PbF_6]$, $Rb[PbF_3]$ and $Cs[PbF_3]$ are formed, when PbF_2 is treated with stoichiometric amounts of the corresponding alkali metal fluorides (Scheme 9.5) [230]. The fluorination of mixed-valent Pb_3O_4 with anhydrous HF yields different products, depending on the reaction temperature. Between −50 and 110 °C, Pb_3F_8 is obtained, whereas Pb(IV) is partially reduced between 110 and 220 °C. Finally, PbF_2 is formed at temperatures up to 500 °C [231]. The mixed-valent Pb_3F_8 is also accessible by the reaction of PbF_2 with BrF_3 (Scheme 9.5). Above 80 °C, it decomposes to PbF_2 and F_2. Structurally, Pb_3F_8 is composed of distorted octahedral and pentagonal bipyramidal units, which are linked by F-bridges to form an infinite, two-dimensional, ladder-like arrangement [232]. Another product of the reaction of PbF_2 with BrF_3 is the fluorido-bromate(III) salt $[PbF][Br_2F_7]$, which loses BrF_3 above 50 °C to leave pure PbF_2 [233]. Lead difluoride has been used as an evaporation additive in the optic industries for the preparation of special glasses.

Tetrafluorides

Silicon tetrafluoride **SiF₄** is a colorless gas that strongly fumes in contact with humidity. It can be prepared from the elements and by reaction of SiO_2 with HF under dehydrating conditions, which may be realized by treatment of a mixture of CaF_2 and SiO_2 with concentrated sulphuric acid (Scheme 9.6). Very pure SiF_4 is accessible by thermolysis of $Ba[SiF_6]$ at 300–350 °C in $vacuo$. The SiF_4 molecule has a tetrahedral structure with a Si-F bond length of 155 pm, which is a value between a typical Si-F

Si $\qquad$ Ba[SiF$_6$] $\qquad$ M$_2$[SiF$_6$]

F_2 — vacuum, 300-350 °C — MF — MOH or M$_2$CO$_3$

$$\text{SiO}_2 \xrightarrow[\text{H}_2\text{SO}_4 \text{ (conc.)}]{\text{CaF}_2} \text{SiF}_4 \xrightarrow{\text{H}_2\text{O}} \text{H}_2\text{SiF}_{6\text{(aq.)}} + \text{SiO}_2$$

H$_2$O, gas phase

evaporation

F$_3$SiOSiF$_3$ + HF $\qquad$ SiF$_4$ + HF

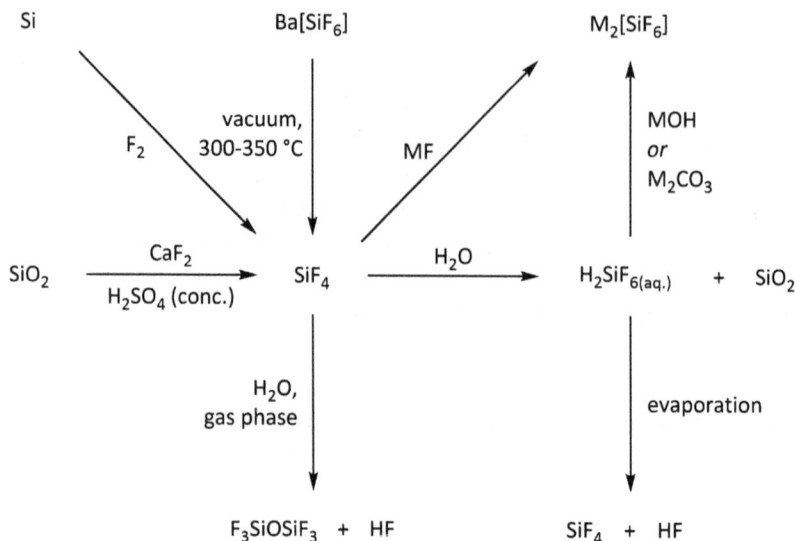

Scheme 9.6: Preparation of SiF$_4$, H$_2$SiF$_6$ and hexafluorosilicate salts.

single- (170 pm) and a double bond (148 pm). This can be explained by the high polarity of the Si-F bond, which leads to a covalent and an electrovalent contribution to the bonding situation. As a consequence, the inversion barrier is rather high, so that an octahedral coordination is not favored in the solid state [172]. Nevertheless, SiF$_4$ is a Lewis acid and prefers the formation of octahedral 1:2-complexes with *cis*-orientation of the base ligands [234, 235]. At low temperature, a 1:1-adduct with pyridine has been described as well. It has a polymeric structure wherein every Si-atom is six-fold coordinated [236]. The ammoniate [SiF$_4$(NH$_3$)$_2$] · 2 NH$_3$ is obtained from the reaction of SiF$_4$ with liquid ammonia. Overall, it shows a large number of N-H- and H-F-bonds, so that a three-dimensional network is formed [122]. Adducts with primary and secondary amines readily split off HF to form amino fluorosilanes F$_3$SiNR$_2$ or fluorosilazanes (RNSiF$_2$)$_n$ [237–240]. Substituted ammonium fluorosilicates with the anions [SiF$_5$]$^-$, [Si$_2$F$_{11}$]$^{3-}$ and [Si$_3$F$_{16}$]$^{4-}$ are formed upon reaction with primary amines in presence of tertiary amines. However, the product formation depends on the structure of the amine and the reaction conditions.

SiF$_4$ is very sensitive toward hydrolysis but rather unreactive if moisture is rigorously excluded. In the gas phase, the reaction with H$_2$O mainly yields hexafluorodisiloxane F$_3$SiOSiF$_3$ [241]. If SiF$_4$ is passed into water, SiO$_2$ and hexafluorosilicic acid H$_2$SiF$_6$ are generated (Scheme 9.6). H$_2$SiF$_6$ has some technical relevance and is also obtained as side product in the reaction of fluorapatite [Ca$_5$(PO$_4$)$_3$F] with SiO$_2$ and concentrated H$_2$SO$_4$. Pure H$_2$SiF$_6$ cannot be isolated from aqueous solutions; it can only be enriched to a mass percentage of $w = 60\%$. If concentrated H$_2$SO$_4$ is added to Ba[SiF$_6$], anhydrous H$_2$SiF$_6$ is indeed liberated, but it immediately decomposes to give SiF_4 and HF. However, in a diluted aqueous environment, HF is not liberated in significant amounts, so that low-concentrated solutions of H_2SiF_6 do not etch glass. The situation changes upon evaporation of aqueous H_2SiF_6 solutions, which leads to the liberation of SiF_4 and HF (Scheme 9.6). The ratio of HF and SiF_4 in the gas-phase over aqueous solutions of hexafluorosilicic acid strongly depends on the concentration of the evaporated solution. At $w = 13.3\%$, two molecules of HF are detected in the gas phase along with one SiF_4 molecule, so that it looks like H_2SiF_6 is distilled "without decomposition." Higher concentrated H_2SiF_6 solutions liberate more SiF_4, in contrast, solutions with a smaller concentration liberate more HF into the gas phase. Thus, evaporation of concentrated H_2SiF_6 solutions results in the enrichment of HF, so that glass will be etched. Even if the free acid is unstable in isolated form, the hydroxonium salt $(H_3O)_2[SiF_6]$ can be isolated from cooled concentrated solutions in form of colourless crystals (mp.: 19 °C).

Hexafluorosilicate salts are generally accessible by the reaction of the strong acid H_2SiF_6 with metal hydroxides and carbonates (Scheme 9.6). They can also be prepared directly from SiF_4 and a metal fluoride. Most of the hexafluorosilicate salts are water-soluble, only $Ba[SiF_6]$ and the alkali metal salts, except $Li_2[SiF_6]$, are sparingly soluble. H_2SiF_6 and its salts are toxic and have been used against bacteria and insects, as for example, $MgSiF_6$ has been used as wood preservative. The hexafluorosilicate anion $[SiF_6]^{2-}$ has an octahedral structure and is much more stable against hydrolysis than SiF_4. In aqueous solutions of SiF_4, one can also find $[SiF_5]^-$ anions, which are in equilibrium with other fluorosilicate species. They can be isolated from their solutions as $[R_4N]^+$ or $[R_4As]^+$-salts (R = Alk, Ph). In contrast to $[SiF_6]^{2-}$-salts, pentafluorosilicates are soluble in organic solvents [242–245]. Based on X-ray structural analysis as well as on NMR- and vibrational spectroscopy, a trigonal-bipyramidal structure has been found for the $[SiF_5]^-$ anion [244, 245]. It has comparable acceptor properties like SiF_4 [245, 246]. A variety of potential uses of silicate salts has been reviewed quite recently. For example, ammonium hexafluorosilicates are used as modifiers in zeolite catalysts, fluorinating agents, ionic liquids, electro-optical materials or even anticaries agents [247]. There are numerous mixed fluorohalosilanes and organofluorosilanes R_nSiF_{4-n} which are derived from SiF_4.

Germanium tetrafluoride **GeF$_4$** is a colorless gas, which shows a number of similarities to its lighter homologue SiF_4. For example, it is also formed by reaction of the elements or by thermolysis of the corresponding barium hexafluorogermanate(IV) salt $Ba[GeF_6]$ at 600 °C (Scheme 9.7). Analogously to SiF_4, it has a tetrahedral structure. Its inversion barrier is still quite high, so that the formation of octahedrally coordinated Ge-centers is not preferred in its solid-state structure [172]. However, GeF_4 is a Lewis acid that forms 1:1- and 1:2-*cis*-coordinated, octahedral adducts with donor molecules [235]. In contact with water, it hydrolyzes to form GeO_2 and H_2GeF_6. Addition of fluorides or hydroxides to H_2GeF_6-solutions yields salts with the octahedral hexafluorogermanate(IV) anion $[GeF_6]^{2-}$ (Scheme 9.7). They are also directly accessible from GeF_4 and fluorides, as for example, $K_2[GeF_6]$ is prepared from GeF_4 and KF [248–250].

$$Ba[GeF_6] \qquad\qquad M_2[GeF_6]$$

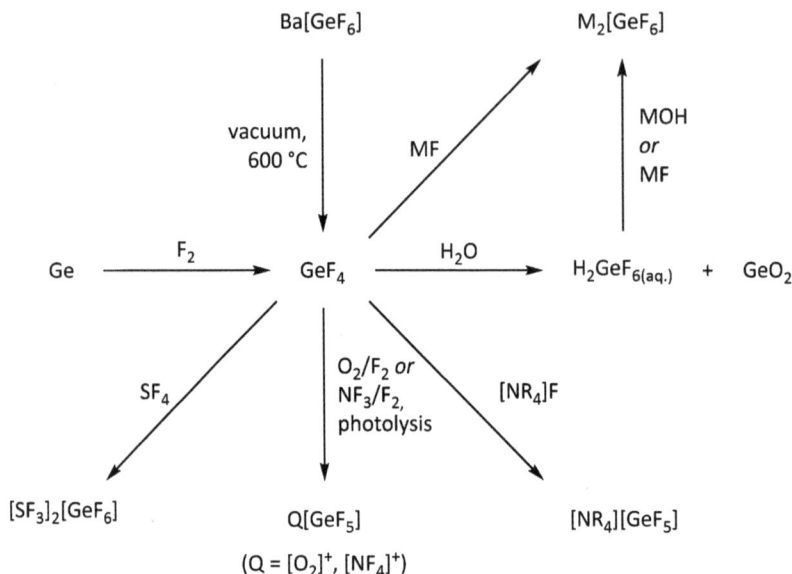

Scheme 9.7: Preparation and selected reactions of GeF_4.

Interestingly, the reaction of the chlorides CsCl and $[NMe_4]Cl$ with GeF_4 also leads to the formation of the respective hexafluorogermanates(IV) $M_2^I[GeF_6]$ (M^I = Cs, NMe_4), along with $GeCl_4$ [251]. SF_4 reacts with GeF_4 to form the complex $[SF_3]_2[GeF_6]$, which contains almost perfectly octahedral $[GeF_6]^{2-}$ anions (Scheme 9.7). The coordination geometry around the sulphur-atoms is made up to a distorted octahedron by three contacts to each one neighboring $[GeF_6]^{2-}$ anion [252]. In analogy to silicon, the pentafluorogermanates $[GeF_5]^-$ can be obtained with large cations like $[(C_3H_7)_4N]^+$, $[(C_4H_9)_4N]^+$ or $[Ph_4As]^+$ (Scheme 9.7) [253]. Vibrational spectra of matrix-isolated $Cs[GeF_5]$ show distinct ion pairs with slightly distorted trigonal-bipyramidal anions [254]. Similarly, a monomeric structure is assigned to $[GeF_5]^-$ in its $[NBu_4]^+$-salt [255]. Photolysis of a mixture O_2-F_2-GeF_4 or NF_3-F_2-GeF_4 at low temperature yields the salts $[O_2][GeF_5]$ and $[NF_4][GeF_5]$, respectively (Scheme 9.7). In these salts, the anions have a polymeric structure, which arises from linkages by cis-F-bridges [256, 257]. Additionally, cis-F-bridging of octahedral $[GeF_6]$-units is observed in $[ClO_2][GeF_5]$, whereas the xenon derivative $[XeF_5][GeF_5]$ contains infinite, $trans$-F-bridged chains of $[GeF_6]$-octahedra [255]. $[GeF_5]^-$ is a strong electron acceptor and reacts, for example, with NH_3 to form the 1:1-complex $[GeF_5(NH_3)]^-$ [253]. GeF_4 vapor reacts with the chlorides $AlCl_3$, $MgCl_2$ or $FeCl_3$ under full halogen substitution, mixed halides cannot be observed. Instead, $GeClF_3$, $GeCl_2F_2$ and $GeCl_3F$ are accessible by fluorination of $GeCl_4$ with SbF_3.

Tin tetrafluoride **SnF$_4$** is an extremely hygroscopic, colorless solid, which is prepared from the elements at about 100 °C or more conveniently by the reaction of $SnCl_4$ with anhydrous HF. SnF_4 can also be made from SnO and F_2 between

200 and 500 °C [258]. Like SiF_4 and GeF_4, the SnF_4 molecule has a tetrahedral structure, but in the solid state it is octahedrally surrounded by six F-atoms. The $[SnF_6]$-units are linked via a common edge to form a layered network [259, 260]. Quantum-chemical calculations have shown, that the inversion-barrier of SnF_4 is rather low, which is in line with the fact, that it easily tends to forms hexacoordinated Sn-species in the solid state [172]. With Lewis bases, for example, pyridine, 2,2'-bipyridine or N,N,N',N'-tetramethylethylenediamine (tmeda), SnF_4 preferably forms octahedral complexes [258]. SnF_4 readily fluorinates other halide compounds and reacts with anhydrous HF to form hexafluorostannic acid H_2SnF_6. Consequently, it also reacts with fluoride donors, so that numerous salts with the octahedral $[SnF_6]^{2-}$ anion have been described [221, 261–264]. Hexafluorostannate(IV) compounds are also accessible in anhydrous liquid ammonia [265]. NMR-investigations on $SnF_4/[SnF_6]^{2-}$-mixtures in SO_2-solution revealed polymeric $[SnF_6]^{2-}$ anions. They consist of *cis*-F-bridged octahedral $[SnF_6]$-units, which are further connected to $[Sn_2F_{11}]^{3-}$ and $[(SnF_5)_n]^{n-}$-fragments [266]. In an alkaline solution, the hexafluorostannate anion is hydrolyzed to $[SnF_5(OH)]^{2-}$, $[SnF_4(OH)_2]^{2-}$, $[SnF_3(OH)_3]^{2-}$, etc., which exist as multiple isomers [267]. In addition, many organic derivatives of tin tetrafluoride as well as a few mixed fluorohalides are known.

Lead tetrafluoride **PbF$_4$** can be obtained by fluorination of PbF_2 at about 300 °C or directly from Pb-powder and F_2 [268, 269]. Above 90 °C, PbF_4 decomposes to PbF_2 and F_2, thus it is a rather strong fluorinating agent [270]. Like the other tetrafluorides of group 14, it has a tetrahedral structure in the gas phase. In the solid state, it forms a layered structure of *cis*-F-bridged $[PbF_6]$-octahedra [259, 260]. Analogously to SnF_4, the inversion barrier of PbF_4 is small, as has been revealed by quantum-chemical calculations. This underlines the fact, that it readily forms hexacoordinated Pb-centers in the solid state [172]. It is very sensitive toward impurities and hydrolysis. PbF_4 forms fluoroplumbates with the compositions $[PbF_5]^-$, $[PbF_6]^{2-}$, $[PbF_7]^{3-}$ and $[PbF_8]^{4-}$ [264, 270, 271]. The most interesting structural behaviour can be observed for the hexafluoroplumbate(IV) salts in the series of alkaline earth metals. $Ca[PbF_6]$ does not contain discrete $[PbF_6]^{2-}$ anions but crystallizes in a structure related to the ReO_3 type. In contrast, $Sr[PbF_6]$ and $Ba[PbF_6]$ contain isolated $[PbF_6]^{2-}$ anions [271].

9.4.3 Study questions

(I)   Which graphite fluorides are known? How are they prepared and in which property do they significantly differ from nonfluorinated graphite?

(II)   What is the ground state of difluorocarbene? Explain this observation.

(III)   What is the final product of the treatment of C_{60} with fluorine, which still contains a fullerene scaffold? Is it possible to prepare a perfluorinated fullerene? Explain.

(IV)   How is SiF_2 prepared? Compare its stability with the carbon species CF_2 and CH_2.

(V)   Which structural similarities are found for gaseous GeF_2, SnF_2 and PbF_2? How do the solid-state structures differ? Briefly describe the coordination geometry of the metal cations.

(VI)   How does the stability of the difluorides GeF_2, SnF_2 and PbF_2 determine their chemical be-
       haviour? Also comment on the stability of their fluorometalates.
(VII) Briefly compare the gas-phase and solid-state structures of the tetrafluorides SiF_4, GeF_4,
       SnF_4 and PbF_4. Which general trend is observed?
(VIII) What happens, if SiF_4 is introduced into water? Can the product be isolated? Are there other
       stable derivatives of it?

9.5 Group 15: Nitrogen, phosphorus, arsenic, antimony, bismuth

9.5.1 Nitrogen

A variety of binary nitrogen fluorine compounds is known; the most important ones include NF_3, N_2F_4, N_2F_2 and FN_3. Apart from that, the oxyfluoride ONF_3 and $[NF_4]^+$-salts
need to be mentioned. In general, all N-F compounds are thermodynamically unsta-
ble, but nevertheless they have been widely explored as they once have been considered for applications as rocket fuels. However, they have never been used for this
purpose. Apart from the binary N-F compounds, there also exist fluoroamines, which
are substituted with organic residues.

NF_3 and its derivatives
Nitrogen trifluoride **NF_3** is a toxic, colorless gas (mp.: −206.8 °C; bp.: −129 °C), which
is formed upon fluorination of NH_3 over Cu (Scheme 9.8). Alternatively, it can be prepared by electrochemical fluorination of molten NH_4F or urea in anhydrous HF [272].
It may also be obtained from NH_4F/nHF and NH_3, using F_2 as fluorinating agent and
ClF_3 as mediator [273]. Analogously to ammonia, NF_3 has a pyramidal structure, but it
does not show the typical umbrella inversion of NH_3. It can be liquified by strong cooling to give a clear, mobile liquid. NF_3 is relatively inert up to 250–300 °C and virtually
insoluble in water or KOH-solution; however, alkaline hydrolysis occurs at temperatures above 100 °C. NF_3 reacts with H_2 under spark ignition, resulting in a sharp bang
and a red-violet luminescence. Analogously, a mixture of NF_3 and H_2O vapour can
be ignited to form HF and nitrous acid HONO accompanied with a blueish luminescence (Scheme 9.8). NF_3 reacts with Al_2Cl_6 at 70 °C to form N_2, Cl_2 and AlF_3. Metals
like copper reduce NF_3 to N_2F_4. Electrical discharge of an O_2-NF_3 mixture yields the
oxyfluoride ONF_3. In contrast to ammonia, NF_3 is only a very weak Lewis base and
does not form adducts with strong fluoride acceptors. In terms of applications, NF_3 is
used as a fluorine source in HF/DF-high energy lasers and as an additive to lightbulb
gas fillings. Furthermore, it has some technical use in the electronic industry for clean-
ing purposes. NF_3 has also been investigated as potential fluorinating and oxidizing
agent for the treatment of used nuclear fuels. A number of metal oxides is transformed
to volatile fluoride- or oxyfluoride compounds, which generally allows a separation of
the different components [274].

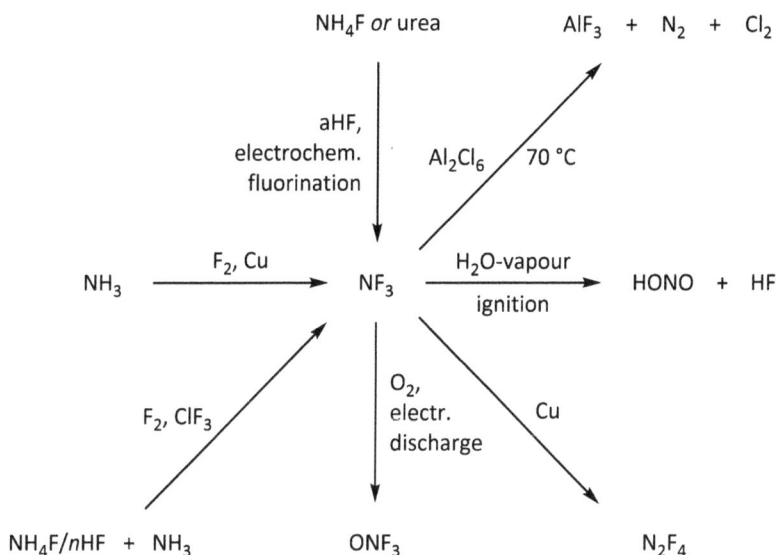

Scheme 9.8: Preparation and selected reactions of NF_3.

An important derivative of nitrogen trifluoride is chlorodifluoroamine **ClNF$_2$** (mp.: < −183 °C; bp.: −67 °C), which may be generally prepared by treatment of HNF_2 with reagents like BCl_3, NaOCl, Cl_2 in presence of KF or RbF as well as chlorine fluorides (Scheme 9.9) [275–278]. It is also accessible by the reaction of NH_4F with ClF_3 or a mixture of NH_4Cl and NaCl with ClF [279–281]. Generally, $ClNF_2$ is in equilibrium with N_2F_4 and Cl_2 [282]. Thus, photolysis of $ClNF_2$ as well as reaction with nucleophilic reagents yields N_2F_4. In turn, N_2F_4 reacts with excess Cl_2 under UV-photolysis to form $ClNF_2$ [283]. $ClNF_2$ further reacts with olefins under addition of NF_2- and Cl-radicals to the double bond [282]. Similarly, oxidative addition to transition metal complexes of the type [M(CO)Cl(PEt$_3$)$_2$] (M = Rh, Ir) occurs [284]. The higher chlorinated dichlorofluoroamine **Cl$_2$NF** is also known, but not that extensively explored. It has been prepared from NaN$_3$ and ClF for the first time via intermediary ClN$_3$ [285]. A safer preparation, that avoids the formation of highly explosive ClN$_3$ includes the fluorination of mixtures of NH_4F and NaCl with F_2 (Scheme 9.9) [281]. Alternatively, NH_4F/nHF and ClF_3 may be used as starting materials [280]. The certainly most convenient synthesis of Cl$_2$NF is described by the low temperature fluorination of FC(O)NCl$_2$ with elemental F_2 [286].

The fluoroamines NH_2F and NHF_2 are unstable, colorless gases. They have pyramidal geometry like NH_3 and NF_3. The preparation of monofluoroamine (fluoramide) **NH$_2$F** requires some synthetic detour, since it is generated by evaporation of fluoroammonium salts [NH$_3$F]$^+$[HF$_2$]$^-$ · nHF in dynamic high vacuum (Scheme 9.9). To remove HF, the gaseous reaction mixture is passed over activated, powdered KF and the purified NH_2F is subsequently condensed at −196 °C. Under these conditions, it is recov-

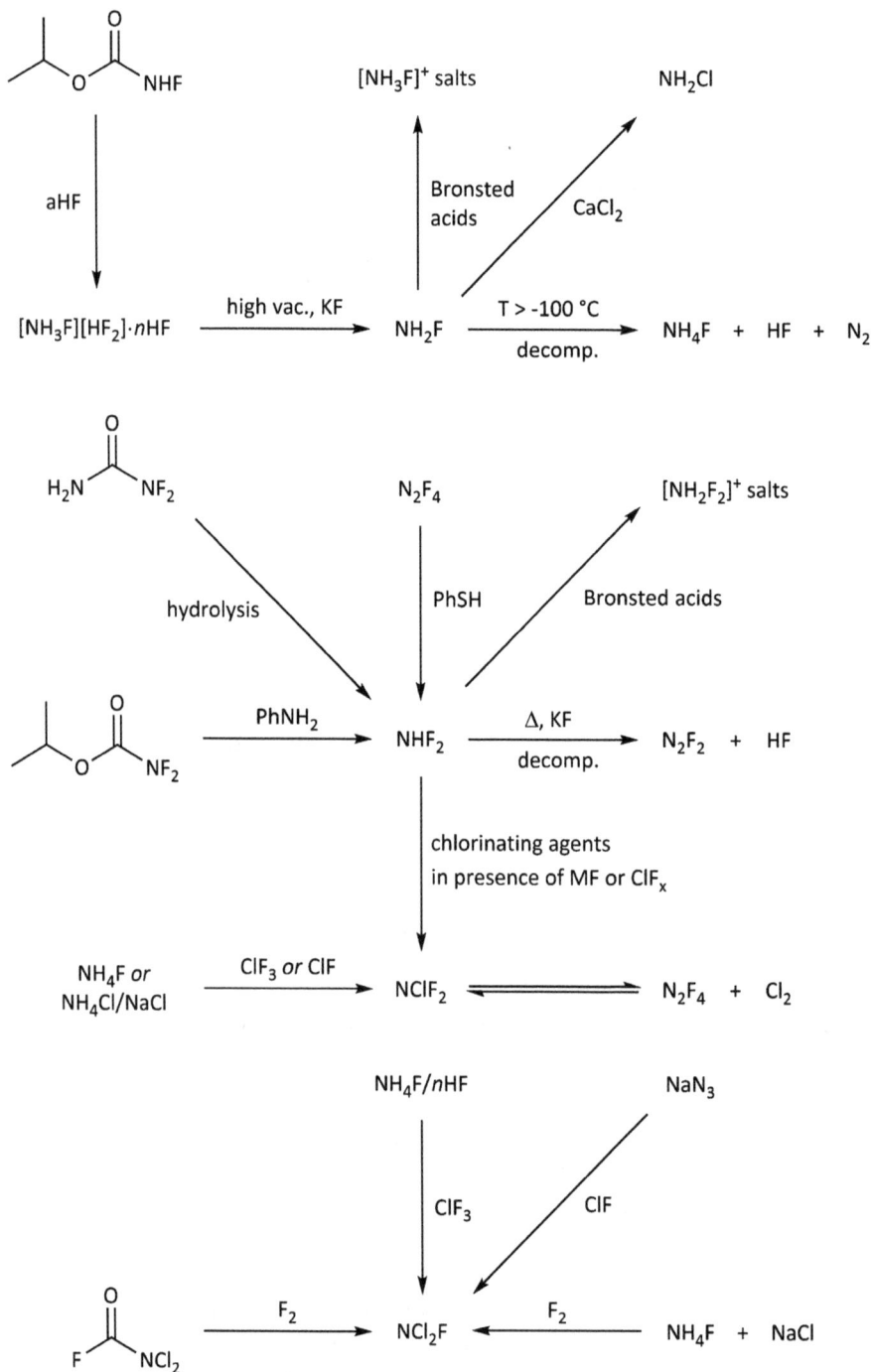

Scheme 9.9: Preparation and selected reactions of the fluoroamines NH_2F and NHF_2 (top) and the chlorofluoroamines NCl_2F and $NClF_2$ (bottom).

ered as an extremely thermolabile, colorless solid that melts at about $-100\,°C$ under complete decomposition into NH_4F, HF and N_2 [287]. Adducts of the type $[NH_3F][HF_2]$ $\cdot nHF$ are prepared by the reaction of anhydrous HF with N-fluoroisopropylcarbamate, which is in turn synthesized by direct fluorination in an aqueous solution [288, 289]. NH_2F is a strong amination agent, as chloroamine NH_2Cl is readily formed, when the fluoroamine is passed through granulated $CaCl_2$ (Scheme 9.9). It is a weak base and forms fluoroammonium salts $[NH_3F]^+$ upon reaction with strong Brønsted acids HX (e. g., $X^- = HF_2^-$, SO_4H^-, SO_3F^-, SO_3Cl^-, ClO_4^-). The anion X^- must not possess nucleophilic properties; otherwise, the fluoride in NH_2F will be substituted.

Difluoroamine (fluorimide) **NHF_2** is a product of the electrolysis of molten ammonium bifluoride $[NH_4]^+[HF_2]^-$, along with many other nitrogen fluorides (NH_2F, NF_3, N_2F_4, N_2F_2) [290]. The same range of compounds is obtained from the direct reaction of NH_3 with F_2 at low temperature [291]. More conveniently, it can be prepared by cleavage of tetrafluoro hydrazine N_2F_4 with thiophenol PhSH at 50 °C (Scheme 9.9). Another approach includes the hydrolysis of N,N-difluoro urea or the treatment of the carbamate $F_2NC(O)O^iC_3H_7$ with aniline [283, 292]. NHF_2 is a colorless, explosive gas (m. p.: $-116.4\,°C$; bp.: $-23.6\,°C$) and may be stored at room temperature. However, it thermally decomposes in presence of KF to yield HF and N_2F_2. The generation of the latter may be explained by intermolecular amination of NHF_2 with NHF_2. The process is supported by H-accepting fluoride ions [293]. The very weak base NHF_2 can form adducts with strong Brønsted or Lewis acids, for example, the difluoro ammonium ion $[NH_2F_2]^+$ (Scheme 9.9). Difluoro amine NHF_2 also acts as acid and reacts with bases like OH^- to generate the difluoroamide anion $[NF_2]^-$. The latter is transformed to N_2F_2 in absence of any other reactant, in presence of Fe^{3+} it is oxidized to N_2F_4 [294].

Tetrafluorohydrazine and difluorodiazene

Tetrafluorohydrazine **N_2F_4** is formed upon reduction of NF_3 with Cu, As, Sb or Bi at 375 °C (Scheme 9.10). The colorless, toxic gas (mp.: $-164.5\,°C$; bp.: $-73\,°C$) is also obtained from electrical discharge of Hg and NF_3 or the oxidation of HNF_2 with hypochlorite in alkaline media (pH = 12) [295]. Its structure is similar to hydrazine N_2H_4, where comparable ratios of the *trans* and the *gauche* conformer are present. The *trans*-conformer is about $2\,kJ\,mol^{-1}$ more stable, the barrier for the internal rotation of the NF_2-units amounts $12.5\,kJ\,mol^{-1}$. *In vacuo* or at elevated temperatures, N_2F_4 is in reversible equilibrium with $2\,NF_2$ (Scheme 9.10). NF_2 is one of the stable fluororadicals, which can be frozen out. It has a dark blue color and is isoelectronic to the radicals $[O_3]^-$, O_2F and ClO_2, and like all of them has an odd valence electron count. It is as "stable" as the radicals NO and NO_2 and has, along with the aforementioned nitrogen oxides, OF_2 and NOF a bent structure.

Due to its easy dissociation, N_2F_4 tends to form NF_2 compounds with many reaction partners, for example, NO, S_2F_{10} or Cl_2, to give products like $ONNF_2$, F_2NSF_5 or

HNF$_2$ NF$_3$ N$_2$F$_2$ + NF$_3$

OCl$^-$, pH = 12   Hg, electr. discharge   Br$_2$, hν

NF$_3$ $\xrightarrow[\text{375 °C}]{\text{Cu, As, Sb or Bi}}$ N$_2$F$_4$ $\xrightleftharpoons[]{Δ, \text{vacuum}}$ 2 ·NF$_2$

S$_2$F$_{10}$ or Cl$_2$ OF$_2$ fluoride acceptors, e.g. SnF$_4$, AsF$_5$ or SbF$_5$

F$_5$SNF$_2$ or ClNF$_2$ NOF + NF$_3$ [N$_2$F$_3$]$^+$[MF$_n$]$^-$

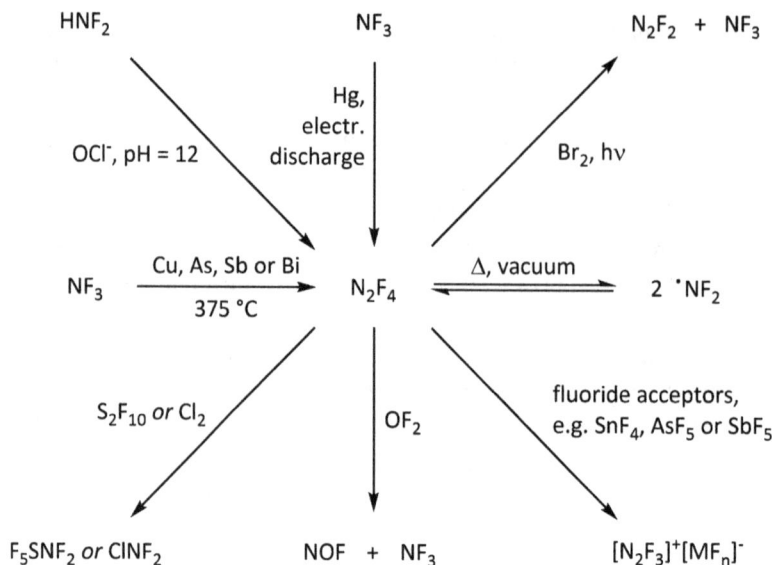

Scheme 9.10: Preparation and selected reactions of N$_2$F$_4$.

ClNF$_2$, respectively (Scheme 9.10) [283]. It can also add to olefins to form organic N,N-difluoroamines. N$_2$F$_4$ has fluorinating properties, for example, sulphur is oxidized to SF$_4$ and F$_2$NSF$_5$ at 110–140 °C. Against OF$_2$, it is a reducing agent and forms NF$_3$ and NOF. Hydrolysis at 60 °C yields HF and NO; at 133 °C nitrate and N$_2$ are obtained. N$_2$F$_4$ explosively reacts with H$_2$ under formation of N$_2$ and HF [296]. Stable complexes of the [N$_2$F$_3$]$^+$ cation are generated from N$_2$F$_4$ and strong fluoride acceptors, for example, [N$_2$F$_3$][SnF$_5$], [N$_2$F$_3$][AsF$_6$] or [N$_2$F$_3$][SbF$_6$] [297–299]. Tetrafluoro hydrazine may also be photolyzed in the presence of Br$_2$ to give N$_2$F$_2$ and NF$_3$ (Scheme 9.10) [300].

Difluorodiazene **N$_2$F$_2$** exists as a planar molecule in two isomeric forms, which are both generated upon thermal decomposition of KF · HNF$_2$ or FN$_3$ as well as upon electric discharge of NF$_3$ in presence of Hg (Scheme 9.11) [293, 295, 301, 302]. Another important synthetic approach is the electrolysis of NH$_4$[HF$_2$] or NH$_4$F in anhydrous HF [290, 303, 304]. Alternatively, N$_2$F$_4$ can be photolyzed in the presence of Br$_2$ [300, 305]. Moreover, a safer approach has been reported, producing N$_2$F$_2$ upon base-mediated decomposition of N,N-difluorourea H$_2$NC(O)NF$_2$ or the carbamate F$_2$NC(O)O^{i}C$_3$H$_7$ between 0-30 °C [292, 306]. At 300 °C, N$_2$F$_2$ decomposes into the elements. It does not react with water or oxygen. Pure $trans$-N$_2$F$_2$ is obtained by the reaction of N$_2$F$_4$ with Al$_2$Cl$_6$ at −80 to −112 °C or by reduction of [N$_2$F$_3$]$^+$[AsF$_6$]$^-$ with NOCl (Scheme 9.11) [307]. The selective reduction of N$_2$F$_4$ to pure $trans$-N$_2$F$_2$ also occurs, if it is reacted with the graphite intercalation compound C$_x$ · AsF$_5$ (x = 10–12) [308, 309]. Both isomers are in a temperature-dependent equilibrium, at 25 °C there are about 90 % cis-N$_2$F$_2$ [309]. The equilibrium is only fast at high temperatures or in presence of catalysts like stainless

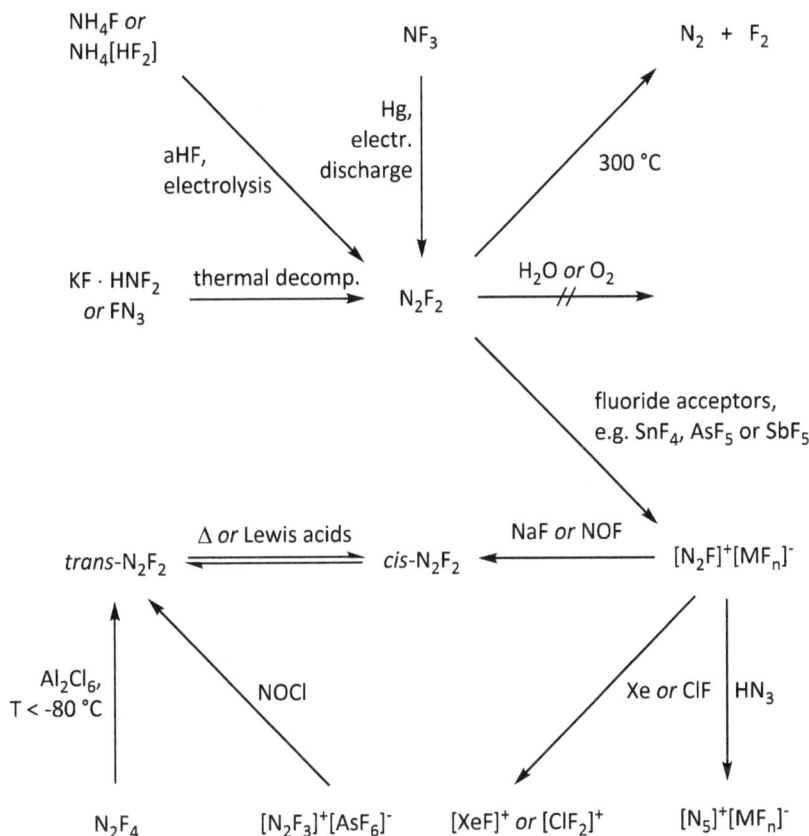

Scheme 9.11: Preparation and selected reactions of N_2F_2.

steel or strong Lewis acids, hence it is possible to separate the isomers by high vacuum distillation or gas chromatography (cis-N_2F_2: mp.: $< -195\,°C$; bp.: $-105.7\,°C$/$trans$-N_2F_2: mp.: $-172\,°C$; bp.: $-111.4\,°C$). The $trans$-isomer is about $12\,kJ\,mol^{-1}$ higher in energy than the cis-form, accordingly, the N=N-bond length in the cis-isomer has been found to be slightly shorter, whereas the N-F-distances are a bit elongated with respect to $trans$-N_2F_2. However, surprisingly cis-N_2F_2 is slightly more reactive [290, 310]. For example, cis-N_2F_2 slowly attacks glass, whereas the $trans$-isomer basically remains unchanged after several weeks. Upon heating or in presence of strong Lewis acids, the $trans$-isomer is transformed into the cis-compound [303, 309]. N_2F_2 reacts with SO_2 under both, photochemical and thermal conditions yielding the same products, being SOF_2, SOF_4, SO_2F_2, $S_2O_5F_2$, N_2O and N_2 [311]. Furthermore, N_2F_2 reacts with strong Lewis acids (AsF_5, SbF_5, SnF_4) in HF to form salts which contain the linear $[N_2F]^+$ cation (Scheme 9.11) [298, 302, 309, 312, 313]. The $[AsF_6]^-$-salt can be sublimed in high vacuum at $75\,°C$ under minor decomposition. $[N_2F]^+$ is isoelectronic to N_2O, $[NO_2]^+$ and CO_2. In its $[AsF_6]^-$-salt, it has been found to have both, a very short N-N-distance

(109.9 pm), comparable to that of N_2, and the shortest N-F bond known (121.7 pm) [313]. It reacts with NaF in HF to regenerate cis-N_2F_2 and $Na[AsF_6]$ (Scheme 9.11). In a related reaction, NOF leads to the formation of $NO[AsF_6]$ and exclusively cis-N_2F_2. The $[N_2F]^+$ cation also oxidizes Xe to $[XeF]^+$ and ClF to $[ClF_2]^+$, however, it does not oxidize ClF_5, BrF_5, IF_5, XeF_4, NF_3 or O_2 [309, 313]. Another interesting reaction of $[N_2F]^+$ with HN_3 yields the highly energetic, planar, V-shaped $[N_5]^+$ cation (Scheme 9.11) [314]. Attempts to prepare a potential N_5F by thermolysis of $[N_5][AsF_6]$ or displacement of $[N_5][SbF_6]$ with CsF have not been successful, due to the kinetic instability of the compound. However, the expected decomposition products NF_3, N_2F_2 and FN_3 have been observed [315]. It has been tried to use the salt $[N_2F][AsF_6]$ as fluorinating agent for arenes, which has not resulted in the formation of satisfying amounts of fluoroaromatics [316]. Neutral N_2F_2 has also been investigated in organic synthesis. For example, it shows the typical reactivity of electrophilic N-F reagents and is thus capable of transforming cyclic ketones into α-fluoroketones via intermediary enamines [317].

Tetrafluoroammonium salts

A covalent nitrogen pentafluoride **NF$_5$** does not exist, even if quantum-chemical calculations predict a local energetic minimum for a potential D_{3h}-symmetric NF_5 molecule [318]. Only a very labile, ionic $NF_4^+F^-$ of comparable energy has been reported, which decomposes to NF_3 and F_2 even at the very low temperature of $-142\,°C$ [319, 320]. However, there are numerous compounds with the tetrahedral tetrafluoroammonium cation $[\mathbf{NF_4}]^+$ [257, 321–328]. Expectedly, the most suitable anions for this species are based on strong fluoride acceptors, such as AsF_5 or SbF_5. For the preparation of $[NF_4]^+$-salts, NF_3, F_2 and the respective element fluoride are reacted under activation by electrical discharge, photolysis or thermal energy at high pressure. The formation mechanism of $[NF_4]^+$ in this reaction system has been investigated by a computational study, revealing the crucial involvement of a radical species $BF_4^.$, $AsF_6^.$ or $SbF_6^.$, which oxidizes NF_3 to the radical cation $[NF_3]^{+\cdot}$. In the next step, the radical cation recombines with a fluorine atom $F^.$ to form $[NF_4]^+$ [329]. Alternative synthetic approaches toward tetrafluoroammonium salts include the oxidation of NF_3 with PtF_6 or more conveniently with $[KrF]^+$-salts [330]. Finally, substitution or metathesis reactions can be used for the generation of $[NF_4]^+$-salts, starting with readily accessible anions like $[BF_4]^-$, $[AsF_6]^-$ or $[SbF_6]^-$. Almost all of the perfluoroammonium salts are colorless, crystalline solids which are very hygroscopic and sensitive toward hydrolysis. They are excellent ionic oxidizers; a special example is $[NF_4]_2[NiF_6]$, a deep-red, hygroscopic solid. It can be generated by the metathesis of $[NF_4][SbF_6]$ with $Cs_2[NiF_6]$ in HF and is stable up to 100 °C. It has been reported as the first compound that is comprised of two extremely oxidizing ions. The thermal decomposition of $[NF_4]_2[NiF_6]$ yields three equivalents of elemental fluorine [327].

Fluoroazide

Fluoroazide **FN_3** is a yellow, very explosive gas (mp.: $-139\,°C$; bp.: $-82\,°C$). It is prepared by fluorination of HN_3 and already starts to decompose to N_2F_2 and N_2 at room
temperature [331, 332]. The decomposition reaction is catalyzed by copper or copper
fluoride. In the solid and liquid phase, FN_3 is extremely sensitive, so that very violent
explosions may result, if it is either cooled to $-196\,°C$ or rapidly vaporized. However, at
$-80\,°C$ and a reduced pressure of 10–20 mbar, it can be stored without decomposition.
In gas-phase reactions at room temperature, FN_3 does not react with H_2O, O_2, XeF_2 or
OF_2, but it transfers intermediary NF to CO, NO and COS [332]. Furthermore, FN_3 forms
stable adducts with the Lewis acids BF_3 and AsF_5 at $-196\,°C$, however, the formation
of $[N_3]^+$ cations is not observed [333]. The molecular structure of FN_3 has been determined experimentally and supported by quantum chemistry, revealing a slightly bent
N_3-unit ($170.9°$) [334, 335].

Oxyfluorides of nitrogen

Apart from the binary nitrogen-fluorine compounds, three mostly covalent oxyfluorides are known. Nitrosyl fluoride **NOF** is the product of the fluorination of NO and can
be considered as the acid fluoride of nitrous acid. It is a colourless gas (mp.: $-132.5\,°C$;
bp.: $-59.9\,°C$) with a bent structure. NOF has fluorinating properties against many elements and forms nitrosyl compounds with fluoride acceptors [336]. Similarly, nitryl
fluoride **NO_2F** is a colorless gas (mp.: $-166\,°C$; bp.: $-72.5\,°C$), which is generated upon
fluorination of NO_2 with CoF_3 at $300\,°C$ or with F_2 at $25\,°C$. It can be considered as the
acid fluoride of nitric acid and is also accessible by treatment of nitryl salts with fluoride donors. NO_2F has a planar structure and a reactivity comparable to that of NOF.
Both of them are sensitive towards hydrolysis [336]. In the low temperature fluorination of NO (8 K) and NO_2 (20 K) in an inert matrix, the isomeric hypofluorites N-O-F
and O-N-O-F are formed as well [337, 338].

Trifluoroamine oxide **ONF_3** is obtained upon electrical discharge of NF_3/O_2 or
$N_2/F_2/O_2$-mixtures at $-196\,°C$ as a colorless, toxic gas (mp.: $-160\,°C$; bp.: $-87.6\,°C$)
[339–341]. Apart from that, it can be prepared by fluorination of NOF with F_2 or IrF_6,
or by oxidative oxygenation of NF_3 using N_2O in presence of SbF_5 (Scheme 9.12).
ONF_3 is stable in glassware at room temperature, resistant to hydrolysis and acts as
oxidizing agent [341–344]. However, even if it is not readily hydrolyzed by water or
aqueous base, it is attacked by H_2SO_4 [345]. ONF_3 has a distorted tetrahedral structure and reacts with strong fluoride acceptors like AsF_5, SbF_5 or AuF_5 to give stable
complexes with the $[ONF_2]^+$ cation (Scheme 9.12) [341, 342, 346–350]. The latter has
a planar, C_{2v}-symmetric geometry and is isoelectronic to carbonyl fluoride COF_2 and
the nitrate anion NO_3 [346, 347, 350]. As the N-O bond in the parent ONF_3 has a high
double-bond character, the structure of it may be better described as $ONF_2^+F^-$. This
indicates that negative hyperconjugation takes place, where the negative charge is
distributed over all three F-atoms [344]. ONF_3 acts as a fluorinating agent against,

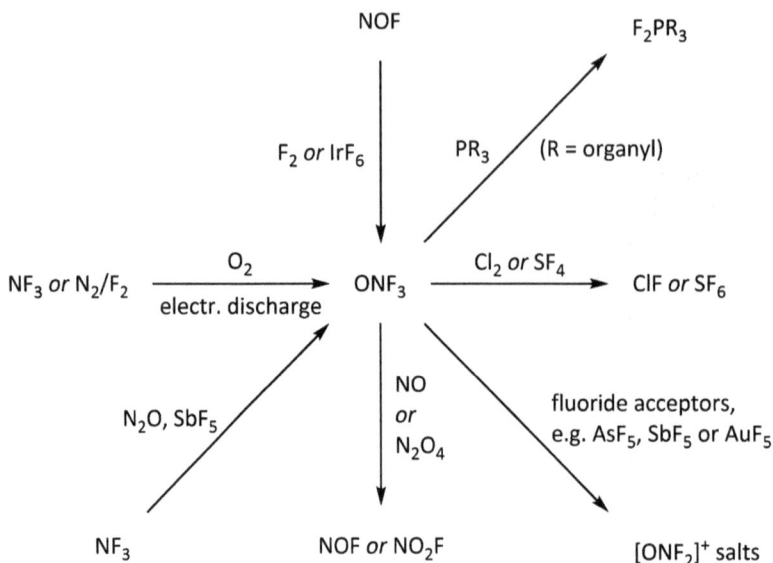

Scheme 9.12: Preparation and selected reactions of ONF_3.

for example, Cl_2, SF_4, NO or N_2O_4 to produce ClF, SF_6, NOF and NO_2F, respectively (Scheme 9.12). Phosphines are readily oxidized to difluorophosphoranes, however, PCl_3 or PF_3 are not oxidized by ONF_3. The outcome of these reactions perfectly indicates that ONF_3 actually acts as fluorinating, but rather not as an oxygenating agent [342, 345, 351, 352]. With the highly fluorinated olefins $CFX=CF_2$ (X = F, Cl, Br), ONF_3 reacts under Lewis acid catalysis to form the respective CF_2X-CF_2-ONF_2 compounds [353]. When solid ONF_3 is radio or photolyzed, the pyramidal $ONF_2^{.}$ radical can be detected by ESR-spectroscopy [341, 354–356].

9.5.2 Phosphorus, arsenic, antimony, bismuth

The heavier group-15 elements all exist as tri- and pentafluorides in binary fluorine compounds. Apart from that, P_2F_4 and many substituted derivatives of the binary compounds are known.

Diphosphorus tetrafluoride

Diphosphorus tetrafluoride $\mathbf{P_2F_4}$ is a stable, colorless gas (mp.: −86.5 °C; bp.: −6.2 °C) that results from the reaction of PF_2I with Hg under reduced pressure [357, 358]. The P-P bond is stronger than the respective N-N bond in N_2F_4, thus $PF_2^{.}$ radicals are only formed at high temperatures and low pressure [358]. The bond dissociation energy of the P-P-bond has been estimated to be in a region of about 240 kJ mol^{-1} [359]. Experimentally, the cleavage starts at about 350 °C and is not completed at 900 °C [360]. If gaseous P_2F_4 is instantaneously cooled from 900 to about −200 °C, the condensate contains PF_3 and P_2F_4 along with ˙PF_2 radicals and liquid, colorless $P(PF_2)_3$ (mp.: −66 °C; decomp.: >10 °C) [360]. In its ground state, the P_2F_4 molecule clearly shows a *trans*-geometry of its two [PF_2]-units [361, 362]. The reactivity of P_2F_4 is generally determined by the homolytic cleavage into ˙PF_2 radicals and follow-up reactions of the latter. Upon reaction with B_2H_6, the relatively stable, gaseous adduct $P_2F_4 \cdot BH_3$ is generated [363]. In contact with water, P_2F_4 is readily cleaved to yield the hydrolysis products PF_2H and F_2POPF_2, the latter of which can also be obtained by the reaction of PF_2I with HgO. In a similar way, the reaction with HI leads to the formation of PF_2H and PF_2I [357]. P_2F_4 may also be added to alkenes and alkynes under irradiative- or thermal generation of ˙PF_2 radicals [364–367]. Preferentially, the olefin substrates should be electron-rich to obtain reasonable yields of addition products, whereas the acetylenes should be substituted by a CF_3-group [367, 368].

Trifluorides

Phosphorus trifluoride **PF_3** is a toxic, colorless and in small concentrations odourless gas (mp.: −151.5 °C; bp.: −101.2 °C), which is produced upon halogen exchange of PCl_3 with HF, AsF_3 or ZnF_2 (Scheme 9.13). It has a trigonal-pyramidal structure with relatively short P-F distances (154.6 pm). This resembles a strong double bond character, when the bond length is compared to typical values of P-F single and double bonds ($d_{(P-F)}$: 166 pm; $d_{(P=F)}$: 152 pm) [369]. PF_3 is only slowly hydrolyzed in water, however, in alkaline solutions hydrolysis occurs rapidly under formation of phosphonic acids or phosphonates, respectively. It reacts with KF at 240 °C to disproportionate to elemental phosphorus and $K[PF_6]$ (Scheme 9.13). PF_3 reacts with O_2 under electrical discharge and forms the anhydride of difluorophosphoric acid $(OPF_2)_2O$ (mp.: 0.1 °C; bp.: 72 °C) along with polymeric $[-PF(O)-O-]_n$. In high pressure reactions, PF_3 is transformed to SPF_3 by H_2S and CS_2; accordingly, OPF_3 is obtained with CO_2 and SO_2 [370, 371]. The chalcogen derivatives EPF_3 (E = O, S, Se) are also accessible by the reaction of PF_3 with O_2, S or Se at about 4000 bar and 300 °C (Scheme 9.13) [372].

PF$_3$ is a very weak Lewis acid, and thus has no pronounced acceptor properties. However, in acetonitrile solution it reacts with anhydrous $[NMe_4]F$ to form colorless $[NMe_4]^+[PF_4]^-$ (Scheme 9.13). $[PF_4]^-$ has a ψ-trigonal-bipyramidal structure and is hydrolyzed in alkaline media to form $[POF_2]^-$, $[HPO_2F]^-$ and finally $[HPO_3]^-$. $[PF_4]^-$ adds H^+ and HF to form HPF_4 and $[HPF_5]^-$, respectively [373]. As a very weak Brønsted base, PF_3 may only be protonated by extremely strong acids like HF/SbF_5. On the other hand, PF_3 is a strong Lewis base and forms adducts, for example, with borane $PF_3 \cdot BH_3$. The reaction of PF_3 with SbF_5 is a bit controversial. In one publication, the formation of the ionic complex $[PF_2][SbF_6]$ has been reported [374]. In other studies, the reactions of the trifluoride with AsF_5 or SbF_5 have been formulated as redox reactions, thus forming PF_5 and the respective trifluoride AsF_3 or SbF_3. Subsequently, excess

P + K[PF$_6$]   EPF$_3$ (E = O, S, Se)

KF, 240 °C

O$_2$, S or Se,
4000 bar, 300 °C

PCl$_3$ $\xrightarrow{\text{HF, AsF}_3 \text{ or ZnF}_2}$ PF$_3$ $\xrightarrow[\text{electr. discharge}]{\text{O}_2}$ (F$_2$PO)$_2$O
+
-[P(O)F-O]$_n^-$

B$_2$H$_6$ [NMe$_4$]F   PtCl$_2$ or PdCl$_2$

PF$_3$ · BH$_3$ [NMe$_4$][PF$_4$]   Pt(PF$_3$)$_4$ or Pd(PF$_3$)$_4$

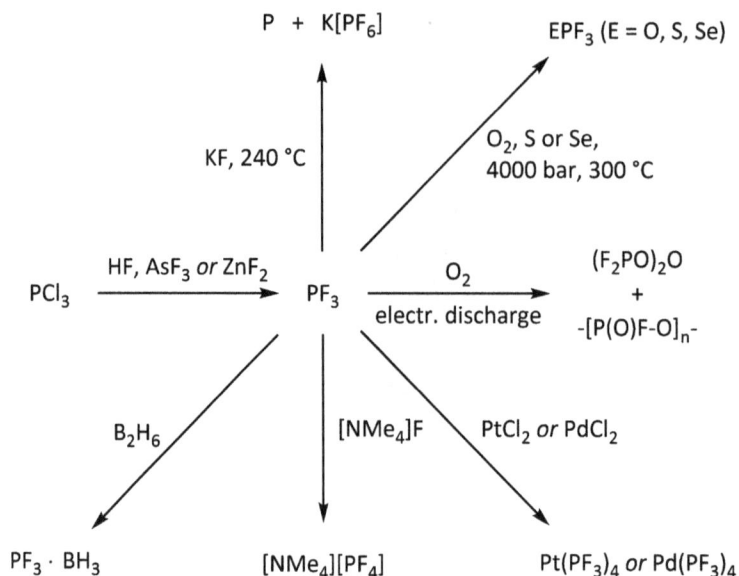

Scheme 9.13: Preparation and selected reactions of PF$_3$.

SbF$_5$ leads to the generation of [PF$_4$][Sb$_3$F$_{16}$] and SbF$_3$ · (SbF$_5$)$_x$ [375]. The most important property of PF$_3$ might be its ligand character, which is comparable to CO. Thus, transition metal complexes of CO and PF$_3$ show a similar behavior [376]. However, PF$_3$ is an even stronger π-acceptor, so that it can replace CO in carbonyl complexes. For example, the carbonyl complexes Pd(CO)$_4$ and Pt(CO)$_4$ can only be isolated at low temperatures under matrix conditions, whereas Pd(PF$_3$)$_4$ and Pt(PF$_3$)$_4$ are comparably stable. They are obtained as volatile liquids by the reaction of the anhydrous metal chlorides with excess PF$_3$, which takes the role of both, reducing agent and ligand (Scheme 9.13) [377–380]. The strong tendency to coordinate to metal centers also explains the toxicity of PF$_3$, as stable haemoglobin complexes are readily formed, which then prevent O$_2$ uptake and consequently lead to suffocation.

There exist many mixed fluorohalides PF$_2$X and PFX$_2$ (X = Cl, Br, I, pseudo halide), out of which PF$_2$I is probably the most relevant derivative. The chlorine- and bromine-containing compounds are typically prepared by partial fluorination of the trihalides PX$_3$ (X = Cl, Br) [381, 382]. In case of PF$_2$I, the synthesis is achieved by treatment of F$_2$PN(CH$_3$)$_2$ or other PF$_2$-containing substrates with HI [383]. PF$_2$I is used for the introduction of PF$_2$ groups into various substrates [384]. If it is reacted with N$_2$F$_4$ at 23 °C, the extremely explosive F$_2$NPF$_2$ is formed [385]. Similarly, the other phosphorus(III) halodifluorides PF$_2$Br and PF$_2$Cl are commonly used for the introduction of PF$_2$-groups into organic molecules. In most cases, this is realized by nucleophilic displacement of the halide using organolithium reagents as well as silyl or stannyl groups [386–388].

For the preparation of monofluoro phosphines with two organic substitutents R_2PF, lithiated substrates are reacted with PCl_2F [389].

Arsenic trifluoride **AsF$_3$** is a colorless liquid (mp.: $-6.0\,°C$; bp.: $62.8\,°C$) and can be generated by fluorination of As_2O_3 with HF or HSO_3F. Like PF_3, AsF_3 has a pyramidal structure [390]. It is readily hydrolyzed and has a wide application as fluorinating agent. Liquid AsF_3 conducts electricity, as it partly dissociates into $[AsF_2]^+$ and $[AsF_4]^-$, where the $[AsF_2]^+$ cation is stabilized by AsF_3 to form adducts of the type $[AsF_2]^+ \cdot nAsF_3$. Arsenic trifluoride has both, acid and base properties, thus it forms salts with the distorted ψ-trigonal bipyramidal anion $[AsF_4]^-$ upon addition of alkali metal or ammonium fluorides, whereas the reaction with fluoride acceptors like SbF_5 leads to the formation of $[AsF_2][SbF_6]$ [391, 392]. When it is reacted with Cl_2 at $0\,°C$, AsF_3 is transformed to a compound of the type $[AsCl_4][AsF_6]$. The latter exists as an ionic compound in polar solvents, but can be converted to molecular $AsCl_2F_3$ upon vacuum sublimation [393, 394]. There are also many derivatives of AsF_3, for example, the chloro compounds $AsClF_2$ and $AsCl_2F$ or many organic variants of the types $RAsF_2$ and R_2AsF, which are typically prepared from the aforementioned chlorinated compounds. AsF_3 can be used as fluorinating agent for the preparation of high-boiling fluorides, as the by-product $AsCl_3$ can be removed from the reaction mixture by distillation.

Antimony trifluoride **SbF$_3$** is formed upon reaction of HF with Sb_2O_3 as colourless, crystalline solid (mp.: $292\,°C$; bp.: ca. $345\,°C$). In the solid state, SbF_3 molecules are linked via F-bridges, which results in a distorted octahedral environment of fluorine atoms for every Sb-center [395]. SbF_3 is, like AsF_3, a moderately strong fluorinating agent and very sensitive toward hydrolysis. In the molten state, electric conductivity is measurable, which means that self-dissociation to $[SbF_2]^+$ and $[SbF_4]^-$ must take place. Like its lighter arsenic homologue, the cation $[SbF_2]^+$ is stabilized by SbF_3 and forms various adducts with different stoichiometry, for example, $[Sb_2F_5]^+$ or $[Sb_3F_7]^{2+}$. However, SbF_3 is only a very weak Lewis base, but a strong Lewis acid. With fluoride donors, salts of the types $M[SbF_4]$ and $M_2[SbF_5]$ are formed. In alkali metal salts, the $[SbF_4]^-$ anion is not monomeric like $[AsF_4]^-$, but forms either polymeric $[SbF_4]_n^{n-}$ or cyclic tetramers of the type $[SbF_4]_4^{4-}$, both being composed of ψ-octahedral $[SbF_5]$-units [396]. However, in presence of a larger cation, for example, 4-amino-1,2,4-triazolium, isolated, ψ-trigonal bipyramidal $[SbF_4]^-$ anions are observed [396]. In $M_2^I[SbF_5]$ (M^I – Na, NH_4), discrete $[SbF_5]^{2-}$ anions have been revealed by X-ray crystallography; they have a distorted square-pyramidal structure, and thus also a ψ-octahedral coordination geometry at the Sb-center [397]. Numerous fluoroantimonate(III) anions are accessible by varying the $SbF_3/[SbF_4]^-$-ratio, like for example, the singly-charged species $[Sb_2F_7]^-$, $[Sb_3F_{10}]^-$ and $[Sb_4F_{13}]^-$ as well as the multiply charged anions $[Sb_2F_9]^{3-}$, $[Sb_3F_{11}]^{2-}$, $[Sb_4F_{15}]^{3-}$ and $[Sb_5F_{19}]^{4-}$ [398, 399]. In general, the structures of the anions may depend on the cation. For example, in $K[Sb_2F_7]$ one can find infinite chains of alternating ψ-trigonal-bipyramidal $[SbF_4]^-$ anions and pyramidal SbF_3 molecules [400]. In contrast, $Cs[Sb_2F_7]$ contains discrete

$[Sb_2F_7]^-$ anions, wherein ψ-trigonal bipyramidal $[SbF_4]$-units are linked by an axial F-bridge [401]. SbF_3 forms covalent 1:1-complexes with AsF_5 and SbF_5 which are connected by F-bridges. Spectra of these compounds do not point to the existence of discrete $[SbF_2]^+$ cations [402]. However, the incomplete fluorination of elemental Sb yields not only SbF_3 but also the binary fluoride $Sb_{11}F_{43}$. Overall, the latter is composed of $[SbF_6]^-$ anions and the polymeric, chain-like cation $[Sb_6F_{13}]_n^{5+}$, in which units of $[SbF_2]^+$ and $[Sb_2F_5]^+$ are found [403, 404]. SbF_3 has some application as fluorinating agent for the synthesis of low-melting fluorides. Here, in contrast to $AsF_3/AsCl_3$, the product can be distilled off, whereas solid $SbCl_3$ typically remains in the reaction vessel.

Bismuth trifluoride **BiF₃** is generated in form of a colourless, crystalline powder (mp.: 649 °C; bp.: 900 °C), when Bi_2O_3 or BiOCl are reacted with HF. It is insoluble in water and is the only fluoride of group 15 to form a lattice of ions. In the β-form, it crystallizes in a distorted UCl_3-structure. The coordination environment of Bi may be described by a threefold capped, trigonal prism. Thus, the Bi-center is surrounded by nine F-atoms, out of which one of the fluorine caps is a bit more distant than the others. This might be due to the stereochemical effect of the lone pair at Bi. As a result, the coordination number of bismuth is written as 8+1. BiF_3 forms 1:1- and 1:3-complexes with AsF_5 and SbF_5, which, however, do not contain a discrete $[BiF_2]^+$ cation, but exist in form of F-bridged, covalent polymers [402]. The oxyfluoride BiOF is formed as crystalline product of the fluorination of Bi_2O_3 with BiF_3 [405].

Pentafluorides and their derivatives
In general, all the pentafluorides of group 15 are strong fluoride acceptors. Especially AsF_5 and SbF_5 are commonly used to abstract fluoride ions and stabilize unusual cationic species as $[MF_6]^-$ salts. However, the structural properties of the pentafluorides significantly change from P to Bi. Whereas PF_5 and AsF_5 are gaseous compounds under standard conditions, SbF_5 is a very viscous liquid and BiF_5 is a solid. These differences have a clear impact on the chemical behavior of the pentafluorides.

Phosphorus pentafluoride **PF₅** is a colourless gas (mp.: −93.7 °C; bp.: −84.5 °C), which can easily be obtained by treatment of PCl_5 with AsF_3 or directly by fluorination of elemental phosphorus (Scheme 9.14). The PF_5 molecule has a trigonal-bipyramidal structure with fluxional behavior. The latter can be visualized by ^{19}F-NMR spectroscopy, as the axial and equatorial nuclei are magnetically indistinguishable at room temperature. The intramolecular exchange of the F nuclei can be explained by means of pseudo rotation. PF_5 is readily hydrolyzed and, therefore, fumes on air, but it is stable in dry glass up to 250 °C. It is a strong Lewis acid and forms the octahedral hexafluorophosphate anion $[PF_6]^-$ upon reaction with fluoride ions (Scheme 9.14). The $[PF_6]^-$-salts can be generated directly from PF_5 and MF or by substitution of PCl_5 and MCl with HF. The free hexafluoro phosphoric acid HPF_6 is not stable in pure form, and thus, solutions of it can only be enriched to a concentration of about

P $\longrightarrow$ HPF_6(aq.) *or* $HPF_6 \cdot 6\,H_2O$

F_2 aqueous HF

PCl_5 $\xrightarrow{\ AsF_3\ }$ PF_5 $\xrightarrow{\ F^-\ }$ $[PF_6]^-$ salts

H^-
gas phase

phosphorous acids $\xrightarrow{\ HF\ }$ HPF_4 *or* H_2PF_3 $\xrightarrow{\ F^-\ }$ $[H_2PF_4]^-$ salts

e.g. NMe_3 decomp. AsF_5

$[HPF_5]^-$ salts $\xleftarrow{\ [HF_2]^-\ }$ $PF_3 + HF$ $[H_2PF_2][AsF_6]$

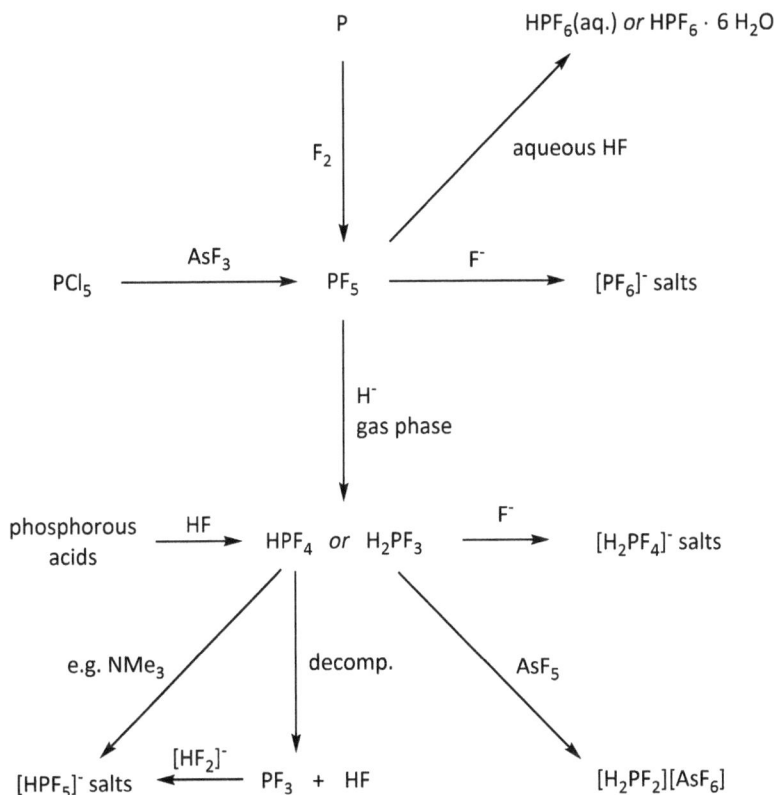

Scheme 9.14: Preparation and selected reactions of PF_5 and the partly hydrogenated derivatives HPF_4 and H_2PF_3.

75 % (Scheme 9.14). It can also be isolated in form of a stable hexahydrate (mp.: 31 °C). With the stronger fluoride acceptor SbF_5, PF_5 yields the unstable 1:3-complex $[PF_4]^+[Sb_3F_{16}]^-$ [375]. Apart from that, various mixed fluorohalides $PF_{5-n}X_n$ (X = Cl, Br) as well as a large number of fluorophosphoranes R_nPF_{5-n} (R = H, organic residue; n = 1–3) are known. Generally, the hydrogen-containing derivatives HPF_4 and H_2PF_3 are accessible by treatment of the respective phosphorous acid with HF or by gas-phase reaction of PF_5 with hydride sources (Scheme 9.14) [406–409]. Difluorophosphorane H_3PF_2 is prepared, if P_2H_4 is reacted with a large excess of HF or upon low-temperature fluorination of PH_3 with XeF_2 [410–412]. The tetrafluorophosphorane HPF_4 is readily cleaved into PF_3 and HF [407]. In turn, the reaction of PF_3 with $[HF_2]^-$ generates the pentafluorohydridophosphate anion $[PF_5H]^-$, which is furthermore accessible by the reaction of PF_4H with N-containing organic bases (Scheme 9.14) [413, 414]. Analogously, salts of the type $M^I[PF_4H_2]$ (M^I = K, Cs) with the D_{4h}-symmetric $[PF_4H_2]^-$ anion are obtained, if PF_3H_2 is reacted with M^IF [414, 415]. Moreover, PF_3H_2 is transformed to the difluorophosphonium cation $[PF_2H_2]^+$ upon treatment with AsF_5

[416]. In all of the fluorophosphoranes HPF_4, H_2PF_3 and H_3PF_2 the equatorial fluorine atoms are replaced by hydrogen, which has been found for the solid, liquid as well as for the gas-phase structures of the molecules [409, 417–419]. All three compounds are fluxional molecules and can act as both, Lewis-acids and bases [420].

Phosphorus oxytrifluoride **POF₃** is obtained by halogen exchange of $POCl_3$ with metal fluorides, for example, ZnF_2, PbF_2 or AgF, or electrical discharge of a PF_3-O_2 mixture [421]. It is a colorless gas (mp.: –39.1 °C (1.05 bar); bp.: –39.7 °C) and has a distorted tetrahedral structure [422]. POF_3 is sensitive toward hydrolysis and attacks glass. It forms adducts with BF_3, AsF_5 and SbF_5, which are coordinated via the oxygen atom [423, 424]. Also, it can add F^- at –140 °C to form the unstable, ψ-trigonal-bipyramidal $[POF_4]^-$ anion. Between –140 and –100 °C, $[POF_4]^-$ reacts with POF_3 to yield $[F_5PO\text{-}POF_2]^-$, which is then transformed to $[PF_6]^-$ and $[PO_2F_2]^-$ in presence of fluoride ions at higher temperatures [425, 426].

Phosphorus thiotrifluoride **PSF₃** can be made in an analogous way like POF_3, when P_2S_5, $SPCl_3$ or $SPBr_3$ are fluorinated [427–430]. The colorless gas (mp.: –148.8 °C; bp.: –52.5 °C) may also be generated via the pressure reaction of PF_3 with CS_2 or COS [370, 371]. As expected, it also has a distorted tetrahedral structure [431].

Arsenic pentafluoride **AsF₅** is a colorless gas (mp.: –79.8 °C; bp.: –53.2 °C) and the product of the fluorination of elemental arsenic or As_2O_3 (Scheme 9.15). Like PF_5, it has a trigonal-bipyramidal structure which is subject to pseudorotational exchange of axial and equatorial positions. Thus, one can only observe one signal in ^{19}F NMR at room temperature. AsF_5 is a strong Lewis acid and a strong fluoride acceptor. With fluorides, it readily forms the octahedral hexafluoroarsenate anion $[AsF_6]^-$ (Scheme 9.15). The bonding situation in $[AsF_6]^-$ salts can vary from strongly ionic to covalent, F-bridged contributions of the ions, depending on the strength of the fluoride-donating base. At –78 °C and in presence of excess AsF_5, the undecafluoro diarsenate anion $[As_2F_{11}]^-$ is detected (Scheme 9.15) [432]. AsF_5 may also form $[AsF_4]^+$ cations upon reaction with very strong fluoride acceptors, as for example, $[AsF_4]^+[PtF_6]^-$ is prepared from AsF_5 and F_2 on a Pt-wire at red heat. Arsenic pentafluoride is capable of intercalating into graphite at room temperature, which results in blue-black $C_{10}AsF_5$. However, upon heating to 60 °C, AsF_5 will be released again [433].

AsF_5 also has oxidizing properties, as it for example, reacts with elemental sulphur to yield $[S_8][AsF_6]_2$ and AsF_3 (Scheme 9.15). Some transition metals are oxidized in SO_2-solution to form compounds of the types $M[AsF_6]_2$, $[MF][AsF_6]$ and their corresponding solvates [434, 435]. An oxidation reaction is also observed for elemental mercury, which is transformed to $[Hg_3][AsF_6]_2$, containing the linear, symmetrical $[Hg_3]^{2+}$ cation (Scheme 9.15) [436]. $[AsF_6]^-$ is not hydrolyzed in alkaline solutions, whereas it forms hexafluoroarsenic acid $HAsF_6$ in strongly acidic media. The latter reacts reversibly with water and generates $H[AsF_5(OH)]$ and HF [437]. The fluorohydroxoarsenates can be isolated as alkali salts. The anions $[AsF_5(OH)]^-$ and $[AsF_4(OH)_2]^-$ rapidly condense to yield the O-bridged dimers $[F_5As\text{-}O\text{-}AsF_5]^{2-}$ and $[(F_4AsO)_2]^{2-}$ with an octahedral coordination environment at the As-centers [438–440]. There are also numer-

As $\xrightarrow{F_2}$ AsF$_5$

As $\xrightarrow{\text{F}_2, \text{Pt-wire (red heat)}}$ [AsF$_4$][PtF$_6$]

As$_2$O$_3$ $\xrightarrow{F_2}$ AsF$_5$ $\xrightarrow{F^-}$ [AsF$_6$]$^-$ salts

AsF$_5$ $\xrightarrow{\text{Hg}}$ [Hg$_3$][AsF$_6$]$_2$

AsF$_5$ $\xrightarrow{\text{sulphur}}$ [S$_8$][AsF$_6$]$_2$

[AsF$_6$]$^-$ salts $\xrightarrow[\text{-78 °C}]{\text{excess AsF}_5,}$ [As$_2$F$_{11}$]$^-$ salts

SbF$_3$ $\xrightarrow{F_2}$ SbF$_5$

SbF$_5$ $\xrightarrow{\text{MF} \ (\text{M = K, Cs, NMe}_4)}$ M$_2$[SbF$_7$]

Sb$_2$O$_3$ $\xrightarrow{F_2}$ SbF$_5$ $\xrightarrow{F^-}$ [Sb$_n$F$_{5n+1}$]$^-$ salts

Bi $\xrightarrow[\text{pressure}]{\text{F}_2, 500\text{-}600 \text{ °C}}$ BiF$_5$ $\xrightarrow{F^-}$ [BiF$_6$]$^-$ salts

BiF$_5$ $\xrightarrow{\text{ClOF}_3 \text{ or } \text{XeF}_6}$ [ClOF$_2$][BiF$_6$] or [XeF$_5$][BiF$_6$]

BiF$_5$ $\xrightarrow{\text{excess F}^-}$ [BiF$_7$]$^{2-}$ salts

BiF$_5$ $\xrightarrow{\text{NF}_3/\text{F}_2 \text{ pyrolysis}}$ [NF$_4$][BiF$_6$]

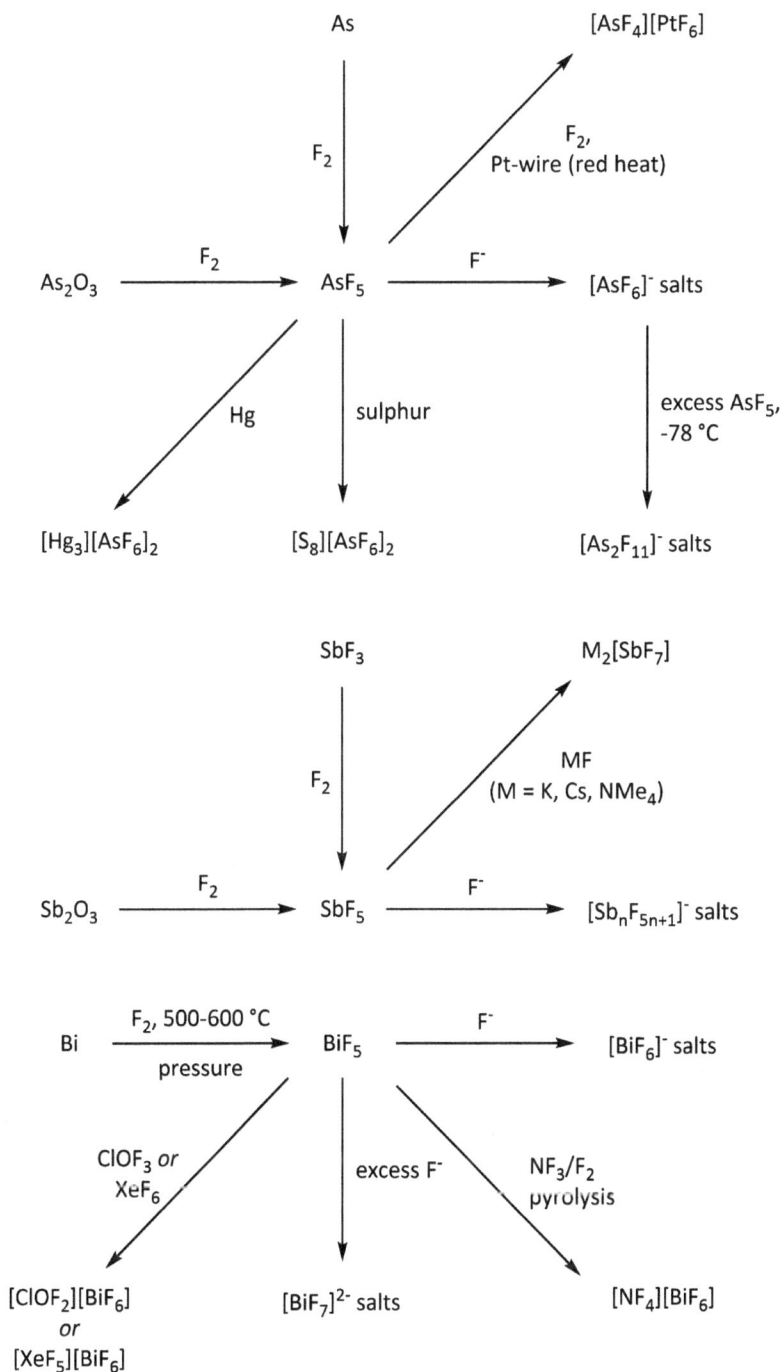

Scheme 9.15: Preparation and selected reactions of AsF$_5$, SbF$_5$ and BiF$_5$.

ous derivatives of AsF_5 and $[AsF_6]^-$, which contain mixed substituents, for example, AsF_3Cl_2, $AsF_{5-n}R_n$ or $[AsF_{6-n}R_n]^-$ (R = organic residue).

Arsenic oxytrifluoride **AsOF₃** is obtained by the thermal decomposition of $AsF_3(ONO_2)_2$ in vacuo. It is a white, hygroscopic solid, which does not dissolve in nonpolar solvents. At 240 °C, $AsOF_3$ decomposes without melting [441]. As expected, the shape of the molecule is derived from a tetrahedron, as has been shown by vibrational spectroscopy under matrix conditions [442].

Antimony pentafluoride **SbF₅** is a viscous liquid (mp.: 8.3 °C; bp.: 141 °C) that is produced by fluorination of Sb_2O_3 (Scheme 9.15). It is associated via F-bridges in the solid, liquid and gaseous state. In the solid state, there are cyclic tetramers with *cis*-F-bridges, so that every Sb atom has a distorted octahedral coordination environment [443]. Both the solid-state structure as well as the molecular structure in the liquid phase are similar to that of NbF_5. In liquid phase and in solution, SbF_5 forms *cis*-F-bridged polymers. The bridges are so strong, that they stay unchanged below the boiling point [444, 445]. Gaseous SbF_5 is still heavily associated, however, above 350 °C one can detect monomeric molecules [446–448].

Antimony pentafluoride is one of the strongest Lewis acids and fluoride acceptors. It can be arbitrarily mixed with liquid HF, which results in the formation of conductive solutions, comprised of the ions $[H_2F]^+$ and $[SbF_6]^-$. The conductivity maximum is achieved, when 3.5 mol-% SbF_5 are added, with higher SbF_5-concentrations, polymeric anions $[Sb_nF_{5n+1}]^-$ are observed [449]. Furthermore, the cation is better written as $[H(HF)_n]^+$, as it is highly hydrogen bonded and polymeric in nature [450]. SbF_5 reacts with graphite to form hydrolytically more stable intercalation compounds, which are additionally better to handle [451]. If SbF_5 is treated with SO_3, the viscous, polymeric adduct $SbF_4(SO_3F)$ is formed, which is a strong acid in HSO_3F-solution. Therefore, SbF_5 is an excellent compound for the enhancement of the acidity of HF and HSO_3F, respectively [452, 453].

With most of the element fluorides, SbF_5 reacts under formation of hexafluoro antimonates $[SbF_6]^-$ or polymeric anions $[Sb_nF_{5n+1}]^-$ (Scheme 9.15). In all of these species, antimony is octahedrally coordinated by six fluorine atoms, in the polymeric anions, this is realized by linkage of the $[SbF_6]$-units via *cis*-F-bridges. SbF_5 is also capable of forming heptafluoro antimonates $[SbF_7]^{2-}$ with a pentagonal bipyramidal structure in salts of the type $M_2[SbF_7]$ (M = K, Cs, NMe_4) [454]. SbF_5 reacts with methyl fluoride in liquid SO_2 and gives rise to the complex salt $[MeOSO]^+[Sb_2F_{11}]^-$ [455–457]. It also forms complexes of the types $[SbF_5 \cdot \text{base}]$ and $[(SbF_5)_2 \cdot \text{base}]$ with many weak bases, for example, SO_2ClF, SOF_2 or SO_2 [458]. This also includes the aggregation of $nSbF_5$ with $[SbF_6]^-$ to form the chain-like anions $[Sb_2F_{11}]^-$, $[Sb_3F_{16}]^-$, etc. The very low basicity of these fluoroantimonates is underlined by the existence of salts like $[O_2]^+[SbF_6]^-$, $[S_4]^{2+}[Sb_2F_{11}]_2^-$, $[Me_3C]^+[SbF_6]^-$ or $[Br_2]^+[Sb_3F_{16}]^-$. Analogously to AsF_5, SbF_5 is an oxidizing agent and stabilizes numerous different and partly unusual cations as fluoroantimonate salts [459]. Another analogy is the reversible equilibrium which is formed upon hydrolysis of $[SbF_6]^-$ in HF solution to form $[SbF_5(OH)]^-$ [460].

Mixed antimony chlorofluorides are generated by careful addition of small amounts $SbCl_5$ to SbF_5. This leads to a decrease of the viscosity as Sb-F-Sb-bridges are cleaved upon F/Cl exchange. In addition, the electric conductivity is increased due to the formation of a species with more ionic character. Mixtures of SbF_5 containing small amounts of $SbCl_5$ are far better fluorinating agents. Upon the fluorination of $SbCl_5$ with AsF_3 as fluorinating agent, the tetrameric $(SbCl_4F)_4$ is formed, with ClF, $SbCl_2F_3$ is obtained. The latter is comprised of $[SbCl_4]^+$ cations and $[F_4ClSb-F-SbClF_4]^-$ anions [461]. $SbCl_4F$ reacts with SbF_5 in liquid SO_2 to yield $SbCl_3F_2$, which is aggregated to cis-F-bridged tetramers, so that every Sb-atom is surrounded by each three Cl- and F-atoms in a distorted octahedral manner [462]. The reaction of SbF_5 with NaCl at $-10\,°C$ in SO_2 yields cis-$Na[SbF_2Cl_4]$ as main product. $Na[SbCl_6]$ is transformed to $Na[SbCl_2F_4]$, when it is treated with excess HF at room temperature. The fully substituted $Na[SbF_6]$ is obtained, when the aforementioned reaction is performed under pressure at $30\,°C$ [463].

Antimony oxytrifluoride **SbOF$_3$** is obtained, if $SbF_3(ONO_2)_2$ is carefully thermolized. Alternatively, it is made from SbF_3Cl_2 and excess Cl_2O. It is a white, very hygroscopic solid and is thermally stable up to $200\,°C$, where it starts to decompose without melting. With pyridine, the formation of a 1:1-adduct is observed. The respective dioxyfluoride **SbO$_2$F** is also known. Accordingly, it is prepared by thermal decomposition of $SbF(ONO_2)_4$ or treatment of $SbFCl_4$ with excess Cl_2O. SbO_2F has comparable solvation and hydrolysis properties like $SbOF_3$ but it is thermally more stable, as its decomposition does not take place below $350\,°C$ [441].

Bismuth pentafluoride **BiF$_5$** is synthesized by fluorination of elemental, molten bismuth at $500-600\,°C$ under pressure (Scheme 9.15) [464]. It is a colorless, crystalline and sublimable solid (mp.: $151.4\,°C$; bp.: $230\,°C$), which exists in form of $trans$-F-bridged chains in the solid state. Every Bi-atom is octahedrally surrounded by six fluorine atoms [465]. In the gas phase, BiF_5 is also associated [446, 466–468]. It forms long, white needles when it is sublimed, decomposes upon heating to form BiF_3 and F_2 and it is a strong oxidant [464]. BiF_5 is a very strong fluorinating agent and Lewis acid and readily forms fluorobismuthates with fluoride donors. For example, UF_5 is fluorinated to UF_6, whereas the reactions with $ClOF_3$ or XeF_6 lead to the formation of the ionic 1:1-adducts $[ClOF_2][BiF_6]$ and $[XeF_5][BiF_6]$, respectively (Scheme 9.15) [469–471]. Interestingly, there also exist mostly oligomeric adducts of SbF_5 and BiF_5 in many different ratios [472]. This might be due to the fact, that both pentafluorides are comparable in their chemical behavior. However, SbF_5 is the slightly stronger Lewis acid, as it is capable of abstracting F^- from $[BiF_6]^-$-salts [473]. If BiF_5 is treated with a mixture of NF_3 and F_2, the adduct $[NF_4][BiF_6] \cdot nBiF_5$ is formed, which can be pyrolyzed to give $[NF_4][BiF_6]$ (Scheme 9.15). The latter is also accessible by the displacement reaction of $[NF_4][BF_4]$ with BiF_5 [325]. Hexafluoro bismuthates(V) of the type $M^I[BiF_6]$ (M^I = Li, Na, K, Cs) are obtained, when BiF_5 is reacted with alkali metal fluorides M^IF [474–476]. The potassium salt $K[BiF_6]$ is comprised of octahedral $[BiF_6]$-units, which can exist in two different forms. One of them is the cubic α-modification, which is stable at room temperature, and has six equivalent Bi-F distances. The other variant is the tetragonal β-form with four shorter and two slightly elongated Bi-F contacts [475]. Apart from that, the pentagonal-bipyramidal $[BiF_7]^{2-}$ anion is also known (Scheme 9.15). It is accessible by treatment of BiF_5 with excess $M^I F$ (M^I = Na, K, Rb, Cs, NO, NMe_4). Salts with the octafluorobismuthate anion $[BiF_8]^{3-}$ have not been prepared so far [454]. In contrast to SbF_5, BiF_5 does not form polyanions of the type $[Bi_nF_{5n+1}]^-$ [473].

9.5.3 Study questions

(I) How is NF_3 prepared? Compare its structural properties and general chemical behaviour with NH_3.

(II) Which are the most important derivatives of NF_3 containing Cl- or H-substituents? How are they accessible and what is their main chemical behavior?

(III) What happens, if N_2F_4 is heated? Which general consequences does this have on its reaction with compounds like S_2F_{10} or Cl_2?

(IV) N_2F_2 exists in form of two isomers. How can one selectively obtain them? Which of them is more stable? Do they differ in their reactivity?

(V) What happens upon treatment of N_2F_2 with strong fluoride acceptors? Which properties does the product have and which ionic nitrogen-species can be prepared from it?

(VI) Comment on the existence of fluorinated pentavalent nitrogen species. How are they prepared?

(VII) What is the main application of the three oxyfluorides NOF, NO_2F and NOF_3? Which species are obtained, if they are treated with a strong fluoride acceptor? Which structures do the corresponding compounds have?

(VIII) How is P_2F_4 prepared? Compare its reactivity with N_2F_4.

(IX) Compare the fluoride donor- and acceptor properties of PF_3, AsF_3, SbF_3 and BiF_3.

(X) Which property of PF_3 leads to a comparable behaviour to CO? Does the use of PF_3 allow the preparation of compounds, which do not exist as stable CO derivatives?

(XI) Why are AsF_3 and SbF_3 common fluorinating agents? In which cases are they preferentially used?

(XII) The pentafluorides PF_5, AsF_5, SbF_5 and BiF_5 are all strong fluoride acceptors. Do they also donate fluoride? Which of them is the strongest acceptor?

(XIII) Which types of fluoroanions are obtained, when the pentafluorides PF_5, AsF_5, SbF_5 and BiF_5 are reacted with fluoride donors?

9.6 Group 16: Oxygen, sulphur, selenium, tellurium, polonium

9.6.1 Oxygen

There are several binary oxygen fluorides known, which mainly includes the radicals $\cdot$OF and $\cdot O_2F$ as well as the compounds OF_2 and O_2F_2. Apart from these species, there exist so called fluoroxy compounds, which are often named hypofluorites, according to the nomenclature of the oxyacids of the higher halogen homologues. All of the oxygen-fluorine compounds are strong oxidants and fluorinating agents and, except OF_2, thermodynamically unstable.

Oxygen fluorides

Oxygen monofluoride ˙**OF** is a radical that is formed upon thermal and photolytic de-composition of OF_2 (Scheme 9.16). It is an intermediate compound in all reactions of the oxygen fluorides [477–479]. It can be detected by infrared-spectroscopy, when OF_2 is photolyzed in a matrix at 4–15 K [480, 481]. Other preparative methods include the photolysis of F_2-NO_2-mixtures at 8 K as well as the simultaneous deposition of OF_2 and a beam of metal atoms (Li, Na, K, Mg) under matrix conditions [482, 483]. Conve-niently, ˙OF can furthermore be generated from F-atoms and O_3 (Scheme 9.16) [484]. The recombination of ˙OF radicals toward O_2F_2 is insignificant, instead, mainly the formation of O_2 and F-atoms is observed [479, 485]. The bond distance in ˙OF (136 pm) has been determined by rotational spectroscopy and can be regarded as intermediate between a single- (ca. 140 pm) and a double bond (ca. 120 pm), which is comparable to the situation in isoelectronic $[O_2]^-$ [483, 484]. The molecular properties of ˙OF have further been predicted by quantum chemical methods [486]. The ˙OF radical has also been discussed in terms of atmospheric chemistry. Since the reaction of fluorine atoms with ozone yields ˙OF, it has been assumed, that this reaction might also play a role in the depletion of the ozone layer. However, based on kinetic investigations, this process has not been found to be of importance [478, 487, 488].

Dioxygen monofluoride ˙O_2F is also a radical species, which is obtained under matrix conditions at 4 K upon photolysis of OF_2-O_2- or F_2-O_2-mixtures (Scheme 9.16) [489, 490]. Apart from that, it can be synthesized by impact of an electrical discharge on a 2:1-mixture of oxygen and fluorine at $-200\,°C$ and a reduced pressure of 10 mbar. ˙O_2F is isoelectronic to $[O_3]^-$ and has a bent structure with a strong, short O-O- and a very weak, long O-F bond [490, 491]. These findings have also been supported by the-oretical calculations [486, 492]. Analogously, the solid, dark red-brown tetraoxygen difluoride O_4F_2 is accessible by the photolytic reaction of a 2:1-mixture of O_2 and F_2 at low temperature or electrical discharge of OF_2 and O_2 (Scheme 9.16) [493–496]. It decomposes above $-185\,°C$ to yield F_2 and O_2. In solution, O_4F_2 is in equilibrium with the radical ˙O_2F. Thus, O_4F_2 can be described as being built from two ˙O_2F molecules, which are weakly linked by an O-O-bridge [497].

˙O_2F is considered as an intermediate in the preparation of dioxygenyl salts from F_2, O_2 and fluoride acceptors [498]. Accordingly, O_4F_2 and OF_2 react with fluoride ac-ceptors like BF_3 to yield dioxygenyl salts of the type $[O_2][MF_6]$ (Scheme 9.16) [499]. In turn, ˙O_2F is generated, if the dioxygenyl salts are treated with fluoride-donating species like alkali metal fluorides or $ClOF_3$. Under appropriate conditions, the ob-tained ˙O_2F can decompose to O_2 and F˙ radicals, thus, being a radiation-free source of atomic fluorine [500, 501]. Both, ˙O_2F and O_4F_2 are stronger oxidizing and fluorinating agents than O_2F_2. Apart from that, O_4F_2 is a better source of the O_2F group than O_2F_2,

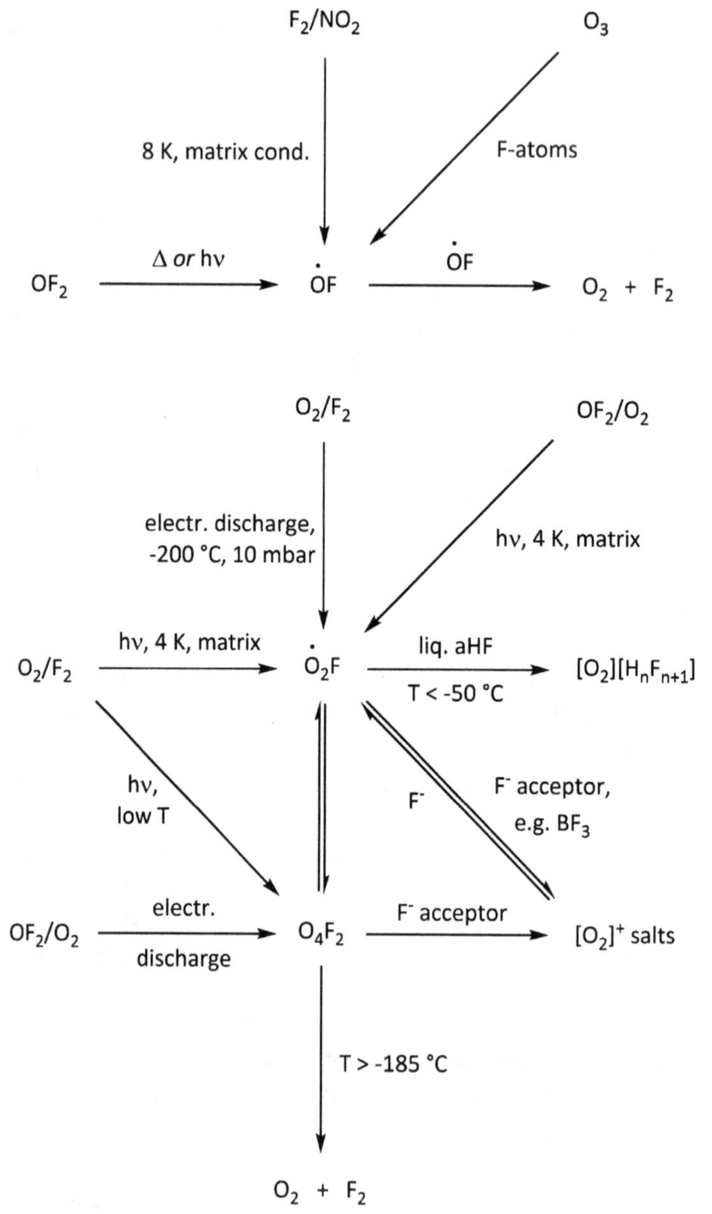

F_2/NO_2 O_3

8 K, matrix cond.

F-atoms

Δ or hν

OF_2 $\overset{\cdot}{O}F$ $\overset{\cdot}{O}F$ $O_2 + F_2$

O_2/F_2 OF_2/O_2

electr. discharge, -200 °C, 10 mbar

hν, 4 K, matrix

hν, 4 K, matrix

O_2/F_2 O_2F liq. aHF, $T < -50\ °C$ $[O_2][H_nF_{n+1}]$

hν, low T

F^- F$^-$ acceptor, e.g. BF_3

OF_2/O_2 electr. discharge O_4F_2 F$^-$ acceptor $[O_2]^+$ salts

$T > -185\ °C$

$O_2 + F_2$

Scheme 9.16: Preparation and selected reactions of the oxygen fluorides OF, O_2F and O_4F_2.

as has been observed in the reaction with SO_2. Still, the main product is SO_2F_2, which is obtained by fluorination, but the yield of FSO_2OOF is significantly increased [502].

O_2F forms loose aggregates of the type $(O_2F)_n$ below $-175\ °C$, which have been found to behave like O_4F_2 [490]. In liquid, anhydrous HF, O_2F is solvated as

$[O_2]^+[H_nF_{n+1}]^-$ and hence is found to be metastable up to −50 °C (Scheme 9.16). Such solutions have strongly oxidizing properties and are capable of transforming Ag(II) to Ag(III), Au(III) to Au(V) or Ni(II) to Ni(IV) in form of $[AgF_4]^-$, $[AuF_6]^-$ and $[NiF_6]^{2-}$, respectively [501]. The radical species O_2F has also been considered to be involved in ozone depletion processes in the stratosphere. However, as for other oxygen-fluoride species, the impact of this compound was found to be negligible, compared to, for example, the corresponding chlorine-containing species [488].

Oxygen difluoride **OF$_2$** is a colourless, very toxic gas (mp.: −223.8 °C; bp.: −145.3 °C), which is generated, when F_2-gas is introduced in water in presence of alkali metal fluorides. Alternatively, F_2 can be passed through a 0.5 M aqueous solution of KOH or NaOH (Scheme 9.17) [503]. It is also possible to synthesize OF_2 by fluorination of wet alkali metal fluorides or electrolysis of an aqueous KHF_2 solution as well as photolysis of a F_2-O_3-mixture [485]. Probably, most of the aforementioned reactions proceed via HOF as intermediate [504]. However, OF_2 is not the anhydride of HOF, as hydrolysis reactions do not result in the formation of hypofluorous acid [505]. In condensed form, OF_2 has an intense yellow color. The pure compound is nonexplosive and thermally stable up to 200–250 °C. The molecule has a bent structure with a F-O-F angle of 98°, where the bonding dissociation energies amount to 178.5 kJ mol^{-1} for the first O-F bond and 201.9 kJ mol^{-1} for the second one, respectively [506, 507]. Thus, OF_2 typically transfers the OF group in reactions. In many reactions, it is a strong oxidizer and fluorinating agent, however, it is less reactive than fluorine due to its higher activation energy. Many nonmetals are partly oxidized, partly fluorinated upon heating with OF_2. Water vapour reacts explosively after ignition, H_2S does so even at room temperature. OF_2 is only sparingly soluble in water and is slowly hydrolyzed to O_2 and HF in basic medium [508]. Aqueous solutions do not have acidic, but strong oxidizing properties as OF_2 reacts, for example, with hydrogen halides HX to form HF, H_2O and the elemental halogen X_2 (Scheme 9.17). In alkaline media, it does not form hypofluorites OF^-, but fluoride and oxygen. It reacts with NF_3 under electrical discharge to form ONF_3 and interacts photolytically with the noble gases Kr and Xe, resulting in KrF_2 and XeF_2. The reaction of OF_2 with SO_3 yields peroxysulphuryl difluoride FSO_2OOF [477, 509]. In presence of CsF, OF_2 transforms carbonyl fluoride COF_2 to the trioxide CF_3OOOCF_3 (mp.: −138 °C; bp.: −16 °C). The latter reaction is assumed to proceed in form of an associative nucleophilic substitution on the oxygen atom [510]. A disassociation into OF^+ and F^- cannot be observed. Instead, treatment of OF_2 with fluoride acceptors, for example, AsF_5, yields dioxygenyl salts of the type $[O_2]^+[AsF_6]^-$ (Scheme 9.17) [511].

Dioxygen difluoride **O$_2$F$_2$** is deposited as orange-yellow compound on the cooled walls of the reaction vessel, when a O_2-F_2 mixture is reacted under high voltage glow discharge at 77–90 K and a reduced pressure of 13–25 mbar or, alternatively, under laser photolysis or radiolysis conditions (Scheme 9.17) [496, 503, 512]. A high-temperature preparation in the liquid phase has also been described [513]. As alternative starting

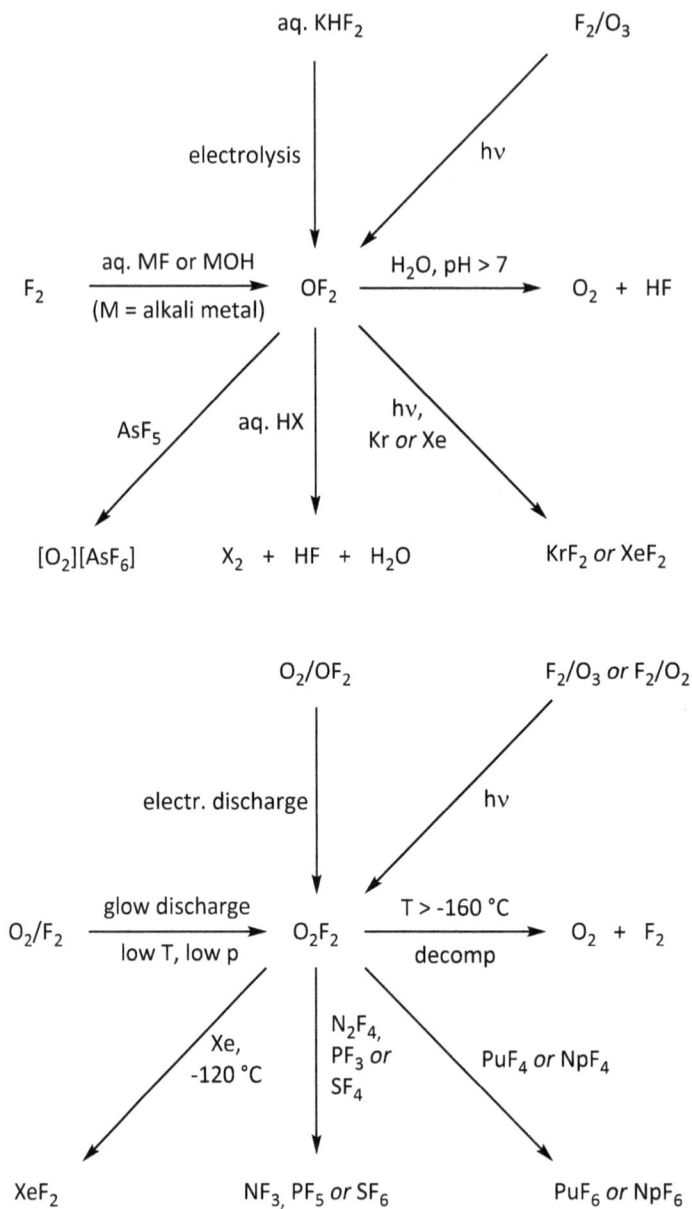

Scheme 9.17: Preparation and selected reactions of the oxygen fluorides OF_2 and O_2F_2.

materials, O_3 and F_2 can be used in a photochemical reaction, as well as an OF_2-O_2-mixture under electric discharge conditions [485, 494]. O_2F_2 melts to give an orange-red liquid (mp.: −163.5 °C; bp. (calc.): −57 °C) that can be dissolved in CCl_3F with yellow color. The gaseous, pale brown O_2F_2 decomposes slowly at −160 °C and rapidly above

$-100\,°C$ to form O_2 and F_2 (Scheme 9.17) [503, 514]. Other preparative accesses include the radiolysis of a liquid O_2-F_2-mixture at 77 K, photolysis of O_3 and F_2 at 120-195 K or the photolysis of liquid O_2 and F_2 at 77 K with near UV-light, which yields quantitative amounts of O_2F_2 [515]. In any case, a strong excitation is required [516]. The structure of O_2F_2 is comparable to H_2O_2, with a very strong O-O and weak O-F bonds (433 vs. 75 kJ mol^{-1}). Compared to hydrogen peroxide, the O-O distance is significantly shorter (121.7 vs. 147.5 pm) and is very similar to molecular O_2 (120.7 pm). Thus, there is a high double bond character, which can be rationalized by "no bond resonance" structures [507, 517–519]. O_2F_2 is a very strong oxidizing and fluorinating agent and reacts violently with organic matter and H-containing compounds below $-120\,°C$. For example, tetrafluoroethylene $F_2C=CF_2$ is mainly converted to COF_2 and CF_4, only minor amounts of C_2F_6, SiF_4 and the peroxide $(CF_3O)_2$ are observed [520]. However, this type of reaction can be moderated, if a less reactive perfluoroolefin is reacted in presence of a solvent. For example, the treatment of perfluoropropene with O_2F_2 at $-183\,°C$ in CCl_3F yields a mixture of 1- and 2-(fluoroperoxy)-perfluoropropane [521]. O_2F_2 often also violently interacts with inorganic substrates, as for example elemental sulphur reacts explosively at $-180\,°C$. At low temperatures, N_2F_4, SF_4 and PF_3 are readily oxidized to NF_3, SF_6 and PF_5, respectively (Scheme 9.17) [522]. Elemental xenon is converted to XeF_2 at $-120\,°C$ [523]. SO_2 is mainly fluorinated by O_2F_2 to give SO_2F_2, a transfer of the fluoroperoxy group O_2F is only observed to a small extent [502]. Accordingly, a number of other sulphur oxides and oxyfluorides is fluorinated [524]. Furthermore, O_2F_2 oxidizes PuF_4 to PuF_6, which is a reaction of some importance in nuclear fuel recycling and actinide purification in general [525]. Similarly, the neptunium(IV) species NpO_2 and NpF_4 are converted to NpF_6 (Scheme 9.17) [526]. Like the other oxygen fluorides, O_2F_2 forms dioxygenyl salts when it is treated with fluoride acceptors (BF_3, PF_5, AsF_5, SbF_5) [499, 512, 527, 528]. In the presence of O_2, CO reacts with O_2F_2 to form *bis*-fluoroformyl trioxide F(O)COOOC(O)F, which readily decomposes below room temperature to give the dioxide F(O)COOC(O)F.

Inorganic hypofluorites

Hypofluorous acid **HOF** (mp.: $-117\,°C$; bp.: 10-20 °C) is prepared by the reaction of F_2 with ice at $-40\,°C$ and low pressure (Scheme 9.18) [529]. Along with HOF, the gaseous product mixture contains HF, O_2, H_2O_2 and H_2O. The latter two can be frozen out at $-50\,°C$, HF is removed at $-78\,°C$ and HOF can be condensed at $-183\,°C$ in mg-amounts as white solid, whereas it has a pale-yellow color in the liquid state. Most likely, it is an intermediate in the preparation of OF_2 [504]. At 25 °C, gaseous HOF has a half-life of about 30 minutes in a PTFE-vessel and slowly decomposes to HF and O_2, presumably via a radical pathway (Scheme 9.18). The same decomposition products are observed in slightly basic, aqueous solution. In neutral or slightly acidic, aqueous solution, the hydrolysis of HOF leads to the formation of H_2O_2 and HF [529, 530]. HOF is stabilized by the formation of a hydrogen-bonded 1:1-complex with acetonitrile, which has been

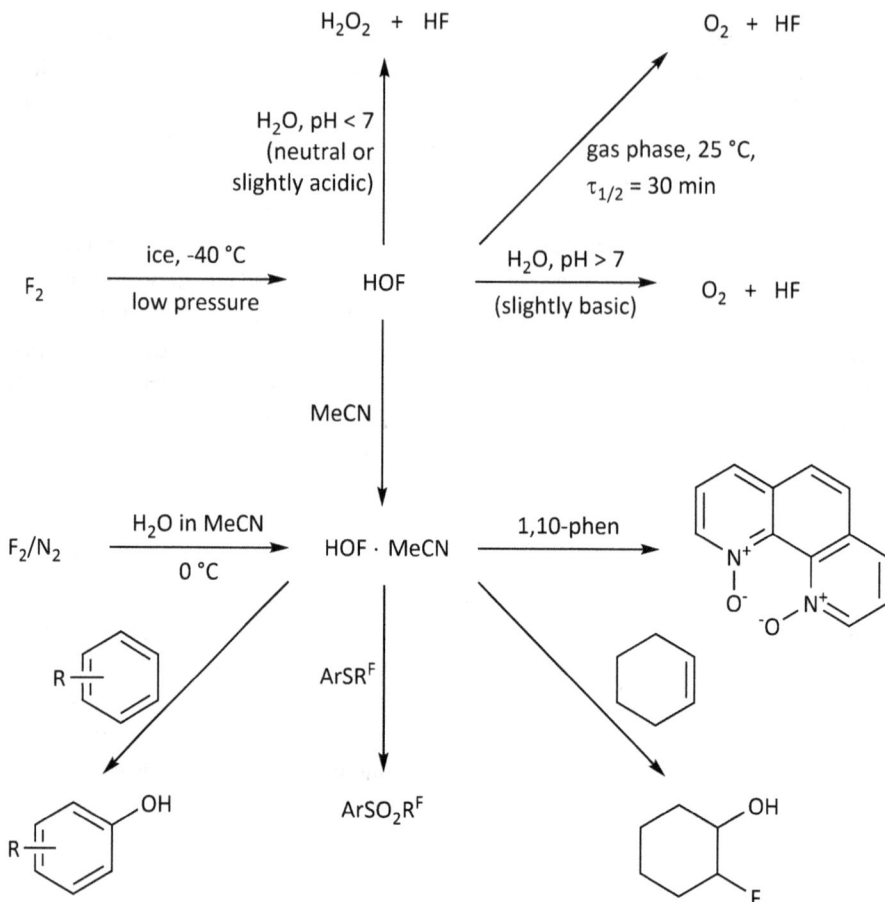

Scheme 9.18: Preparation and selected reactions of HOF.

characterized by X-ray crystallography [531, 532]. Conveniently, this species is directly accessible in solution by reacting dilute F_2 (10 % in N_2) with a low-concentrated mixture of H_2O in MeCN at 0 °C (Scheme 9.18) [533]. With an aqueous iodide solution, HOF yields $[I_3]^-$, OH^- and F^-; with a sulphate solution, peroxosulphate is formed [529]. In the presence of fluorine atoms, HF and OF are generated [534].

The O-H- (96.4 pm) and O-F-bond length (144.2 pm) both correspond to single bonds, the HOF-angle (97.2°) is very small in comparison to OF_2 (103.2°) and OH_2 (104.7°), which is a consequence of the different atomic radii and bond distances in HOF [535]. In the solid state, the molecules are connected by O-H-bridges, to form infinite, bent chains [536]. Quantum-chemical calculations on the acidity and basicity of HOF show that it is a significantly stronger acid and weaker base compared to H_2O and even a slightly stronger acid than HF [537, 538]. Based on matrix-isolation experiments, it has been revealed that HOF is a strong oxidizer, but does not show fluorinating properties against substrates like PF_3, AsF_3 or H_2O [539]. The product formation may be explained by the nucleophilic substitution of the fluoride on the oxygen atom, which is positively polarized. The different outcome of the reaction in alkaline medium could trace back to the hypofluorite ion OF^-, which is formed. OF^- could possibly act as a nucleophile against HOF and form HOOF, which immediately decomposes to O_2 and HF. However, so far, it has not been possible to isolate salts containing OF^-. The synthetic utility of HOF in organic transformations has been known for long times. However, its importance has further been increased with the exploration of the stable acetonitrile adduct [531]. For example, HOF is capable of oxidizing simple arene compounds, such as benzene, p-xylene or naphthalene to the respective phenolic products (Scheme 9.18) [540]. Apart from that, porphyrin derivatives are readily oxidized to N-oxides [541]. HOF also reacts with unsaturated compounds under fluorohydroxylation of the multiple bond, for example, cyclohexene is transformed to 2-fluorocyclohexanol (Scheme 9.18) [542]. Other important reactions include the oxidation of heteroatoms, as well as the generation of epoxides and oxidation of aromatic and aliphatic amines to nitro-compounds. For example, benzo[b]thiophene and electron-deficient sulphides of the type $ArSR^F$ are easily converted to the respective SO_2-derivatives (Scheme 9.18). Epoxidation reactions are observed for steroidal compounds like cholesteryl acetate as well as fluorinated olefins. Finally, the preparation of 1,10-phenanthroline-N,N'-dioxide has been achieved, which is hardly accessible by any other typical oxidizing agent [543–548].

In general, covalent hypofluorites are derivatives of the oxygen fluorides, where the OF group is bound to either inorganic or organic residues. Nitryl hypofluorite **O_2NOF** is an explosive gas (mp.: −175 °C; bp.: −45.9 °C), which is obtained from the reaction of fluorine with diluted or concentrated HNO_3 (Scheme 9.19) [549, 550]. Furthermore, O_2NOF is recovered, when diluted fluorine (F_2/N_2) is passed over anhydrous KNO_3 [551]. A quantitative formation of O_2NOF is achieved, when $[NF_4][SbF_6]$ is treated with alkali metal nitrates at low temperature in MeCN or SO_2 [552]. The molecular structure has been controversially discussed for quite some time. However, finally gas-electron diffraction experiments and theoretical calculations revealed a planar geometry for nitryl hypofluorite [553]. Chemically, O_2NOF behaves like a typical hypofluorite, since it readily adds to double bonds in the sense of a $F^.$ and a $O_2NO^.$-radical (Scheme 9.19) [554].

Fluorosulphuryl hypofluorite **FSO_2OF** (mp.: −158.5 °C; bp.: −31.3 °C) is a colorless, highly reactive gas, which may be prepared by various different methods, for example, the fluorination of $NaOSO_2F$ or SO_3 in presence of AgF_2 at 200 °C (Scheme 9.19) [555]. Instead of the aforementioned thermal preparations, FSO_2OF can also be generated photochemically from SO_3 and F_2 [556]. Another approach is the careful thermal decomposition of solid $[NF_4][SO_3F]$ at 10 °C, which produces a mixture of NF_3 and FSO_2OF in high yield (Scheme 9.19) [557]. The gas-phase structure of the molecule has been determined by gas electron diffraction techniques and quantum-chemical calculations, revealing a *gauche*-conformer as the predominant species [558]. FSO_2OF is a

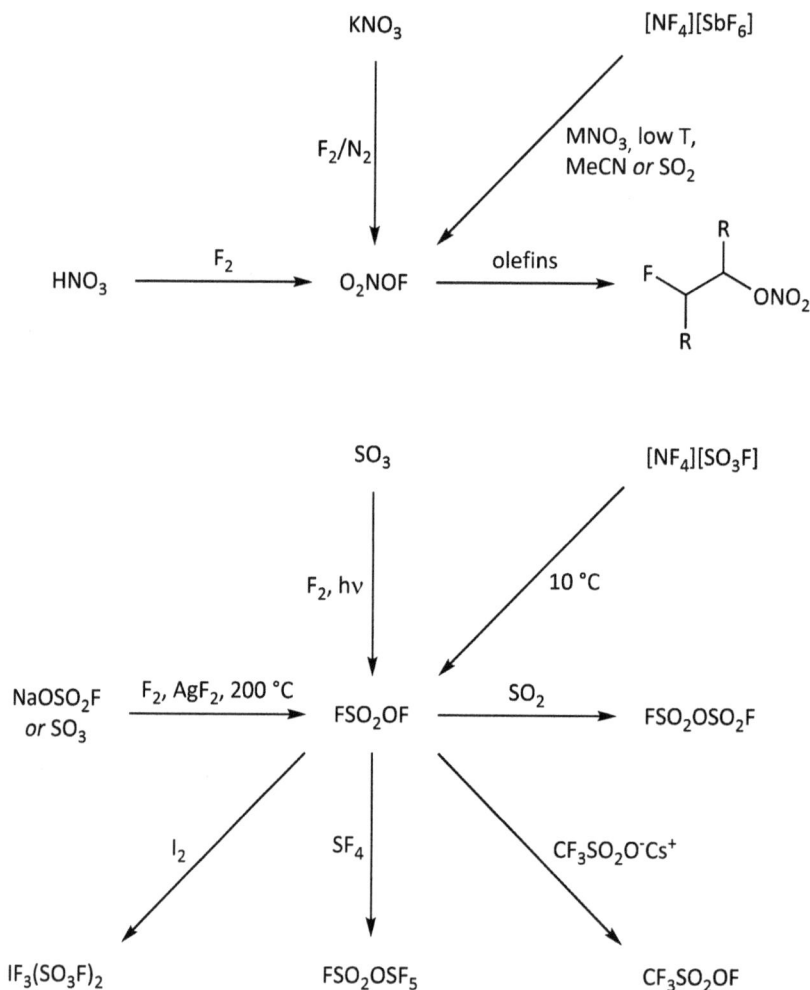

Scheme 9.19: Preparation and selected reactions of the inorganic hypofluorites O_2NOF and FSO_2OF.

strong oxidizer and a dangerous explosive. Above 235 °C, it undergoes thermal decomposition to SO_2F_2 and O_2, which is catalyzed by fluorine [559]. Another observed decomposition pathway leads to the formation of F_2 and FSO_2OOSO_2F [560]. FSO_2OF reacts with SO_2 under formation of pyrosulphuryl fluoride FSO_2OSO_2F, whereas SOF_2 is converted to a mixture of SOF_4 and peroxydisulphuryl difluoride FSO_2OOSO_2F. Apart from that, elemental iodine is transformed to $IF_3(SO_3F)_2$ (Scheme 9.19) [561]. Upon reaction with SF_4, the formation of SF_5OSO_2F is observed [562]. Reactions of FSO_2OF with a few more inorganic fluorides have been explored, where the transfer of F-atoms and OSO_2F-groups is observed in general [560]. The same occurs in reactions with organic substrates, thus FSO_2OF readily adds to olefinic double bonds [562]. In the reaction with CsF at 75 °C, the intermediate CsOF is assumed to be formed [563]. CF_3SO_2OF is a gaseous (mp.: −87 °C; bp.: ca. 0 °C) derivative of FSO_2OF, which is directly accessible from it by treatment with $[CF_3SO_2O]^-Cs^+$ (Scheme 9.19) [564].

If thionyl fluoride SOF_2 is fluorinated in presence of AgF_2 or CsF, pentafluorosulphur hypofluorite **F_5SOF** is generated via intermediary SOF_4 (Scheme 9.20) [563, 565, 566]. It is the most stable inorganic hypofluorite (mp.: −86 °C; bp.: −35.1 °C) and decomposes to O_2 and SF_6 at 210 °C [567]. Photolytic cleavage of F_5SOF leads to the peroxide F_5SOOSF_5 or, in presence of N_2F_4, to F_5SONF_2 [568, 569]. Br_2, I_2, NO_2 and PF_3 are oxidized to yield the respective fluorides BrF_3, IF_5, NO_2F and PF_5, whereas Cl_2 does not react, even at elevated temperatures [570, 571]. F_5SOF readily adds to olefinic double bonds in the sense of a F˙- and a SF_5O˙-radical in a quantitative manner (Scheme 9.20) [567, 572–575]. The sulphur atom of pentafluorosulphur hypofluorite is octahedrally coordinated, with the O-F-group in staggered conformation with respect to the equatorial S-F-bonds [576]. There exists also the colorless selenium analogue **F_5SeOF** (mp.: −54 °C; bp.: −29 °C), which is prepared by fluorination of $KSeO_2F$, SeO_2 or $SeOCl_2$ [577–579]. Additionally, $F_4Se(OF)_2$ is typically generated, where the two OF groups are in *trans*-positions to each other [577]. It is the only non-C-containing *bis*-hypofluorite. An alternative high-yield synthesis of F_5SeOF includes the treatment of the mercury salt $Hg(OSeF_5)_2$ with F_2 [580]. The reactivity of SeF_5OF is comparable to that of the sulphur compound, being characterized by the transfer of F˙- and SeF_5O˙-radicals. However, the selenium-derivatives are less stable, so that the formation of SeF_6 is mostly observed, when the hypofluorite is reacted with olefins. The formation of organic $OSeF_5$-compounds is only achieved with a few substrates, for example, perfluorocyclopentene, but not with common alkenes like ethylene, tetrafluoroethylene or hexafluoropropene [578]. Pentafluorotellurium hypofluorite **F_5TeOF** (mp.: ca. −80 °C) is accessible by the reaction of $Cs[OTeF_5]$ with FSO_2OF at −45 °C [581]. In an improved synthesis, it is prepared in high yield from $B(OTeF_5)_3$ and F_2 [582]. TeF_5OF is a colorless compound, which is stable at room temperature and may be stored in stainless steel vessels [581]. It readily adds to fluorinated olefinic substrates under formation of the respective TeF_5O-containing fluorocarbons [583, 584]. Hexafluorobenzene is converted to the corresponding 1,3-cyclohexadiene derivative or a substituted cyclohexanone-compound [584].

Perchloryl hypofluorite **O_3ClOF** is obtained as colorless and strongly oxidizing gas (mp.: −167.3 °C; bp.: −15.9 °C), when F_2 is passed through $HClO_4$ (w = 70 %) (Scheme 9.20). Furthermore, OF_2 and O_2 are generated as side products [585]. Moreover, O_3ClOF can be liberated from $[NF_4][ClO_4]$ and $[ClF_6][ClO_4]$, where the $[NF_4]^+$-salt is significantly more stable but still decomposes at 25 °C to yield NF_3 and O_3ClOF [586, 587]. Perchloryl hypofluorite has been found to be stable at room temperature and can be stored in PTFE- or passivated steel-containers. The structure of O_3ClOF is described by tetrahedrally coordinated Cl-centers, where the molecules have C_s-symmetry with the O-F group in a staggered conformation with respect to the Cl-O-bonds [588, 589]. Like the other aforementioned hypofluorites, O_3ClOF readily

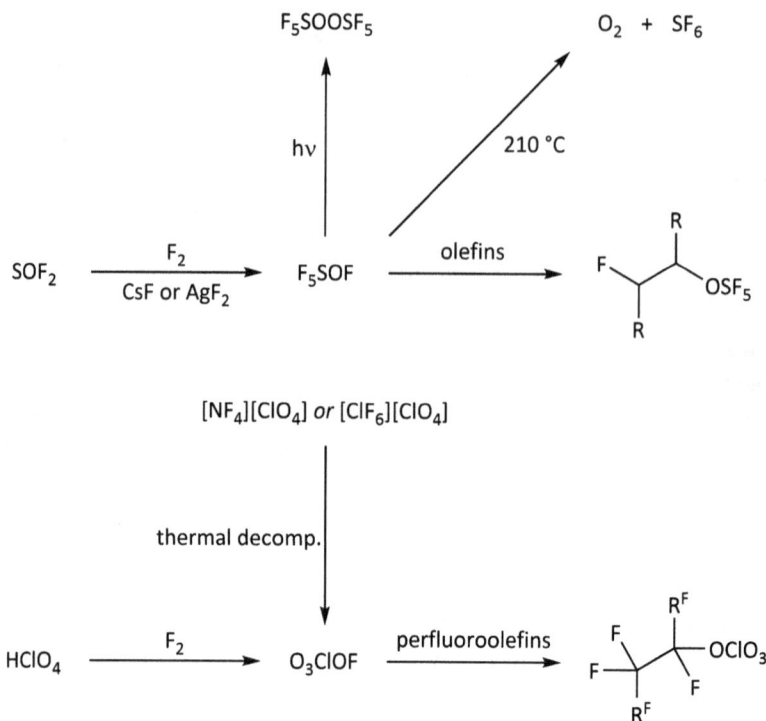

Scheme 9.20: Preparation and selected reactions of the inorganic hypofluorites F_5SOF and O_3ClOF.

adds to perfluorinated olefins, for example, tetrafluoroethylene or hexafluoropropene under formation of explosive, fluorinated alkyl perchlorates (Scheme 9.20) [590].

Organic hypofluorites

Organic hypofluorites can be divided in the two subclasses of explosive perfluoroacyl hypofluorites RC(O)OF and the more stable fluoroxy fluoroalkanes ROF [530]. All of the hypofluorites have an electrophilic fluorine substituent, which is capable of readily adding to electron-rich multiple bonds or other nucleophilic sites [591]. The perfluoroacyl hypofluorites, for example, $CF_3C(O)OF$ (bp.: −21 °C) and $C_2F_5C(O)OF$ (bp. (calc.): 2 °C), are formed upon fluorination of the corresponding perfluorocarboxylic acids or their salts in the presence of water or HF as colorless, highly explosive substances (Scheme 9.21) [592–595]. The stability of this compound class can be increased, if the α-position is substituted with at least one electron-withdrawing group. Apart from that, the incorporation of long alkyl chains should be avoided [591]. The thermal decomposition of the perfluoroacyl hypofluorites yields CO_2 and CF_4 or C_2F_6, respectively [596]. Another important acyl hypofluorite is the corresponding methylated derivative $CH_3CO(OF)$, which is also directly accessible by fluorination of the free acid or its alkali metal salts. Acetyl hypofluorite has proven suitable for a variety of transformations in fluoroorganic chemistry (cf. Section 12.4.1). Fluoroformyl hypofluorite **FC(O)OF** is a gaseous compound (mp.: −158 °C; bp.: −59 °C), which is prepared from *bis*-(fluorocarbonyl) peroxide (FC(O)O)$_2$ and F$_2$ at room temperature under irradiation with UV-light (Scheme 9.21). However, it cannot be directly prepared by fluorination of CO$_2$ [597]. In the gas phase, an equilibrium between *cis*- and *trans*-FC(O)OF is observable by ^{19}F-NMR spectroscopy [598]. Further investigations by means of vibrational spectroscopy and theoretical calculations revealed an energy difference of 4.9 kJ mol^{-1} between the conformers, where *trans*-FC(O)OF is the predominant species [599]. The treatment of FC(O)OF with UV-light gives directly access to difluorodioxirane [600].

The fluoroxy perfluoroalkanes are generally synthesized by fluorination of alcohols and ketones [601, 602]. Mostly, acyl- and diacyl compounds can be transformed quantitatively in presence of alkali metal fluorides. The simple perfluorohypofluorite **CF$_3$OF** (bp.: −95 °C) is in reversible equilibrium with COF$_2$ and F$_2$ between 320-460 °C (Scheme 9.21). When the mixture is treated with COF$_2$ at 280 °C and subsequently quenched, the peroxide CF$_3$OOCF$_3$ (bp.: −37 °C; dec.: >225 °C) is formed. The same reaction occurs under irradiative conditions [603]. Chemically, CF$_3$OF behaves like CF$_3$O$^.$ and F$^.$ but not like CF$_3$ and $^.$OF [530, 604]. The fluorination of trifluoroacetate or oxalate as well as FC(O)OF in the presence of CsF yields *bis*-fluoroxy difluoromethane **CF$_2$(OF)$_2$** (bp.: −64 °C) [597, 605, 606]. It is also directly accessible by treatment of CO$_2$ with excess F$_2$ in the presence of CsF (Scheme 9.21) [607–609]. In general, the reactivity of CF$_2$(OF)$_2$ against fluorinated olefins is comparable to that of other hypofluorites. For example, the reaction with tetrafluoroethylene leads to the formation of perfluorinated diethoxymethane CF$_2$(OC$_2$F$_5$)$_2$ (Scheme 9.21). However, no addition products are observed with inorganic substrates like SO$_2$, SO$_3$ or SF$_4$ [610, 611]. CF$_2$(OF)$_2$ can furthermore add to olefinic double-bonds under formation of 2,2-difluoro-1,3-dioxolane derivatives (Scheme 9.21) [612, 613]. Both, CF$_3$OF and CF$_2$(OF)$_2$ are capable of fluorinating aromatic compounds in an electrophilic fashion (cf. Section 14.3.2). However, activated substrates, such as toluene, xylenes and anisole are not selectively fluorinated but give rise to mixtures of various products [614]. Generally, the hypofluorites can be applied as versatile electrophilic fluorinating agents [605, 615, 616]. If FC(O)OF is treated with ClF instead of F$_2$, the mixed alkyl hypohalite CF$_2$(OCl)(OF) (mp.: −128 °C; bp.: ca. −20 °C) is obtained [617].

Moreover, CF$_2$(OF)$_2$ reacts over several steps with COF$_2$/CsF (= "CF$_3$O$^-$Cs$^+$") to give CF$_3$OF along with the trioxide CF$_3$OOOCF$_3$ (mp.: −138 °C; bp.: −16 °C) and CF$_3$OOOCF$_2$OOCF$_3$ or CF$_3$OOC(O)F, respectively [618, 619]. The trioxide CF$_3$OOOCF$_3$ is alternatively generated from COF$_2$ and OF$_2$ in the presence of CsF [510, 620]. If the reaction conditions are changed, it is possible to obtain fluoroperoxytrifluoromethane CF$_3$OOF, which is also accessible by fluorination of CF$_3$OOH with F$_2$ in the presence of CsF [621, 622]. In general, the *bis*-perfluoroalkyl trioxides R^{F}OOOR^F are surprisingly stable. For example, CF$_3$OOOCF$_3$ has a half-life of about 65 weeks at room temperature, where it slowly decomposes to the peroxide CF$_3$OOCF$_3$ and O$_2$ [623]. The molecular structure of CF$_3$OOF has been determined by gas electron diffraction, revealing a

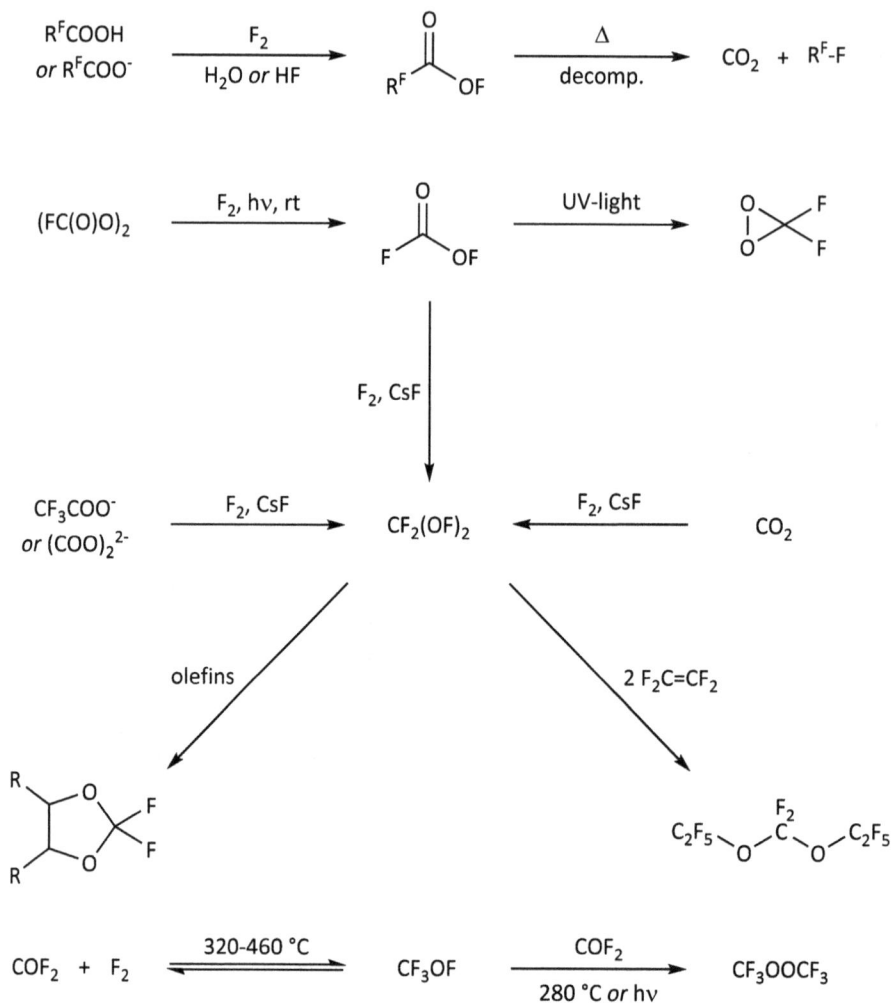

Scheme 9.21: Preparation and selected reactions of organic acyl and alkyl hypofluorites.

situation comparable to O_2F_2 with a short O-O-bond and a dihedral angle of 97.1° [624]. Vibrational spectra of the trioxide CF_3OOOCF_3 indicate a similar bond-order for the oxygen atoms in the O_3-bridge as it is found for the O-O-bond in the peroxide CF_3OOCF_3 [625]. These findings have been confirmed by X-ray crystallography. Apart from that, it has been found that the $[O_3]$-bridge is considerably more bent than in neutral O_3 or the $[O_3]^-$ anion[626]. Both, CF_3OOOCF_3 and CF_3OOF are suitable compounds for the synthesis of other CF_3OO- and OF-derivatives, such as CF_3OONO_2 [627–629]. The addition of CF_3OOOCF_3 to olefins, for example, chlorotrifluoroethylene $CF_2=CFCl$ yields the respective peroxyfluorocarbons $CF_3OOCF_2CFClOCF_3$ and $CF_3OCF_2CFClOOCF_3$ [630, 631].

If all the aforementioned oxygen-fluorine compounds are compared concerning their bonding situation, some general conclusions can be drawn. The O-F bond is stronger in compounds with only one oxygen atom, and the shortest bond distance is observed in the ˙OF radical. Moreover, the O-F bonds are weakened in OF_2 and HOF. The dioxygen compounds generally have very weak O-F and very strong O-O bonds. The O-O distance in ˙O_2F and O_2F_2 is only slightly enlarged in comparison to elemental O_2. This might be explained by the interaction of fluorine with the unpaired π^*-orbitals of oxygen, which results in a three-centered molecular orbital. The latter is antibonding with respect to O-O, but has a bonding character for O-F. Thus, the O-O bond remains almost unaffected, whereas the O-F bonds are weak, compared to bonds that are formed by electron pairs.

9.6.2 Sulphur, selenium, tellurium, polonium

Sulphur, selenium, tellurium and polonium mainly form tetra- and hexafluorides as well as numerous derivatives in the oxidation states +IV and +VI. Several low valent sulphur fluorides and a few binary selenium analogues exist as well, whereas the corresponding tellurium and polonium compounds are unknown in low oxidation states.

Low-valent fluorides
Sulphur difluoride **SF_2** is a very unstable molecule, which is formed upon fluorination of SCl_2-vapors at pressures below 25 mbar with KF (170 °C), HgF_2 (150 °C) or AgF (room temp.) (Scheme 9.22). Along with SF_2, many other sulphur fluorides, for example, S_2F_2 or SF_4 are formed, which have to be separated [632]. Alternatively, the reaction of COS with F_2 yields SF_2 [633]. Along with SF_3-radicals, SF_2 can further be observed under matrix conditions, if SF_4 is photolyzed (Scheme 9.22) [634]. Similarly, UV-photolysis of CF_3SF_3 in an Ar-matrix results in high yields of SF_2 [635]. In the gas phase, the latter has a bent structure and is only stable in diluted form and if no decomposition catalysts, for example, HF, BF_3, PF_5 or other metal fluorides, are present [636]. Even under these conditions, the half-life of SF_2 in stainless steel vessels at a pressure of 13 mbar is no longer than four hours. The decomposition occurs via formation of the di- and trimer of SF_2. The dimer disulphur tetrafluoride **S_2F_4** is formed in a temperature and pressure dependent equilibrium from monomeric SF_2 (Scheme 9.22). It is directly observed upon treatment of SCl_2 with KF, along with the mixed adduct of SF_2 and the intermediary species SFCl. However, the equilibrium between SF_2 and S_2F_4 is also affected by SSF_2 and SF_4, which are rapidly formed upon decomposition of SF_2 [637]. The dimerization of SF_2 has also been explored by quantum-chemical calculations, getting an insight into the mechanism as well as reaction enthalpies and activation energies [638]. Structurally, S_2F_4 can be understood as a derivative of SF_4, wherein

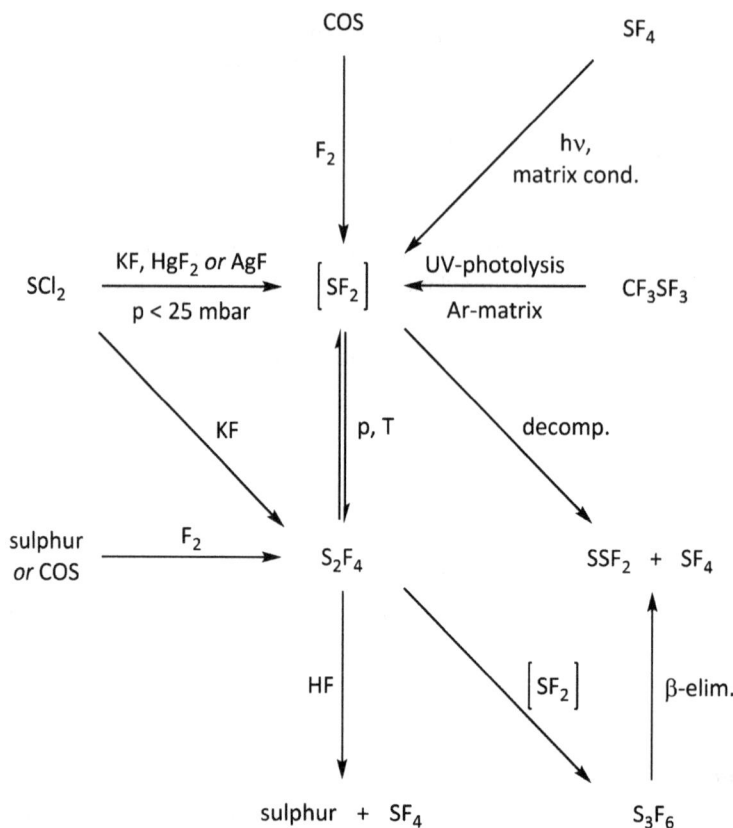

Scheme 9.22: Preparation and selected reactions of the low-valent sulphur fluorides SF_2, S_2F_4 and S_3F_6.

one of the equatorial fluorine atoms has been exchanged by a SF-group (F_3SSF). How-ever, all four S-F-bonds are inequivalently long [638–640]. Besides the dimerization of SF_2, S_2F_4 is also formed by gentle fluorination of sulphur or COS with F_2, along with SSF_2, SF_4 and SF_6 (Scheme 9.22). The side-products can be distilled off and S_2F_4 remains at −78 °C in pure form as a colorless liquid (mp.: −98 °C; bp.: 39 °C) [633]. S_2F_4 is thermally rather stable, but decomposes to sulphur and SF_4, if traces of HF are present [641]. UV-photolysis of a Cl_2/S_2F_4-mixture under matrix conditions gives rise to the mixed sulphur(II) halide SFCl [642]. The dimerization of sulphur difluoride is an α-addition of SF_2 to SF_2, hence the reverse cleavage of S_2F_4 is an α-elimination. The trimer trisulphur hexafluoride $\mathbf{S_3F_6}$ may have the constitution F_3S-S-SF_3 and is formed as an intermediate in the decomposition of SF_2. The molecule is formed by cocondensation of SF_2 and S_2F_4 but readily undergoes β-elimination to yield SF_4 and SSF_2 (Scheme 9.22). The elimination is irreversible, thus, β-addition of SF_4 to SSF_2 is impossible.

Disulphur difluoride **FSSF** is a colorless gas (mp.: −133 °C; bp.: 15 °C) that is prepared by fluorination of molten sulphur with AgF at 125 °C (Scheme 9.23) [643, 644]. The resulting mixture has to be quenched immediately in liquid air and fractionated in high vacuum. Alternatively, it can be made by passing diluted gaseous S_2Cl_2 over AgF at room temperature [645]. In the solid state, FSSF is stable for days when it is stored in glassware, however, it attacks dry glass and quartz at room temperature and is rapidly oxidized by NO_2 to give nitrosyl fluorosulphate $[NO]^+[SO_3F]^-$ (Scheme 9.23) [644, 646]. In the presence of HF or BF_3, FSSF readily isomerizes to SSF_2 [644]. At −50 to −60 °C, a mixture of almost equal amounts of both isomers FSSF and SSF_2 is formed. The mixture is transformed more and more into SSF_2, if it is warmed above −50 °C, at 0 °C only thiothionylfluoride is present. However, the temperature dependent equilibrium is reversible. In Ni-, Au- and Pt-vessels, pure FSSF is stable at room temperature for long times. As soon as there are traces of SSF_2, the isomerization is accelerated [643]. At 22 °C, gaseous FSSF has a half-life of 4.7 h. Its decomposition is catalyzed by HF and BF_3 [644]. If a gaseous mixture of FSSF and SSF_2 is submitted to glow discharge conditions, mainly FSSF is decomposed to S_2F_4, SF_4 and sulphur [641]. FSSF and SSF_2 are oxidized by AgF_2 and CoF_3 to give SF_4 and SF_6 (Scheme 9.23) [644]. Both compounds are attacked by aqueous sodium hydroxide to give F^- and $S_2O_3^{2-}$.

The mixture of FSSF/SSF_2 might also be described as tetrasulphurtetrafluoride $FSSSSF_3$, which is formed by β-addition of FSSF to SSF_2. The possible occurrence of such an addition can be considered as an analogue to the formation of trisulphurtetrafluoride $FSSSF_3$ that is prepared by co-condensation of SF_2 and SSF_2 at low temperatures. $FSSSF_3$ is an unstable, colorless and viscous liquid (mp.: −62 °C; bp.: 94 °C), which decomposes to SF_4 and SSF_2 in the gas phase. It is only stable for longer times, if it is kept at temperatures below −80 °C under exclusion of any traces of moisture [647]. If FSSF is reacted with H_2S, one can observe the difluoropolysulphanes FSSSF and FSSSSF along with F_3SSF and unreacted FSSF (Scheme 9.23). The difluoropolysulphanes generally decompose in glass vessels at temperatures above 0 °C [632, 648, 649]. Difluoro trisulphane FSSSF and difluoro tetrasulphane FSSSSF can also be recovered, if gaseous sulphur reacts with AgF. They are obtained as mixture in form of a light yellow, sparingly volatile oil, which decomposes to sulphur, SSF_2 and FSSF [632, 650].

Thiothionylfluoride **SSF_2** is a colorless gas (mp.: −164.6 °C; bp.: −10.6 °C), which is formed upon fluorination of diluted, gaseous S_2Cl_2 with KF at 140–145 °C or HgF_2 at 20 °C [643, 645, 651]. It is also obtained, if NF_3 is reacted with molten sulphur at 350–400 °C. Further products of this reaction are NSF as well as SF_4, SOF_2 and SF_6 [652]. Apart from the aforementioned methods, SSF_2 can be conveniently prepared by the conversion of a mixture of both S_2F_2-isomers in presence of catalytic amounts HF [644]. SSF_2 has a pyramidal structure. It is sensitive toward hydrolysis and thermally stable up to 250 °C even if the disproportionation toward SF_4 and elemental sulphur is thermodynamically favored. However, in the presence of catalysts like BF_3 or HF, decomposition readily takes place [632, 651]. SSF_2 reacts with electrophilic reagents

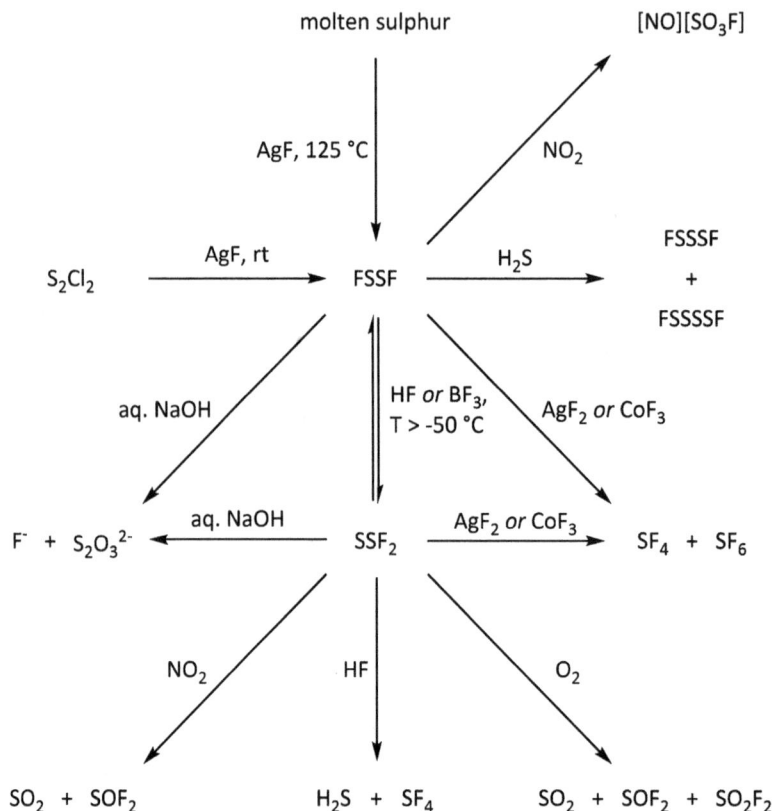

Scheme 9.23: Preparation and selected reactions of the low-valent sulphur fluorides FSSF and SSF$_2$.

like HF or O$_2$ to form SF$_4$ and H$_2$S or SO$_2$, SOF$_2$ and SO$_2$F$_2$, respectively (Scheme 9.23). Apart from that, NO$_2$ oxidizes it to SO$_2$ and SOF$_2$ [646]. It also attacks dry glass and quartz [644]. Photolysis of SSF$_2$ yields FSSF, which is further decomposed to sulphur, SF$_4$ and SF$_2$ or S$_2$F$_4$, respectively [653].

In both of the isomeric S$_2$F$_2$ molecules, the S-S distance (FSSF: 189.0 pm; SSF$_2$: 185.6 pm) clearly points to the presence of a double-bond [654, 655]. This is what has to be expected for thiothionylfluoride, however, in the case of FSSF it can be rationalized by no-bond resonance (Figure 9.3) [646]. The resonance structures also clarify, why the isomerization of FSSF to SSF$_2$ can occur rather easily.

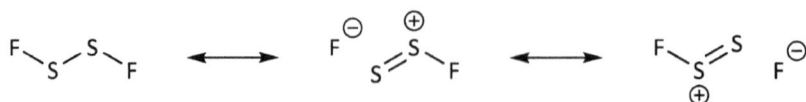

Figure 9.3: Resonance structures of FSSF.

Low-valent fluorides of selenium and tellurium are only scarcely explored, which is due to the fact, that the selenium derivatives are only accessible in the gas phase and under matrix conditions, whereas low-valent tellurium fluorides are completely unknown. Nevertheless, they are rather well explored by quantum chemistry. Selenium difluoride **SeF$_2$**, diselenium difluoride **Se$_2$F$_2$** and SeF$_4$ are observed, if Se-vapor is reacted with highly diluted fluorine gas, followed by condensation of the product mixture on a CsI-window at 12 K in an Ar-matrix [656]. Apart from that, the UV-photolysis of CF$_3$SeF$_3$ results in the formation of SeF$_2$ in high yield [635]. Under standard conditions, the bent SeF$_2$ as well as Se$_2$F$_2$ disproportionate to Se and SeF$_4$. Upon UV-photolysis in a matrix, FSeSeF is partly transformed to selenoseleninylfluoride SeSeF$_2$ [656]. The thermochemical and structural properties of SeF$_2$ and TeF$_2$ have been explored by quantum-chemical methods [657, 658]. Interestingly, theoretical studies on the relative stability of the isomers of Se$_2$F$_2$ and Te$_2$F$_2$ show that the tellurium derivative TeTeF$_2$ is more stable than FTeTeF, like it is found for sulphur, whereas the situation is opposite for SeSeF$_2$ and FSeSeF [659]. Low-valent polonium fluorides have neither been prepared nor explored by theoretical methods.

Tetrafluorides and tetravalent oxyfluorides

Sulphur tetrafluoride **SF$_4$** is a very toxic, colorless gas (mp.: −121.0 °C; bp.: −40.4 °C) with a suffocating smell, that can be prepared by several methods; however, the direct synthesis from the elements is only successful under inert conditions at −78 °C [660]. More conveniently, SF$_4$ is prepared by Cl/F-exchange of mostly low valent sulphur chlorides, as for example, in the reaction of SCl$_2$ with NaF in acetonitrile at 75 °C (Scheme 9.24). Alternatively, SF$_4$ is obtained, if S$_2$Cl$_2$ is treated with F$_2$ at 110–120 °C or upon halogen exchange of SCl$_2$ with pyridine/HF at 45 °C [661, 662]. It is also possible to start from SCl$_2$ and transform it directly to SF$_4$ in a one-pot reaction with Cl$_2$ and NaF in MeCN at 70–80 °C (Scheme 9.24) [661]. SF$_4$ is very reactive and readily hydrolyzes to give HF and SO$_2$. It is not only a good fluorinating agent for organic substrates like alcohols or ketones (cf. Chapters 13.4.2 and 13.4.4), but also a versatile reagent in inorganic fluorine chemistry. In general, oxides, sulphides and carbonyls can be readily transformed into the respective fluorine-containing compounds [663]. For example, SF$_4$ selectively converts P=O groups to PF$_2$ units.

SF$_4$ has a ψ-trigonal-bipyramidal structure with two slightly elongated, axial S-F bonds (164.3 pm (axial) vs. 154.2 pm (equatorial)) and the lone pair in an equatorial position [664]. It mainly exists in monomeric form, only at low temperature in the liquid state, SF$_4$ is found to be associated via the axial F-atoms [665]. It is a fluxional molecule and undergoes rapid intramolecular exchange processes, if the temperature is not too low. SF$_4$ is a weak Lewis acid, thus, stable complexes are formed with strong Lewis bases like pyridine or triethyl amine [666]. Large cations like Cs$^+$ or [NMe$_4$]$^+$ stabilize salts of the type M^I[SF$_5$] (Scheme 9.24) [667–670]. In the square-pyramidal

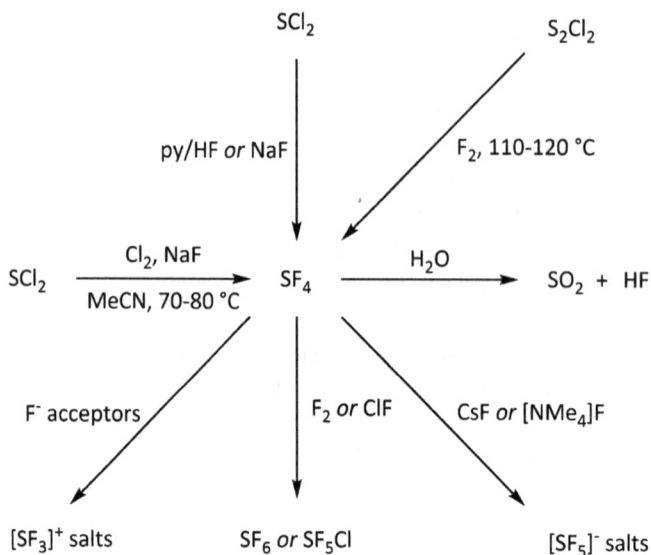

Scheme 9.24: Preparation and selected reactions of SF_4.

$[SF_5]^-$ anion, the sulphur atom is situated at the foot point of the pyramid. Accordingly, ^{18}F-labeled SF_4 is generated, if SF_4 is reacted with dry $Li^{18}F$. The intermediately formed $Li[SF_4^{18}F]$-complex is unstable and decomposes to LiF and labeled SF_4 [671]. With fluoride acceptors like BF_3, PF_5, AsF_5 or SbF_5, SF_4 readily reacts to yield very stable, colorless and crystalline $[SF_3]^+$-salts (Scheme 9.24) [672–674]. Therein the pyramidal cations are linked with the fluoroanions via F-bridges. The sulphur atom occupies the apex of $[SF_3]^+$, which can be formally considered as adduct of "F^+" to SF_2 [125]. The formation of $[SF_3]^+$ species is reversible, if for example, alkali metal fluorides M^IF are added, SF_4 is regenerated. Thus, the formation of, for example, BF_3-adducts can be exploited to separate SF_4 (and also SOF_4) from other sulphur fluoride compounds, such as SOF_2, SO_2F_2 and SF_6 [672, 674]. SF_4 can be oxidized to yield several sulphur(VI) compounds (Scheme 9.24). For example, reactions with F_2 and ClF give SF_6 and SF_5Cl, respectively, whereas oxidation with O_2 in the presence of NO_2 leads to the oxyfluoride SOF_4 [675].

Sulphuroxide difluoride (thionyl fluoride) **SOF$_2$** is a colorless gas (mp.: −129.5 °C; bp.: −43.8 °C), which can be generated by many fluorination reactions with SF_4 or more conveniently by the substitution of thionyl chloride $SOCl_2$ with NaF in acetonitrile. SOF_2 is less reactive than SF_4 and has a distorted pyramidal structure with the sulphur atom in the apical position. SOF_2 is both a weak Lewis acid and base. With $[NMe_4]F$, it reacts to yield the ψ-trigonal-bipyramidal anion $[SOF_3]^-$, with two axial F-atoms and each one O- and F-substituent in equatorial positions. The cesium salt $Cs[SOF_3]$ has been investigated by vibrational spectroscopy under matrix-conditions [676]. SOF_2 is hydrolyzed by water to form HF and SO_2.

Selenium tetrafluoride **SeF$_4$** is a colorless liquid (mp.: −9.5 °C; bp.: 106 °C), which is synthesized by the halogen exchange of SeCl$_4$ with AgF at 50 °C or ClF as well as by fluorination of SeO$_2$ with SF$_4$ at 100 °C or alternatively with excess ClF (Scheme 9.25) [663, 677–680]. It can also be prepared by direct fluorination of elemental selenium or Se$_2$Cl$_2$ at 0 °C under inert conditions as well as by treatment of Se with AgF at 180–200 °C [681–683]. Like SF$_4$, SeF$_4$ is very reactive, and thus is also easily hydrolyzed [682]. SeF$_4$ can be evaporated without decomposition and is stable at relatively high temperatures. Structurally, it is comparable to SF$_4$, with the difference that it is associated via F-bridges in the solid and the gas phase [684]. In the crystalline state, the coordination geometry at the Se centers is found to be a distorted octahedron [685]. In the CH$_3$F solution at −140 °C, monomeric SeF$_4$ is observed, where the exchange of axial and equatorial fluorine substituents is inhibited [686]. SeF$_4$ is a good fluorinating agent and attacks glass [679]. The Lewis base complex with pyridine, SeF$_4$ · py, has been used as convenient fluorinating agent for organic substrates like alcohols, ketones and carboxylic acids [687]. SeF$_4$ is also capable of fluorinating a variety of elements, such as P, As, Sb, Bi and Si [682]. Apart from that, inorganic oxides, such as CrO$_3$, MoO$_3$, WO$_3$ and UO$_3$ are transformed to oxyfluoride compounds of the type M^{VI}OF$_4$ and M^{VI}O$_2$F$_2$ (M^{VI} = Cr, Mo, W, U), which mostly form complexes with excess SeF$_4$ (Scheme 9.25) [688]. It has also been used as a solvent for the preparation of fluorometallates, for example, palladates(IV), when "PdF$_3$" is treated with BrF$_3$ in SeF$_4$. However, SeF$_4$ alone reduces "PdF$_3$" to divalent PdF$_2$ [689]. In liquid SeF$_4$, self-dissociation to pyramidal [SeF$_3$]$^+$ and square-pyramidal [SeF$_5$]$^-$ can be observed to a small extent. However, a more pronounced dissociation ([SeF$_3$]$^+$[HF$_2$]$^-$) is observed in HF solution or even stronger in presence of BF$_3$ ([SeF$_3$]$^+$[BF$_4$]$^-$) [690]. SeF$_4$ reacts with alkali metal fluorides, [NMe$_4$]F and NOF · 3 HF as well as a few other ionic fluorides to form the complexes M^I[SeF$_5$] (M^I = Na, K, Rb, Cs, NMe$_4$, NO), which are more stable than their sulphur analogues (Scheme 9.25). However, at room temperature most of them already dissociate, only the salts with larger cations are stable. The [NMe$_4$]-salt has been characterized by X-ray crystallography, revealing square-pyramidal [SeF$_5$]$^-$ anions, as expected [669, 682, 691, 692]. [SeF$_5$]$^-$ can further be transformed to [SeF$_6$]$^{2-}$, if it is reacted with "naked" F$^-$. The dianion has a distorted octahedral structure, as has been revealed by X-ray crystallography [692, 693]. If SeF$_4$ is reacted with dry Li^{18}F at room temperature, the formation of ^{18}F-labeled SeF$_4$ is observed [671]. Like SF$_4$, SeF$_4$ reacts with fluoride acceptors to form the analogous [SeF$_3$]$^+$-salts, which have been characterized by X-ray crystallography (Scheme 9.25) [672, 674, 694, 695]. Upon treatment with Me$_3$SiN$_3$, SeF$_4$ is substituted by azido ligands, giving rise to neutral Se(N$_3$)$_4$ as well as anionic [Se(N$_3$)$_5$]$^-$ and [Se(N$_3$)$_6$]$^{2-}$ [696]. Accordingly, SeF$_4$ can be reacted with Me$_3$SiCN for the introduction of cyano groups. However, the fully substituted tetracyanide Se(CN)$_4$ is not accessible; the reaction only yields the mixed species SeF$_2$(CN)$_2$ and SeF(CN)$_3$ [697, 698].

Seleniumoxide difluoride **SeOF$_2$** is a colorless, fuming liquid (mp.: 15 °C; bp.: 125 °C), which may be prepared from SeO$_2$ and SeF$_4$, SeO$_2$ and SF$_4$ or SeF$_4$ and TeO$_2$

Se   SeO$_2$
| /
F$_2$ *or* AgF   SF$_4$ *or* ClF
↓   ↙

SeCl$_4$ —— AgF *or* ClF ——→ SeF$_4$ —— F⁻ ——→ [SeF$_5$]⁻ salts

F⁻ acceptors /   MO$_3$ "naked" F⁻
           ↙ | |
                            ↓   ↓

[SeF$_3$]⁺ salts   MOF$_4$ + MO$_2$F$_2$ [SeF$_6$]²⁻ salts

                       (M = Cr, Mo, W, U)

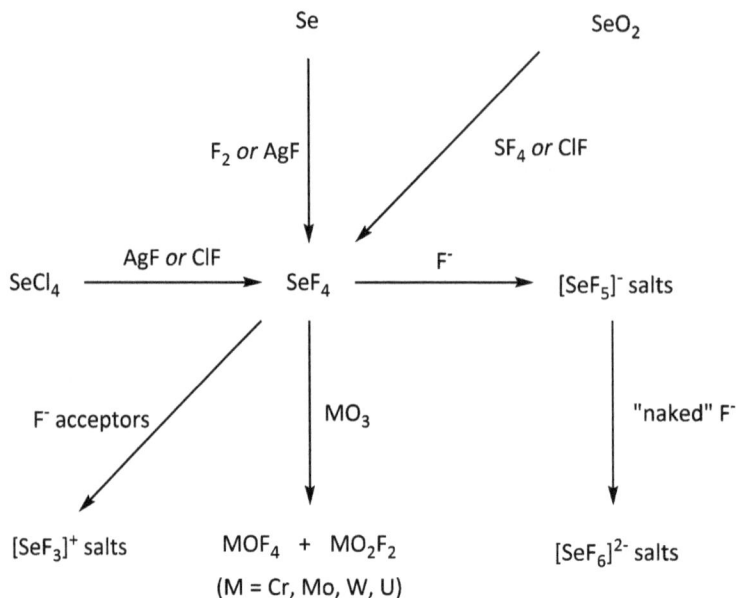

Scheme 9.25: Preparation and selected reactions of SeF$_4$.

[680, 688, 699]. The reaction of SeO$_2$ with ClF, COF$_2$ or diluted F$_2$ also yields SeOF$_2$ [677, 682, 700]. Finally, SeOF$_2$ is accessible by halogen-exchange of SeOCl$_2$ with AgF, NaF or KF, the latter furnishing SeOF$_2$ in high yield [678, 701–703]. Like SOF$_2$, it has a pyramidal structure. In the solid state, SeOF$_2$-molecules are linked by each two O- and one F-bridge, resulting in a distorted octahedral coordination environment at the Se centers. Overall, a layered arrangement is found [704]. SeOF$_2$ rapidly attacks glass and reacts with red phosphorus under ignition [701]. Excess SO$_3$ reacts quanti- tatively with SeOF$_2$ under formation of the *bis*-fluorosulphate SeO(SO$_3$F)$_2$ [705]. With the fluoride acceptor NbF$_5$, the crystalline complex SeOF$_2$ · NbF$_5$ is formed, where the Nb- and Se-atom are bridged by oxygen; in contrast, ionic fluoride donors trans- form SeOF$_2$ to anionic [SeOF$_3$]⁻ [692, 706, 707]. The salts K[SeOF$_3$] and Cs[SeOF$_3$] contain isolated anions with ψ-trigonal-bipyramidal geometry, the lone-pair and oxy- gen atom occupying equatorial positions [707]. The treatment of SeO$_2$ with CsF gives rise to the ionic species Cs[SeO$_2$F] [692]. The reaction of SeO$_2$ with anhydrous HF un- ambiguously yields SeOF$_2$, whereas the analogous telluriumoxide difluoride **TeOF$_2$** is not observed in the system TeO$_2$/HF [708]. However, the anionic derivatives [TeO$_2$F$_2$]²⁻ and [TeOF$_4$]²⁻ have clearly been identified [709]. The parent TeOF$_2$ has been described to be formed in an equimolar mixture of TeO$_2$ and TeF$_4$ at 140 °C. Its structure was determined by powder X-ray crystallography, revealing helical chains of distorted trigonal bipyramidal [TeO$_2$F$_2$]-units. The polyhedra are linked by O-atoms in common corners. In addition, weak Te-F interactions connect the aforementioned chains to a three-dimensional network [710].

Tellurium tetrafluoride **TeF$_4$** is a colorless solid (mp.: 130 °C; bp.: 374 °C), which is accessible by fluorination of TeO$_2$ with SeF$_4$ at 80 °C, with COF$_2$ at 160 °C, with FeF$_3$ at 700 °C or by comproportionation of Te and TeF$_6$ at 200 °C (Scheme 9.26) [688, 699, 700, 711, 712]. Like the other tetrafluoride homologues described above, it may also be prepared from the elements under inert conditions below 0 °C [713]. Pure TeF$_4$ can be generated by thermal decomposition of the fluorocomplexes M^I[TeF$_5$] (M^I = Li, Na, K, Rb, Cs) at 450–900 °C, where the sodium salt is most suitable for this purpose [711, 714, 715]. TeF$_4$ is very reactive and readily hydrolyzes, in glass and quartz vessels it is completely decomposed at 200 °C [716]. It is a fluorinating agent with a comparable efficiency to SF$_4$ and SeF$_4$. At higher temperatures, it reacts with metals like Cu, Ag, Au and Ni under formation of metal fluorides and –tellurides (Scheme 9.26) [716]. It can be evaporated without decomposition, but slowly disproportionates above 190 °C to form Te and TeF$_6$ [716]. Its structure is generally comparable to SF$_4$ and SeF$_4$ [684]. Thus, gaseous TeF$_4$ is described by monomeric ψ-trigonal bipyramidal molecules [717]. In the solid state, it has a ψ-octahedral structure, wherein distorted square-pyramidal [TeF$_5$]-units are linked via cis-F-bridges to form infinite chains [685, 718]. Liquid TeF$_4$ conducts electricity, which is due to its self-dissociation to [TeF$_3$]$^+$ and [TeF$_5$]$^-$. The situation is different in solution, where one can find F-bridged (TeF$_4$)$_n$ oligomers ($n \leq 4$) in unpolar solvents like toluene, benzene or diethyl ether, whereas polar donor solvents (MeCN, EtOH, acetone, tetrahydrofuran, dimethoxyethane, dioxane) rather form weak, monomeric adducts. A Lewis base complex with NMe$_3$ is also known, it is comprised of the ions [(Me$_3$N)$_2$TeF$_3$]$^+$ and [TeF$_5$]$^-$ [717, 719]. Salts with the pentafluorotellurate(IV) anion [TeF$_5$]$^-$ can easily be made by the reaction of TeO$_2$ and CsF in HF (w = 40 %) or SeF$_4$ and are convenient sources of TeF$_4$ [714, 715, 720]. The [TeF$_5$]$^-$ anion is isoelectronic to IF$_5$ and [XeF$_5$]$^+$. In the solid state, one can find distorted square-pyramidal [TeF$_5$]-groups [691, 720, 721]. Hexafluorotellurate(IV) anions [TeF$_6$]$^{2-}$ have been reported to exist in solutions with tetraalkylammonium cations (Scheme 9.26) [722, 723]. If TeF$_4$ is reacted with KF in the solid state in a 2:1-ratio, the fluorotellurate(IV) K[Te$_2$F$_9$] is obtained, which has been found to contain five- and six-fold coordinated Te-centers [724]. Similar to the sulphur and selenium analogues, TeF$_4$ forms [TeF$_3$]$^+$-salts upon reaction with fluoride acceptors like AsF$_5$ or SbF$_5$ [672, 674, 725]. In the same way like SeF$_4$, TeF$_4$ may be substituted by treatment with Me$_3$SiN$_3$, resulting in the formation of the explosive azido-derivative Te(N$_3$)$_4$. Accordingly, [Te(N$_3$)$_5$]$^-$ is accessible from [TeF$_5$]$^-$ and Me$_3$SiN$_3$, whereas the treatment of neutral Te(N$_3$)$_4$ with ionic [PPh$_4$]N$_3$ leads to the formation of the hexaazidotellurate(IV) anion [Te(N$_3$)$_6$]$^{2-}$ [726, 727]. The substitution of TeF$_4$ with Me$_3$SiCN gives rise to the highly pyrophoric and thermally unstable tetracyanide Te(CN)$_4$ [728, 729].

Hexafluorides and hexavalent oxyfluorides
Sulphur hexafluoride **SF$_6$** was first detected by Moissan in 1900 and can be directly prepared from the elements in a very exothermic reaction or by fluorination of SO$_2$

Te   TeO$_2$

TeF$_6$, 200 °C   CsF, aq. HF

TeO$_2$ $\xrightarrow{\text{SeF}_4\text{, COF}_2 \atop \text{or FeF}_3}$ TeF$_4$ $\underset{\Delta}{\overset{\text{F}^-}{\rightleftharpoons}}$ [TeF$_5$]$^-$ salts

F$^-$ acceptors   M, Δ   [NR$_4$]F

[TeF$_3$]$^+$ salts   MF$_n$ + M$_x$Te$_y$ [TeF$_6$]$^{2-}$ salts

(M = Cu, Ag, Au, Ni)

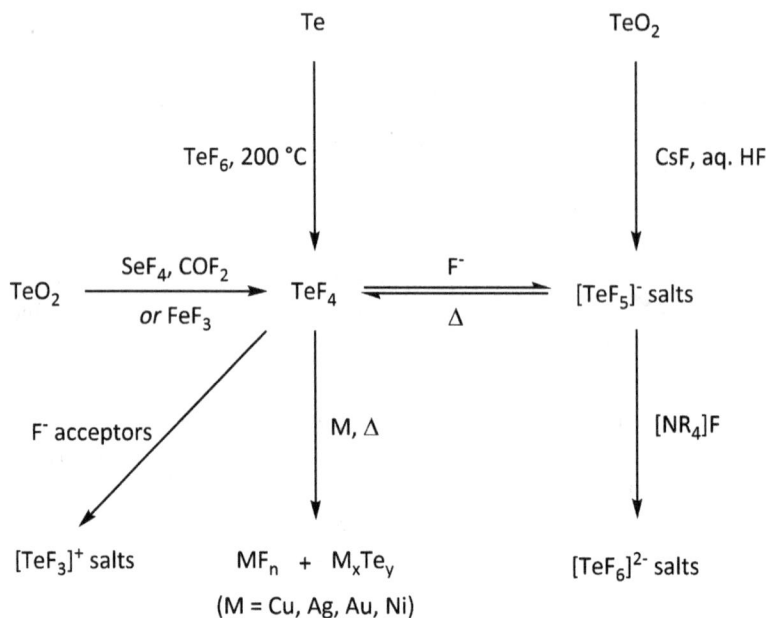

Scheme 9.26: Preparation and selected reactions of TeF$_4$.

[730, 731]. In order to remove other sulphur fluorides from technically produced SF$_6$, the reaction gas is first heated to 400 °C and subsequently washed with alkaline solutions. In the heating step, S$_2$F$_{10}$ is decomposed into SF$_6$ and SF$_4$, the alkaline medium hydrolyzes SF$_4$ to SO$_3^{2-}$ and F$^-$. Finally, SF$_6$ is purified by pressure distillation. It is a color and odorless, volatile gas (mp.: −50.8 °C (pressurized); subl.: −63.8 °C) with octahedral shape and is about five times denser than air. SF$_6$ is nonflammable, nontoxic and does not dissolve in water. It is chemically extremely inert and can be heated with hydrogen without HF formation. Molten alkali hydroxides, hot HCl and overheated water vapor (500 °C) do not attack it. Sulphur hexafluoride can even resist oxygen under electrical discharge conditions. Sodium can be molten in SF$_6$ gas, without showing turbidity on its surface due to NaF formation, only at its boiling point (881.3 °C) it starts to react with SF$_6$. In general, one can conclude, that if SF$_6$ reacts at all, the reactions mostly result in the complete destruction of the molecule. The indifferent behavior can be explained by the perfect steric shielding of the S-atom which especially prevents nucleophilic attacks in the sense of a S$_N$2-reaction and by the full occupation of all valence orbitals of the F-atoms. Thus, a direct attack of Lewis bases is impossible. However, S$_N$1-type reactions should be possible, where F$^-$ is abstracted in the first step under formation of an intermediate [SF$_5$]$^+$ cation, which could readily react with a nucleophile. The F$^-$-abstraction does not occur, even in presence of the strongest Lewis acids, which is a result of the bonding situation. The latter has been of large interest and has also been investigated with the help of computations. The exact bonding situation is still not fully understood, the current knowledge considers multicenter-bonding and clearly neglects the participation of d-orbitals [732]. Reactions of SF_6 are only kinetically inhibited, as soon as one fluoride is substituted by a larger group, like $-CH_3$ or $-Cl$ [733, 734] nucleophilic substitution reactions occur [733–735]. Thermodynamically, SF_6 should be readily hydrolyzed by water in a strongly exothermic reaction to give H_2SO_4 and HF. In liquid ammonia, SF_6 is attacked by metals (Li-Cs, Sr, Ba, Eu, Yb) to form metal fluorides and sulphides, furthermore it reacts with H_2S under formation of S_8 and HF [736]. With electrophilic reagents, like Al_2Cl_6 or SO_3, SF_6 reacts at high temperatures and is decomposed to AlF_3, Cl_2 and low valent sulphur chlorides or SO_2F_2, respectively. Above 500 °C and at high pressure or electrical discharge, SF_6 also reacts with other compounds, like metal oxides, CS_2 or COS [737, 738]. Photolysis in the presence of oxygen yields several different oxides $F_5S-O_n-SF_5$ ($n = 1, 2, 3$), which are formed via intermediary radicals. Due to the chemical inertness and its very low dielectric constant as well as its large dielectric strength, SF_6 is still used on larger scale for electric insulators in high voltage facilities. Apart from that, it is used as protecting gas for metal melts as well as for heat insulation or noise attenuation. The chemical inertness is not only a big advantage for technical applications, but also resulted in environmental concerns, as SF_6 has a global warming potential, which is many thousand times stronger than that of CO_2. In the atmosphere, increasing concentrations of SF_6 have been detected. Therefore, the long-term use of SF_6 is tried to be avoided or at least significantly limited.

Disulphur decafluoride **S_2F_{10}** is an extremely toxic gas (mp.: −52.7 °C; bp.: 26.7 °C), which is formed as side product of the fluorination of sulphur. More conveniently, it can be prepared by photolysis of a SF_5Cl/H_2-mixture or SF_5Br (Scheme 9.27) [739]. It is far more reactive than SF_6 and disproportionates at 150 °C to yield SF_4 and SF_6. S_2F_{10} has D_{4d}-symmetry in its ground state, which means that the two square-pyramidal SF_5 fragments are in a staggered conformation. The S-F bonds (156 pm) are as long as in SF_6, but the S-S distance (221 pm) is relatively long in comparison to usual S-S single bonds (208 pm). This explains why the decay into SF_5 radicals is favored. S_2F_{10} reacts with a number of compounds, as for example, Cl_2 and Br_2 lead to the formation of SF_5Cl and SF_5Br, respectively (Scheme 9.27). Apart from that, $SF_5(SO_2F)$ is obtained with SO_2, whereas N_2F_4 generates SF_5NF_2 and, finally, NH_3 gives rise to NSF_3 [740]. S_2F_{10} is stable toward acidic and alkaline hydrolysis.

Sulphur chloride pentafluoride **SF_5Cl** (mp.: −64 °C; bp.: −19.1 °C) and sulphur bromide pentafluoride **SF_5Br** (mp.: −78.6 °C; bp.: 3.1 °C) are gaseous compounds, which are formed upon halogenation of S_2F_{10} with X_2 (X = Cl, Br) [740]. More conveniently, they are obtained by halofluorination of SF_4 with XF in presence of CsF in a pressure reactor at elevated temperatures (SF_5Cl: 375 °C; SF_5Br: 100 °C) (Scheme 9.27) [668, 741–743]. It is assumed, that the generation of SF_5X proceeds via intermediary $Cs[SF_5]$ which is formed upon addition of CsF to SF_4 [668]. Further substituted derivatives, for example, CF_3SF_4Cl are also accessible by chlorofluorination of the respective sulphur (IV) precursors [744]. Both SF_5Cl and SF_5Br are much more reactive than SF_6, which

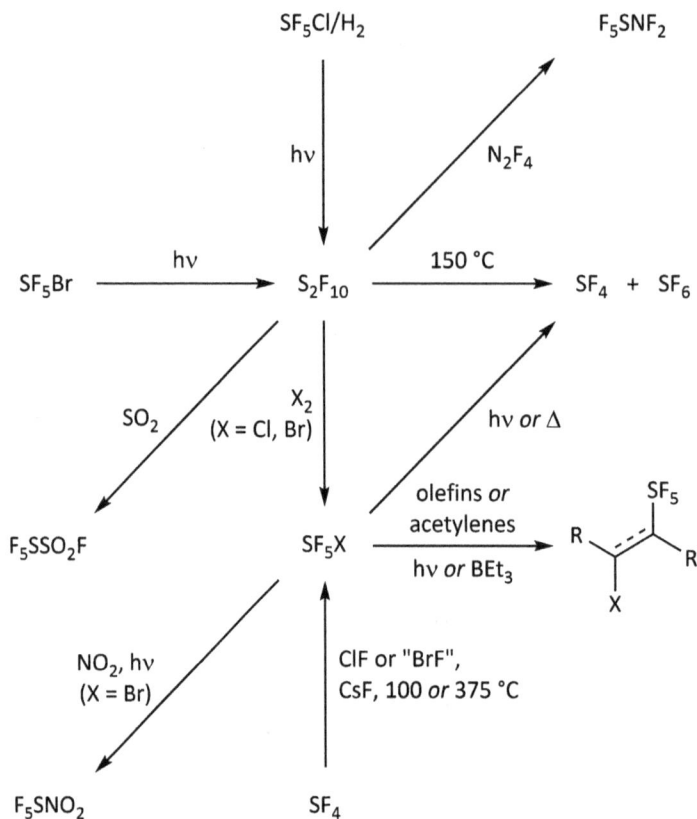

Scheme 9.27: Preparation and selected reactions of SF_5X (X = SF_5, Cl, Br).

is due to the positively polarized Cl- and Br-atom, respectively. Alkaline hydrolysis occurs easily, in the case of SF_5Br decomposition even proceeds with water. At elevated temperatures (SF_5Cl: 400 °C; SF_5Br: 150 °C) or upon irradiation with UV light, SF_5X decompose into halide- and SF_5 radicals which then yield SF_4 and SF_6 (Scheme 9.27). If they are irradiated in the presence of oxygen, several different oxides F_5S-O_n-SF_5 (n = 1, 2, 3) are formed via radical mechanisms. SF_5X (X = Cl, Br) are commonly used as reagents or precursors for organic and inorganic SF_5-derivatives, as they are quite readily added to multiple bonds, for example, in olefins or acetylenes. Especially the organic chemistry of SF_5-containing moieties has largely developed over the last decades. The preparation and utility of this class of compounds has been reviewed elsewhere some time ago [745, 746]. For this reason, the chemistry of SF_5X leading to a large variety of organic compounds will not be discussed herein. The photolysis of SF_5Br and NO_2 gives rise to the nitro derivative SF_5NO_2 (Scheme 9.27) [747, 748]. With strong nucleophiles it is possible to substitute fluorine at the sulphur atom[733, 734]. As expected, the molecular structure of SF_5Cl is described by a distorted octahedron, where the axial S-F-bond is slightly elongated compared to the equatorial ones. Furthermore, the $Cl-S-F_{eq}$ angle is a bit larger than 90° [749, 750]. The situation is inverted for SF_5Br, where the axial S-F-bond is slightly shorter. In addition, the $Br-S-F_{eq}$ angle is a little smaller than 90°[751].

Sulphur oxytetrafluoride **SOF_4**, also known as thionyl tetrafluoride, can be prepared as colorless gas (mp.: −99.6 °C; bp.: −48.5 °C) by the reaction of SF_4 with O_2 in presence of NO_2 (Scheme 9.29) and is sensitive toward hydrolysis. It is also accessible by fluorination of SOF_2 with F_2 above 80 °C or more conveniently by treatment of SOF_2 with ClF_3 or BrF_5 [566, 675, 752, 753]. The treatment of SOF_2 with ClF also produces SOF_4 in high yield, along with Cl_2 [754]. If CsF or AgF_2 are present in the reaction mixture, SOF_4 is further fluorinated to yield the hypofluorite SF_5OF, in absence of F_2, SOF_4 is isomerized to SF_3OF [563, 565, 566]. The SOF_4 molecule has a distorted trigonal-bipyramidal structure, with the oxygen atom in an equatorial position [664, 755]. If SOF_4 is reacted with fluoride donors, for example, CsF, the formation of the distorted octahedral anion $[OSF_5]^-$ is observed (Scheme 9.29) [563, 670, 756–758]. The salts are only stable with a large cation, such as $[(Me_2N)_3S]^+$ or $[NMe_4]^+$, otherwise they tend to decompose back to the starting materials. SOF_4 reacts with fluoride acceptors like AsF_5 or SbF_5 to form the distorted tetrahedral cation $[OSF_3]^+$ [672, 759, 760]. Based on SOF_4, a few organic derivatives have also been described. For example, the reaction of SOF_4 with arylsilylethers $ArOSiR_3$ yields compounds of the type $ArOS(O)F_3$ along with $FSiR_3$ [761]. Similarly, the substitution with amines RNH_2 and azole derivatives AzH gives rise to $RNS(O)F_2$ and $Az_2S(O)F_2$ (Az = pyrazole, imidazole, 1,2,4-triazole), respectively [762, 763]. The iminosulphur oxydifluoride derivatives $RNS(O)F_2$ are valuable linkers for the rather new sulphur(VI) fluoride exchange (SuFEx) click-chemistry, as their fluoride atoms can be readily substituted by further organic groups [762, 764].

Sulphur dioxydifluoride **SO_2F_2**, more commonly known as sulphuryl fluoride, is formed, when $Ba(OSO_2F)$ is heated under anhydrous conditions. SO_2F_2 is also prepared by direct fluorination of SO_2 with F_2. It is a color- and odorless gas (mp.: −120 °C; bp.: −55 °C) and has a distorted tetrahedral molecular structure [765–767]. Unlike sulphuryl chloride, SO_2F_2 is remarkably inert, comparable to the reactivity of SF_6. In a closed tube it may be heated together with water and is not decomposed at 150 °C. It is only slowly attacked by alkaline solutions; sodium may be molten in SO_2F_2 without losing its metallic shine. In contrast to SF_6, sulphuryl fluoride is toxic and has found application in the combat of wood worms. Under matrix-conditions it can be reacted with CsF to yield the anion $[SO_2F_3]^-$, which most likely has a distorted trigonal-bipyramidal structure with the oxygen-substituents occupying equatorial positions [676]. Like SOF_4, SO_2F_2 can be readily functionalized with organic substituents. The treatment with various phenols ArOH yields arylfluorosulphates $ArOSO_2F$, which can be used in the SuFEx click-chemistry (see also SOF_4 above) [762, 768].

Sulphuryl chloride fluoride **SO_2ClF** is a colorless gas (mp.: −124.7 C; bp.: 7.1 °C), which results from incomplete substitution of SO_2Cl_2 for example, with SbF_3 in presence of $SbCl_5$ or with pyridinium poly(hydrogenfluoride) $[py \cdot (HF)_x]$, NaF or PbF_2

as fluorinating agents [662, 769–771]. A high yield synthesis of SO_2ClF comprises the room-temperature reaction of KSO_2F with Cl_2 [772]. Apart from that, it is prepared in quantitative yield by the treatment of SO_2 with ClF [754]. In contact with water, SO_2ClF readily hydrolyzes. The molecular structure of SO_2ClF is described by a distorted tetrahedron [765, 766]. The chemistry of SO_2ClF is mainly determined by the replacement of either one or both of the halide substituents. Moreover, it is also an important solvent in inorganic and organometallic chemistry, as it is for example, used for the generation of carbocations in super acidic media. Furthermore, its low melting point and its relatively inert character make it an attractive solvent for the use in low-temperature transformations.

Selenium hexafluoride **SeF₆** is an extremely poisonous, octahedral, volatile compound (mp.: −46.6 °C; bp.: −34.8 °C), which can be made from the elements. It is slightly more reactive than SF_6 but still does not react with water under standard conditions. Also, there is no indication, that SeF_6 acts as fluoride donor or acceptor under formation of $[SeF_5]^+$ or $[SeF_7]^-$ [732]. It requires 10 % alcoholic NaOH solutions to hydrolyze it. Upon heating, SeF_6 reacts with metals or NH_3 to form elemental Se, HF and N_2. Since it is rather inert, it has even been considered as potential replacement for SF_6 in electric switches [773]. However, due to its reductive instability, which also plays a role in enzymatic degradation in terms of Se-poisoning, this potential use was abandoned. Under ambient conditions, SeF_6 oxidizes aqueous solutions of I^-, even at −78 °C it reacts with phenyl lithium in ether [732]. The latter reaction can be an access point to organoselenium compounds.

Selenium chloride pentafluoride **SeF₅Cl** is a colorless gas (mp.: −19 °C; bp.: 4.5 °C) that is very sensitive toward hydrolysis but can be stored in dry, passivated stainless-steel vessels at ambient temperature. It is formed in the reaction of $Cs[SeF_5]$ with $ClSO_3F$ as well as directly from SeF_4 and ClF [774, 775]. However, the treatment of SeF_4 with BrF does not yield SeF_5Br, but only SeF_6 and Br_2, which might be explained by the fact, that BrF is always at least partly dissociated to Br_2 and BrF_3, where the latter will readily oxidize SeF_4 to SeF_6 [775]. Expectedly, SeF_5Cl molecules have C_{4v}-symmetry[776]. Attempts to add SeF_5Cl to olefins, for example, $CH_2=CF_2$, have not been successful, as SeF_5Cl readily decomposes to ClF and SeF_4, resulting in the exclusive addition of ClF to the double bond [777].

Selenium oxytetrafluoride **SeOF₄** is generated in monomeric form upon vacuum pyrolysis of $Na[OSeF_5]$, where it is recovered as a white solid at −196 °C [758, 778, 779]. Above −100 °C, $SeOF_4$ polymerizes to give, among others, the hydrolytically unstable dimer $(SeOF_4)_2$ in form of a colorless liquid (mp.:12 °C; bp.: 65 °C). The structure is described by two connected $[SeO_2F_4]$ octahedra with a common oxygen edge [779–781]. Above 250 °C, the dimer $(SeOF_4)_2$ decomposes to yield mainly SeF_4 and O_2, along with some $SeOF_2$ and traces of monomeric $SeOF_4$ [782]. Selenium dioxide difluoride ("selenonyl fluoride") **SeO₂F₂** is a colorless gas (mp.: −99.5 °C; bp.: −8.4 °C), which is made from $BaSeO_4$ with excess HSO_3F [783]. It may also be prepared from SeO_3 and SeF_4, PF_5 or AsF_3 and has a distorted tetrahedral structure [766, 767, 784–786]. SeO_2F_2 is hydrolyzed to HF and selenic acid, in general it is far more reactive than the sulphur analogue SO_2F_2, as it violently reacts with gaseous NH_3. However, it does not react with dry glass at room temperature [783]. The fluoride-substituents of SeO_2F_2 can be replaced by silyloxy groups, if it is reacted with hexamethyldisiloxane $(Me_3Si)_2O$, resulting in the formation of $(Me_3SiO)_nSeO_2F_{2-n}$ ($n = 1$ or 2) [787].

Tellurium hexafluoride **TeF$_6$** is the most reactive chalcogen hexafluoride. Analogously to the aforementioned sulphur and selenium hexafluorides, it can be prepared from Te and F_2 as a volatile compound (mp.: $-37.6\,°C$ (at 2 atm., pressurized); bp.: $-38\,°C$). It is furthermore accessible by fluorination of TeO_2 with BrF_3 (Scheme 9.28). At room temperature, TeF_6 is completely hydrolyzed by water after a few hours to finally yield telluric acid $Te(OH)_6$. The partially hydrolyzed compounds $TeF_n(OH)_{6-n}$ ($n = 1–4$) are in equilibrium with each other and have been observed by NMR- and Raman-spectroscopy [788–791]. It is also possible to replace the fluorine atoms by other O- and N-containing ligands [732]. However, upon reaction with tertiary amines, such as Me_3N, TeF_6 is reduced to tetravalent tellurium [691]. A special attention is paid to the pentafluoro orthotelluric acid F_5TeOH, which is conveniently prepared from $Te(OH)_6$ and HSO_3F (Scheme 9.28). It is also obtained, if TeF_6 is reacted with silanols [792]. Numerous alkoxy derivatives $TeF_n(OR)_{6-n}$ are also known and accessible by substitution reactions of TeF_6 with alcohols [793–796]. In contrast to SF_6 and SeF_6, TeF_6 can also act as Lewis acid. For example, the reaction with alkali metal fluorides or AgF yields heptafluorotellurates(VI) with the pentagonal bipyramidal anion $[TeF_7]^-$, with CsF at 250 °C one can even obtain square-antiprismatic $[TeF_8]^{2-}$-species (Scheme 9.28) [797–802].

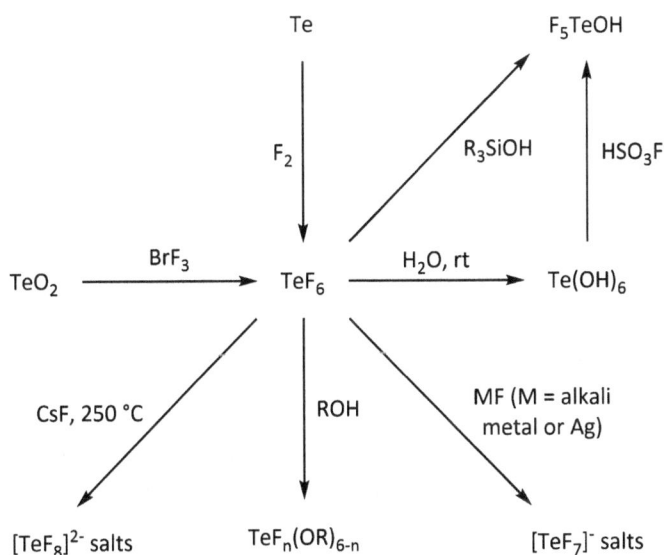

Scheme 9.28: Preparation and selected reactions of TeF_6.

The tellurium compounds **TeF₅Cl** (mp.: −28 °C; bp.: 13.5 °C) and **TeF₅Br** (mp.: −32 °C) are both generated as low-boiling liquids upon fluorination of the tetrahalides $TeCl_4$ or $TeBr_4$ with F_2 or the reaction of TeF_4 with the respective halogenfluoride XF (X = Cl, Br) [803, 804]. Alternatively, TeF_5Cl is made by treatment of TeF_4, $TeCl_4$ or TeO_2 with ClF [805, 806]. TeF_5Br can be stored without noticeable decomposition in dry PTFE-vessels at −78 °C, however, at room temperature Br_2 is liberated after a few days [804]. TeF_5Cl reacts with methanol and its silylether CH_3OSiMe_3 to form a 1:6-mixture of *cis*- and *trans*-$(CH_3O)TeF_4Cl$. However, no reactions are observed with anhydrous HF, SbF_5, F_2 or ClF at room temperature. A slow decomposition reaction is observed with CsF, where TeF_4, TeF_6, Cl_2 and traces of ClF are generated. Thermal decomposition occurs in the range of 70–250 °C to yield TeF_4, TeF_6 and Cl_2 [807]. Upon hydrolysis of TeF_5Cl, species of both series $(HO)_nTeF_{6-n}$ (n = 2–6) and $(HO)_nTeF_{5-n}Cl$ (n = 1–3) are observed [808]. TeF_5Cl is indeed capable of adding to olefins, such as $CH_2=CF_2$. However, the resulting products containing TeF_5-groups are unstable, so that further transformations cannot be performed without decomposition [777, 803].

Monomeric tellurium oxytetrafluoride **TeOF₄** does not exist. Instead, the dimer $(TeOF_4)_2$ is formed, when $Li[OTeF_5]$ is pyrolyzed [758, 779, 780]. The colorless, crystalline **(TeOF₄)₂** (mp.: 28 °C; bp.: 77.5 °C) has a similar structure like the Se-analogue, but it is much more sensitive toward hydrolysis [779–781]. It can also be recovered from thermolysis of $B(OTeF_5)_3$ at 600 °C in a flow system [781, 809]. Above 400 °C, dimeric $(TeOF_4)_2$ decomposes to TeF_4, O_2 and some $TeOF_2$; however, monomeric $TeOF_4$ is not observed [782]. In presence of fluoride ions, $(TeOF_4)_2$ forms the pentagonal-bipyramidal $[TeOF_6]^{2-}$ anion, where oxygen is occupying an axial position, however, salts containing the distorted octahedral pentafluorooxotellurate(VI) anion $[OTeF_5]^-$ are not prepared from $TeOF_4$ but the far more readily accessible $HOTeF_5$ (see below) [810]. Distorted tetrahedral tellurium dioxydifluoride **TeO₂F₂** is prepared by the reaction of $BaTeO_4$ with excess HSO_3F. Anionic $[TeO_2F_4]^{2-}$ is generated upon pyrolysis of $Na[OTeF_5]$ along with dimeric $(TeOF_4)_2$ [779].

Polonium hexafluoride **PoF₆** has not been unambiguously prepared so far. It could be both, unstable or even nonexisting. The most obstructive problem is the radioactivity of all the polonium isotopes. By now, it was impossible to prepare PoF_6 from ²¹⁰Po, which has a half-life of 139 days, most probably due to radioactive decay [811]. However, efforts have been made to react the longer-lived ²⁰⁸Po ($t_{1/2}$ = 2.9 years) with F_2, which resulted in the formation of a volatile compound [812]. Considering the chemistry of TeF_6 and many other molecular hexafluorides, the volatile compound could indeed have been PoF_6.

The pentafluoro orthochalcogenic acids F_5EOH (E = S, Se, Te) have to be prepared following very different methods [758]. Pentafluoro orthosulphuric acid **HOSF₅** is only stable at low temperatures and is generated in a two-step sequence. First, ClF and SOF_4 are reacted to give $ClOSF_5$, which is subsequently treated with HCl to yield $HOSF_5$ and Cl_2. The acid decomposes to HF and SOF_4 at temperatures above −65 °C (Scheme 9.29) [813, 814]. More stable derivatives can be generated, when the

SF_4

$[OSF_5]^-$ salts

$HOSF_5$

O_2, NO_2

F^-

$T > -65\ °C$ / $- HF$

HCl / $- Cl_2$

SOF_2 —— $BrF_5\ or\ ClF_3$ —→ SOF_4 —— ClF —→ $ClOSF_5$

SeO_2F_2 —— HSO_3F, KHF_2 —→ $HOSeF_5$ —— MF_n —→ $M(OSeF_5)_n$

$TeOH_6$ —— HSO_3F —→ $HOTeF_5$ —— XeF_2 —→ $Xe(OTeF_5)_2$

BCl_3

$Te(OTeF_5)_4$

$Cs[B(OTeF_5)_4]$ ←—— $Cs[OTeF_5]$ —— $B(OTeF_5)_3$

$Te(OTeF_5)_6$

UF_6

$U(OTeF_5)_6$

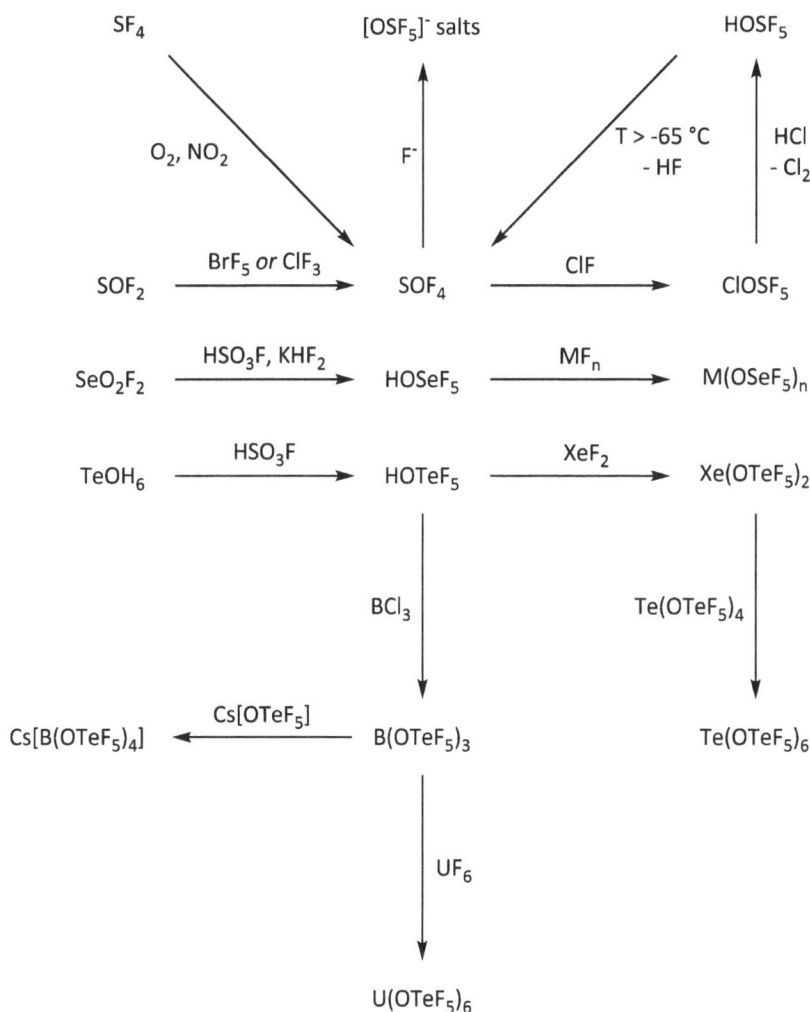

Scheme 9.29: Preparation and selected reactions of EF_5OH (E = S, Se, Te).

hydrogen is replaced by other substituents, which have to be either strongly electron-withdrawing (-F, -OSF_5, -SO_2F) or very electropositive (Cs^+) [814]. A convenient preparation of the respective selenium derivative **$HOSeF_5$** (mp.: 38 °C; bp.: 44 °C) is achieved by the treatment of SeO_2F_2 with HSO_3F and $K[HF_2]$ (Scheme 9.29). $HOSeF_5$ is a strong oxidizer, as it immediately liberates Cl_2 in contact with NaCl. Therefore, the direct transfer of the $OSeF_5$-group is only achieved, when $HOSeF_5$ is reacted with fluorides or oxides (Scheme 9.29). Nevertheless, a few reagents have been explored, such as $Hg(OSeF_5)_2$, $(CH_3)_3SiOSeF_5$ or $ClOSeF_5$, which are capable of transferring this substituent under mild conditions to chloride-containing substrates [758]. The free acid $HOSeF_5$ is corrosive against glass and metals, however, it does not react with PTFE and attacks quartz only slowly [815]. Pentafluoro orthotelluric acid **HOTeF$_5$** is readily accessible by the reaction of BaH$_4$TeO$_6$ with HSO$_3$F in form of a colorless liquid or more conveniently from Te(OH)$_6$ and HSO$_3$F (Scheme 9.29) [816–818]. All the three compounds HOEF$_5$ (E = S, Se, Te) are strong acids with highly electronegative, distorted octahedral OEF$_5$-groups. In case of the selenium- and tellurium-compound, many derivatives are directly accessible from the free acids [670, 810]. The reaction of HOTeF$_5$ with BCl$_3$ leads to the quantitative formation of the important OTeF$_5$-transfer reagent B(OTeF$_5$)$_3$, which can further be reacted with Cs[OTeF$_5$] to result in the borate derivative Cs[B(OTeF$_5$)$_4$] (Scheme 9.29) [809, 818]. Another important reagent for the introduction of OTeF$_5$-substituents is the xenon compound Xe(OTeF$_5$)$_2$, which is prepared from XeF$_2$ and HOTeF$_5$ [818]. There are many more OEF$_5$-derivatives known; the majority of them containing the OTeF$_5$-group [580, 758, 813, 814, 819–823]. For example, U(OTeF$_5$)$_6$ is prepared from UF$_6$ and B(OTeF$_5$)$_3$, whereas Te(OTeF$_5$)$_4$ is transformed to Te(OTeF$_5$)$_6$ upon treatment with Xe(OTeF$_5$)$_2$ (Scheme 9.29) [824–826]. Based on the chemistry of pentafluoroorthotellurates, it is possible to construct weakly coordinating anions, such as the aforementioned borate [B(OTeF$_5$)$_4$]$^-$, the derivatives [M^V(OTeF$_5$)$_6$]$^-$ (M^V = As, Sb, Bi, Nb) or the recent examples of [M^{III}(OTeF$_5$)$_4$]$^-$ (M^{III} = Al, Ga) [827–833]. As further derivatives of the pentafluoro orthochalgogenic acids, the hypofluorites F$_5$EOF (see above), the oxides (F$_5$E)$_2$O and the peroxides (F$_5$EO)$_2$ have to be mentioned. They are formed as side products upon fluorination of the dioxides or oxyhalides [580].

Sulphur-nitrogen-fluorine compounds [834, 835]

Ternary sulphur-nitrogen-fluorine compounds also have a certain relevance. The chemistry of this class of compounds is mainly based on (NSF)$_x$ (x = 1, 3, 4) and NSF$_3$. Thiazyl fluoride **NSF** is an unstable, colorless gas (mp.: −89 °C; bp.: 0.4 °C) with a keen odor that is prepared in good yields upon fluorination of S$_4$N$_4$ with (metal) fluorides, for example, HgF$_2$, AgF$_2$, IF$_5$ or SF$_4$. Furthermore, it can be synthesized by the reaction of SF$_4$ with NH$_3$, NF$_3$ with elemental sulphur as well as MoS$_2$ with NF$_3$ [652]. It is also accessible by thermolysis of Hg(NSF$_2$)$_2$. The NSF molecule has a bent structure, the bond order of the N-S bond amounts to 2.4. In the gas phase *in vacuo*, it decomposes to S$_4$N$_4$ and S$_3$N$_2$F$_2$. In the liquid state, it associates to form the cyclic trimer (NSF)$_3$, whereas it slowly oligomerizes to tetrameric (NSF)$_4$ in the gas phase at pressures above 1 bar. NSF is quite reactive and slowly attacks glass at room temperature. Therefore, it should be stored in PTFE vessels. NSF reacts with AgF$_2$ to yield NSF$_3$, in the reaction with strong fluoride acceptors like AsF$_5$ or SbF$_5$, the salts [NS]$^+$[MF$_6$]$^-$ are generated. It is also possible to treat NSF with fluoride donors, for example, with CsF the compound Cs[NSF$_2$] is formed. The NSF group is also part of the mercury compound Hg(NSF$_2$)$_2$ and the sulphurnitride halides XNSF$_2$. Elemental chlorine transforms NSF into the cyclic trimer (NSCl)$_3$, whereas ClNSF$_2$ is formed, if CsF and Cl$_2$ are present.

The colorless and crystalline (mp.: 74.2 °C; bp.: 92.5 °C), cyclic trithiazyl trifluoride (**NSF**)$_3$ cannot only be obtained as trimerization product of NSF, but also by Cl/F exchange of (NSCl)$_3$ with AgF$_2$ [652]. (NSF)$_3$ is a Lewis base and reacts with BF$_3$ and other fluoride acceptors to form a non-ionic, colorless solid, which is stable up to 60 °C [836]. It is very sensitive toward hydrolysis and decomposes to monomeric NSF, if it is heated to 250 °C. The structure of (NSF)$_3$ is described by a six-membered (NS)$_3$-ring, which adapts a boat-conformation. The fluorine substituents at the sulphur atoms are in axial positions [837, 838]. Tetrathiazyl tetrafluoride (**NSF**)$_4$ (mp.: 153 °C, decomp.) is synthesized by the fluorination of S$_4$N$_4$ with AgF$_2$ in form of colorless crystals. In comparison to S$_4$N$_4$, the ring is planarized and contains localized S-N double- and single bonds. (NSF)$_4$ is hydrolyzed to NH$_4$F and H$_2$SO$_3$, with fluoride acceptors it yields a mixture of complexes with the [N$_3$S$_3$F$_2$]$^+$ and [NS]$^+$ cation. [N$_4$S$_4$F$_3$]$^+$ readily loses NSF and results in [N$_3$S$_3$F$_2$]$^+$, which reacts with remaining fluoride acceptor to finally yield [NS]$^+$ salts [836].

The most relevant of the ternary nitrogen-sulphur-fluorine compounds is the chemically and thermally stable thiazyl trifluoride **NSF$_3$** (mp.: –72.6 °C; bp.: –27.1 °C). It is also a colorless gas with keen odor and one of the products of the fluorination of S$_4$N$_4$ with AgF$_2$. Apart from that, the distorted tetrahedral NSF$_3$ can be made by fluorination of NSF with F$_2$ or AgF$_2$. It can be considered as SF$_6$-derivative, wherein three F atoms have been replaced by one nitrogen atom. Thus, it also has some chemical inertness and is not hydrolyzed by diluted acids and only slowly by water. Further, NSF$_3$ is resistant against Na up to temperatures of about 200 °C. It reacts with alcohols and amines under fluoride substitution and HF liberation. The latter can add to the multiple bond of NSF$_3$ and give rise to F$_5$SNH$_2$ which slowly decomposes back to the starting materials at room temperature. It is also possible to add other molecules to the multiple bond, as for example, ClF will give rise to F$_5$SNCl$_2$ at –78 °C, whereas SF$_4$ in presence of BF$_3$ yields F$_5$SNSF$_2$, which is also obtained in many reactions of N$_2$F$_4$ with sulphur compounds.

9.6.3 Study questions

(I)   Which reactivity is generally observed for the radical species ˙OF and ˙O$_2$F? Do they recombine?

(II) Give some examples for the high oxidation power of the oxygen fluorides OF$_2$ and O$_2$F$_2$. What happens when they are treated with strong fluoride acceptors?

(III) How is HOF prepared? What is it commonly used for in organic chemistry?

(IV) What is the common feature of the inorganic hypofluorites? Which reactivity is observed against unsaturated C-C bonds? Why are they only sparingly used in organic synthesis?

(V)   Which general types of organic hypofluorites are known? How are they prepared and what are they commonly used for?

(VI)   Under which conditions is it possible to observe SF_2? Which are its main decomposition products? Briefly describe the structures of the corresponding molecules and comment on the reversibility of their formation.

(VII)   There are two isomeric compounds with the sum formula S_2F_2. Which molecular structure do they have? How are these compounds prepared? Can they be converted into each other?

(VIII)   How does one prepare SF_4? What happens, if it is reacted with fluoride donors and acceptors? Briefly describe the structures of the corresponding species.

(IX)   How do SF_4 and SeF_4 structurally differ from each other? Does this also have consequences on the chemical properties?

(X)   Which general methods are known for the preparation of TeF_4? Compare its reactivity and stability with those of the lighter homologues SF_4 and SeF_4.

(XI)   Thermodynamically, SF_6 should be readily hydrolyzed by water. Why does such a reaction not occur easily? Explain the inertness of SF_6.

(XII)   Which three compounds are the most important transfer reagents for SF_5-substituents? How are they prepared?

(XIII)   Compare the reactivity of the oxyfluorides SOF_4 and SO_2F_2 with respect to hydrolysis as well as the reaction with fluoride donors and acceptors. Which molecular structures are observed for the corresponding neutral and ionic species?

(XIV)   How does the reactivity of the chalcogen hexafluorides change from sulphur to tellurium? Comment on the behavior against fluoride donors. Which structures do the resulting products have?

(XV)   What are the most convenient methods to prepare inorganic derivatives of the acids $HOSF_5$, $HOSeF_5$ and $HOTeF_5$? Comment on the preparation and the stability of the free acids.

(XVI)   Which molecular structures are observed for the compounds NSF, $(NSF)_3$, $(NSF)_4$ and NSF_3? How are these compounds prepared?

9.7 Group 17: Chlorine, bromine, iodine

In general, the halogen fluorides exist in a variety of binary molecular compounds, but also in form of fluorocations or anions. Their stabilities mainly depend on the halogen and its oxidation state. The halogen fluorides have a number of structural and chemical properties in common, so that they can mostly be discussed together in a more general way. Apart from binary halogen-fluorine compounds, ternary oxygen fluorides exist as well, which will be subject of the second part of this subchapter.

9.7.1 Preparation of the halogen fluorides

Chlorine monofluoride **ClF** is a colorless gas (mp.: −155.6 °C; bp.: −101.1 °C) and is prepared from Cl_2 and F_2, which are passed over activated copper in a monel reactor at 220–250 °C (Scheme 9.30). If the temperature is raised to 300 °C, chlorine trifluoride **ClF$_3$** is formed (mp.: −76.3 °C; bp.: 11.8 C). Chlorine pentafluoride **ClF$_5$** is conveniently synthesized from ClF_3 and F_2 in an autoclave at 350 °C and a pressure of 250 bar. Alternatively, ClF_3 can be fluorinated at −78 °C using O_2F_2 as fluorinating agent, which is

ClF

Cu,
220-250 °C

Cu, 300 °C   F$_2$,
                           350 °C, 250 bar
Cl$_2$ + F$_2$ ⟶ ClF$_3$ ─────────── ClF$_5$
                          or O$_2$F$_2$, -78 °C

[BrF] ⇌ BrF$_3$ + Br$_2$

low T,
inert solv.   pyridine

                  20 °C
Br$_2$ + F$_2$ ⟶ BrF$_3$ BrF · py

200 °C

BrF$_5$

[IF]

-40 °C,
inert solv.   I$_2$

      250-270 °C   -40 °C
IF$_7$ ⟵ I$_2$ + F$_2$ ⟶ [IF$_3$]
                           inert solv.

rt   ozonolysis,
                     aHF

IF$_5$ I$_2$

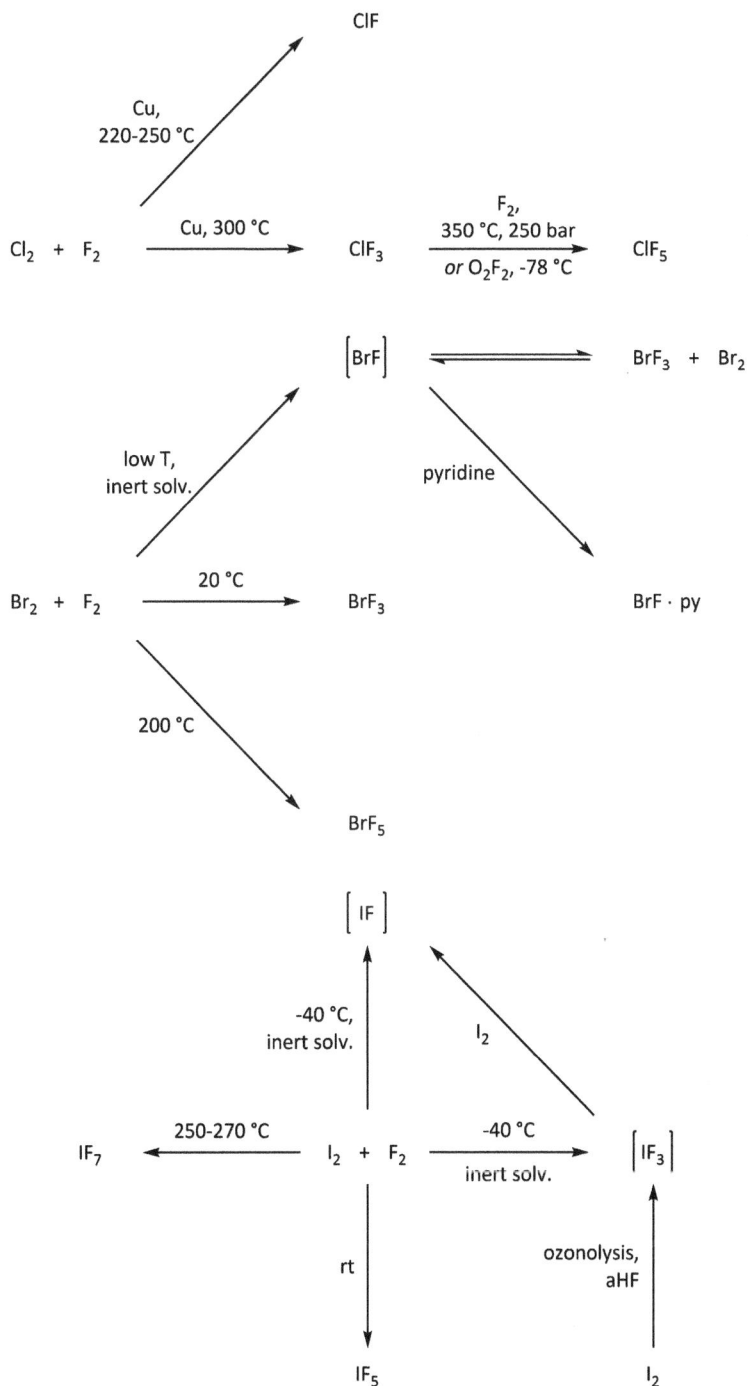

Scheme 9.30: Preparation of the halogen fluorides.

advantageous over the high-temperature preparation, due to the thermal instability of ClF_5 (Scheme 9.30). Like the low-valent chlorine-fluorine compounds, ClF_5 is a colorless gas (mp.: −103 °C; bp.: −13.1 °C) [839, 840]. Chlorine heptafluoride **ClF_7** does not exist. As quantum chemical studies have underlined, it is a highly unstable species and thus immediately splits off F_2 [841].

Similarly, the bromine fluorides are directly prepared from the elements. Bromine monofluoride **BrF** is formed at low temperature in inert solvents as a light-red to orange gas (mp.: −33 °C; bp.: ca. 20 °C). However, it cannot be isolated in pure form, as it is always in equilibrium with Br_2 and BrF_3 (Scheme 9.30) [842]. BrF can be stabilized by pyridine, yielding the adduct BrF · py [843]. Bromine trifluoride **BrF_3** is formed at 20 °C as a colorless liquid (mp.: 8.8 °C; bp.: 125.8 °C), at about 200 °C, bromine pentafluoride **BrF_5** is generated, also in form of a colorless liquid (mp.: −60.5 °C; bp.: 41.3 °C). Similar to ClF_7, bromine heptafluoride **BrF_7** does not exist. Again, theoretical studies revealed a largely exothermic decomposition pathway under the loss of F_2 [841].

Iodine monofluoride **IF** (decomp.: −14 °C) and iodine trifluoride **IF_3** (decomp.: −28 °C) are thermally unstable. They are generated as chocolate brown and yellow to colorless solids, respectively, starting from the elements at −40 °C in inert solvents (Scheme 9.30). IF is also a product of the comproportionation of IF_3 and I_2 [36, 37, 844, 845]. Apart from that, IF_3 is accessible by ozonolysis of I_2 in anhydrous HF [846]. Iodine pentafluoride **IF_5** is a yellow liquid (mp.: 9.4 °C; bp.: 104.5 °C), which is prepared upon fluorination of I_2 at room temperature (Scheme 9.30). Heating the elements to 250–270 °C yields iodine heptafluoride **IF_7** (mp.: 6.5 °C; subl.: 4.8 °C).

9.7.2 Structural properties of the halogen fluorides

The structural properties of all the halogen fluorides may be discussed in a more general way, as there are a lot of similarities. In solid and liquid phase, the halogen fluorides are mainly associated via F-bridges. The trifluorides XF_3 have a T-shape, resulting from a ψ-trigonal bipyramid with two lone pairs in equatorial positions [847, 848]. In the gas phase and under high pressure, ClF_3, BrF_3 and IF_3 also dimerize to a small extent. Furthermore, a fluorine exchange can be observed in the gas phase, which also takes place via dimerization. In the condensed phase, BrF_3 and IF_3 are slightly associated via F-bridges, yielding polymeric $(BrF_3)_x$ and $(IF_3)_x$. In these polymers, Br is surrounded by a distorted square of fluorine atoms, whereas iodine adapts a distorted pentagonal planar coordination geometry, as could be clearly identified by X-ray crystallography [36, 847–850]. The pentafluorides XF_5 still have a lone pair, and thus have the structure of a slightly distorted square-pyramid (ψ-octahedron). In the XF_5 structures, the central atom X is not directly situated in the square plane, but a little bit below. The IF_7 molecule forms a pentagonal bipyramid. The equatorial bonds are longer than the axial ones, in addition, the equatorial substituents dynamically move slightly out of plane.

9.7.3 Stability of the halogen fluorides

The stability of the halogen fluorides can be visualized by a table, wherein a diagonal line is drawn from ClF to IF_5 (Table 9.6). The most stable fluorides, that is, ClF, BrF_3 and IF_5, are on the line. The derivatives above the line, that is, BrF, IF and IF_3 decompose already at low temperatures, liberating the corresponding halogen and the stable fluoride [844]. In acetonitrile solution, IF_3 is in equilibrium with IF and IF_5, where IF also undergoes decomposition processes [851]. The higher fluorides ClF_3, ClF_5, BrF_5 and IF_7 are below the line and only decompose at high temperatures to form equilibria of the stable fluorides and fluorine. Thus, they are strong fluorinating agents. All the halogen fluorides are very reactive compounds. Due to this and their tendency to decompose, they are difficult to obtain and handle in pure form. The reactivity is often increased by impurities, mainly HF, which might also be desirable to achieve sufficiently fast reactions in some cases. The fluorination ability of the halogen fluorides can be ordered in the following way: $ClF_3 > BrF_5 > IF_7 > ClF > BrF_3 > IF_5 > BrF > IF_3 > IF$. A classification of ClF_5 is difficult, as it should be expected to be more reactive than ClF_3 but its actual reactivity is rather slightly lower than that of ClF. This can be explained by kinetic inhibition of many of its reactions. The most important and most established fluorinating agents among the halogen fluorides are ClF, ClF_3, BrF_3 and IF_5. They readily react with many elements and starting materials, often violently, even at room temperature. Halogen fluorides also react with quartz, which is probably initiated by traces of HF. Pure samples only slowly react with SiO_2.

Table 9.6: Stability trends in the series of halogen fluorides.

Most stable	ClF	BrF	IF	Decomposition to X_2 and stable XF_n at low temp.
$\downarrow$	ClF_3	BrF_3	IF_3	$\uparrow$
decomposition to F_2 and stable XF_n at high temp.	ClF_5	BrF_5	IF_5	most stable
	–	–	IF_7	

9.7.4 Reactivity of the halogen fluorides

ClF readily adds to multiple bonds, for example, C=C, N=S, S=O, C=O, the addition of BrF and IF to C=C bonds of halogenated olefins is also known (cf. Section 13.4.9). However, in the latter case, the halofluorination is often realized by more convenient approaches [852]. Another important reaction of ClF is the oxidative chlorofluorination of SF_4 to produce SF_5Cl. Apart from that, it tends to undergo transformations under HF-elimination, as for example, $HONO_2$ is converted to $ClONO_2$ or H_2O reacts to form Cl_2O. In both cases, HF is liberated, too [853].

ClF_3 is one of the most reactive chemicals known, it reacts explosively with many inorganic and organic compounds like water, ammonia, asbestos or wood. Mixtures of ClF_3-NH_3 or ClF_3-N_2H_4 have also been investigated as rocket fuel. Many elements ignite in the presence of ClF_3 under fluoride formation, even xenon or metal chlorides and oxides are readily fluorinated. It also reacts with organic substrates, as it oxidizes perfluorinated alcohols as well as COF_2 to the corresponding peroxides. Such transformations can only be conducted with perfluorinated substrates, as hydrocarbons react extremely violently with ClF_3, often combined with ignition or explosions. ClF_3 is stored in containers made of steel, nickel, copper or Monel, which are passivated by a fluoride layer on the metal surface. ClF_3 has been used for the production of UF_6 and for recycling and regeneration of nuclear fuels.

In general, the higher chlorine fluorides ClF_3 and ClF_5 are already strong oxidizing and fluorinating agents, but they are themselves oxidized by even stronger oxidizers like XeO_3F_2, KrF_2 or PtF_6. Both, ClF_3 and ClF_5 yield the oxyfluoride $FClO_3$ when they are reacted with XeO_3F_2. Upon reaction with KrF_2, ClF_3 is oxidized to ClF_5, whereas ClF_5 is transformed to cationic $[ClF_6]^+$ by PtF_6. The hydrolysis of ClF_5 results in the formation of $FClO_2$, which is also obtained as disproportionation product along with ClF, if ClF_3 is reacted with water [854]. The presence of alkali metal fluorides (KF, RbF, CsF) catalyzes the decomposition of ClF_5 to ClF_3 and F_2 [855].

BrF_3 is less reactive than ClF_3, however, it still reacts explosively with water and acts as strong fluorinating agent toward many elements. Reactions with oxides quantitatively yield fluorides or oxyfluorides, which can be used for the analytical assay of the oxygen content. With pyridine, a 1:1-adduct $[BrF_3 \cdot py]$ is formed, which can be stored at $-35\,°C$ but decomposes slowly, sometimes also explosively, at room temperature [856].

BrF_5 is relatively inert against strong oxidizers, thus it can be used as a solvent, for example, for reactions with PtF_6. The hydrolysis of BrF_5 as well as the reactions with the iodine compounds IO_2F, IOF_3 and I_2O_5 yield BrO_2F, whereas a mixture of BrO_2F and $BrOF_3$ is obtained from the reaction of BrF_5 with IO_2F_3 [857]. Organic BrF_5-derivatives of the type R-BrF_4, where R is an aryl group, are also known [858].

The less reactive iodine fluorides are also capable of undergoing fluorine exchange reactions, as for example the reaction of IF_3 with $(CF_3CO)_2O$ leads to the formation of $(CF_3COO)_3I$ along with CF_3COF. Similarly, IF_5 is substituted by Me_3SiOMe to yield IF_4OMe, where the methoxy-group replaces the axial fluorine atom [847, 859–861]. Derivatives with organic substituents are also accessible by direct fluorination, for example, the treatment of CF_3I with F_2 at $-78\,°C$ yields the IF_3-derived product CF_3IF_2 [862]. In contrast, the perfluoroaryl derivative $C_6F_5IF_2$ is prepared by a replacement reaction starting from IF_3 and $Cd(C_6F_5)_2$ at $-78\,°C$ [863]. Due to its reduced reactivity, IF_5 can be handled in glassware, but always contains traces of IOF_3, if the sample is in contact with oxygen or moisture [864]. IF_5 reacts with a number of elements under ignition, for example, the alkali metals, boron, phosphorus, arsenic, etc. to form the respective fluorides; whereas oxides are mainly transformed to oxyfluoride derivatives

[865]. Attempts to prepare iodine(VII) compounds of the type R^FIF_6 have not been successful by either substitution of IF_7 or fluorination of R^FIF_4 [866].

9.7.5 Fluorinated cations derived from the halogen fluorides

Some of the halogen fluorides show electric conductivity in liquid phase, which can be explained by self-dissociation [867, 868]. As the compounds are mostly associated in condensed phase, a small shift of the fluorine bridges is sufficient to achieve dissociation. However, reactions with fluoride acceptors or -donors also yield salt-like compounds with complex interhalogen ions.

Cationic species are mostly formed upon direct reaction of the halogen fluorides with fluoride acceptors like AsF_5 or SbF_5. The ions in these compounds are coordinatively linked by F-bridges. A difference is observed in case of the monofluorides. ClF reacts with fluoride acceptors to form the associated cation $[Cl_2F]^+$, as Cl^+ itself is not stable (Scheme 9.31). The unsymmetric, bent $[Cl_2F]^+$ is also not remarkably resilient, as it can be obtained at $-78\,°C$ in form of its $[AsF_6]^-$-salt, but at the same temperature it completely disproportionates in SbF_5/HF solution to form $[ClF_2]^+$ and $[Cl_3]^+$ [869, 870]. No fluorobasic properties have been reported for BrF and IF, so that flurocations based on these two monofluorides are still unknown.

Many stable complexes with the cations $[XF_2]^+$ (X = Cl, Br, I) are known to be accessible upon reaction of XF_3 with various Lewis acids (Scheme 9.31) [36, 672, 871–873]. In the $[XF_2]^+$-salts, relatively strong F-bridges between anions and cations are found. For example, the $[MF_6]$-octahedra are distorted in $[ClF_2][SbF_6]$, $[ClF_2][AsF_6]$ and $[BrF_2][SbF_6]$, which clearly would not be the case, if purely ionic interactions were present between the ions [871–873]. The $[XF_2]^+$ cations increase their coordination number in these examples to have a ψ-octahedral environment and a planar orientation of the four F-atoms. Overall, the bonding situation in the aforementioned $[XF_2]^+$-complexes can be summarized as being mainly ionic with covalent contributions. In case of BrF_3, associated cations have also been prepared and characterized, namely $[Br_2F_5]^+$ and $[Br_3F_8]^+$ (Scheme 9.31). In the $[Br_2F_5]^+$ cation, both Br-atoms are coordinated by three F-substituents in a trigonal planar environment, where the $[BrF_2]^+$-fragments are symmetrically bridged. Thus, the compound may also be written as $[F_2Br\text{-}F\text{-}BrF_2]^+$. The structure of $[Br_3F_8]^+$ may be best described as being composed of a central $[BrF_2]^+$-unit, which is coordinated by two BrF_3 molecules. This results in a square planar coordination geometry at the central Br-atom, whereas the other two are again surrounded by a trigonal planar orientation of F-substituents [874].

The halogen pentafluorides do not readily donate a fluoride-ion, this only works with the strongest Lewis acids (Scheme 9.31). A reason for this might be the decrease in symmetry of ψ-octahedral XF_5 upon formation of ψ-trigonal-bipyramidal $[XF_4]^+$ [875–878]. In the fluoroantimonates, strong F-bridges are present, which increase the coordination numbers of the halogens to six (Cl, Br) and eight (I) [877, 878].

$[Cl_2F]^+$ $\xleftarrow{\text{FA}}$ ClF $\xrightarrow{F^-}$ $[ClF_2]^-$

$[Br(OCF_3)_2]^-$ $\xrightarrow[- COF_2]{\Delta}$ $[BrF_2]^-$

IF $\xrightarrow{F^-}$ $[IF_2]^-$

$[XF_2]^+$ $\xleftarrow{\text{FA}}$ XF_3 $\xrightarrow{F^-}$ $[XF_4]^-$ $\xrightarrow[(X = I)]{F^-}$ $[IF_5]^{2-}$

$[Br_2F_5]^+$ or $[Br_3F_8]^+$ $\xleftarrow{\text{FA}}$ BrF_3 $\xrightarrow{F^-}$ $[Br_2F_7]^-$ or $[Br_3F_{10}]^-$

$[XF_4]^+$ $\xleftarrow{\text{SbF}_5}$ XF_5 $\xrightarrow[(X = Br, I)]{F^-}$ $[BrF_6]^-$ or $[I_3F_{16}]^-$

$[IF_6]^+$ $\xleftarrow{\text{FA}}$ IF_7 $\xrightarrow{F^-}$ $[IF_8]^-$

$[XF_6]^+$ $\xleftarrow[\text{SbF}_5]{\text{KrF}_2}$ XF_5

$[ClF_6]^+$ $\xleftarrow{\text{PtF}_6}$ ClF_5 or ClO_2F

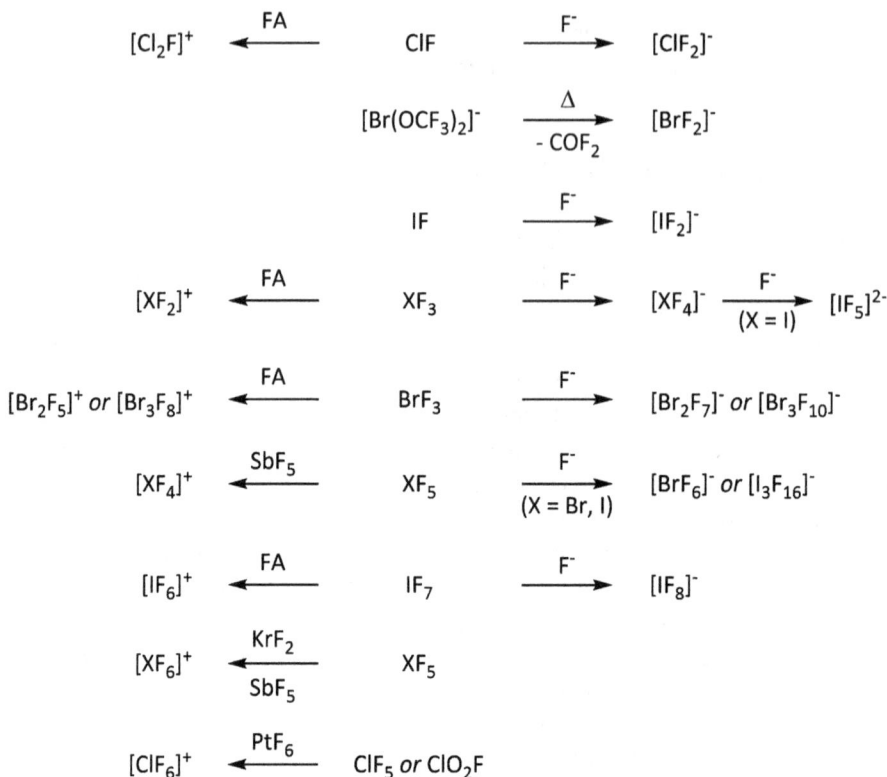

Scheme 9.31: Formation of fluorinated cations and anions based on the halogen fluorides (FA = fluoride acceptor).

The only known halogen heptafluoride IF_7 also reacts with AsF_5 or SbF_5 to form $[IF_6][AsF_6]$ and $[IF_6][Sb_3F_{16}]$, respectively (Scheme 9.31) [672, 879, 880]. Interestingly, the arsenate salt is comprised of two octahedral ions. The $[IF_6]^+$ cation can also be prepared by reaction of IF_5 with $[KrF][Sb_2F_{11}]$ [881]. Even if heptafluorides of chlorine and bromine are unknown and most likely not accessible, it is possible to obtain octahedral $[XF_6]^+$ cations with the halogens in oxidation state +VII (Scheme 9.31) [882]. They are very powerful oxidants which are capable of oxidizing fluoride ions. The reaction of BrF_5 with KrF_2 in presence of AsF_5 or SbF_5 yields colorless, crystalline salts with the octahedral cation $[BrF_6]^+$ [883, 884]. The preparation of $[ClF_6]^+$ can be achieved by oxidation of ClF_5 or ClO_2F with PtF_6 [885–887]. In both cases, a product mixture is obtained, containing the yellow, solid $[ClF_6][PtF_6]$, which is soluble in HF and can be stored at room temperature without decomposition. Efforts to generate ClF_7 by adding NOF only lead to the formation of $[NO][PtF_6]$, ClF_3, ClF_5 and elemental fluorine. Moreover, the oxidation of ClF_5 by KrF_2 in presence of AsF_5 or SbF_5 gives rise to the salts $[ClF_6][AsF_6]$ and $[ClF_6][SbF_6]$, respectively [888].

9.7.6 Fluorohalogenate anions

The fluorohalogenate anions are prepared by reaction of the halogen fluorides with ionic fluorides or, alternatively, by fluorination of halides like KCl, which gives rise to K[ClF$_4$] [889]. The definitely best known fluoroanion of the halogen monofluorides is [ClF$_2$]$^-$, which has a linear, symmetric structure. Recently, this anion has been synthesized by treatment of the ammonium salt [NEt$_3$Me]Cl with ClF in acetonitrile solution and structurally characterized by X-ray crystallography [890, 891]. The preparations of the anions [BrF$_2$]$^-$ and [IF$_2$]$^-$ have been reported quite some time ago, however, their existence has been doubted, due to missing or erroneous spectroscopic data [843, 892, 893]. The definite proof for the existence of [BrF$_2$]$^-$ has been obtained, when a sample of [NMe$_4$][Br(OCF$_3$)$_2$] was thermolyzed between −70 and −10 °C to eliminate COF$_2$ and yield the ammonium salt [NMe$_4$][BrF$_2$] (Scheme 9.31). The linear anion [BrF$_2$]$^-$ has been unambiguously identified by vibrational- as well as NMR-spectroscopy [894]. Similarly, the respective difluoroiodate(I) anion [IF$_2$]$^-$ is accessible in form of its ammonium salt [NEt$_4$][IF$_2$], when IF is directly treated with [NEt$_4$]F (Scheme 9.31). [IF$_2$]$^-$ is also linear, as has been identified by spectroscopic methods [895]. An alternative synthesis of the tetramethylammonium salt [NMe$_4$][IF$_2$] avoids the use of unstable IF and starts from readily available [NMe$_4$]I and XeF$_2$ [896].

The tetrafluoro halogenates [XF$_4$]$^-$ are much more explored and generally accessible by solvolysis of the fluorides M^{I}F (M^I = K, Rb, Cs, NO) in the trifluorides XF$_3$ (X = Cl, Br, I) (Scheme 9.31) [36, 868, 897–900]. An alternative synthesis of the [ClF$_4$]$^-$ anion comprises the treatment of [NEt$_3$Me]Cl with F$_2$ in acetonitrile solution, which gives rise to the corresponding ammonium salt [891]. Recently, the propeller-shaped anion [Cl$_3$F$_{10}$]$^-$ has been reported. It is formed upon reaction of CsF with ClF$_3$ and is structurally described by three T-shaped ClF$_3$ molecules, which are connected by a μ^3-bridging fluoride anion, so that every Cl atom exhibits a distorted square-planar coordination environment [901]. According to the solvolysis method, the barium salt Ba[BrF$_4$]$_2$ is obtained from the reaction of BaF$_2$ with liquid BrF$_3$ [902]. Especially the tetrafluorobromate salts M^I[BrF$_4$] (M^I = K, Rb, Cs) and Ba[BrF$_4$]$_2$ have recently gained attention, as they are discussed as oxidants for noble metals in urban mining processes [903, 904]. In [XF$_4$]$^-$, the F-atoms are oriented in a square around the halogen, which overall has a ψ-octahedral environment with its two additional lone pairs [849, 898–900, 905, 906]. BrF$_3$ also tends to form aggregated anions, such as [Br$_2$F$_7$]$^-$ and [Br$_3$F$_{10}$]$^-$ (Scheme 9.31). For example, the salts Cs[Br$_2$F$_7$] and [PbF][Br$_2$F$_7$] are obtained upon treatment of BrF$_3$ with CsF or PbF$_2$, respectively. The anion [Br$_2$F$_7$]$^-$ is isostructural to the respective heptafluoroaurate [Au$_2$F$_7$]$^-$, being composed of two square-planar [BrF$_4$]-units, which share one common corner [233, 898]. The structure of [Br$_3$F$_{10}$]$^-$ is described by three square planar [BrF$_4$]-units, which are connected at one corner by a μ_3-bridging F-atom [907]. The treatment of [NMe$_4$][IF$_4$] with further [NMe$_4$]F results in the formation of pentagonal planar [IF$_5$]$^{2-}$ (Scheme 9.31) [908]. As a further derivative of IF$_3$, the hexafluoroiodate(III) Cs$_3$[IF$_6$] has also been reported

[845]. However, it has turned out that the latter is a mixture of CsF and Cs[IF$_5$], as [IF$_6$]$^{3-}$ is unstable and spontaneously decomposes upon loss of F$^-$ to result in the fluoroiodate(III) anions [IF$_4$]$^-$ or [IF$_5$]$^{2-}$ [908].

The alkali metal salts M[BrF$_6$] and M[IF$_6$] are the only known fluorohalogenates of the halogenpentafluorides (Scheme 9.31). They are prepared by solvolysis of M^{I}F (M^I = K, Rb, Cs) in the respective fluoride XF$_5$ (X = Br, I) [868]. Surprisingly, [BrF$_6$]$^-$ has the structure of a regular octahedron. Thus, the lone-pair of the bromine atom obviously does not have a steric impact. In contrast, the associate of IF$_5$ with F$^-$ does not yield isolated [IF$_6$]$^-$ in the solid state, but forms aggregates of [I$_3$F$_{16}$]$^-$, where three square-pyramidal IF$_5$ molecules are bridged by a fluoride anion. However, the latter does not occupy the sixth coordination site of the ψ-octahedral IF$_5$-units, but is located aside, due to the steric impact of the lone pair [693, 906, 909]. The influence of the lone pair on the structure of the respective hexafluorohalogenate anions has also been studied by quantum-chemical calculations [910, 911]. Upon reaction of ClF$_5$ with NOF or alkali metal fluorides, the formation of a hexafluorochlorate(V) [ClF$_6$]$^-$ anion cannot be observed. The circumstance that ClF$_5$ reacts with Lewis acids under complex formation, but does not do so with Lewis bases, might allow the conclusion that the highest possible coordination number for chlorine is six, including the five fluorine substituents and the lone pair. This also gives an explanation for the nonexistence of ClF$_7$ as well as the weak association of ClF$_5$ in the liquid state [912].

Octafluoroiodate anions are generated upon reaction of IF$_7$ with CsF, [NMe$_4$]F, NOF and NO$_2$F, respectively [798, 913–915]. Based on vibrational spectroscopy and quantum chemistry, the [IF$_8$]$^-$ anion has been found to possess a square-antiprismatic structure [798, 911]. These findings have been confirmed by X-ray crystallography [916].

Most of the aforementioned compounds are built from ions, which can be clearly deduced from their behavior in solution [868]. For example, [BrF$_2$][SbF$_6$] dissolves in BrF$_3$ under formation of a very well electrically conducting solution. In this solution, [BrF$_2$]$^+$ can be considered as acid which may be quenched by the base [BrF$_4$]$^-$. This behavior can also be transferred to other halogenfluorides. As a consequence, these compounds can be useful reaction media. Analogously to water, redox reactions can also be performed in BrF$_3$, possibly under participation of [BrF$_2$]$^+$ and [BrF$_4$]$^-$. Practically, this allows the oxidative solubilization of even very noble metals (e. g., Ag, Au, Ru) or metal halides (e. g., PdCl$_2$) in liquid BrF$_3$ to form the respective complexes Ag[BrF$_4$], [BrF$_2$][AuF$_4$], [BrF$_2$][RuF$_6$] and [BrF$_2$][PdF$_4$].

The halogen fluorides also exist as ionic compounds, when they are dissolved in anhydrous HF. The trifluorides ClF$_3$ and BrF$_3$ act as fluoride donors to form [XF$_2$]$^+$ and bifluoride [HF$_2$]$^-$. In a similar way, the fluorohalogenates Cs[ClF$_4$] and Cs[BrF$_4$] react to form XF$_3$ and [HF$_2$]$^-$. Fluoride transfer has also been investigated for the cationic compounds, but has only been found for [BrF$_4$][Sb$_2$F$_{11}$], which yields BrF$_5$ and [H$_2$F]$^+$. Interestingly, fluoride donation is also observed in the system ClF$_3$/BrF$_3$, where the ions [ClF$_2$]$^+$ and [BrF$_4$]$^-$ are found in equilibrium, revealing, that BrF$_3$ is the stronger acid [917, 918]. In anhydrous HF, the solvolysis reaction of $[ClF_2]^+$ toward ClF_3 is only very poorly pronounced [919].

9.7.7 Halogen oxyfluorides

There are a lot of chemical and structural relations between the halogen fluorides and halogen oxyfluorides. Interestingly, stable oxyfluoride compounds are only observed, if the central halogen is in the oxidation states +V and +VII, which gives the hint that high oxidation states are preferably stabilized, if oxygen and fluorine substituents are present next to each other. The three-valent derivatives XOF have only been obtained under matrix conditions and are even unstable at very low temperatures.

Trivalent compounds
Chlorosyl fluoride **ClOF** is formed upon partial hydrolysis of ClF_3 in diluted gas phase at low temperatures. Alternatively, it is accessible by the reaction of ClF_3 with HNO_3 or $NaNO_3$ as well as treatment of $ClOF_3$ with SF_4 [854, 855, 920]. ClOF is a thermolabile compound which disproportionates to ClF and ClO_2F with a half-life of ca. 3 min at a pressure of 100 Pa. ClOF cannot even be stabilized by strong Lewis acids. According to vibrational spectra in the gas phase as well as in an Ar-matrix, it has a bent structure [921, 922]. In contrast, the anionic derivative $[ClOF_2]^-$ has been calculated to be stable [841]. However, compounds of the type XOF (X = Cl, Br, I) do not seem to be accessible by standard preparative methods. Thermodynamically, ClOF is more stable than the isomeric hypofluorite. Bromosyl fluoride **BrOF** is even less stable than ClOF and could only be detected in an inert matrix at low temperatures. Similar to the chlorine homologue, quantum chemical calculations predict a stable anion $[BrOF_2]^-$ [841]. Iodosyl fluoride **IOF** has been postulated as an intermediate in the reaction of CF_3IOF_2 with Lewis acids, thermal decomposition of CF_3IO_2 or ozonization of I_2 in HF solution [846].

Pentavalent compounds
Chlorine oxytrifluoride **ClOF₃** is a low boiling liquid (mp.: −42 °C; bp.: 29 °C) that can be made by fluorination of Cl_2O, $NaClO_2$ or $ClONO_2$ at −78 °C in the presence of alkali metal fluorides (Scheme 9.32) [923]. Another preparative access includes the treatment of gaseous mixtures of ClF_3 and OF_2 with ultraviolet light [924]. It is also accessible by UV-irradiation of ClO_2F or ClO_3F in presence of fluorinating agents, such as ClF_5 or F_2, or directly from the elements Cl_2, F_2 and O_2 under photochemical conditions [925]. As has been proven by X-ray crystallography and vibrational spectroscopy, $ClOF_3$ has a ψ-trigonal-bipyramidal structure, where the axial positions are occupied by two fluorine atoms and each one oxygen- and fluorine-atom as well as the lone-pair are found in the equatorial plane [926–928]. $ClOF_3$ is a strong oxidizing and fluorinating agent. It readily reacts with glass and quartz and decomposes above 200 °C to yield

$Cl_2/O_2/F_2$ ClF_3 ClO_2F or ClO_3F

hv

OF_2 hv F_2 or ClF_5 UV-light

ClO_2, $NaClO_2$ or $ClONO_2$ $\xrightarrow{F_2, MF, -78\,°C}$ $ClOF_3$ $\xrightarrow{T > 200\,°C}$ $ClF_3 + O_2$

PtF_6 BF_3, AsF_5 or SbF_5 F^-

$[ClOF_2][PtF_6] + F_2$ $[ClOF_2]^+$ salts $[ClOF_4]^-$ salts

BrF_5

$LiNO_3$

$K[BrOF_4]$ $\xrightarrow[\text{or aHF, -78\,°C}]{[O_2][AsF_6], -78\,°C}$ $BrOF_3$ BrF_5

BF_3, AsF_5 or SbF_5 F^- ⫫ BrO_3^-, BrO_4^- or NO_3^-

$[BrOF_2]^+$ salts $[BrOF_4]^-$ salts

IF_5 IF_5 HIO_4

$KMnO_4$ or $KReO_4$ H_2O F_2/HF

I_2O_5 $\xrightarrow{IF_5}$ IOF_3 IF_5

110 °C F^- ⫫ I_2O_5/F^- or NO_3^-

$IF_5 + IO_2F$ $[IOF_4]^-$ salts

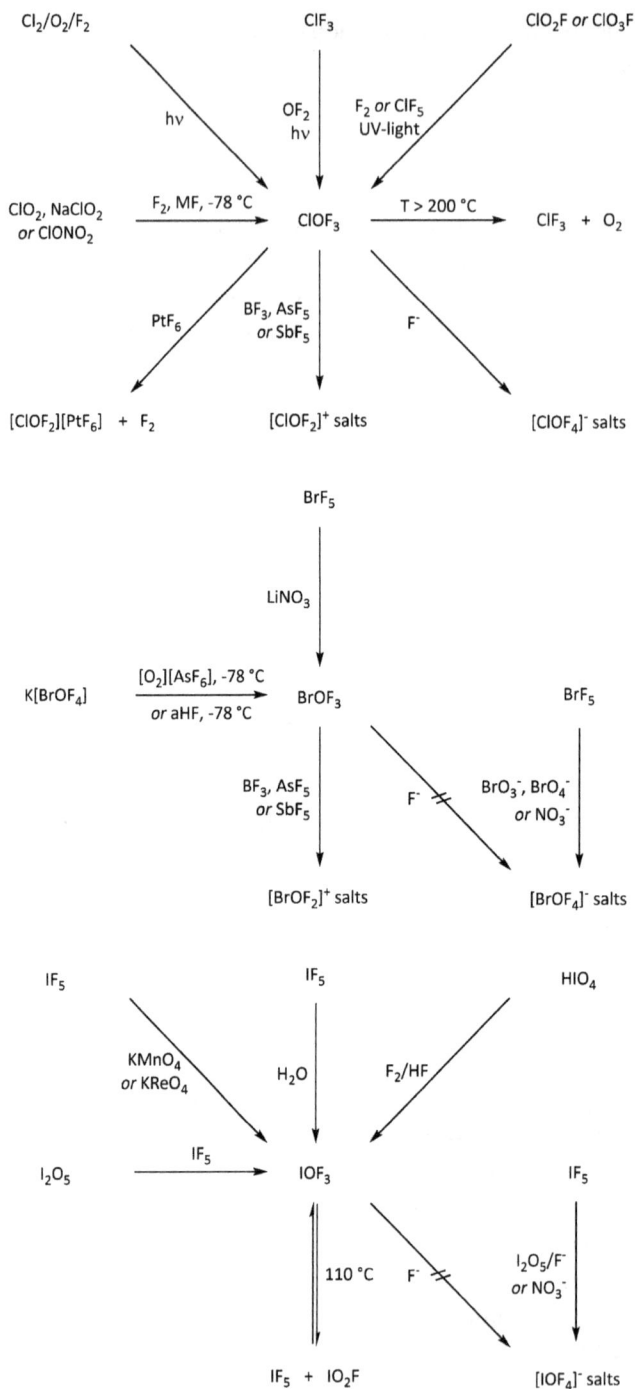

Scheme 9.32: Preparation and selected reactions of the halogen oxytrifluorides XOF_3.

ClF_3 and O_2 [923]. With fluoride acceptors (BF_3, AsF_5, SbF_5), the pyramidal $[ClOF_2]^+$ cation is formed, whereas alkali metal fluorides (KF, RbF, CsF) lead to the formation of distorted, square-pyramidal $[ClOF_4]^-$ anions (Scheme 9.32) [470, 500, 929–931]. The salt $[ClOF_2][PtF_6]$ is obtained from the reaction of $ClOF_3$ with PtF_6, where F_2 is liberated additionally [932]. Bromine oxytrifluoride **$BrOF_3$** (mp.: −5 °C; decomp.: ca. 20 °C) is prepared from $K[BrOF_4]$ and $[O_2][AsF_6]$ or anhydrous HF at −78 °C (Scheme 9.32) [857, 933]. In a more convenient approach, the reaction of excess BrF_5 with $LiNO_3$ also yields pure $BrOF_3$ [934]. Analogously to the chlorine derivative, the structure of $BrOF_3$ is described by a ψ-trigonal bipyramidal geometry, where the lone pair and the oxygen-atom occupy equatorial positions [926]. Furthermore, it is associated in the solid and liquid state, as Raman- and NMR-spectra have shown [935]. The pyramidal $[BrOF_2]^+$ cation is accessible by reaction of $BrOF_3$ with fluoride acceptors like AsF_5 or BF_3 (Scheme 9.32) [936]. However, $BrOF_3$ does not react with fluoride donors, instead the reactions of BrO_3^- or BrO_4^- with BrF_5 yield anionic $[BrOF_4]^-$ (Scheme 9.32) [937]. More conveniently, the reaction of BrF_5 with alkali metal nitrates M^INO_3 (M^I = Na, K, Rb, Cs) leads to the formation of the respective $M^I[BrOF_4]$ salts along with NO_2F [934]. Metathesis of $Cs[BrOF_4]$ with $[NF_4][SbF_6]$ gives rise to the tetrafluoroammonium salt $[NF_4][BrOF_4]$ [938]. The structure of the $[BrOF_4]^-$ anion is ψ-octahedral, with the oxygen atom and the lone-pair occupying the axial positions [926]. It is also possible to obtain $BrOF_3$-derived compounds with organic substituents. They are prepared by hydrolysis of aryl-BrF_4 compounds with $CsNO_3$ [858]. Iodine oxytrifluoride **IOF_3** is a colorless solid, which is formed upon reaction of I_2O_5 with IF_5 or careful hydrolysis of IF_5 [865, 939]. As an oxygen source for the substitution of F-atoms in IF_5, $KMnO_4$ and $KReO_4$ may also be used (Scheme 9.32) [940, 941]. Additionally, the fluorination of HIO_4 with F_2 in HF leads to IOF_3 [942]. The ψ-trigonal-bipyramidal IOF_3 units are linked via weak F-bridges. Again, the oxygen substituent occupies an equatorial position [943, 944]. In contrast to IO_2F, IOF_3 is less prone to hydrolysis and can be stored in glass vessels. Heating to about 110 °C leads to reversible decomposition to IF_5 and IO_2F (Scheme 9.32). In addition, the tendency of IOF_3 to form ionic complexes is less pronounced [942]. Thus, neither cationic nor anionic fluorocomplexes are directly accessible from IOF_3. Similar to the corresponding bromine derivatives, anionic species can be prepared using a synthetic detour. For example, the fluorination of a stoichiometric mixture of I_2O_5 and an alkali metal fluoride with IF_5 yields $M^I[IOF_4]$ [915, 937]. The same products $M^I[IOF_4]$ (M^I = Li, Na, K, Rb, Cs, NO) are obtained, if IF_5 is reacted with M^INO_3 [915]. Expectedly, the $[IOF_4]^-$ anion has C_{4v}-symmetry, as has been verified by vibrational spectroscopy [937, 945]. Heating of stoichiometric amounts of CsF, I_2O_5 and IF_5 gives rise to the dianion $[IOF_5]^{2-}$. Its structure is described by a ψ-pentagonal bipyramid, where oxygen and the lone-pair occupy the axial positions, so that the five F-atoms are oriented in the equatorial plane [946].

Chloryl fluoride **ClO_2F** is a colorless gas (mp.: −115 °C; bp.: −6 °C) with a ψ-tetrahedral structure that is recovered from the reaction of ClF_3 or BrF_3 with $KClO_3$ (Scheme 9.33) [947, 948]. If $NaClO_3$ is used instead of the potassium salt, the amount of

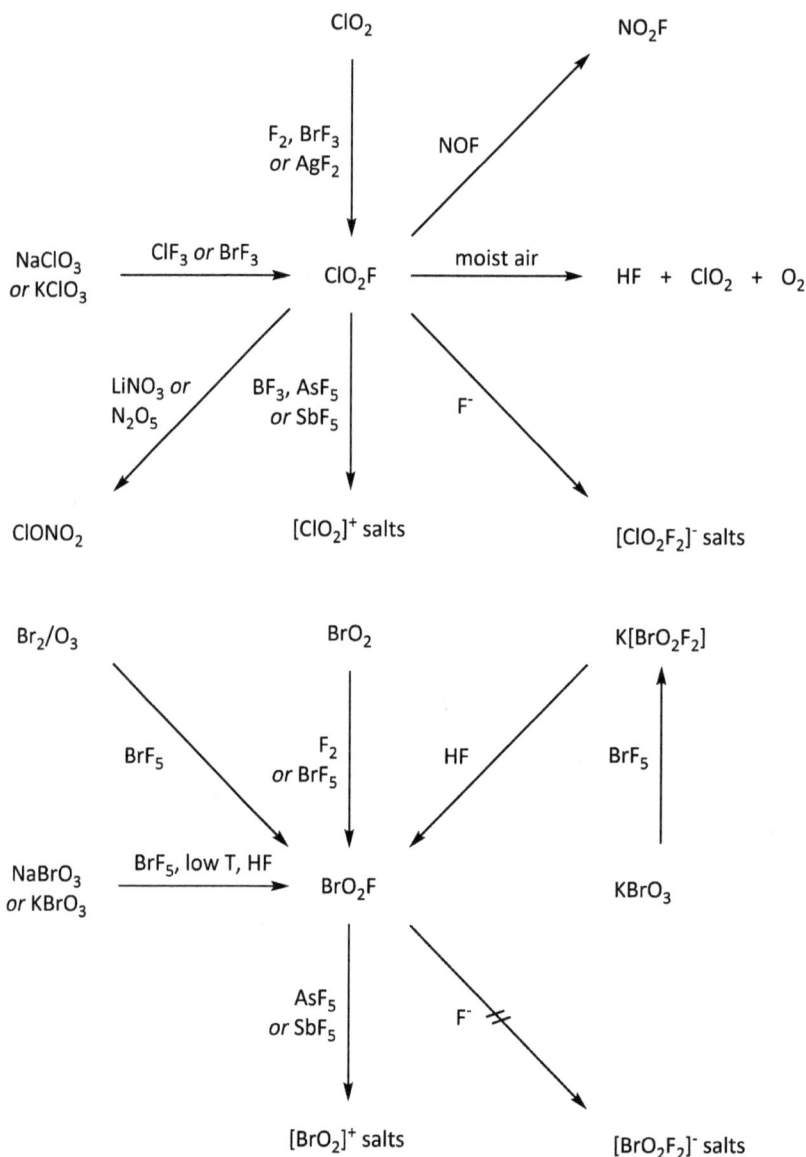

Scheme 9.33: Preparation and selected reactions of the halogen dioxyfluorides ClO_2F and BrO_2F.

required ClF_3 is strongly reduced, as NaF does not form a 1:1-adduct with ClF_3, whereas KF does so [949, 950]. Alternatively, ClO_2F is synthesized by fluorination of ClO_2 with F_2 or better AgF_2 or BrF_3 (Scheme 9.33) [951–954]. In liquid phase, it dissociates into bent $[ClO_2]^+$ and ψ-trigonal-bipyramidal $[ClO_2F_2]^-$, resulting in electric conductivity. Complex compounds with the ions $[ClO_2]^+$ and $[ClO_2F_2]^-$ are also obtained, if ClO_2F is treated with strong Lewis acids (BF_3, PF_5, AsF_5, SbF_5) or fluoride donors (e. g., CsF), respectively (Scheme 9.33) [947, 951, 953, 955–957]. ClO_2F is quite sensitive toward hydrolysis, as it already reacts with moist air under formation of HF, ClO_2 and O_2 (Scheme 9.33). In dry quartz vessels, ClO_2F is relatively stable and storable [952]. It is a strong oxidizer and capable of transforming NOF to NO_2F [958]. In turn, ClO_2F can be oxidized by PtF_6 to yield salts with the cations $[ClO_2]^+$ and $[ClO_2F_2]^+$. The reaction of ClO_2F with $LiNO_3$ or N_2O_5 yields $ClONO_2$ (Scheme 9.33) [855]. Bromyl fluoride **BrO_2F** is formed as a colorless, crystalline compound (mp.: $-9\,°C$; decomp.: $10\,°C$) upon reaction of $KBrO_3$ or $NaBrO_3$ with BrF_5 at low temperature in a HF-solution. Alternatively, it is synthesized by fluorination of BrO_2 with BrF_5 or F_2 as well as upon treatment of Br_2 and O_3 with BrF_5 (Scheme 9.33) [959–961]. BrO_2F is furthermore obtained, if $K[BrO_2F_2]$ is treated with HF or upon careful low-temperature hydrolysis of BrF_5 [857, 962]. Liquid BrO_2F attacks glass, in contact with water or organic compounds violent reactions can occur [960]. Above its melting point, it decomposes often explosively to BrF_3, Br_2 and O_2. Analogously to ClO_2F, BrO_2F has a ψ-tetrahedral structure [857, 962–965]. Upon treatment with KrF_2, BrO_2F is fluorinated to yield first $BrOF_3$ and then BrF_5, however, a Br(VII)-species is not accessible by this method [857]. The reaction with AsF_5 or SbF_5 leads to the bromyl compounds $[BrO_2][AsF_6]$ and $[BrO_2][SbF_6 \cdot xSbF_5]$, respectively (Scheme 9.33) [966, 967]. Upon treatment with AsF_5, the compound $[BrO_2][AsF_6] \cdot 2$ BrO_2F is additionally accessible. The latter decomposes to the mixed-valent species $[Br_3O_4 \cdot Br_2][AsF_6]$, where the cation contains five-, three-, mono- and zero-valent bromine. Similarly, the reaction of excess BrO_2F with CF_3SO_3H gives rise to the mixed-valent cation $[Br_3O_6]^+$, where bromine is in the oxidation states five and three [961]. In the same way like $BrOF_3$, BrO_2F does not directly yield anionic $[BrO_2F_2]^-$ upon reaction with fluoride donors. However, $K[BrO_2F_2]$ is accessible, if BrF_5 and $KBrO_3$ are reacted in a molar 1:1 ratio [968]. It is also generated along with $K[BrOF_4]$, when $K[BrF_6]$ is reacted with $KBrO_3$. $K[BrO_2F_2]$ and $K[BrOF_4]$ can be separated from MeCN solutions, as they have a different solubility [969]. BrO_2F is not oxidized by PtF_6, as it is known for ClO_2F. Instead, the salts $[BrOF_2][PtF_6]$ and $[O_2][PtF_6]$ are obtained together with PtF_5 [967]. Iodyl fluoride **IO_2F** is synthesized by dissolving I_2O_5 in anhydrous HF [942]. It is obtained as polymeric solid, which decomposes at above $200\,°C$ upon melting. IO_2F is also formed by thermal decomposition of IOF_3 at $110\,°C$, in turn, treatment of IO_2F with IF_5 results in the generation of IOF_3 [865, 940, 941]. In the solid-state structure, alternating axial-equatorial linked ψ-trigonal bipyramids are found, where the axial positions are occupied by oxygen and fluorine. The double-bonded oxygen is found in an equatorial position [970]. Analogously to the lighter chlorine and bromine homologues, ionic iodyl- and difluoroiodate(V) compounds are known. However, they are not directly accessible upon treatment of IO_2F with fluoride acceptors or donors, respectively. Instead, synthetic detours are required, as it is also necessary for the ionic derivatives of IOF_3 (see above). As has been proven by several spectroscopic methods, difluoroiodate compounds contain the C_{2v}-symmetric anion $[IO_2F_2]^-$ [945, 971, 972]. A possible route toward salts containing this anion is the treatment of $[NMe_4][IO_3]$

with aqueous HF (w = 48 %) [973]. Further treatment of $[NMe_4][IO_2F_2]$ with $[NMe_4]F$ gives rise to cis-$[IO_2F_3]^{2-}$. Its structure is ψ-octahedral, where the two oxygen-atoms are oriented in relative cis-positions to each other, whereas the three fluorine substituents adapt a mer-geometry. In HF, the $[IO_2F_3]^{2-}$ anion is unstable, if no excess $[NMe_4]F$ is present, which is due to the circumstance, that HF is a stronger fluoride acceptor than $[IO_2F_2]^-$ [974].

Heptavalent compounds

The octahedral chlorine oxypentafluoride **ClOF$_5$** is possibly formed by the UV photolysis of ClF_5 and OF_2 [931]. However, its existence is not unambiguously proven [975]. Theoretical calculations indeed predict an exothermic loss of fluorine, which makes it thermodynamically unstable, but maybe, $ClOF_5$ might have a sufficiently large decomposition barrier to be prepared anyway. The bromine derivative **BrOF$_5$** is unknown, but should be stable and accessible, as has been predicted by quantum chemistry. However, for both the chlorine and the bromine compound, the existence of the ionic derivatives $[XOF_4]^+$ as well as $[XOF_6]^-$ (X = Cl, Br) is predicted to be highly unfavorable [841]. Iodine oxypentafluoride **IOF$_5$** is formed upon hydrolysis of IF_7 or reaction of IF_7 with SiO_2 at 100 °C as a colorless liquid (mp.: 4.5 °C) [976, 977]. More conveniently, two fluoride substituents of IF_7 can be readily replaced using POF_3 as oxygen-donor [978]. IOF_5 is relatively stable against hydrolysis. Expectedly and based on vibrational spectra, the molecular structure of IOF_5 has been found to have C_{4v}-symmetry [979]. Based on experimental data, the structure has additionally been confirmed by theoretical calculations. Interestingly, it has also been found, that both axial and equatorial I-F bonds have about the same length [980]. If IOF_5 is reacted with $[NMe_4]F$, the pentagonal bipyramidal anion $[IOF_6]^-$ is formed, wherein oxygen occupies an axial position [798, 800]. However, an analogous reaction is not observed with LiF or CsF at temperatures up to 60 °C [915]. The $[IOF_6]^-$ anion is a structurally interesting species. The five fluorine atoms in the equatorial plane are puckered in order to decrease steric congestion. The puckering is a fluxional process in solution, as can be observed by ^{19}F-NMR; however, in the solid state the rapid exchange of the F-atoms is inhibited, due to anion-cation interactions [981]. Replacing a fluorine substituent of IOF_5 by oxygen yields a mixture of cis- and trans-$[IO_2F_4]^-$, which is accessible by reacting IO_4^- with HF. Further treatment with HF gives rise to the formation of the respective acids cis- and trans-$HOIOF_4$. Both isomers of $HOIOF_4$ are directly obtained in a 3:1-ratio, if $Ba_3H_4(IO_6)_2$ is treated with HSO_3F (see also IO_2F_3) [982]. The isomeric mixture of $[IO_2F_4]^-$ is transformed to the hypofluorite species cis- and trans- $IOF_4(OF)$ by $[NF_4]^+$, the respective hypochlorite isomers are generated upon reaction with $Cl(OSO_2F)$ [983].

Gaseous chlorine dioxytrifluoride **ClO$_2$F$_3$** (mp.: −81.2 °C; bp.: −21.6 °C) is generated upon reaction of $[ClO_2F_2][PtF_6]$ with NO_2F or NOF at −78 °C [984, 985]. The structure of ClO_2F_3 is described by a C_{2v}-symmetric, trigonal bipyramid, thus the oxygen atoms occupy equatorial positions [986]. It is a fluoride donor and forms thermally stable $[ClO_2F_2]^+$-salts with the fluoride acceptors BF_3 and AsF_5; however, ClO_2F_3 itself does not act as a fluoride acceptor [984, 987]. The stability of the ionic Lewis acid adducts can be explained by a comparison of the structures of the cation and the neutral molecule. The almost tetrahedral $[ClO_2F_2]^+$ cation is expected to be strongly energetically favored over the distorted trigonal-bipyramidal ClO_2F_3 [984]. In contrast, quantum chemical calculations predict an exothermic loss of fluorine for anionic $[ClO_2F_4]^-$, which gives a hint, why this species is not accessible [841]. The bromine species **BrO_2F_3** as well as its anionic derivative $[BrO_2F_4]^-$ have also been investigated by theoretical calculations. It has been found that both compounds should be stable against the loss of fluorine and, therefore, preparatively accessible [841]. However, so far they remain unknown. In contrast, the heavier homologue iodine dioxytrifluoride (**IO_2F_3**)$_2$ is known, however, it does not exist as a monomer. The yellow, oligomeric solid can be sublimed (mp.: 41 °C) and is typically prepared from $Ba_3H_4(IO_6)_2$ and SO_3-containing HSO_3F, which yields intermediary HIO_2F_4. The latter finally reacts with SO_3 to form IO_2F_3 [982]. In condensed state, it is polymeric, up to 100 °C, gaseous IO_2F_3 is associated in form of oxygen-bridged dimers [988–990]. In solution, IO_2F_3 forms cyclic trimers, as has been explored by ^{19}F-NMR spectroscopy [991]. Complex compounds are generated, if $(IO_2F_3)_n$ is reacted with either alkali metal fluorides or strong Lewis acids, such as SbF_5 [990, 992, 993]. The reaction between KIO_4 and IF_5 yields a mixture of $K[IO_2F_4]$ and IO_2F. The potassium salt contains *trans*-$[IO_2F_4]^-$, which has D_{4h}-symmetry. However, in IF_5-solution, it isomerizes to yield a mixture of both, *cis*- and *trans*-$[IO_2F_4]^-$ [993]. Upon further treatment with $[NMe_4]F$, a mixture of *cis*- and *trans*-$[NMe_4][IO_2F_4]$ produces the insoluble *trans*-$[IO_2F_5]^{2-}$, which has a pentagonal-bipyramidal structure. However, *cis*-$[IO_2F_4]^-$ does not accept fluoride ions, and thus remains in solution [994]. The Lewis acid complexes, for example, $IO_2F_3 \cdot SbF_5$, have a polymeric, O-bridged structure [992]. Actually, $(IO_2F_3)_n$ appears to be a stronger fluoride acceptor than SbF_5 [993]. Accordingly, it does not have any fluoride donor properties. It is about as stable against hydrolysis as IOF_5, but it decomposes photolytically to IOF_3 and O_2. Upon reaction of IOF_3 with IO_2F_3, the ions $[IOF_2]^+$ and $[IO_2F_4]^-$ are formed, which has been unambiguously proven by ^{19}F-NMR-spectroscopy [990].

Perchloryl fluoride **ClO_3F** is generated as colorless, toxic and gaseous compound (mp.: −146.5 °C; bp.: −46.7 °C) upon reaction of $KClO_4$ with HSO_3F or HF/SbF_5 as well as upon electrolysis of $NaClO_4$ in HF [995–999]. It can also be prepared by fluorination of $KClO_3$ with F_2 at −20 °C and has a C_{3v}-symmetric molecular structure, which is derived from a tetrahedral geometry [997, 1000–1003]. ClO_3F is a strong oxidizer, stable against hydrolysis and resistant against temperatures of more than 400 °C. It can even be dissolved in water, where it only slowly reacts with the solvent [1004]. Interestingly, ClO_3F is also quite inert in terms of chemical reactivity. Due to its high thermal stability, which is comparable with that of SF_6, it has been used in high voltage facilities as an insulator. As one of only a few reactions, ClO_3F reacts with ammonia to form the amide $ClO_3(NH_2)$. The latter has acidic hydrogen atoms, which can be replaced by metal ions to yield, for example, $K[ClO_3(NH)]$ and $K_2[ClO_3N]$ [997, 1005]. The potassium salts are colorless solids, which are stable up to 300 °C but explode upon shock. They are derived from the perchlorates ClO_4^-, wherein an oxygen substituent has been exchanged by the isoelectronic [NH]- or $[N^-]$-group [1005]. ClO_3F only has weak fluoride accepting properties, so that the formation of anionic $[ClO_3F_2]^-$ is not observed upon reaction with strong fluoride donors [1006]. Similarly, no reactions are observed with the strong Lewis acids AsF_5 and SbF_5 [1007]. ClO_3F has been used as versatile electrophilic fluorinating agent in organic synthesis, since it is capable of substituting hydrogen atoms in activated methylene compounds for example via intermediate carbanion species or enolates (cf. Section 13.5.3) [1008–1013]. This approach has also been used for the preparation of ^{18}F-labeled fluoroaromatics, when lithiated intermediates were treated with $ClO_3^{18}F$ [1014]. Under Friedel–Crafts-type conditions, ClO_3F reacts with aromatic compounds under formation of perchloryl aromatics [1015]. Another useful application of ClO_3F has been found in the synthesis of fluorinated steroids [1016–1018]. Nevertheless, ClO_3F is a hazardous compound due to its oxidizing properties, which especially holds in contact with organic, easily oxidizable materials [1019]. Therefore, its advantageous fluorinating properties are in most cases overcome by safety issues. Apart from that, other common electrophilic fluorinating agents are superior over ClO_3F for the most purposes, due to a higher selectivity and simplicity in their applications [1020]. Perbromyl fluoride **BrO_3F** (mp.: −110 °C; decomp.: ca. 20 °C) is the product of the reaction of $KBrO_4$ with SbF_5 in presence of HF [1021]. It is unstable and sensitive toward hydrolysis, but can be stored at −80 °C in the gas phase. Based on vibrational spectroscopy, the structure of BrO_3F has been found to be similar to ClO_3F [1003, 1022]. In contrast to ClO_3F, BrO_3F reacts with fluoride donors (KF, RbF, CsF, NMe_4F, NOF) under formation of the D_{3h}-symmetric anion $[BrO_3F_2]^-$ [1006]. The reaction with the strong Lewis acids SbF_5 and AsF_5 leads to both, a fluoride abstraction and oxygen elimination, to result in salts of the type $[BrO_2][Sb_nF_{5n+1}]$, where the $[BrO_2]^+$ cation has a bent structure, comparable to $[ClO_2]^+$ [1007]. Periodyl fluoride **IO_3F** is formed upon fluorination of HIO_4 in anhydrous HF. It is a colorless solid, which is rather stable toward hydrolysis, so that it can be stored in glass vessels without decomposition. However, IO_3F thermally decomposes to IO_2F and O_2 at about 100 °C [942]. With fluoride acceptors, it reacts to form $[IO_3]^+$-compounds.

9.7.8 Study questions

(I)   What are the similarities in the preparations of the different halogen fluorides?
(II) Which molecular structures are found for the halogen fluorides?
(III) Which trends are observed regarding the stability of the halogen fluorides? Which are the most stable compounds?

(IV) Which general types of reactions are observed for the halogen fluorides? Is there a trend in the reactivity?
(V) Which fluorocations are accessible from the halogen fluorides? How are they prepared and which structures do they have?
(VI) Which stable fluorohalogenate anions are known? Which structures do they have?
(VII) Describe the molecular structures of the pentavalent oxyfluorides XOF_3 and XO_2F. Which trends are observed regarding their fluoride donating and accepting properties?
(VIII) Comment on the existence of heptavalent halogen oxyfluorides. How are these compounds prepared?

9.8 Group 18: Krypton, xenon, radon

The fluorides are the only noble gas compounds, which are directly accessible from the elements. As the ionization potential of the noble gases decreases with their size, the chemistry of the xenon fluorides is very well explored. Accordingly, the respective radon fluorides can be expected to be more readily accessible, however, since all radon isotopes are radioactive, RnF_n do not play a role in practice. Apart from xenon, krypton-fluorine compounds are known as well, but they are far less stable. For the other noble gases, helium, neon and argon, fluorine compounds are unknown.

9.8.1 Xenon

Xenon difluoride
Xenon difluoride **XeF₂** is formed in an exothermic reaction in high yield, when a 2:1-mixture of xenon and fluorine is passed through a nickel tube at 400 °C and 2 bar pressure (Scheme 9.34). The generated XeF_2 is frozen out at –50 °C [1023]. Another synthetic approach includes the fluorination of elemental Xe with O_2F_2 at –120 °C [523]. More conveniently, XeF_2 is prepared by leaving a Xe/F₂-mixture in the sunlight or submitting it to microwave discharge (Scheme 9.34) [1024]. XeF_2 is a solid (mp.: 129 C), colorless compound that is soluble in H_2O, liquid HF, BrF_5 and even in organic solvents like acetonitrile. The XeF_2 molecule is monomeric in all phases and has a linear structure [1025]. It has a characteristic odor and can be stored in Monel or nickel vessels without decomposition, as long as humidity is excluded. It can easily be sublimed to give large, transparent and shiny, tetragonal crystals. XeF_2 dissolves molecularly in water and dilute acids and hydrolyzes only slowly in aqueous HF with a half-life of 7 h at 0 °C. Interestingly, XeF_2 dissolves without decomposition in dry acetonitrile at room temperature. Apart from that, the solubility and stability of XeF_2 in some other typical solvents has been studied [1026].

XeF_2 is a strong oxidizer, that is capable of transforming Cl^- to Cl_2, IO_3^- to IO_4^-, BrO_3^- to BrO_4^-, Co^{2+} to Co^{3+}, Ce^{3+} to Ce^{4+} and Ag^+ to Ag^{2+} in acidified aqueous solution (Scheme 9.34) [1027]. A similar reactivity is observed in anhydrous HF, where ele-

Xe/F$_2$ Xe   XO$_4^-$

hν *or* microwave discharge

O$_2$F$_2$, −120 °C

XO$_3^-$ (X = Br, I)

H$_2$O, pH < 7

Xe/F$_2$ (2:1-mixture)   400 °C, 2 bar → XeF$_2$ NH$_3$, rt → Xe + HF

PtF$_4$, 140–150 °C

fluoride acceptors

F$^-$ ✗

[Xe$_2$F$_3$][PtF$_6$] + Xe   [XeF]$^+$ salts   [XeF$_3$]$^-$ salts

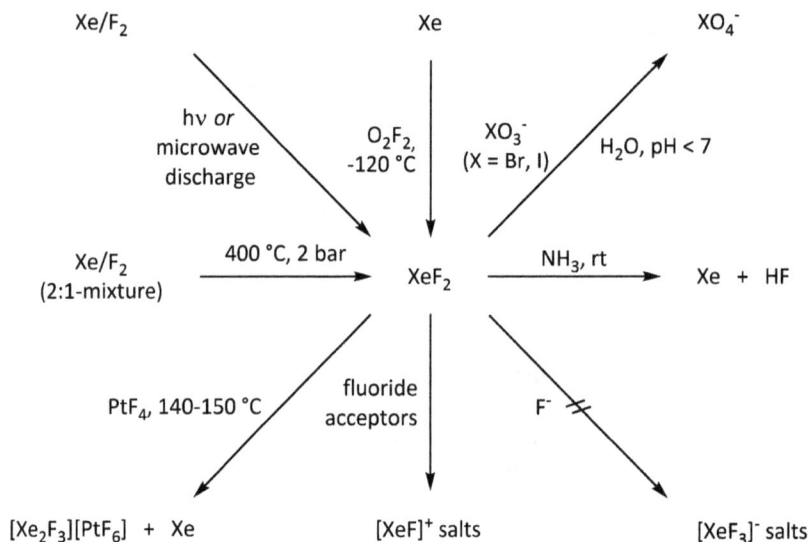

Scheme 9.34: Preparation and selected reactions of XeF$_2$.

ments are oxidized to the respective fluoride compounds. In the same way, low valent fluorides and halometallates are fluorinated, and thus, converted to higher oxidation states. Moreover, fluorination reactions are also observed for many oxides and transition metal carbonyl compounds [39]. Liquid XeF$_2$ furthermore oxidizes PtF$_4$ and Pd$_2$F$_6$ at 140-150 °C (Scheme 9.34) [1028]. In presence of ammonia, XeF$_2$ is reduced at room temperature to yield Xe and HF, a similar reaction occurs with hydrogen at 400 °C. In most cases, the oxidizing effect of XeF$_2$ is accompanied by fluorination, for example, NO$_2$ is oxidatively fluorinated to yield FNO$_2$. Organic molecules also react with XeF$_2$ under formation of fluorinated carbon compounds (cf. Section 12.4.1). In superacidic media, XeF$_2$ becomes an extremely strong oxidiser. Upon treatment with F$_2$, XeF$_2$ is oxidized to higher xenon fluorides. Organoxenon(II) compounds are also known in form of a large variety of xenonium(II) salts, which are commonly prepared from XeF$_2$ and difluoroorganoboranes RBF$_2$ [1029].

XeF$_2$ acts as fluoride donor toward many Lewis-acidic metal fluorides MF$_n$ (n = 3, 4, 5), which gives rise to molecular adducts or complexes containing the [XeF]$^+$ cation (Scheme 9.34) [1030–1032]. These reactions are ideally performed in liquid BrF$_5$, the resulting [XeF]$^+$ salts are crystalline, reactive and very sensitive toward hydrolysis. The [XeF]$^+$ complexes are stronger oxidizers than XeF$_2$ itself, upon contact with water, the Z-shaped cation [FXeOXeFXeF]$^+$ is obtained [1033]. If XeF$_2$ is reacted with an excess of Lewis-acidic metal pentafluoride, the complexes [XeF][M$_2$F$_{11}$] are obtained as well, in contrast, [Xe$_2$F$_3$][MF$_{n+1}$] salts are generated, when excess XeF$_2$ is used [1034]. However, with decreasing fluoride-acceptor strength of the Lewis acid, the covalent contribution to the bonding interaction between cations and anions increases [1035]. In complexes with the weak Lewis acids IF_5, XeF_4, $XeOF_4$ or $[XeF_5]^+$, the bonding situation is almost purely covalent. As a consequence, the solid-state structure of $XeF_2 \cdot IF_5$ shows nearly unchanged molecular units of linear XeF_2 and square-pyramidal IF_5 [1036]. XeF_2 does not have fluoride accepting properties, so that anionic $[XeF_3]^-$ is unknown [1037–1039].

The $[XeF]^+$ cation is stable, as the Xe-F bond in the simple cations is much stronger than in the neutral molecules, for example, the bond length in the pair XeF_2/XeF^+ is decreased from 200 pm to 184 pm, along with an increase of the bond energy from $134 \, kJ \, mol^{-1}$ to $205 \, kJ \, mol^{-1}$. However, it is a strong Lewis acid and exhibits short contacts to the respective anion. It may also add to nitriles like HCN or CF_3CN, where a linear coordination is observed. The associated $[Xe_2F_3]^+$ cation is planar and has a symmetric, V-shaped structure. The bridging F-atom reduces the polarizing effect of $[XeF]^+$, resulting in decreased interactions between the cations and anions in the complexes. If $[XeF]^+[Sb_2F_{11}]^-$ is reacted with excess Xe, the green $[Xe_2]^+$ cation is generated, which under Xe pressure further reacts to give blue $[Xe_4]^+$. The treatment of $[XeF][SbF_6]$ with small amounts of $SbCl_5$ in HF/SbF_5 gives rise to the chloro-cation $[XeCl]^+$ [1040].

In HF/SbF_5, the proton does not attach to the Xe- but the F-atom of XeF_2, which has been observed by the formation of the colorless salt $[F-Xe\cdots F-H]^+[Sb_2F_{11}]^-$. It might be possible that double protonation of XeF_2 occurs in HF/SbF_5 to give the solvated cation $[Xe(FH)_2]^{2+}$, which would then be an extremely strong oxidizer. Analogously to the proton, "naked" metal ions can attach to the F-atom of XeF_2, as has been shown by reactions of XeF_2 with $M[AsF_6]_2$ in an anhydrous liquid HF. In the resulting salts of the type $[M(XeF_2)_m][AsF_6]_2$ (m = 2-6; M = 2 Ag^+, Pb^{2+}, Mg^{2+}, Ca^{2+}, Sr^{2+}, Ba^{2+}, 2/3 Ln^{3+}), M^{2+} shows the coordination environment of an anti-cube or a threefold-capped trigonal prism, where the fluorine ligands are contributed by both, XeF_2 and $[AsF_6]^-$.

Xenon tetrafluoride

Xenon tetrafluoride **XeF_4** is formed upon prolonged heating of xenon with excess fluorine (molar ratio Xe/F_2 = 1/5) at 400 °C and a pressure of 6 bar in a nickel vessel (Scheme 9.35) [1041, 1042]. It can also be obtained by further fluorination of XeF_2 with F_2, or alternatively, upon electric discharge or photochemical reaction of Xe with excess F_2 at room temperature [1043, 1044]. Conveniently, XeF_4 is also accessible by decomposition of NaF-XeF_6 adducts at 350 °C *in vacuo* [1045]. A very mild synthesis at low pressure and low temperatures comprises the treatment of elemental Xe with O_2F_2, resulting in high yields of pure xenon tetrafluoride (Scheme 9.35) [1046]. XeF_4 forms transparent, colorless crystals (mp.: 117 °C), which are stable in absence of moisture. Like the other xenon fluorides, it is soluble in anhydrous HF, but not as much as XeF_2 and XeF_6. Solutions of XeF_4 do only conduct electricity, if a Lewis acid like PF_5, AsF_5 or SbF_5 is added [1047]. Hydrolysis occurs in water as well as in diluted acids

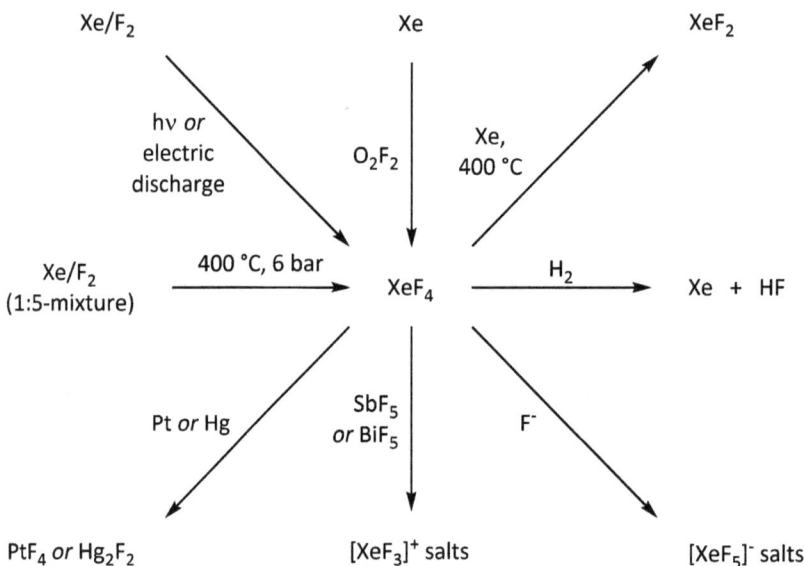

Scheme 9.35: Preparation and selected reactions of XeF_4.

and bases to finally yield Xe and O_2. Under the same conditions, disproportionation to Xe and XeO_3 may occur [1048]. At 0 °C, XeF_4 can be carefully hydrolyzed by water and aqueous H_2SO_4 to yield the dioxide XeO_2 [1049]. The intermediate oxyfluoride $XeOF_2$ is obtained at −80 °C [1050]. XeF_4 has a square-planar structure in both, the crystalline and the gas phase [1025, 1051–1053]. It is a stronger oxidizer and fluorinating agent than XeF_2, as it attacks metallic platinum and mercury under formation of PtF_4 and Hg_2F_2, respectively (Scheme 9.35). XeF_4 reacts violently with ethers like dioxane or THF. The reactions with a few other solvents as well as the solubility in these media have been studied [1026]. The reduction by H_2 proceeds fast at 70 °C and almost immediately at 130 °C to quantitatively yield Xe and HF (Scheme 9.35). XeF_4 reacts with further Xe at 400 °C to give XeF_2, whereas excess F_2 at 300 °C yields XeF_6. Among the xenon fluorides, XeF_4 is the weakest fluoride donor. This can be explained by the fact, that the Xe center in the $[XeF_3]^+$ cation has five electron pairs, which is unfavorable. In addition, XeF_4 already has a ψ-octahedral structure with its two lone pairs, so that the removal of F^- would also decrease the molecular symmetry. Therefore, XeF_4 only reacts with the very strong Lewis-acidic metal fluorides SbF_5 and BiF_5 to form stable adducts of the type $[XeF_3]^+[M_2F_{11}]^-$ containing the T-shaped $[XeF_3]^+$ cation (Scheme 9.35) [1054–1059]. XeF_4 is a stronger fluoride acceptor than XeF_2, but weaker than XeF_6. Upon reaction with alkali metal fluorides, tetra alkyl ammonium fluorides or NOF, pentagonal-planar $[XeF_5]^-$ is generated [1060]. The treatment of XeF_4 with $C_6F_5BF_2$ gives rise to the organoxenon(IV) compound $[C_6F_5XeF_2][BF_4]$, containing a xenonium(IV) cation [1029, 1061, 1062].

Xenon hexafluoride

Xenon hexafluoride $\mathbf{XeF_6}$ is formed in about 90 % yield upon reaction of xenon with excess fluorine (molar ratio $Xe/F_2 = 1/20$) at 300 °C and 60 bar in nickel pressure vessels [1063–1065]. The yield of the reaction is quantitative, if the reaction conditions are changed to 700 °C and 200 bar pressure (Scheme 9.36). More conveniently, the fluorination can be performed in presence of NaF at 50 °C. The resulting $Na_2[XeF_8]$ is formed after one day and can be thermally decomposed to NaF and XeF_6. A low-pressure synthesis utilizing a hot-wire reactor and an excess of F_2 has been described to yield pure XeF_6, avoiding contamination with the lower fluorides [1046]. XeF_6 forms colorless crystals (mp.: 49.4 °C), which are extremely sensitive toward hydrolysis. The molten crystals have a yellow-green color. XeF_6 readily reacts with SiO_2 under formation of highly explosive XeO_3 and SiF_4, so that XeF_6 cannot be handled in glassware (Scheme 9.36). XeF_6 is a strong oxidizer and fluorinating agent, but interestingly, it does not react with NH_4^+, as XeF_6 is electrophilic and will only attack neutral and negatively charged species [1066]. From acetonitrile-solutions, explosive adducts of the compositions $F_6Xe(NCMe)$ and $F_6Xe(NCMe)_2 \cdot MeCN$ can be isolated at low temperature [1067]. It has been shown, that the Xe-N-bonds in these adducts are poorly covalent but reveal large electrostatic character [1068]. XeF_6 has a remarkable solubility in liquid HF, above 30 °C, about 1 mol of XeF_6 dissolves in 2 mol HF. The resulting yellow solutions considerably conduct electricity, presumably due to the formation of $[XeF_5]^+[HF_2]^-$. Crystalline XeF_6 exists in at least eight modifications, which are not all completely characterized [1069]. In most of the structures, $[XeF_5]^+$ and F^- ions are associated to tetrameric rings, but hexamers and lower oligomers can be observed as well. It is expected that a few more modifications are existing under high-pressure conditions. NMR-spectroscopic investigations (^{19}F and ^{129}Xe) of XeF_6-solutions in inert solvents at low temperatures also reveal an association of $(XeF_6)_4$-units with 24 magnetically equivalent fluorine atoms [1070–1072]. In the vapor-phase, XeF_6 is only slightly associated and mainly exists as monomer, as has been shown by mass-spectrometric investigations [1073]. The XeF_6 molecule has a capped octahedral structure, which can also be described as ψ-pentagonal bipyramid, with the lone pair occupying an equatorial position. Experimentally, it has been possible to observe three isomers above room temperature, which exist in equilibrium next to each other and are not very different in their energies [1069, 1074–1076]. These findings are also supported by theoretical calculations [1077]. However, under matrix conditions at 5 K, a single C_{3v}-symmetric isomer is observed, whereas octahedral XeF_6 molecules cannot be found. This might be due to the fact, that the barrier between these isomers is too high to be overcome at a very low temperature [1078]. Nevertheless, XeF_6 belongs to the fluxional molecules under standard conditions, showing fast intramolecular rearrangements. The same dynamic behavior is observed in nonionizing solvents like F_5SOSF_5. As it is also found in the crystal structure, $[XeF_5]^+$-units are bridged by fluoride ions. The xenon atoms are oriented in

Xe

$Na_2[XeF_8]$

$XeO_3 + SiF_4$

F_2, hot-wire reactor

120 °C

SiO_2

Xe/F_2 (1:20-mixture)

700 °C, 200 bar

XeF_6

H_2, HCl or NH_3

Xe + HF

AuF_3 or Hg

fluoride acceptors

F^-

$[Xe_2F_{11}][AuF_6]$ or HgF_2

$[XeF_5]^+$ salts

$[XeF_7]^-$ or $[XeF_8]^-$ salts

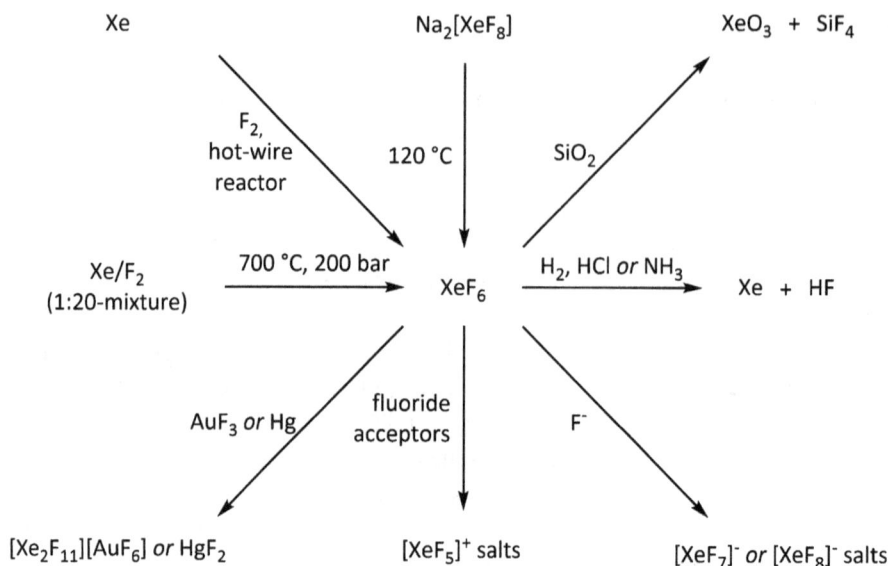

Scheme 9.36: Preparation and selected reactions of XeF_6.

form of a tetrahedron and the fluorine substituents rapidly exchange their bridging and terminal positions around the Xe-centers. In the liquid phase, XeF_6 is comprised of monomers and tetramers. The conductivity points to the self-dissociation of XeF_6 in the molten state under formation of $[XeF_5]^+$, $[Xe_2F_{11}]^+$, $[XeF_7]^-$, $[Xe_2F_{13}]^-$, etc.

Among the xenon fluorides, XeF_6 is the one with the strongest oxidizing and fluorinating properties. It transforms metallic mercury to HgF_2 and oxidizes AuF_3 to the Au(V) compound $[Xe_2F_{11}]^+[AuF_6]^-$ (Scheme 9.36) [1079, 1080]. Upon reaction with hydrogen, XeF_6 is quantitatively reduced to Xe and HF when the mixture is heated. The same is observed for HCl and NH_3, where Cl_2 and N_2 are additionally formed [1081]. Further fluorination of XeF_6 to yield octavalent $[XeF_7]^+$ or XeF_8 has not been successful so far [1082, 1083]. As quantum-chemical calculations have shown, both $[XeF_7]^+$ and XeF_8 are thermodynamically unstable [1084, 1085]. XeF_6 acts as both, fluoride donor and acceptor. It reacts with many fluorides MF_n (PF_5, AsF_5, SbF_5, BiF_5, VF_5, TaF_5, NbF_5, UF_5, AuF_5, GeF_4, SnF_4, PbF_4, TiF_4, ZrF_4, HfF_4, MnF_4, CrF_4, BF_3, AlF_3, GaF_3, InF_3, AgF_3, AuF_3), sometimes indirectly, to form mostly crystalline complexes of the type $[XeF_5]^+[MF_{n+1}]^-$ (Scheme 9.36). In a few cases, molecular adducts are observed [255, 471, 1030–1032, 1066, 1081, 1086–1098]. If an excess of MF_5 is used, $[XeF_5]^+[M_2F_{11}]^-$ is generated, whereas excess XeF_6 yields $[Xe_2F_{11}]^+[MF_6]^-$ [1030, 1031, 1093]. In solution, the square pyramidal $[XeF_5]^+$ cation has clearly been identified by NMR-spectroscopy [1099]. The cation $[Xe_2F_{11}]^+$ is better described as $[F_5Xe\text{-}F\text{-}XeF_5]^+$, wherein the Xe-F-Xe group is bent.

XeF_6 accepts fluoride ions from alkali metal fluorides to result in fluoroxenates(VI) $M^I[XeF_7]$ and $M^I_2[XeF_8]$ (M^I = Na, K, Rb, Cs) (Scheme 9.36) [1100–1102]. The respective perfluoroammonium salts $[NF_4][XeF_7]$ and $[NF_4]_2[XeF_8]$ have also been prepared [1103]. In the series $[XeF_5]^+$, XeF_6, $[XeF_7]^-$ and $[XeF_8]^{2-}$, one can clearly observe the decreasing steric impact of the lone pair, which results from an increase of the crowding. The colorless $Rb_2[XeF_8]$ and $Cs_2[XeF_8]$ cannot be decomposed below 400 °C, and thus are the most stable of the isolated xenon compounds. In contrast the colorless $Rb[XeF_7]$ (decomp.: >20 °C) and the yellow $Cs[XeF_7]$ (decomp.: >50 °C) are degraded rather easily. For the less heavy alkali metals, it is only possible to obtain $K_2[XeF_8]$ and $Na_2[XeF_8]$ at room temperature. The latter decomposes at 120 °C to liberate the starting materials NaF and XeF_6 (Scheme 9.36). Typically, this behavior is exploited for the purification of XeF_6, as XeF_2 and XeF_4 do not react with MF at room temperature. XeF_6 also reacts with NOF and NO_2F to form $[NO]_2[XeF_8]$ as well as $[NO_2][XeF_7]$ and $[NO_2][Xe_2F_{13}]$, respectively. Generally, the $[XeF_8]^{2-}$ anion has the structure of a cubic antiprism, whereas $[XeF_7]^-$ has the shape of a capped octahedron [1100, 1101]. The $[Xe_2F_{13}]^-$ anion consists of capped trigonal-prismatic $[XeF_7]^-$ and XeF_6, which are loosely connected to each other [1101].

General conclusions on the reactivity of xenon fluorides

The different fluoride donor and acceptor properties of the xenon fluorides can be exploited for their separation from mixtures. XeF_4 is by far the weakest fluoride donor among the xenon fluorides, whereas XeF_6 is only slightly more prone to donate F^- than XeF_2. As a consequence, XeF_4 can be separated from a mixture that was treated with excess AsF_5 in BrF_5 at 0 °C in vacuo. XeF_2 and XeF_6 are converted to ionic complexes, which stay in solution, whereas XeF_4 can be driven out in vacuo at 20 °C [1104]. In contrast, XeF_6 is the strongest fluoride acceptor. If a mixture of xenon fluorides is treated with NaF at 50 °C, only XeF_6 will form a nonvolatile compound, that is, $Na_2[XeF_8]$, which can be separated from XeF_2 and XeF_4. Afterwards, The octafluoroxenate(VI) salt can be cleaved at 125 °C in vacuo to liberate pure XeF_6 [1105].

Generally, the xenon fluorides have also been applied for fluorination of organic substrates, such as olefins or other unsaturated, aliphatic compounds but also fluoroaromatics [1106–1108]. Upon reaction with aromatic compounds, XeF_n typically substitute the ring under retention of the aromatic system [1107]. Expectedly, their fluorinating ability increases with their oxidation state, so that the most convenient and easy to handle fluorinating agent is definitely XeF_2. Nevertheless, the higher fluorides have also been used. XeF_6 is a very powerful fluorinating agent for organic substrates. For example, it can transform perfluorocyclopentene C_5F_8 to the respective saturated fluorocarbon C_5F_{10} [1081].

Derivatives of the xenon fluorides

For all the xenon fluorides, it is possible to substitute fluorine atoms by other negative ligands, especially oxygen containing groups. This type of chemistry is mainly restricted to XeF_2, which is carefully treated with the respective acid in an anhydrous medium at low temperatures. Nevertheless, in some cases, the higher fluorides have been substituted as well, as for example, in $XeF_5(OSO_2F)$ [1109]. The substitution occurs stepwise in an equilibrium, which can be driven to the product side by removal of HF. Suitable ligands are typically the anionic species $OClO_3^-$, OSO_2F^- and especially $OSeF_5^-$ and $OTeF_5^-$ [819, 1109–1111]. The latter group is the only substituent, which gives rise to relatively stable derivatives [1112]. The most important of these pentafluorinated chalcogen compounds is certainly $Xe(OTeF_5)_2$ (mp.: 35-40 °C; decomp.: 150 °C), which is only slightly soluble in water and, therefore, hydrolyzes rather slowly. In contrast, it is well soluble in organic solvents like CCl_4 or CH_3CN. Cationic $[XeOTeF_5]^+$ is generated by treatment of neutral $Xe(OTeF_5)_2$ with equimolar amounts of XeF_2 and subsequent fluoride abstraction with AsF_5. Moreover, the $OTeF_5$-group can be substituted by SO_3F, if $[XeOTeF_5][AsF_6]$ is dissolved in HSO_3F [1113]. The reaction of $Xe(OTeF_5)_2$ with $Sb(OTeF_5)_3$ directly yields the strongly oxidizing salt $[XeOTeF_5][Sb(OTeF_5)_6]\cdot SO_2ClF$ [1114]. Starting from XeF_4, an $OTeF_5$-derivative is also accessible. $Xe(OTeF_5)_4$ is a yellow, sublimable solid (mp.: 72 °C) and the only stable xenon(IV) compound apart from XeF_4 [820, 1115]. The formation of $Xe(OTeF_5)_6$, $OXe(OTeF_5)_4$ and $O_2Xe(OTeF_5)_2$ by nucleophilic substitution of XeF_6, $XeOF_4$ and XeO_2F_2 has also been described, however, the XeF_6-derivative is unstable [820, 1116, 1117]. Cationic species are accessible from the compounds $Xe(OTeF_5)_4$ and $OXe(OTeF_5)_4$ by treatment with SbF_5, which leads to the abstraction of $OTeF_5$-substituents and also $OTeF_5$/F-ligand redistribution to some extent [1118].

Xenon oxyfluorides

XeF_2 dissolves molecularly in pure water, in very diluted solutions it only decomposes slowly at room temperature. At higher concentrations or in presence of bases or fluoride acceptors, vigorous reactions occur. In contrast, XeF_4 and XeF_6 react immediately and usually explosively with water [1119]. However, it is possible to perform controlled hydrolysis under certain safety precautions, which gives rise to oxygen-containing xenon compounds. The careful hydrolysis of XeF_4 at −80 °C yields **$XeOF_2$** as a light-yellow solid, which is coordinated via O-bridges and stable up to −25 °C (Scheme 9.37) [1050, 1120, 1121]. Upon warming to −15 °C, it slowly disproportionates to XeF_2 and XeO_2F_2, and at −63 °C it reacts with CsF in HF to give $Cs[XeOF_3]$ (Scheme 9.37) [1120, 1121]. Similarly, the reaction of CsF or $[NMe_4]F$ with $XeOF_2$ in acetonitrile at −45 °C yields $[XeOF_3]^-$ salts, which decompose at room temperature. In the solid state, the ammonium salt shows only little interactions between cations and anions, whereas Cs-O interactions as well as F-bridging between the anions are observed for the alkali metal salt. The structure of $[XeOF_3]^-$ has been calculated and is based on a square-planar geometry, where some small distortions are present [1122]. The formation of cationic $[XeOF]^+$ has never been observed, as reactions of $XeOF_2$ with fluoride acceptors lead either to immediate explosions or, if carefully performed, generate $[XeF]^+$

Cs[XeOF$_3$]   [XeOF$_3$]$^-$ salts

CsF, HF,   CsF or   MeCN,
-63 °C   NMe$_4$F   -45 °C

XeF$_4$ $\xrightarrow{\text{H}_2\text{O, -80 °C}}$ XeOF$_2$ $\xrightarrow{\text{-15 °C}}$ XeF$_2$ + XeO$_2$F$_2$

AsF$_5$,   fluoride
aHF, -78 °C   acceptors

[HOXeF$_2$]$^+$ [XeF]$^+$ salts

XeF$_6$ XeF$_6$ [XeOF$_5$]$^-$ salts

H$_2$O   NO$_3^-$ CsF or NOF

XeF$_6$ $\xrightarrow{\text{POF}_3}$ XeOF$_4$ $\xrightarrow[\text{or SbF}_5]{\text{PF}_5, \text{AsF}_5}$ [XeOF$_3$]$^+$ salts

XeF$_6$ CsNO$_3$

XeO$_3$ $\xrightarrow{\text{XeOF}_4}$ XeO$_2$F$_2$ $\xrightarrow[\text{or SbF}_5]{\text{PF}_5, \text{AsF}_5}$ [XeO$_2$F]$^+$ salts

H$_2$O   CsF or NOF

XeO$_3$ [XeO$_2$F$_3$]$^-$ salts

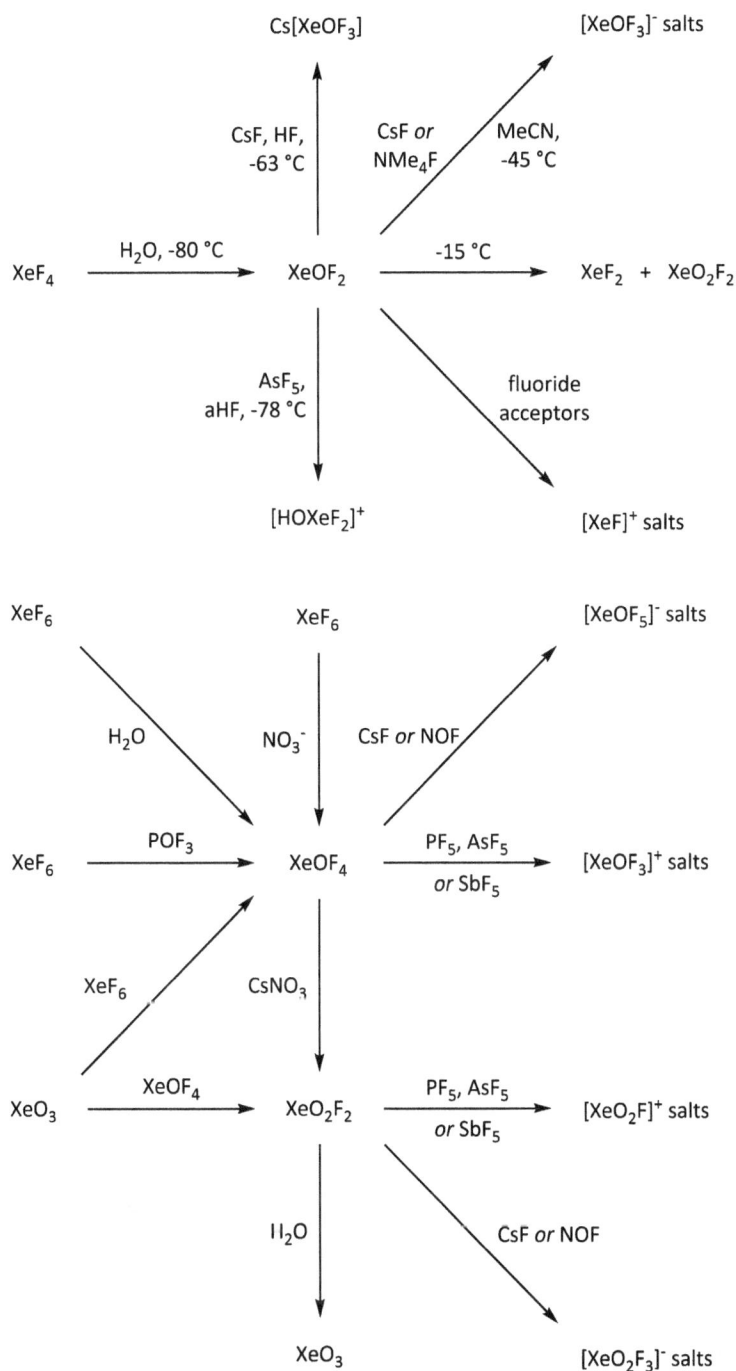

Scheme 9.37: Preparation and selected reactions of xenon oxyfluorides.

salts (Scheme 9.37) [1121]. Nevertheless, a cationic species is accessible, if $XeOF_2 \cdot nHF$ is treated with AsF_5 in anhydrous HF at $-78\,°C$. The resulting product contains $[HOXeF_2]^+$, where the oxygen atom has been protonated. Apart from that, adducts of $XeOF_2$ with cationic species, for example, $[XeF]^+$ have been described [1123]. The solid-state structure of $XeOF_2$ is described by weakly associated, planar monomers. The molecular structure is derived from a trigonal bipyramid, where the two lone-pairs occupy equatorial positions. Thus, $XeOF_2$ has a T-shaped geometry [1124].

XeF_6 is hydrolyzed stepwise via $XeOF_4$ and XeO_2F_2 to finally yield XeO_3. The molecular oxygen-containing derivatives are explosive compounds, whose instability increases with the oxygen content. **XeO_2F_2** exists in form of colorless, hydrolyzable crystals. The molecular structure of it is similar to SF_4 (ψ-trigonal bipyramid), with the fluoride substituents in axial positions. Additionally, Xe-O-contacts are formed to neighboring molecules, resulting in hexacoordinated Xe-centers and a layered solid-state structure of XeO_2F_2 [1125, 1126]. The reaction of excess $XeOF_4$ with $CsNO_3$ leads to the quantitative formation of XeO_2F_2, accompanied by NO_2F (Scheme 9.37) [1127]. XeO_2F_2 is also obtained by vacuum thermolysis of $XeOF_2$ at $-15\,°C$, where the by-product XeF_2 is easily removed by sublimation. At room temperature, XeO_2F_2 slowly decomposes to XeF_2 and O_2, whereas fast heating leads to an explosive decomposition [1120]. Moreover, XeO_2F_2 is accessible by the substitution reaction of $XeOF_4$ with XeO_3, whereas XeF_6 and XeO_3 yield $XeOF_4$ (Scheme 9.37) [1128]. However, upon hydrolysis of XeF_6 or $XeOF_4$, XeO_2F_2 is only poorly observable, as it only seems to be a short-lived intermediate on the way to XeO_3 [1129]. **$XeOF_4$** is a colorless, volatile liquid at standard conditions (mp.: between -46 and $-28\,°C$), which is readily hydrolyzed. It is rapidly formed from XeF_6, so that samples of xenon hexafluoride are often contaminated with $XeOF_4$ [1130]. However, on preparative scale, it is more convenient to avoid the hydrolysis route, which always includes the risk of formation of highly explosive XeO_3. More safely, a slight excess of XeF_6 is reacted with NO_3^- to generate $XeOF_4$ along with NO_2F and $[XeF_7]^-$ (Scheme 9.37) [1127, 1131]. Apart from that, a quantitative formation of $XeOF_4$ is achieved upon reaction of XeF_6 with POF_3 [1132]. The structure of $XeOF_4$ is similar to IF_5, thus it has the shape of a square pyramid, with the O-atom at the apical position and the Xe-atom slightly above the square plane. The Xe-O distance is rather short (170.3 pm) and its force constant quite high, which indicates a considerable double-bond character of the Xe-O bond [1130, 1133]. In anhydrous HF, $XeOF_4$ is dissolved without decomposition [1129]. Attempts to oxidize $XeOF_4$ by PtF_6 or $[KrF]^+$ to yield cationic $[XeOF_5]^+$ have not been successful. Initially, its existence had been claimed, which later turned out to be incorrect, as the product mixtures only contained the adduct $XeOF_4 \cdot [XeF_5]^+$ as well as $[O_2]^+$ salts. Even if the existence of $[XeOF_5]^+$ cannot be excluded, $[XeF_5]^+$ seems to be the far more stable product of these reactions [881, 1034, 1057, 1134].

Both XeO_2F_2 and $XeOF_4$ are quantitatively reduced by hydrogen at 300 °C to yield Xe, HF and H_2O. They react with fluoride acceptors (PF_5, AsF_5, SbF_5) to give the cations

$[XeO_2F]^+$ and $[XeOF_3]^+$, respectively; with fluoride donors (CsF, NOF), the corresponding anions $[XeO_2F_3]^-$ and $[XeOF_5]^-$ are generated (Scheme 9.37) [1056–1058, 1121, 1127, 1135–1138]. Apart from that, the C_3-symmetric anion $[(XeOF_4)_3F]^-$ is known, which is described by a central F-atom, which is coordinated by three $XeOF_4$-units, so that every Xe-center is hexacoordinated [1139]. The geometry of the $[XeOF_3]^+$ cation is derived from a trigonal-bipyramid, where the oxygen and the lone-pair occupy equatorial positions, accordingly, $[XeO_2F]^+$ is distorted trigonal-pyramidal [1055, 1056, 1138]. $[XeOF_5]^-$ has a pentagonal pyramidal structure with the O-atom in apical position, as has been proven by various spectroscopic methods and single crystal X-ray crystallography [1140, 1141]. For the $[XeO_2F_3]^-$-species, two different isomers are possible, which are both derived from a ψ-octahedral geometry [1127]. Furthermore, XeO_2F_2 and $XeOF_4$ can be reacted with $B(OTeF_5)_3$ to yield the derivatives $XeO_2(OTeF_5)_2$ and $XeO(OTeF_5)_4$. XeO_3 reacts with alkali metal halides MX to form the thermally very stable halogenoxenates $M[XeO_3X]$, for example, $Cs[XeO_3F]$ (decomp.: >200 °C). In $M[XeO_3F]$, the XeO_3 units are linked by F-bridges to form infinite chains. The oxidation of strongly basic XeO_3 solutions with O_3 yields the xenon(VIII) compounds $M_4[XeO_6]$ and $M_3H[XeO_6]$. Treatment of the latter two with concentrated H_2SO_4 gives rise to XeO_4, which then can be reacted with XeF_6 to yield oxyfluorides of octavalent xenon. XeO_3F_2 (mp.: –54 °C) is the most stable of these compounds and has the geometry of a trigonal bipyramid, where the axial positions are occupied by fluorine. XeO_2F_4 has also been identified by mass spectrometry, structurally it is most likely described by a D_{4h}-distorted octahedron. Due to its instability, the actual structure of XeO_2F_4 has not been determined so far. The third possible xenon(VIII) oxyfluoride $XeOF_6$ has never been observed. Probably, it is even less stable than the dioxytetrafluoride XeO_2F_4, so that it is immediately decomposed [1142–1144].

9.8.2 Krypton

The only binary compound of krypton is the difluoride **KrF$_2$**, all attempts to further oxidize it have been unsuccessful. KrF_2 is obtained, when a mixture of Kr and F_2 is submitted to electrical discharge at –183 °C and 20 mbar pressure or by UV-irradiation (365–366 nm) of a liquid Kr-F_2-mixture at –196 °C [1145–1149]. Similar to XeF_2, it may also be prepared by exposure of Kr/F_2- or Kr/OF_2-mixtures to sunlight [1150]. It can be purified by sublimation below –10 °C, at higher temperatures it rapidly decomposes. At –78 °C, KrF_2 can be stored for longer times without decomposition. In contrast to XeF_2, KrF_2 is generated endothermally. It forms colorless crystals with linear KrF_2 molecules[1151–1153]. Along with the cation $[KrF]^+$, KrF_2 is a very strong oxidizing agent [883, 1034, 1057]. Even below –78 °C, KrF_2 reacts with water to yield HF, O_2 and Kr, in the same way, Hg, AgF, ClF_3, I_2 and Xe are oxidized to HgF_2, AgF_2, ClF_5, IF_7 and XeF_6, respectively. Analogously to XeF_2, the cations $[KrF]^+$ and $[Kr_2F_3]^+$ are generated upon reaction with strong fluoride acceptors (AsF$_5$, SbF$_5$, BiF$_5$, TaF$_5$,

NbF$_5$) [1057, 1149, 1153–1156]. The structure of [Kr$_2$F$_3$]$^+$ is similar to the respective Xe-derivative, being generally described by a V-shaped geometry. Interestingly, the cation appears to be symmetric in BrF$_5$-solution, but asymmetric in the solid state, which might be due to weak F-bridging to the anion [1153, 1154, 1156]. Expectedly, the SbF$_5$ complexes are the most stable ones which can be handled far more easily than KrF$_2$ itself. Furthermore, they do not decompose below room temperature. The cation [KrF]$^+$ is one of the strongest oxidizing agents known, and transforms IF$_5$ to [IF$_6$]$^+$, BrF$_5$ to [BrF$_6$]$^+$, ClF$_5$ to [ClF$_6$]$^+$, NF$_3$ to [NF$_4$]$^+$ and Au to AuF$_5$ [881, 883, 888, 1034, 1057, 1156]. Generally, the number of KrF$_2$-complexes is rather limited, however, upon reaction with the oxytetrafluorides CrOF$_4$, MoOF$_4$ and WOF$_4$, covalent adducts of the composition KrF$_2 \cdot$ 2 MOF$_4$ (M = Cr, Mo, W) are observed [1157, 1158]. A coordination compound of the composition [BrOF$_2$][AsF$_6$] $\cdot$ 2 KrF$_2$ is also known. As could be shown by X-ray crystallography, the Br-center is six-fold coordinated, where the KrF$_2$ ligands are in *trans*-position to the F-atoms of [BrOF$_2$]$^+$. Thus, [AsF$_6$]$^-$ occupies the position opposite to the oxygen-substituent [1159]. The rare example of a bridging KrF$_2$ ligand is found in the mercury complex Hg(OTeF$_5$)$_2 \cdot$ 1.5 KrF$_2$ [823]. Even a naked metal cation, that is, Mg^{2+}, can be coordinated by krypton difluoride, resulting in the crystalline compound [Mg(KrF$_2$)$_4$][AsF$_6$]$_2$ [1160]. The related mercury compound [Hg(KrF$_2$)$_8$][AsF$_6$]$_2$ $\cdot$ 2 HF contains isolated dications, where the KrF$_2$ ligands terminally coordinate the Hg-center, resulting in a slightly distorted, square-antiprismatic coordination geometry [1161]. Apart from the aforementioned metal centers, KrF$_2$ coordinates HCN under formation of [HCN $\cdot$ KrF]$^+$, which has a linear NKrF group and is stable up to about –50 °C [1162, 1163]. Analogously to the xenon derivatives, the fluoride substituents in KrF$_2$ can be replaced by OTeF$_5$-groups at low temperature (–110 to –90 °C), resulting in thermally unstable Kr(OTeF$_5$)$_2$ [822]. Quantum-chemical calculations on the hypothetic higher krypton fluorides KrF$_4$ and KrF$_6$ have been performed, revealing, that all krypton fluorides, even KrF$_2$, are generally unstable against the loss of F$_2$. Apart from that, the barrier of decomposition is still appreciably high for KrF$_2$ but decreases to less than 4 kJ mol^{-1} in the case of KrF$_6$. As a consequence, KrF$_4$ might still be accessible at moderately low temperature, whereas KrF$_6$ requires very low temperature and is therefore unlikely to be accessible at all [1164].

9.8.3 Radon

Due to the comparably low ionization potential of radon, one could expect, that radon fluorides are readily accessible. However, all radon isotopes are radioactive and the most stable one (^{222}Rn) has a half-life of only 3.8 days, which renders the chemistry of radon compounds very difficult. If ^{222}Rn is fluorinated under various conditions, always the same nonvolatile, ionic RnF$_2$ is formed [880, 1165–1167]. At room temperature, Rn is already oxidized by mild fluorinating agents like ClF$_3$, BrF$_3$ and BrF$_5$ as well as by the solid complexes [ClF$_2$][SbF$_6$], [BrF$_2$][SbF$_6$] and [BrF$_4$][Sb$_2$F$_{11}$]. Probably, RnF$_2$

reacts with fluoride acceptors MF_5 to yield salts of the type $[RnF]^+[MF_6]^-$ (M = Sb, Ta, Bi) [1168]. The higher radon fluorides RnF_4 and RnF_6 are also likely to exist, however, by now it was only possible to detect their hydrolysis product RnO_3. If RnF_6 exists, it is unlikely that an even higher radon fluoride can be formed, due to the impact of relativistic spin-orbit coupling [1169, 1170]. Relativistic effects might also be the reason for a regular octahedral shape of RnF_6 [1168]. The Rn-F bonds are assumed to be rather polar, which results in highly F-bridged structures in the solid phase [1171–1173]. As a consequence, the higher Rn-fluorides should be nonvolatile.

9.8.4 Study questions

(I)   Which xenon fluorides are known? How are they prepared?

(II)   Which molecular structures are observed for the xenon fluorides? Comment on the structure of solid XeF_6.

(III) Comment on the fluoride donor- and acceptor properties of the xenon fluorides. Which structures are observed for the resulting ionic species?

(IV)   All xenon fluorides have oxidizing and fluorinating properties. Give some examples for their reactivity.

(V)   Which xenon oxyfluorides are known? How are they prepared? Comment on their thermal stability.

(VI)   Which binary krypton-fluorine compounds are known? How are they generally prepared? Comment on their stability and give examples for their oxidation power.

10 Transition metal fluorides and rare earth metal fluorides

10.1 Group 3: Scandium, yttrium, lanthanum, actinium

10.1.1 Compounds and properties

Low-valent fluorides

The mono and difluorides of Sc, Y, La and Ac are unknown in condensed phase. Nevertheless, most of these compounds have been observed in the gas phase or under matrix conditions. For example, scandium monofluoride **ScF** is accessible upon heating of solid ScF_3 and Sc metal. However, unreacted ScF_3 and partly reduced ScF_2 are also observed in the mixture of gaseous compounds [1174]. Similarly, elemental Sc and Y can be treated with CaF_2 at high temperatures, which also results in mixtures of the respective mono, di and trifluorides [1175]. Lanthanum monofluoride **LaF** as well as the difluoride **LaF$_2$** are accessible, if a La-plasma is reacted with SF_6 [1176, 1177]. The electronic structure as well as the ground state of ScF and ScF_2 have been determined by both experimental and theoretical investigations [1178, 1179]. Similarly, the electronic structure of YF and its molecular properties have been explored by quantum-chemical calculations [1180, 1181]. Infrared-spectra of matrix-isolated ScF_2-, YF_2- and LaF_2-molecules reveal a C_{2v}-symmetry, which is also supported by theoretical investigations [1182–1184]. In principle, the respective lower fluorides of actinium should also be accessible, however, they have not been described in the literature so far.

Trifluorides

Scandium, yttrium, lanthanum and actinium form the trifluorides **ScF$_3$** (mp.: 1552 °C; bp.: 1607 °C), **YF$_3$** (mp.: 1155 °C; bp.: 2230 °C), **LaF$_3$** (mp.: 1493 °C; bp.: 2330 °C) and **AcF$_3$**. They are all sparingly soluble in water, and thus can be precipitated from aqueous solutions of their salts by addition of hydrofluoric acid or fluorides [1185]. Consequently, the resulting compounds are obtained as hydrates, whereas the anhydrous trifluorides are accessible by treatment of the sesquioxides M_2O_3 with HF or elemental fluorine. Unlike YF_3 and LaF_3, ScF_3 dissolves in presence of excess fluoride under formation of $[ScF_6]^{3-}$. In turn, the ammonium salt $[NH_4]_3[ScF_6]$ can be thermally degraded to yield intermediary $[ScF_4]^-$, which is further decomposed to impure ScF_3 upon heating to 600 °C [1186]. Analogously, one can also observe stable 1:1- and 3:1-complexes in the system NaF/ScF_3, represented by the compounds $Na[ScF_4]$ and $Na_3[ScF_6]$, respectively [1187]. For the other elements of group 3, the complexes $[YF_4]^-$, $[YF_6]^{3-}$, $[LaF_4]^-$ and $[LaF_6]^{3-}$ are also known [1188]. Heating of the hydrates $MF_3 \cdot n$ H_2O leads to the formation of oxyfluorides MOF.

https://doi.org/10.1515/9783110659337-012

All the fluorides of group 3 show a three-dimensional structure in their solid-state. Crystalline ScF_3 has a distorted ReO_3 structure with octahedrally coordinated scandium, whereas molecular ScF_3 is planar [1189]. In contrast, matrix-isolated YF_3- and LaF_3-molecules are clearly nonplanar [1182, 1184]. The solid-state structure of YF_3 represents an own structure type, which is derived from distorted UCl_3. Therein, yttrium has the coordination number nine, leading to a three-fold-capped trigonal prismatic orientation of the ligand atoms [1190]. The coordination environment is further enlarged in the case of LaF_3, where lanthanum has the coordination number eleven. The resulting geometry around La is described by a five-fold-capped trigonal prism [1191]. AcF_3 also crystallises in the LaF_3 lattice [1192]. Most of the fluorometallates $[MF_6]^{3-}$ exhibit a cryolite structure, whereas the majority of the tetrafluorometallates $[MF_4]^-$ is described by a distorted fluorite lattice. LaF_3 and YF_3 react with AsF_5 in anhydrous HF to form the stable compounds $La[AsF_6]_3$ and $YF[AsF_6]_2$. The complexes can be isolated as solids at lower temperatures [1193].

Oxyfluorides

The oxyfluoride derivatives **ScOF**, **YOF**, **LaOF** and **AcOF** are known as well. Generally, they are prepared by hydrolysis of the respective trifluoride, in most cases this requires high temperatures [1194]. LaOF is moreover accessible from mechanochemical synthesis, when La_2O_3 and LaF_3 are ground at room temperature [1195]. In a solid-state synthesis, YOF and LaOF are accessible from their oxides, using PTFE as fluoride source [1196]. Fluorolytic sol-gel synthesis gives rise to nanoscopic YOF and certainly also to the other oxyfluorides [1197]. The solid-state structure of ScOF shows Sc-centers, which are coordinated by four O- and three F-atoms each, so that the coordination polyhedra and the linkage between them very well represent the monoclinic structure type of ZrO_2 [1198]. YOF and LaOF are polymorphic, where both the rhombohedral and the tetragonal phase are distorted from the CaF_2 structure type at lower temperatures [1194, 1199, 1200]. AcOF as well as the high-temperature modification of LaOF crystallize in the cubic fluorite lattice [1192, 1194, 1201]. Matrix-isolated molecules of ScOF, YOF and LaOF have been investigated by vibrational spectroscopy as well as by supporting quantum-chemical calculations. As a consequence, a significantly bent structure has been assigned to the molecules, where the M-O-bonds show pronounced triple-bond-character [1202, 1203]. At temperatures above 280 °C, LaOF reacts with the fluorocarbon CCl_3F under formation of intermediary $LaCl_3$ to finally yield the trifluoride LaF_3 [1204]. Apart from that, LaOF, or more effectively a mixture of LaOF and BaF_2, is capable of catalytically and oxidatively dehydrogenating ethane to ethylene [1205]. LaOF also shows photocatalytic activity in the production of H_2 from methanol-containing solutions [1206]. Its applications in optoelectronics and biotechnology have also been examined [1207]. In general, the oxyfluorides are often doped by other lanthanide ions, in order to achieve interesting material properties, for example, luminescence [1208, 1209].

10.1.2 Study questions

(I)   How are the trifluorides ScF_3, YF_3, LaF_3 and AcF_3 prepared? Can they be obtained as anhydrous compounds?

(II) Which stable fluorometallate anions are observed for the trifluorides of group 3? How do these species arrange in the solid state?

(III) Shortly summarize the synthetic approaches toward the oxyfluorides ScOF, YOF, LaOF and AcOF. What can these compounds be used for?

10.2 The lanthanides

10.2.1 Compounds and properties

All the lanthanides form trifluorides LnF_3, further there exist some tetrafluorides LnF_4 and a few difluorides LnF_2. In some cases, stable tetravalent lanthanide fluorides only exist in form of fluorometallate anions. Apart from binary fluorides, a few oxyfluorides have also been reported.

Monofluorides

The monofluorides LnF are unknown in condensed state. Nevertheless, a few of them, for example, CeF, HoF or TbF, have been investigated by spectroscopic methods in the gas phase and using quantum-chemical calculations [1210–1215]. Along with higher fluorides, PrF has been prepared under matrix conditions by reaction of laser-ablated Pr-atoms with F_2 [1216]. The certainly most explored monofluoride is **YbF**, which might be surprising, as natural Yb consists of seven isotopes. Thus, spectra can get very complicated due to isotopic effects. However, Yb has a similar electron configuration like Ba, so it has been assumed that studying the YbF system will also give further insights into BaF and related molecules [1217–1219]. In general, gaseous lanthanide monofluorides are accessible by either reducing the trifluorides with Ln metal, or by fluorination of elemental Ln with BaF_2 [1220–1222].

Difluorides

Analogously to the monofluorides, most of the difluorides are only known in gaseous state. Accordingly, they are also obtained by fluorination of metallic Ln or reduction of the trifluorides [1220–1222]. PrF_2 has been observed in matrix-experiments and has been identified as one of the products of the fluorination of Pr atoms with F_2 [1216]. In some cases, the lanthanide difluorides have been explored by quantum-chemical calculations, too. For example, CeF_2 has been predicted to have a bent molecular geometry and a triplet ground-state [172]. Nevertheless, stable lanthanide difluorides are known for samarium, europium and ytterbium. They are accessible by reduction of the trifluorides with H_2, Si or the respective metal vapors [1223–1225]. Alternatively, europium difluoride EuF_2 is very simply prepared from $EuSO_4$ and NaF [1226]. At room temperature, **SmF$_2$**, **EuF$_2$** and **YbF$_2$** crystallize in a fluorite lattice, however, an orthorhombic PbCl$_2$ type modification of EuF$_2$ is observed at 400 °C and 114 kbar [1224, 1227]. The lanthanide difluorides are salt-like compounds, which are readily oxidized upon contact with air and react with water under liberation of H$_2$. If they are heated to high temperatures, disproportionation to LnF$_3$ and metallic Ln occurs [1228]. There are various mixed-valent compounds in the system of LnF$_2$ and LnF$_3$. For example, the binary compounds Ln$_3$F$_7$, Ln$_{14}$F$_{33}$ and Ln$_{27}$F$_{64}$ (Ln = Eu, Yb) are known [1229, 1230]. Apart from that, there are also combinations of different lanthanides. Generally, the lanthanide difluorides do not tend to form fluorometallates of the type M^I[LnF$_3$] when starting from LnF$_2$ and M^{I}F. However, upon reduction of LnF$_3$ with alkali metals M, the complexes M[LnF$_3$] (Ln = Eu, Yb; M = Rb, Cs) are obtained [1231]. For ytterbium, the ternary compounds YbBeF$_4$ and YbAlF$_5$ as well as quaternary LiYbAlF$_6$ are known. They are all light-green powders, which are prepared from the respective fluorides at 750 °C. The behavior of Yb^{2+} in the crystalline state can be compared to the alkaline earth dications [1232]. EuF$_2$-nanoparticles have been prepared in ionic liquids and show bright blue luminescence as well as a desirable stability in colloidal solution [1233].

Trifluorides

Similar to the fluorides of group 3, hydrates of the lanthanide trifluorides are prepared by precipitation from aqueous solutions, which is either achieved by the addition of fluoride ions or hydrofluoric acid [1234–1236]. In most cases, the anhydrous salts are generated by treatment with HF or elemental fluorine. However, treatment with F$_2$ is not suitable in case of Ce, Pr and Tb, as these metals would be directly oxidized to the respective tetrafluorides [1237]. In the solid state, **CeF$_3$**, **PrF$_3$**, **NdF$_3$**, **PmF$_3$**, **SmF$_3$** and **EuF$_3$** all have the same hexagonal room structure, with the lanthanide center being eleven-fold coordinated in form of a fully (five-fold)-capped trigonal prism (LaF$_3$ type). The other trifluorides **GdF$_3$**, **TbF$_3$**, **DyF$_3$**, **HoF$_3$**, **ErF$_3$**, **TmF$_3$**, **YbF$_3$** and **LuF$_3$** crystallize in the YF$_3$ structure type, where the metal center is nine-fold coordinated, resulting in a distorted three-capped trigonal prismatic geometry [1190, 1191, 1238, 1239]. Based on experimental data and supported by quantum-chemistry, a pyramidal structure is suggested for all LnF$_3$ molecules [1240]. The structure and bonding of the lanthanide trifluorides has been investigated in more detail by computational chemistry [1241]. The trifluorides are all salt-like, high-melting compounds, which are partly colored. In contact with alkali metal fluorides, they readily form complexes of the type Na[LnF$_4$] and M$_3$[LnF$_6$] [1242]. For the anions [LnF$_4$]$^-$, a three-fold-capped trigonal prismatic coordination geometry is typically found for the metal centers, indicating highly associated structures, whereas [LnF$_6$]$^{3-}$ are found to be mostly octahedral. In anhydrous HF, the lanthanide trifluorides react with the fluoride acceptor AsF$_5$ to form stable complexes of the types LnF[AsF$_6$]$_2$ (Ln = Ce to Er) and Ln$_2$F$_3$[AsF$_6$]$_3$ (Ln = Tm, Yb, Lu), respectively, which can all be isolated at low temperatures [1193].

There are also oxyfluorides of the lanthanides in oxidation state +III. The compounds with the general formula LnOF are typically obtained, if the hydrated trifluorides are heated in vacuo [1235]. They are also accessible from the respective oxide Ln_2O_3 and LnF_3 or solutions of Ln^{3+} which are treated with KF in presence of urea [1243, 1244]. Furthermore, it is possible to prepare the oxyfluorides in solid-state reactions at low temperature, if PTFE is used as fluorinating agent for the lanthanide oxides [1196, 1208]. In case of cerium, the oxyfluoride CeOF is also accessible from a stoichiometric mixture of CeO_2, Ce metal and CeF_3 [1245]. Nanocrystalline oxyfluorides LnOF are accessible, if the trifluoroacetates $Ln(CF_3CO_2)_3$ are carefully fluorinated. The resulting products can have various different shapes, depending on the lanthanide [1246]. The solid-state structures of the lanthanide oxyfluorides are derived from the fluorite type. In most of the cases, the compounds are polymorphic [1247–1249]. TmOF, YbOF and LuOF are isostructural with α-ZrO_2 [1243]. Based on vibrational spectra of matrix-isolated LnOF and supported by quantum-chemical calculations, a bent structure has been assigned to all the oxyfluorides. In each of the compounds, the Ln-F bond has ionic character [1250]. In combination with SrF_2 or BaF_2, SmOF shows catalytic activity in the oxidative dehydrogenation of ethane, as it is capable of activating O_2 [1251]. Apart from that, most of the oxyfluorides possess luminescent properties [1252]. If these properties are combined with controlled structural features, for example, nanowire-shaped materials, this gives rise to potential applications, for example, in displays or other devices [1208, 1253–1255].

Tetrafluorides
The tetrafluorides **CeF$_4$** and **TbF$_4$** are accessible by fluorination of the metals or Ln(III) compounds in form of colorless solids (Scheme 10.1) [268, 1237, 1256–1260]. **PrF$_4$** cannot easily be made by simple fluorination but conveniently results from the treatment of $Na_2[PrF_6]$ with anhydrous HF. The aforementioned complex $Na_2[PrF_6]$ is formed in the reaction of PrF_3 with NaF and F_2 [1257, 1261, 1262]. Nevertheless, the reaction of PrO_2 with KrF_2 directly yields PrF_4 (Scheme 10.1) [1259]. Analogously, PrF_4 and TbF_4 can be prepared in quantitative yield from Pr_6O_{11} and Tb_4O_7 in anhydrous HF at room temperature, if the respective oxide is treated with UV-photolyzed F_2 [1260]. The solid tetrafluorides are isomorphic to UF_4, hence there are eight fluorine atoms coordinated around the metal center with the geometry of a square antiprism [1192, 1257]. The molecular structure of CeF_4 has been explored by matrix-isolation spectroscopy and quantum-chemical methods. It has a tetrahedral geometry with an inversion barrier of about 25 kJ mol^{-1}. Similarly, PrF_4 has been prepared and investigated by vibrational spectroscopy and quantum-chemical methods, revealing a molecular C_{2v}-symmetry. Apart from that, theoretical calculations predict the existence of a stable pentafluoride PrF_5. However, the latter has never been observed in any matrix-experiment [21, 172, 1216]. PrF_4 already decomposes at 90 °C to give PrF_3 and F_2, whereas CeF_4 and TbF_4 are thermally rather stable. However, only CeF_4 can be vaporized without

PrO_2 $Na_2[PrF_6]$ $\xleftarrow{F_2,\ NaF}$ PrF_3

KrF_2 aHF

Ce or Tb $\xrightarrow{F_2}$ LnF_4 $\xrightarrow[\text{(traces)}]{H_2,\ H_2O\ or\ NH_3}$ LnF_3

CeO_2, Δ H_2O F^-
(Ln = Ce)   (Ln = Ce)

$CeF_3 + O_2$ $CeO_2 + HF$ $[LnF_5]^-, [LnF_6]^{2-},$
                                                 $[LnF_7]^{3-}\ or\ [LnF_8]^{4-}$

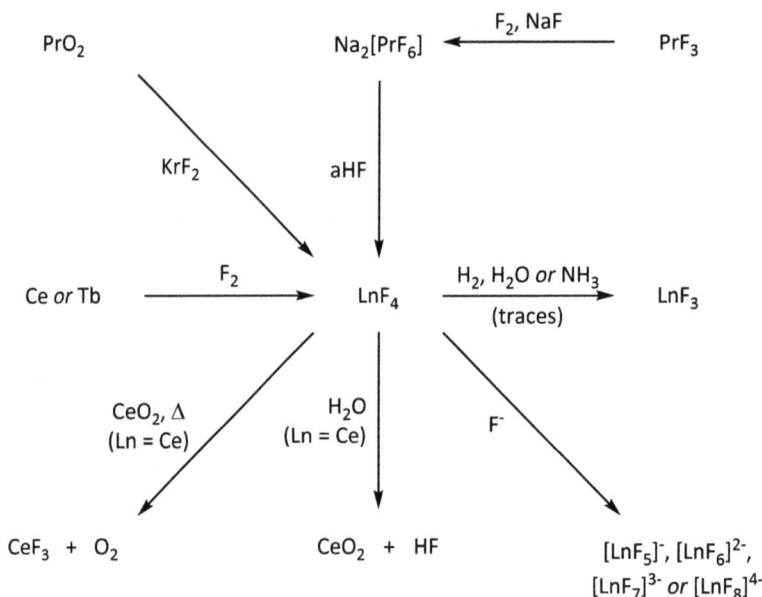

Scheme 10.1: Preparation and selected reactions of lanthanide tetrafluorides LnF_4 (Ln = Ce, Pr, Tb).

substantial decomposition [1263–1265]. The tetrafluorides are reduced to the respective LnF_3 in presence of trace amounts of H_2O, NH_3 or H_2 [1256, 1258]. In case of CeF_4, a monohydrate is known, however, it cannot be dehydrated without decomposition. Hydrolysis of the tetrafluoride yields HF and CeO_2, whereas CeF_4 and CeO_2 react upon heating to form CeF_3 and O_2 [1258]. The lanthanide tetrafluorides readily form fluorometallates, which have the general compositions $M[LnF_5]$, $M_2[LnF_6]$, $M_3[LnF_7]$ or $M_4[LnF_8]$ (Scheme 10.1) [1266]. For example, $(NH_4)_4[CeF_8]$ is accessible by the reaction of CeF_4 with solutions of NH_4F. Therein, the Ce-center has a square antiprismatic coordination environment. $(NH_4)_4[CeF_8]$ can be thermally decomposed, which results mainly in the formation of $(NH_4)_2[CeF_6]$. The $[CeF_6]^{2-}$ anions are linked via F-bridges to form chains, where Ce is again eight-fold coordinated by fluorine substituents in form of a square antiprism [1267]. In turn, $(NH_4)_2[CeF_6]$ reacts with aqueous NH_4F solutions to give $(NH_4)_3[CeF_7] \cdot H_2O$ [1268]. Therein, Ce is eight-fold coordinated and two $[CeF_8]$-dodecahedra are linked by a common edge to give centrosymmetric $[Ce_2F_{14}]^{6-}$-units [1269]. Another octacoordinated Ce-compound is the ammoniate $[CeF_4(NH_3)_4] \cdot NH_3$, which is prepared by the reaction of CeF_4 with liquid ammonia. Here, the geometry around the Ce-center is described by a distorted square antiprism [1270]. Complexes of the general formula $M_3[LnF_7]$ (Ln = Ce, Pr, Nd, Tb, Dy) contain the pentagonal-bipyramidal anion $[LnF_7]^{3-}$. The respective neodymium and dysprosium complexes $[NdF_7]^{3-}$ and $[DyF_7]^{3-}$ are the only stable compounds with Nd and Dy in oxidation state +IV. However, under matrix-conditions, the neutral tetrafluorides NdF_4 and DyF_4 have been identified by vibrational spectroscopy [1271].

Tb$_2$O$_3$ reacts with F$_2$ in presence of alkali metal chlorides to give rise to a large variety of complexes: M[TbF$_5$] (M = K, Rb, Cs), M$_2$[TbF$_6$] (M = K, Rb, Cs), M$_3$[TbF$_7$] (M = Na, K, Rb, Cs), M[Tb$_2$F$_9$] (M = K, Rb), Cs$_3$[Tb$_2$F$_{11}$], Na$_5$[Tb$_2$F$_{13}$], Rb$_2$[Tb$_3$F$_{14}$] as well as M[Tb$_6$F$_{25}$] and M$_7$[Tb$_6$F$_{31}$] [1272]. The barium compound Ba[TbF$_6$] is generated from a stoichiometric mixture of BaF$_2$ and TbF$_4$ under an atmosphere of F$_2$ [1273]. A few more compounds are known in the system M^{II}F$_2$ (M^{II} = Ca, Sr, Ba, Cd) and TbF$_4$, resulting from stoichiometric di- and tetrafluoride ratios of 2:1, 1:1, 1:2 and 1:3 [1274]. Out of the system KF-LnF$_3$-TbF$_4$, the mixed valent lanthanide compounds KLn^{III}[Tb$^{IV}_2$F$_{12}$] (Ln = Ce to Lu) are accessible, including the terbium derivative KTb^{III}[Tb$^{IV}_2$F$_{12}$]. In the solid-state structure, the different terbium centers are perfectly ordered, showing slightly distorted [Tb^{III}F$_8$]$^{5-}$-cubes and dodecahedral [Tb^{IV}F$_8$]$^{4-}$-units [1275]. Several more ternary and quaternary mixed-valent terbium-compounds are accessible upon thermal decomposition of TbF$_4$ in presence of RbCl, AlF$_3$ or KF [1276]. Hydrothermal synthesis gives rise to compounds of the type M$^I_7$[Ce$_6$F$_{31}$] (M^I = Na, K, NH$_4$, Tl) [1277]. The silver compound Ag$_7$[Ce$_6$F$_{31}$] is isostructural to the respective group 4-derivatives Ag$_7$[M$_6$F$_{31}$] (M = Zr, Hf). Thus, the Ce-centers form an octahedron and each of them is surrounded by eight F-atoms in form of a square-antiprism (cf. Section 10.4.2) [1278].

The lanthanide tetrafluorides have oxidizing and fluorinating properties. However, only CeF$_4$ is suitable for synthetic applications, as TbF$_4$ and PrF$_4$ decompose too easily. Therefore, CeF$_4$ has been investigated as potential replacement for CoF$_3$ in fluorocarbon synthesis. As main problem, the high viscosity of CeF$_4$-containing reaction mixtures has been found to be unfavorable for preparative purposes [1279]. Nevertheless, fluorination studies of benzene and benzotrifluoride have been conducted, revealing a mixture of partly fluorinated aromatic derivatives and substituted cyclohexene and cyclohexadiene products [1280, 1281]. CeF$_4$ has also been applied for the fluorination of inorganic compounds, for example, the gas-phase preparation of FeF$_4$ [1282]. Under certain conditions, TbF$_4$ has also been applied as effective fluorinating agent, making use of its instability, and hence, exploiting the easy formation of F$_2$ upon thermal decomposition. For example, TbF$_4$ has been used to generate the compounds CoF$_4$ and CrF$_5$ under mass-spectrometric conditions [1283]. Similarly, CeF$_4$, TbF$_4$ and even PrF$_4$ have been applied to fluorinate the fullerene C$_{60}$ in the gas phase [182, 1284]. Oxyfluorides of the type LnOF$_2$ do not play an important role and have only been observed under matrix conditions [1250].

10.2.2 Study questions

(I) Which stable lanthanide difluorides are known? How are they prepared and how can they be structurally described?

(II) How are the lanthanide trifluorides prepared? What has to be taken into account, when anhydrous CeF$_4$, PrF$_4$ and TbF$_4$ shall be generated?

(III)   Which products are generally obtained, if the lanthanide trifluorides are reacted with alkali metal fluorides? What can one say about their structural behavior?

(IV)   Which tetravalent lanthanide-fluorine compounds are known? How are they prepared? Comment on their thermal stability.

(V)   Which are the most important chemical properties of the lanthanide tetrafluorides? What are they used for?

10.3 The actinides

10.3.1 Compounds and properties

In condensed phase, the actinide fluorides exist in a variety of binary compounds, where the actinide typically adapts an oxidation state in the range from +III to +VI. Apart from those compounds, some mixed-valent fluorides have been described as well as oxyfluorides and a few other derivatives.

Mono- and difluorides

Mono- and difluorides of the actinides are unknown in condensed phase. In some cases, they have been investigated in the gas phase or under matrix conditions, as well as by quantum chemistry. The early actinides Th and U are fairly well investigated, whereas all the others are rather scarcely explored. Thorium monofluoride **ThF** and its difluoride **ThF$_2$** are found in the product mixture, if Th-atoms are reacted with F$_2$ in an Ar-matrix at low temperature. Based on vibrational spectra, a bent structure has been assigned to ThF$_2$ [172, 1285, 1286]. Both thorium compounds have also been identified by mass spectrometry, where they are generated from Th-metal powder and BaF$_2$-vapor [1287]. The low-lying electronic states of ThF have been spectroscopically explored [1288]. In the same way, the low-valent uranium derivatives **UF** and **UF$_2$** are accessible in the gas phase, if UC is heated with BaF$_2$, or upon reduction of UF$_4$ with Ca-metal [1289]. Similar to the aforementioned Th-derivatives, they can also be obtained under matrix conditions [1290]. Spectroscopic and theoretical investigations on UF allowed the identification of its electronic ground state as well as of several excited states [1291]. UF$_2$ is proposed to be an intermediary surface-species in the fluorination of U-metal by NF$_3$ [1292]. If U-atoms are reacted with HF or DF in an argon-matrix, the fluorohydride HUF or fluorodeuteride DUF is formed, respectively [1293, 1294]. The plutonium species **PuF** and **PuF$_2$** are observed in the gas phase, if PuF$_3$ and elemental Pu are heated, or upon fragmentation of PuF$_6$ in a mass-spectrometer [1295, 1296]. The electronic structure of the curium compounds **CmF** and **CmF$_2$** has been investigated by quantum-chemical calculations, where the Cm-F bonds have been predicted to have an ionic character [1297].

Trifluorides and trivalent oxyfluorides

Thorium trifluoride **ThF$_3$** has only been observed under matrix conditions, when Th-atoms were reacted with elemental fluorine. It is a planar compound, as has been investigated by vibrational spectroscopy and supporting quantum-chemical calculations [1285, 1286]. All attempts to prepare ThF$_3$ in condensed phase have been unsuccessful so far [1298]. In contrast, the trifluorides **UF$_3$, NpF$_3$, PuF$_3$, AmF$_3$, CmF$_3$, BkF$_3$** and **CfF$_3$** can be prepared by treatment of aqueous An^{3+} with fluoride ions. The resulting, sparingly soluble actinide trifluorides contain water and can be dehydrated by heating in the presence of HF (Scheme 10.2) [1299–1303]. Other preparative methods include the reaction of the oxides An$_2$O$_3$ with HF at higher temperatures or the reduction of higher fluorides or oxides with a mixture of H$_2$ and HF at 500–700 °C [1304–1310]. UF$_3$ is also obtained from a 3:1-mixture of UF$_4$ and UN at 950 °C [1311]. More recently, it has been found, that UF$_3$ and PuF$_3$ can also be obtained by fluorination of the respective metal under rather mild conditions, using NF$_3$ as source of fluorine (Scheme 10.2). However, at least in case of UF$_3$, the outcome of the reaction depends on the concentration of the fluorinating agent, as higher amounts of NF$_3$ will lead to the formation of UF$_4$ [1292]. UF$_3$ is also accessible by reduction of UF$_4$ with elemental Si [1312]. Generally, actinide trifluorides are poorly volatile compounds with high melting points. In the solid state, all AnF$_3$ have the same structure like LaF$_3$, however, for BkF$_3$ and CfF$_3$ this is only the case at high temperature. Their low-temperature structure is described by the YF$_3$ lattice [1192, 1238, 1313–1315]. The stability of the trifluorides against oxidizing agents is quite different. UF$_3$ is the least stable of the trifluorides, it is already slowly oxidized upon contact with air and water. The fluorination of actinide trifluorides at higher temperatures leads to the formation of either hexa- (U, Np, Pu) or tetrafluorides (Am to Cf). The reduction of CmF$_3$ with stoichiometric amounts of Si at about 1200 °C yields the silicides CmSi, Cm$_2$Si$_3$ or CmSi$_2$, respectively [1316]. Upon contact with fluoride donors, the trifluorides are transformed to anionic fluorocomplexes, however, only very few of them are known (Scheme 10.2). They have the general formula M[AnF$_4$] (M = Na, K), where the actinide is coordinated in form of a three-fold-capped trigonal prism. The complexation of Cm^{3+} by fluoride ions in aqueous solution has been investigated in dependence of the reaction temperature. As a result, the formation of [CmF]$^{2+}$ is favored in a temperature range up to 70 °C, whereas [CmF$_2$]$^+$ is mainly observed at higher temperatures up to 90 °C [1317].

There are also actinide oxyfluorides of the general composition **AnOF** (An = Th, U, Bk, Cf). They are all high-melting, nonvolatile solids, which crystallize in the fluorite lattice [1318, 1319]. **ThOF** is generated by heating Th-metal, ThF$_4$ and ThO$_2$ to 1200 °C. It is a grey-white solid, which is non-hygroscopic and resists oxidation by air at room temperature [1319, 1320]. It is also observed by mass-spectrometry, when oxygen-impurities are present in the thorium-fluorine system [1287]. **UOF** has only briefly been studied by photoelectron spectroscopy and additional quantum-chemical calculations [1321]. **PuOF** has been claimed to be accidentally obtained by attempts to

An$_2$O$_3$ U *or* Pu   UF$_4$

HF, Δ   NF$_3$ Si

An^{3+} $\xrightarrow[\text{then } \Delta,\ aHF]{\text{F}^-,\ \text{H}_2\text{O}}$ AnF$_3$ $\xrightarrow{\text{F}^-}$ [AnF$_4$]$^-$

F$_2$, Δ

AnF$_4$ (An = Am, Cm, Bk, Cf)
or
AnF$_6$ (An = U, Np, Pu)

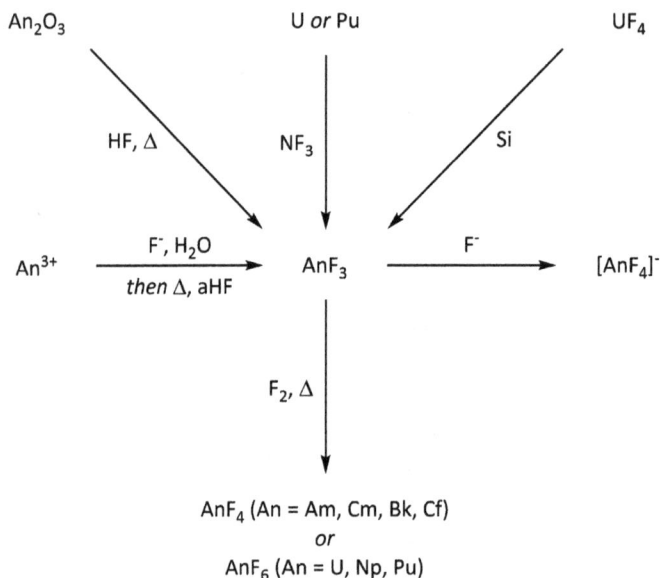

Scheme 10.2: Preparation and selected reactions of actinide trifluorides AnF$_3$ (An = U to Cf).

reduce PuF$_3$ with atomic hydrogen [1322]. However, Investigations on [PuO]$^+$-species in fluoride melts did not confirm the existence of a stable PuOF [1323]. Californium oxyfluoride **CfOF** is the product of high-temperature hydrolysis of the trifluoride CfF$_3$ [1318].

Tetrafluorides and tetravalent oxyfluorides

Generally, the actinide tetrafluorides, especially **ThF$_4$**, **PaF$_4$** and **UF$_4$** are generated by fluorination of the dioxides with SF$_4$, HF or F$_2$ (Scheme 10.3) [1324, 1325]. UF$_4$ is also prepared by reduction of UF$_6$ with S$_8$ [1326]. The other tetrafluorides **NpF$_4$**, **PuF$_4$**, **AmF$_4$**, **CmF$_4$**, **BkF$_4$** and **CfF$_4$** are conveniently prepared by fluorination of the trifluorides with F$_2$ or precipitation from An^{4+}-containing solutions with HF (Scheme 10.3) [1303, 1327–1330]. Einsteinium tetrafluoride **EsF$_4$** has been identified in the gas phase, using thermochromatographic methods [1331]. All the actinide tetrafluorides are chemically quite stable, high-melting solids, which are sparingly soluble in aqueous systems [1303, 1332]. Their structure is described by the UF$_4$ lattice, where eight fluorine atoms are coordinated around the metal center in form of a slightly distorted square antiprism [1330, 1333, 1334]. In the vapor phase, monomeric molecules are present, where ThF$_4$, UF$_4$ and probably also all the other actinide tetrafluorides have a tetrahedral geometry. The molecular structure has been revealed by infrared spectroscopy of matrix-isolated samples and supported by theoretical calculations [1285, 1286, 1335, 1336]. Furthermore, it has been calculated, that ThF$_4$ has a rather small inversion barrier of about 25 kJ mol^{-1} [172].

$$AnF_3$$
$$(An = Np\ to\ Cf)$$

$$AnO_2 \qquad\qquad\qquad\qquad\qquad\qquad UF_6$$

HF, SF$_4$ or F$_2$ $\qquad$ F$_2$ $\qquad$ S$_8$

$$An^{4+} \xrightarrow{\quad HF \quad} AnF_4 \xrightarrow[\text{(An = U)}]{\text{liq. NH}_3} [UF_4(NH_3)_4] \cdot NH_3$$
$$(An = Np\ to\ Cf)$$

F$^-$

$$[AnF_5]^-,\ [AnF_6]^{2-},$$
$$[AnF_7]^{3-}\ or\ [AnF_8]^{4-}$$

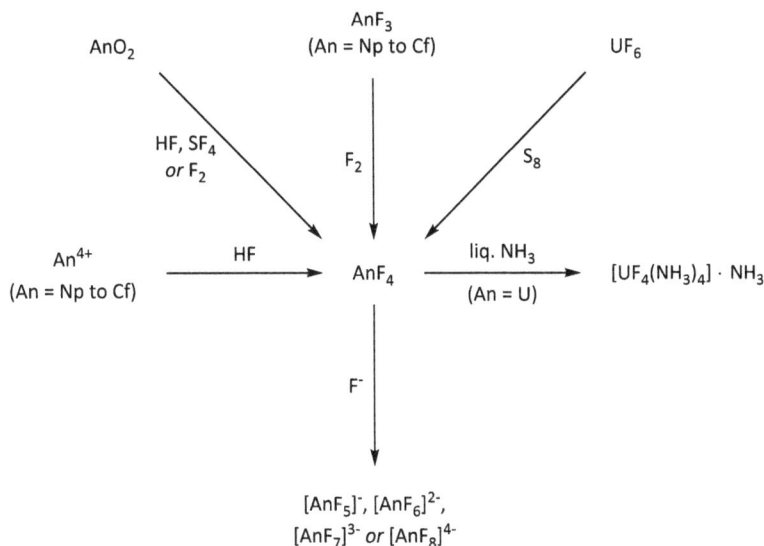

Scheme 10.3: Preparation and selected reactions of actinide tetrafluorides AnF$_4$ (An = Th to Cf).

In the series of actinide tetrafluorides, only PaF$_4$, UF$_4$, NpF$_4$ and PuF$_4$ are oxidized by F$_2$ or O$_2$F$_2$ [526, 1337–1339]. The different behavior against fluorinating agents, such as F$_2$, NF$_3$ or ClF$_3$ can potentially be used for the separation of the different actinides, for example, in nuclear fuel recycling [274, 525]. The actinide tetrafluorides generally tend to form hydrates, which can in some cases be transformed to the anhydrous compounds upon heating. However, heating of UF$_4$ hydrates to 300 °C yields the anhydrous tetrafluoride, whereas PuF$_4 \cdot n$ H$_2$O is reduced to PuF$_3$ under the same conditions [1340]. The actinide tetrafluorides also differ significantly in their volatility. ThF$_4$, UF$_4$, NpF$_4$ and PuF$_4$ can be sublimed, whereas PaF$_4$ is nonvolatile and AmF$_4$ as well as CmF$_4$ decompose to the trifluorides prior to sublimation [1264, 1265, 1341–1344]. In an aqueous solution and in the absence of oxidizable material, AmF$_4$ and CmF$_4$ can be stabilized by high fluoride concentrations, resulting in the formation of fluorometallate complexes [1345]. The pentafluorothorate(IV) anion [ThF$_5$]$^-$ has been identified in the gas phase and under matrix conditions. However, it is believed, that this species might also be obtained in condensed phase, if an appropriate, stabilizing cation is present [1285, 1346]. In aqueous fluoride-containing solutions, mainly ThF$_4$ and [ThF$_3$]$^+$ are observed, at high F$^-$ concentrations, there are also hints for the existence of fluoroanions, such as [ThF$_5$]$^-$ or [ThF$_6$]$^{2-}$ [1347]. However, melting NaF and ThF$_4$ in a 1:2-ratio yields the salt Na[Th$_2$F$_9$], where the structure of the anion in the solid state has been found to be practically identical to that of U$_2$F$_9$ [1348]. If a mixture of stoichiometric amounts of AnF$_4$ and alkali metal- or ammonium fluoride MF is heated, complex salts with the formulas M[AnF$_5$], M$_2$[AnF$_6$], M$_3$[AnF$_7$] and M$_4$[AnF$_8$] are generated, along with several higher aggregates (Scheme 10.3) [1274,

1277, 1349–1353]. In these compounds, the actinides are mainly 8-, 9- and 10-fold coordinated, resulting in partly complicated structures, where anions with a similar composition often have different geometries. For example, thorium is found to be 9-fold coordinated in $(NH_4)_4[ThF_8]$, which contains chains of anionic, F-bridged $[ThF_8]$-units [1354]. In contrast, discrete square-antiprismatic anions are found in $(NH_4)_4[UF_8]$ and $(NH_4)_4[AmF_8]$ [1355]. The ammonium salt $(NH_4)_4[UF_8]$ is prepared by treatment of UO_2 with four equivalents of $NH_4[HF_2]$ [1356]. UF_4 does not show any fluoride-donor properties, as it does not react with the strong fluoride acceptor SbF_5 [1332]. Upon reaction with excess $TiCl_3$ in liquid ammonia, UF_4 reacts to form the complex $[UF(NH_3)_8]Cl_3 \cdot 3.5\ NH_3$. Therein, the geometry around the nine-fold coordinated uranium center is described by a distorted, three-fold capped trigonal prism [1357]. If UF_4 is dissolved in liquid ammonia, the ammoniate $[UF_4(NH_3)_4] \cdot NH_3$ is generated, where the coordination environment of the uranium center is described by a distorted square antiprism (Scheme 10.3) [1326].

In the condensed phase, the only known oxyfluoride with the actinide in oxidation state +IV is **ThOF$_2$**. It is a white solid, which is generated from ThO_2 and ThF_4 at 900 °C [1358]. Upon vaporization, it decomposes back to the starting materials [1359]. In the solid state, the structure of $ThOF_2$ is described by the LaF_3 structure type [1192]. Under matrix conditions, $ThOF_2$ and **UOF$_2$** have been observed, when the respective metal atoms are reacted with OF_2. Based on vibrational spectroscopy, a pyramidal geometry has been assigned to both of the molecules [1360].

Pentafluorides and pentavalent oxyfluorides

Actinide pentafluorides are known for Pa, U and Np, in case of Pu, a stable pentavalent species only exists in form of anionic fluorocomplexes. White **PaF$_5$** and white-blue **UF$_5$** are formed upon reaction of the tetrafluorides with F_2 at 700 °C and 450 °C, respectively (Scheme 10.4) [1338]. The oxidation of UF_4 is also achieved by the xenon fluorides XeF_2, XeF_4 and XeF_6, where the resulting UF_5 is always accompanied by some minor amounts of UF_6 [1092]. The treatment of water-containing Pa_2O_5 with HF vapor at 60 °C results in the formation of the hydrate $PaF_5 \cdot H_2O$, in contrast, the dihydrate $PaF_5 \cdot 2\ H_2O$ is obtained, when a solution of $Pa_2O_5 \cdot nH_2O$ is evaporated from hydrofluoric acid at 110 °C. Thermal decomposition of the hydrates yields oxyfluorides, for example, Pa_2OF_8 [1341, 1361]. PaF_5 is furthermore accessible by fluorination of $PaCl_5$ with F_2 (Scheme 10.4) [1328, 1362]. The certainly best investigated actinide pentafluoride UF_5 can also be prepared by several other methods, mainly reduction of UF_6. The latter can be realized with a mixture of H_2 and HF, silicon powder in anhydrous HF, by UV-irradiation in presence of H_2, by treatment with HBr in HF or upon reaction with Me_3SiX (X = Cl, Br) [1363–1367]. The best yields are obtained upon reduction by CO under UV-irradiation and the reaction of UF_6 with I_2 in IF_5 [1326, 1368–1372]. Both reactions are performed at room temperature; the latter can similarly be used for the generation of **NpF$_5$** (Scheme 10.4) [1328]. Reduction of UF_6 or NpF_6 with PF_3 in anhydrous

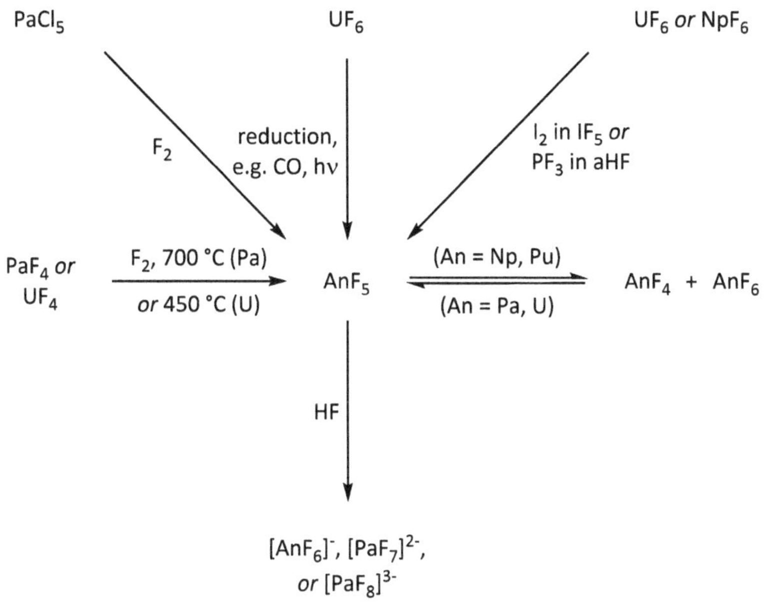

PaCl$_5$ UF$_6$ UF$_6$ *or* NpF$_6$

F$_2$ reduction, e.g. CO, hν I$_2$ in IF$_5$ *or* PF$_3$ in aHF

$$\text{PaF}_4\text{ or UF}_4 \quad \xrightarrow[\text{or 450 °C (U)}]{\text{F}_2, 700\text{ °C (Pa)}} \quad \text{AnF}_5 \quad \underset{(\text{An = Pa, U})}{\overset{(\text{An = Np, Pu})}{\rightleftharpoons}} \quad \text{AnF}_4 + \text{AnF}_6$$

HF

$$[\text{AnF}_6]^-, [\text{PaF}_7]^{2-}, \text{ or } [\text{PaF}_8]^{3-}$$

Scheme 10.4: Preparation and selected reactions of actinide pentafluorides AnF$_5$ (Ln = Pa, U, Np).

HF also gives rise to the pentafluorides, whereas PuF$_6$ is directly transformed to PuF$_4$, which is due to the instability of **PuF$_5$** [1373, 1374]. Another preparative access to NpF$_5$ is the treatment of [NpF$_6$]$^-$-containing solutions with BF$_3$. However, the instability of the pentafluoride with respect to NpF$_4$ and NpF$_6$ renders the preparation of pure NpF$_5$ difficult. This behavior contrasts that of the respective uranium analogues, where the pentafluoride is preferentially formed in UF$_4$/UF$_6$-systems [1375, 1376]. UF$_5$ exists in form of an α- and a β-modification. The latter is conveniently obtained by reduction of UF$_6$ with HBr in anhydrous HF. α-UF$_5$ is generated by heating the β-modification in presence of UF$_6$ [1366]. Low temperature β-UF$_5$ is stable up to 125 °C and forms a three-dimensional structure, where uranium has the coordination number eight. The coordination geometry is described by an intermediate between a square antiprism and a dodecahedron. The α-modification forms chains of *trans*-F-bridged [UF$_6$]-units [1377–1379]. PaF$_5$ is isomorphic to β-UF$_5$, whereas NpF$_5$ adapts the structure of α-UF$_5$ [1328, 1373]. For the ground state of the UF$_5$ molecule, a square-pyramidal structure has been proposed, based on vibrational spectra of the matrix-isolated compound [1380, 1381]. However, quantum-chemical calculations predict a small energetic difference between C$_{4v}$- and D$_{3h}$-geometry, thus proposing a fluxional behavior [1382]. UF$_5$ is well soluble in nitriles and other organic solvents with high polarity, such as dimethyl sulphoxide or N,N-dimethylformamide. In these solvents, the dissociation into [UF$_4$]$^+$ cations and [UF$_6$]$^-$ anions has been observed, further allowing the isolation of solvent-incorporating salts of the type [UF$_4$L$_x$][UF$_6$]. In alcohols, UF$_5$ yields mixed fluoroalkoxides, whereas it degrades ketones, amines and hydrocarbons. However, it is insoluble in solvents like SO_2, CS_2 or fluorocarbons [1383]. In general, the solubility of the actinide pentafluorides is very different. As mentioned before, UF_5 readily dissolves in MeCN, whereas PaF_5 reacts to form a poorly soluble complex. In contrast, NpF_5 does not react or dissolve at all. A similar behavior is found for the reactivity toward Ph_3PO, where UF_5 and PaF_5 form isostructural complexes of the composition $MF_5 \cdot 2 Ph_3PO$, but NpF_5 remains unchanged [1328, 1369]. In case of uranium, complexes are known with XeF_6 and MF_5 (M = As, Sb, Nb, Ta). The pentafluorides VF_5 and BiF_5 do not form adducts, but fluorinate UF_5 to UF_6. The aforementioned coordination-compounds consist of covalently F-bridged species [469, 1092, 1384]. If UF_5 is reacted with anhydrous HCN, a one-dimensional coordination polymer is formed, which has been characterized by X-ray crystallography. The coordination geometry around the U-center is described by a trigonal dodecahedron [1385]. UF_5 is slowly reduced to UF_4 upon reaction with the trifluorides PF_3 or AsF_3, but not with SbF_3 [1332]. In case of PaF_5, reduction to the tetrafluoride only occurs with PF_3, but not with AsF_3 [1362]. In contact with moisture, UF_5 is rapidly hydrolyzed to yield UF_4 and UO_2F_2, however, PaF_5 and UF_5 are very stable against hydrolysis in hydrofluoric acid. UF_5 itself virtually does not dissolve in HF, but it readily forms soluble hydrates, for example, large, blue crystals of $HUF_6 \cdot 2.5 H_2O$, which can be isolated upon cooling to −10 °C. A lower hydrate of the composition $HUF_6 \cdot 1.25 H_2O$ is also known, whereas anhydrous HUF_6 is unknown [1376, 1386]. The uranium(V) fluorosulphate derivatives $UF_{5-n}(SO_3F)_n$ are accessible from UF_6 and SO_3 or UF_5 and $S_2O_6F_2$, respectively [1387, 1388]. A pentafluoride species of thorium with the overall formula "**ThF₅**" has been detected under matrix conditions. However, this species is not a real pentafluoride, but an ionic complex of the composition $[ThF_3]^+[F_2]^-$, as quantum-chemical calculations have shown. Thus, it contains a side-bound $[F_2]^-$-unit [1285, 1286].

Heating of a mixture of UF_4 and UF_6 to 100 °C results in the formation of UF_5 along with the mixed-valent fluorides U_2F_9 (black), U_4F_{17} (black) and U_5F_{22}. The properties of the mixed-valent and thermally unstable U_2F_9 and U_4F_{17} can be compared to the pentafluoride, as they tend to decompose to UF_4 and UF_6 [1332]. Similar mixed valent compounds are formed in case of Pa and Pu, which result in Pa_2F_9 (black), Pa_4F_{17} and Pu_4F_{17} (red). In black U_2F_9, both of the uranium atoms are crystallographically identical, being 9-fold coordinated, where three-fold capped trigonal prismatic polyhedra are symmetrically linked by F-bridges [1379, 1389]. Reactions of PaF_5, UF_5, NpF_5 or PuF_5 with hydrofluoric acid or reduction of the hexafluorides with NO, NO_2, NOF or NO_2F lead to the precipitation of fluorocomplexes $[AnF_6]^-$, $[PaF_7]^{2-}$ and $[PaF_8]^{3-}$ (Scheme 10.4) [1386, 1390, 1391]. The complex anions $[AnF_7]^{2-}$ (An = U, Np, Pu) and $[AnF_8]^{3-}$ (An = U, Np) are furthermore generated upon fluorination of a mixture of alkali metal fluoride MF and AnF_4 with F_2. The aforementioned complexes are also generated by the reaction of UF_6, NpF_6 or PuF_6 with fluorides, however, the Np and Pu compounds are only accessible by solid-phase reactions [1338, 1370, 1371]. Finally, the direct combination of UF_5 with M^IF (M^I = alkali metal, NH_4, Ag, Tl) gives rise to the respective hexafluorouranate(V) complexes, in case of AgF, the octafluorouranate

$[UF_8]^{3-}$ is additionally generated [1392–1396]. Similar to the behavior of the neutral pentafluorides, $[UF_6]^-$ is stable in anhydrous HF, whereas $[NpF_6]^-$ readily decomposes to NpF_4 and NpF_6 [1338]. The coordination geometry of the anions is strongly dictated by the metal and the counter-ion. Interestingly, in $Na_3[AnF_8]$ the actinide (An = Pa, U, Np) is in the center of an almost regular cube of eight F-atoms [1397]. In most of the other salts, except $Cs[UF]_6$, the anions are mainly connected by F-bridges to form chains. A few hydrated complexes of uranium are known as well, that is, $M^{II}[U_2F_{12}]$ (M^{II} = Co, Ni, Cu) [1398]. The $[U_2F_{12}]^{2-}$ anion shows seven-coordinated U-centers, where two mono-capped trigonal prisms are linked by a common edge, as has been shown by X-ray crystallographic investigations on $Sr[U_2F_{12}]$ [1399]. Based on the $[UF_6]^-$ anion, ionic liquids have been investigated and reported. They could possibly be useful for chemical or electrochemical processes in nuclear chemistry as well as in isotope enrichment [1400].

There are several oxyfluorides of actinide compounds in oxidation state +V. Among them, **$NpOF_3$** is the only known oxytrifluoride. It has a green color and is formed upon reaction of Np_2O_5 with anhydrous HF at 40 °C. Initially, a hydrate is obtained, which can be transformed to anhydrous $NpOF_3$ upon heating to 150–200 °C [1401]. $PaOF_3$ has been proposed to be intermediately formed upon fluorination of Pa_2O_5, however, it has never been isolated or characterized, nor has its existence been confirmed by thermochromatography [1342, 1402]. The dioxyfluorides **AnO_2F** (An = Pa, U, Np) are isostructural and can be prepared on several synthetic routes, mostly based on careful hydrolysis of the pentafluorides or incomplete fluorination of the respective oxides. For example, PaO_2F is formed upon reaction of Pa_2O_5-hydrate with HF at 40–60 °C as mono- or dihydrate, which is transformed to Pa_2OF_8 when heated to 140 °C. If Pa_2OF_8 is heated to 250–290 °C on air, it decomposes to PaO_2F. In turn, PaO_2F yields Pa_3OF_7 at 500–600 °C, which finally regenerates the starting material Pa_2O_5 at 650 °C [1361]. UO_2F and U_2OF_8 decompose at 300 °C *in vacuo* to give UF_4, UF_6, and UO_2F_2.

Hexafluorides

The hexafluorides **UF_6** (colorless; mp.: 64 °C (1.5 bar); subl.: 56.5 °C) **NpF_6** (orange; mp.: 54.9 °C; bp.: 55.2 °C) and **PuF_6** (red-brown; mp.: 50.7 °C; bp.: 62.3 °C) are low-melting, volatile solids which are formed upon fluorination of the tri- or tetrafluorides with F_2 at 400–750 °C or with PtF_6 at 25 °C (Scheme 10.5). UF_6 may also be prepared from UF_4 by photosynthesis in liquid F_2 or by fluorination of UO_3 with SF_4 [1324, 1403, 1404]. A mild and efficient room temperature synthesis of pure UF_6 has been described, where O_2F_2 is used to fluorinate U_3O_8 [1405]. On industrial scale, UF_6 is prepared from UO_2 in a two-step sequence. At first, UF_4 is generated by treatment of UO_2 with HF, subsequently, oxidation with F_2 is conducted (Scheme 10.5). A direct reaction of UO_2 or UO_3 with F_2 is inefficient, as it requires a larger amount of elemental F_2 [7]. PuF_6 has been synthesized starting from a variety of plutonium pre-

UO_2 $\qquad\qquad$ UO_3 $\qquad\qquad$ PuF_4 + F_2

HF, *then*
F_2, hv

SF_4

Δ *or*
irradiation

AnF_3 *or*
AnF_4 $\xrightarrow[\text{or PtF}_6,\ 25\ °C]{F_2,\ 450\text{-}700\ °C}$ AnF_6 $\xrightarrow[\text{(An = U)}]{Br_2,\ CH_3CN}$ $[Br(CH_3CN)_3][UF_6]$

NO $\qquad\qquad$ F$^-$ $\qquad\qquad$ F$^-$

$[NO][AnF_6]$ $\qquad$ $[AnF_6]^-$ + F_2 $\qquad$ $[UF_7]^-$, $[UF_8]^{2-}$,
$\qquad\qquad\qquad$ (An = Np, Pu) $\qquad\quad$ $[UF_9]^{3-}$ *or* $[UF_{10}]^{4-}$

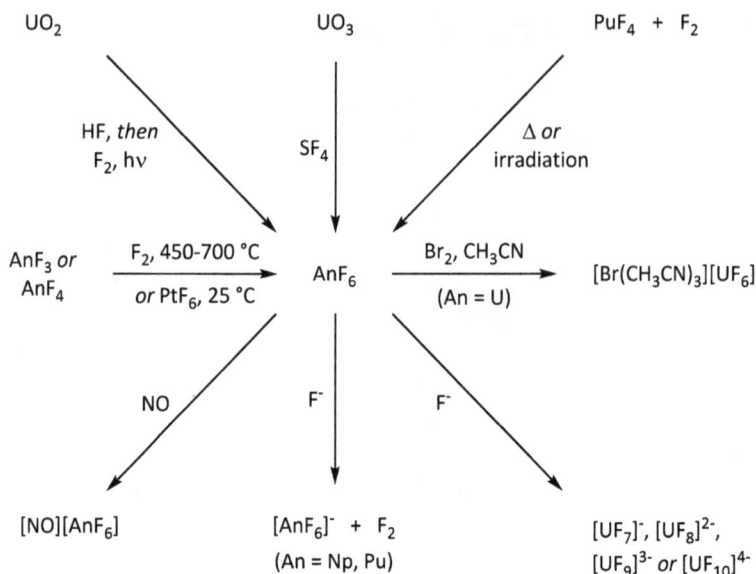

Scheme 10.5: Preparation and selected reactions of actinide hexafluorides AnF_6 (Ln = U, Np, Pu).

cursors, which have been treated with KrF_2 at ambient temperature or with O_2F_2 at low temperature [1406, 1407]. PuF_6 is also accessible from mixtures of PuF_4 and F_2, which either need to be heated or irradiated (Scheme 10.5). In turn, the equilibrium can be shifted, so that it readily decomposes back to the starting materials upon heating or impact of γ-radiation or UV-light [1408–1410]. The three aforementioned hexafluorides are the only ones which have been unambiguously characterized. Clearly, UF_6 is most extensively described, due to its use in nuclear fuel industry. The stability of the hexafluorides strongly decreases from UF_6 to PuF_6, so that the latter one can only be stored at low temperatures. Consequently, AmF_6 or even CmF_6 have not been accessible so far, as the oxidation power of KrF_2 has not been sufficient to produce AmF_6 from AmO_2 [1407]. Furthermore, it has not even been possible to observe AmF_6 by mass-spectrometry in the gas phase [1411]. The actinide hexafluorides have molecular orthorhombic lattices containing regular $[AnF_6]$-octahedra. Due to the f-orbital splitting in an octahedral field, Jahn–Teller distortions do not have to be assumed [732, 1412, 1413]. They are all strong oxidizing and fluorinating agents with increasing oxidation potential from U to Pu. For example, UF_6 and PuF_6 react with SF_4 under formation of the actinide tetrafluorides and SF_6 [1324]. The electron affinities of NpF_6 and PuF_6 are unknown, or at least not described in the open literature [732]. However, experimental data for UF_6 range from 4.0 to 5.3 eV, which is in the same order as for ReF_6 [1414–1417]. Thus, NpF_6 and PuF_6 might have electron affinities which are in the region of IrF_6 and PtF_6, respectively [732].

Despite its volatility and radioactivity, which render it rather difficult in handling, UF_6 is a versatile oxidant. For example, it oxidizes the metals Cu, Cd and Tl at 25 °C in acetonitrile solution to yield Cu(II)-, Cd(II)- and Tl(III)-species, respectively [1418]. In the same way, Br_2 is oxidized by UF_6 in CH_3CN to yield $[Br(CH_3CN)_3]^+[UF_6]^-$, indicating that the oxidation power of UF_6 can be compared to that of OsF_6 (Scheme 10.5) [1419]. Apart from that, the oxidizing properties of UF_6 have also been exploited in organic synthesis [1420, 1421]. In a system of solid UF_5 and gaseous UF_6 it has been found, that fluoride-substituents are readily exchanged. This behavior has been considered to be potentially advantageous for isotope enrichment of uranium, as the valuable isotope ^{235}U can be transferred from solid UF_5 to volatile UF_6, which can then enter the typical purification procedures by gas phase diffusion [1422, 1423]. The hexafluorides NpF_6 and PuF_6 readily react with traces of water, so that the oxyfluorides $AnOF_4$ and AnO_2F_2 (An = Np, Pu) are obtained upon careful hydrolysis [1424–1429]. If NpF_6 or PuF_6 are reacted with alkali metal fluorides, the hexafluorides are transformed to pentavalent, anionic $[AnF_6]^-$ (An = Np, Pu) and F_2 (Scheme 10.5) [1424]. Both, NpF_6 and PuF_6 are not known to form hexavalent fluorometallates of the type $[AnF_7]^-$ or $[AnF_8]^{2-}$. However, UF_6 is capable of forming fluorouranates(VI) of the type $M[UF_7]$, $M_2[UF_8]$, $M_3[UF_9]$ and $M_4[UF_{10}]$ (Scheme 10.5) [1418, 1430–1433]. Unexpectedly, the silver derivative $Ag_2[UF_8]$ has oxidizing properties, as it liberates O_2 upon contact with water [1434]. All three hexafluorides are reduced by NO under formation of the nitrosyl hexafluorometallates(V) $NO[AnF_6]$ (Scheme 10.5) [1435]. The same result is observed for the reactions of NpF_6 and PuF_6 with NOF, but PuF_6 also yields minor amounts of NOF_3 under these conditions [1390]. In the series of the transition metal hexafluorides, this observation can only be made, when IrF_6 is used as oxidant, which indicates, that the oxidation powers of IrF_6 and PuF_6 must be comparable [342]. It is also possible to substitute the fluoride substituents in UF_6 by other ligands like -Cl, $-OTeF_5$ or $-OCH_3$ [732]. UF_6 chemistry can also be performed in anhydrous, liquid ammonia, as for example the reduced uranium(IV) products $[UF_7(NH_3)]^{3-}$ and $UF_4(NH_3)_4$ have been identified and structurally characterized [1436]. In both compounds, the uranium atom is eight-fold coordinated, in the first case as bicapped trigonal-prism, in the latter composition as square antiprism.

There exist oxyfluoride derivatives **AnOF$_4$** (An = U, Np, Pu) and **AnO$_2$F$_2$** (An = U, Np, Pu, Am) for all the three actinide hexafluorides. The oxytetrafluorides UOF_4, $NpOF_4$ and $PuOF_4$ are accessible by careful hydrolysis of the hexafluorides in HF [1424–1428]. Apart from that, UOF_4 also results from the hydrolysis reaction of UF_6 with quartz-wool, treatment of UF_6 with B_2O_3 or fluorination of UO_3 with SeF_4 via intermediary UO_2F_2 [688, 1437, 1438]. In the solid state, the oxytetrafluorides are connected via F-bridges, so that the actinide centers show a pentagonal-bipyramidal coordination geometry. The same structural behavior is observed for the pentafluorides AnF_5. In case of uranium, there are two different modifications, α- and β-UOF_4 [1424, 1425, 1437, 1439–1441]. By quantum-chemical calculations, it has been found, that C_{3v}-symmetry is energetically slightly favored over C_{2v} for the molecular ground-state symmetry of UOF_4. Furthermore, the calculations predicted a triple-bond character for the U-O-bond [1442, 1443]. UOF_4 is a yellow to orange, hygroscopic solid, which, in contrast to many other MOF_4-compounds, is nonvolatile below its degradation point. However, its thermal decomposition is interesting, as it directly forms UO_2F_2 and UF_6 at temperatures of 500 °C, whereas intermediate $U_2O_3F_6$ is observed at lower temperatures of about 290 °C. The latter, yellow compound can be reformulated as $UOF_4 \cdot UO_2F_2$ and is fully converted to UF_6 and UO_2F_2 upon heating to 380 °C. However, the spectroscopic properties and its chemical behavior indicate that it rather represents an own species than being a loose adduct of UOF_4 and UO_2F_2 [1427, 1444]. In general, the stability of the oxytetrafluorides decreases in the series $UOF_4 > NpOF_4 > PuOF_4$ [1425]. Attempts to oxidize $NpOF_4$ by KrF_2 in anhydrous HF have not resulted in the formation of a Np(VII)-species, but partially regenerated NpF_6 [1424]. Nevertheless, thermochromatographic studies indicated that heptavalent compounds of neptunium and plutonium, such as NpF_7, NpO_3F or PuO_3F should be accessible in the gas phase [1402]. Quantum-chemical calculations support the theoretical existence of Np(VII), Pu(VII) and probably even Pu(VIII) species [1445, 1446]. However, it appears to be very challenging to obtain these species. $PuOF_4$ is a dark, chocolate-brown solid, which is immediately hydrolyzed in moist air. However, it is stable at room temperature under inert conditions, as long as there is either some PuF_6 in the sample, or if anhydrous HF is absent. Otherwise, $PuOF_4$ will decompose to PuO_2F_2 and PuF_6 [1425]. UOF_4 has been shown to be a fluoride acceptor, as it forms the pentafluorooxyuranate(VI) complexes $NH_4[UOF_5]$ and $K[UOF_5]$ upon reaction with NH_4F and KF, respectively. The treatment of UOF_4 with other alkali metal fluorides, such as CsF, as well as with NOF does not yield pure or stable products, in case of nitrosyl fluoride, the hexafluorouranate(V) complex $NO[UF_6]$ is finally obtained as decomposition product [1447]. Pentafluorooxyuranate(VI) salts are also accessible by other synthetic methods, as for example, $[Ph_4P][UOF_5]$ is prepared by fluorination of the respective chloro-complex with HF [1448].

The dioxydifluoride compounds **AnO_2F_2** are generated, if $[AnO_2]^{2+}$-containing derivatives are treated with HF or by hydrolysis of the hexafluorides [1322, 1406, 1449–1452]. For the well explored UO_2F_2, a few more synthetic accesses have been found, for example, it can be prepared by treatment of UO_3 with anhydrous HF at 350–500 °C in a Ni-reactor or by oxidation of UF_4 with dry O_2 at 600–900 °C [1376, 1453]. The treatment of uranium oxides with NF_3 also results in intermediary formation of UO_2F_2, however, further fluorination yields UF_6 [1454]. NpO_2F_2 is analogously generated by treatment of $NpO_3 \cdot H_2O$ with anhydrous HF or direct fluorination of $NpO_3 \cdot H_2O$ or Np_2O_5 [1401]. The structure of uranyl fluoride UO_2F_2 and the isomorphous NpO_2F_2, PuO_2F_2 and AmO_2F_2 is rather complicated, consisting of layers of four different types of ordered domains [1449, 1452, 1453, 1455]. At −40 °C, UO_2F_2 reacts with liquid ammonia to form the yellow complex $[UO_2F_2(NH_3)_3]_2 \cdot 2\,NH_3$, which decomposes to yield stable $[UO_2F_2(NH_3)_3]$ at room temperature [1456]. NpO_2F_2 is a pale pink, hygroscopic solid, which is stable on air. Like UO_2F_2 it readily dissolves in water and aqueous mineral acids, however, it is insoluble in anhydrous organic solvents [1401]. Upon sublimation, NpO_2F_2 decomposes to NpO_2, NpF_4 and O_2 [1457]. In aqueous fluoride-containing solutions of $[UO_2]^{2+}$, the ions $[UO_2F]^+$, $[UO_2F_3]^-$ and $[UO_2F_4]^{2-}$ can be observed, along with neutral UO_2F_2 [1458]. The respective Np-complexes are weaker; so far, only cationic $[NpO_2F]^+$ and neutral NpO_2F_2 have been found in fluoride-containing systems of $[NpO_2]^{2+}$ [1451]. The anions $[UO_2F_3]^-$ and $[UO_2F_4]^{2-}$ can further be observed in the gas phase. The latter anion also forms solvates with one or two molecules of acetonitrile or water [1459, 1460]. If U_3O_8 is treated with a 2:1-mixture of $NH_4[HF_2]$ and NH_4F, the complex salt $(NH_4)_3[UO_2F_5] \cdot H_2O$ is generated, which is thermally decomposed to $UO_2F_2 \cdot H_2O$ and, finally, anhydrous UO_2F_2 [1356]. The fluoride substituents of UO_2F_2 can be substituted by azide-groups, resulting in the formation of neutral $UO_2(N_3)_2$ as a mononuclear MeCN adduct or a dinuclear complex with 2,2′-bipyridine (bpy). By further treatment with ionic azides, the azidouranate(VI) anions $[(bpy)UO_2(N_3)_3]^-$ and $[UO_2(N_3)_4]^{2-}$ are also accessible as well as the azido-bridged $[(UO_2)_2(N_3)_8]^{4-}$ [1461].

10.3.2 Study questions

(I)   Which stable actinide trifluorides are known? How are they typically prepared?

(II)   How can one prepare actinide tetrafluorides? Comment on their reactivity toward strongly oxidizing reagents, for example, F_2 or O_2F_2.

(III) For which actinides exist stable pentafluorides? Is there a trend for the stability of pentavalent actinide species? Which consequences does this have on the preparation of AnF_5?

(IV)   Which actinide hexafluorides are known? Are they stable? Which hexafluoride is relevant for nuclear fuel industry and how is it prepared on industrial scale?

(V)   What happens, if the hexafluorides UF_6, NpF_6 and PuF_6 are reacted with alkali metal fluorides? Is there a similar reaction with NO?

10.4 Group 4: Titanium, zirconium, hafnium

Expectedly, the fluorides of group 4 mainly exist in form of tetrafluorides. There are also a few low-valent compounds, which are strong reducing agents. As a general trend, the group 4 fluorides readily form fluorocomplexes.

10.4.1 Titanium

Low-valent fluorides

Binary fluorides of titanium in an oxidation state smaller than +III are rather unusual in condensed state. However, they have been comparably well studied in the gas

phase. Titanium monofluoride **TiF** can be obtained by either flash heating of powdered TiF_3 or TiF_4 samples or by treatment of Ti-metal with SF_6, CF_4 or AlF_3 under
the same conditions [1462, 1463]. It is also observed, when activated helium is reacted
with TiF_4 [1464, 1465]. The absorption and emission spectra of TiF have been mea-
sured and studied [1462–1466]. Apart from that, the low-lying electronic states have
been investigated by means of quantum-chemical calculations, revealing a highly
ionic character of the Ti-F bond [1467, 1468]. The situation is comparable for titanium
difluoride **TiF₂**. It is accessible in the gaseous state, for example, by reduction of
TiF_3 with Ti-metal [1469]. The same reaction can be performed in condensed state at
high pressure (5 GPa). Under these conditions, TiF_2 crystallizes in a distorted fluorite
structure [1470]. It can further be prepared by matrix techniques, when Ti-atoms are
fluorinated by diluted F_2 in argon. In contrast to earlier results, recent vibrational
spectra of matrix-isolated TiF_2 indicate, that it does not have a substantially bent
structure, since it shows bands which are in line with a rather linear geometry. These
findings are also supported by theoretical calculations [172, 1471, 1472]. Apart from the
molecular ground-state geometry, the electronic structure of TiF_2 has been explored
by quantum chemistry [1473]. Solvated TiF_2 is reported to be accessible by alkyla-
tive reduction of TiF_4 in THF-solution, using organometallic reagents, such as ^nBuLi,
$[Al(CH_3)_3]_2$ or Grignard reagents. These reactions yield a black product, which can
probably be described by the formula $[TiF_2(THF)_2]$. However, the initial report on this
compound did not include any comment on its purity or stability [1474].

Titanium trifluoride
Titanium trifluoride **TiF₃** is formed by the reaction of prehydrogenated titanium with
HF at 700 °C and can be purified by sublimation in high vacuum [1475]. The direct treatment of Ti-metal with HF at 200–225 °C also gives rise to TiF_3 (Scheme 10.6) [204]. Apart
from that, the pure compound is generated by thermal decomposition of $NH_4[TiF_4]$
under H_2- or argon-atmosphere at 600–650 °C [1476]. Another synthetic method toward TiF_3 is the heating of CuF_2 in a Ti-vessel at 800 °C for 8 days (Scheme 10.6) [1477].
In the solid state, TiF_3 has a structure similar to VF_3, thus adapting a distorted ReO_3
lattice [1477, 1478]. TiF_3 undergoes a phase transition from a cubic to a rhombohedral structure at about 370 K [1479]. ESR-spectra of matrix-isolated TiF_3 point to the
existence of a planar molecular structure, which is in line with theoretical calculations [1480, 1481]. TiF_3 is a deep-blue to violet, crystalline and paramagnetic solid
(subl.: 930 °C), which is stable on air and resistant against acids and bases. It is insoluble in water and alcohols. Above 950 °C, TiF_3 disproportionates to yield Ti and TiF_4
(Scheme 10.6) [1475]. Upon reduction of titanocene difluoride Cp_2TiF_2 (Cp = cyclopen-
tadienyl, $[C_5H_5]^-$) with activated aluminium, the dimeric Ti(III) derivative $[Cp_2TiF]_2$ is
formed, which is bridged by two F-atoms [1482]. Upon reaction with fluorides M^IF (M^I =
alkali metal, NH_4), TiF_3 forms numerous fluorocomplexes of the general stoichiometry

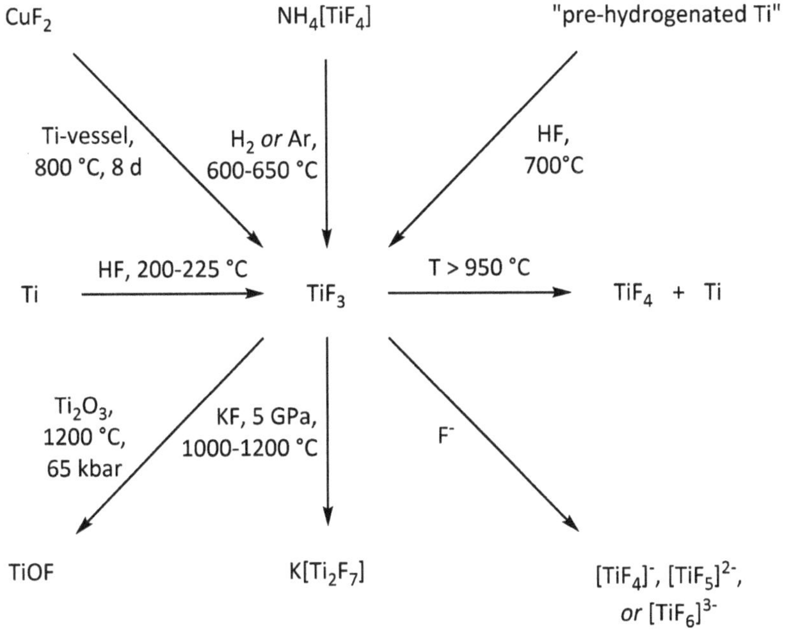

Scheme 10.6: Preparation and selected reactions of TiF_3.

$M^I[TiF_4]$ [1483, 1484]. Apart from the tetrafluorotitanate(III) salts, penta- and hexafluorotitanates are known as well (Scheme 10.6). In $(NH_4)_2[TiF_5]$ and $Ca[TiF_5]$, the anion has a chain structure, where the units are linked by *trans*-F-bridges. Furthermore, the hexafluorotitanates $M_3[TiF_6]$ (M = Li, Na, K, Rb, NH_4, Tl); $M'_2M''[TiF_6]$ (M' = Cs, Rb, K; M'' = Na, K) and $LiM^{II}[TiF_6]$ (M^{II} = Mg, Ca, Sr, Ba, Mn) have been described [1477, 1485–1489]. The lead derivative $Pb_3[TiF_6]_2$ is also known, it is isostructural to the hexafluoroferrate(II) salt $Ba_2[FeF_6]$ [1490]. In the reaction of $(NH_4)_2[TiF_5]$ with zinc in presence of NH_4F in hydrofluoric acid, $NH_4Zn[TiF_6]$ is generated at slightly elevated temperature [1491]. Under high temperature and high-pressure conditions (1000–1200 °C, 5 GPa), interesting species can be observed in the system KF-TiF_3 (Scheme 10.6). If KF is reacted with two equivalents of TiF_3, the salt $K[Ti_2F_7]$ is generated, which is built from pentagonal-bipyramidal $[TiF_7]$-units. The bipyramids share edges with their neighbors, resulting in the formation of ribbons. The latter are linked via F-bridges between the corners of $[TiF_7]$-units to form a three-dimensional network. Under the same preparative conditions, the salt $K_2[TiF_5]$ results from a 2:1-ratio of KF-TiF_3 and contains one-dimensional chains of corner-sharing $[TiF_6]$-octahedra [1492]. Hydrolysis of $K_3[TiF_6]$ yields the oxidation products $K_2[TiF_6]$ and $K_2[TiOF_4]$, along with other compounds. In all fluoro-complexes except $K[Ti_2F_7]$, the Ti-center is octahedrally surrounded by six F-atoms. Complexes of the hexafluorotitanate(III) anion $[TiF_6]^{3-}$ usually have a purple color.

TiF_3 is widely explored as dopant for simple and complex hydrides, for example, MgH_2, $Na[AlH_4]$ or $Li[BH_4]$, as well as other compounds with hydrogen-storage capabilities, like Pt-doped carbon [1493–1497]. There are numerous publications on this field, which is due to the fact, that TiF_3 improves the reversible dehydrogenation of hydrogen storage materials. The latter process is of large interest for the research on renewable energies. Apart from that, TiF_3 has also been investigated as potential electrode material for Li-ion batteries, revealing some promising electrochemical properties. However, the beneficial electrochemical properties are accompanied by irreversible structural changes, since the initial structure of TiF_3 is not regenerated after recharging [1498, 1499]. There is also an oxyfluoride, which is derived from TiF_3. Titanium oxyfluoride **TiOF** is prepared from a stoichiometric mixture of TiF_3 and Ti_2O_3 at 1200 °C and 65 kbar (Scheme 10.6). It is a black crystalline solid which has semiconductor and antiferromagnetic properties. TiOF crystallizes in a rutile structure [1500–1502]. Quantum-chemical calculations predict a bent structure for molecular TiOF [1503].

Titanium tetrafluoride

Titanium tetrafluoride **TiF_4** is a white, hygroscopic solid (subl.: 284 °C) which is obtained upon fluorination of Ti, $TiCl_4$ or TiO_2 (Scheme 10.7) [1475, 1504]. Raman-spectroscopy reveals monomeric, tetrahedral TiF_4 molecules in the gas phase, whereas it is a F-bridged polymer of $[TiF_6]$-octahedra in the solid state [1505]. TiF_4 readily forms 1:2-complexes with N- and O-donors, for example, substituted pyridines or pyridine-1-oxides [1506–1508]. In acetonitrile, the pyridine adducts probably exist in form of $TiF_4(py)_2$, $[TiF_3(py)_3]^+$ and $[TiF_5py]^-$, as has been revealed by ^{19}F-NMR-studies. Similarly, 1:1-adducts with the bidentate N-donor ligands 2,2′-bipyridine and 1,10-phenanthroline are known, where the Ti-center is always six-fold coordinated (Scheme 10.7) [1509]. 1:2-Complexes of TiF_4 with N-heterocyclic carbene ligands have also been prepared and structurally investigated by X-ray crystallography, again revealing hexacoordination of the Ti-center [1510]. TiF_4 also yields stable adducts with very simple donor molecules, such as NH_3. For example, the ammonia complex $TiF_4 \cdot 2\,NH_3$ is stable up to 120 °C, whereas it liberates NH_3 upon further heating. The hydrolysis of the tetrafluoride yields TiO_2 and HF via intermediary $TiOF_2$, however, TiF_4 is only slowly hydrolyzed, so that it is even possible to isolate the dihydrate $TiF_4 \cdot 2\,H_2O$. Furthermore, substitution reactions can be carried out, as for example reactions with secondary amines give rise to derivatives of the type $(R_2N)TiF_3$, $(R_2N)_2TiF_2$ and $(R_2N)_3TiF$ (R = Me, Et) (Scheme 10.7). Among these compounds, $(Et_2N)_3TiF$ is the only monomeric species in solution, whereas all the other derivatives are associated [1511–1514]. The fluoro-substituents of TiF_4 can also be replaced by azide groups, resulting in the formation of either neutral $Ti(N_3)_4$ or the azidotitanate(IV) anions $[Ti(N_3)_5]^-$ and $[Ti(N_3)_6]^{2-}$. The neutral azido-species has been stabilized by the biden-

$[TiF_4(py)_2]$, $[TiF_4(bpy)]$
or $[TiF_4(phen)]$

↑

N-donor, e.g.
py, bpy or phen

TiCl$_4$ or TiO$_2$ $\xrightarrow{\text{HF}}$ TiF$_4$ $\xrightarrow{\text{TiO}_2,\ 550\text{--}800\ °C}$ TiOF$_2$

$[TiF_6]^{2-}$ ↙   HF, F⁻ ↓   640 °C ↑

$[Ti_2F_9]^-$, $[Ti_2F_{10}]^{2-}$, $[TiF_6]^{2-}$ salts $\xrightarrow[\text{then base}]{H_2O_2,\ 20\ °C}$ $[Ti(O_2)F_5]^{3-}$
or $[Ti_2F_{11}]^{3-}$

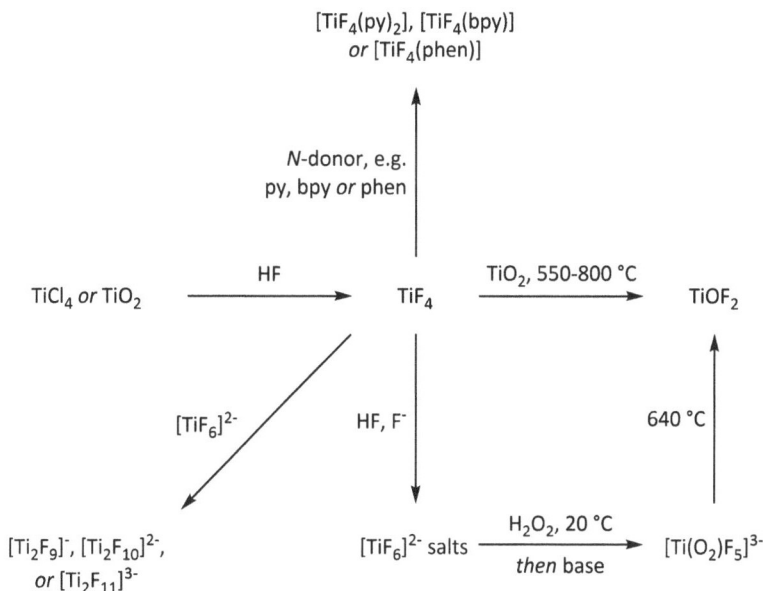

Scheme 10.7: Preparation and selected reactions of TiF$_4$.

tate N-donor ligands 2,2′-bipyridine and 1,10-phenanthroline and characterized by single-crystal X-ray analysis [1515, 1516].

Fluorotitanate(IV)-salts are generally obtained by solubilizing Ti, TiF$_4$ or TiO$_2$ in hydrofluoric acid in presence of the respective fluoride (Scheme 10.7). Moreover, they always contain the octahedral $[TiF_6]^{2-}$ anion [262, 1517–1520]. The alkali metal salts $M_2^I[TiF_6]$ (M^I = K, Rb, Cs) are only poorly soluble in water. Accordingly, they are quite stable against hydrolysis, as for example, K$_2$[TiF$_6$] only slowly reacts with moist air at temperatures up to 700 °C [1521]. Furthermore, the hexafluorotitanates(IV) are very stable in neutral aqueous solution at room temperature, however, they tend to decompose in acidic media [1522]. Upon reaction with Na[HF$_2$], Na$_2$[TiF$_6$] yields the double-salt Na[HF$_2$] · Na$_2$[TiF$_6$] [1523]. The hydrated copper(II) salt Cu[TiF$_6$] · 4 H$_2$O has also been reported [1524]. The substituted ammonium salts (NH$_4$)$_2$[TiF$_6$], $[N^nPr_2H_2]_2[TiF_6]$, $[N^iPr_2H_2]_2[TiF_6]$, $[NEt_2H_2]_2[TiF_6]$ and $[NEt_3H]_2[TiF_6]$ are accessible, if acetone-solutions of TiF$_4$ are saturated with HF and treated with the respective amine [1513, 1514, 1525]. K$_2$[TiF$_6$] is used as starting material for the electrolytic generation of Ti. The reaction of TiF$_4$ with KF and KCl at 500–700 °C yields the colorless chlorofluoro-complex K$_3$[TiF$_6$Cl]. In the TiF$_4$-[TiF$_6$]$^{2-}$ system, the ions [Ti$_2$F$_9$]⁻, [Ti$_2$F$_{10}$]$^{2-}$ and [Ti$_2$F$_{11}$]$^{3-}$ have been detected by ¹⁹F-NMR-spectroscopy in SO$_2$-solution, where [TiF$_6$]-octahedra are bridged by 1,2 or 3 F-atoms, respectively (Scheme 10.7) [266]. Solid complexes of the type M[TiF$_5$ · L] (L = Lewis base) are only generated, if the base is strong enough to prevent the formation of polymeric, F-bridged fluorotitanates [1526]. This is for example the case, if a 1:2-adduct with a N-heterocyclic carbene (NHC) of the type [$TiF_4 \cdot 2\,NHC$] is degraded in solution at room temperature, resulting in the formation of the salt [(NHC)H][$TiF_5 \cdot NHC$] [1510]. Treatment of concentrated solutions of $M_2[TiF_6]$ with H_2O_2 at 20 °C and subsequent addition of a base like NH_3 or KOH yields the very stable, water-soluble peroxo-complexes $M_3^I[Ti(O_2)F_5]$ (M^I = NH_4, K), which are isomorphic to $(NH_4)_3[NbOF_5]$ and $M_3[ZrF_7]$ (Scheme 10.7) [1527]. Structural investigations on single crystals of $(NH_4)_3[Ti(O_2)F_5]$ have revealed a distorted, cubic lattice, where the pentagonal-bipyramidal anions are statistically oriented [1528].

Titanium oxydifluoride

Titanium oxydifluoride **$TiOF_2$** is generated upon hydrolysis of TiF_4; reaction of TiO_2 with hydrofluoric acid (w = 40 %) and subsequent heating; by reaction of $TiCl_2F_2$ with Cl_2O or treatment of TiO_2 with TiF_4 at 550–800 °C (Scheme 10.7). $TiOF_2$ is a colorless powder, which is very air-stable in pure state and is only sparingly hydrolyzed by humidity [1529–1532]. It is also formed by thermal decomposition of the peroxo complex $(NH_4)_3[Ti(O_2)F_5]$ at 640 °C [1533]. Interestingly, $TiOF_2$ is accessible by treatment of the anatase modification of TiO_2 with F_2 or HF. In contrast, fluorination of the rutile phase yields pure TiF_4 [1532, 1534]. The structure of $TiOF_2$ is described by an octahedral coordination of the Ti-atoms with irregular, statistical distribution of O- and F-atoms [1535]. Upon change of the preparation procedure, the structural properties are also affected, for example, if nanostructured, polycrystalline $TiOF_2$ is prepared in a metathetical reaction between SiO_2 and TiF_4 [1536]. Based on theoretical calculations, a planar C_{2v}-symmetric geometry has been proposed for $TiOF_2$ [1503]. Oxofluorotitanates(IV) are formed by the reaction of $M_2^I[TiF_6]$, TiO_2 and M^IF. They have the general formulas $M_3^I[TiOF_5]$ (M^I = Na, K, Rb, NH_4) and $M^{I\prime}{}_2M^{I\prime\prime\prime}[TiOF_5]$ ($M^{I\prime}$ and $M^{I\prime\prime\prime}$ = alkali metal), respectively [1537–1539]. $K_2[TiOF_4]$ is generated by hydrolysis of $K_3[TiF_6]$, whereas $K_3[TiOF_5]$ results from thermal decomposition of the peroxo derivative $K_3[Ti(O_2)F_5]$ [1533]. More conveniently, $K_2[TiOF_4]$ can be directly prepared from KF and TiF_3 in a 1:1-mixture (v/v) of HF and H_2O at 200 °C. It is isostructural to the vanadium analogue $K_2[VOF_4]$ [1540]. The oxotrifluorotitanate(IV) salt $NH_4[TiOF_3]$ is accessible from $(NH_4)_2[TiF_6]$ and H_3BO_3 [1541]. Due to its stability, $TiOF_2$ has been proposed as potential electrode material for Li-ion batteries, as it can reversibly incorporate and release Li [1532, 1534, 1542–1544]. The rate performance of $TiOF_2$ nanoparticles can be further improved by hydrogenation [1545]. More recently, $TiOF_2$ has been prepared in form of nanoscopic single crystals by aerosol-spraying technique [1546]. Nanocubes or ball-flowers of $TiOF_2$ are accessible from $Ti(OR)_4$ (R = $^nC_4H_9$, $^iC_3H_7$), HF and CH_3CO_2H by hydrothermal or solvothermal synthesis [1547–1549]. Under similar conditions, the preparation of nanotubes has also been described, which also show desirable Li-storage properties [1550]. The $TiOF_2$ nanocubes are durable and show high photocatalytic activity in visible light induced degradation of organic pollutants [1551].

10.4.2 Zirconium and hafnium

The chemical properties of zirconium- and hafnium-compounds are generally very similar due to the comparable atomic and ionic radii of the two metals. Analogously to titanium, the heavier homologues of group 4 mainly form tetrafluorides. In case of hafnium, HfF_4 is the only binary Hf-compound, whereas low valent zirconium fluorides exist as well.

Low-valent fluorides

The monofluorides **ZrF** and **HfF** can only be observed in the gas phase. However, they are rather well studied by spectroscopic methods and quantum-chemistry. ZrF is generated by fluorination of a Zr-foil with BaF_2 or SF_6. The mass-spectrometric investigation of ZrF has allowed the evaluation of its bond-dissociation energies [1552]. The electronic ground-state of ZrF has been identified by spectroscopic methods, too [1553, 1554]. Moreover, fundamental properties of the ZrF molecule have been studied by theoretical calculations [1180]. HfF is generated, if HfF_4 is treated with a plasma of electric discharge [1555]. The low-lying electronic states of HfF have been studied by visible and near-infrared laser spectroscopy [1556, 1557].

Zirconium difluoride **ZrF$_2$** is a black solid, which can be generated by reduction of ZrF_4 with atomic hydrogen at 350 °C. In diluted acids and on air, it is readily oxidized to yield tetravalent ZrO_2. Above 800 °C, disproportionation to Zr and ZrF_4 takes place [1558]. ZrF_2 is also obtained, if a stoichiometric mixture of ZrF_4 and Zr-metal is heated in an eutectic mixture composed of each 50 mol-% LiF and KF at 550 °C [1559]. ZrF_2 has also been studied by quantum-chemical calculations, which predict a bent molecular geometry. Hafnium difluoride **HfF$_2$** is unknown, however, calculations predict a bent structure for this compound [172].

Trifluorides

Analogously to TiF_3, Zirconium trifluoride **ZrF$_3$** is formed by the reaction of zirconium hydride with a mixture of H_2 and HF at 750 °C. Less efficiently, impure ZrF_3 is obtained by the reduction of $(NH_4)_2[ZrF_6]$ with H_2 at 650 °C [1298, 1559]. ZrF_3 is a blue-grey, paramagnetic solid, which has a solid-state structure like ReO_3. Thus, the $[ZrF_6]$-octahedra are linked by F-bridges [1298]. ZrF_3 is stable on air at room temperature, but starts to decompose upon heating to 300 °C. It is poorly soluble in hot water, in contrast, it is well soluble in cold, diluted mineral acids, whereas the solubility decreases in warm acids. ZrF_3 is not attacked by alkaline solutions of NaOH or NH_3. If it is heated to more than 850 °C, it disproportionates to Zr and ZrF_4 [1298]. Like the other low-valent hafnium fluorides, **HfF$_3$** is unknown. There are neither experimental nor quantum-chemical studies on this compound. The only small hint on HfF_3 has been found in some mass-spectrometry studies.

Tetrafluorides

Zirconium tetrafluoride **ZrF₄** (mp.: 932 °C) and hafnium tetrafluoride **HfF₄** (m. p.: 1025 °C) are white, crystalline solids, which are generated upon fluorination of the metals or metal compounds [204]. The tetrafluorides can also be prepared by thermal decomposition of the ammonium salts $(NH_4)_2[MF_6]$ (M = Zr, Hf) [1560]. The trihydrate $ZrF_4 \cdot 3\,H_2O$ is obtained upon dissolution of ZrO_2 in aqueous HF and can be dehydrated in high vacuum (Scheme 10.8) [1298]. Similarly, this procedure can be applied for the synthesis of HfF_4. For both tetrafluorides, the dehydration step may also be performed under an atmosphere of HF [1561, 1562]. Purification of the tetrafluorides is typically achieved by sublimation or submission to an ion-exchange method [1563, 1564]. As the Zr-containing minerals in nature typically contain some hafnium (1–3 %), there is the need for efficient separation processes on industrial scale. Due to this, the fluorination of ZrO_2 and HfO_2 with HF and F_2, respectively, has been investigated by thermogravimetry. A working temperature of 525 °C has been identified as being ideal for the conversion of ZrO_2 to ZrF_4. Apart from that, it has been found, that the fluorination process occurs via intermediary oxyfluorides [1565, 1566]. For the subsequent separation of ZrF_4 and HfF_4, the sublimation kinetics of ZrF_4 have been studied, too [1567]. Analogously to TiF_4, ZrF_4 and HfF_4 both exist as tetrahedral molecules in the gas phase [1568]. The barriers for the edge inversion have been calculated by quantum-chemical methods, revealing a value of about 199 kJ mol⁻¹ for ZrF_4, which is slightly higher for HfF_4 (209 kJ mol⁻¹) [172]. In the crystalline phase, ZrF_4 exists in a low temperature α-polymorph below 450 °C, whereas high temperature β-ZrF_4 is present at higher temperatures [1569]. For HfF_4 there are three modifications known, crystallizing in a monoclinic, tetragonal and cubic space group, respectively. In all the aforementioned cases, the metals Zr and Hf are surrounded by eight F-atoms in form of square-antiprisms [1570].

ZrF_4 is very stable on moist air and only poorly soluble in water, mineral acids and alkaline solutions, however, it is slowly attacked by hot bases and acids [1560]. The tetrafluorides react with hydrofluoric acid to form hydrates and oxyfluorides. In contrast to the heavier halides, the hydrates of ZrF_4 and HfF_4 are very stable, especially the mono- and trihydrates are well investigated [1561, 1571, 1572]. The trihydrates $ZrF_4 \cdot 3\,H_2O$ and $HfF_4 \cdot 3\,H_2O$ slightly differ in their structures. In both compounds, the metal center is eight-fold coordinated, but in the zirconium compound one can find pairs of $[ZrF_3(H_2O)_3]$-units, which are bridged by two F-atoms, whereas the hafnium derivative contains chains of $[HfF_4(H_2O)_2]$-units, which are linked by four F-bridges. The remaining third molecule of H_2O occupies free positions in the lattice [1573, 1574]. ZrF_4 and HfF_4 are very hard Lewis acids and, therefore, readily form complexes with Lewis bases, such as N- and O-donor ligands. With monodentate species, such as N,N-dimethylformamide, Ph_3PO, $(CH_3)_3PO$ or Ph_3AsO, the metal center appears to be coordinated by six ligands in total, resulting in an overall formula $[MF_4L_2]$ (Scheme 10.8). The situation is slightly different for dimethyl sulphoxide ligands. Here, the complex is

$(NH_4)_2[MF_6]$

Δ

MO_2 $\xrightarrow[\text{dehydration}]{\text{aq. HF}}$ MF_4 $\xrightarrow[\text{N- or O-donors}]{\text{monodentate}}$ $[MF_4L_2]$

MO_2 (via aq. HF)

MOF_2

F^-, aq. HF

$[MF_6]^{2-}$ salts

F^-, Δ

$[MF_5]^-, [MF_6]^{2-},$
$[MF_7]^{3-} \text{ or } [MF_8]^{4-}$

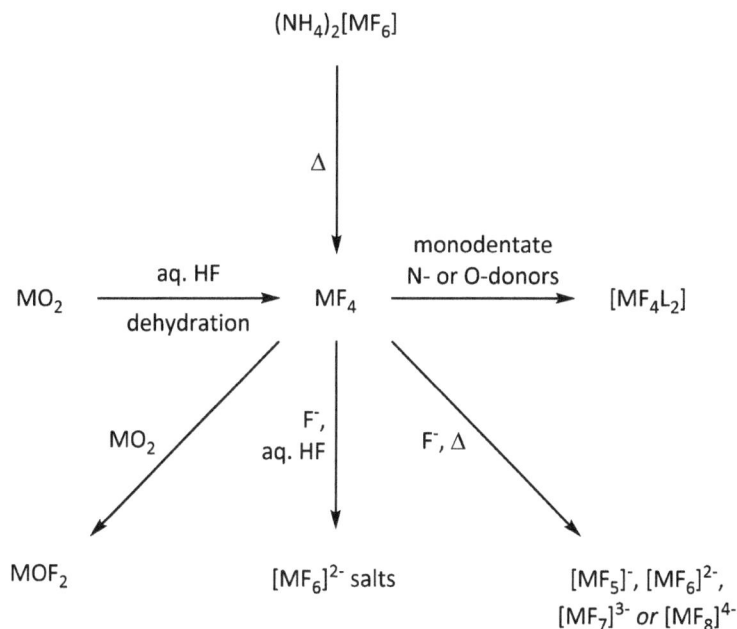

Scheme 10.8: Preparation and selected reactions of the group 4 tetrafluorides MF_4 (M = Zr, Hf).

present in dimerized form, revealing metal-centers, which are bridged by two F-atoms and show a pentagonal-pyramidal coordination geometry each. Moreover, an eight-fold, distorted dodecahedral Zr-coordination is observed in the 1:2-adduct of ZrF_4 with 2,2′-bipyridine or 1,10-phenanthroline. The aforementioned bipyridine complex is stable in methanol solution, but is slowly decomposed upon contact with air [1509, 1575]. Zirconium fluoronitrides of the general formula $ZrN_xF_{4\text{-}3x}$ are generated upon heating of the tetrafluoride in a stream of NH_3 at 580 °C [1576]. The tetrafluorides can also be substituted by azide-groups, however, for the neutral molecules stabilization by a N-donor ligand, that is, 2,2′-bipyridine (bpy) is required. In both cases, the complexes $[Zr(N_3)_4(bpy)_2]_2 \cdot$ bpy and $[Hf(N_3)_4(bpy)_2]_2 \cdot$ bpy show eight-fold-coordinated metal-centers [1516]. If ZrF_4 or HfF_4 are treated with Me_3SiN_3 in presence of two equivalents of ionic $[PPh_4][N_3]$, the hexaazidometallate(IV) anions $[Zr(N_3)_6]^{2-}$ and $[Hf(N_3)_6]^{2-}$ are obtained [1577].

There exist numerous fluorometallates of ZrF_4 and HfF_4. Apart from that, fluorozirconates(IV) and hafnates(IV) are widely explored in various compositions of heavy metal containing glasses. For example, ZrF_4 has been investigated in the ternary systems ZrF_4-BaF_2-ThF_4 and ZrF_4-BaF_2-LnF_3. The glasses resulting from such systems have some unique properties, for example, a small dispersion, a wide area of optical permeability and a reasonable resistance against strong fluorinating agents. Apart from that, they are good amorphous fluoride-ion conductors [1578]. The properties of ZrF_4- and HfF_4-containing melts and glasses have been compared, revealing no significant differences, except for the glass transition temperature [1579]. Generally, the heavy metal fluoride glasses are often discussed with respect to many different applications, for example, in optical components [1580].

Crystalline fluorocomplexes are known with several different $[MF_n]^{(n-4)-}$-units (M = Zr, Hf; n = 5–8) and are typically accessible by thermal treatment of stoichiometric mixtures of alkali metal fluorides and the corresponding tetrafluoride ZrF_4 or HfF_4. Hexafluorometallates are mostly generated from the melts of stoichiometrically composed metal fluorides or by addition of metal fluorides or dioxides to an aqueous HF-solution of ZrF_4 or HfF_4, respectively (Scheme 10.8) [1520, 1581–1584]. The silver(II) salts $Ag[ZrF_6]$ and $Ag[HfF_6]$ are obtained by fluorination of the respective oxychlorides $MOCl_2$ and Ag_2SO_4 with F_2 [262]. Similar to the analogous titanium-compounds, the $[ZrF_6]$-complexes are thermally stable in dry air up to 750 °C, as has been shown for $K_2[ZrF_6]$. However, in moist air, decomposition to HF, ZrO_2 and $K_3[ZrF_7]$ is observed at 700–725 °C. At 750–800 °C, the presence of KF indicates the decomposition of $K_3[ZrF_7]$ [1521]. Octahedral $[ZrF_6]$-units are found in $Li_2[ZrF_6]$ and $[Cu(H_2O)_4][ZrF_6]$, whereas γ-$Na_2[ZrF_6]$ contains irregular $[ZrF_7]$-polyhedra, which are linked by common F-atoms [1585–1588]. The potassium salt $K_2[ZrF_6]$ is built from dodecahedral $[ZrF_8]$-units, which probably contain $[F_4ZrF_4ZrF_4]^{4-}$ anions, as has been investigated by vibrational spectroscopy [1589]. The compounds $M^{II}[ZrF_6]$ (M^{II} = Mg, Ca, Cr, Mn, Fe, Co, Ni, Cu, Zn) crystallize in a cubic ReO_3 lattice [1590]. The same is observed for $V[ZrF_6]$, $V[HfF_6]$ and $Ti[ZrF_6]$, in contrast $Ti[HfF_6]$ exhibits a high-temperature modification of the ReO_3 lattice [1591]. Octahedrally coordinated metal-centers are found in $Cs_2Cu_3MF_{12}$ (M = Zr, Hf), where distorted $[CuF_6]$-octahedra are connected by regular $[MF_6]$-units [1592]. The thermally most stable heptafluorozirconates $[ZrF_7]^{3-}$ also exhibit different structures, which are, however, disordered. $(NH_4)_3[ZrF_7]$ contains pentagonal-bipyramidal $[ZrF_7]$-groups, whereas the anion in $Na_3[ZrF_7]$ is described by a trigonal prismatic geometry, where the seventh fluoride is capping one of the rectangular planes [1593]. Pentagonal-bipyramidal coordination is furthermore observed in the $A^IB^{II}MF_7$-type of compounds (A^I = Rb, Tl; B^{II} = Ca, Cd; M = Zr, Hf) [1594]. $Cu_3[ZrF_7]_2 \cdot 16\,H_2O$ contains $[Zr_2F_{14}]^{6-}$ anions, which are built from two square-antiprisms with a common edge [1595]. The salts $KPd[ZrF_7]$ and $KV[ZrF_7]$ are composed of distorted $[PdF_6]$-/$[VF_6]$-octahedra and pentagonal-bipyramidal $[ZrF_7]$-units, which are connected by a common edge [1591, 1596]. Square-antiprismatic $[ZrF_8]^{4-}$ anions can be found in $[Cu(H_2O)_6]_2[ZrF_8]$, as well as in the triple salt $Li_6[BeF_4][ZrF_8]$ [1597]. A further, interesting example is $Rb_5[Zr_4F_{21}]$, which is comprised of an anionic chain-structure with four crystallographically independent Zr-atoms, all of them having different coordination polyhedra. The different metal centers adapt the coordination numbers seven (pentagonal-bipyramidal), six (octahedral), eight (distorted antiprismatic) and seven (irregular antiprismatic with one vacancy) [1598]. If a mixture of AgF and the respective MF_4 (M = Zr, Hf) is tempered at 450 °C, the ternary compounds $Ag_7Zr_6F_{31}$ and $Ag_7Hf_6F_{31}$ are generated. In the Zr-compound, each of the Zr-centers is surrounded by eight F-atoms in form of a square-antiprism, which are connected by common corners in such a way, that the metal-centers form a $[Zr_6]$-octahedron. The last F-atom is in the middle of the $[Zr_6]$-unit [1278]. In $K_3Ag_2M_4F_{23}$ (M = Zr, Hf), the Zr-centers are again coordinated by each eight F-atoms, resulting approximately in the shape of a dodecahedron [1599]. There are also related ammonium compounds of hafnium with the composition $(NH_4)_6M^IHf_4F_{23}$ (M^I = K, Rb, Cs) [1600]. Furthermore, a dodecahedral coordination geometry is observed in $Rb[HfF_5]$ [1601]. Finally, a distorted cubic-antiprism $[MF_8]$ (M = Zr, Hf) is found in the solid-state structures of the compounds $KCuMF_7$ [1602].

Zirconium oxydifluoride **ZrOF$_2$** and hafnium oxydifluoride **HfOF$_2$** are generated upon heating of a mixture of MO_2 and MF_4 above 550 °C or thermolysis of the trihydrates $MF_4 \cdot 3\,H_2O$ (Scheme 10.8) [1571]. They are also obtained if the oxides are dissolved in aqueous HF. The oxyfluorides are soluble in differently concentrated, aqueous hydrofluoric acid, where $ZrOF_2$ is always better soluble than $HfOF_2$ [1603]. $ZrOF_2$ and $HfOF_2$ form hydrated salts of the type $M'[MOF_3] \cdot 2\,H_2O$, if they are reacted with alkali metal fluorides $M'F$ (M' = K, Rb, Cs) [1604]. Moreover, higher aggregates have been described in the system ZrO_2-ZrF_4, e. g. $Zr_4O_5F_6$, $Zr_7O_9F_{10}$ or $Zr_{10}O_{13}F_{14}$. $Zr_7O_9F_{10}$ is obtained by heating ZrO_2 and ZrF_4 to 800 °C for several days. It has a structure similar to α-UO_3 and contains octahedral and pentagonal-bipyramidal coordinated Zr [1605].

10.4.3 Study questions

(I)   How can one prepare TiF_3? Comment on its stability against oxidation.

(II) Describe the structure of solid and gaseous TiF_4. Which general structural behavior is observed for adducts of TiF_4 with Lewis bases? Give examples for donor molecules.

(III) Which fluorotitanate(IV) anions are known and what is their structure? How can they be prepared? Are they stable against hydrolysis?

(IV) Which product is obtained upon fluorination of TiO_2 with HF or F_2? How does the use of the anatase and rutile modification of TiO_2 influence the outcome of this reaction?

(V)   Are there low valent zirconium and hafnium fluorides? How are they prepared?

(VI) How are the tetrafluorides ZrF_4 and HfF_4 prepared? Comment on their structural differences and similarities with respect to TiF_4.

(VII) Compare the fluorozirconates(IV) and –hafnates(IV) with the corresponding fluorotitanates(IV). Which general trend is observed?

10.5 Group 5: Vanadium, niobium, tantalum

10.5.1 Vanadium

Vanadium forms fluorides in all oxidation states from +II to +V, where the high-valent compounds are more stable. Apart from binary compounds, a number of oxyfluorides and fluorovanadates is known as well.

Low-valent fluorides

Vanadium monofluoride **VF** is unknown in condensed state; however, it has been generated in the gas phase by discharge of VF_4. The electronic states of VF have been investigated by analysis of its emission spectrum [1606]. Apart from that, the ground state as well as the energetically low-lying states have been explored by quantum chemical calculations [1468].

Vanadium difluoride **VF$_2$** is formed upon reduction of VF_3 with a mixture of H_2 and HF at 1150 °C. Alternatively, it can be generated by comproportionation of elemental vanadium with VF_3 at 800 °C or by fluorination of VCl_2 with H_2 and HF at 650 °C [1607–1609]. In a neutral aqueous solution, the blue-violet hydrate $VF_2 \cdot 4\,H_2O$ is generated by electrolytic reduction of VF_5. If the reaction mixture contains free HF, the hydrate also incorporates HF [1610, 1611]. The blue, crystalline VF_2 is antiferromagnetic and crystallizes in a tetragonal rutile lattice [1607, 1612]. Single crystal X-ray diffraction revealed the presence of axially compressed $[VF_6]$-octahedra in the solid state [1613]. The IR-spectrum of matrix-isolated VF_2 points toward a nonlinear molecular structure. The energetically lowest electronic states of VF_2 have been calculated using quantum chemical methods [1614]. VF_2 is a strong reducing agent, so that it will react with excess HF during its preparation, resulting in the formation of VF_3 and H_2. The reaction with alkali metal fluorides M^IF (M^I = Na, K, Rb) yields trifluorovanadates(II) $M^I[VF_3]$ [1609, 1615–1617]. Depending on the molar ratio of M^IF and VF_2, partial or complete disproportionation of VF_2 may also occur. This gives rise to elemental vanadium, hexafluorovanadates(III) $M^I_3[VF_6]$ as well as fluorovanadates with a mixed oxidation state (II and III) [1618, 1619]. The tetrafluorovanadate(II) $Mg[VF_4]$ is a solid solution, which is formed in a high-temperature reaction between VF_2 and MgF_2 [1620]. VF_2-nanobelts have been suggested for the use as floating-type absorbents for wastewater treatment [1621].

Vanadium trifluoride

Vanadium trifluoride **VF$_3$** is formed upon reaction of VCl_3 with HF (600 °C) or NF_3 (300 °C), by disproportionation of VF_4 at 150 °C or by treatment of elemental vanadium with HF at 225–500 °C (Scheme 10.9) [204, 1622–1626]. It can also be obtained by thermal decomposition of $(NH_4)_3[VF_6]$ at 500 °C, which in turn is prepared from V_2O_3 in molten $NH_4[HF_2]$ [1627]. However, in a NH_3-stream at 600 °C, $(NH_4)_3[VF_6]$ is completely degraded to vanadium nitride VN [1628]. VF_3 is a yellow-green, antiferromagnetic and air-stable compound (mp.: ca. 1400 °C), which crystallizes in a rhombohedral lattice with an almost regularly octahedral coordination of vanadium [1629]. The structure may also be described by a distorted, cubic ReO_3 lattice, where the octahedra have been largely rotated around their C_3-axis, so that the F-atoms adapt a hexagonal closest packing in the end [1630]. Quantum-chemical calculations have revealed, that the VF_3 molecule is planar in its ground state, thus having D_{3h}-symmetry [1481]. These findings have been confirmed by vibrational spectra of

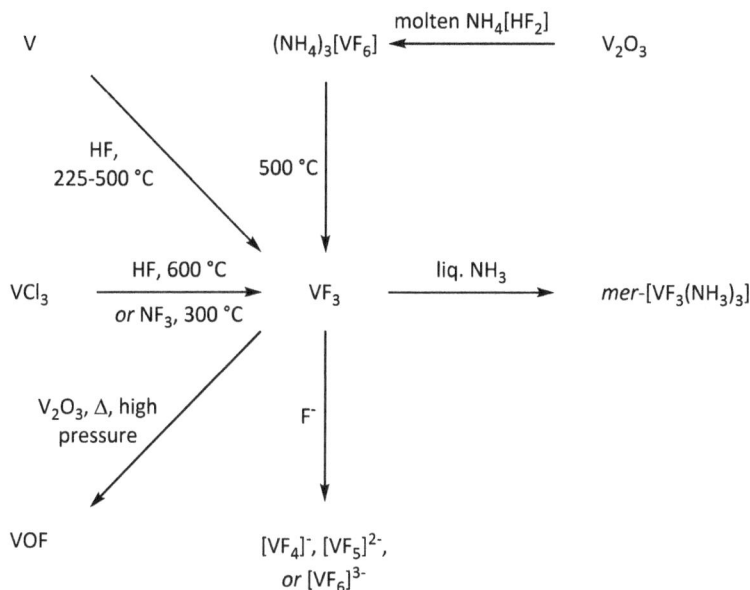

Scheme 10.9: Preparation and selected reactions of VF_3.

matrix-isolated VF_3 [1631]. Moreover, the electronic ground-state of VF_3 has been cal-culated and theoretically studied under consideration of the Jahn–Teller effect [1632]. In typical organic solvents as well as in water, VF_3 is insoluble [1622]. The trihydrate $VF_3 \cdot 3 H_2O$ is obtained from the reaction of V_2O_3 with HF [1633]. In the reaction with anhydrous liquid ammonia, VF_3 forms mer-$[VF_3(NH_3)_3]$, which is composed of dis-crete molecules in the solid state (Scheme 10.9). The $[VF_3(NH_3)_3]$-units, which can be structurally described on the basis of an octahedron, are strongly interconnected by hydrogen-bonding [1634]. VF_3 forms numerous fluorovanadates(III), for example, $M^I[VF_4]$, $M^I_2[VF_5]$ or $M^I_3[VF_6]$ (M^I = Na, K, Rb, Cs, Tl, NH_4) with the green anion $[VF_6]^{3-}$ (Scheme 10.9) [1485, 1627, 1635–1642]. VF_3 reacts with the alkaline earth metal di-fluorides under formation of $M^{II}[VF_5]$ (M^{II} = Ca, Sr, Ba) and $M^{II}_3[VF_6]_2$ (M^{II} = Sr, Ba) [1643]. Analogously, $Pb[VF_5]$ is obtained from PbF_2 and VF_3 [1644]. The strontium compound $Sr[VF_5]$ contains helical chains of cis-F bridged, distorted $[VF_6]$-octahedra and is isostructural to $Sr[FeF_5]$ [1645]. The hydrazinium salt $[N_2H_6][VF_5]$ is synthesized in aqueous HF (w = 40 %) and can be thermally decomposed to liberate VF_3 [1646]. Similarly, $[N_2H_5]_2[VF_5] \cdot H_2O$ is prepared in aqueous solution and can be carefully dehydrated at elevated temperature to yield $[N_2H_5]_2[VF_5]$. The monohydrate contains discrete $[VF_5(H_2O)]^{2-}$ anions, whereas the anhydrous derivative is composed of F-bridged chains of $[VF_5F_{2/2}]$-octahedra [1647]. There are also a few more hydrates of the types $M^I[VF_4 \cdot 2 H_2O]$ and $M^I_2[VF_5 \cdot H_2O]$ [1636]. VF_3 has an appreciably high reversible storage capacity for Li as well as some other beneficial, electrochemical properties, thus making it interesting for potential applications in Li-batteries [1648, 1649]. Furthermore, it can be used as dopant for MgF_2 to enhance its Lewis acidity and surface area [1650]. The derived vanadium oxyfluoride **VOF** is a semiconductor and has a rutile structure. It is commonly prepared in high temperature reactions under high pressure from stoichiometric mixtures of V_2O_3 and VF_3 (Scheme 10.9) [1500, 1501, 1651]. Molecular VOF has a bent structure and can be identified in an argon matrix, if vanadium metal is treated with OF_2 [1503].

Vanadium tetrafluoride

Vanadium tetrafluoride **VF$_4$** is synthesized by fluorination of VCl_4 with HF in CCl_3F at 25 °C or alternatively upon reduction of VF_5 with PCl_3 or elemental vanadium at elevated temperatures (Scheme 10.10) [1622, 1623, 1652]. It is a green, very hygroscopic solid (mp.: 325 °C (decomp.)), which is readily hydrolyzed. The controlled hydrolysis in water yields the hydrate $[VOF_2(H_2O)_2] \cdot H_2O$ [1653]. VF_4 is insoluble in nonpolar solvents and shows only a slight volatility [1622]. In the solid state, VF_4 has a layered structure of F-bridged $[VF_6]$-octahedra, which may be compared to the situation in SnF_4. The main difference is that the $[VF_6]$-units are tilted against each other within a layer, which is not the case in SnF_4. As a consequence, the M-F-M bridges are bent in VF_4, whereas they are linear in SnF_4 [1654]. Above 150 °C, VF_4 starts to disproportionate to VF_3 and VF_5 [1622, 1655]. In contact with fluorinating agents, VF_4 is readily oxidized to VF_5. VF_4 forms adducts with Lewis bases, for example, NH_3, SeF_4 or pyridine [1623]. If 1:1-mixtures of V_2O_5 and V_2O_3 are dissolved in hydrofluoric acid (w = 40 %) together with alkali metal fluorides, the complex, paramagnetic anion $[VF_6]^{2-}$ is formed, which has a very distorted, octahedral structure [1656]. The hexafluorovanadate(IV)-salts $K_2[VF_6]$ and $Cs_2[VF_6]$ are furthermore accessible from 1:2-mixtures of VF_4 and the respective alkali metal fluoride in anhydrous HF, SeF_4 or IF_5 (Scheme 10.10) [1623, 1642]. The analogous lithium compound $Li_2[VF_6]$ can be made in the same way and crystallizes in the trirutile structure type [1657]. The vanadium(IV) chlorofluorides $VClF_3$ and $V_2Cl_3F_5$ are obtained upon reaction of VCl_3 with AsF_3 [1658]. Moreover, the fluoride substituents of VF_4 can be substituted by N_3-groups, using Me_3SiN_3 as azide transfer agent. Starting from $V(N_3)_4$ and ionic azides, the hexaazidovanadate $[V(N_3)_6]^{2-}$ is accessible, too [1659]. The reaction of VF_4 with N-donor-ligands in aqueous solution leads to the formation of VOF_2-complexes [1653].

Free vanadium oxydifluoride **VOF$_2$** is formed as a yellow compound upon reaction of $VOBr_2$ with HF at high temperatures (Scheme 10.10). It is insoluble in common solvents but readily forms hydrates [1622]. In a ligand exchange reaction, TiF_4 converts $VO(acac)_2$ (acac = acetylacetonate) to VOF_2 [1660]. Vanadium oxydifluoride has also been obtained by the reaction of elemental vanadium with OF_2 in an argon matrix at low temperature. Based on vibrational spectroscopy and quantum-chemical calculations, a C_{2v}-symmetry has been assigned to the molecule [1503]. Complex salts can be generated from mixtures of M^IF, VO_2, HF and H_2O in form of $M_2^I[VOF_4]$ (M^I = K,

VF$_5$

[VF$_6$]$^{2-}$ salts

PCl$_3$ or V,
Δ

F$^-$
aHF, SeF$_4$ or IF$_5$

VCl$_4$ $\xrightarrow{\text{HF, CCl}_3\text{F, 25 °C}}$ VF$_4$ $\xrightarrow{\text{T > 150 °C}}$ VF$_3$ + VF$_5$

VOBr$_2$ $\xrightarrow{\text{HF, }\Delta}$ VOF$_2$ $\xrightarrow[\text{aq. HF}]{\text{[M}^{\text{III}}\text{(NH}_3\text{)}_6\text{]Cl}_3}$ [M^{III}(NH$_3$)$_6$][VOF$_5$]
(M^{III} = Cr, Co, Rh)

TiF$_4$

F$^-$
HF, H$_2$O

VO(acac)$_2$

[VOF$_4$]$^{2-}$, [VOF$_5$]$^{3-}$
or [V$_2$O$_2$F$_7$]$^{3-}$

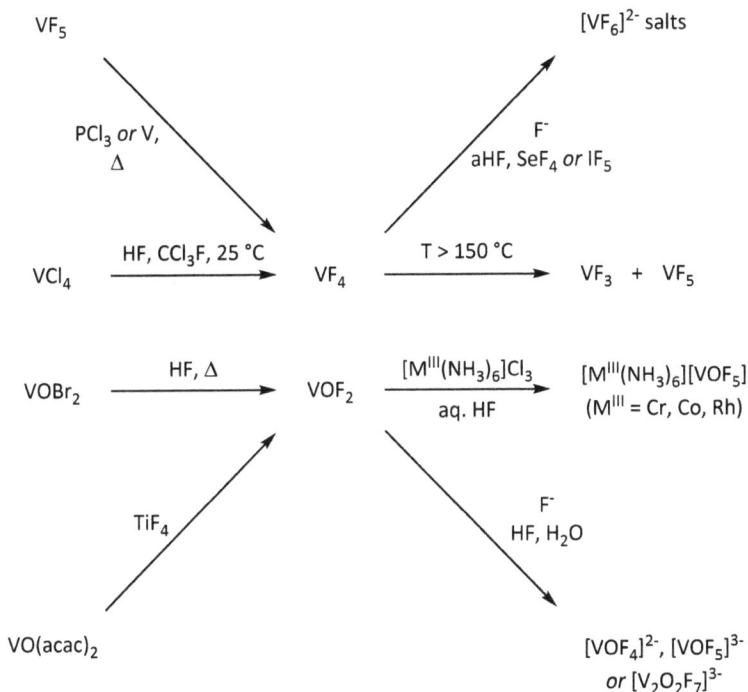

Scheme 10.10: Preparation and selected reactions of VF$_4$ and VOF$_2$.

Rb), Na$_3$[VOF$_5$], Cs$_3$[V$_2$O$_2$F$_7$] and M$^{\text{I}'}_2$M$^{\text{I}''}$[VOF$_5$] (M$^{\text{I}'}$ = K, Rb, Cs; M$^{\text{I}''}$ = Li, Na, K, Rb) [1661]. The salt Na$_3$[VOF$_5$] contains isolated [VOF$_5$]$^{3-}$-octahedra, whereas [VOF$_4$]$^{2-}$ anions exhibit a polymeric chain structure. Therein, the octahedral [VOF$_5$]-units are connected by common corners. The isolated anion [V$_2$O$_2$F$_7$]$^{3-}$ is composed of two [VOF$_5$]-octahedra, which are fused by a common plane [1662]. Accordingly, the dianion [V$_2$O$_2$F$_6$(H$_2$O)$_2$]$^{2-}$ may be rewritten as [VOF$_3$(H$_2$O)]$^{2-}_2$, as it is composed of two distorted [VOF$_4$(H$_2$O)]-octahedra, which share a common edge [1663]. The thermal decomposition of the tetraalkylammonium salts [NR$_4$]$_2$[V$_2$O$_2$F$_6$(H$_2$O)$_2$] (R = Me, Et) results in the formation of impure VF$_3$ [1664, 1665]. Similarly, the ammonium compounds (NH$_4$)$_2$[VOF$_4$], (NH$_4$)$_2$[VOF$_4$ · H$_2$O] and (NH$_4$)$_3$[VOF$_5$] are thermally degraded to mainly result in VF$_3$ [1666]. The reaction of VOF$_2$ with [M^{III}(NH$_3$)$_6$]Cl$_3$ (M^{III} = Cr, Co, Rh) in hydrofluoric acid yields the complexes [M^{III}(NH$_3$)$_6$][VOF$_5$] (Scheme 10.10) [1667].

Vanadium pentafluoride
Vanadium pentafluoride **VF$_5$** is synthesized by fluorination of vanadium metal or VCl$_3$ with F$_2$ or BrF$_3$ at 300 °C. Purification of the crude product mixture is achieved by subsequent low-temperature distillation [1668, 1669]. Alternatively, VF$_5$ is accessible by fluorination of VOF$_3$, V$_2$O$_5$ or VF$_3$ with F$_2$ (Scheme 10.11) [1626, 1670]. VF$_5$ is a colorless, volatile solid (mp.: 19.5 °C; bp.: 48.3 °C), which is transformed into a viscous liq-

VF$_3$, VOF$_3$ or V$_2$O$_5$ BF$_3$ or BiF$_3$ + VCl$_4$ + Cl$_2$

↑ F$_2$ ↗ BCl$_3$ or BiCl$_3$

V or VCl$_3$ $\xrightarrow{\text{F}_2 \text{ or BrF}_3, 300 °C}$ VF$_5$ $\xrightarrow[\text{(M = P, As, Sb)}]{\text{MF}_3}$ VF$_4$ + [VF$_4$][MF$_6$]

ClOF$_3$ ↙   POF$_3$ ↓   F$^-$ ↘

[ClOF$_2$][VF$_6$]   VOF$_3$ + PF$_5$ [VF$_6$]$^-$ salts

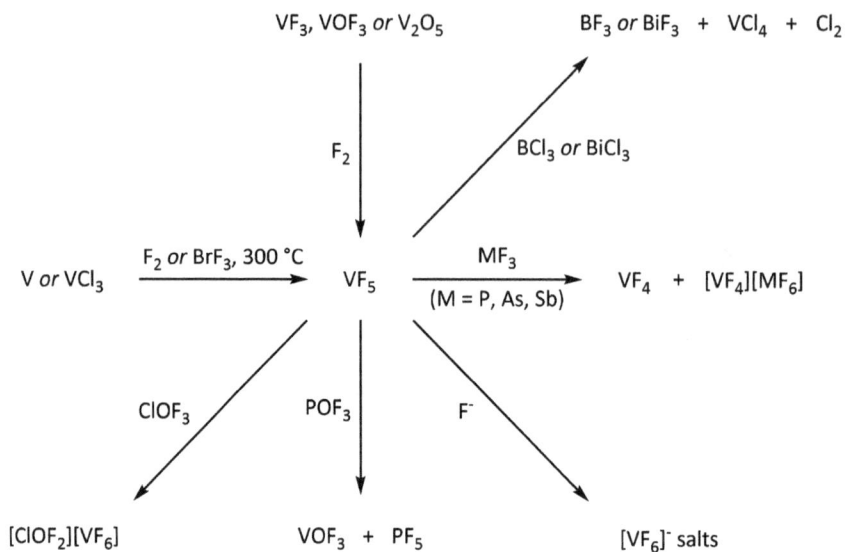

Scheme 10.11: Preparation and selected reactions of VF$_5$.

uid upon melting. After standing for a while, the molten VF$_5$ may change its color to yellow. In the vapor-phase at higher temperatures, VF$_5$ exists as trigonal-bipyramidal monomer, whereas it is mainly associated in the liquid state [468, 1671–1674]. The trigonal-bipyramidal structure of the molecule has also been confirmed by matrix-isolation studies [1675]. The solid-state structure comprises infinite chains of *cis*-F-bridged, slightly distorted [VF$_6$]-octahedra, which are also proposed to be present in the liquid state [1676–1678]. The D$_{3h}$-symmetric molecules in the gas phase are non-rigid, as has been shown by vibrational spectroscopy. Instead, they undergo pseu-dorotation via a pyramidal transition state [1679, 1680]. These investigations have also been supported by quantum-chemical calculations [1681]. Furthermore, it has been shown by theoretical investigations that the D$_{3h}$-symmetry is only due to the ionic character of the vanadium-fluorine bonds, which leads to a slightly distorted spher-ical geometry. In contrast, the covalent derivatives VH$_5$ and VMe$_5$ have a pyramidal, C$_{4v}$-symmetric geometry [1682]. In SO$_2$ClF-solution, the presence of short chains of *cis*-F-bridged [VF$_6$]-units has been revealed [1683]. VF$_5$ is very reactive and easily hy-drolyzed, in water it dissolves with a yellow-red color. It slowly reacts with glass and can be dissolved in anhydrous HF [1655]. Ammonia and pyridine have a reducing ef-fect on VF$_5$ and it also reacts with other oxidizable compounds like PF$_3$, AsF$_3$, SbF$_3$ or I$_2$ to give a mixture of VF$_4$ and the corresponding pentafluoride PF$_5$, AsF$_5$, SbF$_5$ or IF$_5$ (Scheme 10.11). In case of the fluoride acceptors PF$_5$, AsF$_5$ and SbF$_5$ this directly gives rise to the salts [VF$_4$][MF$_6$] (M = P, As, Sb) [1684]. At room temperature, VF$_5$ also fluo-rinates UF$_5$ to UF$_6$ [469]. Upon reaction with BCl$_3$ and BiCl$_3$, the respective trifluorides BF$_3$ and BiF$_3$ are formed along with VCl$_4$ and Cl$_2$ (Scheme 10.11). If the substitution reaction with BCl_3 is carefully performed at $-60\,°C$, the unstable pentachloride VCl_5 is obtained. However, VF_5 does not fluorinate the chalcogen tetrafluorides SeF_4 and SF_4, as the pale pink adduct $SeF_4 \cdot VF_5$ is formed exclusively, whereas SF_4 does not react at all [1685, 1686]. In small quantities, VF_5 and VOF_3 can be adsorbed on NaF, which is of certain relevance for the purification of UF_6 [1687]. The chemistry of VF_5 is often based on fluorine-oxygen exchange reactions, due to the formation of the very stable V=O bond. Thus, VF_5 reacts with POF_3 under formation of VOF_3 and PF_5, accordingly, CF_3COOH is transformed to CF_3COF and HF (Scheme 10.11) [1688]. Upon reaction with fluoride donors, VF_5 yields hexafluorovanadates(V), which are also accessible by the reaction of elemental vanadium with $M^I F$ (M^I = alkali metal, Ag, Tl, NO, NO_2) and BrF_3 [1668, 1689–1692]. $NH_4[VF_6]$ is prepared by the reaction of hydrazinium fluorovanadate(III) with excess XeF_2 [1693]. In the system XeF_6 and VF_5, a 2:1-, 1:1- as well as a 1:2-compound is known. The latter $XeF_6 \cdot 2\,VF_5$ is a purely molecular adduct, whereas $2\,XeF_6 \cdot VF_5$ can be described as ionic $[Xe_2F_{11}][VF_6]$. The 1:1-derivative is intermediate in character [1096]. The pyridinium salt $[C_5H_5NH][VF_6]$ is obtained, if V_2O_5 is reacted with pyridinium poly(hydrogen fluoride) [1694]. VF_5 also reacts with $ClOF_3$ under formation of the ionic 1:1-adduct $[ClOF_2][VF_6]$, as has been identified by vibrational spectroscopy (Scheme 10.11) [470, 1692]. In a low-temperature reaction, polymeric VF_5 and O_2F_2 react to yield $[O_2][V_2F_{11}]$, which rapidly decomposes at room temperature [528]. VF_5 is capable of fluorinating fluoroaromatic compounds, such as hexafluorobenzene, octafluorotoluene, decafluorobiphenyl and octafluoronaphthalene under formation of both, perfluorinated cyclohexene and cyclohexadiene derivatives. However, fully saturated, perfluorinated cyclohexane derivatives are not observed among the reaction products, they are only obtained, if polychlorinated or polybrominated arenes are used as substrates [1695, 1696]. In turn, unsaturated, linear as well as cyclic, polyfluorinated and polychlorinated organic compounds are fluorinated by VF_5 under addition of two fluorine atoms to the C=C-bond [1697]. The same reactivity is observed with respect to trichloroethylene and tetrabromoethylene [1696]. In the reaction with dichloromethane, VF_5 preferentially leads to the exchange of a hydrogen substituent, yielding $CHCl_2F$ as the main product [1698].

Pentavalent vanadium oxyfluorides

Vanadium oxytrifluoride **VOF$_3$** is generated as a yellow solid (mp.: $300\,°C$; bp.: $480\,°C$) upon treatment of VF_3 with O_2 at red heat or by fluorination of V_2O_5 with BrF_3, F_2 or COF_2 [700, 1622, 1699, 1700]. It is also obtained in fluorine-oxygen exchange reactions between VF_5 and excess SO_2 or SO_3 [1692]. In the gas phase, monomeric, C_{3v}-symmetric VOF_3 molecules are observed by vibrational spectroscopy and electron diffraction experiments; however, dimeric $(VOF_3)_2$ molecules are present, too [1701–1705]. Solid, polymeric VOF_3 crystallizes in a layered lattice, where pairs of molecules are linked by di-μ-F-bridges. The pairs are linked to other $[VOF_3]_2$-units by cis-F-bridges, resulting in a distorted octahedral coordination geometry at every V center. Overall, the structure of VOF_3 is closely related to those of VF_4 and CrO_2F_2 [1706, 1707]. VOF_3 forms complexes with N- and O-donor ligands, for example, 2,2′-bipyridine, 1,10-phenanthroline or pyridine N-oxide [1708, 1709]. Moreover, an adduct with a N-heterocyclic carbene ligand has been prepared [1710]. Pure VOF_3 can be dissolved in acetonitrile, however, to a certain extent it will also be reduced to tetravalent vanadium. Upon treatment of such a solution with Me_3SiNMe_2, the fluoride substituents are substituted by dimethylamino-groups. Similarly, Me_3SiOMe leads to the formation of $VO(OMe)_3$, whereas $(Me_3Si)_2O$ yields insoluble VO_2F [1711]. Vanadium dioxyfluoride **VO_2F** is alternatively obtained by fluorination of VO_2Cl or by high-pressure reaction of a 1:1 mixture of V_2O_5 and VOF_3 at 4 GPa and 800 °C as polymeric, orange-brown, crystalline solid (mp.: 350 °C) [1712, 1713]. Based on infrared spectra, it is assumed that it is highly ionic, being composed of discrete $[VO_2]^+$ and F^- ions [1714]. In the solid state, it crystallizes in a trigonal-distorted ReO_3 structure, as has been concluded from powder X-ray diffraction experiments [1715]. VO_2F is insoluble in many organic solvents, such as N,N-dimethylformamide, acetonitrile, tetrahydrofuran, acetone, 1,2-dimethoxyethane or chlorocarbons. However, it dissolves in dimethyl sulphoxide and absolute ethanol, but also slowly reacts with these solvents [1716]. VO_2F is thermally stable as it does not decompose below 300 °C, but it is readily hydrolyzed by water to yield V_2O_5 [1712]. VO_2F-complexes with N-donor ligands (pyridine, 2,2′-bipyridine, 1,10-phenanthroline) are accessible by the reaction of the respective VOF_3-complex with $(Me_3Si)_2O$, but not directly from VO_2F and the free donor ligand [1716].

Oxyfluorovanadates(V) of the type $M^I[VOF_4]$ are generally synthesized from M^IF (M^I = Na, K, Rb, Cs, Tl) and V_2O_5 in HF. The pyramidal $[VOF_4]^-$ anions are present in form of infinite chains, where the single units are linked by cis-F-bridges [1717, 1718]. The same structural feature is observed in the silver(I) salt $Ag[VOF_4]$ and the hydroxylammonium compound $[NH_3OH][VOF_4]$, which are prepared from $AgNO_3$ or $[NH_3OH]Cl$ and VOF_3 in anhydrous HF-solution [1719]. Additionally, the $[NH_3OH]^+$ cations induce an interconnection of the anionic $[VOF_4]$-chains by formation of hydrogen bonds [1720]. The discrete anion $[VOF_4]^-$ is observed in the imidazolium salt $[(dipp)H][VOF_4]$ (dipp = 1,3-bis(2,6-diisopropylphenyl)-1,3-dihydro-$2H$-imidazol-2-ylidene) [1710]. The nitrosyl salt $[NO][VOF_4]$ is obtained, if V_2O_5 is treated with NF_3 [1625]. As has been shown by NMR spectroscopy, the anions rapidly rearrange from a square-pyramidal to a trigonal-bipyramidal coordination geometry [1721]. If $K_2[VOF_5]$ is treated with KOH, it forms the light-yellow, crystalline divanadate complex $K_4[V_2O_3F_8]$ [1722]. A large number of further oxyfluorovanadates is accessible from solutions containing V_2O_5, MF, HF and H_2O but also by solid-state reactions, as for example, $M[VO_2F_2]$, $M_2[VO_2F_3]$, $M_3[VO_2F_4]$, $M'_2M''[VO_2F_4]$ or $M_3[V_2O_4F_5]$ [1723–1726]. The lead derivatives $Pb[VOF_5]$ and $Pb[V_2O_2F_8]$ are obtained from PbF_2 and VOF_3 in anhydrous HF, as a by-product, the mixed-valent $[Pb_3F][V_4O_3F_{18}]$ is generated. The origin of the vanadium(IV) species has been attributed to impurities in the starting material. The $[VOF_5]^{2-}$ salt contains distorted octahedral anions, whereas $[V_2O_2F_8]^{2-}$ can be understood as a dimer of $[VOF_4]^-$, being asymmetrically bridged by two fluorine-substituents. The mixed valent anion $[V_4O_3F_{18}]^{5-}$ is composed of $[V^VOF_4]^-$ and $[V^{IV}F_6]^{2-}$ units, thus it may be reformulated as $[(VOF_4)_3(VF_6)]^{5-}$ [1727]. $[VOF_4]^-$ salts with organic imidazolium and pyridinium cations have also been prepared in ionic liquids. Moreover, the dimeric $[V_2O_2F_8]^{2-}$ has been observed in these reaction media [1728, 1729]. VOF_3 can be treated with tetraalkylammonium salts $[NR_4]X$ (R = Me, Et, n-Bu) to yield complexes with the anions $[VOF_3X]^-$ (X = F, Cl, Br, NO_3, $MeSO_3$) [1708, 1730]. The reaction of VO_2F with $[NMe_4]F$ yields $[NMe_4][VO_2F_2]$ [1716]. The salts $M[VO_2F_2]$ (M = Na, NH_4) are believed to contain polymeric anion chains, whereas enlargement of the cation to $[Ph_4P]^+$ and $[Ph_4As]^+$ leads to monomeric $[VO_2F_2]^-$ anions [1726, 1731, 1732]. In $M_2[VO_2F_3]$, the distorted octahedral $[VO_2F_2F_{2/2}]$-units are linked by cis-F-bridges, forming infinite chains [1733]. Moreover, many structurally diverse oxyfluorovanadate(V) derivatives are accessible by solvothermal or ionothermal synthesis [1734]. The substitution of fluoride by N_3-groups gives rise to explosive $VO(N_3)_3$ and other azide-derivatives, for example, anionic $[VO(N_3)_5]^{2-}$ [1735]. VOF_3 is widely used in organic synthesis, for example, in oxidative biaryl coupling reactions [1736–1743]. Furthermore, it is capable of oxidizing benzylic C-H-bonds [1744]. VO_2F has been found to have a high specific capacity for Li^+ cations, thus making it a potential candidate for cathode-materials in Li-ion batteries [1713, 1715].

10.5.2 Niobium and tantalum

Low-valent fluorides
The monofluorides **NbF** and **TaF** are nonexistent in the condensed phase, but they can be found in the gaseous state under appropriate conditions [1745]. To date, the literature on NbF and TaF is mostly limited to quantum-chemical calculations on their electronic structure or fundamental thermodynamic properties [1180, 1746, 1747]. The electronic transitions of TaF have been studied using laser induced fluorescence spectroscopy supported by theoretical investigations [1748]. The situation does not change a lot for the difluorides **NbF₂** and **TaF₂**, which also have not been observed in a condensed phase. Thus, their existence seems to be limited to gas phase conditions again [1745]. Apart from that, the difluorides have been scarcely investigated by quantum-chemical methods. Recent results predict a slightly bent structure for NbF_2 [1746]. Despite the fact, that a binary niobium(II) fluoride is unknown, a mixed-valent compound, formally having the composition "$NbF_{2.5}$" has been described. It has been synthesized in form of the brown metal cluster fluoride Nb_6F_{15}, which results from the reaction of Nb with NbF_5 at 400 °C. More conveniently, it is accessible by using a temperature gradient from 900–400 °C under argon atmosphere. It crystallizes in a cubic cell, where $[Nb_6F_{12}]^{3+}$ cations are linked by the remaining fluoride ions in a continuous, three-dimensional network. The six Nb-atoms in $[Nb_6F_{12}]^{3+}$ form a regular Nb_6-octahedron, which is surrounded by a cuboctahedron of twelve F-atoms. Nb_6F_{15}

is stable on air and against mineral acids and bases, even at elevated temperatures. However, *in vacuo* above 700 °C, it decomposes to yield Nb and NbF_5 [1749–1751]. Each $[Nb_6F_{12}]^{3+}$-cluster has only 15 valence-electrons, and thus carries an unpaired spin, which leads to antiferromagnetic ordering at temperatures below 6 K [1752]. An analogous tantalum compound with the formula Ta_6F_{15} does not exist, however, the derivatives Ta_6Cl_{15}, Ta_6Br_{15}, Ta_6Br_{14} and Ta_6I_{14} have been described, all of them containing a $[Ta_6X_{12}]$-unit [1753].

Trifluorides

Dark-blue niobium trifluoride **NbF₃** and tantalum trifluoride **TaF₃** are obtained by fluorination of the metals with HF at high temperature and under pressure or, alternatively, upon reduction of the pentafluorides [204, 1754, 1755]. Both trifluorides crystallize in the cubic ReO_3 structure type [1755, 1756]. Quantum-chemical calculations predict a trigonal-planar structure for molecular NbF_3 [1746]. It is stable on air and only decomposes upon heating. Furthermore, it is not attacked by either concentrated mineral acids or alkaline solutions [1755]. NbF_3 and TaF_3 always incorporate oxygen into their crystal lattices for stabilization, thus it is virtually impossible to avoid the formation of mixed oxyfluoride compounds. Consequently, NbF_3 and TaF_3 have never been obtained in pure form by now [1749]. NbF_3 is weakly paramagnetic and has semiconductor properties [1757]. Theoretical investigations also revealed half-metallicity for NbF_3, which makes it interesting as a potential material [1758].

Tetrafluorides

Niobium tetrafluoride **NbF₄** is formed upon reduction of NbF_5, which can be achieved by the use of silicon or niobium powder as well as by other reducing agents [1749, 1759–1761]. It is a black, non-volatile, and very hygroscopic solid with paramagnetic properties [1759]. Its crystal structure is different to other niobium tetrahalides, as it crystallizes tetragonally in a layered structure with two crystallographically different NbF_4 molecules in the unit cell. Like in SnF_4, every Nb-atom is octahedrally surrounded by fluorine-atoms and the $[NbF_6]$-octahedra are connected in a plane via common F-atoms [1749, 1762]. The infrared spectrum of NbF_4 is similar to those of TiF_4, ZrF_4 and VF_4 [1763]. The NbF_4 molecule has D_{2d}-symmetry, as has been investigated by theoretical methods [1746]. On air, NbF_4 is readily hydrolyzed to NbO_2F, in water it forms a brown solution along with a brown precipitate of unknown composition. Moreover, NbF_4 attacks glass vessels, if traces of water are present. Thermally, NbF_4 is significantly more stable than VF_4, as complete and rapid disproportionation is only observed at temperatures above 350 °C [1759]. Hexafluoroniobate(IV) compounds of the composition $M_2^I[NbF_6]$ (M^I = Li, Na, Rb, Cs) are accessible by thermal reactions of NbF_4 and M^IF at 520–650 °C [1764]. Under similar conditions, salts of the type $M^{II}[NbF_6]$ (M^{II} = Mg, Ca, Cr, Mn, Fe, Co, Ni, Zn, Cd) are obtained, too [1765, 1766]. The reduction of a melt of KF and NbF_5 with Nb at 850 °C leads to the formation of cubic $K_3[NbF_7]$, which is isostructural to $NH_4[ZrF_7]$ and $K_3[NbOF_6]$ with the anion having a distorted pentagonal-bipyramidal structure [1767]. Spectra of $K_3[NbF_7]$ in a melt of LiF and BeF_2 at 550 °C have revealed, that Nb(IV) is not coordinated by more than seven atoms, even in a solution [1768]. Tantalum tetrafluoride **TaF$_4$** is still unknown in the condensed phase. It has only been observed in the gas phase by mass-spectrometry [1745].

Pentafluorides

Niobium pentafluoride **NbF$_5$** (mp.: 79 °C; bp.: 234 °C) and tantalum pentafluoride **TaF$_5$** (mp.: 97 °C; bp.: 229 °C) are formed upon fluorination of the metals or their respective compounds, for example, oxides or chlorides (Scheme 10.12) [1759, 1761]. NbF_5 can also be liberated from $K_2[NbF_7]$ at 800 °C using AlF_3 as fluoride acceptor [1749]. The pentafluorides are white, thermally and air-stable, volatile solids, which can be purified by vacuum sublimation. NbF_5 and TaF_5 both crystallize in the NbF_5 lattice, where tetrameric, *cis*-F-bridged $[MF_5]_4$ units are present. The four metal atoms form a square and are each octahedrally coordinated by fluorine. In the tetrameric units, the M-F-M bridges are almost linear (178°) [1769]. The aforementioned structure is retained in the highly viscous melt, however, the obtained Raman-spectra of liquid NbF_5 and TaF_5 depend on the temperature [1677, 1770]. In the vapor phase, an equilibrium between oligomeric $(MF_5)_n$ ($n \leq 4$) and monomeric MF_5 is observed. At lower temperatures, the pentafluorides are associated, whereas they are mainly present as trigonal-bipyramidal monomers at 400 °C [468, 1673, 1771, 1772]. TaF_5 can be dissolved in organic solvents like acetonitrile, dichloromethane or diethyl ether. In acetonitrile, it also forms adducts, which are precipitated upon cooling. Furthermore, NMR-spectroscopic investigations point to the existence of ionic species, such as $[TaF_4(MeCN)_x]^+$ ($x \geq 2$) and $[TaF_6]^-$ [1773]. The pure, solid pentafluorides do not react with dry glass, but in the molten state, they slowly attack it [1749]. Both NbF_5 and TaF_5 are chemically very reactive, however, the reactivity of the pentafluorides in group 15 decreases from VF_5 to TaF_5. 1:1- and 1:2-adducts are readily formed with many neutral and anionic ligands, for example, Et_2O, Me_2S, pyridine or NH_3 [1774]. There are also complexes with the phosphorus(V) containing ligands Ph_3PO, $Ph_2P(O)OEt$ and $Ph_2P(O)SEt$, which are generally described by the formulas $[MF_5L]$, $[M_2F_{10}L]$ as well as $[MF_4L_2]^+$ and $[M_2F_{11}]^-$ [1775]. Recently, complexes of NbF_5 and TaF_5 with a NHC-ligand have also been prepared [1776]. The reaction of NbF_5 and TaF_5 with SO_3 gives rise to the fluorosulphates $NbF_3(SO_3F)_2$ and $TaF_3(SO_3F)_2$, respectively (Scheme 10.12). The latter two compounds are viscous liquids, which fume and decompose on air, in addition they are thermally degraded to liberate SO_2F_2 [1692]. NbF_5 reacts in a 1:1 ratio with SbF_5 to yield an adduct with an infinite chain structure of alternating $[NbF_6]$- and $[SbF_6]$-units, which are bridged by asymmetric *cis*-F-bridges. There are ionic contributions to the bonding situation so that the adduct may also be formulated as $[NbF_4]^+[SbF_6]^-$ (Scheme 10.12) [444, 1777]. Complexes are also formed with SeF_4 to yield $SeF_4 \cdot 2\,NbF_5$, $SeF_4 \cdot NbF_5$ and $SeF_4 \cdot TaF_5$. In contrast to SbF_5, SeF_4 is a weaker

$K_2[NbF_7]$

$[MF_6]^-, [MF_7]^{2-}, [MF_8]^{3-},$
$[NbOF_5]^{2-}$ *or higher aggregates*

$AlF_3,$
800 °C

molten alkali
metal fluorides

metals, oxides
or chlorides

fluorination

MF_5

SbF_5

(M = Nb)

$[NbF_4][SbF_6]$

SeF_4

SO_3

F⁻, H₂O

$[SeF_3][MF_6]$
or $[SeF_3][Nb_2F_{11}]$

$MF_3(SO_3F)_2$

$[MF_6]^-$ *or* $[NbOF_5]^{2-}$

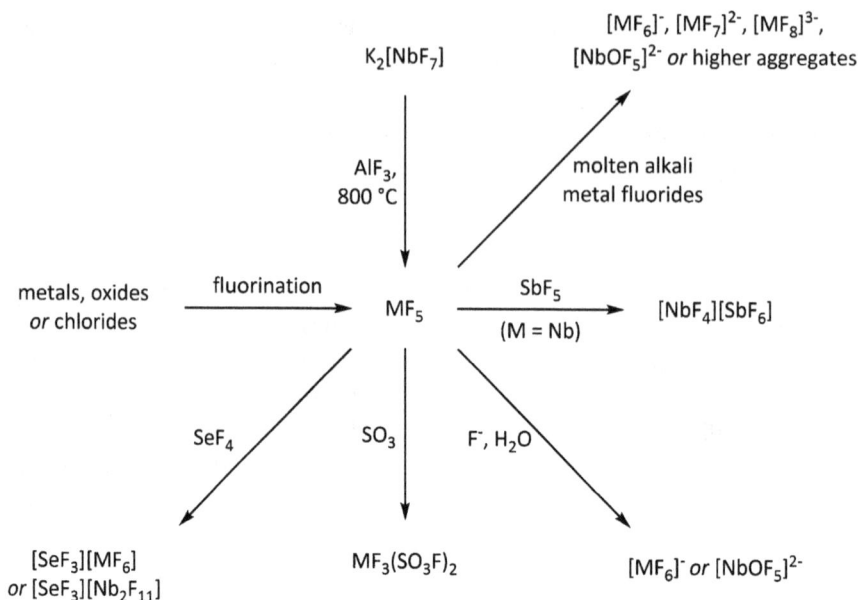

Scheme 10.12: Preparation and selected reactions of the group 5 pentafluorides MF₅ (M = Nb, Ta).

Lewis acid than NbF₅. Thus, the complexes contain $[SeF_3]^+$ cations and $[Nb_2F_{11}]^-$ or $[MF_6]^-$ anions, respectively (Scheme 10.12) [694, 695, 1778]. There are numerous substitution products of NbF₅ and TaF₅ where one or more F-atoms have been replaced by other negatively charged substituents, for example, Cl⁻, Br⁻, MeO⁻ or PhO⁻.

NbF₅ and TaF₅ also form a number of fluorocomplexes (Scheme 10.12). The historically most important salts are certainly K₂[NbOF₅] and K₂[TaF₇], as the first separation of the elements Nb and Ta has been realized by fractional crystallization of these compounds. In the process, the well-soluble Nb-salt remains in aqueous solution, whereas the Ta-salt is heavily soluble in water and precipitates [1779]. Nowadays, the same complexes are separated by fractional extraction with ketones. Fluorocomplexes of NbF₅ and TaF₅ are accessible from melts of the respective fluorides or from reactions with HF. In the latter case, the outcome strongly depends on the concentration of the hydrofluoric acid. Dilute HF-solutions (w < 30 %) yield the anion $[NbOF_5]^{2-}$, whereas high (w = 30–95 %) and very high HF-concentrations lead to the formation of $[NbF_6]^-$ and $[NbF_7]^{2-}$, respectively [1780]. The dependence on the HF-concentration is similar and also significant for the respective fluorocomplexes of tantalum(V). However, in contrast to Nb, the Ta-species do not tend to be hydrolysed, and thus, only binary fluoroanions are found [1781, 1782]. The different types of fluorometallates are $M^I[MF_6]$ (M^I = alkali metal, NH₄, Tl, Ag, Cu, NO, O₂);[1691, 1692] $M_2^I[MF_7]$ (M^I = alkali metal); $M^{II}[MF_7]$ (M^{II} = Mg, Mn, Fe, Co, Ni, Cu, Zn, Cd) and $M_3^I[MF_8]$ (M^I = Li, Na, K, NH₄, Tl) [1783]. The cation of the fluorometallate salts can of course also be organic, as for example, the protonated NHC-moiety in the salts $[(dipp)H][NbF_6]$ and $[(dipp)H][TaF_6]$ (dipp = 1,3-*bis*(2,6-diisopropylphenyl)-1,3-dihydro-2*H*-imidazol-2-ylidene) and their respective solvates [1776]. The hexafluorometallates $M^I[MF_6]$ contain octahedrally coordinated anions, in the systems TlF-NbF_5 and KF-NbF_5 it has been possible to observe the di- and triniobates $M^I[Nb_2F_{11}]$ and $M^I[Nb_3F_{16}]$ [1784]. TaF_5 also yields the fluoroanions $[TaF_6]^-$ and $[TaF_7]^{2-}$ in molten alkali metal fluoride-mixtures. Interestingly, $[TaF_7]^{2-}$ is the exclusive product, if KF is present [1785]. Under similar conditions, NbF_5 yields $[NbF_7]^{2-}$ along with the oxyfluoride derivatives $[NbOF_5]^{2-}$ and $[NbO_2F_4]^{3-}$ [1783, 1786]. In melts of NaF-$K_2[NbF_7]$ as well as LiF-$K_2[NbF_7]$, the octafluoronio-bate(V) anion $[NbF_8]^{3-}$ is observed. If KF is added to the respective mixtures, $[NbF_8]^{3-}$ can also be generated in the solid state [1787]. In contrast, in a diluted aqueous HF-solution, the formation of the hexafluorometallates(V) $[MF_6]^-$ cannot be overcome (Scheme 10.12) [1788]. The solid-state structure of $K_2[NbF_7]$ exhibits isolated anions, which can be described as distorted, monocapped trigonal prisms [1789]. As has been shown by ^{18}F-labeling experiments, the complex anions $[NbF_6]^-$ and $[TaF_6]^-$ are ki-netically labile and exchange their fluorine-substituents in an acetonitrile solution at room temperature [1790]. Apart from that, solutions of the hexafluorometallates $[NbF_6]^-$ and $[TaF_6]^-$ can be treated with BF_3 and PF_5, which results in the forma-tion of complex anions of the type $[M_2F_{11}]^-$ (M = Nb, Ta), along with the respective $[BF_4]^-$ or $[PF_6]^-$ species [1791]. Pyrolytic decomposition of the ammonium complexes $(NH_4)_2[MF_7]$ in a stream of ammonia yields niobium and tantalum nitride compounds [1628]. If NbF_5 and TaF_5 are reacted with $ClOF_3$, they act as fluoride acceptors and lead to the formation of the ionic compounds $[ClOF_2][NbF_6]$ as well as $[ClOF_2][TaF_6]$ [470]. Interestingly, the reaction of $[Sr(HF)_3(TaF_6)_2]$ in liquid ammonia leads to the formation of the neutral, octacoordinated complex $[TaF_5(NH_3)_3]$, indicating, that the fluoride-accepting properties of TaF_5 in liquid NH_3 are not that strong [1792]. The partial hydrolysis of $[Et_4N][TaF_6]$ in MeOH yields $[Et_4N]_2[Ta_2OF_{10}]$ with a linear Ta-O-Ta bond [1793]. In a comparable system, tantalum ethoxyfluoro complexes are generated from TaF_5-$Ta(OEt)_5$ and $Ta(OEt)_5$-HF, respectively. The neutral, anionic as well as cationic species have been identified by ^{19}F-NMR spectroscopy [1794]. The guanidinium salts $[C(NH_2)_3]_2[TaF_7]$ and $[C(NH_2)_3]_3[TaF_8]$ have been investigated by single crystal X-ray diffraction and show isolated mono-capped trigonal prismatic $[TaF_7]^{2-}$ anions and square-antiprismatic $[TaF_8]^{3-}$, respectively [1795].

Pentavalent oxyfluorides
Niobium oxytrifluoride **NbOF$_3$** and tantalum oxytrifluoride **TaOF$_3$** are obtained upon thermal decomposition of $MF_3(SO_3F)_2$ (M = Nb, Ta) at 175–225 °C or partial hydroly-sis of the pentafluorides [1692, 1796]. $TaOF_3$ is also accessible by thermal treatment of TaF_5 with SiO_2 [1797]. In the same way, an oxygen-fluorine exchange reaction of the pentafluorides NbF_5 and TaF_5 with $TiO(acac)_2$ (acac = acetylacetonate) yields the oxytrifluorides $NbOF_3$ and $TaOF_3$ along with $TiF_2(acac)_2$ [1660]. Both $NbOF_3$ and

$TaOF_3$ crystallize in a variant of the SnF_4 structure type, which can be described by sheets of corner-sharing $[MX_6]$-octahedra (X = O, F). The O- and F-atoms are statistically distributed, so that no ordering can be observed. Van der Waals interactions of the sheets lead to a three-dimensional network [1798]. Gas-phase electron diffraction experiments on $NbOF_3$ supported by mass spectrometry have revealed a C_{3v}-symmetric molecular structure [1799]. The oxytrifluorides are sensitive toward moisture and are thus immediately hydrolyzed in water. They are completely soluble in hot, concentrated sulphuric acid and aqueous HF (w = 20 %). Upon heating, $NbOF_3$ (280 °C) and $TaOF_3$ (375 °C) are decomposed to the respective pentafluorides along with the dioxyfluorides MO_2F. However, the system is in a temperature-dependent equilibrium, as the oxytrifluorides are observed as main components at lower temperatures (180–220 °C) [1796, 1798]. Analogously to VOF_3, $NbOF_3$ forms 1:2 complexes with mono- as well as bidentate N- and O-donor ligands, such as 2,2'-bipyridine, 1,10-phenanthroline or different phosphineoxides [1800, 1801]. The fluoride substituents of $NbOF_3$ can be exchanged using Me_3SiN_3, resulting in the formation of explosive niobium oxotriazide $NbO(N_3)_3$, which can further be transformed to N-donor adducts or $[NbO(N_3)_5]^{2-}$-salts [1802].

The dioxyfluorides **NbO_2F** and **TaO_2F** are synthesized from the metals or the metal pentoxides by treatment with hydrofluoric acid (w = 48 %) and subsequent evaporation at 250 °C [1803]. They are also generated upon reaction of the pentafluorides with quartz reaction-vessels [1796]. However, they are hardly obtained in very pure form under aqueous conditions. For this purpose, solid-state reactions have been found to be a suitable alternative [1804]. NbO_2F and TaO_2F crystallize in the cubic ReO_3 structure, where the O- and F-atoms are irregularly distributed over the octahedral positions around the metal centers. As expected, the $[MX_6]$-units (M = O, F) are distorted from ideal octahedral geometry [1804, 1805]. Above 500 °C, NbO_2F decomposes to yield $NbOF_3$, which in turn decays to NbO_2F and NbF_5 upon cooling [1803]. There are also complex oxyfluorometallate salts, mainly niobium compounds, with the general compositions $M^I[MOF_4]$, $M_2^I[MOF_5]$, $M_3^I[MOF_6]$ and $M_3^I[MO_2F_4]$ as well as polymetallates [1533, 1783, 1786, 1806–1814]. The pentafluoroperoxoniobate(V) anion $[Nb(O_2)F_5]^{2-}$ is also known, it has the structure of a monocapped octahedron [1815–1817]. The introduction of Ag^+ into NbO_2F yields a compound, which is photocatalytically active, as it absorbs near ultraviolet light [1818]. NbO_2F has also been proposed as a material for lithium batteries, as it shows appreciable Li-insertion properties. However, for the application as an anode material, it would certainly be too expensive [1542, 1819–1821]. Despite the close similarity of the chemical and physical properties of NbF_5 and TaF_5, the respective oxyfluoride compounds differ in their thermal behavior. Thus, this can be advantageous for the separation of Ta/Nb-mixtures, which otherwise typically requires sophisticated processes. It has been shown that the fluorination of Nb_2O_5/Ta_2O_5-mixtures with HF results in volatile oxyfluorides, which can be separated in temperature ranges below 165 °C or above 650 °C [1822]. Apart from that, fluorination of Nb-compounds is a general method for waste-recycling purposes, as volatile fluorides and oxyfluorides can easily be separated from spent nuclear fuels or other mixtures [1823].

10.5.3 Study questions

(I)   How does one prepare VF_2? Which species is obtained, if it is reacted with alkali metal fluorides? Comment on the stability of divalent vanadium-fluorine compounds.

(II)   Describe the structure of VF_3. How can it be prepared?

(III) To which main group tetrafluoride may the solid-state structure of VF_4 be compared? What is the main difference?

(IV)   How can VOF_2 be prepared? Which fluoroanions does it generate upon reaction with fluoride ions and how can they be structurally described?

(V)   How is VF_5 prepared? Which general reactivity is observed against oxygen-containing compounds, oxidizable materials and unsaturated organic compounds?

(VI)   Which pentavalent vanadium oxyfluorides are known? How can one prepare them? How are they structurally described?

(VII) How are the trifluorides NbF_3 and TaF_3 prepared? What is the main obstacle in the synthesis of the pure compounds?

(VIII) How is NbF_4 synthesized? Comment on its thermal stability and its fluoride accepting properties.

(IX)   Describe the structures of the pentafluorides TaF_5 and NbF_5 in the solid, liquid and gaseous state.

(X)   Which fluoroanions are derived from NbF_5 and TaF_5? Can they be obtained from aqueous systems? If not, how are they prepared?

(XI)   How are the pentavalent oxyfluorides of Ta and Nb prepared? Are they thermally stable?

10.6 Group 6: Chromium, molybdenum, tungsten

The elements of group 6 exist in binary fluorides in oxidation states from +II up to +VI. However, there is a significant change from chromium to tungsten. Chromium is the only metal in group 4, which forms a stable difluoride, in contrast, chromium hexafluoride is unknown. In case of molybdenum and tungsten, the hexafluorides are stable, whereas WF_3 is still unknown in the condensed phase. Comparable to the corresponding fluorides of 2d- and 3d-metals in other groups, for example, Zr/Hf or Nb/Ta, molybdenum and tungsten fluorides have very similar properties.

10.6.1 Chromium

Low-valent fluorides

Chromium monofluoride **CrF** can only be observed in the gas phase. It is generated, if equimolar amounts of elemental Cr and CrF_2 are heated [1824]. The electronic properties of CrF have been investigated by microwave spectroscopy and quantum-chemical calculations [1825, 1826].

Chromium difluoride **CrF$_2$** is a green solid (mp.: 894 °C), which is formed upon reduction of aqueous CrF$_3$ with H$_2$ at 500–600 °C. Alternatively, it is obtained in pure form, when CrF$_3$ or (NH$_4$)$_3$[CrF$_6$] are pyrolyzed at 1100 °C, as Cr(III) is converted to CrF$_2$ and F$_2$ under these conditions (Scheme 10.13). A few more, but less important preparations have also been reported [1827]. For example, electrolytically purified Cr-metal can be dissolved in aqueous HF under formation of green solutions, which indeed contain Cr(II). However, pure CrF$_2$ solutions cannot be prepared by this method [1828]. In contrast, a clean formation of CrF$_2$ is observed, if anhydrous HF is reacted with Cr-metal [204, 1624]. CrF$_2$ crystallizes in a Jahn–Teller distorted rutile lattice with a slightly deformed octahedral coordination geometry at the Cr-centers [1829, 1830]. Gas-phase electron diffraction techniques as well as quantum-chemical calculations have been applied to determine the ground-state structure of the CrF$_2$-molecule. Even if the results remain uncertain, a bent structure is suggested for gaseous CrF$_2$ [1831]. The dihydrate CrF$_2$ · 2 H$_2$O is accessible from aqueous HF-solutions and Cr metal under anaerobic conditions. It is a pale-blue powder, which is well soluble in water. Chemically, CrF$_2$ · 2 H$_2$O is rather inert as it is only slowly oxidized in contact with air to give Cr$_2$O$_3$. CrF$_2$ · 2 H$_2$O can be dehydrated at 100 °C *in vacuo*, resulting in pale-blue, anhydrous CrF$_2$, which is pyrophoric on air, and thus a rather strong reducing agent [1832]. When it is reacted with fluorine, CrF$_2$ is mainly oxidized to CrF$_3$. The treatment with N$_2$ at 650 °C gives rise to chromium(III) nitridefluoride Cr$_2$NF$_3$, which decomposes above 850 °C to result in a mixture of CrF$_3$ and CrN [1833]. In general, CrF$_2$ readily adds water or ammonia to form the respective cations [CrL$_6$]$^{2+}$ (L = H$_2$O, NH$_3$). However, the neutral ammine complex [CrF$_2$(NH$_3$)$_4$] is also known, where the F-atoms occupy *cis*-positions of a distorted octahedron. A few complex salts are derived from CrF$_2$, for example the light-blue, oxidatively very sensitive M^I[CrF$_3$] (M^I = K, Rb, NH$_4$), which have a distorted perovskite structure with tetragonally-compressed [CrF$_6$]-octahedra (Scheme 10.13). In contrast, Na$_2$[CrF$_4$] and Sr[CrF$_4$] are stable on air and contain the square-planar [CrF$_4$]$^{2-}$ anion [1834, 1835].

Upon incomplete reduction of CrF$_3$ or heating of a mixture of CrF$_2$ and CrF$_3$ to 800–900 °C, a mixed fluoride of the approximate composition **Cr$_2$F$_5$** is generated, wherein Cr(II) and Cr(III) are both octahedrally surrounded by F-atoms (Scheme 10.13) [1827]. In the crystalline state, one can find the single structures of CrF$_2$ and CrF$_3$, the octahedral units are linked via common edges and corners [1836]. Cr$_2$F$_5$ is also quite inert and reacts with fluorides of the type M^{I}F to form the chromium(II) complexes Na$_2$[CrF$_4$], K[CrF$_3$] and NH$_4$[CrF$_3$] · 2 H$_2$O. There are also mixed complex salts, for example, K$_x$[Cr^{II}$_x$Cr^{III}$_{1-x}$F$_3$] and M^I[M^{II}Cr^{III}F$_6$] (M^I = Rb, Cs, Tl; M^{II} = Fe, Mg, Ni, Mn) [1837]. The compounds CrMF$_5$ (M = Al, Ti, V) are isostructural with Cr$_2$F$_5$ [1838]. The magnetic properties of the aforementioned compounds have been explored, revealing ferrimagnetism for CrTiF$_5$ and CrVF$_5$, whereas Cr$_2$F$_5$ is antiferromagnetic. In contrast, CrAlF$_5$ is paramagnetic [1836].

CrF$_3$ or (NH$_4$)$_3$[CrF$_6$]   [CrF$_3$]$^-$ or [CrF$_4$]$^{2-}$

1100 °C,
pyrolysis   F$^-$

                   H$_2$, 500-600 °C   CrF$_3$, 800-900 °C
aq. CrF$_3$ $\longrightarrow$ CrF$_2$ $\longrightarrow$ Cr$_2$F$_5$

                   aHF   T > 1100 °C
Cr $\longrightarrow$ CrF$_3$ $\longrightarrow$ CrF$_2$ + F$_2$

Δ, vacuum   F$^-$

(NH$_4$)$_3$[CrF$_6$]   [CrF$_4$]$^-$, [CrF$_5$(H$_2$O)]$^{2-}$
                                              or [CrF$_6$]$^{3-}$

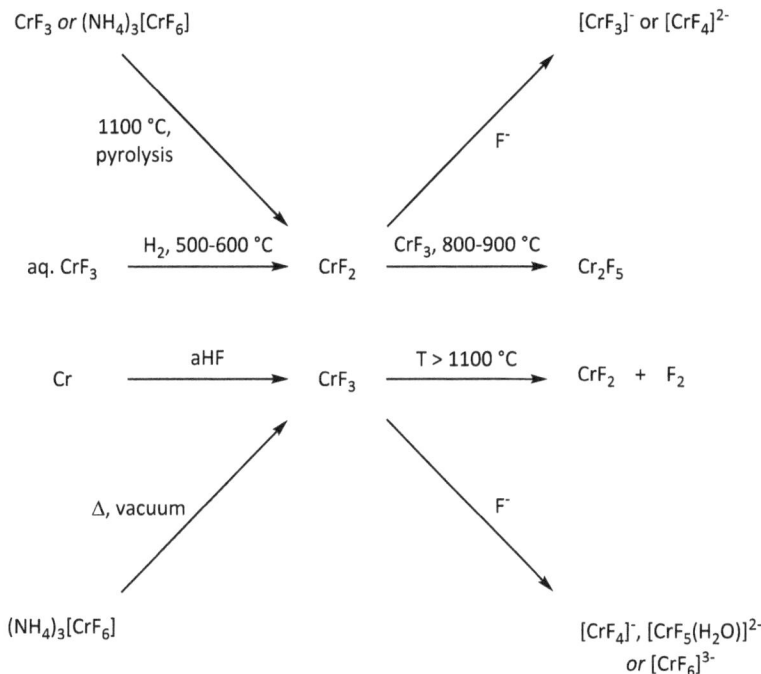

Scheme 10.13: Preparation and selected reactions of CrF$_2$ and CrF$_3$.

Chromium trifluoride

Chromium trifluoride **CrF$_3$** is formed upon reaction of chromium with anhydrous HF or thermal decomposition of (NH$_4$)$_3$[CrF$_6$] *in vacuo* (Scheme 10.13) [204, 1483, 1624, 1827]. Furthermore, CrCl$_3$ or Cr$_2$O$_3$ can be treated with anhydrous HF in a Pt-tube at 1100 °C [1839]. Amorphous CrF$_3$ with a high surface area is accessible, if the hydrazinium complex [N$_2$H$_6$][CrF$_5$] · H$_2$O is treated with F$_2$ in anhydrous HF at room temperature [1840]. CrF$_3$ is a green solid (mp.: 1404 °C), which crystallizes in the VF$_3$ structure type and exists as a monomer in the gas phase [1841]. CrF$_3$ is stable in water and forms hydrates, for example, the violet nonahydrate CrF$_3$ · 9 H$_2$O, which can be reformulated as [Cr(H$_2$O)$_6$]F$_3$ · 3 H$_2$O [1842]. In the solid phase, CrF$_3$ · 3 HF · 3 H$_2$O is generated, which is alternatively written as the complex acid [H$_3$O]$_3$[CrF$_6$]. CrF$_3$ is reduced by strong reducing agents to yield CrF$_2$, at 1100 °C it dissociates to CrF$_2$ and F$_2$ (Scheme 10.13). It also forms a number of fluorochromates, where chromium is always octahedrally coordinated, for example, M^I[CrF$_4$], (M^I = alkali metal, Tl);[1843, 1844] M$_2^I$[CrF$_5$(H$_2$O)] (M^I = K, Rb, NH$_4$, Tl)[1845, 1846] or M$_3^I$[CrF$_6$] (M^I = Li, Na, K, NH$_4$). In the latter case, the compounds can also be composed of different cations [1485, 1845–1850]. The hexafluorochromate(III) anion [CrF$_6$]$^{3-}$ is paramagnetic and has been investigated by EPR-spectroscopy [1851]. For example, [CrF$_6$]$^{3-}$ salts are accessible by treatment of CrCl$_3$ with pyridinium poly(hydrogen fluoride) [1694]. Apart from the aforementioned salts with simple complex anions, the cluster compounds $Na_5[Cr_3F_{14}]$, $Sr_5[Cr_3F_{19}]$[1852] and $K_2[Cr_5F_{17}]$ are known. $Cs[CrF_4]$ shows an interesting structural behavior, as it is composed of linear chains of *trans*-F-bridged $[CrF_6]$-octahedra, which are aggregated to triple strands [1844]. The solid-state structure of $Na_3[CrF_6]$ is described by the cryolite type [1848]. Thermolysis of the ammonium salt $(NH_4)_3[CrF_6]$ at 600 °C exhibits a synthetic route toward chromium nitride CrN [1628].

The oxyfluoride **CrOF** has first been described as dark-green, powdered product of the fluorination of Cr_2O_3 in gaseous HF [1839]. The early preparation of this compound has been doubted, as it has never been confirmed by any other publication. Furthermore, CrOF has only recently been obtained as by-product in an argon matrix, when Cr metal has been reacted with OF_2. Quantum-chemical calculations suggest a bent structure for the oxyfluoride molecule [1853]. CrOF has also been investigated by DFT-methods concerning its behavior in two-dimensional sheets. As could be shown, it has intrinsic ferromagnetic semiconductor properties, making it interesting for potential applications in spintronic devices, that is, storage media [1854].

Chromium tetrafluoride

Chromium tetrafluoride **CrF$_4$** is a green-black solid (subl.: 100 °C; mp.: 277 °C (calc.)), which is formed in its α-modification upon fluorination of chromium or its trihalides CrF_3 and $CrCl_3$ at 300–350 °C (Scheme 10.14) [1855–1857]. In contrast, the dark-violet, crystalline β-CrF_4 is generated, when CrF_5 is submitted to long-term thermolysis (5 months, 130 °C) [1858]. CrF_4 is readily hydrolyzed but can be stored in a dry atmosphere, even above its melting point. Apart from hydrolysis, CrF_4 is relatively inert, as it does not react with NH_3, SO_2, SO_3, SeF_4, halogen fluorides or pyridine at room temperature. However, in boiling BrF_3 or SeF_4, CrF_4 is reduced to lower fluorides, whereas a mixture of BrF_3 and BrF_5 yields the oxyfluoride adduct $CrOF_3 \cdot 0.25\ BrF_3$ [1859]. The CrF_4 molecule has T_d-symmetry, as has been shown by vibrational spectroscopy in a Ne-matrix [1856, 1860]. In the solid state, the chain-like α-modification is composed of dimers of $[CrF_6]$-octahedra, which are connected by common edges. The resulting $[Cr_2F_{10}]$-units are linked by *trans*-F-atoms to form columns [1857]. β-CrF_4 consists of rings of four *cis*-bridged $[CrF_6]$-units, which are connected via common corners. The cyclic tetramers are linked to the neighboring rings via their *trans*-F-atoms, giving rise to the formation of tubes. As a result, two free *cis*-F-atoms remain in every $[CrF_6]$-octahedron [1858]. CrF_4 forms a number of fluorocomplexes of the compositions $M^I[CrF_5]$ (M^I = K, Rb, Cs); $M^I_2[CrF_6]$ (M^I = Li, K, Cs) and $M^{II}[CrF_6]$ (M^{II} = Mg, Ca, Sr, Ba, Zn, Cd, Hg, Ni), which can be obtained by fluorination of mixtures of $CrCl_3$ and metal halides [1859, 1861–1864]. In addition, some of the alkali metal derivatives are generated by thermal decomposition of the respective hexafluorochromate(V) complexes $M^I[CrF_6]$ (M^I = alkali metal) [1865]. The fluorochromate(IV) complexes are also accessible by treatment of CrF_4 with the respective metal chloride in BrF_3-solution (Scheme 10.14) [1859]. The resulting fluorochromates(IV) contain the anions $[CrF_5]^-$

CrF$_5$ [CrF$_5$]$^-$ *or* [CrF$_6$]$^{2-}$

130 °C, 5 months   Cl$^-$, BrF$_3$

Cr, CrF$_3$ or CrCl$_3$ $\xrightarrow{\text{F}_2,\ 300\text{-}350\ °C}$ CrF$_4$ $\xrightarrow{\text{XeF}_6}$ [XeF$_5$][CrF$_5$]

CrF$_3$ or CrO$_2$F$_2$ $\xrightarrow{\text{F}_2}$ CrF$_5$ $\xrightarrow{\text{SbF}_5}$ [CrF$_4$][SbF$_6$]

F$_2$, 400 °C, 200 bar   MF$_3$ (M = P, As, Sb)   CsF, NOF or NO$_2$F

Cr   CrF$_3$ + MF$_5$ [CrF$_6$]$^-$ salts

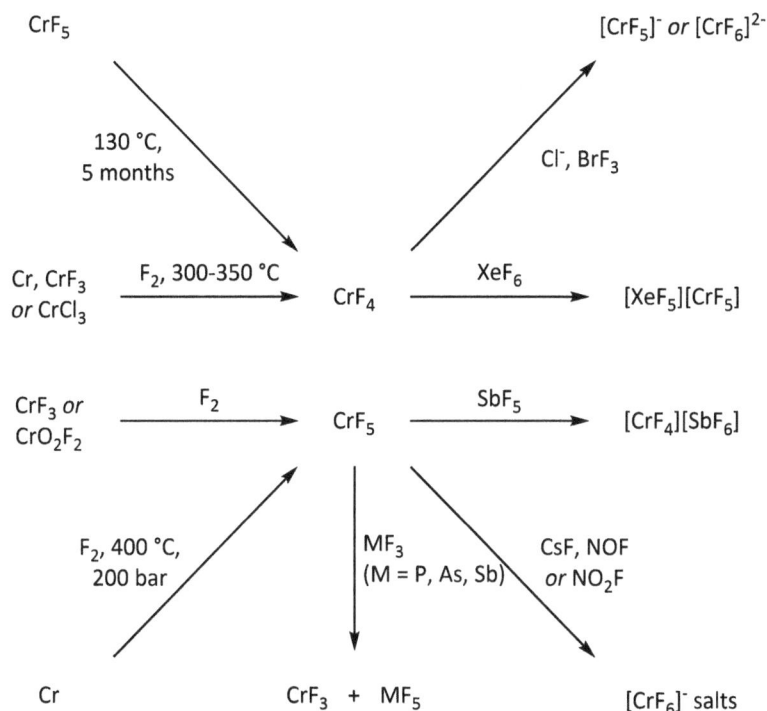

Scheme 10.14: Preparation and selected reactions of CrF$_4$ and CrF$_5$.

or [CrF$_6$]$^{2-}$, where Cr is octahedrally coordinated. In case of [CrF$_5$]$^-$, the *cis*-bridged [CrF$_6$]-octahedra share a common F-atom with their neighbors, resulting in the formation of zig-zag chains. Salts with the [CrF$_6$]$^{2-}$ anion contain discrete, distorted octahedral anions, which show hydrogen-bonding interactions with HF-molecules [1865]. The reaction of elemental chromium with XeF$_2$ and XeF$_6$ leads to the formation of CrF$_2$ and CrF$_3$, along with the complex CrF$_4$ · XeF$_6$ [1866]. The latter is directly accessible by the reaction of CrF$_4$ with XeF$_6$ at room temperature, and can be formulated as [XeF$_5$][CrF$_5$], as it is composed of infinite, anionic chains of distorted [CrF$_6$]-octahedra and discrete [XeF$_5$]$^+$ cations (Scheme 10.14). 1:1- and 2:1-adducts of CrF$_4$ and XeF$_2$ are also well known. The 2:1-compound is forming a three-dimensional network, each Cr-center being distorted-octahedrally coordinated by six F-substituents [1032, 1867]. Another Cr(IV) derivative is the oxydifluoride **CrOF$_2$**, which has been claimed to be formed as a brown-black product upon thermal decomposition of CrO$_2$F$_2$ [1868]. It is also obtained under matrix conditions, if Cr metal is treated with OF$_2$. The molecule has C$_{2v}$-symmetry and strongly ionic Cr-F bonds, as has been shown by infrared spectroscopy and quantum-chemical calculations. Furthermore, the Cr-O-bond has a double-bond character [1853].

Chromium pentafluoride

Chromium pentafluoride **CrF_5** is the only known pentahalide of chromium. It is generated by fluorination of elemental chromium at 400 °C and 200 bar (Scheme 10.14) [1855, 1869, 1870]. More conveniently, it is prepared from CrO_2F_2 and F_2 at rather mild reaction conditions or from CrF_3 and F_2 [1856, 1871]. Generally, the yields of CrF_5 decrease at higher reaction temperatures [1872]. CrF_5 is a volatile, strongly oxidizing, low-melting, purple solid (mp.: 30 °C; bp.: 117 °C). In the gas phase, CrF_5 exists in monomeric form, having C_{2v}-symmetry due to Jahn–Teller distortion, as has been shown by matrix-isolation spectroscopy [468, 1673, 1860, 1872, 1873]. A polymer of cis-F-bridged $[CrF_6]$-octahedra is found in the liquid as well as in the solid state [1856, 1874, 1875]. Under the impact of a mercury arc lamp, CrF_5 is photolyzed to yield CrF_4, however, it shows no tendency to disproportionate to CrF_4 and CrF_6 [1860]. It is very sensitive toward hydrolysis and difficult to purify. Moreover, it already attacks glass at its melting point. In contact with compounds like PF_3, AsF_3 or SbF_3, CrF_5 is readily reduced to CrF_3 (Scheme 10.14). The same behavior is observed with additional halide exchange for the reaction of CrF_5 with PCl_3, which results in $CrCl_3$, PF_3 and PCl_5 [1876]. CrF_5 is also capable of oxidizing Xe and XeF_2 to XeF_2 and XeF_4, respectively, where itself is reduced to CrF_3 [1877]. At slightly elevated temperature, XeF_6 and CrF_5 react to form $[XeF_5][CrF_6]$, which is composed of infinite, anionic chains of distorted, cis-bridged $[CrF_6]$-octahedra and $[XeF_5]^+$ cations [1032]. In the reaction with SO_3, peroxydisulphuryl difluoride $(FSO_2O)_2$ is formed along with chromium(III) fluorosulphate $Cr(SO_3F)_3$ [1878]. CrF_5 also forms 1:1- and 1:2-complexes with SbF_5. In $CrF_5 \cdot SbF_5$, chains of alternating $[CrF_6]$- and $[SbF_6]$-octahedra are found. Interestingly, the bridging-positions at the Cr-centers are in cis-relation to each other, whereas trans-bridges are observed for the $[SbF_6]$-units. The F-bridges are highly asymmetric, so that the 1:1-adduct may also be formulated as $[CrF_4]^+[SbF_6]^-$ (Scheme 10.14) [1875]. Furthermore, an adduct of the composition $CrF_5 \cdot 2\,SbF_5$ is formed, which is a deep red-brown and viscous liquid with strongly oxidizing properties, as it can react with O_2 to yield the dioxygenyl salt $[O_2][CrF_4Sb_2F_{11}]$. It can also oxidize Xe and hexafluorobenzene C_6F_6, however, it is not capable of converting NF_3 to $[NF_4]^+$ [1879, 1880]. The hexafluorochromate(V) complexes $M^I[CrF_6]$ (M^I = alkali metal) are accessible from a mixture of the respective M^IF, F_2 and CrF_3 in anhydrous HF at ambient temperature. The hexafluorochromate(V) complexes readily undergo thermal decomposition to yield tetravalent $M^I[CrF_5]$ and $M_2^I[CrF_6]$, which are then partially solvolyzed in anhydrous HF, resulting in precipitation of neutral CrF_4. From the remaining solutions, several different complexes can be crystallized, among them the potassium salt $K_3Cr_2F_{11} \cdot 2$ HF, which contains a rare example of a $[M_2F_{11}]^{3-}$ anion [1865]. CrF_5 reacts with CsF and NO_2F at 60 and 35 °C, respectively, to yield the red salts $Cs[CrF_6]$ and $[NO_2][CrF_6]$ with the octahedral anion $[CrF_6]^-$ (Scheme 10.14) [1880]. Similarly, NOF gives rise to the complex $[NO][CrF_6]$, which can be transformed to $[NO]_2[CrF_6]$ upon reduction with NO. The controlled pyrolysis of $[NO]_2[CrF_6]$ generates $[NO][CrF_5]$. The reaction of

CrF_5 with excess $[NF_4][HF_2]$ yields the stable salt $[NF_4][CrF_6]$ [1879]. The reaction of CrF_5 with dichloromethane in HF results in an unselective fluorination of the organic molecule [1698]. Under matrix conditions, the dimer **Cr_2F_{10}** has been identified. It consists of two distorted octahedrally coordinated Cr-centers, which are *cis*-bridged by two of the F-substituents [1881].

The purple oxyfluoride **$CrOF_3$** (decomp.: 500 °C) is also a strong oxidizer and sensitive toward hydrolysis. It is obtained upon fluorination of CrO_3 with BrF_3, BrF_5 or ClF_3, but never in pure form, as it always has the formal composition $CrOF_3 \cdot 0.25$ XF_n (X = Cl, n = 3; X = Br, n = 3 or 5) [1882]. If the latter compounds are treated several times with F_2 at 120 °C, the pure, bright purple $CrOF_3$ is obtained, which is polymeric in the solid state, containing three-dimensional networks of corner-sharing $[CrOF_5]$-octahedra [1883, 1884]. More conveniently, crystalline $CrOF_3$ is generated by the reaction of CrO_2F_2 with XeF_2 [1884]. Upon reaction with fluorides, it forms the complex anions $[CrOF_4]^-$ and $[CrOF_5]^{2-}$. The reaction of $M_2^I Cr_2O_7$ (M^I = K, Cs) with BrF_3, BrF_5 or ClF_3 yields the tetrafluorooxychromates(V) $M^I[CrOF_4]$, which are very sensitive toward hydrolysis [1882, 1885]. The potassium salt can also be generated by the reaction of $CrOF_3$ and KF in anhydrous HF [1883].

Chromium hexafluoride and hexavalent oxyfluorides of chromium

For quite some time, chromium hexafluoride **CrF_6** had been believed to be formed along with CrF_5 at the high-pressure fluorination of elemental chromium at 400 °C. It had been described as a lemon yellow, very unstable compound, which already decomposes at −100 to −80 °C under formation of CrF_5 and F_2 [1870]. The fluorination of CrO_3 at 170 °C and a F_2 pressure of 25 atmospheres has also been described [1856, 1886]. However, upon attempts to reproduce the aforementioned experiments, it became clear that CrF_5 is the highest chromium fluoride known [1860, 1872]. Even under matrix conditions, no proof for the existence of CrF_6 has been found to date [1881]. Quantum-chemical calculations revealed that a potential CrF_6 species would have a very high electron-affinity in the order of PtF_6 [1887]. Furthermore, theory has initially suggested, that there is no minimum structure with O_h-symmetry, and thus, a potential CrF_6 would not be octahedral but trigonal-prismatic [1888]. However, when the quality of the quantum-chemical calculations had been further improved, the octahedral geometry was clearly found to be the energetically favored one [1889–1891]. Nevertheless, the activation barrier for a pseudo-rotation from the ground-state octahedral structure to the trigonal-prismatic geometry is predicted to be low [1892].

Despite the nonexistence of CrF_6, hexavalent chromium compounds are well known in form of the oxyfluorides **$CrOF_4$** (deep red; mp.: 55 °C) and **CrO_2F_2**, which are generally synthesized upon fluorination of CrO_3, K_2CrO_4 or $K_2Cr_2O_7$ [688, 700, 865, 1869, 1882, 1893–1897]. The best yields are achieved by treatment of CrO_3 with F_2 at 150 °C (CrO_2F_2) and 220 °C ($CrOF_4$) or the reaction of CrO_3 with excess ClF at 0 °C (Scheme 10.15) [1898–1900]. Chromium oxytetrafluoride is also conveniently prepared from CrO_2F_2 and KrF_2 in anhydrous HF at room temperature or from CrO_2F_2 and F_2 at 200 °C in presence of catalytic amounts of CsF [1871, 1901]. $CrOF_4$ molecules have a square-pyramidal C_{4v}-symmetry, as has been shown by infrared spectra under matrix conditions. In the solid state, it has a F-bridged polymeric structure [1901, 1902]. $CrOF_4$ rapidly hydrolyzes in contact with moist air and reacts violently with elemental S, Se, Si, red P and I_2, mostly accompanied by ignition. With organic solvents such as acetone, acetonitrile or ethanol, it also reacts immediately, however, no reactions are observed with dichloromethane [1902, 1903]. $CrOF_4$ reacts with the fluoride donors CsF and NOF to yield the complexes $Cs[CrOF_5]$ and $[NO][CrOF_5]$ (Scheme 10.15) [1901, 1902]. The latter two compounds are furthermore accessible by the reaction of CrO_2F_2 with elemental F_2 in the presence of the respective fluoride CsF or NOF. The reaction of $CrOF_4$ with dimethyltin(IV) fluoride Me_2SnF_2 gives rise to the complex $[Me_2SnF][CrOF_5]$. In contrast, CrO_2F_2 is transformed to $[Me_2Sn][CrO_2F_4]$ (Scheme 10.15) [1904]. The preparation and X-ray crystallographic investigation of 1:1- and 2:1-adducts of $CrOF_4$ with the noble gas difluorides KrF_2 and XeF_2 has also been described. However, the adducts do not contain discrete anions of the type $[CrOF_5]^-$ or $[Cr_2O_2F_9]^-$, but are weakly coordinated by F-bridges [1158]. No stable products are formed upon reaction with the strong fluoride acceptor AsF_5, indicating that $CrOF_4$ is a weak Lewis base [1901]. Chromyl fluoride CrO_2F_2 (subl.: 29.6 °C; mp.: 31.6 °C (pressurized)) is furthermore accessible from K_2CrO_4 and anhydrous HF as a red-brown gas, which is condensed to a deep violet solid below 30 °C. The fluorination of CrO_2Cl_2 with F_2 represents another synthetic route toward CrO_2F_2, as well as the fluorination of CrO_3 with COF_2 (Scheme 10.15) [1855, 1905]. Similar to $CrOF_4$, CrO_2F_2 is very sensitive toward hydrolysis and has strong oxidizing, but no fluorinating properties. Both oxyfluorides are stable in a dry atmosphere under exclusion of light in aluminium phosphate glass or Kel-F tubes, but not in ordinary laboratory glassware or quartz [1893]. Surprisingly, CrO_2F_2 is thermally very resilient, as it remains unchanged below 500 °C, whereas it undergoes thermal decomposition to O_2 and $CrOF_2$ at higher temperatures. Under the impact of UV-light, it is rather quickly transformed to a polymeric decomposition product [1868]. The structure of CrO_2F_2 has been investigated by Raman-spectroscopy, revealing the presence of tetrahedral monomers in the liquid phase, which are associated in the solid state [1906]. As has been shown by X-ray crystallography, solid CrO_2F_2 forms F-bridged dimers, where the intermolecular forces are moderately strong, meaning that the bonding situation is best described as intermediate between covalent and ionic [1707]. Moreover, the molecular structure of gaseous CrO_2F_2 has been determined by electron diffraction experiments [1907]. The solubility of chromyl fluoride in HF is very low at −78 °C; however, at room temperature it dissolves to form red solutions [1893]. Furthermore, solutions of CrO_2F_2 in anhydrous HF do not appear to undergo autodissociation processes, as they do not conduct electricity, but cationic species are observable, if SbF_5 is added [1908]. CrO_2F_2 reacts with the fluoride donors NOF and NO_2F to form the salts $[NO][CrO_2F_3]$ and $[NO_2][CrO_2F_3]$, with $M^{II}F_2$ (M^{II} = Ca, Mg) and M^IF (M^I = Na, K, Cs), complexes of the

$[Me_2SnF][CrOF_5]$ ←——Me_2SnF_2—— $CrOF_4$ ——CsF or NOF—→ $[CrOF_5]^-$ salts

F_2, 220 °C

KrF$_2$, aHF, rt or
F$_2$, CsF, 200 °C

F_2,
CsF or NOF

CrO_3 ——F_2, 150 °C / or COF$_2$—→ CrO_2F_2 ——Me_2SnF_2—→ $[Me_2Sn][CrO_2F_4]$

TaF_5 or SbF_5

$M^{II}F_2$
or M^IF

NOF or
NO_2F

$[CrO_2F][TaF_6]$ or
$[CrO_2F][Sb_2F_{11}]$

$[CrO_2F_4]^{2-}$ salts

$[CrO_2F_3]^-$ salts

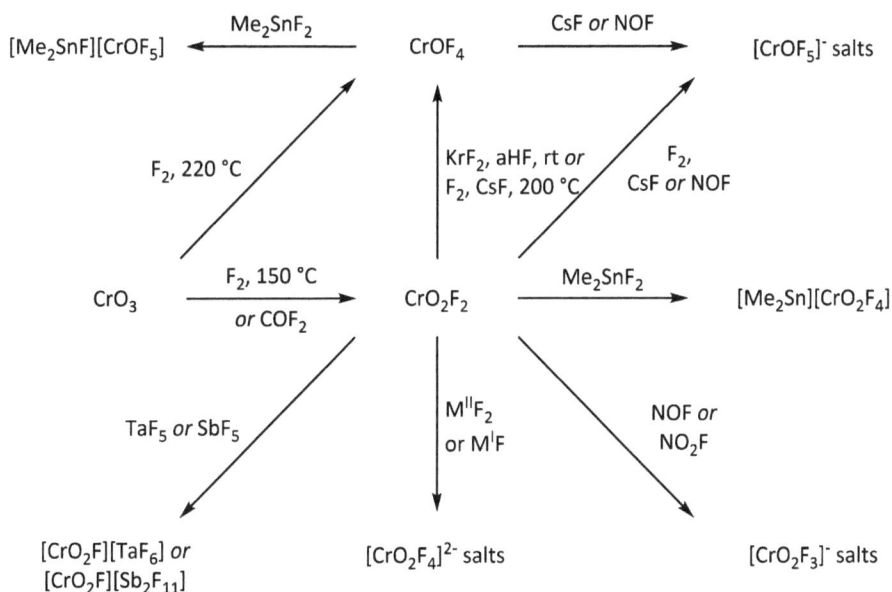

Scheme 10.15: Preparation and selected reactions of $CrOF_4$ and CrO_2F_2.

general compositions $M^{II}[CrO_2F_4]$ and $M_2^I[CrO_2F_4]$ are formed (Scheme 10.15) [1909]. Reactions with Lewis acids are also known, for example, SO_3, TaF_5 and SbF_5 are transformed to $CrO_2(SO_3F)_2$, $[CrO_2F][TaF_6]$ and $[CrO_2F][Sb_2F_{11}]$, respectively (Scheme 10.15) [1900]. Against the predictions of the VSEPR-model, $[CrO_2F_4]^{2-}$ preferentially exists as *cis*-isomer [1682]. If chromyl fluoride is mixed with CrO_2Cl_2, an equilibrium is formed, resulting in the generation of the mixed halide CrO_2ClF [1895]. As another derivative, chromyl nitrate $CrO_2(NO_3)_2$ is prepared from CrO_2F_2 and $NaNO_3$ [1910]. In preparative organic chemistry, applications of CrO_2F_2 have been described in oxidation reactions, for example, of steroidal alkenes [1911].

10.6.2 Molybdenum and tungsten

Low-valent fluorides

Molybdenum monofluoride **MoF** is unknown in condensed state and has only been investigated by mass spectrometric and thermochemical studies as well as by quantum-chemical calculations [1180, 1912]. Similarly, tungsten monofluoride **WF** has only been explored in very special gas-phase chemistry and DFT-calculations so far [1913]. It has not been detected in the gas-phase equilibrium of W metal with SF_6 or WF_6 by mass-spectrometry, however, it has been observable by *in situ* X-ray photoelectron spectroscopy, if elemental tungsten surfaces were exposed to a molecular beam of XeF_2

[1914, 1915]. The situation does not significantly change for the difluorides. Molybdenum difluoride **MoF$_2$** has only been observed by high temperature mass-spectrometry. Based on the obtained data, it has been assumed that the MoF$_2$ molecule probably has a slightly bent structure [1912]. Furthermore, theoretical calculations have been performed on MoF$_2$ [1183]. Tungsten difluoride **WF$_2$** is a high temperature species, which indeed can be observed in the gas-phase equilibrium between W and WF$_6$. Interestingly, it is stable up to 4000 K [1916, 1917]. It is also accessible by the reaction of W metal with SF$_6$ or WF$_6$ or by impact of a molecular XeF$_2$ beam on elemental tungsten surfaces. A linear structure has been assigned to the WF$_2$ molecule [1914, 1915]. Apart from mass-spectrometric investigations, quantum-chemical calculations have been performed, confirming the linear ground state structure [1913].

Trifluorides

Molybdenum trifluoride **MoF$_3$** is most conveniently prepared by reduction of MoF$_6$ or MoF$_5$ with Mo metal at 400–850 °C [1918, 1919]. Alternatively, SbF$_3$ or H$_2$ may be used as reducing agents. If MoBr$_3$ and HF are heated to 600 °C, halogen exchange occurs, thus also resulting in the formation of MoF$_3$ [1668]. It is a yellow-brown, nonvolatile solid, which, comparable to CrF$_3$, crystallizes in a layered VF$_3$ structure [1918, 1920]. The thermochemistry of gaseous MoF$_3$ has been investigated by mass spectrometry [1912]. The geometrical and electronic structure of molecular MoF$_3$ have been studied by quantum chemistry, revealing a planar, D$_{3h}$-symmetric ground state [1921]. MoF$_3$ is insoluble in water and stable up to 600 °C *in vacuo*, however, if MoF$_3$ is heated on air, the trioxide MoO$_3$ is generated. There are also a few complex salts, which typically contain the octahedral anion [MoF$_6$]$^{3-}$, e. g. K$_3$[MoF$_6$]. Compounds with mixed cations of the general composition M$_2^I$M$^{I\prime}$[MoF$_6$] (M^I = K, Rb, Cs, Tl; M$^{I\prime}$ = Na, K, Tl) have also been prepared [1922]. Tungsten trifluoride **WF$_3$** is unknown in the condensed phase. In analogy to WF and WF$_2$, it can be observed in the gas-phase equilibrium of elemental W and SF$_6$ or WF$_6$ [1914]. Furthermore, it is obtained, if tungsten metal surfaces are treated with a molecular beam of XeF$_2$ [1915]. Quantum chemical calculations reveal a D$_{3h}$-ground state structure for the WF$_3$ molecule [1913, 1921]. The oxyfluorides **MoOF** and **WOF** are obtained under matrix conditions, if the corresponding laser-ablated metal is reacted with OF$_2$ in excess argon. They both have a bent structure [1853].

Tetrafluorides

Pale green molybdenum tetrafluoride **MoF$_4$** and red-brown tungsten tetrafluoride **WF$_4$** are non-volatile compounds, which are formed upon treatment of hydrocarbons with the hexafluorides at elevated temperatures [1923]. MoF$_4$ is also accessible by thermal decomposition of Mo$_2$F$_9$ at 170 °C *in vacuo* or upon reduction of MoF$_6$ with Si. It is insoluble in anhydrous HF [1924, 1925]. Consequently, a very convenient room-temperature synthesis starts from the hexafluoromolybdate(IV) salt K$_2$[MoF$_6$], which is treated with SbF_5 in anhydrous HF to precipitate MoF_4. However, if the stronger oxidizing AsF_5 is used instead of SbF_5, the pentafluoride MoF_5 will be obtained [1926]. An alternative synthesis for WF_4 includes the reduction of WCl_6 in acetonitrile with subsequent fluorination by AsF_3 or the disproportionation of WF_5 [1927]. Structurally, MoF_4 and WF_4 can be compared with polymeric NbF_4, showing octahedrally coordinated metal centers in the solid state [1928]. The molecular ground-state structures of MoF_4 and WF_4 have been calculated by quantum-chemical methods, revealing a tetrahedral structure for MoF_4, which has also been verified by gas-electron diffraction. In contrast, the ground state of WF_4 has D_{2d}-symmetry, due to the Jahn–Teller effect. However, square-planar WF_4 (D_{4h}-symmetry) is only slightly higher in energy, so that the molecular structure can be considered as quasiplanar [1921, 1929, 1930]. Both, MoF_4 and WF_4 are sensitive toward hydrolysis and oxidation. At high temperatures, they disproportionate [1923]. WF_4 reacts with quartz under formation of WO_2 and SiF_4 and is practically insoluble in acetonitrile, liquid NH_3, liquid HF and halocarbons. Complexes are formed with Lewis acids like PF_5 or AsF_5 [1927, 1931]. There are also some fluorometallate salts, mainly based on molybdenum. They have the general formulas $M_2^I[MoF_6]$ (M^I = alkali metal, Tl) and $Tl_2M^I[MoF_7]$ (M^I = Na, K) [1932, 1933]. The copper(II) complex $Cu[MoF_6]$ is accessible, if MoF_5 is reacted with Mo metal and CuF_2. The structure of this compound is described by a three-dimensional network of both, $[CuF_6]$- and $[MoF_6]$-octahedra, which are alternatingly connected by common corners [1766]. MoF_4 also reacts with almost all lanthanide trifluorides LnF_3 (Ln = La, Ce, Pr, Nd, Sm, Eu, Gd, Tb, Ho, Er, Tm, Yb, Lu, Y) under formation of ternary compounds with the general formula $LnMoF_7$. All the compounds have been studied by X-ray crystallography and are isotypic to $LnZrF_7$ and $LnNbF_7$ [1934, 1935]. The fluorination of $Mo(CO)_6$ at −75 °C with F_2 yields in a first step the mixed-valent, olive-green **Mo_2F_9**, which decays at 170 °C *in vacuo* to yield volatile MoF_5 and nonvolatile MoF_4 [1924]. The reduction of MoF_6 with silicon in the ratio 2:1 in anhydrous HF also yields a green, polymeric oil of the composition Mo_2F_9, which can be reformulated as ionic compound, comprising the two species $[MoF_3]^+$ and $[MoF_6]^-$. Analogously, the reaction of MoF_6 with chlorides or of $MoCl_5$ with anhydrous HF yields orange-brown $Mo_2Cl_3F_6$, which may be reformulated as $[Mo_3Cl_9][MoF_6]_3$ [1936, 1937]. Tungsten oxydifluoride **WOF_2** is the intermediary product of the fluorination of WO_2 with anhydrous HF at 500 °C. It is a grey solid, which is chemically inert, as it does not react with boiling alkali solutions or concentrated acids [1923]. Both molybdenum oxydifluoride **$MoOF_2$** and WOF_2 can also be prepared in an argon matrix, if the corresponding laser-ablated metal is reacted with OF_2. Highly ionic Mo-F and W-F bonds have been revealed by vibrational spectra in combination with quantum-chemical calculations, whereas the respective metal-oxygen bonds are strongly covalent. The Mo-O-linkage has a double bond character, whereas the W-O-connection is best described by a triple bond. The latter finding may be rationalized by the additional occurrence of a highly delocalized lone-pair at the oxygen atom, which contributes the second π-bond [1853].

Pentafluorides

Molybdenum pentafluoride **MoF$_5$** (mp.: 46 °C) and tungsten pentafluoride **WF$_5$** are yellow, very hydrolytically sensitive solids. There are numerous preparations of MoF$_5$, as for example, the decomposition of Mo$_2$F$_9$ at 150 °C, or the reduction of MoF$_6$ with reducing agents like Mo, PF$_3$, Mo(CO)$_6$, Si or H$_2$ in anhydrous HF (Scheme 10.16) [468, 1372, 1652, 1924, 1925, 1938–1941]. The discovery of WF$_5$ was the result of intensive investigations on halogen lamps. It is formed from WF$_6$ and an electrically heated tungsten wire (500–600 °C) and is deposited at the cooled walls of the reactor (–60 °C) [1652, 1942, 1943]. Both pentafluorides can also be liberated from their hexafluorometallate(V) salts upon treatment with SbF$_5$ in anhydrous HF at relatively low temperatures (Scheme 10.16) [1944]. The mass spectrum of WF$_5$ shows the presence of the oligomers W$_2$F$_{10}$, W$_3$F$_{15}$ and W$_4$F$_{20}$. In the solid state, both MoF$_5$ and WF$_5$ exhibit a tetrameric structure like NbF$_5$ [1938, 1945–1949]. In the temperature range of 0–25 °C, (WF$_5$)$_4$-tetramers are found in the gas phase as well as in nonpolar solvents [1943]. Similarly, the yellow, viscous melts of MoF$_5$ indicate an association of the pentafluoride in the liquid state; moreover, colorless MoF$_5$-vapors still contain oligomeric molecules. The structural behavior of MoF$_5$ has additionally been investigated by quantum-chemical methods, revealing, that the existence of monomers and cyclic trimers is favored in the liquid and the gaseous state [1950]. The structure of isolated MoF$_5$ molecules has been investigated by vibrational spectroscopy under matrix-conditions, initially suggesting a trigonal bipyramidal geometry [1951]. However, MoF$_5$ is slightly distorted from ideal D$_{3h}$-symmetry, which is due to the Jahn–Teller effect. This has been shown by gas-phase electron diffraction in combination with least-square structure refinement calculations and mass spectrometry [1930, 1952]. In practice, the distortion may not always be observed. Recently, the structural properties of MoF$_5$ have been comprehensively revisited [1948]. The pentafluorides both disproportionate irreversibly to the hexa- and tetrafluorides at elevated temperatures of 165 °C (MoF$_5$) and 50–70 °C (WF$_5$), respectively (Scheme 10.16) [1938, 1942, 1943, 1953]. Solid MoF$_5$ is stable in dry glass, however, the liquid pentafluoride attacks it. WF$_5$ can be stored in closed quartz vessels, but disproportionation slowly occurs, even at 0 °C [1942, 1943]. MoF$_5$ is soluble in MoF$_6$ and anhydrous HF and forms 1:1-adducts with NH$_3$, Me$_2$O and Me$_2$S as well as 1:2-adducts with MeCN and pyridine [1925, 1939]. However, it is only a weak fluoride acceptor, since solutions of MoF$_5$ in HF do not contain detectable amounts of the [MoF$_6$]$^-$ anion [1954]. Complexes of MoF$_5$ or [MoF$_6$]$^-$ with the fluoride acceptors BF$_3$, AsF$_5$, TaF$_5$ or NbF$_5$ always contain fluorine bridges [1955]. The hexafluoromolybdates(V) M^I[MoF$_6$] and –tungstates(V) M^I[WF$_6$] are formed upon reduction of the hexafluorides with M^{I}I in liquid SO$_2$ (M^I = Li, Na, K, Rb, Cs); moreover, a few higher fluorometallates are obtained by oxidation of the hexacarbonyls with IF$_5$ in the presence of the respective metal fluorides or iodides (Scheme 10.16). All the fluorometallates are colorless, crystalline solids of the compositions M^I[M^{V}F$_6$], M$_2^I$[M^{V}F$_7$] and M$_3^I$[M^{V}F$_8$] (M^I = Li, Na, K, Rb, Cs, Tl, NO; M^V = Mo, W). The complexes

WF$_6$ [MF$_6$]$^-$ salts MF$_6$ + MF$_4$

W-wire, SbF$_5$, Δ
500–600 °C aHF

reduction
MoF$_6$ ———————→ MF$_5$ MF$_6$
e.g. Mo, PF$_3$, Si

150 °C F$^-$ ≠ I$^-$, liq. SO$_2$

Mo$_2$F$_9$ [MF$_6$]$^-$, [MF$_7$]$^{2-}$ IF$_5$, F$^-$ or I$^-$ M(CO)$_6$
or [MF$_8$]$^{3-}$

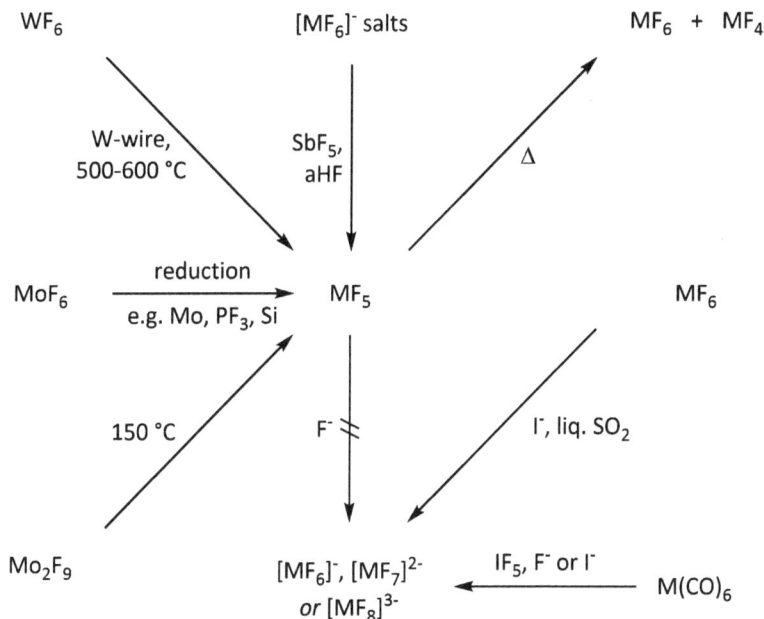

Scheme 10.16: Preparation and selected reactions of the group 6 pentafluorides MF$_5$ (M = Mo, W).

are all very sensitive toward hydrolysis and start to attack glass at temperatures above 250 °C [1691, 1956–1961]. However, the reaction of the pentafluorides with fluoride donors does not give rise to fluorocomplexes. In HF-solution, the [WF$_6$]$^-$ anion is readily oxidized by the cation [NF$_4$]$^+$ to give WF$_6$ [1962]. There are also a few mixed halides of molybdenum(V), as for example, the reaction of MoF$_6$ with MoBr$_4$ yields red-brown MoBrF$_4$ [1939]. Moreover, brown MoCl$_2$F$_3$ is prepared from MoCl$_5$ and AsF$_3$, whereas MoCl$_4$F has a blue-black color [1963].

Molybdenum(V) and tungsten(V) oxyfluorides are only scarcely explored. If MoOF$_4$ is reacted with MoF$_4$ at 200 °C, molybdenum oxytrifluoride **MoOF$_3$** is formed along with MoF$_5$. Both products form complexes with liquid NH$_3$, which have the composition MoF$_5 \cdot m$ NH$_3$ (m = 1–5) and MoOF$_3 \cdot n$ NH$_3$ (n = 1, 2). Even less is known about tungsten oxytrifluoride **WOF$_3$**. However, based on magnetochemical studies, polymeric structures with F-bridges have been claimed for both, MoOF$_3$ and WOF$_3$. *In vacuo*, they disproportionate to MoO$_2$F$_2$ and MoOF$_2$ or WOF$_4$ and WOF$_2$, respectively. The complex anions [M^{V}OF$_5$]$^-$ and [M^{V}OF$_6$]$^{2-}$ have also been observed, mainly in mass-spectrometric investigations.

Hexafluorides

Molybdenum hexafluoride **MoF$_6$** (mp.: 17.4 °C; bp.: 34–35 °C) and tungsten hexafluoride **WF$_6$** (mp.: 2 °C (0.55 bar); bp.: 17.1 °C) are colorless and diamagnetic compounds, which are directly accessible from the elements or by fluorination of Mo- or W-compounds (Scheme 10.17) [1668, 1957, 1964]. In the solid state, they exist in a cubic modification above $-9.6\,°C$ (MoF_6) and $-8.5\,°C$ (WF_6), respectively. Below these temperatures, the hexafluorides crystallize in an orthorhombic lattice [1965]. Both MoF_6 and WF_6 have a regular octahedral structure, however, the analogous permethylated derivatives $Mo(CH_3)_6$ and $W(CH_3)_6$ have the shape of distorted trigonal prisms, which is due to the fact that the methyl group does not act as π-donor, whereas fluorine does so. In addition, the partial negative charges on the fluoride substituents repel each other so that the molecular geometry is pushed toward an octahedron. Calculations have shown, that the trigonal-prismatic geometries of MoF_6 and WF_6 are transition states, which are not much higher in energy compared to the octahedral shape (MoF_6: $\Delta E = 27.7\,kJ\,mol^{-1}$; WF_6: $\Delta E = 45.7\,kJ\,mol^{-1}$) [1966]. The interconversion of trigonal-prismatic structures into octahedral molecules and vice versa is also known as Bailar-twist model [1967]. However, experimentally, this interconversion has only been observed for WF_6, when the derivative C_6F_5-O-WF_5 was investigated [1966]. The experimental and the calculated value for the barrier have been in good agreement. Both MoF_6 and WF_6 are very sensitive toward hydrolysis and reduction. In most of the organic solvents, the hexafluorides are reduced to oxidation state +V, however, at higher temperatures further reduction is also possible. MoF_6 and WF_6 are very well soluble in anhydrous HF and form complexes with Lewis bases [1968–1970]. WF_6 is the weakest oxidizer among the 5d-metal hexafluorides. The oxidation potential of MoF_6 is roughly 1 V larger than that of WF_6 [1971]. However, both MoF_6 and WF_6 are capable of oxidizing metallic Ag, Tl, Pb, Zn, Cd, Hg, Mn, Co, and Ni to solvated cations, whereas they are themselves reduced to $[M^VF_6]^-$. For example, WF_6 oxidizes Ag and Tl to Ag^+ and Tl^+, whereas the stronger oxidizing MoF_6 consequently yields the higher valent species Ag^{2+} and Tl^{3+} (Scheme 10.17) [1972–1974]. Moreover, MoF_6 is capable of oxidizing nitric oxide NO to $[NO]^+$, whereas WF_6 does not react at all [1876]. Upon reaction with ferrocene and cobaltocene derivatives, MoF_6 and WF_6 are reduced to oxidation state +V (ferrocene) and +IV (cobaltocene), respectively [1975]. The reaction of MoF_6 with the tetrabromide $MoBr_4$ results in the formation of the mixed molybdenum(V) halide $MoBrF_4$ [1939].

Fluoromolybdates (VI) and –tungstates (VI) are generated from the hexafluorides and metal fluorides or –iodides in IF_5 or ClF_3 as colorless compounds. Alternatively, they can also be made from the respective fluorides in MeCN (Scheme 10.17) [1976, 1977]. The fluorometallates(VI) have the general compositions $M^I[M^{VI}F_7]$ with the capped octahedral anion $[M^{VI}F_7]^-$ and $M_2^I[M^{VI}F_8]$ where the anion $[M^{VI}F_8]^{2-}$ has a square-antiprismatic structure (M^I = Li, Na, K, Rb, Cs, NH_4, NO, NO_2; M^{VI} = Mo, W) [1876, 1957, 1959, 1978, 1979]. An imidazolium-salt of $[WF_7]^-$ has also been prepared [1980]. Exchange experiments using ^{18}F have shown that both $[MoF_7]^-$ and $[WF_7]^-$ are in equilibrium with the respective hexafluoride in CH_3CN-solution. As a consequence, the fluoride-substituents are simply exchanged, even under heterogeneous conditions [1981]. The reaction of WF_6 with phosphines leads to the formation of 1:1-adducts.

$[NO]^+$ $[(bpy)WF_5]^+$ *or* $[(phen)WF_5]^+$

NO
(M = Mo)

bpy *or* phen,
then $SbF_5 \cdot SO_2$
(M = W)

metals *or* metal compounds   fluorination   → MF_6 Ag *or* Tl   → Ag^+ *or* Tl^+
(M = W)

Ag *or* Tl
(M = Mo)

F^-

Ag^{2+} *or* Tl^{3+} $[MF_7]^-$ *or* $[MF_8]^{2-}$

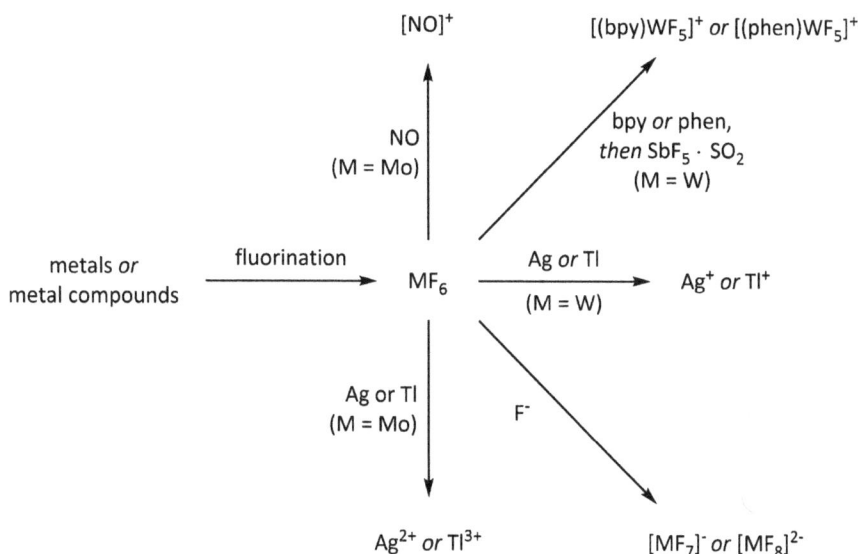

Scheme 10.17: Preparation and selected reactions of the group 6 hexafluorides MF_6 (M = Mo, W).

Depending on the nature of the substituents on the phosphine, the adduct can, for example, adapt the structures of a capped trigonal prism $(P(CH_3)_3)$ or a capped octahedron $(P(CH_3)_2C_6H_5)$ [1982]. WF_6 also reacts with trifluoroacetate ions CF_3COO^- under formation of complex tungsten(VI) mono- and dianions [1983]. Recently, the donor-stabilized $[WF_5]^+$ cation has been prepared, starting from adducts of WF_6, which have subsequently been treated with $SbF_5 \cdot SO_2$ in SO_2-solution (Scheme 10.17). The bidentate ligands 2,2′-bipyridine and 1,10-phenanthroline lead to the formation of hepta-coordinated tungsten centers, adapting the geometry of a capped octahedron [1984]. MoF_6 and WF_6 can be substituted by O- and N-centered nucleophiles, but also others [732, 1985–1990]. In general, the chemistry of both hexafluorides is based on their reduction, as MoF_6 and WF_6 are oxidants with quite high electron affinities. Amongst all the other transition metal hexafluorides, WF_6 has the lowest electron affinity (3.15 eV (calc.)), however, it is not that different from fluorine itself (3.399 eV) [1991].

Tungsten(VI) chlorofluorides $WF_{6-x}Cl_x$ (x = 1–5) are accessible by reaction of WF_6 with chlorides or controlled fluorination of WCl_6. *Cis*- and *trans*-isomers are in equilibrium, in general the chlorofluorides are thermally unstable and decompose to WF_6 and WCl_6 [1876, 1992].

Hexavalent oxyfluorides

Molybdenum oxytetrafluoride **MoOF₄** (mp.: 97.2 °C (33 mbar); bp.: 186 °C) and tungsten oxytetrafluoride **WOF₄** (mp.: 104.7 °C (33 mbar); bp.: 186 °C) are white, volatile solids which are prepared by fluorination of the metals, metal oxides or metal oxytetrachlorides [700, 1964, 1969, 1993, 1994]. More conveniently, WOF_4 is prepared by controlled hydrolysis of WF_6 using SiO_2 [1995]. Both oxytetrafluorides are furthermore accessible by treatment of the hexafluorides with boric acid [1438]. The stoichiometric reaction of MoF_6 with H_2O in HF solution leads to the formation of $MoOF_4$, whereas WF_6 is transformed to the oxonium salt $[H_3O][W_2O_2F_9]$ under the same reaction conditions [1996, 1997]. $MoOF_4$ has a structure comparable to VF_5 and is dimorphic, after a few days, the hexagonal form is converted to the monoclinic one [1998]. The solid-state structure of WOF_4 is comparable to NbF_5, since it is also based on tetrameric $[WOF_4]_4$-units. In both, WOF_4 and $MoOF_4$, the structural units are bridged by fluorine substituents [1999–2003]. In the gas phase, one can find monomeric, square-pyramidal molecules of both oxytetrafluorides, as for example, has been shown by matrix techniques [1994, 2004–2007]. This is against the expectations of the VSEPR model, but can be explained by electronic effects of the strong covalently bonded oxygen-substituents [1682]. Solutions of WOF_4 in acetonitrile, dichloromethane or diethyl ether have been shown to contain monomeric oxyfluoride species [2008]. WOF_4 and $MoOF_4$ form 1:1- and 2:1-complexes with XeF_2 as well as 1:1-adducts with KrF_2, WOF_4-complexes with organic N-donor ligands have been prepared as well [1157, 2009–2014]. However, it does not react with the strong fluoride acceptors BF_3, PF_5 or AsF_5 [1993]. A few complex salts are known for the oxytetrafluorides $MoOF_4$ and WOF_4, where the tungsten compound is the stronger Lewis acid. However, both oxyfluorides are weak fluoride acceptors, as HF-solutions do not contain significant amounts of the respective $[MOF_5]^-$ anion [1954]. The fluorocomplexes have the general compositions $M^I[M^{VI}OF_5]$, $M^I[M^{VI}_2O_2F_9]$ and $M^I_2[M^{VI}OF_6]$ [1625, 1962, 1969, 1993, 2015–2018]. In case of $MoOF_4$ it has been shown that the formation of the fluoroanions strongly depends on the concentration of F^-. With relatively low concentrations of free fluoride, the dimeric anion $[Mo_2O_2F_9]^-$ is preferably formed. If the F^--content is increased, this will first lead to the formation of $[MoOF_5]^-$ and finally $[MoOF_6]^{2-}$. If $MoOF_4$ is treated with M^IF (M^I = alkali metal) in a 1:1-ratio, the respective salts $M^I[Mo_2O_2F_9]$ are obtained, which all contain F-bridged, dimeric $[Mo_2O_2F_9]$-units. Single crystal X-ray analysis has revealed, that the anions have a bent structure with the O-atoms being in *trans*-position with respect to the bridging F-substituents [2019]. Analogously, dimeric $[W_2O_2F_9]^-$ has a bent W-F-W-bridge, as has been confirmed by X-ray crystallography of its oxonium salt [1997]. Moreover, ionic liquids have been described, which are based on 1-ethyl-3-methlimidazolium cations and $[MoOF_5]^-$ or $[WOF_5]^-$ anions, respectively. Interestingly, the tungsten compound is hydrophilic, but stable in aqueous solutions [1728, 2020]. In reactions with boron trihalides BX_3 (X = Cl, Br), the oxytetrafluorides are converted to $MoOCl_4$, $WOCl_4$ and $WOBr_4$ [1438]. Apart from that, the fluoride-substituents in $MoOF_4$ and WOF_4 are fully substituted by azide-ligands upon treatment with Me_3SiN_3. In case of molybdenum this leads to a reduction to Mo(V) and, consequently, the formation of $MoO(N_3)_3$ [2021]. The thio- and selenoderivatives **WSF$_4$** and **WSeF$_4$** are formed upon reaction of WF_6 with Sb_2S_3 or Sb_2Se_3 at 300 or 350 °C, respectively [2022, 2023]. The sulphur derivative is also accessible from WF_6

and $(Me_3Si)_2S$ in a room temperature reaction in dichloromethane or acetonitrile solution. Like its oxygen analogue WOF_4, WSF_4 is monomeric in the aforementioned organic solvents [2008].

Molybdenum dioxydifluoride **MoO₂F₂** (subl.: 120 °C (high vacuum)) is a white solid, which is preferably prepared by pyrolysis of $[Na(H_2O)]_2[cis\text{-}MoO_2F_4]$ at 300 °C *in vacuo* [2024]. Alternatively, MoO_3 can be treated with SeF_4 or IF_5, which results in the formation of the corresponding adducts [688, 1897]. Furthermore, MoO_2F_2 is obtained as intermediate compound, if MoO_3 is reacted with NF_3 at 430 °C. However, the side-product NOF directly transforms MoO_2F_2 to the very hygroscopic nitrosyl salt $[NO][MoO_2F_3]$ [1625]. MoO_2F_2 is very reactive and is immediately reduced in many organic solvents. It also undergoes partial oxygen-fluorine exchange reactions in anhydrous HF. Structurally, MoO_2F_2 may be compared to TiF_4, as it is composed of triangular trimers of double-F-bridged $[MoO_2F_4]$-units which are further connected to columns by oxygen bridges [2024]. The respective tungsten compound **WO₂F₂** is formed upon hydrolysis of WOF_4 in hydrofluoric acid [2025]. It can also be obtained, if $K_2[WO_4]$ is reacted with IF_5 at 120 °C [1897]. At elevated temperatures, it decomposes to WOF_4 and WO_3 [1625]. The structures of the gaseous molecules have been studied by electron diffraction techniques as well as vibrational spectroscopy, revealing monomeric molecules with C_{2v}-symmetry [1994, 2026]. A bipyridine-adduct of WO_2F_2 has been characterized by X-ray crystallography [2013]. The fluoride substituents in both MoO_2F_2 and WO_2F_2 are fully substituted by azide-groups, if the oxyfluorides are treated with Me_3SiN_3 [2027]. There are also complex salts of the dioxydifluorides which have the compositions $M^I[M^{VI}O_2F_3]$, $M^I_2[M^{VI}O_2F_4]$, $M^I_3[M^{VI}O_2F_5]$ and $M^I_3[M^{VI}_2O_4F_7]$ [1625]. The structure of the anions $[MoO_2F_3]^-$ and $[WO_2F_3]^-$ is comparable to the vanadium derivative $[VO_2F_3]^{2-}$, exhibiting distorted octahedrally coordinated metal centers, where the coordination polyhedra are furthermore connected by *cis*-F-bridges to build infinite chains [2028]. The hydrolysis of $[MoF_5]^-$ yields the dimeric anion $[Mo_2O_4F_6]^{2-}$, which has been investigated by single crystal X-ray analysis. It is composed of two distorted octahedral $[MoO_2F_4]$-units, which are bridged by two F-atoms in *cis*-position, thus representing two distorted octahedra with a common edge. The oxygen-substituents occupy the positions *trans* to the bridging F-atoms [1728]. Apart from the aforementioned compounds, salts with the compositions $M^I_2[M^{VI}O_3F_2]$ and $M^I_3[M^{VI}O_3F_3]$ are known.

10.6.3 Study questions

(I)   How is CrF_2 prepared? Comment on its oxidative stability in anhydrous and hydrated form. Which product is obtained, if equimolar amounts of CrF_2 and CrF_3 are heated to 800–900 °C?

(II) Comment on the stability of CrF_3. Give some examples for its preparation. Which structural feature is observed in all fluorochromates(III)?

(III)   Which structural modifications of CrF_4 are known and how are they prepared? Briefly describe their structures.

(IV)   Comment on the stability of pentavalent chromium fluorine compounds. How are these species prepared?

(V)   Are there hexavalent chromium fluorine compounds? How are they prepared? Comment on their general reactivity and stability.

(VI)   How is MoF_3 prepared? Which structures are observed for the gaseous and the solid compound?

(VII)   How can the insolubility of MoF_4 in aHF be exploited for its synthesis? What has to be taken into account?

(VIII) Which ground-state structures are observed for the molecular tetrafluorides MoF_4 and WF_4?

(IX)   Which methods are known for the preparation of the pentafluorides MoF_5 and WF_5? Are they thermally stable? How are pentavalent fluorometallates obtained?

(X)   Which products are observed, when MoF_6 and WF_6 are reacted with Ag, Tl and NO? Give a brief explanation.

(XI)   Which fluorometallate anions are known for hexavalent molybdenum and tungsten? How are these compounds generally prepared and which structures are observed for the anions?

(XII)   Which hexavalent oxyfluorides of molybdenum and tungsten are known and how are they prepared? Are their molecular structures in line with the expectations according to the VSEPR model?

10.7 Group 7: Manganese, technetium, rhenium

The elements of group 7 significantly differ in the formation of their binary fluorides, since manganese forms low valent fluorides up to MnF_4, whereas technetium and rhenium exclusively form high valent fluorides, starting from TcF_5 and ReF_4, respectively. Apart from that, a few oxyfluorides are known as well.

10.7.1 Manganese

Low-valent manganese fluorides

Manganese monofluoride **MnF** is unknown in condensed state. However, it has been investigated in the gas phase by several spectroscopic methods. Furthermore, it has been explored theoretically and by matrix techniques [2029].

Manganese difluoride **MnF₂** is formed upon addition of HF to aqueous Mn(II) solutions. Alternatively, it is obtained by the reaction of HF with $MnCl_2$ or elemental Mn as a pale pink, sparingly soluble, antiferromagnetic salt (mp.: 920 °C). In the solid state, it has the same structure like rutile. MnF_2 forms a number of fluorocomplexes, for example, trifluoromanganates $M^I[MnF_3]$ (M^I = alkali metal, NH_4, Tl), which have an orthorhombic perovskite structure at low temperatures, but exhibit a cubic lattice at higher temperature [2030]. Trifluoromanganates(II) are readily accessible from an aqueous solution, using $MnCl_2$ and M^IF as starting materials. The treatment of $M^I[MnF_3]$ with elemental F_2 results in the formation of pentafluoromanganates(IV) of the general formula $M^I[MnF_5]$ [2031]. The formation of $M^I[MnF_3]$ is also achieved by melting stoichiometric amounts of the fluorides; the same holds for the tetrafluoromanganates $M_2^I[MnF_4]$ (M^I = K, Rb, Cs). In liquid ammonia, the hexafluoromanganate(III) salt $Tl_3[MnF_6]$ is reduced to the fluoroperovskite $Tl[MnF_3]$ [2032]. $Ba[MnF_4]$ has pyroelectric properties at room temperature and is paramagnetic above 30 K [2033–2035]. In all of the $[MnF_4]^{2-}$ complexes, Mn is surrounded by six F-atoms in the form of a distorted octahedron. In $Ba[MnF_4]$, F-bridged $[MnF_6]$-units form a sheet structure, where the terminal F-atoms are in *cis*-orientation. The coordination geometry around the Ba^{2+} cations adapts the structure of a distorted trigonal prism [2033]. Analogously, the anions in the hydroxylammonium salt $[NH_3OH]_2[MnF_4]$ crystallize in a layered structure [2036]. $K[MnF_3]$ undergoes several phase transitions, which do not only lead to structural changes but also affect its magnetic properties [2037–2040]. The same structural behavior is observed, if Li^+ or Na^+ cations are incorporated, yielding the mixed perovskites $K_{1-x}Li_x[MnF_3]$ and $K_{1-x}Na_x[MnF_3]$, respectively [2041, 2042]. The quaternary compound $RbK_2[Mn_2F_7]$ is also known. Structurally, it is described by the perovskite type, wherein the cations are statistically distributed and have the coordination numbers 9 and 12, respectively [2043].

Manganese trifluoride

Manganese trifluoride **MnF$_3$** is formed, if Mn(II) compounds are fluorinated by F_2 or BrF_3. Pure MnF_3 is alternatively accessible upon reaction of MnF_2 with F_2 in anhydrous HF at room temperature [2044]. It is a red-violet, thermally stable, crystalline solid, which is sensitive toward hydrolysis. The structure of solid MnF_3 is derived from VF_3, containing layers of Jahn—Teller distorted $[MnF_6]$-octahedra, which are connected by common corners [2045]. MnF_3 is capable of fluorinating organic compounds, for example benzene, which gives rise to polyfluorinated cyclohexenes [2046]. MnF_3 forms a few fluorocomplexes containing the anions $[MnF_4]^-$, $[MnF_5]^{2-}$ and $[MnF_6]^{3-}$, however, the Mn-center is always hexacoordinated. The intense brown to violet $M^I[MnF_4]$ (M^I = Li, K, Rb, Cs, Tl) are formed upon reduction of the respective $M^I[MnF_5]$ with H_2 [2031, 2047]. $Li[MnF_4]$ and $Na[MnF_4]$ are isostructural, crystallizing in a dirutile lattice [2048, 2049]. The hydrate $Li[MnF_4] \cdot H_2O$ slightly differs from the anhydrous compound, as it crystallizes in a layered structure [2050]. In $Tl[MnF_4]$ one can find layers of $[MnF_6]$-octahedra, which are linked by all four equatorial F-substituents [2051]. The salts $K[MnF_4]$, $Rb[MnF_4]$ and $Cs[MnF_4]$ also crystallize in layered structures, which are derived from the $CsAlF_4$ structure type [2052]. Moreover, layers of corner-sharing $[MnF_6]$-octahedra are found in the compounds $M_2^IM^{I\prime}[Mn_3F_{12}]$ (M^I = Rb, Cs; $M^{I\prime}$ = Li, Na, K), which are prepared by heating mixtures of the binary fluorides [2053]. The hydrates $Cs[MnF_4] \cdot 2 H_2O$ and $Na[MnF_4] \cdot 2 H_2O$ contain discrete tetragonally distorted $[MnF_4(OH_2)_2]^-$-octahedra. The Cs^+ salt is prepared from $Mn(OAc)_3 \cdot 2 H_2O$ and Cs_2CO_3 in aqueous HF (w = 10 %) and can be thermally dehydrated to yield anhydrous $Cs[MnF_4]$, which contains planar, ferromagnetic $[MnF_4]^-$

[2054–2058]. In the same way, the hydrates $Rb[MnF_4(OH_2)]$ and $[NMe_4][MnF_4(OH_2)_2]$ are prepared. In the solid-state structure of the $[NMe_4]^+$ salt, one also finds exclusively isolated $[MnF_4(OH_2)_2]^-$-polyhedra, which are tetragonally elongated. In contrast, the Rb-derivative is alternatingly composed of $[MnF_6]^{3-}$ and $[MnF_4(OH_2)_2]^-$-units [2059]. $K[MnF_4(OH_2)]$ and the one-dimensional antiferromagnet $Tl[MnF_4(OH_2)]$ are also known, their structures are isotypic to the Rb-salt [2060, 2061]. Analogously, $[MnF_4(OH_2)_2]^-$ salts with organic cations have been prepared and structurally investigated. In principal, the structures of the anions are determined by distorted octahedral subunits, which are connected by hydrogen bonding [2062–2064]. Depending on the crystallization conditions, more complex structures are accessible, which contain, for example, tetrameric $[Mn_4F_{18}(OH_2)]^{6-}$ ions [2065]. The two aqualigands in $[MnF_4(OH_2)_2]^{2-}$ can also be replaced by bidentate $H_2PO_4^-$, which then results in a chain structure for $[MnF_4(H_2PO_4)]^{2-}$ [2066]. An interesting compound is the ethylenediammonium complex $[enH_2][MnF_4]$, which is a hybrid layered compound with organic cations between the perovskite layers [2067]. If excess $KMnO_4$ is reacted with aqueous HF (w = 48 %) in pyridine, the complexes $K_2[MnF_5] \cdot H_2O$ and $MnF_4 \cdot py \cdot H_2O$ are obtained. The latter is transformed to the anhydrous manganese(IV) complexes $M^I[MnF_5]$ (M^I = Na, K, Rb) by treatment with HF and $M_2^I CO_3$ [2068]. $KMnO_4$ is reduced by acetylacetone in presence of $M^I[HF_2]$ and gives rise to $M_2^I[MnF_5]$ (M^I = NH_4, Na) and the hydrates $M_2^I[MnF_5] \cdot H_2O$ (M^I = K, Cs) [2069, 2070]. Generally, the pentafluoromanganates $M_2^I[MnF_5]$ often exist as hydrates and contain chains of *trans*-F-bridged $[MnF_6]$-octahedra, where the cations and water molecules (in case of the hydrates) are incorporated between the chains [2071–2076]. BaF_2 and SrF_2 react with solutions of Mn^{3+} in aqueous HF to yield the pale-red compounds $Ba[MnF_5] \cdot H_2O$ and $Sr[MnF_5] \cdot H_2O$, respectively. In both salts, there are kinked chains of *trans*-F-bridged $[MnF_6]$-octahedra [2077]. The same structure is observed for the anions in the ethylenediammonium salt $[enH_2][MnF_5]$ (en = ethylenediamine), whereas each cation is connected to four anion chains by hydrogen bonding. Overall, this results in the formation of a three-dimensional network [2078]. The choice of larger organic cations results in spatial separation of the anionic chains [2079]. If one of the fluoride substituents in $[MnF_5]^{2-}$ is replaced by a bifluoride anion $[HF_2]^-$, the chain structure of the anions is retained, as can be observed in the piperazinium salt $[pipzH_2][MnF_4(HF_2)]$ (pipz = piperazine). The bifluoride anion bridges planar $[MnF_4]$-units, so that strongly elongated $[MnF_6]$-octahedra are formed. Due to the orientation of the HF_2-groups, the anion chain has a zig-zag structure [2080]. In contrast, $[pipzH_2][Mn_2F_8]$ is alternatingly composed of $[pipzH_2]^{2+}$ and $[Mn_2F_8]^{2-}$ layers. The latter consists of edge-sharing $[MnF_6]$-octahedra which are dimerized and linked via a common corner. As there are also strong hydrogen bonds between the cationic and anionic layers, a three-dimensional network is generated [2081]. Treatment of Ni^{2+} or Cu^{2+} solutions with Mn^{3+} in aqueous HF (w = 40 %) leads to the formation of the heptahydrates $Ni[MnF_5] \cdot 7 H_2O$ and $Cu[MnF_5] \cdot 7 H_2O$, respectively, as well as $Cu_3Mn_2F_{12} \cdot 12 H_2O$ in the case of copper(II). Structurally, the latter represents an inverse perovskite, as it can be re-formulated as $[MnF_6][MnF_{6/2}][Cu(H_2O)_4F_{2/2}]_3$ [2082]. The hexafluoromanganates(III) are accessible by treatment of $M_2^I[MnF_5]$ with M^IF or from mixtures of the respec-tive fluorides, which are heated under an inert atmosphere. The range of cations includes the alkali metals as well as NH_4^+ and also 2:1-combinations thereof. All the hexafluoromanganate(III) salts contain isolated, distorted octahedral $[MnF_6]^{3-}$ an-ions [1849, 2083, 2084]. In addition, $Na_3[MnF_6]$ is isostructural to cryolite, $Na_3[AlF_6]$ [2085]. Salts with complex hexamminemetal(III) cations $[M^{III}(NH_3)_6][MnF_6]$ (M^{III} = Rh, Co, Cr) have also been prepared, they are stable on air but decompose upon contact with water or acids [2086]. The hexafluoromanganate(III) salt $Tl_3[MnF_6]$ is reduced in liquid ammonia, resulting in the formation of divalent manganese in $Tl[MnF_3]$ [2032]. Aqua pentafluoromanganates $[MnF_5(OH_2)]^{2-}$ with several nitrogen-containing, organic cations have also been prepared. In all the described compounds, the monomeric, distorted octahedral anions are arranged in either chains or two- and three-dimensional networks by forming hydrogen bonds [2087, 2088].

Manganese tetrafluoride

Manganese tetrafluoride **MnF$_4$** is generated by fluorination of elemental manganese or MnF_3 as well as upon decomposition of hexafluoromanganates(IV) (Scheme 10.18). It is a blue-grey, volatile, very hygroscopic and reactive solid [269, 2047, 2089, 2090]. It is conveniently formed upon treatment of hexafluoromanganates(IV) with AsF_5 in HF solution [2091]. The pure compound is accessible from MnF_2 and KrF_2 in anhydrous HF (Scheme 10.18). The intermediary complex $KrF_2 \cdot MnF_4$ is readily decomposed at room temperature to yield Kr, F_2, KrF_2 and pure MnF_4 [2092]. Similarly, photodissoci-ated F_2 fully oxidizes MnF_2 to MnF_4 in anhydrous HF at room temperature under UV-irradiation [2044]. In the solid state, MnF_4 exists in form of a tetragonal α-modification and a probably rhombohedral β-phase. α-MnF$_4$ consists of tetrameric $[Mn_4F_{20}]^{4-}$ rings, which are connected to bands. The latter are nonplanar and are connected to other bands, giving rise to a three-dimensional network. The structure of β-MnF$_4$ has not been solved yet [2093]. In the gas phase, tetrahedral MnF_4 molecules are found, as has been verified by IR-spectroscopic investigations on the matrix-isolated compound [2094]. MnF_4 is not stable and already slowly decomposes to MnF_3 and F_2 at room temperature (Scheme 10.18). The fluorocomplexes of MnF_4 are slightly more stable and have the general compositions $M^I[MnF_5]$, $M_2^I[MnF_6]$ and $M^{II}[MnF_6]$ (M^I = alkali metal; M^{II} = earth alkali metal, Cu, Ag, Cd, Hg, Ni, Zn). The lithium salt $Li[MnF_5]$ is generated by the reaction of equimolar mixtures of LiF and MnF_2 in a flow of F_2 at 350 °C (Scheme 10.18) [2047]. Alternatively, the treatment of trifluoro-manganate(II) complexes $M^I[MnF_3]$ with F_2 yields pentafluoromanganate(IV) salts $M^I[MnF_5]$ [2031]. Conveniently, the hexafluoromanganate(IV) salts $M_2^I[MnF_6]$ (M^I = al-kali metal) are prepared at room temperature by reacting 2:1-mixtures of M^IF and MnF_2 with F_2 under UV-irradiation in anhydrous HF (Scheme 10.18) [2044]. $Cs_2[MnF_6]$ is also

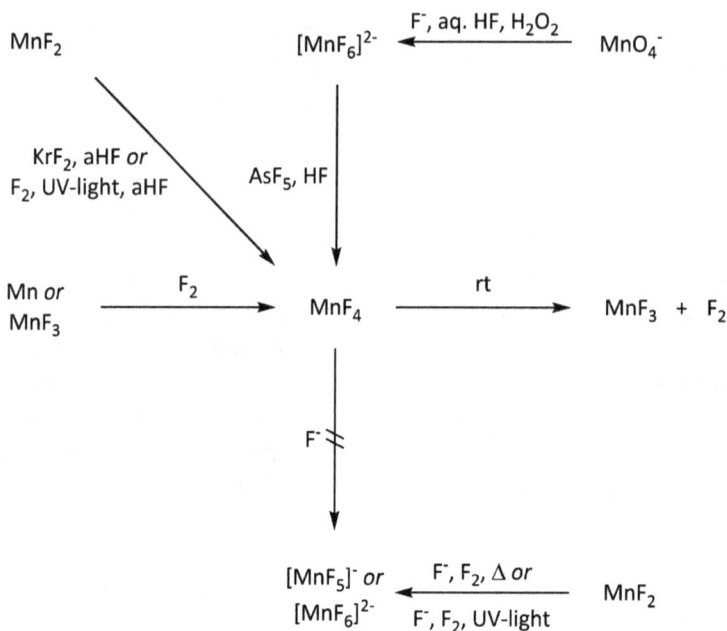

Scheme 10.18: Preparation of MnF$_4$ and fluoromanganate(IV)-salts.

generated from CsF and anhydrous MnCl$_2$, which are reacted in a Monel cylinder at 400 °C for 36 h. The product is obtained as stable, yellow solid and is only slightly decomposed on moist air [2095]. It can be further used as starting material in the metathetical synthesis of the tetrafluoroammonium salt [NF$_4$]$_2$[MnF$_6$] [2017]. Vibrational spectra of the [MnF$_6$]$^{2-}$ anion are in accordance with an octahedral symmetry [2096]. EPR-spectroscopic investigations as well as UV-vis measurements and excitation and luminescence spectroscopy have also been performed [1851, 2097, 2098]. The structures of the yellow hexafluoromanganates M$_2^I$[MnF$_6$] have been investigated, including complexes with mixed cations of the type M'M''[MnF$_6$] (M', M'' = Li, K, Rb, Cs) [2031, 2099–2101]. Compounds of the composition M^{II}[MnF$_6$] (M^{II} = Mg, Ca, Sr, Ba) are also known, they are typically prepared by fluorination of oxocomplexes or equimolar mixtures of the respective chlorides or sulphates [2102, 2103]. The hexafluoromanganate(IV) complexes are very sensitive toward hydrolysis by water, aqueous alkaline solutions and diluted mineral acids. They also rapidly decompose in contact with glass, whereas they are slightly more stable in contact with air [2031]. Historically, K$_2$[MnF$_6$] is certainly the most important hexafluoromanganate(IV) salt, as it has been the first compound, which served as starting material for the purely chemical synthesis of elemental F$_2$. Both, K$_2$[MnF$_6$] and SbF$_5$ can be prepared using HF as fluorinating agent, so that no elemental fluorine is required in the synthetic route toward F$_2$ (Scheme 10.18) [22]. The dioxygenyl-salt [O$_2$]$^+$[Mn$_2$F$_9$]$^-$ is obtained, if MnO$_2$ or MnF$_x$ (x = 2, 3, 4) is reacted with a mixture of F$_2$ and O$_2$ at 350–550 °C and a pressure of

300–3500 atmospheres. It is composed of double-chains of $[MnF_6]$-octahedra, which can be understood as type of layered structure. Between the layers, the $[O_2]^+$ cations are located [2104]. The compounds $[XeF_5]_2[MnF_6]$ and $[ClOF_2]_2[MnF_6]$ have been investigated concerning their thermochemical properties [2105, 2106]. If XeF_2 and MnF_3 are reacted in presence of UV-irradiated F_2 in anhydrous HF, several products are obtained. Among them, $[XeF_5]_2[MnF_6]$ is generated, which is composed of discrete $[XeF_5]^+$ and $[MnF_6]^{2-}$ ions, like it is observed for the isostructural Pd-derivative. In contrast, the adduct $XeF_6 \cdot 2\,MnF_4$ is better described as $[XeF_5]_4[Mn_8F_{36}]$, as it contains octameric anions which are built from $[MnF_6]$-octahedra. Each of these octahedra shares three corners with its neighbors, so that in total a ring-shaped structure is observed for the anion [2107]. Several other adducts of MnF_4 and XeF_2 or XeF_6 have been described [1030, 2108]. If elemental Mn, MnF_2 or MnF_3 is fluorinated in presence of elemental Sb, SbF_3 or SbF_5, adducts of Mn(III) and Mn(IV) are observed, which have the compositions $MnF_3 \cdot SbF_5$, $4\,MnF_4 \cdot SbF_5$ or $MnF_4 \cdot SbF_5$, depending on the reaction conditions, that is, temperature and pressure. However, manganese in higher oxidation states, for example, a salt of the type $[Mn^VF_4]^+[SbF_6]^-$, is not obtained [2109].

Higher binary manganese fluorides than MnF_4 are unknown [2110]. Quantum chemical calculations have been performed for manganese pentafluoride MnF_5, indicating that it is most likely unstable in both the condensed phase and gaseous state. Thus, it is not too surprising that attempts to prepare MnF_5 by matrix techniques have been unsuccessful. The situation becomes even more clear for MnF_6 and MnF_7, where the decomposition pathways under liberation of F_2 are strongly exothermic [2029].

Manganese oxyfluorides

Despite the fact, that high-valent binary manganese fluorides are nonexistent, a heptavalent oxyfluoride is well known. **MnO_3F** is obtained when $KMnO_4$ is reacted with anhydrous HF or HSO_3F. Under milder conditions, the treatment of $KMnO_4$ with IF_5 or HF at room temperature also gives rise to MnO_3F [1897]. It is a dark green liquid (mp.: −38 °C), which is only stable below 0 °C and very sensitive toward humidity. At room temperature, the shock-sensitive compound decomposes explosively into MnO_2, MnF_2 and O_2 [1894, 2111]. MnO_3F is a molecular monomer in the solid state, having an almost tetrahedral structure. It has further been characterized by infrared-, UV-vis- and electronic spectra [2112–2114]. The reaction of MnO_3F with gaseous HCl and HF has been investigated by means of IR-spectroscopy in an Ar-matrix. In the latter case, the formation of a complex $MnO_3F \cdot HF$ is assumed, whereas HCl transforms MnO_3F to MnO_3Cl [2115]. Other manganese oxyfluorides MnO_xF_y have been studied by computational methods, giving an insight into their structures, stabilities and electronic properties [2116]. Manganese(III) oxyfluoride **MnOF** and manganese(IV) oxydifluoride **$MnOF_2$** have been explored by mass spectrometry and infrared spectroscopy under matrix conditions. Supported by theoretical calculations, the ground state of $MnOF_2$

has been found to have C_{2v}-symmetry, whereas MnOF is linear [2117]. Both compounds are also generated under matrix conditions, if laser ablated Mn-atoms are treated with OF_2. Additionally, manganese(V) oxytrifluoride **MnOF$_3$** has been observed, which has a distorted C_{3v}-symmetry [1503].

10.7.2 Technetium and rhenium

Low-valent fluorides

The low-valent fluorides of rhenium and technetium are nonexistent in the condensed phase. However, quantum-chemical calculations have been performed for most of the hypothetic molecules, in some cases they have even been observed in the gas phase. Both rhenium and technetium monofluoride **ReF** and **TcF** have been investigated by calculations, giving an prediction of their fundamental properties like bond lengths, dissociation energies or electron affinities [1180, 2118]. Moreover, ReF has been prepared in a laser-generated plasma and the electronic spectrum of molecular ReF has been investigated [2119]. Rhenium difluoride **ReF$_2$** has been claimed to be formed, if Re metal is electrically exploded in an atmosphere of PF_5 [2120]. Technetium difluoride **TcF$_2$** has been subject of quantum-chemical calculations, which have revealed some properties of the molecule, including the bonding energy [1183].

Trifluorides

Gaseous rhenium trifluoride **ReF$_3$** has been reported to be accessible, if Re-metal is electrically exploded in presence of SF_6 [2120]. Pure ReF$_3$ has never been obtained in the condensed phase. In contrast, the reaction of ReCl$_3$ with anhydrous HF yields Re-metal and some minor amounts of ReF$_6$. It is believed, that ReF$_3$ is generated upon Cl/F-substitution of ReCl$_3$ but immediately decomposes under the reaction conditions [1754]. Apart from that, the only known rhenium-fluorine compound with the metal in oxidation state +III is the octafluorodirhenate(III) anion **[Re$_2$F$_8$]$^{2-}$**. It is prepared by substitution of the respective halide-derivatives [Re$_2$X$_8$]$^{2-}$ (X = Cl, Br, I) with fluoride ions. Interestingly, [Re$_2$F$_8$]$^{2-}$ contains the cluster-ion [Re$_2$]$^{6+}$, which has a Re-Re quadruple bond [2121–2123]. In the solid-state structure, the F-atoms of the [Re$_2$F$_8$]$^{2-}$ anion are oriented in form of a tetragonally distorted cubane-cage around the linear [Re$_2$]$^{6+}$-units [2124]. It is also possible to replace fluoride-substituents by organic, bidentate N,N'- or O,N-donor ligands, resulting in neutral compounds of the general composition [Re$_2$L$_4$F$_2$], (L = hexahydro-2H-pyrimido-[1,2a]-pyrimidinate; diphenyl formamidinate; 2-oxypyridine), where the two F-atoms are on an almost linear F-Re-Re-F-axis [2125, 2126]. Technetium trifluoride **TcF$_3$** is unknown, so that the literature contains neither preparative nor theoretical studies on this compound.

Tetrafluorides

Rhenium tetrafluoride **ReF$_4$** is generated by disproportionation of ReF$_5$ or more conveniently by the reduction of ReF$_5$ or ReF$_6$ with H$_2$, SO$_2$, Re, Zn or Al (Scheme 10.19) [2127–2130]. ReF$_4$ is also accessible from solutions of tetravalent [ReF$_6$]$^{2-}$, which are treated with SbF$_5$ [1926]. It is a blue solid (mp.: 124.5 °C; subl.: ca. 300 °C), which rapidly hydrolyzes to form ReO$_2$ and HF [2129]. In contrast, the green complex salts of the type M$_2^I$[ReF$_6$] are very stable, as they are not even attacked by aqueous alkaline solutions. The hexafluororhenates(IV) are typically obtained by fluorination of M$_2^I$[ReX$_6$] (X = Cl, Br, I) with HF or by reduction of ReF$_6$ with M^{I}I in SO$_2$ at –30 °C (Scheme 10.19). The cation exchange of K$_2$[ReF$_6$] further allows the preparation of the acid H$_2$[ReF$_6$], which can be concentrated from its aqueous solution, where it finally crystallizes as a hydrate. Hexafluororhenate(IV) salts with complex cations of chromium(III) and cobalt(III) are also accessible, as well as the nitrosyl derivative [NO]$_2$[ReF$_6$] [2131]. Technetium tetrafluoride **TcF$_4$** is unknown; it has only been investigated by theoretical calculations. As one of the results, a chain structure has been predicted for solid TcF$_4$ [2132]. However, hydrolytically stable salts of the type M$_2^I$[TcF$_6$] are accessible. One of them is the potassium salt K$_2$[TcF$_6$], which is obtained from a mixture of molten K$_2$[TcBr$_6$] and K[HF$_2$] (Scheme 10.19) [2133]. By a slightly modified protocol, it is also possible to synthesize hexafluorotechnetate(IV) complexes, which include enriched amounts of the radiochemically relevant isotope ^{99m}Tc [2134]. An easier, and thus more convenient preparation of [TcF$_6$]$^{2-}$ is based on the treatment of [TcBr$_6$]$^{2-}$ with concentrated aqueous HF (w = 40 %) [2135]. Starting from K$_2$[TcF$_6$], salts with other cations (Rb$^+$, Cs$^+$, NMe$_4^+$) are accessible by simple metathesis reactions. Alternatively, the reduction of NH$_4$TcO$_4$ with sodium dithionite in aqueous HF and subsequent treatment with M^{I}F generates M$_2^I$[TcF$_6$] (M^I = Na, K, Rb, Cs, NMe$_4$). The ammonium salt (NH$_4$)$_2$[TcF$_6$] is accessible, if NH$_4$TcO$_4$ is reduced by Zn in aqueous HF in presence of NH$_4$F (Scheme 10.19) [2136]. The [TcF$_6$]$^{2-}$ anion is not attacked by dilute aqueous acids, but hydrolyzed by hot, concentrated alkaline solutions [2133]. The solid-state structures of both, hexafluororhenates(IV) [ReF$_6$]$^{2-}$ and –technetates(IV) [TcF$_6$]$^{2-}$ are described by the K$_2$[GeF$_6$] structure type [2137]. The salts contain discrete [MF$_6$]$^{2-}$ anions, which are described by D$_{3d}$-distorted octahedra, as has been shown by X-ray crystallography and Raman spectroscopy [2136]. Mixtures of various chlorofluoro and bromofluororhenates(IV) [ReF$_n$X$_{6-n}$]$^{2-}$ (X = Cl, Br; n = 0–6) can be separated using HPLC-techniques, including the isolation of cis/trans and fac/mer-isomers [2138]. Moreover, the [ReF$_6$]$^{2-}$ anion has been studied concerning its magnetic properties and proposed as building block for the synthesis of new molecule-based magnetic materials [2139].

Pentafluorides

Yellow technetium pentafluoride **TcF$_5$** (mp.: 50 °C; decomp.: 60 °C) and yellow-green rhenium pentafluoride **ReF$_5$** (mp.: 48 °C; bp.: 240 °C) are both solids, which are read-

ReF_6 $[TcBr_6]^{2-}$

I^-, SO_2,
-30 °C HF_2^- or
conc. aq. HF

$[ReX_6]^{2-}$ $\xrightarrow{HF}$ $[MF_6]^{2-}$ $\xleftarrow[\text{aq. HF, then } F^-]{Na_2S_2O_4 \text{ or Zn,}}$ TcO_4^-
(X = Cl, Br, I)

SbF_5
(M = Re)

ReF_5 $\xrightarrow[\text{e.g. } H_2, \text{ Re or Al}]{\text{reduction}}$ ReF_4 $\xrightarrow{\text{hydrolysis}}$ ReO_2 + HF
or ReF_6

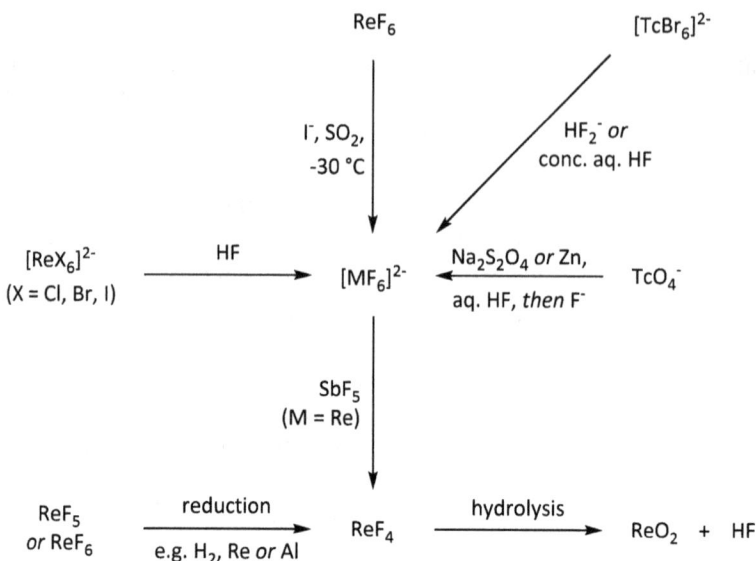

Scheme 10.19: Preparation and selected reactions of tetravalent fluorine compounds of Tc and Re.

ily hydrolyzed [2129]. TcF_5 is formed as one of the products of the fluorination of Tc, or more conveniently by quantitative reduction of TcF_6 with I_2 in IF_5-solution [2140–2142]. Analogously, ReF_5 is prepared by the reduction of ReF_6, where several reducing agents can be applied, for example, metal carbonyls like $W(CO)_6$, W- or Re-wire at 600 °C, Si or H_2 [1652, 2127, 2129]. Both compounds are isostructural to VF_5, existing as trigonal-bipyramidal molecules in the colorless vapor-phase. In the condensed phase, they are associated in chains of $[MF_6]$-octahedra, which are cis-bridged and possess common corners. TcF_5 is more stable than ReF_5, but attacks glass at 60 °C, while itself is decomposed [2140]. At temperatures above 130 °C, ReF_5 irreversibly disproportionates to ReF_4 and ReF_6 [1953, 2129]. The pentafluorides are relatively weak fluoride acceptors and form hexafluorometallates(V), which can either be obtained by direct reaction of the respective fluorides or alternatively, by starting from ReF_6 or TcF_6 [1954, 2140, 2143]. However, from anhydrous HF, ReF_5 does not accept fluoride ions, since significant amounts of $[ReF_6]^-$ cannot be detected [1954]. In the following examples, pentavalent $[ReF_6]^-$ compounds are generated by reactions of the hexafluoride. In SO_2-solution, excess ReF_6 reacts with $M^I I$ to yield antiferromagnetic $M^I[ReF_6]$ (M^I = Na, K, Rb, Cs). Similarly, molecular iodine is oxidized to yield the $[I_2]^+$ cation, using IF_5 as reaction medium. Apart from that, the metals Cd, Cu and Tl are oxidized in acetonitrile to result in white $Cd[ReF_6]_2 \cdot 5$ MeCN, yellow $Cu^I[ReF_6] \cdot 4$ MeCN and pale yellow $Tl^I[ReF_6] \cdot 2$ MeCN, respectively. If IF_5 is added, the complex $Cu^{II}[ReF_6] \cdot 4$ MeCN $\cdot 0.5$ IF_5 is obtained [1369]. The oxyfluorides **$TcOF_3$** and **$ReOF_3$** are formed upon impact of TcF_5 and $ReOF_4$ on glass. $ReOF_3$ is a black, polymeric solid, which is nonvolatile and very hygroscopic. It forms the fluoro complex $[ReOF_4]^-$ upon reaction with fluoride donors [2129, 2144]. Yellow **ReSF₃** results from the reaction of ReF_5 with Sb_2S_3 in anhydrous HF at room temperature. Like the oxygen-analogue, it is very sensitive toward air and moisture [2145].

Hexafluorides

Technetium hexafluoride **TcF₆** (mp.: 37.4 °C; bp.: 55.3 °C) is the final product of the fluorination of elemental Tc at 350–400 °C (Scheme 10.20) [2146]. It can also be directly obtained from commercially available NH_4TcO_4 and elemental F_2 [2147]. In contrast, rhenium hexafluoride **ReF₆** (mp.: 18.5 °C; bp.: 33.7 °C) is formed along with ReF_7 under similar conditions and can be purified by the reaction of the ReF_6/ReF_7 mixture with elemental Re at about 400 °C (Scheme 10.20) [2148]. Both hexafluorides are pale yellow, very volatile compounds, which are dimorphic. At low temperature, they crystallize in an orthorhombic lattice, whereas the high temperature forms have a cubic structure. In the vapor phase, the hexafluorides seem to have octahedral symmetry, however, at lower temperatures, TcF_6 is obviously dynamically distorted from a perfect octahedral shape. This is due to the Jahn–Teller effect and can clearly be observed in vibrational spectra [2149]. The same holds for ReF_6 but as the distortions are very small, it is not surprising that the analysis of the gas-phase structures by X-ray crystallography or electron diffraction yields regular octahedral structures, or more precisely, that the experimental errors are larger than the deviations from O_h-symmetry [1965, 2147, 2150–2156]. Both, TcF_6 and ReF_6, are very sensitive toward hydrolysis, in the first step, the respective oxytetrafluorides $TcOF_4$ and $ReOF_4$ are formed [1997, 2157]. If ReF_6 is partially hydrolyzed in HF, the green $[H_3O]^+[ReOF_5]^-$ has been claimed to be detectable.However, it might rather have the formula $[H_3O]^+[Re_2O_2F_9]^-$, where the anion is composed of two F-bridged $[ReOF_4]$-units [1997, 2158]. ReF_6 is oxidized by $[KrF]^+[Sb_2F_{11}]^-$ to yield the heptavalent cation $[ReF_6]^+$, which is more conveniently accessible by treatment of ReF_7 with SbF_5 (Scheme 10.20) [732, 2159]. However, the hexafluorides themselves are strong oxidizers, where TcF_6 is more easily reduced than ReF_6. Interestingly, the reaction of ReF_6 with a solution of KI in IF_5 yields the pentavalent rhenium salt $K[ReF_6]$, whereas TcF_6 is reduced to tetravalent $[TcF_5]^-$ (Scheme 10.20). Nevertheless, technetium(V) salts of the type $M^I[TcF_6]$ are obtained, if alkali metal chlorides are reacted with TcF_6 in IF_5 [2140, 2143]. Moreover, ReF_6 and TcF_6 both react with NO to yield the nitrosyl salts $[NO][TcF_6]$ and $[NO][ReF_6]$, respectively [2160]. The reaction of PF_3 with ReF_6 results in the formation of PF_5 and ReF_5 [2161]. ReF_6 is also capable of oxidizing decamethylferrocene $[FeCp^*_2]$ to the relatively stable decamethylferrocenium dication $[FeCp^*_2]^{2+}$ [2162]. In SbF_5-solution, ReF_6 is slowly reduced by CO to yield the cationic carbonyl species $[Re(CO)_6]^+$ [2163]. Similarly, ReF_6 reacts with $Re_2(CO)_{10}$ in hydrofluoric acid to yield the complexes $[Re(CO)_5]^+[Re_2F_{11}]^-$ and $[Re(CO)_5]^+[ReF_6]^-$. The latter is also formed upon fluorination of $Re_2(CO)_{10}$ with XeF_2 and may be better

$$\text{NH}_4\text{TcO}_4$$

$$\text{ReF}_6 + \text{ReF}_7 \xleftarrow{\text{F}_2,\ 350\text{-}400\ °C} \text{Re}$$

F$_2$

Re,
400 °C

$$\text{Tc} \xrightarrow{\text{F}_2,\ 350\text{-}400\ °C} \text{MF}_6 \xrightarrow[(M = Re)]{[KrF][Sb_2F_{11}]} [\text{ReF}_6]^+ \text{ salts}$$

H$_2$O, SiO$_2$,
B$_2$O$_3$, etc.

KI, IF$_5$

F⁻

MOF$_4$

K[ReF$_6$] *or* K[TcF$_5$]

[MF$_7$]⁻ *or* [MF$_8$]$^{2-}$

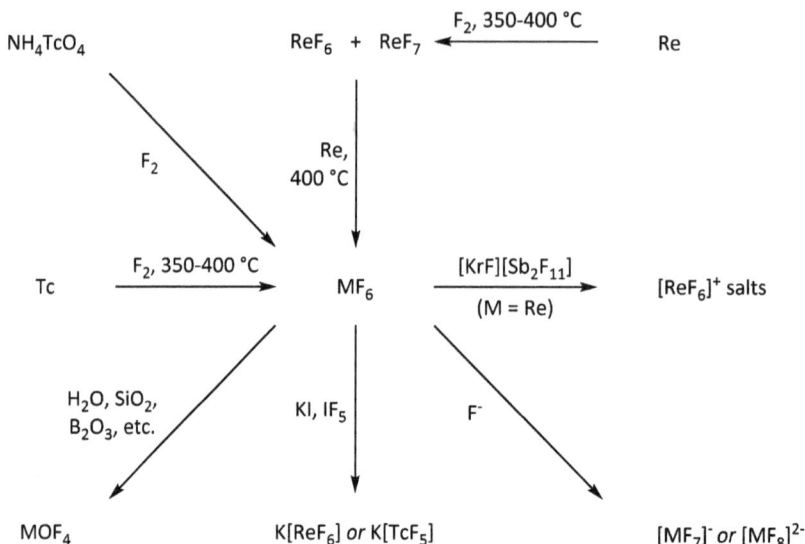

Scheme 10.20: Preparation and selected reactions of the group 7 hexafluorides MF$_6$ (M = Tc, Re).

formulated as [Re(CO)$_5$F]·[ReF$_5$], since its solid-state structure does not reveal equal Re-F bond-lengths, which would be required for a potential [ReF$_6$]⁻ anion. The anion [Re$_2$F$_{11}$]⁻ contains a linear F-Re-F bridge [2164–2166]. Halide exchange of the commercially available chloro-derivative [Re(CO)$_5$Cl] gives rise to the pentacarbonyl-complex [Re(CO)$_5$F], which can be further oxidized by XeF$_2$ to yield trivalent [Re(CO)$_3$F$_3$] [2167]. ReF$_6$ is a weaker oxidizer than PtF$_6$, thus it does not oxidize xenon to Xe⁺, but forms a charge-transfer complex with it [2168]. If ReF$_6$ and TcF$_6$ are reacted with M^{I}F (M^I = NO, NO$_2$, alkali metal), fluorometallates of the types M^I[TcF$_7$], M$^I_2$[TcF$_8$], M^I[ReF$_7$] and M$^I_2$[ReF$_8$] are formed (Scheme 10.20), where the octafluorometallate anions [MF$_8$]$^{2-}$ have a square-antiprismatic structure. The heptafluorometallate anions [MF$_7$]⁻ form pentagonal bipyramids. Interestingly, the heptafluorotechnetate(VI) anion [TcF$_7$]⁻ is unstable, whereas [TcF$_8$]$^{2-}$ is stable [2160]. Apart from their stability, the fluorometallate anions expectedly also differ in their appearance, as the [ReF$_7$]⁻ anion is orange, which is in contrast to violet [ReF$_8$]$^{2-}$ [1100, 2169]. The reaction of ReF$_6$ with an eutectic mixture of LiF, NaF and KF at 600 °C also yields intermediary [ReF$_8$]$^{2-}$, which can then be electrochemically reduced to yield well-crystallized, elemental Re [2170]. It is also possible to substitute the fluoride anions of ReF$_6$ by other halogens, as for example, the reaction with BCl$_3$ yields ReCl$_6$ [1686, 2171, 2172]. Similarly, the reaction with methanol gives rise to Re(OMe)$_6$ [2173]. The fluorinating properties of ReF$_6$ with respect to hydrochlorocarbons have been exemplarily examined for dichloromethane. In contrast to ReF$_7$, ReF$_6$ almost exclusively substitutes one of the Cl-substituents, resulting in the formation of CH$_2$ClF [1698]. Rhenium chloride-pentafluoride **ReClF$_5$** is one of the products of the fluorination of ReCl$_5$ at −30 °C. It is a very unstable, red solid (mp.: $-2\,°C$), which already slowly decomposes at $-30\,°C$ to form ReF_6 and rhenium chlorides [2174]. If ReF_6 is reacted with azidotrimethylsilane Me_3SiN_3 at $-50\,°C$, the nitride fluoride **ReNF₄** is obtained as pale-yellow, crystalline solid. $ReNF_4$ can be fluorinated using ClF_3 to result in a mixture of nitrenes, being composed of the purple $ReF_5(NCl)$ and orange $ReF_5(NF)$ [2175, 2176]. Similarly, $ReF_5(NBr)$ is accessible [2177].

Hexavalent oxyfluorides

There also exist oxyfluorides of technetium(VI) and rhenium(VI). **TcOF₄** (mp.: $133\,°C$; bp.: $165\,°C$) and **ReOF₄** (mp.: $108\,°C$; bp.: $171\,°C$) are generally synthesized from the hexafluorides by treatment with SiO_2, metal carbonyls, ReO_3, B_2O_3 or upon controlled hydrolysis (Scheme 10.20) [1438, 2129, 2144, 2157]. It has been reported that the controlled hydrolysis of ReF_6 with quartz (but not glass or water) does not only yield $ReOF_4$, but also colorless **ReO₂F₂** [864]. The latter has also been claimed to be accessible by the fluorination of Re with oxygen-containing F_2, whereas it hydrolyzes on moist air under formation of a violet color [2128]. However, the existence of ReO_2F_2 has never been confirmed by other research groups and was especially in doubt after the discovery of Re(VII)-oxyfluorides. Thus, the claimed Re(VI)-compound ReO_2F_2 may have been the Re(VII)-compound ReO_2F_3 [2129]. $TcOF_4$ is one of the fluorination products of elemental Tc [2140]. Both compounds, $ReOF_4$ and $TcOF_4$ exist in two different, crystalline shapes. The blue forms are stable at room temperature and have a structure like VF_5, where the molecules are connected to long chains via *cis*-F-bridges [2003, 2178]. The green and light-blue high-temperature modifications consist of cyclic trimers [2179, 2180]. In case of $TcOF_4$, the transition between the two different shapes occurs at $84.5\,°C$. In the gas phase and under matrix conditions, monomeric, C_{4v}-symmetric molecules of $TcOF_4$ and $ReOF_4$ are observed [2004, 2114, 2181, 2182]. Both compounds are very sensitive toward hydrolysis [2129]. Vapors of $ReOF_4$ attack Pyrex-glass at $250\,°C$ under formation of $ReOF_3$, whereas $ReCl_5$ is generated upon reaction with carbon tetrachloride [2129]. $ReOF_4$ forms light-green fluorocomplexes with the $[ReOF_5]^-$ anion, which are accessible by controlled hydrolysis of ReF_6 or heptafluororhenate(VI) anions $[ReF_7]^-$ [2158, 2183]. However, in anhydrous HF, $ReOF_4$ does not yield $[ReOF_5]^-$ in significant amounts [1954]. The sulphur analogue **ReSF₄** is also known. It is a red solid which is obtained, if ReF_6 is reacted with the sulphides Sb_2S_3 or B_2S_3. Similar to the oxygen compound, $ReSF_4$ is very sensitive toward air and moisture [2145].

Heptafluorides

Rhenium heptafluoride **ReF₇** is the final product of the fluorination of Re at $400\,°C$ and 3 bar pressure. It is a yellow solid (mp.: $48.3\,°C$; bp.: $73.7\,°C$), which is soluble in HF [269, 2148, 2184, 2185]. It has also been reported that ReF_7 is accessible, when Re-wire is electrically exploded in a stainless-steel vessel in presence of SF_6 as fluorinating agent [2120]. Along with IF_7, ReF_7 is the only known heptafluoride. Several spectroscopic methods have revealed, that ReF_7 has a nonrigid, pentagonal-bipyramidal structure, which has also been proven by neutron diffraction of powdered samples [2186–2189]. If ReF_7 is hydrolyzed, $HReO_4$ and HF are formed. In general, ReF_7 has a comparable reactivity like ReF_6, the main difference is, that reactions generally occur more vigorously [2190, 2191]. ReF_7 reacts with NOF and NO_2F to form the octafluororhenate(VII)-salts $[NO][ReF_8]$ and $[NO_2][ReF_8]$, respectively, with the yellow, square-antiprismatic anion $[ReF_8]^-$ [2192]. In contrast, the octahedral $[ReF_6]^+$ cation is accessible, when ReF_7 and SbF_5 are heated to 250 °C, which gives rise to $[ReF_6]^+[SbF_6]^-$ and $[ReF_6]^+[Sb_2F_{11}]^-$ [2159]. The vibrational properties of the $[ReF_6]^+$ cation have been explored, too [2193]. The general fluorinating abilities of ReF_7 with respect to organic compounds have been investigated for the representative chlorocarbon dichloromethane. Interestingly, rhenium heptafluoride preferentially substitutes the hydrogen in CH_2Cl_2, whereas the chloro-substituents remain untouched [1698]. Technetium heptafluoride $\mathbf{TcF_7}$ is unknown. Quantum-chemical calculations have revealed that a potential, gaseous TcF_7 molecule should be sufficiently stable against typical decomposition pathways losing F_2 to yield TcF_5 or splitting off a F-atom to result in TcF_6. Furthermore, the cation $[TcF_6]^+$ has also been predicted to be stable against F_2 or F-elimination [2194]. However, the synthetic access to TcF_7 or the closely related $[TcF_6]^+$ is limited, as the fluorination of various technetium(VII) oxyfluorides stops at the stage of $TcOF_5$ (see below). The generation of $[TcF_6]^+$ by treatment of $TcOF_5$ with $[KrF]^+$ turned out to be unsuccessful [2195]. Thus, the preparation of TcF_7 might be restricted to the gas phase or matrix techniques.

Heptavalent oxyfluorides

A few more rhenium(VII) and technetium(VII) compounds are existent in form of the oxyfluorides $ReOF_5$, ReO_2F_3 and ReO_3F as well as the analogous technetium compounds $TcOF_5$, TcO_2F_3 and TcO_3F. Rhenium oxypentafluoride $\mathbf{ReOF_5}$ is conveniently synthesized from Re_2O_7 and ClF_3 [2196]. It is also accessible by fluorination of ReO_2, ReO_3 or $ReOF_4$ at 250–300 °C. It is an off-white, volatile solid (mp.: 41 °C; bp.: 73 °C), which is colorless in the gaseous and molten state and readily hydrolyzes upon contact with air. Expectedly, the $ReOF_5$ molecule has C_{4v}-symmetry [941, 979, 2114, 2144, 2182, 2197, 2198]. $ReOF_5$ acts as fluoride-acceptor against NOF and NO_2F, which results in the formation of ionic $[NO][ReOF_6]$ and $[NO_2][ReOF_6]$. The salts contain the rigid, pentagonal-bipyramidal anion $[ReOF_6]^-$, where the O-atom occupies an axial position [2192]. The analogous sulphur derivative rhenium thiopentafluoride $\mathbf{ReSF_5}$ is also known. It is a maroon solid which is very sensitive toward air and moisture and can be prepared by heating ReF_7 with Sb_2S_3 or B_2S_3. Alternatively, the reaction of ReF_7 with Sb_2S_3 in anhydrous HF also yields $ReSF_5$ [2145]. Technetium oxypentafluoride $\mathbf{TcOF_5}$ is obtained from the room temperature reaction of TcO_2F_3 with KrF_2 in HF as volatile,

orange solid (mp.: 57–58 °C). In the solid state, it is stable for some time at room temperature, as long as it is kept under anhydrous conditions. $TcOF_5$ has a distorted octahedral molecular geometry [2199]. In presence of AsF_5 or SbF_5 in HF, $TcOF_5$ acts as a
fluoride donor, resulting in the formation of salts with the dimeric cation $[Tc_2O_2F_9]^+$.
The latter is composed of two F-bridged, square-pyramidal $[TcOF_4]$-units, where the
bridging F-atom is in a *trans*-position to oxygen. In NMR-spectra of the $[Tc_2O_2F_9]^+$ salts,
one can also identify the $[TcOF_4]^+$ cation. However, in turn, treatment of $TcOF_5$ with
KrF_2 does not yield a potential $[TcF_6]^+$ cation or even the unknown heptafluoride TcF_7
[2195]. Rhenium dioxytrifluoride **ReO_2F_3** is a yellow, polymeric solid (mp.: 95 °C; bp.:
126 °C), which is formed upon reaction of ReO_3Cl with XeF_2 [2200]. Alternatively, it
can be synthesized by fluorination of ReO_2 or $KReO_4$ and is stable in glassware up to
its boiling point [941, 2144, 2198, 2201]. Vibrational spectra have revealed a trigonal-
bipyramidal structure of the monomer, where the O-atoms are occupying equatorial
positions [2114, 2202]. ReO_2F_3 has also been investigated by X-ray crystallography, revealing a polymorphic behavior. The probably thermodynamically most stable conformation has a bent chain-structure; however, cyclic trimers and tetramers have also
been observed. The latter include almost planar six-membered $[Re_3F_3]$-rings and eight-
membered $[Re_4F_4]$-cycles [2200]. The trimers and tetramers also exist in SO_2ClF solution, as has been revealed by NMR spectroscopy [2203]. ReO_2F_3 reacts with fluoride
donors M^IF (M^I = Li, Na, Cs, NMe_4) under formation of the salts $M^I[ReO_2F_4]$. However, the reaction with KF generates $K[Re_2O_4F_7]$, whereas CsF can even lead to the
formation of $Cs[Re_3O_6F_{10}]$ [2201]. Moreover, the fluoride ligands of ReO_2F_3 can be replaced by $OTeF_5$-substituents upon reaction with $B(OTeF_5)_3$. Accordingly, the respective pseudo-octahedral, anionic species $[ReO_2(OTeF_5)_4]^-$ is accessible [2204]. Interestingly, ReO_2F_3 has also fluoride donor properties, as has been found in the reactions
with the fluoride acceptors AsF_5 and SbF_5. In the solid state, the products are not
salt-like, as one can find alternating $[ReO_2F_4]$- and $[SbF_6]$-units in infinite chains of
$ReO_2F_3 \cdot SbF_5$. However, the salt $[ReO_2F_2(MeCN)_2][SbF_6]$ is formed upon dissolving the
adduct $ReO_2F_3 \cdot SbF_5$ in acetonitrile. The $[ReO_2F_2(MeCN)_2]^+$ cation has a distorted octahedral geometry, where the oxo-substituents are oriented in *cis*-position to each other
and in *trans*-orientation to the acetonitrile-ligands. Furthermore, NMR-spectroscopy
revealed that the adduct $ReO_2F_3 \cdot SbF_5$ dissociates in SO_2ClF solution and re-associates
to $[Re_2O_4F_5][Sb_2F_{11}]$, containing the F-bridged $[Re_2O_4F_5]^+$ cation [2205]. The analogous
technetium dioxytrifluoride **TcO_2F_3** is also known. It is prepared as a yellow solid
(mp.: 200 °C) from a 3:1-mixture of XeF_6 and Tc_2O_7 in anhydrous HF. In the solid-state
structure, one can find open chains of F-bridged $[TcO_2F_4]$-units, where the bridging
F-substituent is always in a *trans*-position to oxygen. Interestingly, the ligand atoms
form nearly perfect octahedra, however, the Tc-atoms are not centered, but strongly
displaced toward the O-substituents [2206]. In SO_2ClF-solution, TcO_2F_3 exists in form
of cyclic trimers, as has been explored by NMR-spectroscopy [2203]. Further treatment
of TcO_2F_3 with XeF_6 does not lead to oxygen-fluorine exchange toward $TcOF_5$, but generates the salt $[Xe_2F_{11}][TcO_2F_4]$ [2206]. Like the analogous ReO_2F_3, TcO_2F_3 has also flu-

oride donor properties, as has been proven by the reactions with the fluoride acceptors AsF_5 and SbF_5. Again, the products have to be considered as adducts, which has been shown by X-ray crystallographic analysis of $TcO_2F_3 \cdot SbF_5$ [2205]. Technetium trioxyfluoride (pertechnetyl fluoride) **TcO_3F** is obtained by fluorination of TcO_2 as a yellow, crystalline solid (mp.: 18.3 °C; bp.: ca. 100 °C), which melts to give a yellow liquid and attacks glass at room temperature. Nevertheless, it can be stored in Monel or Ni-vessels at room temperature [2207]. More conveniently, it is obtained, if $K[TcO_4]$ is dissolved in anhydrous HF in presence of BiF_5 [1707]. The TcO_3F molecule has C_{3v}-symmetry, which has been revealed by vibrational spectroscopy [2208, 2209]. In the solid state, double F-bridged dimers are present, which is comparable to the oxyfluorides VOF_3 and CrO_2F_2 [1707]. In HF-solution, TcO_3F is a weak fluoride ion donor, as it is in equilibrium with $[TcO_3]^+$ and $[HF_2]^-$. The reaction with fluoride acceptors, for example, AsF_5, yields stable salts of the type $[TcO_3][AsF_6]$ [2210]. This is not the case, if the aforementioned transformation is performed in HF solution, as this gives rise to derivatives like $[TcO_2F_2][AsF_6]$ [1707]. Rhenium trioxyfluoride (perrhenyl fluoride) **ReO_3F** is usually obtained as a yellow solid upon fluorination of ReO_3Cl with HF, $KReO_4$ with IF_5 as well as Re_2O_7 with ReF_7 or HF [941, 1897, 2111, 2144, 2211, 2212]. Alternatively, the fluorination of ReO_3 yields ReO_3F, if the reaction temperature is kept below 150 °C. ReO_3F melts under formation of a yellow, viscous liquid and decomposes above 200 °C. It crystallizes in a wide temperature range starting at 140 °C. The solid-state structure can be described by spiral chains, which contain almost octahedrally coordinated Re-atoms. Interactions between different chains have been found to be weak [2200]. The structure of ReO_3F has also been investigated by microwave and Raman spectroscopy [2211, 2213]. Moreover, molecular ReO_3F has been investigated by UV-vis spectroscopy in an inert-gas matrix [2114]. ReO_3F is very easily reduced and hydrolyzes immediately upon contact with water, however, it does not attack glass, even at elevated temperatures [2111]. ReO_3F reacts with bidentate bipyridine ligands under formation of stable complexes, which can be handled in air [2214]. Another interesting reaction is observed, if ReO_3F is reacted with XeF_2 in anhydrous HF at −30 °C. The resulting red-orange salt has the composition $[XeOXeOXe][\mu\text{-}F(ReO_2F_3)_2]_2$, containing the planar, zigzag-shaped cation $[XeOXeOXe]^{2+}$. Structurally, the anion is best described as being composed of two distorted $[ReO_2F_4]$-octahedra, which share a common vertex [2215]. A few more derivatives are accessible by derivatization of $[ReO_4]^-$. For example, the reaction of $KReO_4$, KF and IF_5 yields $K_2[ReO_3F_3]$,[2183] whereas $K_2[ReO_4F]$ is generated from $KReO_4$ in hydrofluoric acid [2216].

10.7.3 Study questions

(I)   How is MnF_2 prepared? Which fluoroanions does it form and how are they typically accessible?

(II) Which methods exist for the synthesis of MnF$_3$? What is the common structural feature in virtually all trivalent fluoromanganates?

(III) Comment on the thermal stability of MnF$_4$. For which purpose has it been exploited? How is MnF$_4$ typically prepared?

(IV) Is there a high-valent fluorine-containing manganese compound? How is it prepared? Comment on its stability.

(V) Which is the only existing trivalent rhenium-fluorine compound? How is it prepared? Describe its structure.

(VI) Which neutral and ionic tetravalent fluorine compounds are known for Re and Tc? How are they prepared?

(VII) How are the pentafluorides TcF$_5$ and ReF$_5$ prepared? Are they stable compounds?

(VIII) Why does the procedure for the preparation of TcF$_6$ differ from that of ReF$_6$? Which fluorometallates do the hexafluorides form and how are they structurally described?

(IX) How can ReF$_7$ be prepared? Which products are observed upon reaction with NOF, NO$_2$F and SbF$_5$? Describe the structure of the products.

(X) Which heptavalent oxyfluorides of technetium and rhenium are known? How are these species prepared?

10.8 Iron, cobalt, nickel

10.8.1 Compounds and properties

The 3d-elements of the groups 8 to 10 exclusively form low valent di- and trifluorides. However, fluorometallates with the metal in oxidation state +IV are also known for cobalt and nickel. Due to the similarities of the fluorides of Fe, Co and Ni, these elements are discussed together in this chapter, whereas the platinum metals are treated in the following separate chapter.

Monofluorides

The monofluorides **FeF**, **CoF** and **NiF** are not stable in the condensed phase, but they can be observed in the gaseous state. Apart from that, they have often been subject of quantum-chemical calculations. The electronic states of the FeF molecule have been investigated by rotational spectroscopy, dispersed fluorescence studies and theoretical calculations [2217–2220]. The dissociation energy has also been calculated to have a value of about 4.9 eV [2221]. The total energies, equilibrium bond lengths as well as the binding energies of FeF, CoF and NiF have been determined in another quantum-chemical study and have been in good agreement with experimental data [2222]. Moreover, CoF has been investigated by several different spectroscopic methods in the gaseous state, which, for example, allowed the investigation of electronic states [2223, 2224]. Laser-induced fluorescence spectra have been obtained by excitation in different wavelength regions [2225, 2226]. The permanent electric dipole moment of CoF has also been determined [2227]. In the case of NiF, there is also a large number of publications on various spectroscopic investigations in the gas phase. Among them, near-infrared spectroscopy revealed electronic states and transitions between them [2228, 2229]. The low-lying electronic states have also been investigated by theoretical studies as well as by means of laser spectroscopy [2230, 2231].

Difluorides

White iron difluoride **FeF$_2$** (mp.: 1020 °C), pink cobalt difluoride **CoF$_2$** (mp.: 1200 °C) and yellow nickel difluoride **NiF$_2$** (mp.: 1450 °C) are prepared upon treatment of the metals or metal dichlorides with HF at high temperature (Scheme 10.21). Alternatively, NiF$_2$ is directly accessible from the elements at 550 °C. A compound of the composition NiF$_{2.1}$ is obtained, if Ni(II) salts are treated with F$_2$ at 200 °C [2232]. Thermolysis of the double salts MF$_2 \cdot n$NH$_4$F is conveniently used for the generation of solvate-free MF$_2$. Furthermore, the reduction of FeF$_3$ with elemental iron at 200 °C yields FeF$_2$ (Scheme 10.21). All of the three difluorides crystallize in the rutile lattice with slightly distorted octahedral geometry around the metal centers. NiF$_2$ sublimes above its melting point, in the gas phase one can find linear NiF$_2$ molecules. Generally, the difluorides are only sparingly soluble in water. Upon contact with moisture, FeF$_2$ slowly undergoes hydrolysis, whereas CoF$_2$ only reacts with water vapor at high temperatures to decompose to CoO and HF. However, NiF$_2$ is resilient toward hydrolysis and also quite resistant against concentrated acids. In the presence of hydrogen, CoF$_2$ is reduced at 300 °C, whereas FeF$_2$ does not react below 400 °C. The difluorides readily form hydrates and adducts with ammonia. In aqueous solutions of Co(II) and Ni(II) salts, which also contain fluoride ions, the hydrated complex cations [CoF]$^+_{aq}$ and [NiF]$^+_{aq}$ are detected [2233].

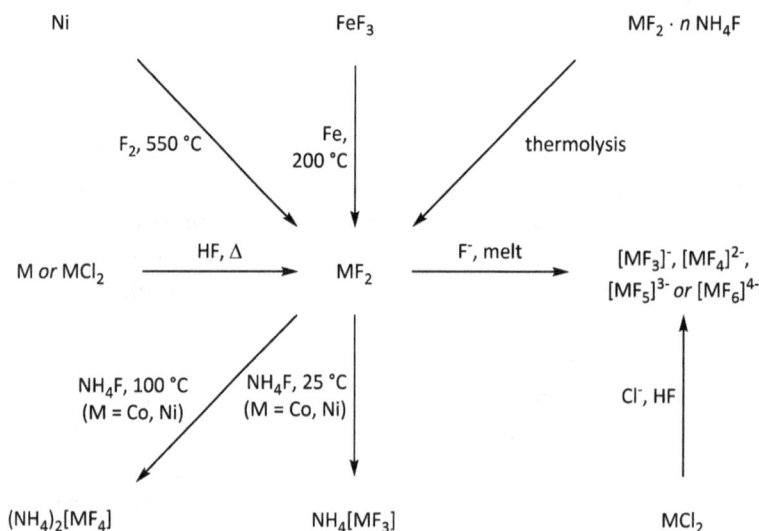

Scheme 10.21: Preparation of the difluorides MF$_2$ (M = Fe, Co, Ni) and their fluorocomplexes.

FeF_2, CoF_2 and NiF_2 form a large number of fluorometallates with the general compositions $M^I[MF_3]$,[2234–2238] $M^I_2[MF_4]$,[2235–2239] $M^{II}[MF_4]$[2043, 2240], $M^{III}[MF_5]$ and $M^{II}_2[MF_6]$ as well as $M^I_3[M_2F_7]$ [2235]. In most of the cases, these complexes are polymorphic, where the metal centers always show a slightly distorted, octahedral coordination geometry. The complex salts are of interest due to their magnetic and optomagnetic properties. Especially the magnetic and structural properties of the Ni compounds have been investigated and categorized [2238]. Generally, the fluorometallates are accessible by melting stoichiometric amounts of the respective fluorides or by reacting mixtures of the metal chlorides with HF (Scheme 10.21). Moreover, the tetrafluoronickelates(II) $M^I_2[NiF_4]$ ($M^I = NH_4$, K, Rb) are accessible, if $[Ni(acac)_2(H_2O)_2]$ is treated with M^IF in aqueous HF (w = 40 %) [2239]. In contrast, $Co[NiF_4]$ is prepared by reduction of the respective hexafluoronickelate(IV) with H_2 at 350 °C. Similarly, $Cu[NiF_4]$ is obtained by the reaction of $Cu[NiF_6]$ with Xe or hexafluoropropene at −20 °C. Both compounds have a pseudo-rutile structure [2240]. The reaction of CoF_2 or NiF_2 with NH_4F at 25 °C yields the 1:1-complexes $NH_4[MF_3]$, whereas heating to 100 °C leads to the formation of the tetrafluorometallate salts $(NH_4)_2[MF_4]$ (Scheme 10.21). The latter derivatives are isostructural with $K_2[NiF_4]$ [2241]. The sodium salts $Na[FeF_3]$, $Na[CoF_3]$ and $Na[NiF_3]$ have an orthorhombic structure similar to $GdFeO_3$, whereas the respective K-, Rb-, Cs-,Tl- and NH_4 salts crystallize in a perovskite lattice [2235, 2242]. $Rb[CoF_3]$ shows ferrimagnetic behavior in a system with $Rb[MgF_3]$, which is due to the magnetic properties of the Co^{2+} cation [2243]. Similarly, $Ba[CoF_4]$ also has paramagnetic and ferroelectric properties at room temperature [2244]. In the antiferromagnetic $Rb[NiF_3]$, the nickel atoms occupy two nonequivalent positions, which are both in the center of an octahedron. 2/3 of the $[NiF_6]$-octahedra are connected over common planes to form $[Ni_2F_9]$-polyhedra, which then are connected via the remaining $[NiF_6]$-octahedra, sharing common corners [2234, 2245]. The compounds $FeCoNiF_6$, Mg_2FeF_6 and $MgFe_2F_6$ have a distorted rutile structure, while $LiFe_2F_6$ contains Fe(II) and Fe(III) in an ordered trirutile structure [2246]. Similarly, di and trivalent Fe centers are both contained in $Fe_2F_5 \cdot 7 H_2O$, where the heptahydrate is isostructural to $FeCoF_5 \cdot 7 H_2O$ and $FeZnF_5 \cdot 7 H_2O$. They all split off water at 75–125 °C and are thus transformed to the dihydrates $FeMF_5 \cdot 2 H_2O$ (M = Fe, Co, Zn). The same behavior is observed for $AlFeF_5 \cdot 7 H_2O$ [2247]. In Na_2NiFeF_7, ferromagnetic chains of $[NiF_6]$-octahedra are present, which are separated from each other by $[FeF_6]$-units [2248, 2249]. Upon crystallization of $K_2[NiF_4]$, the byproduct $K_3[Ni_2F_7]$ is formed, which has intermediate magnetic properties compared to $K[NiF_3]$ and $K_2[NiF_4]$ [2250]. Both, NiF_2 and the complex $K_2[NiF_4]$ are antiferromagnetic. The magnetic behavior of mixtures with the respective magnesium compound has also been investigated [2251]. Structurally, the tetrafluoronickelates(II) consist of $[NiF_6]$-octahedra, which are linked by common corners, comparable to the structure of SnF_4. Modified $K[FeF_3]$ has been investigated for a potential use in batteries. Both, the cubic perovskite structure and the observed electrochemical behavior of the solid solution lead to a desirable reversible capacity, and thus make this compound interesting for applications in sodium-ion batteries [2252].

Trifluorides

Iron trifluoride **FeF$_3$** is formed as a pale-green solid (mp.: 1000 °C) upon fluorination of Fe, FeCl$_2$ or FeCl$_3$ with F$_2$, it may also be prepared from anhydrous FeCl$_3$ and HF at room temperature (Scheme 10.22). If Fe(OH)$_3$ is dissolved in aqueous HF and carefully concentrated, the pale-pink hydrate FeF$_3$ · 4.5 H$_2$O crystallizes out of the reaction mixture. Instead, distillation of the solvent at elevated temperatures *in vacuo* yields the trihydrate FeF$_3$ · 3 H$_2$O. The latter has also been found in small amounts in the volcanic mineral topsøeite on Iceland [2253]. The thermal behavior of the trihydrate has been studied, revealing that both, hydrolysis and dehydration can occur, depending on the reaction conditions. The resulting hydrolysis product with partly exchanged F-substituents has been proposed as a potential electrode material for batteries [2254]. In the solid state, anhydrous FeF$_3$ is isomorphic to AlF$_3$. Moreover, a comparable behavior is also observed in the vapor phase, since gaseous FeF$_3$ exists as a mixture of monomeric and dimeric molecules [1841]. FeF$_3$ is sparingly soluble in water and hydrolyzes only slowly. However, if FeF$_3$ is treated with liquid ammonia in presence of traces of water, the soluble ammoniate *mer*-[FeF$_3$(NH$_3$)$_3$] is obtained (Scheme 10.22). It is stable at low temperatures, but decomposes upon warming. In the solid-state crystal structure, one can find isolated molecules, which are connected by strong H-F-bonds [2255]. Heating of FeF$_3$ in presence of H$_2$ leads to the reduction to FeF$_2$ and Fe (Scheme 10.22). The derived fluorocomplexes of FeF$_3$ are relatively stable in water, however, in an aqueous solution [FeF$_6$]$^{3-}$ is mainly transformed to [FeF$_5$(H$_2$O)]$^{2-}$. If metallic Fe is reacted with hot, concentrated HF, the yellow, mixed-valent fluoride Fe$_2$F$_5$ · 7 H$_2$O is formed, which has been found to be composed of [Fe^{II}(H$_2$O)$_6$]$^{2+}$ and [Fe^{III}F$_5$(H$_2$O)]$^{2-}$ ions. Furthermore, it has been reported that a red dihydrate and finally blue-grey, anhydrous Fe$_2$F$_5$ are obtained upon dehydration [2256, 2257]. Moreover, Fe$_2$F$_5$ · 7 H$_2$O and Fe$_2$F$_5$ · 2 H$_2$O as well as K$_3$[FeF$_6$] are readily available by direct electrochemical synthesis, when a Fe-electrode is oxidized in diluted, aqueous HF [2258].

Iron oxyfluoride **FeOF** is known as well, it is typically prepared by high temperature reaction of FeF$_3$ and Fe$_2$O$_3$ and crystallizes in the rutile structure type (Scheme 10.22) [1500, 1501, 2259, 2260]. Apart from that a polymorph with a layered structure has also been discovered [2261]. Alternatively, FeOF is accessible by treatment of TiO$_2$ with FeF$_3$ [1504]. The reaction of FeOF with FeF$_2$ at 850 °C leads to the formation of solid solutions of the formula Fe$^{II}_{1-x}$Fe$^{III}_x$O$_x$F$_{2-x}$ [2262]. More recently, phase-pure nanorods of FeOF have been prepared, which have shown a high specific capacity. Due to this property, nanostructured FeOF is a potential cathode material for sodium-ion batteries or solid-state Li metal analogues [2263, 2264]. Moreover, it has been shown that mesoporous, amorphous nano-cocoons of FeOF as well as ellipsoidal nanoparticles anchored on reduced graphene oxides perform very well as cathode materials for sodium-ion batteries [2265, 2266]. The structural design of such FeOF nanostructures found its currently best result in the preparation of three-

FeCl₃ FeF₂ + Fe mer-[FeF₃(NH₃)₃]

aHF, rt H₂, Δ liq. NH₃, H₂O traces

Fe, FeCl₂ or FeCl₃ →(F₂) FeF₃ →(F⁻, H₂O) [MF₄]⁻, [MF₅]²⁻, or [MF₆]³⁻

Fe₂O₃ or TiO₂, Δ NH₄F, 180 °C

FeOF NH₄[FeF₄] ←(140 °C) (NH₄)₃[FeF₆]

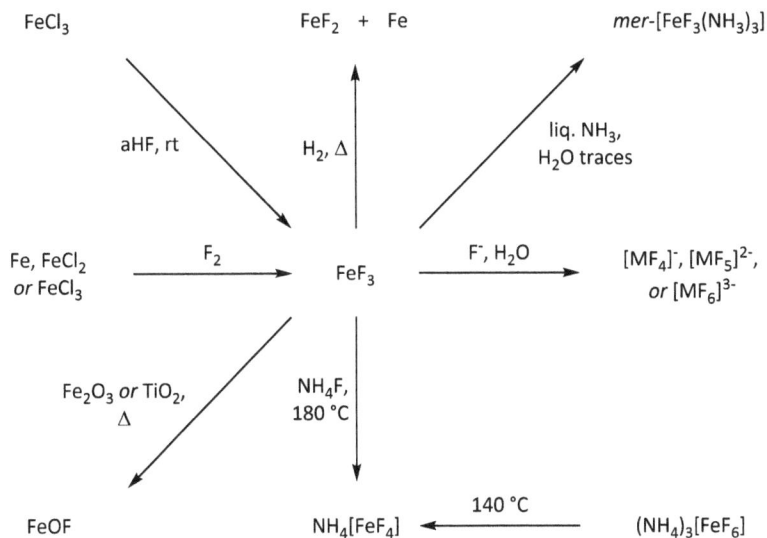

Scheme 10.22: Preparation and selected reactions of FeF₃ and its fluorocomplexes.

dimensional, starfish-like FeOF, which has been attached to reduced graphene oxide sheets. This material has superior properties for the use in lithium-ion batteries, as it shows a high reversible capacity, which only slightly decays per cycle [2267].

Cobalt trifluoride **CoF₃** is a brown solid (decomp.: 350 °C), which is obtained by fluorination of CoCl₂ with F₂ or treatment of elemental Co with ClF₃ or F₂ at 300–400 °C (Scheme 10.23). In the solid state, CoF₃ exhibits the same structure like FeF₃, and thus AlF₃. In their ground state, CoF₃ molecules have D_{3h}-symmetry, as has been shown by vibrational spectroscopy in a noble gas matrix [1631]. Electrolytic oxidation of a saturated solution of CoF₂ in hydrofluoric acid (w = 40 %) yields the green hydrate CoF₃ · 3.5 H₂O. CoF₃ readily undergoes hydrolysis, whereupon it liberates O₂. It is a fluorinating agent and reacts with Al, P, As, S, I₂ as well as other elements. Apart from that, it is an important reagent in organic synthesis, as it transforms hydrocarbons and their halogen-derivatives into poly- and perfluorinated compounds without attacking the carbon skeleton (cf. Section 13.2.3). The thermochemistry of CoF₃ has been investigated, including a comparison of its fluorinating properties with other compounds, that is, FeF₃ or MnF₃. It has turned out that the fluorinating ability rather depends on the enthalpy of formation of the respective di- and trifluorides than on the separation energy of a fluorine atom from MF₃. From this perspective, the Co-compound has weaker fluorinating abilities than, for example, MnF₃, however, CoF₃ is still commonly preferred [2268, 2269]. Thermal decomposition of the ammonia complex [Co(NH₃)₆]F₃ at 125 °C yields the paramagnetic high-spin complex [Co(NH₃)₆][CoF₆] [2270].

Nickel trifluoride **NiF₃** is a black solid, which is synthesized via intermediary NiF₄ by solvolysis of K₂[NiᴵⱽF₆] and AsF₅ in anhydrous HF at 0 °C (Scheme 10.23). Due to

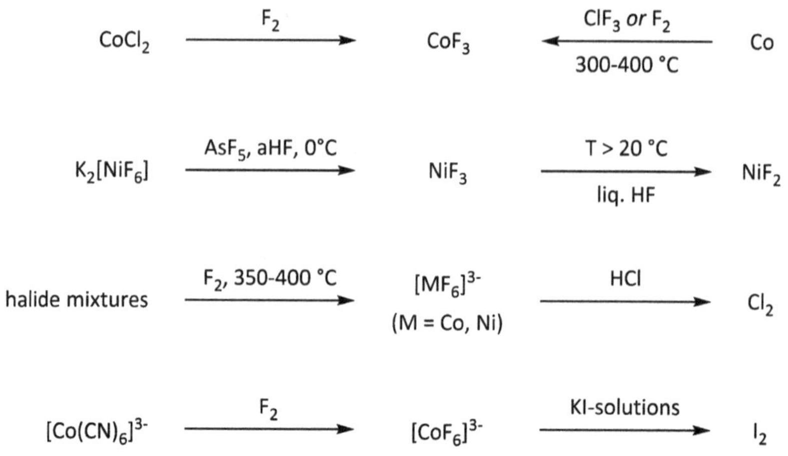

$$CoCl_2 \xrightarrow{F_2} CoF_3 \xleftarrow[\text{300-400 °C}]{ClF_3 \text{ or } F_2} Co$$

$$K_2[NiF_6] \xrightarrow{AsF_5, \text{ aHF, 0°C}} NiF_3 \xrightarrow[\text{liq. HF}]{T > 20 °C} NiF_2$$

$$\text{halide mixtures} \xrightarrow{F_2, 350\text{-}400 °C} \begin{matrix}[MF_6]^{3-}\\(M = Co, Ni)\end{matrix} \xrightarrow{HCl} Cl_2$$

$$[Co(CN)_6]^{3-} \xrightarrow{F_2} [CoF_6]^{3-} \xrightarrow{KI\text{-solutions}} I_2$$

Scheme 10.23: Preparation and selected reactions of CoF_3, NiF_3 and their fluorocomplexes.

its instability, it has never been obtained in pure form [2091, 2232]. Like many other trifluorides of the first row of transition metals, hexagonal NiF_3 has a distorted ReO_3 structure, which is also known as VF_3 structure. Two other slightly different modifications are obtained by thermolysis of NiF_4, containing minor amounts of K^+. In the solid phase, brown-black, rhombohedral NiF_3 is generated, which contains a three-dimensional lattice of $[NiF_6]$-octahedra. The octahedra are linked by common corners and the K^+ cations are situated in voids. In contrast, black hexagonal NiF_3 is obtained, if the thermolysis reaction is performed in liquid HF at 20 °C. It has a hexagonal Na_xWO_3-type structure [2271]. As has been found by X-ray absorption spectroscopy, UV-vis investigations and also by neutron diffraction experiments, NiF_3 is a mixed-valent compound, which is better described as $Ni^{II}[Ni^{IV}F_6]$ [2240, 2272]. The determination of the standard enthalpy of formation of NiF_3 by calorimetric methods leads to the same conclusion [2273]. Above 20 °C in HF solution, NiF_3 decomposes to NiF_2, whereas it is slightly more stable in the solid state. NiF_3 is a strong oxidizer and a powerful fluorinating agent, as it is for example capable of oxidizing Xe to Xe(VI) and fluorinating perfluoropropene to perfluoropropane [2271]. Apart from that, NiF_3 is used for the (per-)fluorination of larger hydrofluorocarbon compounds [2274]. A mixed-valent nickel fluoride of the composition **Ni_2F_5** is also known. It is obtained by thermal decomposition of NiF_3 or reduction of NiF_3 with Xe or XeF_2. Ni_2F_5 reacts with AsF_5 in anhydrous HF under formation of $Ni[AsF_6]$ and F_2. In contrast, disproportionation to NiF_2 and $K_2[NiF_6]$ is observed in basic mixtures of KF in anhydrous HF. Ni_2F_5 is capable of oxidizing Xe to XeF_2 and fluorinating the double bond of perfluoropropene. Apart from that, it is considered as important intermediate in the electrochemical fluorination process (cf. Section 12.2) [2275].

Complexes derived from the trifluorides

There are numerous fluoroferrates(III) and fluorocobaltates(III), however, the only Ni(III) compounds are the hexafluoronickelates(III) [2276]. The structures of the salts $M^I[MF_4]$, $M^I_2[MF_5]$ and $M^{II}[MF_5]$ (M = Fe, Co) are generally based on networks of $[FeF_6]$- and $[CoF_6]$-octahedra, respectively. Typically, the tetrafluorometallates(III) $[MF_4]^-$ exhibit layered structures, whereas the pentafluorometallates(III) $[MF_5]^{2-}$ are composed of *cis*- or *trans*-F-bridged octahedra, which form zig-zag chains [2277–2279]. In the system $FeF_2/FeF_3/M^IF$ (M^I = Li, Na, K, Rb, Tl), several structurally different compounds are observed, depending on the stoichiometry of the starting materials and the reaction conditions. In any case, iron is octahedrally coordinated by six F-atoms. The possible structures include three-dimensional networks of $[FeF_3]_n$, double-layers of $[Fe_2F_7]_n$, single layers of $[FeF_4]_n$, bridged chains of $[Fe_3F_{14}]_n$, isolated chains of $[FeF_5]_n$ or isolated $[FeF_6]$-octahedra [2280].

Thermal decomposition of the ammonium salt $(NH_4)_3[FeF_6]$ at 140 °C gives rise to the antiferromagnetic tetrafluoroferrate(III) $NH_4[FeF_4]$. The latter has a structure, which is a variant of $Tl[AlF_4]$, thus it comprises infinite layers of corner-sharing $[FeF_6]$-octahedra. Moreover, $NH_4[FeF_4]$ is directly accessible from FeF_3 and NH_4F at 180 °C (Scheme 10.22) [2281–2283]. The salts $K[FeF_4]$, $Cs[FeF_4]$ and $Rb[FeF_4]$ belong to the same structure type, in addition, they undergo phase-transitions [2284–2287]. $Li[CoF_4]$ has antiferromagnetic properties and crystallizes in a dirutile lattice [2288]. $K[CoF_4]$ has a certain importance in organic synthesis, as it is a slightly milder fluorinating agent than neutral CoF_3. The latter also forms several adducts with the alkaline earth metal fluorides CaF_2, SrF_2 and BaF_2. The resulting complexes are very hygroscopic and have the compositions $M^{II}[CoF_5]$ (M^{II} = Ca, Sr, Ba); $M^{II}_3[CoF_6]_2$ (M^{II} = Sr, Ba) and $Sr_5[Co_3F_{19}]$ [1852, 2289]. Another structurally interesting example is the tetrafluoroferrate(III) $[NMe_4][FeF_4] \cdot H_2O$, which contains dimeric $[Fe_2F_8(H_2O)_2]^{2-}$ anions, having two F-bridges [2290]. A comparable situation is believed to be found in the oxynonafluoro diferrate(III) anions $[Fe_2OF_9]^{5-}$, where two $[FeF_4]$-fragments are probably bridged by each one O- and F-atom [2291].

As a typical example for the pentafluoroferrates(III), $Ca[FeF_5]$ contains linear zig-zag chains of $[FeF_6]$-octahedra, which are connected to the chains of pentagonal-bipyramidal $[CaF_7]$-units [2292]. Alternatively, *cis*-connected $[FeF_6]$-octahedra can form helical chains, as it is observed for $Sr[FeF_5]$ [2293]. In case of $Ba[FeF_5]$, the chains of $[FeF_6]$-octahedra are branched [2294]. In contrast, the hydrated salts $M_2[FeF_4(H_2O)_2]$ and $M^I[FeF_5(H_2O)]$ contain discrete, octahedral anions [2295]. The same is observed for the heptahydrate $Mn[FeF_5] \cdot 7 H_2O$, where the anionic species $[FeF_6]^{3-}$ and $[FeF_4(H_2O)_2]^-$ are found. They are linked with the $[Mn(H_2O)_6]^{2+}$ cations by hydrogen bonds [2296]. A slightly different situation is observed in $Sr_2Fe_2F_{10} \cdot H_2O$ which is composed of *trans*-chains of $[FeF_5]$-units as well as isolated $[FeF_5(H_2O)]$-octahedra [2297]. A hydrogen-bonded zig-zag chain orientation of isolated $[FeF_5(H_2O)]^{2-}$ anions is found in the salts $K_2[FeF_5(H_2O)]$ and $(NH_4)_2[FeF_5(H_2O)]$. Furthermore, this structural behavior gives rise to one-dimensional, antiferromagnetic properties of the potassium complex [2298, 2299]. The thermolysis of both the ammonium and potassium salt of $[FeF_5(H_2O)]^{2-}$ leads to structural changes, as has been shown by means of EPR spectroscopy. The loss of water ligands results in the aggregation of the anions, forming a layered structure as has been found in $NH_4[FeF_4]$ or, alternatively, chains as in case of $M_2^I[FeF_5]$ (M^I = NH_4, K) [2300–2303]. In $(NH_4)_2[FeF_5]$, strong hydrogen-bonding between one of the ammonium cations and the $[FeF_5]^-$-chain is observed, clearly resulting in distorted NH_4^+ cations [2304]. The iron(III) centers in the hydrazinium salt $[N_2H_6][FeF_5]$ are in a high-spin state and have an almost regular octahedral coordination environment [2305]. A structurally interesting compound is obtained, if the fluorides of sodium, strontium and iron(III) are heated to 650 °C in a ratio of 1:1:2 together with excess NaCl and $ZnCl_2$. The resulting quaternary fluoride $Na_7Sr_2Fe_7F_{32}$ is composed of $[Fe_5F_{26}]^{11-}$ pentamers of $[FeF_6]$-octahedra and infinite, linear chains of $[FeF_5]_n^{2n-}$, containing *trans*-bridged $[FeF_6]$-units. The pentamers are linked with the chains by sharing common corners of $[FeF_6]$-octahedra [2306].

The ammonium salt $(NH_4)_3[FeF_6]$ is composed of isolated $[FeF_6]$-octahedra and ammonium cations, it decomposes upon heating to 140 °C and liberates NH_4F, resulting in the formation of $NH_4[FeF_4]$ (Scheme 10.22). If the thermal decomposition is conducted in a flow of ammonia, FeF_2 is formed at 400 °C, whereas grey iron nitride Fe_2N will be obtained at 600 °C [1628]. The hexafluorometallates(III) $M_3^I[MF_6]$ (M^I = monovalent cation; M = Fe, Co, Ni) typically have a structure, which is derived from cryolite or its variants [2307–2311]. There are also complexes with mixed-valent cations, for example, $CsCu[NiF_6]$, $CsMg[MF_6]$ (M = Co, Fe), $CsZn[MF_6]$ (M = Fe, Co, Ni), $CsM[FeF_6]$ (M = Mn, Co, Ni) and $BaLi[FeF_6]$. The sensitivity of these complexes against hydrolysis by humidity increases from Fe to Ni [2312–2314]. Moreover, hexafluoroferrates(III) and –nickelates(III) with mixed alkali metal cations are known, they have the compositions $Cs_2K[MF_6]$, $Rb_2K[MF_6]$, $Rb_2Na[MF_6]$, $Cs_2Na[MF_6]$ (M = Fe, Co, Ni), $K_2Na[NiF_6]$ and $Cs_2Tl[FeF_6]$, where the Ni-compounds are all obtained as lilac powders. In contrast, the Fe-complexes are colorless and the Co-derivatives typically appear as light-blue, powdered solids. Interestingly, magnetic measurements on $K_3[NiF_6]$ and $Cs_2K[NiF_6]$ have revealed an equilibrium between the respective high- and low-spin species [2315–2318]. The fluoroferrates are readily formed in aqueous media, but the synthesis of the respective cobaltates and nickelates requires more drastic conditions (Schemes 10.22 and 10.23). The best method for their preparation is the fluorination of stoichiometric mixtures of the respective halides with F_2 at 350–400 °C (Scheme 10.23). The salts $M_3^I[CoF_6]$ (M^I = Li, Na, K, Rb, Cs), including $K_2Na[CoF_6]$ and the barium salt $Ba_3[CoF_6]_2$, are conveniently prepared by fluorination of the respective hexacyanocobaltate(III) derivatives $[Co(CN)_6]^{3-}$ with elemental F_2 [2319, 2320]. The hexafluoronickelates $M_3[NiF_6]$ (M = K, Na) are alternatively accessible by fluorination of the respective Ni(II)-salts at 350–400 °C in presence of alkali metal chlorides. Hexafluorocobaltates and –nickelates are very reactive species. However, they are relatively stable when they are exposed to air so that they do not immediately decompose. Their strongly oxidizing properties are underlined by the reaction with H_2O, which liberates O_2. Accordingly, both $[CoF_6]^{3-}$ and $[NiF_6]^{3-}$ react with HCl under formation of Cl_2, in acidic media, the cobalt-complexes evolve I_2 from KI-solutions (Scheme 10.23) [2320]. In liquid HF, the violet, Jahn–Teller-distorted $[NiF_6]^{3-}$ disproportionates to form $K_2[NiF_6]$ and brown NiF_x (x = 2.3–2.5) [2091, 2232].

Tetrafluorides and fluorometallates(IV)

Fluorine-containing compounds of iron(IV) are unknown in the condensed phase. However, iron tetrafluoride **FeF$_4$** has been calculated to be stable in the gas phase and has indeed been prepared at high-temperature and investigated in matrix-isolation experiments. Based on theoretical calculations, the structure of FeF_4 has been expected to be Jahn–Teller distorted. Hence, it has the shape of a flattened tetrahedron, which has been experimentally proven by the observation of additional bands in the infrared spectra of the matrix-experiments [1282, 2321]. The situation of cobalt tetrafluoride **CoF$_4$** is similar to that of FeF_4. It has been prepared in a gas-phase reaction by fluorination of CoF_3, using TbF_4 as fluorinating agent. CoF_4 is relatively stable in the gas phase, as has been identified by mass spectrometry [1283, 2269, 2322]. However, in the condensed state, it has never been observed. In contrast, anionic fluorocomplexes of cobalt(IV) are well known.

Nickel tetrafluoride **NiF$_4$** is also unstable, but in contrast to FeF_4 and CoF_4 it can at least be obtained in condensed state. For this purpose, solutions of $[NiF_6]^{2-}$ are treated with fluoride acceptors like BF_3, AsF_5 or SbF_5 in anhydrous HF at low temperatures (Scheme 10.24). Below –65 °C, NiF_4 is precipitated, however, it immediately starts to decompose to NiF_3 if the temperature is raised. The addition of fluoride donors below –65 °C regenerates the initial hexafluoronickelate(IV) complex. In the presence of excess Lewis acid, NiF_4 dissolves in anhydrous HF to yield cationic $[NiF_3]^+$, which is an extremely powerful oxidizer, as it is capable of converting the pentavalent hexafluorometallate anions $[RuF_6]^-$ and $[PtF_6]^-$ to the neutral hexafluorides MF_6 (Scheme 10.24) [2271, 2323]. Moreover, *in situ* generated NiF_4 is occasionally used for the fluorination of organic substrates at room temperature or below [2274]. The chemistry of NiF_4 and its derivatives is strongly restricted to the use of anhydrous HF as a solvent [34].

The hexafluorocobaltates(IV) $Rb_2[CoF_6]$ and $Cs_2[CoF_6]$ are prepared by fluorination of $M_2^I[CoCl_4]$ with F_2 at 250–300 °C [2320, 2324, 2325]. The yellow-brown $Cs_2[CoF_6]$ is reduced to sky-blue $Cs_3[CoF_6]$ in contact with air, furthermore it reacts violently with water [2320]. The corresponding nickel(IV) salts $M_2^I[NiF_6]$ (M^I = K, Rb, Cs) as well as $CsRb[NiF_6]$ and $RbK[NiF_6]$ are accessible by treatment of the chlorides with F_2 at 350 °C and 350 bar (Scheme 10.24) [2326–2330]. Alternatively, they may be prepared by reaction of $M_2^I[Ni(CN)_4]$ with F_2. Similarly, the bright carmine compounds $Sr[NiF_6]$ and $Ba[NiF_6]$ are prepared by fluorination of the cyano-complexes [2329, 2331]. $K_2[NiF_6]$, $Rb_2[CoF_6]$ and $Cs_2[CoF_6]$ have a cubic $K_2[PtCl_6]$ structure, $Rb_2[NiF_6]$ is described

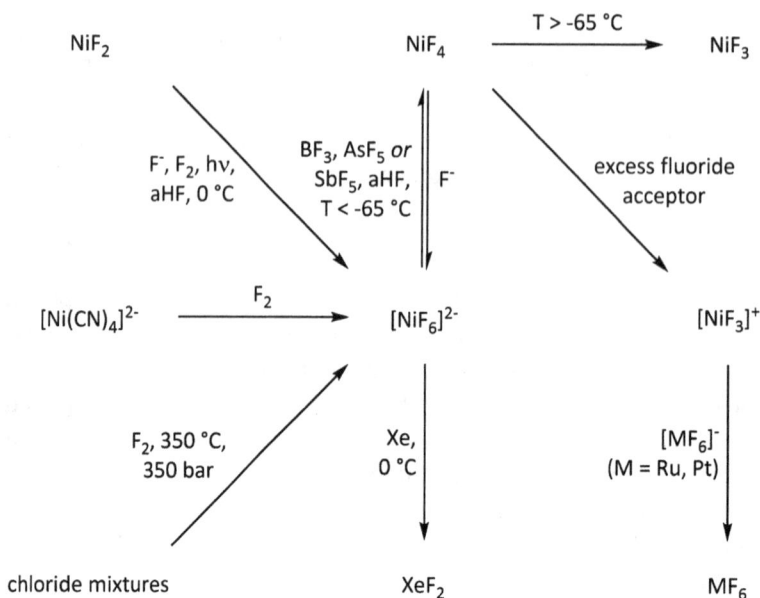

Scheme 10.24: Preparation and selected reactions of NiF_4 and its fluorocomplexes.

by the same structure as hexagonal $Rb_2[MnF_6]$ and $Cs_2[NiF_6]$ crystallizes trigonally like $K_2[GeF_6]$ [2328, 2332]. The sodium salt $Na_2[NiF_6]$ is only accessible at very high pressure and high temperature from a mixture of $NaNiO_2$, Na_2O_2 and F_2. The resulting product contains each 40 % of cubic and hexagonal $Na_2[NiF_6]$, as well as 10 % $Na_3[NiF_6]$ and 10 % $Na[NiF_3]$ [2333]. Apart from that, it has been shown that Ni(IV)-species are conveniently accessible at low temperature, as for example, the treatment of NiF_2 with LiF and photodissociated F_2 in anhydrous HF at 0 °C yields $Li_2[NiF_6]$ (Scheme 10.24) [2334]. Moreover, Ni(III)- and Ni(IV)-complexes are also detected if elemental Ni is anodically oxidized in anhydrous HF. For this reason, fluoronickelates are assumed to be intermediary compounds in the Simons process, the electrofluorination of organic substrates (cf. Section 12.2) [2311]. They have also been discussed as important intermediates in the course of electrochemical fluorination of inorganic compounds [2335]. The hexafluorometallate(IV) salts are extremely reactive oxidizers, as they generate O_2 upon contact with H_2O. In anhydrous HF they liberate F_2, elemental xenon is readily transformed to XeF_2 at 0 °C (Scheme 10.24). The solvolysis of $K_2[NiF_6]$ and AsF_5 in HF yields impure, black NiF_3 [2091, 2232]. Since the mixed-valent NiF_3 is better written as $Ni^{II}[Ni^{IV}F_6]$ (see above), it can also be considered as a hexafluoronickelate(IV) complex. The analogous compounds $M^{II}[NiF_6]$ are accessible from liquid HF by treatment of $K_2[NiF_6]$ with $M^{II}[SbF_6]_2$ (M^{II} = Co, Cu, Zn) or $M^{II}[AsF_6]_2$ (M^{II} = Fe) [2240]. $K_2[NiF_6]$ is also suitable for the preparation of pure F_2 gas. $K_3[NiF_6]$ reversibly absorbs F_2, so that the resulting $K_2[NiF_6]$ splits it off in pure form at elevated temperature [23]. An interesting compound is the tetrafluoroammonium salt $[NF_4]_2[NiF_6]$, which contains both, an oxidizing cation and anion. Due to this, it is a high energy solid propellant oxidizer and, moreover, it is a source of NF_3-F_2 gas mixtures. The preparation is realized by a metathesis reaction of $Cs_2[NiF_6]$ and $[NF_4][SbF_6]$ in anhydrous HF [327]. Similarly, the salt $[ClOF_2]_2[NiF_6]$ is prepared from $[ClOF_2][SbF_6]$ and $Cs_2[NiF_6]$ in HF [2336, 2337]. A complex with $[Xe_2F_{11}]^+$ cations is accessible, if NiF_2 is reacted with XeF_6 and KrF_2, which gives rise to $[Xe_2F_{11}]_2[NiF_6]$ [2338]. The oxidation of nickel and NO_2 with KrF_2 in HF yields the nitryl salt $[NO_2]_2[NiF_6]$ [2339]. Analogously, the calcium salt $Ca[NiF_6]$ has been obtained by the oxidation of nickel in presence of CaF_2 [2340].

10.8.2 Study questions

(I)   How are the difluorides FeF_2, CoF_2 and NiF_2 prepared? Are they stable against hydrolysis?

(II)   Which fluorometallates do the difluorides FeF_2, CoF_2 and NiF_2 form? What is their common structural feature? Give examples for their preparation.

(III)   Comment on the stability of the trifluorides FeF_3, CoF_3 and NiF_3. How are these compounds prepared? Are they all real trifluorides?

(IV)   Which trivalent fluorometallates of Fe, Co and Ni are generally known? How does their preparation differ?

(V)   Which fluorine-containing tetravalent Fe, Co and Ni compounds are accessible in the condensed phase? How are they prepared?

(VI)   How can the strongly oxidizing character of NiF_4 further be increased? Give an example for the reactivity of the resulting species.

10.9 The platinum metals: ruthenium, osmium, rhodium, iridium, platinum, palladium

10.9.1 Compounds and properties

The fluorides of the platinum metals exist in quite a large variety of compounds. There are a few low valent binary fluorides, especially in the case of palladium, but high valent derivatives are also well known, as almost all of the platinum metals form hexafluorides. Apart from binary metal-fluorine compounds, a number of oxyfluorides exists as well, where the metal can even adapt the oxidation state +VIII in case of osmium.

Monofluorides

The monofluorides of the platinum metals are, if at all, only accessible under gasphase conditions. For this reason, they have been mainly investigated by means of quantum chemistry. For the known compounds, some spectroscopic properties have been described. However, due to their limited access and instability, they do not play an important role. Ruthenium monofluoride **RuF** has been reported to be formed along with other fluorides, if Ru powder is vaporized in a stream of IF_5 [2341]. Rhodium monofluoride **RhF** has been generated by the reaction of a Rh plasma with SF_6, which has been doped in He [2342]. Similarly, platinum monofluoride **PtF** has been made from laser-ablated Pt atoms and SF_6 under supersonic free jet conditions or from a Pt-sheet, which has been submitted to a reaction with a glow discharge plasma of CF_4 [2343, 2344].

Difluorides

The only known and stable difluoride in the series of platinum metals is the violet palladium difluoride **PdF$_2$**. It is typically synthesized by reduction of PdF_4 or PdF_3 with SeF_4 or elemental Pd at 930 °C [689, 2345–2348]. Moreover, it is obtained by thermal decomposition of $Pd[GeF_6]$ at 350 °C *in vacuo* [2349]. Similar to the difluorides of Fe, Co and Ni, it typically crystallizes in a rutile lattice, however, at high temperature and under pressure, a second modification is observed, too [2350–2352]. A low temperature form of PdF_2 is recovered, if $Pd[AsF_6]_2$ is treated with CsF in anhydrous HF at room temperature. Its structure is derived from the fluorite type [2353]. The precursor $Pd[AsF_6]_2$ is generated, if Pd metal is dissolved in HF/AsF_5 in presence of F_2 at room temperature [2354]. PdF_2 is paramagnetic and is converted to the diamagnetic ion $[Pd(H_2O)_4]^{2+}$ in aqueous solution. In addition, it has been shown to have weak ferromagnetic properties [2355]. PdF_2 already hydrolyzes in contact with humidity. If PdF_2 is suspended in SeF_4 and subsequently reacted with CsF, the pinkish-brown salt $Cs[PdF_3]$ is generated. In contrast, the respective K- and Rb- salts are accessible by thermal decomposition of the tetravalent hexafluoropalladates $M_2[PdF_6]$ (M = K, Rb) [689, 2352]. Treatment of a 1:2 mixture of Pd and PdF_3 with NaF or KF leads to the formation of the tetrafluoropalladate(II) salts $Na_2[PdF_4]$ and $K_2[PdF_4]$, respectively [2356]. Tetrafluoropalladate(II) salts of the type $M^{II}[PdF_4]$ (M^{II} = Ca, Sr, Ba, Pb, Cd, Hg) are generally accessible by the reaction of PdF_2 with MF_2 at about 930 °C, or by thermal decomposition of hexafluoropalladate(IV) salts $M^{II}[PdF_6]$ [2347, 2348, 2350]. These compounds are slightly colored, diamagnetic and exhibit a tetragonal solid-state structure similar to $K[BrF_4]$. There are also a few pentafluoropalladate(II) compounds of the general compositions $M^I_3[PdF_5]$ (M^I = Rb, Cs) including $Rb_2Cs[PdF_5]$, $K_2Rb[PdF_5]$, $K_2Cs[PdF_5]$ as well as $M^IM^{II}[PdF_5]$ (M^I = K, Rb, Cs; M^{II} = Mg, Zn, Ni, Co, Pd) [2356–2358]. Another class of palladium(II) fluorocomplexes is formed by solid-state reactions of stoichiometric mixtures of alkaline earth metal fluorides with PdF_2. The obtained compounds $M^{II}_2PdF_6$ (M^{II} = Ba, Sr, Pb) and $BaSrPdF_6$ do not contain the isolated anion $[PdF_6]^{4-}$, but separate square-planar $[PdF_4]^{2-}$ units, which aggregate in layers. Overall, antiprismatic $[M^{II}F_8]^{6-}$-polyhedra are oriented in bilayers, interleaving with the $[PdF_4]^{2-}$ units [2359].

Platinum difluoride **PtF$_2$** has been reported several times, however, it has never been isolated. It is believed to be unstable and undergo rapid disproportionation to much more stable PtF_4 and Pt-metal [2360]. Quantum chemical calculations and a few thermodynamic investigations in the gas phase have also been performed on hypothetic PtF_2 [2361]. Ruthenium difluoride **RuF_2** has been identified and thermodynamically characterized by mass spectrometry at high temperatures [2341]. In pure form, it has never been obtained in the condensed phase, which similarly holds for osmium difluoride **OsF_2**. However, by addition of appropriate neutral ligands, that is, CO or PPh_3, it is possible to obtain complexes of RuF_2 as well as of OsF_2, for example, $[MF_2(CO)_2(PPh_3)_2]$ (M = Ru, Os). These complexes are accessible by treatment of $[M(CO)_3(PPh_3)_2]$ with XeF_2, which leads to the intermediate formation of fluoroacyl complexes [2362]. Under elimination of methane, cis-$[Os(CO)_4Me_2]$ reacts with anhydrous HF to yield cis-$[OsF_2(CO)_4]$ [2363]. There is also a number of ruthenium carbonyl fluorides, which are all prepared by oxidative fluorination of $Ru_3(CO)_{12}$ with XeF_2 in anhydrous HF [2364]. Another example for a divalent fluorine-containing ruthenium complex is the bis-phosphine compound $[RuF(dppp)_2]^+[PF_6]^-$, which has been investigated in terms of its ability to fluorinate activated organic bromides [2365]. Both RuF_2- and OsF_2-complexes have been explored with respect to their reversible abilities of cleaving C-F bonds. For this purpose, reactions of mixed carbonyl phosphine complexes with pentafluorophenyl- and trifluoromethyl-substituents have been investigated, revealing the formation of fluorocarbene complexes in the latter case [2366]. This type of C-F-bond activation has further been extended to vinyl fluoride, which gives rise to the OsF_2-vinylidene complex $[OsF_2(C=CH_2)(CO)(P^tBu_2Me)_2]$ [2367]. More recently, RuF_2-complexes with N-heterocyclic carbene ligands have been described as well [2368, 2369]. Rhodium difluoride **RhF_2** and iridium difluoride **IrF_2** are neither known in pure form nor as stabilized complexes.

Trifluorides

Ruthenium trifluoride **RuF_3** is formed as a brown solid (mp.: 650 °C) upon reduction of RuF_5 with I_2 at 500 °C (Scheme 10.25) [2370, 2371]. The pentafluoride may also be reduced by SF_4 under slight pressure at 125 °C to furnish RuF_3 [2372]. Rhodium trifluoride **RhF_3** is a red, very stable solid, which is generated by fluorination of elemental Rh or $RhCl_3$ with F_2 at 500–600 °C (Scheme 10.25) [2373]. The reduction of IrF_6 with Ir yields iridium trifluoride **IrF_3** as a black solid. It is also accessible by reduction of IrF_4 with SF_4, I_2, sulphur or hydrogen [2374]. If HF is passed over palladium sponge at 600 °C, the mixed-valent trifluoride "**PdF_3**" is formed, which is also obtained by fluorination of $PdCl_2$ or $PdBr_2$ with BrF_3 (Scheme 10.25) [2375]. However, since it is not a real trifluoride, PdF_3 is more correctly written as $Pd^{II}[Pd^{IV}F_6]$, where both Pd(II)- and Pd(IV)-centers exhibit a six-fold coordination in the solid state [2349, 2373]. The trifluorides RuF_3, RhF_3 and IrF_3 crystallize in a hexagonal lattice, which is generally derived from the structure of ReO_3 [2272]. RhF_3 describes an own structure type, where the metal ions occupy 1/3 of the octahedral voids in a hexagonal-closest packing of fluoride ions. The Rh centers are surrounded by six fluorine substituents, which are,

$$RuF_5 \xrightarrow[\text{or } SF_4,\ 125\ °C]{I_2,\ 500\ °C} RuF_3$$

$$Rh \text{ or } RhCl_3 \xrightarrow{F_2,\ 500\text{-}600\ °C} RhF_3$$

$$IrF_6 \xrightarrow{Ir} IrF_3 \xleftarrow[\text{or sulphur}]{SF_4,\ I_2,\ H_2} IrF_4$$

$$Pd\text{-sponge} \xrightarrow{aHF,\ 600°C} \text{"}PdF_3\text{"} \xleftarrow[(X = Cl,\ Br)]{BrF_3} PdX_2$$

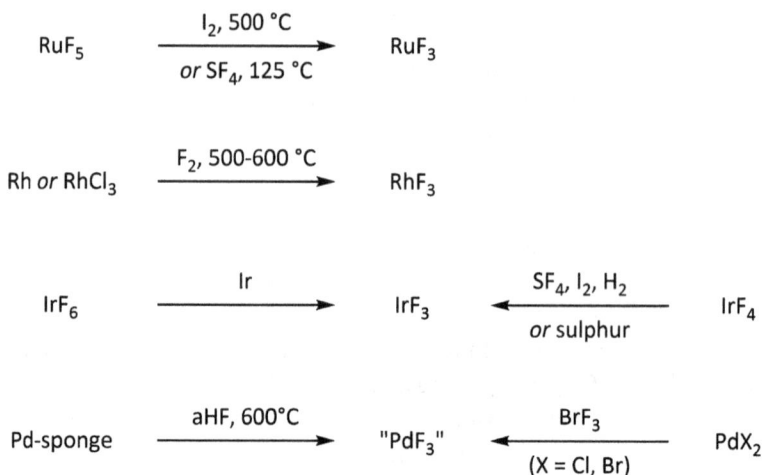

Scheme 10.25: Preparation of the trifluorides RuF_3, RhF_3, IrF_3 and PdF_3.

however, not equally distant from the metal. As a consequence, the coordination environment of Rh is best described by slightly compressed, trigonal antiprisms [2376]. Similarly, the three-dimensional structure of RuF_3 is described by a hexagonal closest packing of fluoride anions, where octahedral voids are occupied by Ru atoms. However, the fluoride packing is not exactly a hexagonal closest one, as the Ru-F-Ru bridging angle (136°) slightly deviates from the ideal value for a closest packing (132°) [2371, 2372]. The structure of PdF_3 may be described by the RhF_3 type, where Pd(II) and Pd(IV) alternatingly occupy 1/3 of the octahedral voids in the hexagonal closest fluoride packing. The Pd(II)-centers are paramagnetic as they have two unpaired electrons, whereas Pd(IV) is diamagnetic. All the trifluorides have also been investigated by UV-vis spectroscopy [2373]. Osmium trifluoride **OsF$_3$** and platinum trifluoride **PtF$_3$** are unknown.

The platinum metal trifluorides are rather unreactive, as for example, IrF_3 is not attacked by water and only slowly reacts with concentrated acids and alkaline solutions [2374]. Dissolving Rh(III)-salts in hydrofluoric acid allows the isolation of the hydrates $RhF_3 \cdot 6\ H_2O$ and $RhF_3 \cdot 9\ H_2O$, which are soluble in water under formation of yellow colored solutions [2377]. Most likely, the solutions contain $[Rh(H_2O)_6]^{3+}$ cations. When $RuCl_3$ is molten in KHF_2, $K_3[RuF_6]$ is generated, which is insoluble in water [2378]. The pentafluoroiridates(III) $[NO]_2[IrF_5]$ and $[NO_2]_2[IrF_5]$ are prepared upon heating of the respective tetravalent hexafluoroiridates [2374]. In a melt of $K_3[Rh(NO_2)_6]$ and KHF_2, the salt $K_3[RhF_6]$ is formed, which further reacts with additional KHF_2 to form $K_2[RhF_5]$ [2377]. There are a few more complexes known, for example, $M^{II}[RhF_5]$ (M = Sr, Ba, Pb), $M^{II}_2[RhF_7]$ (M^{II} = Sr, Pb) [2376, 2379] and $M_3[RhF_6]$ (M_3 = Cs_2K, Rb_2K, Rb_2Na, K_2Na, Tl_2Na) [2380]. In BrF_3 solution, PdF_3 reacts with

K[BrF$_4$] and Ag[BrF$_4$] to form complex salts of the composition M[PdF$_4$] or M$_2$[PdF$_5$] [2375].

Tetrafluorides

All the six platinum metals form binary tetrafluorides. Dark yellow ruthenium tetrafluoride **RuF$_4$** is obtained, if excess RuF$_5$ is reduced by I$_2$ (Scheme 10.26) [2381]. Moreover, it is recovered in form of the adduct RuF$_4$ · AsF$_5$, if RuF$_6$ is reacted with AsF$_3$. RuF$_6$ is also slowly reduced to RuF$_4$, if it is left under an atmosphere of H$_2$ in anhydrous HF [2382]. All of the reduction approaches always yield significant amounts of RuF$_3$ and Ru-metal, which are rather difficult to remove. Thus, a more convenient approach includes the treatment of [RuF$_6$]$^{2-}$ with AsF$_5$ in a solution of anhydrous HF at room temperature, which leads to the quantitative precipitation of polymeric RuF$_4$ in high purity (Scheme 10.26) [1926, 2323, 2372]. The yellow osmium tetrafluoride **OsF$_4$** (mp.: 230 °C; bp.: 280–300 °C) and yellow iridium tetrafluoride **IrF$_4$** (mp.: 106 °C; bp.: 300 °C) are also synthesized by reduction of high-valent compounds [2161, 2374, 2383]. For example, the treatment of OsF$_6$ with W(CO)$_6$, the photolysis of OsF$_6$ or OsF$_5$ in HF as well as the reduction of OsF$_6$ with H$_2$ or Si in HF at room temperature yield OsF$_4$ (Scheme 10.26) [2127, 2383]. Analogously to RuF$_4$, OsF$_4$ is more conveniently prepared by fluoride abstraction from [OsF$_6$]$^{2-}$, using AsF$_5$ as fluoride acceptor [1926, 2323]. IrF$_4$ is accessible by reduction of IrF$_6$ on a W- or Ir-wire at red heat or treatment of the hexafluoride with H$_2$ or Si in HF solution at room temperature (Scheme 10.26) [2127, 2384]. Moreover, it is obtained by photolysis of IrF$_6$ with UV-light or thermal decomposition of IrF$_5$ in presence of Ir [2374, 2385, 2386]. Purple rhodium tetrafluoride **RhF$_4$** and brown platinum tetrafluoride **PtF$_4$** are accessible by fluorination of RhCl$_3$ and Pt or PtCl$_2$ with BrF$_3$, followed by subsequent thermal decomposition of the resulting BrF$_3$-complexes (Scheme 10.26) [2375]. In case of PtF$_4$, the samples are reported to be typically contaminated with Br$_2$, which can be removed by low-temperature fluorination [2360]. Another approach is the repeated thermal decomposition of mixtures of PtF$_6$ and lower platinum fluorides which finally also results in PtF$_4$ [2387]. More conveniently, it can be synthesized from Pt and XeF$_2$ [2388]. RhF$_4$ also results from the fluorination of RhF$_3$ at 250 °C [2386]. Pink to brick-red palladium tetrafluoride **PdF$_4$** is prepared from PdBr$_2$ and BrF$_3$, which yields intermediary "PdF$_3$." The latter is oxidized by F$_2$ at 300 °C and a pressure of 7 bar to yield PdF$_4$ (Scheme 10.26) [2349, 2386]. Like RuF$_4$ and OsF$_4$, PdF$_4$ is also conveniently prepared by precipitation upon treatment of [PdF$_6$]$^{2-}$ with AsF$_5$ [1926].

In the solid-state structures of the isomorphic tetrafluorides, [MF$_2$F$_{4/2}$]-octahedra are found, which are linked by four fluorine atoms each. The distortion of the octahedra is differently pronounced for the various platinum metals. PdF$_4$, PtF$_4$ and IrF$_4$ are isostructural and differ from the structures of SnF$_4$ and VF$_4$ (two unbridged *trans*-F), as two *cis*-fluorine atoms remain unbridged. The resulting space structure

$$RuF_6 \xrightarrow{H_2,\ aHF} RuF_4 \xleftarrow{I_2} RuF_5$$

$$OsF_6\ or\ OsF_5 \xrightarrow{h\nu,\ HF} OsF_4 \xleftarrow[aHF,\ rt]{W(CO)_6,\ H_2\ or\ Si} OsF_6$$

$$RhCl_3 \xrightarrow{BrF_3} RhF_4 \xleftarrow{F_2,\ 250\ °C} RhF_3$$

$$IrF_6 \xrightarrow{H_2\ or\ Si,\ aHF,\ rt} IrF_4 \xleftarrow[red\ heat]{Ir\text{-}\ or\ W\text{-}wire} IrF_6$$

$$Pt\ or\ PtCl_2 \xrightarrow{BrF_3} PtF_4 \xleftarrow{XeF_2} Pt$$

$$PdBr_2 \xrightarrow[then\ F_2,\ 300\ °C,\ 7\ bar]{BrF_3} PdF_4$$

$$[MF_6]^{2-} \xrightarrow[(M = Ru,\ Os,\ Pd)]{AsF_5,\ aHF,\ rt} MF_4$$

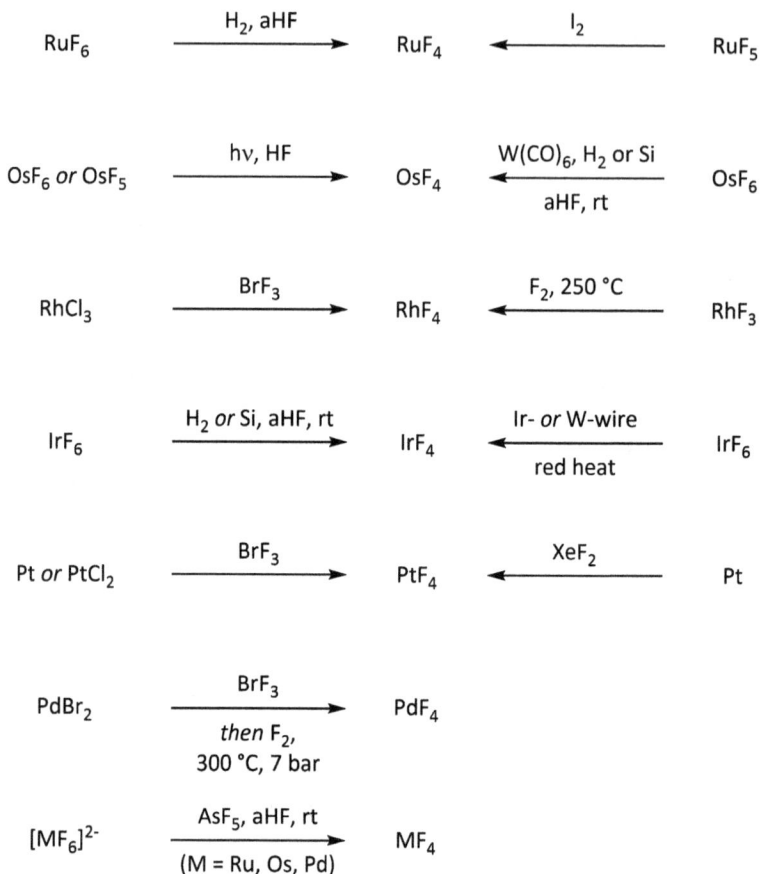

Scheme 10.26: Preparation of the tetrafluorides RuF_4, OsF_4, RhF_4, IrF_4, PdF_4 and PtF_4.

is comparable to rutile, where alternating metal positions of the bands of edge-bridged $[TiO_6]$-octahedra are left unoccupied [2386, 2389]. OsF_4 and RhF_4 are also isostructural but differ significantly from the PdF_4 type, as they are proposed to have a two-dimensional raft structure, which arises from a different linkage of the $[MF_6]$-octahedra [1926, 2389]. RuF_4 crystallizes in a layered structure like VF_4 [2372]. The tetrafluorides are all strong oxidizers and very sensitive toward hydrolysis. IrF_4 is stable up to 400 °C in dynamic vacuum, but disproportionates to iridium tri- and pentafluoride above this temperature. In contrast, RhF_4 is only stable up to 290 °C [2386]. PtF_4 is thermally very stable and can be heated to over 600 °C without decomposition. PdF_4 is easily reduced to $Pd^{II}[Pd^{IV}F_6]$ ("PdF_3"), whereas it is relatively stable *in vacuo* at 200 °C, as it only shows a slight increase in $Pd[PdF_6]$ after 3 hours. However, it is rapidly decomposed above 350 °C [2349]. Hydrolysis of RuF_4 at room temperature and subsequent treatment of the hydrolysis products with F_2 yields RuO_4 [2390]. Upon treatment with CO, IrF_4 is reduced to Ir-metal [2374]. PtF_4 reacts with PF_3

in HF/SbF$_5$ solution to form the stable platinum(II)-complex [Pt(PF$_3$)$_4$][Sb$_2$F$_{11}$]$_2$ · 2 HF, which is very similar to the stable, cationic carbonyl complex [Pt(CO)$_4$][Sb$_2$F$_{11}$]$_2$ [2388, 2391]. Liquid XeF$_2$ oxidizes both PtF$_4$ and mixed-valent Pd^{II}[Pd^{IV}F$_6$] at 140–150 °C under formation of [Pt^{V}F$_6$]$^-$ and XePdF$_6$, the latter being the 1:1 complex of XeF$_2$ and PdF$_4$. XePdF$_6$ is decomposed *in vacuo* at 140–150 °C under elimination of XeF$_2$ to yield the 1:2-complex XeF$_2$ · 2 PdF$_4$. Again, the latter compound is decomposed *in vacuo* at 300 °C, resulting in the formation of mixed-valent Pd$_2$F$_6$ and XeF$_4$ [1028]. This reaction sequence clearly indicates, that PdF$_4$ cannot be oxidized by XeF$_2$, but in turn Pd(IV) is capable of oxidizing XeF$_2$. The species "XeM^{IV}F$_6$" (M^{IV} = Pt, Pd) have been further investigated and transformed to the cesium salts Cs$_2$[M^{IV}F$_6$] by treatment with CsF in anhydrous HF solution [2392].

All the tetrafluorides form fluorocomplexes of the type M$_2^I$[M^{IV}F$_6$] and M^{II}[M^{IV}F$_6$], which are generally prepared by fluorination with F$_2$ or BrF$_3$ as well as by treatment with the respective cation source in molten KHF$_2$ [262, 2377, 2393–2398]. Alternatively, the pentavalent hexafluorometallates(V) M^I[M^{V}F$_6$] can be carefully hydrolyzed to yield air stable M$_2^I$[M^{IV}F$_6$]. (M^I = K, Cs; M$^{V/IV}$ = Ru, Os, Ir) [2399, 2400]. Analogously, Ba[Os^{IV}F$_6$] has been synthesized and structurally investigated [2401]. More recently, the treatment of elemental Ru, Os, Rh, Ir, Pd and Pt with various tetrafluorobromate(III) salts M^I[BrF$_4$] and M^{II}[BrF$_4$]$_2$ (M^I = K, Rb, Cs; M^{II} = Ba) has been investigated. In all cases, the respective salts with the anions [M^{IV}F$_6$]$^{2-}$ (M^{IV} = Rh, Pd, Pt), [M^{V}F$_6$]$^-$ (M^V = Os) or mixtures of them (M = Ru, Ir) are formed, being possibly relevant in future applications in the field of urban mining [903, 904]. Hexafluoroplatinate(IV) salts M$_2^I$[PtF$_6$] (M^I = Li, Na) are accessible by fluorination of a mixture of NaCl or Li$_2$CO$_3$ and (NH$_4$)$_2$[PtCl$_6$] [2393, 2402]. The salts have also been characterized by X-ray crystallography [2393, 2402, 2403]. In addition, the sodium salt Na$_2$[PtF$_6$] reversibly adds ammonia at –40 °C to form the complex [Na(NH$_3$)$_3$]$_2$[PtF$_6$] [2402]. The adduct PdF$_3$ · BrF$_3$ reacts with excess SeF$_4$ to form the salt [SeF$_3$]$_2$[Pd^{IV}F$_6$], which can be transformed to K$_2$[Pd^{IV}F$_6$] upon addition of K$_2$[SeF$_6$] [689]. Similarly to the platinum analogues, hexafluoropalladates(IV) M$_2^I$[Pd^{IV}F$_6$] (M^I = K, Rb, Cs) are accessible by fluorination of the corresponding chloro-compounds M$_2^I$[Pd^{IV}Cl$_6$] with F$_2$ [2404]. Moreover, the hexafluorometallate(IV) complexes M^{II}[M^{IV}F$_6$] (M^{II} = Cu, Ni; M^{IV} = Pd, Pt) have been prepared by solid-phase reactions of NiF$_2$ or CuF$_2$ with PdF$_4$ or PtF$_4$, respectively. Along with the analogous Pd[M^{IV}F$_6$] derivatives, the aforementioned compounds have been investigated in terms of their magnetic properties, revealing ferromagnetic behavior at low temperatures [2405]. A few chlorofluoro- and fluorohydroxy-complexes are also known for Os and Pt [2406–2410]. The isostructural and mixed-valent, ferromagnetic compounds Pd$_2$F$_6$, PdPtF$_6$ and Pt$_2$F$_6$ also belong to the group of hexafluorometallates(IV) [2411].

Pentafluorides

Binary pentafluorides are known for all platinum metals, except palladium. Ruthenium pentafluoride **RuF$_5$** (green; mp.: 86.5 °C; bp.: 227 °C), rhodium pentafluoride **RhF₅** (dark red; mp.: 95.5 °C), iridium pentafluoride **IrF₅** (yellow-green; mp.: 104.5 °C) and platinum pentafluoride **PtF₅** (dark red; mp.: 80 °C) can be synthesized by controlled fluorination of the metals or their low-valent fluorides at 350–400 °C (Scheme 10.27) [468, 2412]. In some cases, pressure has to be applied, too. PtF_5 is also conveniently prepared by fluorination of $PtCl_2$ at a moderately high temperature. It cannot be purified by distillation, as it readily disproportionates to PtF_4 and PtF_6 [2360, 2413]. In the case that the metal forms a stable hexafluoride, the pentafluorides are conveniently accessible by reduction of MF_6 (Scheme 10.27). Thus, the blue-green osmium pentafluoride **OsF₅** (mp.: 70 °C; bp.: 225.7 °C) is prepared by reduction of OsF_6 with either $W(CO)_6$ or I_2 in IF_5 or upon photolysis of the hexafluoride at 25 °C [2383]. Alternatively, treatment of OsF_6 with Si in HF as well as the reduction with W-wire have been described as synthetic methods toward the pentafluoride [1372, 1652]. Finally, it is possible to perform exchange reactions of a fluorometallate salt with a stronger Lewis acid, thus, OsF_5 is obtained by treatment of $Na[OsF_6]$ with SbF_5 (Scheme 10.27) [1944]. Accordingly, IrF_5 may be prepared by reduction of IrF_6 with glass powder, Si or H_2 [2127, 2414]. The reduction of RuF_6 with SbF_3, BiF_3 or Ru metal gives rise to RuF_5 [2382]. The pentafluorides are isomorphic; they form tetrameric units, where the metal atoms occupy the corners of a rhombus. In contrast to NbF_5, the F-bridges are not linear in the RuF_5 structure type, so that the metal atoms are distorted octahedrally coordinated by each six fluorine atoms [2415–2417]. The same association remains in the molten state, which is visualized by an only slight lightening of the color. Even in the gas phase, the pentafluorides are still associated, only at high temperature, colorless vapors containing monomeric, trigonal-bipyramidal molecules are observed [468]. The pentafluorides are sensitive toward hydrolysis and act as strong oxidizers and fluorinating agents. RuF_5 is hydrolyzed at room temperature and can be converted to RuO_4 upon treatment of the hydrolysis products with F_2 [2390]. PtF_5 readily oxidizes H_2O to O_2. At higher temperature, disproportionation of MF_5 occurs, resulting in the tetra- and the hexafluorides. The platinum metal pentafluorides react with halogen fluorides and chalcogen fluorides to yield F-bridged complexes [2378]. In some of those derivatives, the tetrameric structure of the pentafluorides is retained again, as for example, carbonylation of RuF_5 results in the formation of $[Ru(CO)_3F_2]_4$ and fluorination of $Ru_3(CO)_{12}$ gives rise to $[Ru(CO)_3F_2\text{-}RuF_5]_2$ [2418]. OsF_5, RuF_5, RhF_5 and IrF_5 are capable of forming intercalation compounds with graphite in gas-solid transformations or liquid-solid reactions in anhydrous HF [2419]. The pentafluorides also react with XeF_2 in different stoichiometries in BrF_5 solution to yield complexes of the compositions $[XeF]^+[M^VF_6]^-$, $[Xe_2F_3]^+[M^VF_6]^-$ (M^V = Ru, Os, Ir, Pt) and $[XeF]^+[M_2^VF_{11}]^-$ (M^V = Ru, Ir, Pt) (Scheme 10.27) [2420]. The symmetry in the $[M^VF_6]^-$ anions has been proposed to be lowered by interactions with the $[XeF]^+$ cations, as has been indicated by vibrational spectroscopy. Except the osmium complex $[XeF][OsF_6]$, all of the aforementioned fluorometallates are thermally stable at room temperature [2420]. The pentafluorides also react with other fluoride donors to form hexafluorometallates(V) with the general formula $M^I[MF_6]$ and $M^{II}[MF_6]_2$ (M^I = alkali metal, Ag, Tl,

Na[OsF$_6$]   MF$_6$ MF$_4$ + MF$_6$

SbF$_5$

reduction
e.g. Si, H$_2$

Δ

M or   F$_2$, 350-400 °C
low-valent MF$_n$ ————————————   MF$_5$ Pt
                    (M = Ru, Rh, Ir, Pt)

XeF$_2$,   F$^-$ or X$^-$, F$_2$ O$_2$/F$_2$, 450 °C
BrF$_5$ or OF$_2$, 400 °C

[XeF][MF$_6$], [Xe$_2$F$_3$][MF$_6$]   [MF$_6$]$^-$ salts
or [XeF][M$_2$F$_{11}$]
(M = Ru, Ir, Pt)

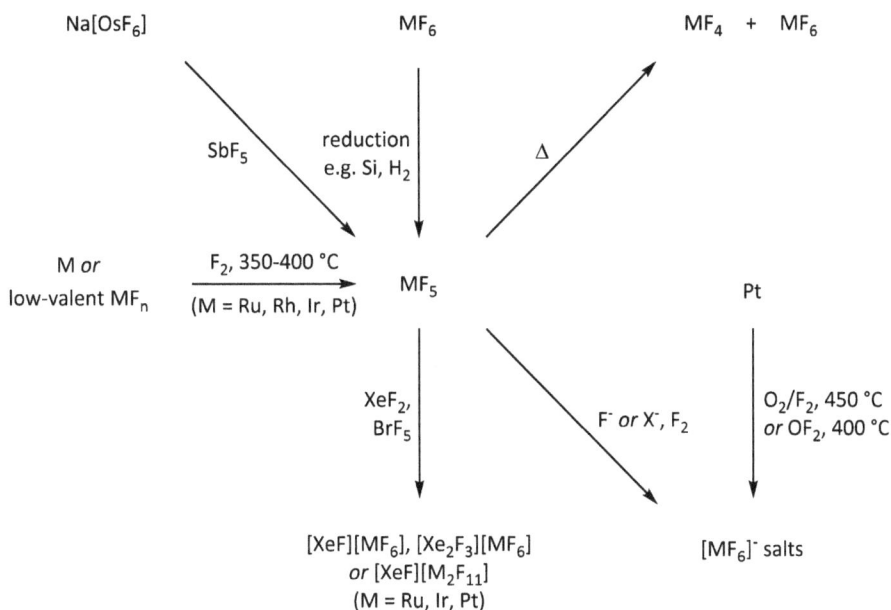

Scheme 10.27: Preparation and selected reactions of the pentafluorides MF$_5$ (M = Ru, Os, Rh, Ir, Pt).

NO, NO$_2$, O$_2$; M^{II} = Ca, Sr, Ba), which are generally accessible by fluorination of the respective halogenide mixtures or, alternatively, by mixing the particular fluorides (Scheme 10.27) [1691, 2399, 2412, 2421, 2422]. In case of Pt and Rh, the hexafluoromet-allate(V) anions are unstable [2423]. Ba[OsF$_6$]$_2$ is accessible by the careful reaction of elemental Os and BaF$_2$ in excess BrF$_3$ at 120 °C. Perspectively, this process might gain relevance in the so-called urban-mining, which is elaborated for recycling of econom-ically relevant, rare noble metals like Os or Ir [2401]. The chloryl salt [ClO$_2$][RuF$_6$] is accessible by fluorination of RuO$_4$ with ClF$_3$ in anhydrous HF at room temperature or by direct treatment of ClO$_2$F with RuF$_5$ [2424]. The crystal structures of [O$_2$]$^+$[RuF$_6$]$^-$, [O$_2$]$^+$[RhF$_6$]$^-$ and [NO]$^+$[RhF$_6$]$^-$ have been investigated and compared [2425]. The plat-inum salts are best prepared via [O$_2$]$^+$[PtF$_6$]$^-$ by treatment with fluorides, as PtF$_5$ is less stable than PtF$_6$. The dioxygenyl-salt [O$_2$][PtF$_6$] is obtained, if platinum sponge is fluorinated in silica-glassware. The initially generated PtF$_6$ provides the required oxygen by attacking the silica. The dioxygenyl salt is a volatile, deep-red solid (subl.: 155 °C; mp.: 260 °C (under pressure)), which is paramagnetic. It reacts violently with water [2413, 2426]. Alternatively, it may also be prepared from platinum sponge and O$_2$/F$_2$ mixtures at 450 °C or direct treatment of Pt with OF$_2$ at 400 °C (Scheme 10.27). It is also possible to fluorinate platinum halides in the presence of oxygen above 350–400 °C [2427]. A more convenient approach, which avoids handling of the extremely reactive PtF$_6$ or O$_2$/F$_2$ mixtures yields [O$_2$][PtF$_6$] by treatment of commercial PtO$_2$ · x H$_2$O with a flow of diluted fluorine gas at 275 °C and atmospheric pressure [2428].

The hydrolytic stability of the complex hexafluorometallate(V) salts is different, as for example, $[RuF_6]^-$ dissolves in H_2O under formation of O_2 and reduction to $[RuF_6]^{2-}$, whereas the same reaction only occurs with $[OsF_6]^-$, if bases are added [2378, 2400]. The reaction of Pd, O_2 and F_2 at 320 °C under high pressure has been claimed to yield the brown salt $[O_2]^+[PdF_6]^-$, representing the first Pd(V)-F-compound [2429]. The hexafluoropalladate(V)-anion has also been reported to be accessible in form of its sodium salt, when KrF_2 is reacted with PdF_4 in presence of liquid HF and solvated NaF [2430]. Later, these findings have been carefully reinvestigated and it has not been possible to obtain any clear hint for the existence of a Pd(V)-species [501, 2431]. Furthermore, matrix experiments rather contradict the existence of such a species, so that the early results clearly have to be doubted [2432]. Quantum-chemical calculations predict the existence of the still unknown $[KrF]^+[IrF_6]^-$, which is assumed to liberate Kr and IrF_7 in an exothermic reaction [2433].

There are only two oxyfluorides of the platinum metals in oxidation state +V. Platinum oxytrifluoride **PtOF$_3$** has been briefly described as pale brown product, which results from the fluorination of PtO_2 with F_2 at 200 °C [2360]. The analogous **RuOF$_3$** has been claimed to arise under similar conditions from RuO_2 [2434]. However, there is no more literature about any further properties or chemistry of this type of compounds.

Hexafluorides

Ruthenium hexafluoride **RuF$_6$** (dark-brown; mp.: 54 °C), rhodium hexafluoride **RhF$_6$** (black; mp.: 70 °C), osmium hexafluoride **OsF$_6$** (yellow; mp.: 34 °C; bp.: 46 °C), iridium hexafluoride **IrF$_6$** (yellow; mp.: 44.8 °C; bp.: 53.6 °C) and platinum hexafluoride **PtF$_6$** (dark-red; mp.: 61.3 °C; bp.: 69.1 °C) are all obtained as volatile products of the reactions between the elements at high temperatures and pressures (Scheme 10.28). Due to their instability, they have to be frozen out of the gaseous reaction mixtures. RuF_6 and RhF_6 are very difficult to prepare, the aforementioned method of direct fluorination of the metals only yields small amounts of products [2382, 2435, 2436]. Similarly, the fluorination of RuF_5 and RhF_5 or elemental Ru and Rh at 300 °C in a flow reaction results in low yields of the hexafluorides [2437]. More conveniently, the reactions of the pentavalent hexafluorometallates $[RuF_6]^-$ and $[RhF_6]^-$ with AgF_3/SbF_5 or AgF_3/BiF_5 in HF result in yields of 50 % (RuF_6) and 18 % (RhF_6), respectively (Scheme 10.28) [2334]. The best synthetic access to PtF_6 is the electric heating of a Pt-wire in an atmosphere of F_2 at liquid nitrogen temperature [2438]. The hexafluorides are isomorphic and crystallize as monomers, having an orthorhombic low-temperature form and a cubic high-temperature modification. In contrast to RhF_6, RuF_6 does not have an ideal octahedral shape [2439]. The reason for this is the Jahn-Teller distortion of the d^2-configured RuF_6 (RhF_6: d^3-configuration). Accordingly, the situation is inverted for the pentavalent hexafluorometallates, since $[RuF_6]^-$ (d^3-configuration) is perfectly octahedral, whereas $[RhF_6]^-$ (d^4-configuration) shows a Jahn–Teller distortion [732].

However, the deviations from a regular octahedral shape are so small, that they cannot be detected by electron diffraction or X-ray crystallography, as the experimental errors are significantly larger [1965, 2147]. The distortions are best observed by vibrational spectroscopy, which works especially fine for OsF_6 [2150]. Based on classical theory, the d^3-configured IrF_6 would be expected to be perfectly octahedral, however, due to relativistic effects, it is not [732]. For the same reason, Jahn–Teller distortions do not play a role in PtF_6. Thus, IrF_6 and PtF_6 represent molecules, which are structurally influenced by relativistic effects [732].

All the hexafluorides are very reactive and extremely powerful oxidizers. In the series of the 5d-transition metals, the stability drops in the direction $WF_6 > ReF_6 > OsF_6 > IrF_6 > PtF_6$. Among the platinum metals, the following additional statements can be made concerning the stability of the hexafluorides: $OsF_6 > RuF_6$, $IrF_6 > RhF_6$ and $RuF_6 > RhF_6$ [2440]. According to the decrease of the stability, the oxidation potentials of the hexafluorides increase. RhF_6 is the least stable hexafluoride, regarding the decay into the elements under F_2 liberation. Rhodium hexafluoride even reacts with carefully dried glass. Both, RuF_6 and RhF_6 as well as PtF_6 readily react with Xe and O_2 at room temperature to form XeF_2 and $[O_2]^+$ (Scheme 10.28) [732]. Upon heating, IrF_6 is also capable of oxidizing Xe [732]. PtF_6 reacts with ReF_6 to form the red complex $[Re^{VII}F_6]^+[Pt^VF_6]^-$ [2159]. The reaction of OsF_6 with Br_2 yields $[Br_2]^+[OsF_6]^-$, whereas Cl_2 is only oxidized by IrF_6 to form $[Cl_4]^+[IrF_6]^-$ (Scheme 10.28) [2441]. Interestingly, deeply colored reaction mixtures are observed in those cases, where no oxidation occurs ($OsF_6 + Cl_2$; $OsF_6 + Xe$; $IrF_6 + Xe$). This is due to charge transfer complexes, which visualize, that the oxidation power is just slightly too small [732]. In the presence of SbF_5, IrF_6 or OsF_6 react with hexafluorobenzene to yield the cation $[C_6F_6]^+$ [2442]. Moreover, IrF_6 is capable of fluorinating SbF_3 to SbF_5, whereas OsF_6 does not do so [2161]. Thermal decomposition of the hexafluorides or reaction with photochemically generated F-atoms at low temperature yields the corresponding pentafluoride and F_2 (Scheme 10.28). Thus, one can conclude, that the first bond-dissociation-energy of the hexafluorides has to be smaller than the dissociation energy of the F_2-molecule. Related suggestions have been made for RuF_6 and PtF_6, which are proposed to add F^- in a sense, that an intermediate species between $[PtF_5-F_2]^-$ and $[PtF_6-F]^-$ is formed [1991, 2443]. This would clearly include some type of preformed F_2, which in turn might explain some reactions, where PtF_6 acts as extreme fluorinating agent [732]. Historically, PtF_6 is of importance, as it has been used for the first oxidation of xenon in 1962. However, to date, the exact fate of the reactants is still under speculation [732]. The hexafluorides generally react violently with water. If the hydrolysis is carefully performed in HF, the oxonium salts $[H_3O]^+[MF_6]^-$ (M = Ru, Ir, Pt) and $[H_3O]_2^+[PtF_6]^{2-}$ are formed. Under the same conditions, OsF_6 reacts to form $OsOF_4$, while RhF_6 yields unstable and noncharacterizable products [2444]. The treatment of OsF_6 with BCl_3 leads to a halogen exchange and yields the pentachloride $OsCl_5$ [732]. OsF_6 can also be transformed to the derivative $OsO(OTeF_5)_4$ [2445]. Reactions of the hexafluorides with F^-

$[RuF_6]^-$ or $[RhF_6]^-$ Pt-wire $[NF_4]^+$ or $[ClF_6]^+$

AgF_3/SbF_5 or AgF_3/BiF_5, HF   electr. heating, F_2, -196 °C NF_3 or ClF_5 (M = Pt)

M F_2, Δ, p MF_6 Cl_2 (M = Ir) $[Cl_4]^+[IrF_6]^-$

Δ or F_2, hv, low T   Xe or O_2, rt (M = Ru, Rh, Pt) Br_2 (M = Os)

$MF_5 + F_2$ XeF_2 or $[O_2]^+$ $[Br_2]^+[OsF_6]^-$

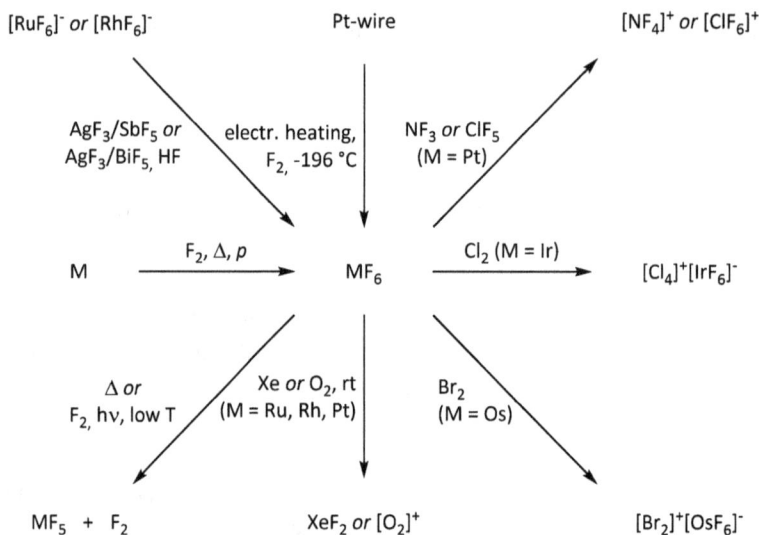

Scheme 10.28: Preparation and selected reactions of the hexafluorides MF_6 (M = Ru, Os, Rh, Ir, Pt).

are only scarcely known, however, $[RuF_7]^-$ has been predicted to behave like a non-classical complex, whereas $[RhF_7]^-$ is supposed to be a normal one [2443]. OsF_6 has been reported to react with nitrosyl fluoride under formation of $[NO]_2[OsF_8]$ [2160]. OsF_6, IrF_6 and PtF_6 react with CO under formation of the cationic carbonyl species $[Os(CO)_6]^{2+}$, $[Ir(CO)_6]^{3+}$ and $[Pt(CO)_4]^{3+}$, respectively. A few other CO-complexes have also been reported [2163, 2446, 2447]. Synthetically, PtF_6 is certainly the most valuable compound due to its extreme oxidizing power [330]. For example, PtF_6 is able to oxidize NF_3 to $[NF_4]^+$ and ClF_5 to $[ClF_6]^+$, however, for the transformation of BrF_5 to $[BrF_6]^+$, $[KrF]^+$ is required as oxidant (Scheme 10.28) [883]. Moreover, the oxidation power of PtF_6 is overcome by NiF_4 and $[NiF_3]^+$, as the latter oxidizes pentavalent $[PtF_6]^-$ to PtF_6 [2334]. Palladium hexafluoride **PdF$_6$** has been calculated by quantum-chemical methods, but is still unknown [2443]. Attempts have been made to prepare it under matrix conditions, however, PdF_6 has rather not been observed, even if its existence cannot be unambiguously excluded [2432].

Hexavalent oxyfluorides

There are also a few oxyfluorides of the platinum metals in oxidation state +VI. Ruthenium oxytetrafluoride **RuOF$_4$** is a very unstable compound, which is formed upon fluorination of RuO_2 with F_2 above 400 °C in form of yellow-green crystals [2448]. Earlier it has also been reported to be the product of the violent reaction of Ru metal with BrF_3 and Br_2 at 20 °C [2449]. Attempts to generate octavalent RuO_2F_4 from RuO_4 and KrF_2 also resulted in the formation of $RuOF_4$ [2450]. It has been investigated by X-ray crystallography, revealing a fluorine-bridged polymeric structure. Interestingly, the bridging F-atom is always in *cis*-position to the double bonded oxygen. The polymer has the shape of helical chains, wherein all Ru-O-bonds point to the same direction [2451]. A second modification of $RuOF_4$ is accessible in form of a yellow sublimate upon treatment of RuO_2 with pure F_2 at 300 °C [2452]. The structure of this modification is compiled of monomeric, tetragonal-pyramidal $RuOF_4$ with only very loose contact to a second $RuOF_4$ molecule, thus forming a very weak dimer. The contact to the second molecule occurs through F-bridging, where fluorine occupies the position opposite to the apical oxygen-substituent [2451]. $RuOF_4$ decomposes under formation of RuF_4 and O_2 [2448]. Analogously to RuF_4 and RuF_5, hydrolysis of $RuOF_4$ and subsequent treatment of the intermediate products with F_2 yields octavalent RuO_4 [2390]. Iridium oxytetrafluoride **IrOF$_4$** has already been described more than 90 years ago. It is a grey-white by-product of the preparation of IrF_6, if the reaction is carried out in glassware. Moreover, it is formed upon reaction of IrF_6 with traces of moisture. Similarly, $IrOF_4$ readily reacts with moisture and glass, however, it is stable in quartz vessels at room temperature. The reaction with water proceeds similar to the hydrolysis of the parent IrF_6, and thus liberates O_2 [2453]. The preparation of platinum oxytetrafluoride **PtOF$_4$** has also been claimed [2413]. However, the initial results of the experiment have been reevaluated and corrected afterwards, so that the obtained product has been identified as dioxygenyl salt $[O_2]^+[PtF_6]^-$ [2426]. Thus, the oxyfluoride $PtOF_4$ remains unknown, even if it might be a presumable intermediate in the controlled hydrolysis of PtF_6. The best synthetic approach toward **OsOF$_4$** (deep-blue; mp.: 100 °C) is the reaction of OsF_6 with OsO_4 in the ratio 5:2 at 175 °C, which also always produces small amounts of OsF_5 [2144]. Crystals of $OsOF_4$ are accessible by either sublimation or recrystallization from HF and exist in two modifications. One of them is isostructural to the aforementioned, polymeric form of $(RuOF_4)_x$. The second one is also deep-blue colored and forms *cis*-F-bridged polymers. The difference is, that the second modification rather builds a sheet than a helix and that it has a slightly smaller molecular volume. Thus, it might be the thermodynamically favored modification [2451].

Heptavalent oxyfluorides

Osmium heptafluoride **OsF$_7$** has been claimed to be formed from the elements at 500–600 °C and 350–400 bar in a Ni-autoclave [2454]. It has been described as a yellow, crystalline compound, which already slowly decomposes at –100 °C. However, the reproduction of OsF_7 under comparable reaction conditions has never been successful [2455]. Instead, very pure OsF_6 has been obtained exclusively and unambiguously identified by Raman spectroscopy. Other synthetic approaches for the generation of a potential OsF_7, for example fluorination of $OsOF_5$ or OsF_6 with KrF_2 have remained unsuccessful, too [2456]. Thus, OsF_7 rather needs to be considered as unknown compound. At least, it has been shown by quantum-chemical calculations, that OsF_7 should be accessible in the gas phase or by matrix isolation techniques [2457].

Nevertheless, there are a few osmium(VII) oxyfluorides, which have been unambiguously identified, namely $OsOF_5$, OsO_2F_3 and $Os_2O_3F_7$. The existence of OsO_3F had also been claimed, however, it became clear, that the sample has obviously been OsO_4 [2455, 2458]. Osmium dioxytrifluoride **OsO_2F_3** is the green product of the reaction of equimolar amounts of OsO_4 and OsF_6 at 150 °C, or $OsOF_4$ and OsO_3F_2 at 100 °C [2144, 2459]. In the solid state, it exists in two different modifications, a F-bridged polymer and a F-bridged tetramer. Interestingly, recrystallization from anhydrous HF typically yields the tetrameric modification, whereas the polymer is obtained by sublimation [2455]. In both structures, the osmium centers differ from each other, so that the compound is more correctly written as mixed-valent adduct $[Os^{VIII}O_3F_2 \cdot Os^{VI}OF_4]$ [2455]. A similar species is generated upon reaction of OsO_4 and a half excess of OsF_6 at 170 °C. The resulting **$Os_2O_3F_7$** may be rewritten as $[OsO_3F_2 \cdot OsF_5]$ and is thus a mixed Os^{VIII}/Os^V-compound [2455]. Osmium oxypentafluoride **$OsOF_5$** is generated upon fluorination of OsO_2 at 250 °C or by static oxyfluorination of elemental osmium at 300 °C and has almost the same green color like OsO_2F_3 [979, 2460, 2461]. Alternatively, it is accessible from a 1:3-mixture of OsO_4 and OsF_6 at 220 °C, resulting always in the formation of $OsOF_4$ as a side-product [2455]. The purification of $OsOF_5$ is realized by sublimation (bp.: 100.6 °C) [2462].

Iridium heptafluoride **IrF_7** is unknown but has been investigated by quantum-chemical calculations and identified as a viable target for gas phase or matrix experiments. A potential access to this compound has been found to be the exothermic decomposition of the ion-pair complex $[KrF]^+[Ir^VF_6]^-$ under liberation of Kr [2433]. To date, this complex is also still unknown, only the analogous $[XeF]^+[Ir^VF_6]^-$ has been reported [2420]. Another interesting target for matrix chemists is iridium oxypentafluoride **$IrOF_5$**, which is less sterically crowded and has therefore been calculated to be even more stable against typical decomposition pathways [2433]. Moreover, a potential platinum heptafluoride **PtF_7** has been explored by quantum-chemical methods. The results clearly indicate that the decomposition of PtF_7 into either PtF_6 or PtF_5 upon loss of F· or F_2, respectively, is strongly exothermic in both cases. Thus, platinum heptafluoride will hardly be accessible [2463].

Octavalent oxyfluorides
Osmium octafluoride **OsF_8** is unknown. Its first preparation has been claimed more than a century ago, however, many years later it turned out that the described sample has clearly not been the octafluoride, but OsF_6 [2464, 2465]. Quantum-chemical calculations indicate that OsF_8 might indeed be accessible under matrix conditions [2457]. Even if the binary osmium(VIII) fluoride is unknown, there are two known oxyfluorides. Osmium trioxydifluoride **OsO_3F_2** (orange; mp.: 170 °C) is the product of the reactions of OsO_4 with BrF_3, Os-metal with a mixture of O_2 and F_2 or OsO_4 and F_2 at 300 °C [2144, 2466]. It is also possible to fluorinate OsO_4 with excess ClF_3 [2467]. In the crystal structure of OsO_3F_2, one can find zig-zag chains of cis-F-bridged

[OsO$_3$F$_3$]-units. All oxygen atoms occupy *cis*-positions relative to each other, which then automatically results in all-*cis*-positioning of the fluorine substituents. This observation can be rationalized by a *trans*-effect [2467]. Upon treatment with fluoride donors M^{I}F (M^I = K, Rb, Cs, Ag), OsO$_3$F$_2$ yields orange salts with the anion [OsO$_3$F$_3$]$^-$, which also has a facial geometry, as expected from the structure of the parent OsO$_3$F$_2$ [2466, 2468, 2469]. In turn, reactions with the strong fluoride acceptors AsF$_5$ and SbF$_5$ in HF lead to the formation of salts with the [OsO$_3$F]$^+$ cation. Under the reaction conditions, the cation is strongly associated with the anions and HF solvent molecules, resulting in cyclic, dimeric structures, where the Os-center exhibits hexacoordination. The situation changes, if OsO$_3$F$_2$ is reacted with neat SbF$_5$, as this leads to the formation of [OsO$_3$F][Sb$_3$F$_{16}$], with isolated, C$_{3v}$-symmetric cations, wherein Os is tetracoordinated [2470]. Furthermore, osmium dioxytetrafluoride **OsO$_2$F$_4$** is known, which is a burgundy-red solid (mp.: 90 °C). It is prepared from OsO$_4$ and KrF$_2$ in HF and has first been described as OsOF$_6$, which has turned out to be incorrect [2471, 2472]. By ^{19}F-NMR- and vibrational spectroscopy it has been possible to reveal *cis*-oxygen positions [2467, 2473]. The analysis of a disordered crystal structure of OsO$_2$F$_4$ points toward the presence of helical chains of [OsO$_2$F$_4$]-octahedra [2467]. *Cis*-OsO$_2$F$_4$ reacts with the strong fluoride acceptors AsF$_5$ and SbF$_5$ under formation of salts with the cation [OsO$_2$F$_3$]$^+$ or the F-bridged *cis*-[μ–F(OsO$_2$F$_3$)$_2$]$^+$ [2474]. The aforementioned osmium oxyhexafluoride **OsOF$_6$** remains unknown, as its initially claimed preparation turned out to be wrong [2471, 2472]. However, it has been explored by quantum-chemical methods, which predict a slightly higher stability for the oxyhexafluoride than for OsF$_8$ [2457]. Thus, OsOF$_6$ might be accessible by matrix-isolation techniques. Iridium octafluoride **IrF$_8$** is also unknown, however, it has recently been calculated to be stable against decomposition into IrF$_6$ and F$_2$ at extreme pressures of more than 39 GPa [2475]. Previously, quantum chemical calculations predicted a strongly exothermic cleavage of F$_2$ due to steric crowding at the Ir-center. Moreover, the derived oxyfluoride IrOF$_6$ has been calculated and revealed slightly less exothermic decomposition pathways, but still remains very unstable [2433]. The question is if it will be ever possible to realize experimental conditions to isolate such unstable species like IrF$_8$ or IrOF$_6$. Finally, quantum-chemical calculations have also been performed for platinum octafluoride **PtF$_8$**. Here, it is quite obvious, that this compound will hardly ever be prepared, as its decomposition into PtF$_6$ and F$_2$ is predicted to be extremely exothermic [2361, 2463].

10.9.2 Study questions

(I)   Which difluorides of the platinum metals can be obtained in condensed phase? Are the binary compounds stable?

(II)   Which general approaches are used for the preparation of the platinum metal trifluorides? Which binary compounds are known? Are they all real trifluorides?

(III) How are the tetrafluorides RuF_4, OsF_4, RhF_4, IrF_4, PdF_4 and PtF_4 prepared? Which tetravalent fluorometallates are known? Comment on their role in urban mining.

(IV) Which structural similarities and differences are observed for the platinum metal pentafluorides and NbF_5? Are the pentafluorides thermally stable?

(V) Which is the most powerful oxidizer among the platinum metal hexafluorides? How can one order them according to their stability? Give examples for their reactivity.

(VI) Which hexavalent oxyfluorides exist in the series of platinum metals? How are they prepared?

(VII) Are there hepta- or even octavalent fluorine-containing compounds of the platinum metals? How are they accessible?

10.10 Group 11: Copper, silver, gold

The binary fluorides of the coinage metals quite differ in their composition, depending of course on the metal itself. Whereas CuF_2 is the only stable binary copper fluoride, silver forms both, a stable mono- and difluoride. In both cases, higher oxidation states are achieved by forming fluorometallate species. In contrast, stable gold fluorides exist for the oxidation states +III and +V, however, low valent AuF has never been isolated. Moreover, oxyfluoride compounds do not play a role in this group of transition metals.

10.10.1 Copper

Copper monofluoride

Pure copper monofluoride **CuF** is unknown. However, it is detected as monomeric, linear molecule (d(Cu-F) = 172 pm) in the vapor phase over molten CuF_2 [2476]. Upon cooling, it rapidly disproportionates to Cu and CuF_2 [2477]. Nevertheless, numerous efforts have been made to prepare binary CuF in the condensed phase, which have been comprehensively summarized elsewhere [2478]. The preparation of CuF has been claimed several times, for example, when Cu_2O is treated with HF or if elemental Cu is heated in a mixture of Cl_2/F_2 at 350 °C [2479, 2480]. However, attempts to reproduce these experiments have never been successful. In addition, some of the results have been found to be based on misinterpretations and especially wrong evaluations of powder X-ray data. The stability and some further properties of CuF have been investigated by quantum chemistry [2222, 2481, 2482]. A close relative of CuF has been prepared in form of the hexafluoroantimonate-salt $Cu[SbF_6]$, which is accessible by treatment of a Cu-wire with SbF_5 in anhydrous HF or reduction of $Cu[SbF_6]_2$ with elemental Cu in anhydrous HF. In the solid-state structure, the Cu^+ cation is octahedrally surrounded by F-atoms of the $[SbF_6]^-$ anions [2483]. Even if pure binary CuF is unknown, the preparation of CuF-complexes has been successful. For example, $[CuF(PPh_3)_3]$ is obtained by treatment of CuF_2 with PPh_3 in refluxing methanol [2484, 2485]. Ammonia complexes of the types $[Cu(NH_3)_3]_2[Cu(NH_3)_4]F_4 \cdot 2\,NH_3$ and $[Cu(NH_3)_2]F \cdot NH_3$ are accessible, if a mixture of Cu and CuF_2 is treated with liquid ammonia and allowed to stand at $-40\,°C$ or room temperature, respectively. Both complexes have been unambiguously charac-terized by single crystal X-ray crystallography [2478]. Moreover, a *bis*-$(\eta^2$-alkyne) co-ordination compound of CuF is known, as well as a T-shaped η^2-acetylene complex, which has been prepared in the gas phase [2486, 2487]. Under matrix conditions, it has been possible to observe the noble-gas compounds NeCuF and ArCuF [2488].

Copper difluoride

Copper difluoride **CuF$_2$** is the only stable, binary copper fluoride. It is generated in a high-temperature reaction of copper and fluorine or by treatment of CuO with HF at $400\,°C$ [2489–2492]. It has also been reported, that copper powder reacts with pyri-dinium poly(hydrogenfluoride) to yield CuF_2 at room temperature [2493]. Anhydrous CuF_2 is a colorless, crystalline solid (mp.: $950\,°C$ (decomp.)), which crystallizes in a tetragonally distorted rutile lattice. In the solid-state structure, it exhibits Jahn–Teller-distorted octahedra with four short (193 pm) and two long Cu-F bonds (227 pm) [2494, 2495]. In the molten state, CuF_2 slowly liberates F_2 under formation of interme-diary "CuF", which of course undergoes immediate disproportionation to CuF_2 and Cu.

In the first row of transition metals, CuF_2 is the compound with the weakest M-F bond, so that it can be used as efficient fluorinating agent at higher temperatures. For example, it transforms Ta to TaF_5 and Mn to MnF_2 [2496]. In presence of H_2O, the lightblue, crystalline dihydrate $CuF_2 \cdot 2\,H_2O$ is obtained, which can also be prepared by dissolving $CuCO_3$ in hydrofluoric acid. In the dihydrate, hydrogen bridges are ob-served between the fluoride- and the water-ligands [2497, 2498]. If $CuF_2 \cdot 2\,H_2O$ is heated, the water molecules are cleaved off. In aqueous solution of CuF_2, the complex cation $[CuF(H_2O)]^+$ is detected [2499]. Upon contact with air, CuF_2 is slowly trans-formed to the green, basic fluoride $CuF_2 \cdot Cu(OH)_2$, which in turn can be converted to the acidic fluoride $CuF_2 \cdot 5\,HF \cdot 6\,H_2O$ *in vacuo* or by addition of sulphuric acid. CuF_2 reacts with the strong fluoride acceptor AsF_5 under formation of 2:3- and 1:1-complexes [2500, 2501]. There are numerous complex fluorocuprate(II) salts, which are derived from CuF_2. They have the general formulas $M^I[CuF_3]$ (M^I = Na, K, Rb, Cs, Ag, Tl, NH_4); $M^I_2[CuF_4]$ (M^I = Na, K, Rb, Cs, NH_4, Tl); $M^{II}[CuF_4]$ (M^{II} = Ca, Sr, Ba) and $M^{II}_2[CuF_6]$ (M^{II} = Sr, Ba, Pb) [1835, 2236, 2241, 2502–2513]. In all of these complexes, the Cu-atom is surrounded by six fluorine atoms in form of a tetragonally distorted octahedron. The compounds CuF_2, $K[CuF_3]$, $Na_2[CuF_4]$ and $K_2[CuF_4]$ have been investigated by ab-sorption spectroscopy. In most of the spectra, the bands are in accordance with the ex-pectations and theoretical values for stretched octahedra of the composition $[CuF_6]^{4-}$ [2514]. Furthermore, quaternary compounds with the composition $M^ICu^{II}M^{III}F_6$ (M^I = K, Rb, Cs; M^{III} = Sc, Al, Ga, In, Tl, Fe, Co, Mn, Rh, Ni, Ti) are known [151, 2312, 2501]. Submitting NaF, CuF_2 and a rare-earth metal fluoride to solid-state reactions, it has also been possible to synthesize rare-earth metal complexes of the general formula $NaM^{III}Cu_2F_8$ (M^{III} = Sm, Eu, Gd, Y, Er, Yb) [2515]. Analogously, the solid-state reaction of CsF and CuF_2 in a molar ratio of 1:2 yields $CsCu_2F_5$. This compound is interesting, as it contains copper centers with the coordination numbers 4, 5 and 6 [2516]. In contrast, octahedrally coordinated copper centers are exclusively found in $K_3Cu_2F_7$. Therein, the $[CuF_6]$-octahedra are F-bridged, and thus connected via a common corner [2508]. More complex derivatives comprised of CuF_2 and several other metal fluorides are also known and have broadly been investigated concerning their structural and magnetic properties [1592, 2517–2523]. A structurally interesting example is $Ba_5Cu_2Al_3F_{23}$, which contains copper centers with both, octahedral and trigonal-prismatic coordination geometry [2524]. The situation gets even more complex in $Ba_{45}Cu_{28}Al_{17}F_{197}$, where $[CuF_6]$-octahedra and $[CuF_6]$-trigonal prisms are found along with $[Cu_2F_{10}]$-bipolyhedral units. The latter consist of either two $[CuF_6]$-octahedra or each one $[CuF_6]$-octahedron and trigonal prism. In both cases, the polyhedra are connected via a common edge [2525]. A further, structurally interesting example is not closely related to the aforementioned CuF_2-based complexes. However, the reaction of $Cu[SbF_6]_2$ with $[XeF_5][SbF_6]$ in anhydrous HF yields the adduct $[XeF_5][Cu[SbF_6]_3]$, which is composed of rings of $[CuF_6]$-octahedra. The $[CuF_6]$-units share apexes with $[SbF_6]$-octahedra, resulting in the formation of an infinite, three-dimensional network. The $[XeF_5]^+$ cations are located in cavities within the aforementioned framework [2526]. In $K_2[CuF_4]$, it is believed, that a high-pressure phase transition at about 9.5 GPa changes the magnetic properties from a ferromagnetic to an antiferromagnetic compound. This transition is induced by a rearrangement of the $[CuF_6]$-octahedra [2527]. Imidazolium salts of the compositions $[H_2Im]_4[CuF_6]$ and $[H_2Im]_2[CuF_4(H_2O)_2]$ have also been reported and are accessible by treatment of $CuO \cdot n\,H_2O$ with $NH_4[HF_2]$ or aqueous HF (w = 40 %) using the corresponding imidazolium-based ionic liquid as solvent [2528]. A mixed copper aluminium fluoride of the composition $CuAl_2F_8$ has been described as highly reactive reagent for the oxidative fluorination of aromatic moieties [2529]. Ammonia complexes of CuF_2 have been known for long time. $CuF_2 \cdot 5\,NH_3$ has been obtained by extraction of $CuF_2 \cdot 2\,H_2O$ with ammonia, the existence of $CuF_2 \cdot 3.33\,NH_3$ as potential decomposition product has also been assumed but not unambiguously clarified [2530]. If $CuF_2 \cdot 2\,H_2O$ is reacted with liquid ammonia, the ammine complex $[Cu(NH_3)_6]F_2 \cdot H_2O$ is formed. X-ray crystal structure analysis revealed the presence of the rare anion $[F(H_2O)F]^{2-}$, which shows strong hydrogen bonding [2531]. In the more recent literature, one can find several publications concerning the application of CuF_2 or derivatives thereof in lithium ion batteries. The use of CuF_2 as potential cathode material has been claimed, when laser-deposited thin films of CuF_2 were electrochemically reacted with lithium [2532]. The mechanism of the conversion of carbon-coated CuF_2 in such a lithium battery has been extensively investigated [2533]. Moreover, the use of nanoscopic CuF_2 and partly hydroxylated species has been explored. However, due to irreversible conversion reactions, copper fluoride based materials have proved to be unsuitable for applications in lithium ion batteries [2534].

Copper trifluoride and tetravalent compounds

Copper trifluoride **CuF$_3$** is only stable at low temperature (−78 °C) and is formed by solvolysis of K$_3$[CuF$_6$] in anhydrous HF. CuF$_3$ is a red solid, which has strong oxidizing and fluorinating properties, as it readily liberates F$_2$ from KHF$_2$-rich solutions of anhydrous HF at −40 °C. It is also capable of oxidizing Xe to XeF$_6$ at −78 °C [2535]. Vibrational spectra of CuF$_3$ have been recorded under matrix conditions [2488]. Apart from binary CuF$_3$, it is possible to synthesize fluorocomplexes, where copper has an oxidation state of +III or even +IV. For example, intense orange Cs[CuF$_4$] is formed, if CsCuCl$_3$ is fluorinated under pressure. The latter is prepared from a mixture of CsCl and CuCl$_2$ · 2 H$_2$O. Cs[CuF$_4$] is very reactive, since it attacks the glassware, if it is stored in a sealed ampoule for a few hours. On contact with air, it decomposes within a few minutes, whereas water immediately and violently hydrolyzes Cs[CuF$_4$] [2536]. The pale-green, crystalline, high-spin (d^8-configuration) complex K$_3$[CuF$_6$] is formed upon fluorination of a KCl/CuCl- or a KCl/CuCl$_2$-mixture [2310, 2326, 2537]. It contains octahedral [CuF$_6$]$^{3-}$ anions and liberates F$_2$, if it is warmed to 0 °C in anhydrous HF [2091]. Cu(III)-compounds of the compositions M$_2^I$Na[CuF$_6$] (M^I = K, Rb, Cs) and Rb$_2$M^I[CuF$_6$] (M^I = Li, K) are generated, if mixtures of CuF$_2$ and the respective alkali metal fluorides are fluorinated [154, 2538]. Cs$_2$K[CuF$_6$] is obtained, if a mixture of CuO, CsCl and KCl is fluorinated with diluted F$_2$ [2539]. K$_2$Li[CuF$_6$] is accessible from a mixture of K$_2$[CuF$_4$] and LiF, which is similarly fluorinated with diluted F$_2$ [2540]. Complexes with divalent cations are known as well, for example, CsMg[CuF$_6$] or CsZn[CuF$_6$], which is prepared by mixing ZnF$_2$ and Cs[CuCl$_3$] and subsequent fluorination [2541, 2542]. In the same class of compounds, a mixed-valent derivative has been described, which has the formula CsCu^{II}[Cu^{III}F$_6$] [2543]. Apart from the aforementioned ternary and quaternary fluorides, there are also other hexafluorocuprates(III), which differ mainly in the composition of the cations. Their syntheses and some of their properties, including magnetic and structural behavior have been summarized in the literature [2544].

The direct fluorination of a pretreated mixture of CsCl and CsCu^{II}Cl$_3$ at 410 °C and 350 bar yields intensive orange-red crystals of Cs$_2$[CuF$_6$] [2545]. Interestingly, this species is also obtained under milder conditions in a flow reactor at 200 °C [2546]. The paramagnetic Cu(IV) complex decomposes violently in water [2545]. Cs$_2$[CuF$_6$] has also been investigated by single crystal X-ray crystallography, revealing slightly elongated [CuF$_6$]-octahedra (d(Cu-F): 4x 175 pm; 2x 177 pm) in a tetragonal unit cell [2547]. Its magnetic behavior indicates that it is a low-spin complex (d^7-configuration) [2548]. Apart from Cs$_2$[CuF$_6$], the respective rubidium and potassium compounds Rb$_2$[CuF$_6$] and K$_2$[CuF$_6$] have been claimed as well, however, due to their instability, they have not been unambiguously characterized [2548, 2549].

10.10.2 Silver

Silver subfluoride

The direct fluorination of silver is a complex process, which is mechanistically not yet fully understood. Depending on pressure and reaction temperature, the fluorides Ag_2F, AgF and AgF_2 can be obtained [2550]. Silver is the only coinage metal, which forms a well-defined subfluoride. Disilver monofluoride **Ag_2F** has been known for long time and is generated upon reaction of finely distributed Ag with AgF in HF or upon electrolysis of AgF at low current density on a silver cathode [2551, 2552]. Ag_2F may also be prepared from Ag and AgF using a mechanochemical reaction [2553]. It crystallizes in form of bronze, hexagonal platelets, wherein two Ag-layers with Ag-Ag metal bonds are alternatingly arranged with a layer of fluoride anions (*anti*-CdI_2-structure) [2554]. Between the Ag- and the F-layers, ionic contacts are present, so that Ag_2F is a good electric conductor [2555, 2556]. Regarding its structure, it is considered as an intermediate between a typical salt-like compound and a metal [2557]. Apart from electrical conductivity, it has some other interesting properties, for example, it has been found that Ag_2F shows superconductivity [2558]. Hydrolysis or thermal decomposition of Ag_2F results in the formation of Ag and AgF [2552, 2559, 2560].

Silver monofluoride

There is a large number of publications about silver monofluoride **AgF**, mainly due to its numerous applications in organic synthesis. It is obtained, if an aqueous solution of Ag_2O or Ag_2CO_3 and HF is evaporated. Upon heating to 40 °C, the resulting dihydrate $AgF \cdot 2\,H_2O$ loses its water ligands [2561]. Alternatively, the thermal decomposition of $Ag[BF_4]$ gives rise to AgF [2562]. It is a yellow solid (mp.: 435 °C; bp.: ca. 1150 °C), which crystallizes in a NaCl type lattice. It conducts electricity and is less light-sensitive than the other silver halides. It is well soluble in H_2O and forms a number of hydrates, such as $AgF \cdot H_2O$, $3\,AgF \cdot 5\,H_2O$, $AgF \cdot 2\,H_2O$ (stable up to 39.5 °C) and $AgF \cdot 4\,H_2O$ (stable from −14 to 18.7 °C). Furthermore, it reacts with HF to form complexes of the compositions $AgF \cdot HF$ and $AgF \cdot 3\,HF$. The most important property of AgF is its ability to gently fluorinate for example I_2 to IF_5 or sulphur to low-valent sulphur fluorides [2378]. As mentioned above, AgF is also used in organic synthesis, as it can for example add to double-bonds of fluorinated olefins [2563]. Apart from that, it is commonly used as fluorinating agent. If an Ag^+ solution is treated with $[AgF_4]^-$ in anhydrous HF, the mixed-valent, red-brown, diamagnetic $Ag^I[Ag^{III}F_4]$ is formed, which exothermally decomposes to paramagnetic AgF_2 below 0 °C [2564]. The reaction of AgF or AgF_2 with liquid ammonia yields the coordination compound $[Ag(NH_3)_2]F \cdot 2\,NH_3$, which already decomposes above −30 °C under liberation of gaseous ammonia [2565]. Another interesting ammine complex of silver(I) fluoride is $[Ag(NH_3)_3]_2[Ag(NH_3)_2]_2[SnF_6]F_2$, which contains two differently coordinated silver(I) centers, showing a linear coordination in $[Ag(NH_3)_2]^+$ and a T-shaped structure in $[Ag(NH_3)_3]^+$. In addition, argentophilic interactions are observed between the complex cations [265].

Silver difluoride

The chemistry of divalent silver is rather rare and, therefore, not that extensively explored [2566]. Silver difluoride **AgF₂** is the only known, stable, binary silver(II) compound; nevertheless, it has been well described already long time ago [864, 2567]. It is prepared from the elements at about 300 °C or by fluorination of silver halides in form of a microcrystalline, brown-black powder (mp.: 690 °C) [864, 2567]. Alternatively, it can also be obtained by electrosynthesis in HF, starting from silver(I)-precursors [2568]. In the solid-state, orthorhombic AgF_2 consists of $[AgF_{4/2}]_\infty$ sheets, where a strongly distorted octahedral coordination of the Ag atoms is achieved by bonding to fluoride substituents of neighboring layers. These findings have been proven by neutron diffraction as well as by X-ray experiments with both, AgF_2 powder and single crystals [2569, 2570]. Moreover, the effect of high pressure on the structural behavior of AgF_2 has been explored by experiments and quantum chemical calculations, revealing transformations to a few high-pressure polymorphs [2571, 2572]. X-ray photoelectron spectroscopy (XPS) has revealed a high degree of covalency in the bonding of Ag(II)-fluoride species [2573]. Below –110 °C, AgF_2 has weak ferromagnetic properties and shows long-range antiferromagnetic ordering [2574]. Thermally, AgF_2 is very stable, at 700 °C its dissociation pressure amounts only 0.1 bar [2575]. However, it is very sensitive toward hydrolysis. AgF_2 is an excellent oxidizer and fluorinating agent, its fluorinating properties are comparable to those of elemental fluorine itself. The main advantage of AgF_2 over F_2 is the purity, as AgF_2 does not contain traces of O_2. Apart from that, the handling of a solid is in most cases more convenient.

Blue solutions of AgF_2 in HF readily oxidize elemental Xe to Xe(II) in presence of BF_3 or AsF_5, whereas solid AgF_2 does not react [2576]. Acidic solutions of Ag(II) are also capable of oxidizing O_2 to $[O_2]^+$ and $[IrF_6]^-$ to IrF_6 [2535, 2577]. In contrast, $[O_2]^+$ oxidizes AgF_2 to $[AgF_4]^-$ in fluorobasic solution [2535]. The reactivity of AgF_2 against many inorganic compounds, for example, oxides, chlorides or oxychlorides has also been explored. As a general conclusion, most of the chloro-derivatives readily react, whereas the oxo-compounds in parts only do so at elevated temperatures, if at all. However, sulphates, nitrates and perchlorates are inert against AgF_2 [2578, 2579]. The catalytic activity of Ag in fluorination reactions may be explained by the intermediary formation of AgF_2. Silver(II) centers are also found in the quaternary fluoride $Ag_2ZnZr_2F_{14}$, which contains $[Ag_2F_7]$-units in form of corner-linked dimers of square-planar $[AgF_4]$-subunits [2580]. $Ag_2ZnZr_2F_{14}$ is derived from the ternary fluorides $Ag_3M^{IV}_2F_{14}$ (M^{IV} = Zr, Hf), which contain two crystallographically different Ag-centers with different coordination environment [2581]. There are also numerous complex salts, which are derived from AgF_2, namely $M^I[AgF_3]$ (M^I = K, Rb, Cs); $M^I_2[AgF_4]$ (M^I = Na, K, Rb, Cs); intensely blue-violet $M^{II}[AgF_4]$ (M^{II} = Ca, Sr, Ba, Cd,

Hg) and $Ba_2[AgF_6]$ [2582–2586]. In all the aforementioned complexes, silver exhibits a Jahn–Teller-distorted, octahedral coordination [2587–2589]. In $Cs[AgF_3]$ one can find stretched $[AgF_6]$-octahedra, which are also observed in $Cs_2[AgF_4]$ [2582]. In the latter case, the $[AgF_6]$-octahedra are two-dimensionally connected, overall resulting in a layered perovskite structure [2590]. $Ba[AgF_4]$ is comprised of isolated $[AgF_4]$-squares, whereas $Ba_2[AgF_6]$ shows again octahedral coordination of the silver centers [2585, 2586, 2591]. AgF_2 also reacts with fluoride acceptors under formation of complex salts with the cations Ag^{2+} and $[AgF]^+$, for example, $Ag^{II}[Ag^{III}F_4]_2$ ($= \mathbf{Ag_3F_8}$); $Ag^{II}[M^{IV}F_6]$ (M^{IV} = Sn, Pb, Zr, Hf, Rh, Pd, Pt); $M^{I}Ag[M^{III}F_6]$ (M^{I} = K, Rb, Cs; M^{III} = Tl, In, Ga, Al, Sc, Fe, Co); $Ag^{II}[SbF_6]_2$; $[Ag^{II}F][Ag^{III}F_4]$ ($= \mathbf{Ag_2F_5}$) and $[Ag^{II}F][SbF_6]$ [150, 262, 2592–2594]. "Ag_2F_5" is obtained as maroon solid upon treatment of a solution of $[AgF][AsF_6]$ with an equimolar amount of $K[AgF_4]$ in anhydrous HF. The addition of AgF_3 to Ag_2F_5 yields "Ag_3F_8", which is also obtained upon decomposition of AgF_3 [2592]. As already mentioned above, silver(II) species are typically coordinated by six fluorine atoms in form of stretched $[AgF_6]$-octahedra. For example, six $[SbF_6]^-$ anions contribute the fluorine ligands in the case of $Ag[SbF_6]_2$ [2594]. In contrast, the $[AgF]^+$ salts are composed of infinite chains, which can differ in their structures, depending on the nature of the anion. The chains are linear in $[AgF][BF_4]$, bent at the F-atoms in $[AgF]_2[AsF_6][AgF_4]$ or even bent at both, the Ag- and F-atoms, as in $[AgF][AsF_6]$ [2564, 2595, 2596]. In the chains, the Ag-centers are weakly coordinated by four further fluorine atoms of the anions, which results in the formation of distorted $[AgF_6]$-octahedra, which have common F-atoms in *trans*-position [2564, 2595]. Structurally, the mixed valent silver fluorides Ag_3F_8 and Ag_2F_5 are interesting as well. In Ag_3F_8 ($= Ag^{II}[Ag^{III}F_4]_2$), there are bands of $[Ag^{II}F_4]$- and $[Ag^{III}F_4]$-squares, which are connected via common corners. The bands are folded in longitudinal direction. In addition, the Ag-centers are weakly coordinated by two F-atoms of parallel bands. Overall, this results in stretched $[AgF_6]$-octahedra [2597]. The situation is different in Ag_2F_5 ($= [Ag^{II}F][Ag^{III}F_4]$), where the $[AgF]^+$ cation forms chains, as discussed above. Additionally, the silver ions in the $[AgF]$-chains are weakly coordinated by four fluorine atoms of four square-planar $[AgF_4]$-units. Hence, the overall coordination of the silver centers is also described as distorted-octahedral [2598]. The structures and potential superconductive properties of many intermediate-valent Ag^I/Ag^{II}- and Ag^{II}/Ag^{III}-fluoride systems have been reviewed and evaluated [2599]. The ternary compound $Ag_3[SbF_6]_4$ also contains mixed-valent Ag(I) and Ag(II) cations and is more precisely written as $Ag_2^I Ag^{II}[SbF_6]_4$. It is prepared from a 2:1-mixture of $Ag[SbF_6]$ and $Ag[SbF_6]_2$ in gaseous HF at 50 °C [2600].

Silver trifluoride

In liquid HF, red silver trifluoride **AgF$_3$** is precipitated from solutions of $[AgF_4]^-$, if BF_3, PF_5, AsF_5 or GeF_4 are added. AgF_3 is isostructural to AuF_3 and thermodynamically unstable, as it is converted to Ag_3F_8 and F_2 at room temperature [2592, 2601].

The chemistry of AgF_3 and its derivatives is strongly restricted to the use of anhydrous HF as solvent [34]. Both AgF_3 and $K[AgF_4]$ have been investigated by UV-vis spectroscopy, which has proven their planar geometry [2602]. Analogously to AgF_2, X-ray photoelectron spectroscopy (XPS) revealed a significant degree of covalent bonding in Ag(III)-fluoride species [2573]. Solutions of AgF_3 in presence of AsF_5, SbF_5 or BiF_5 are capable of oxidizing $[RuF_6]^-$ and $[PtF_6]^-$ to the hexafluorides RuF_6 and PtF_6, respectively. Moreover, O_2 is converted to $[O_2]^+$ under the same conditions [2577]. Thus, Ag(III) is an extremely oxidizing species with an oxidation power comparable to that of Ni(IV). Oxidation of a mixture of M^ICl and AgCl with elemental fluorine in alkaline solution yields the yellow, diamagnetic salts $M^I[AgF_4]$ (M^I = Na, K, Rb, Cs, AgF, XeF_5) as well as $M^{II}[AgF_4]_2$ (M^{II} = Ag, etc.), which are very sensitive toward moisture. Compounds of the type $M^I[AgF_4]$ (M^I = Li, K) are also accessible, if AgF_2 is reacted with LiF or KF in a solution of anhydrous HF, which additionally contains UV-irradiated F_2 [2601]. Apart form that, *in situ* generated O_2F reacts with a suspension of AgF_2 in basic, anhydrous HF to yield $[AgF_4]^-$ [501]. Similar to the complexes $M^{II}[AgF_4]$, the derivatives $M[AgF_4]$ have a structure like $K[BrF_4]$, containing square-planar $[AgF_4]^-$ anions [2603, 2604]. Upon treatment of a mixture of alkali metal chloride and silver nitrate with fluorine at 300 °C, the yellow salt $KCs_2[AgF_6]$ with the octahedral anion $[AgF_6]^{3-}$ is formed [2605].

A binary silver tetrafluoride **AgF₄** is unknown. However, if the fluorination of Cs- and Ag-salts is performed under very high fluorine pressure, the formation of the formal silver(IV) compound $Cs_2[Ag^{IV}F_6]$ has been claimed. More correctly, the oxidation state of silver has been assumed to be an equal mixture of +III and +V, as has been revealed by magnetic measurements, so that the formula of the hexafluoroargentate(IV)-salt is better written as $Cs_2[Ag^{III}_{0.5}Ag^V_{0.5}F_6]$ [2606]. Since it has not been possible to prepare any further compound with silver in an oxidation state higher than +III, these results have to be treated with caution. The potential silverfluorides **AgF₅** and **AgF₆** have been calculated by quantum-chemical methods but are unknown [2443]. Even if it is unlikely that these compounds will ever be prepared, AgF_5 would of course be an extremely powerful Lewis acid and AgF_6 would be an extreme oxidizer, respectively.

10.10.3 Gold

Low-valent gold fluorides

By now, there are only two binary gold fluorides, which have been isolated, namely AuF_3 and AuF_5. However, the formation of gold monofluoride **AuF** has been identified based on vibrational spectra, when Au-films were etched by CF_4/O_2- or SF_6/O_2-glow-discharge plasmas [2607]. Moreover, AuF has been unambiguously identified by mass spectrometry and microwave spectroscopy [2608, 2609]. DFT-calculations revealed a facilitated formation of AuF-species, if xenon is used as mediator [2610]. The complexes XeAuF and ArAuF have been prepared in the gas phase and detected by microwave rotational spectroscopy, too [2611, 2612]. Similarly, the analogous neon compound NeAuF as well as ArAuF have been obtained in the corresponding noble gas matrices and identified by vibrational spectroscopy [2488, 2613]. Further spectroscopic properties of AuF have been investigated by quantum chemical methods [2614]. Some electronic excited states have been explored by laser excitation of gaseous AuF [2615]. So far, it has not been possible to isolate AuF, as it immediately disproportionates to Au and AuF_3. In turn, a high pressure comproportionation of Au and AuF_3 has been proposed as potential synthetic access to AuF [2616]. Attempts to generate a stabilized AuF species by reduction of AuF_3 in presence of AsF_3 in HF/SbF_5 resulted in the formation of $[F_3As-Au]^+[SbF_6]^-$. The solid-state structure of this compound clearly revealed an ionic contact between the gold cation and $[SbF_6]^-$, as the shape of the anion is distorted octahedral. However, this species is still not a real AuF compound [2617].

Matrix experiments also allowed the preparation and characterization of unstable gold difluoride **AuF_2** [2488, 2613]. Quantum-chemical calculations predicted some spectroscopic properties for the AuF_2 molecule [2614]. IR spectra of AuF_2 have been recorded under matrix conditions, too [2488]. However, in the solid state it is still unknown, as it is readily converted to stable AuF_3 [2618]. Nevertheless, it has been possible to synthesize a close relative of AuF_2, as the reaction of Au with F_2 in aHF/SbF_5 at 20 °C gives rise to orange, crystalline $Au[SbF_6]_2$. The structural and magnetic properties of the latter gold(II) compound as well as those of some of its derivatives have also been explored [2619]. The pyrolysis or partial reduction of $Au[SO_3F]_3$ with Au in HSO_3F also yields gold(II) species, which can be identified by means of ESR spectroscopy [2618].

Gold trifluoride

Gold trifluoride **AuF_3** is synthesized upon fluorination of Au_2Cl_6 at about 200 °C or reaction of Au with BrF_3 with subsequent decomposition of the resulting complex $BrF_3 \cdot AuF_3$ [2620, 2621]. The structure of the adduct has been determined by X-ray crystallography, in addition, Raman spectroscopy has revealed a slight ionic character, so that it may be formulated as $[BrF_2]^+[AuF_4]^-$ [2622]. On laboratory scale, AuF_3 is conveniently accessible from the elements in a low-pressure reaction in a quartz reactor [2623, 2624]. AuF_3 is an orange, crystalline solid (decomp.: 500 °C), whose solid-state structure is comparable to that of AgF_2. It forms chains of $[AuF_2F_{2/2}]$, wherein square-planar $[AuF_4]$-units are connected by cis-F-bridges. The chains are polymeric and form a helical-type structure, where parallel coils show weak Au-F-interactions to each other, so that every Au-atom is distorted octahedrally coordinated [2625]. In the gas phase, AuF_3 exists in form of planar, F-bridged Au_2F_6 dimers and T-shaped AuF_3 monomers, as has been explored by electron diffraction studies and quantum-chemical calculations [2626]. AuF_3 is stable up to 500 °C; above this temperature, it decomposes into the elements. It acts as strong fluorinating agent. Yellow complex salts containing the square-planar $[AuF_4]^-$ anion are obtained, when Au or gold chlorides and alkali or alkaline earth metal chlorides are fluorinated [2404, 2603, 2621, 2627]. $K[AuF_4]$ crystallizes in a $K[BrF_4]$-type lattice, where the interactions between the complex anions are very weak [2603, 2628, 2629]. The cesium salt $Cs[AuF_4]$ as well as the parent AuF_3 have been studied by UV-vis spectroscopy [2602]. If the ratio in the reaction of CsF and AuF_3 is changed from about 1:1 to 1:2, the compound $Cs[Au_2F_7]$ with the F-bridged anion $[Au_2F_7]^-$ is obtained [2630]. Higher F-bridged anions of gold have been investigated by computational methods and compared to the respective bromine derivatives. As a result, the $[Au_3F_{10}]^-$ anion has been predicted to be stable and is expected to have a chain-like structure [2631]. There are also several metal salts of the general composition $M^{II}[AuF_4]_2$ (M^{II} = Mg, Ba, Zn, Pd, Cd, Hg, Ni, Ag, Au), which have been synthesized and structurally investigated by X-ray crystallography [2597, 2632–2636]. In case of copper(II), only one of the fluorides is substituted, resulting in the formation of $[CuF][AuF_4]$ [2637]. A number of tetrafluoroaurate(III) complexes of the rare earth metals has been described as well. They are all prepared by treatment of the metal trifluorides with AuF_3 or Au and subsequent fluorination with diluted F_2 [2638–2640]. The early f-block elements yield yellow complexes of the type $[M_2F][AuF_4]_5$ (M = La, Pr, Nd, Sm, Gd), which all crystallize in a tetragonal lattice [2638]. In contrast, the metals Tb, Dy, Ho and Er form complexes with the same color and composition, but crystallize in a triclinic structure [2639]. Finally, the late rare earth metals yield also yellow complexes, however, they have the composition $[MF][AuF_4]_2$ (M = Tm, Yb, Lu) and crystallize in an orthorhombic space group [2640]. The tetrafluoroaurates(III) $[TlF_2][AuF_4]$, $[Bi_2F][AuF_4]_5$, $[Y_2F][AuF_4]_5$, $Sm[AuF_4]_3$ and $La[AuF_4]_3$ have been synthesized and structurally characterized as well [2641, 2642]. For the actinides, a complex derived from ThF_4 has been described, having the formula $[Th_2F_7][AuF_4]$ [2643]. The analogous uranium compound $[U_2F_7][AuF_4]$ has also been prepared and exhibits the same solid-state structure [2630]. The pentafluoroxenon(VI) complex $[XeF_5][AuF_4]$ is generated by a displacement reaction of $BrF_3 \cdot AuF_3$ with XeF_6. The salt is further oxidized by KrF_2 to quantitatively yield the gold(V) complex $[XeF_5][AuF_6]$. In liquid anhydrous HF, $[XeF_5][AuF_4]$ decomposes under liberation of AuF_3 [1098]. The reduction of AuF_3 with elemental xenon has led to the formation, isolation and characterization of the first gold-xenon compound in form of the square-planar cation $[AuXe_4]^{2+}$ [2644]. Gold trifluoride is reduced in superacidic HF/SbF_5 medium, giving rise to orange $[Au_3F_8 \cdot 2\ SbF_5]$, black $[Au_3F_7 \cdot 3\ SbF_5]$ and yellow-green $[Au(HF)_2][SbF_6]_2 \cdot 2\ HF$. In the solid state, all three compounds contain gold(II) centers in square-planar $[Au^{II}F_4]$-units, in addition one can also find square-planar $[Au^{III}F_4]$-motifs in the orange and the black species [2645]. The preparation of the ammonium salts $[NMe_4][AuF_4]$ and $[NEt_4][AuF_4]$ allows the handling of the reactive tetrafluoroaurate(III) species in organic solvents and furnishes a potential starting point for further synthetic transformations. Furthermore, it has been found that AuF_3 itself is complexed by N-donor ligands like MeCN or pyridine, which also allows handling in common organic solvents [2646]. Moreover, AuF_3 is stabilized, if it is complexed with a NHC-ligand. The resulting adduct is accessible from $[AuF_4]^-$ salts or directly from AuF_3 [2647]. The Lewis acidity of such NHC-complexes is tuned by the substitution of one of the fluoride ligands by $-OTeF_5$ or $-Cl$ [2648]. The fluoride substituents of AuF_3 are completely substituted by $OTeF_5$ groups, if the trifluoride is treated with $B(OTeF_5)_3$. Interestingly, the crystal structure analysis of $Au(OTeF_5)_3$ revealed dimers, where $OTeF_5$ has been found to be a bridging ligand [2649]. In anhydrous HF, AuF_3 forms an intercalation compound with graphite in a liquid-solid reaction at room temperature [2419]. The thermally stable but reactive complex anion $trans$-$[AuF_2(CF_3)_2]^-$ is accessible by oxidative fluorination of $[Au(CF_3)_2]^-$ with XeF_2 and undergoes dissociation reactions toward $[Au(CF_3)F_x]^-$ (x = 1–3) in the gas phase [2650]. It has been shown that different stable fluoride complexes of gold(III) react with boronic acids, followed by the formation of C-C bonds upon reductive elimination [2651]. Thus, these types of compounds are operative in coupling reactions.

High-valent gold fluorides

Gold tetrafluoride **AuF₄** is unknown. A compound with the formula $AuF_{3.6}$ has been claimed to be the fluorination product of AuF_3, which is very likely to be the result of a misinterpretation [2652]. However, recently quantum chemical-calculations have indicated, that AuF_4 might indeed be stable under high pressure conditions [2653].

Gold pentafluoride **AuF₅** is obtained if $[O_2][AuF_6]$ or $[KrF][AuF_6]$ are submitted to vacuum thermolysis (Scheme 10.29) [1034, 2654]. It can also be prepared from gold atoms and F_2 under matrix conditions [21]. AuF_5 is a dark red, diamagnetic and crystalline solid, which thermally easily splits off F_2 to yield AuF_3. AuF_5 is the only pentafluoride, which forms dimers of two edge-sharing $[AuF_6]$-octahedra in the solid state [2655]. In the vapor phase, AuF_5 is present as a mixture of dimers and trimers (ca. 4:1), the monomer having a square-pyramidal shape [2656]. A few complex salts are known, where the $[AuF_6]$-group is more or less found to be octahedral. The preparation of the complex salts is achieved by several methods. For example, fluorination of AuF_3 occurs at 400 °C in presence of excess XeF_2 or XeF_6. In the first step, $[Xe_2F_{11}]^+[AuF_6]^-$ is formed, which is then reacted with CsF at 110 °C to yield $Cs[AuF_6]$ [1079, 1080, 2652]. The latter is also accessible by the treatment of $Cs[AuF_4]$ with fluorine and has been investigated by UV-vis spectroscopy [2602, 2652]. Similar to SbF_5, AuF_5 behaves as a strong fluoride acceptor forming, for example, $[NO][AuF_6]$ and $[Xe_2F_3][AuF_6]$, if it is reacted with NOF or XeF_2, respectively (Scheme 10.29) [1034, 2657]. The Raman-spectrum of $[NO][AuF_6]$ has been discussed as general representative for other $[AuF_6]^-$ salts [2658]. In anhydrous HF, a number of hexafluoroaurates(V) $[AuF_6]^-$ is accessible by reaction of a 1:2-mixture of MF_2 (M = Ni, Cu, Ag, Zn, Cd, Hg, Mg, Ca, Sr, Ba) and AuF_3 in presence of KrF_2 or UV-irradiated F_2 [2657]. Analogously, $K[AuF_6]$ is prepared from AuF_3 and KF and has been structurally characterized by X-ray crystallography. A mixed valent gold(III)-gold(V) derivative

Au [AuF$_6$]$^-$ [IF$_6$][SbF$_6$] + AuF$_3$

O_2/F_2, 500 °C, 5 bar or KrF$_2$, aHF, 20 °C

SbF$_5$

SbF$_5$, IF$_5$

[O$_2$][AuF$_6$] or [KrF][AuF$_6$] → vac. thermolysis → AuF$_5$ → thermal decomposition → AuF$_3$ + F$_2$

NOF or XeF$_2$

AuF$_3$ → F$^-$, aHF, KrF$_2$ or F$_2$, hv → [AuF$_6$]$^-$ salts ← O$_2$F or F$_2$, hv ← [AuF$_4$]$^-$

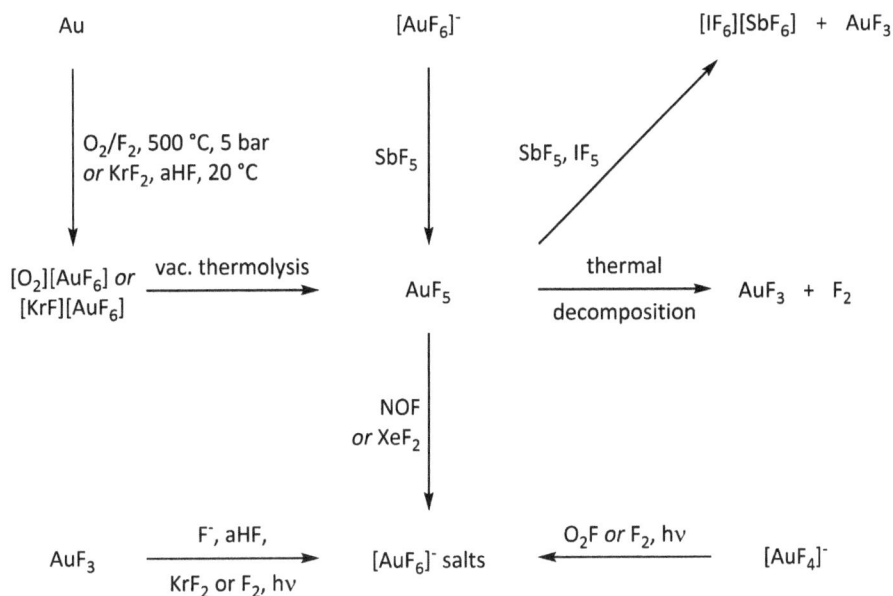

Scheme 10.29: Preparation and selected reactions of pentavalent gold compounds.

of magnesium is also known, having the formula Mg(HF)[AuF$_4$][AuF$_6$] [2659]. Apart from that, the silver(II)-derivative [AgF][AuF$_6$] has been prepared and structurally characterized by X-ray crystallography [2660]. If metallic Au is treated with a mixture of O_2/F_2 (1:4) at 500 °C and a pressure of 5 bar, the dioxygenyl salt [O$_2$][AuF$_6$] is generated (Scheme 10.29) [2661]. [AuF$_6$]$^-$ salts are also accessible by oxidation of [AuF$_4$]$^-$ with O$_2$F or photodissociated F$_2$ [501, 2431]. The krypton salt [KrF][AuF$_6$] is made from KrF$_2$ and Au at 20 °C in anhydrous HF [1034]. Both [O$_2$][AuF$_6$] and [KrF][AuF$_6$] have been characterized by X-ray crystallography, vibrational spectroscopy and electron structure calculations [2662]. All gold (V) compounds are strong oxidizers. In its complex salts, AuF$_5$ is displaced by SbF$_5$ and in a solution of AuF$_5$ and SbF$_5$ in IF$_5$, the salt [IF$_6$][SbF$_6$] is formed along with AuF$_3$ (Scheme 10.29) [2663]. A few AuF$_5$-complexes with nitrogen fluorides (NF$_3$, N$_2$F$_4$) and nitrogen oxyfluorides (NOF$_3$, NOF) are also known and have been characterized by vibrational spectroscopy [350].

AuF$_6$ has been calculated by quantum-chemical methods but has never been prepared [1991]. Analogously to AuF$_4$, high pressure has been proposed as a tool for the preparation of potential AuF$_6$ [2653]. Gold heptafluoride **AuF$_7$** has been claimed to be accessible by the reaction of AuF$_5$ with atomic fluorine and has further been reported to be stable at room temperature, which has been quite surprising [2664, 2665]. However, later the existence of AuF$_7$ has clearly been contradicted by quantum-chemical calculations, as the decomposition into AuF$_5$ and F$_2$ as well as the homolytic cleavage of an Au-F bond to result in also very unstable AuF$_6$ have been found to be very exothermic [2666, 2667].

10.10.4 Study questions

(I)   Are there stable monovalent copper-fluorine compounds in the condensed phase? How are they accessible?

(II) How is CuF_2 prepared? Comment on its thermal and hydrolytic stability.

(III) Which trivalent copper-fluorine compounds are known? How are they prepared? Comment on their stability.

(IV) Which binary silver fluorides are known and how are they prepared? Are they all stable?

(V)   Give examples for the oxidation power of di- and trivalent silver-fluorine species. Under which conditions are these reactions observed and which species are present, respectively?

(VI) How is AuF_3 prepared? Describe its solid-state structure. Which species can be observed in the gas phase?

(VII) How is AuF_5 accessible? Which products are observed, if it is reacted with NOF, XeF_2 or a mixture of SbF_5/IF_5, respectively?

10.11 Group 12: Zinc, cadmium, mercury

The elements of group 12 form mainly difluorides, in the case of mercury, monovalent compounds are known as well. The ionic character of the fluorides decreases from zinc to mercury. Moreover, ZnF_2 and CdF_2 are far more ionic than the other halides, whereas a similar behavior is observed for the mercury chlorides and fluorides.

10.11.1 Zinc

Zinc difluoride **ZnF_2** is generated upon reaction of ZnO with hydrofluoric acid, which results in the intermediary tetrahydrate $[Zn(H_2O)_4]F_2$. The latter can be dehydrated at 100 °C, alternatively, anhydrous ZnF_2 is directly accessible, if HF is passed over elemental Zn at red heat. The reaction of Zn with F_2 also yields ZnF_2, which is a colorless solid (mp.: 872 °C; bp.: 1500 °C) in its anhydrous state and crystallizes in the rutile structure type with a lattice energy of 2970 kJ mol^{-1} [2668, 2669]. Molecular ZnF_2 has a linear structure. The water-free salt is nonhygroscopic and dissolves only sparingly in water, whereas the tetrahydrate is more soluble. ZnF_2 is commonly used as mild fluorinating agent. If it is heated under an H_2-atmosphere, it is reduced to Zn and HF. The treatment of ZnF_2 with aqueous alkali solutions yields the basic fluoride Zn(OH)F, whereas hydrolysis at 500 °C directly yields ZnO and HF. The formation of trifluorozincates $M^I[ZnF_3]$ (M^I = Na, K) from aqueous solutions of M^IF and ZnF_2 has been known for long time. Furthermore, it has been described, that the ammonium salt $(NH_4)_2[ZnF_4] \cdot 2\,H_2O$ is obtained from concentrated aqueous ammonia [2670]. Apart from that, numerous further complex salts are accessible, in most cases by mixing of the respective metal fluorides. Generally, they have the compositions $M^I[ZnF_3]$ (M^I = Na, K, Rb, Cs, NH_4, Ag); $M_2^I[ZnF_4]$ (M^I = K, Rb, Cs, NH_4); $M^{II}[ZnF_4]$ (M^{II} = Ca, Sr, Ba) and

$M^{II}_2[ZnF_6]$ (M^{II} = Ba, Pb) [2239, 2241, 2502, 2671–2678]. Analogously, hydrated derivatives of the composition $M^I_2[ZnF_4] \cdot 2\,H_2O$ (M^I = NH_4, K) are obtained, if zinc salts are treated with NH_4F in aqueous solution [2679]. Moreover, the ammonium complexes $NH_4[ZnF_3]$ and $(NH_4)_2[ZnF_4]$ are formed by corrosion of brass in presence of NH_4F [2680]. The hydroxyl ammonium complex $[NH_3OH]_2[ZnF_4]$ is accessible by dissolving zinc powder in aqueous HF (w = 40 %) and subsequent addition of hydroxylammonium fluoride $[NH_3OH]F$ [2681]. In all of the complex salts, zinc is hexacoordinated. The compounds $M^I[ZnF_3]$ crystallize in a distorted perovskite lattice [2682]. Investigations on a molten mixture of KF and ZnF_2 have shown that the anionic complex is mainly described by $[ZnF_3]^-$ or a polymeric derivative thereof [2683]. $K_2[ZnF_4]$ is isomorphic to $K_2[CoF_4]$.

A structurally interesting feature is observed in the solid-state structure of $Ba_2[ZnF_6]$, which contains a two-dimensional network of distorted $[ZnF_6]$-octahedra. Between these anionic layers, double-layers of $[BaF]^+$ cations are found. Thus, the formula of the compound should not be written as $Ba_2[ZnF_6]$ but better as $[BaF]_2[ZnF_4]$, since only four of the fluoride ligands are used to build up the $[ZnF_6]$-octahedra. The other two F-atoms are contributed by neighboring $[ZnF_6]$-units [2677]. A comparable situation is observed in $Pb_2[ZnF_6]$ [2678]. Quaternary compounds of the composition $Ba_2ZnM^{III}F_9$ (M^{III} = Al, Ga) have also been prepared and investigated by X-ray crystallography. The structurally characteristic feature of these compounds is a statistic distribution of $[ZnF_6]$- and $[M^{III}F_6]$-octahedra, which are connected to double chains, forming stairs of four-rings [2684]. In the quaternary $NaCdZn_2F_7$, both cations Cd^{2+} and Zn^{2+} have a different coordination environment. The Zn-coordination is based on $[ZnF_6]$-octahedra, whereas $[CdF_8]$-cubes are found for Cd [2685, 2686].

10.11.2 Cadmium

Cadmium difluoride **CdF₂** is synthesized in the same way as ZnF_2. The reaction of elemental Cd or its salts CdO, $CdCl_2$ or CdS with F_2 also generates CdF_2 [2687]. Moreover, it is obtained in pure form, if Cd is oxidized by NOF_3 [352]. CdF_2 is a solid compound (mp.: 1078 °C; bp.: 1748 °C), which crystallizes in the fluorite structure type with a large lattice energy of 2770 kj mol^{-1} [2669]. In water, it is slightly better soluble than ZnF_2. However, stable hydrates are unknown, as $CdF_2 \cdot 2\,H_2O$ readily loses water. Furthermore, CdF_2 is not significantly hydrolyzed in aqueous solutions [2688]. If it is treated with aqueous KOH at room temperature, it forms the hydroxyfluoride Cd(OH)F [2689]. CdF_2 reacts with aqueous ammonia under formation of the complex $CdF_2 \cdot 2\,NH_3$, which yields $NH_4[Cd(OH)F_2]$ upon hydrolysis [2690]. Reactions of CdF_2 with liquid ammonia in open reaction vessels at –40 °C lead to the formation of the soluble monohydrate $[Cd(NH_3)_6]F_2 \cdot H_2O$ [2691]. Moreover, the solubility of CdF_2 is improved upon formation of a 1:2-adduct with AsF_5 [2500]. Hydrazine complexes of CdF_2 have also been reported [2692]. Mixing of the respective (alkali) metal fluorides with CdF_2

gives rise to fluorocadmate(II) complexes, which are again comparable to their zinc analogues. Thus, they generally have the compositions $M^I[CdF_3]$ (M^I = K, Rb, Cs, NH_4, Tl) and $M^I_2[CdF_4]$ (M^I = K, Rb) [2241, 2693–2695]. Similarly, hydrated tetrafluorocadmate(II) complexes of the composition $M^I_2[CdF_4] \cdot 2\,H_2O$ (M^I = NH_4, K) are accessible from aqueous solutions of cadmium(II) salts and NH_4F. Furthermore, a mixed tetrahalocadmate(II) complex with the formula $(NH_4)_2[CdClF_3] \cdot 2\,H_2O$ is obtained, if $CdCl_2$ is used as source of Cd(II) [2679]. A ternary hydrate of the composition $Cd_2ZrF_8 \cdot 6\,H_2O$ has been prepared from a mixture of ZrO_2 and $CdCO_3$ in aqueous HF and investigated by EPR-spectroscopy [2696]. Apart from that, the formation of cadmium fluoride complexes in aqueous HF has been explored by means of polarography. As a result, the species $[CdF]^+$, CdF_2 and $[CdF_3]^-$ can be observed, depending on the HF-concentration [2697]. At room temperature, the trifluorocadmate(II) complexes $M^I[CdF_3]$ (M^I = Rb, Cs, Tl) crystallize in a cubic perovskite structure [2694]. The potassium salt $K[CdF_3]$ undergoes three phase transitions between −30 °C and 212 °C. The observed modifications all deviate from the perovskite structure, which is found in ideal form above 212 °C [2698]. CdF_2 may be doped by M^{3+} cations of the rare earth metals, which gives rise to semiconducting properties [2699, 2700]. The reaction of CdF_2 with fluorosulphurylisocyanate FSO_2NCO yields the fluoroformylfluorosulphurylimide salt $Cd[N(SO_2F)COF]_2 \cdot 2\,CH_3CN$ [2701].

10.11.3 Mercury

Dimercury difluoride

Dimercury difluoride Hg_2F_2 is either prepared by the reaction of Hg_2CO_3 with HF or treatment of $Hg_2(NO_3)_2$ with NaF. It is a yellow, crystalline solid (mp.: 570 °C), and is furthermore accessible by the oxidation of elemental Hg with TiF_4 or CdF_2 [2702, 2703]. If it is sublimed, Hg_2F_2 disproportionates to Hg and HgF_2 [2704]. Moreover, it is sensitive toward hydrolysis, as it forms HF, Hg and HgO upon contact with moisture. However, analogously to the silver(I) halides, Hg_2F_2 is significantly more soluble in water than the respective chloride, bromide and iodide. In addition, it is sensitive toward light; at 240 °C Hg_2F_2 also attacks glassware. In the solid-state structure, linear F-Hg-Hg-F molecules are found [2705]. Quantum chemical calculations have shown, that the stability of $[Hg_2]^{2+}$, and thus Hg_2F_2, is not due to relativistic effects but most likely results from condensed-phase interactions. Conditions for the observation of the lighter $[Zn_2]^{2+}$ and $[Cd_2]^{2+}$ homologues have also been proposed, requiring electronegative substituents and systems, where the respective M^{2+} is disfavored over $[M_2]^{2+}$. In the gas phase, all three M_2F_2 have been calculated to be stable against decomposition [2706, 2707]. Moreover, the influence of the ligand electronegativity on the Hg-Hg bond length and its stability has been examined in more detail [2708, 2709]. In molten $HgBr_2$ (242 °C), Hg_2F_2 undergoes solvolysis, resulting in the formation of stable Hg_2Br_2 along with HgBrF and HgF_2, which are unstable under the reaction conditions [2710]. Similarly, the mixed Hg(II) halides HgBrF and HgFI are formed upon treatment of Hg_2F_2 with Br_2 or I_2 in solid-state reactions. Interestingly, HgFI is in a photochemical equilibrium and reversibly dissociates into Hg_2F_2 and I_2 under the impact of light [2711]. If Hg_2F_2 is reacted with an excess of sulphur or selenium, $Hg_3S_2F_2$ or $Hg_3Se_2F_2$ are obtained, respectively [2712]. The reaction of Hg_2F_2 with the n-perfluoropropyl tetracarbonyliron(II) complex $[(^nC_3F_7)Fe(CO)_4]I$ leads to the formation of the mercury complexes $Hg[Fe(CO)_4(^nC_3F_7)]_2$ as well as $Hg[Fe(CO)_4(^nC_3F_7)](^nC_3F_7)$. The complexes are decomposed by sunlight as well as in presence of excess Hg_2F_2 [2713]. Due to its light sensitivity, Hg_2F_2 has also been used as initiator for radical reactions in organic chemistry, as it is, for example, capable of inducing the decarboxylation of arylacetic acids in photoreactions [2714].

Mercury difluoride

Mercury difluoride **HgF_2** is prepared by fluorination of elemental Hg with ClF_3 [2715]. Apart from that, it is obtained, if Hg(II) compounds are treated with HF or by disproportionation of Hg_2F_2 at 450 °C (Scheme 10.30) [2716, 2717]. HgF_2 is a white solid (decomp.: 645 °C), which crystallizes in a fluorite structure with a lattice energy of $2740\ kJ\ mol^{-1}$ [2669, 2718]. At high pressures of about 4.7 GPa, a phase transition from the cubic fluorite type to an orthorhombic lattice is observed, which is present up to 63 GPa [2719]. Analogously to CdF_2, the solubility of HgF_2 in HF is improved upon formation of an adduct with 2 equivalents of AsF_5 [2500]. HgF_2 is volatile and readily decomposes in aqueous solution to yield HgO and HF. It also reacts with ammonia to form $HgF_2 \cdot 2\ NH_3$, which readily hydrolyzes and finally yields $Hg(NH_2)F$ [2690]. HgF_2 reacts with dry Br_2 (105 °C) and Cl_2 (120 °C), and forms the mixed halides HgBrF and HgClF, respectively. Complex fluoromercurate salts of the type $M^I[HgF_3]$ (M^I = K, Rb, Cs) or $M_2^I[HgF_4]$ (M^I = Rb, Cs) are accessible from mixtures of the corresponding fluorides (Scheme 10.30) [2705, 2720]. The compounds $M^I[HgF_3]$ (M^I = Rb, Cs) have a cubic crystal structure like perovskite, whereas the potassium salt (M^I = K) crystallizes in an orthorhombic lattice [2720]. A variety of related mercury(II) *ortho*-pentafluorotellurate(VI) salts has been prepared and characterized by X-ray crystallography. The main difference to the aforementioned fluoromercurates(II) is the absence of F-bridging in the solid-state structures [2721]. Treatment of a 1:2 mixture of HgF_2 and AuF_3 with KrF_2 yields the hexafluoroaurate(V) complex $Hg[AuF_6]_2$. Reduction of the hexafluoroantimonate compound $Hg[SbF_6]_2$ with H_2 results in the formation of $[Hg_2][Sb_2F_{11}]_2$ [2657]. In anhydrous HF, the analogous hexafluoroarsenate $[Hg_2][AsF_6]_2$ is oxidized by F_2 to yield $[HgF][AsF_6]$, which contains infinite $-[Hg-F-Hg]$-zig-zag chains [2722]. Structurally interesting mercury polycations are accessible, if HgF_2 is reacted with the fluoride acceptors TaF_5 and NbF_5. In SO_2-solution, the intermediary salts of the composition $Hg[MF_6]_2$ (M = Ta, Nb) are reduced to $[Hg_3](MF_5)_2SO_4$ (M = Ta, Nb) by the addition of elemental mercury (Scheme 10.30). Therein, the $[Hg_3]^{2+}$

Hg^{2+} Hg_2F_2 $[HgXe]^{2+}$

HF   450 °C   excess SbF_5, Xe

ClF_3

Hg —————→ HgF_2 CS_2, 100–250 °C   → $Hg(SCF_3)_2$

SO_3 F^- TaF_5 or NbF_5 then Hg, SO_2

SO_2F_2 $[HgF_3]^-$ or $[HgF_4]^{2-}$ $[Hg_3]^{2+}$ or $[Hg_4]^{2+}$

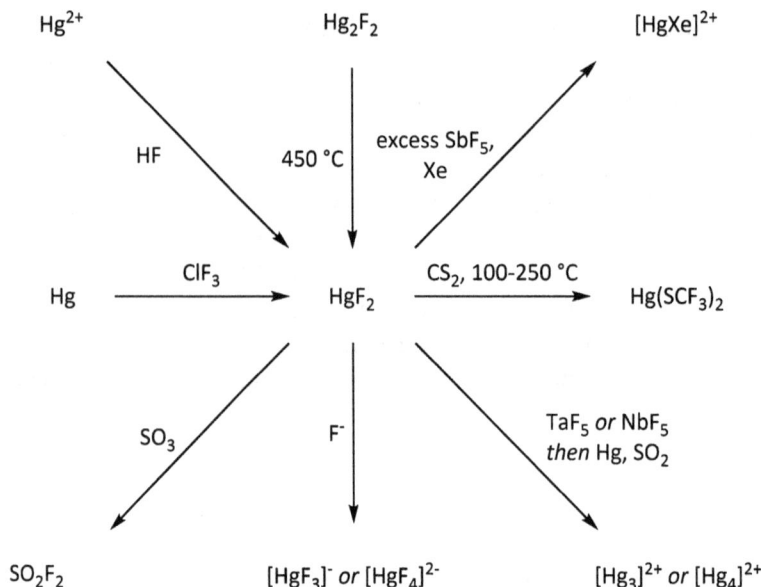

Scheme 10.30: Preparation and selected reactions of HgF_2.

cation is almost linear. Apart from that, the $[Hg_4]^{2+}$ cation also has an almost linear structure and is obtained, if $Hg[Ta_2F_{11}]_2$ is reduced by Hg in SO_2, resulting in the formation of $[Hg_4][Ta_2F_{11}]_2$ [2723]. The reaction of HgF_2 with excess SbF_5 in the presence of Xe yields $[XeHg][SbF_6][Sb_2F_{11}]$, containing the mercury-xenon cation $[HgXe]^{2+}$ (Scheme 10.30) [2724].

HgF_2 is a good and mild fluorinating agent for both organic and inorganic substrates. It fluorinates elements like, for example, Cu, Pb, Sn, Mg, Cr, As or S. If HgF_2 is reacted with sulphur trioxide SO_3, the latter is partly fluorinated to yield sulphuryl fluoride SO_2F_2 (Scheme 10.30) [2725]. The halogen exchange reactions with chlorocarbons, that is, CH_2Cl_2 and $CHCl_3$ have been described long time ago [2716]. Similarly, the reaction of HgF_2 with tetraiodomethane CI_4 gives rise to CF_2I_2 [2726]. Moreover, HgF_2 has been proposed as a potential substitute for CoF_3 in the industrial preparation of fluorocarbons. However, due to its sensitivity toward hydrolysis and its comparably weak fluorinating properties, it has never been used for this purpose [1279]. Many reactions of HgF_2 with unsaturated compounds have been described, showing that it is formally capable of transferring "F_2" to acetonitrile, ethylene, carbon monoxide and dimethyl sulphide [2715]. Accordingly, the systems HgF_2/Cl_2 and HgF_2/Br_2 transfer "ClF" and "BrF" to CN- and SN-triple bonds [2727, 2728]. The addition of "BrF" to alkenes using the HgF_2/Br_2 system has been reported as well [2729]. Furthermore, HgF_2 promotes the addition of CF_3SCl to the triple bonds of chlorine cyanide, ClCN, as well as trichloro- and trifluoroacetonitrile CX_3CN (X = Cl, F) [2730]. Similar to CdF_2, HgF_2 reacts with fluorosulphurylisocyanate FSO_2NCO and yields the fluoroformylfluorosulphurylimide salts $Hg[N(SO_2F)COF]_2 \cdot x\ CH_3CN$ (x = 1, 2), which are more stable than their cadmium analogues [2701]. The chemistry of NSF-building blocks is also strongly connected to HgF_2 [2731]. HgF_2 is also capable of undergoing addition reactions to the double bond of fluoroolefins. The resulting organomercury compounds may be further used as reagents in organic synthesis [2732–2735]. Alternatively, fluoroorganic mercury compounds are accessible, if perfluoroalkyl carbanions are treated with HgF_2 [2736, 2737]. A reaction with cyanogen and cyanogen chloride has also been described. In both reactions, a mixture of perfluoro-2-azapropene and the related amide $Hg[N(CF_3)_3]_2$ has been observed [2738]. HgF_2 also adds to CN-double bonds in $EF_5N{=}CCl_2$ (E = S, Te), resulting in mercury amido-salts of the composition $Hg[N(CF_3)EF_5]_2$ [2739]. The reaction of HgF_2 with CS_2 in a temperature range between 100–250 °C yields the trifluoromethanethiolate salt $Hg(SCF_3)_2$ (Scheme 10.30). The latter is readily transmetalated to yield the copper salt $CuSCF_3$, which is a valuable reagent for the introduction of SCF_3-groups into organic molecules [2740]. HgF_2 has also been used in photochemical fluorination reactions of organic molecules [2741]. Starting from a 1:3-mixture of HgO and HgF_2, the oxyfluoride Hg_4OF_6 is prepared in a solid-state reaction at 380 °C. It has a deep red color and is relatively stable toward air [2742].

10.11.4 Zinc, cadmium and mercury in higher oxidation states

The existence of high-valent group 12 fluorides has been investigated by computational chemistry for quite a long time. The trifluorides of zinc, cadmium and mercury have all been calculated to be unstable. It has been predicted that a stabilization of HgF_3 is achieved by dimerization, however, Hg_2F_6 is still unstable, since quantum chemistry revealed exothermic decomposition pathways. The formation of anionic $[MF_4]^-$ and $[MF_5]^{2-}$ (M = Zn, Cd, Hg) has been identified as the most promising approach toward oxidation state +III in group 12 fluorides, since these species are stabilized with respect to neutral MF_3 [2743]. In another quantum-chemical study, it has been noted that the choice of an appropriate counter ion for such fluorometallates might be challenging. The anion $[ZnF_4]^-$ has been calculated to be stable in the gas phase, however the formation of the potassium salt $K_2[ZnF_4]$ with zinc in oxidation state +II would always be favored over the generation of trivalent $K[ZnF_4]$. In contrast to ZnF_3, the even more exotic $Zn[AuF_6]_3$ has been predicted to be a stable compound, clearly containing zinc in oxidation state +III [2744]. Alternatively, high pressure has been proposed as a tool to stabilize higher oxidation states. For example, HgF_3 has been calculated to be stable at pressures above 73 GPa, whereas the more stable HgF_4 (see below) has already been found to be stable in a pressure range from 38 to 73 GPa [2745].

The situation is comparable for the tetrafluorides of zinc and cadmium. Quantum-chemical calculations have revealed, that the elimination of F_2 from ZnF_4 and CdF_4

is strongly exothermic [2746]. Thus, it became quite clear that Zn and Cd will most likely not be able to extend oxidation state +II, whereas a potential mercury tetrafluoride **HgF$_4$** has been calculated to be thermodynamically stable in the gas phase [2746, 2747]. In fact, it has been possible to generate square-planar HgF$_4$ molecules by treatment of mercury atoms with fluorine at 4 K in a matrix experiment [2748]. DFT-calculations on the superheavy, artificial element copernicium predict a significantly larger stability for CnF$_4$ compared to HgF$_4$ [2749]. However, as the most stable Cn-isotope has a half-life of far less than a minute and is only accessible on a small scale of a few atoms, it is unlikely that the (fluorine) chemistry of it will ever play a role.

10.11.5 Study questions

(I)   How are the fluorides ZnF$_2$ and CdF$_2$ prepared? Which structural behavior do they exhibit in the solid state?

(II)   Which fluorometallates are based on ZnF$_2$ and CdF$_2$? How are they prepared? Is there a common structural feature?

(III) Which stable mercury fluorides are known and how are they synthesized? Which solid-state structures are observed for these compounds?

(IV)   Which products are observed, if HgF$_2$ is reacted with fluoride donors? What happens upon reaction of HgF$_2$ with fluoride acceptors, such as TaF$_5$ or NbF$_5$? Is there a similar reaction with SbF$_5$ in presence of Xe?

Part C: **Organic fluorine chemistry**

In the third section of this manuscript, the chemistry of fluoroorganic compounds is presented with a focus on general properties and applications as well as of course preparative methods to access them. This includes an overview over common fluorinating agents for the main approaches of radical, nucleophilic and electrophilic fluorination. Consequently, the preparative chapters are partitioned after principal types of fluorine introduction. In addition, transformations are presented, which do not introduce but selectively remove fluorine substituents from a highly fluorinated molecule. A few oxidative and reductive conversions are discussed, too.

https://doi.org/10.1515/9783110659337-013

11 Fluoroorganic compounds – unusual properties and versatile applications

11.1 Physical properties of fluoroorganic compounds [2750]

The physical properties of fluorinated organic compounds are in many cases very different to those observed for nonfluorinated molecules. In a very simple way, the majority of these changes can be rationalized by the atomic properties of the fluorine atom. For example, it has a high ionization potential and a low polarizability, which consequently leads to very weak interactions between perfluorinated moieties. Due to the electronegativity of fluorine, C-F bonds are always strongly polarized; nevertheless, perfluorocarbons belong to the most nonpolar compounds known. This is explained by the fact, that the dipole moments in a perfluorocarbon molecule cancel each other. Consequently, partly fluorinated compounds can have a significant polar character. Overall, the physical properties of hydrofluorocarbons are often very different from those of both, perfluorocarbons and hydrocarbons. Finally, the properties of fluoroorganic compounds are also determined by the excellent match of the corresponding 2s and 2p orbitals of carbon and fluorine.

Since fluorine is a larger substituent than hydrogen, the structural behavior of perfluorocarbons differs from that of their nonfluorinated analogues. Typically, linear hydrocarbons adapt a linear zig-zag conformation, which is not the case for the corresponding perfluorocarbons. The steric repulsion of fluorine substituents in relative 1,3-positions renders a zig-zag conformation very unfavorable, so that nonbranched perfluorocarbons have a helical structure. Expectedly, helical chains are less flexible than zig-zag conformations, so that perfluorocarbon molecules occur as rigid rods.

11.1.1 Boiling points and dipole moments

An interesting behavior is observed for the physical properties of organofluorine compounds, for example, boiling point, density or dielectric constant. In general, the densities of perfluorocarbons, such as C_6F_{14} ($\rho = 1.67\,\text{g cm}^{-3}$), are up to 2.5 times higher than those of the corresponding nonfluorinated analogues (C_6H_{14}: $\rho = 0.66\,\text{g cm}^{-3}$). In contrast, the low polarizability and the nonpolar character of saturated perfluorocarbons are perfectly represented by significantly lower boiling points and dielectric constants (C_6F_{14}: bp. = 57 °C; $\varepsilon = 1.69$), compared to hydrocarbons (C_6H_{14}: bp. = 69 °C; $\varepsilon = 1.89$) [2751]. Especially the low boiling point is remarkable, since the perfluorocarbon is almost four times as heavy as C_6H_{14}. In another comparison, the boiling points of almost equally heavy CF_4 (MW = 88 g mol^{-1}; bp. = –128 °C) and C_6H_{14} (MW = 86 g mol^{-1}; bp. = 69 °C) dramatically differ by nearly 200 °C. Apart from that, branching does not significantly influence the boiling points of fluorocarbons, since

https://doi.org/10.1515/9783110659337-014

all three isomeric perfluoropentane derivatives boil at 29 to 30 °C. This is in large contrast to the corresponding hydrocarbons, where the boiling points range from 9.5 °C (neopentane) to 36 °C (*n*-pentane). The same general observation is made for branched ethers and ketones. In some aspects, perfluorocarbons may be well compared to the noble gases. Having similar molecular masses, the fluorinated molecules show boiling points, which are only about 25–30 °C higher. Apart from that, the low reactivity of perfluorocarbons reminds on the chemical behavior of the noble gases.

As already implied above, the physical properties of partly fluorinated molecules are strongly determined by the polar character of the compounds. This is nicely underlined by the behavior of the boiling points of the methane derivatives $CH_{4-n}F_n$ (n = 0–4), where the maximum is found for CH_2F_2 (bp.: –52 °C), whereas a decrease is observed for both, lower and higher fluorine contents (Table 11.1). This is in clear contrast to the other series of halomethanes $CH_{4-n}X_n$ (X = Cl, Br), where the boiling points increase with halogen content (Table 11.1). Accordingly, the dipole moments of fluorinated methane behave the same way, which means that CH_2F_2 is the most polar compound. Some general correlations may be made between the polarity and boiling points of fluorinated aliphatic compounds, however, there are also many exceptions. For example, CH_3CF_3 (μ = 2.32D) and CH_3CHF_2 (μ = 2.27D) have a comparable dipole moment, but the boiling point of 1,1-difluoroethane (bp.: –25 °C) is more than 20 °C higher than that of 1,1,1-trifluoroethane (bp.: –47 °C). A rather irregular behavior is also observed for fluorinated ethylene derivatives, where CH_2 = CF_2 has a lower boiling point than tetrafluoroethylene (–83 °C vs. –76 °C). Nevertheless, a reliable correlation is found for (*E*)/(*Z*)-isomers, where the compound with the higher dipole moment typically boils at a higher temperature. Other physical properties, such as the heat of vaporization are only weakly affected by fluorination. If so, the changes are only poorly predictable.

Table 11.1: Boiling points of halogenated methanes.

| | Boiling point [°C] | | | | |
	n = 0	*n* = 1	*n* = 2	*n* = 3	*n* = 4
$CH_{4-n}F_n$	−161	−79	−52	−82	−128
$CH_{4-n}Cl_n$		−24	40	61	77
$CH_{4-n}Br_n$		4	98	150	190

11.1.2 Solvent properties

Organofluorine compounds show interesting solvent properties, too. Due to their extreme non-polarity, perfluorocarbons are generally poor solvents, however, the situation changes for compounds with very low cohesive energies, such as gases, other highly fluorinated organic molecules, but also molecular inorganic fluorides like WF_6.

The high solubility of O_2 in perfluorinated solvents has led to applications as oxygen carriers, for example, in artificial blood [2752, 2753]. Apart from that, perfluorinated compounds are virtually insoluble in water and HF and mostly only sparingly soluble in hydrocarbons, so that a third phase is typically observed, when perfluorocarbons are mixed with water and organic solvents. This miscibility gap is exploited in fluorous chemistry, which has gained a lot of attraction and importance in the past decades.

11.1.3 Surface tension and surface energy

Liquid perfluoroalkanes have very low surface tensions, which is due to their weak intermolecular interactions. Therefore, they are capable of wetting virtually all surfaces. In the same way, the anti-stick properties of solid perfluorocarbon materials, especially PTFE, result from their low surface energies. This property is exploited in the preparation of frying pans, but is also beneficial for many other applications. However, the surface energies of solid perfluorocarbons, including fluoropolymers, strongly depend on the fluorine content, as they significantly increase from $-(CF_2CF_2)_n-$ to $-(CH_2CHF)_n-$ and finally nonfluorinated polyethylene. Accordingly, polarizable substituents, such as chlorine, also significantly increase the surface energy of the polymer, since $-(CF_2CFCl)_n-$ has about the same surface energy like polyethylene. If a highly fluorinated carbon chain is attached to a hydrophilic group, fluorosurfactants are obtained, which significantly reduce the surface tension of water. This effect is more pronounced than for non-fluorinated hydrocarbon-based surfactants.

11.1.4 Lipophilicity

The introduction of fluorine substituents also changes the lipophilicity of a molecule, which is often a desirable effect in the design of biologically active compounds. Typically, mono and trifluoromethylation of saturated, aliphatic moieties lead to a decreased lipophilicity, whereas the contrary effect is observed for fluoroaromatics as well as compounds, which are fluorinated adjacent to functions with π-electrons, including C=C bonds. For α-fluorinated carbonyl compounds, the situation is rather complicated, so that a precise generalization is impossible to make. Similar to the boiling points, the lipophilicity of partly fluorinated molecules, especially hydrofluorocarbons, may be correlated with the dipole moment. Consequently, more polar compounds tend to be less lipophilic, so that CH_2F_2 has the highest hydrophilicity in the series of fluorinated methanes $CH_{4-n}F_n$ ($n = 0-4$) [2754].

11.1.5 Acidity and basicity

Due to the inductively, strongly electron-withdrawing effect of fluorine substituents and fluorinated groups, the acidities of corresponding alcohols, acids and amides are significantly increased compared to the nonfluorinated compounds (Table 11.2). For example, perfluoro *tert*-butanol is more than thirteen orders of magnitude more acidic than ^{t}BuOH. The acidity is especially increased by the presence of the strongly electron-withdrawing, fluorinated sulphonyl groups -SO_2F and -SO_2CF_3, as for example, $(CF_3SO_2)_2CH_2$ ($pK_a = -1$) is more acidic than CF_3COOH and $(CF_3SO_2)_2NH$ is a very acidic amide. An extreme acidity is furthermore observed for the Brønsted superacids CF_3SO_3H ($H_0 \approx -14$), FSO_3H ($H_0 \approx -15$) as well as mixtures of the latter with SbF_5 ("magic acid", $H_0 \approx -20$).

Table 11.2: Acidities and basicities of selected fluoroorganic compounds and their nonfluorinated analogues.

Acids	pK_a ($X = F$)	pK_a ($X = H$)
CX_3COOH	0.52	4.76
C_6X_5COOH	1.75	4.21
CX_3CH_2OH	12.4	15.9
$(CX_3)_2CHOH$	9.3	16.1
$(CX_3)_3COH$	5.4	19.0
C_6X_5OH	5.5	10.0
Bases	pK_b ($X = F$)	pK_b ($X = H$)
$CX_3CH_2NH_2$	8.1	3.3
$C_6X_5NH_2$	14.4	9.4

Fluorination does not only lead to an increased acidity of the aforementioned compound classes, but also results in a decreased basicity of amines, ethers and carbonyl compounds (Table 11.2). The basicities of $CF_3CH_2NH_2$ and $C_6F_5NH_2$ are about five orders of magnitude lower than those of the corresponding nonfluorinated molecules. Consequently, perfluorinated secondary amines of the type $(R^F)_2NH$ and perfluoropyridine do not react with HCl or BF_3, in the same way, no Lewis-basic properties are observed for tertiary amines $(R^F)_3N$. In general, perfluorinated ethers $(R^F)_2O$ and amines $(R^F)_3N$ behave more like perfluorocarbons, so that their chemical inertness has led to some special commercial applications. However, the introduction of fluorine does not significantly affect the nucleophilicity of amines.

11.1.6 Study questions

(I)   By which fundamental properties is the extraordinary behavior of many fluoroorganic compounds typically rationalized?

(II)   How does the molecular structure of nonbranched perfluorocarbons differ from that of the corresponding nonfluorinated analogues and which consequences does this have on the flexibility of the molecules? Briefly explain this observation.

(III) How do the boiling points change from a non-fluorinated hydrocarbon toward a perfluoro-carbon with the same number of carbon atoms? Is there a comparable behavior for chlori-nated and brominated compounds?

(IV) Why do perfluorocarbons exhibit extraordinary solvent properties? What can they be used for?

(V) What is the reason for the very low surface tensions and energies of liquid perfluorocarbons and solid compounds, respectively? What can these properties be used for?

(VI) How does the introduction of fluorine substituents influence the acidity and basicity of or-ganic compounds? Give some examples.

11.2 Chemical properties and stability of fluoroorganic compounds and reactive intermediates [2750]

The introduction of fluorine into organic moieties typically influences their chemical behavior, that is, the stability of molecules or reactive intermediates. Apart from that, fluorination may give rise to steric interactions, which might be either stabilizing or destabilizing in nature, depending on the substrate, of course.

11.2.1 Fluorine at sp^3-carbons

In saturated systems, the C-F bond is the strongest carbon-element single bond. It is typically by about 105 kJ mol^{-1} more stable than a C-Cl bond, thus alkyl fluorides are 10^2 to 10^6 times less reactive in S_N1- and S_N2-reactions. However, this reactivity can be significantly increased by acid-catalysis, when the fluoride displacement is assisted by H-bonding. The introduction of fluorine also increases the stability of adjacent C-F and C-O bonds, as for example, the C-F bond energy increases from about 450 kJ mol^{-1} in CH$_3$F to almost 550 kJ mol^{-1} in CF$_4$. In the same way, the stabilization is represented by decreasing C-F bond-lengths in the series CH$_3$F (140 pm) > CH$_2$F$_2$ (137 pm) > CHF$_3$ (135 pm) > CF$_4$ (130 pm). This effect may be best described by very similar 2s- and 2p-orbital sizes of carbon and fluorine atoms, leading to a very good overlap and hence back-donation of electron density from the lone-pairs of the fluorine atoms. Conse-quently, several dipolar resonance structures may be formulated for highly fluorinated molecules. Thus, this resonance stabilization is also one of the reasons for the prefer-ential accumulation of fluorine substituents in partly fluorinated compounds, which is observed upon rearrangement (cf. Section 13.1.1).

However, α-fluorination has only a small impact on the strength of C-H or C-X bonds (X = Cl, Br). In contrast, β-fluorination significantly stabilizes C-H bonds, whereas only small effects are observed for C-F bonds. For example, the C-H bond in (CF$_3$)$_3$CH is about 63 kJ mol^{-1} stronger than that in the corresponding nonfluo-rinated compound. In most cases, the introduction of fluorine slightly strengthens aliphatic C-C bonds, except for those substrates, where strain energies are increased, that is, cyclopropanes and epoxides.

The reactivity of saturated substrates toward nucleophilic substitution reactions is reduced by both, α- and β-fluorination. Comparing CF_3CH_2Br and CH_3CH_2Br, the fluorinated compound reacts by about five orders of magnitude slower with NaI in acetone. Under similar conditions, no halide exchange is observed for substrates of the type RCF_2Br. The decreased reactivity of the aforementioned substrates may be explained by steric shielding of the carbon center on the one hand, but also by inductive effects. Finally, the strong $C(sp^3)$-F bond contributes to the stability of CF_3- and CF_2H-groups toward nucleophilic substitution reactions and hydrolysis. Therefore, fluoride displacement in polyfluoroalkanes requires rather drastic conditions and one- or two-electron transfer processes, involving, for example, radical-ion mechanisms or carbene-mediated ion-chain processes.

The thermal stability of perfluorocarbons is determined by the presence of very strong C-F bonds. The only stability limit is given by the strength of the C-C bonds, which generally decreases with increasing chain length, but also for branched molecules. Consequently, CF_4 is the most stable perfluorocarbon, which shows C-F bond homolysis only above 2000 °C. The perfluoroalkanes C_2F_6 and n-C_3F_8 are already pyrolyzed at about 1000 °C, whereas poly(tetrafluoroethylene) readily decomposes above 500 °C. Accordingly, perfluorinated compounds with tertiary C-C bonds as well as highly branched molecules degrade even more easily; sometimes they are less stable than their corresponding hydrocarbon analogues. The decomposition tendency of perfluorinated cycloalkanes strongly depends on the ring size, as c-C_3F_6 liberates:CF_2 at around 170 °C, whereas the cyclobutene homologue c-C_4F_8 only shows slow degradation at 500 °C. Hydrofluorocarbons are thermally less stable than perfluorinated compounds, which is due to their preference to undergo HF-elimination instead of C-C bond cleavage. This decomposition pathway is significantly accelerated by the presence of bases.

Perfluorocarbons are also kinetically stabilized, which is due to the steric and electrostatic shielding of the rather electrophilic carbon-centers by fluorine substituents and their each three lone-pairs. Therefore, nucleophilic attacks are efficiently prevented and perfluorocarbons are mostly inert, for example, against basic hydrolysis. However, due to their rather high electron-affinities, they react with reducing agents, such as alkali metals. A single electron transfer will give rise to an intermediate perfluoroalkyl radical anion, which then decomposes to a fluoride anion and a perfluoroalkyl radical which then undergoes further decomposition. Overall, the formation of metal fluorides and carbon is typically observed under these conditions.

11.2.2 Fluorine at sp²-and sp-carbons

Still, C-F bonds are rather strong in unsaturated compounds, having bond energies of about 485 (fluoroethylene) to 525 kJ mol^{-1} (fluorobenzene). However, the π-bond strength of C=C strongly depends on the degree of fluorination. Interestingly, the π-bond dissociation energy is very similar for the non- and partially fluorinated olefins $CH_2 = CH_2$ (ca. $270\,kJ\,mol^{-1}$) and $CH_2 = CF_2$ (ca. $263\,kJ\,mol^{-1}$), respectively. A significant decrease is observed for the π-bond strength of $CF_2 = CF_2$, which has a bond dissociation energy of roughly $222\,kJ\,mol^{-1}$. In general, monofluorination of olefins leads to a stabilization of the π-bond, whereas vicinal difluorination as well as trifluorination have a destabilizing effect. Fluorinated alkynes are generally destabi-lized, since mono- and difluoroacetylene are both explosive compounds and perfluoro 2-butyne is a very reactive dienophile.

Due to repulsive interactions between the lone-pairs of fluorine and the π-system of the olefin, fluorinated alkenes are destabilized to a certain extent. Consequently, fluorine substituents at unsaturated carbon atoms are generally less favored than at sp^3-carbons and partly fluorinated olefins will tend to rearrange to isomers with a re-duced number of $C(sp^2)$-F bonds (cf. Section 13.1.2).

The introduction of fluorine also influences the strength of carbon-heteroatom π-bonds. For example, the π-bond in carbonyl fluoride COF_2 is significantly stronger than in formaldehyde. In the same way, C=O bonds are stabilized in acid fluo-rides by π-backdonation, which can also be formulated by resonance structures. Therefore, acid fluorides are chemically more stable than acid chlorides. However, the substitution with a fluoroalkyl group, for example, CF_3 destabilizes the C=O bond, which is, for example, observed for hexafluoroacetone. Its C=O bond is sig-nificantly weaker than in acetone, and thus, $(CF_3)_2CO$ readily undergoes addition reactions.

11.2.3 Fluorinated carbocations

The presence of a fluorine substituent in α position stabilizes carbocations by reso-nance, which is more pronounced than the opposing destabilizing inductive effect (Table 11.3). Therefore, the gas-phase stability of methyl cations increases in the or-der $[CH_3]^+ < [CF_3]^+ < [CH_2F]^+ < [CHF_2]^+$. In solution, the α-stabilizing effect of a fluorine substituent is more pronounced than those of the other halogens. Neverthe-less, α-fluorinated carbocations are quite unstable, and alkyl chains are still far bet-ter groups for the stabilization of cationic species. If there is a fluorine substituent in β-position, the carbocation is strongly destabilized by inductive effects, due to which simple β-fluorinated carbocations have neither been observed in the gas-phase, nor in solution (Table 11.3).

In general, fluorinated carbocations are observed as intermediates in electrophilic addition reactions to fluoroolefins or electrophilic aromatic substitutions. However, highly fluorinated olefins are resistant to electrophilic attacks, especially if they are substituted with one or more perfluoroalkyl substituents. This resistance can be ratio-nalized by β-fluorine destabilization of the respective cationic intermediates. For ex-

Table 11.3: Stabilization and destabilization of fluorinated carbocations and carbanions by inductive- and resonance effects.

	Inductive effects	Resonance effects
α-fluoro carbocations	destabilize	stabilize
β-fluoro carbocations	strongly destabilize	–
α-fluoro carbanions	slightly stabilize	destabilize
β-fluoro carbanions	stabilize	stabilize

ample, both tetrafluoroethylene and hexafluoropropene do not react with HF/FSO_3H, and even 3,3,3-trifluoropropene is not protonated by FSO_3H. Indeed, hydrofluoro- and halofluoroethylene derivatives have been shown to react with a variety of electrophiles, however, their reactivity decreases with increasing fluorine content. The same holds for the electrophilic substitution of fluoroaromatics, where pentafluorobenzene still reacts with some electrophiles under forcing conditions, but hexafluorobenzene is resistant to electrophilic attacks, since this would require the formal elimination of "F^+". As a general rule, one can conclude from both α- and β-fluorine effects, that electrophiles will in most cases add regioselectively to fluoroolefins, so that a carbocationic intermediate with a minimized number of β-fluorine substituents is generated.

11.2.4 Fluorinated carbanions

Fluorine substituents in α- and β-position also have an impact on the stability of carbanions, however, the effect is essentially opposite to that described above for carbocations (Table 11.3). In comparison to chlorine- and bromine-substituents, α-fluorine only slightly stabilizes carbanions, which is nicely represented by the acidity in the haloform series $CHBr_3$ (pK_a 22.7) $\approx$ $CHCl_3$ (pK_a 22.4) > CHF_3 (pK_a 30.5). The reason for this observation is the electron-pair repulsion of the fluorine substituent. This process dominates over the inductive stabilization of the carbanion. In general, α-fluorocarbanions prefer to adapt a pyramidal structure, as the π-repulsion is maximized in planar species. Consequently, the pyramidal $[CF_3]^-$ anion has been calculated to have a large inversion barrier of more than $400\,kJ\,mol^{-1}$, whereas the inversion of the methyl anion occurs virtually barrierless. Despite its rather small stabilizing effect compared to the other halogens, α-fluorine substituents are still far more beneficial for the stabilization of a carbanion than a hydrogen substituent. This becomes clear, as the acidity of CF_3H is about 40 orders of magnitude stronger than that of methane. However, the situation might change for planar carbanions, where an α-fluorine substituent can have a destabilizing effect, as it is observed for fluorene, which is more acidic than 9-fluorofluorene.

The presence of a fluorine substituent in β-position stabilizes carbanions by inductive but also by resonance effects, that is, negative hyperconjugation. The stabilizing effect of β-fluorine is rationalized by the comparison of the acidities of selected hydrofluorocarbons in methanol solution. Fluoroform CF_3H, which does not possess a β-fluorine and $1H$-perfluoroheptane $CF_3(CF_2)_5CF_2H$, which only has two of these substituents have a rather low acidity ($pK_a \approx 30$). Instead, the acidity significantly increases for $2H$-perfluoropropane ($pK_a \approx 20$) and $2H$-perfluoro-2-methylpropane ($pK_a \approx 11$), having six and nine β-fluorine substituents, respectively. A very impressive example is 1,2,3,4,5-*pentakis*(trifluoromethyl)-1,3-cyclopentadiene, which is about 18 orders of magnitude more acidic than 1,3-cyclopentadiene itself. In contrast, 1,2,3,4,5-pentafluoro-1,3-cyclopentadiene is only slightly more acidic than the parent, nonfluorinated molecule. These examples clearly underline that the stabilizing effect of β-fluorine substituents is far more pronounced than that of α-fluorine, regardless of the geometry of the carbanion. To date, a number of stable poly- and perfluoroalkyl salts is known and has been isolated, also including several fluorinated alkoxides and thiolates. Many of them are used for preparative purposes to transfer the respective fluoroalkyl groups.

The stabilization of fluorinated carbanions by both α- and more so by β-fluorine substituents helps to explain, why fluorinated olefins as well as fluoroaromatics are prone to undergo attacks by nucleophiles. In addition, fluorine substituents increase the electrophilicity of the double bond, of course. In general, fluoroalkenes react with nucleophiles in a way that the number of β-fluorine atoms is maximized in the intermediate carbanionic species, which therefore gives rise to a general regioselectivity rule. Consequently, the reactivity of unsaturated perfluorocarbons toward a nucleophilic attack increases in the order $CF_2 = CF_2 < CF_2 = CF(CF_3) < CF_2 = (CF_3)_2$ as well as $CF_2 = CF_2 < CF_2 = CFCl < CF_2 = CFBr$. For unsymmetrical olefins, the nucleophilic attack exclusively occurs at the difluoromethylene site. However, even if its regioselectivity might be easily anticipated, the actual product formation is rather unpredictable. Depending on the nature of the nucleophile, the substrate and the reaction conditions, the reaction can either follow an addition- or an addition-elimination pathway. The C-H acidity and the stability of carbanions also determine the regiochemistry of base-promoted eliminations of HX from polyfluoroalkanes. For example, the reaction of $CF_3CH_2CHBrCH_3$ with KOH in EtOH gives rise to $CF_3CH=CHCH_3$ and $CCl_2HCCl_2CHF_2$ produces $CCl_2 = CClCHF_2$.

Overall, one can conclude, that negatively charged intermediates have an increased importance in fluorocarbon chemistry, which is apparently due to the stability of fluorinated carbanions. This is in contrast to hydrocarbon chemistry, where electrophilic processes appear to be more important in terms of isomerizations, oligomerizations but also addition- or Friedel–Crafts reactions. In fluoroorganic chemistry, such transformations are preferentially promoted by F^-, and hence involve carbanions.

11.2.5 Fluorinated radicals

On a thermodynamic viewpoint, fluorination only leads to small changes in the stability of radicals, since fluorinated species are not more stable than their nonfluorinated analogues, in contrast, they are even more destabilized, often. However, kinetic effects can lead to a significant stabilization of fluorinated radicals. Thus, the presence of fluorine can significantly influence reactions that proceed via radical intermediates. Structurally, carbon-centered radicals, which have a fluorine substituent in α-position prefer to adapt a pyramidal geometry to minimize repulsion. However, this effect is less pronounced than in carbanions. If fluorine is attached in β-position, the geometry of the radical is only slightly affected.

The inductive effect of fluorine leads to a deactivation of the C-H abstraction by electrophilic radicals, such as $Cl^{\cdot}$, $Br^{\cdot}$ or $HO^{\cdot}$. This is rationalized by the relative reaction rates (k_{rel}) in the tertiary C-H abstraction of $(CH_2X)CH(CH_3)_2$ (X = F, Cl, Br, CH_3) by $Br^{\cdot}$, where the reactivity significantly increases in the order F (k_{rel} = 1) < Cl (11) < CH_3 (27.5) < Br (220). In the presence of multiple deactivating fluorine substituents, the radical C-H abstraction can even be more impressively decelerated. For example, the H-abstraction of CH_3CH_3 with $Cl^{\cdot}$ is more than four orders of magnitude faster than in case of CH_3CF_3. The situation changes, if the C-H abstraction occurs with nucleophilic radicals, for example, $CH_3^{\cdot}$, since fluorine can have an activating effect under these conditions. For example, the reaction of CF_3H with electrophilic $Br^{\cdot}$ is slower than that of CH_4, however, upon reaction with nucleophilic $CH_3^{\cdot}$, fluoroform proves to be more reactive than methane. A similar behavior is observed for free-radical additions to olefinic double bonds. The electron-deficient tetrafluoroethylene reacts about 100 times faster with nucleophilic $CH_3^{\cdot}$ than with strongly electrophilic $CF_3^{\cdot}$. Consequently, electrophilic radicals like $CF_3^{\cdot}$ preferentially react with electron-rich alkenes. Apart from their electrophilic character, fluorinated radicals also differ from their hydrocarbon analogues regarding their tendency to undergo migrations or disproportionation reactions. In a few rare examples, fluorine-atom migration is observed, for example, it has been found that $CH_2FCF_2^{\cdot}$ can rearrange to $^{\cdot}CH_2CF_3$. However, a disproportionation reaction of fluorinated radicals by F-transfer is never observed. Thus, the reaction of perfluoroalkyl radicals $R^FCF_2CF_2^{\cdot}$ with alkyl radicals like $CH_3CH_2^{\cdot}$ will give rise to $R^FCF_2CF_2H$ and $CH_2 = CH_2$, but not to $R^FCF = CF_2$ and CH_3CH_2F. Consequently, two perfluoroalkyl radicals $R^FCF_2CF_2^{\cdot}$ will not disproportionate to give a saturated perfluorocarbon $R^FCF_2CF_3$ and a perfluoroolefin $R^FCF = CF_2$. The latter behavior is important for the production of fluoropolymers, since the free-radical polymerization cannot be terminated by a disproportionation step, and thus allows the preparation of high molecular weight products.

Apart from preparative chemistry, fluorinated radical intermediates also play an important role in the atmosphere. Typically, hydrofluorocarbons and hydrochlorofluorocarbons are degraded upon initial reaction with $HO^{\cdot}$ radicals. Therefore, refrigerants and other commercially used, volatile compounds should ideally react fast with hydroxyl radicals to have a low impact on global warming (cf. Section 7.2).

11.2.6 Fluorinated carbenes

The presence of fluorine substituents also has an impact on the stability and the structure of carbenes. Most importantly and in contrast to their hydrocarbon analogues, all α-fluorinated carbenes have a singlet ground-state. This is due to resonance stabilization, which is comparable to that in α-fluorinated carbocations. Consequently, $:CF_2$ is the least electrophilic species among the singlet carbenes $:CX_2$ (X = F, Cl, Br, H). The electrophilicity within the halogenated carbenes increases in the order F < Cl < Br, where $:CH_2$ is still more electrophilic than $:CBr_2$. Thus, $:CHF$ is more electrophilic than difluorocarbene, but less so than $:CH_2$. In general, α-fluorinated carbenes show the same reactivity like other singlet carbenes, that is, they stereospecifically add to simple alkenes to form fluorinated cyclopropanes, but they do not insert into C-H bonds. Due to the resonance stabilization by fluorine, $:CF_2$ is the most selective electrophilic carbene known. In contrast, $(CF_3)_2C:$ is a very electrophilic triplet-species, which both adds to olefins and inserts into C-H-bonds. Without any doubt, the preparatively most important fluorocarbene is $:CF_2$, which plays an important role in the industrial production of fluoroolefins, especially $CF_2 = CF_2$ and hexafluoropropene. In turn, the thermal decomposition of perfluorocarbons can also involve fluorinated carbenes.

11.2.7 Study questions

(I)   How strong are C-F bonds in saturated systems? Which consequences does this have on their reactivity in nucleophilic substitution reactions, if they are compared with chlorinated compounds?

(II)   How does the presence of fluorine substituents in α- or β-position of saturated substrates influence the reactivity of other carbon-heteroatom bonds, for example, C-Br?

(III)   Comment on the thermal stability of perfluorocarbons and hydrofluorocarbons. Which limits do exist?

(IV)   Why are perfluorocarbons virtually inert against nucleophilic attacks? With which type of reagents do they react?

(V)   Which effect does the presence of fluorine substituents have on the stability of olefins and carbonyl compounds? What is the main reason for this behavior?

(VI)   How does α- and β-fluorination influence the stability of carbocations? Where are they principally observed?

(VII)   Explain the stabilization of carbanions by α- and β-fluorine substituents. By which property is the stabilizing effect of β-fluorination easily represented?

(VIII) How does the stability of fluorinated carbanions help to predict the regioselectivity of nucleophilic attacks toward fluorinated olefins?

(IX)   Are fluorinated radicals generally more stable than their nonfluorinated analogues? Which behavior of perfluoroalkyl radicals is beneficial for the preparation of fluoropolymers?

(X)   Which electronic ground-state do all α-fluorinated carbenes exhibit? Briefly explain this observation. Which reactions do the carbenes undergo?

11.3 Applications of organofluorine compounds [2755]

11.3.1 Versatile uses of halo- and hydrofluorocarbons

Based on their high volatility and their low chemical reactivity, chlorofluorocarbons (CFCs) have been the first fluoroorganic compounds, which have been prepared on technical scale. They are also known as "freons" and have especially been applied as refrigerants, but they also found uses as propellants or foaming agents. Further benefits of chlorofluorocarbons include that they are non-flammable and also non-toxic, however, they significantly contribute to the ozone depletion in the atmosphere. Therefore, their use and production are regulated nowadays (cf. Section 7.1). At the peak of their use, chlorofluorocarbons have been produced in a worldwide scale of about 1 million tons per year.

In former times, bromofluorocarbons, also known as "halons," have been widely used as efficient fire extinguishing agents, as they are both, nontoxic and non-flammable. The fire-extinguishing properties of halons arise from the homolytic C-Br bond dissociation energy of about 270 kJ mol^{-1}, so that the cleavage of C-Br bonds withdraws heat from a fire. Nowadays, this application is prohibited for environmental reasons (cf. Section 7.1). Another interesting application of halons is found in the field of medicine, as bromofluorocarbons are nontoxic and proved to be suitable for inhalative narcotization.

With the Montreal protocol, the use of ozone-depleting compounds has been strongly restricted since 1987. However, due to their outstanding and unique properties, many halofluorocarbons, especially CFCs, could not be replaced from one day to another. As a general strategy, efforts have been made to substitute harmful compounds by non-ozone-depleting agents with shorter atmospheric lifetimes. In many cases, hydrofluorocarbons (HFCs) as well as fluorinated ethers have been found and established as efficient replacements for refrigeration and foam-blowing processes.

11.3.2 Fluoropolymers

A very large amount of fluoroorganic compounds is synthesized for the preparation of fluoropolymers. The most important one is certainly polytetrafluoroethylene (PTFE), which is extremely stable against even the most aggressive chemicals, such as F_2, UF_6, molten alkali metal hydroxides or hot mineral acids. It can be used for a large variety of applications from temperatures near absolute zero up to 260 °C [2756]. However, at higher temperatures it will decompose thermally (cf. Section 11.2.1). Apart from its chemical stability, PTFE also has other interesting and useful attributes, such as a very low surface energy, which results in low friction and antistick properties. Therefore, one of the first applications of PTFE was the antistick coating of frying pans. Another application of PTFE is the processing to fibers, which can be used for the production of functional clothes, for example, waterproof jackets. Polytetrafluoroethylene with a lower molecular mass (3.000 to 50.000 Da) is a very efficient lubricant, which is widely used and shows an extremely high resistance against chemical degradation. However, PTFE has also some significant drawbacks, for example, a very high melt viscosity. Thus, PTFE devices cannot be prepared by extrusion, but require sintering of PTFE powder under pressure. As a consequence, PTFE components, for example, labware, are always opaque.

If tetrafluoroethylene is copolymerized with trifluorovinyl ethers, polymers with elastomeric properties are formed. In general, these copolymers show both, thermal and chemical resistance and can be processed by extrusion and molding. Typically, they find applications in coatings and seals. Another important fluoropolymer is poly-chlorotrifluoroethylene (PCTFE), which is typically used as thermoplastic or lubricant. Compared to PTFE, PCTFE has the advantage that it can be processed by extrusion at 250–300 °C. Polyvinylidene difluoride (PVDF) is produced in high quantities and is used, for example, in stable mechanical components, but also as material for transparent films. Furthermore, oriented films of PVDF have piezoelectric properties and are therefore widely used in electronic devices.

Finally, perfluoropolyethers (PFAs) play an important role, especially in the preparation of transparent labware, which is resistant to aggressive chemicals. Oligomers of perfluoroethers find application in lubricants, since they have low surface energies and friction coefficients combined with a high chemical stability. It is also possible to derive functional materials from PFAs, for example, membranes. An important example for this is Nafion™, which is based on PFA, which has been derivatized by sulphonic acid functions.

11.3.3 Fluorinated dyes

Fluorine-containing dyes have been known for many decades. Based on the electron-withdrawing properties of fluorine substituents as well as fluorinated groups, the electronic levels and colors of dyes can be tuned. The majority of commercialized fluorinated dyes belong to the class of azo-compounds, but anthraquinone-based compounds are also known. Apart from that, fluorinated polymethine cyanines and azomethines have found some applications, for example, in photography. The introduction of perfluoroalkyl- or perfluoroalkylsulphonyl substituents leads to a high hyperpolarizability in intramolecular push-pull chromophores. In addition, the substitution of the dyes with perfluoroalkyl groups typically enhances the lipophilicity, and thus the solubility in organic solvents.

11.3.4 Fluorinated liquid crystals

Liquid crystals play an important role in our everyday life, as they are used for displays in many electronic devices. Currently, most of the liquid crystalline materials contain fluorine, which can either be a part of a polar group or the mesogenic core structure. In the latter case, lateral fluorination typically leads to a decrease of the clearing point of the liquid crystal. This behavior can be rationalized by a decreased length-to-breadth ratio of the corresponding molecules, since a fluorine substituent is larger than a proton. The steric effect of lateral fluorination generally amounts to a reduction of the clearing temperature by 30–40 K per fluorine substituent. In contrast, the fluorination of polar groups gives advantageous effects by the electronegativity of fluorine, and thus the polarity of C-F bonds. In the simplest case, a terminal fluorine substituent on an aromatic moiety can serve as a polar group, leading to an increased dielectric anisotropy. This effect is enhanced, if one or even two further fluorine substituents are attached in *ortho*-position. Moreover, a larger dielectric anisotropy is achieved, if more complex fluorinated groups are attached to the liquid crystalline molecule, such as $-CF_3$, $-OCHF_2$ or $-OCF_3$, but also more exotic functions like $-SF_5$. The latter is currently the most effective group for the induction of polarity in liquid crystalline materials. Fluorinated bridges, that is, $-CF_2-CF_2-$ and $-OCF_2-$ units have also proven to be suitable for the tuning of the properties of liquid crystalline molecules.

However, fluorination does not only improve the properties of the compounds, but also leads to an improved reliability of liquid crystalline materials. Fluorinated liquid crystals interact significantly less with ionic impurities than for example cyano-substituted materials. Thus, the fluorinated derivatives are more reliable and their capacitor behavior is maintained at a higher level for longer times, which results, for example, in reduced flickering of the display.

11.3.5 Organic electronics

In the last decades, fluorinated compounds have also gained importance in the field of organic electronics. This includes several applications, such as organic light-emitting diodes (OLEDs) and organic field-effect transistors (OFETs) but also organic photovoltaics. In organic semiconductors, the introduction of fluorine or fluorinated groups principally decreases the energies of the highest occupied molecular orbitals (HOMOs) as well as the corresponding lowest unoccupied species (LUMOs). This is caused by the negative inductive effect of fluorine. Another effect is observed in aromatic molecules, where the substitution with fluorine gives rise to a resonance effect, meaning that fluorine can donate electron density by π-backdonation. This can lead to a change of the band gap of the semiconductor, as for example, the p-type semiconductor pentacene is converted to a n-type semiconducting species upon fluorination or introduction of perfluoroalkyl substituents. Furthermore, the presence of fluorine also increases the stability of the materials against water or oxygen. Finally, the fluorination of semiconductor materials influences the intermolecular interactions between neighboring molecules. Typically, fluorinated regions of polyaromatics prefer to orient to more electron-rich areas of their neighbors in the solid state. Consequently, a shorter edge-to-edge distance is observed, often. In addition, the orientation of the molecules is influenced positively in many cases, so that nontilted π-stacks are obtained. Of course, the latter is beneficial for an optimal charge-transport. In general, fluorinated organic semiconductors have the benefit, that they are rather stable against chemical as well as photochemical impact.

11.3.6 Biologically active compounds

The introduction of fluorine into biologically active compounds typically has dramatic effects on their properties. For example, it is possible to modulate the lipophilicity, inhibit metabolic pathways but also modify the interactions between the drug and the target structure. In general, fluorinated drugs often have a better bioavailability, an increased selectivity for the target organs, and finally, they need to be administered in lower effective doses. Depending on the actual compound, the properties of the drug can be changed by introduction of a single fluorine atom, but also by substitution with more complex fluorinated groups, such as $-CF_3$, $-CHF_2$ but also heteroatom-based functionalities like $-OCF_3$ or $-SCF_3$. Overall, the beneficial properties of fluorinated drugs and agrochemicals have led to a dramatic increase of the number of fluorine-containing products on the market over the last decades [43, 2757–2762]. Currently, fluorine is found in roughly 25 % of blockbuster drugs; however, this share will certainly further increase in future [2763]. Often, fluorinated pharmaceuticals are based on natural products, such as carbohydrates, steroids, alkaloids or amino acids [2764]. However, a large variety of fluorinated small molecules has also been used for therapeutic applications [2765]. Interestingly, fluorine does not play an important role in the field of flavors and fragrances [2766]. This has been rationalized by the fact, that the odor of a compound is mainly affected by the shape of a molecule. Consequently, the replacement of hydrogen by a fluorine substituent does not significantly change the shape and the size of a compound. However, the replacement of carbonyl-groups by $-CF_2$- in macrocyclic musk molecules has been shown to influence the conformation, and hence the odor of the respective compounds [2767, 2768]. Apart from that, fluorinated compounds might become of relevance in the production of pheromones in the future, as they have the potential to be used in field traps and, under certain circumstances, for the monitoring of insect populations [2766].

11.3.7 Medicinal diagnostics

Fluorinated compounds are also used in medicinal diagnostics, in most cases this applies to radiopharmaceuticals, which are labeled with ^{18}F. This application arises from the ability that fluorine can mimic several other elements in pharmaceuticals, such as hydrogen or oxygen. In addition, the majority of fluorinated compounds and metabolites behave similarly to their non-fluorinated analogues in terms of transport pathways and metabolism. The artificial isotope ^{18}F is a positron (β^+) emitter and has a half-life of about 110 minutes. Upon contact with electrons, the positrons emit two energetic γ-photons with a characteristic energy of 511 keV. Therefore, ^{18}F-radiopharmaceuticals are almost exclusively used for positron emission tomography (PET), which is a valuable method for the diagnosis of, for example, cancer or brain diseases, such as Parkinson's or Alzheimer's disease. For the visualization of areas with a high glucose metabolism, ^{18}F-labeled fluorodeoxyglucose is commonly used as a tracer. In contrast, Parkinson's disease is mainly diagnosed by the use of ^{18}F-dopamine derivatives. Furthermore, ^{18}F-radiolabeling can also serve to trace metabolic pathways of newly developed drugs in the human body.

Apart from radiopharmaceuticals, it is also possible to use nonlabeled organofluorine compounds as contrast agents for medicinal diagnostics. For example, perfluorocarbons with a heavy-atom substituent, for example, $C_8F_{17}Br$, can be used for the X-ray imaging of soft tissues, such as the lungs. In some cases, diagnostics are also based on ^{19}F-NMR, using compounds like perfluorooctyl bromide as contrast agents. In terms of medicinal diagnostics, this method is also called ^{19}F-magnetic resonance imaging (MRI). Developments on more sophisticated tracers and applications have been reviewed some years ago [2769]. A more recent technique relies on ultrasound imaging, which is realized by the application of rather small perfluorocarbons with relatively low boiling points as contrast agents.

11.3.8 Further medical applications

The use of fluorinated compounds in medicine is not restricted to fluorinated drugs and radiopharmaceuticals. For example, small molecules, in particular halofluorocarbons and highly fluorinated ethers, have been used as inhalation anaesthetics for many years. Their main advantage is their nonflammability and of course their chemical inertness under physiological conditions. Apart from that, the property of perfluorocarbons to dissolve gases, especially oxygen, has been exploited by the design of artificial blood substitutes (cf. Section 11.1.2). Typically, stable emulsions of perfluorocarbons, perfluorinated tertiary amines, phospholipids, electrolytes and water are administered to the patient during surgery. However, afterwards the artificial blood needs to be replaced by reperfusion of the original blood of the patient, since the artificial substitute is of course not capable of adapting any other function than oxygen transport. In most cases, the perfluorocarbon is excreted by the lungs after some time, depending of course on the volatility of the fluorinated compound.

11.3.9 Study questions

(I)   Give some examples for applications of halofluorocarbons. Why are these compounds often problematic?

(II)   What is the major benefit of fluoropolymers? Which types of applications arise from it?

(III) How can fluorination influence the properties of liquid crystals?

(IV)   In which types of organic electronics does fluorine typically play a role? Which properties of fluorinated compounds are exploited for this purpose?

(V)   What are typical advantages of fluorinated drugs and agrochemicals over their nonfluorinated analogues? How popular are fluorine-containing drugs?

(VI)   What is the role of ^{18}F-labeled compounds in medicinal diagnostics? Which derivatives are commonly used for the visualization of cancer and brain diseases?

(VII) What is the role of artificial blood and of which compounds is it composed? Why does it need to be replaced by original blood after surgery?

12 General reaction types and reagents for the introduction of fluorine

In the following section, the most important reaction types for the preparation of fluoroorganic compounds are discussed. This also includes an introduction to common reagents as well as a general evaluation of their suitability in terms of accessibility, reactivity and selectivity.

12.1 Direct fluorination and radical reactions

Direct fluorination techniques typically include very drastic conditions, as they are performed with strongly fluorinating and oxidizing reagents, for example, elemental F_2 itself or high-valent, and thus very reactive metal or halogen fluorides. As a consequence, direct fluorination techniques show a low selectivity and, therefore, the products are often perfluorinated or contain at least a large number of fluorine substituents. Due to this, sensitive substrates are not suitable for this approach, especially when they contain functional groups. Furthermore, one has to keep in mind that multiple bonds as well as aromatic systems are likely to become saturated. Nevertheless, many methods have been developed to use very reactive fluorinating agents under controlled conditions, in order to obtain selectively fluorinated compounds. For the preparation of simple compounds and highly fluorinated building blocks on industrial scale, direct fluorination is still the most important approach without any serious alternative.

12.1.1 Elemental fluorine and high-valent metal fluorides

As elemental F_2 is readily split into two F˙ radicals (cf. Section 2.1), it will mostly substitute substrates following a radical-chain mechanism. This leads to the formation of equimolar amounts of HF and the fluorinated product. However, if the reactions are not carefully controlled, for example, by high dilution of F_2, organic compounds will finally be decomposed to CF_4, HF and other stable products, depending on the composition of the substrate. Other major drawbacks of using elemental F_2 as a radical fluorinating agent in fluoroorganic chemistry are the requirements for specially equipped laboratories and sophisticated reaction apparatus. Nevertheless, under appropriate conditions, elemental fluorine is a very efficient and powerful reagent, which sometimes allows the preparation of compounds and intermediates that are hardly accessible on any other synthetic route.

Some high-valent transition metal fluorides can also serve as donors for fluoride radicals, for example, CoF_3, which is widely used for the preparation of fluorocarbons

https://doi.org/10.1515/9783110659337-015

from hydrocarbons. Upon reaction, it is reduced to CoF_2, but can be readily recycled to CoF_3 by treatment with F_2. Along with CoF_2 and the fluorine-substituted organic molecule, HF is formed as by-product.

12.1.2 Radical fluorination using electrophilic fluorinating reagents

More recently, techniques evolved, which allow selective radical fluorination reactions. For this, radical initiators or photocatalysts are combined with electrophilic fluorinating reagents (cf. Section 12.4). Some time ago, a quantum-chemical study has been published, evaluating the N-F bond strength, and thus the fluorine-donating ability of commonly used, electrophilic fluoraza reagents [2770]. The results give a general idea about the relative reactivities of electrophilic fluorinating agents in radical reactions. In this mild strategy of radical functionalization, regioselectivity is achieved by the presence of directing groups and also by the stability of the intermediate carbon-centered radicals. In most examples, the substrates are fluorinated in tertiary or benzylic positions. More recently, N-fluoro-N-arylsulphonamides (NFASs) have been introduced as a new generation of radical fluorinating reagents. In these compounds, the N-F bond dissociation energies are significantly lower than those of typical electrophilic fluoraza reagents. Therefore, clean radical transformations are observed with this class of reagents, suppressing undesired side-reactions. The NFASs reagents are prepared by fluorination of the corresponding amides using electrophilic NFSI in presence of Cs_2CO_3 [2771].

12.1.3 Study questions

(I)   Which substrates are generally suitable for direct and radical fluorination approaches and which types of products will be obtained?

(II) What is the prerequisite for using F_2 in direct fluorination reactions of organic substrates?

(III) What is the benefit of using CoF_3 for fluorination reactions?

(IV) Which general type of reagents has recently been used for mild radical fluorination? How is it possible to achieve selectivity?

12.2 Electrochemical fluorination

Fluorine chemistry has always been related to electrochemical processes, which is not only due to the preparation of F_2 by electrolysis of HF/KF-mixtures (cf. Section 2.3). In fluoroorganic chemistry, electrochemical fluorination has been developed as a valuable tool for the preparation of highly fluorinated compounds, however, under less forcing conditions than in direct fluorinations using F_2 or high valent metal fluorides. The most important variant of this general approach is known as Simons process and has been developed in the 1940s [2772]. Since then, numerous publications have appeared about the design of the electrolysis cell, experimental protocols, but also mechanistic investigations [2773]. Therefore, this method will only be briefly discussed herein.

12.2.1 Prerequisites

The main advantage of electrochemical fluorination is the use of anhydrous HF and electric current as rather cheap and readily available reagents. Apart from that, the electrolysis cell may be constructed in a rather simple fashion, as fluorination reactions are typically carried out at atmospheric pressure and moderate temperatures. For preparations on larger scale, a number of small cells can easily be combined. As a prerequisite for the electrode-setup, the anode must be made of nickel with a purity of at least 99.6. The material for the cathode may also be iron, however, often Ni is used, too. Some requirements also have to be met by the starting material, which should be soluble in anhydrous HF. Often, this is realized by the presence of heteroatoms, such as O, N, S or P. In addition, a polar group is required to allow protonation by HF and, therefore, the formation of an electroconductive solution. Since the perfluorination of a substrate may be considered as a multi-step procedure, yields are generally higher for smaller molecules with short carbon chains and those that have comparably less C-H bonds. Ideally, the perfluorinated product should be insoluble in anhydrous HF to avoid further fluorination, resulting in the cleavage of C-C or C-hetero bonds.

12.2.2 Mechanistical aspects

The full mechanism of electrochemical fluorination is still not fully understood; however, it is generally accepted, that high-valent nickel fluorides are involved, which are generated by oxidation at the anode, whereas H_2 is liberated at the cathode. This is supported by experiments, which show that the chemically prepared fluorides NiF_3 and NiF_4 are indeed capable of efficiently perfluorinating organic substrates [2274]. Upon electrolysis, the nickel anode is covered with a film of NiF_3, which is probably doped with some NiF_4. Organic molecules are adsorbed on this $NiF_3(NiF_4)$-surface and undergo fluorination, generating HF and NiF_2 as by-products. The latter remains at the anode and is reoxidized to the trifluoride. As a general trend, the substrate will adsorb to the surface of the anode with the part of the molecule that has the highest electron-density, which typically corresponds to the most hydrogenated area. Since the electrochemical fluorination in anhydrous HF is a radical process, the intermediate carbon-centered radical species may undergo side-reactions, such as rearrangements or cyclizations.

12.2.3 Study questions

(I)   What are the major advantages of electrochemical fluorination over direct fluorination using F_2 or CoF_3?

(II)   Which experimental prerequisites have to be fulfilled by the electrolysis cell and the substrates?

(III) Briefly describe the mechanism of electrochemical fluorination.

12.3 Nucleophilic fluorination

Due to the electronegative nature of fluorine, the number of nucleophilic fluorinating reagents is quite large, including both covalent and ionic inorganic fluorides as well as organic compounds. However, as pointed out in Part A (cf. Section 3.5), the fluoride anion is not only a very good nucleophile, but also a rather strong Lewis base. Therefore, substitution reactions must be carried out under anhydrous conditions in aprotic solvents in most cases. Generally, the substitution with F^- works well for sp^3-hybridized carbon, but is not that easily achieved for sp^2-centers.

12.3.1 HF and its complexes

The simplest and cheapest source of fluoride is certainly hydrogen fluoride. Since HF is only a weak acid in aqueous systems (cf. Section 3.3), fluorination reactions are mostly carried out using anhydrous HF as both, reagent and solvent. Under these conditions, anionic $[(HF)_nF]^-$ species are present, which can serve as fluoride source. However, due to the aggressive and hazardous nature of anhydrous HF, transformations with this compound are rather inconvenient and require very special handling along with strict safety precautions. Therefore, a number of HF-complexes has been developed over the years, allowing a safer use of hydrogen fluoride. The most prominent among them is the adduct with pyridine, also very well known as Olah's reagent. It contains about nine molar equivalents of HF per pyridine molecule, which is equivalent to a weigh-content of about 70 % HF. The reagent is also referred to as pyridinium poly(hydrogenfluoride), which corresponds to the formula $[pyH][(HF)_nF]$. It is an acidic complex, which etches glass and must therefore be stored and handled in plastic vessels. Up to 50 °C, Olah's reagent does not release HF in significant amounts. A further development of the liquid pyridine/HF-reagent led to the attachment of the pyridine moieties to a polymeric support, therefore allowing easier handling.

 Another important HF-complex is formed with triethylamine, having the composition $NEt_3 \cdot 3\,HF$. It is also known as triethylamine trihydrofluoride (TREAT-HF). In contrast to the pyridine adduct, it has neutral, or even slightly basic properties, and consequently does not etch glassware. In addition, it does not show any HF-vapour pressure [2774, 2775]. Due to its stability, $NEt_3 \cdot 3\,HF$ is commercially available. The corresponding triethylamine-adducts with two and one HF-molecules are also known, but less important [2776]. HF-complexes are also accessible upon reaction with ammonium salts, for example, $[NBu_4]F$, which then leads to the formation of $[NBu_4][H_2F_3]$. Other important bases for the formation of HF-adducts include urea derivatives, such as cyclic N,N'-dimethylpropylene urea (DMPU). Typically, this reagent has a weight-content of 65 % HF. It can be stored in PTFE vessels at room temperature, but continuously loses HF above 50 °C [2777, 2778]. Generally, the hydrogen bonding in the aforementioned HF-complexes leads to fine-tuning of their nucleophilicity and basicity. In contrast to Olah's reagent and $NEt_3 \cdot 3\,HF$, the DMPU adduct has slightly different properties, which is mainly due to the weak basicity of the urea derivative and its increased hydrogen-bond accepting properties [2779].

12.3.2 Ionic fluorides

Metal fluorides

Other very popular examples for nucleophilic fluorinating reagents are the alkali metal fluorides $M^I F$ (M^I = Na, K, Cs), which are widely used in synthesis, however, their reactivity is somewhat limited by their poor solubility in organic solvents, so that highly polar media and high reaction temperatures are often required to achieve a reasonable conversion of the starting material. Alternatively, additives, such as crown ethers or cryptands are commonly used to solubilize the inorganic salts. Often, phase-transfer conditions are helpful, too. The main advantages of these simple fluoride sources are their availability and their low costs. Less often, other ionic metal fluorides, such as ZnF_2, CuF_2 or NiF_2 are used as nucleophilic fluorinating reagents. A rather expensive but also widely used fluoride source is AgF, which often has additional advantages, as for example, the removal of other halides by the formation of insoluble silver(I) salts.

Tetraalkylammonium fluorides

To overcome the limitation of low solubility, many ionic fluorides with larger cations have been developed, which are soluble in organic solvents, especially the tetraalkylammonium salts $[NR_4]F$ (R = Me, ^nBu). Tetramethylammonium fluoride can be prepared from $[NMe_4]Cl$ and KF, however, the product of this salt metathesis will always contain at least traces of chloride. More conveniently, $[NMe_4]OH$ is neutralized by titration with aqueous HF. Removal of the solvent *in vacuo* and subsequent drying at 150 °C yields the monohydrate $[NMe_4]F \cdot H_2O$. Heating to higher temperatures will result in the decomposition of the ammonium salt to NMe_3 and methyl fluoride. The pure and anhydrous salt is obtained by recrystallization from anhydrous isopropanol, which initially leads to the formation of an alcohol adduct. The latter is decomposed at 80 °C in dynamic vacuum to result in strongly hygroscopic $[NMe_4]F$ (Scheme 12.1). It is a very reactive compound, which leads to the dimerization of CH_3CN [2780]. Anhydrous tetramethylammonium fluoride is also accessible by the salt metathesis of [NMe$_4$][BF$_4$] with KF in dry methanol. The side-product K[BF$_4$] is removed by filtration and the intermediate methanol adducts [NMe$_4$]F · n CH$_3$OH are liberated from the alcohol by heating to 80–140 °C *in vacuo* (Scheme 12.1) [2781]. The anhydrous salt as well as several hydrates of [NMe$_4$]F are commercially available.

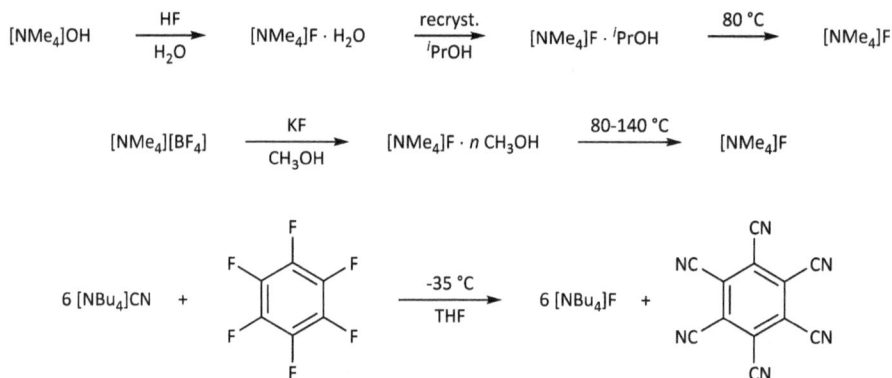

Scheme 12.1: Preparation of anhydrous tetraalkylammonium fluorides.

Analogously to the tetramethylammonium salt, hydrated salts of the type [NBu$_4$]F · n H$_2$O are readily accessible by salt metathesis in aqueous media. However, the preparation of anhydrous tetrabutylammonium fluoride cannot be achieved by drying of the commercially available hydrates in high vacuum, which is due to a far higher tendency of the [NBu$_4$]$^+$ cation to undergo E2-elimination. Therefore, anhydrous [NBu$_4$]F is prepared in another way, using nonaqueous conditions. For this, hexafluorobenzene is reacted with [NBu$_4$]CN in tetrahydrofuran at low temperature to undergo a nucleophilic substitution, yielding hexacyanobenzene and anhydrous tetrabutylammonium fluoride, which can be precipitated from the reaction mixture (Scheme 12.1). The solid is stable under inert atmosphere at –35 °C for weeks, whereas it slowly undergoes E2-elimination above 0 °C. The same decomposition behavior is observed in THF solutions [2782]. A more stable and by far less hygroscopic version of [NBu$_4$]F is obtained, when the hydrated salt is recrystallized from a 4:1-mixture of *tert*-butanol and *n*-hexane. The resulting solvate [NBu$_4$]F · (^{t}BuOH)$_4$ shows comparable fluorinating abilities to the anhydrous compound with a decreased basicity, but can be more easily handled and stored [2783]. Another variant is [NBu$_4$][Ph$_3$SiF$_2$], which is prepared by treatment of the ammonium fluoride with Ph$_3$SiF. It is a stable solid, which is both, less basic and less nucleophilic than the parent [NBu$_4$]F [2784].

A few less important relatives of the aforementioned tetraalkylammonium salts can also serve as sources of "naked" fluoride. For example, the preparation of 1,1,3,3, 5,5-hexamethylpiperidinium fluoride has been described, which shows a very high fluoride activity toward inorganic substrates [693]. Apart from that, anhydrous 1-methyl-hexamethylenetetramine fluoride is prepared by halogen exchange of the corresponding iodide compound with AgF or directly by mixing methylamine, HF, formaldehyde and NH_3 in aqueous solution, in both cases with subsequent drying in high vacuum [2785]. Another example of a "naked" fluoride source is anhydrous tetramethylphosphonium fluoride. It is prepared in a four-step sequence from triphenyl phosphite via the intermediate stages of PMe_3, $[PMe_4]Br$ and $Me_3P = CH_2$. It is a volatile solid, which can be purified by sublimation *in vacuo*. As a consequence, it has a molecular structure in the gas phase as well as in nonpolar solvents. The fluorine substituent is only weakly bound to phosphorous in an axial position of a distorted trigonal bipyramid [2786].

Other salt-like sources of fluoride

There are some more salts, which serve as sources of "naked" fluoride and are of relevance in preparative fluoroorganic chemistry. One of them is *tris*(dimethylamino)sulphonium difluorotrimethylsilicate $[(Me_2N)_3S][SiMe_3F_2]$, often abbreviated as TASF. It is prepared by the reaction of SF_4 with Me_2NSiMe_3 in Et_2O, $CFCl_3$ or *n*-pentane (Scheme 12.2) [2787, 2788]. In a similar way, the hexamethylguanidinium derivative $[(Me_2N)_3C][SiMe_3F_2]$ is accessible by chlorination of tetramethylurea and treatment of the resulting tetramethylchloroformamidinium chloride with anhydrous $[NMe_4]F$ to give *bis*(dimethylamino)difluoromethane. The subsequent reaction with Me_2NSiMe_3 yields the hypervalent silicate salt in a clean reaction at 0 °C (Scheme 12.2) [2789–2791]. In contrast to TASF, which does not yield the "naked" fluoride $[(Me_2N)_3S]F$ upon loss of Me_3SiF, hexamethylguanidinium fluoride has been shown to exist in dynamic equilibrium with the silicate in a solvent mixture of diethyl ether and dichloromethane. Both species have been isolated and structurally characterized by X-ray crystallography. Hexamethylguanidinium fluoride crystallizes as CH_2Cl_2-solvate, where the fluoride ion is octahedrally coordinated by six solvent molecules via hydrogen bridges (see also the cover picture) [2791]. Apart from the two aforementioned silicate-based fluoride sources, a few more compounds have been described, which are also based on hypervalent main-group element anions, such as stannates. However, in practice, they are rather less popular.

12.3.3 Molecular fluorides

Inorganic main-group element fluorides

Analogously to metal salts, molecular main-group fluorides can serve as nucleophilic fluoride sources, too. Despite the fact, that their use has become rather unpopular nowadays, compounds like SbF_3 or SeF_4 have proven to be suitable fluorinating agents several decades ago. However, one of the most important main-group fluorides in flu-

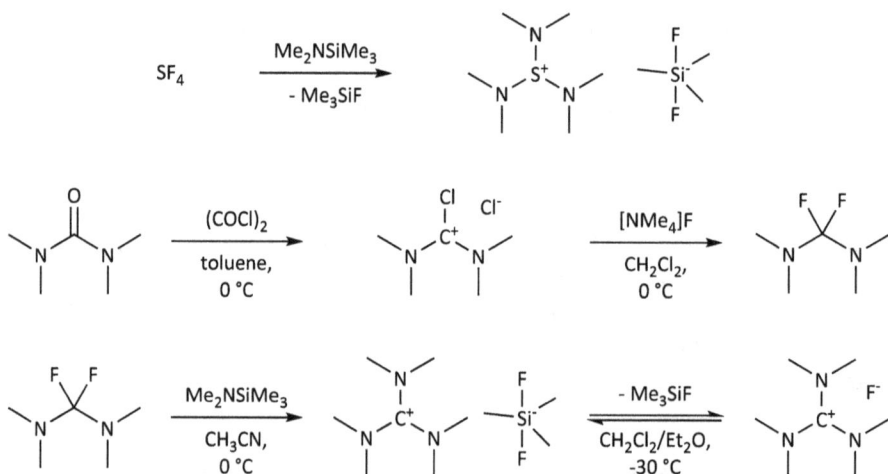

Scheme 12.2: Preparation of *tris*(dimethylamino)sulphonium difluorotrimethylsilicate (top) and hexamethylguanidinium fluoride (bottom).

oroorganic synthesis is sulphur tetrafluoride. Still until today, a large and important class of nucleophilic fluorinating agents is based on SF_4 and its derivatives [2792]. Generally, these reagents are applied in deoxyfluorination reactions of hydroxy-, carbonyl- and carboxyl-compounds. Due to the fact that it is a hazardous gas, SF_4 itself is only sparingly used in fluoroorganic synthesis nowadays. Slightly safer and more convenient liquid alternatives have been developed, which are mostly based on the replacement of a fluoride substituent by a dialkylamino-group, which is achieved by the reaction of SF_4 with the corresponding silylamine R_2NSiMe_3 (Scheme 12.3) [2793, 2794]. The most important representative of this family of R_2NSF_3-compounds is certainly diethylaminosulphur trifluoride, commonly abbreviated as DAST. The major drawback of this type of reagents is the instability of the S-N-bonds, which might result in violent and explosive decomposition reactions at elevated temperature. For example, DAST is only stable up to about 50 °C [2793].

Other dialkylaminosulphuranes, such as diisopropylamino-, dimethylamino-, piperidino- and pyrrolidinosulphur trifluoride show a similar behavior, as they only slightly differ in their substituents at the nitrogen atom. Therefore, some safer alternatives, in particular morpholinosulphur trifluoride (MOST) and *bis*(methoxyethyl)-aminosulphur trifluoride (BAST, Deoxo-Fluor™) have been developed (Scheme 12.3) [2794, 2795]. They still decompose upon heating, but do not tend to explode or evolve large amounts of heat. Further derivatives of DAST and MOST are generated by treatment with BF_3 (Scheme 12.4) [2796, 2797]. The resulting sulphiminium compounds XtalFluor-E™ and XtalFluor-M™ are air-stable solids and show a clearly reduced tendency to degrade. The decomposition of S-F-reagents can be completely avoided, if the reagent does not contain any S-N bonds, which is the case in Fluolead™,

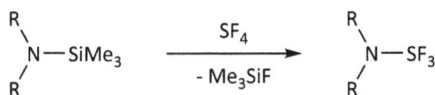

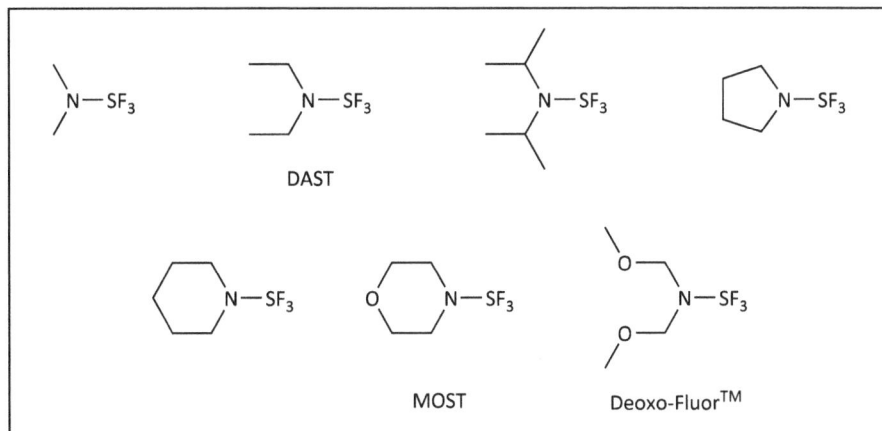

Scheme 12.3: Preparation of dialkylamino sulphurtrifluoride reagents.

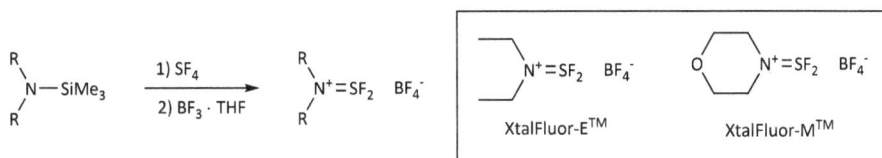

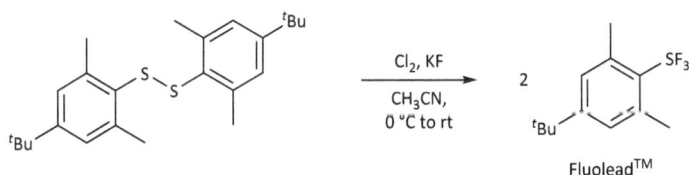

Scheme 12.4: Preparation of sulphiminium- and aryl sulphurtrifluoride reagents.

where the SF$_3$-group is directly attached to a sterically demanding 4-*tert*-butyl-2,6-dimethylphenyl moiety. Apart from that, it is remarkably resilient toward aqueous hydrolysis. Fluolead™ is prepared by chlorination of the corresponding disulphide in presence of excess KF (Scheme 12.4) [2798]. Aryl sulphur trifluorides may also be generated *in situ* by reduction of ArSF$_4$Cl with a reducing agent like pyridine or the corresponding disulphides ArSSAr [2799].

Another important main-group fluoride in preparative fluoroorganic chemistry is represented by boron trifluoride and its derivatives, since nucleophilic fluorination of sp^3- as well as sp^2-carbons is also generally achieved by the use of BF$_3$ or [BF$_4$]$^-$. The versatility of these boron compounds has been demonstrated in a comprehensive review article some time ago [2800]. They are not only used in typical substitution reactions of substrates with a leaving group but are especially useful for the ring-opening of strained systems, such as epoxides or cyclopropanes. Apart from that, BF_3 and $[BF_4]^-$ salts play a role in fluorofunctionalization reactions of unsaturated compounds as well as in fluorinative cyclization approaches.

α,α-Difluoroalkylamine reagents

The family of α,α-difluoroalkylamine compounds can be somewhat considered as milder organic analogues to the aforementioned class of $R_2N\text{-}SF_3$ reagents. Hence, fluoroalkylamines are almost exclusively used in deoxyfluorination reactions of alcohols. The historically first compound of this class that has been extensively applied for fluorodehydroxylation, is commonly known as Yarovenko's reagent and is prepared by the addition of diethyl amine to chlorotrifluoroethylene (Scheme 12.5) [2801]. Upon reaction with water, the $\alpha\text{-}CF_2$-group of the compound is easily hydrolyzed, resulting in the formation of HF and the corresponding amide [2802]. The main disadvantage of Et_2NCF_2CFClH is its short shelf-life, which significantly limits its availability. The stability is significantly improved, if the chloro-substituent is replaced by a CF_3-group, which is readily achieved, when $HNEt_2$ is added to perfluoropropene instead of $CF_2=CFCl$, giving rise to Ishikawa's reagent (Scheme 12.5) [2803]. This fluoroalkylamine compound generally shows a low tendency to decompose, however, often it is contaminated with minor amounts of the HF-elimination product, that is, $Et_2NCF = CFCF_3$. However, in presence of HF, it can regenerate the amine. In the same way, a mixture of fluoroalkylamines and corresponding enamines is obtained, when secondary amines are added to 1,1,3,3,3-pentafluoropropene (Scheme 12.5). Again, the enamine is converted to the active, fluorinating fluoroalkylamine species upon addition of HF [2804]. Interestingly, the electron-rich fluoroenamine $Me_2C = CFNMe_2$ is capable of transforming alcohols to fluorides under neutral conditions [2805]. It is prepared by chlorination of the corresponding amide and subsequent deprotonation of the intermediate iminium species. The halogen exchange is achieved by treatment with KF (Scheme 12.5) [2806, 2807]. Another variant of fluorinating reagents are α,α-difluorobenzylamines, especially N,N-diethyl-α,α-difluoro-(m-methylbenzyl)amine (DFMBA), which is also prepared in two steps via chlorination of the benzamide (Scheme 12.5) [2808]. The reactivity of α-fluoroalkylamine reagents is further increased, if they are based on cyclic imidazolidine compounds. A very convenient fluorinating agent is 2,2-difluoro-1,3-dimethylimidazolidine (DFI), which is readily prepared by chlorination of the corresponding cyclic urea derivative and subsequent halogen exchange with KF [2809, 2810]. The big advantages of DFI are that it is made from a rather cheap urea starting material, which can additionally be recycled, since N,N'-dimethylethylene urea is regenerated after deoxofluorination reactions. A more sophisticated and unsaturated variant of DFI is based on an imidazoline-scaffold and contains 2,6-diisopropylphenyl-groups instead of the methyl

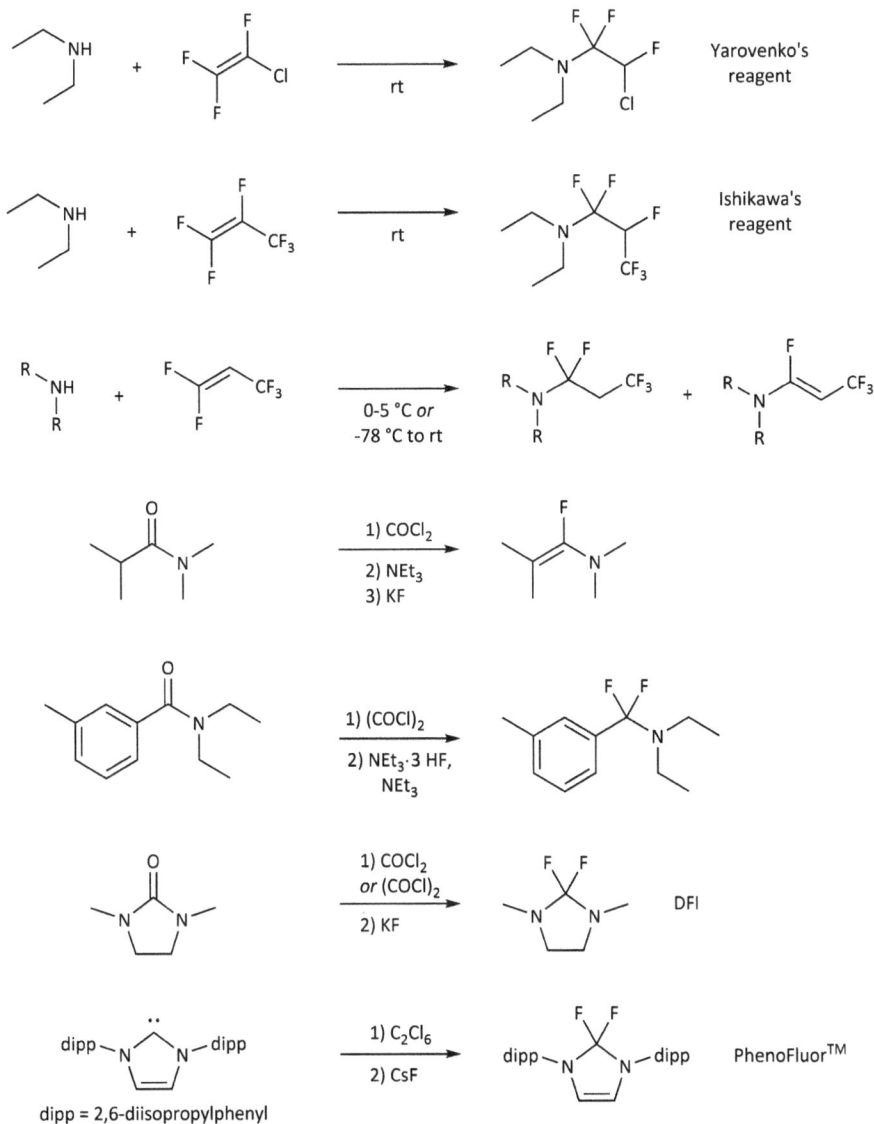

Scheme 12.5: Preparation of fluoroalkylamines as well as imidazolidine- and imidazoline-based fluorinating agents.

substituents (Scheme 12.5). It has been commercialized as PhenoFluor™ and is capable of converting phenols to aryl fluorides. In contrast to DFI, its preparation is not that straightforward, since the corresponding urea derivative is not commercially available, at least not at reasonable costs. Therefore, it is prepared by chlorination of the corresponding N-heterocyclic carbene and subsequent halogen exchange with CsF [2811, 2812].

12.3.4 Enzymatic fluorination

A very special and still developing field of nucleophilic fluorination is given by enzymatic techniques, using fluorinase [2813]. As already mentioned in Part A of this manuscript, the number of naturally-occurring fluorinated molecules is comparably small. However, over the years, more and more insights have been gained on the biosynthesis of these compounds and the involved enzymes. Therefore, it has also been possible to make use of the fluorinating abilities of fluorinases for the preparation of artificial bioactive, fluorinated compounds, which are, for example, based on carbohydrates, nucleosides or peptides. Enzymatic techniques have especially been used for the preparation of ^{18}F-labeled molecules, making use of the high efficiency and site-selectivity of such transformations. Certainly, enzymatic fluorination will further gain importance in preparative organofluorine chemistry in the future. Since its applications are still limited to a few substrates to date, this technique will not be discussed in detail within this manuscript.

12.3.5 Study questions

(I)   What is Olah's reagent? What needs to be taken into account, when using and handling it?

(II)   Are there other synthetically relevant adducts of HF with nitrogen bases? How does the base influence the properties of the adducts?

(III)   What is the major problem of using ionic inorganic fluorides as nucleophilic fluorinating agents in fluoroorganic synthesis? How can this problem be overcome?

(IV)   How are the tetraalkylammonium salts $[NMe_4]F$ and $[NBu_4]F$ prepared? What needs to be taken into account, when the anhydrous compounds are required?

(V)   What is the difference between the hypervalent silicate salts $[(NMe_2)_3S][Me_3SiF_2]$ and $[(NMe_2)_3C][Me_3SiF_2]$? Which species are observable in solution? How are the silicate salts prepared?

(VI)   From which main group fluoride originates a large number of nucleophilic fluorinating agents, such as DAST or MOST? What is their main advantage over their parent fluoride compound and how are they prepared? Can DAST and MOST be converted to other fluorinating reagents?

(VII)   How are α,α-difluoroalkylamine reagents typically prepared? Which decomposition reactions do they show and how can they be reactivated?

(VIII) What is the major advantage of DFI? How is it prepared?

12.4 Electrophilic fluorination

The term "electrophilic" might sound very conflicting when talking about the introduction of the most electronegative element into organic compounds. Indeed, one should always be aware of the fact, that there is of course no "F⁺." Nevertheless, a number of reagents has been explored, which are able to donate fluorine as if it would be positively charged. Some time ago, a quantum-chemical study has been published, evaluating the N-F bond strength, and thus the fluorine-donating ability of commonly used, electrophilic fluoraza reagents [2770]. The results give a general idea about the relative reactivities of electrophilic fluorinating agents in radical reactions.

12.4.1 Fluoroxy compounds, perchloryl fluoride and xenon difluoride

The simplest electrophilic fluorinating agent is elemental fluorine itself, which reacts under several preconditions with hydrocarbon substrates at low temperature and high dilution. However, forcing reaction conditions have to be strictly avoided, as F_2 will react in radical fashion otherwise. Some more compounds are known, which have been widely used as sources of electrophilic fluorine in the past. As a prerequisite, fluorine needs to be attached to an element, which is also strongly electronegative, for example, oxygen. The oxygen fluorides OF_2 and O_2F_2 have only scarcely been used as electrophilic fluoride sources, however, reagents like $CsSO_4F$ are more popular. More importantly, hypofluorites, especially CF_3OF have been used as sources of electrophilic fluorine. The preparation of all these compounds is described in the inorganic part of this manuscript (cf. Section 9.6.1). However, as a general conclusion, all these compounds are highly reactive and quite hazardous, as they can react violently with organic matter. Further members of the class of fluorooxy compounds are given by acyl hypofluorites, which are accessible by the reaction of the corresponding carboxylates $CH_3COO^-Na^+$ and $CF_3COO^-Na^+$ with F_2 at low temperature [2814, 2815]. Typically, they are freshly prepared and used directly in solution without isolation or purification.

Another inorganic source of electrophilic fluorine is perchloryl fluoride, which has widely been applied in fluoroorganic synthesis in former times. Finally, XeF_2 necessarily needs to be mentioned among the group of inorganic electrophilic fluorinating reagents. It has the main advantages that it is comparably easy to handle and acts as a rather mild fluorinating agent. As described in Part B, it is readily available from Xe and F_2, however, it is rather expensive and, therefore, not really suitable for large-scale preparations.

12.4.2 Fluoraza reagents

For practical uses, electrophilic fluorination reactions almost exclusively rely on the use of fluoraza reagents today. Generally, these compounds are far more stable and less hazardous than the aforementioned fluorooxy compounds as well as ClO_3F, in most cases they are even commercially available. They can be divided in two subcategories, namely neutral N-fluoroamine compounds and cationic N-fluoroammonium derivatives. Neutral N-F reagents are mainly represented by sulphonimide and sultam compounds, the most prominent one certainly being *N*-fluorobenzenesulphonimide (NFSI). It is a commercially available solid, which is prepared by direct fluorination of the sulphonimide with F_2/N_2 in presence of NaF (Scheme 12.6) [2816]. In the same way, *N*-fluoro-*o*-benzenedisulphonimide is generated [2817]. A variety of derivatives having perfluoroalkyl substituents instead of the phenyl groups has also been described. However, these compounds are preparatively less important [2818]. Similar to the sulphonimide reagents, *N*-fluorosultams are generated by direct fluorination of the corresponding N-H compounds (Scheme 12.6). Two prominent examples are based on a saccharin- and camphor-scaffold, respectively, where the latter one has been used in the first asymmetric preparation of α-fluorocarbonyl compounds [2819, 2820].

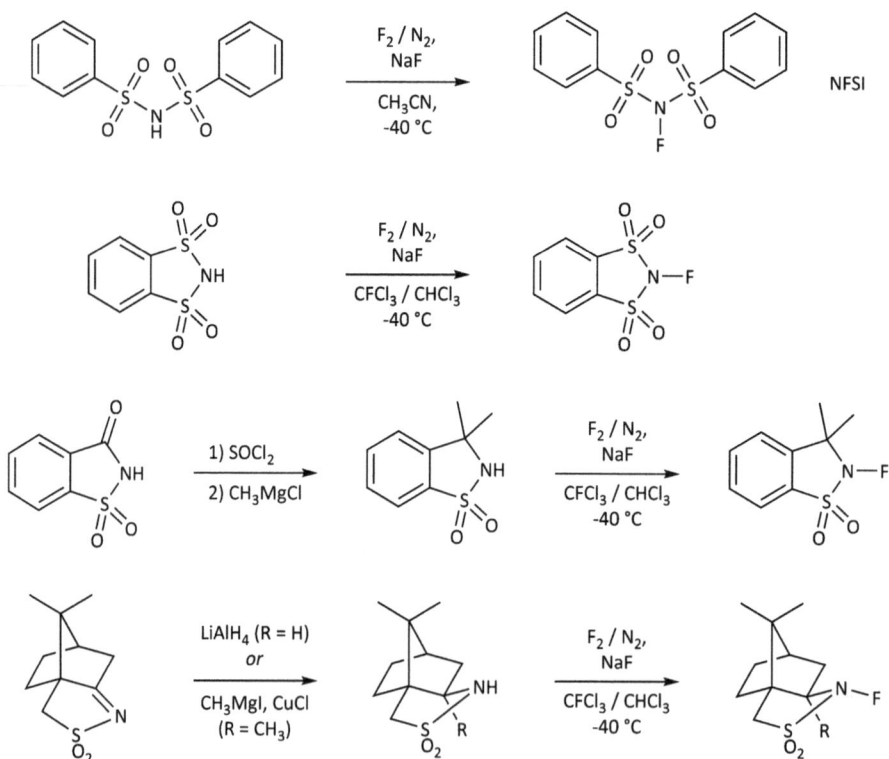

Scheme 12.6: Preparation of *N*-fluorosulphonimide- and *N*-fluorosultam reagents.

The cationic *N*-fluoroammonium reagents are mainly represented by *N*-fluoropyridinium salts and the different Selectfluor™-type reagents, which are based on a 1,4-diazabicyclo[2.2.2]octane moiety. The preparation of *N*-fluoropyridinium salts is achieved by treatment of the corresponding heteroarene with F_2 at low temperature. However, the replacement of the nucleophilic fluoride anion by other counter-ions, such as OTf⁻, [BF₄]⁻, [PF₆]⁻ or [SbF₆]⁻ is required to obtain stable fluorinating reagents. Apart from that, the introduction of additional substituents is helpful. Unsubstituted pyridine indeed reacts with elemental fluorine to generate [pyF]F, however, the adduct readily decomposes above –2 °C [2821]. The introduction of methyl groups in *ortho-* or *ortho-* and *para*-position has proven suitable for the generation of stable fluorinating reagents (Scheme 12.7). Chlorine substituents in *meta*-position also enhance the stability of *N*-fluoropyridinium salts. The anion exchange is achieved by the reaction of the intermediate *N*-fluoropyridinium fluorides with Lewis or Brønsted acids. Alternatively, the triflate anion is introduced by the addition of its sodium salt or the silane reagent Me₃SiOTf [2822].

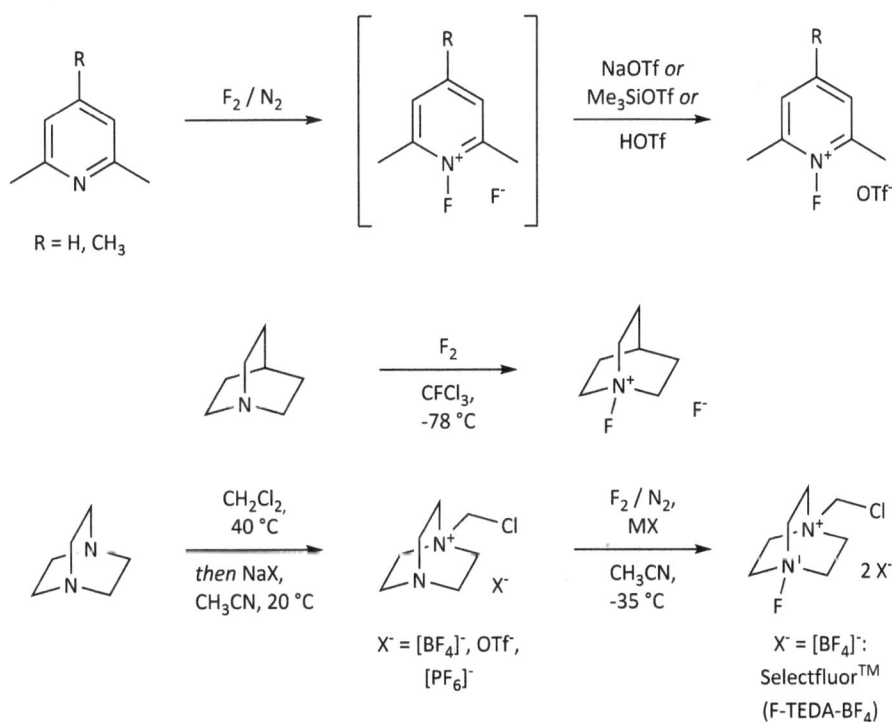

Scheme 12.7: Preparation of *N*-fluoropyridinium-, *N*-fluoroquinuclidinium- and Selectfluor™-type reagents.

In a similar way, heterobicyclic compounds are directly fluorinated by F₂ to give cationic *N*-fluoro species. In contrast to the aforementioned pyridine, the fluorination of quinuclidine yields stable *N*-fluoroquinuclidinium fluoride, which is a stable solid (Scheme 12.7). However, its hygroscopic properties render it rather inconvenient for the application as fluorinating agent [2823]. A significant improvement has been made when a triethylenediamine-moiety (TEDA) was converted to a fluorinating agent. Initially, the amine is alkylated at one of the nitrogen substituents, to give a type of tetraalkylammonium salt. Fluorination of the latter with elemental F_2 gives rise to dicationic N-fluoroammonium salts. The most important member of this family contains a chloromethyl-substituent at one of the heteroatoms and two tetrafluoroborate counter-ions (Scheme 12.7). It is commercialized as Selectfluor™, but is also commonly known as F-TEDA-BF$_4$. It is a bench-stable, nonhygroscopic solid. However, a large variety of reagents has been created, for example, by exchanging the CH$_2$Cl-group by another alkyl substituent, or by replacement of the [BF$_4$]$^-$ anions by, for example, OTf$^-$ or [PF$_6$]$^-$, which might lead to an improved solubility of the dicationic reagent [2824, 2825]. In contrast to the standard F-TEDA-BF$_4$ reagent, Selectfluor™ II does not contain a chloromethyl-substituent, but is substituted with a CH$_3$-group. Another commercialized fluorinating agent very similar to Selectfluor™ is Accufluor™, which bears a hydroxy group instead of the CH$_2$Cl-function. Accufluor™ is conveniently used for the fluorofunctionalization of olefinic substrates but also converts aromatic alcohols to α,α-difluoroketones [2826, 2827]. More sophisticated derivatives of F-TEDA-BF$_4$ have been developed in terms of asymmetric electrophilic fluorination and will be briefly discussed in Section 13.5.5.

12.4.3 Study questions

(I)   Which inorganic compounds have been used as electrophilic fluorinating agents? What is the main problem of their application in organic synthesis?

(II)   What are major advantages and disadvantages of using XeF$_2$ as electrophilic fluorinating agent?

(III)   What is the benefit of fluoraza reagents over fluoroxy compounds? In which general classes are they typically divided?

(IV)   What is NFSI? How is it prepared? Are there related derivatives?

(V)   How can the stability of N-fluoropyridinium salts be tuned? How are these compounds prepared?

(VI)   What is Selectfluor™? How is it prepared? Are there important derivatives?

12.5 Metal-mediated reactions, including photocatalysis

Metal-mediated reactions can be very powerful tools for the preparation of certain fluorinated scaffolds in organic chemistry, as they allow the introduction of fluorine substituents to otherwise hardly accessible positions. In addition, the presence of a metal catalyst often increases the regioselectivity, in presence of appropriate ligands, asymmetric fluorination may be achieved, too. A special relevance is assigned to direct C-H fluorination reactions, which are very important methods for the late-stage functionalization of rather complex molecules, for example, pharmaceuticals. Therefore, a vast amount of metal-mediated C-H fluorination techniques has been described in the last decades.

In general, these methods rely on the application of the typical transition metals, especially palladium, but also copper, silver and a few more. However, for catalytic reactions, the combination with a stoichiometric fluorinating agent is required. Virtually all direct C-H-transformations need to be carried out in the presence of electrophilic fluorinating reagents, mainly NFSI or Selectfluor™. Their role is not only defined by donating the fluorine substituent, but they also work as oxidizers. Consequently, C-H-fluorination reactions may also be carried out using nucleophilic fluoride sources, but will then require the presence of an additional oxidant. In case of substrates with a leaving group, such as a halide or triflate, metal-mediated fluorination reactions can be carried out with nucleophilic fluorinating agents. Often, AgF is used for this purpose.

There are also reactions, where the metal is not directly involved in the formation of C-F bonds, but generates a reactive fluorinating species. This is, for example, the case in photocatalytic reactions, where the excited metal complex activates the fluorinating reagent, mostly by a single electron transfer (SET). Apart from that, a few more examples are known, where the C-F bond forming event is not related to a classical reductive-elimination-type reaction.

12.6 Hydrodefluorination and defluorofunctionalization

In some cases, it appears to be more attractive to apply the inverse synthetic approach toward certain fluorinated moieties, that is, abstracting fluorine substituents from a higher fluorinated compound instead of fluorinating a substrate with a smaller fluorine-content. Of course, the success of this technique strongly depends on the target structure and the starting materials. Selective hydrodefluorination, or more general defluorofunctionalization is a very suitable method for the preparation of fluorinated arenes, but is not that easily achieved with aliphatic substrates. This is not surprising, keeping in mind that the $C(sp^3)$-F bond is very strong, whereas $C(sp^2)$-F bonds are significantly weaker (cf. Section 11.2).

There is no classical set of reagents for this type of transformation; however, some general conclusions can be made for hydrodefluorination reactions. Most importantly, the hydride source should form a strong element-fluorine bond, so that there is a thermodynamic driving force, which is conveniently achieved by using boranes, alanes or silanes as stoichiometric reagents. Moreover, hydrodefluorination can be facilitated, if an organocatalyst or transition-metal complex is used. In the latter case, the majority of reactions have been reported using early transition metals, especially titanium.

Of course, the removal of a fluorine substituent is not necessarily connected with the introduction of a hydrogen substituent. Instead, other nucleophiles can be introduced, too, which is especially the case in the reactions of fluoroarenes. Using selectively fluorinated substrates, it is also possible to perform C-C bond-forming reactions. However, this technique, leading to mostly nonfluorinated products, will not be covered in this manuscript.

13 Fluorination of aliphatic substrates

13.1 General trends in the synthesis of aliphatic organofluorine compounds

13.1.1 The general fluorine effect

Prior to discussing detailed synthetic approaches toward aliphatic organofluorine compounds, two general trends need to be pointed out. Many of the fluorination methods in the following subchapters include side-reactions, especially migrations and rearrangements, as for example, inter- and intramolecular halogen transfer. In general, halogen migration is preferentially observed under drastic reaction conditions and in presence of strong Lewis acids like $SbCl_5$, $AlCl_3$ or $AlBr_3$ (Scheme 13.1) [2828]. If possible, fluorine substituents clearly tend to accumulate at one carbon atom, rendering trifluoromethyl- and geminal difluoromethyl-derivatives the thermodynamically favored products.

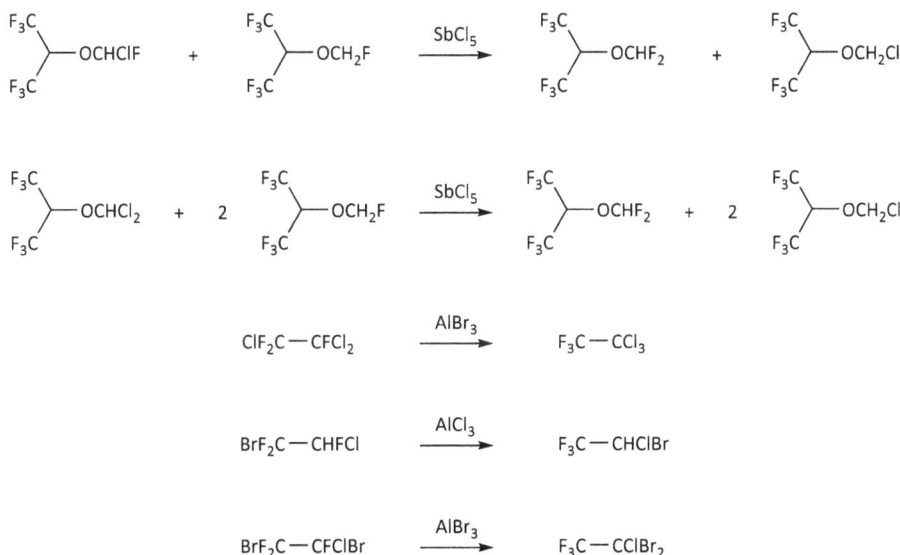

Scheme 13.1: Inter- and intramolecular halogen migration catalyzed by Lewis acids.

The occurrence of this behavior, often referred to as "general fluorine effect," can be rationalized by energetic stabilization due to resonance structures (Figure 13.1). At electron-poor sp^3-carbon centers, the lone-pairs of the fluorine substituents can shift some electron-density toward the positively polarized carbon atom by π-back-donation. This results in negative hyperconjugation and the formation of ionic resonance structures.

https://doi.org/10.1515/9783110659337-016

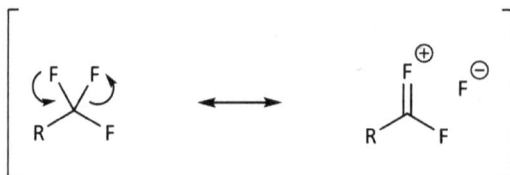

Figure 13.1: Stabilization of fluorinated sp^3-carbon centers by resonance effects.

13.1.2 The special fluorine effect

Another type of rearrangement occurs in unsaturated, fluorinated molecules and is called the "special fluorine effect." The connection of fluorine substituents to sp^2-carbons generally includes repulsive interactions between the π-system and the lone-pairs of fluorine. Thus, the linkage of fluoride substituents to sp^3-carbon centers is energetically preferred, as there are less steric repulsions. The conversion of C(sp^2)-F to C(sp^3)-F-bonds is often observed upon irradiation with UV-light or treatment with F$^-$. For example, irradiation of 1,2,3,4,5-pentafluorocyclopenta-1,3-diene leads to a rearrangement to the respective difluoromethylene compound [2829]. Analogously, UV-light converts hexafluorobenzene to perfluorinated, bicyclic cyclohexa-2,5-diene (Dewar benzene) (Scheme 13.2) [2830]. In a similar way, ω-H-perfluoro-1-butene is converted to various internal olefins with a reduced number of C(sp^2)-F bonds upon impact of CsF at elevated temperature. The stereochemistry of the observed products is largely influenced by H-F bridges [2831]. The "special fluorine effect" also leads to surprising behavior in terms of the acidity of C-H bonds in fluoroorganic compounds. As the deprotonation of pentafluorocyclopentadiene would result in the formation of an additional C(sp^2)-F bond, the generation of the planar [C$_5$F$_5$]$^-$ is not much favored. As a consequence, the acidity of the proton in C$_5$F$_5$H (pK$_a$ about 14) is comparable to that of the non-fluorinated analogue C$_5$H$_6$ (pK$_a$ 15.5). However, replacement of the fluorosubstituents by CF$_3$-groups in C$_5$(CF$_3$)$_5$H renders the proton more acidic by sixteen orders of magnitude (pK$_a$ − 2) [2832, 2833].

13.1.3 Study questions

(I)   What is the general fluorine effect and under which conditions is it typically observed?
(II)   Which are the generally preferred rearrangement products and how can this be rationalized?
(III) What is the special fluorine effect? For which types of compounds and under which typical conditions does it play a role?
(IV)   Compare the C-H acidity of 1,2,3,4,5-pentafluorocyclopenta-1,3-diene with that of 1,2,3,4,5-*pentakis*(trifluoromethyl)cyclopenta-1,3-diene. How can the difference be explained by the special fluorine effect and what happens if the former compound is irradiated with UV-light?

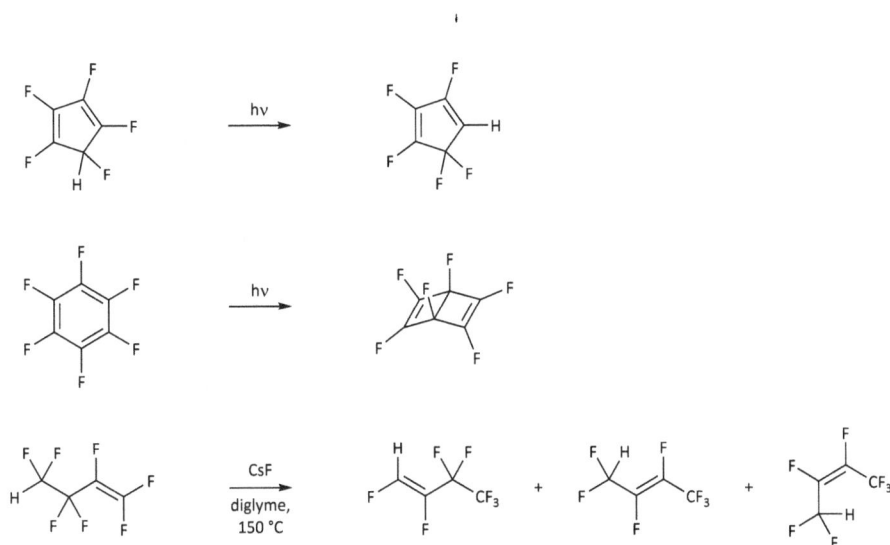

Scheme 13.2: Rearrangements of fluorinated aromatics and olefins induced by UV-light or fluoride ions.

13.2 Radical reactions and direct fluorination

The simplest and probably most efficient methods for the preparation of highly or even perfluorinated aliphatic substrates are direct fluorination techniques using elemental F_2 or other highly reactive reagents. However, one has to keep in mind that exhaustive, uncontrolled fluorination of organic matter will always end up with the formation of CF_4 and other stable products, if conditions are not appropriately chosen. Apart from that, reaction temperature and dilution can have major effects on the nature and the mechanism of the fluorination. For example, F_2 acts as an electrophilic fluorinating agent in high dilution at low temperatures, whereas it clearly initiates radical-chain conditions under more forcing circumstances. Apart from rather traditional approaches of fluorination using reactive reagents like F_2 or high-valent metal fluorides, more recent methods will also be presented, which rely on the use of significantly more selective reagents and milder conditions.

13.2.1 General remarks on direct fluorination techniques

In principle, the preparation of perfluorinated compounds from hydrocarbons requires at least two very contrary sets of conditions. In the beginning, when the substrate contains a lot of substitutable C-H bonds, a too rapid attack of fluorine has to be avoided. Otherwise, extensive fragmentation will occur. With increasing fluorine content, the substitution reactions at the substrate will generally slow down, so that finally more vigorous conditions will be required to achieve perfluorination.

Alternatively, a constantly slow fluorination under mild reaction conditions has to be performed for a longer period of time. The use of an HF-scavenger is furthermore beneficial to limit the number of acid-promoted rearrangements; in most cases, alkali metal fluorides are used for this purpose. For the preparation of perfluorinated compounds, especially in industrial processes, batch- and flow-techniques have to be distinguished. In principle, batch-fluorinations are controlled by the flow of elemental F_2 during the reaction, which is mostly regulated by adjustment of the inert diluent, for example, He or N_2. As a consequence, reaction control is achieved and run-away reactions are avoided, since F_2 can only be consumed as fast, as it is added to the reactor. In contrast, the main advantage of flow processes is the reduced reaction time in the order of a few minutes. As a general principle, both F_2 concentration and reaction temperature increase over the length of the reactor. On the final stage, the reaction mixture is photolyzed to obtain highly fluorinated compounds and decrease the residual hydrogen to a negligible amount.

For many substrates, a clean formation of perfluorocarbons is observed upon direct fluorination. However, in sterically hindered molecules like 2,2,4,4-tetramethylpentane, the generation of the fully fluorinated compound can be difficult (Scheme 13.3). This effect may be explained by the fact that fluorination of peripheral C-H bonds occurs faster than that of internal ones, resulting in the rather rapid formation of perfluorinated *tert*-butyl groups. This increases the shielding of the methylene group and leads to a mixture of products with CF_2-, CFH- or CH_2- units, respectively. Apart from steric impact, electronic effects can also influence the fluorination of certain carbon-centers. For example, a reduced electron-density is found at bridgehead hydrogen substituents in fluorinated bicyclic compounds, such as norbornane or norbornadiene. Consequently, the more acidic protons are rather resistant to the attack of F_2, which occurs in an electrophilic fashion at low temperature (Scheme 13.3) [2834]. Typically, the fluorination of alkanes, ethers, acid halides, esters, alkyl chlorides, most ketones, ketals and orthoesters yields the respective perfluorinated molecules. Primary chlorine substituents are rather stable against fluorination and are typically retained. For example, neopentyl chloride is cleanly converted to perfluoroneopentyl chloride. However, the same behavior is not observed for the respective bromocompound, which is transformed to perfluoroisopentane. The rearrangement is believed to involve carbocationic intermediates, which are formed upon disproportionation of initially generated bromine fluorides $RBrF_2$ and $RBrF_4$, respectively [2835]. Under radical-chain conditions, rearrangements are usually not observed. Nevertheless, in some cases, alkyl chlorides can undergo a 1,2-shift to generate more stable radicals. This behavior has been studied for the direct fluorination of different mono- and polychloroalkanes. Primary alkyl chlorides did not show any rearranged product, whereas tertiary substrates resulted in perfluorinated products, which have at least undergone one 1,2-chlorine shift. Secondary alkyl chlorides are only partially rearranged and yield mixtures of primary and secondary perfluoroalkylchlorides [2836]. In substrates, which do not contain any chloro- or bromo-substituent, rearrangements are scarcely

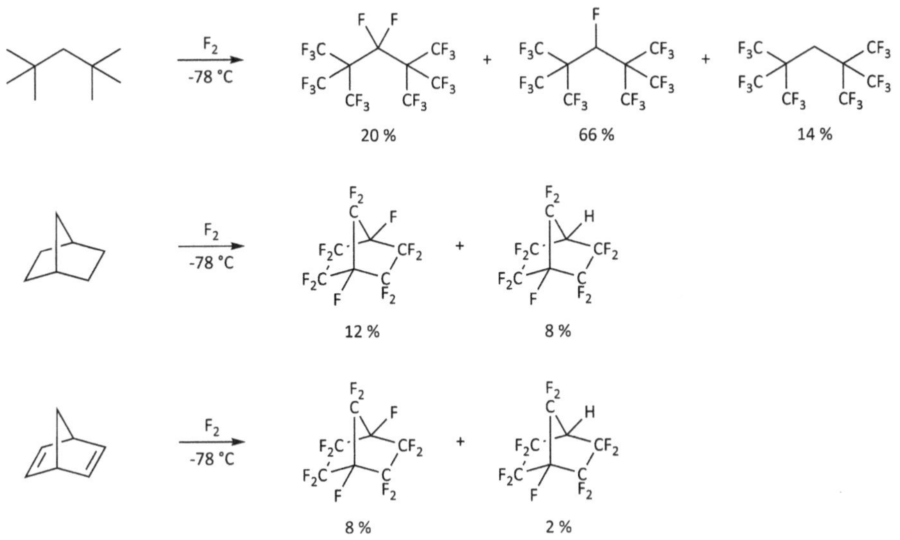

F_2

−78 °C

20 % 66 % 14 %

F_2

−78 °C

12 % 8 %

F_2

−78 °C

8 % 2 %

Scheme 13.3: The impact of steric and electronic effects on the perfluorination of hydrocarbons.

observed. One example is the fluorination of di-*tert*-butyl ketone, which exclusively yields perfluorinated *tert*-butyl isobutyl ketone, clearly showing a 1,2-acyl shift.

13.2.2 Direct fluorination techniques using elemental F_2

Small molecules are typically fluorinated using vapor-phase techniques, such as "jet fluorination" or "porous tube fluorination." For example, perfluoropropane is accessible by either fluorinating the saturated hydrocarbon C_3H_8 or unsaturated propene. In principle, it is also possible to fluorinate acetone to obtain hexafluoroacetone, however, on industrial scale it is prepared from hexafluoropropene or by halogen exchange of hexachloroacetone with HF. For more complex molecules, it is suitable to perform the perfluorination reactions by using a solid-phase low-temperature method, which is named after its inventors Lagow and Margrave. The LaMar process requires low conversion rates and furnishes the fluorinated products in varying yields. For the preparation of liquid perfluoropolyethers, this process has been scaled up to produce the respective compounds on a "gallons per hour"-scale. The LaMar process is not only applied for the generation of perfluoropolyethers from polyethers, but can, for example, also be used for the preparation of perfluorinated diisopropylether or heterocycles like perfluoro-1,2,6-trifluoromethylpiperidine. Based on the LaMar process, a flow version has been established, which is known as aerosol direct fluorination. This method may be applied to prepare compounds like $C(CF_3)_4$ or $CF_3COC(CF_3)_3$ from their nonfluorinated precursors $C(CH_3)_4$ and $CH_3COC(CH_3)_3$, respectively. The LaMar technique is also suitable for fluorinating finely divided, polymeric hydrocarbons, such as polypropylene, to obtain their perfluorinated analogues in a type of surface fluorination reaction. Mechanistically, all the aforementioned reactions follow radical-chain steps (Scheme 13.4). In order to prevent the cleavage of C-C bonds as well as to ensure safe reactions, it is essential to control F_2 concentrations and conversion rates. Furthermore, the evolving heat of reaction has to be efficiently dissipated. Like in every other free radical mechanism, the first step is the initiation, which is either achieved by homolytic cleavage of F_2 into radicals or the direct interaction of F_2 with a hydrocarbon molecule (Scheme 13.4). Afterwards, a fluoride radical can interact with a hydrocarbon molecule or a carbon-centered radical cleaves a F_2 molecule, however, both steps will contribute to the propagation. Finally, termination is achieved by recombination of either an alkyl and a fluoride radical or two carbon-centered radicals. In the overall reaction, F_2 transforms a hydrocarbon R-H into a fluorocarbon R-F, along with the formation of HF.

initiation:

propagation:

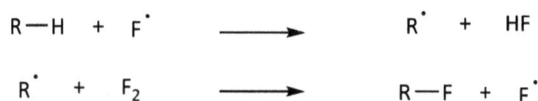

termination:

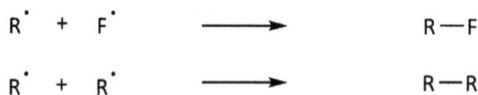

overall reaction:

Scheme 13.4: Reaction steps in the radical fluorination of aliphatic hydrocarbons.

13.2.3 Fluorination using high-valent metal fluorides

Instead of elemental F_2, high-valent metal fluorides are often used for the preparation of highly fluorinated, organic compounds, the most important one being CoF_3. There are also derived tetrafluorocobaltate(III) complexes, such as $K[CoF_4]$ and $Cs[CoF_4]$, which are slightly milder reagents than the parent trifluoride. Other metal fluorides and fluoride complexes used for the fluorination of organic substrates include, for example, AgF_2, MnF_3, CeF_4, PbF_4, BiF_5, $K_2[AgF_4]$, $K_3[NiF_6]$ and $K_2[PtF_6]$. Silver difluoride has shown to be a suitable reagent in liquid-phase fluorinations, whereas CoF_3 is typically applied in vapor phase techniques. Using these reactive, high-valent metal fluorides, one has to be aware of the fact that they basically all require F_2 for their preparation and that they have to be handled under solvent-free conditions in the gaseous or condensed state. Alternatively, fluorination reactions can be carried out in anhydrous HF, which is the only suitable solvent for this purpose. In addition, they often yield complex mixtures of poly- and perfluorinated compounds, accompanied by degradation products and rearranged species, which renders high-valent metal fluorides rather unattractive for preparative purposes. Nevertheless, they are the method of choice for the conversion of certain substrates. For example, cyclic ethers with a fluorinated substituent in α-position are remarkably stable toward CoF_3 and are rather cleanly converted to their perfluorinated analogues at 440 °C. As a consequence, the choice of appropriate reaction conditions, including the selection of a suitable metal fluoride, is essential. In some cases, fluorinated substrates may be used to obtain products, where the hydrogen substituents are not or only partially substituted. A typical example for this is the preparation of CF_3CH_2F from CHF_2CH_3 or $CF_2 = CH_2$. However, typically, complete substitution will occur, since CoF_3 is readily capable of transforming hydrocarbons to perfluorocarbons. For example, C_5H_{12} is perfluorinated to yield mainly C_5F_{12}. The occurring rearrangements are exemplified for the conversion of n-hexane, which also mainly yields the perfluorinated n-hexane derivative. However, branched compounds are obtained, too (Scheme 13.5). Apart from aliphatic hydrocarbons, aromatic moieties are transformed under formation of the fully fluorinated, saturated compounds, for example, the treatment of toluene with CoF_3 yields perfluorinated methylcyclohexane. The treatment of naphthalene with $K[CoF_4]$ results in the almost complete saturation of the polyaromatic system, only one double bond is retained in the resulting hexadecafluoro-1,2,3,4,5,6,7,8-octahydronaphthalene. In contrast, the treatment of naphthalene or tetrahydronaphthalene with CoF_3 at 250 °C yields a complex mixture of nine different regio- and stereoisomers of highly fluorinated octa- and decahydronaphthalene.

Mechanistically, this type of fluorination is believed to proceed via radical-cation intermediates (Scheme 13.5). Initially, CoF_3 oxidizes the hydrocarbon substrate to a radical cation, whereas itself is reduced to a trifluorocobaltate(II) anion. Under evolution of HF and CoF_2, a carbon-centered radical is formed, which in the next step can interact again with CoF_3 to be oxidized to a carbocation. The latter captures a fluoride anion from $[CoF_3]^-$, and the C-F-bond is generated. Upon treatment with elemental F_2, CoF_3 is easily recovered from CoF_2. In principle, this mechanism is similar to the electrochemical fluorination of arenes (cf. Scheme 14.1).

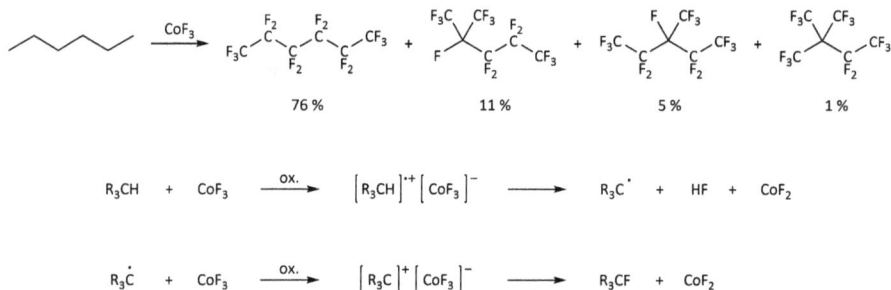

76 %   11 %   5 %   1 %

$R_3CH + CoF_3 \xrightarrow{\text{ox.}} \left[R_3CH\right]^{\cdot +}\left[CoF_3\right]^{-} \longrightarrow R_3C^{\cdot} + HF + CoF_2$

$R_3\overset{\cdot}{C} + CoF_3 \xrightarrow{\text{ox.}} \left[R_3C\right]^{+}\left[CoF_3\right]^{-} \longrightarrow R_3CF + CoF_2$

Scheme 13.5: Perfluorination of *n*-hexane (top) and proposed reaction steps in the oxidative fluorination using CoF$_3$ (bottom).

13.2.4 Selective radical fluorination with electrophilic fluorinating agents

Techniques using catalytic and stoichiometric radical initiators

A rather mild variety of radical fluorination has been reported, where a number of primary, secondary, tertiary, benzylic as well as heteroatom-stabilized radicals is reacted with an electrophilic fluorination agent. The fluorinated products are obtained, when perester-substrates are decarboxylated in presence of excess NFSI in acetonitrile at elevated temperatures (Scheme 13.6) [2837]. Similarly, benzylic as well as secondary and tertiary alkyl fluorides are generated upon treatment of the substrate with *N,N'*-dihydroxypyromellitimide (NDHPI) as radical initiator and Selectfluor™ (Scheme 13.6). NDHPI readily generates *N*-oxy radicals, which are capable of abstracting hydrogen radicals from an electron-rich sp^3-carbon atom. Afterwards, the generated alkyl or benzyl radical is trapped by the fluorinating agent to yield the fluorinated product and an amino radical cation. The latter regenerates *N*-oxy radicals upon reaction with NDHPI, and thus closes the cycle. Interestingly, this method fails, if NFSI or *N*-fluoropyridinium salts are used as fluorinating agents [2838]. An advancement of an initially silver-mediated method (cf. Section 13.6.1) has been developed for the preparation of secondary and tertiary alkyl fluorides, using K$_2$S$_2$O$_8$ as a radical initiator in conjunction with the hexafluorophosphate-salt of Selectfluor™ II in a solvent-mixture of water and acetonitrile (Scheme 13.6). It is assumed that homolysis of the persulphate anion generates sulphate radical anions, which then abstract hydrogen-radicals from the substrate. This approach has proved to be remarkably site-selective, since it exclusively led to the substitution of a tertiary C-H bond of a taxol derivative, leaving all other functionalities of the natural product intact, including a double-bond and a free hydroxy-group. Thus, this method appears to be suitable for late-stage fluorination of complex molecules [2839]. In a related approach, benzylic fluorides are accessible by radical C-H fluorination using K$_2$S$_2$O$_8$ and Selectfluor™ (Scheme 13.6). Selective mono- or difluorination is achieved by varying the amounts of reagents, however, the reaction needs to be carried out in a mixture of water and

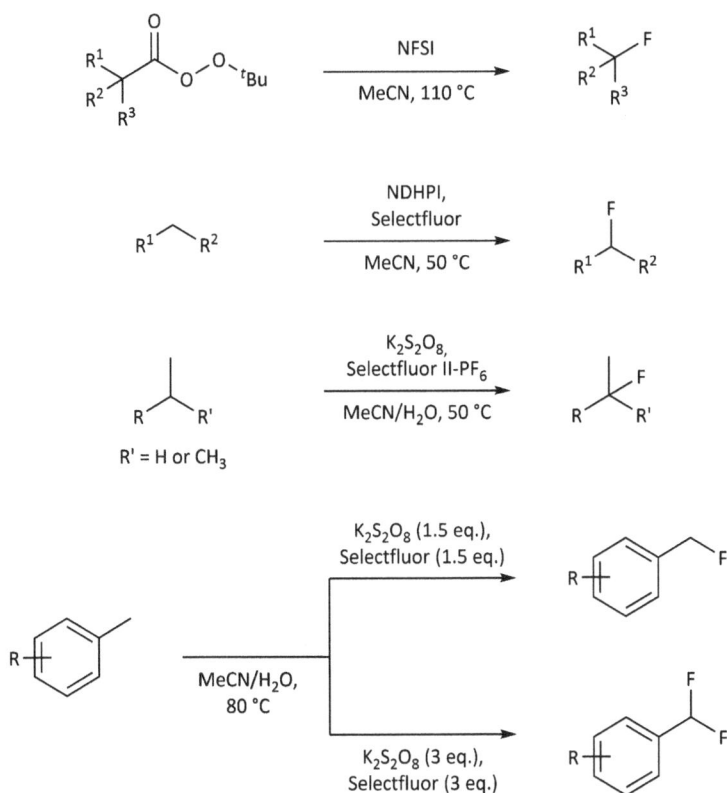

Scheme 13.6: Radical fluorination of perester substrates (top), preparation of secondary and tertiary alkyl fluorides (middle) and synthesis of benzylic fluorides (bottom).

acetonitrile at 80 °C. At a lower temperature, benzaldehyde products are preferentially obtained, which result from oxidation and cannot be further converted to the fluorinated products using the $K_2S_2O_8$/Selectfluor™ system [2840]. In combination with a nucleophilic ^{18}F-radiofluorination, radical fluorination of benzylic positions with $Na_2S_2O_8$ and Selectfluor™ has been reported to give access to ^{18}F-labeled difluoromethyl arenes [2841]. Radical fluorination reactions of benzylic, secondary and tertiary carbon-centers with Selectfluor™ can also be initiated by a combination of BEt_3 and O_2 [2842]. The radical fluorination of heterobenzylic substrates with Selectfluor™ proceeds at room temperature in absence of a separate radical initiator. It is assumed, that the fluorinating agent initially forms a charge-transfer complex with the substrate, which then either undergoes stepwise electron and proton transfer, or proceeds via a concerted, proton-coupled transfer of the electron, in both cases generating a heterobenzylic radical. As usual, the radical is then trapped by Selectfluor™ to produce a heterobenzyl fluoride, which may be derived from various *N*-heterocycles, such as pyridines, pyrimidines or purines [2843].

Photocatalysis

Benzylic mono- and difluorination is also achieved by a photocatalytic approach using xanthone or 9-fluorenone as photosensitizers in conjunction with Selectfluor™ or Selectfluor™-II and visible light. Depending on the choice of the photocatalyst and the fluorinating agent, either mono- or difluorobenzyl products are selectively accessible (Scheme 13.7). 9-Fluorenone in conjunction with Selectfluor™ typically yields monofluorinated products, whereas the more electron-rich xanthone and Selectfluor™-II lead to the formation of geminal difluorides. The method tolerates a variety of functional groups, for example, tertiary alcohols, carboxylic acids or aryl halides. It is proposed that the excited photocatalyst-species abstracts a hydrogen radical from the benzylic position of the substrate to generate an α-hydroxy radical and a benzyl radical, the latter of them being trapped by the fluorinating agent. The intermediately generated amine radical cation afterwards abstracts a hydrogen radical from the α-hydroxy radical and regenerates the photocatalyst [2844]. The addition of an external photosensitizer is not required, if such a function is already included in the substrate. For example, aliphatic C-H bonds are fluorinated upon treatment with Selectfluor™ and 365 nm irradiation, if a phthalimide function is present (Scheme 13.7) [2845]. Apart from the aforementioned systems, a variety of other metal-free photosensitizers may be used, such as tetracyanobenzene, anthraquinone or acetophenone. Virtually all of them are used in conjunction with Selectfluor™ as a fluorinating agent [2846–2849].

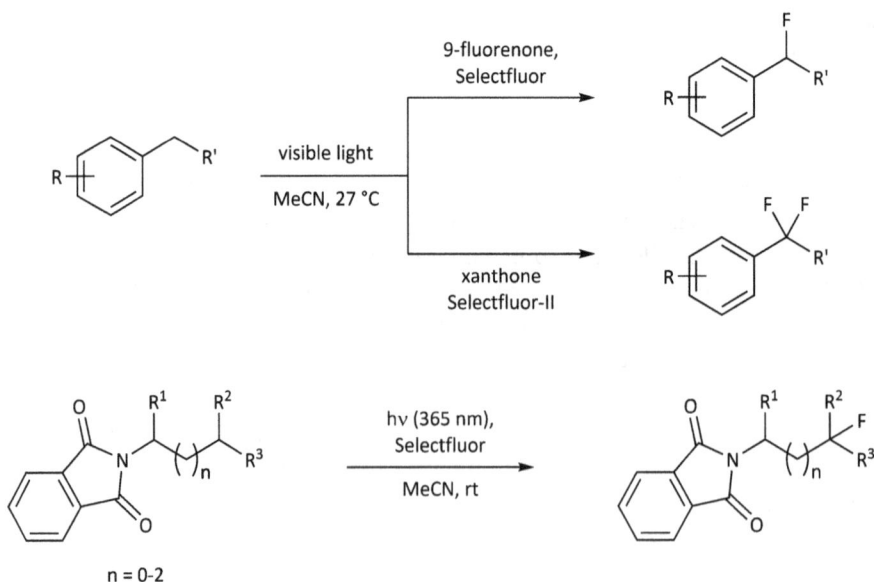

Scheme 13.7: Photocatalytic preparation of benzylic mono- and difluorides (top) and synthesis of alkyl fluorides using a phthalimide-moiety as internal photosensitizer (bottom).

Selective radical fluorination is not limited to substrates with reactive C-H bonds, for example, in benzylic position. It can also be performed in presence of directing groups, as for example, photoexcitable polycyclic enones are selectively fluorinated in β- or γ-position, using Selectfluor™ under UV-irradiation (Scheme 13.8). This method can also be applied for the preparation of bioactive polycycles and steroids, which are site-selectively functionalized, even in presence of up to 65 different $C(sp^3)$-H bonds [2850]. Similarly, β- and γ-fluorination is achieved, when cyclic ketones serve as directing groups in the reaction with Selectfluor™ under visible-light catalysis. Additionally, the presence of benzil is required, which acts as photosensitizer (Scheme 13.8). Interestingly, this method works best for complex, polycyclic substrates, but is not site-selective for linear molecules. Mechanistically, this reaction most likely does not involve photoexcitation of the ketone, but rather proceeds via stepwise electron transfer/proton transfer or concerted, proton-coupled electron transfer [2851].

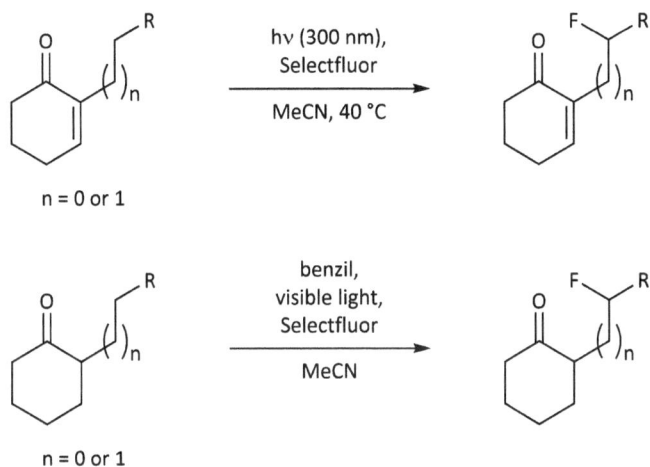

Scheme 13.8: Photocatalytic C-H fluorination directed by cyclic enones (top) and ketones (bottom).

13.2.5 Study questions

(I)   What needs to be taken into account, when preparing perfluorocarbons from hydrocarbons using direct fluorination techniques? Which major approaches are used to meet these requirements?

(II)   How can steric and electronic effects influence the outcome of perfluorination reactions? For which types of substrates does this typically play a role? Briefly explain and give examples.

(III) What is observed upon fluorination of primary, secondary and tertiary alkyl chlorides? Is there a similar behavior of the respective bromo compounds?

(IV)   Which types of products are accessible by the LaMar process? Which measures are typically taken to control the reactions? Why are they necessary?

(V)   Which mechanism does the direct fluorination of hydrocarbon substrates mostly follow? Briefly describe the elementary steps.

(VI)   Which high-valent metal fluorides are typically used for the preparation of perfluorocarbons? Under which conditions are such reactions conducted? Comment on general advantages and disadvantages of the use of high-valent metal fluorides in fluorination reactions.

(VII) Briefly explain the mechanism that has been proposed for fluorination reactions using CoF_3.

(VIII) Which substrates are suitable for radical fluorination using electrophilic fluorinating agents? Do they require an initiator? How can the selectivity of such transformations be rationalized?

(IX)   How can radical C-H fluorination be achieved in photocatalytic approaches? What are typical photocatalysts?

(X)   How can one selectively fluorinate substrates by radical methods in presence of directing groups? Which reagents and directing groups are typically used for this purpose?

13.3 Electrochemical fluorination techniques

Similar to the direct fluorination techniques discussed in the previous chapter, electrochemical fluorination methods are suitable for the preparation of highly or even perfluorinated molecules. Thus, the application of these techniques is very useful in synthetic fluoroorganic chemistry, especially in the production of commercial compounds. Over the years, this topic has often been reviewed, covering preparative procedures, the design of electrochemical cells but also mechanistic aspects [2773]. As a consequence, electrochemical fluorination is only briefly presented herein, focusing on rather general aspects.

13.3.1 Conditions and substrates

The main advantage of electrochemical fluorination is the use of readily available anhydrous HF as fluorine source and solvent at the same time. Fluorination reactions are typically performed at 0 °C in a potential range between 4.5 and 6.0 V. The archetype of these methods is represented by the Simons process. Another profit over the direct fluorination using F_2 or CoF_3 is the tolerance against some important functional groups, such as ethers, amines and carboxylic as well as sulphonic acids. Especially for COOH-groups, the stability arises from immediate conversion to acid fluorides, as free carboxylic acids as well as anhydrides decarboxylate upon electrolysis. In turn, polar functionalities are required to provide a reasonably dissolved starting material. Another significant benefit in this context is the solubility-difference between substrates and products in HF, since the amount of dissolved compound usually decreases with increasing degree of fluorination. The highly fluorinated derivatives often become even immiscible with HF and form a separate, denser phase, which may rather easily be removed. However, the fluorinated product mixtures still contain side-products, which are formed upon skeletal rearrangements, fragmentations or cyclizations [2852]. Moreover, unsaturated substrates will become saturated upon electrochemical fluorination, for example, pyridine is transformed to give mainly perfluoropiperidine and NF_3.

A variety of electrochemical fluorination is the Phillips method. Therein, gaseous substrates like ethane or vapors of liquid compounds are fluorinated at porous carbon anodes in a molten $KF \cdot 2\ HF$ electrolyte. Thus, the reactions are conducted in a fluorine-generator at moderate temperatures of about 100 °C.

13.3.2 Cyclization of carbonyl compounds

Sometimes, originally unintended reactions can lead to valuable products, as in the case of octanoic acid, where it is possible to obtain cyclic perfluoroethers in reasonable yields (Scheme 13.9). The amount of cyclic products decreases with increasing substrate concentration and the cyclization is believed to proceed via hydrogen-bonded onium polycations [2853]. Mixtures of cyclic perfluoroethers and acid fluorides are also obtained from primary alcohols and aldehydes with short carbon chains of four to eight atoms. Cyclic ethers are typically transformed to the respective perfluorinated compounds; however, fluorination is accompanied by rearrangements, typically causing ring contraction or expansion as well as ring scission in some cases. Electrochemical fluorination of α- and β-cycloalkyl-substituted esters yields perfluorinated bicyclic and monospiro ethers, respectively (Scheme 13.9) [2854]. Esters with α-cyclohexenyl-moieties are converted to bicyclic perfluoroethers with and without additional trifluoromethoxy substituents, resulting from the methoxy group of the ester (Scheme 13.9) [2855].

13.3.3 Electrochemical fluorination of heteroatom-containing substrates

Other examples for electrochemical fluorination reactions include the preparation of perfluorinated triethylamine from NEt_3 or the fluorination of polyethers (Scheme 13.10). As some general trend in the electrofluorination of chloroalkyl ethers and amines, chloro-substituents in α-position are readily cleaved, whereas β-chlorine is retained. It is also possible to fluorinate CS_2 under retention of one of the C-S bonds and formation of CF_3SF_5. The Simons process is also used for the fluorination of sulphonyl compounds (Scheme 13.10) [2856, 2857]. In addition, *tris*(perfluoroalkyl)difluorophosphoranes are readily accessible, which may serve as starting materials for ionic liquids, conducting salts and strong Brønsted acids [2858].

13.3.4 Study questions

(I) What are the main advantages of electrochemical fluorination over direct fluorination using F_2 or high-valent metal fluorides?

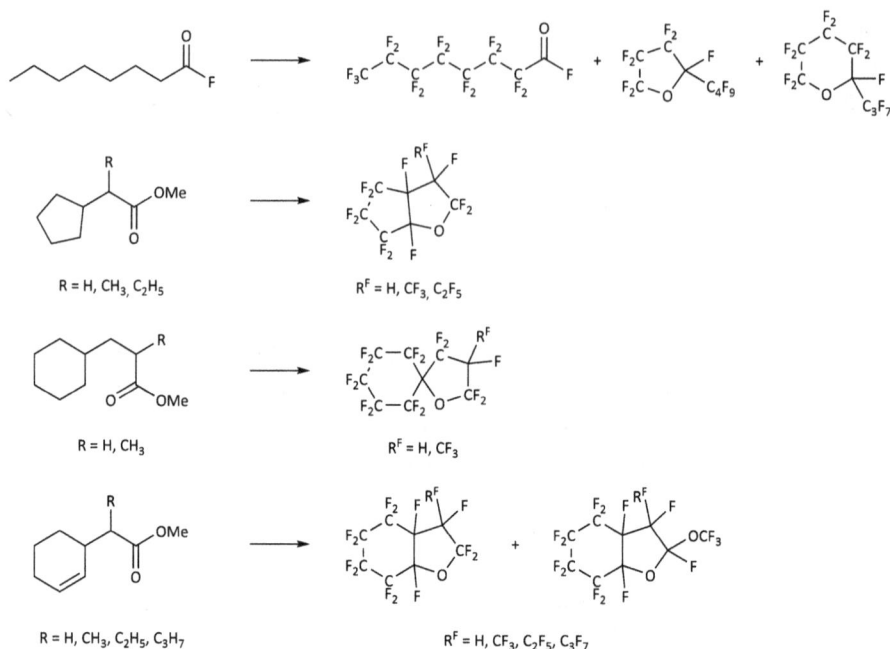

Scheme 13.9: Formation of cyclic perfluoroethers in electrochemical fluorination reactions.

(II)   Which types of compounds are preferentially accessible by electrochemical fluorination? What are the limits of the substrate scope?

(III) Which major side-reaction is observed upon electrochemical fluorination of carbonyl compounds?

13.4 Nucleophilic substitution

13.4.1 Halogen exchange

Halogen fluorides as fluorinating agents

In principal, halogen exchange reactions can be performed with a large variety of reagents, including elemental fluorine or the halogen fluorides BrF_3 and ClF. However, F_2 is not especially suitable for this purpose, as it preferentially reacts with the hydrogen-substituents of the substrate. Nevertheless, F_2 can selectively exchange bromo- and iodo-substituents in a few saturated compounds, for example, 1- and 2-haloadamantanes, in good yields at low temperature. The exchange reactions are believed to proceed via oxidation of the halogen and subsequent replacement of the XF_n-unit by a fluoride ion [2859]. Bromine trifluoride is slightly more useful for halogen replacement reactions, especially in chlorinated substrates. It is very reactive and can exchange chloro-substituents in positions, which are mostly not attacked

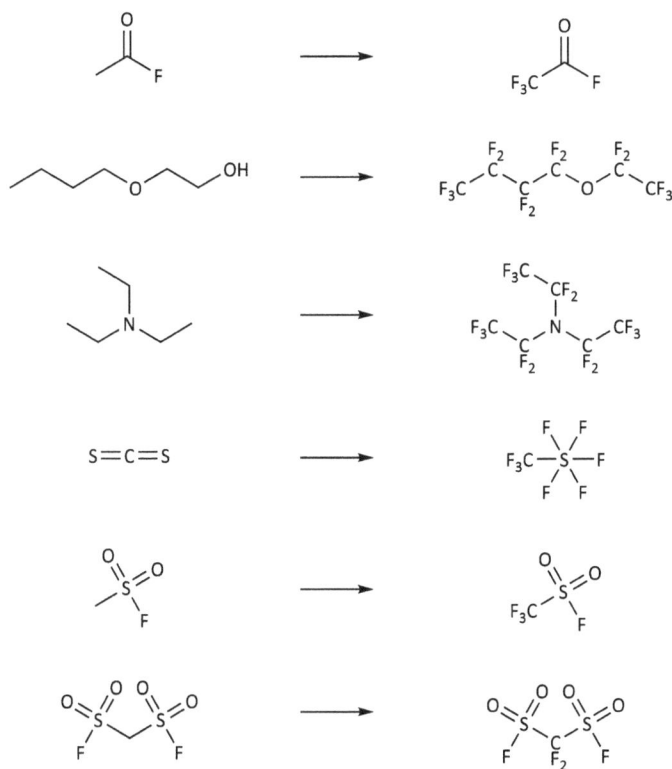

Scheme 13.10: Examples for electrochemical fluorination of oxygen-, nitrogen- and sulphur-containing compounds.

by other reagents, for example, in adjacent location to CF_3-groups (Scheme 13.11). However, due to its high reactivity, it typically generates product mixtures with low selectivity, including the replacement of C-H bonds. In some cases, the addition of a catalyst like $SnCl_4$ can be beneficial to control the product formation. For example, a mixture of BrF_3 and 1% $SnCl_4$ converts 1,3-dichloro-2-fluoropropane exclusively to 1,2,3-trifluoropropane (Scheme 13.11). In 1,1,1,3-tetrachloropropane, the substitution of tertiary chloro-substituents is favored over the replacement of the primary chlorine (Scheme 13.11). Apart from the catalyst, the solvent significantly influences the product formation. If the aforementioned reaction of 1,3-dichloro-2-fluoropropane with BrF_3 is stopped at 40 % conversion, contrary results are observed in CCl_2F-CF_2Cl (Freon 113) and HF. In the chlorofluorocarbon solvent, a clean halogen exchange of one of the chloro-substituents in CH_2Cl-CHF-CH_2Cl occurs, whereas HF additionally promotes the cleavage of C-H bonds, resulting in the formation of CH_2Cl-CF_2-CH_3 and CHF_2-CHF-CH_2Cl.

The use of ClF in halogen exchange reactions has been studied in more detail. In general, vicinal halogens slow down the exchange of the leaving halogen,

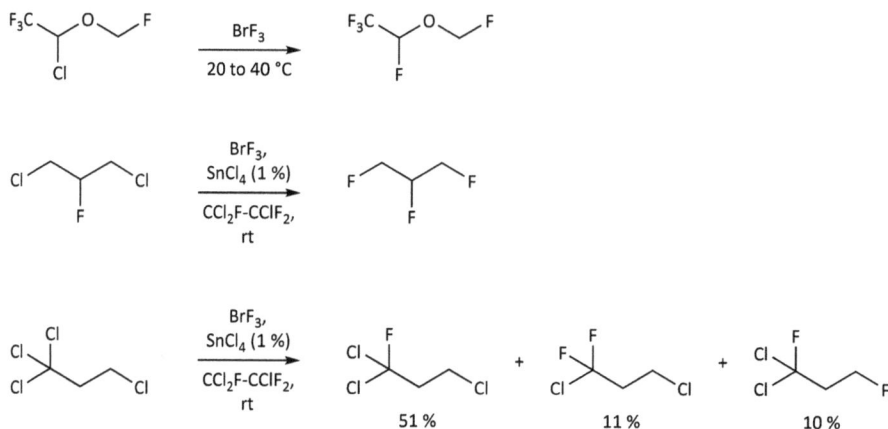

Scheme 13.11: Halogen exchange of alkyl chlorides using BrF$_3$.

whereas geminal halogen substituents do not show a significant impact. Additionally, rearrangements likely occur, if primary alkyl bromides with vicinal chloro- or bromo-substituents are fluorinated. Often, hydride shifts are observed, if stable carbocations can be formed. The reaction of 1,2,3-trichloropropane with one equivalent of ClF in HF between –30 and 0 °C initially leads to the exchange of the secondary chloro-substituent, however, a second equivalent of ClF yields the rearranged product 1-chloro-2,2-difluoropropane (Scheme 13.12). The addition of a third equivalent of chlorine monofluoride does not lead to any further transformation. This behavior can be changed upon addition of SbCl$_5$, which increases the reactivity of ClF against 1,2,3-trichloropropane, so that an exchange of the primary chloro-substituents can occur, too. Nevertheless, the product mixture mostly contains the secondary fluoride. The successive halogen exchange of 1,1,1,2-tetrachloroethane with ClF initially replaces two chloro-substituents of the CCl$_3$-group. Only in the third step, a beginning substitution of the CH$_2$Cl-group is observed. Finally, upon addition of the fourth equivalent of ClF, 1,1,1,2-tetrafluoroethane is obtained. The regioselectivity of the chlorine exchange can also be influenced by substituents. In methyl 2,3-dichloro-3-fluoro-2-methyl-propionate, the primary chlorine substituent is replaced prior to the tertiary one. In contrast, the primary chloro-substituent of methyl 2,2,3-trichloropropionate remains untouched upon fluorination (Scheme 13.12).

Fluoride salts, HF-reagents or antimony fluorides as fluorinating agents
The simplest method for the preparation of alkyl fluorides is certainly the halogen exchange of respective chlorides, bromides or iodides using readily available nucleophilic reagents like HF or alkali metal fluorides. For industrial products, chloride starting materials are conveniently used. The halide exchange of polychloroalkanes is an important method to access hydrofluorocarbons (HFCs), hydrochlorofluoro-

ClF, -30 to 0 °C   ClF, -30 to 0 °C

ClF, aHF, -60 to 0 °C   ClF, aHF, -60 to 0 °C

ClF, aHF, -60 to 0 °C   ClF, aHF, -60 to 0 °C

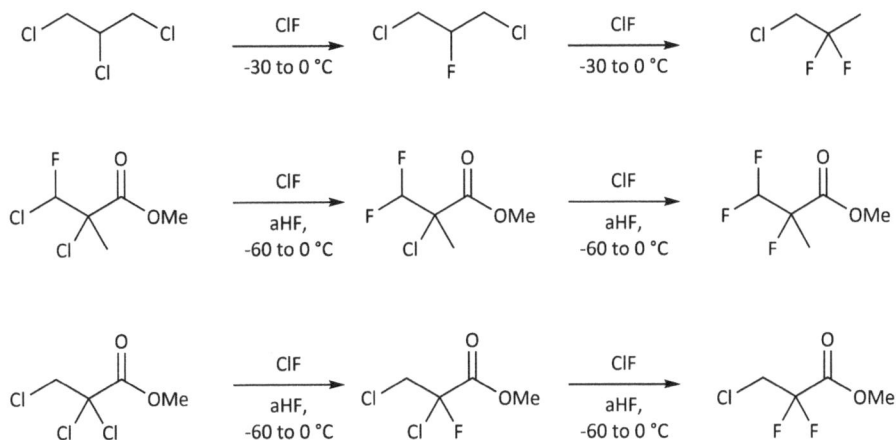

Scheme 13.12: Halogen exchange of alkyl chlorides using ClF.

carbons (HCFCs) and chlorofluorocarbons (CFCs). Typically, anhydrous HF or SbF_3 are used as fluorinating agents, in addition, antimony chlorofluorides $SbCl_xF_{5-x}$, chromium-based catalysts or other Lewis acids are often required to activate the leaving groups. The Lewis acid assisted fluorination is also often referred to as Swarts-fluorination. The outcome of these transformations is generally influenced by the reaction conditions. However, complete substitution is mostly not observed, as for example, CCl_4 is transformed to CF_xCl_{4-x} ($x = 1$–3). The treatment of chloroform with $SbCl_xF_{5-x}$ in anhydrous HF yields the mono- and disubstituted $CHFCl_2$ and CHF_2Cl. In contrast, 1,1,1-trichloroethane is converted to CF_3CH_3. The gas-phase fluorination of perchloroethane CCl_3CCl_3 with anhydrous HF in presence of a Cr-catalyst yields a mixture of five products with one to five fluorine substituents. The reaction of 1,1,1,3-tetrachloropropane $CCl_3CH_2CH_2Cl$ with $SbF_3/SbCl_5$ yields 1 chloro 3,3,3 trifluoropropane $CF_3CH_2CH_2Cl$. Swarts fluorination can also be applied in carbohydrate chemistry, if significantly milder conditions are applied. For example, AgF or ZnF_2 are used in CH_3CN solution at 20 or 70 °C, respectively. Based on this variation, glycosyl fluorides are accessible from the corresponding bromides.

A slightly different approach is used for the halide exchange in geminal di-halogeno-alkanes and -cycloalkanes. In this case, $Ag[BF_4]$, SbF_3 or HgF_2 serve as fluorinating agents, the latter may also be generated *in situ* by treatment of HgO with anhydrous HF. The reaction of these reagents with the 1,2,2-trihalopropanes $CH_2XCX_2CH_3$ (X = Cl, Br) gives rise to the respective compounds $CH_2XCF_2CH_3$. Analogously, 1,1-dichlorocyclohexane is transformed to 1,1-difluorocyclohexane.

Common reagents for the substitution of alkyl monohalides are KF, AgF, "CuF" (*in situ* generation from Cu_2O and anhydrous HF) and tetraalkylammonium salts, especially $[NBu_4]F$. Over the years, a few more special sources of F^- have been reported, which are capable of replacing alkyl halides [2860]. For better results, nucleophilic fluorinating agents should be applied under anhydrous conditions. However, one has to be aware of the fact, that nonsolvated, "naked" fluoride is a rather strong base, which will readily induce elimination reactions. This especially holds for secondary and even more so for tertiary substrates. Typically, anhydrous KF is used in polar aprotic solvents, for increasing its solubility, crown ethers or cryptands are added. Furthermore, spray-dried KF is conveniently used, as it has a larger surface area and is thus a more efficient fluoride source. Metal fluorides can also be applied under solvent-free, supercritical conditions, where the overheated reactant dissolves KF and NaF. Using this method, chloromethyl ethers can be substituted (Scheme 13.13). Similarly, a secondary chloride is substituted under supercritical conditions in difluoromethyl 2-chloro-1,1,1-trifluoroethyl ether. However, better yields are achieved, if the latter reaction is carried out in highly polar solvents like sulpholane or DMF in the presence of a phase-transfer catalyst, for example, [NMe$_4$]Cl, or a crown-ether (Scheme 13.13). Instead of quaternary ammonium salts, the respective phosphonium analogues may be used.

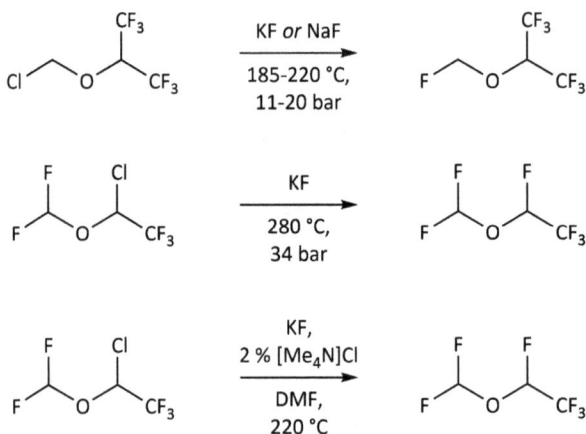

Scheme 13.13: Halogen exchange under supercritical- and phase-transfer conditions.

A general advantage of halogen exchange reactions using nonvolatile reagents is the possibility that the generated, often volatile fluoroorganic compounds can be simply distilled out of the reaction mixture. Silver fluoride and Ag[BF$_4$] can be used under milder reaction conditions and have the additional benefit that the halogen exchange is accompanied by the formation of insoluble AgX. However, a quite high price has to be paid for this complementary thermodynamic driving force, as silver salts are rather expensive. Active copper-fluorine species for halogen exchange reactions are typically generated *in situ* from the respective copper oxide and HF. In conjunction with bidentate ligands, for example, 2,2'-bipyridine, they can effectively displace halogen substituents in primary alkyl halides.

Halogen exchange reactions in activated substrates, that is, allylic and benzylic positions, are also suitable for the generation of fluoroorganic compounds. In general, chlorine in allylic positions is exchanged more readily than in polychloroalkanes. In contrast, substitution of vinylic Cl-substituents is hardly observed. Suitable reagents for allylic fluorinations are again the abovementioned combinations of anhydrous HF or SbF_3 with $SbCl_5$, in some cases, no catalyst is required. The substitution of allylic and especially benzylic halides can also be achieved under mild conditions by the use of ammonium or phosphonium salts, such as $[NBu_4][HF_2]$, $[Et_3PMe]F$, $[Bu_3PMe]F$ or $[PPh_4][HF_2]$. Interestingly, perchloropropene $CCl_3CCl=CCl_2$ is fluorinated by anhydrous HF to give 1,1,2-trichloro-3,3,3-trifluoro-2-propene, whereas treatment of perchlorobuta-1,3-diene with $SbF_3/SbCl_5$ yields 2,3-dichloro-1,1,1,4,4,4-hexafluoro-2-butene. The latter reaction can be explained by a 1,4-addition of chlorine, which occurs prior to halogen exchange. As a typical example for benzylic fluorination reactions, benzotrichloride $PhCCl_3$ is cleanly converted to $PhCF_3$ upon treatment with anhydrous HF. Instead, the reaction of the respective perchloroethyl derivative $PhCCl_2CCl_3$ yields a mixture of $PhCF_2CCl_3$, $PhCF_2CFCl_2$ and $PhCF_2CF_2Cl$, but no $PhCF_2CF_3$, which would have a fully fluorinated side-chain. If an additional chlorine substituent is attached to the aromatic moiety, as for example in $p\text{-}ClC_6H_4CCl_3$, it is not affected by the substitution reaction with SbF_3, which only yields $p\text{-}ClC_6H_4CF_3$. However, this behavior might change upon presence of an increased amount of halogen substituents at the aromatic scaffold (cf. Section 14.2.1). Di- and trichloromethyl arenes are furthermore fluorinated by HF in presence of basic catalysts or solvents, for example, pyridine, triethylamine or dioxane. In addition, the presence of catalytic amounts of SbF_3 or $SbCl_5$ has been found to be beneficial [2861–2863].

Halogen exchange can also take place in α-position of aliphatic carbonyl compounds. Typically, KF is used for this purpose, as it for example, converts methyl chloroacetate $CH_2ClCOOCH_3$ to methyl fluoroacetate $CH_2FCOOCH_3$. Under milder conditions, AgF is capable of substituting α-bromo carbonyl compounds like $C_6H_{13}CHBrCOOCH_3$. More recently, tetrabutylammonium-salts such as $[NBu_4][HF_2]$ or $[NBu_4]F \cdot 3\,H_2O$, have been used for the fluorination of α-bromoketones, sometimes in conjunction with metal fluorides like KF or ZnF_2 [2864–2866]. Furthermore, KF transforms carbonyl chlorides, such as benzoyl chloride and trichloroacetyl chloride, to the respective carbonyl fluorides C_6H_5COF and CCl_3COF. In the latter case, the chloro-substituents in the acetyl-moiety remain unaffected upon reaction with KF, however, if anhydrous HF is used as fluorinating agent in presence of a catalyst, all chloro-substituents are substituted to give trifluoroacetylfluoride CF_3COF. Analogously, anhydrous HF converts perchloro acetone $CCl_3C(O)CCl_3$ to the perfluorinated compound $CF_3C(O)CF_3$. A variant of nucleophilic α-fluorination of 1,3-dicarbonyl compounds has been described using aqueous HF (w = 55 %) and a hypervalent iodine reagent ArIO. Initially, a carbon-iodine bond is formed, which is afterwards replaced by F^- [2867, 2868]. The same method can be adapted for enantioselective α-fluorination, if a chiral iodoarene, that is, 2,2'-diiodo-1,1'-binaphthyl, is used together with a peracid-oxidant and pyridine/HF [2869]. Apart from that, α-fluorinated β-ketoesters are accessible by treatment of the corresponding enantioenriched α-chloro-compounds with CsF and 18-crown-6. The halogen exchange proceeds under full inversion following an S_N2-type mechanism and retains the enantiopurity of the starting material [2870].

13.4.2 Deoxyfluorination of alcohols

HF-reagents and other acidic fluoride sources

A very convenient approach for the preparation of aliphatic fluorides is the fluorodehydroxylation of alcohols. For tertiary substrates, this can directly be realized upon treatment with anhydrous HF or other acidic fluoride sources. These transformations generally follow an S_N1-mechanism and will thus lead to product mixtures, unless thermodynamic control favors or disfavors one stereoisomer. For leaving groups other than -OH, activation is achieved by the use of amine-HF complexes, for example, pyridine/HF. The reactivity of Olah's reagent toward alcohol substrates depends on the position of the hydroxy group. Tertiary alcohols, such as *tert*-butanol or 1-adamantanol react already below 0 °C, whereas secondary alcohols like 2-butanol require slightly higher reaction temperatures in the range of 20–50 °C. Primary alcohols do not react well with pyridine/HF, they also need additional fluoride ions to be converted to primary fluorides. For example, 1-octanol or neopentyl alcohol are reacted with Olah's reagent and NaF to form 1-fluorooctane or neopentyl fluoride, respectively. In benzylic position, hydroxy-groups are readily replaced by fluorine using Olah's reagent. This works especially well for 1-chloro-2-hydroxy-2-phenylethanes, which are readily fluorinated at room temperature. In principle, the conversion of the alcohols is nearly quantitative, however, the yield of fluorinated products depends on their stability. The reaction of benzylic α,β-aminoalcohols with pyridine/HF yields α,β-fluoroamines. However, cyclopropyl methanols undergo fluorination followed by a rearrangement toward homoallylic fluorides, when they are treated with a mixture of HF, pyridine, KHF_2 and $^i Pr_2NH$.

Fluoroalkylamine reagents

Other very common reagents for fluorodehydroxylation include the fluoroalkylamine reagents (FAR) $(C_2H_5)_2NCF_2CHFCl$ (Yarovenko's reagent) and $(C_2H_5)_2NCF_2CHFCF_3$ (Ishikawa's reagent) [2803, 2871]. Similarly, the fluoroenamine reagent $(CH_3)_2C= CFN(CH(CH_3)_2)_2$ is capable of fluorinating primary alcohols, but in a neutral reaction medium [2805]. In contrast to pyridine/HF, Yarovenko's reagent stereospecifically fluorinates benzylic α,β-aminoalcohols to give α,β-fluoroamines with retention of the configuration. Under similar conditions, ethyl 2-hydroxy-2-arylacetates are converted to 2-fluoro-2-arylacetates with moderate stereoselectivity, resulting in an enantiomeric excess of 46 %. The fluorination of bridgehead hydroxy-groups, for example, in bi-cyclo[2.2.2]octanols is achieved by heating the substrate together with Yarovenko's reagent under solvent-free conditions. Fluorinated carboxylic acids are converted to the respective acid fluorides, which do not require any initiation for short-chain derivatives. In contrast, longer chain acids do not react exothermically and need to be heated. In practice, Yarovenko's reagent has only scarcely been used, which is mainly due to its short shelf-life and thus its limited commercial availability. A more stable variant has been developed, embedding the reagent in a styrene polymer. The polymer-supported fluoroamine is not significantly less active and can be stored for months.

The stability problem is overcome in Ishikawa's reagent, which does not de-compose, so that it can still be used months after its preparation. Often, the ac-tive fluoroalkylamine reagent $(C_2H_5)_2NCF_2CHFCF_3$ is contaminated with a small amount of $(C_2H_5)_2NCF=CFCF_3$. The latter olefinic species does not possess any flu-orinating capability, however, upon reaction with HF, generated as a side prod-uct from the conversion of hydroxyl-functions, it is converted to the active com-pound. Ishikawa's reagent shows good fluorinating properties; however, consider-able amounts of by-products are typically obtained (Scheme 13.14). The fluorination of lower aliphatic primary and benzylic alcohols yields exclusively alkyl fluorides. The conversion of higher primary alcohols is mainly accompanied by the formation of 2,3,3,3-tetrafluoropropionates. Unsaturated alcohols as well as saturated substrates, which are branched in C2-position, exclusively give rise to the ester product. Sec-ondary and tertiary alcohols tend to be transformed to the corresponding fluorides, mostly accompanied by significant amounts of alkenes or dialkyl ethers. In con-trast, Ishikawa's reagent efficiently converts secondary benzylic hydroxy esters to the respective secondary benzyl fluorides (Scheme 13.14). This reaction shows a high de-gree of stereospecificity and proceeds with inversion of the configuration. For tertiary benzylic hydroxy esters, lower yields of fluorinated products are observed, as a sub-stantial amount of substrate is dehydrated to give 2-aryl acrylates. Ishikawa's reagent also introduces fluorine substituents into secondary aliphatic α- and β-hydroxy esters (Scheme 13.14). The β-substituted substrates are mostly converted to the respective fluorides, whereas the α-hydroxy compounds preferentially give rise to tetrafluoro-propionates.

Apart from Yarovenko's and Ishikawa's reagent, a few more α,α-difluoroamine reagents have been described, for example, $(C_2H_5)_2NCF_3$ and $(CH_3)_2NCF_2C_6H_5$. They are both quite stable compounds, which are capable of exchanging hydroxyl-groups in alcohols and carboxylic acids. In general, their fluorinating ability can be compared with that of the other fluoroalkylamine reagents. $(C_2H_5)_2NCF_3$ is suitable for the fluo-rination of secondary and tertiary alcohols as well as small carboxylic acids, but it does not fluorinate primary alcohols and higher acids. In contrast, the difluorobenzy-lamine reagent $(CH_3)_2NCF_2C_6H_5$ shows good results in the transformation of primary

$Et_2NCF_2CHFCF_3$

CH_2Cl_2, rt

50-75 %

$Et_2NCF_2CHFCF_3$

CH_2Cl_2, rt

52-60 %   15-20 %

$Et_2NCF_2CHFCF_3$

CH_2Cl_2, rt

R = CH$_3$ (38/0), C$_2$H$_5$ (50/3),
C$_5$H$_{11}$ (81/10), C$_{11}$H$_{23}$ (41/20)

$Et_2NCF_2CHFCF_3$

CH_2Cl_2, rt

R = H (10/31), CH$_3$ (13/42),
C$_4$H$_9$ (3/10), C$_6$H$_{11}$ (1/3)

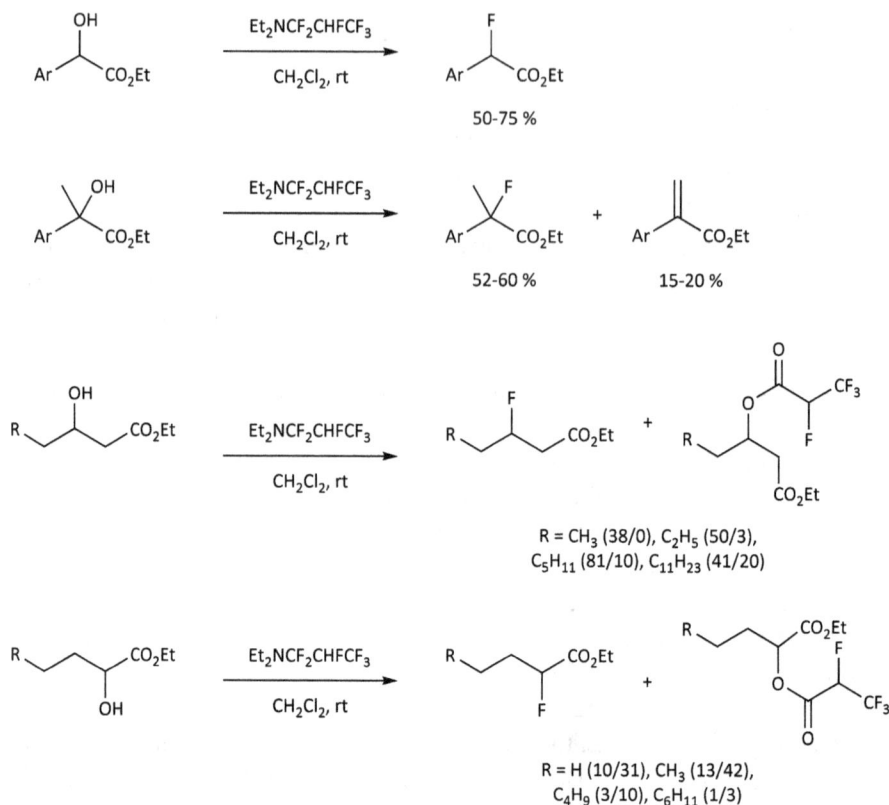

Scheme 13.14: Fluorodehydroxylation using Ishikawa's reagent.

and secondary alcohols but only moderate yields for *tert*-butanol. More recently, α,α-difluorobenzylamine derivatives have been described, which contain a substituted arene-ring. *N,N*-diethyl-α,α-difluoro-(*m*-methylbenzyl)amine selectively deoxyfluorinates alcohols to generate the corresponding fluorides [2872]. Interestingly, the reaction with 1,2- and 1,3-diols stops at the monofluoride stage and results in the formation of fluorinated esters, which are believed to be formed through a cyclic intermediate [2873]. Similarly, another derivative of difluorobenzyl amines, carrying fluorinated $CH_2CH_2C_8F_{17}$-side-chains on the arene-system efficiently deoxyfluorinates monoalcohols and converts diols to the corresponding fluorinated esters [2874]. The addition of several secondary amines to 1,1,3,3-pentafluoropropene results in the formation of compounds of the type $R_2NCF_2CH_2CF_3$, which are also capable of fluorinating primary, secondary and even tertiary alcohols [2804].

Another versatile fluorinating agent in the α-fluoroalkylamine family is 2,2-difluoro-1,3-dimethylimidazolidine (DFI). Due to the presence of two stabilizing nitrogen-atoms in close proximity of the active difluoromethylene-unit, it is even more reactive than the other reagents discussed above. It can also be used for the conversion of carbonyl-groups to CF_2-units [2809]. PhenoFluor™ is a more complex variant of DFI, where the methyl-substituents have been replaced by bulky 2,6-diisopropylphenyl-groups. Initially, this reagent has been developed for the deoxyfluorination of phenols and heteroaromatic alcohols (cf. Section 14.2.4), but also proved its utility for the selective late-stage preparation of complex alkyl fluorides, for example, in natural products or pharmaceuticals [2811, 2812, 2875]. A noncyclic relative of DFI, that is, tetramethylformamidinium hexafluorophosphate has been used for deoxyfluorination reactions under mild conditions and produces alkyl, benzyl and allylic fluorides. However, for satisfying yields it requires the presence of an additional fluoride source, such as $Et_3N \cdot 3\ HF$ [2876].

In order to explain the reactivity of all the α,α-difluoroalkylamine reagents, one has to consider the resonance stabilization of the neutral amine by forming an iminium cation and a corresponding fluoride anion (Scheme 13.15). The iminium-species is very susceptible to a nucleophilic attack of the hydroxyl group of the substrate. Afterwards, the remaining α-fluorine substituent leaves the electron-rich reagent-substrate complex and substitutes the C-O bond of the starting material. Upon fluorination of the substrate, the reagents are transformed to amides, as depicted in Scheme 13.15. In addition, one equivalent of HF is always generated. In general, the fluoroalkylamine reagents, except DFI, are rather tolerant against other functional groups, as for example alkyl bromides or carbonyl functions remain unaffected.

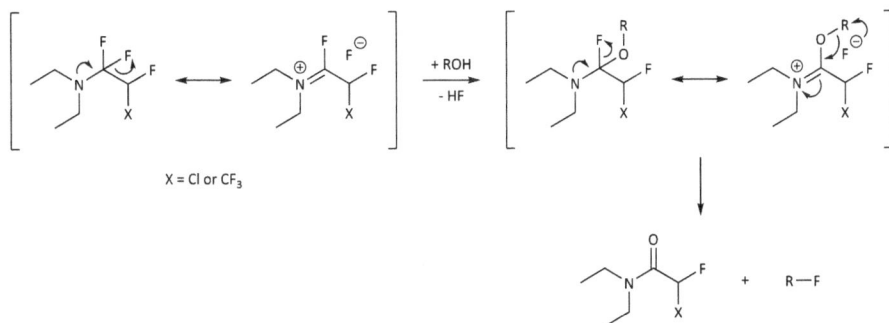

Scheme 13.15: General mechanism for fluorodehydroxylation reactions with fluoroalkylamine reagents.

Sulphur(IV)-based deoxyfluorinating agents

A large and important class of reagents for fluorodehydroxylation is based on SF_4 and its derivatives. Upon reaction with alcohols R-OH, SF_4 is converted to SOF_2, along with the fluorinated product R-F, one equivalent of HF is formed. The use of sulphur tetrafluoride itself has some major disadvantages, as it is a toxic gas, which requires metal autoclaves as reaction vessels. Despite its rather difficult handling, SF_4 is an excellent reagent for the transformation of alcohols to fluorides. However, direct reactions without any additives usually do not yield fluorinated products, which is due to decomposition or polymerization side-reactions. In presence of HF-scavengers, for example, NEt_3 or pyridine, even sensitive benzylic alcohols can be fluorinated, as well as 2-phenylethanol or 2-furylmethanol. In reactions with 1,2-diols, SF_4 initially generates fluorosulphites, which are converted to monofluoro alcohols upon hydrolysis. A selective fluorination of α-hydroxy ketones toward α-fluoro ketones is observed in Et_2O, which probably acts as HF-scavenger. In contrast, the presence of HF is strongly required for the fluorination of hydroxyamines and hydroxyamino acids, where it serves as a protecting agent for the NH_2-functions, as a catalyst and finally also takes the role of the solvent. Predominantly, the replacement of the hydroxy group occurs under inversion of configuration. The treatment of highly fluorinated tertiary alcohols with SF_4 typically results in elimination reactions. Nevertheless, a few examples are known, where fluorination takes place, one of them being represented by hexafluoro-2-aryl-2-propanols, which are converted to heptafluoroisopropyl-substituted arenes. A similar reaction takes place in lower yield, when the hexafluoro-2-propanol moiety is attached to a saturated cyclohexyl ring.

As already implied above (cf. Section 12.3.3), the application of SF_4 itself is rather limited in fluorodehydroxylation reactions, which can mostly be attributed to its hazardous properties. Hence, its liquid substitutes of the type R_2NSF_3, especially DAST (R = Et), are commonly used for this purpose. In the same way, the sulphiminium compounds XtalFluor-E™ and XtalFluor-M™ can be used for fluorodehydroxylation. Upon reaction with alcohols, they do not generate HF, however, the addition of amine-HF complexes is often necessary to obtain satisfying amounts of deoxyfluorinated products [2796, 2797]. Aryl sulphurtrifluorides, such as Fluolead™, convert a variety of different alcohols to the corresponding fluorinated compounds, including alkyl, benzyl and glycosyl substrates as well as diols [2798]. Apart from hydroxy compounds, they are also capable of transforming carboxylic acids to acyl fluorides [2799].

Compared to the parent SF_4, the fluorinating ability of all the aforementioned reagents is only slightly reduced, for example, $(C_2H_5)_2NSF_3$ converts 1-octanol to 1-fluorooctane and 2-bromoethanol to 1-bromo-2-fluoroethane, respectively. Apart from that, tertiary alcohols and benzylic hydroxy groups are transformed, but carboxylic acids remain unchanged, as observed in the reaction of $C_6H_5CH(OH)COOH$ toward $C_6H_5CHFCOOH$ [2793]. Allylic and propargylic fluorides are also very commonly prepared by deoxofluorination reactions [2877]. Typically, fluorinations with DAST are carried out in slightly polar, aprotic solvents, such as CH_2Cl_2, $CFCl_3$ or mono- and diglyme, however, the use of nonpolar solvents, for example, benzene or toluene has also been reported. More recently, ionic liquids have been applied as recyclable reaction media [2878]. In most cases, the reactions are started at $-78\,°C$ and allowed to warm to room temperature; sometimes they are directly performed at $0\,°C$ or room temperature. Due to their mild reaction conditions, fluorinations with DAST show a good chemoselectivity, as many other functional groups remain intact. This includes phenolic hydroxy groups and a number of carbonyl-compounds, since hydroxy esters, ketones, lactones, lactams and nitriles are only substituted at their OH-group(s) to yield the respective fluorinated compounds (Scheme 13.16).

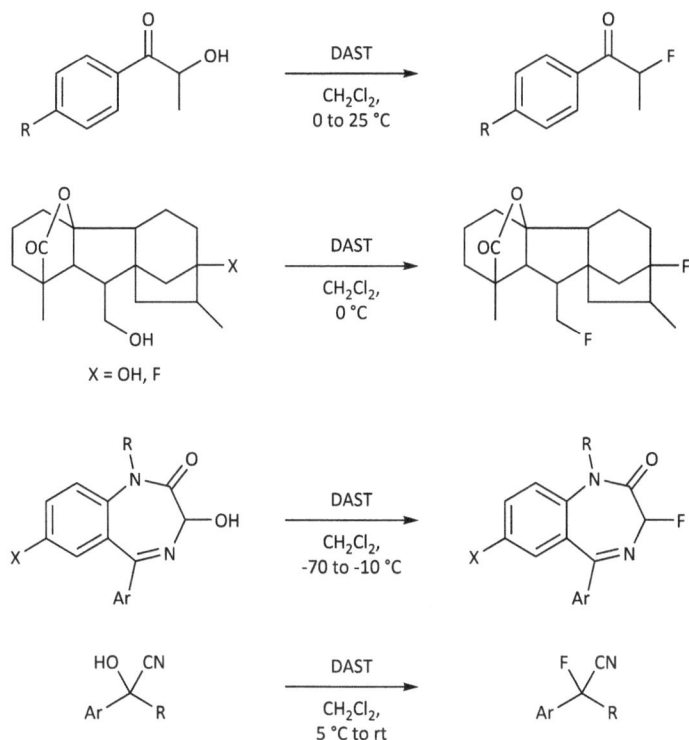

Scheme 13.16: Selective fluorodehydroxylation of hydroxy ketones, -lactones, -lactams and -nitriles.

Fluorodehydroxylation reactions with DAST and its derivatives typically follow an S_N2-mechanism, as depicted in Scheme 13.17, however, elimination products are often observed, especially in case of steroidal substrates. In addition, diphenylmethanol derivatives typically tend to undergo intermolecular dehydration, resulting in the formation of ether by-products. Furthermore, rearrangements can occur with some secondary alcohols, since carbocationic intermediates may sometimes be involved in the reaction mechanism [2793]. For example, isobutanol is converted to a mixture of iso- and *tert*-butylfluoride. Rearrangements are also observed upon reaction of DAST with allylic alcohols, as both crotyl alcohol and 3-butenol yield mixtures of crotyl fluoride and 3-fluorobutene. However, the secondary allyl alcohol is rearranged to a far smaller extent than the primary one. Furthermore, the product ratio is slightly influenced by the choice of the solvent (Scheme 13.18). A few substrates, especially

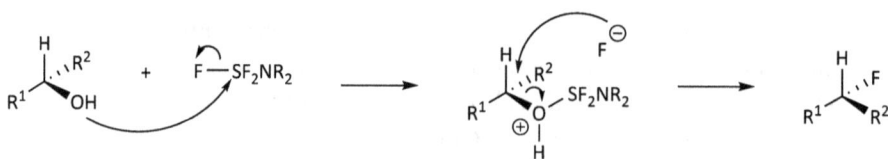

Scheme 13.17: General mechanism of fluorodehydroxylation reactions with DAST-derivatives.

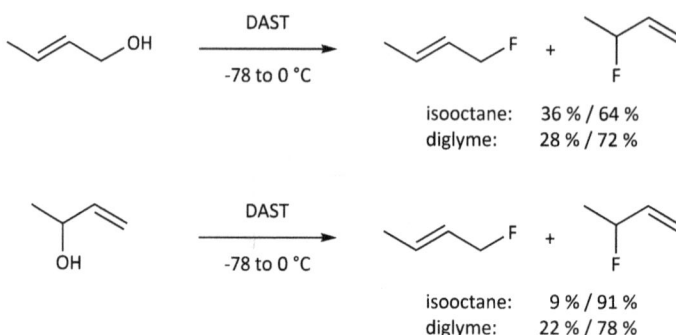

isooctane:	36 % / 64 %
diglyme:	28 % / 72 %

isooctane:	9 % / 91 %
diglyme:	22 % / 78 %

Scheme 13.18: Reactions of crotyl alcohol and 3-butenol with DAST in different solvents.

benzylic alcohols, are prone to undergo S_N1-type substitutions upon deoxofluorination, which leads to two major problems. On the one hand, elimination reactions can occur to yield styrene derivatives, on the other hand, the fluoride substitution of a benzyl-cation will lead to a loss of stereopurity. These problems have been overcome, when silylamines, such as Et_2NSiMe_3 or N-trimethysilylmorpholine, were added to the reaction mixture, which suppress the S_N1-pathway and, therefore, lead to stereospecific deoxyfluorination. Applying the silylamine additive in conjunction with DAST, Fluolead™ or DeoxoFluor™ converts enantiopure benzylic alcohols to fluorides under inversion of the configuration, whereas almost racemic product mixtures are obtained, if the reactions are carried out without the additive [2879–2882].

In some cyclic alcohols, fluorodehydroxylation can also lead to skeletal rearrangements, resulting in ring contraction or -expansion [2883, 2884]. In most cases, reactions with DAST proceed under complete inversion of configuration, which is especially useful for the preparation of fluorinated carbohydrates and steroids. Nevertheless, neighboring groups can have a major impact on the stereoselectivity of fluorodehydroxylations, which may result in partial or even complete retention of configuration in rather complex substrates [2885, 2886]. In polyhydroxyl compounds, especially carbohydrates, a maximum of two OH-groups may be replaced, a higher degree of substitution is not achieved, even if a large excess of DAST is used. Another drawback arises from the structural properties of the fluorodehydroxylating reagent. Due to the increased steric crowding of the nitrogen-based substituents, it is sometimes difficult to access sterically shielded carbon-atoms. Attempts have been made to develop chiral DAST-type reagents for the asymmetric substitution of racemic alcohols. However, the results have been rather poor, as the generated fluorocompounds show low enantiomeric excess [2887].

Another sulphur-based deoxyfluorinating agent for the conversion of primary and secondary alcohols under basic conditions is pyridine-2-sulphonyl fluoride (PyFluor), which typically avoids the formation of elimination products. Along with PyFluor, stoichiometric amounts of sterically hindered bases, such as DBU are required. The reagent can also be used for the introduction of ^{18}F-substituents [2888].

Fluorodehydroxylation of activated alcohol functions

A two-step version of fluorodehydroxylation reactions of alcohols is described by the reaction of activated substrates with fluoride ions. Prior to the fluoride substitution, the hydroxy group is transformed to a sulphonic acid ester, that is, a tosylate, mesylate or triflate, which are far better leaving groups. Subsequently, the addition of an ionic fluoride leads to the formation of the C-F bond. This method works very well for the preparation of primary and secondary alkyl fluorides. Depending on the stability of the substrate, KF may be used as a fluorinating agent in hot polar-aprotic solvents. If milder reaction conditions are required, more active fluoride sources like CsF or tetraalkylammonium fluorides are suitable fluoride donors to be used in CH_3CN or tetrahydrofuran. A mild fluorination of alkyl triflates has, for example, been described using $[NBu_4][HF_2]$ as a fluorinating agent [2889]. It has been reported that ionic liquids enhance the reactivity of alkali metal fluorides in substitution reactions of alkyl sulphonates [2890, 2891]. In the same way, oligoethylene glycols can improve the nucleophilic fluorination by activating the electrophile and reducing the basicity of the fluoride anion [2892, 2893]. The activation of hydroxy groups is of course not restricted to sulphonic acid esters. In a recent example, a variety of primary and secondary alcohols has been converted to O-alkylisourea derivatives, which are subsequently reacted with CuF_2 to give the corresponding deoxyfluorinated products. The probably chelate-driven fluorination step requires the presence of water and clearly follows an S_N2-mechanism [2894].

Furthermore, it has been found, that alcohols are converted to alkyl fluorides upon treatment with KF and PPh_3 under mild conditions. This method tolerates a large number of functional groups and furnishes primary, secondary but also tertiary fluorides and is furthermore applicable for the synthesis of difluorides. Moreover, propargylic fluorides are accessible by this method [2895].

Apart from that, alkyl fluorides are accessible by thermal fluorodecarboxylation reactions of halogenoformates. However, this strategy requires some additional synthetic steps in many cases. The most straightforward approach toward alkylfluoroformates ROC(O)F is the transformation of the respective chloro-compounds ROC(O)Cl with KF. Starting from aliphatic carbamates $ROC(O)NH_2$, $NaNO_2$ and Olah's reagent are necessary to obtain the fluoroformates. It is also possible to start directly from the alcohol and treat it with FC(O)Cl in presence of NBu$_3$. Alternatively, the treatment of alkyl 1,2,2,2-tetrachloroethyl carbonates with a complex of KF and 18-crown-6 generates the respective alkyl fluoroformate. The decarboxylation of the fluoroformates ROC(O)F can be either achieved thermally or by reaction with BF$_3$ · OEt$_2$. In general, the fluorodecarboxylation strategy works well for the generation of primary and secondary alkyl fluorides, such as C$_2$H$_5$F, (CH$_3$)$_2$CHF or *cyclo*-C$_6$H$_{11}$F.

13.4.3 Cleavage of ethers and epoxides

Linear and cyclic α-haloethers are rather easily cleaved by anhydrous HF, which is typically not observed for non-halogenated ethers. For example, *bis*(1,1-difluoroalkyl) ethers are transformed to 1,1,1-trifluoroalkanes and acid fluorides upon treatment with HF (Scheme 13.19). The ease of the cleavage depends on the further substitution of the alkyl residues, as ethers without any electron-withdrawing substituents apart from α-fluorine substituents are already cleaved below room temperature. However, an additional halogen substituent in 3-position increases the required temperature to about 70 °C, whereas 2-halo-1,1-difluoroethers are virtually stable against the attack of HF. Analogously to their aliphatic relatives, cyclic α,α,α',α'-tetrafluoroethers are opened by anhydrous HF to yield acid fluorides with a terminal CF$_3$-group (Scheme 13.19). As a general trend, six-membered rings are more readily cleaved than the respective furan derivatives. The presence of halogen substituents in β- and β'-position as well as a double bond in the cyclic backbone increase the stability toward cleavage. A synthetically relevant example for the cleavage of cyclic α,α,α',α'-tetrafluoroethers is given by the reaction of 1,1,3,3-tetrafluoro-1,3-dihydroisobenzofuran with anhydrous HF, which initially generates 2-(trifluoromethyl)benzoyl fluoride. Subsequent hydrolysis gives rise to the respective benzoic acid. The cleavage of perfluorinated small-ring heterocycles requires superacidic conditions, such as HF/SbF$_5$ or HF/AsF$_5$ mixtures. The ring opening occurs regioselectively upon HF-addition, so that oxetanes are converted to linear perfluoro alcohols, while 1,2-oxazetidines yield fluorinated O-alkylated hydroxyl amines (Scheme 13.19). Interestingly, the typically labile N-O bond stays intact in this case, whereas the C-N bond is cleaved.

    A convenient approach toward 1,2-substituted fluoro-hydroxy compounds ("fluorohydrins") is described by the ring-opening of epoxides with HF-containing reagents. Typically, the amine adducts (^{i}C$_3$H$_7$)$_2$N · 3 HF or (C$_2$H$_5$)$_3$N · 3 HF are used for this purpose, Olah's reagent also serves well for the opening of epoxide-rings. The regioselectivity of the fluoride attack depends on the substitution of the epoxide ring, but also on the nature of the HF-complex [2896, 2897]. The use of basic or neutral HF-reagents leads to a direct nucleophilic, S$_N$2-like attack on the more electropositive carbon center of the epoxide ring. According to the example in Scheme 13.20, this means that substituent R^1 has to be stronger electron-withdrawing than R^2 [2898]. In contrast, acidic HF-complexes initially protonate the oxygen atom. As a consequence, the epoxide ring

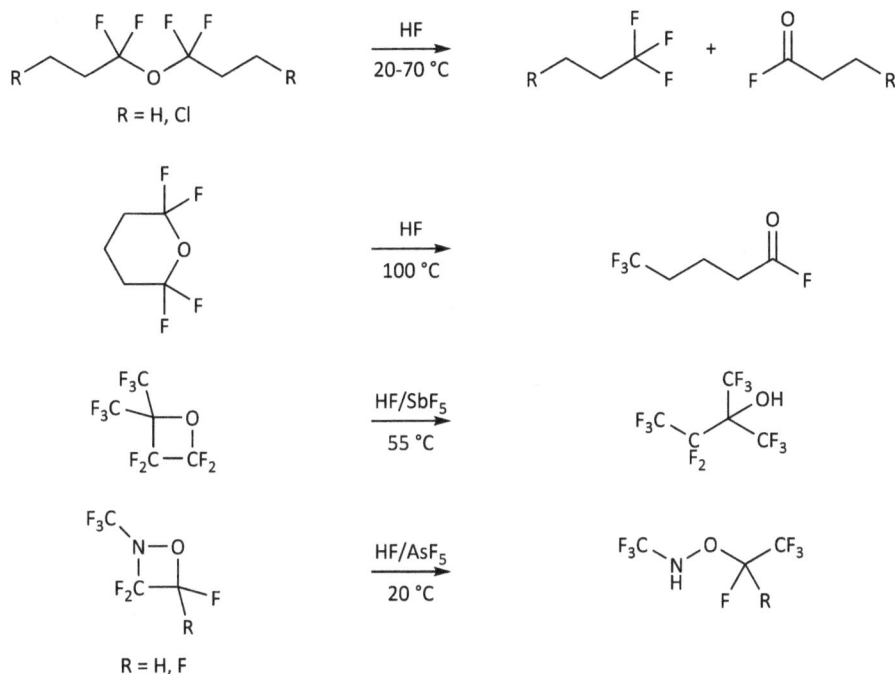

Scheme 13.19: Cleavage of linear and cyclic α,α-difluoroethers using anhydrous HF.

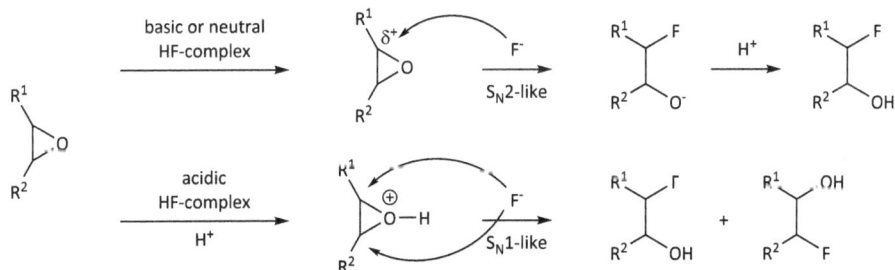

Scheme 13.20: Possible boundary mechanisms for epoxide ring-opening using HF-reagents.

will be opened in a way that generally follows an S_N1-mechanism (Scheme 13.20). However, as long as the substituents R^1 or R^2 do not significantly influence the attack of the fluoride ion, this pathway will lead to the formation of product mixtures. Apart from HF-complexes, the diethyl ether adducts $BF_3 \cdot OEt_2$ and $HBF_4 \cdot OEt_2$ are very commonly used Lewis acids for the ring opening fluorination of a large variety of different epoxide substrates [2800].

The ring-opening of ethylene oxide with anhydrous HF is a convenient approach toward 2-fluoroethanol. In a similar way, perfluoroisobutylene oxide is converted to perfluoro-*tert*-butanol in presence of SbF_5. If the epoxide, for example, ethylene oxide, is attacked by a fluorinated alcoholate, such as $C_3F_7CH_2O^-$, a hydroxy ether is generated, which may be further functionalized [2899].

A high regio- and stereoselectivity has been observed for the reaction of alkyl phenyl 2,3-epoxycarboxylates and similar compounds with Olah's reagent at room temperature (Scheme 13.21). The regioselectivity of the oxirane-cleavage is caused by the presence of the phenyl-substituent, allowing the generation of a stabilized benzyl-cation, which is subsequently attacked by a fluoride anion. For ester- and amide-substituted substrates, the ring opening preferentially generates *trans*-products, whereas a nitrile-group slightly favors the respective *cis*-isomer. The different behavior of acidic and nucleophilic HF-complexes in the reaction with epoxides is nicely visualized in the treatment of the monoepoxide of (*Z*,*Z*)-1,5-cyclooctadiene with pyridine/HF and $Et_3N \cdot 3$ HF. The nucleophilic triethylamine-complex selectively opens the epoxide under quantitative formation of the *trans*-fluorohydrin. In contrast, acidic pyridine/HF leads to transannular cyclization, which is due to the occurrence of intermediate carbocations (Scheme 13.21). A mild, ionic liquid-based HF-source has also been used for the opening of epoxides [2900].

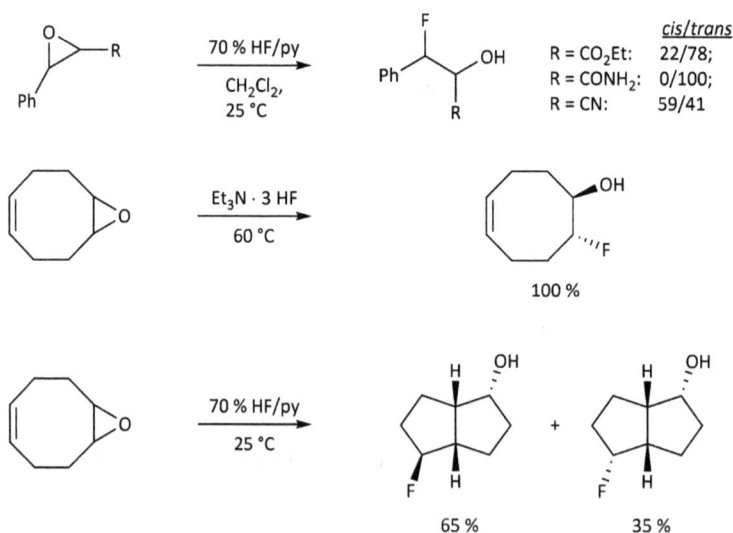

Scheme 13.21: Regioselective opening of alkyl phenyl 2,3-epoxycarboxylates and derivatives using Olah's reagent (top) and treatment of monoepoxy 1,5-cyclooctadiene with different HF-complexes (bottom).

The stereochemistry of α-substituents in cyclic epoxides can also strongly influence the cleavage of the oxirane, as it is, for example, observed for the *cis*- and *trans*-epoxide of isophorol (Scheme 13.22). Upon treatment with $Me_3N \cdot 2$ HF, the *cis*-substrate gives rise to the 3-fluoro-1,2-diol product, whereas the *trans*-compound

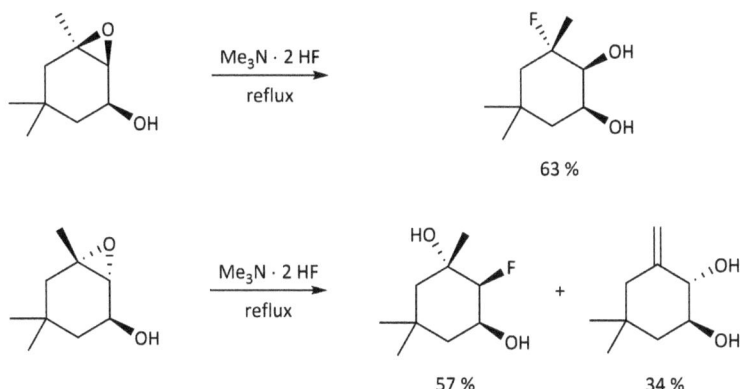

Scheme 13.22: Influence of α-substituents on the cleavage of epoxides in cyclic systems.

yields a mixture of the 2-fluoro-1,3-diol and a nonfluorinated elimination product. The different outcome of these transformations is explained by the influence of the α-substituent on the transition-state conformations [2901]. Instead of HF-complexes, other fluoride sources are also suitable for the generation of fluorohydrins from epoxides. For example, the phosphonium salts $[Bu_4P]F$, $[Bu_4P][HF_2]$ and $[Bu_4P][H_2F_3]$ serve well for the cleavage of simple, aliphatic oxiranes. The same applies to mixtures of $M[HF_2]$ ($M = Na$, K, NH_4) and porous AlF_3 in dimethoxyethane solutions under ultrasound-conditions. The use of diethylaminosulphur trifluoride (DAST) has also been described, however, its reaction with epoxides yields rather complicated mixtures of vicinal- and geminal difluorides as well as *bis*(2-fluoroalkyl) ethers. Another selective approach for the cleavage of epoxides is described by the use of SiF_4 along with a Lewis base, typically $^i Pr_2NEt$. This method uses mild conditions and appears to be highly chemoselective, as it leaves alkenes, ethers, long-chain internal epoxides and C-Si-bonds intact. Moreover, the stereoselectivity of the epoxide-cleavage depends on the use of additives. Upon addition of H_2O or $[NR_4]I^-$, the oxirane-opening occurs in an *anti*-fashion, thus leading to *trans*-products. In the absence of these additives, *cis*-fluorohydrins are obtained [2902, 2903]. Substituted aryl epoxides are converted to *syn*-fluorohydrins upon reaction with $BF_3 \cdot OEt_2$ at −20 °C. Mechanistically, this stereoselective transformation follows an S_N1-mechanism [2904]. In carbohydrate chemistry, KHF_2 is commonly used for ring-opening, whereas a mixture of anhydrous HF and $BF_3 \cdot OEt_2$ is very useful for the preparation of steroidal fluorohydrins. As an extension, the opening of epoxides may be combined with a subsequent fluorodehydroxylation, which gives rise to vicinal difluorides. Depending on the substrate, the latter may be obtained in diastereomerically or even enantiomerically pure fashion, if selective reagents, such as $NEt_3 \cdot 3$ HF or DAST are used [2905].

For a selective epoxide-opening, reagent combinations, for example, KHF_2 and 18-crown-6 are commonly applied. However, since such reagents are typically less reactive, their fluorination reactions are also significantly slower. An acceleration is achieved by using electrophilic transition metal complexes, which might also lead to the enantioselective preparation of fluorohydrins, if the catalyst bears a chiral auxiliary (Scheme 13.23). The halogen-substituent of the chlorohydrin by-product originates from the catalyst [2906]. The method has also been applied for a variety of other *meso-* or racemic epoxides to give fluorohydrins with good enantioselectivity. As an improvement, the formation of chlorohydrins has been suppressed, when AgF was used as fluoride source [2907]. The same salen-ligand has been combined with a cobalt(II)-center and a chiral isothiourea additive to convert *meso*-epoxides to fluorohydrins in enantioselective fashion. In this transformation, the required HF is generated *in situ* from benzoyl fluoride and 1,1,1,3,3,3-hexafluoroisopropanol [2908]. The ring opening of oxabicyclic alkenes is furthermore enantioselectively achieved with a Rh-catalyst and $Et_3N \cdot 3 HF$ to give *trans*-fluorohydrins [2909].

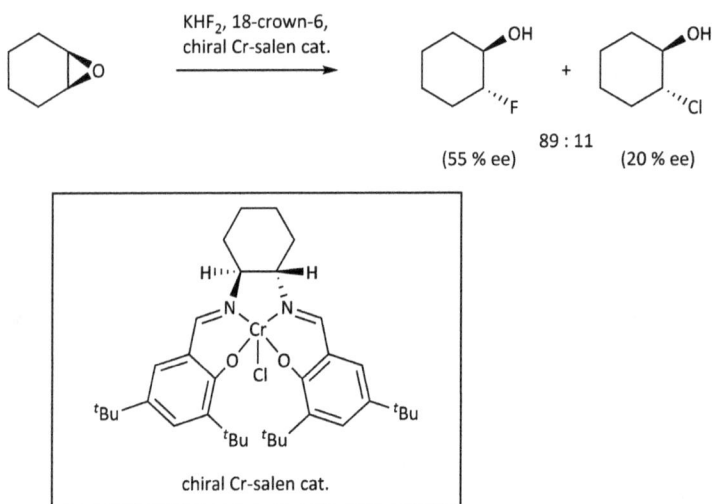

Scheme 13.23: Epoxide-opening with $K[HF_2]$ in the presence of a chiral chromium catalyst.

13.4.4 Deoxyfluorination of carbonyl compounds

Deoxyfluorination reactions are not restricted to alcohol- and epoxide-derivatives but can also be performed with carbonyl compounds. However, the products will not just bear a single fluorine substituent in most cases, but contain CF_2-units or even CF_3-groups, if the starting material is a carboxylic acid. SF_4 and its derivatives are the most commonly used reagents for such transformations; typically, catalytic amounts of HF or BF_3 are required, too. The transformation of aliphatic and aromatic aldehydes gives rise to CF_2H-groups, whereas keto-functions are converted to difluoromethylene-units. Fluoroformates are transformed to the respective trifluoromethoxy derivatives, and carboxylic acids yield trifluoromethyl-substituents. A general reactivity scale of carbonyl compounds with respect to fluorine substitution may be formulated as follows: aldehydes ≈ ketones > carboxylic acids ≈ amides > esters > anhydrides. However, alcohols are the most reactive substrates for deoxyflorination. The mechanism of the conversion of carbonyl compounds by SF_4 is rationalized by the initial attack of an $[SF_3]^+$ cation, which results from solvolysis in anhydrous HF (Scheme 13.24). The SF_3-group adds to the carbonyl-oxygen atom, making the carbonyl-carbon very electrophilic, so that a fluoride ion is intramolecularly transferred from the OSF_3-group. Subsequently, SOF_2 is eliminated and the resulting resonance-stabilized carbocation adds another fluoride ion, which originates from an $[HF_2]^-$ anion. Since the reaction proceeds via a carbocationic species, the main side-products are expectedly formed by rearrangements [2910].

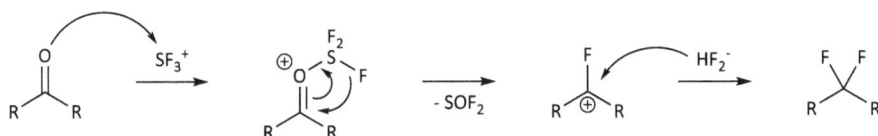

Scheme 13.24: Mechanism for the deoxyfluorination of carbonyl compounds using SF_4.

The majority of straight-chain aliphatic aldehydes is cleanly converted to the corresponding 1,1-difluoroalkanes using SF_4 in a temperature range between −20 and 40 °C. However, some substrates tend to be transformed to *bis*(1-fluoroalkyl) ethers. For example, butyraldehyde is cleanly converted to 1,1-difluorobutane at 20 °C, but yields significant amounts of *bis*(1-fluorobutyl) ether, when it is reacted below 0 °C. In the same way, the reactions of trichloro- and trifluoroacetaldehyde lead to the formation of substantial amounts of ether side-products. In case of chloral, a fluorine-chlorine rearrangement additionally occurs, resulting in the formation of 1,1,2-trichloro-1,2-difluoroethane as major product (Scheme 13.25). Aldehydes typically undergo fluorine-halogen migrations, when they are branched in α-position to the formyl group. As a consequence, reactions of such aldehydes with SF_4 result in the formation of product mixtures containing 1,1-difluoroalkanes, 1,2-difluoroalkanes and, in some cases, again ethers. A skeletal rearrangement also occurs upon treatment of pivaloyl aldehyde with SF_4. Therein, a methyl group exchanges its place with a fluorine substituent (Scheme 13.25). In addition, a certain amount of starting material is again converted to the respective *bis*(1-fluoroalkyl) ether. The product formation can also be influenced by the solvent, as SF_4 cleanly converts *p*-nitrobenzaldehyde to the respective difluoromethylated arene in benzene solution. In contrast, solvent-free conditions exclusively generate *bis*(*p*-nitro-α-fluorobenzyl) ether. The transformation of unsaturated aldehydes requires the presence of a HF-scavenger to avoid saturation. The

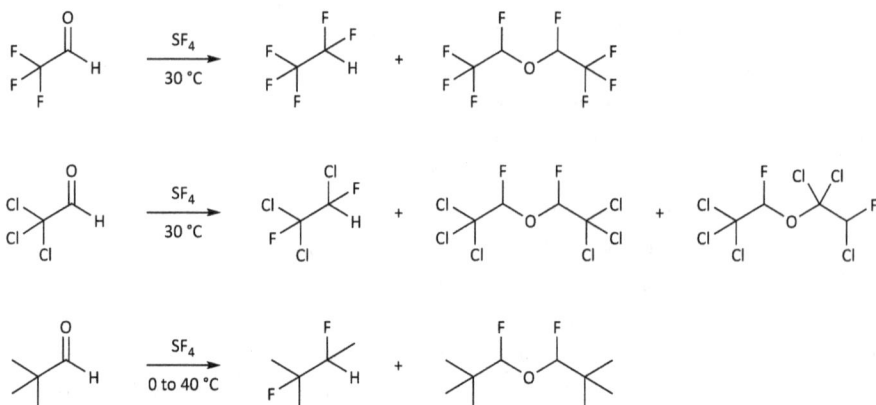

Scheme 13.25: Deoxyfluorination of trihaloacetaldehydes and pivaloyl aldehyde with SF$_4$.

treatment of such substrates with SF$_4$ in the presence of KF cleanly gives rise to the corresponding difluoromethyl compounds.

SF$_4$ reacts with both α- and β-diketones to yield tetrafluoroalkanes, along with substantial amounts of alkenes, ethers and sulphites. However, the formation of the side-products can be suppressed by conducting the reactions in excess anhydrous HF. In some cases, diketones can exclusively yield cyclic ethers, as it is the case for 1,5-diphenylperfluoropentan-1,5-dione. The fluorination of β-keto esters can lead to a variety of products, depending on the substitution of the α-carbon and the presence of anhydrous HF (Scheme 13.26). In total absence of HF and α-substituents, extensive elimination will take place to give mostly the acetylenic ester product. Instead, allene-derivatives are formed, when a carbon-chain is attached to the α-position. In contrast, elimination is completely avoided, if either both α-hydrogens are replaced by alkyl-substituents or when the reactions are carried out in presence of anhydrous HF. Treatment of aliphatic α-hydroxyketones with SF$_4$ gives rise to mixtures of α,β,β-trifluoroalkanes, difluoroalkyl sulphites and bis(difluoroalkyl)sulphates.

The use of SF$_4$-derivatives in deoxyfluorination reactions slightly changes the outcome of the reactions. Due to its milder character, DAST can be more readily applied in the transformation of more sensitive compounds, such as steroids or carbohydrates. The same holds for the XtalFluor™- and Fluolead™-reagents. Typically, all these fluorinating agents are used without any catalyst. As a consequence, the general substrate scope is rather limited, as only aldehydes and some ketones are readily converted. The conversion of carbonyl compounds with DAST requires elevated temperatures of about 80 °C, thus, hydroxyaldehydes and -ketones are exclusively converted to fluorocarbonyl compounds at lower temperature (cf. Scheme 13.16). Apart from that, aldehydes react more readily with aminosulphuranes than ketones, so that the choice of appropriate reaction conditions allows the selective fluorination of the formyl group. Analogously to the deoxyfluorination reactions of alcohols, the transformations of

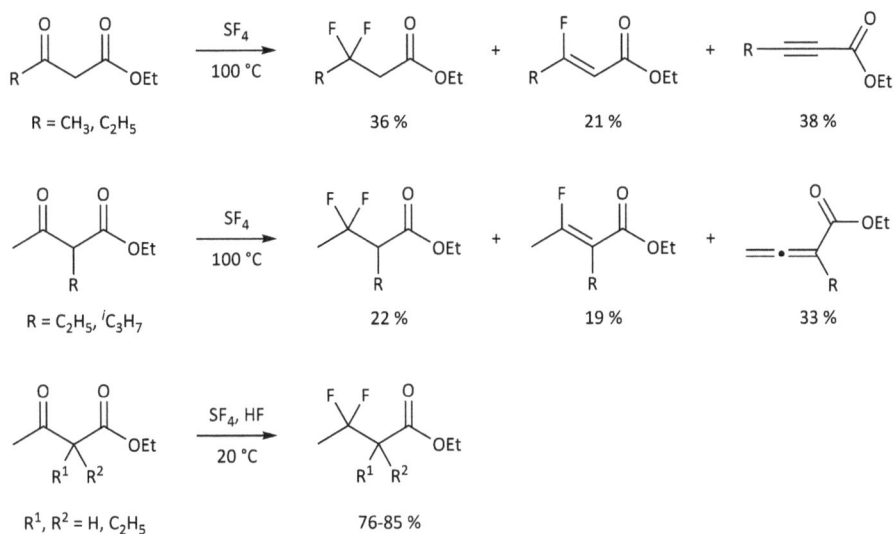

R = CH₃, C₂H₅ ... 36 % ... 21 % ... 38 %

R = C₂H₅, ⁱC₃H₇ ... 22 % ... 19 % ... 33 %

R¹, R² = H, C₂H₅ ... 76–85 %

Scheme 13.26: Fluorination of β-ketoesters with SF$_4$.

carbonyl compounds are mainly accompanied by eliminations and rearrangements. However, the side-reactions can efficiently be suppressed by the choice of appropriate nonpolar solvents, since ionic intermediates are preferentially stabilized in polar media. Thus, polar solvents support the occurrence of rearrangements and eliminations.

Some representative deoxyfluorination reactions with DAST include the conversion of pivaloyl aldehyde to the difluoromethyl compound $(CH_3)_3CCHF_2$, the transformation of the α-ketoester $PhC(O)CO_2Et$ to the α,α-difluoroester $PhCF_2CO_2Et$ as well as the reaction of p-benzoylbenzaldehyde $PhC(O)C_6H_4CHO$ to p-difluoromethylbenzophenone $PhC(O)C_6H_4CHF_2$. In the latter example, the aldehyde function is exclusively converted, whereas the keto-group remains unchanged. Similar to SF$_4$, DAST converts haloacetaldehydes predominantly to bis(1-fluorohaloethyl) ethers. Under suitable reaction conditions, aliphatic, aromatic and heterocyclic ketones are converted to the corresponding difluoromethylene compounds. However, especially the fluorination of cycloaliphatic ketones is very likely to give substantial amounts of fluoroalkenes as dehydrofluorination products (Scheme 13.27). The choice of the solvent has an impact on the product ratio, as compared to dichloromethane a slightly higher amount of fluoroolefins is obtained in 1,2-dimethoxyethane (glyme). The treatment of 4-ketocarboxylic acids with DAST yields γ-fluorolactones in almost quantitative amounts (Scheme 13.27). Upon reaction with β-keto esters, the behavior of DAST is quite different to that of SF$_4$ (cf. Scheme 13.26). Sulphur tetrafluoride gives rise to the corresponding difluoromethylene-compound and its elimination products, whereas the aminosulphurane converts compounds like ethyl acetoacetate to an equal mixture of the respective (E)- and (Z)-vicinal difluoroolefins (Scheme 13.27). Furthermore, the

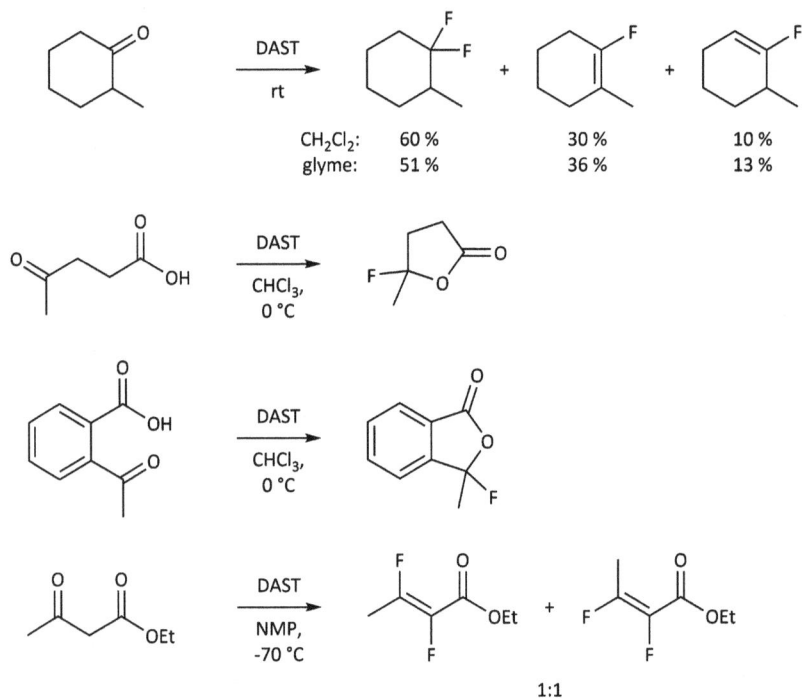

Scheme 13.27: Reactions of 2-methylcyclohexanone, 4-ketocarboxylic acids and ethyl acetoacetate with DAST.

deoxyfluorination reactions with DAST and related reagents do not work for sterically hindered ketone substrates, as well as for ester- and anhydride-functions. Another important property of DAST and similar SF_3-compounds is their capability of smoothly converting carboxylic acids to the respective acid fluorides, but not to CF_3-groups.

Moreover, cyclic and straight chain aliphatic ketones are efficiently transformed to difluoromethylene-compounds, when the carbonyl-group is initially converted to an imine or azine and subsequently fluorinated using BrF_3 in $CFCl_3$ solution [2911].

13.4.5 Deoxyfluorination of carboxylic acids and their derivatives

The conversion of carboxylic acids to CF_3-groups is more or less exclusively achieved by the use of SF_4, most conveniently, such transformations are conducted in anhydrous HF. However, aryl sulphur trifluorides, which are *in situ* generated from $ArSF_4Cl$, have also been found to be suitable reagents for this transformation [2799]. The first step, the conversion of the carboxylic acid to an acid fluoride, can also be realized by the use of many other fluorinating reagents like DAST or the fluoroenamine compounds. The use of cyanuric fluoride has also been reported for this purpose [2912].

However, for the second step, an electrophilic $[SF_3]^+$ cation is required, which attacks the carbonyl-oxygen following the same mechanism that has been described for the deoxyfluorination of carbonyl compounds above (cf. Scheme 13.24). Occasionally, the formation of *bis(α,α*-difluoroalkyl) ethers is observed, which can be rationalized by the side-reaction of cationic intermediates with acid fluorides (Scheme 13.28). The yields of the ether by-products depend on both, the nature of the acid and the reaction conditions. For simple, unsubstituted acids, ether formation will only be observed at low temperatures, whereas cycloaliphatic acids and those with longer alkyl chains require higher temperatures to generate the corresponding ether. For example, SF_4 cleanly converts acetic acid to 1,1,1-trifluoroethane at 4 °C, whereas substantial amounts of *bis*(1,1-difluoroethyl) ether are obtained at –10 °C. With cyclohexanecarboxylic acid, full conversion to the CF_3-compound is observed at 60 °C, however, at 40 °C, a mixture of the ether and the trifluoromethyl product is formed. Carboxylic acids with halogen substituents generally require increased reaction temperatures to react with SF_4.

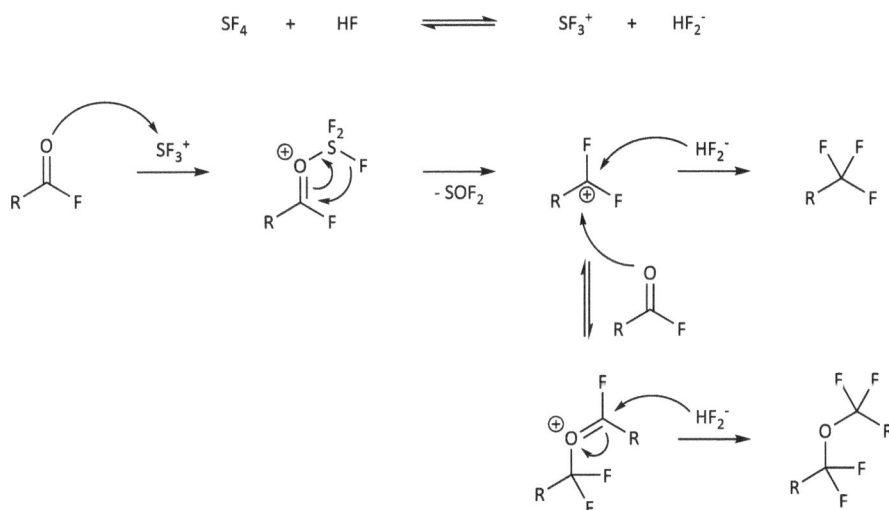

Scheme 13.28: Dissociation of SF_4 in HF (top) and mechanism of the transformation of acid fluorides to CF_3-groups, including the side-reaction, which results in the formation of *bis(α,α*-difluoroalkyl) ethers.

The reaction of carboxylic acids, acid chlorides and *tert*-butyl esters with BrF_3 at 0 °C displays another method for the simple but effective preparation of acyl fluorides [2913]. The latter are furthermore accessible upon reaction of carboxylic acids with trichloroacetonitrile and triphenylphosphine, followed by the treatment with $[NBu_4]F \cdot (^tBuOH)_4$ [2914]. In a more exotic approach, UF_6 may be used for the conversion of carboxylic acids to acid fluorides, too [2915]. Apart from that, the halogen exchange of acyl halides can be achieved with nucleophilic reagents like [NBu₄]F or in a more simple way with alkali metal fluorides [2782, 2916]. The use of a C,N-chelated di-*n*-butyl tin fluoride has also been reported for the preparation of acyl fluorides as well as fluoroformates and fluorophosgene. The advantages of this reagent comprise its high stability and its solubility in organic solvents [2917]. In an oxidative approach, primary aliphatic aldehydes are converted to acyl fluorides, using aryl difluorobromanes as fluorinating reagents. In contrast, aromatic aldehydes are converted to aryl difluoromethylethers due to an occurring 1,2-aryl shift [2918].

Reactions of SF₄ with α- or β-hydroxyacids rather yield complicated product mixtures. In most cases, the COOH-function is cleanly converted to a trifluoromethyl group, however, the hydroxy-substituent can undergo different reactions, leading to fluorination, fluorosulphination, esterification or dehydration (Scheme 13.29). Depending on the substitution of the alkyl residue, the formation of some of these products can be suppressed. Product mixtures are also generally obtained upon treatment of alkanedicarboxylic acids with SF₄, typically they contain *bis*(trifluoromethyl) alkanes, cyclic α,α,α′,α′-tetrafluoroethers, linear *bis*(pentafluoroethyl) ethers as well as polyfluoroethers. The conversion of 1,2- and 1,3-dicarboxylic acids yields cyclic tetrafluoroethers as major products, which is not the case for 1,1- or 1,4-diacids. If possible, the formation of a five-membered ring is preferred, as exemplified in case of propane-1,2,3-tricarboxylic acid (Scheme 13.29). Analogously, cyclohexane-*cis*-1,2-dicarboxylic acid is mainly transformed to its corresponding cyclic tetrafluoroether, accompanied by only small amounts of the *bis*(trifluoromethyl) product.

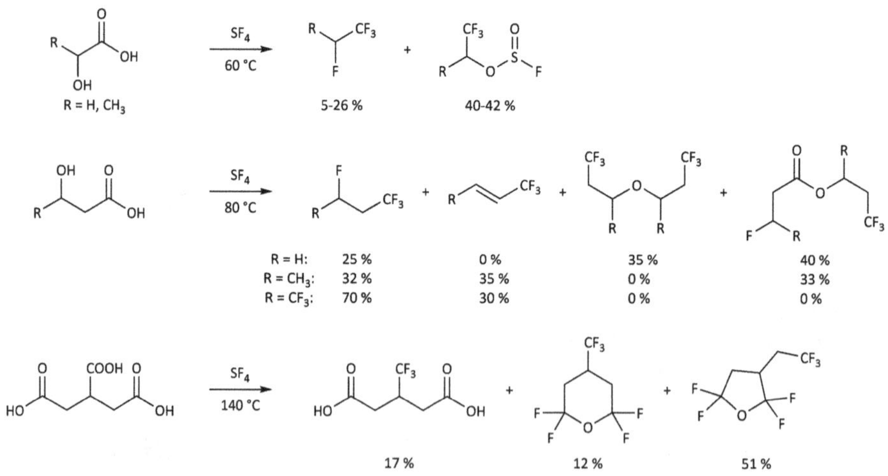

Scheme 13.29: Reactions of α- and β-hydroxyacids and a tricarboxylic acid with SF₄.

Analogously to their aliphatic relatives, benzene carboxylic acids react with SF₄ to give benzotrifluorides. However, the yields strongly depend on the substituents on the aromatic ring. For example, electron-donating groups in *p*-position lead to rather poor results, whereas a *p*-nitro-substituent drastically increases the yield of the CF_3-arene. Under rather mild conditions in HF-solution, hydroxybenzoic acids react with SF_4 to form hydroxybenzotrifluorides. The conversion of benzene dicarboxylic acids again depends on the substitution pattern. Derivatives of 1,2-diacids, such as tetrachloroph-thalic acid or its anhydride are mostly converted to cyclic $\alpha,\alpha,\alpha',\alpha'$-tetrafluoroethers. Accordingly, 1,2,4,5-benzenetetracarboxylic acids, for example, dichloropyromel-litic acid, are converted to give products with two $\alpha,\alpha,\alpha',\alpha'$-tetrafluoroether rings (Scheme 13.30). A similar behavior is observed for naphthalene-1,8-dicarboxylic acid derivatives as well as the respective 1,4,5,8-tetracarboxylic acid compounds (Scheme 13.30). However, the cyclic $\alpha,\alpha,\alpha',\alpha'$-tetrafluoroethers are found to be part of six-membered rings in this case. Under mild conditions, SF_4 initially dehydrates the carboxylic acid functions, yielding the cyclic acid anhydrides in quantitative amounts. Further conversion of the anhydrides is achieved by heating them with SF_4 to about 150–250 °C.

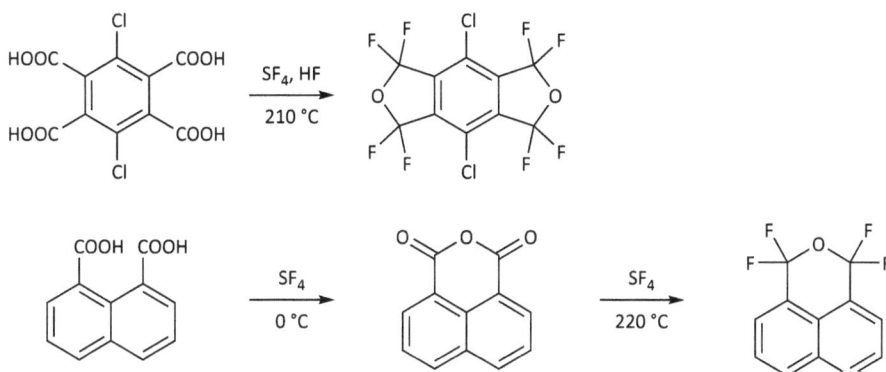

Scheme 13.30: Reactions of benzene- and naphthalene polycarboxylic acids with SF_4.

Interestingly, furan-2-carboxylic acid cannot be transformed to the CF_3-compound. The reaction with SF_4 at 0 °C stops at the stage of the acid fluoride. More forcing con-ditions only result in resin formation. However, the presence of a second COOH-group or another electron-withdrawing substituent stabilizes the furan system and enables the preparation of poly(trifluoromethyl)furans. The treatment of furantetracarboxylic acid with SF_4 converts the acid functions in 2- and 5-position to CF_3-groups. The remaining two carboxylic acid substituents are transformed to a cyclic $\alpha,\alpha,\alpha',\alpha'$-tetrafluoroether. In presence of excess HF, both 2,4- and 2,5-furandicarboxylic acid yield the corresponding *bis*-CF_3-compound, where additionally two fluorine sub-stituents are added in α- and α'-position to the oxygen. The latter reactivity is not observed for the analogous sulphur compound, as thiophene-2,5-dicarboxylic acid is exclusively converted to *bis*(trifluoromethyl)thiophene. The reactions of imidazole mono- and dicarboxylic acids with SF_4 also give rise to the corresponding CF_3-compounds.

Reactions of SF_4 with carboxylic acid derivatives, such as amides and esters, can also lead to the formation of CF_3-groups. In presence of HF, nonfluorinated amides are cleaved to initially generate acyl fluorides, which are further converted to trifluoromethyl substituents. Using SF_4 along with KF as an HF-scavenger, the carbonyl-group is fluorinated. This behavior is, for example, observed for *N,N*-dialkylbenzamides, which are transformed to *N,N*-dialkyl-*α,α*-difluorobenzylamines. Perfluorinated tertiary amides as well as cyclic imides react with a mixture of SF_4 and HF to give tertiary perfluoroamines. An interesting behavior is furthermore observed for formamides, where the formyl group is directly converted to a CF_3-substituent, using SF_4 and KF at elevated temperature (Scheme 13.31). Interestingly, both, the carbonyl-oxygen as well as the hydrogen-substituent are replaced to give rise to *N*-(trifluoromethyl)amines. In absence of HF, esters of carboxylic acids are fairly resistant toward SF_4, even at elevated temperatures. Upon impact of excess hydrogen fluoride, esters are converted to trifluoromethyl compounds. This is indeed not the case, if esters of highly fluorinated acids or alcohols are reacted with SF_4 in HF-solution under mild conditions, where the ester-bond remains intact so that only the carbonyl-group is fluorinated to yield *α,α*-difluoroethers (Scheme 13.31). The treatment of lactones with SF_4 can yield different products, depending on the ring size (Scheme 13.31). For example, *γ*-butyrolactone is initially cleaved to give *γ*-fluorobutyryl fluoride, which will be further fluorinated to result in 1,1,1,4-tetrafluorobutane. In contrast, the six-membered ring of 1,4-dioxane-2,5-diones is not opened, instead fluorination of the carbonyl-groups occurs exclusively.

13.4.6 Sulphur replacement

Sulphur can generally be considered as synthetic precursor for fluorine, since organo-sulphur compounds are rather easily converted to valuable fluoroorganic molecules. Typical substrates for such transformations are thiocarbonyl compounds, thianylium salts but also dithiolanes and dithianes. Even if these sulphur-containing substrates as such may not be largely important, other groups are readily converted to them, especially carbonyl functions.

The probably most important reaction in terms of sulphur replacement is the oxidative fluorodesulphuration, which generally requires stoichiometric amounts of a soft, electrophilic oxidant ("X^+" (X = F, Cl, Br, I), NO^+) and a nucleophilic fluoride source, such as amine-HF complexes [2919–2925]. The oxidation step can also be performed electrochemically [2926, 2927]. For example, benzophenone is initially reacted with 1,2-ethanedithiol to generate a 1,3-dithiolane derivative (Scheme 13.32). In the next step, the thiophilic oxidant, formally Br^+ in the present example, adds to one of

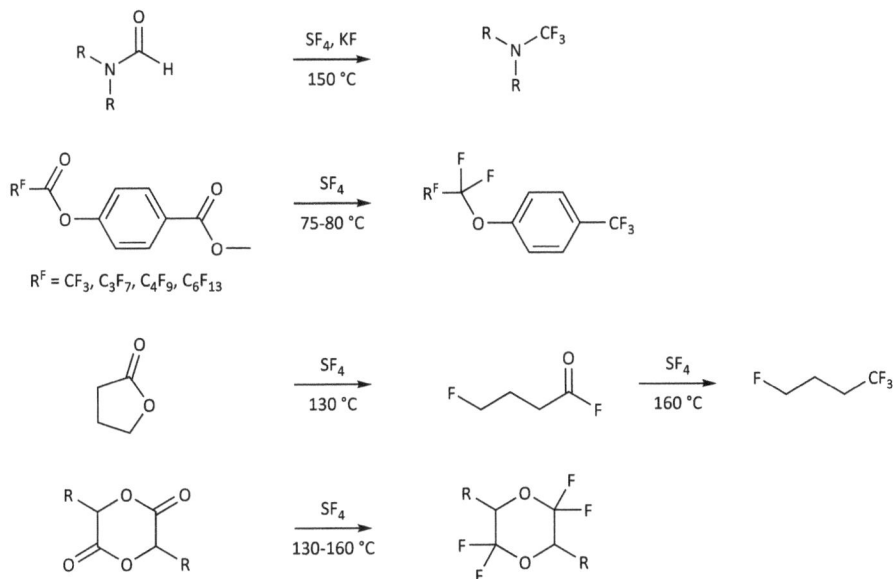

Scheme 13.31: Deoxyfluorination of amides and esters with SF_4.

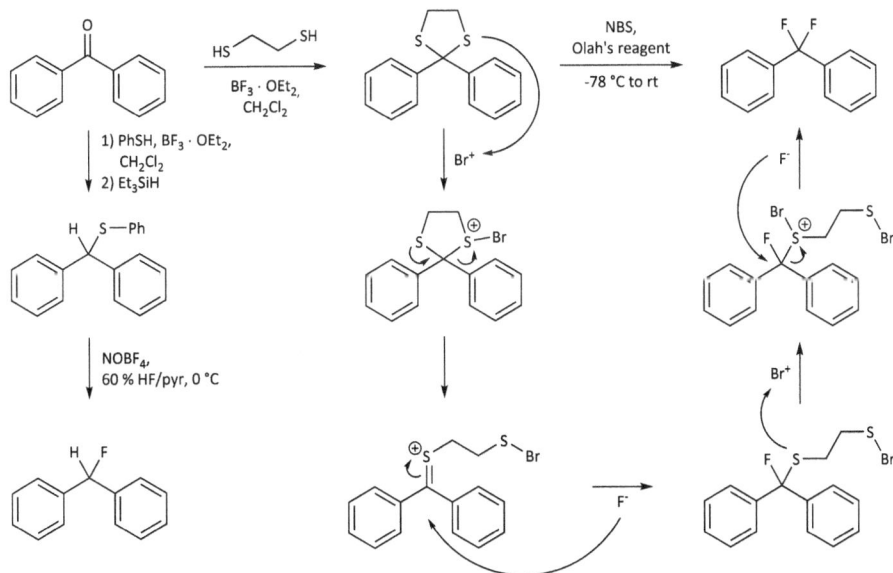

Scheme 13.32: Mechanism for the oxidative fluorodesulphuration of benzophenone showing also the variant with an additional reductive step (left side).

the sulphur atoms and leads to the opening of the dithiolane ring. As a result, the former carbonyl-carbon is now electron-deficient so that a fluoride ion can readily add to this position. Subsequently, the other S-atom is halogenated and a second fluoride ion is introduced. Upon elimination of the rather unstable dihalogenated sulphur derivative, the difluoromethylene product is obtained [2919]. The conversion of ketone-based dithiolanes readily takes place at −78 °C, whereas aldehyde-derived substrates require higher temperatures of about 0 °C. For acid-sensitive substrates, the amine-HF complexes are substituted by less acidic reagents, such as [NBu$_4$][H$_2$F$_3$]. Moreover, the generation of difluoromethylene-units from 1,3-dithianes or -dithiolanes is directly achieved by reaction with BrF$_3$ [2928]. The major drawback of the fluorodesulphuration approach is the generation of relatively large amounts of waste, since the sulphur auxiliary cannot be regenerated, once it is cleaved off.

There are many variants of fluorodesulphuration reactions of protected or activated carbonyl-substrates. The introduction of an additional reductive step, for example, will lead to the formation of CHF-groups instead of difluoromethylene units (Scheme 13.32). Furthermore, the conversion of thioesters and xanthogenates strongly depends on the choice of the oxidant and the fluoride source. The selection of a rather mild oxidant, such as N-iodosuccinimide (NIS) in conjunction with neutral to slightly acidic [NBu$_4$][H$_2$F$_3$] will readily convert the thiocarbonyl group to a CF$_2$-unit. However, for the transformation of the resulting α,α-difluorothioether, the use of strongly acidic pyridine/HF and the more powerful oxidant 1,3-dibromo-5,5-dimethylhydantoin (DBH) is required (Scheme 13.33). The reaction of dithiocarbamates with "X$^+$" and an HF-amine complex almost quantitatively yields N-trifluoromethylamines. The aforementioned conversions of thiocarbonyl compounds can also be performed with BrF$_3$, which, for example, generates α,α-difluoroethers from thioesters within a few seconds at 0 °C. Accordingly, dithioesters are transformed to CF$_3$-groups whereas xanthogenates will give rise to trifluoromethylethers. In the same way the corresponding chlorodifluoromethylethers are accessible from chlorothioformates [2929–2932]. Apart from that, xanthogenates, thioesters as well as dithianes also react with ArSF$_3$-reagents to give the corresponding fluorinated compounds [2799].

Fluorodesulphuration is also a suitable method in carbohydrate chemistry, where it is applied for the preparation of glycosyl fluorides from the respective thioethers. The treatment of aryl thioglycosides with NBS and either DAST or pyridine/HF at −15 °C yields the corresponding glycosyl fluorides with retention of configuration. In contrast, full inversion occurs, if the alkyl or aryl thioglycoside is reacted with [(CH$_3$)$_2$SSCH$_3$]$^+$[BF$_4$]$^-$. The advantage of all three aforementioned reagents is their tolerance against glycosidic linkages and sensitive protecting groups. Fluorodesulphuration gives rise to a variety of aliphatic compounds, including trifluoromethyl-, α,α-difluoroalkyl- and also perfluoroalkylethers [2933, 2934]. Aromatic thioorthoesters are converted to CF$_3$-arenes using pyridine/HF in conjunction with either DBH or NBS. The same holds for aliphatic derivatives, which are easily converted to trifluoromethylated molecules upon reaction with BrF$_3$ [2935].

Another approach in fluorodesulphuration chemistry is the route via stable dithianylium salts, which are generally easily accessible from carboxylic acids or acid

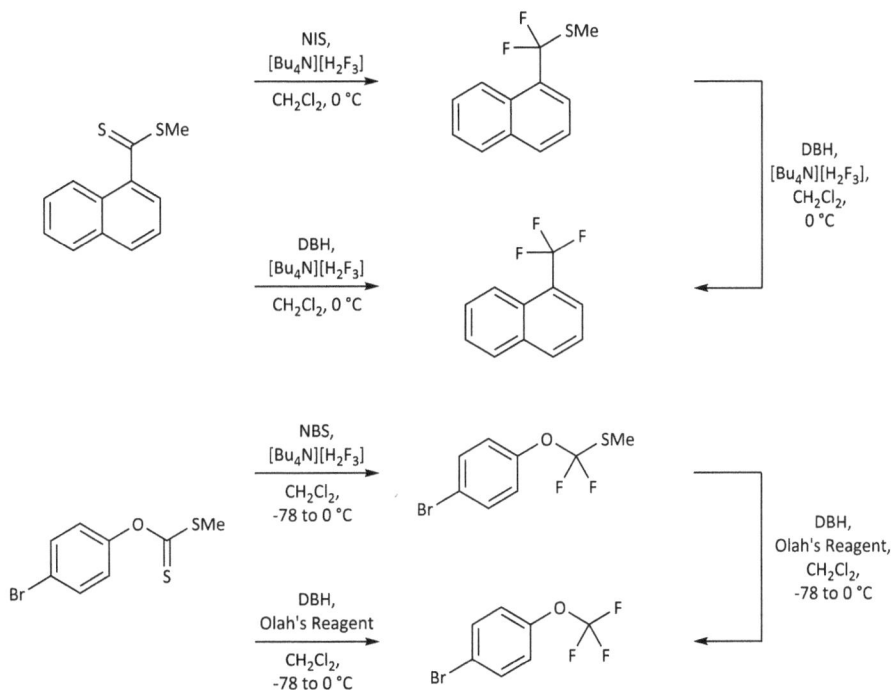

Scheme 13.33: Fluorodesulphuration of dithioesters and xanthogenates.

chlorides and mercaptans, for example, 1,3-propanedithiol. For the synthesis of α,α-difluoroethers, the dithianylium salt is attacked by an alcoholate, leading to a dithioorthoester intermediate, which may even be isolated at room temperature, if it is substituted by a strongly electron-withdrawing perfluoroalkyl group. In contrast, dithioorthoesters derived from alkyl- or aryl dithianylium salts are unstable and decompose above −50 °C. The intermediate dithioorthoester is submitted to a "standard" fluorodesulphuration process, following the mechanism depicted in Scheme 13.32. The dithianylium salts can also be reacted with other O- and N-centered nucleophiles, such as ^{t}BuOOH, Me$_3$SiN$_3$ or imidazole [2936]. Apart from its preparative relevance, fluorodesulphuration has also been found to occur in nature in a non-oxidative variant. So far, this is the only known enzymatic process, which is capable of introducing inorganic fluoride ions. The sulphur-based key-intermediate in the enzymatic reaction-sequence is a trialkylsulphonium ion, which is then substituted by fluoride [2937, 2938].

Apart from oxidative fluorodesulphuration, a few other methods have been described to convert sulphur-containing molecules to fluorinated compounds. This works especially well for thiocarbonyl derivatives, as they are more readily transformed to the corresponding difluoromethylene derivatives than their oxygen analogues. In a very simple way, trithiocarbonates and thiuramsulphides are directly reacted with SF$_4$ at 100–120 °C. For example, ethylene trithiocarbonate is converted to 2,2-difluoro-1,3-dithiolane, whereas thiuramsulphides give rise to N,N-dialkyl-N-trifluoromethylamines. The main difference between the reactions of SF$_4$ with C=S- and C=O-bonds is the redox-chemistry of the sulphur atoms that occurs for both the substrate and the fluorinating reagent. Thus, elemental sulphur is generated as a by-product and all four fluorine atoms of SF$_4$ are used. The increased reactivity of thiocarbonyl-groups is also beneficial for the conversion of ester functions, as thioesters are readily transformed to α,α-difluoroethers upon treatment with DAST at room temperature. Interestingly, this method even allows the conversion of silyl esters, leaving the O-Si bond intact and giving rise to α,α-difluorosilylethers. However, the formation of a difluoromethylene-group fails for lactone-substrates.

13.4.7 Nitrogen replacement

Analogously to the oxygen- and sulphur-containing substrates in the previous sections, amino-groups can serve as leaving groups for deaminative fluorination. Typically, this procedure is used for the generation of arylfluorides (cf. Section 14.2.5), however, a few examples of aliphatic fluorides have also been prepared by this method. α-Amino acids are readily converted to α-fluorocarboxylic acids via intermediary diazo-species. For this transformation, NaNO$_2$ is required as diazotization agent along with a fluoride source, which is typically Olah's reagent or another HF-complex (Scheme 13.34). A big advantage of this method is the availability of α-amino acids. The major drawback is indeed, that in some cases rearrangements of the intermediary cationic species likely occur, leading to the formation of β-fluorocarboxylic acids. However, rearrangements can be suppressed by using 1:1-mixtures of pyridine and HF.

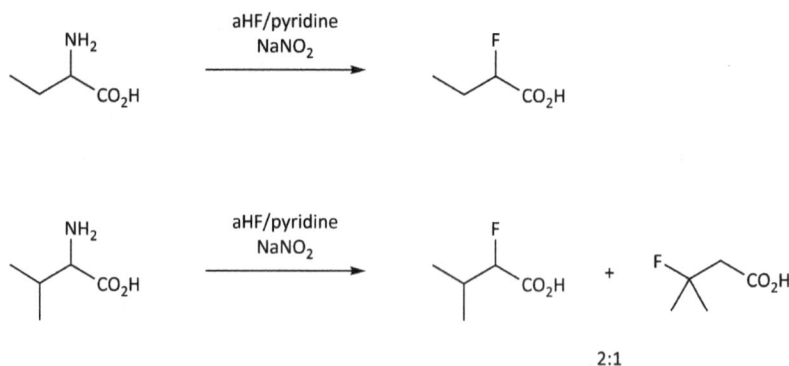

Scheme 13.34: Deaminative Fluorination of α-amino acids.

Another family of nitrogen-containing substrates includes diazoalkanes, -ketones and -esters. Depending on the reagent, diazoalkanes can be transformed to a variety of fluorinated compounds. In the simplest way, the reaction with an HF-source, for example, Olah's reagent or HBF_4 leads to hydrofluorination reactions, whereas direct treatment with F_2 at low temperatures generates geminal difluorides. Accordingly, the treatment of diazoalkanes with fluoroformates produces α-fluoroesters, whereas acid fluorides give rise to α-fluoroketones. In a similar manner, diazoketones can be reacted with anhydrous HF or pyridine/HF to yield fluoromethylketones. Using a reagent combination of HF/pyridine and N-halosuccinimide, α-fluoro-α-haloketones are accessible from diazoketones. At low temperatures, elemental F_2 can convert diazoketones, for example, 2-diazocyclohexanone into the corresponding α,α-difluoroketone. Similarly, hydrofluorination, halofluorination and geminal difluorination can be realized with diazoester substrates.

The ring opening of cyclic nitrogen-containing systems, that is, aziridines and azirines gives rise to α-fluoroamines. Again, these reactions can be generally realized by the application of Olah's reagent. However, other reagents may be used, which might also influence the outcome of the reaction in terms of regio- and stereochemistry. Of course, the selectivity also strongly depends on the substrate structure. For example, 2,2-diphenylaziridine is converted to 2-fluoro-2,2-diphenylethan-1-amine using pyridine/HF. The other possible product, 1-fluoro-2,2-diphenylethan-1-amine, is not observed, as it is unstable toward α-elimination of HF. Activated aziridines, which are substituted with an electron-withdrawing group at the nitrogen-atom, are selectively converted to the corresponding *anti*-fluoroamines using $NiF_2 \cdot n\,H_2O$ ($n < 4$) in the presence of $[NBu_4]F$, however, unsymmetric substrates lead to the formation of regioisomers [2939]. In the same way, *trans*-β-fluoroamines are obtained upon ring-opening of aziridines in presence of catalytic amounts of a bicyclic nitrogen-base, which generates an amine-HF reagent *in situ*. As HF-source, a mixture of benzoyl fluoride and hexafluoroisopropanol is used [2940]. An asymmetric ring-opening is observed when chiral aziridine substrates are treated with $Et_3N \cdot 3$ HF, however, a mixture of regioisomers is obtained [2941]. The enantioselective preparation of *trans*-β-fluoroamines is furthermore achieved by metal-catalysis using a latent HF-source. However, the yields and enantiopurities of the products strongly depend on the substrate [2942]. A substrate-dependence is furthermore observed for the ring-opening of azirine derivatives, which can yield β,β-difluoroamines, α-fluoroketones or even pyrazines (Scheme 13.35). The reaction of benzylic α-hydroxyaziridines with Olah's reagent gives access to α-fluoroaziridines and α,γ-difluoroamines.

Despite not involving a nitrogen-replacement reaction, the treatment of methyl and phenyl isocyanate with anhydrous HF or Olah's reagent is worth being mentioned. In this transformation, the HF-addition occurs at the C-N-bond, so that carbamyl fluorides are formed. It has been shown, that the use of anhydrous HF is by far more efficient for this type of reaction.

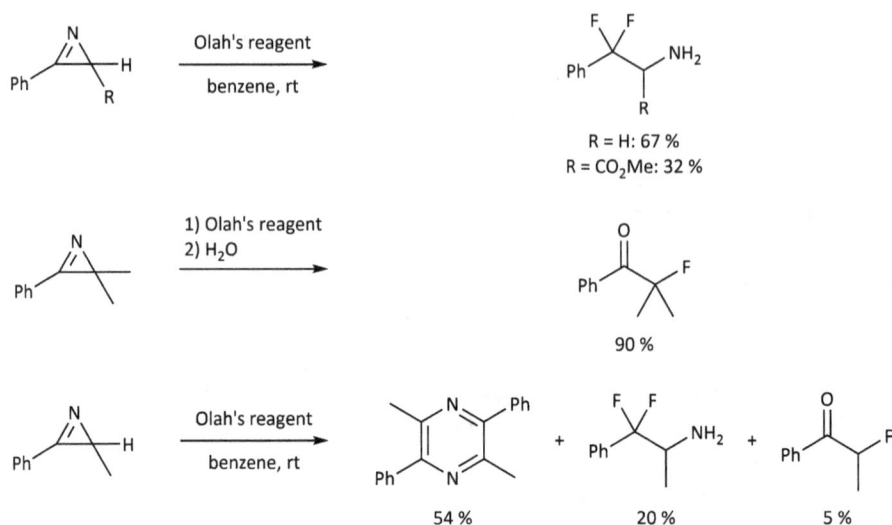

Scheme 13.35: Ring opening of different azirine derivatives using Olah's reagent.

13.4.8 Hydrofluorination of unsaturated C-C bonds

Aliphatic fluorine compounds can also be obtained by the reaction of fluorine-containing reagents with carbon-carbon double- and triple-bonds. This is often re-alized by acid-catalyzed reactions with anhydrous HF, but less reactive HF-reagents may also be used. In principle, electron-rich alkenes are very susceptible toward an HF-addition, so that they do not require the addition of any catalyst. For example, *tert*-butyl fluoride is readily generated from isobutene. However, the product yields depend on the substrate and the reaction conditions. In general, the addition of HF to unsaturated organic compounds has to be carried out at low temperatures to avoid cationic polymerization. In contrast, more forcing conditions are required for elec-tronically deactivated substrates, such as trichloroethylene, where the HF addition needs to be catalyzed by $SbCl_5$ or other Lewis acids. The same holds for acetylenic substrates, which can in principle be directly reacted with anhydrous HF, but will only give poor yields of hydrofluorinated products. Therefore, the use of catalysts or additives is beneficial. Depending on the choice of the reagents, the hydrofluori-nation of acetylene will either stop at the vinylfluoride stage or cleanly lead to the formation of 1,1-difluoroethane. Full conversion is especially achieved, if FSO_3H is added to the reaction mixture, whereas Hg- and Cd-compounds, for example, Ph_2Hg, $Hg(NO_3)_2$, $Cd(NO_3)_2$ or $Cd(BF_4)_2$, tend to stop the hydrofluorination on the stage of the olefin. The treatment of alkylacetylenes with Olah's reagent leads to the formation of geminal difluorides. In a similar way, phenylacetylene is transformed to 1-phenyl-1,1-difluoroethane, however, a large excess of HF is required. Alternatively, the reaction can be performed in the gas-phase over an HgO-catalyst. Acetylenes with electron-withdrawing substituents do not react under any of the aforementioned conditions, as they will add a fluoride ion first and subsequently undergo protonation. Thus, they require the presence of fluoride ions and a proton source (Scheme 13.37).

As a drawback of forcing reaction conditions, initially generated chlorine-containing products likely undergo subsequent halogen exchange. Treating trichloroethylene with HF in presence of SbCl₅, 1,1,2-trichloro-1-fluoroethane will not be isolated, but further converted to 1-chloro-2,2,2-trifluoroethane. The product formation of HF-addition reactions is furthermore influenced by the choice of appropriate solvents. For example, propene is cleanly transformed to 2-fluoropropane by using Olah's reagent in THF at 20 °C. In contrast, reactions in dichloromethane proceed more selectively (Scheme 13.36). At room temperature, the two internal double bonds of an unsaturated bi(cyclohexyl)-molecule are hydrofluorinated to give axially fluorinated cyclohexane moieties. At slightly lower temperature, the same reactivity is observed in a similar system; however, therein, the terminal olefinic bonds remain untouched (Scheme 13.36) [2943, 2944].

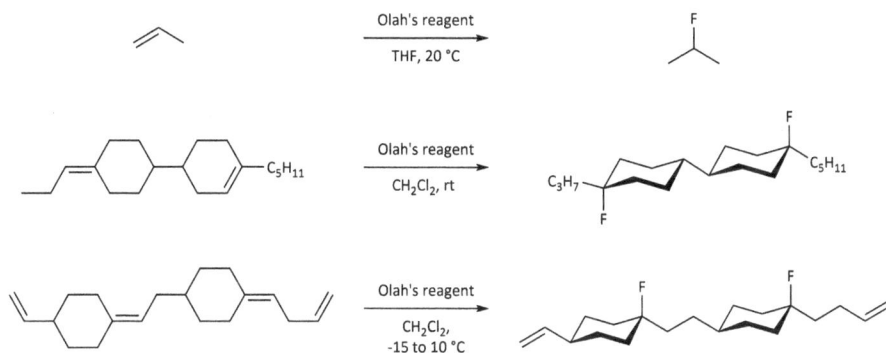

Scheme 13.36: Hydrofluorination of olefins with Olah's reagent in different solvents.

The addition of hydrogen fluoride also works well for more sensitive unsaturated substrates, such as steroids or terpenes, if the reactions are carried out at sufficiently low temperatures in inert solvents. Mechanistically, the hydrofluorination of an unsaturated C-C bond is rationalized by an initial electrophilic attack of a proton to the multiple bond, followed by F⁻ addition to the intermediary carbocation, thus yielding Markovnikov products. In more complex systems, the hydrofluorination of unsaturated C-C-bonds is often accompanied by rearrangements of the carbocationic species. In very electrophilic alkenes, the HF-addition typically occurs in another way, as the fluoride anion is added first, before protonation of the carbanion occurs (Scheme 13.37). Some more details on the chemistry of highly fluorinated olefins are discussed in an own subchapter (cf. Section 13.4.11).

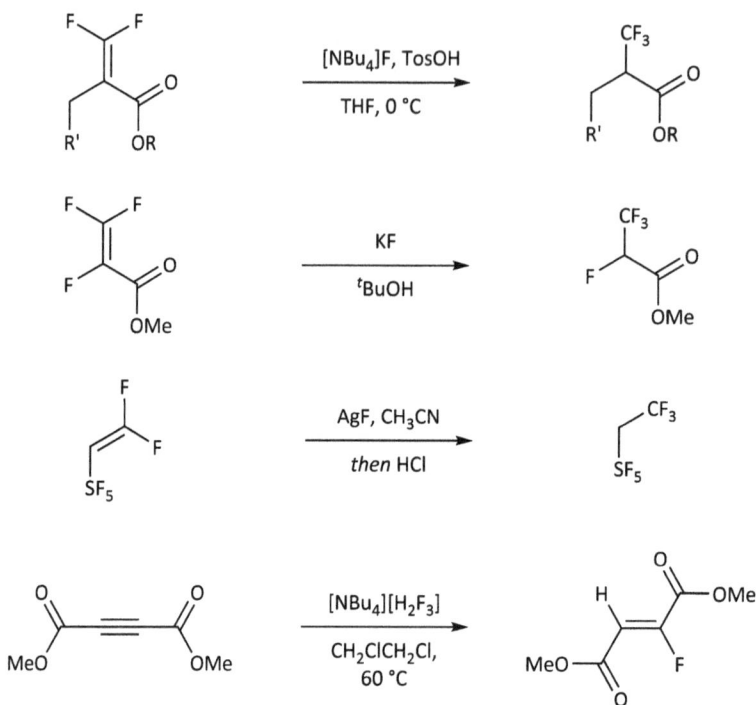

Scheme 13.37: Hydrofluorination of highly electrophilic olefins and acetylenes.

The HF-addition to unsaturated carbon-scaffolds is not restricted to the use of anhydrous HF and its amine complexes. Similarly, [NO][BF$_4$] mediates the fluorination of diarylacetylene derivatives by Olah's reagent. Initially, the electrophilic nitrosonium cation adds to the triple bond, which is followed by an attack of F$^-$. After a series of alternating protonations and further fluoride additions, the acetylene function is finally converted to a tetrafluoroethylene unit [2945].

Based on partly fluorinated olefins, fluoroalkyl ethers can be prepared, too. For example, 1,1-difluoroethene reacts with formaldehyde in anhydrous HF to give the terminally fluorinated dipropyl ether (CF$_3$CH$_2$CH$_2$)$_2$O. Another important functionalization of olefins includes the inter- or intramolecular aminofluorination, leading to β-fluoroamines, which are important motifs in pharmaceutical industry. Therefore, a number of methods, including asymmetric ones, has been described for their preparation [2860].

13.4.9 Halofluorination of unsaturated C-C bonds

Instead of HF, halogen monofluorides XF (X = Cl, Br, I) can formally be added to olefins, which gives rise to vicinal halofluoroalkanes. Typically, the regioselectivity of the addition follows Markovnikov's rule; apart from that the products are formed with *trans*-stereochemistry. However, in some complex molecules, especially steroids and carbohydrates, the results may differ. Additionally, Wagner–Meerwein rearrangements are likely to occur in some cyclic substrates. For electron-rich starting materials, the product formation can be rationalized by the involvement of halonium intermediates in a polar mechanism.

In most halofluorination reactions, the halogen monofluorides XF are not directly used for this purpose. ClF is a stable compound and can indeed be used for the chlorofluorination of halogenated olefins, acrylates, dienes or styrenes. However, due to its high reactivity it is rather inconvenient to use, especially with sensitive substrates. BrF and IF are unstable, so that they are rather difficult to use in fluoroorganic synthesis. More conveniently, an electrophilic halogenation reagent, such as a *N*-halosuccinimide, is used in conjunction with a fluoride source, often HF, one of its amine-complexes or fluoroalkylamines, like Yarovenko's reagent. The positively charged halogen-species may also originate from the respective halogens, hypochlorites, hypobromites, 1,3-dibromo-5,5-dimethylhydantoin (DBH) or iodine-pyridine complexes. A number of further reagent-combinations has also been reported [2946]. More recently, *N*-halosuccinimides have been used in conjunction with a poly(hydrogen fluoride)-based ionic liquid, which serves as fluoride source [2900, 2947]. The combination of 4-iodotoluene difluoride and I_2 represents a source of IF and converts a variety of alkenes and styrenes in Markovnikov fashion to the corresponding iodofluorinated products [2948]. Another variant is the use of trihaloisocyanuric acids together with pyridine/HF, which has been reported to show good results with cyclohexene and styrene substrates [2949]. The bromo- and iodofluorination of alkenes and styrenes is furthermore achieved, when adducts of 1,1,1,3,3-pentafluoropropene and dialkylamines are used as fluoride source in combination with either DBH or *N*-iodosuccinimide [2950]. The fluorocyanation of enamine substrates has been achieved combining an electrophilic fluorinating agent with a source of CN⁻. Therein, β-fluoroiminium intermediates are trapped by cyanide nucleophiles [2951].

Typically, halofluorination reactions are carried out in nonpolar solvents like CCl_4, CH_2Cl_2 or Et_2O at room temperature or below. Exemplarily, the reaction of 1-hexene with pyridine/HF and *N*-bromosuccinimide (NBS) yields 1-bromo-2-fluorohexane (Scheme 13.38). Using the same reagent system, diphenylacetylene is converted to 1-bromo-2-fluoro-1,2-diphenylethene and acrylic acid undergoes bromofluorination to give 2-bromo-3-fluoropropionic acid. The latter is easily transformed to an α-amino acid, if it is treated with ammonia. For the addition of "BrF" to 1,1-diphenylethylene, NBS is used in conjunction with polymeric pyridine/HF (Scheme 13.38). As expected, the product shows the typical Markovnikov regiochemistry. Furthermore, the *trans*-stereoselectivity is illustrated in the iodofluorination of 1,2-dihydronaphthalene [2952]. As already mentioned above, cyclic systems, such as (*1Z,5Z*)-1,5-cyclooctadiene often undergo rearrangements upon halofluorination, however, the product formation also strongly depends on the choice of the solvent and the reagents.

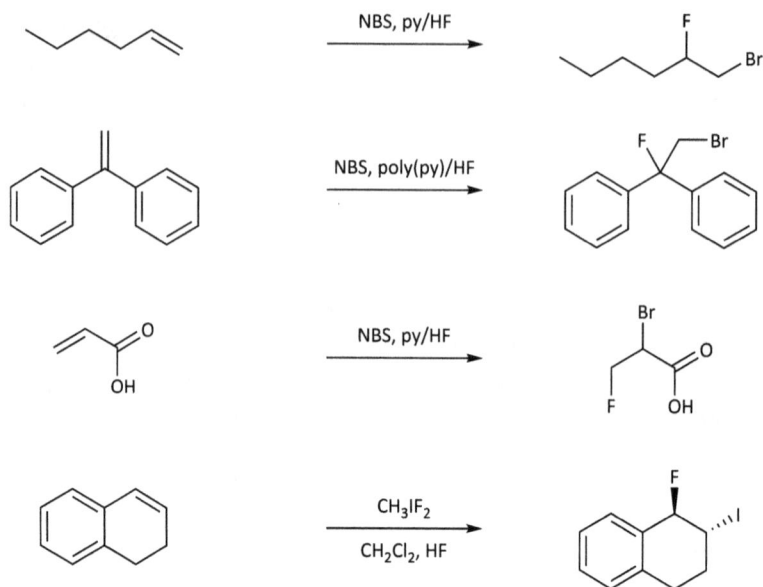

Scheme 13.38: Halofluorination of selected olefins.

The use of NBS with pyridine/HF in ethereal solution does exclusively yield bicyclic rearrangement products, whereas NBS and $Et_3N \cdot 3\ HF$ in CH_2Cl_2 cleanly lead to the bromofluorination of one double bond (Scheme 13.39). The treatment of (1Z,5E)-1,5-cyclodecadiene with NBS and $Et_3N \cdot 3\ HF$ in Et_2O again yields exclusively one rearranged product. Accordingly, the monoepoxide of (1Z,5Z)-1,5-cyclooctadiene is halofluorinated, accompanied by the cleavage of the oxirane ring. However, a new oxygen bridge is formed across the molecule, resulting in the formation of 9-oxabi-cyclononane-derivatives (Scheme 13.39). A variety of rearranged halofluorination products is also observed for the bicyclic substrates norbornene and norbornadiene [2946]. The iodofluorination of allenes is achieved by treatment with N-iodosuccin-imide and $NEt_3 \cdot 3\ HF$, yielding benzylic and allylic fluorides, which contain a vinylic iodo-substituent [2953].

It is also possible to formally add F_2 to unsaturated systems under mild condi-tions, if the XF-adduct, which is obtained in the first step is further converted by halo-gen exchange. For example, cyclohexene reacts with N-bromo or N-iodosuccinimide in presence of Olah's reagent to give 1-bromo-2-fluorocyclohexane or 1-fluoro-2-iodo-cyclohexane, respectively. Further treatment with AgF and pyridine/HF results in halogen exchange and the generation of 1,2-difluorocyclohexane. Similarly, 1,2-di-phenylethene may be converted to 1,2-difluoro-1,2-diphenylethane.

Halofluorination reactions of acetylenes are only scarcely described in the lit-erature. However, dialkyl- and diphenyl alkynes are converted to vicinal halofluo-roalkenes using N-halosuccinimides in conjunction with Olah's reagent. Acetylenes

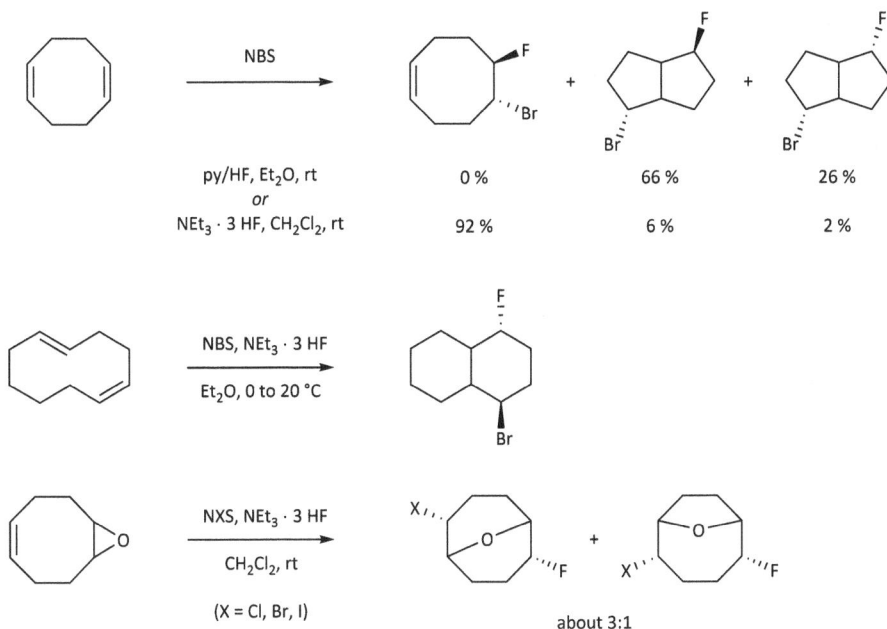

	py/HF, Et$_2$O, rt *or* NEt$_3$ · 3 HF, CH$_2$Cl$_2$, rt			
		0 %	66 %	26 %
		92 %	6 %	2 %

Scheme 13.39: Halofluorination of cyclic substrates, resulting in rearranged products.

can also be directly treated with "BrF" or "IF" at low temperatures, however, they are mainly converted to saturated compounds, especially in presence of an excess of reagents.

13.4.10 Fluorocarbene addition to unsaturated C-C bonds

Another important reaction type for the derivatization of unsaturated compounds is the addition of fluorocarbenes to olefins, which leads to the formation of fluorinated cyclopropane rings [2954, 2955]. The most prominent and important example is difluorocarbene :CF$_2$, a relatively stable and moderately electrophilic species. The stability is explained by π-donation of the fluorine substituents, which counters the destabilization by inductive effects. Interestingly, :CF$_2$ does not insert into C-H bonds, even though this behavior is observed with the heavier chlorine homologue :CCl$_2$.

The preparation of difluorocarbene can be achieved by several strategies, such as the decomposition of heavy-metal trifluoromethyl compounds, like Me$_3$SnCF$_3$ or PhHgCF$_3$. For example, cyclohexene is reacted with Hg(CF$_3$)$_2$ and NaI to give 7,7-difluoronorcarane [2956–2958]. Alternatively, :CF$_2$ is obtained by thermal decomposition of halodifluoroacetate salts, which requires quite drastic reaction conditions [2959–2961]. Under significantly milder conditions, derivatives of (fluorosulphonyl)difluoroacetic acid are decomposed in base- or fluoride-induced transformations

[2962, 2963]. The treatment of dihalodifluoromethanes with different reducing agents also gives rise to difluorocarbene, as well as the thermal decomposition of perfluoro-cyclopropane and other perfluorocarbons [2964, 2965]. Industrially, difluorocarbene is prepared by thermal elimination of HCl from chlorodifluoromethane and serves as important starting material for the synthesis of tetrafluoroethylene and other perfluorocarbons [2954].

As a general trend, a poor reactivity is observed when difluorocarbene is reacted with electron-deficient alkenes, whereas electron-rich olefins are rather easily converted to difluorocyclopropane-derivatives [2966]. The transformation of relatively electron-poor butyl acrylate to the corresponding difluorocyclopropane has been achieved, using $(SO_2F)CF_2CO_2SiMe_3$ as difluorocarbene source [2967]. However, the substrate scope of this transformation is limited. In many cases, it is helpful, to protect electron-withdrawing carbonyl-functions as ketals or acetals [2968]. A rather mild difluorocarbene-transfer toward alkenes and alkynes is achieved, when CF_3SiMe_3 is treated with NaI, giving rise to difluorocyclopropanes and -propenes, respectively [2969].

13.4.11 Halo- and hydrofluorination of highly fluorinated olefins

The reactivity of perfluoroolefins is inverted to that of unsaturated hydrocarbons, as they are very susceptible to nucleophilic attacks. Hence, the addition of HF or XF to perfluoroolefins proceeds via initial attack of F^- to result in the formation of perfluorinated carbanions, which are subsequently protonated or halogenated. However, with large cations like Cs^+, Ag^+ or $[(NMe_2)_3S]^+$, tertiary fluorinated carbanions can even be isolated. A typical reagent combination for the transfer of HF is that of KF and urea, which is for example used to convert hexafluoropropene to 1,1,1,2,3,3,3-heptafluoropropane. Alternatively, AgF can be used to generate the perfluoroisopropyl anion, which is afterwards protonated by the addition of water. The preparation of perfluorohaloalkanes $C_nF_{2n+1}X$ (X = Br, I) is very important in industry, as those compounds are useful precursors for many other fluorochemicals. Due to their instability, the halogen fluorides BrF and IF have to be prepared *in situ* at elevated temperatures by adding stoichiometric mixtures of BrF_3/Br_2 and IF_5/I_2, respectively. Generally, the addition of "IF" and "BrF" to highly electrophilic olefins requires comparably higher reaction temperatures than the halofluorination of hydrocarbons.

The treatment of perfluoropropene with a mixture of IF_5 and I_2 results in the addition of "IF" to the olefin, giving rise to 2-iodoheptafluoropropane (Scheme 13.40). The same result is obtained under milder conditions, if the fluoroolefin is reacted with KF and iodine in acetonitrile. As further examples for iodofluorinations, tetrafluoroethylene is converted to pentafluoroiodoethane and 1,2-difluoroethene analogously yields 2,2,2-trifluoro-1-iodoethane. Similar to the aforementioned transfer of

I$_2$, IF$_5$

150 °C

KF, I$_2$

CH$_3$CN, 110 °C

F$_3$C   CF$_3$

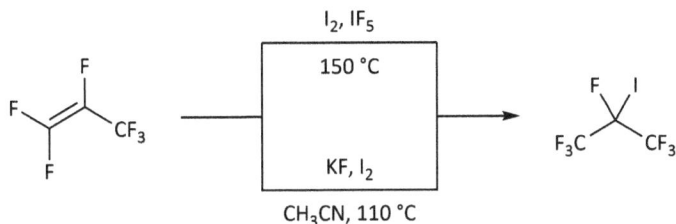

Scheme 13.40: Different approaches to the iodofluorination of perfluoropropene.

"IF," the reaction of perfluoropropene with BrF gives rise to 2-bromo-1,1,1,2,3,3,3-heptafluoropropane. The mechanism of these transformations is not fully discovered, but most likely proceeds via formation and subsequent capture of perfluorocarbanionic species. In a few cases, bromination- and iodination-reactions of stable tertiary perfluorocarbanions are reversible, especially in sterically demanding systems (Scheme 13.41). Sometimes, steric effects might also be responsible for the formation of unexpected halogenation products. For example, the treatment of a sterically crowded perfluorocarbanion with ICl does not result in the transfer of the formally more positively charged iodine, but the smaller chlorine-substituent (Scheme 13.41).

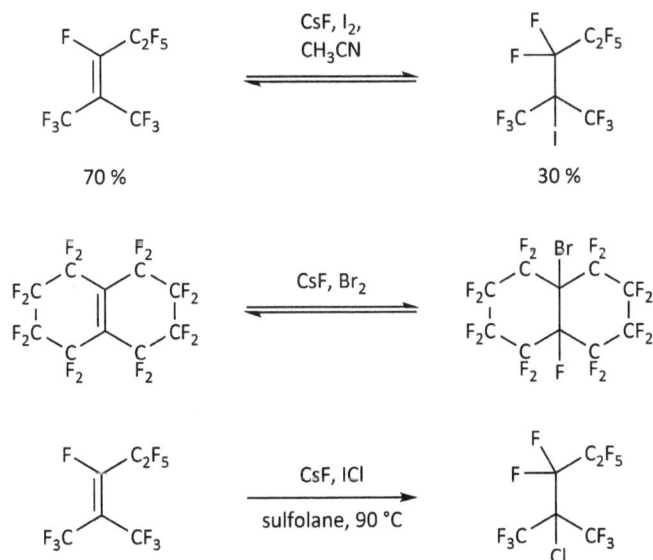

CsF, I$_2$, CH$_3$CN

70 %   30 %

CsF, Br$_2$

CsF, ICl

sulfolane, 90 °C

Scheme 13.41: Steric impact in bromo- and iodofluorination reactions of bulky fluoroolefins.

The high reactivity of halogen monofluorides furthermore allows the addition to highly fluorinated aromatic substrates, such as hexafluorobenzene or octafluoronaphthalene. In the sense of a 1,4-addition, a mixture of Br$_2$ and BrF$_3$ will initially transfer F_2 to the aromatic system. Afterwards, BrF is added to one of the remaining double-bonds. Halofluorinations are not limited to unsaturated C-C moieties, but can also occur across C=O-, C=N- or C≡N-bonds. The reaction of fluorinated carbonyl-compounds with alkali metal fluorides and ClF gives rise to hypochlorites, as for example, COF_2 is transformed to CF_3OCl. Analogously, fluorinated imines are converted to N-chloro- or N-bromoamines, whereas nitriles are transformed to N-dihaloamines or N-haloimines, depending on the addition of either one or two equivalents of halogenfluoride.

13.4.12 Highly fluorinated olefins as building blocks

Of course, the derivatization of fluorinated olefins is not restricted to the attachment of further halide substituents or protons. However, in most cases, such substitution reactions proceed via the initial addition of a fluoride ion, which gives rise to perfluorinated carbanions. Afterwards, these anions can be treated with a number of electrophilic species. The extremely toxic and electrophilic perfluoroisobutene easily adds any type of nucleophile, as it is for example readily hydrolyzed in a mixture of acetone and water to yield α-hydroperfluoroisobutyric acid. In the presence of CsF, it gives rise to the perfluorinated *tert*-butylcarbanion, which subsequently can undergo a nucleophilic attack toward alkyl-, benzyl- or allyl halides, respectively (Scheme 13.42). Apart from that, dimerization or oligomerization of the fluoroolefin can also occur in presence of fluoride ions. If different unsaturated compounds are reacted together, mixed oligomers are formed (Scheme 13.42). Perfluoroalkyl anions can also be added to perfluorinated (hetero-)aromatic compounds and other activated arenes, giving rise to the respective perfluoroalkylated derivatives.

Perfluorinated anions can further be functionalized with carbon dioxide, carbonyl fluoride, acyl halides, fluoroformates as well as hexafluorothioacetone. Accordingly, the treatment of a mixture of perfluoroisobutylene, phosgene and CsF results in the formation of perfluoropivaloyl fluoride (Scheme 13.43). Highly fluorinated ketones and acid fluorides are easily converted to the respective secondary and primary fluoroalkoxides, if they are treated with a fluoride source. The intermediate fluoroalcoholates can then be reacted with a number of electrophiles, for example, alkyl- and allyl halides, α-haloketones or alkyl sulphates to result in the formation of fluoroethers (Scheme 13.43). Treatment of alkoxides with a mixture of tetrafluoroethylene and iodine gives rise to 2-iodo-1,1,2,2-tetrafluoroethyl ethers. Apart from that, acid fluorides can intramolecularly be transformed to ether derivatives. Perfluoroethers are also obtained, if the corresponding fluorinated alkoxides are reacted with epoxides like hexafluoropropene oxide (HFPO), which results in ring opening and the formation of a new alkoxide or acid fluoride species, respectively (Scheme 13.43). Based on this method, the formation of higher oligomers or perfluoropolyethers can be achieved.

CsF, RX

diglyme

R = Me, Bu, Bn, allyl

CsF

sulpholane

KF

CH$_3$CN

E/Z = 85:15

CsF,
CF$_2$=CF$_2$

CH$_3$CN

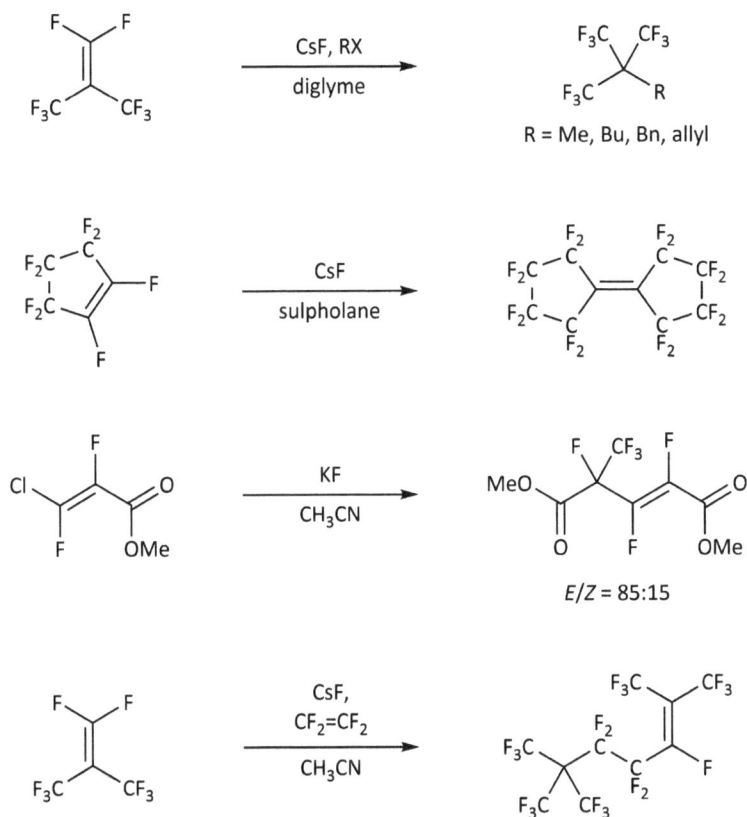

Scheme 13.42: Derivatization of perfluoroolefins under formation of C-C bonds.

Fluoride ions are capable of opening lactone rings as well, again the generated alkoxide species may be trapped by electrophiles (Scheme 13.43). Moreover, fluoride ions add to C=N- and C≡N-bonds in highly fluorinated, electron-deficient compounds. For example, pentafluoro-2-azapropene is converted to an aza-anion, which gives rise to *N,N-bis*(trifluoromethyl)amine compounds upon alkylation (Scheme 13.43). Nitriles and imines react with carbonyl fluoride in the presence of a fluoride source to result in the formation of fluorinated isocyanates.

Fluorinated olefins are also useful starting materials for the synthesis of fluorine-containing sulphur- and selenium compounds. For example, fully fluorinated *tert*-butyl mercaptan is prepared via intermediary K[C(CF$_3$)$_3$], which is initially prepared from perfluoro isobutene and KF. The potassium compound is reacted with Hg(CF$_3$COO)$_2$ and subsequent reaction of the Hg-salt with sulphur and KF gives rise to the respective mercury thiolate, from which perfluorinated *tert*-butyl thiol is liberated upon acidification with HCl. The derived disulphide is directly accessible from the reaction of perfluoro isobutene with sulphur and SbF$_5$. Analogously, many other perfluoroalkyl substituted sulphur compounds are accessible if the fluorinated carban-

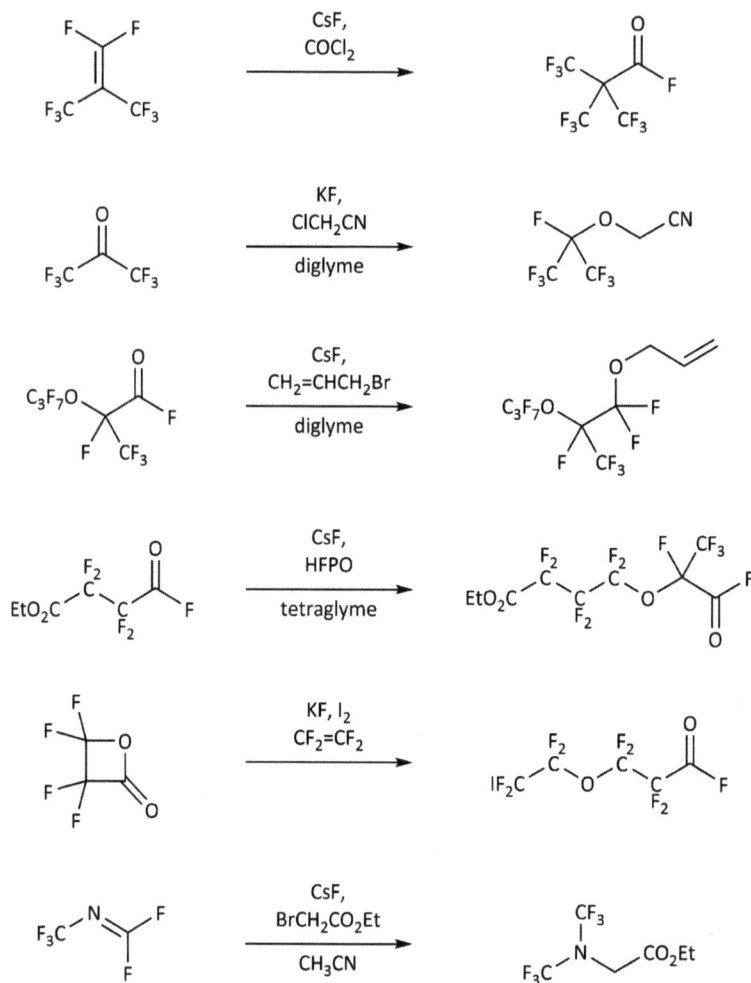

Scheme 13.43: Derivatization of unsaturated perfluorinated compounds.

ion is treated with electrophilic species like sulphenyl halides, dialkyl disulphides, thionyl fluoride or sulphuryl fluoride. This synthetic route also works for the preparation of selenium compounds, as for example, tertiary perfluoroalkyl selenoethers are synthesized from the respective perfluorinated carbanions and selenenyl halides, for example, PhSeCl.

Addition reactions to unsaturated fluorocarbons are furthermore suitable for the preparation of amines with partially fluorinated alkyl chains. The treatment of 1,1-difluoroethene with HNO_3 in anhydrous HF initially yields 2,2,2-trifluoronitroethane, which is subsequently reduced by Fe/HCl to give the corresponding trifluoroethylamine hydrochloride. Instead of HNO_3/HF, NO_2F or $[NO_2][BF_4]$ can alternatively be used for the simultaneous introduction of a fluoro- and a nitro-substituent. Analogously, nitroso-derivatives are obtained, if perfluorocarbanions are treated with NOCl, or directly from the fluoroolefin and NOF. Tertiary perfluoronitroso compounds are useful precursors of perfluoro alcohols. Finally, fluorinated amines are directly accessible from perfluoroolefins, if for example tetrafluoroethylene is reacted with hexamethylenetetramine in anhydrous HF, which gives rise to 2,2,3,3,3-pentafluoropropylamine.

Depending on the nature of the nucleophile and the reaction conditions, the obtained product will not always be saturated. For example, tetrafluoroethylene is substituted by sodium ethoxide to give ethyltrifluorovinyl ether. However, if traces of ethanol are still present in C_2H_5ONa, the saturated addition product $CF_2HCF_2OC_2H_5$ is generated in significant amounts. The same observation is made for the thiolates $NaSC_4H_9$ and $NaSC_2H_5$ in reactions with tetrafluoroethylene and perfluoroisobutylene, which mainly lead to the formation of the respective perfluorovinyl- and perfluoroisobutenyl thioethers.

13.4.13 Study questions

(I)   What is the advantage of using BrF_3 in halogen exchange reactions? How can one improve its selectivity? How does the solvent influence the outcome of the reaction?

(II)   Which general trend is observed in the reactions of polychloroalkanes with ClF? How can this behavior be influenced?

(III)   What is Swarts fluorination? To which types of substrates is it mostly applied and which products are generally obtained?

(IV)   Which fluoride sources are typically used for the halogen exchange of alkyl monohalides? What needs to be taken into account when using naked fluoride?

(V)   Which products are obtained upon halogen exchange of perchloropropene, benzotrichloride and pentachloroethylbenzene?

(VI)   Which reagents are typically used to substitute halogen substituents in α-position of carbonyl groups and in acid chlorides? Is there a different result when CCl_3COCl is reacted with KF and aHF?

(VII)   Which are the advantages and disadvantages of using pyridine/HF or aHF in fluorodehydroxylation reactions?

(VIII) Which general reactivity and selectivity is observed for fluoroalkylamine reagents in fluorodehydroxylation reactions? Which are commonly the major side-products?

(IX)   How can one explain the reactivity of the fluoroalkylamine reagents? Is there a general mechanism?

(X)   Why is SF_4 only sparingly used in fluorodehydroxylation reactions? Which products are obtained upon reaction with 1,2-diols and α-hydroxyketones?

(XI)   How does the reactivity of DAST and other R_2NSF_3 reagents differ from that of parent SF_4? Under which conditions are they typically used? Comment on their chemoselectivity.

(XII)   Which general mechanism do fluorodehydroxylation reactions with DAST typically follow? Which major problems are observed with secondary substrates, especially allylic and benzylic alcohols?

(XIII)   Which functional groups and reagents are suitable to activate alcohols for a replacement by fluoride?

(XIV)   Which products are obtained upon reaction of linear and cyclic α-fluoroethers with aHF? How is the reactivity influenced by further substituents in β- and γ-position?

(XV)   What is a common method for the preparation of fluorohydrins? How can the reagent influence the outcome of the reaction?

(XVI)   How can the stereoselectivity of epoxide-cleavage toward α-fluorohydroxy compounds be influenced?

(XVII)   Which reagents are required to deoxyfluorinate carbonyl compounds? What types of products are commonly obtained? What is the general mechanism of this transformation?

(XVIII)   Which side-products are observed upon deoxyfluorination of aldehydes? Does this behavior depend on the reaction conditions and the substrate?

(XIX)   Which products are obtained, when α- and β-diketones are submitted to deoxyfluorination with SF_4? What are the products of the reaction of β-ketoesters with SF_4? Which side-products are observed and how can their formation be suppressed?

(XX)   Which are the main differences, if SF_4-derivatives are used for deoxyfluorination reactions of carbonyl compounds instead of the parent tetrafluoride?

(XXI)   Which are the main side-products of the reaction of carboxylic acids with SF_4/HF? How can one rationalize their formation? Which products are obtained upon fluorination of 1,1-; 1,2-; 1,3- and 1,4-dicarboxylic acids?

(XXII)   Which products are obtained, when esters and amides are reacted with SF_4? Give examples.

(XXIII)   Which reagents are generally needed for the oxidative fluorodesulphuration? To which substrates is this technique typically applied and which conditions does it require?

(XXIV)   How does the product formation in the fluorodesulphuration of dithioesters and xanthogenates depend on the choice of reagents? Which products are obtained?

(XXV)   How can nitrogen-containing substrates, such as α-aminoacids, diazocompounds or small heterocycles serve as starting materials for the preparation of fluorinated compounds? With which reagents do they need to be reacted and which products are generally obtained?

(XXVI)   Under which conditions does HF add to unsaturated carbon-carbon bonds? How can one rationalize the regioselectivity of the product formation?

(XXVII)   Which reagents are typically used for the halofluorination of electron-rich olefins? How does the choice of reagents influence the product formation?

(XXVIII) Which types of products are obtained, if fluorocarbenes, for example, :CF_2, are reacted with olefins? How is :CF_2 prepared?

(XXIX)   How does the halofluorination of highly fluorinated olefins differ from that of hydrocarbons? Which reagents are used for this type of transformation?

(XXX)   How are perfluorinated tert-butyl and perfluoroisopropyl compounds accessible? Which species are obtained upon treatment of perfluoroketones and perfluorinated acid fluorides with fluoride? What are they used for?

13.5 Electrophilic fluorination

A variety of reagents may be used for the electrophilic introduction of fluorine into aliphatic substrates. Depending on structure and sensitivity of the compound, this can in some cases also be achieved by the use of F_2 at low temperatures. However, more conveniently, fluorine is attached to a group of atoms, which serves as better leaving group than F⁻, thus donating some sort of "F⁺." The latter is believed to be transferred in halophilic S_N2-type manner or by the involvement of radical intermediates, however, the actual mechanism is still not fully understood. Typically, electrophilic reagents are applied for the fluorination of activated alkenes, for example, enol ethers, enol acetates, silyl enol ethers or enamines, yielding mostly α-fluorocarbonyl derivatives. Apart from that, activated C-H bonds can be substituted as well as organometallic species. Electrophilic fluorination techniques can furthermore serve for the preparation of ^{18}F-labeled compounds.

13.5.1 Elemental fluorine as electrophilic fluorinating agent

The simplest reagent for electrophilic fluorination of aliphatic systems is elemental fluorine itself. However, special reaction conditions are required to suppress a radical reaction mode (cf. Section 12.1.1). To enable electrophilic fluorination with F_2, it needs to be highly diluted by inert gases, such as N_2, Ar or He. Typically, mixtures containing 1–10 % F_2 in N_2 are used for this purpose. The dilution principle is also mandatory for the substrate, which is generally dissolved in $CFCl_3$, mixtures of $CHCl_3$ and $CFCl_3$ or also CH_3CN. The advantage of $CHCl_3/CFCl_3$-mixtures is their property to efficiently scavenge radicals, and thus inhibiting radical chain mechanisms. Finally, reaction control is achieved by working in a low temperature range between −40 and −78 °C. Mechanistically, it is clear that no actual "F⁺" is responsible for the introduction of the fluorine substituent. Instead, the transformation of C-H to C-F bonds might be explained by a three-centered transition state and the coordination-assistance of the solvent (Scheme 13.44). Electrophilic fluorination reactions with elemental F_2 work well for sp^3-hybridized, anionic or strongly nucleophilic carbon centers, especially tertiary ones in steroids or other complex molecules (Scheme 13.44). Apart from that, fluorination readily occurs in α-position of electron-withdrawing functions, for example, nitro- or carboxylato-groups. Accordingly, enolates, enols and enol silyl ethers are converted to α-fluoro ketones.

13.5.2 Electrophilic fluorination using fluoroxy compounds

Rather simple but hazardous variants of electrophilic fluorination are represented by the addition of OF-compounds to unsaturated systems. Typical reagents for this purpose include the oxygen fluorides OF_2 and O_2F_2 and more conveniently hypofluorites, such as CF_3OF. The addition of the latter to electron-rich olefins, especially steroids and carbohydrates, occurs under formation of cis-Markovnikov products, involving polarization of the O-F bond. In contrast, electrophilic, halogenated substrates are

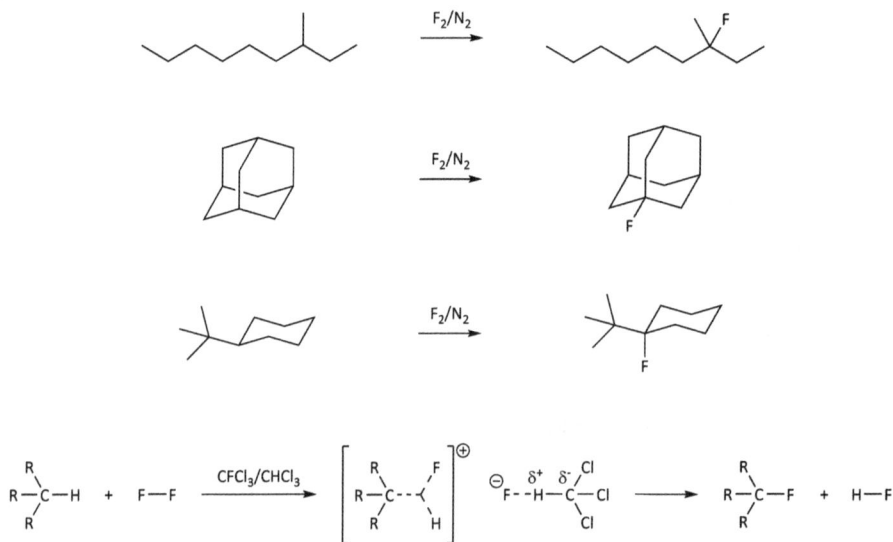

Scheme 13.44: Electrophilic Fluorination of substrates with tertiary C-H bonds using elemental F_2 (top) and proposed reaction mechanism (bottom).

transformed by radical processes, which can either lead to the formation of fluoro-trifluoromethoxylated molecules or oligomeric products. Perfluoroalkoxyfluorina-tions are also observed for perfluorinated aromatics like hexafluorobenzene, which add the respective perfluoroalkyl hypofluorite in the sense of a 1,2- or 1,4-addition (Scheme 13.45). Another very useful reagent for electrophilic fluorination is acetyl hy-pofluorite, which adds to olefins in a polar mechanism to yield vicinal fluoroacetates with Markovnikov selectivity. Apart from that, *syn*-stereochemistry is observed in most cases, as for example in the reaction with (*E*)-stilbene (Scheme 13.45). CH_3COOF is es-pecially useful for the preparation of fluorinated carbohydrates and can also serve for ^{18}F-labeling, as it is rapidly prepared from acetate salts and F_2. The respective perfluo-rinated acetyl hypofluorite CF_3COOF has also been used for electrophilic fluorination reactions, as it is for example capable of converting enol acetates to α-fluoroketones. However, due to its high reactivity it is not commonly used as fluorinating agent.

The addition of fluoroxy compounds to unsaturated C-C bonds is not restricted to organic hypofluorites, but can also occur with inorganic compounds, such as $CsSO_4F$. If it is reacted with styrene or cyclohexene in aprotic solvents like CH_3CN or CH_2Cl_2, the respective vicinal fluoroalkyl sulphates are generated. However, both, the regio- and stereoselectivity of these transformations are poor. In protic solvents like CH_3OH or CH_3COOH, vicinal fluoroalkyl methyl ethers or acetates are obtained, respectively, showing predominantly *syn*-Markovnikov selectivity. Accordingly, the reaction of phenyl acetylene with two equivalents of $CsSO_4F$ in methanol yields a mixture of difluoroacetophenone and its dimethyl acetal (Scheme 13.46). Apart from

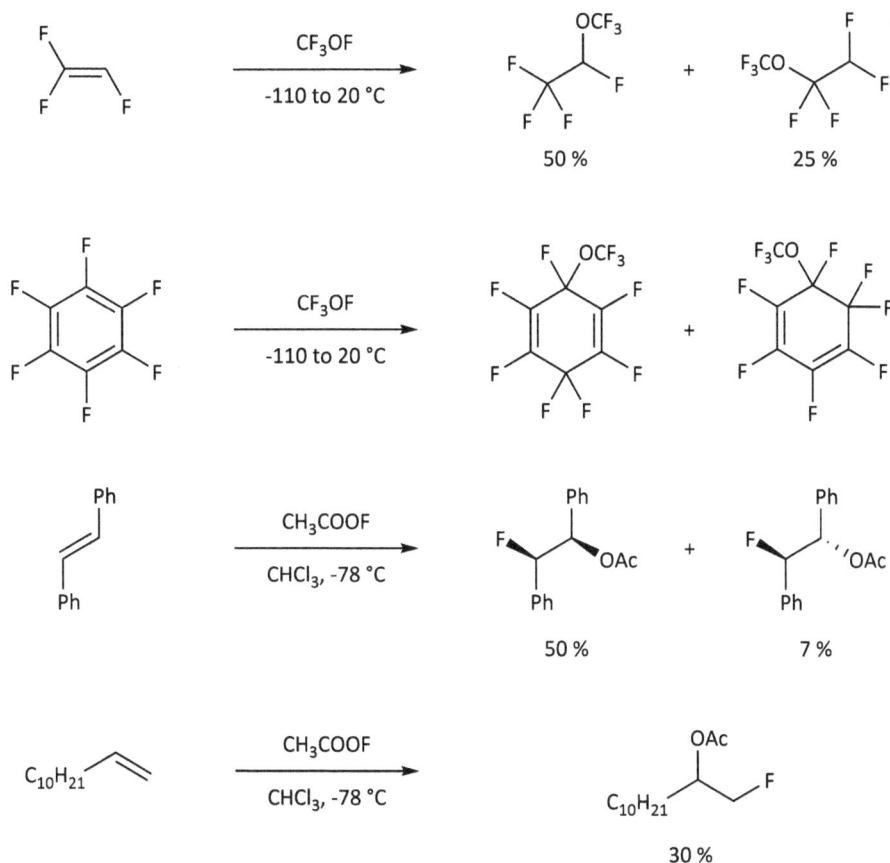

Scheme 13.45: Addition of organic hypofluorites to unsaturated C-C bonds.

fluoroxysulphates, inorganic sources of electrophilic fluorine include fluoroxenonium reagents, which are formed by reaction of XeF_2 with CF_3SO_3H, FSO_3H or HNO_3. They are stable below $-10\,°C$ and react with olefins to give vicinal adducts with *syn*-selectivity (Scheme 13.46). Another inorganic reagent, that has initially been used for electrophilic fluorination is the extremely unstable perchloryl hypofluorite ClO_4F. Analogously to the other fluorooxy compounds discussed above, it can add to olefins, for example, perfluoropropene, to give a mixture of perfluoropropyl perchlorate regioisomers (Scheme 13.46).

The reactions of CF_3OF with enol ethers, enol acetates and enamines are especially useful for the synthesis of steroidal α-fluoroketones. Depending on the structure of the substrate, the fluorine substituent may even be introduced with stereoselectivity. Silyl enol ethers are suitable starting materials for the generation of α-fluorinated esters, carboxylic acids or amides in reactions with CF_3OF at $-70\,°C$. Upon conversion of trimethylsilyl ethers, COF_2 and $(CH_3)_3SiF$ are liberated, so that a subsequent hydrol-

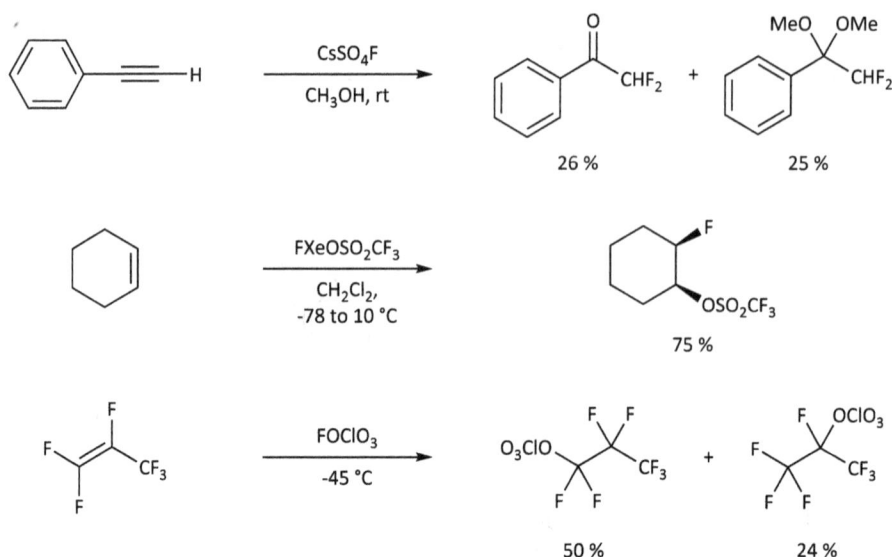

Scheme 13.46: Addition of inorganic fluoroxy compounds to unsaturated C-C bonds.

ysis of the reaction mixture is not required. Analogously, acyl hypofluorites are used for the fluorination of enols, enol acetates, vinyl ethers and enolates. A smooth fluorination is observed for enol acetates in steroidal scaffolds. For the fluorination of vinyl acetates, mixtures of acyl- and perfluoroalkyl hypofluorites also work well. At room temperature, $CsSO_4F$ is typically applied in CH_3CN- or CH_2Cl_2-solution to fluorinate enols and enol acetates.

The capability of fluorinating tertiary C-H bonds is not restricted to elemental F_2 but can also be performed using fluoroxy reagents. In principle, both secondary and tertiary C-H bonds can be replaced, however, tertiary protons are more likely substituted. A typical model-substrate for such type of transformation is adamantane, which is rather easily converted to 1-fluoroadamantane by any electrophilic fluorination agent. The best yields are observed for reactions with CF_3OF and F_2 at −25 °C, whereas other common agents like XeF_2 or $CsSO_4F$ give rather poor yields. Again, CF_3OF is suitable for the derivatization of complex molecules with biological applications, for example, amino acids or steroids.

13.5.3 Electrophilic fluorination using perchloryl fluoride

Even if its preparative relevance might be negligible nowadays, ClO_3F clearly needs to be mentioned among the list of electrophilic fluorinating agents. Historically, it has been widely applied in electrophilic fluorination reactions, however, today its use is rather unattractive, as it is hazardous and rather difficult to handle. Apart from that, it often reacts under formation of chlorinated by-products. Nevertheless, it is a useful reagent for the fluorination of highly nucleophilic anions and enamines. For example, anionic species derived from malonaldehyde and organophosphonates are readily converted to fluorinated analogues. Accordingly, perchloryl fluoride has been used for the preparation of diethyl 2-ethyl-2-fluoromalonate, starting from the sodium salt of the carbanion. For very electron-deficient systems, like mono- or dinitro enolates, the fluorination with ClO_3F proceeds at room temperature in the presence of KF.

13.5.4 Fluoraza compounds as electrophilic fluorinating agents

The preparative use of the aforementioned fluoroxy compounds as well as ClO_3F is closely associated with a variety of drawbacks, the most important one certainly being the safety aspect. Practically, all hypofluorites are unstable compounds, which can spontaneously decompose, often this happens violently. Perchloryl fluoride is a strong oxidizer and can also react violently with organic matter. Thus, it is impossible to perform fluorination reactions with such reagents on a larger scale. Furthermore, most of the reagents require special handling and reaction vessels. Their high reactivity additionally renders them nonsuitable for the conversion of sensitive compounds. Thus, electrophilic fluorination agents have been developed, which are capable of selectively transferring fluorine-substituents to electron-rich substrates under rather mild conditions.

In practice, the reagents for the transformation of C-H acidic compounds or (masked) carbanions, for example, enol acetates, silyl enol ethers and enolates, are virtually all N-F reagents, which are generally divided into the two classes of neutral N-fluoroamine compounds and ionic N-fluoroammonium reagents. Under appropriate reaction conditions, they can also fluorinate thioethers, sulphoxides and sulphones in α-position [2970, 2971]. For example, $(CF_3SO_2)_2NF$ converts the activated 1,3-diketone 3-methyl 2,4-pentanedione to 3-fluoro-3-methyl-2,4-pentanedione. Similarly, diethyl (1-fluoroethyl) phosphonate is accessible by treatment of the corresponding potassium salt with NFSI. In case of the fluorination of β-diketones and β-ketoesters with $(CF_3SO_2)_2NF$, reaction control may be achieved by the addition of water (Scheme 13.47). Under anhydrous conditions, the activated methylene group is converted to a CF_2-unit, whereas the presence of H_2O inhibits the second fluorination, as it prevents further enolization of the monofluorinated product [2972]. Ester- and amide enolates, generated at low temperatures by the use of lithium diisopropylamide (LDA), as well as neutral dicarbonyl compounds are also efficiently fluorinated by $(CF_3SO_2)_2NF$ (Scheme 13.47). Analogously, α-fluorinated-β-dicarbonyl compounds are prepared using Selectfluor™ in acetonitrile under microwave conditions [2973]. With the same class of substrates, both NFSI and Selectfluor™ can be used as reagents for α-monofluorination under solvent-free conditions or in water (Scheme 13.47) [2974, 2975]. The α-fluorination of β-ketoesters has also been

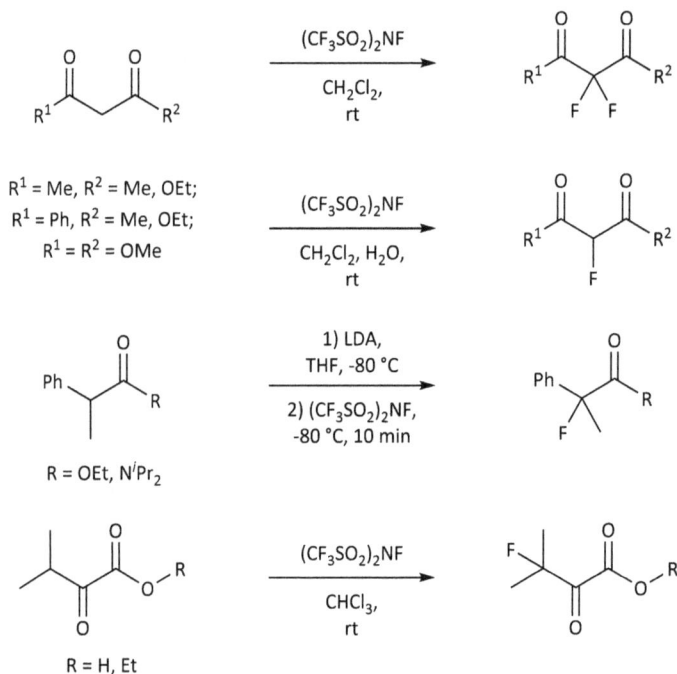

R^1 = Me, R^2 = Me, OEt;
R^1 = Ph, R^2 = Me, OEt;
R^1 = R^2 = OMe

R = OEt, N^{i}Pr$_2$

R = H, Et

Scheme 13.47: Electrophilic fluorination of enols and enolates using (CF$_3$SO$_2$)$_2$NF.

achieved using HF/pyridine in presence of a peracid-oxidant and catalytic amounts of 4-iodotoluene. The reaction proceeds through intermediary ArIF$_2$ species and works also well for β-ketoamides and β-ketosulphones [2869].

Analogously to activated β-ketocarbonyls, the α-fluorination of enolizable ketones can be achieved without pre-functionalization, if the carbonyl-substrate, for example, acetophenone, is directly treated with Selectfluor™. This can be realized under several different sets of conditions; however, the use of a polar solvent is required. For example, α-fluorination occurs upon reaction of the starting material with the fluorinating reagent in refluxing methanol or under microwave irradiation; furthermore, electrophilic fluorination reactions can be carried out in aqueous micellar systems or in presence of catalytic amounts of H$_2$SO$_4$ in methanol [2864, 2976–2978]. Alternatively, the α-fluorination of acetophenone-derivatives as well as silyl enol ethers can also be achieved by the use of nucleophilic fluoride sources, such as complexes of Et$_3$N and HF, if a hypervalent iodine-reagent, for example, ArIO or ArIF$_2$, is added [2979, 2980]. The treatment of styrene derivatives with Selectfluor™ and an oxidant generates α-fluorinated acetophenones as well [2981].

The formation of mono- and difluorinated products can further be controlled in the fluorination of enolates using N-fluorosultam reagents. If a slight excess of base (1.2 eq.) and fluorinating reagent (1.3 eq.) is used, the monofluorinated deriva-

tive will preferentially be generated, whereas larger excess of both, base (3.6 eq.) and
N-fluorosultam (3.6 eq.) mostly leads to the formation of difluorinated compounds.
The electrophilic fluorination of C-H acidic compounds can also proceed via enamines, as for example in the case of cyclohexanone, which is initially activated by
the use of morpholine. Upon reaction with Selectfluor™, 2-fluorocyclohexanone is
generated. In principle, it is also possible to react organometallic species, for example, Grignard compounds or organosilanes with N-F-reagents and obtain fluorinated
molecules. However, this approach has only received limited attention.

Aliphatic hydrocarbons can also be directly substituted by electrophilic fluorination using Selectfluor™. However, the reaction conditions strongly influence the
type of products formed, being either the respective alkylfluorides or, alternatively,
acetamides. The latter result from the reaction with the solvent and are especially observed after prolonged heating of the substrate and the fluorinating agent in acetonitrile. The presence of Lewis acid catalysts, such as $BF_3 \cdot OEt_2$, further supports this reaction pathway [2982]. More recently, the selective C-H fluorination of heterobenzylic
substrates has been reported. Selective mono- and difluorination is achieved at elevated temperature (60 °C), when NFSI is used in conjunction with Li_2CO_3. The transformation is believed to proceed via initial formation of an N-sulphonylpyridinium
salt, which increases the acidity of the benzylic protons in its proximity. Thus, deprotonation can occur rather easily and is followed by electrophilic fluorination with
NFSI. If the reaction is carried out at higher temperature (75 °C) and in presence of increased equivalents of the reagents, difluorination is achieved. Apart from pyridines,
pyrimidines can also be fluorinated in benzylic position; however, their transformation needs to be carried out at significantly higher temperatures (150 °C) [2983, 2984].
The treatment of styrene derivatives with Selectfluor™ gives rise to aryl allylic fluorides and proceeds through the formation of benzylic cations [2985]. Similarly,
N-substituted 4-aryl-1,2,5,6-tetrahydropyridines are converted to the corresponding
allylic fluorides using Selectfluor™ in the presence of water [2986]. In ionic liquids,
an electrophilic fluorination is observed upon impact of Selectfluor™ on methylated
phenols, for example, 3,4,5-trimethylphenol, which is converted to the corresponding cyclohexadienone derivative, bearing a fluorine substituent in 4-position [2987].
Analogously, N-tosylaniline substrates are converted to 4-fluorocyclohexadienimine
products upon reaction with Olah's reagent and a hypervalent iodine reagent [2988].
Fluorinated dienones are also obtained when polymethylated benzene derivatives are
reacted with Selectfluor™ in a mixture of acetonitrile and water [2989].

The actual mechanism of all the aforementioned reactions depends on the substrate and the reaction conditions, but can be categorized in the whole range between
a halophilic S_N2-replacement and a single-electron-transfer (SET) (Scheme 13.48).
In the depicted example, a N-F reagent is shown, however, the mechanisms are,
with slight deviations, also applicable for most of the other electrophilic fluorination
reagents.

Nu-   F—NR$_2$ $\xrightarrow{\text{S}_\text{N}2}$ Nu—F   NR$_2^-$

Nu$^-$ F—NR$_2$ $\xrightarrow{\text{SET}}$ Nu$^\cdot$ $\left[\text{F—NR}_2\right]^{\cdot-}$

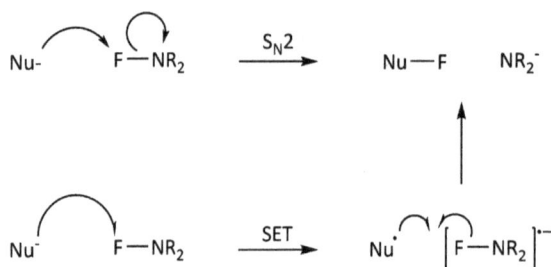

Scheme 13.48: General mechanism of a halophilic S$_\text{N}$2-substitution reaction (top) and a single electron transfer (SET, bottom).

13.5.5 Asymmetric fluorination using fluoraza reagents

In many cases, it is not only required to introduce a fluorine substituent into organic compounds, but also, to do so in a stereoselective fashion. In the last decades, many important developments have been made in the field of enantioselective fluorination, especially including approaches, which use electrophilic fluorinating agents. Due to this, a number of detailed review articles have appeared on this topic [2990–2993]. In general, three major approaches toward chiral C-F compounds can be differentiated, which are based on chiral fluorinating agents, substrates with given stereoinformation or alternatively, the application of suitable auxiliaries. Activated, chiral substrates are often generated by organocatalysis, that is, the treatment of carbonyl compounds with appropriate amines, such as proline-derivatives. The stereoinformation of the intermediary chiral enamine species then induces a stereoselective attack of the electrophilic fluorinating agent. Phase-transfer conditions with chiral anions have also proven suitable in conjunction with cationic fluorinating reagents. Apart from that, it is possible to functionalize fluorine-containing starting materials by, for example, asymmetric alkylation. Overall, the asymmetric introduction of fluorine substituents is mostly restricted to activated carbonyl substrates, including rather simple moieties like ketones or ketoesters, but also more complex (hetero-)cyclic starting materials.

A number of chiral N-F reagents has been developed over the years, however, due to readily occurring racemization of unstable intermediates, the generation of reasonable *ee*-values is not that easily achieved. The first approaches toward chiral fluorinating reagents included the generation of N-F reagents derived from camphorsultam or other *N*-fluorosultams (Figure 13.2) [2994]. In later developments, the chiral fluorinating species has been derived from cinchona alkaloids (Figure 13.2). Using this method, acceptable *ee*-values of about 90 % can be achieved [2995, 2996]. Cinchona-based fluorinating agents can also be prepared *in situ* during the fluorination reaction by treatment with a source of "F$^+$." In many cases, this can be realized in catalytic fashion, reducing the necessary amounts of alkaloids. The attachment of a fluorous tag furthermore enables simple recovery and reuse of the cinchona-alkaloid [2997]. It

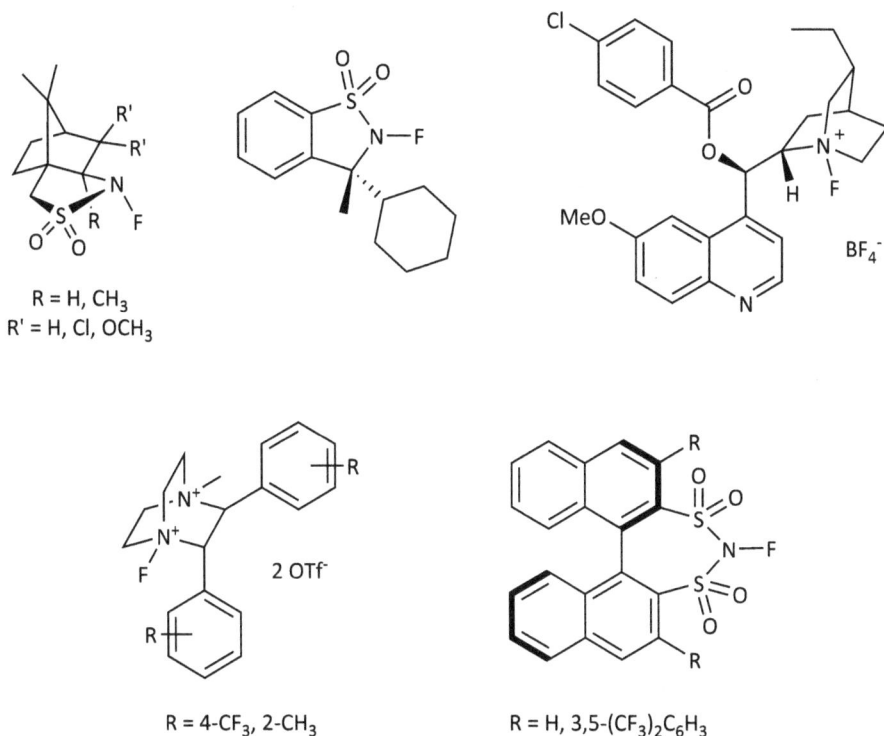

R = H, CH$_3$
R' = H, Cl, OCH$_3$

R = 4-CF$_3$, 2-CH$_3$

R = H, 3,5-(CF$_3$)$_2$C$_6$H$_3$

Figure 13.2: Chiral N-F-reagents for electrophilic fluorination based on N-fluorosultams (top left and middle), cinchona alkaloids (top right) as well as commercial Selectfluor™ (bottom left) and NFSI reagents (bottom right).

has been found that chiral thiourea-derivatives can also be used as catalysts for enantioselective fluorination. Interestingly, the α-fluorination of cyclic β-ketoesters only yields enantioenriched products, if NFSI is used as fluorinating agent, whereas Selectfluor™ leads to the formation of a racemic mixture. This finding has been rationalized by interactions between NFSI and the thiourea-moiety [2998, 2999]. Apart from natural-product inspired molecules, a few other variants of chiral fluorinating agents have been developed over the years. For example, the enantioselective α-fluorination of cyclic β-ketoesters has been achieved with chiral derivatives of NFSI (Figure 13.1) [3000]. Other methods include chiral Selectfluor™ reagents (Figure 13.2) or the combination of achiral, dicationic Selectfluor™ with a chiral phosphate anion and give rise to reasonable enantioselectivities [3001, 3002].

The enantioselective preparation of α-fluorinated cyclic ketones is readily achieved, when activated substrates, such as acyl- or silyl enol ethers are treated with Selectfluor™ or NFSI in presence of catalytic amounts of various cinchona alkaloids, so that the chiral fluorinating agent is generated *in situ* [3003, 3004]. The treatment of phenols with Selectfluor™ and a chiral phosphoric acid derivative under phase-transfer-catalysis conditions leads to the enantioselective formation of α-fluorinated cyclohexadienones in a fluorinative dearomatization approach [3005]. Similarly, α-fluorinated aldehydes are prepared in enantioselective fashion by treatment with NFSI, when catalytic amounts of a cinchona alkaloid are present [3006]. Alternatively, enamine catalysis with chiral prolinol or imidazolidinone derivatives in conjunction with electrophilic fluorinating agents also yields stereoselectively fluorinated products [3007–3010]. In the same way, the enantioselective α-fluorination of ketones is achieved by applying enamine catalysis with proline derivatives [3011]. Cinchona alkaloids with a primary amine function have also been used and studied in terms of a stereoselectivity model [3012–3014]. Apart from that, enamine catalysis has been combined with the use of a chiral anion in phase-transfer catalysis [3015].

In case of chiral substrates, such as steroidal vinyl esters, ethers or similar compounds, both regio- and stereoselectivity may be induced by the structure of the starting material. Accordingly, stereoselectivity in electrophilic fluorination reactions can also be achieved by the attachment of chiral auxiliaries onto the substrate, for example Evans' chiral oxazolidinone groups or menthol-derived substituents [3016–3020]. Typically, N-acyloxazolidinone derivatives are deprotonated in α-position at low temperature and subsequently reacted with NFSI [3021–3023].

The enantioselective preparation of β-fluoroalcohols is achieved by the organocatalytic α-fluorination of aliphatic and α-chloroaliphatic aldehydes, using NFSI in presence of a chiral prolinol or imidazolidinone-derivative, which initially generates α-fluoro- and α-chloro-α-fluoro aldehydes, respectively. Subsequent reduction of the carbonyl group with $NaBH_4$ yields the corresponding β-fluoroalcohols [3024, 3025]. In a similar one-pot-approach, β-fluoroamines are directly accessible in enantioselective fashion, if the organocatalytic α-fluorination of aldehydes is combined with a reductive amination step, using $Na[BH(OAc)_3]$ as reducing agent. Analogously, β,β-difluoroamines are generated, if two equivalents of the fluorinating agent are used in the first step [3024, 3026]. The organocatalytic, enantioselective α-fluorination of aldehydes has also been combined with a Wittig-olefination, which gives rise to chiral allylic fluorides [3027].

Allylic alcohols are enantioselectively fluorinated by Selectfluor™ in presence of a chiral phosphoric acid derivative and a boronic acid to give α-fluorinated homoallylalcohols. The boronic acid generates a directing monoester *in situ* and thus significantly supports the enantiocontrol in this approach [3028]. A rather important route toward allylic fluorides is given by fluorodesilylation of allylsilanes. For example, acyclic substrates, bearing a chiral oxazolidinone-auxiliary, are asymmetrically converted to branched allylic fluorides using Selectfluor™. However, the diastereoselectivity of this approach is rather poor [3029]. In the same way, chiral N-Boc-aminoallylsilanes are converted to allylic fluorides in high yield but with virtually no stereoselectivity, yielding a 1:1-mixture of diastereomers [3030]. Far better *syn/anti*-ratios have been observed, when chiral, branched (*E*)-allylsilanes were used as starting materials [3031, 3032]. Apart from that, the procedure has been used for the preparation of cyclic allylfluorides and fluorinated oxazines, too [3033–3036]. In another interesting approach, the fluorodesilylation of allylsilanes has been combined with fluorous chemistry, where the removal of the fluorous tag is realized by the treatment with Selectfluor™ and thus generates allylic fluorides [3037]. The enantioselective preparation of allylic fluorides is furthermore achieved in catalytic fashion, if achiral allylsilane-substrates are reacted with NFSI in presence of catalytic amounts of *bis*-cinchona alkaloids. However, this method requires low reaction temperatures and quite long reaction times of up to 5 days [3003]. Fluorodesilylation reactions have also been reported for allenylsilanes, which give rise to propargylic fluorides. If enantiopure starting materials are reacted with Selectfluor™, the chirality is transferred to the products via a highly stereospecific *anti*-S_E2' mechanism, thus yielding propargylic fluorides with high enantiomeric excess [3038, 3039].

13.5.6 Xenon difluoride

A special role in electrophilic fluorination is attributed to XeF_2, which is also capable of mildly adding F_2 to unsaturated C-C-bonds, but rather operates in an oxidative fashion under radical conditions. Typically, it is used in CH_3CN-solutions without catalysts or in CH_2Cl_2 in conjunction with HF- or BF_3-reagents at temperatures below 25 °C. It is believed, that in a first step the olefin is oxidized to a radical cation, whereas XeF_2 is reduced to a XeF' radical, the second fluoride substituent being transferred to the Lewis acid HF or BF_3. Subsequently, the organic radical cation is fluorinated by XeF' to result in a carbocation, which then adds F^- from the respective $[HF_2]^-$ or $[BF_4]^-$ anion (Scheme 13.49). For example, 1,1-diphenylethene is converted to 1,2-difluoro-1,1-diphenylethane. Moreover, XeF_2 is conveniently used for the fluorination of unsaturated C-C bonds in carbohydrate chemistry.

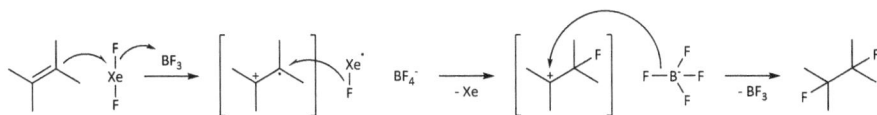

Scheme 13.49: Proposed mechanism for the fluorination of unsaturated C-C bonds using XeF_2.

Apart from alkenes, XeF_2 and its complexes with dialkyl sulphides as well as its graphite intercalate fluorinate enols, enolates and carboxylic acids. In the latter case, conversion proceeds via unstable xenon esters, which readily decarboxylate to form free radical intermediates. Depending on the reaction conditions, the radical either captures a fluorine substituent to generate an alkyl fluoride or it abstracts a hydrogen radical from the solvent. Similarly, perfluorinated acids decarboxylate in presence of XeF_2. Upon reaction with arene-substituted acids, this gives rise to perfluoroalkylated aromatics. In contrast to their aliphatic relatives, aromatic and vinylic acids are not decarboxylated by xenon difluoride.

XeF$_2$ efficiently fluorinates electron-deficient enolates and silyl enol ethers. Especially the latter conversion is a versatile method toward fluorinated complex molecules, for example, steroids or peptides. In tertiary positions, activated hydrocarbons like adamantane are directly fluorinated by XeF$_2$, however, the yields for such transformations are generally rather poor. The presence of a sulphur-atom in α-position also leads to the activation of C-H bonds, which can then readily be fluorinated by XeF$_2$. For example, methyl thioethers are easily fluorinated to give fluoromethyl thioethers via intermediate S-F species.

13.5.7 Study questions

(I)   Which general conditions have to be met to use F$_2$ as electrophilic fluorinating agent? What are typical substrates and how can one generally describe the mechanism of such transformations?

(II)   Which general types of products are observed upon reaction of olefins or acetylenes with fluoroxy compounds? How does the mechanism differ for electron-rich and electron-deficient substrates and which consequences does this have on the stereochemistry of the products?

(III) How can fluoroxycompounds serve for the preparation of α-fluorocarbonyl compounds and direct C-H fluorination? Which substrates are generally suitable and what is the advantage of using CF$_3$OF for the conversion of silyl enol ethers?

(IV)   Which aliphatic substrates are generally suitable for an electrophilic fluorination with fluoraza reagents? How can one influence the product formation? Is it necessary to add a base?

(V)   By which general mechanisms may the electrophilic fluorination with fluoraza reagents be described?

(VI)   Which major approaches can be differentiated in terms of stereoselective fluorination?

(VII) Give examples for common chiral, electrophilic fluorinating reagents. Which types of compounds are they based on?

(VIII) Which are the most important auxiliaries and organocatalysts for the enantioselective preparation of α-fluorinated carbonyl compounds? How can one enantioselectively prepare β-fluoroalcohols and β-fluoroamines?

(IX)   Why is XeF$_2$ a special electrophilic fluorinating agent? Under which conditions is it typically used in organic reactions? How can one mechanistically describe the reaction of XeF$_2$ with olefins?

13.6 Metal-mediated reactions, including photocatalysis

In general, the formation of alkyl-fluorides is easily achieved by a number of methods, which have been shown in the previous sections. However, a variety of transition-metal mediated methods has also been reported [3040–3042]. Often, these approaches lead to more selective formations of carbon-fluorine bonds, in many cases even in asymmetric fashion [2990–2993]. Furthermore, the fluorination of otherwise poorly accessible substrates and positions may be enabled.

13.6.1 C-H fluorination

Directing-group approaches

The selective transformation of C-H bonds requires the presence of a directing group in most cases. For example, the presence of strongly coordinating quinoline-moieties enables the selective monofluorination, which can either be achieved with a catalytic amount of Pd(OAc)$_2$ and stoichiometric quantities of N-fluoro-2,4,6-trimethylpyridinium salts under microwave conditions or, alternatively, by the use of Pd(OAc)$_2$ in conjunction with AgF and an oxidizing hypervalent iodine reagent (Scheme 13.50). Both variants are proposed to proceed via a catalytic Pd^{II}/Pd^{IV}-cycle [3043, 3044]. Apart from that, the β-selective C-H-fluorination of amino acid derivatives is achieved by Pd-catalysis but requires a directing amide-function in the substrate as well as a highly optimized quinoline ligand for the metal-complex. In addition, stoichiometric amounts of Selectfluor™ and Ag$_2$CO$_3$ are necessary (Scheme 13.50). In general, this procedure works better for benzylic than for aliphatic substrates [3045]. Similarly, β-selective C-H fluorination of amino acid derivatives is realized, if the substrate contains a chelating 2-(pyridine-2-yl) isopropyl amine group. The reaction proceeds with catalytic amounts of Pd(OAc)$_2$ and stoichiometric quantities of Selectfluor™ and selectively generates *anti*-fluorinated products without an additional ligand (Scheme 13.50). Again, benzylic substrates give better results than aliphatic ones [3046]. The latter transformation is also suitable for slightly modified substrates, if Fe(OAc)$_2$ and Ag$_2$CO$_3$ are added [3047].

Another interesting directing group is 8-aminoquinoline which can be attached to carboxylic acid derivatives to enable β-selective fluorination by Pd-catalysis. Both, secondary alkyl and benzyl fluorides can be accessed by this approach, which requires the additives Ag$_2$O and ^{t}BuCOOH as well as NFSI as fluorinating agent (Scheme 13.51) [3048]. Benzylic and aliphatic alcohol derivatives are fluorinated in β-position upon treatment with Pd(OAc)$_2$ and NFSI, if a N,N-bidentate oxime moiety is present (Scheme 13.51). After the fluorination step, the directing group can be split off with Mo(CO)$_6$ to yield the free β-fluoroalcohol [3049].

Using the approach of transient directing groups, *ortho*-alkylated benzaldehyde derivatives are converted to enantioenriched benzyl fluorides, when they are treated with a chiral α-amino amide auxiliary (Scheme 13.52). The transformation is mediated by a Pd^{II}-catalyst and requires a N-fluoro-2,4,6-trimethylpyridinium salt as fluorinating oxidant. The reaction works well, when an electron-withdrawing substituent is attached to the phenyl ring, however, it fails with a methoxy-group in *meta*-position. Furthermore, the aryl ring will become substituted in *ortho*-position, if there is no blocking function [3050].

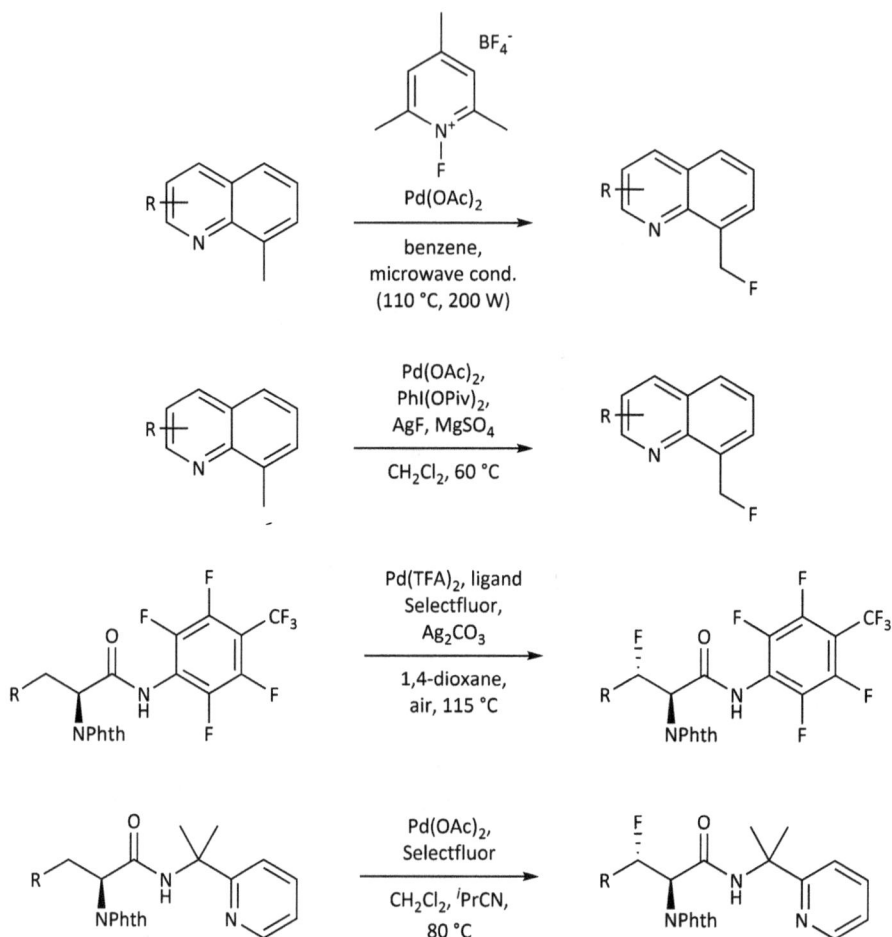

Scheme 13.50: Benzylic fluorination of quinoline-substrates (top), and directed C-H functionalization in presence of different amide functions (middle and bottom).

Fluorine substituents can also be introduced in allylic position without prefunctionalization, for example, by oxidative treatment of olefins with a Pd^{II}-complex and a Cr-salen co-catalyst in presence of a *bis*(sulphoxide) ligand, stoichiometric amounts of Et$_3$N · 3 HF as fluoride source and *p*-benzoquinone as oxidant (Scheme 13.52). In this transformation, the C-F bond is not formed in a reductive elimination step from a high-valent Pd-species, instead fluorination is enabled by the generation of a palladium(II) π-allyl complex. The regioselectivity of this method is only moderate and yields product mixtures, where the formation of branched allylic fluorides is generally favored. However, various functional groups are tolerated by this approach, including esters and amides, as well as alkyl bromides; another advantage is the use of an inexpensive nucleophilic fluoride source [3051].

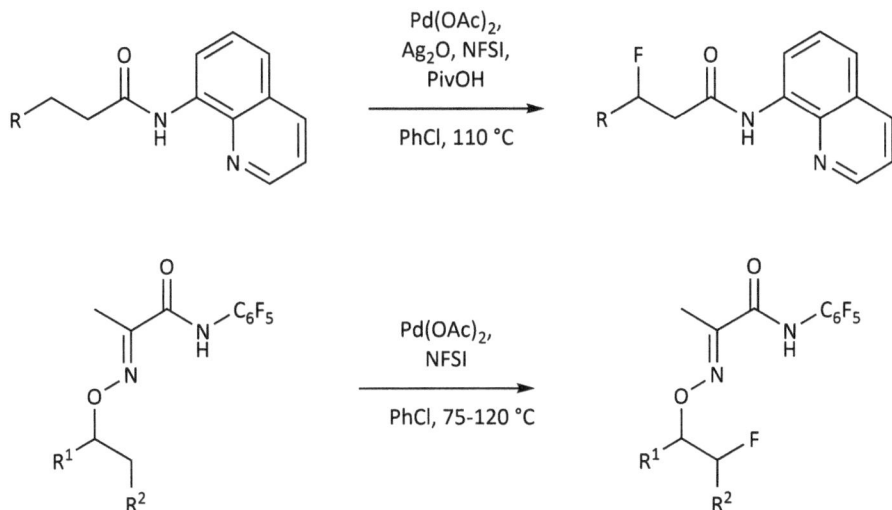

Scheme 13.51: Pd-mediated C-H fluorination in presence of a directing aminoquinoline moiety (top) and a bidentate oxime function (bottom).

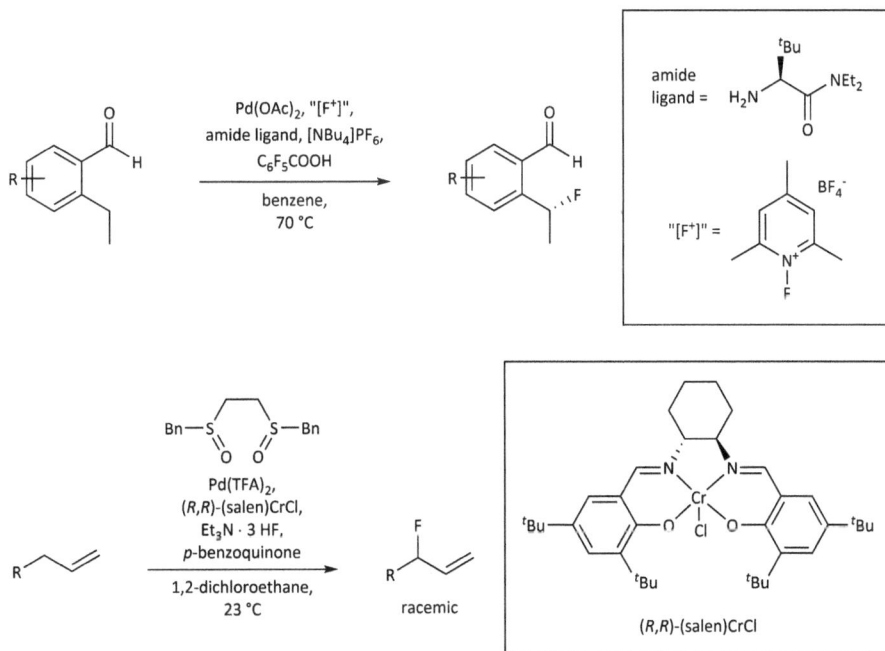

Scheme 13.52: Stereoselective, benzylic C-H fluorination using a chiral amide as transient directing group (top) and preparation of allylic fluorides in presence of Pd(II) and a Cr(III)-salen complex (bottom).

Enantioselective C-H fluorination of activated substrates

In case that the substrate contains activated C-H bonds, for example in proximity of electron-withdrawing groups, regioselective fluorination is observed without the presence of an additional directing group. Thus, β-ketoesters are selectively fluorinated in α-position upon treatment with NFSI and a Pd(II)-catalyst. This type of reaction has been extensively studied, using a chiral biphosphine-ligand at the Pd-center, which yields enantioenriched products. The transformation is proposed to proceed via a chiral Pd-enolate species, which then directs the attack of the fluorine substituent [3052–3056]. The enantioselective fluorination of β-ketoesters is furthermore achieved by the use of electrophilic fluorinating agents in conjunction with a chiral titanium(IV) complex. Again, the metal-catalyst coordinates to the ketoester and generates an enolate, which is then susceptible to electrophilic attacks. The chiral ligand blocks one of the faces of the enolate, so that the fluoride addition occurs stereoselectively [3057, 3058]. The aforementioned method has also been extended to other 1,3-dicarbonyl compounds, such as 1,3-diketones or β-ketoamides [3059]. In addition, chiral Co(II)-salen complexes as well as nickel(II)-systems with *bis*(oxazoline) or *N,N,N*-tridentate ligands have been used together with NFSI for the enantioselective α-fluorination of β-ketoesters [3060–3063]. The enantioselective α-fluorination of α-chloro-β-ketoesters has been achieved with a Pd(II)-biphosphine complex and NFSI [3064]. An improved enantioselectivity has been observed for the same reaction, when a chiral Ni(II)-catalyst was used in conjunction with Selectfluor™ [3065]. Analogously, the enantioselective α-fluorination of 1,3-dicarbonyl compounds is achieved by copper(II) catalysis, using a variety of chiral ligands, for example, amino sulphoximines, bidentate bioxazolines or tridentate, amine-linked *bis*(thiazolines) [3066–3068]. The sequential treatment of β-ketoesters with *N*-chlorosuccinimide and NFSI in presence of Cu(II) and a chiral spiro pyridyl oxazoline ligand gives rise to α-chloro-α-fluoro-β-ketoesters with high enantioselectivity. Excluding the chlorination step, the latter method has also been adapted to the monofluorination of cyclic substrates [3069]. However, a reasonable stereoselectivity is only observed, if the substrate contains a *tert*-butyl ester function, but not for ethyl esters [3070]. The reaction of α-substituted malonates with NFSI in presence of Zn(OTf)$_2$ and a chiral dibenzofuran-linked *bis*(oxazoline) ligand yields the corresponding α-fluorinated products in high yields and enantioselectivities [3071].

The enantioselective fluorination of α-ketoesters has also been reported. Using a dimeric Pd-complex with chiral bisphosphine ligands and NFSI as fluorinating agent, the fluorine substituent is introduced in proximity of the keto-group in high yields and with controlled stereochemistry. Subsequent reduction of the carbonyl group gives rise to fluorohydrins with high enantioselectivity; however, the outcome depends on the reducing agent. Diisobutylaluminiumhydride (DIBAL) gives rise to *anti*-fluorohydrins, whereas *L*-selectride selectively generates *syn*-products [3072].

The enantioselective α-fluorination of a variety of other functionalized ester substrates, such as α-cyanoesters or α-nitroesters has also been described [3073–3075]. Apart from that, various activated substrates are substituted in proximity of the carbonyl function. For example, chiral N-acyloxazolidinone substrates are asymmetrically fluorinated upon reaction with NFSI, Et_3N and $TiCl_4$. Cleavage of the oxazolidinone-moiety, which is a widely used auxiliary in asymmetric synthesis, gives access to enantioenriched, fluorinated building blocks [3076]. In the same way, achiral N-(arylacetyl)oxazolidinones and -thiazolidinones are enantioselectively fluorinated in α-position, when they are treated with NFSI, a chiral Ni(II)-catalyst and 2,6-lutidine [3077–3079].

Apart from the examples above, the metal-catalyzed, asymmetric preparation of several more α-fluorocarbonyl compounds is well known. Since many of them are part of rather complex molecules, especially heterocycles, they are not discussed herein, but have been reviewed elsewhere [2860].

Radical fluorination using electrophilic fluorinating agents

Apart from the aforementioned, mostly enantioselective transformations, allylic and benzylic as well as secondary alkyl fluorides can be accessed in a Cu-mediated radical fluorination, which uses Selectfluor™ as fluoride source (Scheme 13.53). In addition, a few additives are required. Mechanistically, the reaction is believed to proceed via reduction of the fluorinating agent by a Cu(I)-species, which yields an amine radical cation. The latter abstracts a hydrogen radical from the substrate and the resulting carbon-radical adds a fluoride radical from Selectfluor™, and thus regenerates an amine radical cation. The product yields of this synthetic method are strongly time-dependent, since the products are getting depleted after prolonged reaction [3080]. Double benzylic C-H fluorination occurs, if the substrate is reacted with catalytic amounts of $AgNO_3$ in conjunction with $Na_2S_2O_8$ as oxidizer and Selectfluor™ and generates difluorobenzyl-products (Scheme 13.53). This reaction is proposed to proceed via an Ag(II)-species, which is generated upon oxidation of Ag^I by sodium persulphate. Afterwards, Ag^{II} oxidizes a benzylic C-H bond to give a benzyl radical, which is then trapped by Selectfluor™. Interestingly, this method works best in a solvent mixture of water and acetonitrile, but not under anhydrous conditions [3081]. A direct C-H functionalization is furthermore achieved in benzylic position, if the substrate is treated with Selectfluor™ in presence of a Fe(II)-catalyst (Scheme 13.53) [3082]. A few more reactions have been described, where Selectfluor™ serves as source of fluoride radicals in C-H fluorination reactions [3041]. In principal, all these methods only differ in the nature of the radical initiator or catalyst, which can for example be a mixture of $AgNO_3$ and glycine but also V_2O_3 [3083, 3084].

Photocatalytic fluorination

Another important type of metal-mediated radical fluorination is given by photocatalytic reactions. Both, aliphatic and benzylic substrates are fluorinated in an UV-light

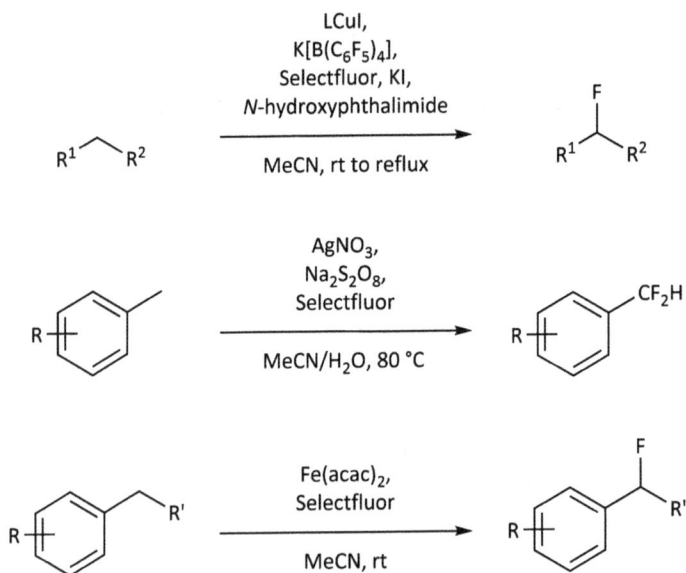

Scheme 13.53: Cu-mediated radical C-H fluorination of allylic, benzylic and secondary sp^3-carbon atoms (top), double fluorination of benzylic substrates in presence of Ag(I) (middle) and Fe(II)-catalyzed benzylic fluorination (bottom).

promoted reaction, using the decatungstate salt [NBu$_4$]$_4$[W$_{10}$O$_{32}$] as photocatalyst and NFSI as fluorine source (Scheme 13.54). For benzylic substrates, NaHCO$_3$ needs to be added to avoid side-product formation. This method is also versatile for the preparation of more complex fluorinated compounds, as it selectively fluorinates the branched positions of amino acid derivatives or specific sites in natural products [3085, 3086]. Furthermore, the aforementioned procedure has been adapted to the decagram-scale, high-yield preparation of fluorinated leucine methyl ester in a flow process (Scheme 13.54). The product is a drug-precursor, which is otherwise only available in multi-step procedures, thus, this example clearly underlines the utility of photocatalytic fluorination techniques [3087]. Apart from that, ^{18}F-labeled amino acids are available upon treatment of the free acids with [^{18}F]-NFSI, sodium decatungstate and 365 nm irradiation. Again, fluorination site-selectively occurs at tertiary carbon-atoms [3088]. A variant of photocatalytic fluorination has also been reported using an Ir-complex as photosensitizer and NEt$_3$ · 3 HF as nucleophilic fluorine source. In a decarboxylative approach, a variety of N-hydroxyphthalimide esters has been converted to primary, secondary as well as tertiary aliphatic and benzylic fluorides. Using this method, fluorination is enabled by the irradiation with blue LEDs. Mechanistically, it is proposed, that the activated Ir(III)*-photocatalyst induces the decarboxylation of the substrate in a single electron transfer process, leading to the formation of an alkyl radical and an Ir(IV)-species. In a radical-polar crossover, the

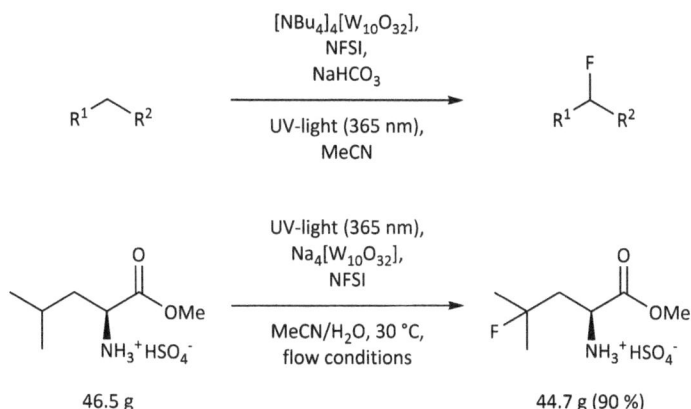

Scheme 13.54: Photocatalytic C-H fluorination using decatungstate as photosensitizer (top) and adaption of the method for multigram-scale preparation of a fluorinated amino acid ester (bottom).

latter oxidizes the alkyl radical to a carbocation, which is then trapped by a fluoride anion [3089].

Radical fluorination in presence of directing groups

Radical fluorinations can also be performed in presence of directing groups and generally proceed by the formation of a heteroatom-centered radical, which subsequently undergoes a 1,5-hydrogen atom transfer (HAT). The carbon radical is then trapped by the fluorinating agent [3041]. For example, N-tert-butyl-N-fluoroamides react with Fe(OTf)$_2$ to give a fluorinated Fe(III)-species and a N-centered radical. The latter undergoes a 1,5-hydrogen atom transfer and generates a carbon-radical, which is then fluorinated by the fluoroiron(III)-complex. Thus, the fluorinated product is formed and the Fe(OTf)$_2$ catalyst is regenerated. This method works well for the transformation of primary, secondary and tertiary benzylic C-H bonds and shows excellent site-selectivity (Scheme 13.55) [3090]. The radical fluorination of a variety of secondary and tertiary carbon-centers can also be directed by hydroperoxy-functions (Scheme 13.55). Initially, an alkoxy radical is generated upon reaction of the peroxy substrate with a Fe(II)-porphyrin catalyst. The alkoxy radical then abstracts a hydrogen atom in a 1,5-HAT to generate a carbon-centered radical, which is then fluorinated by NFSI. Additionally, the presence of LiBH$_4$ is required for the regeneration of the iron-catalyst [3091].

Radical fluorination using nucleophilic fluorinating agents

Radical C-H fluorination can also be achieved by the use of nucleophilic fluoride sources. For example, benzylic fluorides are accessible by a Mn-mediated reaction with Et$_3$N · 3 HF and iodosylbenzene PhIO (Scheme 13.56) [3092]. With some

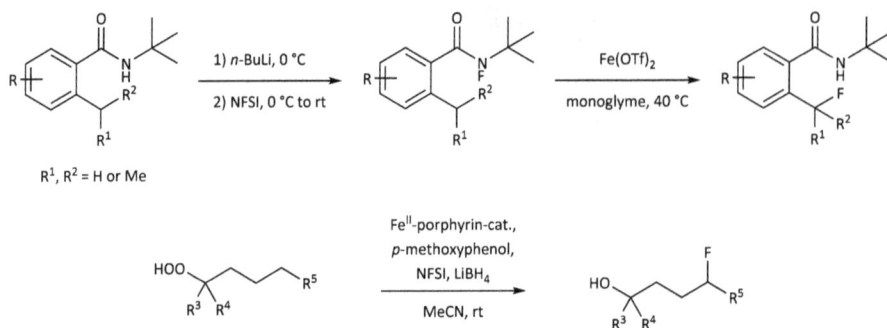

Scheme 13.55: Fe(II)-catalyzed C-H fluorination directed by a fluoroamide (top) and a hydroperoxy-function (bottom).

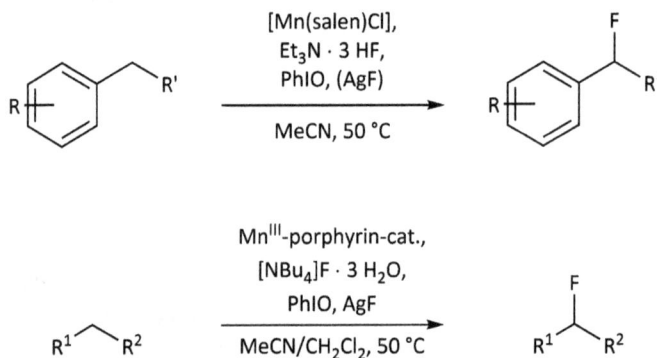

Scheme 13.56: Mn(III)-mediated C-H-fluorination of benzylic (top) and aliphatic substrates (bottom) using nucleophilic fluorine sources.

slight changes in the synthetic procedure, this approach has also been used for ^{18}F-radiofluorination of simple and more complex substrates, proving a general suitability for site-selective late-stage functionalization [3093]. The aforementioned method is furthermore applicable to the preparation of aliphatic fluorides and shows excellent site-selectivity with natural products and steroids. It appeared, that Mn-porphyrin catalysts in conjunction with [NBu$_4$]F and AgF work best for aliphatic fluorination, whereas the respective benzylic fluorides are obtained, when Mn-salen complexes are used. In case of electron-rich benzylic substrates, an efficient fluorination is observed with Et$_3$N · 3 HF, whereas electron-poor starting materials additionally require the use of AgF. Typically, the manganese-catalysts are available as chloromanganese(III)-species, which are converted to the corresponding fluorocomplex and oxidized to the active oxo-Mn^V-F catalyst. The latter abstracts a hydrogen atom from the substrate to form a carbon-centered radical and a Mn^{IV}OH-species. The Mn(IV)-hydroxo-complex undergoes a ligand exchange to give a Mn(IV)-difluoro-derivative, which transfers a fluoride radical to the substrate and thus generates the product. The polar effect induces the site-selectivity and leads to the homolytic cleavage of the most hydridic C-H bond. In addition, the selectivity is influenced by the steric bulk of the Mn-catalyst. The formation of both, alcohol and ketone by-products is generally observed and can be rationalized by radical recombination of the Mn^{IV}-OH-complex with a carbon-centered radical [3094, 3095].

13.6.2 Fluorofunctionalization of olefins

Allylic amines are transformed to benzyl fluorides upon 1,1-arylfluorination, using aryl boronic acid nucleophiles in conjunction with electrophilic Selectfluor™ in presence of a Pd-catalyst (Scheme 13.57). The choice of an appropriate chiral ligand further-more allows the enantioselective preparation of the products [3096]. Apart from that, Pd-catalyzed intramolecular aminofluorination-reactions of inactivated olefins have been described, which lead to cyclic, β-fluorinated amines. This method requires AgF as fluoride source and a hypervalent iodine reagent as oxidant (Scheme 13.57) [3097]. Recently, an adaption of the aforementioned method has been reported leading to the formation of β-fluoro piperidines. Similarly, the conversion of δ,ε-unsaturated amine substrates is catalyzed by a Pd(II)-complex and requires a chiral quinoline-oxazoline ligand. In this approach, [NEt$_4$]F · 3 HF serves as fluoride source and is used in conjunc-tion with bis(pivaloyloxy) iodobenzene as oxidant. The enantioselectivity is further improved by the addition of CsOCF$_3$ [3098]. In general, aminofluorination reactions of olefins are rather well explored, as the resulting products are important structural motifs in pharmaceuticals. Therefore, asymmetric methods have been reported as well [2860]. Another example for a Pd^{II}/Pd^{IV}-catalyzed halofunctionalization is the fluoro-sulphonylation of styrene derivatives, which regioselectively yields β-fluorosulphones (Scheme 13.57). Again, an electrophilic fluorinating agent is applied in conjunction with a Pd^{II}-complex and a bidentate, 1,10-phenanthroline-based ligand. This trans-formation is believed to proceed via radical intermediates and shows a high diastere-oselectivity, if internal (E)- or (Z)-alkenes are used as substrates [3099].

Benzylic fluorides are furthermore accessible in fluorosulphonimidation reac-tions by treatment of styrenes with a Cu(I)-catalyst in presence of catalytic amounts of a boron-reagent and AgF, using NFSI as fluorine- and sulphonimide source [3100]. Al-ternatively, a hypervalent iodine reagent may be used for the sulphonimide transfer, whereas the fluoride substituent is provided by Et$_3$N · 3 HF [3101]. The regioselec-tive and syn-specific hydrofluorination of styrenes is achieved with Selectfluor™ and Et$_3$SiH by Pd(II)-catalysis under very mild conditions (Scheme 13.58) [3102]. In proximity of a directing group (N-quinolin-8-yl amide), styrene-substrates react with Selectfluor™ and aryl boronic acids in a Pd-mediated fluoroarylation reaction (Scheme 13.58). The use of a chiral pyridyl-oxazoline ligand leads to asymmetric product formation [3103].

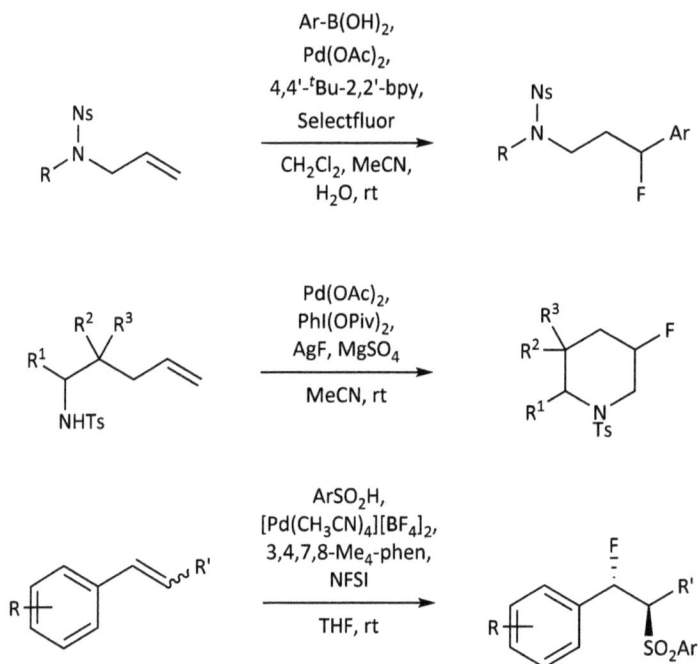

Scheme 13.57: Fluorofunctionalization of allyl- and styrene substrates leading to benzylic fluorides (top), β-fluorinated cyclic amines (middle) and *anti*-β-fluorosulphones (bottom).

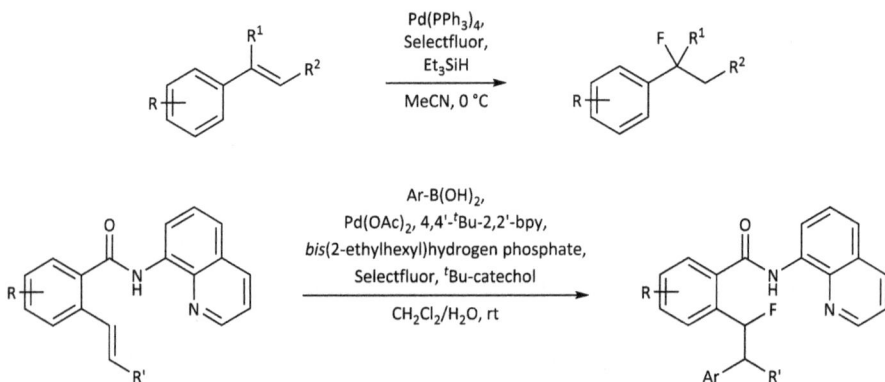

Scheme 13.58: Pd-catalyzed hydrofluorination (top) and fluoroarylation (bottom) of styrene derivatives.

A mixture of branched and linear allylic fluorides is obtained upon Pd-mediated carbofluorination of allenes with aryl- or vinyl iodides, using AgF as fluoride source [3104].

13.6.3 Allylic, benzylic and propargylic substrates with a leaving group

The reaction of racemic cyclic allylic chlorides with a chiral Pd-catalyst and stoichio-metric amounts of AgF yields enantiomerically enriched allylic fluorides (Scheme 13.59) [3105]. In a related approach, linear allylic chlorides are preferentially transformed to branched allylic fluorides, again with a reasonable enantioselectivity (Scheme 13.59) [3106]. A variety of allyl p-nitrobenzoate substrates has been selectively converted to the corresponding linear allylic fluorides, using a Pd-catalyzed oxida-tive approach with p-benzoquinone and $[NBu_4]F \cdot (^tBuOH)_4$ (Scheme 13.59) [3107]. Phosphorothioate esters can also serve as substrates for the Pd-mediated synthesis of allylic fluorides, when they are treated with AgF [3108].

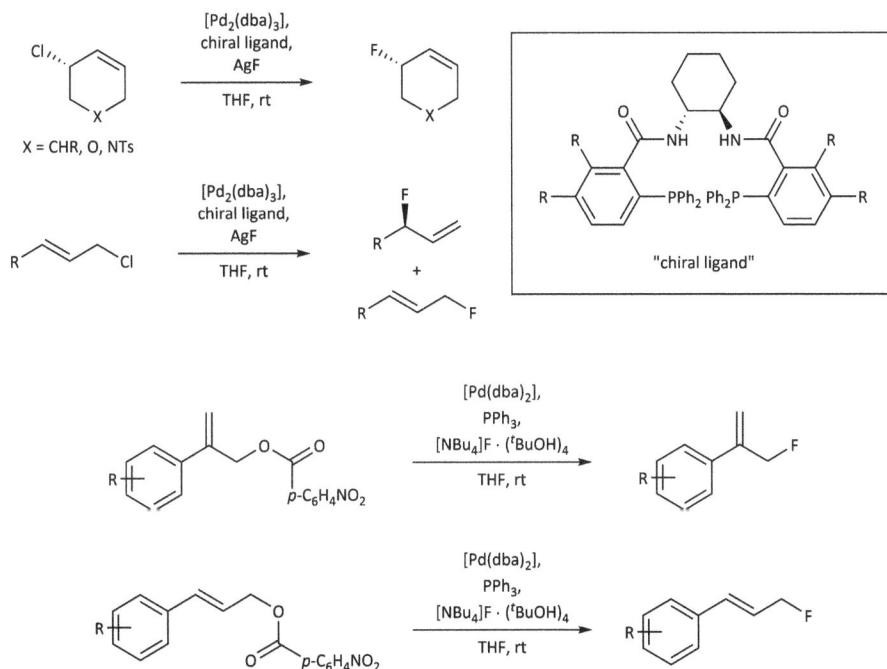

Scheme 13.59: Pd-catalyzed preparation of allylic fluorides from chlorinated substrates (top) and p-nitrobenzoates (bottom).

Secondary as well as tertiary allylic fluorides are accessible by a Rh- or Ir-catalyzed fluorination of allylic trichloroacetimidates using $Et_3N \cdot 3$ HF as fluoride source (Scheme 13.60) [3109]. A number of branched and also linear (E)- and (Z)-allylic fluorides has been prepared from the corresponding carbonate substrates by an Ir-mediated method using $[NBu_4]F \cdot (^tBuOH)_4$ as nucleophilic fluoride source (Scheme 13.60). Both the position and the configuration of the double bond are re-

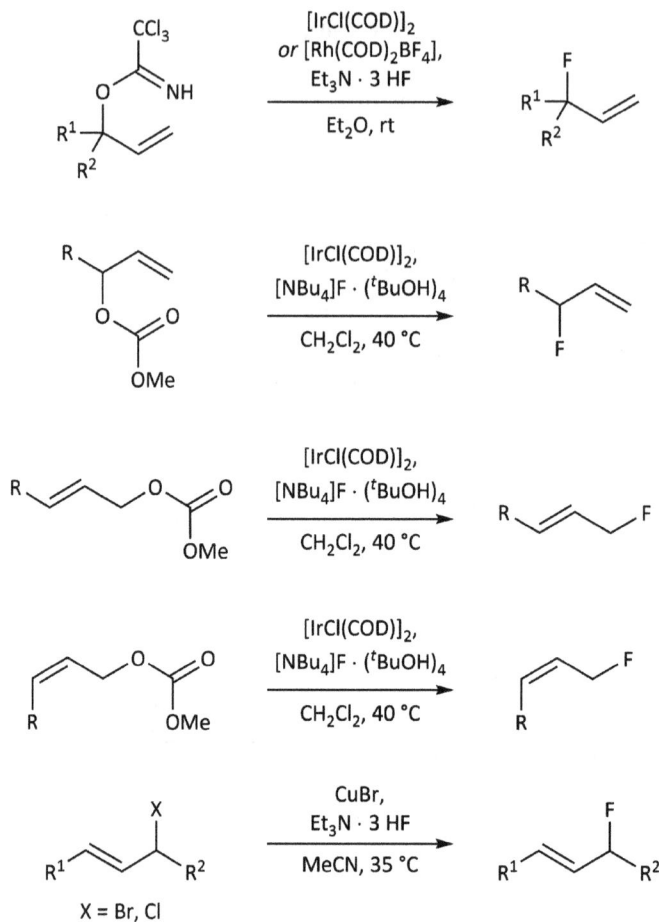

Scheme 13.60: Rh- and Ir-mediated synthesis of allylic fluorides from trichloroacetimidates (top), regioselective and configurationally retentive fluorination of allylic carbonates (middle) and Cu-catalyzed fluorination of allylic chlorides and bromides (bottom).

tained in this approach [3110]. Allylic chlorides and -bromides are regioselectively converted to the corresponding fluorides by $Et_3N \cdot 3\,HF$ in presence of catalytic CuBr, if the substrate contains a heteroatom function in R^1 or R^2 (Scheme 13.60) [3111]. The phosphine reagent $[(PPh_3)_3CuF]$ has been reported to substitute allylic chlorides and bromides in stoichiometric fashion to give almost exclusively linear allylic fluorides [3112]. Silver(I)-salts catalyze decarboxylative fluorination reactions in benzylic-, allylic- and other secondary- and tertiary alkyl substrates, where Selectfluor™ serves as fluorinating agent [3113]. If the benzylic α-position of the carboxylic acid contains one or two fluorine-substituents, the fluorodecarboxylation approach will give rise to CF_2H- and CF_3-substituted arenes, respectively [3114]. Benzylic as well as alkyl fluorides are furthermore accessible in a radical process when alkylboronates or boronic acids are treated with Selectfluor™ in presence of $AgNO_3$ in aqueous medium [3115].

Propargylic fluorides are mostly prepared by deoxyfluorination of corresponding alcohol substrates, which generally does not require the presence of a metal catalyst (cf. Section 13.4.2). However, it has been found, that they are obtained with some stereoselectivity, if a mixture of diastereoisomeric starting materials is transformed to cobalt-carbonyl complexes. The subsequent deoxofluorination with DAST favors the formation of the *anti*-products [3116].

13.6.4 Study questions

(I)   Give examples for directing groups, which are typically used in metal-mediated C-H fluorinations. Which substrates are generally most suitable for this approach and which metals and reagents are used?

(II)   What is the role of metal complexes in the stereoselective α-fluorination of activated C-H bonds? Which substrates are generally used and which type of fluorinating reagent is required?

(III) How does the metal-mediated radical C-H fluorination with electrophilic fluorinating reagents generally proceed? To which substrates is it applied and what is the role of the metal?

(IV) How can directing groups lead to a site-selective, radical C-H fluorination? What is the role of the metal and which reagents are used?

(V)   Briefly describe the elementary steps of the Mn(III)-mediated radical C-H fluorination using nucleophilic fluorinating reagents. Which reagents are needed and which role do they play?

(VI) Give examples for metal-mediated fluorofunctionalizations of unsaturated compounds. Which substrates are used, which reagents are required and which products are obtained?

(VII) Which substrates are typically used for the metal-mediated preparation of allylic fluorides? Which fluoride sources are used and what can one say about the selectivity of the fluorination methods?

13.7 Selected transformations of fluorinated compounds

Sometimes the preparation of fluorinated molecules requires the oxidation or reduction of a given, often easily accessible substrate. Despite not including any fluorination technique as such, this general approach is discussed here, since it may be important to rationalize the origin of the one or other important fluorinated building block. In addition, halogenation reactions of fluorinated substrates are briefly discussed. However, further transformation, such as perfluoroalkylations or cyclizations as well as general organometallic chemistry of fluorinated compounds will not be covered in this manuscript.

13.7.1 Reductive transformations

Due to their strength, carbon-fluorine bonds in alkyl fluorides are typically not cleaved by reducing agents. There are only a few examples, where the replacement of C-F bonds has been achieved using potassium metal or sodium naphthalide as reducing agents. Highly fluorinated compounds are also reduced by alkali metals, as for example, lithium-amalgam converts PTFE to a monolayered carbon polymer, which consists of six-membered rings. However, the presence of functional groups, especially carbonyl-substituents can activate C-F bonds in its proximity. Typically, this reactivity leads to the loss of α-fluorine atoms, as can be observed, when hexafluoroacetone is reacted with magnesium in tetrahydrofuran. Initially, an enolate salt is generated, which then reacts with another molecule of hexafluoroacetone, yielding a perfluorinated β-hydroxyketone. Under the reaction conditions, the latter is also partly reduced to the corresponding perfluorinated α-hydro-β-hydroxyketone (Scheme 13.61). In a similar way, the reaction proceeds with aluminium, however, upon acidification the reduction can be stopped on the enolate-stage, yielding the isolable enol of pentafluoroacetone. In a related approach, trifluoromethyl ketones and –imines are converted to difluoro enol silyl ethers and N-silylated difluoroenamines, respectively, when they are treated with Mg metal in presence of Me$_3$SiCl (Scheme 13.61) [3117–3119]. The aforementioned reactions of trifluoromethylated ketones and imines can also be performed by electroreduction. If the reduction of hexafluoroacetone with Mg is carried out in more polar N,N-dimethylformamide, perfluoropinacol is obtained (Scheme 13.61). Carbon-fluorine bonds are typically resistant to reductive cleavage by zinc, but C-Cl bonds are not, which gives the opportunity of a selective reduction of chlorodifluoromethylketones. Activated C-F bonds in esters and ketones are also cleaved by photochemical or electrochemical methods.

In contrast to fluorine-substituents at saturated sp^3-carbon centers, those in vinylic or allylic positions are rather easily cleaved by appropriate reducing agents. This may, for example, be realized by the use of H$_2$ in conjunction with catalysts. Furthermore, hydrides like LiAlH$_4$ can be used for selective displacement of vinylic and allylic fluorine substituents. Moreover, the reductive cleavage of olefinic fluorine substituents is achieved by the use of phosphines like PBu$_3$. A few examples are known, where aryl fluorides are reduced by electrochemical methods, giving rise to hydrodefluorinated arenes.

Depending on the reaction conditions, catalytic hydrogenation of chlorofluorocarbons can be used for the selective generation of hydrochlorofluorocarbons or hydrofluorocarbons. Chloro-substituents may also be replaced upon reaction with LiAlH$_4$, tin- or silicon hydrides. Furthermore, chlorine abstraction occurs with active metals like Na or Zn in protic solvents, as well as by reaction with UV-light in presence of a hydrogen source, often isopropanol.

Unsaturated organofluorine compounds can furthermore undergo reductive coupling. For example, the treatment of perfluoro 1-methylcyclopentene with PPh$_3$ leads

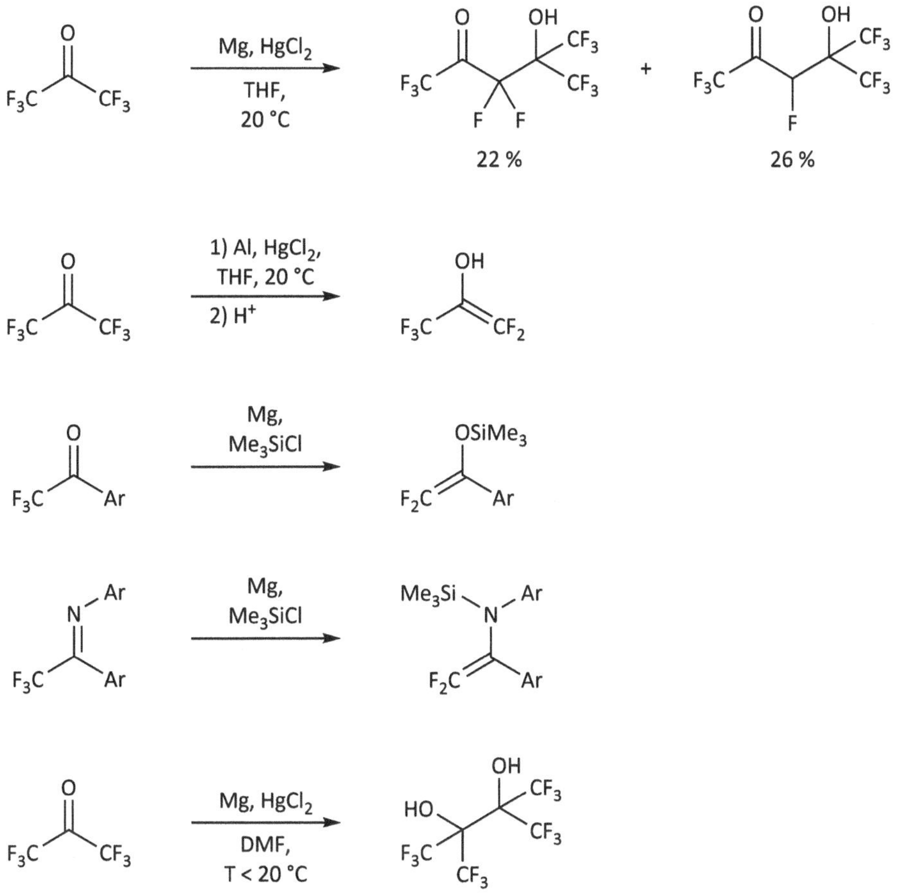

Scheme 13.61: Reduction of hexafluoroacetone and trifluoromethyl ketones and -imines.

to the cleavage of the vinylic fluorine substituent and the formation of dimers. In a similar way, biphenyl and bipyridyl compounds are generated upon electrochemical reduction of fluorinated arenes like pentafluoronitrobenzene or pentafluoropyridine (Scheme 13.62).

Apart from C-F bonds, reduction reactions can of course also affect other functional groups of fluorine-containing substrates, as for example, activated C-O bonds are readily cleaved. Some fluoroallylic alcohols are attacked by $NaBH_4$ in S_N2'-fashion, leading to a displacement of the hydroxy group, however, vinylic fluorine substituents are not replaced. The reaction of perfluorinated epoxides with $LiAlH_4$ results in ring-opening and thus the formation of alcohols. The reductive cleavage of 1,1-difluorocyclopropanes can yield interesting compounds. Depending on the reagents and conditions, the two fluorine-substituents are partially or completely retained. For example, catalytic hydrogenation of 2-phenyl-1,1-difluorocyclopropane

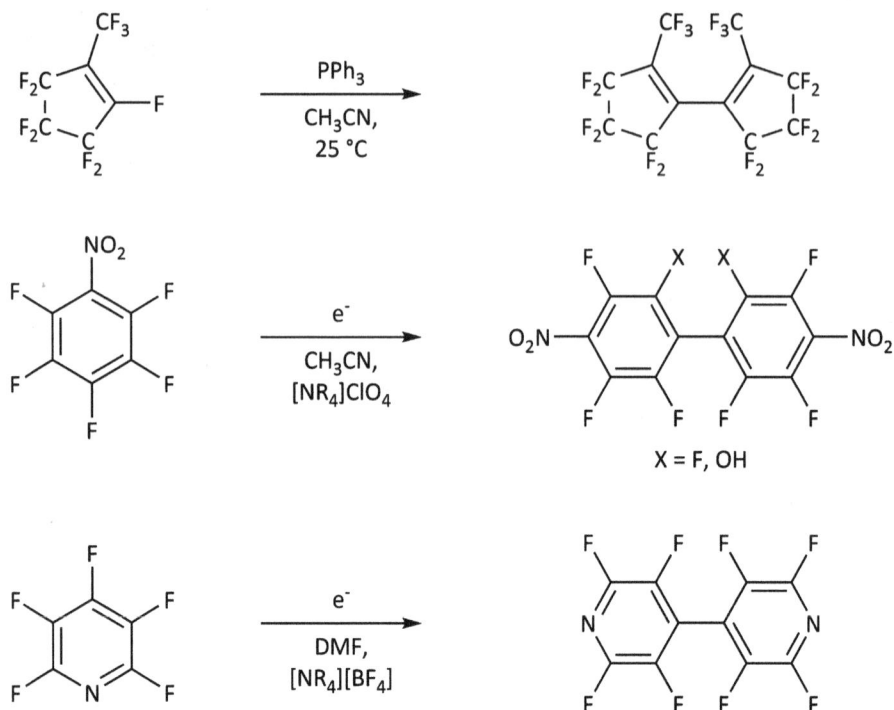

Scheme 13.62: Reductive coupling of fluoroolefins and -aromatics.

results in the cleavage of the C2–C3 bond and gives rise to 2,2-difluoro-1-phenylpropane and the respective monofluoride in a ratio of about 1:1. If the difluorocyclopropane ring is substituted with an acetoxy-group, one fluoride substituent is split off upon reaction with LiAlH$_4$, resulting in the generation of β-fluoroallyl alcohols. Under free radical conditions, difluorocyclopropane derivatives are converted to allylidene difluorides, if Bu$_3$SnH is used as reducing agent.

Aliphatic internal fluoroolefins are catalytically hydrogenated by H$_2$/Pd to give the corresponding dihydrogenated compounds (Scheme 13.63). However, as soon as branching occurs at the double bond, the loss of fluoride is observed, resulting in dihydro- and trihydro-derivatives. In case of tetrasubstituted olefins, the initial product may eliminate HF to result in the overall exchange of one H- against a F-atom and the shift of the double bond (Scheme 13.63). Fluorinated olefins, which contain vinylic bromine or chlorine substituents are typically dehalogenated prior to saturation of the double bond, giving finally rise to tetrahydro derivatives. Formation of the latter is enhanced, if the hydrogenation is carried out in the liquid phase in the presence of a base, for example, NaOH. Hydroformylation reactions are not observed, if the alkene substrate contains vinylic fluorine substituents, however, they can be realized with

H₂, Pd/C or
H₂, Pd/Al₂O₃

50 °C

87 %

H₂, Pd/C or
H₂, Pd/Al₂O₃

50 °C

11 %   72 %

H₂, Pd/Al₂O₃

H₂, CO,
Rh₆(CO)₁₆

80 °C

RF = CF₃, C₆F₅

94-97 %

H₂, CO,
Co₂(CO)₈

90-100 °C

RF = CF₃, C₆F₅

44-88 %

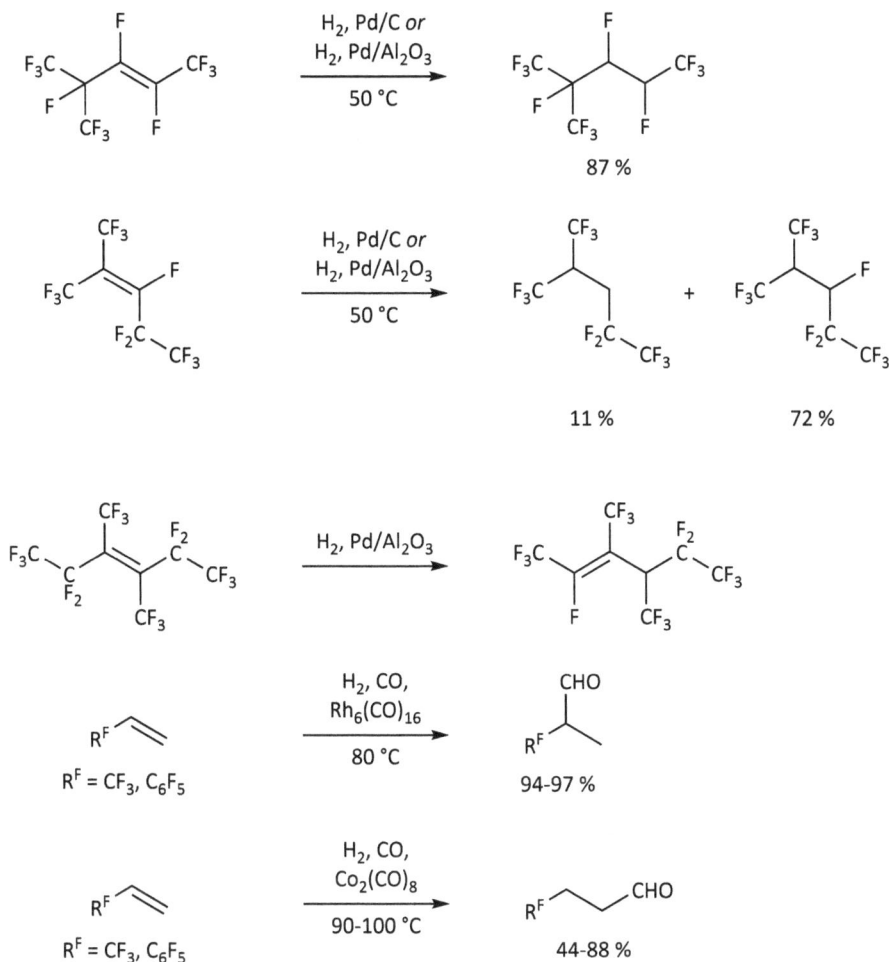

Scheme 13.63: Reduction of perfluoroolefins and hydroformylation of alkenes with fluorinated substituents.

fluoroalkyl- and fluoroaryl olefins. The application of rhodium and cobalt catalysts gives rise to branched and linear aldehydes, respectively (Scheme 13.63).

The synthetically most important class of substrates for reduction reactions are certainly carbonyl-compounds. In a very simple fashion, these transformations are used to generate fluorinated alcohols like 2,2,2-trifluoroethanol or 2-hydroperfluoro-isopropanol, which are accessible by reduction of trifluoroacetates or hexafluoroacetone, respectively. In both cases, LiAlH₄ is commonly used as reducing agent at slightly elevated temperatures. However, the choice of suitable reagents and possible additives will further allow the regio- and stereoselective preparation of more complex compounds. The reaction of perfluorinated diketones with LiAlH₄ leads to the reduction of both keto-groups, resulting in the formation of the corresponding diol, however, the treatment with the borohydrides NaBH$_4$ or KBH$_4$ only affects one carbonyl function and produces a hydroxyketone (Scheme 13.64). Similarly, LiAlH$_4$ reduces a perfluorinated ketoacid fluoride to the respective diol, whereas the hydroxyacid is obtained upon treatment with KBH$_4$. For the stereoselective reduction of carbonyl compounds, the presence of an additive can be helpful. For example, trifluoroacetophenone is reduced to the corresponding (S)-alcohol using a BH$_3$-complex with pyridine in presence of β-cyclodextrin. However, the enantiomeric excess of the product is rather poor. Apart from that, enantiomerically pure α,α-difluoro-β-hydroxy-ketones are predominantly transformed to *erythro*-2,2-difluoro-1,3-diols using a mixture of diisobutylaluminiumhydride (DIBAL), zinc chloride and N,N,N',N'-tetramethyl ethylenediamine [3120]. Another example for a highly stereoselective reduction of a carbonyl group is given by the application of baker's yeast, which works well for fluoromethyl- or 2,2,2-trifluoroethyl ketones (Scheme 13.64). The reduction of fluorinated ketones may also be exploited for C-C couplings, as for example, the treatment of hexafluoroacetone with triethylphosphite gives rise to perfluorinated pinacol. Alternatively, the latter can be prepared by hydrodimerization in isopropanol under UV-irradiation.

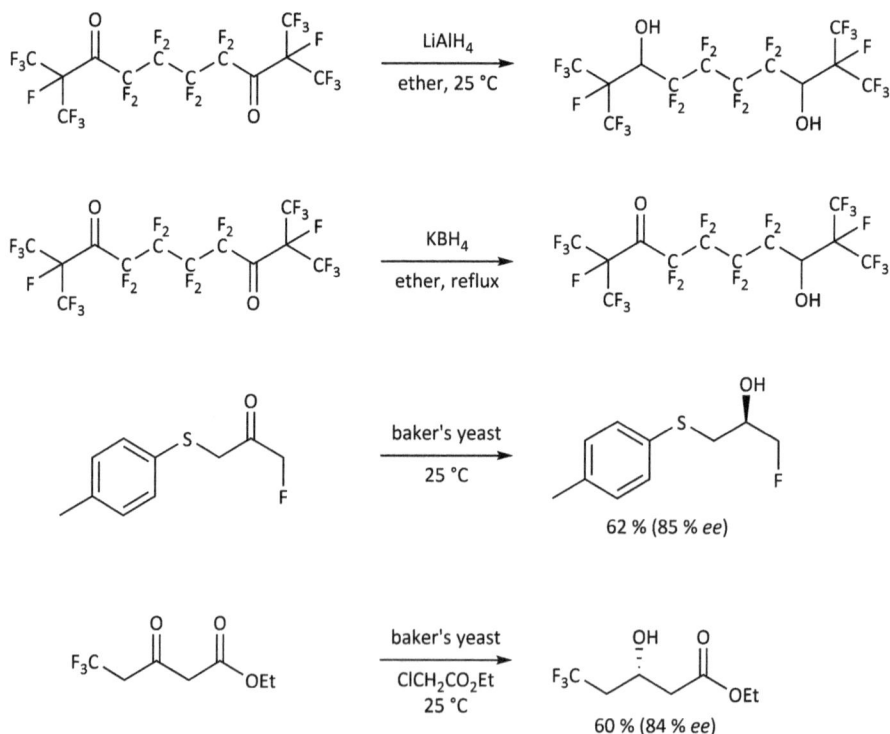

Scheme 13.64: Selective reduction of fluorinated carbonyl compounds.

Moreover, fluorinated carboxylic acids and esters are valuable starting materials for less-oxidized functionalities. However, the catalytic hydrogenation of COOH-groups works only well for trifluoroacetic acid. Fluorinated alcohols are readily accessible by reduction of the corresponding esters with KBH_4 in methanol or with $LiAlH_4$ in diethyl ether. The use of $Na[AlH_2(OCH_2CH_2OCH_3)_2]$ at low temperature stops the reduction of the fluoroester at the aldehyde stage [3121]. In contrast, trifluoroacetaldehyde is typically prepared by treatment of CF_3COOH with $LiAlH_4$ at -10 to -5 °C. Trifluoromethyl ketones are accessible by reaction of trifluoroacetic acid with Grignard reagents, for example, CH_3MgI, which leads to the formation of trifluoroacetone $CF_3C(O)CH_3$. Isolated fluorine atoms in esters and nitriles are mostly cleaved by $LiAlH_4$, if they are in α-position of the activating group. As the borohydrides are milder reducing agents, they do not affect fluorine substituents in α-fluorocarboxylate esters. Furthermore, $NaBH_4$ selectively reduces the ester group of polychlorofluoro carboxylates, but does not attack the chloro-substituents in the alkyl chain.

In standard organic chemistry, aliphatic and aromatic nitro compounds are important intermediates for the preparation of amines and other functional groups based on nitrogen, the same holds for fluorinated derivatives. Similar to their non-fluorinated analogues, aromatic polynitro-compounds are selectively reduced under appropriate reaction conditions. For example, the treatment of 2-fluoro-1,3-dinitro-benzene with H_2 and Pd/C leads to the diamino-product, whereas the use of Fe/HCl only reduces one of the nitro-substituents. The reduction of *m*-nitrobenzotrifluoride is also achieved by treatment with alkali metal hydroxides, however, full conversion to the aniline derivative additionally requires Na_2S. Thus, azoxy derivatives are obtained, if the nitro compound is only treated with KOH.

13.7.2 Oxidative transformations

An important class of compounds in preparative fluoroorganic chemistry is given by epoxides. Sometimes, they are generated by reaction of fluoroolefins, such as trifluoro acrylonitrile, with O_2 at elevated pressure and temperature. Another approach uses substituted peroxybenzoic acids to generate oxiranes from 2-substituted pentafluoro-propene derivatives bearing ester, amide or phosphonic ester functions. Some simple epoxides are furthermore accessible using high-valent transition metal oxides in acidic media. An example for this is the treatment of hexafluoropropene with a solution of CrO_3 in FSO_3H, which gives rise to hexafluoropropene oxide. If a mixture of CrO_3 and Cr_2O_3 is used, pentafluoroacetonyl fluorosulphate will be formed instead. Hexafluoropropene oxide and perfluoroisobutylene oxide are furthermore prepared by oxidation of the fluoroalkenes in aqueous or alcoholic, alkaline solution using H_2O_2. Alternatively, transition metal oxidizers, such as $KMnO_4$, may be used in anhydrous HF. However, the use of different oxidants can influence the outcome of the reaction, as for example, H_2O_2 converts perfluoroisobutene to its oxide, but $KMnO_4$

and perfluoropropene yield hexafluoroacetone hydrate, which can be dehydrated to liberate hexafluoroacetone. Apart from that, the latter is accessible by SbF_5-mediated isomerization of hexafluoropropene oxide. Accordingly, perfluorinated thioacetone is prepared by oxidation, however, in this case, sulphur is used as oxidant in presence of KF. At room temperature, hexafluorothioacetone exists as a dimer, but it can be converted to its monomeric form upon heating to 600 °C.

A very convenient reagent for the synthesis of perfluorinated oxiranes is NaOCl, which may be used in aqueous acetonitrile or other aprotic solvents (Scheme 13.65). Typically, epoxidation reactions of (E)- and (Z)-olefins proceed with retention of configuration. If the substrate contains allylic chlorine or other substituents, which are susceptible to nucleophilic replacement, the oxidation in alkaline media needs to be carried out at sufficiently low temperature. Epoxidation reactions with NaOCl in aqueous CH_3CN also work well for cyclic and bicyclic perfluoroolefins. The situation changes for non-fluorinated olefins with perfluoroalkyl groups, which are typically not oxidized by m-chloroperbenzoic acid (mCPBA). A very efficient reagent for such transformations is HOF · CH_3CN, which commonly reacts with perfluoroalkyl ethenes at 0 °C in a few minutes (Scheme 13.65). For some less reactive substrates, reactions are completed at room temperature within 2–3 hours [547]. However, HOF fails to oxidize strongly electron-deficient double-bonds and those, which are electron-poor and sterically hindered at the same time, for example, in 1,2-bis(perfluorobutyl)ethene or 2-(vinyl)perfluoroalkanes. If the perfluorinated substituent contains a trifluorovinyl-unit, it will not be oxidized, so that partially fluorinated, terminal dienes cleanly yield monoepoxides at the hydrocarbon site (Scheme 13.65) [547]. In general, epoxidation of dienes is realized under similar conditions as those of substrates with only one C=C-bond. The selectivity of the oxidation depends on the structure of the starting material and on the reagent. For example, nonconjugated perfluoro 1,4-cycloheptadiene is converted to the bis-epoxide using NaOBr. Under the same conditions, conjugated perfluoro 1,3-cycloheptadiene yields a mixture of the monoepoxide and a bridged bis-epoxide (Scheme 13.65).

A second major approach toward functionalized fluoroolefins is dihydroxylation, which sometimes proceeds via intermediary epoxides. For example, terminal ω-fluoroalkenes are easily converted to diols using a mixture of H_2O_2 and HCOOH. In absence of vinylic fluorine-substituents, perfluoroolefins yield vicinal diols upon reaction with $KMnO_4$ (Scheme 13.66). Typically, the dihydroxylation occurs stereospecifically under formation of syn-diols. In contrast, α-hydroxyketones are obtained, if the double-bond is substituted with one fluorine atom; the presence of two vicinal vinylic fluorine-substituents leads to the formation of α-diketones. Accordingly, the reaction of 1-alkoxy-1-fluoroalkenes with $KMnO_4$ yields α-hydroxyesters (Scheme 13.66). Another interesting example is given by the treatment of tetrakis(trifluoromethyl)allene with $KMnO_4$. Initially, dihydroxylation of one of the double bonds occurs to give an intermediary hydroxy enol, which subsequently tautomerizes to yield an α-hydroxyketone (Scheme 13.66).

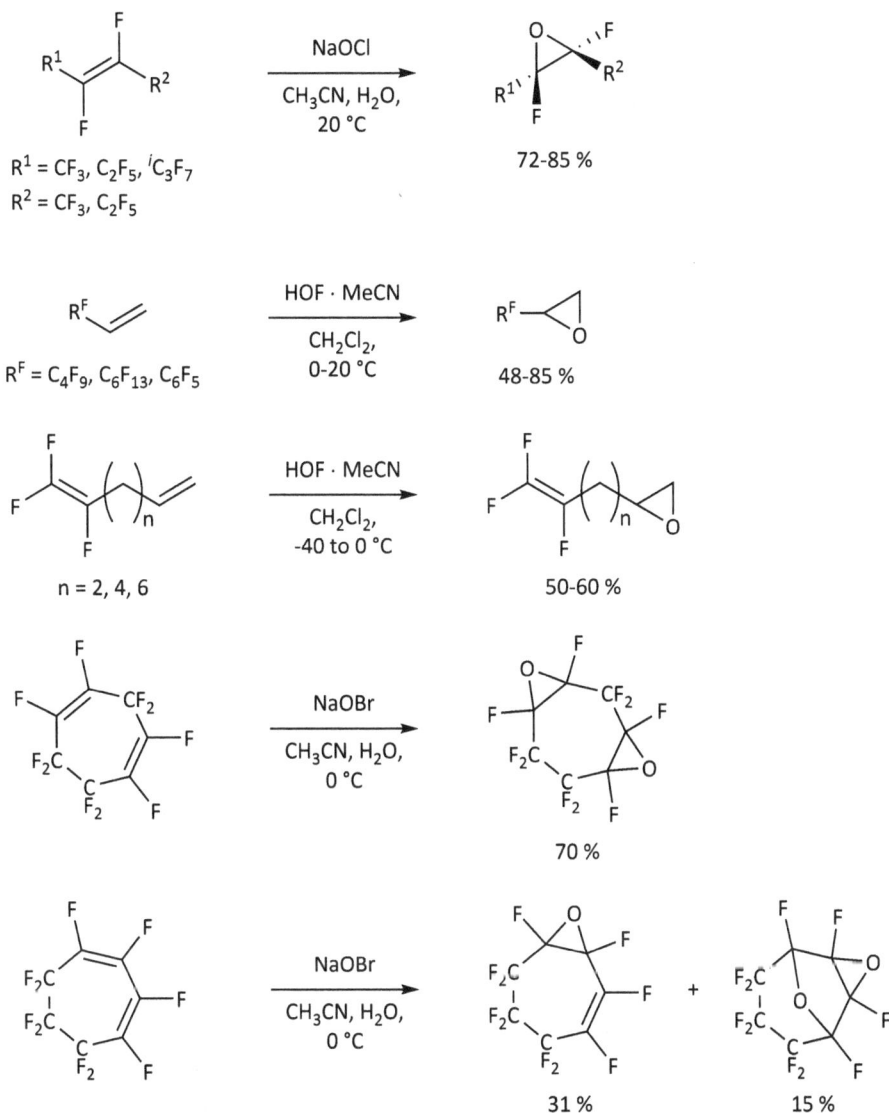

R^1 = CF$_3$, C$_2$F$_5$, ^{i}C$_3$F$_7$
R^2 = CF$_3$, C$_2$F$_5$

NaOCl

CH$_3$CN, H$_2$O,
20 °C

72-85 %

R^F = C$_4$F$_9$, C$_6$F$_{13}$, C$_6$F$_5$

HOF · MeCN

CH$_2$Cl$_2$,
0-20 °C

48-85 %

n = 2, 4, 6

HOF · MeCN

CH$_2$Cl$_2$,
-40 to 0 °C

50-60 %

NaOBr

CH$_3$CN, H$_2$O,
0 °C

70 %

NaOBr

CH$_3$CN, H$_2$O,
0 °C

31 %   15 %

Scheme 13.65: Epoxidation of fluorinated alkenes and dienes.

Fluoroolefins can further be treated to form halohydrins, which are also useful synthetic intermediates. A common reagent for the preparation of chlorohydrins is *tert*-butylhypochlorite, which tends to yield 1,2-dichloro derivatives as side-products in aqueous solution. Chlorohydrin acetates are generated, if the solvent is changed to acetic acid. In a similar way, bromohydrin acetates are obtained upon reaction of perfluoroalkylated alkenes with Br$_2$ and Hg(OAc)$_2$ in CH$_3$COOH. The same system gives rise to a mixture of bromohydrin acetate regioisomers and the corresponding dibromo-

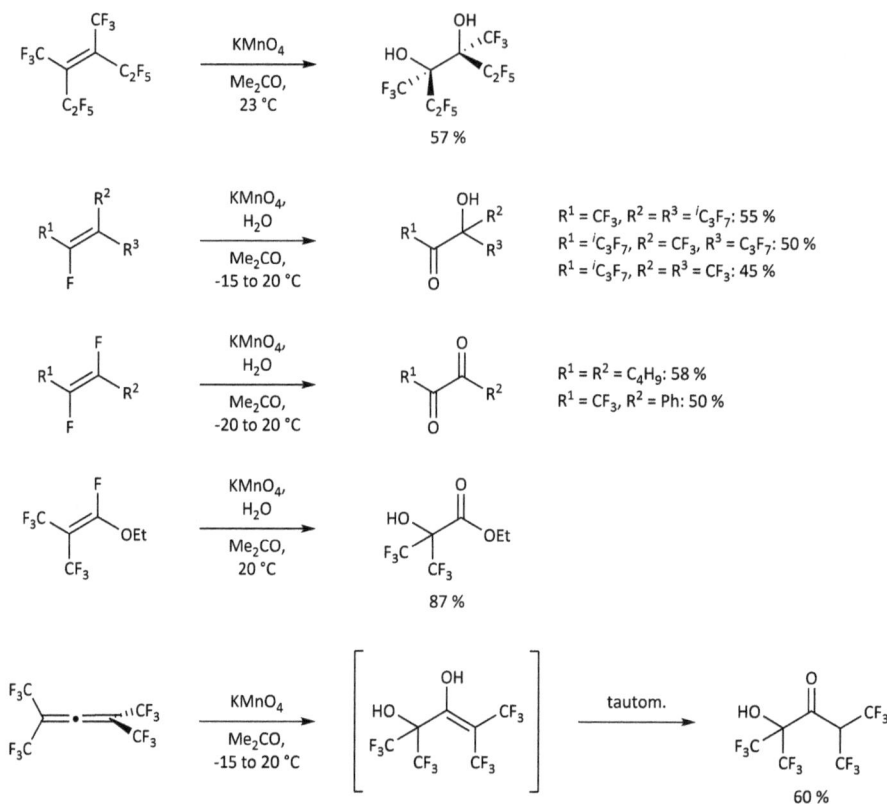

Scheme 13.66: Dihydroxylation of fluorinated alkenes.

derivative, if the vinylic substrate contains a branched perfluoroalkyl chain in allylic position.

The oxidation of fluorinated alkenes can also lead to the cleavage of the double bond, where the actual products strongly depend on the substrate structure and the used reagents. For example, ozonization of perfluoroalkenes in CF_3COOH gives rise to acid fluorides. The oxidative treatment of perfluorocyclobutene with $KMnO_4$ results in the formation of perfluorosuccinic acid $COOH-CF_2-CF_2-COOH$. Similarly, internal perfluoroalkenes are converted to carboxylic acids using RuO_4, which is in most cases more effective than $KMnO_4$. In a related approach, perfluorooctylethene is transformed to perfluorononanoic acid when it is treated with RuO_2 in the presence of an oxidant, for example, NaOCl or HIO_4.

Upon reaction with aqueous $KMnO_4$ under phase-transfer conditions, hydroxymethyl groups of primary alcohols, for example, $C_3F_7C(CF_3)_2(CH_2)_2CH_2OH$, are oxidized to carboxylic acids. Accordingly, fluorinated secondary alcohols are readily oxidized to the corresponding ketones. This transformation can be achieved by a variety of reagents. For example, a hydroxy-group in proximity of a fluoroalkyl chain may be oxidized by $Na_2Cr_2O_7$ in acidic media. The same system is capable of generating ketones from fluoroalkyl as well as chloroalkyl fluoromethyl alcohols under phase-transfer conditions. The oxidation of 4-(2-hydroxyperfluoroisopropyl) cyclohexanol with pyridinium chlorochromate (PCC) yields the cyclohexanone derivative, however, cleavage to a dicarboxylic acid is observed with HNO_3 and NH_4VO_3 (Scheme 13.67). A number of secondary alcohols with neighboring, terminal CF_3-groups are oxidized by the Dess–Martin periodinane reagent. Under Sharpless-conditions using $Ti(O^iPr)_4$, *tert*-butyl hydroperoxide and $(+)$-(L)-diisopropyltartrate, asymmetric epoxidation of racemic α,β-unsaturated fluoroalcohols is achieved (Scheme 13.67). In the present example, the unreacted (S)-alcohol is recovered in good enantiomeric excess, thus this method can be exploited for the kinetic resolution of enantiomers [3122]. Primary unsaturated alcohols are oxidized by $KMnO_4$ under cleavage of the double bond, resulting in the formation of dicarboxylic acids. A perfluorinated ketene derivative has been epoxidized by NaOCl to give a rare example of a stable α-lactone (Scheme 13.67) [3123]. Apart from that, chlorofluoroalkyl ketones are converted to esters by Baeyer–Villiger oxidation using trifluoroperoxyacetic acid.

Similar to Kolbe reactions, carboxylic acids and their salts are electrolyzed to result in decarboxylative coupling. For example, electrolysis of 3,3,3-trifluoro-2-trifluoro-methylpropanoic acid in presence of its potassium salt yields the corresponding fluoroalkane 1,1,1,4,4,4-hexafluoro-2,3-*bis*(trifluoromethyl)butane. It is also possible to electrochemically oxidize mixtures of acids under specific conditions to get single products. For example, electrolysis of mixtures of trifluoroacetic or pentafluoropropionic acid with trideuteroacetic acid gives rise to trifluoroethane-d_3 and pentafluoropropane-d_3, respectively [3124]. As expected, Kolbe electrolysis of two different perfluorinated acids R-COOH and R′-COOH generally gives rise to a mixture of the three products R-R, R-R′ and R′-R′.

Moreover, oxidation reactions are suitable for the preparation of functionalized aromatic compounds. The treatment of hexafluorobenzene with concentrated H_2O_2 at 140 °C yields pentafluorophenol. Under the same conditions, pentafluorobiphenyl C_6F_5-C_6H_5 is monohydroxylated on the nonfluorinated ring. In contrast, the reaction of perfluoronaphthalene with H_2O_2 at 100 °C yields a complex mixture, which is comprised of naphthol derivatives, naphthoquinone but also ring degradation products. Fluorinated phenols are mostly oxidized to the corresponding fluorine-containing cyclohexadienone and benzoquinone derivatives, in some cases, rearrangement and degradation of the ring occur as well. If a CF_3-group is attached to the aromatic system, $K_2S_2O_8$ hydroxylates the ring, whereas chlorous acid gives rise to a mixture of 2-trifluoromethyl-1,4-benzoquinone and a ring contracted, chlorinated 1,3-cyclopentadienone derivative [3125]. The use of $Na_2Cr_2O_7$ and H_2SO_4 leads to the oxidation of C_2F_5O-substituted *p*-aminophenols to *p*-benzoquinones. Mechanistically, the oxidation of pentafluorophenol typically proceeds via pentafluorophenoxy radicals, which can intermolecularly attack the aromatic ring to yield di-

88 %

58 %

R = C$_6$H$_{13}$, R^F = CF$_3$:   50 %   46 % (60 % *ee*)
R = C$_5$H$_{11}$, R^F = CHF$_2$:   52 %   39 % (98 % *ee*)
R = C$_5$H$_{11}$, R^F = CH$_2$F: 56 %   43 % (98 % *ee*)

85 %

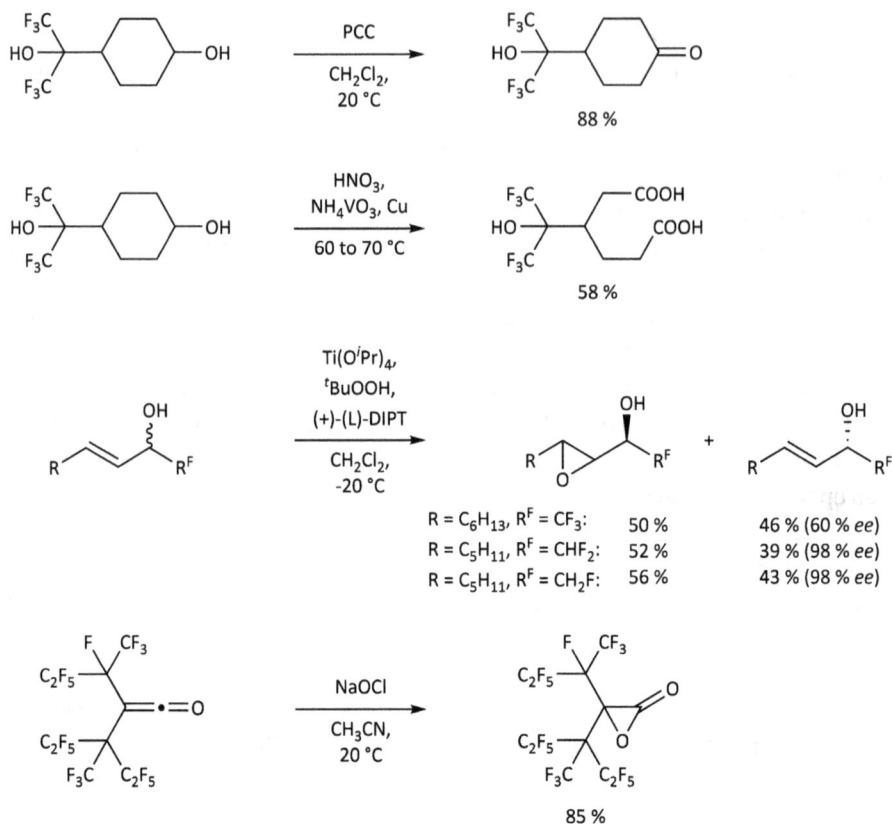

Scheme 13.67: Oxidation of secondary fluoroalcohols and fluorinated ketene.

and trimeric products. Accordingly, the oxidation of C$_6$F$_5$OH with H$_2$O$_2$ strongly depends on the concentration of the oxidant. Diluted reaction mixtures will give rise to *tetrakis*(pentafluorophenoxy)-*p*-benzoquinone, whereas high concentrations lead to the destruction of the ring, yielding (Z)-2,3-difluoromaleic acid. The latter product is also obtained, if perfluoro-1,4-benzoquinone is reacted with peroxyacetic acid. Treatment of pentafluorophenol with Pb(OAc)$_4$ produces perfluorinated 2,5-cyclohexadien-1-one in good yield.

13.7.3 Halogenations

The addition of elemental halogens to unsaturated fluorocompounds including trifluorovinyl ethers and fluorinated styrene substrates is rather easily achieved. Depending on the substrate and the reaction conditions, the halogenation may proceed via ionic or free-radical processes. Compared to nonfluorinated substrates, halogen-addition to fluoroalkenes is only poorly stereoselective, as the addition can occur in *syn*- or *anti*-fashion. The actual product formation strongly depends on steric and electronic factors, as can be observed for the ionic bromination of fluorinated styrene derivatives [3126]. The bromination of (*E*)- and (*Z*)-1-fluoropropene under ionic conditions in the dark gives rise to the respective *anti*-addition products in high yields. Apart from that, *trans*-1,4-adducts are mostly obtained upon halogenation of conjugated, cross-conjugated and homoconjugated fluoroolefins. In the same way, interhalogens, for example, ICl, are added to fluorine-containing olefins. Again, the formation of 1,4-adducts is preferred for conjugated fluorinated dienes, however, mixtures of regioisomers are obtained in most cases. Moreover, halogen or interhalogen addition also occurs with acetylenic substrates. For example, *syn*-addition of I_2 or ICl to terminal perfluoroalkyl acetylenes leads to the formation of olefins with an iodo-substituent on the terminal carbon-atom.

Halogen substituents are not exclusively introduced in addition reactions with unsaturated carbon substrates. Another important approach is the replacement of hydrogen-atoms in saturated starting materials, which can generally take place under thermal or photochemical conditions. For example, fluoroform is catalytically brominated to CF_3Br at 600–650 °C and perfluoro *tert*-butyl chloride is generated upon hydrogen-replacement with ICl in the presence of KF. In fluoroaromatic compounds, the hydrogen-replacement proceeds via an ionic mechanism and follows the typical directing effects. Thus, iodination of fluorobenzene mostly yields the *para*-substituted compound and only minor amounts of the *meta*-isomer. In the same way, 2-fluoroanisole is brominated to give 4-bromo-2-fluoroanisole. Highly fluorinated arenes, such as 2,3,5,6-tetrafluorophenol, are also halogenated by X_2 (X = Cl, Br, I) under strongly acidic conditions. The preparation of perfluorinated 1-bromonaphthalene is realized at room temperature and requires the presence of iron. In general, hydrogen-substituents in heterocycles can also be substituted by halogens. Hydrogen-replacement furthermore readily occurs in *α*-position of activated aliphatic substrates, such as fluorinated ethers, amines, aldehydes or nitriles. Sometimes, an interesting regioselectivity is observed with different halogens, for example, upon treatment of 2,2,3,4,4,4-hexafluorobutyl methyl ether. Photochemical bromination with Br_2 mostly affects the *α*-position of the butyl-chain, whereas Cl_2 leads to the substitution of the methyl group under identical conditions.

The treatment of alkali metal salts of highly fluorinated carboxylic acids with a halogen yields fluoroalkyl halides. For example, the decarboxylative treatment of CF_3CO_2K with I_2 represents an efficient method for the preparation of CF_3I. This synthetic approach can also be extended to the synthesis of fluoroiodoalkyl ethers and fluorovinyl iodides. Similarly, treatment of fluorinated carboxylic acids or their alkali metal salts with fluorosulphonyl hypohalites also yields perfluoroalkyl halides. A hydrogen-substituent in the polyfluoroalkyl chain of a ketone or acyl fluoride is efficiently replaced by chlorine or bromine, if it is reacted with PCl_5 or PBr_5, respectively. A mixture of SO_2Cl_2 and *N*,*N*-dimethylaniline has been used to replace the

α-hydrogen of polyfluoronitrile compounds. The treatment of perfluoroketones with PCl$_5$ leads to the replacement of the carbonyl-group and the formation of a geminal dichloride. Analogously, α-keto esters are converted to tetrachlorinated ethers, however 1,4-dicarbonyl compounds give rise to cyclic α,α'-dichloroethers. Furthermore, PCl$_5$ converts perfluoroacyl halides and carboxylic acids to the corresponding trichloromethyl compounds. An efficient reagent for the replacement of secondary and tertiary fluoro-substituents by bromine is BBr$_3$, which, however, does not attack CF$_3$-groups. In a related approach, difluoromethylamine functions are substituted by BCl$_3$. The generation of 1,1,1-trihaloalkanes is achieved by the treatment of primary fluoroalkyl iodides with AlCl$_3$ or AlBr$_3$, respectively. Analogously, benzylic and vinylic fluorine-substituents are replaced by chlorine, when the starting material is reacted with AlCl$_3$. The latter is also used to replace the α-fluorine atoms in polyfluoroethers by chlorine substituents.

13.7.4 Study questions

(I)   Which products are obtained by reduction of hexafluoroacetone with Mg metal? How does the solvent influence the outcome of the reaction? Which compound can be isolated, if the reduction is carried out with Al and stopped by acidification?

(II)   How can chloro-substituents in chlorofluorocarbons be selectively replaced? Which methods and reagents are typically used?

(III) Which products are generally observed upon catalytic hydrogenation of fluorinated olefins? Does branching at the double bond influence the product formation? What happens, if the substrate contains vinylic chloro- or bromo-substituents?

(IV)   Which reagents may be used to convert fluorinated diketones to diols or hydroxyketones, respectively? Which products are obtained, if these reagents are applied for the reduction of ketoacid fluorides?

(V)   Which reagents are commonly used to prepare fluorinated epoxides? Give examples.

(VI)   Which products are generally obtained, if fluoroolefins are reacted with KMnO$_4$? Comment on the role of vinylic fluorine substituents. What happens, if fluorinated enol ethers are used as substrate?

(VII) How are fluorinated olefins converted to carboxylic acids? Which substrates and reagents are typically used?

(VIII) Which types of products are obtained upon electrolysis of fluorinated carboxylic acid/carboxylate mixtures?

(IX)   How can one oxidatively prepare pentafluorophenol from hexafluorobenzene? What happens upon further oxidation with the same reagent?

(X)   Is it possible to halogenate fluoroolefins by treatment with the elemental halogens X$_2$? Which stereoselectivity is generally observed? Does it depend on the reaction conditions?

(XI)   By which methods may (per-)fluoroalkyl halides be prepared? What happens upon treatment of fluorinated carbonyl compounds with PCl$_5$?

14 Fluorination of (hetero-) aromatic compounds

Analogously to their saturated relatives, fluoroarenes may also be prepared by a large number of methods, including more classical substitution reactions as well as modern metal-catalyzed approaches. However, in contrast to the functionalization of sp^3-carbon atoms, strongly oxidizing reagents are in most cases not suitable for the fluorination of arenes, since they will rather generate saturated fluoroorganic molecules. In the past years, several review articles have been published, covering new developments and trends in the preparation of aryl fluorides, including the synthesis of ^{18}F-labeled compounds [3127–3129]. In general, radiofluorination reactions are very similar to those, which are used for the introduction of nonlabeled fluorine substituents, despite the fact, that they need to be carried out more rapidly and sometimes with slightly different fluorinating agents. In addition, late-stage functionalizations are preferred for this purpose. However, since most of the techniques for the introduction of ^{18}F-substituents do not significantly differ from standard fluorination, they are not explicitly covered in the following chapter.

14.1 Electrochemical fluorination techniques

Generally, electrochemical fluorination techniques are versatile methods for the introduction of fluorine into robust starting materials. However, one has to be aware of the fact that this approach is rather not suitable, if the substrate contains functional groups or sensitive carbon scaffolds, since the reaction conditions are typically not very mild.

The anodic fluorination of arenes using amine-HF adducts or tetraalkylammonium fluorides as fluorine sources is also known as the Knunyants–Rozhkov method. Therein, the substrate is typically dissolved in CH_3CN in an electrolysis cell containing Pt-anodes. For example, chlorobenzene is converted to a mixture of o- and p-chlorofluorobenzene, however, the dearomatization product 3-chloro-3,6,6-trifluoro-1,4-cyclohexadiene is also obtained. The electrochemical fluorination of naphthalene results in the formation of a mixture of the corresponding 1-fluoroarene and the 1,4-difluorinated product. Mechanistically, this electrochemical oxidation may be described by a four-step sequence, starting with the formation of an aryl radical cation [ArH]$^{+\cdot}$, which is subsequently transformed to a radical species [ArFH]$^{\cdot}$ upon addition of a fluoride anion (Scheme 14.1). Afterwards, this radical is again electrochemically oxidized, giving rise to cationic [ArFH]$^+$. The final step can proceed in two different ways, where either the fluoroarene is formed by elimination of a proton, or the addition of another fluoride anion yields a dearomatized, difluorinated compound.

https://doi.org/10.1515/9783110659337-017

$$ArH \xrightarrow[- e^-]{ox.} \left[ArH \right]^{\bullet +} \xrightarrow{+ F^-} \left[ArFH \right]^{\bullet} \xrightarrow[- e^-]{ox.} \left[ArFH \right]^{+} \begin{array}{c} \xrightarrow{+ F^-} ArF_2H \\ \\ \xrightarrow{- H^+} ArF \end{array}$$

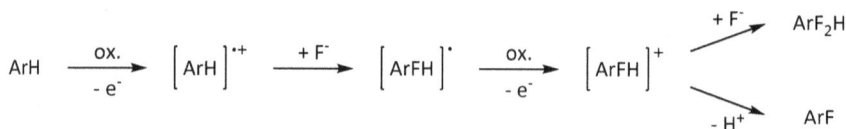

Scheme 14.1: General reaction steps in electrochemical fluorination reactions of arenes.

14.2 Nucleophilic substitution

14.2.1 Halogen exchange

Although it might not appear too obvious to prepare fluorinated arenes and heteroarenes by halogen exchange, it is a suitable approach for the synthesis of a number of activated substrates. In general, activation is achieved by the introduction of electron-withdrawing groups, typically -NO$_2$, -CN, -CF$_3$, -CHO or multiple -Cl substituents. Pyridine rings also belong to the activated substrates. However, one has to be aware of the fact that C-F bonds are rather easily cleaved in aromatic compounds, which can even occur under very mild conditions, where the instability of C(sp^2)-F bonds results from repulsive interactions of the π-system and the lone-pairs of the fluoride substituents (cf. Section 13.1.2). The stability of the other carbon-halogen bonds increases from chlorine to iodine. For highly fluorinated arenes and heteroaromatics, such as hexafluorobenzene or pentafluoropyridine, no further activation is required and they are easily substituted by a variety of nucleophiles, resulting in the cleavage of C-F bonds. A typical example for this reactivity is the reaction of C$_6$F$_6$ with thiolates, such as NaSPh, which will lead to the complete replacement of the fluorine substituents at room temperature, generating the respective thioether functions [3130]. Nucleophilic aromatic substitution of hexafluorobenzene with [NBu$_4$]CN yields hexacyanobenzene and, more importantly, enables the convenient preparation of anhydrous [NBu$_4$]F, which is highly hygroscopic and cannot be obtained by drying [2782]. Pentafluoropyridine is very susceptible to nucleophilic substitution in 4-position, which can be beneficial for preparative purposes. For example, the synthesis of 4-(perfluoro-*tert*-butyl) tetrafluoropyridine is achieved using a mixture of C$_5$F$_5$N, perfluoroisobutene and CsF. Generally, the tendency of highly fluorinated arenes to undergo nucleophilic substitution reactions gives rise to the inverse synthetic approach of selectively preparing fluoroaromatic compounds by C-F-bond activation, to which a separate subchapter is dedicated (see Section 14.6).

Nevertheless, it is possible to prepare fluoroarenes by nucleophilic aromatic substitutions. In many cases, spray-dried KF is used as fluorinating agent in suspensions of hot polar-aprotic solvents. Sometimes, a few modifications help to improve the conversion of the substrates, for example, the addition of some CsF increases the reactivity of KF. The application of phase-transfer reagents or solvent-free high temperature conditions may also help. As a typical example, 1,3-dichloro-4-nitrobenzene is converted to 1,3-difluoro-4-nitrobenzene, using a KF-suspension in a polar solvent. Similarly, tetrachloroisophtalonitrile is fluorinated by KF in sulpholane at 220 °C to give tetrafluoroisophtalonitrile in good yield. However, the fluorination of pentachlorobenzonitrile with KF affords the identical yield of pentafluorobenzonitrile, no matter if the reaction is carried out in benzonitrile or under solvent-free conditions. In the same way, the fluorination of perchlorobenzene C_6Cl_6 or -pyridine C_5Cl_5N is achieved at 450–500 °C without any solvent, resulting in the formation of a mixture of chlorofluorobenzenes $C_6F_{6-x}Cl_x$ ($x = 0, 1, 2, 3$) and pentafluoropyridine C_5F_5N, respectively. For highly reactive substrates, such as 2,4,6-trichlorotriazine, a mixture of NaF and HF can also be used to substitute the chloro-substituents. More recently, significantly milder conditions have been reported, using anhydrous $[NBu_4]F$ as fluorinating agent at room temperature. Using this method, a number of chlorinated benzene derivatives, 2-chloropyridines as well as other nitrogen heteroaromatics has been converted to the corresponding fluorinated compounds [3131, 3132]. Apart from that, the halogenexchange of a variety of chlorine-containing pyridine and -quinoline derivatives has been realized using *in situ* generated $[NBu_4]F$ [3133]. Nonactivated bromoarenes can also serve as substrates for nucleophilic fluorination; however, the treatment with anhydrous $[NMe_4]F$ generally leads to the formation of regioisomers, since the reaction proceeds via aryne intermediates [3134].

14.2.2 Fluorodenitration and fluorodesulphonylation

Fluoroarenes can also be prepared by the substitution of nitro-groups in fluorodenitration reactions, which are typically performed using KF at about 200 °C in highly polar solvents like sulpholane (Scheme 14.2). In addition, the presence of stoichiometric amounts of phthaloyl dichloride is required to decompose the liberated nitrite anions, which would otherwise induce side-reactions, especially the formation of phenolates and consequently diphenylethers [3135, 3136]. Fluorodenitration reactions are not limited to *m*-polynitro aromatics, but can also be performed in the presence of other functional groups, for example, chloro- or cyano-substituents. Apart from that, fluoroaromatic compounds are also formed, if *m*-sulphonyl fluorides Ar-SO$_2$F are reacted with KF (Scheme 14.2). Nevertheless, nitro-substituents are more prone to substitution, which is clearly observed when a mixed nitro-fluorosulphonyl substrate is submitted to a nucleophilic substitution reaction with KF [3137]. A significantly milder method has been reported, using anhydrous $[NBu_4]F$ in fluorodenitration reactions at room temperature. This approach cleanly yields fluoroarenes and does not require any additive, however, the substrate must contain another strongly electron-withdrawing group [3131, 3132]. The fluorodenitration method can also be applied for the preparation of fluorinated heterocyclic compounds, such as pyridines, triazoles or purines. In case of the latter two, anhydrous HF may serve as fluorinating agent. The fluorodenitration of nitropyridines has also been achieved using $[NBu_4]F$ as fluoride source in THF solutions, where the presence of water does not have to be strictly excluded.

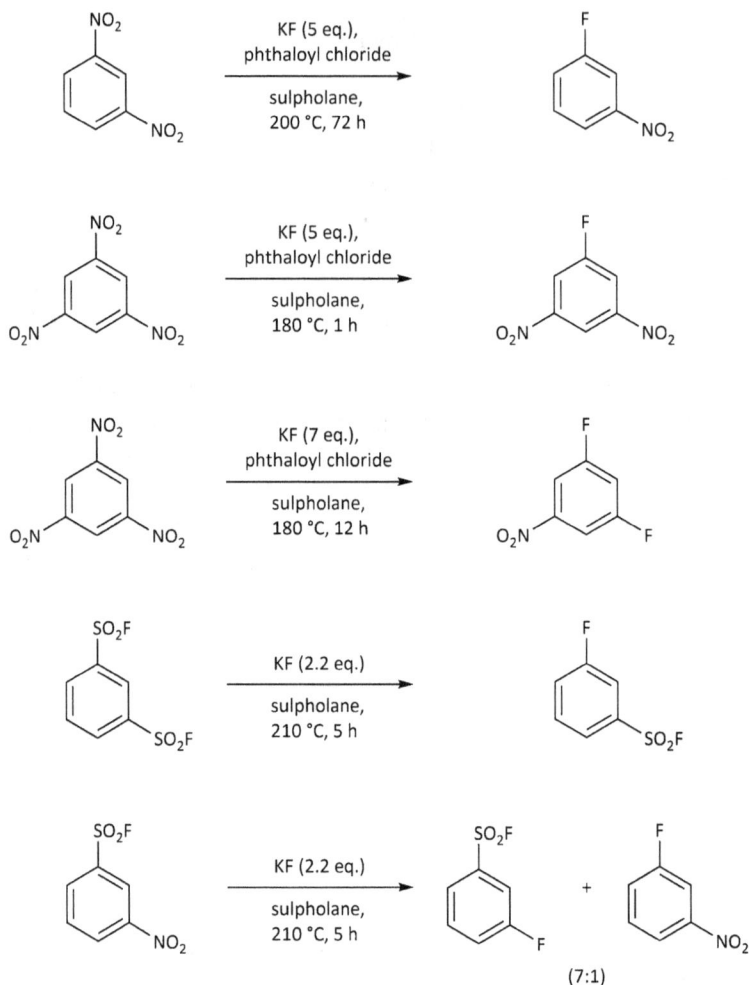

Scheme 14.2: Preparation of fluoroarenes using fluorodenitration and fluorodesulphonylation reactions.

The substitution reaction works well with 2- and 4-nitropyridines, but requires an additional electron-withdrawing group in case of 3-nitropyridines [3138]. Fluorodenitration reactions of several different arene substrates have also been conducted under microwave conditions, furnishing the corresponding fluoroaromatics after a short reaction time [3139].

14.2.3 Preparation of pentafluorophenyl compounds

The pentafluorophenyl-substituent is an important moiety in fluoroorganic chemistry, which will be discussed in a little more detail in the following section, although it is not closely related to the introduction of fluorine. Typically, the C_6F_5-group is introduced via building-block approaches, since it is by far more convenient, to substitute a highly fluorinated arene ring instead of performing an exhaustive fluorination of potentially sensitive substrates on a late stage. For this purpose, a few compounds may be used, which often can be traced back to pentafluorochlorobenzene, which is the major side-product in the halogen exchange reaction of hexachlorobenzene toward the perfluorinated compound (Scheme 14.3). An important transfer reagent for C_6F_5-moieties is the Grignard compound C_6F_5MgCl, which is stable in ethereal solution at room temperature. Starting from the organomagnesium compound, a few derivatives are accessible, for example, pentafluorobenzene is obtained by protonation, whereas the respective iodo-derivative is generated upon treatment with I_2 (Scheme 14.3). Bromopentafluorobenzene is typically prepared from C_6F_5H in an electrophilic bromination reaction, using a mixture of Br_2 and $AlBr_3$ (Scheme 14.3).

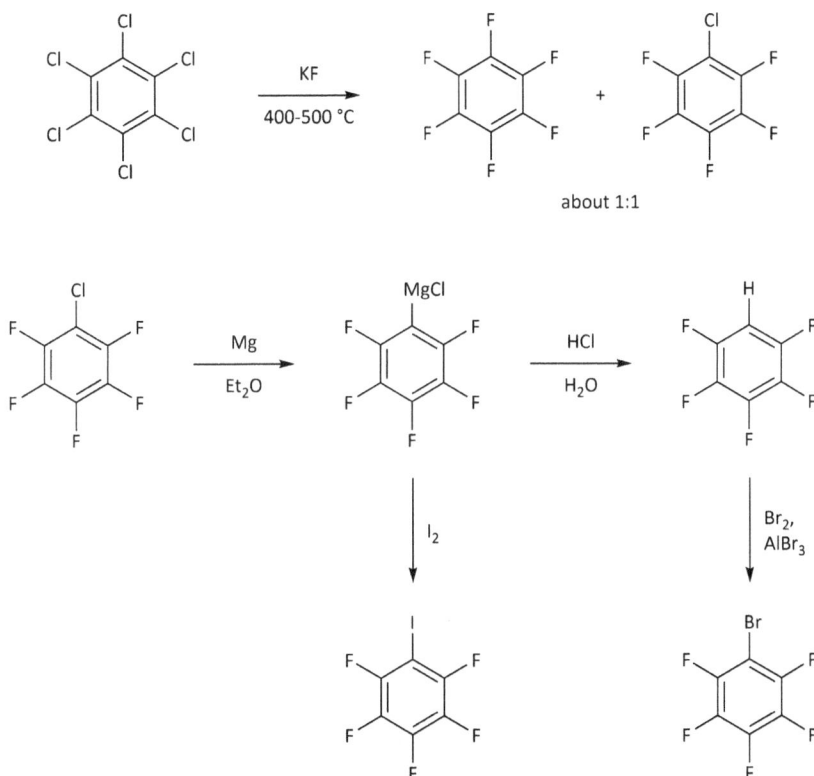

Scheme 14.3: Halogen exchange of hexachlorobenzene (top) and preparation of pentafluorophenyl compounds using C_6F_5MgCl (bottom).

Pentafluorophenol is obtained upon reaction of hexafluorobenzene with aqueous KOH, whereas ethylene glycol solutions of KSH or NaSH lead to the formation of pentafluorothiophenol C_6F_5SH (Scheme 14.4). Hexafluorobenzene is also directly substituted by CH_3ONa and aqueous NH_3 to yield pentafluoroanisole and pentafluoroaniline, respectively. $C_6F_5NH_2$ is further oxidized to the nitroso- or nitro-compound by the peracids HCO_3H and CF_3CO_3H. Apart from that, the explosive azide $C_6F_5N_3$ is generated from perfluorobenzene and NaN_3 (Scheme 14.4). For the synthesis of pentafluorotoluene, C_6F_6 is reacted with methyllithium, whereas pentafluorobenzyl alcohol arises from the reaction of C_6F_5MgCl with formaldehyde. If acetaldehyde is used as electrophile, pentafluorostyrene is generated after subsequent dehydration with P_4O_{10}. In a similar way, pentafluorophenylacetylene is synthesized, if C_6F_5MgCl is reacted with trichloroacetaldehyde. After chlorination of the hydroxy-group, the chlorine substituents are fully abstracted in two steps using $AlCl_3$ and Mg, respectively, giving rise to C_6F_5-substituted acetylene. In contrast, perfluorinated styrene is prepared from C_6F_5H, which is initially treated with CCl_2F-$CClF_2$ and SbF_5, leading to the formation of the saturated intermediate C_6F_5-CF_2-CF_2Cl. Dehalogenation of the latter compound by treatment with iron at 650 °C results in the formation of perfluorostyrene. Moreover, pentafluorobenzaldehyde is formed in the reaction of C_6F_5MgCl with N-methyl- or N-ethylformanilide. Analogously, treatment of the organomagnesium compound with CO_2 leads to the formation of pentafluorobenzoic acid, which is also accessible by other synthetic routes, for example, the reaction of C_6F_5H with SbF_5 and phosgene or hydrolysis of perfluorotoluene in presence of SbF_5. Starting from C_6F_5COOH, other derivatives, such as pentafluorobenzonitrile, are accessible upon treatment with NH_3 and subsequent dehydration of the intermediate benzamide $C_6F_5CONH_2$ using P_4O_{10}.

14.2.4 Deoxyfluorination

The fluoroformate process
Starting from phenols, fluoroarenes are accessible by a thermal fluorodecarboxylation method. For this, the phenolic substrate has to be converted to a fluoroformate $ArOC(O)F$ using gaseous $FC(O)Cl$ in presence of NBu_3. Typically, the decarboxylation step is achieved in presence of Pt/Al_2O_3 or anhydrous HF and AlF_3. In general, this method may not be very suitable for the application on laboratory scale, since it requires the use of toxic $FC(O)Cl$. However, it is an important industrial procedure for the generation of aryl fluorides. Therein, chloroformates may be directly used as starting materials in the system of anhydrous HF and AlF_3, as the halogen exchange from $ArOC(O)Cl$ toward $ArOC(O)F$ readily takes place *in situ*, followed by the elimination of CO_2. It is also possible to prepare the intermediate fluoroformate compounds by a catalytic reaction of the corresponding phenol with CO_2 in HF-solution. Furthermore, the costly Pt-catalyst in the decarboxylation step is not necessarily required and may generally be replaced by Al-based compounds. The major advantage of the fluorofor-

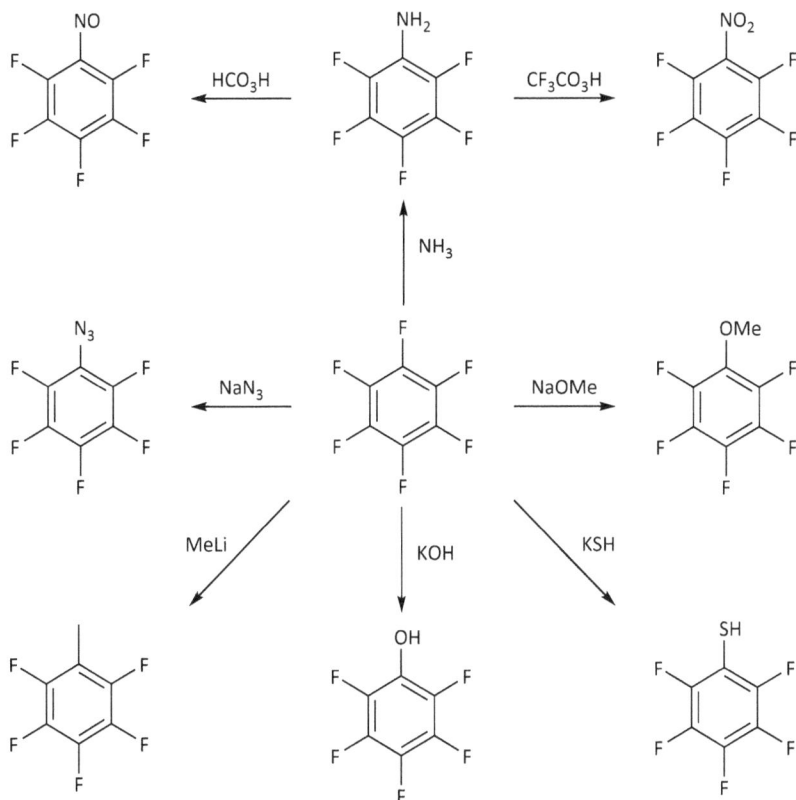

Scheme 14.4: Direct functionalization of C_6F_6 using nucleophilic aromatic substitution reactions.

mate process is given by the high, nearly quantitative yields, which are typically obtained.

Direct deoxyfluorination of phenols

Generally, the direct deoxofluorination of aromatic alcohols is by far more challenging than that of aliphatic substrates. Nevertheless, the conversion of phenols to aryl fluorides has been achieved, using the α,α-difluoroamine-derived PhenoFluor™ reagent in presence of stoichiometric amounts of CsF. This approach tolerates a large number of ring-substituents, such as halides, carbonyl functions or amines. In addition, it can be applied to heteroaromatic substrates but also steroids or other rather complex natural products [2811]. Catechol substrates are deoxofluorinated, when they are initially oxidized to o-quinones using a hypervalent iodine reagent. The subsequent treatment with a R_2NSF_3-reagent, that is, DAST or Deoxo-Fluor™, followed by reduction with NaBH$_4$ yields regioisomeric mixtures of o-fluorophenols [3140]. This procedure has also been adapted to the deoxofluorination of bridged biphenyl substrates [3141].

14.2.5 Nitrogen replacement

Decompositon of aryl diazonium salts

A very convenient method for the selective preparation of aryl fluorides is represented by the substitution of aryl diazonium salts with fluoride. Initially, aromatic amines are treated with $NaNO_2$ and thus converted to diazonium salts, which are subsequently reacted with anhydrous HF, aqueous HBF_4 or $NaBF_4$, giving rise to the intermediates $[Ar\text{-}N_2]^+F^-$ or $[Ar\text{-}N_2]^+[BF_4]^-$, respectively. In the final step, the diazonium salts are heated to liberate N_2 and the aryl fluoride (Scheme 14.5). The main difference between these transformations is the stability of the diazonium salts, as the tetrafluoroborates may be isolated prior to decomposition, whereas the diazonium fluorides need to be prepared *in situ*. The Balz–Schiemann reaction via aryldiazonium tetrafluoroborates is one of only a few named reactions in fluoroorganic chemistry. The stability of the diazonium salts also has a direct impact on the laboratory equipment, as the Balz–Schiemann reaction is conveniently performed in standard glassware, whereas decomposition reactions of arenediazonium fluorides essentially need to be carried out in metal vessels.

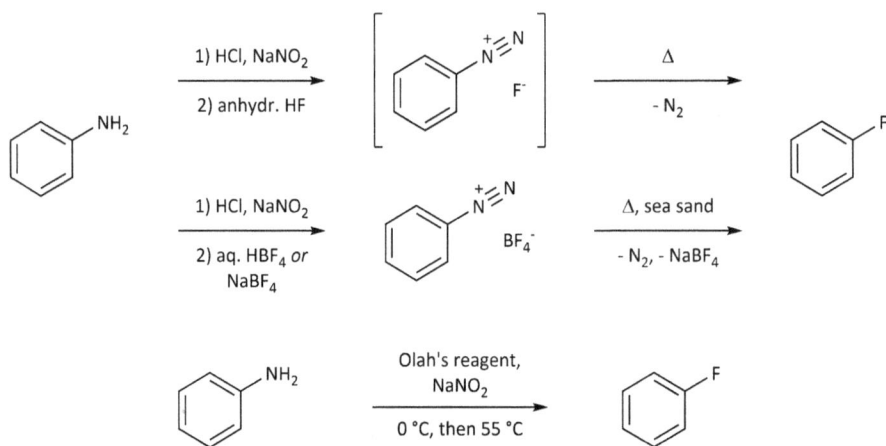

Scheme 14.5: Preparation of arylfluorides by thermal decomposition of aryldiazonium salts.

Over the years, a few modifications have been made to the Balz–Schiemann reaction, mainly focusing on the replacement of reagents and the development of less hazardous synthetic methodologies. For example, *tert*-butyl nitrite has been used as replacement for aqueous $NaNO_2$-solutions. In combination with $BF_3 \cdot Et_2O$, the fluorodediazoniation of anilines is achieved in a one-pot procedure in *o*-dichlorobenzene, thus avoiding the use of Brønsted acids [3142]. Apart from that, $[PF_6]^-$ anions have been used as replacement for tetrafluoroborate. The main benefit of using hexafluorophosphate salts is their limited water-solubility; in addition, they sometimes lead to increased yields of fluoroarenes. Other fluoroanions, such as $[AsF_6]^-$, $[SbF_6]^-$ or $[SiF_6]^{2-}$ play a less important role and are only scarcely used.

As another alternative, anilines can be directly diazotized with $[NO][BF_4]$ and subsequently decomposed in organic solvents. The decomposition of arenediazonium tetrafluoroborate salts is furthermore achieved by treatment of their water- or acetone-solutions with elemental copper or its halides. For example, in a one-pot procedure, 2-isopropyl-6-methylaniline is converted to the corresponding aryl fluoride using aqueous $NaNO_2$, HBF_4 and copper powder at room temperature. Apart from thermal or copper-mediated decomposition, aryldiazonium salts furnish aryl fluorides upon photochemical treatment or impact of ultrasound. Both techniques generally proceed under relatively mild reaction conditions. The photochemical method is furthermore used for the preparation of fluorinated heterocycles, for example, imidazoles or pyrazoles [3143].

In practice, the Balz–Schiemann method is one of the most frequently used techniques for the preparation of simple fluoroarenes. However, even the relatively stable $[ArN_2][BF_4]$ are still hazardous compounds, so that a large-scale application of this method is strongly inadvizable. On laboratory scale, the large amount of reaction energy is commonly controlled by diluting the aryldiazonium tetrafluoroborates with sea-sand or other inert, solid materials. Another major drawback of the Balz–Schiemann method is its low reproducibility, as reaction yields can strongly vary in a number of experiments, which are carried out under identical conditions. In more recent examples, the $[ArN_2][BF_4]$ salts are also not isolated but prepared *in situ*. Subsequently, the product solutions are directly thermolyzed in a temperature range of about 55–160 °C (Scheme 14.5). This avoids the isolation of the hazardous diazonium salts and consequently allows the preparation of aryl fluorides on larger scale. Moreover, Balz Schiemann reactions have also been carried out in continuous flow [3144].

Decomposition of aryl triazenes

Another long-known approach toward aryl fluorides has been based on fluorodediazoniation of arenediazonium piperidides in aqueous HF. Such reactions, involving intermediary aryl triazenes, are also known as Wallach fluorination. Over the years, the synthetic procedure has been modified, for example, piperidine has been replaced by aqueous solutions of Me_2NH, yielding the respective dimethyltriazene intermediates. For the subsequent fluorination, a number of conditions have been examined, for example, the use of neat anhydrous HF or mixtures with co-solvents like CH_2Cl_2, benzene or tetrahydrofuran. Furthermore, other fluorine sources have been tested, such as acidic pyridine/HF, or the fluoride salts CsF and $[NBu_4]F$ in conjunction with strong acids. Wallach-reactions have also been performed under fluorous conditions in solution-phase or using solid-supported triazene starting materials. The major advantages of this method are given by a reduced occurrence of side-reactions and comparably mild reaction conditions [3145]. In the same way, an ionic liquid has been reported to be a suitable reaction medium for the decomposition of triazene intermediates [3146]. For example, the preparation of 1-amino-8-fluoronaphthalene is achieved by the treatment of the corresponding cyclic triazene with HF/pyridine under mild conditions [3147].

Fluorodequaternization

An alternative approach toward aryl fluorides is described by the method of fluorodequaternization. For this, aryltrimethylammonium salts are treated with CsF in polar solvents like dimethyl sulphoxide or acetonitrile. In most cases, perchlorate, or more safely triflate anions are used. The ring fluorination works best, when a strongly electron-withdrawing substituent, such as a nitro-group, is attached in *para*-position. The driving force for the replacement of the trimethylammonium group is its high nucleofugacity. The fluorodequaternization method can also be applied to the synthesis of fluorinated heterocycles, for example, pyrimidines and purines.

14.2.6 Miscellaneous transformations

Diaryliodonium salts can also serve as precursors for aryl fluorides, when they are treated with nucleophilic fluorine sources, such as [NMe$_4$]F. Typically, symmetric hexafluorophosphate- or triflate salts are used as substrates to avoid the formation of side-products. The reaction proceeds via initial formation of iodonium fluoride intermediates, which are readily generated in acetonitrile solution. However, in the same solvent, their thermal degradation to aryl fluorides and iodoarenes is not that easily achieved. In addition, increasing concentrations of the by-products [NMe$_4$]OTf and [NMe$_4$][PF$_6$] decelerate the formation of fluoroarenes. Therefore, it is beneficial to remove the [NMe$_4$]-salts and change the solvent to benzene prior to the decomposition step [3148]. Another problem of this type of chemistry includes fluoride promoted ligand-exchange reactions of the iodonium salts [3149].

Recently, a site-selective functionalization of aromatic C-H-bonds has been described, which proceeds via aryl sulphonium salts. Initially, the arene is reacted with a derivative of dibenzothiophene *S*-oxide to give an intermediary salt, which can be isolated and purified. The transformation to an aryl fluoride is achieved upon treatment with a source of fluoride ions, for example, [NMe$_4$]F. In general, this method tolerates a large variety of functional groups, works well with electron-rich arenes and shows mostly *p*-selectivity. Thus, this approach is suitable for late-stage functionalization and can also be used for the introduction of ^{18}F-substituents [3150]. The selective *para*-fluorination of pivaloyl protected anilines is achieved by a hypervalent iodine reagent in combination with pyridine/HF. It has been proposed that the anilide function is attacked by the iodine reagent to form ionic nitrenium intermediates. The latter are stabilized by the phenyl ring, thus charge-delocalized resonance-structures with an activated *para*-carbon exist. Finally, trapping of these species with HF gives rise to *para*-fluorinated products [3151].

14.2.7 Study questions

(I)   Which substrates are generally suitable for the preparation of fluoroarenes by halogen exchange? What needs to be considered regarding the stability of fluoroaromatic compounds?

(II)   What are typical reagents and conditions for the preparation of fluorinated aromatic compounds by halogen exchange?

(III)   How can aromatic nitro-compounds and fluorosulphonyl-derivatives serve as precursors for fluoroarenes? Which reagents and conditions are required? Are there additional requirements on the substrate?

(IV)   How is C_6F_5Cl prepared and why is it an important building-block for the preparation of pentafluorophenyl compounds? Give examples.

(V)   Give examples for the direct functionalization of hexafluorobenzene and comment on the required reagents and conditions.

(VI)   Which methods may be used for the conversion of phenolic substrates to aryl fluorides? Comment on their importance and suitability.

(VII)   What is the Balz–Schiemann reaction? Which are its major benefits and disadvantages and which improvements can help to overcome the drawbacks?

(VIII)   Aryl fluorides may also be prepared involving diaryl iodonium or aryl sulphonium intermediates. Comment on major advantages and problems of these approaches.

14.3 Electrophilic fluorination

14.3.1 Elemental fluorine as electrophilic fluorinating agent

In principle, electrophilic fluorination of aromatic systems can be achieved using F_2 in high dilution. However, this approach is rather inconvenient, as the regioselectivity is rather low, resulting in the formation of product mixtures. For example, the electrophilic fluorination of nitrobenzene with F_2/N_2 results in the formation of a 1:9:1.5-mixture of the respective *ortho*-, *meta*- and *para*-fluoronitrobenzenes. Apart from that, substrates containing OH- or NH_2-substituents have to be fluorinated in anhydrous HF to prevent formation of OF- or NF_2-functions. Again, the regioselectivity of these transformations is low, even if some of the positions on the aromatic ring are blocked by other substituents. The fluorination of 4-hydroxybenzoic acid results in the formation of both, 4-hydroxy-3-fluorobenzoic acid and the *o*-difluorinated 4-hydroxy-3,5-difluorobenzoic acid. Nevertheless, the reaction of 1,4-disubstituted benzene derivatives with elemental F_2 may result in selective fluorination, if both substituents direct fluorine to the same site. For example, the treatment of 4-nitroanisole with F_2/N_2 exclusively yields 2-fluoro-4-nitroanisole [3152]. It has been found that acidic solvents, such as sulphuric or formic acid can serve as reaction media for electrophilic fluorination reactions using F_2 [3153]. The fluorination of aniline-derivatives in CF_3SO_3H mainly yields *meta*-substituted products. The presence of an electron-donating substituent in *para*-position further enhances the regioselectivity [3154]. The fluorination of benzaldehyde derivatives gives rise to mixtures of fluorobenzalde-hydes and benzoyl fluorides. The product formation also depends on the substituents on the arene-ring, since electron-donating groups preferentially lead to the substitution of the aromatic moiety, whereas electron-poor substrates are mainly converted to acid fluorides [3155]. The direct fluorination of arene substrates is not restricted to substituted benzenes, but has also been extended to more complex molecules, such as coumarins [3156]. Apart from that, direct fluorination has also been reported for acetyl-protected purine nucleosides [3157]. Generally, it can be concluded, that other methods are certainly more suitable for the electrophilic fluorination of arenes, as they are often more selective and additionally do not require the use of special laboratory equipment for handling elemental fluorine.

14.3.2 Electrophilic fluorination using fluorooxy reagents

A few very reactive reagents for electrophilic fluorination of arenes contain O-F bonds, such as CF_3OF, $Cs^+[SO_3OF]^-$ or the acetyl hypofluorites $CH_3C(O)OF$ and $CF_3C(O)OF$. Typically, they are prepared using F_2 and are also very hazardous reagents, which renders them rather inconvenient for fluorination reactions of organic sub-strates, especially if the preparation of larger amounts of products is required. Due to their high reactivity, these reagents generally show a rather poor selectivity, as for example, the reaction of benzene with CF_3OF yields a mixture of fluoroben-zene, *p*-difluorobenzene and trifluoromethoxybenzene. However, if some positions of the aromatic ring are blocked and the arene-moiety is deactivated, as in the case of *p*-trifluoromethylacetanilide, selective fluorination can be achieved. In the lat-ter example, monofluorination in *ortho*-position to the nitrogen-substituent is ob-served. The fluorination of activated substrates, such as *m*-dimethoxybenzene or 2,6-dimethoxyacetophenone, is achieved upon treatment with CF_3OF, which gives rise to 4-fluororesorcinol (Scheme 14.6). Moreover, a mixture of 2- and 4-fluororesorcinol is obtained, if the free alcohol is treated with $CsSO_4F$ in presence of BF_3, however, this method shows a rather poor regioselectivity and the yield is significantly lower. As a general trend, *ortho*-substitution is preferentially achieved upon fluorination. The same is observed for polycyclic aromatics which are to some extent addition-ally or even exclusively converted to geminal difluoro products (Scheme 14.6). Their formation may be rationalized by the occurrence of an addition-elimination mecha-nism.

 An alternative and selective approach toward fluoroaromatic molecules is the treatment of organometallic species with sources of electrophilic fluorine. For this purpose, CH_3COOF is a suitable reagent for the replacement of silyl-, germyl- or

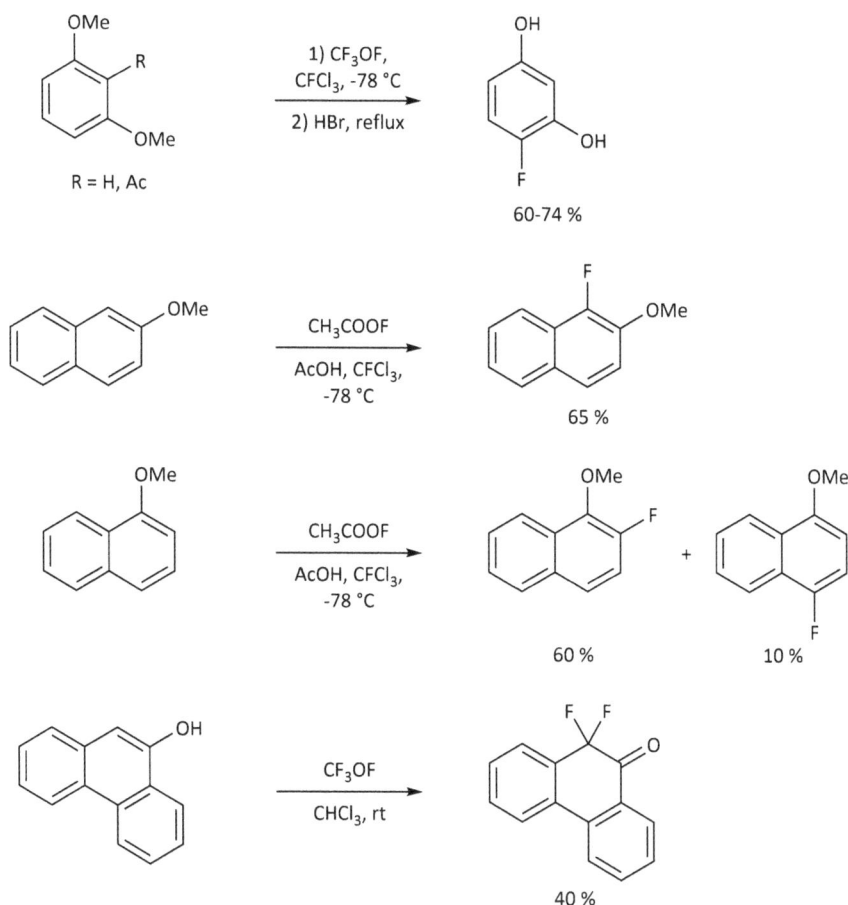

Scheme 14.6: Electrophilic fluorination of aromatic systems using fluoroxy reagents.

mercury-containing groups. However, the yields for these transformations are typically low; in addition, competitive substitution of aromatic protons can occur, especially with aryl silicon compounds. A high yield of p-fluoroanisole is obtained upon reaction of p-anisyl mercuric acetate p-$(CH_3O)C_6H_4HgOAc$ with CF_3OF. Furthermore, high yields of fluorinated products are achieved upon treatment of aryl tin compounds with CH_3COOF, $CsSO_4F$ or F_2. Electron-deficient aryl boronic acids or boronate esters are converted to aryl fluorides using AcOF. In most cases, moderate to high yields and a high regioselectivity are observed, however, some substrates give rise to a mixture of regioisomeric aryl fluorides [3158]. Apart from that, aryl boronic esters are fluorinated in moderate yield, if $CsSO_4F$ is used as fluorinating agent.

Perchloryl fluoride is a reagent, which is only scarcely used for the fluorination of arenes, however, it rapidly and efficiently converts aryl lithium species to the respective fluoroarenes.

14.3.3 Xenon difluoride as a source of electrophilic fluorine

XeF_2 is a rather mild reagent for electrophilic fluorination, which can be used up to room temperature, either without any catalyst in CH_3CN or in presence of catalytic amounts of HF or $BF_3 \cdot OEt_2$ in CH_2Cl_2 or $CHCl_3$. Other advantages, at least for small-scale preparations, are its commercial availability and that it can be principally handled in standard laboratory glassware. The reaction of mesitylene with XeF_2/HF yields the monofluorinated product, whereas a 72:4-mixture of *meta-* and *para*-monofluorinated aromatics is obtained in the reaction with trifluoromethylbenzene. XeF_2 may not only be used in pure form but also as a graphite intercalate. This is especially useful to control the fluorination of polycyclic arenes, such as anthracene or dibenz[*a,h*]anthracene. The treatment of anthracene with XeF_2 at -196 °C and subsequent slow warming to 12 °C results in the formation of all three possible monofluoroanthracene regioisomers in a ratio of about 5:1:3 [3159]. Instead, dibenz[*a,h*]anthracene is selectively fluorinated by the graphite intercalate $C_{19}XeF_6$ under similar conditions to yield exclusively 7-fluorodibenz-[*a,h*]anthracene. If the same transformation is carried out with pure XeF_2, a second, difluorinated compound is generated, too [3160]. XeF_2 is furthermore capable of fluorinating *para*-substituted aryl trimethylsilane substrates. However, reasonable product yields are only obtained if the reaction is carried out in C_6F_6; in addition, substrates with an electron-withdrawing group are not fluorinated [3161].

14.3.4 Electrophilic fluorination using fluoraza reagents

The N-F reagents can also be applied in substitution reactions of aromatic substrates. However, the selectivity of the fluorination reactions again strongly depends on the substrate structure and the activation or deactivation of the arene. Activated aromatic compounds are fluorinated by both, neutral and cationic fluoraza reagents. For example, the treatment of anisole with NFSI yields a mixture of all three possible monofluorinated regioisomers in an *ortho:meta:para*-ratio of 58:5:37, whereas acetanilide is fluorinated by Selectfluor™ to give the *ortho-* and *para*-fluorinated derivatives in a ratio of 62:38. The actual ratios of course strongly depend on the choice of reagents and reaction conditions. As already indicated by the ratios of regioisomers given above, *ortho*-fluorinated isomers are preferentially formed upon electrophilic fluorination of substituted arenes. The rather poor selectivity of this type of reaction is furthermore observed when 4-alkylphenols or 4-alkylphenylethers are reacted with a source of "F⁺". Typically, a mixture of 2-fluorinated arenes, 4-fluoro-2,5-cyclohexadienones and fluorodealkylated 4-fluoroarenes is obtained. The actual ratio of the products depends on the reaction conditions and the structure of the substrate [2860]. The fluorination of β-naphthol and N-(2-naphthyl)-acetamide with one equivalent of Selectfluor™ in ionic liquids yields the corresponding arenes with a fluorine substituent in 1-position. In contrast, 1,1-difluoronaphthalene-2(1*H*)-one is preferentially obtained, if the amount of fluorinating reagent is doubled. In both cases, the corresponding 3-fluorinated naphthalene derivatives are only generated in traces [3162]. Fluorinated pyrrole and pyrazole derivatives are accessible upon reaction with Selectfluor™ under microwave conditions [3163, 3164].

Fluoraza reagents are also suitable for the selective introduction of fluorine substituents into aromatic substituents of amino acids and steroidal structures, such as tyrosine and estradiol. *N*-fluoropyridinium salts do not only serve as fluorinating agents but can themselves be transformed to 2-fluoropyridines upon treatment with excess base, for example, NEt₃. This procedure is believed to proceed via a cyclic, N-F containing carbene species, which is then substituted by a fluorine atom from the counter anions of the pyridinium reagent, usually $[BF_4]^-$ or $[PF_6]^-$ [3165].

Moreover, fluoroarenes are accessible by the treatment of aryl lithium intermediates with fluoraza reagents. For example, the reaction of iodonaphthalene substrates with *tert*-butyllithium initially leads to halogen lithium exchange and subsequent trapping of the organometallic intermediate with NFSI gives rise to the corresponding fluorinated naphthalene [3166]. As a major drawback, this methodology cannot be applied to molecules with electrophilic substituents, such as carbonyl, cyano or nitro groups. Starting from a variety of aryl bromides and iodides, the fluorination of aryl lithium compounds has also been achieved in flow-microreactors [3167]. In general, the involvement of organolithium intermediates is especially popular for the fluorination of heteroaromatic substrates, such as thiophenes [3168–3172]. In some cases, the presence of a halogen substituent is not even required, and the substrate is selectively deprotonated by the alkyllithium reagent. Again, treatment with NFSI or other electrophilic fluorinating agents leads to the formation of fluorinated heteroarenes [3169, 3173, 3174]. Analogously, aryl bromides can serve as starting materials, when they are treated with either elemental Mg or Grignard reagents. As usual, NFSI is used as fluorinating reagent in a solvent mixture of dichloromethane and perfluorodecaline. The perfluorinated solvent is required to suppress radical pathways leading to the formation of side-products. Overall, a variety of functionalized aryl and heteroaryl substrates, including five-membered heterocycles, is converted to the corresponding fluorides [3175, 3176]. In a slightly different approach, a *N*-fluoro-2,4,6-trimethylpyridinium salt has been used as electrophilic fluorinating agent [3177]. The preparation of 2-arylfluoroarenes has been achieved, when aryl Grignard species were reacted with an aryne and subsequent trapping of the intermediary 2-(aryl)arylmagnesium compounds with an electrophilic source of fluorine [3178]. Fluoraza reagents may also be applied for the conversion of other organometallic species like aryl silanes. Aryl boronic acids or trifluoroborate derivatives are converted to aryl fluorides upon reaction with Selectfluor™. However, for most of the substrates, protodeboronated side-products occur in significant amounts [3179].

The direct C-H fluorination of arenes has been achieved by deprotonation with lithium alkylmagnesate bases Bu₃MgLi or Bu₄MgLi₂ and subsequent treatment of the aryl magnesium intermediates with NFSI. However, the presence of an *ortho*-directing group, for example, an amide-function, is strongly required [3180].

14.3.5 Study questions

(I)   Comment on the suitability of using elemental F_2 for the electrophilic fluorination of arenes. Which substrates may be fluorinated and which problems do typically occur?

(II) Which prerequisites need to be fulfilled by the substrate to become efficiently fluorinated by fluoroxy compounds? Are there general trends regarding the product formation?

(III) What are the major advantages of using XeF_2 for the electrophilic fluorination of arenes? How can its reactivity be controlled?

(IV) Do fluoraza reagents show a better selectivity in electrophilic fluorination reactions than for example F_2 or CF_3OF? Which factors generally influence the regioselectivity and which general trend is observed?

(V)   How can organometallic aryl species be exploited for the synthesis of fluoroarenes? Which substrates and reagents are typically used and which limits exist regarding the compatibility of functional groups?

14.4 Reductive aromatization

A convenient approach toward perfluorinated aromatic compounds is the reductive aromatization of the respective cyclic, fluorinated compounds using iron or iron oxide in a temperature range between 500 and 750 °C (Scheme 14.7) [3181]. This method is also applied on industrial scale and has the advantage that the starting materials are quite readily available and that the iron surface can be easily regenerated by treatment with H_2. Upon replacement of the iron-based materials by other reducing agents, the preparation of partly aromatized polycyclic compounds as well as partially reduced fluoroorganic compounds can be achieved. In addition, the reaction conditions for reductive aromatization are significantly milder, when other reductants are used [3181].

14.5 Metal-mediated reactions

Usually, the preparation of fluoroarenes requires rather drastic reaction conditions, which are in most cases incompatible with sensitive organic scaffolds or functional groups. Thus, the development of mild, selective and transition metal catalyzed fluorinations of, for example, aryl bromides, -triflates, -boronic acids or -tin organyls has been a great achievement, especially in terms of late-stage fluorination. Another important approach is the direct C-H-functionalization, which generally requires the presence of directing groups. In the last decades, a large number of methods has been described and reviewed, out of which a selection will be presented in the following

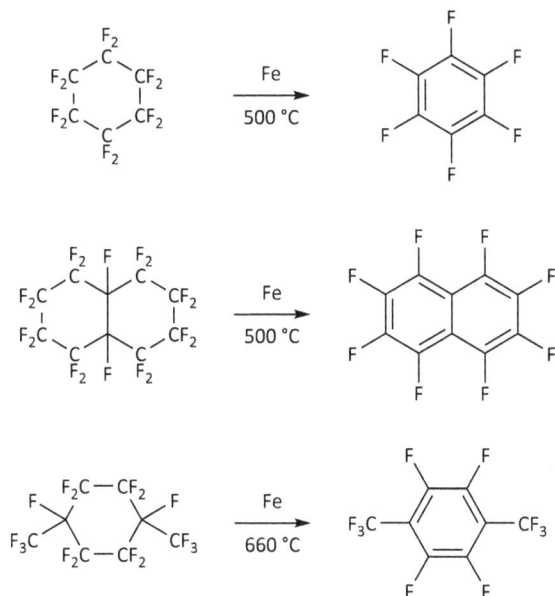

Scheme 14.7: Reductive aromatization of cyclic, perfluorinated compounds.

section [3040, 3041]. In general, fluorination reactions are always mediated by late transition metals, which is rationalized by the fact that the crucial C-F-bond-forming reductive elimination step does not take place with the early d-block elements due to high M-F bonding energies.

14.5.1 Reactions mediated by palladium

General aspects about Pd-mediated fluorination reactions

Before discussing specific metal-mediated approaches toward fluoroarenes, a few general aspects of this type of chemistry should be pointed out. The treatment of a bisphosphine Pd^{II}ArI complex with electrophilic N-fluoro-2,4,6-trimethylpyridinium reagents preferentially generates the aryl iodide upon reductive elimination; the corresponding fluoroarene is only formed in minor amounts (Scheme 14.8). However, arylpalladium(II) species, which may be generated from the corresponding Pd-acetate complex and an aryl boronic acid, have been shown to undergo fluorination with Selectfluor™, giving rise to the corresponding aryl fluoride and an unreactive Pd-complex in a stoichiometric reaction (Scheme 14.8) [3182]. Apart from that, the treatment of an N,N'-donor-stabilized Pd^{II}ArF-complex with XeF$_2$ produces aryl fluorides. Upon treatment of the Pd(II)-complex with XeF$_2$, an intermediate Pd(IV)-species has been isolated and characterized, which only produces the aryl-fluoride after further treatment with an electrophilic fluorinating agent. Thermal treatment of the Pd(IV)

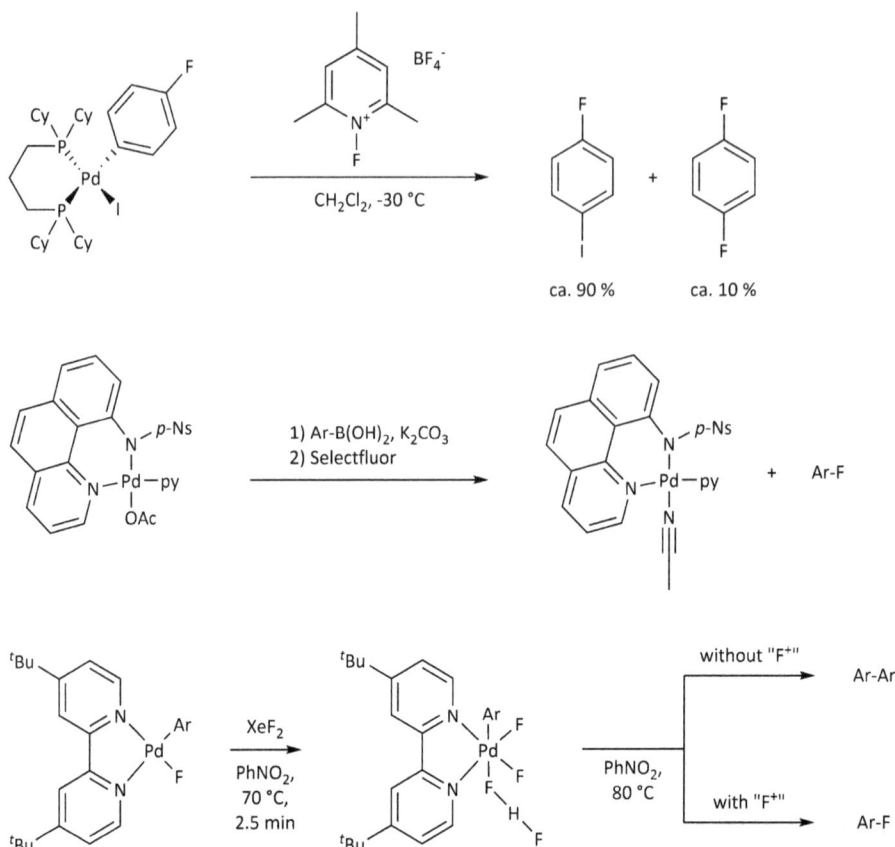

Scheme 14.8: Preferential formation of aryl iodides compared to fluorides from Pd^{II}-aryl species (top), stoichiometric generation of fluoroarenes from a Pd(II)-complex (middle) and investigation of the reductive elimination of aryl fluorides from an Pd(IV)-intermediate (bottom).

complex only liberates traces of ArF and mainly gives rise to biphenyl-derivatives (Scheme 14.8). Thus, the reductive elimination of aryl fluorides from Pd(IV)-complexes appears to be unfavorable. Hence, these findings give a hint, that Pd(IV)-species may be generally involved in the formation of C-F bonds, but they most likely do not release the fluorinated product upon classic reductive elimination [3183].

The majority of aromatic fluorination reactions, especially in terms of directed C-H functionalization, involves high-valent Pd(IV)-species. This type of reaction generally requires the use of an electrophilic fluorinating agent, which is capable of oxidizing Pd^{II} to Pd^{IV}. In addition, the defluorinated reagent subsequently serves as a base and takes up the liberated proton. A general catalytic cycle is depicted in Scheme 14.9 [3041]. Furthermore, C-F bonds are also generated, when arylfluorides are reductively eliminated from low-valent Pd^{II}-complexes. As a prerequisite for this pathway, a bulky ligand needs to be used, since the Pd(II)-fluoride complexes tend to form unreactive

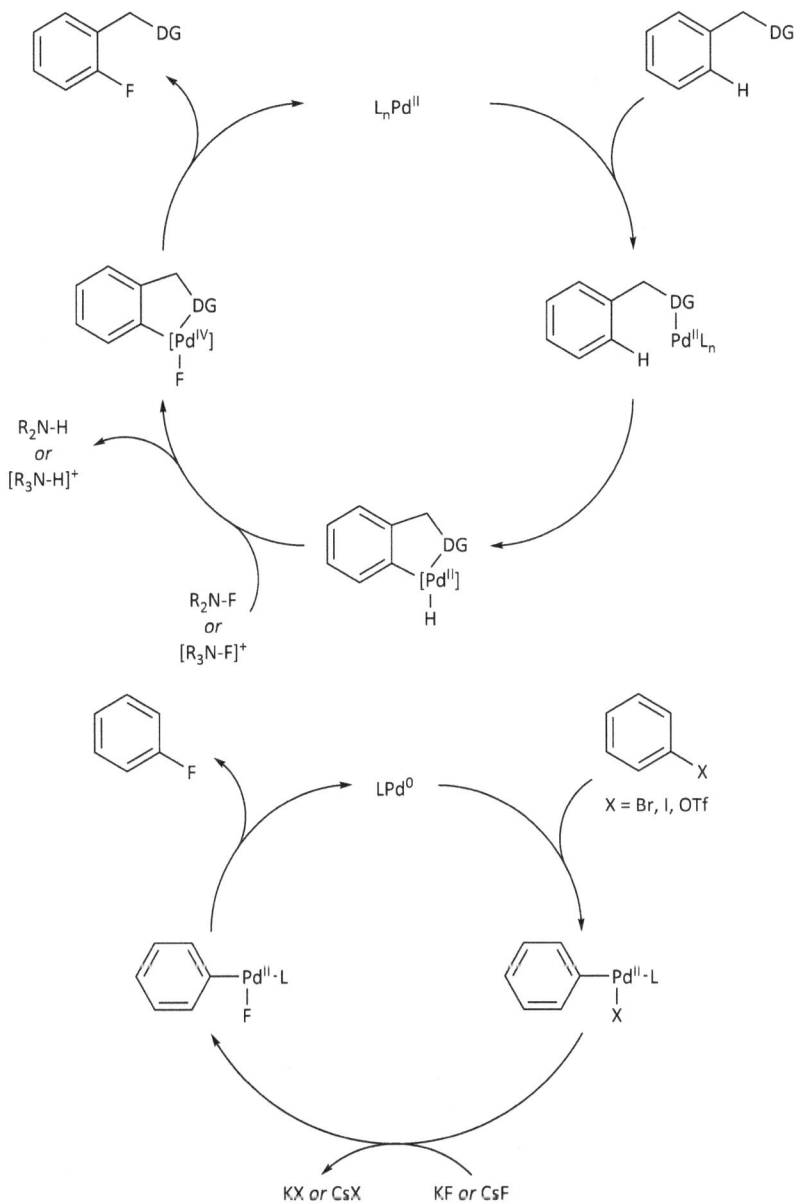

Scheme 14.9: General catalytic cycles for the Pd^{II}/Pd^{IV}-mediated C-H fluorination of arenes (top) and the corresponding Pd^{0}/Pd^{II}-mediated conversion of pre-functionalized substrates (bottom).

dimers otherwise. However, T-shaped, three-coordinate Pd-complexes have proven to be suitable intermediates, which undergo reductive elimination under formation of a C-F bond. In contrast to the aforementioned approach, this synthetic method cannot be applied for direct C-H fluorination of arenes, but requires the presence of a leaving group, typically Br⁻, I⁻ or OTf⁻. However, as a major advantage, more readily available nucleophilic fluoride sources, for example, CsF or KF can be used (Scheme 14.9) [3041]. Finally, a few Pd-mediated reactions have been described, where the fluorinated product is not formed in a reductive elimination step.

Pd(II)-mediated C-H fluorination in presence of directing groups

Palladium is certainly the most important transition metal for the direct introduction of fluorine into aromatic hydrocarbons. The selective fluorination of C-H bonds generally requires the presence of a directing group. For example, 2-aryl pyridines are converted to 2-(2-fluoroaryl) pyridines upon catalytic reaction with Pd(OAc)$_2$ and stoichiometric amounts of N-fluoro-2,4,6-trimethylpyridinium tetrafluoroborate under microwave-irradiation conditions (Scheme 14.10) [3043]. Analogously, other nitrogen heterocycles, such as quinoxalines, oxazoles, pyrazines and pyrazoles can serve to direct the Pd-catalyzed fluorination of arenes (Scheme 14.10). Monofluorination in *ortho*-position has efficiently been achieved using 1.5 equivalents of NFSI as fluorinating agent, whereas *ortho*-difluorination has been observed with the double amount of the reagent [3184]. A few more nitrogen heterocycles can serve as directing groups in the Pd-mediated C-H fluorination of arenes by electrophilic fluorinating reagents [3041].

However, unless they are not part of the target structure, heterocyclic substituents are rather inconvenient directing groups, as they can neither be easily removed nor converted to other functional groups. Thus, a more useful approach has been described, using a triflamide group as directing auxiliary, which can easily be converted to other functionalities, including aldehydes, nitriles and amines. Aromatic triflamide substrates are fluorinated in *ortho*-position, when they are treated with Pd(OTf)$_2$ · 2 H$_2$O, stoichiometric amounts of N-fluoro-2,4,6-trimethylpyridinium triflate and N-methyl-2-pyrrolidone (NMP) as additive (Scheme 14.11). However, without blocking one of the *ortho*-positions, a mixture of mono- and difluorinated products is obtained [3185]. This problem can be overcome when benzamide-substrates are used, which again react with a Pd(II)-salt, that is, Pd^{II}(OTf)$_2$(MeCN)$_4$ and N-fluoro-2,4,6-trimethylpyridinium triflate in presence of NMP (Scheme 14.11). Depending on the reaction conditions, selective mono- or difluorination is possible without any further structural requirements on the substrate. Performing the reaction in polar acetonitrile yields the *ortho*-monofluorinated benzamides, whereas difluorination is achieved when the reaction is carried out in nonpolar benzotrifluoride with a higher amount of fluorinating agent. Subsequent cleavage of the amide function gives access to *o*-fluorinated benzoic acid derivatives [3186]. In the same way, oxalylamide protected benzylamines can be reacted with Pd(OAc)$_2$ and NFSI to yield *ortho*-fluorinated products. Again, selective mono- or difluorination is achieved upon variation of the reaction conditions [3187]. Apart from that, some more C-H fluorinations have been reported, using other amide-variants as directing groups [3041].

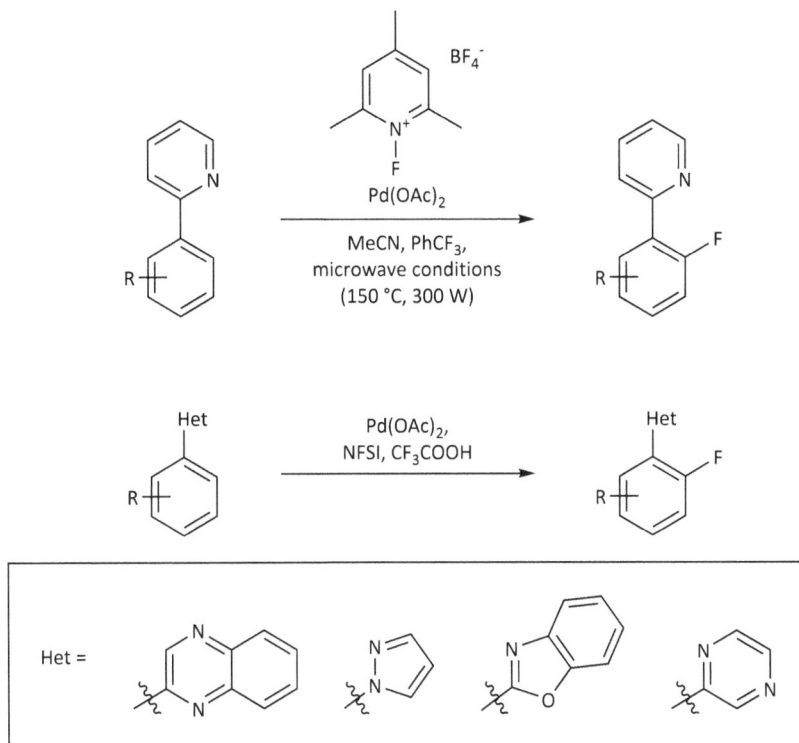

Scheme 14.10: Directed C-H fluorination of 2-pyridyl arenes (top) and other heterocycle-substituted substrates (bottom).

Selective *ortho*-monofluorination is furthermore achieved when a cleavable *O*-methyl oxime ether function serves as directing group, which is easily accessible from ketones or aldehydes (Scheme 14.12). The transformation of the oxime ether substrate is believed to be mediated by a cationic $[Pd(NO_3)]^+$ species and proceeds under mild conditions at 25 °C. In addition, this approach tolerates a number of functional groups [3188]. *Ortho*-monofluorinated phenol-derivatives are obtained when a cleavable 2-pyridyloxy-function is used as directing group. The treatment of 2-phenoxy pyridines with a source of Pd^0 and NFSI selectively introduces one fluorine-substituent into the aryl ring (Scheme 14.12). After the fluorination step, the directing 2-pyridyl group can be cleaved upon reaction with methyl triflate and subsequent treatment with sodium in methanol [3189]. In a more elegant way, the directing group does not need to be installed and removed separately, but is linked to the substrate *in situ*. This method has been used for the selective *ortho*-fluorination of a number of *ortho*- and *meta*-substituted benzaldehydes, using aniline-2,4-disulphonic acid as transient directing group (Scheme 14.12). The generated imine is fluorinated by *N*-fluoro-2,4,6-trimethylpyridinium triflate in a Pd^{II}-catalyzed reaction and releases the fluorinated carbonyl product upon hydrolysis. The method works well for electron-deficient, but

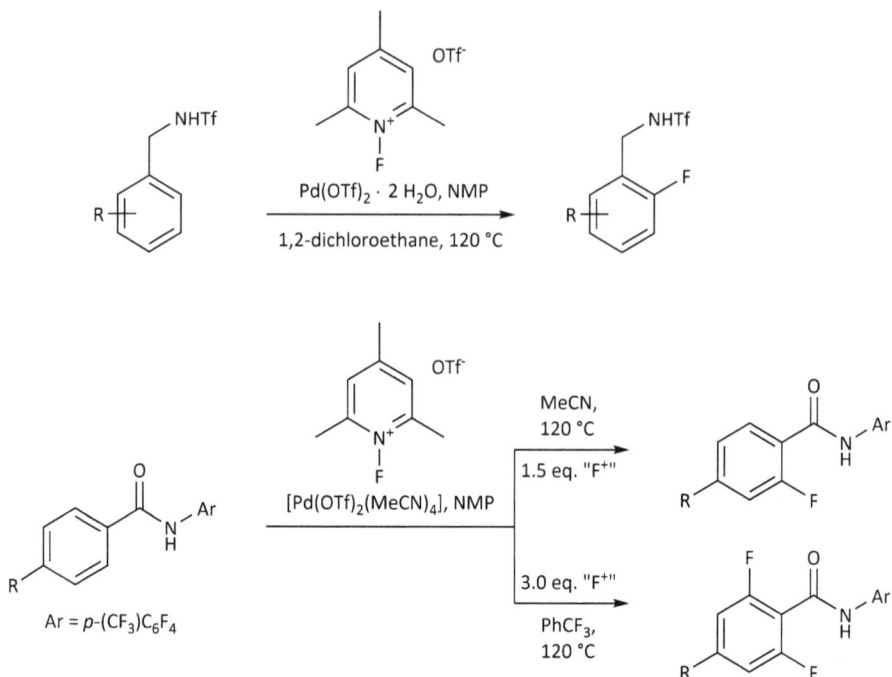

Scheme 14.11: Triflamide (top) and benzamide (bottom) directed C-H fluorination of arenes.

not for electron-rich substrates. Since the directing group contains sulphonic acid functions, the additional use of an acid is not required for the attachment and detachment of the directing group [3190].

Pd(0)-mediated fluorination

The formation of aryl fluorides can also be achieved by reductive elimination from Pd(II)-species. This method works well for a number of aryl triflate substrates using CsF as fluoride source. The Pd^0-catalyst is generated from [(cinnamyl)PdCl]$_2$ and a bulky biaryl monophosphine ligand L (L = BrettPhos or ^{t}BuBrettPhos). Upon treatment of the triflate with the Pd^0-species PdL, oxidative addition occurs to give LPd^{II}Ar(OTf), which is then converted to LPd^{II}Ar(F) by anion exchange with CsF. Finally, reductive elimination regenerates LPd^0 and liberates the fluorinated arene ArF (cf. Scheme 14.9) [3191]. A few variants of this approach have been reported, for example, the transfer from a batch reaction to flow conditions. The use of a reactor with a packed-bed of CsF provides a solution for both, the solubility issues of CsF in unpolar solvents as well as the waste of the rather costly alkali metal fluoride, which needs to be used in excess [3192]. Apart from that, the Pd^0-precursor in the batch reactions has been replaced by [(LPd)$_2$(cod)], which avoids the formation of ArCl side-products. In addition, the adamantyl-substituted AdBrettPhos is used instead of the ^{t}BuBrettPhos-

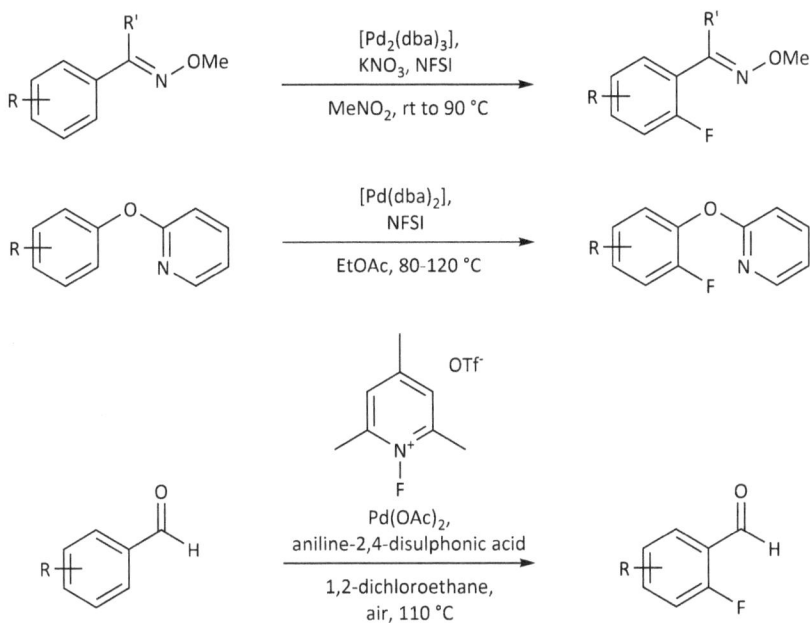

Scheme 14.12: Selective C-H fluorination using directing *O*-methyl oxime ether functions (top), 2-pyridyloxy substrates (middle) and a transient directing group (bottom).

ligand to enable a better reactivity toward highly electron-rich and heteroaryl substrates (Scheme 14.13) [3193]. The same catalyst can be used for the conversion of aryl bromides and iodides (Scheme 14.13). Further slight changes on the BrettPhos ligand have led to the extension of the substrate scope to aryl- and heteroaryl bromides and -iodides. In addition, the stoichiometric use of AgF is required for halide substrates to assist in the transmetalation step from LPd^{II}Ar(X) toward LPd^{II}Ar(F) [3194]. The latter procedure has also been used for the fluorination of five-membered heteroaryl bromides, however, the phosphine ligand needs to be adapted again [3195]. Finally, the optimized ligand allowed the conversion of aryl triflates and -bromides under mild conditions at room temperature, which diminishes side reactions and provides the fluorinated products with high regioselectivity [3196]. In a two-step one-pot procedure, a variety of aromatic alcohols has been converted to fluoroarenes via intermediary aryl nonaflates (Scheme 14.13). Initially, the alcohol is treated with a mixture of $C_4F_9SO_2F$ and CsF to generate $ArOSO_2C_4F_9$, which then oxidatively adds to a Pd^0-catalyst. Replacement of the nonaflate anion by F$^-$ is followed by reductive elimination of the aryl fluoride product [3197].

Miscellaneous variants of Pd-catalyzed fluorination

In both, general organic chemistry and also fluoroorganic synthesis, transformations involving Pd(III)-species are very rare. However, an example has been reported,

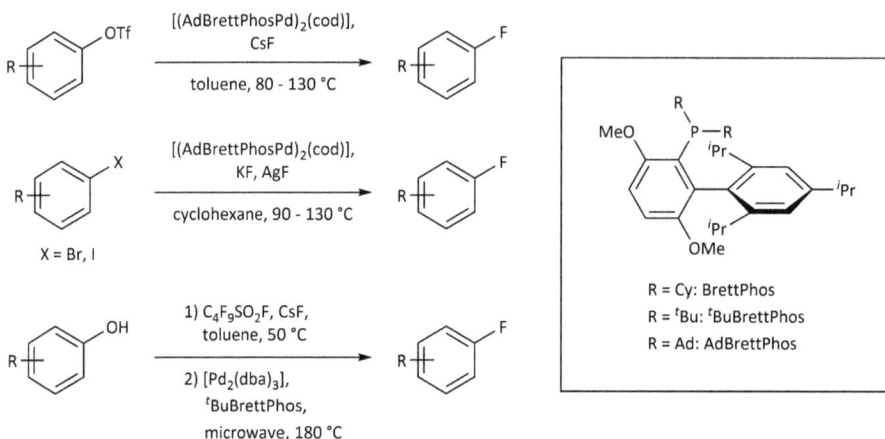

Scheme 14.13: Pd-mediated selective fluorination of aryl triflates (top), bromo- and iodoarenes (middle) as well as phenols (bottom).

where aryltrifluoroborates are converted to fluoroarenes using a palladium(II) terpyridine complex in conjunction with Selectfluor™ (Scheme 14.14). The reaction is proposed to start with a turnover-limiting oxidation of the Pd(II)-species by the oxidant. Afterwards, a fluoride radical is transferred from the fluorinating agent to the ipso-carbon of the aryl trifluoroborate substrate. Finally, single-electron transfer (SET) from the aryl radical to the metal-center regenerates the Pd^{II}-terpyridine complex and yields the fluoroarene upon liberation of BF_3 [3198]. Another approach has been described more recently and includes the nondirected C-H fluorination of arenes with a catalyst composed of a Pd(II)-center, a terpyridine-ligand (terpy) as well as 2-chlorophenanthroline (2-Cl-phen) (Scheme 14.14). Again, Selectfluor™ serves as fluorinating agent. The reaction proceeds without the formation of organometallic intermediates and typically generates regioisomers. However, substituents on the arene ring can influence the product distribution. Mechanistically, this transformation is believed to be initiated by an oxidation of the square planar $[(terpy)Pd(2\text{-}Cl\text{-}phen)]^{2+}$-complex by the fluorinating agent to give an octahedral $[(terpy)PdF(2\text{-}Cl\text{-}phen)]^{3+}$-species. The latter compound is a stronger electrophilic fluorination agent and substitutes even weakly nucleophilic arenes. Thus, this method contrasts the non-catalyzed electrophilic fluorination using NFSI or Selectfluor™, which only works for electron-rich substrates (cf. Section 14.3.4) [3199].

14.5.2 Reactions mediated by copper

An important method for the preparation of fluoroarenes, which is also applied on industrial scale, is the copper catalyzed fluorination in presence of O_2, where HF is used as fluorine source (Scheme 14.15). Upon reaction of arenes with CuF_2 at 450-550 °C,

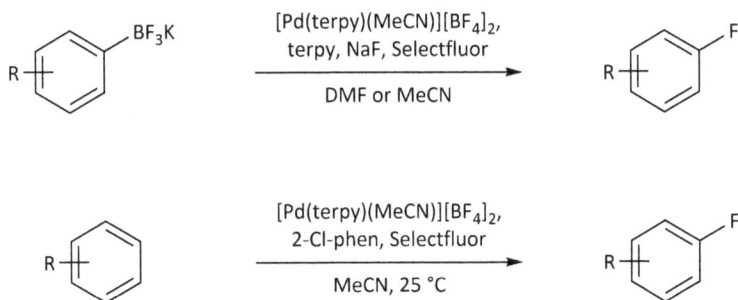

Scheme 14.14: Pd-mediated fluorination of aryltrifluoroborates (top) and nondirected C-H fluorination of arenes (bottom).

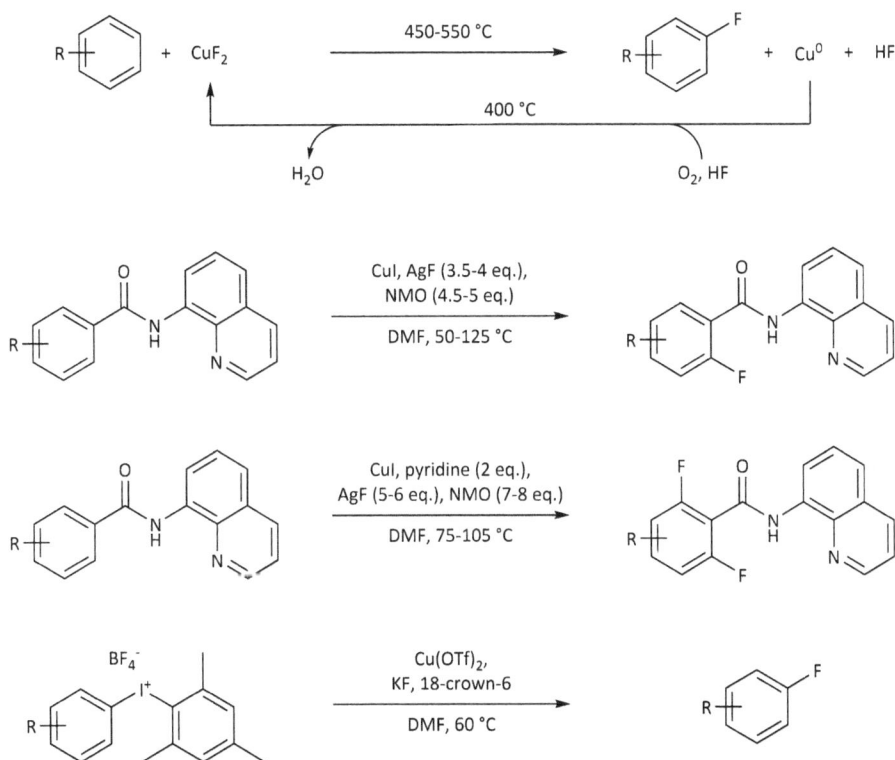

Scheme 14.15: CuF₂-catalyzed fluorination of simple arenes (top), directed mono and difluorination of benzamide-derivatives (middle) and fluorination of diaryliodonium salts (bottom).

the metal fluoride is reduced to elemental copper, generating the fluorinated aromatic compound and HF. The treatment of Cu with HF and O_2 at 400 °C regenerates CuF_2 under formation of H_2O, which is overall the only stoichiometric by-product. This method can generally be applied for the synthesis of fluorobenzenes, -toluenes but also diflu-

orobenzenes. The major advantages of this procedure include the retention of the activity of the copper catalyst after the reaction-regeneration cycles as well as the waste-
efficiency, since only H_2O is evolved [3200]. A more sophisticated approach toward
ortho-mono- or difluorinated derivatives of N-(8-quinolinyl) benzamide has been described using catalytic amounts of CuI in conjunction with stoichiometric quantities
of pyridine, AgF and N-methylmorpholine-N-oxide as oxidant (Scheme 14.15). The reaction temperature and of course the reagent equivalents determine, if mono- or difluorinated products are obtained. In addition, the directing group can be removed by alkaline hydrolysis, giving rise to the corresponding fluorinated benzoic acid derivatives
[3201]. Recently, this method has been modified for ^{18}F-radiofluorination, too [3202].

Furthermore, a Cu-catalyzed fluorination takes place, when diaryliodonium salts
are reacted with $Cu(OTf)_2$ in presence of KF and 18-crown-6 (Scheme 14.15). This
method gives rise to fluoroarenes in high yield and is proposed to proceed via a
Cu^I/Cu^{III}-catalytic cycle. Indeed, the fluorination of the substrate takes also place in
absence of the copper-catalyst (cf. Section 14.2.6), however, a mixture of fluoroarenes
is obtained under these conditions. The selectivity of the metal-catalyzed reaction is
rationalized by the oxidative insertion of Cu(I) into the less crowded carbon-iodine
bond followed by reductive elimination from a Cu(III)-fluoride species [3203]. Aryl
boronate esters can also serve as starting materials for a copper-mediated fluorination,
when they are additionally treated with AgF and an electrophilic N-fluoropyridinium
reagent. Moreover, the fluorination step has been combined with a preliminary bo-
rylation of aromatic C-H bonds or aryl bromides, applying Ir- or Pd-catalysis, respectively. However, this fluorination method requires an increased amount of copper- and
silver-salts [3204]. In a comparable approach, aryl stannanes and aryl trifluoroborate
substrates are converted to aryl fluorides, when they are treated with a Cu(I)-catalyst
and an electrophilic N-fluoropyridinium reagent [3205]. Further improvements on
the conversion of aryl trifluoroborate substrates have led to the replacement of the
electrophilic fluorinating reagent by nucleophilic KF, but in turn required the use of a
larger amount of Cu(II)-salt [3206]. The halogen exchange of aryl chlorides, bromides
and iodides toward fluorides is rather easily achieved in macrocyclic substrates in
presence of stoichiometric or catalytic amounts of Cu(I) and AgF as fluorinating agent
[3207]. In a similar system, direct C-H fluorination has been achieved using $Cu(ClO_4)_2$
and KF [3208]. Apart from that, aryl iodides are converted to fluorides, when they are
reacted with stoichiometric amounts of a Cu(I)-salt and AgF [3209]. A directed halogen exchange has been described for 2-pyridyl-2′-bromophenyl substrates, which are
fluorinated in presence of a Cu(I)-catalyst and AgF [3210]. Recently, a directed fluorination of aryl halides with KF has been described, which is mediated by a N-heterocyclic
carbene-copper(I) complex and requires a pyridyl- or another N-donor function in *ortho*-position of the halide. Furthermore, 4-dimethylaminopyridine has been used as
additive to enhance product yields. This method is also suitable for the preparation
of ^{18}F-labeled arenes [3211].

A few other Cu-mediated methods have also been described, leading to more complex fluorinated products. For example, aryl enynes are transformed to polycyclic fluorinated compounds using a cyclization/fluorination approach. The reaction is believed to proceed via Cu^0/Cu^{II}-catalysis and requires stoichiometric amounts of Selectfluor™ as oxidant and fluorine source [3212].

14.5.3 Reactions mediated by silver

Aryl fluorides are accessible by a silver(I)-mediated fluorination of arylstannane substrates, using stoichiometric amounts of AgOTf and the PF_6-salt of Selectfluor™ (Scheme 14.16). This method is believed to proceed via arylsilver intermediates, which are oxidatively fluorinated by Selectfluor™. Upon reductive elimination, the fluoroarene is formed and an inactive silver(I) species is recovered [3213]. This method has been further modified, so that the silver(I)-source Ag_2O is only required in catalytic amounts. It has been shown, that late-stage fluorination of complex small molecules can be achieved using this fluorination approach [3214].

In a similar way, arylboronic acids and aryl trifluoroborates are converted to arylfluorides using a silver(I)-salt and Selectfluor™ as fluorinating agent (Scheme 14.16) [3215, 3216]. Aryl trialkoxysilanes can serve as prescursors for aryl fluorides, too, since they are fluorinated using Ag_2O and Selectfluor™ in presence of the additive BaO. This approach tolerates a larger number of functional groups but completely fails with trialkylsilyl-substituted starting materials [3217].

A direct, *ortho*-selective C-H fluorination of pyridine and pyrimidine heterocycles has been described using AgF_2 in a reaction at room temperature (Scheme 14.16). Mechanistically, this approach is proposed to proceed via coordination of AgF_2 to the Lewis-basic donor-atom of the substrate. Afterwards, [Ag]-F adds to the π-system and generates a silver amido complex, which is converted to the fluorinated heteroarene upon removal of a hydrogen atom. Along with the product, two equivalents of AgF and one molecule HF are liberated (Scheme 14.16). This method selectively produces monofluorinated compounds and tolerates substrates, which are substituted in 2-, 3- or 4-position [3218].

A few reactions have been reported, which proceed by gold(I)- or silver(I)-mediated intramolecular aminofluorination of alkynes and thus yield fluorinated, nitrogen-containing heterocycles, such as pyrazoles and isoquinolines [3219, 3220].

14.5.4 Reactions mediated by bismuth

Only recently, the conversion of aryl boronic acid derivatives to fluoroarenes by a Bi(III)/Bi(V)-redox couple has been described. Interestingly, the main group metal bismuth is capable of undergoing the same organometallic reactions like transition

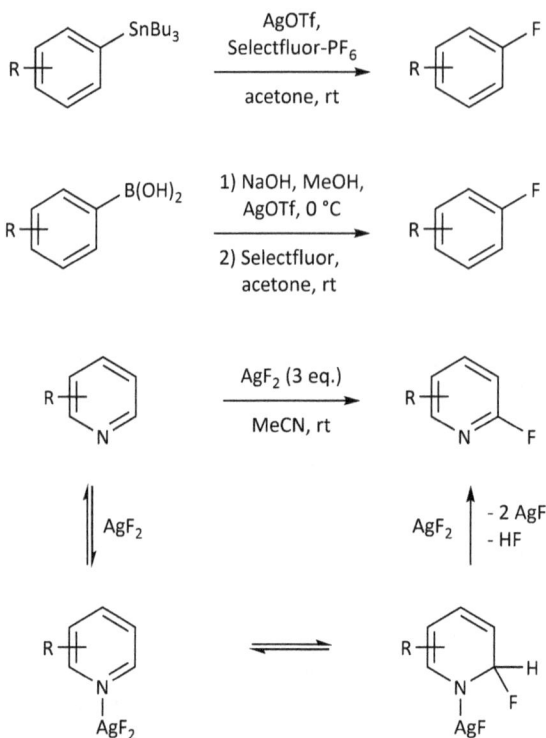

Scheme 14.16: Silver-mediated fluorination of aryl stannanes and aryl boronic acids (top) and fluorination of pyridine derivatives using AgF_2 with proposed mechanism (bottom).

metals, that is, transmetallation, oxidative addition and reductive elimination. Likewise, an electrophilic fluorinating agent is required for the oxidation of Bi(III) to Bi(V) as it is also necessary, when Cu-, Ag- or Pd-redox couples are used (see above). In general, a large variety of aryl boronic acids and corresponding pinacol boronic esters has been fluorinated by this approach using either stoichiometric or catalytic amounts of the catalytically active bismine (Scheme 14.17) [3221].

14.5.5 Study questions

(I)   Which two major approaches need to be distinguished in terms of Pd-mediated synthesis of fluoroarenes? Which substrates and reagents are generally used? Are there other important things, which one needs to be aware of when preparing aryl fluorides by Pd-catalysis?

(II)   Which functions are typically used as directing groups for the direct C-H fluorination of arenes? Comment on their major advantages and disadvantages.

(III) What are the main benefits and drawbacks of the Pd-mediated fluorination of aryl bromides, iodides and triflates? Comment on the availability of the required Pd-catalysts and reagents.

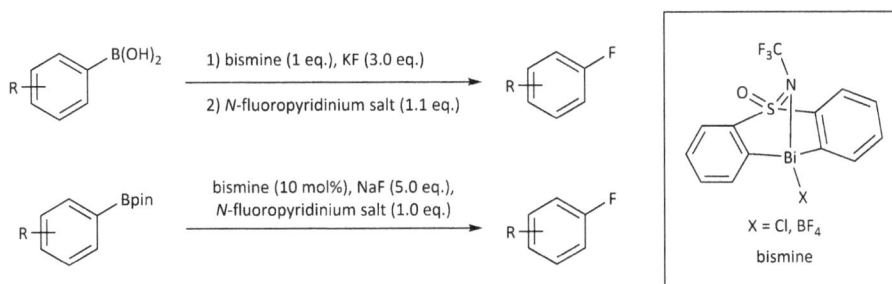

Scheme 14.17: Bi-mediated stoichiometric (top) and catalytic fluorination (bottom) of aryl boron substrates.

(IV)   What is the major advantage of the direct fluorination of arenes by CuF_2? To which substrates is it typically applied and which reagents and conditions are used?

(V)   Which substrates are generally suitable for a Cu-mediated fluorination? Which reagents are required for these approaches?

(VI)   What is most likely the role of Ag(I) in the fluorination of aryl stannanes? Which fluorinating reagent is required? Is there a similar reaction with other substrates?

(VII)   What are the most surprising findings about the bismuth-mediated fluorination of aryl boronic acid derivatives? Which fluorinating reagent is used?

14.6 C-F bond activation

An important feature of perfluorinated aromatic compounds, for example, C_6F_6, is the orthogonal reactivity with respect to hydrocarbons like benzene. As the charge-distribution in highly fluorinated arenes is inverted, meaning that the most positive charge is found in the center of the π-system, these molecules are generally susceptible toward nucleophilic attacks but not to electrophilic substitution. In contrast, benzene is readily substituted by electrophiles but not by nucleophilic species, as the center of its aromatic system is negatively charged. Apart from the charge-density in perfluoroaromatics, steric repulsions play a role in the increased reactivity toward nucleophiles, for example, fluoride ions. In the aromatic system, the p-orbital of the sp^2-carbon atom repulsively interacts with the lone-pairs of the fluorine substituents. This interaction is reduced, if the hybridization of the carbon-atom is changed toward sp^3, which is realized by a nucleophilic attack. The obtained difluoromethylene unit is furthermore stabilized by negative hyperconjugation effects.

14.6.1 Replacement of fluoride substituents in highly fluorinated aromatics

For the preparation of fluoroaromatics with a certain substitution pattern, it is often suitable to start from highly- or even perfluorinated substrates and selectively replace fluoride substituents by other functionalities. For the replacement of fluoride sub-stituents, some general trends in reactivity and selectivity can be observed, as for example, the second nucleophilic substitution of C_6F_6 always takes place in *para*-position. Hexafluorobenzene is readily substituted by a dimethylamino-group upon treatment with a mixture of the amine in isopropanol. In the next step, a nucleophile, which is represented by the methylthiolate anion in the present case, selectively at-tacks the *para*-position of the dimethylamino-substituted fluoroarene (Scheme 14.18) [3222]. This selectivity can be rationalized in terms of stabilization effects, as the ini-tially generated fluoroarene-nucleophile adduct is best stabilized by *o*-fluorine sub-stituents, followed by fluorine in the *meta*-position. Thus, the attack in *para*-position yields an intermediate, which is stabilized by two *ortho*- and two *meta*-fluorine sub-stituents. In case of pentafluoropyridine, some other trends in reactivity and selectiv-ity are observed (Scheme 14.18) [3223]. Bromination with $HBr/AlBr_3$ initially replaces the fluorine substituents in *ortho*- and *para*-positions, whereas the *meta*-fluorine substituents remain unchanged. Starting from the intermediate 2,4,6-tribromo-3,5-difluoropyridine, two different synthetic routes can be followed in terms of further nucleophilic substitutions. The addition of a soft nucleophile, for example, thio-phenolate PhS⁻ leads to the replacement of the *para*-bromine exclusively. Instead, hard nucleophiles, such as methoxide, will substitute the fluorine substituents in *meta*-positions. Depending on the stoichiometry, either one or both of them are re-placed. It is also possible to remove the bromo-substituents in 2,4,6-tribromo-3,5-difluoropyridine by hydrogenation, which then gives rise to 3,5-difluoropyridine.

In principle, it is also possible to replace fluorine in aromatic compounds that are less active and bear fewer fluorine atoms. Often, such substrates are used for C-C coupling reactions, but of course also for the generation of carbon-heteroatom bonds. However, these substitution reactions generally require higher reaction temperatures or the application of transition-metal catalysts. Since this approach typically yields nonfluorinated molecules, it will not be covered in this manuscript.

14.6.2 Fluoride replacement by main-group organometallics

Fluoroaromatic compounds are also activated by the use of organometallic reagents, especially lithium organyls, which are sometimes used in conjunction with potas-sium alcoholates [3224, 3225]. In case of monofluorinated substrates, directed *or-tho*-lithiation is the main reaction pathway at low temperatures, which may be used for coupling reactions but also for further functionalization [3226]. In general, the acidity of aromatic C-H bonds is increased, if electron-withdrawing substituents are present. The site selectivity can further be explained by stabilization, which occurs by electron-donating interactions of the *ortho*-substituent toward the metal atom. As a consequence, very efficient *ortho*-directing groups include, for example, alkoxy- and dialkylamido-groups, but also fluorine-substituents [3227, 3228]. However, one

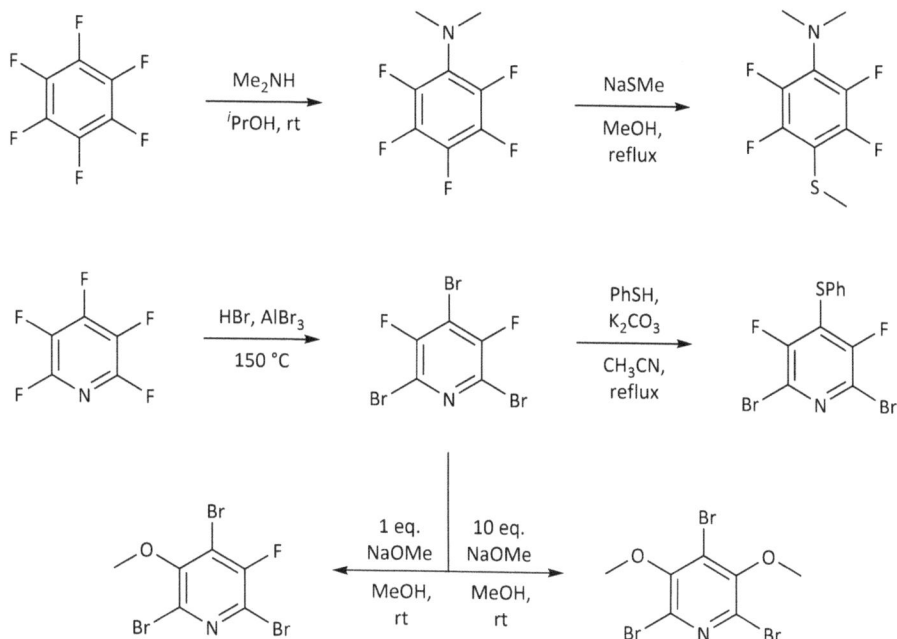

Scheme 14.18: Nucleophilic aromatic substitution of perfluorobenzene and -pyridine.

has to be aware of the fact, that *ortho*-metalation needs to be performed at low-temperature to obtain kinetic intermediates. Upon warming above −30 °C, the thermodynamically unstable derivatives will eliminate the metal fluoride and generate reactive aryne compounds, which immediately polymerize, if they are not trapped by other reagents (Scheme 14.19) [3229]. Nevertheless, the controlled defluorination approach via transient aryne-intermediates can give rise to a variety of interesting transformations, including intramolecular cyclizations [3119]. For example, lithiation of highly fluorinated substrates like C_6F_5Cl gives access to partly fluorinated naphthalene (Scheme 14.19) [3230].

Ortho-metalation is an important tool for the functionalization of partly fluorinated arenes, however, the outcome of the reaction strongly depends on the choice of the base but also on the solvent (Scheme 14.20). As shown in the example, the stronger base butyllithium will rather lead to the halogen-metal exchange of aryl bromides, whereas lithium diisopropylamide (LDA) preferentially results in *ortho*-lithiation and thus the abstraction of protons [3228]. The other representative transformation demonstrates the effect of the solvent, as the same substrate leads to different products in Et_2O and THF. This can be rationalized by a transmetalation, which occurs in THF, but not in less coordinating Et_2O [3231, 3232]. The tendency of lithiated fluoroarenes to undergo transmetalation may be explained by a gain of stability, which is achieved by two stabilizing *ortho*-substituents in the present example [3233]. In some cases, it is

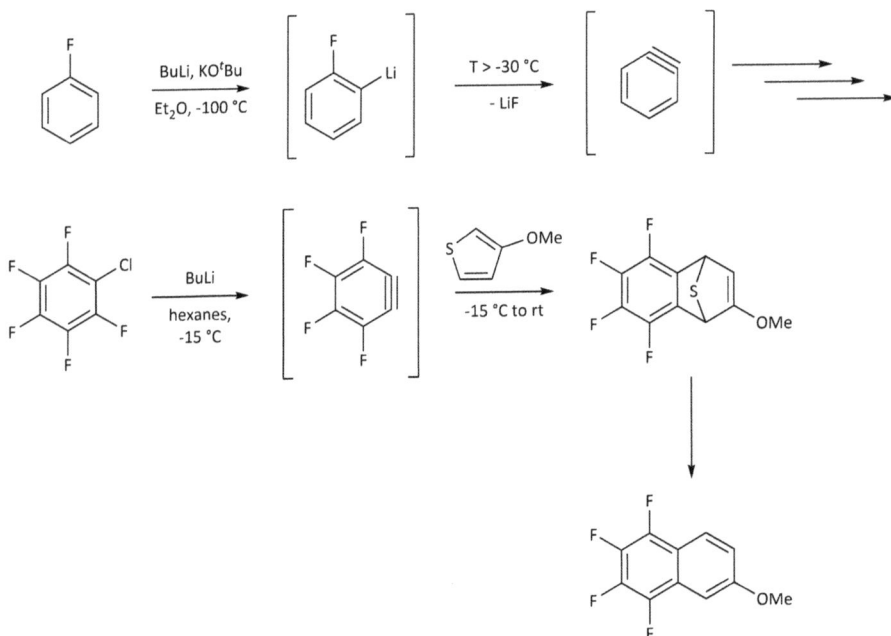

Scheme 14.19: Derivatization of fluoroarenes by *ortho*-lithiation.

necessary to block certain reactive positions of the aromatic ring by the introduction of protective groups, as desired substitution patterns might otherwise not be accessible.

14.6.3 C-F bond activation by transition metal complexes

Fluoroaromatic substrates are also activated by transition-metal complexes. This plays again a big role in the field of C-C coupling reactions, but can furthermore serve for the selective preparation of fluorinated arenes, which are hardly accessible by other methods. In general, highly fluorinated substrates, such as perfluorobenzene, -toluene or -pyridine are far more susceptible to C-F bond activation, whereas the transformation of mono- or difluorinated arenes is more difficult. Of course, catalytic C-F bond activation reactions require stoichiometric amounts of additives, which regenerate the catalytically active transition metal complex. These reagents are often based on silicon, aluminium or other elements that exhibit strong element-fluorine bonds. The certainly most widely explored approach in transition-metal mediated C-F bond activations are hydrodefluorination reactions, which typically proceed in presence of early transition metals, for example, titanium, zirconium, scandium or the rare earth metals. In contrast, the late transition metals, such as nickel, platinum, palladium or rhodium are commonly used in transformations, which either lead to C-C coupling or the formation of carbon-heteroatom connections. In general, transition-

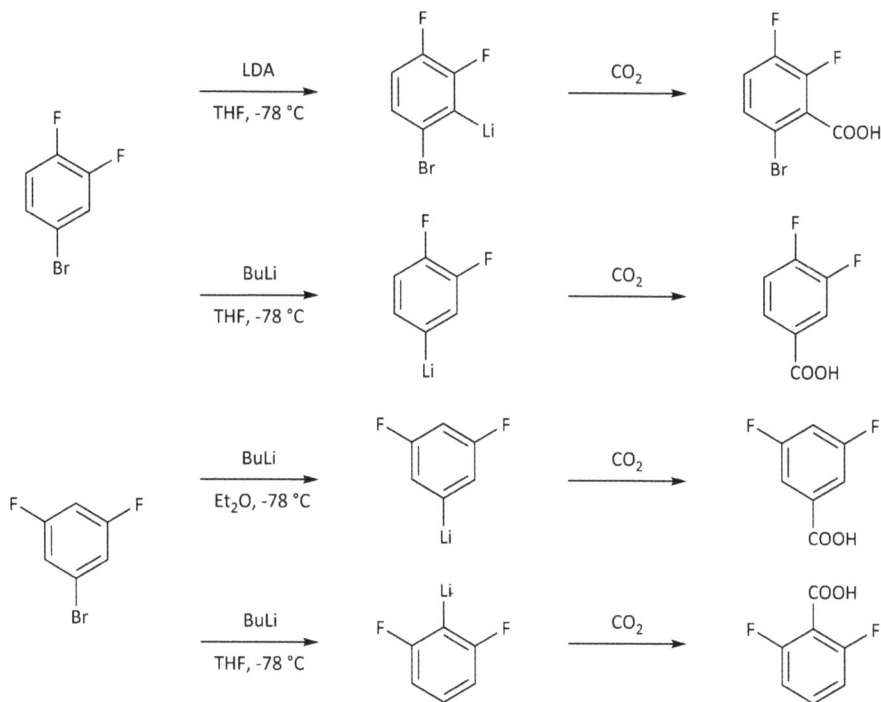

Scheme 14.20: Lithiation and carboxylation of fluoroarenes under different conditions.

metal mediated C-F bond activation has been extensively studied and reviewed over the past decades [3119, 3234]. However, it still develops rather quickly, especially in terms of the conversion of fluorinated greenhouse-gases and other potentially harmful waste-products into valuable fluorinated building blocks.

Nevertheless, the selective C-F bond activation can be challenging, especially in partly fluorinated arenes, which contain both C-H- and C-F bonds. This topic has been reviewed some time ago, giving also a lot of details on mechanisms and energetics, which have been partly obtained by quantum-chemical methods [3235]. A few general tendencies for the preference of either C-H- or C-F activation may be summarized as follows. In general, the preference for the one or other activation depends on the metal complex and, of course, the substrate. In partly fluorinated arenes, the strength of the C-F bonds decreases with the number of *ortho*-fluorine substituents, whereas C-H-bonds are getting stronger with fluorine in direct proximity. Early transition-metals with mostly d^0-configuration as well as lanthanides usually tend to give C-F-activation products. The situation gets more complex, when metals with partially or completely filled d-shells are regarded. In the first row of transition metals, C-F activation is mainly observed in presence of Co- and Ni-centers, whereas Rh is the only well-studied metal in the second row, which rather tends to activate C-H bonds, but occasionally also does so with C-F bonds. The C-F activation by third-row transition metals is only scarcely described. A few general trends are also observed in terms of regioselectivity, as for example *ortho*-fluorine substituents favor the selective activation of C-H bonds. This might be surprising, since this type of C-H-bond is the strongest in fluoroarenes. However, the reason for this can be found in the even larger strength of the metal-carbon bond in the activation product [3235].

Hydrodefluorination

For the synthesis of partly fluorinated arenes, it is often suitable to start from highly or even perfluorinated substrates and selectively replace fluorine substituents in hydrodefluorination reactions [3236, 3237]. This can be achieved by rather simple approaches as described above or by the application of more sophisticated, transition-metal mediated variants. For example, the treatment of polyfluoroarenes with the zirconium complex $Cp_2^*ZrH_2$ (Cp^* = pentamethylcyclopentadienyl, $[C_5(CH_3)_5]^-$) leads to the formation of hydrodefluorinated arenes [3238]. However, since most of the transition metals are rather expensive, catalytic transformations are desirable for both, economic and ecological reasons. Regarding the aforementioned example involving a zirconium hydride species, a catalytic reaction is observed, when stoichiometric amounts of an alane, that is, diisobutylaluminiumhydride (DIBAL) are added to a mixture of a zirconocene-derived hydride and a fluorinated arene [3239]. To date, numerous varieties of catalytic hydrodefluorinations have been described, using a number of different transition metals. For example, an *N*-heterocyclic carbene complex of ruthenium has been applied in conjunction with triethylsilane to selectively defluorinate C_6F_6, C_6F_5H or C_5F_5N [3240]. Catalytic C-F bond activation with subsequent introduction of a hydrogen-substituent is also achieved with Au-, Cu-, Co- or Fe-based complexes [3241–3245]. The reaction of several perfluorinated arenes with catalytic amounts of titanocene difluoride and stoichiometric Ph_2SiH_2 leads preferentially to *para*-hydrodefluorinated products [3246]. The same system can be used for the selective hydrodefluorination of more complex 4-amino-substituted fluoropyridines [3247, 3248]. Treatment of pentafluorobenzene with the rhodium catalyst $[RhC_6F_5(PMe_3)_3]$ and stoichiometric amounts of $(EtO)_3SiH$ as hydride source selectively yields 1,2,4,5-tetrafluorobenzene [3249]. With the same rhodium complex, C_6F_6 is converted to C_6F_5H in presence of a base under elevated H_2-pressure [3250]. A modification of the catalyst allows the synthesis of pentafluorobenzene under atmospheric H_2-pressure [3251]. More recently it has been found that the presence of a transition-metal complex is not necessarily required for the C-F bond activation. In the presence of an organocatalyst, that is, *O*- or *N*-donor molecules, the hydrodefluorination of polyfluoroarenes readily proceeds with aluminium and gallium hydrides [3252, 3253].

Moreover, selective hydrodefluorination reactions can proceed under reductive conditions, using stoichiometric amounts of a reductant, such as zinc, in conjunction with a transition metal complex. For example, the treatment of pentafluorobenzoic acid with Zn in presence of catalytic amounts of $NiCl_2$ and a *N*-donor ligand like

2,2′-bipyridine preferentially leads to a fluoride exchange in *ortho*-position, yielding either the mono- or disubstituted product, depending on the amount of catalyst [3254]. The selective fluoride abstraction from *N*-acetyl protected polyfluorinated aromatic amines does not require a catalyst, if the reductions are carried out with Zn in aqueous ammonia. The amine-function directs the fluoride abstraction to occur almost exclusively in *ortho*-position [3255]. In a similar approach, C_6F_5COOH, $C_6F_5CH_2OH$, C_5F_5N and heptafluoro-2-naphthoic acid are substituted in *para*-position, when they are treated with Zn in aqueous NH_3-solution [3256, 3257].

Formation of carbon-heteroatom bonds

The C-F bond activation of fluorinated aromatic and heteroaromatic compounds with late transition metals is typically used in C-C-coupling reactions [3234]. The selective derivatization of polyfluoroarenes with other aryl or alkyl substituents is a valuable tool for the preparation of interesting, functional molecules. Of course, this approach also works for substrates with fewer fluorine substituents, which will then mostly lead to nonfluorinated products. Often, the selectivity of the activation is additionally enhanced by the presence of directing groups, such as imines, ketones, or pyridines. The coupling reactions include mostly Kumada–Tamao, Suzuki–Miyaura and Negishi-type couplings and will not be discussed herein. However, there are a few cases, where the selective arylation or alkylation of fluoroarenes has been extended by the introduction of another functional group. For example, pentafluoropyridine has been activated and functionalized to give acylated products. The treatment of C_6F_5N with a zerovalent Ni-compound and PEt_3 leads to the activation of a fluoride-substituent in 2-position. Subsequent methylation of the complex and treatment with CO gives rise to 2-acetyltetrafluoropyridine (Scheme 14.21) [3258]. In contrast, the activation with a rhodium-catalyst selectively activates the perfluoropyridine-substrate in 4-position, and thus gives rise to the corresponding 4-acylated product after treatment with CO and a methylating agent (Scheme 14.21) [3259]. However, the mechanisms of these reactions are quite different, as in the case of the Ni-activation, the product is liberated from a tetracoordinated complex upon treatment with CO. The Rh-mediated reaction proceeds via a ligand exchange, a subsequent oxidative addition of methyl iodide to give a six-coordinated metal-center and finally reductive elimination of the acetylated product in presence of CO.

    Transition-metal complexes are also used to improve the selectivity of substitution-reactions, which would otherwise give completely different products. Indeed, the reaction of hexafluorobenzene with aryl thiolates will proceed without any transition-metal catalyst to give the corresponding hexasubstituted aryl thioethers [3130]. However, in presence of a Rh-catalyst and PPh_3, the fluoride replacement using diaryl disulphides proceeds more selectively. Depending on the stoichiometry of the reactants, different degrees of substitution are accessible. As a general trend, the remaining fluorine-substituents will preferentially occupy *para*-positions with respect

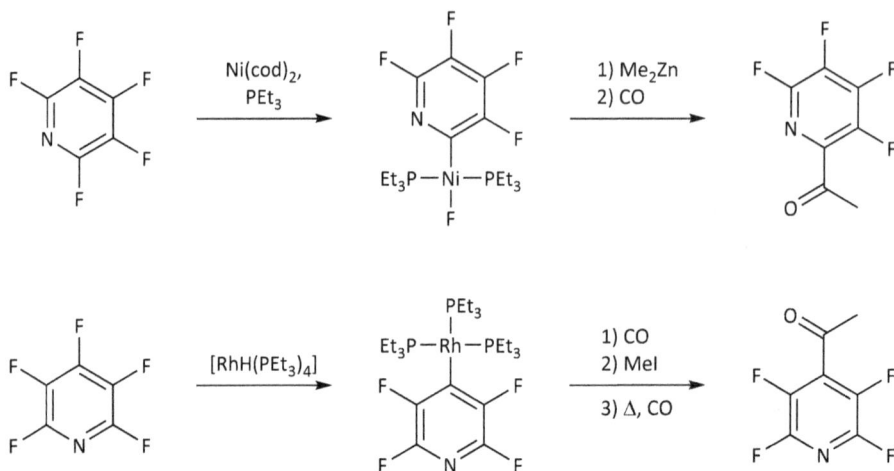

Scheme 14.21: Selective acetylation of pentafluoropyridine with Ni- and Rh-catalysts.

to each-other. The same is observed for minor fluorinated substrates and substituted pentafluorophenyl compounds [3260, 3261]. The formation of aryl- and benzyl thioethers has also been achieved starting from polyfluorobenzenes and the respective thioacetates in presence of CuBr and proline or by using aryl thiols or disulphides with CuBr, 1,10-phenanthroline and DDQ (2,3-dichloro-5,6-dicyano-1,4-benzoquinone) as oxidant. However, in the latter case the outcome strongly depends on the substitution of the arylsulphur compound [3262, 3263].

The activation of fluoroarenes may also be coupled with a formation of C-O bonds in terms of methoxylation reactions. This works well in a Pt-catalyzed reaction with $Si(OMe)_4$, if the substrate contains a directing group, such as an imine, imidazole or oxazoline. Other halogen substituents on the aromatic ring are tolerated, when they are not in adjacent position to the directing group [3264]. Apart from that, aryl-ethers are generated from highly fluorinated arenes, aryl boronic acids and oxygen in presence of nickel(II) acetylacetonate. However, this method fails with minor fluorinated substrates like C_6H_5F, p-$C_6H_4F_2$ and $1,3,5$-$C_6H_3F_3$ [3265]. Aryl ethers are also obtained, if pentafluorobenzene is reacted with various phenols in presence of catalytic amounts of $Pd(OAc)_2$ and stoichiometric $AgNO_3$. The substitution of the fluoroarene occurs chemo- and regioselectively in p-position to the hydrogen-substituent [3266]. The functionalization of C-F-activated substrates with other heteroatoms than sulphur or oxygen is only scarcely explored. However, examples are known, where C_6F_5-substituted cyclopentadiene- and indene-derivatives have been reacted with the dimethylamido complex $[Ti(NMe_2)_4]$, which leads to the selective exchange of the *ortho*-substituents of the pentafluorophenyl moiety, and thus allows the conversion to double-aminated products [3267]. In the presence of a directing groups, for example, a keto-function in 2,6-difluoroacetophenones, one of the adjacent fluorine substituents is replaced by a silyl-group, if the substrate is reacted with catalytic amounts of [Rh(cod)$_2$][BF$_4$] and Me$_3$Si-SiMe$_3$ [3268].

A selective activation, and thus functionalization can also be achieved for other fluorinated heterocycles than pyridine. For example, the reaction of 5-chloro-2,4,6-trifluoropyrimidine with Ni(cod)$_2$ and PCy$_3$ followed by subsequent treatment with iodine selectively gives rise to 5-chloro-2,6-difluoro-4-iodopyrimidine (Scheme 14.22). However, the use of a less sterically demanding phosphine, for example, PEt$_3$, will lead to the activation of the chloro-substituent [3269]. In a similar approach, 2,4,6-trifluoropyrimidine is activated by the addition of Ni(cod)$_2$ and PEt$_3$. The reaction of the intermediate nickelfluorido-complex with an excess of the trifluoropyrimidine starting material and CsOH · H$_2$O, followed by addition of HCl yields a 4-pyrimidinone-derivative (Scheme 14.22) [3270]. Other versatile methods for the functionalization of nitrogen heterocycles include the formation of C-O, C-B and C-Si bonds. For example, the treatment of pentafluoropyridine with the palladium complex [Pd(Me)$_2$(tmeda)] (tmeda = N,N,N',N'-tetramethylethylenediamine) in presence of water and NEt$_3$ leads to the activation of the fluorine-substituent in 4-position and the replacement by oxygen. Subsequent treatment of the intermediate complex with Me$_3$SiCl gives rise to the corresponding pyridyl silyl ether (Scheme 14.22). The catalytically active Pd-complex is then regenerated using methyl lithium [3271]. In a similar way, activated C$_5$F$_5$N is reacted with the disilane FMe$_2$SiSiMe$_2$F to give the corresponding tetrafluoropyridine derivative with a silyl-group in 4-position [3272]. However, the reaction with a tertiary silane like HSiPh$_3$ gives rise to the hydrodefluorination-product. The reaction of pentafluoropyridine with the rhodium-complex [Rh(SiPh$_3$)(PMe$_3$)$_3$] activates the heteroarene in both, 2- and 4- position in a ratio of about 3:1. Treatment of the intermediate mixture with *bis*(catecholato)diboron (B$_2$cat$_2$) affords pyridyl boronate esters (Scheme 14.22). However, their formation has not been achieved in a catalytic fashion so far [3273]. In contrast to the aforementioned example, the rhodium boryl complex [Rh(Bpin)(PMe$_3$)$_3$] selectively activates pentafluoropyridine in 2-position (Scheme 14.22). In the presence of B$_2$pin$_2$, this reaction proceeds catalytically and generates the respective pinacolboronate ester [3274]. As a general trend, the C-F activation of pentafluoropyridine with Ni-complexes preferentially occurs adjacent to the nitrogen, whereas the corresponding Pd-, Pt- and Rh-compounds mostly lead to an activation in 4-position [3119].

14.6.4 Study questions

(I)   Perfluorocarbons exhibit an orthogonal reactivity to hydrocarbons. What does this mean?

(II)   Which product is obtained, if hexafluorobenzene is first reacted with Me$_2$NH and then with MeS$^-$? How can one rationalize the formation of this product?

(III)   Which general reactivity is observed, if partly fluorinated arenes are treated with organolithium reagents? Why is it important to control the reaction temperature?

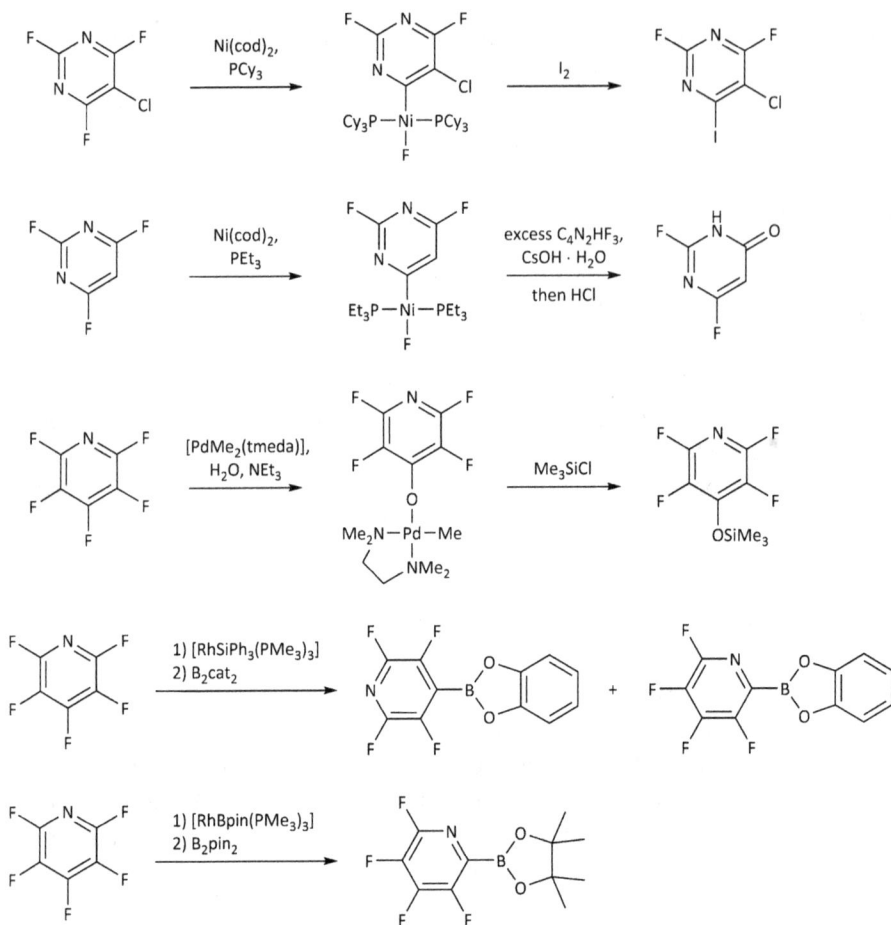

Scheme 14.22: Functionalization of fluorinated pyrimidines and pyridines.

(IV) What happens, if halofluoroarenes are reacted with lithium organyls? Comment on the influence of the organometallic reagent and the solvent on the product formation.

(V) Which general trends are observed for the transition-metal mediated C-F bond activation of fluoroarenes? Which metals are typically used? Is there any selectivity for C-H or C-F activation, when transition metal complexes are reacted with aromatic hydrofluorocarbons?

(VI) What is the advantage of using hydrodefluorination techniques? Which types of compounds are typically prepared by this method and which conditions need to be met by the reagent system to allow an efficient catalytic transformation?

(VII) Does the C-F activation of pentafluoropyridine with a Ni⁰- or a Rhᴵ-complex, followed by a subsequent acetylation step yield different products? How do the mechanisms of the acetylation sequences differ?

(VIII) Which reagents and transition metal catalysts may be used for the introduction of heteroatoms into polyfluoroarenes? Give examples.

15 Preparation of fluorinated olefins and acetylenes

Unsaturated fluorinated compounds have a certain importance, especially as monomeric building blocks for the preparation of polymeric functional materials. Apart from that, they are important synthetic intermediates, as they are typically susceptible to a variety of further derivatizations [3275]. Finally, they may introduce interesting structural properties to molecules, rendering them suitable for applications. Due to this, a number of efficient synthetic methods has been developed and reviewed over the last decades [2899, 3275–3279]. In addition, a comprehensive summary of preparative strategies toward di-, tri- and tetrasubstituted monofluoroalkenes has been published some years ago [3280]. In general, synthetic approaches toward fluoroolefins and -acetylenes are comparable to standard hydrocarbon chemistry, where the most important preparative methods are certainly based on elimination reactions. Fluorine substituents may also be directly introduced to vinylic positions, when a leaving group is replaced by nucleophilic or electrophilic fluorinating reagents. Furthermore, a few coupling procedures play a role, which are somehow also comparable to the synthesis of nonfluorinated molecules, especially in terms of Wittig- or Horner–Wadsworth–Emmons reactions, but also for Pd-catalyzed cross-coupling of fluorovinyl-fragments. In the same way, C-F-bonds of fluorinated olefins may be selectively activated by transition metal catalysts, and subsequently coupled to a variety of other carbon fragments [3234]. However, the focus in the following section will be on the transfer of fluorine substituents to unsaturated carbon-centers and the formation of unsaturated C-C-bonds with fluorinated alkylidene species. Thus, transition metal mediated coupling-reactions of complete, fluorinated vinyl-units will not be included herein. In contrast to hydrocarbon chemistry, transition-metal catalyzed metathesis techniques only play a minor role in organofluorine chemistry. Nevertheless, a number of substrates with a fluorinated backbone or side-chain may be used in this type of transformation, which has been reviewed some time ago [3281]. The selective transition-metal catalyzed hydrodefluorination of highly fluorinated olefins can also be a valuable tool for the preparation of partly fluorinated molecules and will therefore also be briefly discussed in this chapter.

15.1 Elimination reactions

The most important group of reactions toward unsaturated, aliphatic organofluorine compounds is certainly represented by dehydrohalogenations or dehalogenations of saturated derivatives. Apart from that, the elimination of other leaving groups, such as sulphonyl- and sulphinyl-functions, has also widely been described and reviewed [2792, 3275, 3277].

https://doi.org/10.1515/9783110659337-018

15.1.1 Elimination of HX

Perfluoroalkylated, terminal alkenes are accessible by treatment of perfluoroalkyl-ethyl iodides with strong bases, which leads to the elimination of HI. However, with one more methylene-group, that is, in perfluoroalkylpropyl iodides, nucleophilic substitution of the iodine-substituent predominates over elimination reactions. In a similar way, iodotrifluoroethene is prepared upon treatment of CF_2Cl-$CHFI$ with aqueous KOH. The starting material for this transformation is initially generated by addition of ICl to trifluoroethylene. Accordingly, controlled halofluorination of alkynes gives rise to halofluoroolefins. For example, a formal IF-transfer is achieved by the use of a combination of 4-iodotoluene difluoride and I_2 [2948]. For many compounds, different synthetic routes have been described, as for example 1,1,3,3,3-pentafluoropropene can be synthesized by halogen exchange of 3-bromo-1,1,3,3-tetrafluoropropene with HF, dehydrohalogenation of saturated 1-iodo-1,1,3,3,3-pentafluoropropane but also by decomposition of fluorinated potassium isobutyrate (Scheme 15.1). Fluorinated vinylic iodides are converted to acetylenes, when they are reacted with KF in presence of dicyclohexyl-18-crown-6.

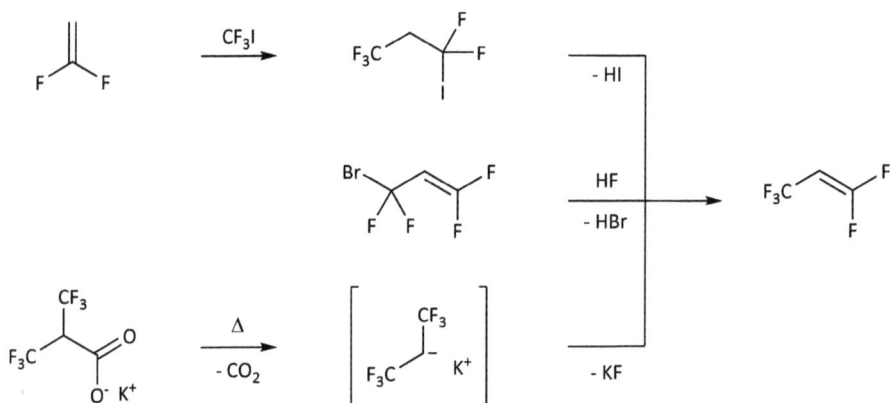

Scheme 15.1: Different synthetic routes toward 1,1,3,3,3-pentafluoropropene.

15.1.2 Elimination of HF

In general, HF-elimination is far less readily achieved than removal of HCl or HBr, as the reaction rates for break-away of the latter two in bimolecular reactions are about two to five orders of magnitude higher. However, the presence of a sufficient amount of β-fluorides may render protons acidic enough to enable elimination following an E1cb-mechanism. For example, treatment of 6,7-difluorododecane with KO^tBu in tetrahydrofuran leads to the stereospecific elimination of HF. Depending on the configuration of the substrate, either (E)- or (Z)-6-fluoro-6-dodecene is generated. The conversion of geminal difluorides to vinylic fluorides typically requires a strong base, however, some aluminium oxides can efficiently dehydrofluorinate such type of substrate under mild conditions.

Dehydrofluorination is also readily induced by aqueous or alcoholic solutions of alkali metal hydroxides, if a single hydrogen is surrounded by a large number of fluorine atoms. Most likely, this type of elimination proceeds via an E1cb-mechanism, too. Similarly, polyfluorocyclohexanes are converted to polyfluorocyclohexenes and polyfluorocyclohexadienes upon reaction with KOH. Under the same conditions, 1,3-dihydrododecafluorocycloheptanes yield a mixture of monohydroundecafluorocycloheptenes and decafluorocycloheptadienes (Scheme 15.2). This type of reaction is not restricted to carbocycles, but also occurs with heterocycles, such as oxetanes and 1,4-dioxanes (Scheme 15.2). Relatively weak bases are sufficient to abstract HF from highly fluorinated compounds, which is especially beneficial for the conversion of substrates with rather sensitive functional groups. For example, methyl 3,3-difluoro-2-(trifluoromethyl) acrylate is readily accessible from the saturated 2-hydro-hexafluoroisopropyl-precursor and $Et_3N \cdot BF_3$. The elimination of HF from vinylic fluorides gives rise to acetylenes or allenes, respectively (Scheme 15.2). The HF-abstraction from 1,1,3,3-pentafluoropropene requires a strong base and gives rise to tetrafluoroallene [3282].

15.1.3 Elimination of halogens

Elimination reactions are not restricted to the abstraction of hydrogenhalides but can also include the removal of halogens. For example, sodium amalgam converts perfluorinated 3,4-dimethyl-3-hexene to the corresponding 2,4-hexadiene under formal loss of F_2. In some cases, defluorination also occurs with NaF at elevated temperature, as it is for example observed for the conversion of 2,2,5,5-tetrafluoro-2,5-dihydrothiophene to 2,5-difluorothiophene at 530 °C. Interestingly, dehydrofluorination does not occur under these conditions. Product mixtures may be obtained from substrates, where more than one type of halogen is present. However, in most cases, fluorine will be more stable toward elimination. For example, symmetric dichlorodifluoroethene CFCl =CFCl is prepared by chlorine-abstraction from $CFCl_2$-$CFCl_2$, using zinc in boiling ethanol or dioxane. Another good example for selective dechlorination is given by the treatment of perfluoroalkyl-substituted pentachloroethane derivatives with zinc in N,N-dimethylformamide, which results in the formation of perfluoroalkyl acetylenes. The situation changes, if chlorine and fluorine are more dispersed in a polychlorofluorocarbon, where both halogens may be susceptible to removal by zinc. Nevertheless, selective dechlorination can be achieved, if the substrate is treated with PPh_3 in dioxane solution. Formal removal of "BrF" occurs, when 2-bromo-1,1-difluoroethene is initially reacted with Mg metal to form a Grignard compound, followed by β-elimination

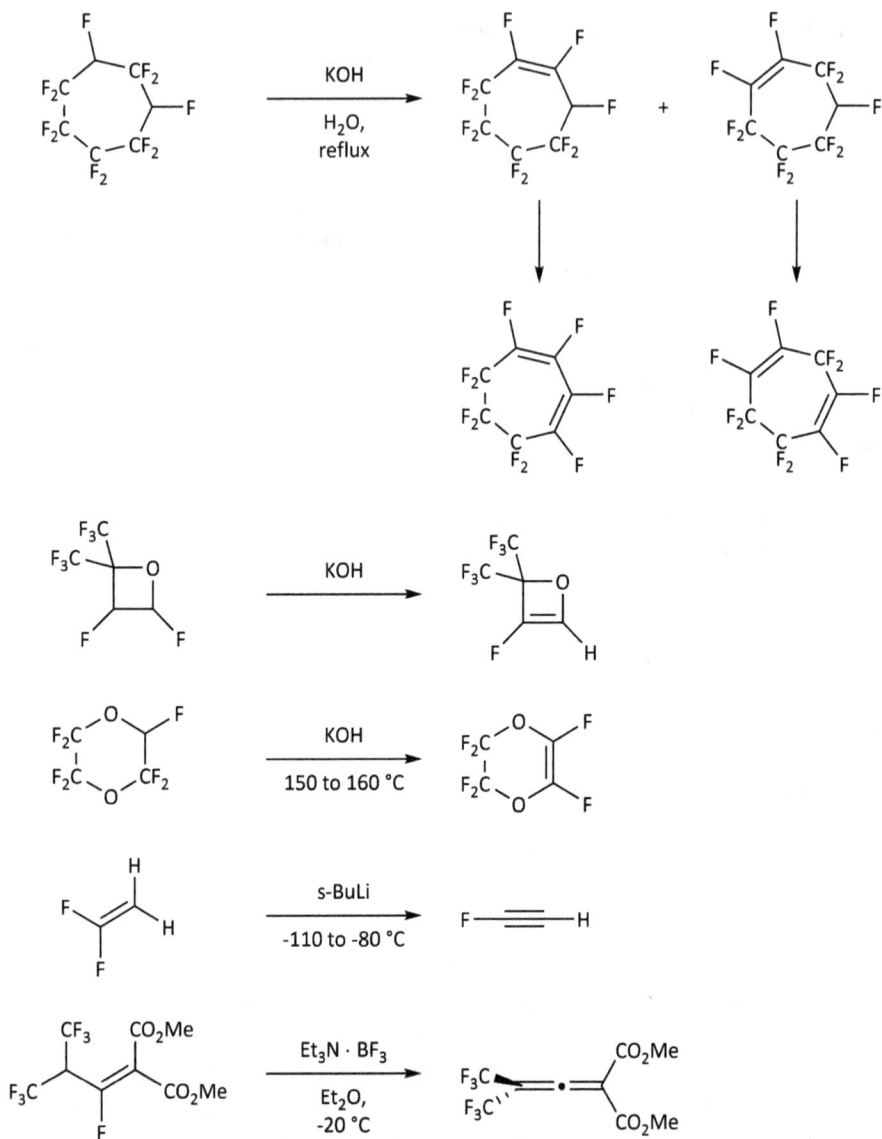

Scheme 15.2: HF-elimination from fluorinated substrates.

at about 60 °C to give monofluoroacetylene. In a related approach, perfluoroallenes are accessible, if the starting materials are reacted with copper powder at 200 °C under reduced pressure. However, the allene products readily dimerize, and thus give rise to cyclobutene derivatives. Zinc-copper couples are commonly used for the preparation of perfluoroolefins from perfluoroalkyl iodides in highly polar solvents, such as dimethyl sulphoxide or N,N-dimethylformamide. Mostly, the addition of a nucleophile is crucial to avoid the formation of complex product mixtures. For example, the conversion of C_4F_9I mainly generates perfluoro 2-butene and nonafluorobutane, if NaOAc or sodium bisulphite are present, whereas KSCN leads to the preferential formation of the terminal perfluoroolefin, along with perfluorinated 2-butene and a small amount of C_4F_9H.

15.1.4 Elimination of other small molecules

Elimination reactions may also be used to prepare unsaturated compounds from alcohols. Typically, this is achieved by conversion of the hydroxy function to a sulphonate or other good leaving group and subsequent heating in the presence of KO^tBu. Highly fluorinated alcohols are only dehydrated by P_2O_5 or H_2SO_4, however, hexafluoroisobutanol is converted to hexafluoroisobutylene under mild conditions using KOH at 20–50 °C. A mixture of $POCl_3$ and pyridine is capable of dehydrating trifluoromethyl homoallyl alcohols, but fails to eliminate water from the corresponding saturated substrates.

Sometimes, elimination reactions proceed by the removal of larger fragments or multiple small molecules. For example, pyrolysis of perhalogenated acid chlorides leads to the simultaneous liberation of CO and ClF, resulting in the formation of olefins (Scheme 15.3). Apart from that, some fluorinated ethers are capable of eliminating alkyl fluorides upon reaction with Lewis acids, such as SbF_5 or $BF_3·Et_2O$ (Scheme 15.3). Fluorinated olefins are also accessible upon desilylative defluorination, which is catalyzed by fluoride ions [3119]. The formal elimination of fluorosilanes can occur in several fashions, depending on the relative positions of the substituents. For example, the treatment of 1-(3′-chlorophenyl)-1-trimethylsilyl-1,2,2,2-tetrafluoroethane with $[NBu_4][Ph_3SiF_2]$ results in a 1,2-desilylative defluorination and gives rise to a trifluorostyrene derivative (Scheme 15.3) [3283].

15.1.5 Pyrolysis of fluorinated compounds

Some other fluorinated olefins are preferentially prepared by pyrolysis of fluorinated compounds. A typical example for this is perfluoroisobutene, which results from pyrolysis of tetrafluoroethene, hexafluoropropene, poly(tetrafluoroethene) or perfluorocyclobutane. However, it is also accessible by dechlorination of the respective saturated chlorofluorocarbon. Alkenes, which are substituted with a perfluorinated *tert*-butyl group in general also arise from octafluoroisobutene. As the latter is extremely electrophilic, it readily adds any type of nucleophile. Thus, the treatment with CsF results in the formation of intermediary $Cs[C(CF_3)_3]$, which can be attached to other molecules by nucleophilic substitution. The perfluorinated *tert*-butyl anion may be reacted with ethyleneoxide, where the epoxide ring will be opened to give an alcoholate. Subsequent elimination results in the formation of a (perfluoro-*tert*-butyl)

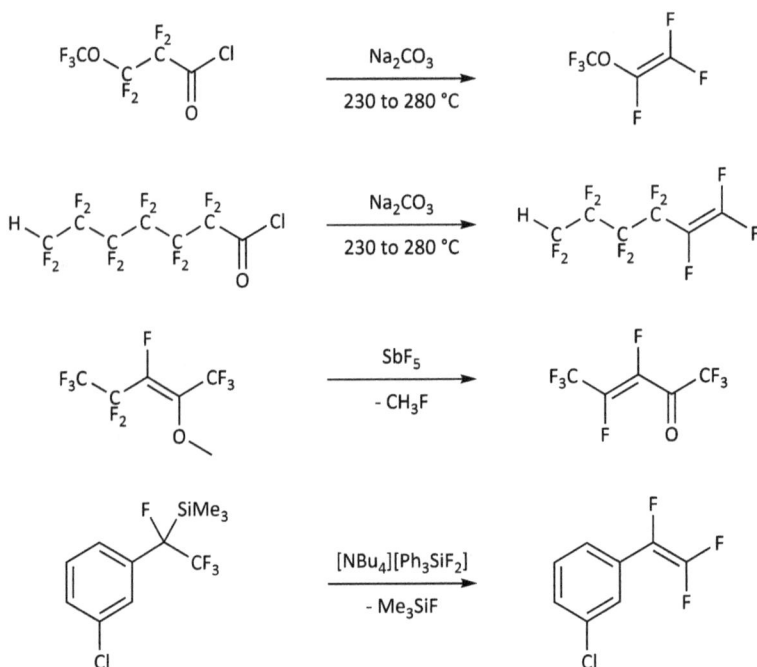

Scheme 15.3: Preparation of olefins by elimination of larger fragments and alkyl fluorides.

ethylene derivative. Alternatively, the perfluoro-*tert*-butyl anion may be reacted with 2-bromoethyl acetate to give 2-(perfluoro-*tert*-butyl)ethyl acetate, which eliminates acetic acid upon pyrolysis and thus generates the substituted olefin.

15.1.6 Preparation of fluorine-containing polyenes

The preparation of fluorinated compounds with more than one unsaturated C-C bond generally follows the same principles as described above for rather simple olefins. For example, perfluorobuta-1,3-diene is accessible in a two-step procedure from CF_2Cl-$CFICl$, which is initially submitted to a C-C coupling with elemental Zn (Scheme 15.4). Afterwards, the intermediary 1,2,3,4-tetrachlorohexafluorobutane CF_2Cl-$CFCl$-$CFCl$-CF_2Cl is dehalogenated by elemental Zn in ethanol to yield perfluorobutadiene (Scheme 15.4). In a slightly different fashion, perfluorinated 2,3-dimethyl-1,3-butadiene is prepared by initial electrolysis of perfluoroisobutyric acid, resulting in the coupling of two $(CF_3)_2CH$-units under liberation of CO_2. Subsequently, the dihydro-compound $(CF_3)_2CHCH(CF_3)_2$ is dehydrofluorinated by an aqueous suspension of NaOH and Al_2O_3 to liberate the unsaturated compound (Scheme 15.4).

It is also possible to prepare derivatives like 1,4-divinylperfluorobutane, where terminal vinyl-groups are attached to perfluoroalkyl linkers. Starting from tetrafluo-

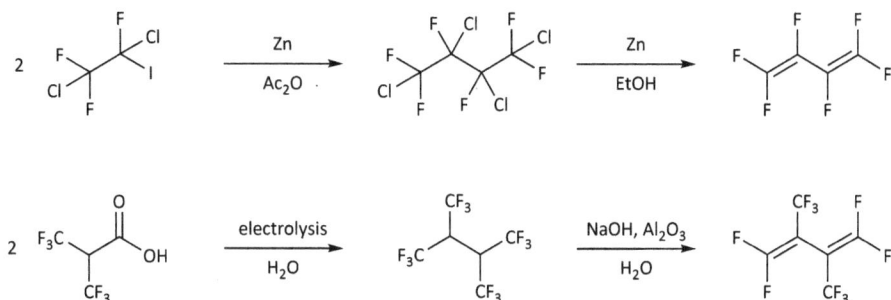

Scheme 15.4: Preparation of fluorinated butadiene-derivatives.

roethylene, I_2 is added in a first step to generate CF_2I-CF_2I, which is thermally converted to 1,4-diiodoperfluorobutane. The latter adds to two equivalents of ethylene and thus gives rise to a saturated compound with two terminal iodo substituents. Finally, 1,4-divinylperfluorobutane is obtained by elimination of HI, using an alcoholic solution of KOH (Scheme 15.5). The fully fluorinated analogue of 1,4-divinylperfluorobutane, 1,7-perfluorooctadiene is also accessible in several steps, following a similar synthetic strategy as for the preparation of perfluorobutadiene. The major steps include the addition of a perhalogenated alkyl iodide to tetrafluoroethylene and a subsequent Zn-mediated coupling. Finally, dechlorination is achieved by the use of elemental Zn in a mixture of AcOH and Ac_2O (Scheme 15.5).

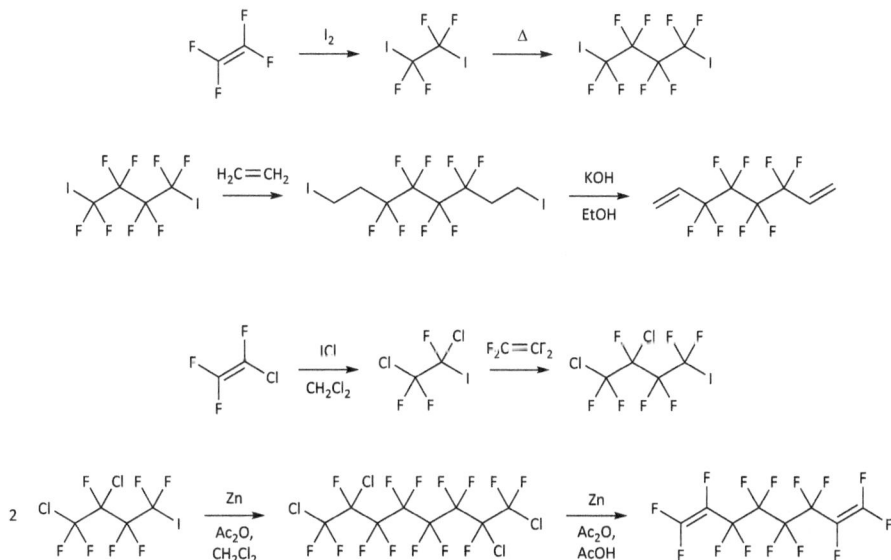

Scheme 15.5: Preparation of fully- and partially fluorinated 1,7-octadiene-derivatives.

15.1.7 Fluorinated acetylenes

Fluoroacetylenes, that is, compounds with a fluorine substituent directly attached to a sp-hybridized carbon are extremely reactive and, therefore, unstable. A long time ago, HCCF has been prepared by pyrolysis of fluoromaleic anhydride. It is a colorless gas, which spontaneously trimerizes to 1,2,4-trifluorobenzene. It has been characterized as "treacherously explosive" in the liquid state [3284]. Alternatively, fluoroacetylene is also obtained by the reaction of 1,1-difluoorethylene with *sec*-butyllithium at low temperature [3285]. In the same way, fluorophenylacetylene PhCCF is generated by the reaction of 2-chloro-1,2-difluorostyrene with organolithium reagents or by the elimination of fluorotrimethylsilane from the corresponding silyl alkene [3286, 3287]. The gas-phase preparation of the halofluoroacetylenes ClCCF and BrCCF by dehalogenation methods has also been reported [3288]. Starting from 1,1-difluoroethylene, fluorosilylacetylenes are accessible, which have been shown to form stable Dewarbenzene derivatives upon spontaneous trimerization [3289, 3290]. The latter observation has also been made for the oligomerization of *tert*-butylfluoroacetylene, which additionally generates other valence-bond isomers of benzene [3291]. Furthermore, fluoroacetylenes readily undergo cycloaddition reactions [3292]. Fluoro(silyl)- as well as fluoro(stannyl)acetylenes have been found to undergo radical addition reactions in tetrahydrofuran solution at −78 °C, leading to the formation of fluorovinyl-substituted cyclic ethers [3293].

Nevertheless, alkyne derivatives with fluorinated groups, such as CF_3, are well known and accessible. In principle, they are prepared by the same methods, which have been described for the synthesis of fluoroolefins in the previous section. For example, 3,3,3-trifluoropropyne is accessible by dehydroiodination of 1-iodo-3,3,3-trifluoropropene, which in turn is prepared by the addition of CF_3I to acetylene at 220 °C and elevated pressure (Scheme 15.6). Similarly, hexafluoro-2-butyne is generated by Zn-mediated chlorine-abstraction from the respective butene-precursor, the latter being accessible from perchloro-1,3-butadiene and antimonychlorofluoride SbF_3Cl_2 (Scheme 15.6). Addition of the perfluoro-*tert*-butyl anion to 1-chloro-3,3,3-trifluoropropyne gives rise to perfluorinated 4,4-dimethyl-2-pentyne. The chloroacetylene is prepared upon dehydrochlorination of 1,1,1-trichloro-3,3,3-trifluoropropane in the presence of KOH (Scheme 15.6).

15.1.8 Study questions

(I)   How are terminal perfluoroalkyl-substituted alkenes typically prepared? Is there a similar access to 1,1,3,3,3-pentafluoropropene?

(II) Does the elimination of HF from saturated compounds occur as readily as the cleavage of HCl or HBr? Which structural features help to enhance the tendency for a HF elimination?

H≡H → (CF₃I, 220 °C) → F₃C∕∖I → (KOH) → F₃C≡H

(CCl₂ diene structure) → (SbF₃Cl₂) → (fluorinated structure) → (Zn, Ac₂O) → F₃C≡CF₃

F₃C∕∖Cl → (Cl₂) → F₃C∕C(Cl)₂∕CHCl → (KOH) → F₃C≡Cl

(F₂C=C(CF₃)₂) + F₃C≡Cl → (CsF) → (F₃C)₃C≡CF₃

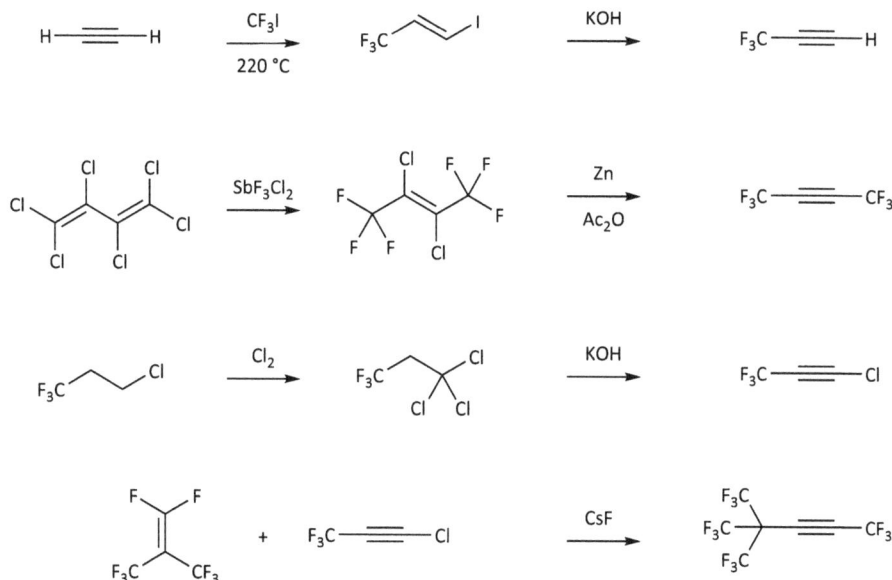

Scheme 15.6: Synthesis of fluorinated acetylenes.

(III)   Unsaturated compounds are also accessible by the formal elimination of halogen or inter-halogen molecules. Which reagents and conditions are typically required to observe such transformations and which products are obtained? Give examples.
(IV)   How can fluoroolefins be prepared from fluorinated alcohols and perfluoroacyl chlorides? Which reagents and conditions are required?
(V)   Which compound is an intermediate in the synthesis of both perfluorinated 1,3-butadiene and perfluoro 1,7-octadiene? Briefly describe the synthesis of the two dienes.
(VI)   Are fluoroacetylenes stable compounds? How can they be accessed? Briefly describe the synthesis of *mono-* and *bis*-trifluoromethyl acetylene.

15.2 Nucleophilic substitution

15.2.1 Nucleophilic attack on CF₃-substituted vinyl compounds

Direct nucleophilic substitution reactions of vinylic leaving groups are rather scarcely observed. However, 1-trifluoromethylvinyl compounds, such as α-trifluoromethylsty-rene, are easily attacked by nucleophiles. The nature of the nucleophile and the solvent strongly influence, if an addition or addition-elimination reaction occurs. The latter is preferentially observed with organometallic species in aprotic solvents, whereas protic nucleophiles, such as amines or alcohols, in protic media mostly lead to addition products. Addition-elimination reactions with CF₃-substituted vinyl substrates give access to 1,1-difluoroalkenes (Scheme 15.7). Typical nucleophiles are

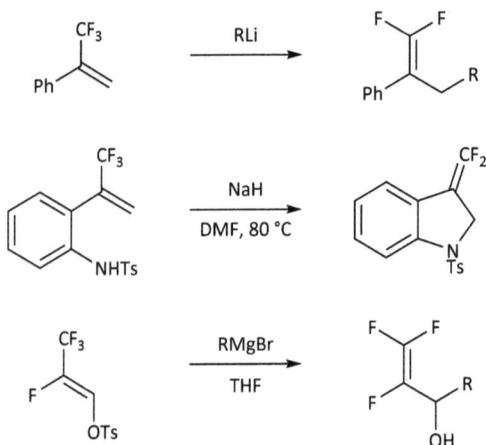

Scheme 15.7: Preparation of difluorovinyl-compounds from trifluoromethylated olefins.

organolithium compounds, Grignard reagents, but also N-lithiated amines or ester enolates. The nucleophilic attack can also occur intramolecularly, giving rise to cyclic, difluoromethylene-substituted compounds, for example, 3-difluoromethyleneindoline derivatives (Scheme 15.7) [3119, 3294]. Trifluorovinylated molecules are accessible by S_N2'-type reactions, if for example, β-fluoro-β-trifluoromethyl enol- or enamine species are reacted with Grignard reagents [3295, 3296]. S_N2'-type reactions of fluorinated imino and ditihioesters may lead to a substitution at the heteroatom. Depending on the length of the fluorinated alkyl chain in the substrate, the product contains either a geminal difluoroolefin function, or alternatively, one of the fluorine atoms is replaced by a perfluoroalkyl group [3297, 3298].

15.2.2 Addition-elimination reactions of fluorinated olefins

On the other hand, geminal difluoroolefins can serve as starting materials, as they are typically attacked by a number of carbon and heteroatom nucleophiles in an addition-elimination mechanism, to give monofluoroalkenes [3119]. The nucleophilic attack can also proceed intramolecularly, giving rise to cyclic vinyl fluorides or partly unsaturated, fluorinated heterocycles [3119, 3299]. It is also possible to replace both fluorine substituents; however, this will lead to nonfluorinated products, which are not discussed herein.

Reactions with nucleophiles are also observed for perfluoroolefins and their cyclic analogues, as they are very electrophilic species. However, often, the addition of nucleophiles leads to saturated compounds, which is discussed in Sections 13.4.11 and 13.4.12. Nevertheless, there are also examples, where an addition-elimination mechanism is operative, resulting in the formation of partly defluorinated species.

For example, perfluorocyclopentene reacts with carbanions under substitution of either one or both of the vinylic fluorine substituents [3119, 3300, 3301]. Depending on the nature of the nucleophile and the stability of the product, the second addition may proceed via an S_N2'-type mechanism, affording unsaturated products, where the double bond has been shifted [3302]. An interesting reaction is furthermore observed, if CF_3-substituted acetylenes with a sterically demanding substituent on the other side of the triple bond, for example, $SiPh_3$ or CPh_3, are reacted with *tert*-butyl lithium. In principle, an addition-elimination mechanism is observed for this type of transformation, however, it further includes a cyclization-step, and thus generates geminal difluoropropene-derivatives [3303].

15.2.3 Miscellaneous reactions toward fluoroolefins

Fluorinated olefins are furthermore accessible by iodofluorination of alkynes, using a combination of a hypervalent iodine fluoride and I_2, which generate "IF" *in situ*. However, this approach requires reduced temperatures to avoid the full saturation of the multiple bond. For internal acetylene substrates, an (*E*)-selective product formation has been observed [2948]. In a related strategy, allene substrates are reacted with nucleophilic fluoride to give fluoroolefins. For example, the treatment of allenyl ketones with [NBu_4]F in an aqueous system yields β-fluoroenones, which are preferentially (*E*)-configured. However, the corresponding (*Z*)-isomers are formed as well [3304].

Some sort of nucleophilic fluorination toward fluoroolefins is represented by the thermal decomposition of vinyliodonium tetrafluoroborate salts. This method works well for substrates derived from propene and 1-decene, which are selectively transformed to the (*Z*)-configured vinyl fluorides in high yield; however, more complex substrates readily undergo side-reactions, especially rearrangements [3305]. The hydrofluorination of alkynyl iodonium tetrafluoroborate salts using either aqueous HF or CsF in presence of water yields the corresponding (*Z*)-(2-fluoroalkenyl)iodonium compounds [3306, 3307]. Apart from that, (*Z*)-fluoroalkenyliodonium compounds are obtained by the reaction of potassium alkynyltrifluoroborate salts with a hypervalent aryliodonium difluoride species and subsequent treatment with aqueous HF [3308]. The analogous (*E*)-configured derivatives are obtained upon treatment of terminal alkynes with PhIO and pyridine/HF, followed by the addition of $BF_3 \cdot OEt_2$. Instead of PhIO, PhI may also be used in conjunction with *m*-chloroperbenzoic acid (*m*CPBA) [3309]. Moreover, the reaction of vinyliodonium salts with CsF has been shown to lead to the formation of fluoroalkenes [3310].

Fluorine-containing unsaturated compounds are also accessible by deoxyfluorination. In a very basic example, 3,3,3-trifluoropropyne is prepared by the reaction of acetylene carboxylic acid with SF_4, which is performed under pressure at 120 °C. Under milder conditions, β-diketones are deoxyfluorinated using an α,α-difluoroamine reagent, which gives rise to β-fluoro-α,β-unsaturated ketones. The stereoselectivity of this transformation strongly depends on the substrate, so that both (*E*)- and (*Z*)-products are generally accessible. For unsymmetrical diketones, a regioselective deoxyfluorination is observed [3311].

15.2.4 Study questions

(I)   How can nucleophilic attacks on trifluoromethyl-substituted alkenes give rise to geminal difluoroolefiles? Which nucleophiles and conditions are generally suitable for this type of transformation?

(II)   Which unsaturated products are typically obtained, if geminal difluoroolefins and perfluoroolefins are reacted with nucleophiles? Which mechanism is observed?

(III)   How can vinyliodonium salts serve as synthetic intermediates toward vinyl fluorides? How are they prepared and which major problem arises from their use?

15.3 Electrophilic fluorination

15.3.1 Fluorination of vinyl lithium intermediates

Similar to fluoroarenes, fluoroolefins are accessible, if for example, an intermediary vinyl lithium species is treated with an electrophilic fluorinating reagent. Typically, the organolithium compound is generated from vinyl iodides at low temperature. In general, the organometallic intermediates are stereoselectively fluorinated under retention of the configuration. As major side-products, the respective protonated olefins are formed [3312]. The lithiation approach can also be applied for the introduction of vinylic fluorine substituents into partly hydrogenated heterocycles. For example, treatment of a 1,2-dihydropyridine derivative with butyllithium and subsequent addition of NFSI yields the corresponding 6-fluoro-1,2-dihydropyridine [3313].

15.3.2 Electrophilic fluorination of substituted cyclopropanes and allenes

Another class of substrates, which is converted to monofluoroolefins upon impact of electrophilic fluorinating agents is described by strained methylenecyclopropanes. The reaction of these compounds with NFSI leads to the cleavage of the cyclopropane ring and selectively generates monofluorinated alkenes, where the $N(SO_2Ph)_2$-moiety is attached to the terminal position [3314]. In general, the ring opening proceeds faster and more efficiently, if an electron-rich arene is attached to the C-C double bond. In contrast, a complex product mixture is obtained, if dialkyl-substituted methylenecyclopropanes are reacted. The treatment of diarylmethylene cyclopropanes with Selectfluor™ in nitrile solvents again leads to the ring opening and generates monofluoroolefins. In this case, an amide function is transferred to the terminal carbon atom [3315]. In a similar fashion, aryl-substituted allenes can be submitted to an electrophilic fluorination by Selectfluor™ in aqueous acetonitrile, giving rise to fluorohydroxylated compounds, that is, 2-fluoroalken-3-ols. However, the presence of at least one arene-substituent on the allene-moiety is mandatory, most likely for stabilization of the corresponding cationic allylic intermediate [3316]. The treatment of 2,3-allenoic acids with Selectfluor™ in aqueous acetonitrile or water results in an electrophilic fluorocyclization, and thus gives rise to fluorinated furanone products [3317]. Similarly, 4,5-allenoic acids may be converted to γ-lactones with a fluorovinyl substituent, whereas 4,5-allenoic tosylamides yield pyrrolidine derivatives [3318]. The electrophilic fluorodesilylation of allenyl silanes gives rise to 2-fluoro-1,3-dienes, which are generally rather poorly accessible [3319].

15.3.3 Miscellaneous transformations

Fluoroolefins are also accessible from vinylic substrates with a leaving group. For example, the fluorodesilylation of alkenylsilanes with Selectfluor™ or NFSI under ultrasound irradiation yields monofluoroalkenes, which are, however, formed as mixtures of stereoisomers [3320]. When the fluorodesilylation of 1-silyl-3-boryl-2-alkenes is combined with an allylboration of aldehydes, fluoroalkenes are obtained with high (Z)-selectivity. The use of either pure (E)- or (Z)-configured starting material does not change the outcome of the reaction [3321].

Finally, the preparation of vinyl fluorides is also achieved by direct C-H fluorination. For example, the treatment of 1,1-diphenylethene with Selectfluor™ and water in an ionic liquid yields the corresponding monofluorinated olefin [2987]. In the same way, cyclic 1,1-diarylethylenes as well as 1,1-diaryldienes are fluorinated upon treatment with Selectfluor™ in a mixture of acetonitrile and water. However, this method is limited to a small number of suitable substrates [3322].

15.3.4 Study questions

(I)   How are vinylfluorides prepared from vinyllithium intermediates? Are these transformations selective? Which side-products are typically obtained?

(II)   How can methylenecyclopropanes and allenes serve as substrates for vinyl fluorides? Under which conditions are they fluorinated and which products are typically obtained?

(III)   Which type of olefinic substrate can generally be submitted to direct C-H fluorination? What is the fluorinating agent?

15.4 Transition metal-free introduction of fluorinated fragments

Another important approach toward fluorinated olefins comprises the introduction of not only single vinylic fluorine substituents but the transfer of fluorinated fragments. However, this manuscript will only cover approaches, where the double-bond is generated upon coupling with a fluorinated fragment.

15.4.1 Wittig olefination

The most prominent example for olefinations in organic chemistry is probably given by the Wittig reaction and its variants, which can also be used in organofluorine chemistry. Thus, carbonyl-compounds are converted to geminal difluoroolefins upon treatment with a phosphine and a source of difluorocarbene, which together give rise to an active ylide. For example, a variety of aliphatic, alicyclic and aromatic ketones and aldehydes is readily converted to 1,1-difluoroalkenes, when they are reacted with PPh_3 or $P(NMe_2)_3$ and CF_2Br_2 in triglyme-solution. However, this method strongly depends on the purity of the solvent, which can lead to poorly reproducible results. This problem is overcome when zinc dust and dimethylacetamide are added to the reaction mixture. Phosphonium salts of the type $[Ph_3PCF_2Br]^+Br^-$ are further used in other transformations, which yield fluoroolefins, such as chain extension under formation of dienes or conversion of diketones to the corresponding dienes (Scheme 15.8). In a similar approach, geminal difluoroalkenes are obtained, if nonstabilized phosphorous ylides are reacted with CF_2HCl (Scheme 15.8). In more recent approaches, less problematic sources of difluorocarbene have been used, such as fluorosulphonyl difluoroacetates, Me_3SiCF_2Cl or transition metal complexes containing CF_3-anions [3275].

Starting from PPh_3 and $CHFI_2$, monofluoroalkenes are prepared from a number of aromatic ketones and aldehydes. The reactions are conducted in N,N-dimethylformamide in presence of a Zn/Cu-couple. If fluorodihalomethanes or fluorotrihalomethanes are used in the olefination reaction, halofluoroalkenes may be prepared, too. For example, the use of $CFCl_3$ gives rise to a number of chlorofluoroalkenes, if the corresponding phosphonium reagent is reacted with aromatic or aliphatic ketones in benzonitrile. When fluorotrihalomethanes are treated with two equivalents of the corresponding phosphane, fluorinated phosphoranium salts are generated, which are capable of transferring CHF-units to aliphatic and aromatic aldehydes as well as polyfluoroacyl fluorides (Scheme 15.9).

The transfer of larger fluorinated alkylidene moieties is not that easily achieved, since the preparation of the corresponding ylide-species is often accompanied by β-fluoride elimination. Nevertheless, perfluoroisopropylidene-units can be attached to a large variety of aldehydes, when the latter are reacted with PPh_3 and 2,2-dichlorohexafluoropropane or *tetrakis*(trifluoromethyl)-1,3-dithietane, the dimer of hexafluorothioacetone (Scheme 15.9).

CF_2Br_2 + PPh_3 $\longrightarrow$ Ph_3P^+ ... Br^- Br

Ph_3P^+ ... Br^- Br + F_2C—CF_3 ... Ph $\xrightarrow[CH_3CN]{Hg}$ F_2C—CF_3 ... Ph_3P^+—C ... Ph Br^- $\xrightarrow{H_2O}$ ... Ph

$\xrightarrow[CF_2Br_2]{PPh_3,}$

2 Ph_3P ... R^1 R^2 $\xrightarrow[{- [Ph_3PCHR^1R^2]^+Cl^-}]{\underset{- PPh_3}{HCF_2Cl}}$ ... R^1 R^2

R^1 = H, Me, Ph
R^2 = alkyl or aryl

n = 2, 3, 4

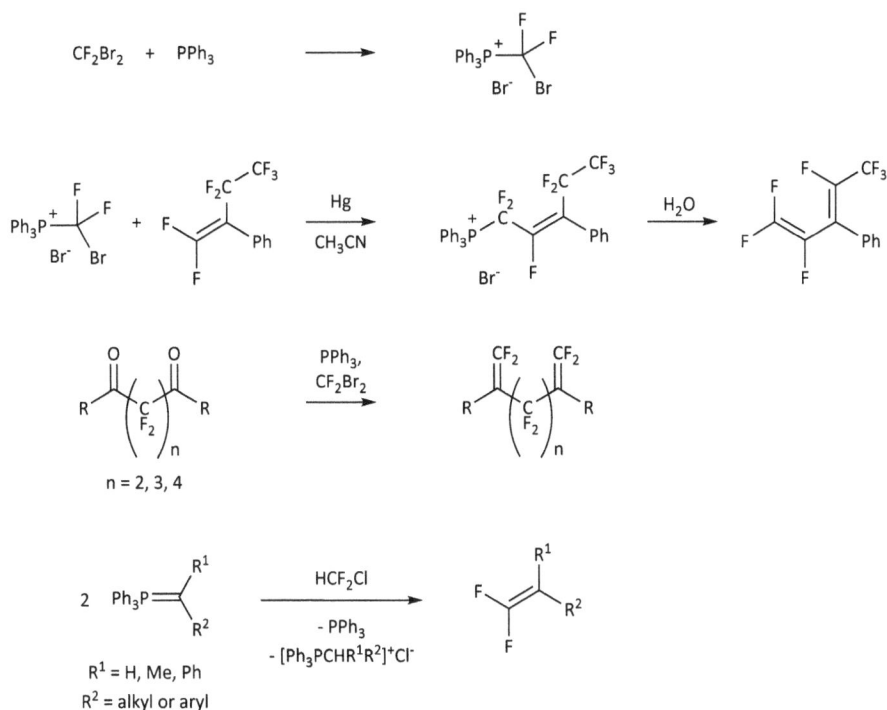

Scheme 15.8: Generation and transfer of fluorinated alkylidene units.

The reaction of isopropylidenetriphenylphosphine ylide with anhydrides of perfluorinated carboxylic acids yields fluorinated acyl phosphonium salts. Upon reaction with $Ph_3P = CH_2$ and PhLi, a new ylide is formed, which can be submitted to a Wittig reaction with an aldehyde, resulting in the formation of a perfluoroalkylated, conjugated diene. Alternatively, the acyl intermediate can be treated with a nucleophile, for example, an acetylide, which gives rise to a perfluoroalkylated enyne [3323, 3324]. In a similar approach, perfluoroalkylated alkynes are accessible, if perfluoroacyl chlorides are reacted with two equivalents of a phosphorous ylide (Scheme 15.10). In the final step, the acetylene is obtained by vacuum-pyrolysis at 200–260 °C, however, when substituent R contains an aldehyde or a keto-group in direct proximity of the ylide, product mixtures are obtained [3325–3327]. Pentafluorophenyl acetylenes are accessible, when hexafluorobenzene is initially reacted with $Ph_3P = CH_2$. Subsequent treatment of the obtained ylide with either an acid fluoride or anhydride followed by pyrolysis finally yields C_6F_5-substituted alkynes. Wittig-reactions of (1-fluoroacetyl) methylenetriphenylphosphorane with aldehydes give rise to α,β-unsaturated fluoromethylketones. The preparation of trifluoromethyl-substituted alkenes is also achieved by Wittig reactions. For this, triphenylphosphorane-derivatives are reacted

$3\ PBu_3\ +\ CFX_3\ \longrightarrow\ Bu_3P{=}\substack{F\\X^-\\+\\PBu_3}\ +\ Bu_3PX_2$

$R^F\text{-COF}$ $R\text{-CHO}$

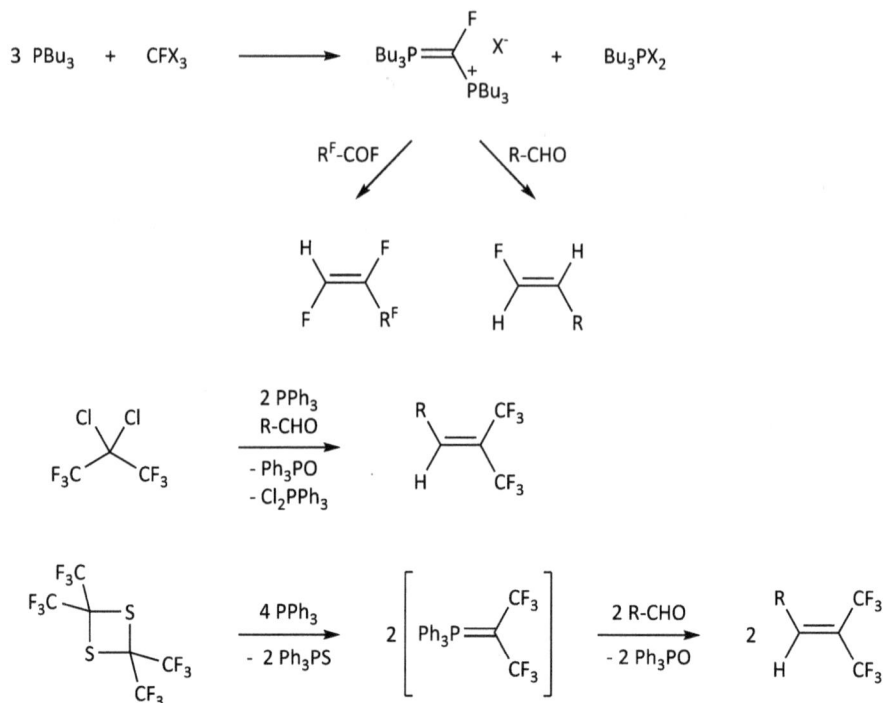

Scheme 15.9: Transfer of fluoromethylene- and hexafluoroisopropylidene units.

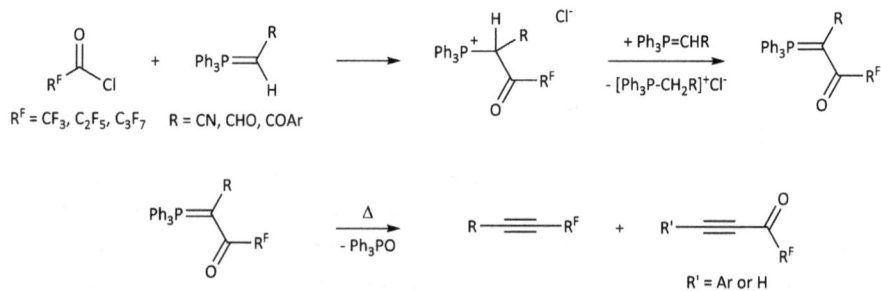

$R^F = CF_3, C_2F_5, C_3F_7$ R = CN, CHO, COAr

$+ Ph_3P{=}CHR$

$- [Ph_3P\text{-}CH_2R]^+Cl^-$

Δ

$- Ph_3PO$

R' = Ar or H

Scheme 15.10: Preparation of perfluoroalkylated acetylenes by Wittig-type reactions.

with trifluoromethylketones. The latter can also be used in Peterson-type olefination, which will give rise to trifluoromethylated olefins [3328]. The use of fluorinated esters and amides in Wittig reactions gives rise to fluorine-containing enol ethers and enamines [3329, 3330]. Fluorinated α,β-unsaturated aldehydes are suitable starting materials for the generation of conjugated dienes, which contain vinylic fluorine-substituents.

15.4.2 Julia- and Julia–Kocienski-type olefinations

Initially, a Julia-type difluoromethylenation of aldehydes has been described, using $PhSO_2CF_2Br$ in conjunction with *tetrakis*(dimethylamino)ethene (TDAE) as electron-transfer agent. The intermediate (benzenesulphonyl)difluoromethyl alcohols are then reductively cleaved by sodium amalgam to liberate the corresponding 1,1-difluoroolefin [3331]. However, this method comprises a multistep sequence and requires rather drastic reaction conditions, which do not tolerate sensitive functional groups. The use of difluoromethyl 2-pyridylsulphone, also known and commercially available as Hu-reagent, avoids the use of strong reducing agents, but still needs to be conducted at low temperature and in presence of the strong base KO^tBu. Nevertheless, this Julia–Kocienski-type olefination can be applied for a large variety of carbonyl substrates and shows a good tolerance against functional groups. Further improvements on this approach have been made by using the potassium-salt of 2-pyridyl sulphonyldifluoroacetic acid in a decarboxylative Julia–Kocienski reaction, as this reagent does not require an additional base and converts aldehydes to 1,1-difluoroolefins under mild conditions [3332].

15.4.3 Horner–Wadsworth–Emmons olefination

Fluorinated olefins are also accessible by the common variants of the Wittig reaction, for example, Horner–Wadsworth–Emmons olefination. In the simplest way, difluoromethylphosphonates are deprotonated by a base like lithium diisopropylamide (LDA) and subsequently reacted with an aromatic aldehyde or ketone. Thermolysis of the adduct gives then rise to 1,1-difluoroolefins [3333]. The use of difluoromethyldiphenylphosphine oxide enlarges the substrate scope and allows the conversion of both, aliphatic and aromatic carbonyl compounds [3334].

Upon reaction with alkyl diethylphosphonofluoroacetates, a variety of aliphatic aldehydes is converted to mostly (E)-configured α-fluoro-α,β-unsaturated esters with high stereoselectivity. In a more elegant way, ester starting materials may be used instead of sensitive aldehydes, if they are *in situ* reduced by diisobutylaluminium hydride. Horner–Wadsworth–Emmons reactions may also be carried out with diethylphosphonofluoroacetonitrile, which will give rise to the corresponding cyano-substituted fluoroolefins. For the preparation of alkenes, fluoroalkylphosphonates can also be deprotonated prior to the reaction with aldehydes. Partly fluorinated dienes are accessible by treatment of electrophilic fluoroalkenol phosphates with ylides. Trifluoromethylketones can also be used in Horner–Wadsworth–Emmons reactions to give rise to α,β-unsaturated-β-trifluoromethyl esters with an E/Z-selectivity of 9:1. However, the products readily rearrange to finally yield non-conjugated β,γ-unsaturated esters [3335].

15.4.4 Study questions

(I)   How are geminal difluoroolefins prepared by Wittig-type reactions? Which reagents are re-
      quired?
(II) Is it possible to transfer other fluorinated fragments than CF_2 to carbonyl compounds?
      Which reagents are used for these transformations?
(III) Which substrates and reagents are required to prepare fluorinated (conjugated) dienes,
      enynes and acetylenes by Wittig-type transformations?
(IV) Which reagents are used to convert carbonyl compounds to 1,1-difluoroolefins by Julia or
      Julia–Kocienski olefination? What are the main advantages of this method?
(V)   How can Horner–Wadsworth–Emmons reactions be used to prepare fluoroolefins with other
      electron-withdrawing substituents, such as ester or nitrile functions?

15.5 Olefin metathesis

15.5.1 General aspects about metathesis reactions of fluorinated olefins

In hydrocarbon-chemistry, metathesis reactions are versatile methods for the formation of carbon-carbon double bonds. In parts, these techniques can also be transferred
to fluoroorganic chemistry, however, like other electron-deficient alkenes, fluorinated
substrates are generally less reactive in such transformations. The use of olefins with a
fluorinated backbone, fluoroalkyl side-chains or even vinylic fluorine substituents has
been reviewed some time ago [3281]. Apart from that, metathesis reactions can be performed under fluorous conditions or in fluorinated solvents. However, in the present
manuscript, the focus will be on reactions, where fluorine or a fluorinated substituent
is directly attached to the olefinic bond, which undergoes metathesis.

Initial attempts to react 1,1-difluoroethylene with a ruthenium alkylidene complex (2nd generation Grubbs catalyst) have indeed resulted in the cleavage of the
fluoroolefin, but only gave rise to a stable ruthenium difluorocarbene compound
along with the respective methylene-derivative, β,β-difluorostyrene as well as styrene
(Scheme 15.11) [3336]. Thus, the metathesis reaction proceeded only stoichiometrically, but not catalytically, as it would be required for successful metathesis of
olefinic substrates. So far, catalytic reactions have only been described for mono-
substituted systems, which is certainly due to the fact, that intermediary generated
difluorocarbene-complexes of the transition metal, mostly ruthenium, are catalytically inactive. In practice, cross metathesis of two different olefins is only scarcely
described, as for example, 3,3,3-trifluoropropene and nonafluorobutylethylene have
been submitted to a reaction with several other terminal alkenes in presence of either
Grubbs-II or the second generation of Grubbs–Hoveyda catalyst, the latter generating
better yields at milder conditions (Scheme 15.11). As a general trend, all the reactions proceed with (E)-selectivity and require a large excess of the electron-deficient
fluorinated olefin [3337].

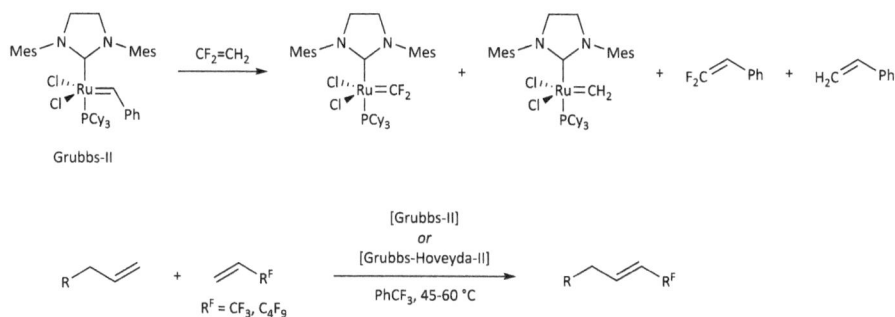

Scheme 15.11: Reaction of 1,1-difluoroethylene with Grubbs-II catalyst and cross metathesis involving perfluoroalkyl olefins.

15.5.2 Preparation of fluorinated carbocycles

The majority of metathesis reactions involving fluorinated olefins comprise ring-closing reactions, which are in most cases favored by entropy. Thus, ring-closing metathesis (RCM) of 2-fluoroallyl or related fluorovinyl substrates furnishes a versatile route toward fluorinated cycloalkenes and a number of unsaturated heterocycles. In the same way, 2-trifluoromethylallyl compounds can serve as starting materials for the respective cyclic trifluoromethylated products. As an example for the formation of carbocycles, diethyl 2-(2-fluoroallyl)-2-alkenyl malonates are readily converted to six- and seven-membered rings, whereas the formation of the respective cyclopentene-derivative is not observed (Scheme 15.12) [3338]. This is probably due to steric reasons and increased ring-strain in the transient metallabicycle species. Interestingly, the respective CF_3-substituted cyclopentene-derivatives are accessible by RCM (Scheme 15.12) [3339].

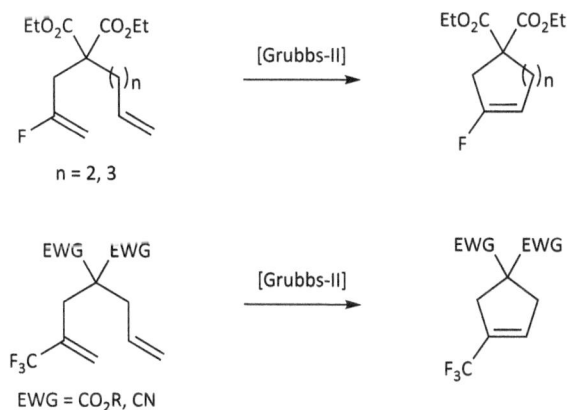

Scheme 15.12: Synthesis of fluorinated unsaturated carbocycles using ring-closing metathesis.

15.5.3 Preparation of fluorinated heterocycles

Ring-closing metathesis of various 2-fluoroallyl- or 2-(trifluoromethyl)allyl-substituted amines gives rise to a number of partially hydrated nitrogen heterocycles, such as tetrahydropyridines, pyrrolines, tetrahydro-1H-azepines, tetrahydropyridazines and tetrahydro-1,2-diazepines. Apart from that, fluorine-containing lactones and lactams are accessible (Scheme 15.13) [3339–3345]. The preparation of fluorinated tetrahydropyridines is readily achieved with a number of different protecting groups at the nitrogen atom, however, the best yield for the fluorine-substituted derivative is achieved with a p-nitrobenzenesulphonyl-group (nosyl, Ns), whereas the corresponding CF_3-compound is best prepared from the tosyl-protected precursor [3344]. Interestingly, the formation of N-protected 3-fluoropyrroline-derivatives is not observed, if the corresponding allyl (2-fluoroallyl)amine is treated with the second generation Grubbs catalyst. Instead, deallylation occurs exclusively. However, the presence of an additional methyl group in neighboring position of the amine function or the replacement of the vinylic fluorine substituent by CF_3 allow the preparation of fluorinated pyrrolines [3344]. Fluorinated tetrahydro-1H-azepines are readily obtained from the corresponding diene-precursors, however, larger heterocyclic systems are only scarcely obtained [3344]. The use of appropriate hydrazine-substrates gives rise to fluorinated cyclic hydrazines, for example, tetrahydropyridazines or tetrahydro-1,2-diazepines, if the basicity of the nitrogen atoms is efficiently reduced by the attachment of suitable protecting groups [3340]. Moreover, the synthesis of 3-fluorocoumarin is possible, if the respective 2-vinylphenyl 2-fluoroacrylate is used as starting material, however, significantly increased product yields are obtained, if the substrate contains a propenylinstead of a vinyl-group [3341, 3343]. Apart from lactones, lactams are accessible, for example, if tosyl-protected N-alkenyl-N-(2-fluoroacryl)amides are submitted to ring-closing metathesis reactions. The formation of γ- and δ-lactams proceeds with moderate to good yields, whereas ε-lactams are accompanied by ring-contraction products. The formation of larger lactam-cycles is typically not observed [3343]. The synthesis of 2-phenyl-4-(trifluoromethyl)-2,5-dihydrofuran is a rare example of a fluorinated oxygen-heterocycle, that has been prepared by a metathesis reaction [3339].

15.5.4 Study questions

(I)   Which types of fluorinated olefins are generally suitable to be used in metathesis reactions? How can this behavior be explained?

(II)   Which substrates are used for the metathetical preparation of unsaturated carbocycles, which contain vinylic fluorides or trifluoromethyl groups? Which ring sizes are preferentially obtained?

(III) Which types of partly saturated, fluorine containing heterocycles are accessible by ring-closing metathesis? Which preconditions have to be met by the substrates?

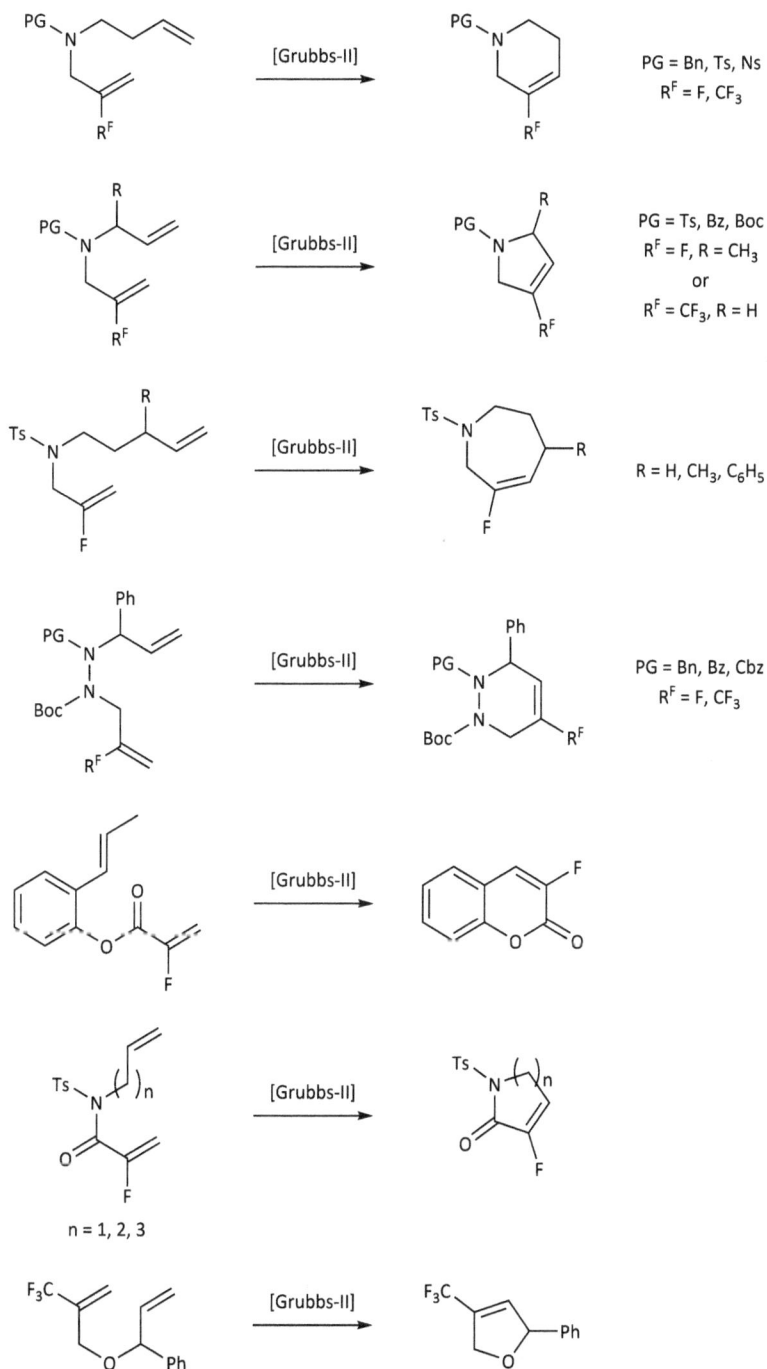

Scheme 15.13: Preparation of fluorinated- and trifluoromethylated heterocycles by metathesis.

15.6 Metal-mediated reactions, including photocatalysis

15.6.1 Hydrofluorination of alkynes

Compared to the number of methods, which have been described for the synthesis of aromatic or alkyl fluorides, transition-metal mediated fluorination techniques leading to vinyl fluorides are only sparingly represented in literature [3040]. Nevertheless, they can be very helpful for the preparation of fluorinated olefins.

In general, the hydrofluorination of alkynes is a rather simple preparative method; however, upon direct conversion of acetylenes with HF, completely saturated derivatives will mostly be obtained. Using a gold(I)-catalyst with a bulky N-heterocyclic carbene ligand and $Et_3N \cdot 3$ HF, vinyl fluorides are selectively generated from acetylenes. This method further requires acidic additives. Overall, the hydrofluorination proceeds *trans*-selectively and works well for both, symmetrical and unsymmetrical alkyne substrates. In the latter case, a moderate to excellent regioselectivity is observed [3346]. The latter is further improved, if a directing group is present, as for example, in N-troc protected (troc = 2,2,2-trichloroethyl carbamate) propargylic amines [3347]. Branched vinylic fluorides are selectively prepared upon treatment of terminal alkynes with DMPU/HF in presence of a gold(I)-catalyst [2777]. In another approach, a N-heterocyclic carbene gold(I)-bifluoride complex has been used for the hydrofluorination of alkynes. Starting from symmetric diarylacetylenes, the corresponding (Z)-stilbenes have been generated exclusively. Under slightly modified conditions, that is, lower temperatures and longer reaction times, unsymmetric substrates have been regioselectively hydrofluorinated. In both cases, $NEt_3 \cdot 3$ HF and $NH_4[BF_4]$ have been used in stoichiometric amounts [3348]. The reaction of *in situ* generated bromoalkynes with AgF in a mixture of acetonitrile and water yields the corresponding bromofluoroalkenes with (Z)-selectivity. In the same way, electron-deficient acetylenes, bearing an acyl or ester substituent, are converted to (Z)-fluoroolefins using AgF [3349].

15.6.2 Directed C-H fluorination of vinylic substrates

The fluorination of vinylic C-H bonds in proximity of directing O-methyl oxime ethers can be achieved by the use of electrophilic NFSI in presence of a cationic $[Pd(NO_3)]^+$ catalyst, which is efficiently generated from a Pd^0-source and KNO_3. This method gives access to a number of different vinylfluorides, mostly α-fluorostyrenes, if a trisubstituted alkene substrate is used. Moreover, this approach works for the preparation of fluoroarenes, too (cf. Section 14.5.1) [3188].

15.6.3 Fluorination of vinyl boronic acids and vinyl stannanes

Fluorinated olefins can also be prepared from substrates with a leaving group. For example, a selective method is given by the silver(I)-mediated reaction of vinylboronic acids with Selectfluor™. Using this method, fluorovinyl compounds are accessible in reasonable yields with complete stereocontrol [3215]. Similarly, vinyl stannanes are transformed to fluorinated alkenes when they are reacted with XeF_2 in presence of Ag(I) [3350–3352].

15.6.4 Miscellaneous transformations toward monofluoroolefins

Apart from the approaches presented above, there are only very few reports about transition-metal catalyzed preparations of fluoroolefins, which are in addition rather exotic in nature. For example, the reaction of propargyl acetates with a gold(I)-catalyst and stoichiometric amounts of Selectfluor™ generates α-fluoroenones [3353]. A very similar reaction has been described, using an additional AgOTf catalyst [3354]. Fluorinated dihydropyrrole-derivatives are obtained upon Ag(I)-mediated fluorocyclization of allenic amines, using NSFI as fluorinating agent [3355]. A silver(I)-mediated radical aminofluorination has been reported, when terminal allenes are treated with NFSI in presence of catalytic amounts of AgF [3356].

15.6.5 Preparation of geminal difluoroolefins from CF₃-substituted alkenes

Trifluoromethyl-substituted alkenes, especially styrenes, are versatile starting materials for the preparation of geminal difluoroolefins. The simplest derivatization approach is a hydrodefluorination reaction, which can, for example, easily be achieved by a Ni-catalyzed method using stoichiometric amounts of $PhSiH_3$ as hydride-source (Scheme 15.14) [3357]. A similar approach to the S_N2'-type nucleophilic substitution of trifluoromethyl-substituted alkenes, which generates geminal difluoroolefins (cf. Section 15.2.1), proceeds via radical intermediates and may be performed under photoredox-conditions or using transition-metal catalysis. Initially, an alkyl or aryl radical is transferred to the olefinic starting material to generate a tertiary, carbon-centered radical. In case of photocatalysis, the latter undergoes a single electron reduction by the reduced form of the catalyst and is transformed to a carbanion, which then eliminates a fluoride ion to yield the 1,1-difluoroolefin. Similarly, β-fluoride elimination of an intermediate transition metal complex will lead to the same product [3358–3365]. Indole heterocycles and other electron-rich aromatic substrates may be reacted with trifluoromethyl-substituted diazoalkanes in a Pd-catalyzed C-H functionalization approach to yield the corresponding geminal difluoroolefins [3366]. 1,1-Difluoro-1,4-dienes are accessible by a Ni-catalyzed defluorinative coupling of trifluoromethylalkenes with acetylenes, whereas a Cu-catalyzed reaction of trifluo-

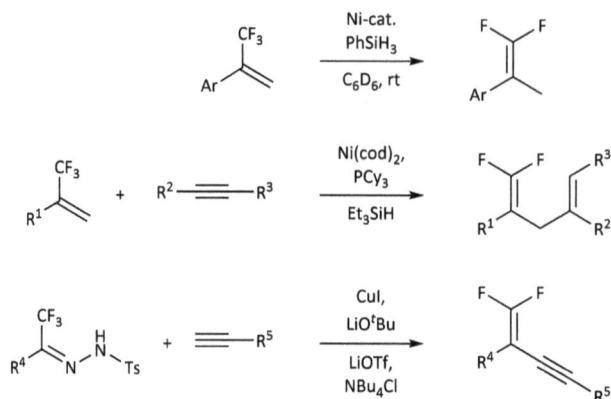

Scheme 15.14: Transition-metal mediated hydrodefluorination and coupling of trifluoromethylated substrates with acetylenes.

romethyl ketone *N*-tosylhydrazones with terminal alkynes gives rise to 1,1-difluoro-1,3-enynes (Scheme 15.14) [3367, 3368].

15.6.6 Preparation of geminal difluoroolefins from diazocompounds

Furthermore, 1,1-difluoroolefins are accessible from the corresponding diazo-sub-strates upon treatment with sources of difluorocarbene in a copper-mediated reac-tion. Further improvements of this method have been described, using Me_3SiCF_3 of Me_3SiCF_2Br as difluorocarbene-reagents without initiation by a transition metal. Instead, activation is achieved by compounds like $[NBu_4]Br$ or NaI. A few other variants have been described involving, for example, intermediate boronic acid in-termediates, which undergo β-fluoride elimination [3369]. Recently, perfluorinated isopropylidene fragments have been transferred to α-diazoesters, yielding geminal *bis*(trifluoromethyl)olefins. This has been realized by Cu-catalysis using Me_3SiCF_3 as only fluorocarbon-source. The transformation is believed to proceed via four suc-cessive steps, starting with a $:CF_2$ migratory insertion to yield an α-CF_3-substituted organocopper species. Afterwards, β-fluoride elimination gives rise to a geminal di-fluoroolefin, which then undergoes double addition-elimination with $[CF_3]^-$, finally resulting in the replacement of the vinylic fluorine substituents by trifluoromethyl groups [3370].

15.6.7 Study questions

(I)   Why are transition metal catalysts needed for the selective preparation of fluoroolefins by hydrofluorination of acetylenes? Which catalysts and reagents are used? Give examples.

(II)   How can vinyl fluorides be prepared by metal-mediated direct C-H fluorination and replacement reactions of vinyl boronic acids and -stannanes? Which catalysts and fluorinating agents are required?

(III)   Which products are obtained, if trifluoromethyl-substituted, terminal olefins and N-tosylhydrazone-protected trifluoromethylketones are reacted with acetylenes? Which transition metals are required for this transformation?

(IV)   How can diazo-compounds serve as substrates for the preparation of fluorinated olefins? Which reagents and catalysts are typically required and which types of products are generated?

15.7 Selective hydrodefluorination of highly fluorinated olefins

In most cases, perfluorinated olefins are more easily accessible than partly fluorinated ones, especially, if a certain substitution pattern is required. Thus, the controlled removal of fluorine substituents is a valuable method for the preparation of such compounds from highly fluorinated substrates. However, the subject of transition-metal catalyzed, selective hydrodefluorination has been reviewed elsewhere, and will hence only be discussed herein very briefly [3236].

In the literature, there are various methods for the selective F-abstraction from fluoroolefins, most of them requiring the presence of a transition-metal complex, sometimes even in stoichiometric amounts. As a prerequisite, the generated metal-fluorine bond needs to be rather strong, in addition a stoichiometric hydride source is required for catalytic transformations, typically a silane or alane. For example, fluoroolefins are hydrodefluorinated by the zirconium hydride compound $Cp_2^*ZrH_2$ ($Cp^* =$ pentamethylcyclopentadienyl; $[C_5(CH_3)_5]^-$) [3238]. The latter complex is also capable of detaching allylic fluorine-substituents, for example, in 3,3,3-trifluoropropene to generate 1,1-difluoropropene via an insertion/β-fluoride elimination mechanism [3371]. In general, vinylic and allylic fluorine substituents appear to be more reactive toward initial hydrodefluorination than those in fluoroarenes, as has been shown in the reaction of perfluoroallylbenzene with titanocene difluoride and Ph_2SiH_2. However, in the subsequent activation step, the arene will be defluorinated, too [3372]. The same titanium/silane system efficiently hydrodefluorinates fluorinated propenes in a catalytic fashion, such as hexafluoropropene, 1,1,3,3,3-pentafluoropropene or 3,3,3-trifluoropropene [3373, 3374]. More recently, it has been found that fluoroolefins are also hydrodefluorinated in an organocatalytic approach, using N- or O-donor molecules in conjunction with stoichiometric amounts of an aluminium or gallium hydride [3252, 3253].

16 Solutions to study questions

1.4

(I)   Fluorinated compounds can be found in the atmosphere as well as of course in large amounts in minerals in the earth's crust and the oceans. The most important fluoride minerals are fluorspar (CaF_2), cryolite $Na_3[AlF_6]$ and fluoroapatite [3 $Ca_3(PO_4)_2 \cdot CaF_2$], whereas volatile compounds, such as HF or the man-made fluorocarbons are found in the atmosphere.

(II)   The majority of natural fluorine compounds in the atmosphere is contributed by volcanoes, which mainly liberate HF in amounts of up to 6 million tons per year. However, large amounts of fluorinated compounds in the atmosphere are man-made, which especially holds for chlorofluorocarbons, hydrofluorocarbons and perfluorocarbons. Apart from that, some other persistent compounds, for example, SF_6, are originating from anthropogenic sources.

(III) Elemental fluorine can be found in trace amounts in some varieties of fluorite minerals, for example, in antozonite, which is found in Germany. It is liberated by grinding or shattering the mineral and can be detected by its smell. Its formation is explained by radiolytic cleavage of CaF_2, which is caused by accompanying radioactive components in the mineral.

(IV)   Fluoroacetate is found in some higher plants in tropical and subtropical regions. Since it is very toxic to animals, it is used as defence mechanism against them. Fluorocitrate is a metabolite of fluoroacetate, and thus also sometimes found in plants, which produce fluoroacetate. The seeds of some plants in Western Africa contain toxic ω-fluorooleic acid, which again serves as protection against animals.

(V)   Fluorinase is capable of connecting inorganic fluoride to biomolecules, thus forming C-F bonds. In combination with the generally high selectivity and efficiency of enzymatic reactions, this could potentially be exploited in synthetic approaches toward fluorinated molecules.

2.5

(I)   Elemental fluorine reacts with virtually every inorganic compound or element, unless it is already a fluoride in its highest oxidation state. Mostly, these reactions are very violent and often accompanied by explosions or ignitions. For example, F_2 reacts violently with sulphur or phosphorous even at −185 °C. Similarly, organic materials are often ignited, since fluorine has a very high affinity to hydrogen.

(II)   The extreme reactivity of fluorine is mainly explained by three facts, the low dissociation energy of the fluorine molecule, the high bond-energy of many element fluorine bonds, and finally, the small size of both the F-atom and the fluoride anion.

(III) The isotope ^{19}F is the only natural occurring isotope of fluorine. However, the artificial ^{18}F has a high relevance in medicinal diagnostics as positron emitter in PET-imaging.

(IV)   So far, the linear trifluoride anion $[F_3]^-$ and the V-shaped pentafluoride anion $[F_5]^-$ have been characterized by gas-phase and matrix isolation techniques.

(V)   In a purely chemical synthesis, F_2 is accessible by treatment of $K_2[MnF_6]$ with SbF_5. On an industrial scale, elemental fluorine is prepared by the electrolysis of HF/KF-mixtures in Monel- or steel cells at 70–130 °C. The cell material serves as cathode, whereas the anodes are typically made of graphite-free petroleum coke. During the electrolysis process, the evolved gases H_2 and F_2 strictly need to be kept separated from each other.

https://doi.org/10.1515/9783110659337-019

(VI)   Commercially available F_2 typically has a purity of 99.0 to 99.5 %, which is achieved by low temperature purification. Very pure fluorine gas (>99.7 %) is accessible, if the crude F_2 is reacted with NiF_2 and KF and subsequent thermal decomposition of the resulting nickel complex. Similarly, O_2 is removed by treatment with SbF_5. A final low-temperature distillation yields spectroscopically pure F_2.

(VII)   The choice of appropriate reaction vessels strongly depends on the reactants and the reaction conditions. Fluorinated compounds, which are nonvolatile and do not react with glass can generally be handled in standard laboratory glassware or quartz vessels. For all fluorides, which readily react with SiO_2, perfluorinated plastics may be used as materials for reaction vessels. The latter are also suitable materials, if F_2 is used in diluted form at low temperature under nonpressurized conditions. Very reactive materials and pure F_2 necessarily require metal reaction vessels, which are made of passivated stainless steel, copper or mostly nickel or Monel. However, no material is indefinitely stable against elemental F_2 or other highly reactive compounds, especially at elevated pressures and temperatures.

3.7

(I)   On an industrial scale, anhydrous HF is prepared by the treatment of fluorspar (CaF_2) with concentrated sulphuric acid at elevated temperatures. Multiple distillation processes furnish aHF with a purity of about 99.5 %. On laboratory scale, aHF may also be conveniently prepared by thermolysis of acidic fluorides MF · HF.

(II)   In the solid-state, HF exists as a polymer in form of unsymmetrically bridged, planar zig-zag chains. Upon melting, the polymeric chains break up to shorter fragments, which are built of about seven HF-molecules. The heptamers are typically not branched but interact with each other via H-bridges. Gaseous HF is comprised of monomeric HF and hexameric $(HF)_6$-molecules, which have a chair-like structure. At temperatures above 90 °C monomeric HF is observed exclusively.

(III)   Liquid anhydrous HF is highly acidic, which is due to its self-dissociation, yielding stable associated anions. The acidity is further increased, if a fluoride acceptor is added, typically a strong Lewis acid, such as AsF_5 or SbF_5. In contrast, aqueous solutions of HF are only weakly acidic, since the high H-F affinity leads to the formation of ion pairs $[H_3O]^+ \cdot F^-$. Consequently, no significant amount of free hydronium ions is available.

(IV)   An acidic fluoride salt contains one or more molecules of HF and forms a stable solvate. A very prominent example is KF · HF ($K^+[HF_2]^-$), apart from that KF can also incorporate two, three or even four molecules of HF. In all these salts, the anions are described by HF molecules, which coordinate to F^- forming very strong H-bridges.

(V)   A "naked" fluoride anion as such does not exist in condensed phase, since there are always at least small interactions with counterion, solvent, etc. In practice, ionic fluorides with weakly coordinating cations, mostly Cs^+ or $[N(CH_3)_4]^+$, are used as sources of naked fluoride. These compounds are strong fluoride donors, where F^- has strongly nucleophilic but also Lewis-basic properties.

(VI)   In basic aHF, one can find anions of the type $[F(HF)_n]^-$, in most cases bifluoride $[HF_2]^-$. Such media are obtained, if ionic fluorides, especially alkali metal fluorides, are dissolved in aHF. In contrast, the dissolution of many covalent fluorides, especially the strongly Lewis acidic pentafluorides PF_5, AsF_5 and SbF_5 yields acidic aHF. Therein, fluoronium cations $[H_2F]^+$ are found.

(VII)   The fluoroacidity of a reaction medium strongly influences the species that may be found in it. A typical example is the chemistry of Ag(III). In basic aHF, $[AgF_4]^-$ is obtained by oxidation of AgF_2 with $[O_2]^+$; upon acidification with AsF_5, AgF_3 is precipitated under neutral conditions. Finally, strongly oxidizing, cationic Ag(III)-species are obtained in acidic aHF.

(VIII)   Generally, element fluorides with a low oxidation state dissolve better in aHF than high-valent ones, which is especially pronounced for the corresponding Ag(I)/Ag(II) and Tl(I)/Tl(III) couples. Apart from that, the solubility of the element fluorides significantly increases with the size of the cation, which is especially underlined for the compounds in groups 1 (LiF vs. CsF) and 2 (BeF$_2$ vs. BaF$_2$).

4.4

(I)   Fluorides in the highest oxidation states are typically accessible by gas-phase reactions with strongly oxidizing reagents like F_2 or the higher halogen fluorides. The reactions can either be conducted in flow or under pressure. Typical substrates are the elements as well as oxides or low-valent fluorides.

(II)   Fluorides in medium oxidation states are conveniently accessible by reduction, as long as a stable high-valent species is readily accessible. Typical reducing agents are silicon, CO, H_2 or the respective element.

(III)   Solid-state reactions are mostly used for the preparation of polynary fluorides with a certain stoichiometry. In some cases, these techniques are also applied to prepare anhydrous fluorides from decomposable starting materials.

(IV)   In most cases, fluorination reactions in chloride flux can be conducted at significantly lower temperature than in solid state. Moreover, for the preparation of fluorometallate complexes, a flux may be used, which is the fluorinating agent at the same time, for example, KHF_2.

(V)   Nonhydrolyzable fluorides can be prepared in aqueous solution, where they are in most cases precipitated from chlorides, hydroxides, carbonates or oxides upon addition of aqueous HF or some other fluoride source, typically an alkali metal fluoride. In some cases, the resulting fluoride is obtained as hydrate. In contrast, hydrolyzable fluorides can of course not be prepared in aqueous systems. In some rare cases, they may be precipitated from organic solvents, which are problematic, as both a soluble starting material and a soluble fluoride salt are required. More conveniently, the starting material is submitted to halogen exchange in inert solvents, mostly aHF or acetonitrile, using aHF, alkali metal fluorides or even organic tetraalkylammonium salts as fluoride source. This method is applicable to both, organic and inorganic syntheses.

(VI)   Since they are conducted with oxidizing reagents, such as F_2 or XeF_2, fluorination reactions need to be carried out in inert solvents, often aHF, anhydrous MeCN or a chlorofluorocarbon, for example, CCl_3F. As another important prerequisite, the transformations have to be carried out at sufficiently low temperature and under high dilution conditions, which especially holds for the use of F_2.

5.3

(I)   In the rutile type, the cations are octahedrally coordinated by fluorine, whereas fluorine has a trigonal-planar coordination environment. In the fluorite structure, the cations are surrounded by eight fluoride anions, forming a cube, whereas the fluoride anions are tetrahedrally coordinated by the metal cations. The occurrence of these two structure types mainly depends on the size-ratio of the cation toward the fluoride anion. Smaller cations typically prefer the rutile type, as it is observed for MgF_2 or many difluorides of the 3d-metals, whereas difluorides with larger cations tend to adapt a fluorite structure, for example, the heavier earth alkaline fluorides SrF_2 and BaF_2.

(II) In all three structure types, the cations are octahedrally surrounded by fluorine, thus forming $[MF_6]$-units, which are sharing all six corners with neighboring octahedra. The differences arise from the M-F-M angle, and thus the twist of neighboring octahedra. In the ReO_3 type, the M-F-M bridges are linear, whereas a slight twist is observed in the VF_3 type, which is even more pronounced in RhF_3.

(III) Trifluorides with large cations often prefer either the LaF_3 or the YF_3 structure type. In both cases, the coordination environment around the cation is based on a trigonal prism. In the LaF_3 type, the trigonal prism is fully capped, however, the fluorides on the top and bottom side of the prism are more distant, so that the geometry is often described as 9+2. The YF_3 type contains threefold-capped trigonal prisms, where one of the fluorides is a little more distant, resulting in an 8+1 coordination. In both structure types, the coordination environment around fluorine is trigonal planar.

(IV) Both the SnF_4 and the PdF_4 type are composed of $[MF_6]$-octahedra, which are connected by four common corners. The difference is the relative position of the two nonbridging fluorine substituents, which are in *trans*-position in the SnF_4 type but in *cis*-position in PdF_4. As a consequence, SnF_4 forms two-dimensional layers, whereas PdF_4 exhibits a three-dimensional network.

(V) The α-UF_5 type contains linear chains of $[MF_6]$-octahedra, which are connected via common corners. In contrast, the M-F-M bridges are not linear in the VF_5 type, so that the octahedral units are twisted against each other. A similar behavior is observed for the cyclic tetramers in the NbF_5 and RuF_5 type. In the NbF_5 structure, the M-F-M bridges are linear, whereas they are bent in RuF_5.

7.4

(I) Since F_2 is detected by its smell, even in very low concentrations, poisoning hardly occurs. However, the contact with skin and tissue leads to severe skin burns, mostly accompanied by fluoride poisoning, which is caused by the generated HF and other fluoride compounds.

(II) Typically, the lethal dose of fluoride for adults amounts to 30-60 mg per kg bodyweight. Skin contact with HF does not only result in very painful skin burns, but additionally causes fluoride poisoning. Depending on the affected skin area and the amount of HF, and thus its concentration, the skin contact may be lethal, too.

(III) Both, fluoroacetate and perfluoroisobutene are extremely toxic compounds, since they inhibit important processes in the human body. However, fluoroacetate is not volatile under standard conditions and is therefore by far less easily incorporated than gaseous perfluoroisobutene.

(IV) The most "famous" persistent compounds are probably perfluorooctanoic acid (PFOA) and perfluorooctyl sulphonates (PFOS), which have been used as impregnations for textiles and in fire-extinguishing foams.

(V) The Montreal Protocol has been established in 1987 to regulate the use and limit the liberation of ozone-depleting chlorofluorocarbons.

(VI) Hydrofluorocarbons do not decompose under the liberation of ozone-depleting Cl^- radicals. However, they are long-lasting in the atmosphere and significantly contribute to the greenhouse-effect. To avoid this, replacements are developed, which are less long-lasting and/or have a smaller GWP. In addition, attempts are made to convert volatile hydro- and chlorofluorocarbon waste into valuable starting materials.

(VII) "Up-Cycling" is a process, which converts fluorinated polymers back into valuable fluorinated monomers, which can be reused in production. Thus, the amount of waste is reduced at the same time as resources are saved.

(VIII) In general, fluorous chemistry makes use of the miscibility behavior of highly fluorinated compounds with nonfluorinated organic solvents and aqueous phases. Upon warming, fluorous solvents typically mix with organic phases, but they will separate again when cooled down. This, for example, allows the use of catalysts, which are substituted with perfluoroalkyl chains. The substrate and the product typically dissolve in the organic phase. After the reaction, simple phase separation gives rise to a reusable, fluorous phase, containing the catalyst. Overall, this saves resources and produces less waste. In addition, perfluorocarbons can take up a rather large amount of gases, especially O_2, due to which fluorous chemistry is extremely valuable for oxidation reactions.

8.4

(I) ^{19}F-NMR is a versatile analytical tool. Due to the high sensitivity of the only natural occurring isotope ^{19}F, it allows for example the online reaction control, as long as fluorinated groups are present in the substrate and/or the product. Another very beneficial aspect is the high dispersity of fluorine chemical shifts, so that changes in the molecular structure often lead to significant changes in δ, which easily allows a fast conclusion, if a reaction was successful or not.

(II) Nowadays, $CFCl_3$ is almost exclusively used as reference substance, in former times, other substances, for example, trifluoroacetic acid, have been used, too. The chemical shift of a certain nuclei depends on a few experimental parameters, such as the nature of the solvent, the temperature, but in many cases also the concentration of the sample. Therefore, the use of an internal standard with a well-known chemical shift under the respective measurement conditions is very beneficial for an exact determination of $\delta(^{19}F)$ for an unknown compound.

(III) In binary fluorides, the chemical shift typically increases with the electronegativity of the element, this is, for example, observed for the pentafluorides SbF_5 and AsF_5, but also for the hexafluorides TeF_6 and SeF_6. Moreover, this effect is observed for organic compounds, where the α-fluorine atoms of ethers have a higher chemical shift than those of the corresponding thioether derivatives.

(IV) SF_4 has a ψ-trigonal-bipyramidal molecular structure, where the lone-pair occupies one of the equatorial positions. Accordingly, the two axial and the two equatorial fluorine atoms are not equivalent. However, at room temperature, a single resonance is observed for all four fluorine substituents, which is due to the fluxional behavior of the molecule. If the temperature is sufficiently high, the barrier for the interconversion of the different fluorine nuclei is overcome and a fast exchange process is observed, making the fluorine nuclei equivalent in the NMR-spectrum. At low temperature, this interconversion process can be slowed down and finally inhibited, so that in case of SF_4 two triplet resonances will be observed.

(V) For both types of compounds, the chemical shift of the fluorine nuclei strongly depends on the nature of the substituents R. If R is another halide, the chemical shift will be slightly positive or around zero, whereas the proton-substituted compounds are found in strongly negative shift regions. Depending on the shielding, other residues, for example, organic groups will move the chemical shifts closer to the one or other extreme.

(VI)   The chemical shifts of fluoroarenes may also be found in a rather wide region of the ^{19}F-NMR spectrum. For example, fluorobenzene appears at about −116 ppm, whereas the shift values generally decrease for higher fluorinated aromatics. However, the most negative shift values are typically observed for pentafluorophenyl groups, and not for hexafluorobenzene. For di-, tri- and tetrafluorobenzene derivatives, the chemical shift also depends on the substitution pattern, as for example the shift values decrease in the series $\delta(meta\text{-}C_6H_4F_2) > \delta(para\text{-}C_6H_4F_2) > \delta(ortho\text{-}C_6H_4F_2)$.

(VII) The coupling constant can give information, where a fluorine substituent is bound, which especially plays a role in inorganic compounds, for example, derivatives of transition metal fluorides. For example, very high ^1J coupling constants are observed for platinum and other transition metal fluorides. Especially in organic compounds, the coupling constants can also provide structural information, as for example the vicinal ^3J coupling constants depend on the dihedral angle. Moreover, the ^3J coupling constants in fluoroolefins provide information about the E/Z-geometry.

9.1.2

(I)   Generally, the alkali metal fluorides are prepared by treatment of hydroxides or carbonates with aqueous HF. Some of the fluorides form stable hydrates, which can be dried by heating in high vacuum. For the lighter homologues, the sensitivity toward hydrolysis has to be taken into account.

(II)   All the alkali metal fluorides crystallize in a NaCl lattice, where both cations and fluoride anions exhibit an octahedral coordination environment. However, since Cs^+ is far bigger than F^-, CsF adapts an inverted NaCl structure.

(III) LiF is only poorly soluble in water, however, with increasing size of the cation, the alkali metal fluorides are better soluble. In contrast, the sensitivity toward hydrolysis decreases from LiF to CsF. For example, NaF is hydrolyzed by humidity upon heating.

9.2.3

(I) BeF_2 is prepared by treatment of the oxide BeO with aHF at 220 °C or by thermal decomposition of the ammonium beryllate $(NH_4)_2[BeF_4]$. It cannot be isolated from aqueous solution, because the hydrate $[BeF_2(H_2O)_2]$ undergoes hydrolysis before dehydration takes place.

(II)   In general, BeF_2 has some structural relationship to quartz. Thus, it shows tetrahedral $[BeF_4]$-units in the solid state, and undergoes several temperature-dependent phase transitions. In the gas phase, BeF_2 exists as linear monomer, the existence of dimeric structures has also been identified by matrix experiments.

(III) BeF_2 readily accepts fluoride ions under formation of several different fluoroberyllate anions. The $[BeF_4]^{2-}$ anion is readily accessible from aqueous solutions, if elemental beryllium is treated with HF or NH_4F.

(IV)   All the compounds are typically prepared by treatment of the carbonates with HF. They are resistant toward hydrolysis.

(V) MgF_2 differs from the other compounds, since it crystallizes in a rutile lattice (octahedral coordination of Mg^{2+}; trigonal planar coordination of F^-). The others are described by a fluorite lattice (cubic coordination of M^{2+}; tetrahedral coordination of F^-).

9.3.3

(I)   Boron subfluorides include all binary boron fluorine compounds, where boron has an oxidation state smaller than +III. They are all more or less unstable and decompose already at low temperatures (BF, B_3F_5, B_4F_6, B_8F_{12}) or latest at room temperature (B_2F_4). They are typically accessible by low temperature condensation reactions, the thermally more stable B_2F_4 may also be prepared by halogen exchange of B_2Cl_4 or B_2Br_4.

(II) The structure of B_8F_{12} is generally related to that of diborane B_2H_6, thus B_8F_{12} may be rewritten as $B_2(BF_2)_6$. Consequently, four of the BF_2-groups are terminal, whereas the other two are bound in a bridging mode. In contrast to the $[BH_2B]$-fragment in B_2H_6, the central $[B(BF_2)_2B]$-unit is not planar in B_8F_{12}. The structure of $B_{10}F_{12}$ is described by a distorted B_4-tetrahedron, where four terminal BF_2-unit are connected to each of the four corners. The remaining two BF_2-fragments are attached in bridging mode on opposite edges of the tetrahedron. One can reformulate $B_{10}F_{12}$ as $B_4(BF_2)_6$, however in this case a corresponding hydrogen derivative, that is, $closo$-B_4H_6, is nonexisting.

(III) On technical scale, BF_3 is prepared by the reaction of B_2O_3, $B(OH)_3$ or $Na_2B_4O_7$ with CaF_2 and oleum. The reaction of the latter two components yields HF, which then replaces the oxygen substituents of the boron substrate. The dehydrating properties of oleum are exploited to remove the liberated water. BF_3 is a strong Lewis acid, thus a large number of stable adducts with Lewis bases is formed. One of the most relevant compounds is the commercially available ether adduct $BF_3 \cdot OEt_2$, which significantly simplifies handling of the toxic and corrosive BF_3.

(IV) In contrast to BH_3, BF_3 does not tend to oligomerize and thus exists as trigonal planar monomer. Some weak intermolecular interactions are only found in solid BF_3. Due to π-backdonation of the fluorine substituents, the B-F bonds are comparably short compared to typical B-F single bonds.

(V) $[BF_4]^-$ is a very weak Lewis base and thus belongs to the family of weakly coordinating anions. Typically, anhydrous tetrafluoroborate salts are obtained by the treatment of metal oxides, hydroxides or carbonates with concentrated solutions of HBF_4. In some cases, they may also be prepared by treatment of a metal chloride or fluoride with BF_3.

(VI) For the group 13 metals, the stability of the trifluorides decreases from aluminium to thallium, whereas the stability of the monofluorides increases. Consequently, AlF_3, GaF_3, InF_3 and TlF are the most stable fluorides of these elements. This observation is explained by the relativistic inert-pair effect.

(VII) TlF is prepared by the treatment of thallium salts, such as Tl_2CO_3, with HF. The chemistry of TlF is comparable to those of the alkali metal fluorides and AgF, which is due to the similar ionic radii of the cations.

(VIII) In the solid state, AlF_3 crystallizes in a distorted ReO_3 lattice. In the vapor phase one can find both, trigonal planar AlF_3 monomers and $(AlF_3)_2$ dimers, which are in equilibrium. AlF_3 is obtained by high temperature treatment of the metal or oxide with HF.

(IX) Cryolite is a very important additive in the fused salt synthesis of aluminium metal. For the synthesis of $Na_3[AlF_6]$, Al_2O_3 is treated with HF, which initially yields $H_3[AlF_6]$. Neutralization of the latter with Na_2CO_3 yields synthetic cryolite. Alternatively, it may also be prepared by treatment of NaOH and $Al(OH)_3$ with NH_4F.

(X)   With fluoride donors, AlF_3 readily forms fluoroaluminates of the general composition $M[AlF_4]$, $M_2[AlF_5]$ and $M_3[AlF_6]$. In most of them, aluminium is octahedrally coordinated by six fluorine substituents. Thus, isolated octahedral $[AlF_6]^{3-}$ anions are found in $M_3[AlF_6]$, whereas a connection of the $[MF_6]$-units is found in $[AlF_5]_n^{n-}$ and $[AlF_4]_n^{2n-}$, which form one-dimensional chains and two-dimensional sheets, respectively. However, isolated tetra-hedral $[AlF_4]^-$ and trigonal-bipyramidal $[AlF_5]^{2-}$ anions are observed in the tetramethylam-monium salts.

(XI) The thermal decomposition of $(NH_4)_3[GaF_6]$ and $(NH_4)_3[InF_6]$ at 170 °C mainly yields the tetrafluorometallate complexes $NH_4[GaF_4]$ and $NH_4[InF_4]$. Further heating of the tetrafluo-roindate species to 300 °C gives rise to neutral InF_3.

(XII) TlF_3 is commonly obtained by treatment of the oxide Tl_2O_3 with F_2, BrF_3 or SF_4 at 300 °C. Its solid-state structure is described by the YF_3 type; thus, the Tl centers are surrounded by each nine fluorine substituents. Six of them build a distorted trigonal prism around the metal; the other three are capping the rectangular faces, where one of them is a little more distant than the other two.

9.4.3

(I) The graphite fluorides can be divided in the two general groups of graphite fluoride $(CF_x)_n$ and graphite subfluoride $(C_yF)_n$. The actual composition of these compounds depends on their preparation; however, the parameters x and y always range around 1 and 4, respec-tively. They are both prepared by treatment of graphite with fluorine in presence of HF. For the preparation of $(C_yF)_n$, the reaction is performed at room temperature, whereas $(CF_x)_n$ re-quires higher temperatures of 400–600 °C. Both graphite fluorides significantly differ from nonfluorinated graphite with respect to their conductivity, since they do not $((CF_x)_n)$ or only slightly $((C_yF)_n)$ conduct electricity.

(II) Unlike CH_2, difluorocarbene is a singlet carbene, meaning that the free electrons occupy the same orbital. This behavior is explained by the fact, that the p-orbitals of carbon and fluorine are similar in size, which leads to a good overlap. Thus, the fluorine substituents can donate electron density to the empty orbital of carbon by π-backdonation, which leads to a stabilization of the carbene. This would indeed be impossible, if CF_2 was a triplet species. In addition, difluorocarbene has a bent structure, which is in contrast to the almost linear triplet carbene CH_2.

(III) The fluorination of C_{60} will finally yield a fullerene fluoride of the composition $C_{60}F_{48}$. Further fluorination does not yield higher fluorinated fullerenes, such as the hypothetical perfluori-nated $C_{60}F_{60}$, but results in fragmentation of the carbon cage. This is explained by the fact that the six remaining double bonds in $C_{60}F_{48}$ are perfectly shielded by the fluorine sub-stituents.

(IV) SiF_2 is typically prepared by comproportionation of SiF_4 and elemental Si or thermal decom-position of Si_2F_6. In both cases, the reactions need to be carried out at high temperature and *in vacuo*. As the carbenes CF_2 and CH_2, difluorosilylene is only stable in the gas-phase or if it is frozen out in an inert gas matrix. However, it is comparably far more stable than CF_2 and even more so in comparison to CH_2, which is represented by the respective half-lives of about 150, 1 and 0.1 s.

(V) In the gas-phase, all three difluorides are found in monomeric form, at lower temperatures, GeF_2 and SnF_2 may also exist as di- or trimers. In their ground-state, all the monomers are singlet species, which therefore have a bent molecular structure. In the solid state, all the metal fluorides show a different coordination environment, thus they also crystallize in dif-ferent lattices. GeF_2 is surrounded by four fluorine substituents, which results in a ψ-trigonal bipyramidal environment, also described as see-saw shape. In contrast, a distorted octahe-dral environment is found for every Sn-atom in solid SnF_2. The coordination is further increased for lead difluoride, where every Pb-center is surrounded by nine F-substituents in form of a distorted, three-fold capped trigonal prism.

(VI)   Due to the inert-pair effect, the stability of the difluorides significantly increases from GeF_2 to PbF_2. Accordingly, GeF_2 is a strong reducing agent, which readily converts SeF_4 to elemental Se. In the same way, $[GeF_3]^-$ salts are oxidized upon contact with air. This behavior is also observed for the corresponding fluorostannates(II). In contrast, fluoroplumbates(II) are stable against air.

(VII)   In the vapor phase, all the tetrafluorides exist as tetrahedral molecules. The inversion barrier of the tetrahedron is rather high for SiF_4 and GeF_4, but decreases for SnF_4 and PbF_4. Thus, the former two do not tend to form octahedrally coordinated metal centers in the solid state, which is in contrast clearly observed for SnF_4 and PbF_4.

(VIII) SiF_4 is very sensitive toward hydrolysis, thus, it is converted to SiO_2 and H_2SiF_6, when it is passed into water. H_2SiF_6 cannot be isolated, but aqueous solutions with a concentration of 60 % may be prepared. Stable derivatives, for example hexafluorosilicate salts, are obtained, if H_2SiF_6 is treated with metal hydroxides or carbonates.

9.5.3

(I) NF_3 is prepared by the fluorination of ammonia with F_2 in presence of copper. Alternatively, it is also accessible by treatment of mixtures of NH_3, NH_4F and HF with F_2 or electrochemical fluorination of urea or NH_4F in aHF. Like NH_3, NF_3 has a trigonal pyramidal structure, but it does not show the umbrella inversion of ammonia. A significant difference is also observed in the chemical behavior, since NF_3 is only a very weak Lewis base and consequently virtually insoluble in water.

(II)   The fluoroamines NH_2F and NHF_2 as well as the chloro compounds $NClF_2$ and NCl_2F are certainly the most important derivatives of NF_3. NH_2F is prepared by the decomposition of $[NH_3F]^+$ salts in vacuo. It is a strong amination agent and a weak base. NHF_2 is conveniently prepared by the hydrolytic cleavage of -C(O)NF_2 groups from difluorourea or corresponding carbamates. The cleavage of N_2F_4 with PhSH also yields difluoroamine, which is a very weak base. With strong acids, it generates $[NH_2F_2]^+$, whereas it acts as an acid toward strong bases like OH^-, giving rise to $[NF_2]^-$. $NClF_2$ is generally prepared by chlorination of NHF_2. Since it is in equilibrium with N_2F_4 and Cl_2, it is a source of the corresponding radicals and will also add to unsaturated C-C bonds. NCl_2F is less extensively explored and is most safely prepared by low temperature fluorination of $FC(O)NCl_2$ with F_2.

(III)   At elevated temperatures, N_2F_4 dissociates reversibly to NF_2 radicals. Thus, it yields products like SF_5NF_2 or $ClNF_2$ upon reaction with S_2F_{10} or Cl_2, respectively.

(IV) N_2F_2 exists as a cis- and a trans-isomer. They can be separated by gas chromatography or high vacuum distillation, apart from that they are also selectively accessible by certain synthetic methods. For example, the trans-isomer is obtained by reduction of N_2F_4 with Al_2Cl_6 or $[N_2F_3][AsF_6]$ with NOCl. In contrast, cis-N_2F_2 is exclusively obtained by treatment of $[N_2F][AsF_6]$ with NaF or NOF. Interestingly, the cis-isomer is $12 \, kJ \cdot mol^{-1}$ lower in energy but at the same time more reactive than the trans-species, as it, for example, slowly attacks glass.

(V)   The treatment of N_2F_2 with AsF_5 or SbF_5 yields the linear cation $[N_2F]^+$. It is a strong oxidizer and converts Xe to $[XeF]^+$ and ClF to $[ClF_2]^+$. Upon reaction of $[N_2F]^+$ salts with HN_3, the planar, V-shaped cation $[N_5]^+$ is obtained.

(VI)   A covalent NF_5 does not exist, only an ionic $NF_4^+F^-$ has been reported, which decomposes to NF_3 and F_2 at very low temperature. Nevertheless, numerous pentavalent salts with the tetrahedral $[NF_4]^+$ cation are known. Typically, they are prepared upon treatment of NF_3 with F_2 and a fluoride acceptor, for example, AsF_5 or SbF_5.

(VII) The three nitrogen oxyfluorides are mainly used as fluorinating agents. In all cases, cationic species are obtained upon treatment with strong fluoride acceptors. The bent NOF is converted to linear $[NO]^+$, planar NO_2F gives rise to the bent nitryl cation $[NO_2]^+$ and finally, fluoride abstraction from distorted tetrahedral NOF_3 results in the formation of the planar, C_{2v}-symmetric cation $[NOF_2]^+$.

(VIII) P_2F_4 is accessible by the reductive treatment of PF_2I with Hg. The P-P bond is significantly stronger than the N-N bond in N_2F_4. Consequently, the homolytic cleavage of P_2F_2 into $^{\cdot}PF_2$ radicals requires more energy. However, the general chemical behavior is comparable; thus, the transfer of $^{\cdot}PF_2$ radicals is mainly observed, for example, to unsaturated C-C bonds.

(IX) PF_3 is a very weak Lewis acid and does only accept a fluoride ion, when it is reacted with anhydrous $[NMe_4]F$. In the same way, it does not show clear fluoride donating properties upon reaction with AsF_5 or SbF_5. In contrast, AsF_3 reacts with both fluoride donors and acceptors. SbF_3 is a strong Lewis acid but only a weak Lewis base, so that it preferentially acts as fluoride acceptor, giving rise to mostly poly- and oligomeric anions. BiF_3 is converted to covalent polymeric species upon treatment with AsF_5 or SbF_5.

(X) Like CO, PF_3 has π-acceptor properties and may thus be used as ligand. Compared to CO, this property is even more pronounced for PF_3, so that for example the complexes $Pd(PF_3)_4$ and $Pt(PF_3)_4$ are relatively stable, whereas the corresponding carbonyl complexes may only be isolated at very low temperatures.

(XI) AsF_3 and SbF_3 are conveniently used in halogen exchange reactions of chlorides. The treatment of a chloride substrate with AsF_3 yields $AsCl_3$ as by-product, which is volatile and can be removed by distillation. Thus, AsF_3 is preferred for the preparation of high-boiling fluorides. In contrast, SbF_3 is suitable for the preparation of volatile fluorides, since $SbCl_3$ is a solid, which typically remains in the reaction vessel when the reaction mixture is distilled.

(XII) If PF_5 is reacted with SbF_5, it donates a fluoride ion and gives rise to the unstable complex $[PF_4][Sb_3F_{16}]$. In the same way, a fluoride ion is abstracted from AsF_5, when it reacts with F_2 at a Pt-wire, which leads to the formation of $[AsF_4][PtF_6]$. The other two pentafluorides SbF_5 and BiF_5 do not show any fluoride donor properties, since they are the strongest acceptors. In the end, SbF_5 is the slightly stronger acid, as it is capable of abstracting fluoride from $[BiF_6]^-$.

(XIII) Upon contact with fluoride donors, PF_5 exclusively yields $[PF_6]^-$. The same is observed for AsF_5, with the extension, that $[As_2F_{11}]^-$ may be additionally obtained at low temperature and in presence of excess AsF_5. The situation completely changes for SbF_5, which not only forms $[SbF_6]^-$, but also higher aggregates like $[Sb_2F_{11}]^-$ or $[Sb_3F_{16}]^-$. Moreover, the dianion $[SbF_7]^{2-}$ is known. Finally, BiF_5 does not form polyanions, but only accepts fluoride ions to form $[BiF_6]^-$ and, in a few cases, $[BiF_7]^{2-}$.

9.6.3

(I) Both oxygen fluorides are unstable compounds, which possess strongly oxidizing and fluorinating properties. In some cases, $^{\cdot}O_2F$ can be decomposed to O_2 and $^{\cdot}F$, which makes it a convenient and radiation-free source of atomic fluorine. Apart from that, it can also serve as reagent for the transfer of O_2F-moieties. $^{\cdot}OF$ does not show a pronounced tendency to recombine to O_2F_2, as it mainly yields O_2 and F-atoms. In contrast, $^{\cdot}O_2F$ can dimerize to O_4F_2, which exhibits a weak O-O bridge.

(II) Both OF_2 and O_2F_2 are very strong oxidizing reagents. For example, they are both capable of oxidizing Xe to XeF_2, in case of O_2F_2 this occurs at very low temperature (-120 °C). O_2F_2

also converts tetravalent Np- and Pu-compounds to the hexafluorides NpF_6 and PuF_6, respectively. Due to its thermal instability, its use is limited to reactions at low temperature. In contrast, OF_2 is stable up to 200 °C and can thus be applied in reactions at higher temperatures, which, for example, allows the photolytic oxidation of Kr to KrF_2. In aqueous solution it reacts with HX (X = Cl, Br, I) and liberates HF, H_2O and X_2. Upon treatment with strong fluoride acceptors, both compounds are converted to dioxygenyl salts.

(III)   Pure HOF is prepared by the reaction of F_2 with ice at –40 °C. However, it is a rather unstable gas, which has a half-life of about 30 minutes at room temperature. More conveniently, the stable adduct HOF · MeCN is obtained, if F_2 is reacted with a low-concentrated mixture of water in acetonitrile at 0 °C. The adduct is a very powerful oxidant for organic substrates, as it is for example used for the synthesis of epoxides, N-oxides or other heteroatom oxides, which are often hardly accessible by other common oxidizing agents.

(IV)   All the inorganic hypofluorites ROF possess an electrophilic fluorine substituent, thus they can be used as electrophilic fluorinating agents. In reactions with unsaturated compounds, they typically add a RO˙ and a F˙ radical to the organic substrate. However, in contact with organic matter the hypofluorites can react very violently, so that they cannot be safely used in organic synthesis.

(V)   The organic hypofluorites can be divided into the two sub-classes of perfluoroacyl- and fluoroalkyl hypofluorites. The acyl compounds are conveniently prepared by the treatment of the perfluorocarboxylic acids or their salts with F_2 in the presence of HF or water. The fluoroalkyl fluoroxy compounds are similarly obtained by the fluorination of alcohols or ketones. Both the acyl and the alkyl derivatives are used as sources of electrophilic fluorine in organic synthesis. Apart from that, the fluoroalkyl fluoroxy compounds, especially CF_3OF and $CF_2(OF)_2$ may be added to olefinic substrates in the sense of RO˙ and F˙ radicals.

(VI) SF_2 is an unstable molecule, which may only be observed in absence of decomposition catalysts in diluted gas phase. It is a bent molecule and mainly decomposes via di- and trimerization processes. The dimerization is reversible and yields F_3SSF, which may be described as derivative of SF_4, where one of the fluorine substituents has been replaced by an SF-group. The trimerization is irreversible and may be formally described as addition of SF_2 to the dimer S_2F_4, thus yielding a compound of the constitution F_3SSSF_3. The latter decomposes irreversibly to SF_4 and SSF_2. Its structure may be derived from two SF_4 molecules, where each one fluorine substituent is replaced by a common, bridging sulphur substituent.

(VII)   Disulphur difluoride FSSF can be described by two SF-fragments, which are connected by a comparably short S-S single bond. The small bonding distance between the sulphur atoms can be explained by a partial double-bond character, which is enabled by no-bond resonance of the fluorine substituents. In contrast, thiothionylfluoride $S=SF_2$ has a pyramidal structure. FSSF is typically prepared by fluorination of molten sulphur with AgF at 125 °C or alternatively by halogen exchange of S_2Cl_2, again using AgF as fluorinating agent. In a similar way, SSF_2 may be prepared by treatment of gaseous S_2Cl_2 with KF at 145 °C or with HgF_2 at room temperature, alternatively molten sulphur may be fluorinated by NF_3 at elevated temperatures. Both compounds are in equilibrium with each other, especially in presence of HF or BF_3, FSSF is readily isomerized to SSF_2. Moreover, the isomerization of FSSF is accelerated, as soon as traces of SSF_2 are present. However, the equilibrium is reversible and temperature dependent, thus, FSSF is preferentially formed upon cooling.

(VIII) SF_4 may be prepared by several different methods, however in any case a low-valent sulphur chloride is used as starting material, that is, S_2Cl_2 or SCl_2. For example, the reaction of SCl_2 with NaF or pyridine/HF yields sulphur tetrafluoride via intermediary, unstable SF_2. In the same way, the treatment of SCl_2 with Cl_2 and NaF is a suitable method to prepare SF_4. Sulphur tetrafluoride can act as both, fluoride acceptor and donor. Thus, stable $[SF_5]^-$ salts are obtained, if SF_4 is reacted with fluoride donors with large cations, such as CsF or $[NMe_4]F$. In contrast, the treatment with fluoride acceptors, for example, BF_3 or SbF_5, yields very stable $[SF_3]^+$ salts. Neutral SF_4 has a ψ-trigonal bipyramidal shape with each two axial and equatorial fluorine substituents, often also referred to as see-saw structure. The $[SF_5]^-$ anion forms a square pyramid with the sulphur atom at the foot point. The $[SF_3]^+$ cation has a trigonal pyramidal structure, where sulphur occupies the apex of the pyramid.

(IX) The molecular structures of both tetrafluorides are very similar, showing see-saw shaped structures. However, SF_4 is monomeric and does not tend to associate, whereas SeF_4 does so in the solid and the gaseous state. For example, it exhibits F-bridges in the solid state, which lead to a distorted octahedral coordination geometry around the Se-centers. Both compounds are good fluorinating agents for organic as well as inorganic compounds, so in general, their chemical reactivity and behavior does not differ a lot. This also holds for the fluoride accepting properties, since both fluorides can be converted to the pentafluorochalcogenate(IV) anions $[SF_5]^-$ and $[SeF_5]^-$, respectively. However, $[SeF_5]^-$ can additionally react with naked fluoride donors to obtain the distorted octahedral dianion $[SeF_6]^{2-}$, which is indeed not possible for the sulphur analogue.

(X) TeF_4 is typically prepared by fluorination of TeO_2, for example, using SeF_4 or FeF_3 as fluorinating agents. In the same way, it can be obtained by a comproportionation reaction of elemental Te and TeF_6. Moreover, pure TeF_4 is accessible by thermal decomposition of $[TeF_5]^-$ salts. The chemical reactivity of TeF_4 is comparable with those of SF_4 and SeF_4, thus it is also a good fluorinating agent. However, it is significantly less stable and starts to disproportionate above 190 °C. Like the other tetrafluorides, it is very sensitive toward hydrolysis. TeF_4 completely decomposes in glass and quartz vessels, if it is heated to 200 °C.

(XI) Based on thermodynamics, the aqueous hydrolysis of SF_6 toward HF and H_2SO_4 should occur readily in an exothermic reaction. However, such a reaction is not observed, which is due to the kinetic stability of the molecule. A nucleophilic attack is efficiently inhibited by the perfect shielding of the electrophilic sulphur atom by the six fluorine substituents; thus, an S_N2-reaction cannot occur. On the other hand, the bonding situation prevents the abstraction of a fluoride ion by strong acceptors like SbF_5 to yield $[SF_5]^+$, so that S_N1-reactions cannot take place either. In general, SF_6 is only attacked under extreme conditions, for example, by metals in liquid ammonia or in high-temperature reactions with electrophilic reagents. However, as soon as one of the fluoride substituents is replaced by another atom or group, the corresponding molecule becomes susceptible to nucleophilic attacks.

(XII) To date, the only relevant SF_5-reagents are S_2F_{10}, SF_5Cl and SF_5Br. S_2F_{10} is most conveniently prepared by photolysis of a SF_5Cl/H_2 mixture, whereas the halides SF_5X are accessible by treatment of SF_4 with XF in the presence of CsF. Under UV-irradiation or heating, all the reagents liberate SF_5 radicals, which may be transferred to appropriate substrates, for example, unsaturated organic compounds. In some cases, the activation of the reagents may also be achieved by the use of reagents like BEt_3.

(XIII) The oxyfluorides of hexavalent sulphur significantly differ in their reactivity. SOF_4 is readily hydrolyzed by traces of water, whereas sulphuryl fluoride is relatively inert, as it is not decomposed, when it is reacted with water at 150 °C. A similar reactivity is observed in the reactions with fluoride donors and acceptors. The distorted trigonal bipyramidal SOF_4 quite easily accepts fluoride to form distorted octahedral $[OSF_5]^-$. The analogous reaction of SO_2F_2 has only been observed at low temperature in a matrix, most likely, because the transformation of the distorted tetrahedral SO_2F_2 toward probably distorted trigonal bipyramidal $[SO_2F_3]^-$ is not favored. Moreover, the treatment of SOF_4 with strong fluoride acceptors gives rise to distorted tetrahedral $[SOF_3]^+$, whereas a similar reaction has not been described for SO_2F_2.

(XIV) The reactivity of the hexafluorides generally increases from sulphur to tellurium. SeF_6 is already slightly more reactive than the very inert SF_6, however, it still does not react with water under standard conditions. Consequently, it also does not react with fluoride donors and acceptors. The situation is different for TeF_6, which is completely hydrolyzed by water at room temperature. The increased reactivity of TeF_6 is further represented by the reaction with fluoride donors, as it is capable of forming pentagonal bipyramidal $[TeF_7]^-$ and square antiprismatic $[TeF_8]^{2-}$ anions.

(XV) The free acids $HOSF_5$, $HOSeF_5$ and $HOTeF_5$ significantly differ regarding their synthetic accessibility and stability. The sulphur derivative $HOSF_5$ decomposes above $-65\,°C$ and is prepared by the reaction of $ClOSF_5$ with HCl. Derivatives containing the OSF_5-group are therefore most conveniently prepared by the reaction of SOF_4 with fluoride donors. The acids of the heavier homologues are stable at room temperature and can thus be used as source of the $OSeF_5$- and $OTeF_5$-group, respectively. However, the oxidation potential of hexavalent selenium needs to be taken into account, so that $OSeF_5$-derivatives are only obtained upon reaction of $HOSeF_5$ with oxides and fluorides, but not with chlorides, or other readily oxidizable anions. This synthetic problem does not occur with $HOTeF_5$. $HOSeF_5$ is prepared by the reaction of SeO_2F_2 with HSO_3F and $K[HF_2]$, whereas the tellurium analogue is conveniently generated by the reaction of $Te(OH)_6$ with HSO_3F.

(XVI) Thiazyl fluoride is an unstable molecule with a bent structure. It is, for example, obtained by fluorination of S_4N_4. Trimeric $(NSF)_3$ and $(NSF)_4$ are obtained upon oligomerization of the monomer as well as upon halogen exchange of $(NSCl)_3$ or fluorination of N_4S_4, respectively. $(NSF)_3$ is composed of a six-membered $(NS)_3$ ring, which shows a boat conformation. The fluorine substituents at the sulphur atoms are in axial positions. In contrast, tetrameric $(NSF)_4$ contains a planarized, eight-membered $(NS)_4$ ring. NSF_3 has a distorted tetrahedral structure and is for example accessible by the reaction of NSF with F_2 or AgF_2.

9.7.8

(I) In general, all halogen fluorides can be directly prepared from the elements. The actual outcome depends on the reaction conditions, especially the temperature, thus, the preparation of low-valent halogen fluorides is only achieved under relatively mild conditions and low temperatures, whereas high-valent fluorides, such as BrF_5 or IF_5 require high temperatures and pressures.

(II) The structures of the halogen trifluorides are derived from a trigonal-bipyramid, where two equatorial positions are occupied by lone pairs. Thus, the molecules have a T-shape. The structure of the pentafluorides is described by a ψ-octahedral geometry, resulting in distorted square-pyramidal molecules. The only heptafluoride IF_7 has a pentagonal-bipyramidal structure.

(III) The most stable halogen fluorides are ClF, BrF_3 and IF_5. The low-valent fluorides BrF, IF and IF_3 are unstable and readily decompose to the stable fluoride and the corresponding halogen X_2. In contrast, the high-valent compounds ClF_3, ClF_5, BrF_5 and IF_7 only decompose at higher temperatures to yield F_2 and the respective stable halogen fluoride.

(IV) All halogen fluorides have principally oxidizing and fluorinating properties. However, the low-valent derivatives, especially the monofluorides, are used for halofluorination reactions of both, unsaturated organic compounds but also inorganic substrates like SF_4. The high-valent halogen fluorides liberate F_2 upon heating, thus they are often used as strong fluorinating agents. However, they require very special laboratory equipment and react violently with organic matter, so that they may only be applied for the preparation of very stable compounds. For example, ClF_3 is extremely reactive and often results in explosions and ignitions in contact with any substrate, even asbestos. In some cases, BrF_5, which is relatively stable against further oxidation, is also used as solvent in reactions with even stronger oxidants. In some general reactivity scale, the halogen fluorides can be ordered in the sense $ClF_3 > BrF_5 > IF_7 > ClF > BrF_3 > IF_5 > BrF > IF_3 > IF$. Thus, the iodine fluorides are less reactive than the corresponding chlorine and bromine compounds. Finally, the less reactive halogen fluorides, such as IF_3 or IF_5 can also be substituted by other functional groups, and hence serve as starting point for the preparation of other high-valent halogen compounds.

(V)   In general, fluorocations are obtained by the treatment of the neutral fluorides with strong fluoride acceptors, such as AsF_5 or SbF_5. Among the monofluorides, ClF is the only compound that may be converted to a cationic species. However, it does not have the composition Cl^+ but aggregates to form $[Cl_2F]^+$, which has an unsymmetric and bent structure. The trifluorides XF_3 all react with fluoride acceptors to form species of the general type $[XF_2]^+$. However, these cations always show a rather strong association to the corresponding anion $[MF_6]^-$, resulting in a square-planar coordination environment around the halogen centers. In case of bromine, aggregation may also occur with BrF_3 molecules to form the cations $[Br_2F_5]^+$ and $[Br_3F_8]^+$, respectively. The structure of $[Br_2F_5]^+$ is described by two trigonal-planar $[BrF_3]$-units, which share one fluorine substituent, so that they are symmetrically bridged. In $[Br_3F_8]^+$, a $[BrF_2]^+$ cation is associated to two BrF_3 molecules, in a way, that the central bromine atom shows a square-planar coordination environment, whereas the other two exhibit a trigonal-planar geometry. The pentafluorides XF_5 can only be converted to cationic $[XF_4]^+$, if they are reacted with the strongest fluoride acceptors. The ionic species have a see-saw shape and show strong F-bridges to the corresponding anions, resulting in an increase of the coordination numbers. The treatment of IF_7 with AsF_5 or SbF_5 gives rise to octahedral $[IF_6]^+$. Since ClF_7 and BrF_7 are nonexistent, the corresponding heptavalent fluorocations cannot be prepared by fluoride abstraction. In contrast, they are obtained by oxidation of the pentafluorides with KrF_2 or PtF_6.

(VI)   Typically, the reaction of the halogen fluorides with fluoride donors results in the formation of fluorohalogenate anions. Based on the halogen monofluorides, the linear anions $[XF_2]^-$ can be obtained, whereas the trifluorides are typically converted to square-planar $[XF_4]^-$. Depending on the stoichiometry of the halogen fluoride and the fluoride donor, it is also possible to obtain larger fluoroanions, such as the propeller-shaped $[Cl_3F_{10}]^-$, which is best described by three T-shaped ClF_3 molecules, which are μ^3-bridged by a fluoride anion. Additionally, $[BrF_4]^-$ also forms higher aggregates with BrF_3, that is, $[Br_2F_7]^-$ and $[Br_3F_{10}]^-$. The structure of $[Br_2F_7]^-$ can be described by two square-planar $[BrF_4]$-units, which share a common corner. In $[Br_3F_{10}]^-$, three square-planar $[BrF_4]$-groups share one corner, thus containing a μ^3-bridging fluorine substituent. In case of IF_3, it is also possible to obtain pentagonal planar $[IF_5]^{2-}$. In the series of the halogen pentafluorides only BrF_5 and IF_5 can be converted to fluoroanions which, however, differ in their structures. In case of bromine, isolated, regular octahedral $[BrF_6]^-$ is observed, whereas the iodine compound aggregates to form $[I_3F_{16}]^-$, which is described by three square-pyramidal $[IF_5]$-units, which are bridged by a fluorine substituent. Due to the steric impact of the lone pairs, the overall coordination geometry at the iodine centers is distorted octahedral. A fluoroanion is also derived from IF_7, as its reaction with fluoride donors results in the formation of square-antiprismatic $[IF_8]^-$.

(VII)   The halogen oxytrifluorides XOF_3 all show a ψ-trigonal-bipyramidal structure with the lone pair and the oxygen substituent occupying equatorial positions. The dioxyfluorides XO_2F (X = Cl, Br) have a ψ-tetrahedral structure. The iodine compounds are associated, since they show weak F-bridges in IOF_3, whereas linked ψ-trigonal-bipyramidal units are observed in polymeric IO_2F. The trigonal-bipyramids are alternatingly connected by axial and equatorial substituents. The treatment of the chlorine compounds $ClOF_3$ and ClO_2F with fluoride donors and acceptors gives rise to the corresponding ionic species $[ClOF_4]^-$, $[ClOF_2]^+$, $[ClO_2F_2]^-$ and $[ClO_2]^+$, respectively. The bromine compounds $BrOF_3$ and BrO_2F react with strong fluoride acceptors to yield the corresponding cations $[BrOF_2]^+$ and $[BrO_2]^+$, but they do not accept fluoride ions. Instead, the anions $[BrOF_4]^-$ and $[BrO_2F_2]^-$ are obtained by treatment of, for example, BrO_3^- with BrF_5 or $[BrF_6]^-$. The iodine compounds IOF_3 and IO_2F do not react with either fluoride donors or acceptors. However, the fluoroanions $[IOF_4]^-$ and $[IO_2F_2]^-$ are accessible by treatment of IF_5 with NO_3^- and IO_3^- with aqueous HF, respectively.

(VIII) In the series of halogen oxypentafluorides, only IOF_5 is existing. It is typically prepared by the careful reaction of IF_7 with compounds like H_2O, SiO_2 or POF_3. Among the dioxytrifluorides, the bromine compound BrO_2F_3 is unknown, whereas ClO_2F_3 is typically accessible by the reaction of its fluorocation $[ClO_2F_2]^+$ with fluoride donors. IO_2F_3 does not exist in monomeric form, however, its oligomers are accessible by the reaction of $Ba_3H_4(IO_6)_2$ with SO_3-containing HSO_3F. The perhalogenyl fluorides XO_3F are all known and, for example, accessible by treatment of the perhalogenates XO_4^- (X = Cl, Br) with HF/SbF_5 or fluorination of $HIOF_4$ in aHF. However, BrO_3F is thermally rather unstable and can only be stored at low temperatures.

9.8.4

(I)   Xenon forms the stable fluorides XeF_2, XeF_4 and XeF_6, which are in principle all accessible by the reaction of the elements. However, the stoichiometry of Xe and F_2 significantly determines the outcome of the reaction, along with applied pressure and temperatures. Most conveniently, XeF_2 is prepared by the photolytic reaction of a mixture of Xe and F_2 in a glass vessel. Starting from the elements, the preparation of XeF_4 already requires five molar equivalents of fluorine as well as a reaction temperature of 400 °C at a pressure of 6 bar. Alternatively, a low pressure and low temperature reaction of Xe with O_2F_2 yields pure XeF_4 in high yields. In contrast, the preparation of XeF_6 requires high temperatures and pressures as well as a large excess of F_2. A quantitative formation of XeF_6 is observed, when Xe is reacted with 20 molar equivalents of F_2 in nickel vessels at 700 °C and 200 bar pressure.

(II) XeF_2 is a linear molecule, which can be considered as ψ-trigonal-bipyramid with the three lone pairs occupying all equatorial positions. Accordingly, XeF_4 has two lone pairs and is thus ψ-octahedral resulting in a square-planar structure. XeF_6 still possesses a lone pair, so that it cannot be perfectly octahedral. In contrast, its shape may be described by a ψ-pentagonal-bipyramid, where the lone pair occupies one of the equatorial positions. Alternatively, the resulting structure may be described as capped octahedron. In the solid-state, XeF_6 mainly associates to form tetrameric rings, which are alternatingly composed of $[XeF_5]^+$ and F^-. Some lower oligomers as well as hexameric structures may be observed as well.

(III) XeF_2 acts as fluoride donor toward strong fluoride acceptors, resulting in the formation of $[XeF]^+$ cations. In presence of excess XeF_2, aggregation to the planar, V-shaped $[Xe_2F_3]^+$ is observed. XeF_2 does not possess any fluoride acceptor properties, so that anionic $[XeF_3]^-$ does not exist. Among the xenon fluorides, XeF_4 is the weakest fluoride donor, so that the T-shaped $[XeF_3]^+$ is only obtained upon reaction with the very strong acceptors SbF_5 and BiF_5. In contrast to XeF_2, XeF_4 is capable of accepting fluoride, so that the reaction with fluoride donors, yields pentagonal-planar $[XeF_5]^-$. Both the fluoride donor and -acceptor properties of XeF_6 are relatively pronounced. It reacts with a large number of acceptors to form square-pyramidal $[XeF_5]^+$, in presence of excess XeF_6, aggregation to $[Xe_2F_{11}]^+$ is observed. It is comprised of two $[XeF_5]$-units, which are connected by a F-bridge. Therein, the Xe-F-Xe group has a bent geometry. XeF_6 is the strongest fluoride acceptor among the xenon fluorides, thus, it is readily converted to capped-octahedral $[XeF_7]^-$ and cubic-antiprismatic $[XeF_8]^{2-}$. In addition, the aggregate $[Xe_2F_{13}]^-$ is accessible, which is comprised of a capped trigonal-prismatic $[XeF_7]^-$ and a loosely connected XeF_6.

(IV) The oxidation power of the xenon fluorides increases from XeF_2 to XeF_6. For example, XeF_2 oxidizes Cl^- to Cl_2, XO_3^- to XO_4^- (X = Br, I) or Ag^+ to Ag^{2+} in acidified aqueous solution. An increased oxidation power is observed for $[XeF]^+$ salts. XeF_4 converts metallic platinum and mercury to the fluorides PtF_4 and Hg_2F_2, respectively. XeF_6 is even more powerful, as it converts AuF_3 to the pentavalent species $[AuF_6]^-$ and oxidizes elemental mercury to HgF_2.

(V) $XeOF_2$ is prepared by careful hydrolysis of XeF_4 at $-80\,°C$. It is a moderately stable compound, as it starts to decompose above $-25\,°C$ to yield divalent XeF_2 and hexavalent XeO_2F_2. In principle, the stepwise hydrolysis of XeF_6 gives rise to the oxyfluorides $XeOF_4$ and XeO_2F_2. More conveniently, they are prepared by the reaction of $CsNO_3$ with excess XeF_6 or $XeOF_4$, respectively. At room temperature, XeO_2F_2 decomposes to XeF_2 and O_2, whereas $XeOF_4$ appears to be stable under standard conditions. Octavalent xenon oxyfluorides are obtained, if XeO_4 is reacted with XeF_6. The most stable of them is XeO_3F_2, however, it is only existing at low temperature. Due to its instability, XeO_2F_4 has only been observed by mass-spectrometry, whereas $XeOF_6$ remains completely unknown.

(VI) The only stable neutral krypton fluoride is KrF_2, which is generally prepared from the elements. However, an additional initiation is always required, such as electrical discharge or UV-irradiation. The treatment of KrF_2 with strong fluoride acceptors, for example, AsF_5 or SbF_5, gives rise to the cationic species $[KrF]^+$ and $[Kr_2F_3]^+$. Neutral KrF_2 is stable up to about $-10\,°C$, above it rapidly decomposes, whereas the cationic species do not decompose below room temperature. KrF_2 is a very strong oxidizing agent and, for example, converts ClF_3 to ClF_5, I_2 to IF_7 and Xe to XeF_6. The oxidation power is further increased for $[KrF]^+$, which for example oxidizes NF_3 to $[NF_4]^+$, Au to AuF_5 and the halogen pentafluorides ClF_5 and BrF_5 to the corresponding heptavalent species $[XF_6]^+$.

10.1.2

(I) The trifluorides ScF_3, YF_3, LaF_3 and AcF_3 are typically accessible from aqueous solutions of the corresponding metal. The addition of aqueous HF or fluoride salts results in the precipitation of the trifluorides, mostly in form of hydrates. The anhydrous trifluorides are obtained, if the oxides M_2O_3 are reacted with anhydrous HF or F_2. However, heating of the hydrates $MF_3 \cdot n\,H_2O$ does not yield the anhydrous compounds but the oxyfluorides MOF.

(II) The trifluorides react with fluoride salts to form stable 1:1- and 1:3- adducts, which are the fluoroanions $[MF_4]^-$ and $[MF_6]^{3-}$. In case of scandium, hexafluoroscandate(III) compounds are obtained upon dissolution of ScF_3 in presence of excess fluoride. The same reaction does not occur with YF_3 and LaF_3. However, tetra- and hexafluorometallate salts of the trifluorides are accessible by solid-state reactions of stoichiometric mixtures of MF_3 and a fluoride source, for example, NaF. Most likely, the same reactions will occur with AcF_3, however, they have not been described in the literature. The solid-state structure of the $[MF_4]^-$ salts is in most cases described by a distorted fluorite lattice, whereas a cryolite structure is observed for the hexafluorometallates.

(III) The oxyfluorides ScOF, YOF, LaOF and AcOF are commonly prepared by hydrolysis of the trifluorides, which typically requires high temperatures. A mechanochemical synthesis of LaOF has also been described, where LaF_3 is ground together with a stoichiometric amount of La_2O_3. Other preparative methods for the generation of the oxyfluorides include the solid-state synthesis of YOF and LaOF from their oxides and poly(tetrafluoroethylene) as fluoride source. Finally, nanoscopic YOF has been prepared by fluorolytic sol-gel synthesis. Potential applications have especially been described for LaOF, which catalytically dehydrogenates ethane to ethylene and also shows a photocatalytic activity in the liberation of H_2 from methanol-solutions. Potential applications of LaOF in optoelectronics and biotechnology have also been explored.

10.2.2

(I)   Stable lanthanide difluorides are only known for Sm, Yb and Eu. They are typically prepared by reduction of the trifluorides with H_2, Si or the corresponding metal vapors. At room temperature, the solid-state structure of SmF_2, EuF_2 and YbF_2 is described by a fluorite lattice.

(II)   The treatment of aqueous lanthanide solutions with HF or fluoride typically leads to the precipitation of the trifluorides LnF_3. However, this synthetic method mostly furnishes hydrates. The anhydrous trifluorides are obtained, if the starting materials, for example, oxides or chlorides are treated with anhydrous HF or F_2, in the same way, these reagents may be used for the dehydration of $LnF_3 \cdot n\,H_2O$. For the preparation of CeF_3, PrF_3 and TbF_3 the treatment with F_2 is not suitable, since this would give rise to the tetravalent compounds.

(III)   The lanthanide trifluorides LnF_3 readily react with alkali metal fluorides to form salts of the compositions $M[LnF_4]$ and $M_3[LnF_6]$. In the solid-state structures of these compounds, a high association is observed for the tetrafluorometallate anions, whereas the hexafluorometallate anions typically exist as isolated octahedral species.

(IV)   There are only three stable lanthanide tetrafluorides, namely CeF_4, TbF_4 and PrF_4. Apart from that, the heptafluorometallates $[NdF_7]^{3-}$ and $[DyF_7]^{3-}$ contain tetravalent lanthanide centers. The neutral NdF_4 and DyF_4 are very unstable and have never been obtained by preparative methods; they have only been observed at very low temperatures in an inert gas matrix. The tetrafluorides CeF_4 and TbF_4 are conveniently prepared by treatment of the metals or the trifluorides with F_2. In case of PrF_4, a synthetic detour is required, as it is accessed by intermediary $Na_2[PrF_6]$, which in turn is obtained by the treatment of PrF_3 with F_2 in presence of NaF. The reaction of the $[PrF_6]^{2-}$ salt with anhydrous HF finally liberates neutral PrF_4. Alternatively, it is also obtained by treatment of PrO_2 with KrF_2. PrF_4 only has a limited thermal stability, as it starts to decompose to PrF_3 and F_2 at 90 °C. The stability increases for TbF_4 and CeF_4, the latter being the most stable lanthanide tetrafluoride. Similar to $[PrF_6]^{2-}$, the tetravalent Nd- and Dy-complexes $[MF_7]^{3-}$ are prepared by fluorination of low-valent species in presence of a fluoride source.

(V)   The lanthanide tetrafluorides are oxidizing and fluorinating agents. Thus, they have been used as fluorinating agents for both inorganic and organic substrates, especially in gas-phase reactions. This application is based on their thermal instability, as they tend to liberate F_2 upon heating. Exploiting this property, FeF_4, CoF_4 and CrF_5 have been prepared in the gas-phase. The tetrafluorides CeF_4, TbF_4 and PrF_4 have also been used for fluorination reactions of C_{60}.

10.3.2

(I)   Stable trifluorides are known for the actinides from U to Cf. ThF_3 has only been observed under matrix conditions and PaF_3 is completely unknown. Most commonly, the actinide trifluorides are generated by precipitation from aqueous solutions of An^{3+}, which is realized by the addition of fluoride ions. The resulting hydrates are then dried by thermal treatment in presence of HF. Alternatively, AnF_3 are obtained by the reaction of the oxides An_2O_3 with HF at elevated temperatures or reduction of higher fluorides by a mixture of H_2 and HF. In case of UF_3 and PuF_3, the trifluorides may also be obtained by mild fluorination of the corresponding metal with NF_3.

(II)   For the preparation of ThF_4, PaF_4 and UF_4, it is convenient to start from the dioxides AnO_2 and treat them with fluorinating agents, for example, SF_4, HF or F_2. The other tetrafluorides NpF_4, PuF_4, AmF_4, CmF_4, BkF_4 and CfF_4 are typically obtained by fluorination of the trifluorides with F_2, or by precipitation from An^{4+} solutions using HF as fluoride source. However, one has to pay attention, when PaF_4, UF_4, NpF_4 or PuF_4 are prepared by the use of strong oxidizers like F_2, since these tetrafluorides can be further oxidized. This especially occurs, if the reactions are carried out at comparably forcing conditions, that is, high temperatures and high concentrations of the oxidizer.

(III)   Stable pentafluorides are only known for Pa, U and Np. In general, the stability of pentavalent fluorine compounds of the actinides decreases from Pa to Pu, since PuF_5 immediately disproportionates to PuF_4 and PuF_6. Similarly, NpF_5 slowly decomposes to NpF_4 and NpF_6, so that it is difficult to obtain pure samples of it. In contrast, heating of a mixture of UF_4 and UF_6 generates UF_5, so that the pentafluoride is the preferred species in this system. Accordingly, UF_5 is accessible by a number of synthetic approaches, such as treatment of the tetrafluoride with F_2 at 450 °C or a large variety of reductions of UF_6. PaF_5 is best obtained by fluorination of $PaCl_5$ or also by oxidation of the tetrafluoride, in this case at 700 °C. NpF_5 can be generated by reduction of NpF_6 or treatment of $[NpF_6]^-$ with a fluoride acceptor, but as mentioned above, pure samples are hardly obtained due to its instability. PuF_5 is only stable in its anionic fluorocomplex $[PuF_6]^-$, which is obtained by reduction of PuF_6 with NO, NO_2 or similar reagents. Alternatively, $[PuF_7]^{2-}$ is generated upon oxidation of PuF_4 with F_2 in presence of fluoride.

(IV)   Actinide hexafluorides only exist for U, Np and Pu, where their stability strongly decreases from U to Pu. Thus, PuF_6 can only be stored at low temperature, whereas UF_6 is stable. However, they are all strongly oxidizing species. UF_6 also plays an important role in the nuclear fuel industry, where it is used for isotope enrichment. On industrial scale, it is prepared in a two-step sequence, at first, UO_2 is converted to UF_4 using HF as fluorinating agent. Subsequently, UF_4 is oxidized by F_2 to yield UF_6.

(V)   Upon reaction with alkali metal fluorides, the hexafluorides behave differently. UF_6 forms hexavalent fluorouranates of the compositions $[UF_7]^-$, $[UF_8]^{2-}$, $[UF_9]^{3-}$ and $[UF_{10}]^{4-}$, whereas NpF_6 and PuF_6 are reduced to pentavalent $[NpF_6]^-$ and $[PuF_6]^-$, respectively. However, when the hexafluorides are reacted with NO, they are all reduced to pentavalent $NO[AnF_6]$.

10.4.3

(I) TiF_3 is typically prepared by the direct reaction of Ti metal with HF at moderately high temperature of 200–225 °C. Other synthetic methods include the heating of CuF_2 in a Ti-vessel at 800 °C or, for the preparation of very pure TiF_3, the thermal decomposition of $NH_4[TiF_4]$ under an atmosphere of Ar or H_2. In general, TiF_3 is quite stable toward oxidation, as it is not oxidized by air at room temperature. However, it will be oxidized upon heating. Moreover, it disproportionates to Ti and TiF_4, if it is heated above 950 °C under an inert atmosphere.

(II)   The molecular structure of gaseous TiF_4 is described by a tetrahedron. In contrast, it is polymeric in the solid state, forming a network of $[TiF_6]$-octahedra. The tendency of forming octahedral species is also clearly pronounced for the formation of Lewis base adducts. In general, suitable ligands are *N*- and *O*-donor ligands, such as pyridines or pyridine-1-oxides. The ligands may also be bidentate *N,N*-donor ligands, such as 2,2'-bipyridine or 1,10-phenanthroline, apart from that, adducts with *N*-heterocyclic carbenes are also known. Finally, a stable 1:2 complex of TiF_4 is even obtained with ammonia.

(III) Upon reaction with fluoride donors in HF solution, TiF_4 exclusively yields octahedral $[TiF_6]^{2-}$ salts. They are also conveniently prepared by the reaction of Ti or TiO_2 with a fluoride source in HF. In general, hexafluorotitanate(IV) salts are only poorly soluble in water, thus, they are also rather stable against hydrolysis. However, in acidic aqueous media, they are readily decomposed.

(IV)   The treatment of TiO_2 with HF or F_2 is certainly one of the most convenient methods to prepare TiF_4. However, the modification of the TiO_2 starting material significantly determines the outcome of the reaction. Interestingly, only the rutile modification gives rise to TiF_4, whereas the fluorination of anatase results in the formation of the oxyfluoride $TiOF_2$.

(V)   Indeed, the low-valent zirconium fluorides ZrF_2 and ZrF_3 are known, whereas comparable hafnium compounds do not exist. ZrF_2 is typically prepared by the reduction of ZrF_4 with atomic hydrogen or the comproportionation of stoichiometric mixtures of Zr metal and ZrF_4 in an eutectic mixture of each 50 mol-% LiF and KF at 550 °C. ZrF_3 is obtained, if prehydrogenated Zr metal ("Zr hydride") is reacted with H_2 and HF at 750 °C.

(VI)   The tetrafluorides ZrF_4 and HfF_4 may be prepared by fluorination of the metals or their compounds. For example, the treatment of the dioxides ZrO_2 and HfO_2 in aqueous HF yields hydrates, which can be dehydrated in high vacuum or an atmosphere of HF. Alternatively, ZrF_4 and HfF_4 can also be liberated by thermal decomposition of the ammonium complexes $(NH_4)_2[MF_6]$. Like TiF_4, ZrF_4 and HfF_4 are tetrahedral molecules in the gas phase. However, in contrast to TiF_4, they are surrounded by eight fluorine substituents in the solid state, resulting in square-antiprismatic coordination environments.

(VII) Reactions of TiF_4 with fluoride donors exclusively generate hexafluorotitanates(IV) $[TiF_6]^{2-}$, which are in most cases present as isolated, octahedral anions. In contrast, ZrF_4 and HfF_4 are readily forming higher coordinated fluorometallate species with seven or even eight fluorine substituents around the metal center. Nevertheless, $[MF_5]$- and $[MF_6]$-units are observed, too. In many cases, the actual coordination geometry of the zirconium and hafnium species strongly depends on the nature of the cations. In addition, the fluorometallates are mostly present in F-bridged networks and are only scarcely observed as isolated anions.

10.5.3

(I)   In general, VF_2 is accessible by reduction of higher fluorides, for example by treatment of VF_3 with a mixture of H_2/HF or vanadium metal. It is also accessible by electrolytic reduction of VF_5 in aqueous solution, where it will be obtained as hydrate $VF_2 \cdot 4\,H_2O$. Finally, it is accessible by fluorination of VCl_2 with HF under an atmosphere of H_2. The treatment of VF_2 with alkali metal fluorides gives rise to the trifluorovanadates(II) $M[VF_3]$. VF_2 is a strong reducing agent, so that it will be oxidized by excess HF during its preparation. Accordingly, disproportionation to vanadium metal and trivalent $[VF_6]^{3-}$ may occur during the synthesis of trifluorovanadates, depending on the molar ratio of MF and VF_2.

(II) VF_3 molecules have a trigonal-planar geometry. In the solid state, it may be described by a distorted ReO_3 lattice, where the octahedral $[VF_6]$-units have been rotated around their

C_3-axis. Overall, this results in a hexagonal closest packing of the F-atoms. VF_3 is prepared by fluorination of VCl_3 using HF or NF_3 or by reduction of VF_4 with vanadium metal. Alternatively, it may also be liberated by thermal decomposition of the ammonium complex $(NH_4)_3[VF_6]$.

(III) The structure of solid VF_4 is comparable to that of SnF_4, thus it also forms layers of $[MF_6]$-octahedra. However, the octahedral $[VF_6]$-units are tilted against each other, so that the V-F-V bridges are bent. In SnF_4, the $[SnF_6]$-octahedra are connected by linear Sn-F-Sn bridges.

(IV) Careful hydrolysis of VF_4 in water yields the trihydrate $VOF_2 \cdot 3\,H_2O$. The anhydrous compound is accessible by the treatment of $VOBr_2$ with aHF at high temperatures or a ligand exchange reaction of $VO(acac)_2$ with TiF_4. VOF_2 reacts with fluoride donors to form the complex anions $[VOF_4]^{2-}$, $[VOF_5]^{3-}$ and $[V_2O_2F_7]^{3-}$, most conveniently, these species are generated by the reaction of VO_2 and a fluoride source in aqueous HF. All three anions are built from octahedral $[VOF_5]$-units, which are isolated in $[VOF_5]^{3-}$ and connected to polymeric chains in $[VOF_4]^{2-}$. In $[V_2O_2F_7]^{3-}$, two $[VOF_5]$-octahedra share a common edge.

(V) VF_5 is accessible by treatment of vanadium metal or low-valent compounds, such as VCl_3, with fluorinating agents like F_2 or BrF_3. It can also be prepared by fluorination of VOF_3 or V_2O_5. Due to the stability of V=O-bonds, VF_5 typically reacts with oxygen containing substrates like POF_3 to undergo an oxygen-fluorine exchange; in the present example PF_5 and VOF_3 are generated. A similar reactivity is observed against organic substrates. Oxidizable materials, such as PF_3 or I_2 are readily oxidized to the pentafluorides PF_5 and IF_5, whereas VF_5 is reduced to VF_4. VF_5 also fluorinates fluoroaromatic compounds as well as polyfluoroolefins. However, the reaction with arenes gives only rise to fluorinated cyclohexene and cyclohexadiene derivatives, but not to the fully saturated products.

(VI) In oxidation state +V, vanadium forms the oxyfluorides VOF_3 and VO_2F. VOF_3 is typically prepared by oxygen fluorine exchange, starting from VF_5 and excess SO_2 or SO_3 or from the oxide V_2O_5, which is partly fluorinated by BrF_3, F_2 or COF_2. Alternatively, VF_3 may be oxidized by O_2 at red heat. VO_2F is either obtained by fluorination of VO_2Cl or high-pressure reaction of a 1:1-mixture of V_2O_5 and VOF_3. In the gas phase, monomeric VOF_3 molecules exhibit a C_{3v} symmetry, but dimeric $(VOF_3)_2$ can be observed, too. In the solid state, it shows a layered structure where pairs of VOF_3 are connected by F-bridges, so that the V centers overall have a distorted octahedral coordination geometry. VO_2F is a highly ionic compound and crystallizes in a distorted ReO_3 lattice.

(VII) In general, the trifluorides NbF_3 and TaF_3 are prepared by reduction of the pentafluorides or fluorination of the metals with HF at high pressures and temperatures. However, they both have a high oxygen affinity, so that it is virtually impossible to obtain oxygen-free NbF_3 and TaF_3. To date, the preparation of the pure trifluorides has not been successful.

(VIII) NbF_4 is typically prepared by reduction of the pentafluoride, which can be achieved by the use of reductants like Si or Nb-powder. It is thermally far more stable than the lighter homologue VF_4, since complete disproportionation is only observed above 350 °C. Upon heating with fluoride donors, NbF_4 forms hexafluoroniobates(IV) $[NbF_6]^{2-}$. Distorted pentagonal-bipyramidal $[NbF_7]^{3-}$ is also known, however, it is not directly prepared from NbF_4 but a mixture of NbF_5, Nb and KF at 850 °C.

(IX) In general, the structures of TaF_5 and NbF_5 are similar in all states. In the solid state, they form cyclic tetramers of $[MF_6]$-octahedra, containing almost perfectly linear M-F-M bridges. The same structural behavior is mainly retained in the highly-viscous melts, however, the actual situation also depends on the temperature. The same holds for the gaseous pentafluorides, where oligomeric $(MF_5)_n$ ($n \leq 4$) are in equilibrium with monomers of MF_5. The latter have a trigonal-bipyramidal structure and are mostly observed at higher temperatures of about 400 °C.

(X) The pentafluorides TaF_5 and NbF_5 accept fluoride ions to form fluoroanions of the composi-
     tions $[MF_6]^-$, $[MF_7]^{2-}$ and $[MF_8]^{3-}$. As a general difference, the Nb-compounds tend to be
     hydrolyzed, which is not observed for the Ta-based anions. Thus, the oxygen-containing an-
     ions $[NbOF_5]^{2-}$ and $[NbO_2F_4]^{3-}$ are also known. However, in aqueous systems, only the hex-
     acoordinated species $[MF_6]^-$ and $[NbOF_5]^{2-}$ are accessible. The higher fluoroanions require
     either highly concentrated, or better anhydrous, HF media. More conveniently, they are pre-
     pared from melts of the corresponding fluoride mixtures or in solid-state reactions. Finally,
     the aggregates $[M_2F_{11}]^-$ are obtained upon treatment of $[MF_6]^-$ with a fluoride acceptor, for
     example, BF_3 or PF_5.

(XI) The oxytrifluorides $TaOF_3$ and $NbOF_3$ are obtained by several methods, for example, the
     fluorine-oxygen exchange of the pentafluorides, which can be achieved by careful hydrol-
     ysis or reaction with $TiO(acac)_2$. In contrast, the dioxyfluorides TaO_2F and NbO_2F are rather
     prepared by the inverse approach, for example, treatment of the oxides M_2O_5 with aqueous
     HF. For the preparation of pure samples, solid-state reactions have proven to be superior.
     The oxytrifluorides MOF_3 decompose upon heating (280–375 °C) to yield a mixture of MF_5
     and MO_2F, however, these compounds are in a reversible equilibrium, so that the oxytriflu-
     orides are regenerated at lower temperature (180–220 °C). Upon heating to 500 °C, NbO_2F
     decomposes to $NbOF_3$, which is then degraded to NbO_2F and NbF_5. Thus, the thermal de-
     composition of NbO_2F is irreversible.

10.6.3

(I) CrF_2 is prepared by the reduction of aqueous CrF_3 with H_2 at 500–600 °C or the reaction of
     chromium metal with aHF. Pyrolysis of $(NH_4)_3[CrF_6]$ or directly CrF_3 at 1100 °C also yields the
     difluoride, since CrF_3 liberates CrF_2 and F_2 at high temperatures. The hydrate $CrF_2 \cdot 2 H_2O$ is
     obtained, if Cr metal is reacted with aqueous HF under anaerobic conditions. The dihydrate is
     relatively stable toward oxidation, as it is only slowly oxidized to Cr_2O_3 upon contact with air.
     However, anhydrous CrF_2 is a pyrophoric, strong reducing agent and accessible by heating
     of the dihydrate to 100 °C *in vacuo*. Heating of equimolar amounts of CrF_2 and CrF_3 results
     in the formation of mixed-valent Cr_2F_5, which contains both, octahedrally coordinated Cr(II)
     and Cr(III) centers in the solid-state.

(II) In general, oxidation +III is favorable for chromium compounds, this also holds for CrF_3.
     Thus, it is stable on air under standard conditions; however, strong oxidizers or reducing
     agents will oxidize or reduce it, respectively. CrF_3 is also stable in aqueous systems, where
     it forms hydrates, such as $CrF_3 \cdot 9 H_2O$, which may be reformulated as hexaaquacomplex
     $[Cr(H_2O)_6]F_3 \cdot 3 H_2O$. Upon heating to 1100 °C, CrF_3 splits off F_2 and is thus reduced to CrF_2.
     CrF_3 is typically prepared by the treatment of chromium metal or trivalent precursors, for
     example, $CrCl_3$ or Cr_2O_3 with aHF. CrF_3 reacts with fluoride sources to form a number of dif-
     ferent fluorochromate anions, which may also contain water ligands. However, as a common
     structural feature, the coordination geometry around the Cr centers in these anionic species
     is always based on an octahedral arrangement.

(III) CrF_4 exists in form of an α- and β-modification. α-CrF_4 is obtained by fluorination of Cr metal
     or the trihalides CrF_3 or $CrCl_3$ at 300–350 °C. The β-modification is accessible, when CrF_5 is
     thermolyzed at 130 °C over a period of 5 months. The solid-state structure of α-CrF_4 is de-
     scribed by pairs of $[CrF_6]$-octahedra, which share a common edge. The dimeric $[Cr_2F_{10}]$-units
     are further connected by *trans*-F-bridges, resulting in the formation of columns. In β-CrF_4,
     the $[CrF_6]$-octahedra are linked to cyclic tetramers, which are again connected by *trans*-F-
     bridges, in this case giving rise to tubes.

(IV)   The thermal stability of CrF_5 is rather limited at higher temperatures, which is perfectly represented by decreasing yields in high-temperature preparations, for example, the fluorination of Cr metal or CrF_3 using F_2. The corresponding $[CrF_6]^-$ salts are even less stable, since they readily undergo thermal decomposition to tetravalent $[CrF_5]^-$ and $[CrF_6]^{2-}$, respectively. Pentavalent $[CrF_6]^-$ salts are accessible by treatment of CrF_5 with fluoride donors, for example, NOF, NO_2F or CsF, as well as by reaction of CrF_3 with F_2 in presence of F^- in aHF at ambient temperature.

(V)   A binary chromium hexafluoride does not exist, however, there are the hexavalent oxyfluorides $CrOF_4$ and CrO_2F_2, which are, for example, obtained by the treatment of CrO_3 with F_2 at 220 °C and 150 °C, respectively. Alternatively, other hexavalent oxygen-containing substrates like CrO_4^{2-} or $Cr_2O_7^{2-}$ can be fluorinated to yield the oxyfluorides, in addition, CrO_2F_2 is also obtained by halogen exchange of CrO_2Cl_2 using F_2 as fluorinating agent. Both oxyfluorides are very sensitive toward hydrolysis and act as strong oxidizing agents, however, they do not possess fluorinating properties. CrO_2F_2 is thermally stable up to about 500 °C, above this temperature it decomposes to tetravalent $CrOF_2$ and O_2. Under exclusion of light and in a dry atmosphere, both $CrOF_4$ and CrO_2F_2 are stable in materials like Kel-F, however, they decompose in standard laboratory glassware and quartz vessels.

(VI)   Most conveniently, MoF_3 is prepared by reduction of the high-valent fluorides MoF_5 or MoF_6 using Mo metal, Si or H_2 as reducing agents. Alternatively, it is also obtained by halogen exchange of $MoBr_3$ with HF. For gaseous MoF_3, a trigonal-planar geometry has been proposed based on quantum-chemical calculations, whereas the solid state is described by the VF_3 structure type.

(VII)   The insolubility of MoF_4 in aHF allows its precipitation, when hexafluoromolybdates $[MoF_6]^{2-}$ are treated with the fluoride acceptor SbF_5. However, if AsF_5 is used for this purpose, molybdenum will be oxidized, so that MoF_5 is obtained.

(VIII)   In its ground state, molecular MoF_4 has a tetrahedral structure. The situation is different for gaseous WF_4, as it shows Jahn–Teller distortion, thus, decreasing the molecular symmetry from T_d to D_{2d}. However, since the square-planar, D_{4h}-symmetric structure of WF_4 is only slightly higher in energy, the molecule is considered to be quasiplanar.

(IX)   The preparation of MoF_5 may be achieved by several different methods, for example, the decomposition of Mo_2F_9 or reduction of MoF_6. In the same way, WF_5 is obtained by the reduction of WF_6 at an electrically heated tungsten wire. More conveniently, both pentafluorides are liberated upon treatment of the hexafluorometallates $[MF_6]^-$ with SbF_5 under rather mild conditions. The thermal stability of the pentafluorides is rather limited, as they irreversibly disproportionate to the tetra- and the hexafluoride at <70 °C (WF_5) and 165 °C (MoF_5). Fluorometallates of MoF_5 and WF_5 cannot be obtained by the treatment of MF_5 with F^-, instead, hexafluorometallates $[MF_6]^-$ are obtained by reduction of the hexafluorides with iodide. The corresponding hepta- and octafluorometallates $[MF_7]^{2-}$ and $[MF_8]^{3-}$ are typically prepared by oxidation of the hexacarbonyls $M(CO)_6$ with IF_5 in presence of fluorides or iodides.

(X)   Both, MoF_6 and WF_6 are strongly oxidizing agents, thus metallic Ag and Tl as well as NO will be oxidized. The reaction with MoF_6 yields Ag^{2+}, Tl^{3+} and $[NO]^+$, whereas WF_6 is a significantly weaker oxidizer, and thus only yields Ag^+ and Tl^+. In contrast to MoF_6, it does not react with NO.

(XI)   Hexavalent fluorometallates are obtained, when MoF_6 or WF_6 are reacted with fluorides or iodides in an oxidizing medium, for example, IF_5 or ClF_3. They typically have the composition $[MF_7]^-$ or $[MF_8]^{2-}$, where the heptafluorometallate typically exhibits a capped-octahedral structure, whereas $[MF_8]^{2-}$ is described by a square-antiprism.

(XII)   Both tungsten and molybdenum form oxyfluorides of the compositions MOF_4 and MO_2F_2. The oxytetrafluorides are typically obtained by controlled oxygen-fluorine exchange of the hexafluorides, for example, careful hydrolysis ($MoOF_4$), or reaction with other oxygen-donors, such as SiO_2 (WOF_4) or boric acid ($MoOF_4$, WOF_4). Alternatively, they are accessible by fluorination of the metals, metal oxides or the oxytetrachlorides $MOCl_4$. The preparation of the dioxydifluorides is a little bit more problematic. MoO_2F_2 is generated upon partial fluorination of MoO_3 by SeF_5, IF_5 or NF_3. However, it is very reactive and easily undergoes reduction or adduct formation. Thus, it is most conveniently obtained by pyrolysis of an-ionic cis-$[MoO_2F_2]^{2-}$ in vacuo. WO_2F_2 is accessible by controlled hydrolysis of WOF_4 or partial fluorination of WO_4^{2-} with IF_5. According to the VSEPR-model, a distorted trigonal-bipyramidal geometry has to be expected for the molecular oxytetrafluorides. However, due to electronic effects of the strongly covalent M-O bonds, it appears that $MoOF_4$ and WOF_4 exhibit a square-pyramidal shape. In contrast, the dioxydifluorides MoO_2F_2 and WO_2F_2 have a distorted tetrahedral geometry, as it is expected.

10.7.3

(I) MnF_2 is obtained by the reaction of aqueous Mn^{2+} solutions or $MnCl_2$ with HF. In the same way, it is accessible from manganese metal and HF. MnF_2 readily forms the fluoroman-ganates(II) $[MnF_3]^-$ and $[MnF_4]^{2-}$. The trifluoromanganates are accessible from aqueous solution, for example, by treatment of $MnCl_2$ with fluorides. Apart from that, both $[MnF_3]^-$ and $[MnF_4]^{2-}$ compounds are obtained upon melting of stoichiometric amounts of MnF_2 and a fluoride.

(II) The fluorination of divalent manganese with F_2 or BrF_3 yields MnF_3, which is also accessible by oxidation of MnF_2 with F_2 in aHF at room temperature. MnF_3 forms a number of anionic fluorocomplexes, however, in virtually all of them, the coordination environment around the Mn centers is based on an octahedral geometry. Depending on the composition of the flu-oromanganates, their actual shape may be represented by a rather perfect octahedron, for example, in $[MnF_6]^{3-}$, or it can show deviations, as for example, in tetragonally distorted $[MnF_4(H_2O)_2]^-$.

(III) MnF_4 is thermally unstable, since it already slowly decomposes to MnF_3 and F_2 at room tem-perature. This decomposition has been exploited in the first purely chemical synthesis of F_2, which started from $K_2[MnF_6]$ and SbF_5, both of which are accessible without the use of ele-mental fluorine. Apart from the liberation of its $[MnF_6]^{2-}$ salts by the addition of SbF_5, MnF_4 can also be made by fluorination of manganese metal or MnF_3. Moreover, the oxidation of MnF_2 with KrF_2 or photodissociated F_2 in aHF gives rise to the tetrafluoride MnF_4.

(IV) Binary manganese fluorides do not exist in high oxidation states, however, the heptavalent oxyfluoride MnO_3F is well known. It is typically prepared by the reaction of $KMnO_4$ with aHF, HSO_3F or IF_5. It is only stable below 0 °C and explosively decomposes at room temperature upon shock-impact to yield MnO_2, MnF_2 and O_2.

(V)   Neutral ReF_3 does not exist in condensed phase. The reaction of $ReCl_3$ with aHF results in the formation of Re metal and ReF_6, indicating, that ReF_3 is most likely highly unstable, and thus immediately disproportionates. However, trivalent rhenium is found in anionic $[Re_2F_8]^{2-}$, which is most conveniently prepared by halogen exchange reactions of $[Re_2X_8]^{2-}$. The di-anion contains linear units of $[Re_2]^{6+}$, which exhibit a quadruple bond between the metal centers. The fluoride atoms are oriented around this linear unit in form of a tetragonally dis-torted cubane cage.

(VI) Tetravalent rhenium is found in the neutral fluoride ReF_4 as well as in the corresponding an-ionic $[ReF_6]^{2-}$. In contrast, neutral TcF_4 is unknown, whereas $[TcF_6]^{2-}$ is accessible. ReF_4 is typically prepared by reduction of the high-valent fluorides ReF_5 or ReF_6, which may be realized by the use of H_2 or Re metal as reducing agent. More conveniently, the tetrafluoride is liberated from $[ReF_6]^{2-}$ by treatment with SbF_5. Hexafluororhenate(IV) salts are commonly prepared by halogen exchange of $[ReX_6]^{2-}$ with HF or reduction of ReF_6 with iodides in SO_2 solution. Similarly, $[TcF_6]^{2-}$ is mostly accessed by halogen exchange of $K_2[TcBr_6]$ with different fluoride sources, for example, concentrated aqueous HF.

(VII)   Both pentafluorides are typically obtained by reduction of the corresponding hexafluoride. Thus, TcF_5 is synthesized by quantitative reduction of TcF_6 by I_2 in IF_5. For the reduction of ReF_6, several reducing agents can be used, for example, Re wire, H_2 or Si. In general, the pentafluorides are stable under inert conditions at room temperature, if they are not stored in glassware. TcF_5 is more stable than ReF_5, the latter irreversibly disproportionating to ReF_4 and ReF_6 above 130 °C.

(VIII) Since there is no stable TcF_7, TcF_6 is the final product of the fluorination of Tc metal but also of TcO_4^-. In contrast, a mixture of ReF_6 and ReF_7 is obtained under similar conditions, when Re-compounds are used as starting materials. Thus, the ReF_6/ReF_7 mixture needs to be submitted to an additional reduction step with Re metal at 400 °C to generate pure ReF_6. In principle, both hexafluorides form heptafluorometallates $[MF_7]^-$ and octafluorometallates $[MF_8]^{2-}$. However, among these species, $[TcF_7]^-$ is unstable. The $[MF_7]^-$ anions have a pentagonal-bipyramidal structure, whereas $[MF_8]^{2-}$ are described by square-antiprisms.

(IX) ReF_7 is the final product of the fluorination of Re metal, which is obtained at 400 °C and a pressure of 3 bar. It exhibits both fluoride accepting and donating properties, thus the square-antiprismatic $[ReF_8]^-$ is generated upon reaction with NOF and NO_2F. In contrast, octahedral $[ReF_6]^+$ is obtained, if ReF_7 is reacted with SbF_5.

(X)   For the heptavalent metals Tc and Re all six possible oxyfluorides MOF_5, MO_2F_3 and MO_3F are existing. $ReOF_5$ is typically prepared by the treatment of Re_2O_7 with ClF_3 or fluorination of an oxide or oxyfluoride in a lower oxidation state, such as ReO_2, ReO_3 or $ReOF_4$. The preparation of the corresponding $TcOF_5$ is more complicated, as it is generated by the reaction of TcO_2F_3 with KrF_2 at room temperature. ReO_2F_3 is synthesized by the treatment of ReO_3Cl with XeF_2 or fluorination of ReO_2 or ReO_4^-. Again, the preparation of the Tc derivative TcO_2F_3 is slightly different, as it is made by the reaction of Tc_2O_7 with three equivalents of XeF_6. ReO_3F can be made by several approaches, for example, reaction of ReF_7 with Re_2O_7 or fluorination of ReO_4^- or ReO_3. It is also accessible by halogen exchange of ReO_3Cl, using HF as fluorinating agent. Finally, TcO_3F is synthesized by fluorination of TcO_2.

10.8.2

(I)   The difluorides FeF_2, CoF_2 and NiF_2 are typically prepared by treatment of the metals or metal dichlorides with HF at high temperature. Alternatively, NiF_2 is directly accessible from Ni metal and F_2 at 550 °C and FeF_2 may be prepared by the reduction of FeF_3 with iron metal at 200 °C. The stability against hydrolysis increases from FeF_2 to NiF_2. FeF_2 is slowly hydrolyzed by moisture at room temperature, whereas CoF_2 only reacts with water vapor at high temperatures. Finally, NiF_2 is stable against hydrolysis and partly also resists concentrated acids.

(II)   The difluorides form a variety of fluorometallates, the most important ones containing the anions $[MF_3]^-$, $[MF_4]^{2-}$, $[MF_5]^{3-}$ and $[MF_6]^{4-}$. In all these species, the coordination environment of the metal center is based on an octahedron, which may be more or less distorted, depending on the composition of the anion. Consequently, most of these anions do not exist as discrete octahedra, but form one-, two- or three-dimensional networks. Commonly, the fluorometallates are prepared by melting of stoichiometric mixtures of fluorides, but they are mostly also accessible by treatment of the corresponding chlorides with HF. Especially in case of Ni, the divalent fluoronickelates are also often prepared by reduction of tetravalent $[NiF_6]^{2-}$, for example, with H_2.

(III) The stability of the trifluorides significantly decreases from FeF_3 to NiF_3. FeF_3 is only reduced, when it is heated in presence of H_2, in contrast, CoF_3 is stable, but a strong oxidizer, as it reacts with a number of elements and also (per-)fluorinates hydrocarbons. NiF_3 is unstable, since it is already decomposed to NiF_2 in HF at 20 °C. FeF_3 is conveniently prepared by treatment of Fe metal or its chlorides $FeCl_2$ and $FeCl_3$ with F_2, alternatively, it is generated upon halogen exchange of anhydrous $FeCl_3$ with HF. CoF_3 is obtained, if Co metal or $CoCl_2$ are fluorinated by F_2 or ClF_3. The unstable NiF_3 is obtained, if $K_2[NiF_6]$ is treated with AsF_5 in aHF at 0 °C, which leads to the precipitation of intermediary NiF_4, which is even less stable and thus decomposes to NiF_3. NiF_3 is not a real trifluoride, since it is composed of di- and tetravalent Ni centers, according to the formula $Ni^{II}[Ni^{IV}F_6]$.

(IV) Based on FeF_3 and CoF_3, tetra-, penta- and hexafluorometallates $[MF_4]^-$, $[MF_5]^{2-}$ and $[MF_6]^{3-}$ are known, however, NiF_3 only forms hexafluoronickelates $[NiF_6]^{3-}$. According to the (in-)stability of the neutral trifluorides, the preparation of the cobaltates and nickelates is not achieved in aqueous media, whereas this perfectly works for the iron compounds. In contrast, the Co- and Ni-species are, for example, obtained by fluorination of chloride mixtures with F_2 at 350–400 °C.

(V) Tetravalent iron-fluorine compounds do not exist in condensed state, the same holds for neutral CoF_4. However, anionic hexafluorocobaltates(IV) are accessible, as well as the corresponding Ni species and the unstable NiF_4. $[CoF_6]^{2-}$ is typically obtained by fluorination of $[CoCl_4]^-$ with F_2 at 250–300 °C. Similarly, $[NiF_6]^{2-}$ is generated by fluorination of $[Ni(CN)_4]^{2-}$ with F_2. It is also possible to fluorinate chloride mixtures under high-temperature and –pressure conditions. More conveniently, the treatment of NiF_2 and a fluoride source with photodissociated F_2 in aHF at 0 °C also yields hexafluoronickelate(IV) salts. NiF_4 is liberated from $[NiF_6]^{2-}$ salts by treatment with fluoride acceptors below −65 °C.

(VI) Upon treatment with excess fluoride acceptor, NiF_4 is converted to cationic $[NiF_3]^+$, which is an extremely powerful oxidizer. For example, it oxidizes the pentavalent anions $[RuF_6]^-$ and $[PtF_6]^-$ to the neutral hexafluorides RuF_6 and PtF_6, respectively, both of which are extreme oxidizers, too.

10.9.2

(I) The only stable difluoride of the platinum metals is PdF_2. Apart from that, RuF_2 and OsF_2 can be stabilized by neutral ligands, such as CO or PPh_3. However, in pure form they are not accessible in condensed phase.

(II) The trifluorides are typically prepared by two different approaches, which are either reduction of a high-valent fluoride, as in case of RuF_3 and IrF_3, or fluorination of low-valent species. The latter method is used for the preparation of RhF_3 and PdF_3, where the latter does not contain trivalent Pd, but is composed of di- and tetravalent metal, according to the formula $Pd^{II}[Pd^{IV}F_6]$. The other two platinum metals Pt and Os do not form trifluorides.

(III) According to their preparation, the tetrafluorides can again be divided in two major groups. RuF_4, OsF_4 and IrF_4 are commonly made by reduction of the penta- or hexafluorides with reducing agents like H_2 or Si. In contrast, RhF_4, PdF_4 and PtF_4 are synthesized by fluorination of the metals or low-valent species, such as the di- or trihalides. Moreover, all the tetrafluorides form hexafluorometallates $[MF_6]^{2-}$, which may also serve as source for the preparation of RuF_4, OsF_4 and PdF_4, when they are treated with AsF_5. Moreover, the hexafluorometallates play an important role in urban mining processes, as they are obtained upon treatment of the metals with $[BrF_4]^-$. By exploiting this behavior, the precious metals may be isolated from solid waste.

(IV)   The pentafluorides RuF_5, OsF_5, RhF_5, IrF_5 and PtF_5 are isomorphic and exist as tetramers in the solid-state. In contrast to the NbF_5 structure type, the platinum metals exhibit nonlinear M-F-M bridges, so that the octahedral $[MF_6]$ units are twisted against each-other. Like NbF_5, the platinum metal pentafluorides are highly associated in the liquid and also in the gas phase, so that monomeric molecules are only observed at high temperatures. The pentafluorides are not indefinitely stable, as they disproportionate to the tetra- and hexafluorides at higher temperatures.

(V)   The most powerful and certainly most commonly used oxidizer among the platinum metal hexafluorides is PtF_6. Accordingly, the stability of the hexafluorides decreases in the series $OsF_6 > IrF_6 > RhF_6 > PtF_6$. The hexafluorides RuF_6, RhF_6 and PtF_6 readily oxidize O_2 to $[O_2]^+$ and Xe to XeF_2. Moreover, OsF_6 converts Br_2 to $[Br_2]^+$, whereas IrF_6 oxidizes Cl_2 to $[Cl_4]^+$. PtF_6 also reacts with ReF_6, NF_3 and ClF_5 to generate $[ReF_6]^+$, $[NF_4]^+$ and $[ClF_6]^+$. However, its oxidation power is not sufficient to oxidize BrF_5 to $[BrF_5]^+$.

(VI)   The platinum metals form the hexavalent oxyfluorides $RuOF_4$, $IrOF_4$ and $OsOF_4$. Other compounds have been claimed but their existence has not been confirmed. $RuOF_4$ is prepared by the reaction of RuO_2 with F_2 above 400 °C. Alternatively, it is obtained, if RuO_4 is treated with KrF_2. In contrast, $IrOF_4$ is generated by the reaction of IrF_6 with traces of moisture, which may also occur during the preparation of the hexafluoride in glassware. $OsOF_4$ is best prepared by the stoichiometric reaction of five equivalents OsF_6 with two equivalents OsO_4 at 175 °C.

(VII)   Osmium is the only platinum metal that forms hepta- as well as octavalent oxyfluorides. However, binary hepta- or even octafluorides do not exist for any of the metals. The heptavalent osmium oxyfluorides include the compounds OsO_2F_3, $Os_2O_3F_7$ and $OsOF_5$. Actually, OsO_2F_3 is a mixed-valent compound, which is prepared by the reaction of OsO_4 with OsF_6 or $OsOF_4$ with OsO_3F_2 and more correctly written as $[OsO_3F_2 \cdot OsOF_4]$. In the same way, $Os_2O_3F_7$ is a mixed-valent compound, and also written as $[OsO_3F_2 \cdot OsF_5]$. It is obtained by the treatment of OsO_4 with a half excess of OsF_6. Finally, $OsOF_5$ is generated upon fluorination of OsO_2 or oxyfluorination of Os metal. The known octavalent compounds are OsO_3F_2 and OsO_2F_4. The former is for example obtained by the reaction of OsO_4 with BrF_3, ClF_3 or F_2. OsO_2F_4 is synthesized by treatment of OsO_4 with KrF_2.

10.10.4

(I)   Pure, nonstabilized CuF is highly unstable and thus unknown in condensed state. However, it is possible to prepare complexes, such as $[CuF(PPh_3)_3]$ or $[Cu(NH_3)_2]F \cdot NH_3$ by refluxing CuF_2 and PPh_3 in methanol or reacting Cu and CuF_2 in liquid ammonia.

(II)   Anhydrous CuF_2 is obtained, if Cu metal is reacted with F_2 or by treatment of CuO with aHF at 400 °C. It is also accessible in aqueous media, if for example $CuCO_3$ is dissolved in hydrofluoric acid. The obtained dihydrate $CuF_2 \cdot 2 H_2O$ cleaves off the water ligands if it is heated, thus, CuF_2 is stable against hydrolysis. However, upon contact with air, the anhydrous compound is slowly converted to $CuF_2 \cdot Cu(OH)_2$. Thermally, copper difluoride is a stable compound up to its melting point, but liquid CuF_2 slowly liberates F_2.

(III)   Neutral CuF_3 is only stable below −78 °C and typically obtained by solvolysis of $K_3[CuF_6]$ in anhydrous HF. Moreover, the trivalent fluorocuprates $[CuF_4]^-$ and $[CuF_6]^{3-}$ are known.

The cesium salt $Cs[CuF_4]$ is prepared by fluorination of $Cs[CuCl_3]$ under pressure, whereas $K_3[CuF_6]$ is obtained by fluorination of $KCl/CuCl$- or $KCl/CuCl_2$-mixtures. The neutral trifluoride is a strong oxidizer since it is capable of oxidizing Xe to XeF_6 at $-78\ °C$. At higher temperatures it readily decomposes under liberation of F_2. The cuprates are slightly more stable, however, $[CuF_6]^{3-}$ also evolves F_2, if it is warmed to $0\ °C$.

(IV)   Silver forms the stable binary fluorides Ag_2F, AgF, AgF_2 and unstable AgF_3. Ag_2F is prepared from Ag and AgF upon electrolysis in HF or by a mechanochemical synthesis. AgF is typically obtained in aqueous media by the treatment of silver(I) compounds, such as Ag_2O or Ag_2CO_3, with HF, giving rise to the hydrate $AgF \cdot 2\ H_2O$, which loses its water ligands upon slight heating. Alternatively, anhydrous AgF is generated upon thermal decomposition of $Ag[BF_4]$. The preparation of AgF_2 typically requires higher temperatures and is achieved by treatment of Ag metal with F_2 or fluorination of silver(I) halides. AgF_3 is not directly accessible by fluorination of low-valent species, but is liberated by treatment of anionic $[AgF_4]^-$ with fluoride acceptors, such as BF_3, PF_5 or AsF_5. However, salts of the composition $M[AgF_4]$ are accessible by fluorination of mixtures of MCl and AgCl with F_2. AgF_3 is the only neutral silver-fluorine compound which is thermodynamically unstable at room temperature, and thus decomposes under liberation of F_2.

(V) AgF_2 is a strong oxidizer, whereas trivalent silver even exhibits an extreme oxidation potential. However, in both cases, the oxidizing properties are especially obtained in fluoroacidic media, that is, in HF solutions, which contain strong fluoride acceptors, such as AsF_5 or SbF_5. Consequently, the active species under these conditions are $[AgF]^+$ and a corresponding cationic trivalent species. For example, divalent $[AgF]^+$ oxidizes Xe, O_2 and $[IrF_6]^-$ to XeF_2, $[O_2]^+$ and IrF_6, respectively. Cationic Ag(III) species are even capable of liberating the neutral hexafluorides RuF_6 and PtF_6 from the pentavalent anions $[MF_6]^-$. A similar oxidation power is only observed for tetravalent Ni compounds.

(VI) AuF_3 can be prepared by fluorination of Au_2Cl_6 at $200\ °C$ or dissolution of elemental gold in BrF_3, followed by subsequent decomposition of the adduct $AuF_3 \cdot BrF_3$. On laboratory scale, it may also be generated in a low-pressure reaction of Au and F_2 in a quartz-reactor. In its solid-state structure, AuF_3 exhibits polymeric chains of square-planar $[AuF_4]$-units, which are connected by *cis*-F-bridges. Moreover, the chains adapt a helical structure, where parallel coils show weak Au-F interactions, so that every Au center is distorted octahedrally coordinated. In the gas-phase, both T-shaped AuF_3 monomers as well as planar, F-bridged Au_2F_6 dimers are found.

(VII) AuF_5 is not directly accessible, but is liberated from anionic $[O_2][AuF_6]$ or $[KrF][AuF_6]$ upon vacuum thermolysis. The pentavalent hexafluoroaurates are, for example, obtained by oxidation of metallic gold with either a 1:4-mixture of O_2/F_2 at $500\ °C$ and a pressure of 5 bar, or by the room temperature reaction with KrF_2 in aHF, respectively. AuF_5 is a strong oxidizer and also a strong fluoride acceptor, so that it reacts with XeF_2 and NOF to form $[Xe_2F_3][AuF_6]$ and $[NO][AuF_6]$, respectively. However, it is a weaker fluoride acceptor than SbF_5, so that it gives rise to $[IF_6][SbF_6]$ and AuF_3, if it is reacted with a solution of SbF_5 in IF_5.

10.11.5

(I)   The preparation of the fluorides ZnF_2 and CdF_2 is quite similar. The reaction of ZnO with aqueous HF yields the tetrahydrate $[Zn(H_2O)_4]F_2$, which can be dehydrated upon heating. The anhydrous compound is also accessible by treatment of Zn metal with HF at red heat or the direct reaction of the elements. The same approaches are suitable for the preparation of CdF_2; moreover, the fluorination of CdO, $CdCl_2$ or CdS with F_2 also yields CdF_2. The solid-state structure of ZnF_2 is described by a rutile lattice, whereas the bigger size of Cd^{2+} results in a fluorite-type structure for CdF_2.

(II) Both ZnF_2 and CdF_2 form fluorometallates $[MF_3]^-$ and $[MF_4]^{2-}$ upon reaction with alkali metal fluorides or similar fluoride donors like TlF or NH_4F. In most cases, these species are also accessible in hydrated form, if the reactions are carried out in an aqueous medium. In all of these compounds, the metals are hexacoordinated in form of distorted octahedra. The reaction of two equivalents BaF_2 or PbF_2 with ZnF_2 formally gives rise to $M_2[ZnF_6]$. However, structural investigations on the Ba salt have shown that these species are more correctly written as tetrafluorozincates $[MF]_2[ZnF_4]$, and thus do not contain isolated, octahedral $[ZnF_6]^{4-}$ anions.

(III) Mercury forms the stable fluorides Hg_2F_2 and HgF_2. Hg_2F_2 is prepared by the reaction of Hg_2CO_3 with HF or treatment of $Hg_2(NO_3)_2$ with NaF. In contrast, HgF_2 is obtained, if Hg(II) compounds are treated with HF, upon fluorination of Hg with ClF_3, but also by disproportionation of Hg_2F_2 at 450 °C. In the solid-state structure of Hg_2F_2, linear molecules are observed, whereas HgF_2 crystallizes in a fluorite lattice.

(IV) Similar to ZnF_2 and CdF_2, HgF_2 yields the fluoromercurates $[HgF_3]^-$ and $[HgF_4]^{2-}$ upon reaction with fluoride donors. The reaction with fluoride acceptors, such as NbF_5 or TaF_5, initially converts HgF_2 to salts of the type $Hg[MF_6]_2$ or $Hg[Ta_2F_{11}]_2$, respectively. Upon reduction of these species with Hg in SO_2, the almost linear polycations $[Hg_3]^{2+}$ and $[Hg_4]^{2+}$ are accessible. If HgF_2 is reacted with excess SbF_5 in presence of Xe, the cation $[XeHg]^{2+}$ is formed.

11.1.6

(I) In many aspects, fluorinated organic compounds significantly differ from their nonfluorinated analogues. Often, these differences are explained by the outstanding properties of the fluorine atom itself, that is, its high ionization potential and its low polarizability, which both lead to weak interactions between fluorinated moieties. Moreover, the high electronegativity of fluorine leads to highly polarized C-F bonds, however, in perfluorocarbons the dipole moments cancel, so that these compounds belong to the most unpolar compounds known. Finally, the C-F bonds in fluoroorganic compounds are strengthened by a good match of the corresponding 2s and 2p orbitals of carbon and fluorine.

(II) Linear hydrocarbons typically adapt a linear zig-zag conformation, which is clearly not observed for nonbranched perfluorocarbons. The latter preferentially adapt a helical conformation, which is due to the fact, that sterical repulsions between fluorine substituents in relative 1,3-position render a zig-zag conformation highly unfavorable. Since helical chains are far less flexible than zig-zag chains, perfluorocarbon molecules mostly behave like rigid rods.

(III) An interesting behavior is observed for the boiling points of fluoroorganic compounds. For example, the highest boiling point in the series of methanes $CH_{4-n}F_n$ is observed for CH_2F_2, whereas the values decrease for higher and lower fluorine content. Thus, the dipole moments of the molecules seem to play an important role. However, a similar behavior is not observed for the analogous chlorinated and brominated methane compounds, as their boiling points increase with increasing halogen content. More impressively, the hydrocarbon C_6H_{14} and the perfluorinated analogue C_6F_{14} exhibit similar boiling points (57 vs. 69 °C), despite the fact, that the perfluorocarbon is almost four times as heavy as the hydrocarbon.

(IV) The outstanding solvent properties of perfluorocarbons arise from their extreme nonpolarity. Thus, they are generally very poor solvents for other organic compounds and of course also for hydrophilic substances. However, they are very good solvents for compounds with very low cohesive energies, such as gases, molecular inorganic fluorides like WF_6 and of course other organic compounds with a high fluorine content. Thus, they are for example used as oxygen carrier in artificial blood and very commonly as solvents in fluorous chemistry.

(V) Due to the very weak intermolecular interactions, liquid perfluorocarbons exhibit a very low surface tension and are thus capable of wetting virtually all surfaces. In the same way, solid perfluorocarbons show very high surface energies and consequently have antistick properties. The latter is especially exploited in the preparation of frying pans, whereas the low surface tensions of liquid, highly fluorinated compounds find for example applications in the preparation of fluorosurfactants.

(VI) The introduction of fluorine typically increases the acidity of órganic compounds, such as alcohols and carboxylic acids. For example, trifluoroacetic acid is by four orders of magnitude stronger than CH_3COOH. Even more pronounced is the difference between the couples hexafluoroisopropanol and isopropanol as well as perfluoro *tert*-butanol and *tert*-butanol, where the pKa values differ by about 7 and 13.5, respectively. In the same way, fluorination decreases the basicity of amines, ethers and carbonyl compounds. For example, pentafluoroaniline is about five orders of magnitude less basic than nonfluorinated aniline.

11.2.7

(I) In saturated systems, the C-F bond energy typically amounts to $450–550\ kJ \cdot mol^{-1}$, and thus represents the strongest carbon-element bond. Consequently, the reactivity of C-F bonds in S_N1- and S_N2-reactions is low, which is nicely underlined by the fact, that C-Cl bonds in comparable systems are by about two to six orders of magnitude more reactive.

(II) Both, α- and β-fluorination reduce the reactivity of saturated substrates toward nucleophilic substitution. For example, the replacement of the Br substituent in CF_3CH_2Br by NaI in acetone is by far slower than in nonfluorinated CH_3CH_2Br. The deactivating effect is even more pronounced for α-fluorination, since the bromo substituent of $-CF_2Br$ groups cannot be replaced using NaI in acetone.

(III) Perfluorocarbons are thermally rather stable, which is due to the strength of the C-F bonds. Thus, the weakest bonds in perfluorocarbons are found within the carbon skeleton, so that CF_4 is extremely stable, but the susceptibility to thermal degradation significantly increases with the number of carbon atoms. This effect is even more pronounced for nonlinear, highly branched perfluorocarbons. For hydrogen-containing compounds, the thermal stability is drastically decreased, which is due to the fact, that hydrofluorocarbons can readily eliminate HF upon thermal degradation.

(IV) Perfluorocarbons are kinetically inert, since the fluorine substituents both sterically and electrostatically protect the electrophilic carbon skeleton from nucleophilic attacks, for example, basic hydrolysis. However, perfluorocarbons react with strong reducing agents, such as alkali metals, which will initially give rise to perfluoroalkyl radical anions. The latter split off a fluoride anion and the perfluoroalkyl radicals typically undergo further decomposition.

(V) The presence of fluorine substituents can lead to both, stabilization or destabilization of olefinic double bonds. For example, monofluorination leads to a stabilization of the C=C bond, whereas the presence of two vicinal or even three fluorine substituents destabilizes the double bond. In carbonyl compounds, the presence of fluorine always stabilizes the C=O bond, but only if it is directly connected to the carbonyl carbon. The main reason for these observations is the good overlap of the 2s and 2p orbitals of carbon and fluorine. The fluorine substituents back-donate some electron density into the π-system, which leads to a strengthening of the bond. This is best achieved, if there is no interfering group, such as other fluorine substituents.

(VI)   In general, fluorinated carbocations are rather unstable and only scarcely observed. A fluorine substituent in α-position stabilizes the carbocation by resonance, typically, this behavior is more pronounced than the destabilization by inductive effects. However, the latter lead to strong destabilization, if there are fluorine substituents in β-position of the cation. In most cases, fluorinated carbocations are obtained by electrophilic addition reactions to unsaturated substrates. Since the stability of the intermediate cations decreases with increasing fluorine content, this also gives an explanation, why highly fluorinated olefins are commonly not attacked by electrophiles.

(VII) Carbanions are stabilized by both α- and β-fluorination. However, the stabilization due to the inductive effect of α-fluorine substituents is only weakly pronounced. Thus, β-fluorination leads to significantly higher stabilization of carbanions by both, inductive and resonance effects. The influence of β-fluorine is nicely represented by the acidity of C-H bonds. Thus, fluoroform and other 1H-perfluorocarbons are only poorly acidic, since they have no, or at maximum two β-fluorine substituents. Consequently, the acidity is drastically increased for secondary and tertiary C-H bonds, for example, in 2H-perfluoropropane and 2H-perfluoro-2-methylpropane.

(VIII) The nucleophilic attack on fluoroolefins always gives rise to fluorinated carbanions. The regioselectivity of the nucleophilic attack can generally be predicted, since the intermediate anion tends to have a maximum amount of stabilizing β-fluorine atoms. For example, the nucleophilic attack on unsymmetric fluoroolefins always occurs at the difluoromethylene site. Nevertheless, the formation of the final product depends on many more factors and is thus hardly predictable.

(IX)   On a thermodynamical viewpoint, fluorinated radicals are not generally more stable than their nonfluorinated analogues, in contrast, they are often even slightly destabilized. However, in some cases kinetic stabilization occurs, so that the presence of fluorine can indeed influence the outcome of radical reactions. Interestingly, perfluoroalkyl radicals cannot undergo disproportionation reactions, so that it is impossible to generate a perfluorocarbon and a perfluoroolefin from two perfluoroalkyl radicals. This behavior is very beneficial for the preparation of fluoropolymers, since an important termination reaction does not occur. Thus, this behavior allows the preparation of high molecular weight products.

(X)   All α-fluorinated carbenes have a singlet ground-state, which is due to resonance stabilization. The latter does not allow the free electrons of the carbon atom to occupy different orbitals. Like other singlet carbenes, the fluorinated derivatives typically add to unsaturated systems, for example, :CF$_2$ reacts with olefins under formation of difluorocyclopropane moieties. However, the fluorinated carbenes do not insert into C-H bonds.

11.3.9

(I)   Halofluorocarbons, especially chlorofluorocarbons, have been widely used as refrigerants, but also as propellants and foam-blowing agents. In general, their use is based on their high volatility and low reactivity, in conjunction with nonflammability and nontoxicity. Moreover, non-toxic and non-flammable bromofluorocarbons have been used as fire extinguishing agents and as inhalative narcotics. However, the chloro- and bromofluorocarbons significantly contribute to the ozone-depletion in the atmosphere, so that their applications have been banned or at least strongly restricted. Therefore, hydrofluorocarbons have been developed to replace the halofluorocarbons, as they are not ozone-depleting. Nevertheless, they mostly show a rather long atmospheric lifetime and are tried to be replaced by less problematic compounds with short atmospheric lifetimes.

(II) The main advantage of fluoropolymers is their resistance against aggressive chemicals. Apart from that, especially the anti-stick properties of PTFE play an important role, so that it is used as coating in frying pans, but also as fiber in waterproof functional clothes. For the chemist, the most popular uses of fluoropolymers are certainly described by the preparation of labware, seals and coatings, which are resistant to aggressive chemicals like HF and even F_2 under appropriate conditions.

(III) The lateral fluorination of liquid crystal molecules typically influences the length-breadth ratio of the compound, which results in decreased clearing temperatures. In contrast, the introduction of fluorine or fluorinated substituents in polar groups increases the dielectric anisotropy. In addition, fluorinated liquid crystal molecules are more reliable, since they do not that easily interfere with ionic impurities as it is, for example, the case for cyano-substituted materials. Thus, a high level of performance is retained for longer times.

(IV) Fluorine also improves the performance and properties of organic electronics, such as OLEDs, OFETs but also photovoltaics. In organic semiconductors, fluorine decreases the HOMO and LUMO levels, which is due to its electronegativity and thus its negative inductive effect. In contrast, the resonance effects, and hence the π-backdonation of fluorine substituents in aromatic moieties can change the band gap of the semiconductor. Finally, fluorination of organic semiconductor materials also has an impact on the intermolecular interactions. Since fluorinated regions generally tend to orient to more electron-rich areas of their neighboring molecules, this can result in shorter edge-to-edge distances and sometimes also in better, nontilted π-stacking.

(V) The introduction of fluorine into biologically active compounds often has a large number of advantages. Common benefits include a modulated lipophilicity, the inhibition of metabolic pathways and also modified interactions with the target. Overall, these changes typically lead to an increased bioavailability, an increased selectivity as well as decreased effective doses. Due to the beneficial properties, fluorinated drugs are increasing on the market. To date, about every fourth blockbuster drug contains fluorine. Perspectively, this value will further increase in future.

(VI) The artificial isotope ^{18}F is a positron emitter with a half-life of about 110 minutes and is commonly used in PET imaging. Depending on the target organ, several ^{18}F-labeled molecules are used as tracers. For example, ^{18}F-fluorodeoxyglucose serves for the diagnosis of cancer, since it is preferentially transported to areas with an increased glucose metabolism. In contrast, ^{18}F-dopamine derivatives are applied for the diagnosis of brain diseases.

(VII) Artificial blood is mostly a stable emulsion of perfluorocarbons, perfluorinated tertiary amines, phospholipids, electrolytes and water. Its most important property is the capability of dissolving large amounts of oxygen, so that it is administered to the patient during surgery. However, since it can only serve as adequate replacement in terms of oxygen transport, it needs to be replaced by original blood after the surgery. All the other important functions of original blood cannot be taken by artificial blood.

12.1.3

(I) Direct and radical fluorinations exhibit quite drastic reaction conditions; thus, they are mostly not suitable for sensitive substrates, especially if they contain functional groups.

Consequently, these techniques are commonly used for the preparation of highly or even perfluorinated molecules.

(II) Elemental fluorine is one the simplest but still a very versatile reagent for the preparation of fluorinated compounds. However, careful reaction control is required, which is typically achieved by dilution and cooling. Apart from that, a suitable laboratory and appropriate reaction vessels are required, since fluorinations with F_2 cannot be conducted in standard organic laboratories, of course.

(III) CoF_3 is certainly the most frequently used high-valent transition metal fluoride for the preparation of fluoroorganic molecules. It is a very reactive and efficient fluorinating agent for the preparation of highly fluorinated compounds. Other benefits include its easier handling than F_2 and its recyclability by treatment with F_2, so that HF is the only waste product that is generated during the reaction.

(IV) As very mild alternatives, fluoraza reagents have been used in radical fluorination reactions. If they are activated by a photocatalyst or some other radical initiator, the N-F bond is homolytically cleaved to yield a nitrogen-centered radical and F˙. The former abstracts a H˙ radical from the substrate, where the regioselectivity of the abstraction is determined by either the presence of a directing group or often the stability of the intermediary carbon-centered radical species. Thus, this method of radical fluorination typically yields tertiary or benzylic fluorides.

12.2.3

(I) The main benefit of electrochemical fluorination over direct fluorination is given by significantly less forcing reaction conditions. In addition, the applied reagents HF and electrical current are by far cheaper than F_2, they also allow slightly easier handling, since electrochemical fluorination is typically conducted at atmospheric pressure. But still, electrochemical fluorination belongs to the rather drastic methods of fluorination, and is thus not suitable for the preparation of sensitive compounds.

(II) The most important parameter for the setup of an electrolysis cell is the anode, which has to be of Ni metal with high purity. For the cathode material, iron might be used, however, in most cases they are also made of Ni. The general setup of the cell can be rather simple, as electrofluorinations are mostly conducted at atmospheric pressure and moderate temperatures. For large-scale preparations, multiple electrolysis cells can be combined. The substrates have to meet some requirements, too, since they need to be soluble in aHF and additionally contain a polar group, which is needed for the formation of an electroconductive solution. Ideally, the fluorinated compound should be less soluble or better insoluble in aHF, so that it can be easily separated. As a general trend, best yields are observed for short-chained substrates with comparatively less C-H bonds.

(III) The complete mechanism of electrochemical fluorination processes is still not fully understood. However, films of high-valent nickel fluorides are generated upon oxidation of HF at the anode, whereas H_2 is liberated at the cathode. Most likely, the anode is then covered with NiF_3, which also contains some NiF_4 as dopant. The organic substrate is then adsorbed at the nickel fluoride surface and undergoes fluorination, which additionally yields HF and NiF_2. The latter remains at the anode surface and is reoxidized to high-valent fluorides. As a general trend, the area of the substrates with the highest electron-density adsorbs best on the anode. The electrochemical fluorination in aHF is a radical process, so that side-reactions of the carbon centered radicals may occur, such as rearrangements or cyclizations.

12.3.5

(I)   Olah's reagent is a complex of HF with pyridine, also known as pyridinium poly(hydrogen-fluoride). The adduct contains about nine molar equivalents of HF per pyridine molecule, which equals a mass percentage of roughly 70 %. When handling it, it needs to be taken into account, that it has acidic properties and will thus etch glassware. Consequently, it has to be stored and used in plastic vessels. In addition, it releases HF, if it is heated above 50 °C.

(II)   Other important complexes of HF with organic nitrogen bases are $NEt_3 \cdot 3\,HF$ (TREAT-HF) and the adduct with N,N'-dimethlypropyleneurea (DMPU). The nature of the base determines the properties of the complex, for example, TREAT-HF has a neutral or slightly basic character, so that it does not etch glassware as it is observed for Olah's reagent. Finally, the weak basicity of DMPU and its increased hydrogen-bond accepting properties lead to a slightly different behavior of its HF complex. Overall, the use of appropriate bases allows fine-tuning of the basicity and nucleophilicity of F^- in the aforementioned HF complexes.

(III)   The major problem of the use of ionic fluorides in organic synthesis is their limited solubility in organic solvents. This issue can be overcome by using additives, such as crown ethers or cryptands. The latter will complex cations like K^+, so that F^- is brought into solution as well. The solubility problem is sometimes also overcome by using CsF as fluorinating agent, since it is better soluble than the lighter alkali metal fluorides.

(IV)   In hydrated form, the tetraalkylammonium salts $[NMe_4]F$ and $[NBu_4]F$ are readily available by salt metathesis in aqueous medium. However, the anhydrous compounds cannot be obtained by thermal treatment of the hydrates, since the cations, especially $[NBu_4]^+$, will decompose under these conditions. Therefore, the anhydrous fluorides directly need to be prepared in anhydrous state, in case of $[NBu_4]F$ this is conveniently achieved by the reaction of $[NBu_4]CN$ with hexafluorobenzene. The corresponding anhydrous methyl derivative $[NMe_4]F$ is commonly prepared by salt metathesis of $[NMe_4][BF_4]$ with KF in methanol. The resulting alcohol adducts are then decomposed in vacuo to yield pure and anhydrous $[NMe_4]F$.

(V)   The hypervalent silicate anion $[Me_3SiF_2]^-$ is found in both, the sulphonium compound TASF as well as the corresponding guanidinium salt. However, in solution, only the hexamethyl-guanidinium derivative shows a dynamic equilibrium between F^- and Me_3SiF. TASF is prepared by the reaction of SF_4 with Me_2NSiMe_3, whereas the guanidinium salt is accessible by the reaction of Me_2NSiMe_3 with the difluoromethane derivative $(Me_2N)_2CF_2$. The latter is prepared by chlorination of tetramethylurea and subsequent halogen exchange with $[NMe_4]F$.

(VI)   A large number of fluorinating agents is derived from SF_4, such as DAST or MOST. Their main advantage over the gaseous and highly toxic SF_4 is their simplified handling, as they are liquid compounds. However, they are thermally not very stable, so that they tend to decompose explosively upon heating. Their preparation is rather simple, since they are obtained from the reaction of SF_4 with Me_3Si-substituted, secondary amines. Treatment of DAST or MOST with BF_3 yields the sulphiminium compounds XtalFluor-E™ and XtalFluor-M™, respectively, which are air-stable solids with a reduced tendency to decompose and thus, safer fluorinating agents.

(VII) α,α-Difluoroalkylamine reagents are commonly prepared by the addition of secondary amines, mostly Et_2NH to fluorinated olefins, such as chlorotrifluoroethylene, perfluoropropene or 1,1,3,3,3-pentafluoropropene. However, the difluoroalkylamine reagents tend to eliminate HF, which then gives rise to α-fluoroenamines. In most cases, the latter are no active fluorinating agents, but they can add HF again to be transformed back to the original difluoroalkylamine, and thus the active compound.

(VIII)   DFI is a very reactive and convenient fluorinating agent. Its main advantage is that it is pre-pared from a readily available urea derivative, which also allows recycling. For the prepara-tion of DFI, *N,N'*-dimethylethylenurea is initially chlorinated to yield an intermediate ami-dinium chloride, which is then converted to DFI upon halogen exchange with KF.

12.4.3

(I)   In principle, electrophilic fluorinating agents are present, if F-substituents are connected to other highly electronegative elements, such as oxygen. Thus, the majority of inorganic reagents comprises fluoroxy compounds, in the simplest fashion oxygen fluorides like OF_2 or O_2F_2. Apart from that, anionic fluoroxysulphates $[FOSO_3]^-$ have been used as electrophilic fluorinating agents, as well as simple organic compounds like CF_3OF or the acyl derivatives CH_3COOF and CF_3COOF. Another important compound that needs to be mentioned is per-chloryl fluoride ClO_3F. The problem of all these reagents is their extreme reactivity in contact with organic matter, which can easily result in violent reactions. Thus, they all have been replaced by safer alternatives and are not commonly used in organic synthesis.

(II)   In contrast to the fluoroxy compounds, XeF_2 is a rather mild fluorinating agent. It is quite easy to handle and can be stored in PTFE vessels, so that it is also commercially available. However, it is rather expensive, so that it is not a suitable reagent for large scale prepara-tions.

(III)   The fluoraza reagents are generally more stable and less hazardous than the oxygen-based compounds. Therefore, they are in most cases commercially available. The reagents are divided into two major groups, being neutral *N*-fluoroamine compounds and cationic *N*-fluoroammonium reagents.

(IV)   NFSI (*N*-fluorobenzenesulphonimide) belongs to the N-fluoroamine compounds and is cer-tainly the most important derivative of this subclass of fluoraza reagents. It is prepared by the reaction of the sulphonimide $(PhSO_2)_2NH$ with F_2/N_2 in presence of NaF. There are also derivatives of NFSI, where the phenyl groups are replaced by other substituents, for exam-ple, CF_3. However, these reagents are by far less important.

(V)   *N*-fluoropyridinium salts are readily accessible by fluorination of the corresponding het-eroarene with F_2 at low temperature. However, in order to obtain stable reagents, the arene needs to be substituted, for example, by methyl groups in *ortho*- or *ortho*- and *para*-positions. A stabilization is also achieved, if chlorine substituents are attached in *meta*-positions. Apart from that, it is necessary to replace the nucleophilic fluoride anion by other, less reactive anions, such as $[BF_4]^-$, $[PF_6]^-$, $[SbF_6]^-$ or OTf$^-$. This is conveniently achieved by treatment of the intermediary *N*-fluoropyridinium fluoride with the corresponding Lewis acids BF_3, PF_5 and SbF_5 or by treatment with the silane reagent Me_3SiOTf, respectively.

(VI)   Selectfluor™ is the commercial name of a very common electrophilic fluorinating reagent, which is based on a dicationic and bicyclic triethylendiammonium moiety. It is prepared by alkylation of triethylenediamine (TEDA) with CH_2Cl_2 and subsequent fluorination with di-luted F_2 at low temperature. During its preparation, the anions Cl$^-$ and F$^-$ have to be re-placed by other, non-nucleophilic species, typically $[BF_4]^-$. The commercially available Se-lectfluor™ contains a chloromethyl group at one of the nitrogen centers and two $[BF_4]^-$ an-ions. However, there are also derivatives with other anions, for example, $[PF_6]^-$ or OTf$^-$, which typically show an improved solubility. Another derivative is Selectfluor™ II, which contains a methyl group instead of the CH_2Cl-unit. The latter may also be replaced by a hy-droxy function, which then gives rise to commercially available Accufluor™.

13.1.3

(I)   The general fluorine effect describes the preferential accumulation of fluorine substituents at saturated carbon centers, which typically occurs under drastic reaction conditions and in presence of strong Lewis acids. This behavior can result in both inter- and intramolecular halogen transfer as well as in other migrations and rearrangements.

(II)   The general fluorine effect leads to the accumulation of fluorine substituents at saturated carbon centers. Consequently, products containing CF_3-groups and geminal CF_2-units are the preferred ones. This can be rationalized by resonance stabilization of the electron-deficient carbon centers by π-backbonding from the lone pairs of the fluorine substituents. This behavior results in negative hyperconjugation, and thus gives rise to ionic resonance structures.

(III)   The special fluorine effect is observed for unsaturated compounds and is caused by repulsive interactions between the π-system of sp^2-carbons and the lone pairs of fluorine substituents. Thus, $C(sp^3)$-F bonds are generally preferred over $C(sp^2)$-F, since they exhibit less steric repulsion. Consequently, unsaturated fluoroorganic compounds tend to minimize their number of $C(sp^2)$-F bonds, which especially occurs under UV-irradiation or in presence of fluoride ions.

(IV)   The special fluorine effect may also be visualized by the C-H acidity of fluoroorganic compounds, such as 1,2,3,4,5-pentafluorocyclopenta-1,3-diene. The latter contains 4 $C(sp^2)$-F bonds and one $C(sp^3)$-bond, upon deprotonation, it increases the amount of $C(sp^2)$-F to 5. Thus, deprotonation is not very favorable, since the inductive stabilization of the negatively charged $[C_5F_5]^-$ is less pronounced than the destabilization by an additional $C(sp^2)$-F bond. Consequently, 1,2,3,4,5-pentafluorocyclopenta-1,3-diene is only slightly more acidic (pKa about 14) than its nonfluorinated analogue (pKa 15.5). In contrast, the replacement of the fluoro-substituents by CF_3-groups results in a highly acidic compound (pKa −2). In this case, deprotonation does not lead to unfavorable $C(sp^2)$-F bonds. Moreover, the special fluorine effect, and thus the reduction of the number of $C(sp^2)$-F bonds is observed, if 1,2,3,4,5-pentafluorocyclopenta-1,3-diene is irradiated with UV-light, as this will lead to a rearrangement to 1,2,3,5,5-pentafluorocyclopenta-1,3-diene.

13.2.5

(I)   When hydrocarbons are fluorinated, it is important that the concentration of the fluorinating agent, in most cases F_2, is not too high in the beginning, when there are many substitutable C H bonds. Otherwise, undesired reactions may occur, resulting in increased fragmentation of the substrate. To fulfil these requirements, two general approaches may be chosen. Either, the F_2 concentration is gradually increased during the reaction by decreasing the amount of the inert diluent, since the final C-H substitutions will require more forcing conditions. Or, alternatively, the reaction may be run at constantly low F_2 flow, which will require longer reaction times to achieve perfluorination.

(II)   Both steric and electronic effects can influence the outcome of perfluorination reactions. Steric effects especially play a role in substrates containing *tert*-butyl groups or other sterically demanding substituents. For example, it is difficult to achieve perfluorination of 2,2,4,4-tetramethylpentane, since the 18 C-H bonds of the terminal methyl groups undergo faster fluorination than those of the internal CH_2-unit. As a consequence, the perfluorinated *tert*-butyl groups further increase the shielding of the methylene unit, so that a mixture of $((CF_3)_3C)_2CF_2$, $((CF_3)_3C)_2CHF$ and $((CF_3)_3C)_2CH_2$ is obtained. In the same way, electronic effects can deactivate C-H bonds, which is, for example, observed in bicyclic substrates, such as norbornane or norbornadiene. Therein, the CH_2-groups are faster fluorinated than the tertiary C-H bonds in bridgehead positions. The latter are then further deactivated by the electron-withdrawing effect of the fluorinated bicyclic system. The rather acidic protons are then more resistant to the attack of F_2.

(III) The fluorination of primary alkyl chlorides typically yields the corresponding primary perfluoroalkyl chlorides, since rearrangements do not take place. In contrast, 1,2-chlorine shifts can be observed in secondary and tertiary substrates. In secondary alkyl chlorides the rearrangements are mostly not very pronounced, so that a mixture of rearranged and nonrearranged products will be obtained. The situation changes for tertiary alkyl chlorides, which readily undergo rearrangements upon fluorination. In general, alkyl bromides are less stable and will mostly result in rearrangements within the carbon skeleton, giving rise to nonbrominated perfluorocarbons.

(IV) The LaMar process is one of the fundamental methods for the preparation of perfluorinated molecules, which relies on the low-temperature fluorination in the solid-phase. Over the time, several varieties have been developed to adapt the fluorination process to other classes of substrates. Overall, the LaMar process and its varieties have been used for the preparation of liquid perfluoropolyethers, simple *bis*(perfluoroalkyl) ethers and fluorinated heterocycles, such as piperidine derivatives. Moreover, finely divided, solid hydrocarbon polymers, such as polypropylene are converted to the respective perfluorinated polymers. In order to ensure a safe fluorination process, the F_2 concentrations in the reactions generally need to be controlled, in addition, an efficient heat dissipation is required to avoid the cleavage of C-C bonds.

(V) Mechanistically, the direct fluorination approaches are described by radical-chain processes. Thus, initiation is typically achieved by the homolytic cleavage of F_2 or its direct reaction with the hydrocarbon substrate, which gives rise to HF as well as F˙ and a carbon-centered radical. Propagation is achieved by reaction of F˙ with the substrate, which will give rise to HF and a carbon-centered radical. The latter can also react with F_2 to form a fluorinated product and F˙. Finally, termination occurs upon recombination of two carbon-centered radicals or a carbon-centered radical and F˙. Overall, F_2 reacts with the hydrocarbon substrate under formation of a fluorinated product and HF.

(VI) Instead of F_2, a number of high-valent metal fluorides can be used as fluorinating agents, such as the neutral compounds AgF_2, MnF_3, CeF_4, PbF_4, BiF_5, but also anionic species like $[AgF_4]^{2-}$, $[CoF_4]^{-}$, $[NiF_6]^{3-}$ or $[PtF_6]^{2-}$. The most important metal fluoride in this context is CoF_3, as it has been widely used for the preparation of highly fluorinated organic compounds. All the aforementioned compounds are relatively strong oxidizers, so that they can only be handled as pure substances or in HF, but not in any other solvent. Consequently, high-valent metal fluorides are typically applied to fluorinate hydrocarbon substrates in the vapor-phase. For selected substrates, these fluorinating agents show excellent results, however, they also exhibit significant drawbacks. First of all, the preparation of all the aforementioned metal fluorides is more or less closely associated to elemental fluorine, so that they cannot be considered as real alternatives to avoid the use of F_2. Moreover, the reaction conditions typically result in complex product mixtures of poly- and perfluorinated compounds, also including degradation products and rearranged species.

(VII) Mechanistically, the fluorination with CoF_3 is believed to proceed via radical-cation intermediates. In the first step, CoF_3 oxidizes the hydrocarbon to a radical-cation, where itself is reduced to $[CoF_3]^{-}$. The radical-cation loses a proton, which reacts with $[CoF_3]^{-}$ to liberate

CoF$_2$ and HF. In the next step, the carbon-centered radical is oxidized by CoF$_3$ to result in a carbocation and [CoF$_3$]$^-$. The cation abstracts F$^-$ from the latter anion, so that the C-F bond is formed along with another molecule of CoF$_2$.

(VIII) Apart from the drastic techniques using F$_2$ or metal fluorides, it is also possible to use mild radical conditions for the preparation of more sensitive compounds. In some cases, it is not even necessary to use an additional initiator, for example, when radicals are generated by decarboxylation of perester substrates. In the same way, the use of heterobenzylic substrates does not require the addition of an initiator. Other substrates with reactive C-H bonds in secondary, tertiary and especially benzylic positions need to be activated by radical initiators, mostly compounds like K$_2$S$_2$O$_8$, a combination of BEt$_3$ and O$_2$ but also organic initiators like NDHPI. The selectivity of the direct C-H fluorinations arises from the stability of the intermediately generated carbon-centered radicals, which are generated by H˙ abstraction by the initiator. Thus, most of the techniques will lead to a selective fluorination in benzylic or tertiary position, depending on the actual substrate. The intermediate carbon-centered radicals are capable of abstracting F˙ from fluoraza reagents, in most cases Selectfluor™, to form C-F bonds. The resulting nitrogen-centered radical often works as base and abstracts H˙ to regenerate the initiator.

(IX) Direct C-H fluorination of reactive positions is also achieved under photocatalytic conditions, where the activated photosensitizer is believed to abstract H˙ from the substrate. The carbon-centered radical then abstracts F˙ from the fluoraza reagent, which in turn regenerates the photosensitizer by uptake of H˙. Typical photocatalysts for this purpose are 9-fluorenone or xanthone, however, others like tetracyanobenzene, anthraquinone or acetophenone may also be used. In some special cases, it is not required to use an additional photosensitizer, when there is already a photosensitive group in the molecule, such as a phthalimide-moiety.

(X) A selective, radical C-H fluorination is not only achieved for activated substrates containing benzylic or tertiary hydrogen substituents. In presence of photoexcitable, directing groups, such as polycyclic enones, it is also possible to selectively fluorinate in β- or γ-position of the directing function. A similar behavior is observed, if cyclic ketones serve as directing groups, which requires the use of additional benzil as photosensitizer. In both cases, Selectfluor™ serves as electrophilic fluorinating agent.

13.3.4

(I) In contrast to direct fluorination techniques, electrochemical fluorination tolerates several important functional groups, such as ether or amine functions. More correctly, electrochemical fluorination even requires some polar functionalities, in order to make the substrates soluble in aHF. Another advantage is the use of aHF as both solvent and cheap fluorinating agent under typically less forcing reaction conditions. Overall, the reaction conditions of electrochemical fluorination are milder than in direct fluorination reactions, however, side-reactions, such as rearrangements and especially cyclizations can readily occur.

(II) Electrochemical fluorination is especially suitable for the preparation of perfluorinated ethers, amines, sulphonyl fluorides and carbonyl fluorides. It is also possible to fluorinate gaseous substrates or vapors of liquid starting materials. However, unsaturated compounds like pyridine will become saturated. Moreover, chlorine substituents in α-position of heteroatoms are readily cleaved off.

(III) The electrochemical fluorination of carbonyl compounds, especially carboxylic acids, often leads to cyclization reactions. Initially, the carboxylic acids are converted to carbonyl fluorides, which then undergo cyclization, probably via hydrogen-bonded onium polycations.

This side-reaction gives rise to mixtures of perfluorinated acid fluorides and cyclic ethers, which are also obtained from primary alcohols and aldehydes. Both, α- and β-cycloalkyl-substituted esters as well as α-cyclohexenyl-substituted derivatives are transformed to cyclic ethers, too.

13.4.13

(I) BrF_3 is a very reactive compound and thus only suitable for halogen exchange reactions in very robust species. However, some unreactive substrates, for example, alkyl chlorides with adjacent CF_3-groups are efficiently substituted by BrF_3. In general, substitution reactions with BrF_3 occur rather unselectively and thus generate complicated product mixtures. The addition of a catalyst like $SnCl_4$ can help to control the product formation, as for example the addition of 1% $SnCl_4$ leads to a clean conversion of CH_2Cl-CHF-CH_2Cl to CH_2F-CHF-CH_2F. In addition, the solvent plays a role, since rather clean halogen exchange is observed in chlorofluorocarbon solvents, whereas the use of HF typically leads to additional cleavage of C-H bonds, and thus rearrangements.

(II)   Generally, ClF shows a rather poor reactivity against primary chloro-substituents in alkyl chlorides. Thus, the secondary chloride is replaced first, when 1,2,3-trichloropropane is treated with ClF. However, if the primary chlorine is not part of a CH_2Cl-group, it is more easily replaced by fluorine, which is for example observed, if 1,1,1,2-tetrachloroethane is treated with ClF. Initially, two chlorine substituents of the CCl_3-group are replaced, before an exchange is observed at the CH_2Cl-unit. Typically, a primary chlorine is also substituted, if a fluoride is attached to the same carbon atom, or if $SbCl_5$ is added to the reaction mixture.

(III)   Swarts fluorination is a method, which is especially applied for the industrial preparation of hydrofluorocarbons, chlorofluorocarbons and hydrochlorofluorocarbons. It uses simple fluorinating agents, such as aHF, which are combined with Lewis acid catalysts in most cases antimony chlorofluorides $SbCl_xF_{5-x}$. The Lewis acid assisted fluorination is commonly applied to poly- and perchlorinated substrates, however, in most cases not all chlorine substituents are replaced. For example, CCl_4 is converted to CCl_3F and $CHCl_3$ yields a mixture of $CHCl_2F$ and $CHClF_2$ upon reaction with aHF in presence of $SbCl_xF_{5-x}$. Nevertheless, 1,1,1-trichloroethane is cleanly converted to 1,1,1-trifluoroethane under similar conditions.

(IV)   The halogen exchange of alkyl monohalides is commonly achieved by the use of alkali metal fluorides, mostly KF. Less often, AgF or the tetraalkylammonium salts $[NR_4]F$ are used for this purpose. In general, better results are achieved, when the reactions are carried out under anhydrous conditions in polar aprotic solvents. However, one has to take into account that naked, nonsolvated fluoride is a rather strong base and will thus induce elimination side-reactions, especially upon treatment of secondary and tertiary substrates.

(V)   In general, allylic and benzylic halogen substituents are more reactive toward halogen exchange. Thus, perchloropropene is converted to CF_3-CCl=CCl_2 upon reaction with HF. However, the vinylic chlorine-substituents remain unchanged. Similarly, all benzylic chlorine substituents are replaced in $PhCCl_3$ to give benzotrifluoride. In contrast, the halogen exchange of $PhCCl_2CCl_3$ does not yield the pentafluoroethyl derivative, but a mixture of difluorobenzyl products, that is, $PhCF_2CCl_3$, $PhCF_2CCl_2F$ and $PhCF_2CClF_2$.

(VI)   In most cases, exchange reactions in α-halogenated carbonyl compounds are achieved by the use of KF or other metal salts, for example, AgF. Tetraalkylammonium salts are also generally suitable for this purpose. Furthermore, KF is used for the halogen exchange in acid chlorides, such as PhCOCl or CCl_3COCl. In the latter case, KF will exclusively exchange the acid chloride function, but does not attack the CCl_3-group. However, if anhydrous HF is used as fluorinating agent, CCl_3COCl is converted to the perfluorinated derivative CF_3COF.

(VII)   Tertiary alcohols generally undergo fluorodehydroxylation reactions, when they are treated with aHF or other acidic fluorides. However, since these transformations follow a S_N1-mechanism, they are mostly not stereoselective and generate product mixtures. Alcohols can also be dehydroxylated using pyridine/HF as fluorinating agent, where the reactivity depends on the position of the hydroxy groups. Thus, tertiary alcohols are readily transformed, whereas secondary hydroxy groups already require slight heating. Finally, primary alcohols do not react well with pyridine/HF. In summary, HF and its pyridine complex are cheap and valuable reagents for fluorodehydroxylation, however, they are not very selective and also require activated substrates.

(VIII)   In general, fluoroalkylamine reagents are capable of converting all types of aliphatic alcohols, including primary, secondary and tertiary substrates. However, the actual reactivity depends on the reagent, so that some of them will preferentially convert tertiary substrates, whereas others will give better results with primary alcohols. The fluorodehydroxylation reactions proceed under inversion, thus they generally show a rather good stereoselectivity. Nevertheless, side-products are often observed, such as esters, alkenes or ethers.

(IX)   The reactivity of all the fluoroalkylamine reagents may be explained by a general mechanism (cf. Scheme 13.15). The fluoroalkylamines are resonance-stabilized, so that an iminium fluoride may also be formulated. The iminium cation is very susceptible to nucleophilic attacks, so that the hydroxy function readily adds to it. Along with the adduct, HF is formed. The intermediate is also resonance-stabilized and may be formulated as another iminium fluoride, where the fluoride anion then substitutes the C-O bond, giving rise to the fluorodehydroxylated product and an amide.

(X)   In general, SF_4 is a very efficient reagent for fluorodehydroxylation reactions. However, it has some major drawbacks, as it is a very toxic and corrosive gas, which thus requires special handling and metal reaction vessels. Apart from that, it often leads to the formation of significant amounts of side-products, which is due to decomposition and polymerization reactions, which are induced by the presence of HF. However, the addition of HF-scavengers, such as NEt_3 or Et_2O can improve the formation of alkylfluorides, even for sensitive benzylic substrates. Interestingly, the treatment of 1,2-diols with SF_4 does not yield vicinal difluorides but α-fluoroalcohols. Similarly, the keto group in α-hydroxyketones remains unchanged if the reaction is conducted in Et_2O, so that such substrates are cleanly converted to α-fluoroketones.

(XI)   For practical purposes, dialkylaminosulphur trifluorides R_2NSF_3, such as DAST are mostly superior over SF_4, since they are more easily handled. In addition, they do not liberate HF upon fluorodehydroxylation, therefore, the addition of amine/HF complexes is often required to achieve reasonable conversion of the alcohol substrates. Typically, dialkylaminosulphur trifluoride reagents are used in slightly polar, aprotic solvents, such as CH_2Cl_2, $CFCl_3$ or monoglyme, in some cases, fluorodehydroxylation reactions with these reagents have also been carried out in unpolar solvents, for example, benzene or toluene. Mostly the reactions are started at −78 °C and are warmed to room temperature, sometimes, cooling to 0 °C is sufficient, too. In addition, fluorodehydroxylation may also be directly carried out at room temperature. Due to their generally rather mild reaction conditions, dialkylaminosulphur trifluorides show a good chemoselectivity and do not react with phenolic hydroxy groups as well as carbonyl groups and nitrile functions in corresponding hydroxyester, -ketone, -lactone, -lactam and -nitrile substrates.

(XII)   Fluorodehydroxylation reactions using DAST and similar reagents follow a S_N2-mechanism, however, for some secondary substrates carbocationic intermediates are involved, which can lead to rearrangements and eliminations. Apart from that, allylic alcohols readily undergo rearrangements, which especially holds for primary substrates, such as crotyl alcohol, which mainly yields rearranged 3-fluorobutene. In contrast, rearrangements are less pronounced, when secondary 3-butenol is used as starting material, but still, primary crotyl fluoride is observed as reaction product. A similar problem is observed for benzylic substrates, which often follow a S_N1-mechanism upon fluorodehydroxylation. Consequently, this leads to two major problems, as on the one hand the intermediary cationic species can undergo elimination toward styrene derivatives and on the other hand, fluorination of a benzyl cation leads to a loss of stereopurity.

(XIII)   Apart from direct fluorodehydroxylation, alcohols can be converted to alkyl fluorides in a two-step procedure. Thus, the alcohols are first activated by formation of esters or other good leaving groups. Very commonly, sulphonyl esters of the type -OSO_2R serve as leaving groups, but the use of O-alkylisoureas has also been reported. PPh_3 can also serve as activating reagent for the fluorodehydroxylation of alcohols. In all the aforementioned cases, nucleophilic fluoride sources, such as KF are used as fluorinating agents in hot, polar aprotic solvents. Under milder conditions, better soluble reagents like CsF or tetraalkylammonium salts may also be applied. Another type of activation is achieved, if the alcohol substrate is converted to fluoroformates ROC(O)F, which yield alkyl fluorides upon decarboxylation.

(XIV)   The reaction of α-fluoroethers, especially *bis*(1,1-difluoroalkyl) ethers with aHF leads to the cleavage of the ether and gives rise to a CF_3 group and an acid fluoride, in case of cyclic ethers, both functions will be part of one molecule. In linear substrates, the stability toward the ether cleavage is increased by the presence of an additional halogen substituent in γ-position, whereas β-halogens virtually render the compounds stable against cleavage. In cyclic compounds, six-membered rings are more readily cleaved than five-membered ones; moreover, the presence of β-halogens again increases the stability of the ether function against cleavage.

(XV)   The preparation of 1,2-fluorohydroxy compounds is conveniently achieved by the cleavage of epoxides with HF. However, in most cases, HF is used in form of one of its complexes, such as pyridine/HF or $NEt_3 \cdot 3$ HF. The nature of the HF complex influences the mechanism of the epoxide cleavage and thus the product formation. Acidic reagents, such as pyridine/HF initially protonate the oxygen, so that the oxirane is opened in a S_N1-type fashion, giving rise to product mixtures. In contrast, basic HF reagents attack the most electropositive carbon of the epoxide in S_N2-fashion, thus generally allowing a more selective product formation. However, in both cases, the substituents at the oxirane ring also influence the attack of the fluoride ion and thus the nature of the product.

(XVI)   In general, the cleavage of simple epoxides by HF-reagents will give rise to *trans*-fluorohydrins. However, the stereochemistry of the product is of course influenced by the substrate and the presence of directing groups. In some cases, the outcome is also influenced by the choice of reagents, for example, the use of $SiF_4/^iPr_2NEt$ instead of HF-complexes typically yields *anti*-products, if additives like H_2O or tetraalkylammonium fluorides are present. In absence of the latter, *syn*-products will be obtained. For selective oxirane-opening it is also suitable to use chiral transition metal complexes in conjunction with fluoride sources like $K[HF_2]$ and crown ethers to increase the solubility of the fluorinating agent.

(XVII)   The deoxyfluorination of carbonyl compounds does not occur as easily as fluorodehydroxylation, thus these transformations are in most cases achieved by the use of SF_4 in presence of a Lewis acid, mostly HF or BF_3. This system is capable of converting aldehyde functions to CF_2H-groups and ketones to difluoromethylene compounds, respectively. Among all carbonyl compounds, these compound classes are the most reactive ones, reactions with esters, amides or anhydrides are less readily achieved. In presence of HF, SF_4 converts carboxylic acids to CF_3 groups. Mechanistically, SF_4 is solvolyzed in HF to give $[SF_3]^+$ cations, which are attacked by the carbonyl-oxygen of the substrate. The intermediate cationic species intramolecularly transfers F^- to the carbonyl carbon and SOF_2 is cleaved off. In the last step, the fluorinated carbocation adds another fluoride anion and the CF_2-product is obtained.

(XVIII) Since the deoxyfluorination of aldehydes involves carbocationic intermediates, side-reactions, such as rearrangements or ether formation are likely to occur. For example, butyraldehyde cleanly yields 1,1-difluorobutane, when it is reacted with SF_4 at 20 °C, however, at 0 °C significant amounts of *bis*(1-fluorobutyl) ether are obtained, too. Moreover, fluorine-halogen migrations are observed, especially when the aldehyde is branched in α-position. Thus, the reaction of CCl_3CHO with SF_4 does not yield CCl_3-CF_2H but rearranged CCl_2F-$CFClH$, again along with ether products. A similar behavior is observed for the transformation of pivaloyl aldehyde, where a methyl group exchanges its position with one fluorine substituent.

(XIX) The reaction of α- as well as β-diketones with SF_4 gives rise to tetrafluoroalkanes. However, these transformations are accompanied by a large number of side-products, including alkenes and ethers. In the same way, the conversion of β-ketoesters does not yield clean products in most cases. Instead of β,β-difluoroesters, substantial amounts of elimination products are obtained, depending on the substitution of the α-position. If there is no other α-substituent than hydrogen, the system will preferentially eliminate two equivalents of HF to give acetylenic esters. However, an alkyl substituent in α-position does not allow the formation of a triple bond, but leads to allenic esters. A clean formation of β,β-difluoroesters is only achieved, if there are two alkyl groups attached to the α-position, or if the reaction is carried out in aHF.

(XX) In general, the dialkylaminosulphur trifluoride reagents are less reactive in the deoxyfluorination of carbonyl compounds. Thus, reagents like DAST only react with aldehydes and some ketones. Moreover, these transformations require higher reaction temperatures and still yield significant amounts of side-products by eliminations and rearrangements. As an advantage, the SF_3 reagents can be more readily used in organic solvents, so that the side-reactions may be suppressed by the use of less-polar reaction media. Overall, DAST and related reagents are especially suitable for the deoxyfluorination of rather sensitive substrates like steroids or carbohydrates, however, they do not react with sterically hindered ketones.

(XXI) The reaction of carboxylic acids with SF_4 in aHF generally exhibits a method for the preparation of CF_3-substituted compounds. As a major side-reaction, the formation of *bis*(α,α-difluoroalkyl) ethers is observed, which can be explained by the reaction of intermediate carbocationic species with acid fluorides. In the same way, dicarboxylic acids are preferentially converted to cyclic $\alpha,\alpha,\alpha',\alpha'$-tetrafluoroethers, where 1,2-dicarboxylic acids yield five-membered rings and 1,3-dicarboxylic acids give rise to six-membered cycles, respectively. This observation is also made for 1,2- and 1,3-dicarboxylic acids, which are attached to a cyclohexane moiety or aromatic systems. However, 1,1- and 1,4-dicarboxylic acids are not converted to cyclic compounds.

(XXII) Esters and amides also react with SF_4, however, the actual product formation strongly depends on the substrate and the reaction conditions. In presence of HF, nonfluorinated amides are cleaved and finally converted to CF_3 groups. In contrast, fluorination of the carbonyl group exclusively occurs, if SF_4 is used with KF as, for example, N,N-dialkylbenzamides are converted to α,α-difluorobenzyl-N,N-dialkylamines. The reaction of perfluorinated tertiary amides with SF_4/HF gives rise to tertiary perfluoroamines, whereas treatment of formamides with SF_4/KF yields N-trifluoromethylamines. In absence of HF, esters are virtually stable against SF_4, however excess HF will cleave them, and thus enable the formation of CF_3-groups. Esters of fluorinated acids or alcohols are converted to α,α-difluoroethers, if the reactions are carried out with SF_4 in HF-solution under mild conditions.

(XXIII) Oxidative fluorodesulphuration is a common method for the preparation of difluoromethylene compounds from ketone and aldehyde substrates, which are typically converted to 1,3-dithianes by the reaction with 1,2-ethanethiol. This method requires stoichiometric amounts of an electrophilic oxidant, such as "Br^+" or "I^+", and of course a nucleophilic fluoride source, in most cases amine-HF reagents. As a general trend, ketone-based substrates react more readily than those, which are derived from aldehydes. Thus, reactions with the latter are conducted at 0 °C, whereas the ketone derivatives are mostly reacted at −78 °C.

(XXIV) The choice of the reagents significantly influences the outcome of the fluorodesulphuration reactions of dithioesters and xanthogenates. The treatment of dithioesters with N-iodosuccinimide as rather mild oxidant yields α,α-difluorothioethers, the corresponding CF_3-groups are only obtained upon use of the stronger oxidant 1,3-dibromo-5,5-dimethylhydantoin (DBH). In both cases, the reactions are conducted at 0 °C in CH_2Cl_2, using $[H_2F_3]^-$ as source of fluoride. Similarly, N-bromosuccinimide and $[H_2F_3]^-$ only attack the thiocarbonyl group of xanthogenates, but do not cleave the thioether. For the latter transformation, the use of both, the stronger oxidant DBH as well as the more acidic pyridine/HF is required to give the trifluoromethoxy-group. Again, solvent and reaction temperature do not change, since both variants are conducted in CH_2Cl_2 in a temperature range from −78 to 0 °C.

(XXV) Despite not being that popular, nitrogen-containing substrates can serve as starting materials for the preparation of fluorinated compounds. The diazotization of α-amino acids with $NaNO_2$ and subsequent reaction with pyridine/HF yields α-fluorocarboxylic acids. In contrast, diazoalkanes, diazoketones and diazoesters undergo hydrofluorination, halofluorination or geminal difluorination, if they are reacted with sources of HF, "XF" or F_2, respectively. The diazo-substrates may also be reacted with other reagents, such as acid fluorides or fluoroformates, which will give rise to α-fluoroketones and α-fluoroesters, respectively. In contrast, the ring opening of cyclic nitrogen-containing systems, such as aziridines with HF-reagents yields α- or β-fluoroamines.

(XXVI) The hydrofluorination of unsaturated carbon-carbon bonds represents a method for the preparation of saturated hydrofluorocarbons. In most cases, electron-rich alkenes readily add HF, so that the presence of a catalyst is not required. In contrast, electron-deficient olefins, for example, trichloroethylene do not react with HF alone, but need the presence of a Lewis acid, such as $SbCl_5$. In the same way, acetylene does not react well with HF alone, however, the presence of a catalyst also determines the final product. For example, HSO_3F promotes the full saturation of the triple bond to give 1,1-difluoroethane, whereas the presence of Hg(II)- or Cd(II)- compounds will result in the formation of vinyl fluoride. Electron-rich acetylenes with alkyl- or phenyl-substituents rather easily add HF to become fully saturated. In contrast, the substitution of electron-deficient alkynes is only achieved, if a source of fluoride ions and protons is present. Mechanistically, the hydrofluorination of unsaturated C-C bonds in electron-rich systems proceeds via an initial protonation step and the subsequent addition of a fluoride anion to the intermediate carbocation, thus giving rise to Markovnikov products in most cases. In contrast, electron-deficient π-systems first add a fluoride anion, followed by protonation of the intermediate carbanion.

(XXVII) In the simplest fashion, halofluorination of electron-rich olefins is achieved by treatment of the unsaturated substrate with halogen monofluorides XF. However, since ClF is very reactive and BrF and IF are unstable, it is more convenient to use reagent combinations, which are composed of an electrophilic halogenating agent and a fluoride source. In most cases, this comprises the use of a N-halosuccinimide with a complex of HF. The choice of the reagents strongly influences the outcome of the halofluorination reactions. For example, the use of acidic pyridine/HF as fluoride source typically leads to the formation of carbocationic intermediates, so that rearrangements are likely to occur. In contrast, $NEt_3 \cdot$ 3 HF mostly leads to clean halofluorination of olefinic bonds.

(XXVIII) Reactions of olefins with fluorocarbenes, such as $:CF_2$, lead to the formation of cyclopropyl compounds. For electron-rich alkenes, this transformation is readily achieved, whereas the carbene addition to electron-deficient olefins does not occur that easily. $:CF_2$ is always generated *in situ*, for example, by decomposition of CF_3 compounds or thermolysis of difluoroacetate salts.

(XXIX) In contrast to hydrocarbons, highly- and perfluorinated olefins are not attacked by electrophiles, thus, the formal addition of "XF" cannot proceed via carbocationic species but involves perfluoroalkyl anions. Consequently, the unsaturated substrates are initially reacted with a fluoride source like CsF or KF and the resulting anionic species are subsequently treated with a source of "X^+." In most cases, the elemental halogens X_2 can take this role.

(XXX) Due to their electrophilicity, perfluoroolefins readily add fluoride anions and are converted to perfluoroalkyl anions. Accordingly, the treatment of perfluoroisobutene with fluoride yields perfluorinated *tert*-butyl anions, which may be trapped by any electrophile. In the same way, hexafluoropropene is converted to perfluorinated isopropyl anions upon reaction with F^-. Similarly, fluoride is readily added to perfluoroketones and perfluorinated acid fluorides, which gives rise to secondary and primary perfluoroalkoxides, respectively. Fluorinated alkoxides especially play a role in the preparation of fluorinated ethers.

13.5.7

(I) Elemental fluorine can serve as electrophilic fluorinating agent, if it is highly diluted by inert gases. In addition, the substrate needs to be highly diluted in solvents like $CHCl_3$, $CFCl_3$ or CH_3CN. Finally, the fluorination reaction needs to be carried out at low temperatures between $-78\,°C$ and $-40\,°C$. If the aforementioned criteria are not fulfilled, F_2 will react in a radical manner. Typical substrates for electrophilic fluorination using elemental fluorine are electron-rich hydrocarbons, especially those with tertiary C-H bonds, one of the prime examples certainly being adamantane. Hydrogen substituents in α-position of electron-withdrawing groups are also readily replaced. Mechanistically, the electrophilic fluorination with F_2 is believed to proceed via a three-centered transition state, where coordination assistance of the solvent also plays an important role.

(II) Reactions of fluoroxy compounds ROF with unsaturated carbon-carbon bonds generally lead to the addition of the electrophilic fluorine substituent and the corresponding RO-group to the C-C multiple bond. However, for electron-rich substrates, the addition of ROF involves polarization and occurs in *syn*-fashion. In contrast, electron-deficient starting materials add

ROF in a radical fashion, so that the products do not show a clear stereochemistry. In addition, the radical intermediates may undergo poly- or oligomerization side reactions.

(III) Enolizable carbonyl compounds are readily fluorinated in α-position, if enol acetates, enol esters, enamines or other reactive intermediates are treated with fluoroxycompounds. The advantage of using CF_3OF for the conversion of trimethylsilyl enol ethers to α-fluoorocarbonyl products is the formation of the volatile by-products COF_2 and Me_3SiF, which simplifies the workup of the reaction. Fluorooxy compounds are also capable of directly fluorinating C-H bonds in electron-rich tertiary and sometimes even secondary positions. Under appropriate conditions, this approach is especially suitable for the electrophilic fluorination of complex molecules, for example, steroids.

(IV) In general, C-H acidic substrates, such as β-dicarbonyl compounds are especially suitable for electrophilic fluorination reactions with fluoraza reagents. In most cases, it is not required to add a base, since the acidic C-H compounds are deprotonated by anionic R_2N^-, resulting from the fluorinating agent. However, less acidic systems may initially be converted to reactive carbanions, for example, by the use of bases like LDA. Reaction control can be achieved by the addition of water, since it inhibits the enolization of α-monofluorinated carbonyl compounds. Thus, α,α-difluorocarbonyl products are often obtained without the use of additives. With some reagents, it is also possible to control the degree of fluorination in α-position by the amount of added fluorinating agent and base, if required.

(V) Mechanistically, there are two boundary situations that may be used to rationalize electrophilic fluorination using fluoraza reagents. On the one hand, a halophilic S_N2-type substitution includes a nucleophilic attack of the deprotonated substrate Nu^- on the N-F bond of the reagent R_2NF, giving rise to the fluorinated product NuF and anionic R_2N^-, which serves as base. The second possibility involves a single electron transfer (SET) between Nu^- and R_2NF, which proceeds via an intermediary radical $Nu^{\cdot}$ and a radical anion $[R_2NF]^{\cdot-}$, in the end also leading to the formation of NuF and R_2N^-. The actual mechanism for the fluorination of the one or other substrate depends on the starting material and of course also on reagents and conditions. However, it is most likely somewhere in between halophilic S_N2-substitution and SET for the most cases.

(VI) The stereoselective preparation of organic compounds necessarily requires stereoinformation, which may originate from different sources. In the simplest case, the substrate itself contains a chiral carbon-center, which is capable of directing the introduction of further substituents, such as fluorine. Alternatively, chiral fluorinating reagents may be used. Especially for the asymmetric α-fluorination of carbonyl compounds it is suitable to use chiral organocatalysts. In some cases, it is also helpful to introduce chiral auxiliaries, which then direct the attack of the fluorinating reagent.

(VII) Chiral fluorinating reagents are typically based on small natural-occurring molecules or other, readily accessible motifs. For example, neutral fluoraza reagents have been prepared, which are based on camphorsultam. In contrast, cationic derivatives have been prepared, which originate from cinchona alkaloids. They also enable asymmetric fluorination, if the cinchona derivative is only added in catalytic amounts. Apart from the aforementioned examples, chiral derivatives of the commercial reagents NFSI and Selectfluor™ have also been described. In case of cationic reagents, it is also suitable to combine a commercial fluorinating reagent, mostly Selectfluor™, with a chiral anion, which is, for example, based on chiral phosphoric acids. Typically, the latter approach is conducted under phase-transfer conditions.

(VIII) Especially α-fluorinated carbonyl compounds are commonly prepared in stereoselective manner by using organocatalytic approaches. Typically, such chiral organocatalysts are based on proline or imidazolidinone derivatives. In the same way, stereoselectivity is achieved, when chiral auxiliaries are attached to the substrate, such as oxazolidinones (Evans' auxiliaries) or menthol derivatives. The stereoselective synthesis of β-fluoroalcohols and -amines is achieved by the initial asymmetric α-fluorination of aldehydes and subsequent reduction or reductive amination, respectively.

(IX) XeF_2 belongs to the electrophilic fluorinating agents, however, it rather transfers its fluorine substituents in an oxidative fashion. In most cases, it is used below room temperature in CH_3CN or CH_2Cl_2, in the latter case often in conjunction with HF or BF_3. The oxidative fluorination of olefins is believed to proceed via initial oxidation of the substrate to a radical cation, whereas XeF_2 is reduced to $XeF\cdot$ and transfers one of its fluoride substituents to the corresponding Lewis acid HF or BF_3. $XeF\cdot$ then fluorinates the organic radical cation under liberation of Xe and formation of a fluorinated carbocation, which subsequently abstracts a fluoride anion from anionic $[HF_2]^-$ or $[BF_4]^-$, respectively.

13.6.4

(I)   The direct C-H fluorination can be directed by several functional groups, which mostly contain N-donor atoms. For example, quinoline based substrates are fluorinated in benzylic position, if they are treated with a Pd(II)-catalyst and N-fluoropyridinium reagents or a combination of a hypervalent iodine reagent and AgF. The selective β-fluorination of α-amino acids is achieved, if the substrate is converted to a directing amide prior to the fluorination step. The latter is again realized by using Pd(II)-catalysis in combination with Selectfluor™ as fluorinating reagent. In some cases, additional ligands and metal salts, such as Ag_2CO_3 are required. Another example for a directing group is 8-aminoquinoline, which is also attached to carboxylic acids to generate amides. Reacting the latter with Pd(II)-salts, NFSI and some additives gives rise to β-fluorinated products. Similarly, β-fluorinated products are accessible, if a N,N-bidentate oxime moiety is attached to alcohol substrates. Again, fluorination is achieved by the use of Pd(II) with NFSI. Overall, most of the aforementioned transformations show best results, when the fluorinated C-H bond is a benzylic one.

(II)   In activated substrates, C-H bonds may be stereoselectively fluorinated in presence of suitable metal complexes. Typically, such substrates are 1,3-dicarbonyl compounds, especially β-ketoesters, which are fluorinated in α-position. Indeed, such transformations generally proceed without a metal complex, however, the latter can help to induce stereoselectivity, for example, by the use of chiral ligands. In a very general and simple picture, the metal center forms an enolate with the substrate, where the chiral ligand shields one face against the attack of the fluorine substituent. Consequently, this approach generally requires the use of electrophilic fluorinating agents. In case that the substrate already contains stereoinformation, for example, by the presence of an auxiliary, a chiral ligand is not needed at the metal center. This is for example the case, if chiral N-acyloxazolidinone substrates are asymmetrically fluorinated upon reaction with NFSI, Et_3N and $TiCl_4$.

(III)   In general, radical C-H fluorination does not necessarily require the presence of metal compounds, if there are other activating agents. However, metal compounds can serve for the generation of radical intermediates as, for example, a Cu(I) species has been reported to reduce the electrophilic fluorinating agent to a nitrogen-centered radical, which is then capable of abstracting a hydrogen radical from the substrate. In a slightly different approach, Ag(I) is oxidized by stoichiometric amounts of $Na_2S_2O_8$ and the resulting Ag(II)-species abstracts a hydrogen radical from the substrate. In both cases, the carbon-centered radical abstracts F⁻ from the fluorinating agent. This method is generally suitable for the fluorination of activated C-H bonds in benzylic, allylic or sometimes secondary alkyl substrates.

(IV) In presence of directing groups, metal mediated C-H fluorination can be very site-selective. For example, *N*-fluoro-*N*-*tert*-butylbenzamides are selectively fluorinated in benzylic position of *ortho*-alkyl substituents. Upon reaction with a Fe(II)-catalyst, the N-F bond is cleaved, leading to a fluorinated Fe(III)-species and a nitrogen-centered radical. The latter undergoes a 1,5-hydrogen atom transfer (HAT), so that a benzylic radical is formed, which then abstracts F⁻ from the fluoroiron(III) complex, thus regenerating the Fe(II)-catalyst and liberating the benzyl fluoride product. Since the starting material already contains electrophilic fluorine, the addition of a fluorinating reagent is not needed. In a similar way, hydroperoxy-functions can serve as directing groups, also inducing 1,5-HAT upon reaction with a Fe(II)-complex, which initially leads to the formation of an alkoxy radical. However, in this case, NFSI is required as fluorinating agent as well as LiBH$_4$ for the regeneration of the Fe(II)-catalyst.

(V) Mn(III)-salen and Mn(III)-porphyrine complexes are capable of mediating the C-H fluorination of benzylic and aliphatic substrates, respectively. Along with the metal-catalyst, a nucleophilic fluorinating agent, that is, NEt$_3$ · 3 HF or [NBu$_4$]F · 3 H$_2$O, as well as the oxidant PhIO is required. Initially, the trivalent fluorocomplexes are obtained upon halogen exchange of commercial chloro-derivatives and subsequently oxidized by PhIO to yield pentavalent manganeseoxo species. The latter abstract a hydrogen radical from the most hydridic C-H bond of the substrate and are thus reduced to tetravalent Mn-hydroxy complexes, which then undergo ligand exchanges with the nucleophilic fluoride source, so that a tetravalent Mn-difluoride-species is obtained. The carbon-centered radical finally abstracts a fluoride radical from the aforementioned tetravalent species, thus forming the fluorinated product and regenerating the Mn(III)-fluoro catalyst.

(VI) The metal-mediated fluorofunctionalization of unsaturated substrates can be achieved by a large variety of reagent combinations. In general, electrophilic fluorinating reagents are combined with nucleophilic sources of other functionalities or alternatively, nucleophilic sources of fluorine can be used, when the other function is introduced in electrophilic fashion. For example, the treatment of allylic amines with a nucleophilic aryl boronic acid and Selectfluor™ in presence of a Pd-catalyst yields benzyl fluorides. Another very common functionalization of olefinic substrates includes aminofluorination, which is often conducted in intramolecular fashion, giving rise to *β*-fluorinated, cyclic amines. Again, this transformation is catalyzed by a Pd-complex, however, in this case nucleophilic AgF is used in conjunction with a hypervalent iodine oxidant. Moreover, the fluorosulphonylation of styrene derivatives is achieved by the Pd-mediated reaction with NFSI and ArSO$_2$H. This approach selectively yields *β*-fluorosulphones, typically even in high diastereoselectivity. Benzylic fluorides are furthermore obtained upon Cu-mediated fluorosulphonimidation of styrenes, where NFSI serves as source of both, fluorine and the sulphonimide group.

(VII) The preparation of allylic fluorides is achieved by several different metal-mediated methods, however, they all require the presence of a leaving group and virtually all use nucleophilic fluorinating agents. The treatment of racemic, cyclic allylic chlorides with a chiral Pd-catalyst and AgF gives rise to enantiomerically enriched, cyclic allylic fluorides. The same catalyst and reagent convert linear allylic chlorides to allylic fluorides; however, a mixture of linear and branched products is obtained, the latter being preferentially generated. Starting from a variety of allyl *p*-nitrobenzoates, linear allylic fluorides are selectively obtained under Pd-catalysis, using [NBu$_4$]F · (ᵗBuOH)$_4$ as fluorinating agent. A few variants have been described using Rh- or Ir-catalysts for the conversion of allyl trichloroacetimidates and allyl carbonates. In all of these approaches, the position and the configuration of the double-bond are retained. Allylic chlorides and bromides are converted to the corresponding fluorides, when they contain a heteroatom. For this, they are reacted with catalytic amounts of CuBr and $NEt_3 \cdot 3\,HF$.

13.7.4

(I)   The reduction of hexafluoroacetone with Mg metal can yield different products, depending on the polarity of the solvent. In highly polar *N,N*-dimethylformamide, two molecules are reductively coupled to yield perfluoropinacol. In contrast, the formation of a β-hydroxyketone is observed, if the reaction is carried out in less polar THF. In the latter case, an intermediate enolate species reacts with another molecule of hexafluoroacetone. If the reduction is carried out with Al, it can be stopped by acidification to yield the isolable enol of pentafluoroacetone.

(II) In general, chloro-substituents at sp^3-carbons are far more easily cleaved than the corresponding C-F bonds. Thus, there are several techniques for a selective hydrodechlorination of chlorine- and fluorine containing substrates. For example, C-Cl may be reductively cleaved by the use of metals like Zn or Na in protic solvents. Other suitable reducing agents include $LiAlH_4$ as well as silicon or tin hydrides. In some cases, catalytic hydrogenation may be applied, photolytic approaches have been described, too.

(III) The catalytic hydrogenation of internal fluoroolefins generally gives rise to dehydrogenated products. However, as soon as there is a third carbon substituent at the double bond, the former vinylic fluorine substituent will be partly split off, thus giving rise to a mixture of di- and trihydrogenated molecules. The outcome of the catalytic hydrogenation changes again, when all four substituents at the double bond are fluorinated alkyl residues. In this case, elimination of HF may occur, hence giving rise to trihydrogenated products, which additionally show a shift of the double bond. Fluorinated olefin substrates with vinylic chloro- or bromo-substituents are typically completely dehalogenated upon catalytic hydrogenation, thus giving rise to tetrahydrogenated compounds.

(IV) Of course, the outcome of reductive transformations does not only depend on the substrate but also on the reducing agent. For example, both carbonyl groups in fluorinated diketones are reduced by $LiAlH_4$, giving rise to diols. In contrast, $NaBH_4$ or KBH_4 only reduce one of the keto-functions, which then leads to the formation of hydroxy ketones. Similarly, fluorinated ketoacid fluorides are converted to diols, when they are treated with $LiAlH_4$, whereas $[BH_4]^-$ reagents yield hydroxyacids.

(V) Fluorinated epoxides are valuable building blocks, which are generally prepared by oxidation of fluoroolefins. However, these transformations may be realized using a number of different reagents. The simplest oxygen source is O_2 itself, which may be used for epoxidation reactions at elevated pressures and temperatures. Other typical oxidants include peroxybenzoic acids, H_2O_2 and hypohalites, such as NaOCl or NaOBr. Of course, high-valent metal compounds, for example, CrO_3 or $KMnO_4$ may also be used, however, in some cases, they give rise to other products. For example, the oxidation of hexafluoropropene with $KMnO_4$ does not yield the epoxide but the hydrate of hexafluoroacetone, which results from dihydroxylation of the double bond. Another very efficient reagent for the epoxidation of perfluoroalkyl-substituted olefins is $HOF \cdot MeCN$.

(VI) The reaction of fluorinated olefins with $KMnO_4$ generally results in dihydroxylation. However, depending on the substrate, other products than 1,2-diols may be obtained. This is the case, if the starting material contains vinylic fluorine substituents, which will then lead to an elimination of HF. Thus, the presence of one vinylic fluorine in the substrate gives finally rise to α-hydroxyketones, whereas two vicinal fluoro-substituents lead to α-diketones. Accordingly, the dihydroxylation of α-fluorinated enol ethers produces α-hydroxyesters.

(VII) Fluorinated olefins may be oxidized to carboxylic acids, this is especially observed for strained, cyclic substrates, such as perfluorocyclobutene, upon reaction with strong metal oxidizers, for example, $KMnO_4$ or RuO_4. The oxidation of the olefin leads to a cleavage of the double-bond, giving rise to dicarboxylic acids in case of cyclic starting materials. However, linear fluoroolefins are also oxidatively cleaved to furnish carboxylic acids. Alternatively, ozonization of fluoroolefins yields acid fluorides, which may also be hydrolyzed to give the corresponding acid.

(VIII) Electrolysis of mixtures of fluorinated carboxylic acids and their carboxylates leads to decarboxylative coupling, similar to Kolbe reactions. The use of a single substrate gives rise to symmetric fluoroalkanes, whereas mixed compounds are also accessible, if different acid starting materials are used. However, in the latter case, this will typically lead to the formation of three products.

(IX) The oxidation of hexafluorobenzene with concentrated H_2O_2 at 140 °C yields pentafluorophenol. The phenol product may be further oxidized by H_2O_2, where the product formation depends on the concentration of the reaction mixture. Low concentrated mixtures will give rise to *tetrakis*(pentafluorophenoxy)-*p*-benzoquinone, whereas higher concentrations lead to the cleavage of the ring and thus generate (Z)-2,3-difluoromaleic acid.

(X) Elemental halogens are rather easily added to unsaturated fluorinated compounds, giving rise to the corresponding dihalogenated products. However, in most cases, such halogenations are not stereoselective, since the reaction can proceed via ionic but also via radical intermediates, thus halogen addition can occur in *syn*- or *anti*-fashion. Of course, the stereoselectivity depends on the steric and electronic situation of the starting material. For example, *anti*-bromination is observed, if (*E*)- and (*Z*)-1-fluoropropene are reacted in the dark under ionic conditions.

(XI) Fluoroalkyl halides and especially perfluoroalkyl iodides are important compounds in synthetic fluoroorganic chemistry. For example, they are obtained upon halogenation of hydrogen-containing starting materials, such as CF_3H or $^tC_4F_9H$. CF_3Br is prepared upon bromination of fluoroform at 600–650 °C in presence of a catalyst, in the same way, perfluorinated *tert*-butyl chloride is accessible by the reaction with ICl, if KF is added. Another important approach toward fluoroalkyl halides is the decarboxylative halogenation of fluorinated alkali metal carboxylates. The halogenation is achieved by the use of the elemental halogen as, for example, the reaction of CF_3CO_2K with I_2 yields CF_3I. Apart from that, fluorosulphonyl hypohalites may serve as halogenating agents. In contrast, geminal dichlorides are obtained, if carbonyl-compounds are reacted with PCl_5. This transformation may also be extended to dicarbonyl compounds, such as α-ketoesters, which are transformed to tetrachloroethers. However, 1,4-dicarbonyl compounds undergo cyclization upon chlorination and are thus converted to α,α'-dichloroethers. Finally, perfluorinated acyl halides and carboxylic acids are transformed to CCl_3-groups, when they are treated with PCl_5.

14.2.7

(I) In general, fluoroarenes are accessible by halogen exchange, however, the substrate needs to be activated by the presence of an electron-withdrawing substituent. Activation of the starting material is also achieved, when a heteroatom is present, that is, in pyridine derivatives. The problem with halogen exchange reactions is the comparably low stability of aromatic $C(sp^2)$-F bonds, so that the latter are readily cleaved by nucleophiles, sometimes even under mild conditions. In contrast, the stability of the other carbon-halogen bonds in arenes increases from chlorine to iodine. Especially perfluorinated molecules, such as C_6F_6 or C_5NF_5 are very susceptible to substitution reactions.

(II)   The halogen exchange of activated aromatic substrates is typically realized by the use of fluoride salts, mostly spray-dried KF, in hot, polar-aprotic solvents, such as sulpholane. In some cases, the reactivity of KF is increased by the addition of CsF, moreover the halogen exchange can occur under phase-transfer conditions or in absence of any solvent at high temperatures. The introduction of fluorine substituents occurs under significantly milder conditions, if anhydrous $[NBu_4]F$ or $[NMe_4]F$ are used as fluoride sources, in some cases they may even be generated *in situ*.

(III) Fluorodenitration is typically realized by the treatment of aromatic nitro-compounds with KF in highly polar solvents at high temperature. In order to suppress the formation of phenolates and diphenylesters, phthaloyl dichloride needs to be added in stoichiometric amounts. Fluorosulphonyl groups can also be replaced under similar conditions, however, they are less prone to nucleophilic attacks. In some cases, especially for the fluorination of heteroaromatic molecules it might be more suitable to use $[NBu_4]F$ as fluorinating agent, which can be applied under mild conditions at room temperature. As a general requirement, all substrates need to contain another strongly electron-withdrawing group in *m*-position, that is, a NO_2-group in fluorodenitration approaches and a SO_2F-substituent for successful fluorodesulphonylation.

(IV)   Pentafluorochlorobenzene is the major side-product of the halogen exchange of C_6Cl_6 with KF at 400–500 °C. It is an important synthetic intermediate, since it can be converted to the Grignard-compound C_6F_5MgCl, which is stable in solution at room temperature. Starting from the organomagnesium species, a variety of functionalizations can be achieved as, for example, C_6F_5I is prepared by treatment with I_2 or protonation with HCl results in the formation of C_6F_5H. In turn, the latter undergoes electrophilic bromination with $Br_2/AlBr_3$, giving rise to C_6F_5Br.

(V)   Hexafluorobenzene readily undergoes substitution reactions with a number of nucleophiles. For example, the reaction with KOH leads to the formation of the phenol C_6F_5OH, whereas the corresponding hydrosulphides yield the analogous thiophenol. Moreover, direct substitution is achieved with NH_3 and NaOMe, generating pentafluoroaniline and pentafluoroanisole, respectively. Remarkably, the reactions with NH_3 and KOH occur under very mild conditions in aqueous media, which again underlines the high reactivity of perfluorobenzene toward nucleophilic attacks.

(VI)   In general, there are two major approaches, which may be used for the preparation of aryl fluorides from phenols. The first one includes the conversion of the substrate to a fluoroformate, which is subsequently decarboxylated to yield the fluorinated arene. This technique is scarcely used in laboratories, but is an important process for the preparation of fluorinated aromatics on industrial scale. The second approach is similar to the deoxyfluorination of aliphatic hydroxy groups and uses an α,α-difluoroamine reagent, commercialized as PhenoFluor™. The latter method is far more easily realized in a standard laboratory and is also significantly milder, so that it may be used for the conversion of sensitive substrates, including steroids or other natural products. However, due to the rather expensive reagent, it is not very suitable for deoxyfluorination on industrial scale.

(VII) The Balz–Schiemann reaction is used for the conversion of aniline-derivatives to the corresponding arylfluorides via intermediate aryldiazonium tetrafluoroborate salts. The major advantage of this approach is the availability of aniline starting materials and the rather simple reagents and reaction conditions, however, Balz–Schiemann reactions typically suffer from a poor reproducibility and safety issues, since the diazonium salts may generally decompose violently. Thus, large-scale preparations in bulk are not advisable. A number of improvements have been described, making the reaction procedure safer and more reliable, for example, the *in situ* formation and decomposition of the aryldiazonium tetrafluoroborate intermediates, avoiding the isolation of these hazardous species. Moreover, the decomposition of the diazonium salts may also be achieved under photochemical or ultrasound conditions, which are both milder than thermal fluorodediazoniation.

(VIII) The reaction of diaryliodonium salts with fluoride sources, for example, $[NMe_4]F$ yields intermediary iodonium fluorides, which are thermally decomposed to yield fluoroarenes. The main benefit of this procedure is certainly the synthetic availability of diaryliodonium compounds from aryl iodides. The drawbacks of this method include the thermal decomposition of the intermediates, which is often not that easily achieved. In addition, symmetric diaryliodonium salts should be used to avoid the formation of side-products, however, in the end only one of the aryl rings will be converted to the product. Finally, fluoride-promoted rearrangements of the iodonium salts may occur. The C-H fluorination of electron-rich arenes can also be achieved via intermediary aryl sulphonium salts, which are generated upon reaction of the substrate with dibenzothiophene *S*-oxide. Treatment of the sulphonium salt with a fluoride source, such as $[NMe_4]F$ gives rise to mostly *p*-fluorinated products. In addition, this method shows a high tolerance against functional groups, therefore, it may also be applied for late-stage functionalization.

14.3.5

(I) If it is used in high dilution, elemental F_2 can serve as electrophilic fluorinating reagent for the preparation of fluoroarenes. However, due to its reactivity it is generally poorly selective, so that mixtures of regioisomeric products are usually obtained. However, in presence of directing substituents as well as blocking functions, a better regioselectivity may be achieved. In general, only a rather small range of functional groups is tolerated, such as nitro-, carboxy-, hydroxy- or amino-functions, for the latter two the fluorination needs to be carried out in aHF to prevent the formation of OF- and NF_2-derivatives. Therefore, the electrophilic fluorination of arenes using F_2 is only suitable for robust substrates.

(II) Similar to F_2, fluoroxy reagents are very reactive, so that they generally exhibit poor selectivity in fluorination reactions of arene substrates. However, if the starting material is deactivated and additionally blocked in some positions by the attachment of appropriate substituents, a selective fluorination may be achieved. For example, the fluorination of *p*-CF_3-acetanilide selectively yields the corresponding *o*-monofluorinated product. A selective fluorination is also observed for activated substrates, such as 1,3-dimethoxybenzene, when the substituents clearly direct the fluorine substituent to a certain position. In the present example, the methoxy groups are both *ortho*- and *para*-directing, so that 4-fluororesorcinol is obtained. As a general trend, *ortho*-fluorinated products are preferentially obtained upon fluorination of activated arenes with fluoroxy compounds.

(III) XeF_2 is a rather mild fluorinating reagent, which can therefore be used in standard laboratory glassware and in organic solvents, typically CH_2Cl_2 or CH_3CN. In addition, its commercial availability is beneficial for the use in small scale transformations in the laboratory. Apart from the pure compound, graphite intercalates of XeF_2 may also be used for the electrophilic fluorination of arenes, which is especially useful to control the reactivity in the fluorination of polyaromatic substrates.

(IV)   Despite being far less reactive, fluoraza reagents do not generally exhibit a significantly increased regioselectivity in electrophilic fluorination reaction of arenes. Again, the selectivity is by far more influenced by the substrate itself and the presence of activating or deactivating groups. Moreover, the blocking of certain positions at the aromatic ring will increase the regioselectivity of the fluorine attack. However, as a general trend, substituted substrates are preferentially converted to o-fluorinated products, if possible.

(V)   Electrophilic fluoraza reagents are also used to prepare fluoroarenes from intermediary organometallic species, especially aryl lithium and corresponding organomagnesium species. Typically, aryl bromides or iodides are reacted with lithium organyls, such as $^t C_4 H_9 Li$ to undergo halogen-lithium exchange, and thus generate aryl lithium intermediates, which are then trapped by "F^+", mostly using NFSI as fluorinating reagent. In the same way, aryl magnesium intermediates are prepared from aryl bromides and Mg metal or Grignard reagents and subsequently fluorinated by NFSI or other reagents. In some cases, it is also suitable to directly deprotonate the arene by the use of organolithium reagents, so that the presence of a halogen substituent is not required. As a major disadvantage, the synthetic approach via organometallic intermediates cannot be applied to substrates with electrophilic groups, such as carbonyl-, cyano- or nitro-functions.

14.5.5

(I)   The application of Pd-catalysis can be very helpful for the preparation of aryl fluorides, however, one has to distinguish the approaches using Pd^{II}/Pd^{IV}-couples from those, which rely on Pd^0/Pd^{II}-catalysis, as this significantly influences the choice of the substrates and the reagents. In general, direct C-H fluorination is achieved by the use of Pd(II)-catalysts, when the substrate contains a directing group. However, this approach necessarily requires electrophilic fluorinating reagents, which serve as both, oxidizer for Pd^{II} and fluorine source. Using this approach, one needs to be aware of the fact, that a standard reductive elimination step of the fluoroarene does not readily occur from the intermediate Pd(IV)-species. The second approach involves the Pd^0/Pd^{II}-mediated fluorination of aryl bromides, iodides or triflates Ar-X, and thus requires a leaving group. Since the zerovalent Pd-catalyst oxidatively inserts into the Ar-X bond, an additional oxidizer is not required, and thus simple nucleophilic fluorinating reagents, such as CsF or AgF may be used. Furthermore, the C-F bond is formed by a classic reductive elimination step from the intermediate Pd(II)-species, giving rise to the aryl fluoride and regenerating the zerovalent Pd-catalyst. Again, some details need to be considered, when using this synthetic method, as for example the use of stoichiometric amounts of AgF is typically necessary to achieve ligand exchange from Ar-$[Pd^{II}]$-X to Ar-$[Pd^{II}]$-F. Even more importantly, a large monodentate phosphine ligand is required to allow the formation of a T-shaped, monomeric $LPd^{II}Ar(X)$ intermediate, and thus avoid the formation of unreactive dimers.

(II)   The choice of the directing group for direct C-H fluorination mainly depends on the target structure, and thus the question, if the directing group needs to be removed or converted to another functionality. For example, the fluorination can be directed by simple heterocyclic substituents, such as pyridine, pyrazine or pyrazole, which will give rise to o-fluorinated arenes. However, in most cases, these substituents cannot be cleaved off or converted to other functions, so that their use is rather limited. Cleavable directing groups, such as triflamide substituents can be converted to aldehyde, nitrile or amine functions after the fluorination step, in the same way, benzamides may be transformed to benzoic acids. The drawback of the triflamide substituent is the necessity to block one of the *ortho*-positions, otherwise, a mixture of *o*-mono- and *o*-difluorinated products will be obtained. In contrast, mono- or difluorination of benzamide substrates is easily controlled by the amount of fluorinating reagent and the reaction conditions, especially the choice of the solvent. Other cleavable directing groups include *O*-methyl oxime ethers and 2-pyridyloxy functions, which finally give rise to o-fluorinated phenylketone and benzaldehyde derivatives or o-fluorophenols, respectively. Especially the oxime approach can be conducted under very mild conditions at room temperature, so that it tolerates a number of functional groups. In a very elegant way, benzaldehyde derivatives are converted to directing imines by the *in situ* reaction with aniline-2,4-disulphonic acid. This method works well for electron-deficient, but not for electron-rich arenes and selectively gives rise to o-monofluorinated benzaldehydes.

(III) In general, aryl bromides, iodides as well as triflates are widely used starting materials in organic synthesis, therefore it is certainly beneficial to use them for the selective preparation of aryl fluorides. However, for large scale applications, these substrates might be too expensive, especially if they are no simple and readily available building blocks. Another important problem is the necessity to use very sophisticated ligands of the BrettPhos family, which of course leads to excellent results in small-scale reactions, but will definitely render upscaling difficult, due to the availability of these phosphine ligands. Finally, the use of aryl bromides and iodoarenes as starting materials requires the use of stoichiometric amounts of AgF to enable ligand exchange at the Pd(II)-center, which is of course a rather costly reagent. Apart from that, the fluoride exchange in the analogous triflate systems is usually achieved by the use of CsF, which is also a rather costly fluorinating agent.

(IV) The treatment of arenes with CuF_2 at 450–550 °C leads to direct C-H fluorination and gives thus rise to a mixture of aryl fluoride, HF and zerovalent copper. The main advantage of this reaction, which is applied on industrial scale, is the fact, that CuF_2 is regenerated by the reaction of Cu^0 with HF and O_2 at 400 °C. Upon regeneration, water is formed, which is actually the only waste from the overall process. This fluorination technique is generally suitable for the preparation of rather simple fluoroarenes, such as mono- and difluorobenzenes or fluorinated toluenes.

(V) The preparation of fluoroarenes is also achieved, if diaryliodonium-salts are reacted with Cu(II), KF and 18-crown-6 as fluorinating agent. Moreover, aryl boronate esters can serve as starting materials, however, in this case, stoichiometric amounts of AgF as well as an electrophilic *N*-fluoropyridinium reagent are required. Both, aryl stannanes and aryl trifluoroborate salts are converted to fluoroarenes, if they are reacted with a Cu(I) catalyst and an electrophilic *N*-fluoropyridinium reagent. The halogen exchange of aryl iodides is achieved, if they are reacted with stoichiometric amounts of Cu(I) and AgF. A directed approach has been described, which includes the directed exchange of aryl halides by the presence of N-donor functions in o-position of the halide. In this case, KF is the stoichiometric fluorinating reagent, whereas a *N*-heterocyclic carbene-Cu(I) complex serves as mediator.

(VI) In the Ag(I)-mediated fluorination of aryl stannanes, it is believed that intermediate arylsilver species are formed during the reaction. The latter are then oxidatively fluorinated, thus the use of an electrophilic fluorinating reagent, that is, Selectfluor, is required. Probably, the Ag(I)-mediated fluorination of aryl boronic acids proceeds in a comparable fashion.

(VII) The preparation of fluoroarenes from aryl boronic acid derivatives may also be mediated by a Bi^{III}/Bi^V-couple, which undergoes the typical organometallic elemental reactions, such as oxidative addition and reductive elimination. Thus, bismuth behaves like a tran-

sition metal and consequently requires an electrophilic fluorinating reagent, that is, a
N-fluoropyridinium reagent, for the oxidation step from Bi^{III} to Bi^{V}.

14.6.4

(I)   Perfluorinated aromatics show an inverted charge distribution and thus exhibit an orthogo-
      nal reactivity with respect to hydrocarbons. In fluorinated arenes, the most positive charge
      is found in the center of the π-system, so that these molecules are generally attacked by nu-
      cleophiles, but not by electrophiles. In contrast, the most negative charge is located in the
      center of the π-system of hydrocarbons, so that they are readily attacked by electrophiles,
      but not by nucleophiles.

(II) The reaction of C_6F_6 with dimethylamine initially leads to the formation of the monosub-
      stituted product $C_6F_5NMe_2$. However, the further treatment with methylthiolate selectively
      gives rise to the *para*-substituted product. This selectivity can be explained by the stabiliza-
      tion of the fluoroarene-nucleophile adduct by *ortho*- and *meta*-fluorine substituents.

(III) If aromatic hydrofluorocarbons, such as fluorobenzene, are reacted with organolithium com-
      pounds at low temperature, *ortho*-lithiation is typically observed, leading to the formation
      of kinetic products. In some cases, potassium alcoholates are additionally required to en-
      able this reactivity. However, if the reaction mixture is warmed over −30 °C, the thermody-
      namically unstable intermediates eliminate LiF and generate reactive arynes, which tend to
      oligomerize, if they are not trapped.

(IV) The outcome of the treatment of aromatic halofluorocarbons with lithium organyls depends
      on the choice of the organolithium compound. For example, halogen-metal exchange is typ-
      ically observed with the strong base butyllithium, whereas lithium diisopropylamide (LDA)
      rather results in *ortho*-lithiation. In addition, the choice of the solvent plays a role as, for ex-
      ample, *trans*-metalation can occur in coordinating solvents like THF, but not in diethyl ether.

(V)   In general, aromatic C-F bonds are activated by a large number of transition metals, however,
      the choice of the metal significantly influences the outcome of the activation and also has an
      impact on subsequent functionalizations of the arene. For example, early transition metals,
      like titanium or zirconium, are preferentially used in hydrodefluorination reactions, whereas
      late transition metals, for example, nickel, platinum or rhodium, are useful for the generation
      of C-C and carbon-heteroatom bonds. In addition, the selective activation of C-H or C-F bonds
      in aromatic hydrofluorocarbons is often challenging. Carbon-hydrogen as well as carbon-
      fluorine bonds also influence the strength of their neighbors, as for example the strength of
      C-F bonds decreases with the number of *ortho*-fluorine substituents. In the same way, *ortho*-
      fluorine strengthens adjacent C-H bonds. Early, mostly d^0-configured metal centers tend to
      activate C-F bonds, the same is typically observed for Co- and Ni-centers. In contrast, the
      4d-metal rhodium rather activates C-H bonds, but can also do so with C-F bonds, depending
      on the actual substrate and the conditions.

(VI) Hydrodefluorination reactions are typically used for the preparation of partly fluorinated aro-
      matic hydrocarbons from readily accessible highly or even perfluorinated substrates. Often,
      this approach is far more easily realized than a partial and selective fluorination of a nonflu-
      orinated substrate. In some cases, hydrodefluorination techniques might also give access
      to substitution patterns, which cannot be prepared by other synthetic routes. For efficient
      metal-catalyzed reactions, some general requirements need to be fulfilled. On the one hand,
      the transition-metal catalyst should selectively activate C-F bonds and replace the fluorine
      substituents by hydrogen. On the other hand, a stoichiometric hydride source is required

to regenerate the active transition metal hydride from the fluorocomplexes. Therefore, the hydride source should form strong element-fluorine bonds to efficiently abstract fluoride; in most cases, this includes boranes, silanes and alanes.

(VII) The reaction of pentafluoropyridine with Ni(cod)$_2$ and PEt$_3$ results in the activation of a C-F bond adjacent to the nitrogen atom, whereas the treatment with [RhH(PEt$_3$)$_4$] activates perfluoropyridine in 4-position. The subsequent acetylation thus gives rise to 2-acetyltetrafluoropyridine and the corresponding 4-acetylated derivative. Based on the different metal centers, the acetylation of the intermediate pyridine complexes follows different pathways. In case of the Ni-mediated reaction, the acetylated product is liberated from a tetracoordinated metal center, whereas the analogous reaction sequence at the Rh-center also involves hexacoordinated intermediates.

(VIII) Apart from carbon-substituents and hydrogen, a few heteroatoms may also be connected to fluorinated arenes after C-F bond activation. For example, methoxy-substituents are transferred from Si(OMe)$_4$ in a Pt-catalyzed approach, if the substrate contains a directing group. Moreover, aryl ethers are generated upon C-F bond activation with a Ni(II)-complex and subsequent treatment with an aryl boronic acid and O$_2$. Dimethylamino functions are transferred to pentafluorophenyl compounds, if the latter are reacted with [Ti(NMe$_2$)$_4$]. Pentafluoropyridine is converted to a silyl ether, if it is activated by a Pd(II)-catalyst in presence of water and subsequently trapped with Me$_3$SiCl. Finally, the activation of pentafluoropyridine with Rh-complexes and subsequent treatment with reagents like B$_2$pin$_2$ or B$_2$cat$_2$ gives rise to boronate esters.

15.1.8

(I) Terminal perfluoroalkylated olefins are generally accessible from ethylene and the corresponding perfluoroalkyl iodide, which are initially reacted to give a saturated compound of the composition R^F-CH$_2$-CH$_2$I. Treatment of the latter intermediate with a strong base results in the elimination of HI and generates a terminal olefin R^F-CH=CH$_2$. In the same way, the addition of CF$_3$I to 1,1-difluoroethylene gives rise to CF$_3$-CH$_2$-CF$_2$I, which also eliminates HI under basic conditions to form 1,1,3,3-pentafluoropropene.

(II) In general, reaction rates for the elimination of HF in bimolecular reactions are by two to five orders of magnitude smaller than those for dehydrochlorination and dehydrobromination, respectively. However, HF-elimination readily occurs, if the substrate contains acidic protons in proximity of a large number of fluorine atoms. This behavior is observed for acyclic as well as for cyclic polyfluorocarbons and heterocyclic substrates.

(III) Unsaturated fluorinated compounds are also accessible by formal elimination of halogen molecules, for example the treatment of perfluorinated compounds, such as 3,4-dimethyl-3-hexene, with sodium amalgam leads to the formal cleavage of F$_2$ and gives rise to the corresponding 2,4-hexadiene. More frequently, Cl$_2$ is eliminated from chlorofluorocarbons by the use of Zn, which is applied under a variety of conditions. For example, treatment of CFCl$_2$-CFCl$_2$ with Zn in hot ethanol or dioxane yields the chlorofluoroolefin CFCl=CFCl. Similarly, perfluoroalkylated pentachloroethanes are converted to perfluoroalkylacetylenes by the use of Zn in N,N-dimethylformamide. A selective removal of chlorine substituents from polychlorofluorocarbons is achieved by the use of PPh$_3$ in dioxane. A formal elimination of BrF occurs, if CHBr=CF$_2$ is reacted with Mg, which finally gives rise to monofluoroacetylene. Analogously, IF is typically removed from perfluoroalkyliodides by the use of Cu/Zn-couples in highly polar solvents, which gives rise to perfluoroolefins. However, the actual product formation of the latter transformation depends on the presence of nucleophilic additives.

(IV) Fluorinated olefins are also accessible from alcohols and acyl chlorides. The treatment of highly fluorinated hydroxy compounds with the strong dehydrating agents P_2O_5 or H_2SO_4 results in the elimination of water, and thus gives rise to the corresponding olefins. However, for some activated substrates, for example, hexafluoroisobutanol or trifluoromethylated homoallyl alcohols, dehydration occurs under significantly milder conditions. Alternatively, alcohol substrates are transformed to olefins, if the hydroxy-function is converted to a good leaving group, such as a sulphonate, and subsequently eliminated by heating. In a similar way, pyrolysis of perfluorinated acyl chlorides leads to the elimination of CO and ClF, and thus also generates fluorinated olefins.

(V) Both perfluorinated dienes may be prepared from intermediate $CF_2Cl\text{-}CFICl$. For the synthesis of fluorinated 1,3-butadiene, the aforementioned C2-building block is initially submitted to a C-C coupling reaction, which is realized by the treatment with Zn in acetic anhydride. In this step, the C-I bonds are cleaved, so that the resulting product only contains chloro- and fluoro-substituents. Upon treatment with Zn in ethanol, the chlorine-atoms are split off and perfluoro 1,3-butadiene is obtained. The preparation of fluorinated 1,7-octadiene requires four additional carbon atoms, thus, the C2-fragment $CF_2Cl\text{-}CFICl$ is reacted with $CF_2=CF_2$ prior to the coupling step, giving rise to the C4-unit $CF_2Cl\text{-}CFCl\text{-}CF_2\text{-}CF_2I$. Treatment of the latter with Zn in acetic anhydride leads to the cleavage of the C-I bonds and coupling of two C4-fragments. Finally, the chloro-substituents are removed by treatment with Zn in a mixture of acetic anhydride and acetic acid, giving rise to the two terminal trifluorovinyl units of fully fluorinated 1,7-octadiene.

(VI) Fluoride substituents, which are directly attached to a sp-carbon are very reactive, therefore, fluoroacetylenes are generally unstable. The preparation of monofluoroacetylene HCCF is for example realized by the treatment of 1,1-difluoroethylene with *sec*-butyllithium at low temperatures or pyrolysis of fluoromaleic anhydride. Due to its reactivity, it spontaneously trimerizes to give 1,2,4-trifluorobenzene. However, trifluoromethylated acetylenes, for example CF_3CCH and CF_3CCCF_3 are stable compounds, which are interesting building blocks in fluoroorganic chemistry. 3,3,3-Trifluoropropyne is prepared in a two-step reaction, where CF_3I is initially added to acetylene to form $CF_3CH=CHI$, which is then treated with KOH to eliminate HI. Hexafluoro-2-butyne is, for example, generated, if perchlorinated 1,3-butadiene is fluorinated by SbF_3Cl_2, which converts the terminal CCl_2-units to CF_3-groups, and thus leads to the formation of the symmetric olefin $CF_3CCl=CClCF_3$. The formal removal of Cl_2 upon treatment with Zn generates *bis*(trifluoromethyl) acetylene.

15.2.4

(I) The nucleophilic attack on CF_3-substituted alkenes, especially styrene derivatives, can be exploited for the preparation of 1,1-difluoroolefins. However, this transformation does not occur with protic nucleophiles, such as amines or alcohols, which typically undergo an addition to the double bond. In contrast, an S_N2'-type addition-elimination mechanism is observed with organometallic nucleophiles, especially organolithium compounds and Grignard reagents.

(II) The reactions of geminal difluoroolefins as well as perfluoroolefins with nucleophiles give rise to unsaturated, partly defluorinated products, if the transformation follows an addition-elimination mechanism. Otherwise, saturated products will be obtained, of course. Typically, 1,1-difluoroalkenes are converted to vinyl fluorides, which can also be part of cyclic moieties, if the nucleophilic attack occurs intramolecularly. Accordingly, partly unsaturated heterocycles are accessible by this method, too. In addition-elimination reactions, fully halogenated olefins, such as perfluorocyclopentene, react with nucleophiles under replacement of one or both vinylic fluorine substituents.

(III) Vinyliodonium salts can be converted to vinyl fluorides, if they are treated with CsF, moreover, the thermal decomposition of tetrafluoroborate salts gives rise to fluroolefins. Typically, the vinyliodonium species are prepared by hydrofluorination of alkynyliodonium salts or iodofluorination of alkynyltrifluoroborates using, for example, hypervalent aryliodonium difluorides. The main problem of this approach is the involvement of cationic species, which especially leads to rearranged side-products, when complex substrates are used.

15.3.4

(I) The treatment of vinyllithium intermediates with electrophilic fluorinating agents, such as NFSI, gives rise to vinyl fluorides. In most cases, the vinyllithium intermediates are generated by low-temperature treatment of vinyl iodides with organolithium compounds. In general, the fluorination occurs stereoselectively under retention of the configuration of the double bond. Major side-products are hydrodefunctionalized olefins, which are formed upon protonation of the reactive vinyllithium intermediates.

(II) The reaction of strained methylenecyclopropanes with NFSI leads to the cleavage of the cyclopropane ring and gives rise to monofluorinated olefins. In general, this transformation proceeds faster, if electron-rich arenes are attached to the double bond. The treatment of diaryl-substituted methylenecyclopropanes with electrophilic Selectfluor in nitrile solvents also gives rise to monofluoroolefins. In the same way, electrophilic fluorination of aryl-substituted allenes is achieved by the use of Selectfluor in aqueous MeCN, which gives rise to fluorohydroxylated products. Moreover, lactones and pyrrolidines with vinylic fluoride substituents are generated upon fluorocyclization of allenoic acids and allenoic tosylamides, respectively.

(III) So far, only a very few substrates have been successfully submitted to a direct C-H fluorination in vinylic position. This transformation has been observed for 1,1-diaryl-substituted olefins, such as 1,1-diphenylethene, 1,1-diaryldienes or cyclic 1,1-diarylethylenes. In all cases, Selectfluor has been used as electrophilic fluorinating agent in presence of water in polar solvents.

15.4.4

(I) The Wittig-reaction is used for the conversion of aliphatic, alicyclic as well as aromatic ketones and aldehydes to 1,1-difluoroolefins. Like the standard protocol, which is used for the transfer of non-fluorinated fragments, the preparation of geminal difluoroolefins requires a phosphine, mostly PPh_3, which is used in conjunction with a difluorocarbene source, such as CF_2Br_2. Alternatively, it is also possible to react nonstabilized phosphorous ylides with CHF_2Cl, which also gives rise to 1,1-difluoroolefins.

(II) Apart from geminal difluoroolefins, other fluorinated alkenes are accessible by Wittig-type reactions. For example, monofluoroolefins are obtained, if $CHFI_2$ is used as reagent. In the same way, fluorodihalomethanes and fluorotrihalomethanes, such as $CFCl_3$, give rise to 1,1-chlorofluooralkenes. The treatment of fluorotrihalomethanes with two equivalents of the phosphine yields intermediary phosphoranium salts, which are capable of converting aromatic aldehydes and polyfluoroacyl halides to monofluoroalkenes. Moreover, perfluorinated isopropylidene-units are transferred to aldehydes, if 2,2-dichloro-1,1,1,3,3,3-hexafluoropropane or *tetrakis*(trifluoromethyl)-1,3-dithietane are reacted with PPh$_3$ and subsequently treated with the carbonyl-compound.

(III)   The reaction of Ph$_3$P=C(CH$_3$)$_2$ with fluorinated acid anhydrides of the type (R^{F}CO)$_2$O yields fluorinated acylphosphonium salts, which are further converted to a new ylide upon treatment with Ph$_3$P=CH$_2$ and PhLi. This ylide can be submitted to a Wittig reaction with aldehydes, giving rise to perfluoroalkylated, conjugated dienes. In contrast, perfluoroalkylated enynes are obtained, if the acylphosphonium intermediate is treated with an acetylide. Perfluoroalkylated acetylenes are accessible from perfluoroacyl chlorides and two equivalents of a phosphorus ylide. Conjugated dienes with vinylic fluorine substituents are generated, if fluorinated α,β-unsaturated aldehydes are submitted to a Wittig olefination.

(IV)   In contrast to the Wittig approach, which uses fluorinated methane derivatives as reagents, Julia– and Julia–Kocienski olefination rely on the use of difluoromethylsulphonyl compounds, such as difluoromethyl 2-(pyridyl)sulphone or 2-pyridylsulphonyl difluoroacetic acid. The advantage of this type of olefination are its suitability to a number of carbonyl substrates, its mild reaction conditions and its tolerance against functional groups.

(V)   Fluorinated olefins are also accessible from aromatic aldehydes and ketones by Horner-Wadsworth-Emmons reactions using difluoromethylphosphonates, which finally give rise to 1,1-difluoroolefins. This method may also be applied to aliphatic carbonyl substrates, if CF$_2$H-P(O)Ph$_2$ is used as source of the CF$_2$-unit. Apart from the formation of geminal difluoroolefins, Horner—Wadsworth—Emmons reactions can also be applied for the preparation of fluorinated olefins with other electron-withdrawing groups. For example, the treatment of aliphatic aldehydes with diethylphosphonofluoroacetates gives rise to α-fluorinated α,β-unsaturated esters. In the same way, the reaction with diethylphosphonofluoroacetonitrile yields cyano-substituted fluoroolefins. The fluorinated substituent can also be part of the substrate and is not necessarily attached to the phosphorous reagent as, for example, the treatment of trifluoromethylketones with diethylphosphonoacetates generates β-trifluoromethylated α,β-unsaturated esters. However, the latter often undergo rearrangements to nonconjugated β,γ-unsaturated esters.

15.5.4

(I)   Olefin metathesis of fluorinated substrates is restricted to starting materials with a single vinylic fluorine substituent. The reason for this is probably found in the catalytic inactivity of transition metal-difluorocarbene complexes, so that the transfer of CF$_2$-units cannot be achieved in a catalytic fashion. In practice, the synthetic approach of olefin metathesis is mostly applied to the preparation of cyclic products, that is, using ring-closing metathesis (RCM), which is typically favored by entropy.

(II)   For the preparation of unsaturated carbocycles, which contain vinylic F- or CF$_3$-substituents, the corresponding 2-substituted allenyl substrates are typically submitted to ring-closing metathesis. Typical substrates include for example malonates, which are α-functionalized by both, a 2-substituted allyl-group and a nonfluorinated alkenyl substituent. Metathesis of these substrates gives rise to six- and seven-membered rings with a vinylic fluorine substituent, whereas the corresponding cyclopentene derivative is not accessible. However, it can be prepared with a CF$_3$-group instead of the vinylic fluorine substituent.

(III)   Ring-closing metathesis of a variety of 2-fluoroallyl- and 2-(trifluoromethyl)allyl-substituted amines gives rise to a number of partly saturated nitrogen heterocycles, such as tetrahydropyridines, pyrrolines, tetrahydro-1*H*-azepines, tetrahydropyridazines and tetrahydro-1,2-diazepines. In the same way, tetrahydropyridazines are accessible from hydrazine-derivatives, whereas fluorinated lactones and lactams are prepared from appropriately substituted 2-fluoroacrylic esters and amides, respectively. However, in case of the nitrogen-containing substrates, it is necessary to reduce their basicity by the attachment of suitable protecting groups to the heteroatom.

15.6.7

(I)   In general, hydrofluorination of acetylenes is a rather simple transformation, which does not necessarily require the presence of a transition metal catalyst. However, if this trans-formation needs to be stopped at the olefinic stage, a metal complex is indeed required, since otherwise fully saturated products are preferentially generated. In some cases, the presence of a transition metal catalyst can also help to improve the regio- and stereoselec-tivity of the HF-addition. For example, a *trans*-selective addition occurs, if a gold(I)-catalyst with a *N*-heterocyclic carbene ligand is used in conjunction with $Et_3N \cdot 3$ HF. In another approach, DMPU/HF has been used as fluorinating agent, using again a gold(I) complex. The treatment of diarylacetylenes with a NHC-stabilized gold(I)-bifluoride complex gives rise to (*Z*)-stilbenes, if it is additionally treated with $NEt_3 \cdot 3$ HF or $NH_4[BF_4]$. Alternatively, electron-deficient alkynes as well as bromoacetylenes are converted to (*Z*)-fluoroolefins and (*Z*)-bromofluoroolefins, respectively, if they are treated with AgF.

(II) The direct C-H fluorination of vinylic substrates is achieved, if trisubstituted substrates are treated with $[Pd(NO_3)]^+$ and electrophilic NFSI. This approach additionally requires the presence of directing *O*-methyloxime ethers and works best for the preparation of α-fluorostyrenes. In contrast, the substitution reaction of vinyl boronic acids and vinyl stan-nanes is catalyzed by Ag(I) and requires the electrophilic fluorinating agents Selectfluor and XeF_2, respectively.

(III) The treatment of 2-CF_3-substituted olefins with acetylenes in presence of a Ni-catalyst and a hydride source leads to a defluorinative coupling, which follows a mechanism similar to a S_N2'-reaction, thus leading to the formation of 1,1-difluoro-1,4-dienes. In contrast, the re-action of *N*-tosylhydrazone-protected trifluoromethylketones with terminal alkynes in pres-ence of a Cu(I)-catalyst and a base gives rise to 1,1-difluoro-1,3-enynes. The product is gen-erated upon attack of an acetylide species on the hydrazone carbon-atom, which leads to the elimination of a fluoride anion from the neighboring CF_3-group, and thus generates a geminal difluoroolefin.

(IV) Diazocompounds are transformed to fluorinated olefins, if they are reacted with some type of fluorinated carbene species. For example, the treatment with a source of :CF_2 in presence of a copper-catalyst gives rise to 1,1-difluoroolefins. However, upon use of the silane reagents Me_3SiCF_3 or Me_3SiCF_2Br, the metal catalyst is not required and can also be replaced by other initiators. Interestingly, the reaction of α-diazoesters with Me_3SiCF_3 and a Cu-catalyst re-sults in the formation of geminal *bis*(trifluoromethyl)olefins, which is proposed to proceed via the formation of an intermediary 1,1-difluoroolefin and subsequent replacement of the vinylic fluorine substituents by CF_3-groups.

Bibliography

[1]   S. J. Tavener, J. H. Clark. Chapter 5 Fluorine: Friend or Foe? A Green Chemist's Perspective. In: A. Tressaud (ed.) Advances in Fluorine Science: Elsevier; 2006:177–202.

[2]   G. M. Sawyer, C. Oppenheimer. Chapter 5: Volcanic Fluorine Emissions: Observations by Fourier Transform Infrared Spectroscopy. In: A. Tressaud (ed.) Advances in Fluorine Science: Elsevier; 2006:165–85.

[3]   J. Schmedt auf der Günne, M. Mangstl, F. Kraus, *Angew. Chem., Int. Ed. Engl.* **2012**, 51, 7847–9.

[4]   V. R. Celinski, M. Ditter, F. Kraus, F. Fujara, J. Schmedt auf der Günne, *Chem. Eur. J.* **2016**, 22, 18388–93.

[5]   K. K. J. Chan, D. O'Hagan. Chapter Eleven – The Rare Fluorinated Natural Products and Biotechnological Prospects for Fluorine Enzymology. In: D. A. Hopwood (ed.) Methods Enzymol: Academic Press; 2012:219–35.

[6]   J. Moschner, V. Stulberg, R. Fernandes, S. Huhmann, J. Leppkes, B. Koksch, *Chem. Rev.* **2019**, 119, 10718–801.

[7]   F. Kraus, *Nachr. Chem.* **2019**, 67, 54–8.

[8]   S. I. Ivlev, A. J. Karttunen, M. Hoelzel, M. Conrad, F. Kraus, *Chem. Eur. J.* **2019**, 25, 3310–7.

[9]   P. Politzer, *J. Am. Chem. Soc.* **1969**, 91, 6235–7.

[10]   P. Politzer, *Inorg. Chem.* **1977**, 16, 3350–1.

[11]   G. L. Caldow, C. A. Coulson, *J. Chem. Soc. Faraday Trans.* **1962**, 58, 633–41.

[12]   R. S. Mulliken, *J. Am. Chem. Soc.* **1955**, 77, 884–7.

[13]   A. A. Woolf, *J. Fluorine Chem.* **1978**, 11, 307–15.

[14]   K. Sonnenberg, L. Mann, F. A. Redeker, B. Schmidt, S. Riedel, *Angew. Chem., Int. Ed. Engl.* **2020**, 59, 5464–93.

[15]   Y. Ogawa, O. Takahashi, O. Kikuchi, *Comput. Theor. Chem.* **1998**, 424, 285–92.

[16]   A. A. Tuinman, A. A. Gakh, R. J. Hinde, R. N. Compton, *J. Am. Chem. Soc.* **1999**, 121, 8397–8.

[17]   B. Braïda, P. C. Hiberty, *J. Am. Chem. Soc.* **2004**, 126, 14890–8.

[18]   S. Riedel, T. Köchner, X. Wang, L. Andrews, *Inorg. Chem.* **2010**, 49, 7156–64.

[19]   F. A. Redeker, H. Beckers, S. Riedel, *RSC Adv.* **2015**, 5, 106568–73.

[20]   T. Vent-Schmidt, F. Brosi, J. Metzger, et al., *Angew. Chem., Int. Ed. Engl.* **2015**, 54, 8279–83.

[21]   F. Brosi, T. Vent-Schmidt, S. Kieninger, T. Schlöder, H. Beckers, S. Riedel, *Chem. Eur. J.* **2015**, 21, 16455–62.

[22]   K. O. Christe, *Inorg. Chem.* **1986**, 25, 3721–2.

[23]   L. B. Asprey, *J. Fluorine Chem.* **1976**, 7, 359–61.

[24]   E. Jacob, K. O. Christe, *J. Fluorine Chem.* **1977**, 10, 169–72.

[25]   J. Janzen, L. S. Bartell, *J. Chem. Phys.* **1969**, 50, 3611–8.

[26]   G. A. Olah, R. H. Schlosberg, *J. Am. Chem. Soc.* **1968**, 90, 2726–7.

[27]   A. W. Jache, G. H. Cady, *J. Phys. Chem.* **1952**, 56, 1106–9.

[28]   J. D. Forrester, M. E. Senko, A. Zalkin, D. Templeton, *Acta Crystallogr.* **1963**, 16, 58–62.

[29]   M. Tramšek, E. Goreshnik, M. Lozinšek, B. Žemva, *J. Fluorine Chem.* **2009**, 130, 1093–8.

[30]   D. Mootz, K. Bartmann, *Angew. Chem.* **1988**, 100, 424–5.

[31]   T. A. O'Donnell, *J. Fluorine Chem.* **1978**, 11, 467–80.

[32]   A. F. Clifford, H. C. Beachell, W. M. Jack, *J. Inorg. Nucl. Chem.* **1957**, 5, 57–70.

[33]   A. F. Clifford, W. D. Pardieck, M. W. Wadley, *J. Phys. Chem.* **1966**, 70, 3241–5.

[34]   B. Zemva, *C. R. Acad. Sci., Ser. IIc: Chim.* **1998**, 1, 151–6.

[35]   T. Schlöder, F. Kraus, S. Riedel. Fluorides: Solid-State Chemistry. In: R. A. Scott (ed.) Encyclopedia of Inorganic and Bioinorganic Chemistry. 2014:1–30.

https://doi.org/10.1515/9783110659337-020

[36]   M. Schmeißer, W. Ludovici, D. Naumann, P. Sartori, E. Scharf, *Chem. Ber.* **1968**, 101, 4214–20.

[37]   M. Schmeißer, P. Sartori, D. Naumann, *Chem. Ber.* **1970**, 103, 880–4.

[38]   L. A. Shimp, R. J. Lagow, *Inorg. Chem.* **1977**, 16, 2974–5.

[39]   R. C. Burns, I. D. MacLeod, T. A. O'Donnell, T. E. Peel, K. A. Phillips, A. B. Waugh, *J. Inorg. Nucl. Chem.* **1977**, 39, 1737–9.

[40]   H. C. Hodge, F. A. Smith, *J. Occup. Environ. Med.* **1977**, 19, 12–39.

[41]   A. W. Davison, L. H. Weinstein. Chapter 8: Some Problems Relating to Fluorides in the Environment: Effects on Plants and Animals. In: A. Tressaud (ed.) Advances in Fluorine Science: Elsevier; 2006:251–98.

[42]   A. F. Lailey, L. Hill, I. W. Lawston, D. Stanton, D. G. Upshall, *Biochem. Pharmacol.* **1991**, 42, S47–S54.

[43]   G. Theodoridis. Chapter 4 Fluorine-Containing Agrochemicals: An Overview of Recent Developments. In: A. Tressaud (ed.) Advances in Fluorine Science: Elsevier; 2006:121–75.

[44]   M. S. Onyango, H. Matsuda. Chapter 1 Fluoride Removal from Water Using Adsorption Technique. In: A. Tressaud (ed.) Advances in Fluorine Science: Elsevier; 2006:1–48.

[45]   M. Pontié, C. Diawara, A. Lhassani, et al., Chapter 2 Water defluoridation processes: a review. Application: nanofiltration (NF) for future large-scale pilot plants. In: A. Tressaud (ed.) Advances in Fluorine Science: Elsevier; 2006:49–80.

[46]   A. F. Danil de Namor, I. Abbas. Chapter 3 Calixpyrrole–Fluoride Interactions: From Fundamental Research to Applications in the Environmental Field. In: A. Tressaud (ed.) Advances in Fluorine Science: Elsevier; 2006:81–119.

[47]   R. Verduzco, M. S. Wong, *ACS Cent. Sci.* **2020**, 6, 453–5.

[48]   E. Kumarasamy, I. M. Manning, L. B. Collins, O. Coronell, F. A. Leibfarth, *ACS Cent. Sci.* **2020**, 6, 487–92.

[49]   A. Sekiya, M. Yamabe, K. Tokuhashi, Y. Hibino, R. Imasu, H. Okamoto. Chapter 2: Evaluation and Selection of CFC Alternatives. In: A. Tressaud (ed.) Advances in Fluorine Science: Elsevier; 2006:33–87.

[50]   R. P. Tuckett. Chapter 3: Trifluoromethyl Sulphur Pentafluoride, SF5CF3: Atmospheric Chemistry and Its Environmental Importance via the Greenhouse Effect. In: A. Tressaud (ed.) Advances in Fluorine Science: Elsevier; 2006:89–129.

[51]   H. Nishiumi, K. Sato, R. Kato. Chapter 4: Production of Second- or Third-Generation Fluorine-based Refrigerants from (Photo)-Dechlorination of Fluorocarbon Wastes. In: A. Tressaud (ed.) Advances in Fluorine Science: Elsevier; 2006:131–64.

[52]   Up-Cycling. Closing the Loop. 3M Dyneon Fluoropolymers, 2016 (Accessed August 24, 2020, at https://multimedia.3m.com/mws/media/9073230/up-cycling-fluoropolymers-brochure.pdf?fn=Up-Cycling_Brochure_EN.pdf).

[53]   C. Pettinari, G. Rafaiani. NMR Applications, Solution state 19F. In: J. C. Lindon, G. E. Tranter, D. W. Koppenaal (eds.) Encyclopedia of Spectroscopy and Spectrometry (Third Edition): Oxford: Academic Press; 2017:117–24.

[54]   J. C. Lindon. Nuclear Magnetic Resonance Spectroscopy | Fluorine-19∗. In: P. Worsfold, C. Poole, A. Townshend, M. Miró (eds.) Encyclopedia of Analytical Science (Third Edition): Oxford: Academic Press; 2016:141–51.

[55]   C. P. Rosenau, B. J. Jelier, A. D. Gossert, A. Togni, *Angew. Chem.* **2018**, 130, 9672–7.

[56]   J. W. Emsley, L. Phillips, *Prog. Nucl. Magn. Reson. Spectrosc.* **1971**, 7, 1–520.

[57]   H. Koroniak, K. W. Palmer, W. R. Dolbier Jr, H.-Q. Zhang, *Magn. Reson. Chem.* **1993**, 31, 748–51.

[58]   C. L. Liotta, H. P. Harris, *J. Am. Chem. Soc.* **1974**, 96, 2250–2.

[59]   Z. K. Ismail, R. H. Hauge, J. L. Margrave, *J. Inorg. Nucl. Chem.* **1973**, 35, 3201–6.

[60]   W. F. Howard, L. Andrews, *Inorg. Chem.* **1975**, 14, 409–13.
[61]   S. I. Ivlev, T. Soltner, A. J. Karttunen, M. J. Mühlbauer, A. J. Kornath, F. Kraus, *Z. Anorg. Allg. Chem.* **2017**, 643, 1436–43.
[62]   N. von Aspern, G.-V. Röschenthaler, M. Winter, I. Cekic-Laskovic, *Angew. Chem., Int. Ed. Engl.* **2019**, 58, 15978–6000.
[63]   J.-C. Tedenac, L. Cot, *C. R. Acad. Sci., Ser. IIc: Chim.* **1969**, 268C, 1687–90.
[64]   A. S. Quist, J. B. Bates, G. E. Boyd, *Spectrochim. Acta, Part A* **1972**, 28, 1103–5.
[65]   A. H. Narten, *J. Chem. Phys.* **1972**, 56, 1905-9.
[66]   W. Yu, L. Andrews, X. Wang, *J. Phys. Chem. A* **2017**, 121, 8843–55.
[67]   R. A. Kovar, G. L. Morgan, *J. Am. Chem. Soc.* **1970**, 92, 5067–72.
[68]   M. G. Hogben, K. Radley, L. W. Reeves, *Can. J. Chem.* **1970**, 48, 2960–4.
[69]   M. Schmidt, H. Schmidbaur, *Z. Naturforsch. B* **1998**, 53, 1294.
[70]   A. K. Baral, H. K. Saha, N. N. Ray, *J. Inorg. Nucl. Chem.* **1967**, 29, 2490–4.
[71]   F. Kraus, M. B. Fichtl, S. A. Baer, *Z. Naturforsch. B* **2009**, 64, 257.
[72]   F. Kraus, S. A. Baer, M. R. Buchner, A. J. Karttunen, *Chem. Eur. J.* **2012**, 18, 2131–42.
[73]   C. J. Hardy, D. Scargill, *J. Chem. Soc.* **1961**, 2658–63.
[74]   E. Thilo, H. Budzinski, *Z. Anorg. Allg. Chem.* **1963**, 320, 177–82.
[75]   H. Budzinski, E. Thilo, *Z. Anorg. Allg. Chem.* **1964**, 327, 8–14.
[76]   S. Deganello, *J. Am. Ceram. Soc.* **1972**, 55, 584.
[77]   S. Deganello, *Acta Crystallogr. B* **1973**, 29, 2593–7.
[78]   J. H. Burns, E. K. Gordon, *Acta Crystallogr.* **1966**, 20, 135–8.
[79]   A. D. van Ingen Schenau, G. C. Verschoor, R. A. G. de Graaff, *Acta Crystallogr. B* **1976**, 32, 1127–32.
[80]   M. R. Anderson, S. Vilminot, I. D. Brown, *Acta Crystallogr. B* **1973**, 29, 2961–2.
[81]   D. M. Roy, R. Roy, E. F. Osborn, *J. Am. Ceram. Soc.* **1954**, 37, 300–5.
[82]   R. E. Thoma, H. Insley, H. A. Friedman, G. M. Hebert, *J. Nucl. Mater.* **1968**, 27, 166–80.
[83]   A. Snelson, B. N. Cyvin, S. J. Cyvin, *J. Mol. Struct.* **1975**, 24, 165–76.
[84]   G. Boucherle, S. Aléonard, *Mater. Res. Bull.* **1971**, 6, 525–35.
[85]   L. M. Toth, J. B. Bates, G. E. Boyd, *J. Phys. Chem.* **1973**, 77, 216–21.
[86]   H. Cho, W. Wang, R. Kim, et al., *Proc. Natl. Acad. Sci. USA* **2001**, 98, 8525–30.
[87]   F. Weigel, A. Trinkl, *Radiochim. Acta* **1968**, 9, 36.
[88]   E. P. F. Lee, P. Soldán, T. G. Wright, *Inorg. Chem.* **2001**, 40, 5979–84.
[89]   M. Vasiliu, J. G. Hill, K. A. Peterson, D. A. Dixon, *J. Phys. Chem. A* **2018**, 122, 316–27.
[90]   P. L. Timms, *J. Am. Chem. Soc.* **1967**, 89, 1629–32.
[91]   A. K. Holliday, F. B. Taylor, *J. Chem. Soc.* **1964**, 2731–4.
[92]   R. W. Kirk, D. L. Smith, W. Airey, P. L. Timms, *J. Chem. Soc., Dalton Trans.* **1972**, 1392–6.
[93]   A. Finch, H. I. Schlesinger, *J. Am. Chem. Soc.* **1958**, 80, 3573–4.
[94]   M. Arrowsmith, J. Böhnke, H. Braunschweig, et al., *Chem. Commun.* **2017**, 53, 8265–7.
[95]   R. J. Brotherton, A. L. McCloskey, H. M. Manasevit, *Inorg. Chem.* **1963**, 2, 41–3.
[96]   D. A. Saulys, J. Castillo, J. A. Morrison, *Inorg. Chem.* **1989**, 28, 1619–24.
[97]   J. R. Durig, J. W. Thompson, J. D. Witt, J. D. Odom, *J. Chem. Phys.* **1973**, 58, 5339–43.
[98]   L. Trefonas, W. N. Lipscomb, *J. Chem. Phys.* **1958**, 28, 54–5.
[99]   D. D. Danielson, J. V. Patton, K. Hedberg, *J. Am. Chem. Soc.* **1977**, 99, 6484–7.
[100] L. A. Nimon, K. S. Seshadri, R. C. Taylor, D. White, *J. Chem. Phys.* **1970**, 53, 2416–27.
[101] A. Szabó, A. Kovács, G. Frenking, *Z. Anorg. Allg. Chem.* **2005**, 631, 1803–9.
[102] C. R. Watts, J. K. Badenhoop, *J. Chem. Phys.* **2008**, 129, 104307.
[103] A. K. Holliday, F. B. Taylor, *J. Chem. Soc.* **1962**, 2767–71.
[104] J. J. Ritter, T. D. Coyle, *J. Chem. Soc. A* **1970**, 1303–4.
[105] B. W. C. Ashcroft, A. K. Holliday, *J. Chem. Soc. A* **1971**, 2581–3.

[106] J. J. Ritter, T. D. Coyle, J. M. Bellama, *J. Organomet. Chem.* **1971**, 29, 175–84.

[107] W. Haubold, K. Stanzl, *Chem. Ber.* **1978**, 111, 2108–12.

[108] W. Haubold, K. Stanzl, *J. Organomet. Chem.* **1979**, 174, 141–7.

[109] C. Pubill-Ulldemolins, E. Fernánez, C. Bo, J. M. Brown, *Org. Biomol. Chem.* **2015**, 13, 9619–28.

[110] A. Kerr, N. C. Norman, A. Guy Orpen, et al., *Chem. Commun.* **1998**, 319–20.

[111] N. Lu, N. C. Norman, A. G. Orpen, M. J. Quayle, P. L. Timms, G. R. Whittell, *J. Chem. Soc., Dalton Trans.* **2000**, 4032–7.

[112] J. H. Muessig, D. Prieschl, A. Deißenberger, et al., *J. Am. Chem. Soc.* **2018**, 140, 13056–63.

[113] G. Bélanger-Chabot, H. Braunschweig, *Angew. Chem., Int. Ed. Engl.* **2019**, 58, 14270–4.

[114] J. A. J. Pardoe, N. C. Norman, P. L. Timms, et al., *Angew. Chem., Int. Ed. Engl.* **2003**, 42, 571–3.

[115] P. L. Timms, N. C. Norman, J. A. J. Pardoe, et al., *Dalton Trans.* **2005**, 607–16.

[116] B. G. DeBoer, A. Zalkin, D. H. Templeton, *Inorg. Chem.* **1969**, 8, 836–41.

[117] J. S. Hartman, P. L. Timms, *J. Chem. Soc., Dalton Trans.* **1975**, 1373–6.

[118] N. N. Greenwood, R. L. Martin, *J. Chem. Soc.* **1953**, 1427–32.

[119] N. N. Greenwood, *J. Inorg. Nucl. Chem.* **1958**, 5, 224–8.

[120] N. N. Greenwood, *J. Inorg. Nucl. Chem.* **1958**, 5, 229–36.

[121] C. Gascard, G. Mascherpa, *J. Chim. Phys.* **1973**, 70, 1040–7.

[122] F. Kraus, S. A. Baer, *Z. Anorg. Allg. Chem.* **2010**, 636, 414–22.

[123] D. F. Wolfe, G. L. Humphrey, *J. Mol. Struct.* **1969**, 3, 293–303.

[124] R. E. Mesmer, A. C. Rutenberg, *Inorg. Chem.* **1973**, 12, 699–702.

[125] D. D. Gibler, C. J. Adams, M. Fischer, A. Zalkin, N. Bartlett, *Inorg. Chem.* **1972**, 11, 2325–9.

[126] J. S. Hartman, P. Stilbs, *J. Chem. Soc., Chem. Commun.* **1975**, 566–7.

[127] J. S. Hartman, G. J. Schrobilgen, P. Stilbs, *Can. J. Chem.* **1976**, 54, 1121–9.

[128] K. Duda, A. Himmelspach, J. Landmann, F. Kraus, M. Finze, *Chem. Commun.* **2016**, 52, 13241–4.

[129] J. Hoeft, F. J. Lovas, E. Tiemann, T. Törring, *Z. Naturforsch. B* **1970**, 25, 901.

[130] D. Welti, R. F. Barrow, *Nature* **1951**, 168, 161.

[131] J. Hoeft, F. J. Lovas, E. Tiemann, T. Törring, *Z. Naturforsch. A* **1970**, 25, 1029.

[132] H. Schnöckel, H. J. Gocke, *J. Mol. Struct.* **1978**, 50, 281–4.

[133] N. W. Alcock, H. D. B. Jenkins, *J. Chem. Soc., Dalton Trans.* **1974**, 1907–11.

[134] C. W. F. T. Pistorius, J. B. Clark, *Phys. Rev.* **1968**, 173, 692–9.

[135] A. Ruoff, J. Weidlein, *Z. Anorg. Allg. Chem.* **1969**, 370, 113–8.

[136] F. J. Keneshea, D. Cubicciotti, *J. Phys. Chem.* **1967**, 71, 1958-60.

[137] M. L. Lesiecki, J. W. Nibler, *J. Chem. Phys.* **1975**, 63, 3452–61.

[138] J. Grannec, J. Portier, *C. R. Acad. Sci., Ser. IIc: Chim.* **1971**, 272C, 942–4.

[139] J. Grannec, L. Lozano, J. Portier, P. Hagenmuller, *Z. Anorg. Allg. Chem.* **1971**, 385, 26–32.

[140] G. Shanmugasundaram, G. Nagarajan, *Z. Phys. Chem.* **1969**, 2400, 363.

[141] N. A. Matwiyoff, W. E. Wageman, *Inorg. Chem.* **1970**, 9, 1031–6.

[142] J. W. Akitt, N. N. Greenwood, G. D. Lester, *J. Chem. Soc. A* **1971**, 2450–7.

[143] V. Sćepanović, S. Radosavljević, J. Misović, *J. Fluorine Chem.* **1974**, 3, 403–8.

[144] E. E. Flagg, D. L. Schmidt, *J. Inorg. Nucl. Chem.* **1969**, 31, 2329–41.

[145] M.-C. Kervarec, C. P. Marshall, T. Braun, E. Kemnitz, *J. Fluorine Chem.* **2019**, 221, 61–5.

[146] Z. Mazej, E. Goreshnik, *New J. Chem.* **2010**, 34, 2806–12.

[147] D. H. Feather, A. Buechler, *J. Phys. Chem.* **1973**, 77, 1599–600.

[148] E. Vajda, I. Hargittai, S. Tremmel, *Inorg. Chim. Acta* **1977**, 25, L143–L5.

[149] M. J. Reisfeld, *Spectrochim. Acta, Part A* **1973**, 29, 1923–6.

[150] B. Müller, K. Hoppe, *Z. Anorg. Allg. Chem.* **1973**, 395, 239–48.

[151] R. Hoppe, R. Jesse, *Z. Anorg. Allg. Chem.* **1973**, 402, 29–38.

[152] J. Setter, R. Hoppe, *Z. Anorg. Allg. Chem.* **1976**, 423, 133–43.

[153] Y. A. Buslaev, S. P. Petrosyants, V. P. Tarasov, *J. Struct. Chem.* **1974**, 15, 187–90.

[154] S. Schneider, R. Hoppe, *Z. Anorg. Allg. Chem.* **1970**, 376, 268–76.

[155] H. Bode, E. Voss, *Z. Anorg. Allg. Chem.* **1957**, 290, 1–16.

[156] G. Mermant, C. Belinski, F. Lalau-Keraly, *C. R. Acad. Sci., Ser. IIc: Chim.* **1967**, 265C, 238–41.

[157] R. Haiges, J. A. Boatz, J. M. Williams, K. O. Christe, *Angew. Chem., Int. Ed. Engl.* **2011**, 50, 8828–33.

[158] R. Haiges, P. Deokar, M. Vasiliu, T. H. Stein, D. A. Dixon, K. O. Christe, *Chem. Eur. J.* **2017**, 23, 9054–66.

[159] C. Hebecker, *Z. Anorg. Allg. Chem.* **1972**, 393, 223–9.

[160] C. Hebecker, R. Hoppe, *Sci. Nat.* **1966**, 53, 104.

[161] C. Hebecker, *Sci. Nat.* **1971**, 58, 361.

[162] C. Hebecker, *Z. Naturforsch. B* **1975**, 30, 305.

[163] C. Hebecker, *Z. Anorg. Allg. Chem.* **1975**, 412, 37–46.

[164] J. Koch, C. Hebecker, *Sci. Nat.* **1985**, 72, 431–2.

[165] O. Ruff, O. Bretschneider, *Z. Anorg. Allg. Chem.* **1934**, 217, 1–18.

[166] W. Rüdorff, G. Rüdorff, *Z. Anorg. Allg. Chem.* **1947**, 253, 281–96.

[167] W. Rüdorff, G. Rüdorff, *Chem. Ber.* **1947**, 80, 417–23.

[168] R. J. Lagow, R. B. Badachhape, J. L. Wood, J. L. Margrave, *J. Am. Chem. Soc.* **1974**, 96, 2628–9.

[169] R. J. Lagow, L. A. Shimp, D. K. Lam, R. F. Baddour, *Inorg. Chem.* **1972**, 11, 2568–70.

[170] R. J. Lagow, R. B. Badachhape, P. Ficalora, J. L. Wood, J. L. Margrave, *Synth. React. Inorg. Met.-Org. Chem.* **1972**, 2, 145–9.

[171] F. X. Powell, D. R. Lide Jr., *J. Chem. Phys.* **1966**, 45, 1067–8.

[172] K. S. Thanthiriwatte, M. Vasiliu, S. R. Battey, et al., *J. Phys. Chem. A* **2015**, 119, 5790–803.

[173] A. A. Tuinman, P. Mukherjee, J. L. Adcock, R. L. Hettich, R. N. Compton, *J. Phys. Chem.* **1992**, 96, 7584–9.

[174] A. A. Gakh, A. A. Tuinman, J. L. Adcock, R. A. Sachleben, R. N. Compton, *J. Am. Chem. Soc.* **1994**, 116, 819–20.

[175] O. V. Boltalina, L. N. Sidorov, V. F. Bagryantsev, et al., *J. Chem. Soc., Perkin Trans. 2* **1996**, 2275–8.

[176] O. V. Boltalina, N. A. Galeva, *Russ. Chem. Rev.* **2000**, 69, 609–21.

[177] A. V. Kepman, V. F. Sukhoverkhov, A. Tressaud, et al., *J. Fluorine Chem.* **2006**, 127, 832–6.

[178] A. A. Gakh, A. A. Tuinman, *Tetrahedron Lett.* **2001**, 42, 7137–9.

[179] O. V. Boltalina, A. a. K. Abdul-Sada, R. Taylor, *J. Chem. Soc., Perkin Trans. 2* **1995**, 981–5.

[180] S. J. Austin, P. W. Fowler, J. P. B. Sandall, F. Zerbetto, *J. Chem. Soc., Perkin Trans. 2* **1996**, 155–7.

[181] R. W. Clare, D. L. Kepert, *Comput. Theor. Chem.* **1997**, 389, 97–103.

[182] O. V. Boltalina, A. Y. Lukonin, V. K. Pavlovich, L. N. Sidorov, R. Taylor, A. K. Abdul-Sada, *Fullerene Sci. Technol.* **1998**, 6, 469–79.

[183] S. I. Troyanov, P. A. Troshin, O. V. Boltalina, I. N. Ioffe, L. N. Sidorov, E. Kemnitz, *Angew. Chem., Int. Ed. Engl.* **2001**, 40, 2285–7.

[184] L. Hedberg, K. Hedberg, O. V. Boltalina, N. A. Galeva, A. S. Zapolskii, V. F. Bagryantsev, *J. Phys. Chem. A* **2004**, 108, 4731–6.

[185] R. J. Papoular, H. Allouchi, A. V. Dzyabchenko, et al., *Fuller. Nanotub. Carbon Nanostructures* **2006**, 14, 279–85.

[186] A. A. Gakh, A. A. Tuinman, *J. Phys. Chem. A* **2000**, 104, 5888–91.

[187] P. A. Troshin, Y. A. Mackeyev, N. V. Chelovskaya, Y. L. Slovokhotov, O. V. Boltalina, L. N. Sidorov, *Fullerene Sci. Technol.* **2000**, 8, 501–17.

[188]   J. Szala-Bilnik, M. F. Costa Gomes, A. A. H. Pádua, *J. Phys. Chem. C* **2016**, 120, 19396–408.

[189]   A. A. Gakh, A. A. Tuinman, J. L. Adcock, R. N. Compton, *Tetrahedron Lett.* **1993**, 34, 7167–70.

[190]   P. Strobel, J. Ristein, L. Ley, K. Seppelt, I. V. Goldt, O. Boltalina, *Diam. Relat. Mater.* **2006**, 15, 720–4.

[191]   J. L. Margrave, P. W. Wilson, *Acc. Chem. Res.* **1971**, 4, 145–52.

[192]   M. Schmeißer, K.-P. Ehlers, *Angew. Chem.* **1964**, 76, 781–2.

[193]   V. M. Rao, R. F. Curl Jr., P. L. Timms, J. L. Margrave, *J. Chem. Phys.* **1965**, 43, 2557–8.

[194]   P. L. Timms, R. A. Kent, T. C. Ehlert, J. L. Margrave, *J. Am. Chem. Soc.* **1965**, 87, 2824–8.

[195]   P. L. Timms, T. C. Ehlert, J. L. Margrave, F. E. Brinckman, T. C. Farrar, T. D. Coyle, *J. Am. Chem. Soc.* **1965**, 87, 3819–23.

[196]   K. G. Sharp, J. F. Bald, *Inorg. Chem.* **1975**, 14, 2553–6.

[197]   H. P. Hopkins, J. C. Thompson, J. L. Margrave, *J. Am. Chem. Soc.* **1968**, 90, 901–2.

[198]   Y. N. Tang, G. P. Gennaro, Y. Y. Su, *J. Am. Chem. Soc.* **1972**, 94, 4355–7.

[199]   E. E. Siefert, R. A. Ferrieri, O. F. Zeck, Y. N. Tang, *Inorg. Chem.* **1978**, 17, 2802–9.

[200]   J. C. Thompson, J. L. Margrave, *Inorg. Chem.* **1972**, 11, 913–4.

[201]   M. Schmeisser, *Angew. Chem.* **1954**, 66, 713–4.

[202]   H. Oberhammer, *J. Mol. Struct.* **1976**, 31, 237–45.

[203]   E. L. Muetterties, *Inorg. Chem.* **1962**, 1, 342–5.

[204]   E. L. Muetterties, J. E. Castle, *J. Inorg. Nucl. Chem.* **1961**, 18, 148–53.

[205]   N. Bartlett, K. C. Yu, *Can. J. Chem.* **1961**, 39, 80–6.

[206]   P. Riviere, A. Castel, J. Satge, C. Abdennadher, *Organometallics* **1991**, 10, 1227–8.

[207]   J. Trotter, M. Akhtar, N. Bartlett, *J. Chem. Soc. A* **1966**, 30–3.

[208]   K. Zmbov, J. W. Hastie, J. L. Margrave, *J. Chem. Soc. Faraday Trans.* **1968**, 64, 861–7.

[209]   H. Takeo, R. F. Curl, *J. Mol. Spectrosc.* **1972**, 43, 21–30.

[210]   H. Huber, E. P. Kündig, G. A. Ozin, A. V. Voet, *Can. J. Chem.* **1974**, 52, 95–9.

[211]   P. Riviere, J. Satge, A. Boy, *J. Organomet. Chem.* **1975**, 96, 25–40.

[212]   A. Castel, P. Riviere, J. Satge, J. J. E. Moreau, R. J. P. Corriu, *Organometallics* **1983**, 2, 1498–502.

[213]   P. Rivière, J. Satgé, A. Castel, *C. R. Acad. Sci., Ser. IIc: Chim.* **1976**, 282C, 971–4.

[214]   P. Rivière, J. Satgé, A. Castel, *C. R. Acad. Sci., Ser. IIc: Chim.* **1977**, 284C, 395–8.

[215]   P. Rivière, A. Castel, J. Satgé, D. Guyot, *J. Organomet. Chem.* **1986**, 315, 157–64.

[216]   A. L. Hector, A. Jolleys, W. Levason, D. Pugh, G. Reid, *Dalton Trans.* **2014**, 43, 14514–6.

[217]   J. D. Donaldson, R. Oteng, *Inorg. Nucl. Chem. Lett.* **1967**, 3, 163–4.

[218]   A. Lari Lavassani, L. Cot, C. Geneys, C. Avinens, *C. R. Acad. Sci., Ser. IIc: Chim.* **1975**, 280C, 1211–4.

[219]   R. R. McDonald, A. C. Larson, D. T. Cromer, *Acta Crystallogr.* **1964**, 17, 1104–8.

[220]   A. Lari Lavassani, G. Jourdan, C. Avinens, L. Cot, *C. R. Acad. Sci., Ser. IIc: Chim.* **1974**, 307C, 307–10.

[221]   H. Kriegsmann, G. Kessler, *Z. Anorg. Allg. Chem.* **1962**, 318, 266–76.

[222]   J. D. Donaldson, J. D. O'Donogue, R. Oteng, *J. Chem. Soc.* **1965**, 3876–9.

[223]   G. Bergerhoff, L. Goost, *Acta Crystallogr. B* **1973**, 29, 632–3.

[224]   E. Acker, K. Recker, S. Haussühl, *J. Cryst. Growth* **1976**, 35, 165–8.

[225]   G. Bergerhoff, L. Goost, *Acta Crystallogr. B* **1970**, 26, 19–23.

[226]   G. Denès, J. Pannetier, J. Lucas, *C. R. Acad. Sci., Ser. IIc: Chim.* **1975**, 280C, 831–4.

[227]   T. Birchall, P. A. W. Dean, R. J. Gillespie, *J. Chem. Soc. A* **1971**, 1777–82.

[228]   M. F. A. Dove, R. King, T. J. King, *J. Chem. Soc., Chem. Commun.* **1973**, 944–5.

[229]   R. Sabatier, A.-M. Hebrard, J.-C. Cousseins, *C. R. Acad. Sci., Ser. IIc: Chim.* **1974**, 279C, 1121–3.

[230] V. Dracopoulos, D. T. Kastrissios, G. N. Papatheodorou, *J. Non-Cryst. Solids* **2005**, 351, 640–9.

[231] M. Bannert, G. Blumenthal, H. Sattler, M. Schönherr, H. Wittrich, *Z. Chem.* **1972**, 12, 191–2.

[232] H. L. Deubner, M. Sachs, J. Bandemehr, et al., *Chem. Eur. J.* **2019**, 25, 15656–61.

[233] J. Bandemehr, M. Sachs, S. I. Ivlev, A. J. Karttunen, F. Kraus, *Eur. J. Inorg. Chem.* **2020**, 2020, 64–70.

[234] A. P. Hagen, B. W. Callaway, *J. Inorg. Nucl. Chem.* **1972**, 34, 487–90.

[235] A. D. Adley, P. H. Bird, A. R. Fraser, M. Onyszchuk, *Inorg. Chem.* **1972**, 11, 1402–9.

[236] B. J. Aylett, I. A. Ellis, C. J. Porritt, *J. Chem. Soc., Dalton Trans.* **1972**, 1953–8.

[237] J. J. Harris, B. Rudner, *J. Inorg. Nucl. Chem.* **1972**, 34, 75–87.

[238] B. J. Aylett, I. A. Ellis, C. J. Porritt, *J. Chem. Soc., Dalton Trans.* **1973**, 83–7.

[239] U. Klingebiel, D. Fischer, A. Meller, *Monatsh. Chem.* **1975**, 106, 459–71.

[240] U. Klingebiel, D. Enterling, A. Meller, *J. Organomet. Chem.* **1975**, 101, 45–56.

[241] J. L. Margrave, K. G. Sharp, P. W. Wilson, *J. Am. Chem. Soc.* **1970**, 92, 1530–2.

[242] H. C. Clark, P. W. R. Corfield, K. R. Dixon, J. A. Ibers, *J. Am. Chem. Soc.* **1967**, 89, 3360–1.

[243] K. Behrends, G. Kiel, *Sci. Nat.* **1967**, 54, 537.

[244] F. Klanberg, E. L. Muetterties, *Inorg. Chem.* **1968**, 7, 155–60.

[245] H. C. Clark, K. R. Dixon, J. G. Nicolson, *Inorg. Chem.* **1969**, 8, 450–3.

[246] K. Kleboth, *Monatsh. Chem.* **1970**, 101, 357–61.

[247] V. O. Gelmboldt, V. C. Kravtsov, M. S. Fonari, *J. Fluorine Chem.* **2019**, 221, 91–102.

[248] B. Hájek, F. Benda, *Collect. Czechoslov. Chem. Commun.* **1972**, 37, 2534–6.

[249] B. Hájek, F. Benda, *Z. Chem.* **1974**, 14, 365–6.

[250] A. E. Gebala, M. M. Jones, *J. Inorg. Nucl. Chem.* **1967**, 29, 2301–5.

[251] F. Klanberg, *Z. Naturforsch. B* **1963**, 18, 845.

[252] T. E. Mallouk, G. L. Rosenthal, G. Mueller, R. Brusasco, N. Bartlett, *Inorg. Chem.* **1984**, 23, 3167–73.

[253] I. Wharf, M. Onyszchuk, *Can. J. Chem.* **1970**, 48, 2250–6.

[254] A. M. McNair, B. S. Ault, *Inorg. Chem.* **1982**, 21, 2603–5.

[255] T. E. Mallouk, B. Desbat, N. Bartlett, *Inorg. Chem.* **1984**, 23, 3160–6.

[256] K. O. Christe, R. D. Wilson, I. B. Goldberg, *Inorg. Chem.* **1976**, 15, 1271–4.

[257] K. O. Christe, C. J. Schack, R. D. Wilson, *Inorg. Chem.* **1976**, 15, 1275–82.

[258] C. J. Wilkins, H. M. Haendler, *J. Chem. Soc.* **1965**, 3174–9.

[259] R. Hoppe, W. Dähne, *Sci. Nat.* **1962**, 49, 254–5.

[260] M. Bork, R. Hoppe, *Z. Anorg. Allg. Chem.* **1996**, 622, 1557–63.

[261] R. Hoppe, V. Wilhelm, B. Müller, *Z. Anorg. Allg. Chem.* **1972**, 392, 1–9.

[262] B. Müller, R. Hoppe, *Z. Anorg. Allg. Chem.* **1972**, 392, 37–41.

[263] M. J. Durand, J. L. Galigne, A. Lari-Lavassani, *J. Solid State Chem.* **1976**, 16, 157–60.

[264] C. Hebecker, R. Hoppe, *Sci. Nat.* **1966**, 53, 106.

[265] F. Kraus, M. Fichtl, S. Baer, *Acta Crystallogr., Sect. E* **2016**, 72, 1860–3.

[266] P. A. W. Dean, *Can. J. Chem.* **1973**, 51, 4024–30.

[267] P. A. W. Dean, D. F. Evans, *J. Chem. Soc. A* **1968**, 1154–66.

[268] H. von Wartenberg, *Z. Anorg. Allg. Chem.* **1940**, 244, 337–47.

[269] H. W. Roesky, O. Glemser, K.-H. Hellberg, *Chem. Ber.* **1965**, 98, 2046–8.

[270] U. Reusch, E. Schweda, *Mater. Sci. Forum* **2001**, 378–381, 326–30.

[271] J. Bandemehr, H. L. Deubner, M. Sachs, F. Kraus, *Z. Anorg. Allg. Chem.* **2018**, 644, 1721–6.

[272] A. Tasaka, T. Osada, T. Kawagoe, et al., *J. Fluorine Chem.* **1998**, 87, 163–71.

[273] T. Miyazaki, I. Mori, T. Umezaki, S. Yonezawa, *J. Fluorine Chem.* **2019**, 219, 55–61.

[274] R. Scheele, B. McNamara, A. M. Casella, A. Kozelisky, *J. Nucl. Mater.* **2012**, 424, 224–36.

[275] D. Pilipovich, C. J. Schack, *Inorg. Chem.* **1968**, 7, 386–7.

[276]   K. O. Christe, *Inorg. Chem.* **1969**, 8, 1539.

[277]   R. C. Petry, *J. Am. Chem. Soc.* **1960**, 82, 2400–1.

[278]   W. C. Firth, *Inorg. Chem.* **1965**, 4, 254–5.

[279]   D. M. Gardner, W. W. Knipe, C. J. Mackley, *Inorg. Chem.* **1963**, 2, 413–4.

[280]   T. Miyazaki, I. Mori, T. Umezaki, S. Yonezawa, *J. Fluorine Chem.* **2018**, 210, 126–31.

[281]   J. V. Gilbert, R. A. Conklin, R. D. Wilson, K. O. Christe, *J. Fluorine Chem.* **1990**, 48, 361–6.

[282]   R. C. Petry, *J. Am. Chem. Soc.* **1967**, 89, 4600–4.

[283]   R. C. Petry, J. P. Freeman, *J. Am. Chem. Soc.* **1961**, 83, 3912.

[284]   R. W. Cockman, E. A. V. Ebsworth, J. H. Holloway, *J. Chem. Soc., Chem. Commun.* **1986**, 1622–3.

[285]   B. Sukornick, R. F. Stahl, J. Gordon, *Inorg. Chem.* **1963**, 2, 875.

[286]   J. S. Thrasher, D. D. Desmarteau, *J. Fluorine Chem.* **1991**, 52, 51–5.

[287]   R. Minkwitz, R. Naß, *Z. Naturforsch. B* **1988**, 43, 1478.

[288]   R. Minkwitz, A. Liedtke, R. Nass, *J. Fluorine Chem.* **1987**, 35, 307–15.

[289]   V. Grakauskas, K. Baum, *J. Am. Chem. Soc.* **1969**, 91, 1679–83.

[290]   M. Schmeisser, P. Sartori, *Z. Naturforsch. B* **1966**, 21, 314.

[291]   J. Jander, V. Münch, *Z. Anorg. Allg. Chem.* **1978**, 446, 193–207.

[292]   D. L. Klopotek, B. G. Hobrock, *Inorg. Chem.* **1967**, 6, 1750–1.

[293]   E. A. Lawton, D. Pilipovich, R. D. Wilson, *Inorg. Chem.* **1965**, 4, 118–9.

[294]   K. J. Martin, *J. Am. Chem. Soc.* **1965**, 87, 394–5.

[295]   J. W. Frazer, *J. Inorg. Nucl. Chem.* **1959**, 11, 166–7.

[296]   L. P. Kuhn, C. Wellman, *Inorg. Chem.* **1970**, 9, 602–7.

[297]   J. K. Ruff, *J. Am. Chem. Soc.* **1965**, 87, 1140–1.

[298]   J. K. Ruff, *Inorg. Chem.* **1966**, 5, 1791–4.

[299]   K. O. Christe, C. J. Schack, *Inorg. Chem.* **1978**, 17, 2749–54.

[300]   L. M. Zaborowski, K. E. Pullen, J. n. M. Shreeve, *Inorg. Chem.* **1969**, 8, 2005–7.

[301]   H. W. Roesky, O. Glemser, D. Bormann, *Angew. Chem.* **1964**, 76, 713–4.

[302]   H. W. Roesky, O. Glemser, D. Bormann, *Chem. Ber.* **1966**, 99, 1589–93.

[303]   C. B. Colburn, F. A. Johnson, A. Kennedy, K. McCallum, L. C. Metzger, C. O. Parker, *J. Am. Chem. Soc.* **1959**, 81, 6397–8.

[304]   M. Schmeisser, P. Sartori, *Angew. Chem.* **1959**, 71, 523.

[305]   L. M. Zabrowski, R. A. De Marco, J. M. Shreeve, M. Lustig, *Inorg. Synth.* **1973**, 14, 34–9.

[306]   F. A. Johnson, *Inorg. Chem.* **1966**, 5, 149–50.

[307]   G. L. Hurst, S. I. Khayat, *J. Am. Chem. Soc.* **1965**, 87, 1620.

[308]   V. Munch, H. Selig, *J. Fluorine Chem.* **1980**, 15, 253–6.

[309]   K. O. Christe, D. A. Dixon, D. J. Grant, et al., *Inorg. Chem.* **2010**, 49, 6823–33.

[310]   R. K. Bohn, S. H. Bauer, *Inorg. Chem.* **1967**, 6, 309–12.

[311]   M. Lustig, *Inorg. Nucl. Chem. Lett.* **1969**, 5, 723–4.

[312]   K. O. Christe, R. D. Wilson, W. Sawodny, *J. Mol. Struct.* **1971**, 8, 245–53.

[313]   K. O. Christe, R. D. Wilson, W. W. Wilson, R. Bau, S. Sukumar, D. A. Dixon, *J. Am. Chem. Soc.* **1991**, 113, 3795–800.

[314]   K. O. Christe, W. W. Wilson, J. A. Sheehy, J. A. Boatz, *Angew. Chem., Int. Ed. Engl.* **1999**, 38, 2004–9.

[315]   H. M. Netzloff, M. S. Gordon, K. Christe, et al., *J. Phys. Chem. A* **2003**, 107, 6638–47.

[316]   G. A. Olah, K. Laali, M. Farnia, J. Shih, B. P. Singh, K. O. Christe, *J. Org. Chem.* **1985**, 50, 1338–9.

[317]   J. Bensoam, F. Mathey, *Tetrahedron Lett.* **1977**, 18, 2797–800.

[318]   H. F. Bettinger, P. v. R. Schleyer, H. F. Schaefer, *J. Am. Chem. Soc.* **1998**, 120, 11439–48.

[319]   C. T. Goetschel, V. A. Campanile, R. M. Curtis, K. R. Loos, C. D. Wagner, J. N. Wilson, *Inorg. Chem.* **1972**, 11, 1696–701.

[320]   K. O. Christe, W. W. Wilson, *J. Am. Chem. Soc.* **1992**, 114, 9934–6.

[321]   W. E. Tolberg, R. T. Rewick, R. S. Stringham, M. E. Hill, *Inorg. Chem.* **1967**, 6, 1156–9.

[322]   K. O. Christe, R. D. Wilson, A. E. Axworthy, *Inorg. Chem.* **1973**, 12, 2478–81.

[323]   K. O. Christe, C. J. Schack, R. D. Wilson, *J. Fluorine Chem.* **1976**, 8, 541–4.

[324]   K. O. Christe, C. J. Schack, R. D. Wilson, *Inorg. Chem.* **1977**, 16, 849–54.

[325]   K. O. Christe, R. D. Wilson, C. J. Schack, *Inorg. Chem.* **1977**, 16, 937–40.

[326]   K. O. Christe, C. J. Schack, *Inorg. Chem.* **1977**, 16, 353–9.

[327]   K. O. Christe, *Inorg. Chem.* **1977**, 16, 2238–41.

[328]   W. W. Wilson, K. O. Christe, *J. Fluorine Chem.* **1982**, 19, 253–62.

[329]   K. O. Christe, R. Haiges, M. Vasiliu, D. A. Dixon, *Angew. Chem., Int. Ed. Engl.* **2017**, 56, 7924–9.

[330]   K. O. Christe, W. W. Wilson, R. D. Wilson, *Inorg. Chem.* **1984**, 23, 2058–63.

[331]   K. Dehnicke, *Angew. Chem.* **1967**, 79, 253–9.

[332]   K. Gholivand, G. Schatte, H. Willner, *Inorg. Chem.* **1987**, 26, 2137–40.

[333]   G. Schatte, H. Willner, *Z. Naturforsch. B* **1991**, 46, 483.

[334]   N. J. S. Peters, L. C. Allen, R. A. Firestone, *Inorg. Chem.* **1988**, 27, 755–8.

[335]   D. Christen, H. G. Mack, G. Schatte, H. Willner, *J. Am. Chem. Soc.* **1988**, 110, 707–12.

[336]   R. Schmutzler, *Angew. Chem., Int. Ed. Engl.* **1968**, 7, 440–55.

[337]   R. R. Smardzewski, W. B. Fox, *J. Chem. Soc., Chem. Commun.* **1974**, 241–2.

[338]   R. R. Smardzewski, W. B. Fox, *J. Am. Chem. Soc.* **1974**, 96, 304–6.

[339]   W. B. Fox, J. S. MacKenzie, E. R. McCarthy, J. R. Holmes, R. F. Stahl, R. Juurik, *Inorg. Chem.* **1968**, 7, 2064–7.

[340]   P. J. Bassett, D. R. Lloyd, *J. Chem. Soc. A* **1971**, 3377–9.

[341]   W. B. Fox, J. S. MacKenzie, N. Vanderkooi, et al., *J. Am. Chem. Soc.* **1966**, 88, 2604–5.

[342]   N. Bartlett, J. Passmore, E. J. Wells, *Chem. Commun.* **1966**, 213–4.

[343]   W. B. Fox, J. S. MacKenzie, R. Vitek, *Inorg. Nucl. Chem. Lett.* **1970**, 6, 177–9.

[344]   K. O. Christe, *J. Am. Chem. Soc.* **1995**, 117, 6136–7.

[345]   W. B. Fox, C. A. Wamser, R. Eibeck, D. K. Huggins, J. S. MacKenzie, R. Juurik, *Inorg. Chem.* **1969**, 8, 1247–9.

[346]   K. O. Christe, W. Maya, *Inorg. Chem.* **1969**, 8, 1253–7.

[347]   C. A. Wamser, W. B. Fox, B. Sukornick, et al., *Inorg. Chem.* **1969**, 8, 1249–53.

[348]   E. C. Curtis, D. Pilipovich, W. H. Moberly, *J. Chem. Phys.* **1967**, 46, 2904–6.

[349]   V. Plato, W. D. Hartford, K. Hedberg, *J. Chem. Phys.* **1970**, 53, 3488–94.

[350]   S. S. Nabiev, V. B. Sokolov, B. B. Chaivanov, *Russ. Chem. Bull.* **2012**, 61, 497–505.

[351]   S. A. Kinkead, J. M. Shreeve, *Inorg. Chem.* **1984**, 23, 4174–7.

[352]   O. D. Gupta, R. L. Kirchmeier, J. M. Shreeve, *Inorg. Chem.* **1990**, 29, 573–4.

[353]   R. D. Wilson, W. Maya, D. Pilipovich, K. O. Christe, *Inorg. Chem.* **1983**, 22, 1355–8.

[354]   N. Vanderkooi, J. S. Mackenzie, W. B. Fox, *J. Fluorine Chem.* **1976**, 7, 415–20.

[355]   K. Nishikida, F. Williams, *J. Am. Chem. Soc.* **1975**, 97, 7166–8.

[356]   R. R. Smardzewski, W. B. Fox, *J. Chem. Phys.* **1974**, 60, 2193–4.

[357]   R. W. Rudolph, R. C. Taylor, R. W. Parry, *J. Am. Chem. Soc.* **1966**, 88, 3729–34.

[358]   M. Lustig, J. K. Ruff, C. B. Colburn, *J. Am. Chem. Soc.* **1966**, 88, 3875.

[359]   C. R. S. Dean, A. Finch, P. J. Gardner, D. W. Payling, *J. Chem. Soc., Faraday Trans.* **1974**, 70, 1921–5.

[360]   D. Solan, P. L. Timms, *Chem. Commun.* **1968**, 1540–1.

[361]   H. L. Hodges, L. S. Su, L. S. Bartell, *Inorg. Chem.* **1975**, 14, 599–603.

[362]   K. H. Rhee, A. M. Snider, F. A. Miller, *Spectrochim. Acta, Part A* **1973**, 29, 1029–35.

[363]   K. W. Morse, R. W. Parry, *J. Am. Chem. Soc.* **1967**, 89, 172–3.

[364]   K. W. Morse, J. G. Morse, *J. Am. Chem. Soc.* **1973**, 95, 8469–70.

[365]   J. G. Morse, K. W. Morse, *Inorg. Chem.* **1975**, 14, 565–9.

[366]   E. R. Falardeau, K. W. Morse, J. G. Morse, *Inorg. Chem.* **1974**, 13, 2333–7.

[367]   J. G. Morse, J. J. Mielcarek, *J. Fluorine Chem.* **1988**, 40, 41–9.

[368]   W. K. Glanville, K. W. Morse, J. G. Morse, *J. Fluorine Chem.* **1976**, 7, 153–8.

[369]   Y. Morino, K. Kuchitsu, T. Moritani, *Inorg. Chem.* **1969**, 8, 867–71.

[370]   A. P. Hagen, B. W. Callaway, *Inorg. Chem.* **1975**, 14, 1622–4.

[371]   A. P. Hagen, B. W. Callaway, *Inorg. Chem.* **1978**, 17, 554–5.

[372]   A. P. Hagen, E. A. Elphingstone, *Inorg. Chem.* **1973**, 12, 478–80.

[373]   K. O. Christe, D. A. Dixon, J. C. P. Sanders, G. J. Schrobilgen, W. W. Wilson, *Inorg. Chem.* **1994**, 33, 4911–9.

[374]   R. D. W. Kemmitt, R. M. McRae, R. D. Peacock, I. L. Wilson, *J. Inorg. Nucl. Chem.* **1969**, 31, 3674–6.

[375]   G. S. H. Chen, J. Passmore, *J. Chem. Soc., Chem. Commun.* **1973**, 559.

[376]   T. Kruck, *Angew. Chem.* **1967**, 79, 27–43.

[377]   T. Kruck, K. Baur, *Angew. Chem., Int. Ed. Engl.* **1965**, 4, 521.

[378]   G. F. Svatos, E. E. Flagg, *Inorg. Chem.* **1965**, 4, 422–3.

[379]   T. Kruck, K. Baur, W. Lang, *Chem. Ber.* **1968**, 101, 138–42.

[380]   T. Kruck, K. Baur, *Z. Anorg. Allg. Chem.* **1969**, 364, 192–208.

[381]   H. S. Booth, S. G. Frary, *J. Am. Chem. Soc.* **1939**, 61, 2934–7.

[382]   H. S. Booth, A. R. Bozarth, *J. Am. Chem. Soc.* **1939**, 61, 2927–34.

[383]   R. W. Rudolph, J. G. Morse, R. W. Parry, *Inorg. Chem.* **1966**, 5, 1464–6.

[384]   G. G. Flaskerud, K. E. Pullen, J. n. M. Shreeve, *Inorg. Chem.* **1969**, 8, 728–30.

[385]   K. Cohn, J. E. Smith, R. Steen, *J. Am. Chem. Soc.* **1970**, 92, 6359–60.

[386]   L. Heuer, K. Bode Ulrike, R. Jones Peter, R. Schmutzler, *Z. Naturforsch. B* **1989**, 44, 1082.

[387]   W. Krüger, R. Schmutzler, H.-M. Schiebel, V. Wray, *Polyhedron* **1989**, 8, 293–300.

[388]   L. Heuer, P. G. Jones, R. Schmutzler, *J. Fluorine Chem.* **1990**, 46, 243–54.

[389]   G. Meyer Thomas, G. Jones Peter, R. Schmutzler, *Z. Naturforsch. B* **1993**, 48, 875.

[390]   T. Chikaraishi, E. Hirota, *Bull. Chem. Soc. Jpn.* **1973**, 46, 2314–6.

[391]   A. J. Edwards, R. J. C. Sills, *J. Chem. Soc. A* **1971**, 942–5.

[392]   P. Klampfer, P. Benkič, A. Lesar, B. Volavšek, M. Ponikvar, A. Jesih, *Collect. Czechoslov. Chem. Commun.* **2004**, 69, 339–50.

[393]   L. Kolditz, *Z. Anorg. Allg. Chem.* **1955**, 280, 313–20.

[394]   H. Preiss, *Z. Anorg. Allg. Chem.* **1971**, 380, 45–50.

[395]   A. J. Edwards, *J. Chem. Soc. A* **1970**, 2751–3.

[396]   A. A. Udovenko, L. A. Zemnukhova, E. V. Kovaleva, Y. E. Gorbunova, Y. N. Mikhailov, *Russ. J. Coord. Chem.* **2004**, 30, 550–4.

[397]   R. R. Ryan, D. T. Cromer, *Inorg. Chem.* **1972**, 11, 2322–4.

[398]   D. R. Schroeder, R. A. Jacobson, *Inorg. Chem.* **1973**, 12, 515–8.

[399]   B. Ducourant, R. Fourcade, *C. R. Acad. Sci., Ser. IIc: Chim.* **1976**, 282C, 741–4.

[400]   R. R. Ryan, S. H. Mastin, *Inorg. Chem.* **1971**, 10, 1757–60.

[401]   R. R. Ryan, S. H. Mastin, A. C. Larson, *Inorg. Chem.* **1971**, 10, 2793–5.

[402]   T. Birchall, P. A. W. Dean, B. D. Valle, R. J. Gillespie, *Can. J. Chem.* **1973**, 51, 667–73.

[403]   A. J. Edwards, D. R. Slim, *J. Chem. Soc., Chem. Commun.* **1974**, 178–9.

[404]   A. J. Hewitt, J. H. Holloway, B. Frlec, *J. Fluorine Chem.* **1975**, 5, 169–71.

[405]   A. Rulmont, *Spectrochim. Acta, Part A* **1972**, 28, 1287–96.

[406]   R. R. Holmes, R. N. Storey, *Inorg. Chem.* **1966**, 5, 2146–50.

[407]   B. Blaser, K.-H. Worms, *Z. Anorg. Allg. Chem.* **1968**, 361, 15–21.

[408]   P. M. Treichel, R. A. Goodrich, S. B. Pierce, *J. Am. Chem. Soc.* **1967**, 89, 2017–22.

[409]   A. H. Cowley, R. W. Braun, *Inorg. Chem.* **1973**, 12, 491–2.

[410]   F. Seel, K. Velleman, *Z. Anorg. Allg. Chem.* **1971**, 385, 123–30.

[411]   R. Minkwitz, A. Liedtke, *Inorg. Chem.* **1989**, 28, 4238–42.

[412]   H. Beckers, *Z. Anorg. Allg. Chem.* **1993**, 619, 1869–79.

[413]   J. F. Nixon, J. R. Swain, *Inorg. Nucl. Chem. Lett.* **1969**, 5, 295–9.

[414]   A. H. Cowley, P. J. Wisian, M. Sanchez, *Inorg. Chem.* **1977**, 16, 1451–5.

[415]   K. O. Christe, C. J. Schack, E. C. Curtis, *Inorg. Chem.* **1976**, 15, 843–8.

[416]   H. W. Roesky, K.-L. Weber, J. Schimkowiak, *Angew. Chem.* **1981**, 93, 1017.

[417]   F. Klapdor Martin, H. Beckers, W. Poll, D. Mootz, *Z. Naturforsch. B* **1997**, 52, 1051.

[418]   R. R. Holmes, C. J. Hora, *Inorg. Chem.* **1972**, 11, 2506–15.

[419]   A. J. Downs, G. S. McGrady, E. A. Barnfield, D. W. H. Rankin, *J. Chem. Soc., Dalton Trans.* **1989**, 545–50.

[420]   J. W. Gilje, R. W. Braun, A. H. Cowley, *J. Chem. Soc., Chem. Commun.* **1974**, 15b–7.

[421]   U. Wannagat, J. Rademachers, *Z. Anorg. Allg. Chem.* **1957**, 289, 66–79.

[422]   K. Kuchitsu, T. Moritani, Y. Morino, *Inorg. Chem.* **1971**, 10, 344–50.

[423]   H. Selig, N. Aminadav, *Inorg. Nucl. Chem. Lett.* **1970**, 6, 595–7.

[424]   E. Lork, B. Görtler, C. Knapp, R. Mews, *Solid State Sci.* **2002**, 4, 1403–11.

[425]   H. Selig, N. Aminadav, *J. Inorg. Nucl. Chem.* **1973**, 35, 3371–3.

[426]   K. O. Christe, D. A. Dixon, G. J. Schrobilgen, W. W. Wilson, *J. Am. Chem. Soc.* **1997**, 119, 3918–28.

[427]   T. E. Thorpe, J. W. Rodger, *J. Chem. Soc. Trans.* **1888**, 53, 766–7.

[428]   H. S. Booth, M. C. Cassidy, *J. Am. Chem. Soc.* **1940**, 62, 2369–72.

[429]   H. S. Booth, C. A. Seabright, *J. Am. Chem. Soc.* **1943**, 65, 1834–5.

[430]   F. Seel, K. Ballreich, R. Schmutzler, *Chem. Ber.* **1962**, 95, 199–202.

[431]   K.-I. Karakida, K. Kuchitsu, *Inorg. Chim. Acta* **1976**, 16, 29–34.

[432]   P. A. W. Dean, R. J. Gillespie, R. Hulme, D. A. Humphreys, *J. Chem. Soc. A* **1971**, 341–6.

[433]   L. Chun-Hsu, H. Selig, M. Rabinovitz, I. Agranat, S. Sarig, *Inorg. Nucl. Chem. Lett.* **1975**, 11, 601–3.

[434]   P. A. W. Dean, *J. Fluorine Chem* **1975**, 5, 499–507.

[435]   C. David Desjardins, J. Passmore, *J. Fluorine Chem.* **1975**, 6, 379–88.

[436]   B. D. Cutforth, C. G. Davies, P. A. W. Dean, R. J. Gillespie, P. R. Ireland, P. K. Ummat, *Inorg. Chem.* **1973**, 12, 1343–7.

[437]   W. L. Lockhart, M. M. Jones, D. O. Johnston, *J. Inorg. Nucl. Chem.* **1969**, 31, 407–14.

[438]   L. Kolditz, M. Gitter, *Z. Chem.* **1967**, 7, 202–3.

[439]   L. Kolditz, M. Gitter, *Z. Chem.* **1967**, 7, 240–1.

[440]   L. Kolditz, M. Gitter, *Z. Anorg. Allg. Chem.* **1967**, 354, 15–22.

[441]   K. Dehnicke, J. Weidlein, *Z. Anorg. Allg. Chem.* **1966**, 342, 225–32.

[442]   A. J. Downs, G. P. Gaskill, S. B. Saville, *Inorg. Chem.* **1982**, 21, 3385–93.

[443]   A. J. Edwards, P. Taylor, *J. Chem. Soc., Chem. Commun.* **1971**, 1376–7.

[444]   T. K. Davies, K. C. Moss, *J. Chem. Soc. A* **1970**, 1054–8.

[445]   J. Bacon, P. A. W. Dean, R. J. Gillespie, *Can. J. Chem.* **1970**, 48, 3413–24.

[446]   E. W. Lawless, *Inorg. Chem.* **1971**, 10, 2084–6.

[447]   L. E. Alexander, *Inorg. Nucl. Chem. Lett.* **1971**, 7, 1053–6.

[448]   L. E. Alexander, I. R. Beattie, *J. Chem. Phys.* **1972**, 56, 5829–31.

[449]   R. J. Gillespie, K. C. Moss, *J. Chem. Soc. A* **1966**, 1170–5.

[450]   J.-C. Culmann, M. Fauconet, R. Jost, J. Sommer, *New J. Chem.* **1999**, 23, 863–7.

[451]   J. M. Lalancette, J. Lafontaine, *J. Chem. Soc., Chem. Commun.* **1973**, 815.

[452]   A. Commeyras, G. A. Olah, *J. Am. Chem. Soc.* **1969**, 91, 2929–42.

[453]   J. Sommer, S. Schwartz, P. Rimmelin, P. Canivet, *J. Am. Chem. Soc.* **1978**, 100, 2576–7.

[454]   G. W. Drake, D. A. Dixon, J. A. Sheehy, J. A. Boatz, K. O. Christe, *J. Am. Chem. Soc.* **1998**, 120, 8392–400.

[455]   J.-Y. Calves, R. J. Gillespie, *J. Chem. Soc., Chem. Commun.* **1976**, 506–7.

[456]   J. Y. Calves, R. J. Gillespie, *J. Am. Chem. Soc.* **1977**, 99, 1788–92.

[457]   R. J. Gillespie, F. G. Riddell, D. R. Slim, *J. Am. Chem. Soc.* **1976**, 98, 8069–72.

[458]   P. A. W. Dean, R. J. Gillespie, *J. Am. Chem. Soc.* **1969**, 91, 7260–4.

[459]   F. Aubke. Stabilization of unusual cations by very weakly basic fluoro anions. In: Inorganic Fluorine Chemistry: American Chemical Society; 1994:350–65.

[460]   J. E. Griffiths, G. E. Walrafen, *Inorg. Chem.* **1972**, 11, 427–9.

[461]   H. Preiss, *Z. Anorg. Allg. Chem.* **1972**, 389, 254–62.

[462]   J. G. Ballard, T. Birchall, D. R. Slim, *J. Chem. Soc., Dalton Trans.* **1977**, 1469–72.

[463]   W. Schmidt, D. Steinborn, L. Kolditz, *Z. Chem.* **1970**, 10, 440.

[464]   J. Fischer, E. Rudzitis, *J. Am. Chem. Soc.* **1959**, 81, 6375–7.

[465]   C. Hebecker, *Z. Anorg. Allg. Chem.* **1971**, 384, 111–4.

[466]   M. J. Vasile, G. R. Jones, W. E. Falconer, *Int. J. Mass Spectrom.* **1973**, 10, 457–69.

[467]   M. J. Vasile, W. E. Falconer, *Inorg. Chem.* **1972**, 11, 2282–3.

[468]   W. E. Falconer, G. R. Jones, W. A. Sunder, et al., *J. Fluorine Chem.* **1974**, 4, 213–34.

[469]   J. H. Holloway, G. M. Staunton, K. Rediess, R. Bougon, D. Brown, *J. Chem. Soc., Dalton Trans.* **1984**, 2163–6.

[470]   R. Bougon, T. Bui Huy, A. Cadet, P. Charpin, R. Rousson, *Inorg. Chem.* **1974**, 13, 690–5.

[471]   B. Družina, B. Žemva, *J. Fluorine Chem.* **1988**, 39, 309–15.

[472]   G. S. H. Chen, J. Passmore, P. Taylor, T. K. Whidden, P. S. White, *J. Chem. Soc., Dalton Trans.* **1985**, 9–16.

[473]   A. I. Popov, A. V. Scharabarin, V. F. Sukhoverkhov, N. A. Tchumaevsky, *Z. Anorg. Allg. Chem.* **1989**, 576, 242–54.

[474]   C. Hebecker, *Z. Anorg. Allg. Chem.* **1970**, 376, 236–44.

[475]   C. Hebecker, *Z. Anorg. Allg. Chem.* **1971**, 384, 12–8.

[476]   K. O. Christe, W. W. Wilson, C. J. Schack, *J. Fluorine Chem.* **1978**, 11, 71–85.

[477]   I. J. Solomon, A. J. Kacmarek, J. Raney, *J. Phys. Chem.* **1968**, 72, 2262–3.

[478]   R. P. Wayne, G. Poulet, P. Biggs, et al., *Atmos. Environ.* **1995**, 29, 2677–881.

[479]   I. V. Nikitin, *Russ. Chem. Rev.* **2008**, 77, 739–49.

[480]   A. Arkell, R. R. Reinhard, L. P. Larson, *J. Am. Chem. Soc.* **1965**, 87, 1016–20.

[481]   L. Andrews, *J. Chem. Phys.* **1972**, 57, 51–5.

[482]   R. R. Smardzewski, W. B. Fox, *J. Chem. Phys.* **1974**, 61, 4933–4.

[483]   L. Andrews, J. I. Raymond, *J. Chem. Phys.* **1971**, 55, 3078–86.

[484]   F. Tamassia, J. M. Brown, S. Saito, *J. Chem. Phys.* **2000**, 112, 5523–6.

[485]   A. D. Kirshenbaum, *Inorg. Nucl. Chem. Lett.* **1965**, 1, 121–3.

[486]   D. Feller, K. A. Peterson, D. A. Dixon, *J. Phys. Chem. A* **2010**, 114, 613–23.

[487]   A. J. Colussi, M. A. Grela, *Chem. Phys. Lett.* **1994**, 229, 134–8.

[488]   J. Sehested, K. Sehested, O. J. Nielsen, T. J. Wallington, *J. Phys. Chem.* **1994**, 98, 6731–9.

[489]   A. Arkell, *J. Am. Chem. Soc.* **1965**, 87, 4057–62.

[490]   R. D. Spratley, J. J. Turner, G. C. Pimentel, *J. Chem. Phys.* **1966**, 44, 2063–8.

[491]   Z. Singh, G. Nagarajan, *Monatsh. Chem.* **1974**, 105, 91–104.

[492]   B. S. Jursic, *Comput. Theor. Chem.* **1996**, 366, 97–101.

[493]   A. V. Grosse, A. G. Streng, A. D. Kirshenbaum, *J. Am. Chem. Soc.* **1961**, 83, 1004–5.

[494]   A. G. Streng, L. V. Streng, *Inorg. Nucl. Chem. Lett.* **1966**, 2, 107–10.

[495]   A. G. Streng, *Can. J. Chem.* **1966**, 44, 1476–9.

[496]   M. Al-Mukhtar, J. H. Holloway, E. G. Hope, *J. Fluorine Chem.* **1992**, 59, 1–4.

[497]   D. J. Gardiner, J. J. Turner, *J. Fluorine Chem.* **1972**, 1, 373–5.

[498]   Z. Mazej, M. Ponikvar-Svet, J. F. Liebman, J. Passmore, H. D. B. Jenkins, *J. Fluorine Chem.* **2009**, 130, 788–91.

[499]   J. N. Keith, I. J. Solomon, I. Sheft, H. H. Hyman, *Inorg. Chem.* **1968**, 7, 230–4.

[500]   K. O. Christe, R. D. Wilson, I. B. Goldberg, *J. Fluorine Chem.* **1976**, 7, 543–9.

[501]   G. M. Lucier, C. Shen, S. H. Elder, N. Bartlett, *Inorg. Chem.* **1998**, 37, 3829–34.

[502]   I. J. Solomon, A. J. Kacmarek, *J. Fluorine Chem.* **1971**, 1, 255–6.

[503]   O. Ruff, W. Menzel, *Z. Anorg. Allg. Chem.* **1933**, 211, 204–8.

[504]   E. H. Appelman, A. W. Jache, *J. Am. Chem. Soc.* **1987**, 109, 1754–7.

[505]   I. V. Nikitin, *Russ. Chem. Rev.* **2004**, 73, 609–19.

[506]   M. C. Lin, S. H. Bauer, *J. Am. Chem. Soc.* **1969**, 91, 7737–42.

[507]   R. Marx, K. Seppelt, *Dalton Trans.* **2015**, 44, 19659–62.

[508]   S. N. Misra, G. H. Cady, *Inorg. Chem.* **1972**, 11, 1132–4.

[509]   F. Neumayr, N. Vanderkooi, *Inorg. Chem.* **1965**, 4, 1234–7.

[510]   L. R. Anderson, W. B. Fox, *J. Am. Chem. Soc.* **1967**, 89, 4313–5.

[511]   A. Šmalc, K. Lutar, *J. Fluorine Chem.* **1977**, 9, 399–408.

[512]   C. T. Goetschel, V. A. Campanile, C. D. Wagner, J. N. Wilson, *J. Am. Chem. Soc.* **1969**, 91, 4702–7.

[513]   T. R. Mills, *J. Fluorine Chem.* **1991**, 52, 267–76.

[514]   P. Frisch, H.-J. Schumacher, *Z. Anorg. Allg. Chem.* **1936**, 229, 423–8.

[515]   A. Šmalc, K. Lutar, J. Slivnik, *J. Fluorine Chem.* **1975**, 6, 287–9.

[516]   R. H. Krech, G. J. Diebold, D. L. McFadden, *J. Am. Chem. Soc.* **1977**, 99, 4605–8.

[517]   J. K. Burdett, D. J. Gardiner, J. J. Turner, R. D. Spratley, P. Tchir, *J. Chem. Soc., Dalton Trans.* **1973**, 1928–32.

[518]   K. R. Loos, C. T. Goetschel, V. A. Campanile, *Chem. Commun.* **1968**, 1633–4.

[519]   L. Hedberg, K. Hedberg, P. G. Eller, R. R. Ryan, *Inorg. Chem.* **1988**, 27, 232–5.

[520]   R. T. Holzmann, M. S. Cohen, *Inorg. Chem.* **1962**, 1, 972–3.

[521]   I. J. Solomon, A. J. Kacmarek, J. N. Keith, J. K. Raney, *J. Am. Chem. Soc.* **1968**, 90, 6557–9.

[522]   A. G. Streng, *J. Am. Chem. Soc.* **1963**, 85, 1380–5.

[523]   S. I. Morrow, A. R. Young, *Inorg. Chem.* **1965**, 4, 759–60.

[524]   I. J. Solomon, A. J. Kacmarek, J. M. McDonlugh, *J. Chem. Eng. Data* **1968**, 13, 529–31.

[525]   T. R. Mills, L. W. Reese, *J. Alloys Compd.* **1994**, 213–214, 360–2.

[526]   P. G. Eller, L. B. Asprey, S. A. Kinkead, B. I. Swanson, R. J. Kissane, *J. Alloys Compd.* **1998**, 269, 63–6.

[527]   A. R. Young, T. Hirata, S. I. Morrow, *J. Am. Chem. Soc.* **1964**, 86, 20–2.

[528]   J. E. Griffiths, A. J. Edwards, W. A. Sunder, W. E. Falconer, *J. Fluorine Chem.* **1978**, 11, 119–42.

[529]   M. H Studier, E. H. Appelman, *J. Am. Chem. Soc.* **1971**, 93, 2349–51.

[530]   J. N. M. Shreeve. Fluorinated Hypofluorites and Hypochlorites. In: H. J. Emeléus, A. G. Sharpe (eds.) Adv. Inorg. Chem: Academic Press; 1983:119–68.

[531]   E. H. Appelman, O. Dunkelberg, M. Kol, *J. Fluorine Chem.* **1992**, 56, 199–213.

[532]   O. Dunkelberg, A. Haas, M. F. Klapdor, D. Mootz, W. Poll, E. H. Appelman, *Chem. Ber.* **1994**, 127, 1871–5.

[533]   S. Rozen, M. Brand, *Angew. Chem., Int. Ed. Engl.* **1986**, 25, 554–5.

[534]   E. H. Appelman, M. A. A. Clyne, *J. Chem. Soc. Faraday Trans.* **1975**, 71, 2072–84.

[535]   H. Kim, E. F. Pearson, E. H. Appelman, *J. Chem. Phys.* **1972**, 56, 1–3.

[536]   W. Poll, G. Pawelke, D. Mootz, E. H. Appelman, *Angew. Chem.* **1988**, 100, 425–6.

[537]   P. Burk, I. A. Koppel, A. Rummel, A. Trummal, *J. Phys. Chem.* **1995**, 99, 1432–5.

[538]   M. N. Glukhovtsev, A. Pross, L. Radom, *J. Phys. Chem.* **1996**, 100, 3498–503.

[539]   E. H. Appelman, A. J. Downs, C. J. Gardner, *J. Phys. Chem.* **1989**, 93, 598–608.

[540]   E. H. Appelman, R. Bonnett, B. Mateen, *Tetrahedron* **1977**, 33, 2119–22.

[541]   R. Bonnett, R. J. Ridge, E. H. Appelman, *J. Chem. Soc., Chem. Commun.* **1978**, 310–1.

[542]   K. G. Migliorese, E. H. Appelman, M. N. Tsangaris, *J. Org. Chem.* **1979**, 44, 1711–4.

[543]   L. E. Andrews, R. Bonnett, E. H. Appelman, *Tetrahedron* **1985**, 41, 781–4.

[544]   R. Beckerbauer, B. E. Smart, Y. Bareket, S. Rozen, *J. Org. Chem.* **1995**, 60, 6186–7.

[545]   M. Kol, S. Rozen, *J. Chem. Soc., Chem. Commun.* **1991**, 567–8.

[546]   S. Rozen, M. Kol, *J. Org. Chem.* **1992**, 57, 7342–4.

[547]   M. H. Hung, B. E. Smart, A. E. Feiring, S. Rozen, *J. Org. Chem.* **1991**, 56, 3187–9.

[548]   S. Rozen, S. Dayan, *Angew. Chem., Int. Ed. Engl.* **1999**, 38, 3471–3.

[549]   G. H. Cady, *J. Am. Chem. Soc.* **1934**, 56, 2635–7.

[550]   O. Ruff, W. Kwasnik, *Angew. Chem.* **1935**, 48, 238–40.

[551]   D. M. Yost, A. Beerbower, *J. Am. Chem. Soc.* **1935**, 57, 782.

[552]   B. Hoge, K. O. Christe, *J. Fluorine Chem.* **2001**, 110, 87–8.

[553]   B. Casper, D. A. Dixon, H.-G. Mack, S. E. Ulic, H. Willner, H. Oberhammer, *J. Am. Chem. Soc.* **1994**, 116, 8317–21.

[554]   B. Tittle, G. H. Cady, *Inorg. Chem.* **1965**, 4, 259–60.

[555]   F. B. Dudley, G. H. Cady, D. F. Eggers, *J. Am. Chem. Soc.* **1956**, 78, 290–2.

[556]   M. Gambaruto, J. E. Sicre, H. J. Schumacher, *J. Fluorine Chem.* **1975**, 5, 175–9.

[557]   K. O. Christe, R. D. Wilson, C. J. Schack, *Inorg. Chem.* **1980**, 19, 3046–9.

[558]   F. Aubke, B. Casper, H. S. P. Müller, H. Oberhammer, H. Willner, *J. Mol. Struct.* **1995**, 346, 111–20.

[559]   F. B. Dudley, *J. Chem. Soc.* **1963**, 1406–14.

[560]   H. Willner, F. Mistry, F. Aubke, *J. Fluorine Chem.* **1992**, 59, 333–49.

[561]   J. E. Roberts, G. H. Cady, *J. Am. Chem. Soc.* **1960**, 82, 354–5.

[562]   W. P. Gilbreath, G. H. Cady, *Inorg. Chem.* **1963**, 2, 496–9.

[563]   J. K. Ruff, M. Lustig, *Inorg. Chem.* **1964**, 3, 1422–5.

[564]   E. H. Appelman, A. W. Jache, *J. Am. Chem. Soc.* **1993**, 115, 1376–8.

[565]   F. B. Dudley, G. H. Cady, D. F. Eggers, *J. Am. Chem. Soc.* **1956**, 78, 1553–7.

[566]   J. K. Ruff, R. Czerepinski, *Inorg. Synth.* **1968**, 11, 131–7.

[567]   S. M. Williamson, G. H. Cady, *Inorg. Chem.* **1962**, 1, 673–7.

[568]   C. I. Merrill, G. H. Cady, *J. Am. Chem. Soc.* **1961**, 83, 298–300.

[569]   W. H. Hale, S. M. Williamson, *Inorg. Chem.* **1965**, 4, 1342–6.

[570]   B. W. Tattershall, G. H. Cady, *J. Inorg. Nucl. Chem.* **1967**, 29, 3003–5.

[571]   A. J. Colussi, H. J. Schumacher, *J. Inorg. Nucl. Chem.* **1971**, 33, 2680–2.

[572]   S. M. Williamson, *Inorg. Chem.* **1963**, 2, 421–2.

[573]   J. R. Case, G. Pass, *J. Chem. Soc.* **1964**, 946–8.

[574]   C. J. Schack, R. D. Wilson, K. O. Christe, *J. Fluorine Chem.* **1989**, 45, 283–91.

[575]   L. Du, B. Elliott, L. Echegoyen, D. D. DesMarteau, *Angew. Chem., Int. Ed. Engl.* **2007**, 46, 6626–8.

[576]   R. A. Crawford, F. B. Dudley, K. Hedberg, *J. Am. Chem. Soc.* **1959**, 81, 5287–8.

[577]   J. E. Smith, G. H. Cady, *Inorg. Chem.* **1970**, 9, 1293–4.

[578]   J. E. Smith, G. H. Cady, *Inorg. Chem.* **1970**, 9, 1442–5.

[579]   G. Mitra, G. H. Cady, *J. Am. Chem. Soc.* **1959**, 81, 2646–8.

[580]   K. Seppelt, *Chem. Ber.* **1973**, 106, 157–64.

[581]   C. J. Schack, W. W. Wilson, K. O. Christe, *Inorg. Chem.* **1983**, 22, 18–21.

[582]   C. J. Schack, K. O. Christe, *Inorg. Chem.* **1984**, 23, 2922.

[583]   C. J. Shack, K. O. Christe, *J. Fluorine Chem.* **1984**, 24, 467–76.

[584]   C. J. Schack, K. O. Christe, *J. Fluorine Chem.* **1988**, 39, 153–62.

[585]   G. H. Rohrback, G. H. Cady, *J. Am. Chem. Soc.* **1947**, 69, 677–8.

[586]   K. O. Christe, W. W. Wilson, R. D. Wilson, *Inorg. Chem.* **1980**, 19, 1494–8.

[587]   K. O. Christe, W. W. Wilson, *J. Fluorine Chem.* **1984**, 26, 257–61.

[588]   K. O. Christe, E. C. Curtis, *Inorg. Chem.* **1982**, 21, 2938–45.

[589]   B. Casper, H.-G. Mack, H. S. P. Mueller, H. Willner, H. Oberhammer, *J. Phys. Chem.* **1994**, 98, 8339–42.

[590]   C. J. Schack, K. O. Christe, *Inorg. Chem.* **1979**, 18, 2619–20.

[591]   S. Rozen, D. Hebel, *J. Org. Chem.* **1990**, 55, 2621–3.

[592]   G. H. Cady, K. B. Kellogg, *J. Am. Chem. Soc.* **1953**, 75, 2501–2.

[593]   S. Rozen, O. Lerman, *J. Am. Chem. Soc.* **1979**, 101, 2782–4.

[594]   S. Rozen, O. Lerman, *J. Org. Chem.* **1980**, 45, 672–8.

[595]   A. Menefee, G. H. Cady, *J. Am. Chem. Soc.* **1954**, 76, 2020–1.

[596]   R. D. Stewart, G. H. Cady, *J. Am. Chem. Soc.* **1955**, 77, 6110–4.

[597]   R. L. Cauble, G. H. Cady, *J. Am. Chem. Soc.* **1967**, 89, 5161–2.

[598]   G. A. Argueello, G. Balzer-Joellenbeck, B. Juelicher, H. Willner, *Inorg. Chem.* **1995**, 34, 603–6.

[599]   G. A. Arguello, B. Juelicher, S. E. Ulic, et al., *Inorg. Chem.* **1995**, 34, 2089–94.

[600]   K. L. Bierbrauer, J. Codnia, M. L. Azcárate, G. A. Argüello, *Inorg. Chem.* **2001**, 40, 2922–4.

[601]   J. H. Prager, *J. Org. Chem.* **1966**, 31, 392–4.

[602]   J. K. Ruff, A. R. Pitochelli, M. Lustig, *J. Am. Chem. Soc.* **1966**, 88, 4531–2.

[603]   M. Wechsberg, G. H. Cady, *J. Am. Chem. Soc.* **1969**, 91, 4432–6.

[604]   G. Pass, H. L. Roberts, *Inorg. Chem.* **1963**, 2, 1016–9.

[605]   G. K. Mulholland, R. E. Ehrenkaufer, *J. Org. Chem.* **1986**, 51, 1482–9.

[606]   P. G. Thompson, *J. Am. Chem. Soc.* **1967**, 89, 1811–3.

[607]   F. A. Hohorst, J. n. M. Shreeve, *J. Am. Chem. Soc.* **1967**, 89, 1809–10.

[608]   R. L. Cauble, G. H. Cady, *J. Am. Chem. Soc.* **1967**, 89, 1962.

[609]   F. A. Hohorst, J. M. Shreeve, S. M. Williamson, *Inorg. Synth.* **1968**, 11, 143–7.

[610]   F. A. Hohorst, J. n. M. Shreeve, *Inorg. Chem.* **1968**, 7, 624–6.

[611]   W. Navarrini, S. Corti, *J. Fluorine Chem.* **2004**, 125, 189–97.

[612]   W. Navarrini, L. Bragante, S. Fontana, V. Tortelli, A. Zedda, *J. Fluorine Chem.* **1995**, 71, 111–7.

[613]   A. Russo, W. Navarrini, *J. Fluorine Chem.* **2004**, 125, 73–8.

[614]   M. J. Fifolt, R. T. Olczak, R. F. Mundhenke, J. F. Bieron, *J. Org. Chem.* **1985**, 50, 4576–82.

[615]   W. H. Bailey III, W. J. Casteel Jr., R. G. Syvret, *Collect. Czechoslov. Chem. Commun.* **2002**, 67, 1416–20.

[616]   R. G. Syvret, W. J. Casteel, G. S. Lal, J. S. Goudar, *J. Fluorine Chem.* **2004**, 125, 33–5.

[617]   A. Russo, D. D. DesMarteau, *Inorg. Chem.* **1995**, 34, 6221–2.

[618]   D. D. Des Marteau, *Inorg. Chem.* **1970**, 9, 2179–81.

[619]   L. R. Anderson, W. B. Fox, *Inorg. Chem.* **1970**, 9, 2182–3.

[620]   L. R. Anderson, D. E. Gould, W. B. Fox, J. K. Ruff, *Inorg. Synth.* **1970**, 12, 312–5.

[621]   I. J. Solomon, A. J. Kacmarek, W. K. Sumida, J. K. Raney, *Inorg. Chem.* **1972**, 11, 195–6.

[622]   D. D. DesMarteau, *Inorg. Chem.* **1972**, 11, 193–5.

[623]   P. G. Thompson, *J. Am. Chem. Soc.* **1967**, 89, 4316–9.

[624]   C. J. Marsden, DesMarteau, L. S. Bartell, *Inorg. Chem.* **1977**, 16, 2359–66.

[625]   J. D. Witt, J. R. Durig, D. Des Marteau, R. M. Hammaker, *Inorg. Chem.* **1973**, 12, 807–10.

[626]   K. I. Gobbato, H. Oberhammer, M. F. Klapdor, et al., *Angew. Chem., Int. Ed. Engl.* **1995**, 34, 2244–5.

[627]   F. A. Hohorst, D. D. DesMarteau, L. R. Anderson, D. E. Gould, W. B. Fox, *J. Am. Chem. Soc.* **1973**, 95, 3866–9.

[628]   F. A. Hohorst, D. D. DesMarteau, *Inorg. Chem.* **1974**, 13, 715–9.

[629]   R. A. DeMarco, W. B. Fox, *Inorg. Nucl. Chem. Lett.* **1974**, 10, 965–71.

[630]   F. A. Hohorst, D. D. DesMarteau, *J. Chem. Soc., Chem. Commun.* **1973**, 386b–7.

[631]   F. A. Hohorst, J. V. Paukstelis, D. D. DesMarteau, *J. Org. Chem.* **1974**, 39, 1298–302.

[632]   F. Seel. Lower Sulfur Fluorides. In: H. J. Eméleus, A. G. Sharpe (eds.) Advances in Inorganic Chemistry and Radiochemistry: Academic Press; 1974:297–333.

[633]   A. Haas, H. Willner, *Z. Anorg. Allg. Chem.* **1980**, 462, 57–60.

[634]   R. R. Smardzewski, W. B. Fox, *J. Fluorine Chem.* **1976**, 7, 353–5.

[635]   H. Willner, *Z. Anorg. Allg. Chem.* **1981**, 481, 117–25.

[636]   D. R. Johnson, F. X. Powell, *Science* **1969**, 164, 950.

[637]   W. Gombler, A. Haas, H. Willner, *Z. Anorg. Allg. Chem.* **1980**, 469, 135–48.

[638]   Y. Steudel, R. Steudel, Ming W. Wong, D. Lentz, *Eur. J. Inorg. Chem.* **2001**, 2001, 2543–8.

[639]   F. Seel, R. Budenz, W. Gombler, *Chem. Ber.* **1970**, 103, 1701–8.

[640]   M. V. Carlowitz, H. Oberhammer, H. Willner, J. E. Boggs, *J. Mol. Struct.* **1983**, 100, 161–77.

[641]   F. Seel, R. Stein, *J. Fluorine Chem.* **1979**, 14, 339–46.

[642]   H. Willner, *Z. Naturforsch. B* **1984**, 39, 314.

[643]   F. Seel, R. Budenz, D. Werner, *Chem. Ber.* **1964**, 97, 1369–72.

[644]   R. Brown, G. Pez, *Aust. J. Chem.* **1967**, 20, 2305–13.

[645]   A. Haas, H. Willner, *Spectrochim. Acta, Part A* **1979**, 35, 953–9.

[646]   F. Seel, R. Budenz, *Chem. Ber.* **1965**, 98, 251–8.

[647]   H. Willner, *Z. Anorg. Allg. Chem.* **1984**, 514, 171–8.

[648]   K.-P. Wanczek, C. Bliefert, R. Budenz, *Z. Naturforsch. A* **1975**, 30, 1156.

[649]   B. Meyer, T. V. Oommen, B. Gotthardt, T. R. Hooper, *Inorg. Chem.* **1971**, 10, 1632–5.

[650]   F. Seel, R. Budenz, W. Gombler, H. Seitter, *Z. Anorg. Allg. Chem.* **1971**, 380, 262–6.

[651]   F. Seel, H. D. Gölitz, *Z. Anorg. Allg. Chem.* **1964**, 327, 32–50.

[652]   O. Glemser, U. Biermann, J. Knaak, A. Haas, *Chem. Ber.* **1965**, 98, 446–50.

[653]   F. Seel, K. P. Wanczek, *Z. Phys. Chem.* **1970**, 72, 109.

[654]   R. L. Kuczkowski, *J. Am. Chem. Soc.* **1964**, 86, 3617–21.

[655]   C. J. Marsden, H. Oberhammer, O. Lösking, H. Willner, *J. Mol. Struct.* **1989**, 193, 233–45.

[656]   A. Haas, H. Willner, *Z. Anorg. Allg. Chem.* **1979**, 454, 17–23.

[657]   V. E. Jackson, D. A. Dixon, K. O. Christe, *Inorg. Chem.* **2012**, 51, 2472–85.

[658]   S. A. Shlykov, H. Oberhammer, A. V. Titov, N. I. Giricheva, G. V. Girichev, *Eur. J. Inorg. Chem.* **2008**, 2008, 5220–7.

[659]   M. El-Hamdi, J. Poater, F. M. Bickelhaupt, M. Solà, *Inorg. Chem.* **2013**, 52, 2458–65.

[660]   D. Naumann, D. K. Padma, *Z. Anorg. Allg. Chem.* **1973**, 401, 53–6.

[661]   C. W. Tullock, F. S. Fawcett, W. C. Smith, D. D. Coffman, *J. Am. Chem. Soc.* **1960**, 82, 539–42.

[662]   G. A. Olah, M. R. Bruce, J. Welch, *Inorg. Chem.* **1977**, 16, 2637.

[663]   A. L. Oppegard, W. C. Smith, E. L. Muetterties, V. A. Engelhardt, *J. Am. Chem. Soc.* **1960**, 82, 3835–8.

[664]   K. Kimura, S. H. Bauer, *J. Chem. Phys.* **1963**, 39, 3172–8.

[665]   W. Gombler, F. Seel, *J. Fluorine Chem.* **1974**, 4, 333–9.

[666]   D. K. Padma, *J. Fluorine Chem.* **1974**, 4, 441–3.

[667]   R. Tunder, B. Siegel, *J. Inorg. Nucl. Chem.* **1963**, 25, 1097–8.

[668]   C. W. Tullock, D. D. Coffman, E. L. Muetterties, *J. Am. Chem. Soc.* **1964**, 86, 357–61.

[669]   K. O. Christe, E. C. Curtis, C. J. Schack, D. Pilipovich, *Inorg. Chem.* **1972**, 11, 1679–82.

[670]   W. Heilemann, R. Mews, S. Pohl, W. Saak, *Chem. Ber.* **1989**, 122, 427–32.

[671]   M. Azeem, R. J. Gillespie, *J. Inorg. Nucl. Chem.* **1966**, 28, 1791–3.

[672]   F. Seel, O. Detmer, *Z. Anorg. Allg. Chem.* **1959**, 301, 113–36.

[673]   M. Azeem, M. Brownstein, R. J. Gillespie, *Can. J. Chem.* **1969**, 47, 4159–67.

[674]   N. Bartlett, P. L. Robinson, *J. Chem. Soc.* **1961**, 3417–25.

[675]   W. C. Smith, V. A. Engelhardt, *J. Am. Chem. Soc.* **1960**, 82, 3838–40.

[676] K. Garber, B. S. Ault, *Inorg. Chem.* **1983**, 22, 2509–13.
[677] C. Lau, J. Passmore, *J. Fluorine Chem.* **1975**, 6, 77–81.
[678] E. B. R. Prideaux, C. B. Cox, *J. Chem. Soc.* **1927**, 928–9.
[679] E. B. R. Prideaux, C. B. Cox, *J. Chem. Soc.* **1928**, 1603–7.
[680] K. Seppelt, D. Lentz, G. Klöter, C. J. Schack, *Inorg. Synth.* **1986**, 24, 27–31.
[681] O. Glemser, F. Meyer, A. Haas, *Sci. Nat.* **1965**, 52, 130.
[682] E. E. Aynsley, R. D. Peacock, P. L. Robinson, *J. Chem. Soc.* **1952**, 1231–4.
[683] P. L. Goggin, *J. Inorg. Nucl. Chem.* **1966**, 28, 661–2.
[684] C. J. Adams, A. J. Downs, *Spectrochim. Acta, Part A* **1972**, 28, 1841–54.
[685] R. Kniep, L. Korte, R. Kryschi, W. Poll, *Angew. Chem.* **1984**, 96, 351–2.
[686] K. Seppelt, *Z. Anorg. Allg. Chem.* **1975**, 416, 12–8.
[687] G. A. Olah, M. Nojima, I. Kerekes, *J. Am. Chem. Soc.* **1974**, 96, 925–7.
[688] N. Bartlett, P. L. Robinson, *J. Chem. Soc.* **1961**, 3549–50.
[689] N. Bartlett, J. W. Quail, *J. Chem. Soc.* **1961**, 3728–32.
[690] M. Brownstein, R. J. Gillespie, *J. Chem. Soc., Dalton Trans.* **1973**, 67–70.
[691] A. R. Mahjoub, D. Leopold, K. Seppelt, *Z. Anorg. Allg. Chem.* **1992**, 618, 83–8.
[692] K. O. Christe, D. A. Dixon, R. Haiges, et al., *J. Fluorine Chem.* **2010**, 131, 791–9.
[693] A. R. Mahjoub, X. Zhang, K. Seppelt, *Chem. Eur. J.* **1995**, 1, 261–5.
[694] R. J. Gillespie, A. Whitla, *Can. J. Chem.* **1970**, 48, 657–63.
[695] A. J. Edwards, G. R. Jones, *J. Chem. Soc. A* **1970**, 1891–4.
[696] T. M. Klapötke, B. Krumm, M. Scherr, R. Haiges, K. O. Christe, *Angew. Chem., Int. Ed. Engl.* **2007**, 46, 8686–90.
[697] S. Fritz, D. Lentz, M. Szwak, *Eur. J. Inorg. Chem.* **2008**, 2008, 4683–6.
[698] S. Fritz, C. Ehm, D. Lentz, *Inorg. Chem.* **2015**, 54, 5220–31.
[699] R. Campbell, P. L. Robinson, *J. Chem. Soc.* **1956**, 776–97.
[700] S. P. Mallela, O. D. Gupta, J. M. Shreeve, *Inorg. Chem.* **1988**, 27, 208–9.
[701] E. B. R. Prideaux, C. B. Cox, *J. Chem. Soc.* **1928**, 739–45.
[702] C. W. Tullock, D. D. Coffman, *J. Org. Chem.* **1960**, 25, 2016–9.
[703] R. Paetzold, K. Aurich, *Z. Anorg. Allg. Chem.* **1962**, 315, 72–8.
[704] J. C. Dewan, A. J. Edwards, *J. Chem. Soc., Dalton Trans.* **1976**, 2433–5.
[705] M. Černík, K. Dostal, *Z. Chem.* **1984**, 24, 192–3.
[706] A. J. Edwards, G. R. Jones, *J. Chem. Soc. A* **1969**, 2858–61.
[707] J. Milne, P. Lahaie, *Inorg. Chem.* **1983**, 22, 2425–8.
[708] H. Selig, U. El-gad, *J. Inorg. Nucl. Chem.* **1976**, 28, 233–4.
[709] J. B. Milne, D. Moffett, *Inorg. Chem.* **1973**, 12, 2240–4.
[710] L. Guillet, A. Ider, J. P. Laval, B. Frit, *J. Fluorine Chem.* **1999**, 93, 33–8.
[711] J. H. Moss, R. Ottie, J. B. Wilford, *J. Fluorine Chem.* **1974**, 3, 317–22.
[712] G. A. R. Hartley, T. H. Henry, R. Whytlaw-Gray, *Nature* **1938**, 142, 952.
[713] J. Carre, P. Claudy, M. Kollmannsberger, J. Bousquet, E. Garnier, P. Barberi, *J. Fluorine Chem.* **1978**, 11, 613–27.
[714] J. Carre, P. Germain, J. Thourey, G. Perachon, *J. Fluorine Chem.* **1986**, 31, 241–5.
[715] P. Germain, G. Perachon, J. M. Letoffe, P. Claudy, *Thermochim. Acta* **1987**, 119, 243–8.
[716] J. H. Junkins, H. A. Bernhardt, E. J. Barber, *J. Am. Chem. Soc.* **1952**, 74, 5749–51.
[717] S. A. Shlykov, N. I. Giricheva, A. V. Titov, M. Szwak, D. Lentz, G. V. Girichev, *Dalton Trans.* **2010**, 39, 3245–55.
[718] A. J. Edwards, F. I. Hewaidy, *J. Chem. Soc. A* **1968**, 2977–80.
[719] N. N. Greenwood, A. C. Sarma, B. P. Straughan, *J. Chem. Soc. A* **1968**, 1561–3.
[720] J. C. Jumas, M. Maurin, E. Philippot, *J. Fluorine Chem.* **1977**, 10, 219–30.
[721] A. Haas, M. Pryka, *J. Chem. Soc., Chem. Commun.* **1993**, 993–4.

[722]   L. B. Asprey, N. A. Matwiyoff, *J. Inorg. Nucl. Chem.* **1976**, 28, 123–5.

[723]   J. Thourey, S. Bendaoud, G. Perachon, *J. Fluorine Chem.* **1987**, 36, 395–405.

[724]   L. Guillet, J. P. Laval, B. Frit, *J. Fluorine Chem.* **1997**, 85, 177–81.

[725]   A. J. Edwards, P. Taylor, *J. Chem. Soc., Dalton Trans.* **1973**, 2150–3.

[726]   T. M. Klapötke, B. Krumm, P. Mayer, I. Schwab, *Angew. Chem., Int. Ed. Engl.* **2003**, 42, 5843–6.

[727]   R. Haiges, J. A. Boatz, A. Vij, et al., *Angew. Chem., Int. Ed. Engl.* **2003**, 42, 5847–51.

[728]   T. M. Klapötke, B. Krumm, J. C. Gálvez-Ruiz, H. Nöth, I. Schwab, *Eur. J. Inorg. Chem.* **2004**, 2004, 4764–9.

[729]   D. Lentz, M. Szwak, *Angew. Chem., Int. Ed. Engl.* **2005**, 44, 5079–82.

[730]   H. Moissan, P. Lebeau, *C. R. Acad. Sci., Ser. IIc: Chim.* **1900**, 130C, 865–71.

[731]   A. A. Opalovskii, E. U. Lobkov, *Russ. Chem. Rev.* **1975**, 44, 97–109.

[732]   K. Seppelt, *Chem. Rev.* **2015**, 115, 1296–306.

[733]   T. Kitazume, J. n. M. Shreeve, *J. Chem. Soc., Chem. Commun.* **1976**, 982–3.

[734]   T. Kitazume, J. n. M. Shreeve, *J. Am. Chem. Soc.* **1977**, 99, 3690–5.

[735]   J. Wessel, G. Kleemann, K. Seppelt, *Chem. Ber.* **1983**, 116, 2399–407.

[736]   H. L. Deubner, F. Kraus, *Inorganics* **2017**, 5, 68.

[737]   A. P. Hagen, D. J. Jones, S. R. Ruttman, *J. Inorg. Nucl. Chem.* **1974**, 36, 1217–9.

[738]   A. P. Hagen, B. W. Callaway, *Inorg. Chem.* **1975**, 14, 2825–7.

[739]   R. Winter, P. G. Nixon, G. L. Gard, *J. Fluorine Chem.* **1998**, 87, 85–6.

[740]   B. Cohen, A. G. MacDiarmid, *Inorg. Chem.* **1965**, 4, 1782–5.

[741]   C. J. Schack, R. D. Wilson, M. G. Warner, *J. Chem. Soc., Chem. Commun.* **1969**, 1110a–a.

[742]   T. A. Kovacina, A. D. Berry, W. B. Fox, *J. Fluorine Chem.* **1976**, 7, 430–2.

[743]   R. Winter, R. J. Terjeson, G. L. Gard, *J. Fluorine Chem.* **1998**, 89, 105–6.

[744]   J. I. Darragh, D. W. A. Sharp, *J. Chem. Soc., Chem. Commun.* **1969**, 864b–5.

[745]   P. R. Savoie, J. T. Welch, *Chem. Rev.* **2015**, 115, 1130–90.

[746]   S. Altomonte, M. Zanda, *J. Fluorine Chem.* **2012**, 143, 57–93.

[747]   N. Lu, H. P. S. Kumar, J. L. Fye, et al., *Angew. Chem., Int. Ed. Engl.* **2006**, 45, 938–40.

[748]   N. Lu, J. S. Thrasher, S. von Ahsen, H. Willner, D. Hnyk, H. Oberhammer, *Inorg. Chem.* **2006**, 45, 1783–8.

[749]   C. J. Marsden, L. S. Bartell, *Inorg. Chem.* **1976**, 15, 3004–9.

[750]   R. E. Noftle, R. R. Smardzewski, W. B. Fox, *Inorg. Chem.* **1977**, 16, 3380–1.

[751]   D. Christen, H. G. Mack, H. Oberhammer, *J. Chem. Phys.* **1987**, 87, 2001–5.

[752]   K. Seppelt, *Z. Anorg. Allg. Chem.* **1971**, 386, 229–31.

[753]   H. Jonas, *Z. Anorg. Allg. Chem.* **1951**, 265, 273–83.

[754]   C. J. Schack, R. D. Wilson, *Inorg. Chem.* **1970**, 9, 311–4.

[755]   G. Gundersen, K. Hedberg, *J. Chem. Phys.* **1969**, 51, 2500–7.

[756]   M. Lustig, J. K. Ruff, *Inorg. Chem.* **1967**, 6, 2115–7.

[757]   K. O. Christe, C. J. Schack, D. Pilipovich, E. C. Curtis, W. Sawodny, *Inorg. Chem.* **1973**, 12, 620–2.

[758]   K. Seppelt, *Angew. Chem.* **1982**, 94, 890–901.

[759]   M. Brownstein, P. A. W. Dean, R. J. Gillespie, *J. Chem. Soc., Chem. Commun.* **1970**, 9.

[760]   C. Lau, H. Lynton, J. Passmore, P.-Y. Siew, *J. Chem. Soc., Dalton Trans.* **1973**, 2535–8.

[761]   D. S. Ross, D. W. A. Sharp, *J. Chem. Soc., Dalton Trans.* **1972**, 34–7.

[762]   S. Li, P. Wu, J. E. Moses, K. B. Sharpless, *Angew. Chem., Int. Ed. Engl.* **2017**, 56, 2903–8.

[763]   E. Lork, M. Müller, J. Wessel, R. Mews, T. Borrmann, W. D. Stohrer, *J. Fluorine Chem.* **2001**, 112, 247–58.

[764]   B. Gao, S. Li, P. Wu, J. E. Moses, K. B. Sharpless, *Angew. Chem., Int. Ed. Engl.* **2018**, 57, 1939–43.

[765]   D. Mootz, A. Merschenz-Quack, *Acta Crystallogr. Sect. C: Cryst. Struct. Commun.* **1988**, 44, 924–5.

[766]   T. Birchall, R. J. Gillespie, *Spectrochim. Acta* **1966**, 22, 681–8.

[767]   K. Hagen, V. R. Cross, K. Hedberg, *J. Mol. Struct.* **1978**, 44, 187–93.

[768]   Z. Liu, J. Li, S. Li, G. Li, K. B. Sharpless, P. Wu, *J. Am. Chem. Soc.* **2018**, 140, 2919–25.

[769]   H. S. Booth, C. V. Herrmann, *J. Am. Chem. Soc.* **1936**, 58, 63–6.

[770]   H. J. Eméleus, J. F. Wood, *J. Chem. Soc.* **1948**, 2183–8.

[771]   D. K. Padma, V. Subrahmanya Bhat, A. R. Vasudevamurthy, *J. Fluorine Chem.* **1978**, 11, 187–90.

[772]   F. Seel, L. C. Duncan, R. G. Czerepinski, G. H. Cady, *Inorg. Synth.* **1967**, 9, 111–3.

[773]   J. A. D. Stockdale, R. N. Compton, H. C. Schweinler, *J. Chem. Phys.* **1970**, 53, 1502–7.

[774]   C. J. Schack, R. D. Wilson, J. F. Hon, *Inorg. Chem.* **1972**, 11, 208–9.

[775]   S. Colton, J. L. Margrave, P. W. Wilson, *Synth. React. Inorg. Met.-Org. Chem.* **1971**, 1, 149–54.

[776]   K. O. Christe, C. J. Schack, E. C. Curtis, *Inorg. Chem.* **1972**, 11, 583–6.

[777]   T. Grelbig, T. Krügerke, K. Seppelt, *Z. Anorg. Allg. Chem.* **1987**, 544, 74–80.

[778]   K. Seppelt, *Angew. Chem.* **1974**, 86, 103.

[779]   K. Seppelt, *Z. Anorg. Allg. Chem.* **1974**, 406, 287–98.

[780]   K. Seppelt, *Angew. Chem.* **1974**, 86, 104.

[781]   H. Oberhammer, K. Seppelt, *Inorg. Chem.* **1979**, 18, 2226–9.

[782]   M. J. Vasile, F. A. Stevie, K. Seppelt, *J. Fluorine Chem.* **1979**, 13, 487–99.

[783]   A. Engelbrecht, B. Stoll, *Z. Anorg. Allg. Chem.* **1957**, 292, 20–4.

[784]   J. Toužín, L. Mitáček, *Z. Chem.* **1980**, 20, 32–3.

[785]   H.-G. Jerschkewitz, *Angew. Chem.* **1957**, 69, 562.

[786]   A. J. Blake, D. Bevilacqua, M. Černík, Z. Žák, *J. Chem. Soc., Dalton Trans.* **1995**, 689–93.

[787]   K. Dostál, J. Šikola, *Z. Chem.* **1983**, 23, 185.

[788]   L. Kolditz, I. Fitz, *Z. Anorg. Allg. Chem.* **1967**, 349, 175–83.

[789]   G. W. Fraser, G. D. Meikle, *J. Chem. Soc., Chem. Commun.* **1974**, 624–5.

[790]   U. Elgad, H. Selig, *Inorg. Chem.* **1975**, 14, 140–5.

[791]   W. Tötsch, F. Sladky, *Z. Naturforsch. B* **1983**, 38, 1025.

[792]   G. W. Fraser, J. B. Millar, *J. Chem. Soc., Chem. Commun.* **1972**, 1113.

[793]   G. W. Fraser, J. B. Millar, *J. Chem. Soc., Dalton Trans.* **1974**, 2029–31.

[794]   G. W. Fraser, G. D. Meikle, *J. Chem. Soc., Dalton Trans.* **1975**, 1033–6.

[795]   G. W. Fraser, G. D. Meikle, *J. Chem. Soc., Dalton Trans.* **1977**, 1985–7.

[796]   G. W. Fraser, G. D. Meikle, *J. Chem. Soc., Perkin Trans. 2* **1975**, 312–4.

[797]   K. O. Christe, D. A. Dixon, J. C. P. Sanders, G. J. Schrobilgen, W. W. Wilson, *J. Am. Chem. Soc.* **1993**, 115, 9461–7.

[798]   K. O. Christe, J. C. P. Sanders, G. J. Schrobilgen, W. W. Wilson, *J. Chem. Soc., Chem. Commun.* **1991**, 837–40.

[799]   A.-R. Mahjoub, T. Drews, K. Seppelt, *Angew. Chem., Int. Ed. Engl.* **1992**, 31, 1036–9.

[800]   A.-R. Mahjoub, K. Seppelt, *J. Chem. Soc., Chem. Commun.* **1991**, 840–1.

[801]   X. Zhang, K. Seppelt, *Z. Anorg. Allg. Chem.* **1997**, 623, 491–500.

[802]   H. Selig, S. Sarig, S. Abramowitz, *Inorg. Chem.* **1974**, 13, 1508–11.

[803]   G. W. Fraser, R. D. Peacock, P. M. Watkins, *Chem. Commun.* **1968**, 1257a–a.

[804]   L. Lawlor, J. Passmore, *Inorg. Chem.* **1979**, 18, 2921–3.

[805]   C. Lau, J. Passmore, *Inorg. Chem.* **1974**, 13, 2278–9.

[806]   M. Murchie, J. Passmore, K. Seppelt, *Inorg. Synth.* **1986**, 24, 31–3.

[807]   L. J. Lawlor, J. Passmore, *Can. J. Chem.* **1984**, 62, 1477–82.

[808]   L. J. Lawlor, A. Martin, M. P. Murchie, J. Passmore, J. C. P. Sanders, *Can. J. Chem.* **1989**, 67, 1501–5.

[809]   H. Kropshofer, O. Leitzke, P. Peringer, F. Sladky, *Chem. Ber.* **1981**, 114, 2644–8.

[810]   P. K. Miller, K. D. Abney, A. K. Rappe, O. P. Anderson, S. H. Strauss, *Inorg. Chem.* **1988**, 27, 2255–61.

[811]   O. Redlich, E. A. Coulson, E. F. G. Herington, et al., *J. Chem. Soc.* **1948**, 1987–95.

[812]   B. Weinstock, C. L. Chernick, *J. Am. Chem. Soc.* **1960**, 82, 4116–7.

[813]   K. Seppelt, *Angew. Chem.* **1976**, 88, 56–7.

[814]   K. Seppelt, *Z. Anorg. Allg. Chem.* **1977**, 428, 35–42.

[815]   K. Seppelt, *Angew. Chem.* **1972**, 84, 212.

[816]   A. Engelbrecht, F. Sladky, *Angew. Chem.* **1964**, 76, 379–80.

[817]   A. Engelbrecht, F. Sladky, *Monatsh. Chem.* **1965**, 96, 159–68.

[818]   F. Sladky, C. J. Schack, *Inorg. Synth.* **1986**, 24, 33–7.

[819]   K. Seppelt, *Angew. Chem.* **1972**, 84, 715–6.

[820]   E. Jacob, D. Lentz, K. Seppelt, A. Simon, *Z. Anorg. Allg. Chem.* **1981**, 472, 7–25.

[821]   C. J. Schack, K. O. Christe, *J. Fluorine Chem.* **1982**, 21, 393–6.

[822]   J. C. P. Sanders, G. J. Schrobilgen, *J. Chem. Soc., Chem. Commun.* **1989**, 1576–8.

[823]   J. R. DeBackere, H. P. A. Mercier, G. J. Schrobilgen, *J. Am. Chem. Soc.* **2014**, 136, 3888–903.

[824]   K. Seppelt, *Chem. Ber.* **1976**, 109, 1046–52.

[825]   L. K. Templeton, D. H. Templeton, N. Bartlett, K. Seppelt, *Inorg. Chem.* **1976**, 15, 2720–2.

[826]   D. Lentz, H. Pritzkow, K. Seppelt, *Angew. Chem.* **1977**, 89, 741.

[827]   M. D. Noirot, O. P. Anderson, S. H. Strauss, *Inorg. Chem.* **1987**, 26, 2216–23.

[828]   D. M. Van Seggen, P. K. Hurlburt, M. D. Noirot, O. P. Anderson, S. H. Strauss, *Inorg. Chem.* **1992**, 31, 1423–30.

[829]   H. P. A. Mercier, J. C. P. Sanders, G. J. Schrobilgen, *J. Am. Chem. Soc.* **1994**, 116, 2921–37.

[830]   D. M. Van Seggen, P. K. Hurlburt, O. P. Anderson, S. H. Strauss, *Inorg. Chem.* **1995**, 34, 3453–64.

[831]   A. Wiesner, T. W. Gries, S. Steinhauer, H. Beckers, S. Riedel, *Angew. Chem., Int. Ed. Engl.* **2017**, 56, 8263–6.

[832]   S. Hämmerling, G. Thiele, S. Steinhauer, H. Beckers, C. Müller, S. Riedel, *Angew. Chem., Int. Ed. Engl.* **2019**, 58, 9807–10.

[833]   A. Wiesner, L. Fischer, S. Steinhauer, H. Beckers, S. Riedel, *Chem. Eur. J.* **2019**, 25, 10441–9.

[834]   O. Glemser, R. Mews. Sulfur-Nitrogen-Fluorine Compounds. In: H. J. Emeléus, A. G. Sharpe (eds.) Advances in Inorganic Chemistry and Radiochemistry: Academic Press; 1972:333–90.

[835]   R. Mews. Nitrogen-Sulfur-Fluorine Ions. In: H. J. Emeléus, A. G. Sharpe (eds.) Advances in Inorganic Chemistry and Radiochemistry: Academic Press; 1976:185–237.

[836]   R. Mews, D.-L. Wagner, O. Glemser, *Z. Anorg. Allg. Chem.* **1975**, 412, 148–54.

[837]   B. Krebs, S. Pohl, O. Glemser, *J. Chem. Soc., Chem. Commun.* **1972**, 548.

[838]   B. Krebs, S. Pohl, *Chem. Ber.* **1973**, 106, 1069–75.

[839]   D. F. Smith, *Science* **1963**, 141, 1039.

[840]   S. A. Kinkead, L. B. Asprey, P. G. Eller, *J. Fluorine Chem.* **1985**, 29, 459–62.

[841]   K. S. Thanthiriwatte, M. Vasiliu, D. A. Dixon, K. O. Christe, *Inorg. Chem.* **2012**, 51, 10966–82.

[842]   O. Ruff, A. Braida, *Z. Anorg. Allg. Chem.* **1933**, 214, 81–90.

[843]   D. Naumann, E. Lehmann, *J. Fluorine Chem.* **1975**, 5, 307–21.

[844]   M. Schmeisser, E. Scharf, *Angew. Chem.* **1960**, 72, 324.

[845]   M. Schmeißer, P. Sartori, D. Naumann, *Chem. Ber.* **1970**, 103, 590–3.

[846]   D. Naumann, E. Renk, E. Lehmann, *J. Fluorine Chem.* **1977**, 10, 395–403.

[847]   M. Schmeisser, D. Naumann, R. Scheele, *J. Fluorine Chem.* **1972**, 1, 369–72.

[848]   M. Schmeisser, D. Naumann, E. Lehmann, *J. Fluorine Chem.* **1974**, 3, 441–4.

[849]   H. H. Hyman, T. Surles, L. A. Quarterman, A. I. Popov, *Inorg. Chem.* **1970**, 9, 2726–30.

[850]   S. Hoyer, K. Seppelt, *Angew. Chem., Int. Ed. Engl.* **2000**, 39, 1448–9.

[851]   E. Lehmann, D. Naumann, M. Schmeisser, *J. Fluorine Chem.* **1976**, 7, 33–42.

[852]   P. Sartori, A. J. Lehnen, *Chem. Ber.* **1971**, 104, 2813–20.

[853]   C. J. Schack, *Inorg. Chem.* **1967**, 6, 1938–9.

[854]   K. O. Christe, *Inorg. Chem.* **1972**, 11, 1220–2.

[855]   K. O. Christe, W. W. Wilson, R. D. Wilson, *Inorg. Chem.* **1989**, 28, 675–7.

[856]   S. I. Ivlev, M. R. Buchner, A. J. Karttunen, F. Kraus, *J. Fluorine Chem.* **2018**, 215, 17–24.

[857]   R. J. Gillespie, P. H. Spekkens, *J. Chem. Soc., Dalton Trans.* **1977**, 1539–46.

[858]   W. Breuer, H. J. Frohn, *Z. Anorg. Allg. Chem.* **1993**, 619, 209–14.

[859]   M. Schmeißer, P. Sartor, D. Naumann, *Chem. Ber.* **1970**, 103, 312.

[860]   G. Oates, J. M. Winfield, *Inorg. Nucl. Chem. Lett.* **1972**, 8, 1093–5.

[861]   G. Oates, J. M. Winfield, O. R. Chambers, *J. Chem. Soc., Dalton Trans.* **1974**, 1380–4.

[862]   E. Lehmann, D. Naumann, W. Stopschinski, *Spectrochim. Acta, Part A* **1975**, 31, 1905–11.

[863]   H. J. Frohn, K. Schrinner, *Z. Anorg. Allg. Chem.* **1997**, 623, 1847–9.

[864]   O. Ruff, *Angew. Chem.* **1934**, 47, 480.

[865]   E. E. Aynsley, R. Nichols, P. L. Robinson, *J. Chem. Soc.* **1953**, 623–6.

[866]   H.-J. Frohn, V. V. Bardin, *J. Fluorine Chem.* **2010**, 131, 1000–6.

[867]   T. Surles, L. A. Quarterman, H. H. Hyman, *J. Fluorine Chem.* **1974**, 3, 453–6.

[868]   E. D. Whitney, R. O. MacLaren, C. E. Fogle, T. J. Hurley, *J. Am. Chem. Soc.* **1964**, 86, 2583–6.

[869]   K. O. Christe, W. Sawodny, *Inorg. Chem.* **1969**, 8, 212–9.

[870]   R. J. Gillespie, M. J. Morton, *Inorg. Chem.* **1970**, 9, 811–4.

[871]   H. Lynton, J. Passmore, *Can. J. Chem.* **1971**, 49, 2539–43.

[872]   A. J. Edwards, R. J. C. Sills, *J. Chem. Soc. A* **1970**, 2697–9.

[873]   A. J. Edwards, G. R. Jones, *J. Chem. Soc. A* **1969**, 1467–70.

[874]   S. I. Ivlev, A. J. Karttunen, M. R. Buchner, M. Conrad, F. Kraus, *Angew. Chem., Int. Ed. Engl.* **2018**, 57, 14640–4.

[875]   K. O. Christe, E. C. Curtis, C. J. Schack, S. J. Cyvin, J. Brunvoll, W. Sawodny, *Spectrochim. Acta, Part A* **1976**, 32, 1141–7.

[876]   E. J. Baran, *J. Fluorine Chem.* **1981**, 17, 543–8.

[877]   K. O. Christe, X. Zhang, J. A. Sheehy, R. Bau, *J. Am. Chem. Soc.* **2001**, 123, 6338–48.

[878]   A. Vij, F. S. Tham, V. Vij, W. W. Wilson, K. O. Christe, *Inorg. Chem.* **2002**, 41, 6397–403.

[879]   K. O. Christe, W. Sawodny, *Inorg. Chem.* **1967**, 6, 1783–8.

[880]   F. A. Hohorst, L. Stein, E. Gebert, *Inorg. Chem.* **1975**, 14, 2233–6.

[881]   J. F. Liebman, *J. Fluorine Chem.* **1977**, 9, 147–51.

[882]   J. F. Lehmann, G. J. Schrobilgen, K. O. Christe, A. Kornath, R. J. Suontamo, *Inorg. Chem.* **2004**, 43, 6905–21.

[883]   R. J. Gillespie, G. J. Schrobilgen, *Inorg. Chem.* **1974**, 13, 1230–5.

[884]   K. O. Christe, R. D. Wilson, *Inorg. Chem.* **1975**, 14, 694–6.

[885]   F. Q. Roberto, *Inorg. Nucl. Chem. Lett.* **1972**, 8, 737–40.

[886]   K. O. Christe, *Inorg. Nucl. Chem. Lett.* **1972**, 8, 741–5.

[887]   K. O. Christe, *Inorg. Chem.* **1973**, 12, 1580–7.

[888]   K. O. Christe, W. W. Wilson, E. C. Curtis, *Inorg. Chem.* **1983**, 22, 3056–60.

[889]   J. Shamir, *Isr. J. Chem.* **1978**, 17, 37–47.

[890]   K. O. Christe, W. Sawodny, J. P. Guertin, *Inorg. Chem.* **1967**, 6, 1159–62.

[891]   P. Pröhm, J. R. Schmid, K. Sonnenberg, et al., *Angew. Chem., Int. Ed. Engl.* **2020**, 59, 16002–6.

[892]   T. Surles, L. A. Quarterman, H. H. Hyman, *J. Inorg. Nucl. Chem.* **1973**, 35, 668–70.

[893]   H. Meinert, *Z. Chem.* **1967**, 7, 41–57.

[894]   R. Minkwitz, R. Bröchler, R. Ludwig, *Inorg. Chem.* **1997**, 36, 4280–3.

[895]   D. Naumann, A. Meurer, *J. Fluorine Chem.* **1995**, 70, 83–4.

[896] K. O. Christe, W. W. Wilson, G. W. Drake, M. A.Petrie, J. A. Boatz, *J. Fluorine Chem.* **1998**, 88, 185–9.

[897] E. D. Whitney, R. O. MacLaren, T. J. Hurley, C. E. Fogle, *J. Am. Chem. Soc.* **1964**, 86, 4340–2.

[898] S. Ivlev, P. Woidy, V. Sobolev, I. Gerin, R. Ostvald, F. Kraus, *Z. Anorg. Allg. Chem.* **2013**, 639, 2846–50.

[899] S. I. Ivlev, A. J. Karttunen, R. Ostvald, F. Kraus, *Z. Anorg. Allg. Chem.* **2015**, 641, 2593–8.

[900] S. I. Ivlev, R. V. Ostvald, F. Kraus, *Monatsh. Chem.* **2016**, 147, 1661–8.

[901] B. Scheibe, A. J. Karttunen, U. Müller, F. Kraus, *Angew. Chem., Int. Ed. Engl.* **2020**, 59, 18116–9.

[902] S. Ivlev, V. Sobolev, M. Hoelzel, et al., *Eur. J. Inorg. Chem.* **2014**, 2014, 6261–7.

[903] S. Ivlev, P. Woidy, F. Kraus, I. Gerin, R. Ostvald, *Eur. J. Inorg. Chem.* **2013**, 2013, 4984–7.

[904] S. I. Ivlev, A. V. Malin, A. J. Karttunen, R. V. Ostvald, F. Kraus, *J. Fluorine Chem.* **2019**, 218, 11–20.

[905] K. O. Christe, J. P. Guertin, *Inorg. Chem.* **1966**, 5, 473–6.

[906] A. R. Mahjoub, A. Hoser, J. Fuchs, K. Seppelt, *Angew. Chem.* **1989**, 101, 1528–9.

[907] S. I. Ivlev, A. J. Karttunen, R. V. Ostvald, F. Kraus, *Chem. Commun.* **2016**, 52, 12040–3.

[908] K. O. Christe, W. W. Wilson, G. W. Drake, D. A. Dixon, J. A. Boatz, R. Z. Gnann, *J. Am. Chem. Soc.* **1998**, 120, 4711–6.

[909] K. O. Christe, W. W. Wilson, *Inorg. Chem.* **1989**, 28, 3275–7.

[910] J. Pilmé, E. A. Robinson, R. J. Gillespie, *Inorg. Chem.* **2006**, 45, 6198–204.

[911] D. A. Dixon, D. J. Grant, K. O. Christe, K. A. Peterson, *Inorg. Chem.* **2008**, 47, 5485–94.

[912] K. O. Christe, D. Pilipovich, *Inorg. Chem.* **1969**, 8, 391–3.

[913] C. J. Adams, *Inorg. Nucl. Chem. Lett.* **1974**, 10, 831–5.

[914] F. Seel, M. Pimpl, *J. Fluorine Chem.* **1977**, 10, 413–30.

[915] K. O. Christe, W. W. Wilson, R. D. Wilson, *Inorg. Chem.* **1989**, 28, 904–8.

[916] A.-R. Mahjoub, K. Seppelt, *Angew. Chem.* **1991**, 103, 844–5.

[917] T. Surles, L. A. Quarterman, H. H. Hyman, *J. Fluorine Chem.* **1974**, 3, 293–306.

[918] H. H. Hyman, T. Surles, L. A. Quarterman, A. I. Popov, *Inorg. Chem.* **1971**, 10, 913–6.

[919] M. Brownstein, J. Shamir, *Can. J. Chem.* **1972**, 50, 3409–10.

[920] R. Bougon, M. Carles, J. Aubert, *C. R. Acad. Sci., Ser. IIc: Chim.* **1967**, 265C, 179–82.

[921] T. D. Cooper, F. N. Dost, C. H. Wang, *J. Inorg. Nucl. Chem.* **1972**, 34, 3564–7.

[922] L. Andrews, F. K. Chi, A. Arkell, *J. Am. Chem. Soc.* **1974**, 96, 1997–2000.

[923] D. Pilipovich, C. B. Lindahl, C. J. Schack, R. D. Wilson, K. O. Christe, *Inorg. Chem.* **1972**, 11, 2189–92.

[924] R. Bougon, J. Isabey, P. Plurien, *C. R. Acad. Sci., Ser. IIc: Chim.* **1970**, 271C, 1366–9.

[925] D. Pilipovich, H. H. Rogers, R. D. Wilson, *Inorg. Chem.* **1972**, 11, 2192–5.

[926] A. Ellern, J. A. Boatz, K. O. Christe, T. Drews, K. Seppelt, *Z. Anorg. Allg. Chem.* **2002**, 628, 1991–9.

[927] K. O. Christe, E. C. Curtis, *Inorg. Chem.* **1972**, 11, 2196–201.

[928] H. Oberhammer, K. O. Christe, *Inorg. Chem.* **1982**, 21, 273–5.

[929] K. O. Christe, E. C. Curtis, C. J. Schack, *Inorg. Chem.* **1972**, 11, 2212–5.

[930] K. O. Christe, C. J. Schack, D. Pilipovich, *Inorg. Chem.* **1972**, 11, 2205–8.

[931] K. Züchner, O. Glemser, *Angew. Chem.* **1972**, 84, 1147–9.

[932] C. J. Schack, C. B. Lindahl, D. Pilipovich, K. O. Christe, *Inorg. Chem.* **1972**, 11, 2201–5.

[933] R. Bougon, T. Bui Huy, *C. R. Acad. Sci., Ser. IIc: Chim.* **1976**, 283C, 461–3.

[934] W. W. Wilson, K. O. Christe, *Inorg. Chem.* **1987**, 26, 916–9.

[935] K. O. Christe, E. C. Curtis, R. Bougon, *Inorg. Chem.* **1978**, 17, 1533–9.

[936] R. Bougon, T. B. Huy, P. Charpin, R. J. Gillespie, P. H. Spekkens, *J. Chem. Soc., Dalton Trans.* **1979**, 6–12.

[937] K. O. Christe, R. D. Wilson, E. C. Curtis, W. Kuhlmann, W. Sawodny, *Inorg. Chem.* **1978**, 17, 533–8.

[938] K. O. Christe, W. W. Wilson, *Inorg. Chem.* **1986**, 25, 1904–6.

[939] O. Ruff, A. Braida, *Z. Anorg. Allg. Chem.* **1934**, 220, 43–8.

[940] E. E. Aynsley, *J. Chem. Soc.* **1958**, 2425–6.

[941] E. E. Aynsley, M. L. Hair, *J. Chem. Soc.* **1958**, 3747–8.

[942] M. Schmeisser, K. Lang, *Angew. Chem.* **1955**, 67, 156.

[943] J. W. Viers, H. W. Baird, *Chem. Commun.* **1967**, 1093–4.

[944] A. J. Edwards, P. Taylor, *J. Fluorine Chem.* **1974**, 4, 173–9.

[945] J. B. Milne, D. M. Moffett, *Inorg. Chem.* **1976**, 15, 2165–9.

[946] K. O. Christe, W. W. Wilson, D. A. Dixon, J. A. Boatz, *J. Am. Chem. Soc.* **1999**, 121, 3382–5.

[947] A. A. Woolf, *J. Chem. Soc.* **1954**, 4113–6.

[948] G. Tantot, P. Joubert, R. Bougon, *Can. J. Chem.* **1978**, 56, 1634–7.

[949] K. O. Christe, R. D. Wilson, C. J. Schack, *Inorg. Nucl. Chem. Lett.* **1975**, 11, 161–3.

[950] K. O. Christe, R. D. Wilson, C. J. Schack, D. D. DesMarteau, *Inorg. Synth.* **1986**, 3–6.

[951] M. Schmeisser, W. Fink, *Angew. Chem.* **1957**, 69, 780.

[952] H. Schmitz, H. J. Schumacher, *Z. Anorg. Allg. Chem.* **1942**, 249, 238–44.

[953] M. Schmeisser, F. L. Ebenhöch, *Angew. Chem.* **1954**, 66, 230–1.

[954] J. E. Sicre, H. J. Schumacher, *Z. Anorg. Allg. Chem.* **1956**, 286, 232–6.

[955] H. A. Carter, W. M. Johnson, F. Aubke, *Can. J. Chem.* **1969**, 47, 4619–25.

[956] K. O. Christe, E. C. Curtis, *Inorg. Chem.* **1972**, 11, 35–9.

[957] D. K. Huggins, W. B. Fox, *Inorg. Nucl. Chem. Lett.* **1970**, 6, 337–9.

[958] R. H. Toeniskoetter, F. P. Gortsema, *J. Inorg. Nucl. Chem.* **1966**, 28, 2205–8.

[959] M. Schmeisser, E. Pammer, *Angew. Chem.* **1957**, 69, 781.

[960] M. Schmeisser, E. Pammer, *Angew. Chem.* **1955**, 67, 156.

[961] K. Seppelt, *Angew. Chem., Int. Ed. Engl.* **2019**, 58, 18928–30.

[962] E. Jacob, *Z. Anorg. Allg. Chem.* **1977**, 433, 255–60.

[963] R. J. Gillespie, P. Spekkens, *J. Chem. Soc., Chem. Commun.* **1975**, 314–6.

[964] K. O. Christe, E. C. Curtis, E. Jacob, *Inorg. Chem.* **1978**, 17, 2744–9.

[965] I.-C. Hwang, R. Kuschel, K. Seppelt, *Z. Anorg. Allg. Chem.* **1997**, 623, 379–83.

[966] E. Jacob, *Angew. Chem.* **1976**, 88, 109–90.

[967] M. Adelhelm, E. Jacob, *Angew. Chem.* **1977**, 89, 476–7.

[968] G. Tantot, R. Bougon, *C. R. Acad. Sci., Ser. IIc: Chim.* **1975**, 281C, 271–3.

[969] R. J. Gillespie, P. Spekkens, *J. Chem. Soc., Dalton Trans.* **1976**, 2391–6.

[970] R. Minkwitz, M. Berkei, R. Ludwig, *Inorg. Chem.* **2001**, 40, 6493–5.

[971] A. Finch, P. N. Gates, M. A. Jenkinson, *J. Fluorine Chem.* **1972**, 2, 111–2.

[972] J. B. Milne, D. Moffett, *Inorg. Chem.* **1975**, 14, 1077–81.

[973] M. Gerken, J. P. Mack, G. J. Schrobilgen, R. J. Suontamo, *J. Fluorine Chem.* **2004**, 125, 1663–70.

[974] J. P. Mack, J. A. Boatz, M. Gerken, *Inorg. Chem.* **2008**, 47, 3243–7.

[975] K. O. Christe, C. J. Schack. Chlorine Oxyfluorides. In: H. J. Emeléus, A. G. Sharpe (eds.) Advances in Inorganic Chemistry and Radiochemistry: Academic Press; 1976:319–98.

[976] R. J. Gillespie, J. W. Quail, *Proc. Chem. Soc.* **1963**, 278.

[977] N. Bartlett, L. E. Levchuk, *Proc. Chem. Soc.* **1963**, 342–3.

[978] C. J. Schack, K. O. Christe, *J. Fluorine Chem.* **1990**, 49, 167–9.

[979] J. H. Holloway, H. Selig, H. H. Claassen, *J. Chem. Phys.* **1971**, 54, 4305–11.

[980] K. O. Christe, E. C. Curtis, D. A. Dixon, *J. Am. Chem. Soc.* **1993**, 115, 9655–8.

[981] K. O. Christe, D. A. Dixon, A. R. Mahjoub, et al., *J. Am. Chem. Soc.* **1993**, 115, 2696–706.

[982] A. Engelbrecht, P. Peterfy, *Angew. Chem.* **1969**, 81, 753.

[983]   K. O. Christe, R. D. Wilson, C. J. Schack, *Inorg. Chem.* **1981**, 20, 2104–14.

[984]   K. O. Christe, R. D. Wilson, *Inorg. Chem.* **1973**, 12, 1356–7.

[985]   K. O. Christe, *Inorg. Nucl. Chem. Lett.* **1972**, 8, 457–9.

[986]   K. O. Christe, E. C. Curtis, *Inorg. Chem.* **1973**, 12, 2245–51.

[987]   K. O. Christe, R. D. Wilson, E. C. Curtis, *Inorg. Chem.* **1973**, 12, 1358–62.

[988]   I. Beattie, R. Crocombe, A. German, et al., *J. Chem. Soc., Dalton Trans.* **1976**, 1380–7.

[989]   M. J. Vasile, W. E. Falconer, F. A. Stevie, I. R. Beattie, *J. Chem. Soc., Dalton Trans.* **1977**, 1233–7.

[990]   A. Engelbrecht, O. Mayr, G. Ziller, E. Schandara, *Monatsh. Chem.* **1974**, 105, 796–806.

[991]   R. J. Gillespie, J. P. Krasznai, *Inorg. Chem.* **1976**, 15, 1251–6.

[992]   H. A. Carter, J. N. Ruddick, J. R. Sams, F. Aubke, *Inorg. Nucl. Chem. Lett.* **1975**, 11, 29–34.

[993]   R. J. Gillespie, J. P. Krasznai, *Inorg. Chem.* **1977**, 16, 1384–92.

[994]   J. A. Boatz, K. O. Christe, D. A. Dixon, et al., *Inorg. Chem.* **2003**, 42, 5282–92.

[995]   G. Barth-Wehrenalp, *J. Inorg. Nucl. Chem.* **1956**, 2, 266.

[996]   A. Engelbrecht, H. Atzwanger, *Monatsh. Chem.* **1952**, 83, 1087–9.

[997]   A. Engelbrecht, H. Atzwanger, *J. Inorg. Nucl. Chem.* **1956**, 2, 348–57.

[998]   C. A. Wamser, W. B. Fox, D. Gould, B. Sukornick, *Inorg. Chem.* **1968**, 7, 1933–5.

[999]   C. A. Wamser, B. Sukornick, W. B. Fox, D. Gould, A. F. Clifford, J. W. Thompson, *Inorg. Synth.* **1973**, 29–33.

[1000] H. Bode, E. Klesper, *Z. Anorg. Allg. Chem.* **1951**, 266, 275–80.

[1001] J. E. Sicre, H. J. Schumacher, *Angew. Chem.* **1957**, 69, 266.

[1002] A. H. Clark, B. Beagley, D. W. J. Cruickshank, *Chem. Commun.* **1968**, 14–5.

[1003] H. H. Claassen, E. H. Appelman, *Inorg. Chem.* **1970**, 9, 622–4.

[1004] G. H. Cady, *J. Fluorine Chem.* **1978**, 11, 225–41.

[1005] H. C. Mandell, G. Barth-Wehrenalp, *J. Inorg. Nucl. Chem.* **1959**, 12, 90–4.

[1006] J. F. Lehmann, G. J. Schrobilgen, *J. Am. Chem. Soc.* **2005**, 127, 9416–27.

[1007] J. F. Lehmann, S. Riedel, G. J. Schrobilgen, *Inorg. Chem.* **2008**, 47, 8343–56.

[1008] C. E. Inman, R. E. Oesterling, E. A. Tyczkowski, *J. Am. Chem. Soc.* **1958**, 80, 6533–5.

[1009] H. Gershon, J. A. A. Renwick, W. K. Wynn, R. D'Ascoli, *J. Org. Chem.* **1966**, 31, 916–8.

[1010] M. Schlosser, G. Heinz, *Chem. Ber.* **1969**, 102, 1944–53.

[1011] W. J. Gensler, Q. A. Ahmed, M. V. Leeding, *J. Org. Chem.* **1968**, 33, 4279–81.

[1012] W. A. Sheppard, *Tetrahedron Lett.* **1969**, 10, 83–4.

[1013] G. H. Posner, L. L. Frye, *J. Fluorine Chem.* **1985**, 28, 151–9.

[1014] R. E. Ehrenkaufer, R. R. MacGregor, *Int. J. Appl. Radiat. Isot.* **1983**, 34, 613–5.

[1015] C. E. Inman, R. E. Oesterling, E. A. Tyczkowski, *J. Am. Chem. Soc.* **1958**, 80, 5286–8.

[1016] H. M. Kissman, A. M. Small, M. J. Weiss, *J. Am. Chem. Soc.* **1960**, 82, 2312–7.

[1017] G. R. Allen, N. A. Austin, *J. Org. Chem.* **1961**, 26, 5245–7.

[1018] S. Nakanishi, E. V. Jensen, *J. Org. Chem.* **1962**, 27, 702–3.

[1019] *Chem. Eng. News* **1960**, 38, 62–4.

[1020] M. S. Marma, L. A. Khawli, V. Harutunian, B. A. Kashemirov, C. E. McKenna, *J. Fluorine Chem.* **2005**, 126, 1467–75.

[1021] E. H. Appleman, M. H. Studier, *J. Am. Chem. Soc.* **1969**, 91, 4561–2.

[1022] J. Baran Enrique, J. Aymonino Pedro, *Z. Naturforsch. B* **1972**, 27, 1568.

[1023] P. A. Agron, G. M. Begun, H. A. Levy, A. A. Mason, C. G. Jones, D. F. Smith, *Science* **1963**, 139, 842.

[1024] L. V. Streng, A. G. Streng, *Inorg. Chem.* **1965**, 4, 1370–1.

[1025] P. Tsao, C. C. Cobb, H. H. Claassen, *J. Chem. Phys.* **1971**, 54, 5247–53.

[1026] H. Meinert, G. Kauschka, S. Büdiger, *Z. Chem.* **1967**, 7, 111–2.

[1027] E. H. Appleman, *J. Am. Chem. Soc.* **1968**, 90, 1900–1.

[1028] N. Bartlett, B. Žemva, L. Graham, *J. Fluorine Chem.* **1976**, 7, 301–20.

[1029] H. J. Frohn, V. V. Bardin. Organoxenonium Salts: Synthesis by "Xenodeborylation", Reactivities, and NMR Spectroscopic Properties. Recent Developments in Carbocation and Onium Ion Chemistry: American Chemical Society; 2007:428–57.

[1030] M. Bohinc, J. Grannec, J. Slivnik, B. Žemva, *J. Inorg. Nucl. Chem.* **1976**, 38, 75–6.

[1031] B. Žemva, J. Slivnik, M. Bohinc, *J. Inorg. Nucl. Chem.* **1976**, 38, 73–4.

[1032] K. Lutar, H. Borrmann, B. Žemva, *Inorg. Chem.* **1998**, 37, 3002–6.

[1033] M. Gerken, M. D. Moran, H. P. A. Mercier, et al., *J. Am. Chem. Soc.* **2009**, 131, 13474–89.

[1034] J. H. Holloway, G. J. Schrobilgen, *J. Chem. Soc., Chem. Commun.* **1975**, 623–4.

[1035] B. Frlec, J. H. Holloway, *J. Chem. Soc., Dalton Trans.* **1975**, 535–40.

[1036] G. R. Jones, R. D. Burbank, N. Bartlett, *Inorg. Chem.* **1970**, 9, 2264–8.

[1037] J. F. Liebman, *J. Fluorine Chem.* **1976**, 7, 531–5.

[1038] N. Vasdev, M. D. Moran, H. M. Tuononen, et al., *Inorg. Chem.* **2010**, 49, 8997–9004.

[1039] V. I. Spitzin, Y. M. Kiselev, N. E. Fadeeva, A. I. Popov, N. A. Tchumaevsky, *Z. Anorg. Allg. Chem.* **1988**, 559, 171–81.

[1040] S. Seidel, K. Seppelt, *Angew. Chem., Int. Ed. Engl.* **2001**, 40, 4225–7.

[1041] J. H. Holloway, R. D. Peacock, *Proc. Chem. Soc.* **1962**, 389–90.

[1042] H. H. Claassen, H. Selig, J. G. Malm, *J. Am. Chem. Soc.* **1962**, 84, 3593.

[1043] A. Šmalc, K. Lutar, J. Slivnik, *J. Fluorine Chem.* **1976**, 8, 95–6.

[1044] A. D. Kirshenbaum, L. V. Streng, A. G. Streng, A. V. Grosse, *J. Am. Chem. Soc.* **1963**, 85, 360–1.

[1045] K. Lutar, A. Smalc, B. Zemva, S. A. Kinkead, *Inorg. Synth.* **1992**, 4–6.

[1046] J. B. Nielsen, S. A. Kinkead, J. D. Purson, P. G. Eller, *Inorg. Chem.* **1990**, 29, 1779–80.

[1047] D. Martin, *C. R. Acad. Sci., Ser. IIc: Chim.* **1969**, 268C, 1145–8.

[1048] S. M. Williamson, C. W. Koch, *Science* **1963**, 139, 1046.

[1049] D. S. Brock, G. J. Schrobilgen, *J. Am. Chem. Soc.* **2011**, 133, 6265–9.

[1050] J. S. Ogden, J. J. Turner, *Chem. Commun.* **1966**, 693–4.

[1051] J. A. Ibers, W. C. Hamilton, *Science* **1963**, 139, 106.

[1052] D. H. Templeton, A. Zalkin, J. D. Forrester, S. M. Williamson, *J. Am. Chem. Soc.* **1963**, 85, 242.

[1053] H. H. Claassen, C. L. Chernick, J. G. Malm, *J. Am. Chem. Soc.* **1963**, 85, 1927–8.

[1054] P. Boldrini, R. J. Gillespie, P. R. Ireland, G. J. Schrobilgen, *Inorg. Chem.* **1974**, 13, 1690–4.

[1055] R. J. Gillespie, B. Landa, G. J. Schrobilgen, *Inorg. Chem.* **1976**, 15, 1256–63.

[1056] D. E. McKee, C. J. Adams, N. Bartlett, *Inorg. Chem.* **1973**, 12, 1722–5.

[1057] D. E. McKee, C. J. Adams, A. Zalkin, N. Bartlett, *J. Chem. Soc., Chem. Commun.* **1973**, 26–8.

[1058] R. J. Gillespie, G. J. Schrobilgen, *Inorg. Chem.* **1974**, 13, 2370–4.

[1059] D. E. McKee, A. Zalkin, N. Bartlett, *Inorg. Chem.* **1973**, 12, 1713–7.

[1060] K. O. Christe, E. C. Curtis, D. A. Dixon, H. P. Mercier, J. C. P. Sanders, G. J. Schrobilgen, *J. Am. Chem. Soc.* **1991**, 113, 3351–61.

[1061] H.-J. Frohn, N. LeBlond, K. Lutar, B. Žemva, *Angew. Chem., Int. Ed. Engl.* **2000**, 39, 391–3.

[1062] K. Koppe, J. Haner, H. P. A. Mercier, H.-J. Frohn, G. J. Schrobilgen, *Inorg. Chem.* **2014**, 53, 11640–61.

[1063] E. E. Weaver, B. Weinstock, C. P. Knop, *J. Am. Chem. Soc.* **1963**, 85, 111–2.

[1064] J. G. Malm, I. Sheft, C. L. Chernick, *J. Am. Chem. Soc.* **1963**, 85, 110–1.

[1065] C. L. Chernic, J. G. Malm, S. M. Williamson, *Inorg. Synth.* **1966**, 8, 258–60.

[1066] B. Žemva, J. Slivnik, *J. Fluorine Chem.* **1981**, 17, 375–9.

[1067] K. Matsumoto, J. Haner, H. P. A. Mercier, G. J. Schrobilgen, *Angew. Chem., Int. Ed. Engl.* **2015**, 54, 14169–73.

[1068] J. Haner, K. Matsumoto, H. P. A. Mercier, G. J. Schrobilgen, *Chem. Eur. J.* **2016**, 22, 4833–42.

[1069] S. Hoyer, T. Emmler, K. Seppelt, *J. Fluorine Chem.* **2006**, 127, 1415–22.

[1070] H. Rupp, K. Seppelt, *Angew. Chem., Int. Ed. Engl.* **1974**, 13, 612–3.

[1071] K. Seppelt, N. Bartlett, *Z. Anorg. Allg. Chem.* **1977**, 436, 122–6.

[1072] G. J. Schrobilgen, J. H. Holloway, P. Granger, C. Brevard, *Inorg. Chem.* **1978**, 17, 980–7.

[1073] W. E. Falconer, M. J. Vasile, F. A. Stevie, *J. Chem. Phys.* **1977**, 66, 5335–8.

[1074] L. S. Bartell, R. M. Gavin Jr., *J. Chem. Phys.* **1968**, 48, 2466–83.

[1075] G. L. Goodman, *J. Chem. Phys.* **1972**, 56, 5038–41.

[1076] H. H. Claassen, G. L. Goodman, H. Kim, *J. Chem. Phys.* **1972**, 56, 5042–53.

[1077] D. A. Dixon, W. A. de Jong, K. A. Peterson, K. O. Christe, G. J. Schrobilgen, *J. Am. Chem. Soc.* **2005**, 127, 8627–34.

[1078] M. Gawrilow, H. Beckers, S. Riedel, L. Cheng, *J. Phys. Chem. A* **2018**, 122, 119–29.

[1079] K. Leary, A. Zalkin, N. Bartlett, *Inorg. Chem.* **1974**, 13, 775–9.

[1080] K. Leary, A. Zalkin, N. Bartlett, *J. Chem. Soc., Chem. Commun.* **1973**, 131–2.

[1081] G. L. Gard, G. H. Cady, *Inorg. Chem.* **1964**, 3, 1745–7.

[1082] B. Weinstock, E. E. Weaver, C. P. Knop, *Inorg. Chem.* **1966**, 5, 2189–203.

[1083] T. Schroer, K. O. Christe, *Inorg. Chem.* **2001**, 40, 2415–9.

[1084] B. Frlec, J. H. Holloway, J. Slivnik, A. Šmalc, B. Volavšek, A. Zemljič, *J. Inorg. Nucl. Chem.* **1970**, 32, 2521–7.

[1085] D. J. Grant, T.-H. Wang, D. A. Dixon, K. O. Christe, *Inorg. Chem.* **2010**, 49, 261–70.

[1086] N. Bartlett, B. G. DeBoer, F. J. Hollander, F. O. Sladky, D. H. Templeton, A. Zalkin, *Inorg. Chem.* **1974**, 13, 780–5.

[1087] K. O. Christe, E. C. Curtis, R. D. Wilson, *J. Inorg. Nucl. Chem.* **1976**, 28, 159–65.

[1088] K. E. Pullen, G. H. Cady, *Inorg. Chem.* **1966**, 5, 2057–9.

[1089] K. E. Pullen, G. H. Cady, *Inorg. Chem.* **1967**, 6, 1300–2.

[1090] K. E. Pullen, G. H. Cady, *Inorg. Chem.* **1967**, 6, 2267–8.

[1091] G. H. Cady, J. Aubert, *Inorg. Chem.* **1970**, 9, 2600–2.

[1092] M. Bohinc, B. Frlec, *J. Inorg. Nucl. Chem.* **1972**, 34, 2942–6.

[1093] B. Žemva, J. Slivnik, *J. Fluorine Chem.* **1976**, 8, 369–71.

[1094] B. Žemva, S. Milićev, J. Slivnik, *J. Fluorine Chem.* **1978**, 11, 519–26.

[1095] B. Žemva, S. Milićev, J. Slivnik, *J. Fluorine Chem.* **1978**, 11, 545–53.

[1096] A. Jesih, B. Žemva, J. Slivnik, *J. Fluorine Chem.* **1982**, 19, 221–6.

[1097] B. Žemva, A. Jesih, *J. Fluorine Chem.* **1984**, 24, 281–9.

[1098] K. Lutar, A. Jesih, I. Leban, B. Zemva, N. Bartlett, *Inorg. Chem.* **1989**, 28, 3467–71.

[1099] R. J. Gillespie, G. J. Schrobilgen, *Inorg. Chem.* **1974**, 13, 765–70.

[1100] S. W. Peterson, J. H. Holloway, B. A. Coyle, J. M. Williams, *Science* **1971**, 173, 1238–9.

[1101] A. Ellern, A.-R. Mahjoub, K. Seppelt, *Angew. Chem., Int. Ed. Engl.* **1996**, 35, 1123–5.

[1102] R. D. Peacock, H. Selig, I. Sheft, *J. Inorg. Nucl. Chem.* **1966**, 28, 2561–7.

[1103] K. O. Christe, W. W. Wilson, *Inorg. Chem.* **1982**, 21, 4113–7.

[1104] N. Bartlett, F. O. Sladky, *J. Am. Chem. Soc.* **1968**, 90, 5316–7.

[1105] I. Sheft, T. M. Spittler, F. H. Martin, *Science* **1964**, 145, 701.

[1106] T.-C. Shieh, N. C. Yang, C. L. Chernick, *J. Am. Chem. Soc.* **1964**, 86, 5021–2.

[1107] N.-C. Yang, T.-C. Shieh, E. D. Feit, C. L. Chernick, *J. Org. Chem.* **1970**, 35, 4020–4.

[1108] R. Filler, *Isr. J. Chem.* **1978**, 17, 71–9.

[1109] D. D. Desmarteau, M. Eisenberg, *Inorg. Chem.* **1972**, 11, 2641–4.

[1110] F. Sladky, *Angew. Chem.* **1969**, 81, 330.

[1111] F. Sladky, *Angew. Chem.* **1969**, 81, 536–7.

[1112] K. Seppelt, D. Nothe, *Inorg. Chem.* **1973**, 12, 2727–30.

[1113] N. Keller, G. J. Schrobilgen, *Inorg. Chem.* **1981**, 20, 2118–29.

[1114]   H. P. A. Mercier, M. D. Moran, J. C. P. Sanders, G. J. Schrobilgen, R. J. Suontamo, *Inorg. Chem.* **2005**, 44, 49–60.

[1115]   D. Lentz, K. Seppelt, *Angew. Chem.* **1978**, 90, 391–2.

[1116]   D. Lentz, K. Seppelt, *Angew. Chem., Int. Ed. Engl.* **1979**, 18, 66–7.

[1117]   G. A. Schumacher, G. J. Schrobilgen, *Inorg. Chem.* **1984**, 23, 2923–9.

[1118]   R. G. Syvret, K. M. Mitchell, J. C. P. Sanders, G. J. Schrobilgen, *Inorg. Chem.* **1992**, 31, 3381–5.

[1119]   E. H. Appelman, J. G. Malm, *J. Am. Chem. Soc.* **1964**, 86, 2141–8.

[1120]   E. Jacob, R. Opferkuch, *Angew. Chem.* **1976**, 88, 190.

[1121]   R. J. Gillespie, G. J. Schrobilgen, *J. Chem. Soc., Chem. Commun.* **1977**, 595–7.

[1122]   D. S. Brock, H. P. A. Mercier, G. J. Schrobilgen, *J. Am. Chem. Soc.* **2010**, 132, 10935–43.

[1123]   D. S. Brock, H. P. A. Mercier, G. J. Schrobilgen, *J. Am. Chem. Soc.* **2013**, 135, 5089–104.

[1124]   D. S. Brock, V. Bilir, H. P. A. Mercier, G. J. Schrobilgen, *J. Am. Chem. Soc.* **2007**, 129, 3598–611.

[1125]   H. H. Claassen, E. L. Gasner, H. Kim, J. L. Huston, *J. Chem. Phys.* **1968**, 49, 253–7.

[1126]   S. W. Peterson, R. D. Willett, J. L. Huston, *J. Chem. Phys.* **1973**, 59, 453–9.

[1127]   K. O. Christe, W. W. Wilson, *Inorg. Chem.* **1988**, 27, 3763–8.

[1128]   J. L. Huston, *J. Phys. Chem.* **1967**, 71, 3339–41.

[1129]   H. H. Selig, L. A. Quarterman, H. H. Hyman, *J. Inorg. Nucl. Chem.* **1966**, 28, 2063–5.

[1130]   D. F. Smith, *Science* **1963**, 140, 899.

[1131]   K. O. Christe, W. W. Wilson, *Inorg. Chem.* **1988**, 27, 1296–7.

[1132]   J. B. Nielsen, S. A. Kinkead, P. G. Eller, *Inorg. Chem.* **1990**, 29, 3621–2.

[1133]   G. M. Begun, W. H. Fletcher, D. F. Smith, *J. Chem. Phys.* **1965**, 42, 2236–42.

[1134]   K. O. Christe, R. D. Wilson, *J. Fluorine Chem.* **1976**, 7, 356–8.

[1135]   H. Selig, *Inorg. Chem.* **1966**, 5, 183–6.

[1136]   M. C. Waldman, H. Selig, *J. Inorg. Nucl. Chem.* **1973**, 35, 2173–82.

[1137]   K. O. Christe, W. W. Wilson, *Inorg. Chem.* **1988**, 27, 2714–8.

[1138]   B. E. Pointner, R. J. Suontamo, G. J. Schrobilgen, *Inorg. Chem.* **2006**, 45, 1517–34.

[1139]   J. H. Holloway, V. Kaucic, D. Martin-Rovet, D. R. Russell, G. J. Schrobilgen, H. Selig, *Inorg. Chem.* **1985**, 24, 678–83.

[1140]   A. Ellern, K. Seppelt, *Angew. Chem., Int. Ed. Engl.* **1995**, 34, 1586–7.

[1141]   K. O. Christe, D. A. Dixon, J. C. P. Sanders, G. J. Schrobilgen, S. S. Tsai, W. W. Wilson, *Inorg. Chem.* **1995**, 34, 1868–74.

[1142]   J. L. Huston, *J. Am. Chem. Soc.* **1971**, 93, 5255–6.

[1143]   H. H. Claassen, J. L. Huston, *J. Chem. Phys.* **1971**, 55, 1505–7.

[1144]   J. L. Huston, *Inorg. Chem.* **1982**, 21, 685–8.

[1145]   J. Slivnik, A. Šmalc, K. Lutar, B. Žemva, B. Frlec, *J. Fluorine Chem.* **1975**, 5, 273–4.

[1146]   F. Schreiner, J. G. Malm, J. C. Hindman, *J. Am. Chem. Soc.* **1965**, 87, 25–8.

[1147]   A. Smalc, K. Lutar, B. Zemva, S. A. Kinkead, *Inorg. Synth.* **1992**, 11–5.

[1148]   S. A. Kinkead, J. R. FitzPatrick, J. Foropoulos, R. J. Kissane, J. D. Purson. Photochemical and Thermal Dissociation Synthesis of Krypton Difluoride. Inorganic Fluorine Chemistry: American Chemical Society; 1994:40–55.

[1149]   M. Al-Mukhtar, J. H. Holloway, E. G. Hope, G. J. Schrobilgen, *J. Chem. Soc., Dalton Trans.* **1991**, 2831–4.

[1150]   L. V. Streng, A. G. Streng, *Inorg. Chem.* **1966**, 5, 328–9.

[1151]   R. D. Burbank, W. E. Falconer, W. A. Sunder, *Science* **1972**, 178, 1285.

[1152]   H. H. Claassen, G. L. Goodman, J. G. Malm, F. Schreiner, *J. Chem. Phys.* **1965**, 42, 1229–32.

[1153]   J. F. Lehmann, D. A. Dixon, G. J. Schrobilgen, *Inorg. Chem.* **2001**, 40, 3002–17.

[1154]   R. J. Gillespie, G. J. Schrobilgen, *Inorg. Chem.* **1976**, 15, 22–31.

[1155]   B. Frlec, J. H. Holloway, *Inorg. Chem.* **1976**, 15, 1263–70.

[1156] R. J. Gillespie, G. J. Schrobilgen, *J. Chem. Soc., Chem. Commun.* **1974**, 90–2.

[1157] J. H. Holloway, G. J. Schrobilgen, *Inorg. Chem.* **1981**, 20, 3363–8.

[1158] H. P. A. Mercier, U. Breddemann, D. S. Brock, M. R. Bortolus, G. J. Schrobilgen, *Chem. Eur. J.* **2019**, 25, 12105–19.

[1159] D. S. Brock, J. J. Casalis de Pury, H. P. A. Mercier, G. J. Schrobilgen, B. Silvi, *J. Am. Chem. Soc.* **2010**, 132, 3533–42.

[1160] M. Lozinšek, H. P. A. Mercier, D. S. Brock, B. Žemva, G. J. Schrobilgen, *Angew. Chem., Int. Ed. Engl.* **2017**, 56, 6251–4.

[1161] J. R. DeBackere, G. J. Schrobilgen, *Angew. Chem., Int. Ed. Engl.* **2018**, 57, 13167–71.

[1162] P. J. MacDougall, G. J. Schrobilgen, R. F. W. Bader, *Inorg. Chem.* **1989**, 28, 763–9.

[1163] G. J. Schrobilgen, *J. Chem. Soc., Chem. Commun.* **1988**, 863–5.

[1164] D. A. Dixon, T.-H. Wang, D. J. Grant, K. A. Peterson, K. O. Christe, G. J. Schrobilgen, *Inorg. Chem.* **2007**, 46, 10016–21.

[1165] L. Stein, *Science* **1970**, 168, 362.

[1166] L. Stein, *J. Inorg. Nucl. Chem.* **1973**, 35, 39–43.

[1167] C. L. Chernick, H. H. Claassen, P. R. Fields, et al., *Science* **1962**, 138, 136.

[1168] L. Stein, *Inorg. Chem.* **1984**, 23, 3670–1.

[1169] M. Dolg, W. Küchle, H. Stoll, H. Preuss, P. Schwerdtfeger, *Mol. Phys.* **1991**, 74, 1265–85.

[1170] J. F. Liebman, *Inorg. Nucl. Chem. Lett.* **1975**, 11, 683–5.

[1171] M. Kaupp, C. van Wüllen, R. Franke, F. Schmitz, W. Kutzelnigg, *J. Am. Chem. Soc.* **1996**, 118, 11939–50.

[1172] G. L. Malli, *Comput. Theor. Chem.* **2001**, 537, 71–7.

[1173] M. Filatov, D. Cremer, *PCCP, Phys. Chem. Chem. Phys.* **2003**, 5, 1103–5.

[1174] D. McLeod, W. Weltner, *J. Phys. Chem.* **1966**, 70, 3293–300.

[1175] K. F. Zmbov, J. L. Margrave, *J. Chem. Phys.* **1967**, 47, 3122–5.

[1176] B. Simard, A. M. James, *J. Chem. Phys.* **1992**, 97, 4669–78.

[1177] D. L. Hildenbrand, K. H. Lau, *J. Chem. Phys.* **1995**, 102, 3769–75.

[1178] K. D. Carlson, C. Moser, *J. Chem. Phys.* **1967**, 46, 35–46.

[1179] V. G. Solomonik, A. A. Mukhanov, *J. Struct. Chem.* **2012**, 53, 28–34.

[1180] L. Cheng, M. Y. Wang, Z. J. Wu, Z. M. Su, *J. Comput. Chem.* **2007**, 28, 2190–202.

[1181] T. Ma, G. S. M. Tong, A. S. C. Cheung, *Mol. Phys.* **2012**, 110, 1407–14.

[1182] R. D. Wesley, C. W. DeKock, *J. Phys. Chem.* **1973**, 77, 466–8.

[1183] P. E. M. Siegbahn, *Theor. Chim. Acta* **1994**, 87, 441–52.

[1184] X. Wang, L. Andrews, *J. Phys. Chem. A* **2010**, 114, 2293–9.

[1185] J. G. Stites, M. L. Salutsky, B. D. Stone, *J. Am. Chem. Soc.* **1955**, 77, 237–40.

[1186] B. Hájek, *Z. Chem.* **1965**, 5, 341.

[1187] R. E. Thoma, R. H. Karraker, *Inorg. Chem.* **1966**, 5, 1933–7.

[1188] V. Dracopoulos, B. Gilbert, B. Brrensen, G. M. Photiadis, G. N. Papatheodorou, *J. Chem. Soc. Faraday Trans.* **1997**, 93, 3081–8.

[1189] R. H. Hauge, J. W. Hastie, J. L. Margrave, *J. Less-Common Met.* **1971**, 23, 359–65.

[1190] A. Zalkin, D. H. Templeton, *J. Am. Chem. Soc.* **1953**, 75, 2453–8.

[1191] A. Zalkin, D. H. Templeton, T. E. Hopkins, *Inorg. Chem.* **1966**, 5, 1466–8.

[1192] W. Zachariasen, *Acta Crystallogr.* **1949**, 2, 388–90.

[1193] M. Fele-Beuermann, K. Lutar, Z. Mazej, S. Milićev, B. Žemva, *J. Fluorine Chem.* **1998**, 89, 83–9.

[1194] W. Zachariasen, *Acta Crystallogr.* **1951**, 4, 231–6.

[1195] J. Lee, Q. Zhang, F. Saito, *J. Am. Ceram. Soc.* **2001**, 84, 863–5.

[1196] S. E. Dutton, D. Hirai, R. J. Cava, *Mater. Res. Bull.* **2012**, 47, 714–8.

[1197] G. Scholz, M. Dreger, R. Bertram, E. Kemnitz, *Dalton Trans.* **2015**, 44, 13522–9.

[1198]   B. Holmberg, *Acta Chem. Scand.* **1966**, 20, 1082–8.
[1199]   A. W. Mann, D. J. M. Bevan, *Acta Crystallogr. B* **1970**, 26, 2129–31.
[1200]   J. W. Fergus, *J. Mater. Sci. Lett.* **1997**, 16, 267–9.
[1201]   J. W. Fergus, H. P. Chen, *J. Electrochem. Soc.* **2000**, 147, 4696–704.
[1202]   Y. Gong, L. Andrews, C. W. Bauschlicher Jr., *Chem. Eur. J.* **2012**, 18, 12446–51.
[1203]   Y. Gong, L. Andrews, C. W. Bauschlicher, *J. Phys. Chem. A* **2012**, 116, 10115–21.
[1204]   M. Bannert, G. Blumenthal, H. Sattler, M. Schönherr, H. Wittrich, *Z. Anorg. Allg. Chem.* **1978**, 446, 251–6.
[1205]   X. P. Zhou, Z. S. Chao, J. Z. Luo, H. L. Wan, K. R. Tsai, *Appl. Catal. A* **1995**, 133, 263–8.
[1206]   Q. Xie, Y. Wang, B. Pan, H. Wang, W. Su, X. Wang, *Catal. Commun.* **2012**, 27, 21–5.
[1207]   C. Suresh, H. Nagabhushana, G. P. Darshan, et al., *Arab. J. Chem.* **2018**, 11, 196–213.
[1208]   Y. Wang, T. Wen, H. Zhang, et al., *J. Phys. Chem. C* **2014**, 118, 10314–20.
[1209]   V. P. Smagin, A. P. Khudyakov, *Inorg. Mater.* **2019**, 55, 64–76.
[1210]   D. J. W. Robbins, R. F. Barrow, *J. Phys. B* **1974**, 7, L234–L5.
[1211]   D. J. W. Lumley, R. F. Barrow, *J. Mol. Spectrosc.* **1978**, 69, 494–5.
[1212]   R. M. Clements, R. F. Barrow, *J. Mol. Spectrosc.* **1984**, 107, 119–23.
[1213]   A. L. Kaledin, M. C. Heaven, R. W. Field, L. A. Kaledin, *J. Mol. Spectrosc.* **1996**, 179, 310–9.
[1214]   L. A. Kaledin, M. C. Heaven, R. W. Field, *J. Mol. Spectrosc.* **1999**, 193, 285–92.
[1215]   S. A. Mucklejohn, *J. Phys. D, Appl. Phys.* **2011**, 44, 224010.
[1216]   T. Vent-Schmidt, S. Riedel, *Inorg. Chem.* **2015**, 54, 11114–20.
[1217]   R. F. Barrow, A. H. Chojnicki, *J. Chem. Soc. Faraday Trans.* **1975**, 71, 728–35.
[1218]   C. S. Dickinson, J. A. Coxon, N. R. Walker, M. C. L. Gerry, *J. Chem. Phys.* **2001**, 115, 6979–89.
[1219]   V. L. Ryabov, V. F. Ezhov, M. A. Kartasheva, A. V. Loginov, A. E. Pirogov, *Opt. Spectrosc.* **2007**, 102, 371–5.
[1220]   K. F. Zmbov, J. L. Margrave, *J. Chem. Phys.* **1966**, 45, 3167–70.
[1221]   K. F. Zmbov, J. L. Margrave, *J. Phys. Chem.* **1966**, 70, 3379–82.
[1222]   P. D. Kleinschmidt, K. H. Lau, D. L. Hildenbrand, *J. Chem. Phys.* **1981**, 74, 653–60.
[1223]   T. Petzel, O. Greis, *Z. Anorg. Allg. Chem.* **1973**, 396, 95–102.
[1224]   G. Beck, W. Nowacki, *Sci. Nat.* **1938**, 26, 495–6.
[1225]   A. Gin-ya, N. Toru, U. Katumi, S. Jiro, *Chem. Lett.* **1976**, 5, 189–90.
[1226]   H. Pink, *Z. Anorg. Allg. Chem.* **1968**, 356, 319–20.
[1227]   H. P. Beck, *Z. Anorg. Allg. Chem.* **1979**, 459, 72–80.
[1228]   T. Petzel, O. Greis, *J. Less-Common Met.* **1976**, 46, 197–207.
[1229]   O. Greis, *Z. Anorg. Allg. Chem.* **1977**, 430, 175–98.
[1230]   O. Greis, *Z. Anorg. Allg. Chem.* **1978**, 441, 39–46.
[1231]   G. Wu, R. Hoppe, *Z. Anorg. Allg. Chem.* **1984**, 514, 92–8.
[1232]   J. Köhler, *Solid State Sci.* **1999**, 1, 545–53.
[1233]   N. v. Prondzinski, J. Cybinska, A.-V. Mudring, *Chem. Commun.* **2010**, 46, 4393–5.
[1234]   S. Kern, P. M. Raccah, *J. Phys. Chem. Solids* **1965**, 26, 1625–8.
[1235]   F. Weigel, V. Scherer, *Radiochim. Acta* **1967**, 7, 40.
[1236]   O. Greis, T. Petzel, *Z. Anorg. Allg. Chem.* **1974**, 403, 1–22.
[1237]   J.-H. Kim, S. Yonezawa, M. Takashima, *J. Fluorine Chem.* **2003**, 120, 111–6.
[1238]   J. R. Peterson, B. B. Cunningham, *J. Inorg. Nucl. Chem.* **1968**, 30, 1775–84.
[1239]   S. Cotton. Binary Compounds of the Actinides. Lanthanide and Actinide Chemistry: John Wiley & Sons, Ltd; 2006:155–72.
[1240]   Z. Akdeniz, M. Gaune-Escard, M. P. Tosi, *Z. Naturforsch. A* **2001**, 56, 381.
[1241]   W. Xu, W.-X. Ji, Y.-X. Qiu, W. H. E. Schwarz, S.-G. Wang, *PCCP, Phys. Chem. Chem. Phys.* **2013**, 15, 7839–47.
[1242]   R. E. Thoma, H. Insley, G. M. Hebert, *Inorg. Chem.* **1966**, 5, 1222–9.

[1243] A. Taoudi, J. P. Laval, B. Frit, *Mater. Res. Bull.* **1994**, 29, 1137–47.

[1244] Y. Zhang, X. Li, Z. Hou, J. Lin, *Nanoscale* **2014**, 6, 6763–71.

[1245] J. Pannetier, J. Lucas, *C. R. Acad. Sci., Ser. IIc: Chim.* **1969**, 268C, 604–7.

[1246] X. Sun, Y.-W. Zhang, Y.-P. Du, et al., *Chem. Eur. J.* **2007**, 13, 2320–32.

[1247] O. Greis, R. Ziel, T. Petzel, B. Breidenstein, A. Haase, *Mater. Sci. Forum* **1994**, 166–169, 677–82.

[1248] J. H. Müller, T. Petzel, *J. Alloys Compd.* **1995**, 224, 18–21.

[1249] J. W. Fergus, *Mater. Res. Bull.* **1996**, 31, 1317–23.

[1250] T. Mikulas, M. Chen, D. A. Dixon, K. A. Peterson, Y. Gong, L. Andrews, *Inorg. Chem.* **2014**, 53, 446–56.

[1251] C.-T. Au, X. Zhou, *J. Chem. Soc. Faraday Trans.* **1997**, 93, 485–91.

[1252] Y.-P. Du, Y.-W. Zhang, L.-D. Sun, C.-H. Yan, *J. Phys. Chem. C* **2008**, 112, 405–15.

[1253] Y.-P. Du, Y.-W. Zhang, Z.-G. Yan, L.-D. Sun, C.-H. Yan, *J. Am. Chem. Soc.* **2009**, 131, 16364–5.

[1254] T. Wen, X. Li, D. Ning, J. Yao, B. Yang, Y. Wang, *J. Mater. Chem. C* **2018**, 6, 11007–14.

[1255] D. Kim, J. R. Jeong, Y. Jang, et al., *PCCP, Phys. Chem. Chem. Phys.* **2019**, 21, 1737–49.

[1256] W. Klemm, P. Henkel, *Z. Anorg. Allg. Chem.* **1934**, 220, 180–2.

[1257] B. B. Cunningham, D. C. Feay, M. A. Rollier, *J. Am. Chem. Soc.* **1954**, 76, 3361–3.

[1258] W. Asker, A. Wylie, *Aust. J. Chem.* **1965**, 18, 959–68.

[1259] V. I. Spitzin, L. I. Martynenko, J. U. M. Kiselew, *Z. Anorg. Allg. Chem.* **1982**, 495, 39–51.

[1260] Z. Mazej, *J. Fluorine Chem.* **2002**, 118, 127–9.

[1261] T. P. Perros, T. R. Munson, C. R. Naeser, *J. Chem. Educ.* **1953**, 30, 402.

[1262] J. Soriano, M. Givon, J. Shamir, *Inorg. Nucl. Chem. Lett.* **1966**, 2, 13–4.

[1263] E. W. Kaiser, W. A. Sunder, W. E. Falconer, *J. Less-Common Met.* **1972**, 27, 383–7.

[1264] J. K. Gibson, R. G. Haire, *J. Solid State Chem.* **1988**, 73, 524–30.

[1265] J. K. Gibson, R. G. Haire, *J. Less-Common Met.* **1988**, 144, 123–31.

[1266] M. Josse, M. El-Ghozzi, D. Avignant, G. André, F. Bourée, *J. Solid State Chem.* **2007**, 180, 1623–35.

[1267] R. R. Ryan, A. C. Larson, F. H. Kruse, *Inorg. Chem.* **1969**, 8, 33–6.

[1268] R. A. Penneman, A. Rosenzweig, *Inorg. Chem.* **1969**, 8, 627–30.

[1269] R. R. Ryan, R. A. Penneman, *Acta Crystallogr. B* **1971**, 27, 1939–43.

[1270] F. Kraus, S. A. Baer, *Z. Naturforsch. B* **2011**, 66, 868.

[1271] T. Vent-Schmidt, Z. Fang, Z. Lee, D. Dixon, S. Riedel, *Chem. Eur. J.* **2016**, 22, 2406–16.

[1272] D. Avignant, J.-C. Cousseins, *C. R. Acad. Sci., Ser. IIc: Chim.* **1974**, 278C, 613–6.

[1273] E. Largeau, V. Gaumet, M. El-Ghozzi, D. Avignant, J. C. Cousseins, *J. Mater. Chem.* **1997**, 7, 1881–5.

[1274] E. Largeau, M. El-Ghozzi, *J. Fluorine Chem.* **1998**, 89, 223–8.

[1275] E. Largeau, M. El-Ghozzi, D. Avignant, *J. Solid State Chem.* **1998**, 139, 248–58.

[1276] M. Josse, M. Dubois, M. El-Ghozzi, D. Avignant, *J. Alloys Compd.* **2004**, 374, 213–8.

[1277] C. C. Underwood, C. D. McMillen, J. W. Kolis, *J. Chem. Crystallogr.* **2014**, 44, 493–500.

[1278] D. Koller, B. G. Müller, *Z. Anorg. Allg. Chem.* **2000**, 626, 1426–8.

[1279] R. Fowler, H. Anderson, J. J. Hamilton, et al., *Ind. Eng. Chem.* **1947**, 39, 343–5.

[1280] A. G. Hudson, A. E. Pedler, J. C. Tatlow, *Tetrahedron* **1969**, 25, 4371–4.

[1281] A. G. Hudson, A. E. Pedler, *Tetrahedron* **1970**, 26, 3435–40.

[1282] J. V. Rau, S. Nunziante Cesaro, N. S. Chilingarov, M. S. Leskiv, G. Balducci, L. N. Sidorov, *Inorg. Chem. Commun.* **2003**, 6, 643–5.

[1283] M. V. Korobov, L. N. Savinova, L. N. Sidorov, *J. Chem. Thermodyn.* **1993**, 25, 1161–8.

[1284] A. A. Goryunkov, Z. Mazej, B. Žemva, S. H. Strauss, O. V. Boltalina, *Mendeleev Commun.* **2006**, 16, 159–61.

[1285] L. Andrews, K. S. Thanthiriwatte, X. Wang, D. A. Dixon, *Inorg. Chem.* **2013**, 52, 8228–33.

[1286] K. S. Thanthiriwatte, X. Wang, L. Andrews, et al., *J. Phys. Chem. A* **2014**, 118, 2107–19.

[1287] K. H. Lau, R. D. Brittain, D. L. Hildenbrand, *J. Chem. Phys.* **1989**, 90, 1158–64.

[1288] B. J. Barker, I. O. Antonov, M. C. Heaven, K. A. Peterson, *J. Chem. Phys.* **2012**, 136, 104305.

[1289] K. H. Lau, D. L. Hildenbrand, *J. Chem. Phys.* **1982**, 76, 2646–52.

[1290] R. D. Hunt, C. Thompson, P. Hassanzadeh, L. Andrews, *Inorg. Chem.* **1994**, 33, 388–91.

[1291] I. O. Antonov, M. C. Heaven, *J. Phys. Chem. A* **2013**, 117, 9684–94.

[1292] B. K. McNamara, R. Scheele, A. M. Casella, A. E. Kozelisky, D. Neiner, *MRS Proc.* **2010**, 1264, 1264-Z06-03.

[1293] T. Vent-Schmidt, R. D. Hunt, L. Andrews, S. Riedel, *Chem. Commun.* **2013**, 49, 3863–5.

[1294] T. Vent-Schmidt, L. Andrews, S. Riedel, *J. Phys. Chem. A* **2015**, 119, 2253–61.

[1295] R. A. Kent, *J. Am. Chem. Soc.* **1968**, 90, 5657–9.

[1296] P. D. Kleinschmidt, J. W. Ward, *J. Less-Common Met.* **1986**, 121, 61–6.

[1297] Y. Mochizuki, H. Tatewaki, *J. Chem. Phys.* **2003**, 118, 9201–7.

[1298] P. Ehrlich, F. Plöger, E. Koch, G. Kaupa, *Z. Anorg. Allg. Chem.* **1964**, 333, 209–15.

[1299] J. C. Wallmann, W. W. T. Crane, B. B. Cunningham, *J. Am. Chem. Soc.* **1951**, 73, 493–4.

[1300] G. A. Burney, F. W. Tober, *Ind. Eng. Chem. Process Des. Dev.* **1965**, 4, 28–32.

[1301] W. R. Wilmarth, G. M. Begun, R. G. Haire, J. R. Peterson, *J. Chem. Phys.* **1988**, 89, 4666–70.

[1302] A. A. Lizin, S. V. Tomilin, O. E. Gnevashov, et al., *At. Energy* **2013**, 115, 11–7.

[1303] C. J. Mandleberg, K. E. Francis, R. Smith, *J. Chem. Soc.* **1961**, 2464–8.

[1304] E. F. Westrum, J. C. Wallmann, *J. Am. Chem. Soc.* **1951**, 73, 3530–1.

[1305] O. J. C. Runnalls, *Can. J. Chem.* **1953**, 31, 694–6.

[1306] J. L. Burnett, *J. Inorg. Nucl. Chem.* **1966**, 28, 2454–6.

[1307] J. D. Moseley, H. N. Robinson, *J. Inorg. Nucl. Chem.* **1968**, 30, 2277–9.

[1308] H. A. Friedman, C. F. Weaver, W. R. Grimes, *J. Inorg. Nucl. Chem.* **1970**, 32, 3131–3.

[1309] U. Berndt, B. Erdmann, *Radiochim. Acta* **1973**, 19, 45.

[1310] B. Claux, O. Beneš, E. Capelli, P. Souček, R. Meier, *J. Fluorine Chem.* **2016**, 183, 10–3.

[1311] H. Tagawa, *J. Inorg. Nucl. Chem.* **1975**, 37, 731–3.

[1312] S. S. Rudel, H. L. Deubner, B. Scheibe, M. Conrad, F. Kraus, *Z. Anorg. Allg. Chem.* **2018**, 644, 323–9.

[1313] J. N. Stevenson, J. R. Peterson, *J. Inorg. Nucl. Chem.* **1973**, 35, 3481–6.

[1314] D. H. Templeton, C. H. Dauben, *J. Am. Chem. Soc.* **1953**, 75, 4560–2.

[1315] L. B. Asprey, T. K. Keenan, F. H. Kruse, *Inorg. Chem.* **1965**, 4, 985–6.

[1316] F. Weigel, R. Marquart, *J. Less-Common Met.* **1983**, 90, 283–90.

[1317] A. Skerencak, P. J. Panak, V. Neck, et al., *J. Phys. Chem. B* **2010**, 114, 15626–34.

[1318] J. R. Peterson, J. H. Burns, *J. Inorg. Nucl. Chem.* **1968**, 30, 2955–8.

[1319] J. P. Rannou, J. Lucas, *Mater. Res. Bull.* **1969**, 4, 443–50.

[1320] J. Lucas, J.-P. Rannou, *C. R. Acad. Sci., Ser. IIc: Chim.* **1968**, 266C, 1056–8.

[1321] S. K. Roy, T. Jian, G. V. Lopez, et al., *J. Chem. Phys.* **2016**, 144, 084309.

[1322] J. Fuger. Thermodynamics of Plutonium Halides and Halogeno Complexes in the Solid State and in Aqueous Media. Plutonium Chemistry: American Chemical Society; 1983:75–98.

[1323] C. E. Bamberger, R. G. Ross, C. F. Baes, J. P. Young, *J. Inorg. Nucl. Chem.* **1971**, 33, 3591–4.

[1324] C. E. Johnson, J. Fischer, M. J. Steindler, *J. Am. Chem. Soc.* **1961**, 83, 1620–2.

[1325] W. V. Conner, *J. Less-Common Met.* **1971**, 25, 379–84.

[1326] S. S. Rudel, S. A. Baer, P. Woidy, et al., *Z. Kristallogr.* **2018**, 233, 817.

[1327] R. G. Haire, L. B. Asprey, *Inorg. Nucl. Chem. Lett.* **1973**, 9, 869–74.

[1328] D. Brown, B. Whittaker, J. A. Berry, J. H. Holloway, *J. Less-Common Met.* **1982**, 86, 75–84.

[1329] L. B. Asprey, F. H. Ellinger, S. Fried, W. H. Zachariasen, *J. Am. Chem. Soc.* **1957**, 79, 5825.

[1330] T. K. Keenan, L. B. Asprey, *Inorg. Chem.* **1969**, 8, 235–8.

[1331] G. Bouissières, B. Jouniaux, Y. Legoux, J. Merinis, *C. R. Acad. Sci., Ser. IIc: Chim.* **1980**, 290C, 381–2.

[1332] T. O'Donnell, P. Wilson, *Aust. J. Chem.* **1969**, 22, 1877–82.

[1333] L. B. Asprey, R. G. Haire, *Inorg. Nucl. Chem. Lett.* **1973**, 9, 1121–8.

[1334] H. O. Haug, R. D. Baybarz, *Inorg. Nucl. Chem. Lett.* **1975**, 11, 847–55.

[1335] R. J. M. Konings, D. L. Hildenbrand, *J. Alloys Compd.* **1998**, 271–273, 583–6.

[1336] K. R. Kunze, R. H. Hauge, D. Hamill, J. L. Margrave, *J. Phys. Chem.* **1977**, 81, 1664–7.

[1337] M. J. Steindler, D. V. Steidl, R. K. Steunenberg, *Nucl. Sci. Eng.* **1959**, 6, 333–40.

[1338] L. B. Asprey, R. A. Penneman, *J. Am. Chem. Soc.* **1967**, 89, 172.

[1339] L. E. Trevorrow, T. J. Gerding, M. J. Steindler, *J. Inorg. Nucl. Chem.* **1968**, 30, 2671–7.

[1340] J. K. Dawson, R. W. M. D'Eye, A. E. Truswell, *J. Chem. Soc.* **1954**, 3922–9.

[1341] L. Stein, *Inorg. Chem.* **1964**, 3, 995–8.

[1342] M. Laser, E. Merz, *J. Inorg. Nucl. Chem.* **1969**, 31, 349–60.

[1343] C. J. Mandleberg, D. Davies, *J. Chem. Soc.* **1961**, 2031–7.

[1344] P. D. Kleinschmidt, K. H. Lau, D. L. Hildenbrand, *J. Chem. Phys.* **1992**, 97, 1950–3.

[1345] T. K. Keenan, *J. Am. Chem. Soc.* **1961**, 83, 3719–20.

[1346] L. N. Sidorov, L. V. Zhuravleva, M. V. Varkov, et al., *Int. J. Mass Spectrom.* **1983**, 51, 291–311.

[1347] R. Felmy Andrew, D. Rai, J. M. Batteile Marvin, *Radiochim. Acta* **1993**, 62, 133.

[1348] W. Zachariasen, *Acta Crystallogr.* **1949**, 2, 390–3.

[1349] R. Benz, R. M. Douglass, F. H. Kruse, R. A. Penneman, *Inorg. Chem.* **1963**, 2, 799–803.

[1350] L. B. Asprey, F. H. Kruse, R. A. Penneman, *Inorg. Chem.* **1967**, 6, 544–8.

[1351] D. Brown, B. Whittaker, N. Edelstein, *Inorg. Chem.* **1974**, 13, 1805–8.

[1352] H. Abazli, J. Jové, M. Pagès, *C. R. Acad. Sci., Ser. IIc: Chim.* **1979**, 288C, 157–9.

[1353] R. N. Singh, I. P. Saraswat, S. K. Suri, *Synth. React. Inorg. Met.-Org. Chem.* **1983**, 13, 21–8.

[1354] R. R. Ryan, R. A. Penneman, A. Rosenzweig, *Acta Crystallogr. B* **1969**, 25, 1958–62.

[1355] A. Rosenzweig, D. T. Cromer, *Acta Crystallogr. B* **1970**, 26, 38–44.

[1356] B. N. Wani, S. J. Patwe, U. R. K. Rao, K. S. Venkateswarlu, *J. Fluorine Chem.* **1989**, 44, 177–85.

[1357] P. Woidy, A. J. Karttunen, S. S. Rudel, F. Kraus, *Chem. Commun.* **2015**, 51, 11826–9.

[1358] R. W. M. D'Eye, *J. Chem. Soc.* **1958**, 196–9.

[1359] K. H. Lau, D. L. Hildenbrand, *J. Phys. Chem.* **1989**, 93, 3768–70.

[1360] Y. Gong, X. Wang, L. Andrews, T. Schlöder, S. Riedel, *Inorg. Chem.* **2012**, 51, 6983–91.

[1361] D. Brown, J. F. Easey, *J. Chem. Soc. A* **1970**, 3378–81.

[1362] T. A. O'Donnell, A. B. Waugh, C. H. Randall, *J. Inorg. Nucl. Chem.* **1977**, 39, 1597–600.

[1363] J. Levy, P. Wilson, *Aust. J. Chem.* **1973**, 26, 2711–4.

[1364] L. B. Asprey, R. T. Paine, *J. Chem. Soc., Chem. Commun.* **1973**, 920–1.

[1365] M. R. Bennett, L. M. Ferris, *J. Inorg. Nucl. Chem.* **1974**, 36, 1285–90.

[1366] E. Jacob, *Z. Anorg. Allg. Chem.* **1973**, 400, 45–50.

[1367] D. Brown, J. A. Berry, J. H. Holloway, G. M. Staunton, *J. Less-Common Met.* **1983**, 92, 149–53.

[1368] G. W. Halstead, P. G. Eller, L. B. Asprey, K. V. Salazar, *Inorg. Chem.* **1978**, 17, 2967–9.

[1369] J. A. Berry, A. Prescott, D. W. A. Sharp, J. M. Winfield, *J. Fluorine Chem.* **1977**, 10, 247–54.

[1370] D. K. Sanyal, D. W. A. Sharp, J. M. Winfield, *J. Fluorine Chem.* **1981**, 19, 55–60.

[1371] G. W. Halstead, P. G. Eller, R. T. Paine, *Inorg. Synth.* **1982**, 162–7.

[1372] R. T. Paine, L. B. Asprey, L. Graham, N. Bartlett, *Inorg. Synth.* **1979**.

[1373] M. Baluka, S. Yeh, R. Banks, N. Edelstein, *Inorg. Nucl. Chem. Lett.* **1980**, 16, 75–7.

[1374] T. A. O'Donnell, R. Rietz, S. Yeh, *J. Fluorine Chem.* **1983**, 23, 97–102.

[1375] J. G. Malm, C. W. Williams, L. Soderholm, L. R. Morss, *J. Alloys Compd.* **1993**, 194, 133–7.

[1376] L. M. Ferris, *J. Am. Chem. Soc.* **1957**, 79, 5419–21.

[1377] W. Zachariasen, *Acta Crystallogr.* **1949**, 2, 296–8.

[1378] R. R. Ryan, R. A. Penneman, L. B. Asprey, R. T. Paine, *Acta Crystallogr. B* **1976**, 32, 3311–3.

[1379] P. Gary Eller, A. C. Larson, J. R. Peterson, D. D. Ensor, J. P. Young, *Inorg. Chim. Acta* **1979**, 37, 129–33.

[1380] B. J. Krohn, W. B. Person, J. Overend, *J. Chem. Phys.* **1976**, 65, 969–76.

[1381] L. H. Jones, S. Ekberg, *J. Chem. Phys.* **1977**, 67, 2591–5.

[1382] J. Onoe, H. Nakamatsu, T. Mukoyama, R. Sekine, H. Adachi, K. Takeuchi, *Inorg. Chem.* **1997**, 36, 1934–8.

[1383] G. W. Halstead, P. G. Eller, M. P. Eastman, *Inorg. Chem.* **1979**, 18, 2867–72.

[1384] R. Bougon, P. Cearpin, *J. Fluorine Chem.* **1979**, 14, 235–41.

[1385] B. Scheibe, S. S. Rudel, M. R. Buchner, A. J. Karttunen, F. Kraus, *Chem. Eur. J.* **2017**, 23, 291–5.

[1386] L. B. Asprey, R. A. Penneman, *Inorg. Chem.* **1964**, 3, 727–9.

[1387] W. W. Wilson, C. Naulin, R. Bougon, *Inorg. Chem.* **1977**, 16, 2252–7.

[1388] J. P. Masson, C. Naulin, P. Charpin, R. Bougon, *Inorg. Chem.* **1978**, 17, 1858–61.

[1389] J. C. Taylor, *Inorg. Nucl. Chem. Lett.* **1976**, 12, 725–8.

[1390] P. G. Eller, J. G. Malm, B. I. Swanson, L. R. Morss, *J. Alloys Compd.* **1998**, 269, 50–6.

[1391] M. N. Bukhsh, J. Flegenheimer, F. M. Hall, A. G. Maddock, C. F. de Miranda, *J. Inorg. Nucl. Chem.* **1966**, 28, 421–31.

[1392] R. A. Penneman, L. B. Asprey, G. Sturgeon, *J. Am. Chem. Soc.* **1962**, 84, 4608–9.

[1393] R. A. Penneman, G. D. Sturgeon, L. B. Asprey, *Inorg. Chem.* **1964**, 3, 126–9.

[1394] G. D. Sturgeon, R. A. Penneman, F. H. Kruse, L. B. Asprey, *Inorg. Chem.* **1965**, 4, 748–50.

[1395] P. Charpin, *C. R. Acad. Sci., Ser. IIc: Chim.* **1965**, 260C, 1914–6.

[1396] R. Bougon, P. Plurien, *C. R. Acad. Sci., Ser. IIc: Chim.* **1965**, 260C, 4217–8.

[1397] D. Brown, J. F. Easey, C. E. F. Rickard, *J. Chem. Soc. A* **1969**, 1161–4.

[1398] F. Montoloy, P. Plurien, *C. R. Acad. Sci., Ser. IIc: Chim.* **1968**, 267C, 1036–8.

[1399] B. Scheibe, C. Pietzonka, O. Mustonen, et al., *Angew. Chem., Int. Ed. Engl.* **2018**, 57, 2914–8.

[1400] T. Kanatani, K. Matsumoto, R. Hagiwara, *Chem. Lett.* **2009**, 38, 714–5.

[1401] K. W. Bagnall, D. Brown, J. F. Easey, *J. Chem. Soc. A* **1968**, 2223–7.

[1402] M. Fargeas, R. Fremont-Lamouranne, Y. Legouv, J. Merini, *J. Less-Common Met.* **1986**, 121, 439–44.

[1403] J. Slivnik, K. Lutar, A. Šmalc, *J. Fluorine Chem.* **1977**, 9, 255–6.

[1404] J. Slivnik, K. Lutar, *J. Fluorine Chem.* **1978**, 11, 643–6.

[1405] L. B. Asprey, S. A. Kinkead, P. G. Eller, *Nucl. Technol.* **1986**, 73, 69–71.

[1406] J. G. Malm, P. G. Eller, L. B. Asprey, *J. Am. Chem. Soc.* **1984**, 106, 2726–7.

[1407] L. B. Asprey, P. G Eller, S. A. Kinkead, *Inorg. Chem.* **1986**, 25, 670–2.

[1408] L. E. Trevorrow, W. A. Shinn, R. K. Steunenberg, *J. Phys. Chem.* **1961**, 65, 398–403.

[1409] M. J. Steindler, D. V. Steidl, J. Fischer, *J. Inorg. Nucl. Chem.* **1964**, 26, 1869–78.

[1410] L. E. Trevorrow, T. J. Gerding, M. J. Steindler, *Inorg. Nucl. Chem. Lett.* **1969**, 5, 837–9.

[1411] J. K. Gibson, R. G. Haire, *J. Alloys Compd.* **1992**, 181, 23–32.

[1412] M. Kimura, V. Schomaker, D. W. Smith, B. Weinstock, *J. Chem. Phys.* **1968**, 48, 4001–12.

[1413] J. H. Levy, J. C. Taylor, P. W. Wilson, *J. Chem. Soc., Dalton Trans.* **1976**, 219–24.

[1414] M. Boring, *Chem. Phys. Lett.* **1977**, 46, 242–4.

[1415] B. P. Mathur, E. W. Rothe, G. P. Reck, *J. Chem. Phys.* **1977**, 67, 377–81.

[1416] R. N. Compton, *J. Chem. Phys.* **1977**, 66, 4478–85.

[1417] J. L. Beauchamp, *J. Chem. Phys.* **1976**, 64, 929–35.

[1418] J. A. Berry, R. T. Poole, A. Prescott, D. W. A. Sharp, J. M. Winfield, *J. Chem. Soc., Dalton Trans.* **1976**, 272–4.

[1419] L. McGhee, D. S. Rycroft, J. M. Winfield, *J. Fluorine Chem.* **1987**, 36, 351–9.

[1420] G. A. Olah, J. Welch, T.-L. Ho, *J. Am. Chem. Soc.* **1976**, 98, 6717–8.

[1421] G. A. Olah, J. Welch, *J. Am. Chem. Soc.* **1978**, 100, 5396–402.

[1422]   Y. Yato, H. Funasaka, *J. Nucl. Sci. Technol.* **1992**, 29, 296–9.

[1423]   Y. Yato, O. Suto, H. Funasaka, *J. Nucl. Sci. Technol.* **1995**, 32, 430–8.

[1424]   R. D. Peacock, N. Edelstein, *J. Inorg. Nucl. Chem.* **1976**, 38, 771–3.

[1425]   R. C. Burns, T. A. O'Donnell, *Inorg. Nucl. Chem. Lett.* **1977**, 13, 657–60.

[1426]   P. W. Wilson, *J. Chem. Soc., Chem. Commun.* **1972**, 1241.

[1427]   P. W. Wilson, *J. Inorg. Nucl. Chem.* **1974**, 36, 303–5.

[1428]   E. Jacob, W. Polligkeit, *Z. Naturforsch. B* **1973**, 28, 120.

[1429]   D. Brown. Halides of the Lanthanides and Actinides: Wiley-Interscience (1968); 1968.

[1430]   R. Bougon, P. Charpin, J. P. Desmoulin, J. G. Malm, *Inorg. Chem.* **1976**, 15, 2532–40.

[1431]   M. Iwasaki, N. Ishikawa, K. Ohwada, *J. Inorg. Nucl. Chem.* **1977**, 39, 2191–2.

[1432]   M. Iwasaki, N. Ishikawa, K. Ohwada, *J. Inorg. Nucl. Chem.* **1978**, 40, 503–5.

[1433]   J. G. Malm, H. Selig, S. Siegel, *Inorg. Chem.* **1966**, 5, 130–2.

[1434]   J. G. Malm, *J. Fluorine Chem.* **1983**, 23, 267–82.

[1435]   B. Scheibe, S. Lippert, S. S. Rudel, et al., *Chem. Eur. J.* **2016**, 22, 12145–53.

[1436]   F. Kraus, S. A. Baer, *Chem. Eur. J.* **2009**, 15, 8269–74.

[1437]   R. T. Paine, R. R. Ryan, L. B. Asprey, *Inorg. Chem.* **1975**, 14, 1113–7.

[1438]   R. C. Burns, T. A. O'Donnell, A. B. Waugh, *J. Fluorine Chem.* **1978**, 12, 505–17.

[1439]   J. C. Taylor, P. W. Wilson, *Acta Crystallogr. B* **1974**, 30, 1701–5.

[1440]   J. C. Taylor, P. W. Wilson, *J. Chem. Soc., Chem. Commun.* **1974**, 232–3.

[1441]   J. H. Levy, J. C. Taylor, P. W. Wilson, *J. Inorg. Nucl. Chem.* **1977**, 39, 1989–91.

[1442]   A. Kovács, R. J. M. Konings, *ChemPhysChem* **2006**, 7, 455–62.

[1443]   G. A. Shamov, G. Schreckenbach, T. N. Vo, *Chem. Eur. J.* **2007**, 13, 4932–47.

[1444]   P. W. Wilson, *J. Inorg. Nucl. Chem.* **1974**, 36, 1783–5.

[1445]   M. Straka, K. G. Dyall, P. Pyykkö, *Theor. Chem. Acc.* **2001**, 106, 393–403.

[1446]   W. Huang, P. Pyykkö, J. Li, *Inorg. Chem.* **2015**, 54, 8825–31.

[1447]   P. Joubert, R. Bougon, *C. R. Acad. Sci., Ser. IIc: Chim.* **1975**, 280C, 193–5.

[1448]   K. W. Bagnall, J. G. H. du Preez, B. J. Gellatly, J. H. Holloway, *J. Chem. Soc., Dalton Trans.* **1975**, 1963–8.

[1449]   C. J. Mandleberg, H. K. Rae, R. Hurst, G. Long, D. Davies, K. E. Francis, *J. Inorg. Nucl. Chem.* **1956**, 2, 358–67.

[1450]   A. E. Florin, I. R. Tannenbaum, J. F. Lemons, *J. Inorg. Nucl. Chem.* **1956**, 2, 368–79.

[1451]   S. Ahrland, L. Brandt, *Acta Chem. Scand.* **1968**, 22, 106–14.

[1452]   T. K. Keenan, *Inorg. Nucl. Chem. Lett.* **1968**, 4, 381–4.

[1453]   M. Atoji, M. J. McDermott, *Acta Crystallogr. B* **1970**, 26, 1540–4.

[1454]   B. McNamara, R. Scheele, A. Kozelisky, M. Edwards, *J. Nucl. Mater.* **2009**, 394, 166–73.

[1455]   W. Zachariasen, *Acta Crystallogr.* **1948**, 1, 277–81.

[1456]   P. Woidy, A. J. Karttunen, F. Kraus, *Z. Anorg. Allg. Chem.* **2012**, 638, 2044–52.

[1457]   P. D. Kleinschmidt, K. H. Lau, D. L. Hildenbrand, *J. Chem. Phys.* **1992**, 97, 2417–21.

[1458]   G. Tian, L. Rao, *Inorg. Chem.* **2009**, 48, 6748–54.

[1459]   J. Su, P. D. Dau, Y.-H. Qiu, et al., *Inorg. Chem.* **2013**, 52, 6617–26.

[1460]   P. D. Dau, J. Su, H.-T. Liu, et al., *Chem. Sci.* **2012**, 3, 1137–46.

[1461]   R. Haiges, M. Vasiliu, D. A. Dixon, K. O. Christe, *Chem. Eur. J.* **2017**, 23, 652–64.

[1462]   R. L. Diebner, J. G. Kay, *J. Chem. Phys.* **1969**, 51, 3547–54.

[1463]   R. S. Ram, J. R. D. Peers, Y. Teng, et al., *J. Mol. Spectrosc.* **1997**, 184, 186–201.

[1464]   A. Chatalic, P. Deschamps, G. Pannetier, *C. R. Acad. Sci., Ser. IIc: Chim.* **1970**, 270C, 146–9.

[1465]   R. S. Ram, P. F. Bernath, *J. Mol. Spectrosc.* **2005**, 231, 165–70.

[1466]   E. A. Shenyavskaya, V. M. Dubov, *J. Mol. Spectrosc.* **1985**, 113, 85–92.

[1467]   A. I. Boldyrev, J. Simons, *J. Mol. Spectrosc.* **1998**, 188, 138–41.

[1468]   C. Koukounas, S. Kardahakis, A. Mavridis, *J. Chem. Phys.* **2004**, 120, 11500–21.

[1469] J. W. Hastie, R. H. Hauge, J. L. Margrave, *J. Chem. Phys.* **1969**, 51, 2648–56.
[1470] N. Morita, T. Endo, T. Sato, M. Shimada, *J. Mater. Sci. Lett.* **1987**, 6, 859–61.
[1471] A. V. Wilson, A. J. Roberts, N. A. Young, *Angew. Chem., Int. Ed. Engl.* **2008**, 47, 1774–6.
[1472] S. G. Wang, W. H. E. Schwarz, *J. Chem. Phys.* **1998**, 109, 7252–62.
[1473] M. Vogel, W. Wenzel, *Chem. Phys. Lett.* **2005**, 413, 42–6.
[1474] J. J. Eisch, J. N. Gitua, P. O. Otieno, X. Shi, *J. Organomet. Chem.* **2001**, 624, 229–38.
[1475] P. Ehrlich, G. Pietzka, *Z. Anorg. Allg. Chem.* **1954**, 275, 121–40.
[1476] K. Koyama, Y. Hashimoto, *Nippon Kagaku Kaishi* **1973**, 1973, 195–7.
[1477] R. Hoppe, S. Becker, *Z. Anorg. Allg. Chem.* **1989**, 568, 126–35.
[1478] S. Siegel, *Acta Crystallogr.* **1956**, 9, 684.
[1479] B. J. Kennedy, T. Vogt, *Mater. Res. Bull.* **2002**, 37, 77–83.
[1480] T. C. De Vore, W. Weltner, *J. Am. Chem. Soc.* **1977**, 99, 4700–3.
[1481] J. H. Yates, R. M. Pitzer, *J. Chem. Phys.* **1979**, 70, 4049–55.
[1482] R. S. P. Coutts, P. C. Wailes, R. L. Martin, *J. Organomet. Chem.* **1973**, 47, 375–82.
[1483] P. Bukovec, J. Šiftar, *Monatsh. Chem.* **1974**, 105, 510–6.
[1484] J. Omaly, P. Batail, D. Grandjean, D. Avignant, J.-C. Cousseins, *Acta Crystallogr. B* **1976**, 32, 2106–10.
[1485] W. Massa, W. Rüdorff, *Z. Naturforsch. B* **1971**, 26, 1216.
[1486] W. E. Hatfield, P. J. Nassiff, T. W. Couch, J. F. Villa, *Inorg. Chem.* **1971**, 10, 368–73.
[1487] E. Alter, R. Hoppe, *Z. Anorg. Allg. Chem.* **1974**, 403, 127–36.
[1488] P. Burkert, H. P. Fritz, G. Stefaniak, *Z. Naturforsch. B* **1968**, 23, 872.
[1489] J. Gaile, W. Rüdorff, W. Viebahn, *Z. Anorg. Allg. Chem.* **1977**, 430, 161–74.
[1490] J. Ravez, M. Vassiliadis, *C. R. Acad. Sci., Ser. IIc: Chim.* **1970**, 270C, 219–21.
[1491] L. Herman, G. Mitra, *J. Fluorine Chem.* **1972**, 1, 498–9.
[1492] S. Yamanaka, A. Yasuda, H. Miyata, *J. Solid State Chem.* **2010**, 183, 256–61.
[1493] P. Wang, X. D. Kang, H. M. Cheng, *ChemPhysChem* **2005**, 6, 2488–91.
[1494] A. Grzech, U. Lafont, P. C. M. M. Magusin, F. M. Mulder, *J. Phys. Chem. C* **2012**, 116, 26027–35.
[1495] Z. Z. Fang, L. P. Ma, X. D. Kang, P. J. Wang, P. Wang, H. M. Cheng, *Appl. Phys. Lett.* **2009**, 94, 044104.
[1496] Y. H. Guo, X. B. Yu, L. Gao, G. L. Xia, Z. P. Guo, H. K. Liu, *Energy Environ. Sci.* **2010**, 3, 464–9.
[1497] H. Chen, R. T. Yang, *Langmuir* **2010**, 26, 15394–8.
[1498] A. Kitajou, I. Tanaka, Y. Tanaka, et al., *Electrochemistry* **2017**, 85, 472–7.
[1499] A. Kitajou, K. Eguchi, Y. Ishado, H. Setoyama, T. Okajima, S. Okada, *J. Power Sources* **2019**, 419, 1–5.
[1500] B. L. Chamberland, A. W. Sleight, *Solid State Commun.* **1967**, 5, 765–7.
[1501] B. L. Chamberland, A. W. Sleight, W. H. Cloud, *J. Solid State Chem.* **1970**, 2, 49–54.
[1502] J. Cumby, M. B. Burchell, J. P. Attfield, *Solid State Sci.* **2018**, 80, 35–8.
[1503] R. Wei, Z. Fang, M. Vasiliu, D. A. Dixon, L. Andrews, Y. Gong, *Inorg. Chem.* **2019**.
[1504] B. Leng, J. H. Moss, *J. Fluorine Chem.* **1975**, 5, 93–8.
[1505] L. E. Alexander, I. R. Beattie, *J. Chem. Soc., Dalton Trans.* **1972**, 1745–50.
[1506] D. S. Dyer, R. O. Ragsdale, *Inorg. Chem.* **1967**, 6, 8–11.
[1507] J. W. Downing, R. O. Ragsdale, *Inorg. Chem.* **1968**, 7, 1675–7.
[1508] C. E. Michelson, D. S. Dyer, R. O. Ragsdale, *J. Chem. Soc. A* **1970**, 2296–8.
[1509] R. J. H. Clark, W. Errington, *J. Chem. Soc. A* **1967**, 258–61.
[1510] G. B. Nikiforov, H. W. Roesky, P. G. Jones, J. Magull, A. Ringe, R. B. Oswald, *Inorg. Chem.* **2008**, 47, 2171–9.
[1511] H. Bürger, K. Wiegel, *Z. Anorg. Allg. Chem.* **1973**, 398, 257–72.
[1512] W. S. Sheldrick, *J. Fluorine Chem.* **1974**, 4, 415–21.

[1513]   J. A. Chandler, R. S. Drago, *Inorg. Chem.* **1962**, 1, 356–8.

[1514]   J. A. Chandler, J. E. Wuller, R. S. Drago, *Inorg. Chem.* **1962**, 1, 65–9.

[1515]   R. Haiges, J. A. Boatz, S. Schneider, T. Schroer, M. Yousufuddin, K. O. Christe, *Angew. Chem., Int. Ed. Engl.* **2004**, 43, 3148–52.

[1516]   T. Saal, P. Deokar, K. O. Christe, R. Haiges, *Eur. J. Inorg. Chem.* **2019**, 2019, 2388–91.

[1517]   J. Fischer, G. Keib, R. Weiss, *Acta Crystallogr.* **1967**, 22, 338–40.

[1518]   A. Decian, J. Fischer, R. Weiss, *Acta Crystallogr.* **1967**, 22, 340–3.

[1519]   B. Kojic-Prodic, B. Matkovic, S. Scavnicar, *Acta Crystallogr. B* **1971**, 27, 635–7.

[1520]   I. W. Forrest, A. P. Lane, *Inorg. Chem.* **1976**, 15, 265–9.

[1521]   D. L. Deadmore, J. S. Machin, A. W. Allen, *J. Am. Ceram. Soc.* **1962**, 45, 120–2.

[1522]   Y. A. Buslaev, D. S. Dyer, R. O. Ragsdale, *Inorg. Chem.* **1967**, 6, 2208–12.

[1523]   R. Weiss, J. Fischer, B. Chevrier, *Acta Crystallogr.* **1966**, 20, 534–7.

[1524]   R. Weiss, J. Fischer, G. Keib, *C. R. Acad. Sci., Ser. IIc: Chim.* **1964**, 259C, 1125–7.

[1525]   J. A. Chandler, R. S. Drago, R. Latham, *J. Inorg. Nucl. Chem.* **1961**, 21, 283–6.

[1526]   H. G. Lee, D. S. Dyer, R. O. Ragsdale, *J. Chem. Soc., Dalton Trans.* **1976**, 1325–9.

[1527]   W. P. Griffith, *J. Chem. Soc.* **1964**, 5248–53.

[1528]   R. Stomberg, I.-B. Svensson, *Acta Chem. Scand.* **1977**, 31 A, 635–7.

[1529]   K. S. Vorres, F. B. Dutton, *J. Am. Chem. Soc.* **1955**, 77, 2019.

[1530]   K. Dehnicke, *Sci. Nat.* **1965**, 52, 660.

[1531]   J. H. Moss, A. Wright, *J. Fluorine Chem.* **1975**, 5, 163–7.

[1532]   N. Louvain, Z. Karkar, M. El-Ghozzi, P. Bonnet, K. Guérin, P. Willmann, *J. Mater. Chem. A* **2014**, 2, 15308–15.

[1533]   J. K. Ghosh, G. V. Jere, *J. Fluorine Chem.* **1987**, 35, 669–76.

[1534]   S.-T. Myung, M. Kikuchi, C. S. Yoon, H. Yashiro, Y.-K. Sun, *J. Power Sources* **2015**, 288, 376–83.

[1535]   K. Vorres, J. Donohue, *Acta Crystallogr.* **1955**, 8, 25–6.

[1536]   S. Shian, K. H. Sandhage, *J. Appl. Crystallogr.* **2010**, 43, 757–61.

[1537]   G. Pausewang, W. Rüdorff, *Z. Anorg. Allg. Chem.* **1969**, 364, 69–87.

[1538]   V. Ya. Kavun, S. G. Kozlova, I. A. Tkachenko, S. P. Gabuda, *J. Struct. Chem.* **2010**, 51, 463–70.

[1539]   G. Blasse, G. J. Dirksen, G. J. Pausewang, R. Schmidt, *J. Solid State Chem.* **1990**, 88, 586–9.

[1540]   J. B. Felder, J. Yeon, H.-C. zur Loye, *Solid State Sci.* **2015**, 48, 212–7.

[1541]   S. Deki, Y. Aoi, O. Hiroi, A. Kajinami, *Chem. Lett.* **1996**, 25, 433–4.

[1542]   M. V. Reddy, S. Madhavi, G. V. Subba Rao, B. V. R. Chowdari, *J. Power Sources* **2006**, 162, 1312–21.

[1543]   B. Li, D. Wang, Y. Wang, et al., *Electrochim. Acta* **2015**, 180, 894–901.

[1544]   D. Deng, *ChemNanoMat.* **2017**, 3, 146–59.

[1545]   M. He, Z. Wang, X. Yan, L. Tian, G. Liu, X. Chen, *J. Power Sources* **2016**, 306, 309–16.

[1546]   J. Zhu, D. Zhang, Z. Bian, et al., *Chem. Commun.* **2009**, 5394–6.

[1547]   Z. Wang, B. Huang, Y. Dai, et al., *CrystEngComm* **2012**, 14, 4578–81.

[1548]   L. Chen, L. Shen, P. Nie, X. Zhang, H. Li, *Electrochim. Acta* **2012**, 62, 408–15.

[1549]   F. Li, Z. P. Fu, Y. L. Lu, *Adv. Mater. Res.* **2013**, 634–638, 2297–300.

[1550]   Y. Zeng, W. Zhang, C. Xu, et al., *Chem. Eur. J.* **2012**, 18, 4026–30.

[1551]   J. Wang, F. Cao, Z. Bian, M. K. H. Leung, H. Li, *Nanoscale* **2014**, 6, 897–902.

[1552]   D. L. Hildenbrand, K. H. Lau, *J. Chem. Phys.* **1997**, 107, 6349–52.

[1553]   S. Soorkia, N. Shafizadeh, J. Liévin, et al., *J. Phys. Chem. A* **2011**, 115, 9620–32.

[1554]   A. Martinez, M. D. Morse, *J. Chem. Phys.* **2011**, 135, 024308.

[1555]   E. N. Moskvitina, Y. Y. Kuzyakov, *Spectrosc. Lett.* **1999**, 32, 719–28.

[1556]   A. G. Adam, W. S. Hopkins, D. W. Tokaryk, *J. Mol. Spectrosc.* **2004**, 225, 1–7.

[1557]   M. Grau, A. E. Leanhardt, H. Loh, et al., *J. Mol. Spectrosc.* **2012**, 272, 32–5.

[1558]   F. McTaggart, A. Turnbull, *Aust. J. Chem.* **1964**, 17, 727–30.

[1559]   F. Basile, E. Chassaing, G. Lorthioir, *J. Less-Common Met.* **1984**, 98, 1–10.

[1560]   G. v. Hevesy, W. Dullenkopf, *Z. Anorg. Allg. Chem.* **1934**, 221, 161–6.

[1561]   C. E. F. Rickard, T. N. Waters, *J. Inorg. Nucl. Chem.* **1964**, 26, 925–30.

[1562]   O. N. Carlson, F. A. Schmidt, H. A. Wilhelm, *J. Electrochem. Soc.* **1957**, 104, 51–6.

[1563]   M. F. Churbanov, N. K. Rudnevsky, A. M. Tumanova, V. I. Zvereva, Y. V. Maslov, *Mater. Sci. Forum* **1985**, 5–6, 73–5.

[1564]   K. Tanaka, Y. Kita, H. Kawamoto, S. Yoshikwa, *Mater. Sci. Forum* **1991**, 67–68, 63–8.

[1565]   B. M. Vilakazi, O. S. Monnahela, J. B. Wagener, P. A. B. Carstens, T. Ntsoane, *J. Fluorine Chem.* **2012**, 141, 64–8.

[1566]   O. S. Monnahela, B. M. Vilakazi, J. B. Wagener, A. Roodt, P. A. B. Carstens, W. L. Retief, *J. Fluorine Chem.* **2012**, 135, 246–9.

[1567]   C. J. Pretorius, A. D. Pienaar, P. L. Crouse, H. F. Niemand, *Adv. Mater. Res.* **2014**, 1019, 398–405.

[1568]   A. Büchler, J. B. Berkowitz-Mattuck, D. H. Dugre, *J. Chem. Phys.* **1961**, 34, 2202–3.

[1569]   A. Chrétien, B. Gaudreau, *C. R. Acad. Sci., Ser. IIc: Chim.* **1959**, 248C, 2878–9.

[1570]   G. Benner, B. G. Müller, *Z. Anorg. Allg. Chem.* **1990**, 588, 33–42.

[1571]   T. N. Waters, *J. Inorg. Nucl. Chem.* **1960**, 15, 320–8.

[1572]   L. Kolditz, A. Feltz, *Z. Anorg. Allg. Chem.* **1961**, 310, 217–24.

[1573]   D. Hall, C. E. F. Rickard, T. N. Waters, *Nature* **1965**, 207, 405–6.

[1574]   D. Hall, C. E. F. Rickard, T. N. Waters, *J. Inorg. Nucl. Chem.* **1971**, 33, 2395–401.

[1575]   S. L. Benjamin, W. Levason, D. Pugh, G. Reid, W. Zhang, *Dalton Trans.* **2012**, 41, 12548–57.

[1576]   W. Jung, R. Juza, *Z. Anorg. Allg. Chem.* **1973**, 399, 129–47.

[1577]   P. Deokar, M. Vasiliu, D. A. Dixon, K. O. Christe, R. Haiges, *Angew. Chem., Int. Ed. Engl.* **2016**, 55, 14350–4.

[1578]   D. Leroy, D. Ravaine, *C. R. Acad. Sci., Ser. IIc: Chim.* **1978**, 286C, 413–6.

[1579]   W. C. Hasz, J. H. Whang, C. T. Moynihan, *J. Non-Cryst. Solids* **1993**, 161, 127–32.

[1580]   V. Nazabal, M. Poulain, M. Olivier, et al., *J. Fluorine Chem.* **2012**, 134, 18–23.

[1581]   B. Gaudreau, *C. R. Acad. Sci., Ser. IIc: Chim.* **1966**, 263C, 67–70.

[1582]   D. Avignant, J.-C. Cousseins, *C. R. Acad. Sci., Ser. IIc: Chim.* **1972**, 274C, 631–4.

[1583]   I. Ban, B. Volavšek, L. Goliš, *Z. Anorg. Allg. Chem.* **2002**, 628, 695–8.

[1584]   G. D. Robbins, R. E. Thoma, H. Insley, *J. Inorg. Nucl. Chem.* **1965**, 27, 559–68.

[1585]   R. Hoppe, W. Dähne, *Sci. Nat.* **1960**, 47, 397.

[1586]   G. Brunton, *Acta Crystallogr. B* **1973**, 29, 2294–6.

[1587]   J. Fischer, R. Weiss, *Acta Crystallogr. B* **1973**, 29, 1955–7.

[1588]   G. Brunton, *Acta Crystallogr. B* **1969**, 25, 2164–6.

[1589]   A. P. Lane, D. W. A. Sharp, *J. Chem. Soc. A* **1969**, 2942–5.

[1590]   M. Poulain, J. Lucas, *C. R. Acad. Sci., Ser. IIc: Chim.* **1970**, 271C.

[1591]   R. Schmidt, M. Kraus, B. G. Müller, *Z. Anorg. Allg. Chem.* **2001**, 627, 2344–50.

[1592]   M. Müller, B. G. Müller, *Z. Anorg. Allg. Chem.* **1995**, 621, 993–1000.

[1593]   H. J. Hurst, J. C. Taylor, *Acta Crystallogr. B* **1970**, 26, 2136–7.

[1594]   L. Meddar, M. El-Ghozzi, D. Avignant, *Z. Anorg. Allg. Chem.* **2008**, 634, 565–70.

[1595]   J. Fischer, R. Weiss, *Acta Crystallogr. B* **1973**, 29, 1963–7.

[1596]   M. Müller, B. G. Müller, *Z. Anorg. Allg. Chem.* **1995**, 621, 1047–52.

[1597]   J. Fischer, R. Elchinger, R. Weiss, *Acta Crystallogr. B* **1973**, 29, 1967–71.

[1598]   G. Brunton, *Acta Crystallogr. B* **1971**, 27, 1944–8.

[1599]   D. Koller, B. G. Müller, *Z. Anorg. Allg. Chem.* **2000**, 626, 1429–33.

[1600]   T. F. Antokhina, T. A. Kaidalova, N. N. Savchenko, L. N. Ignat'eva, *Russ. J. Inorg. Chem.* **2012**, 57, 1535–9.

[1601] D. Koller, B. G. Müller, *Z. Anorg. Allg. Chem.* **2002**, 628, 575–9.

[1602] M. Kraus, B. G. Müller, *Z. Anorg. Allg. Chem.* **2000**, 626, 1929–33.

[1603] G. v. Hevesy, O. H. Wagner, *Z. Anorg. Allg. Chem.* **1930**, 191, 194–200.

[1604] G. Petit, C. Bourlange, *C. R. Acad. Sci., Ser. IIc: Chim.* **1969**, 269C, 657–60.

[1605] B. Holmberg, *Acta Crystallogr. B* **1970**, 26, 830–5.

[1606] W. E. Jones, G. Krishnamurty, *J. Phys. B* **1980**, 13, 3375–82.

[1607] J. W. Stout, W. O. J. Boo, *J. Appl. Phys.* **1966**, 37, 966–7.

[1608] C. Cros, R. Feurer, M. Pouchard, *J. Fluorine Chem.* **1975**, 5, 457–66.

[1609] M. W. Shafer, *Mater. Res. Bull.* **1969**, 4, 905–12.

[1610] H. J. Seifert, B. Gerstenberg, *Angew. Chem.* **1961**, 73, 657.

[1611] H.-J. Seifert, B. Gerstenberg, *Z. Anorg. Allg. Chem.* **1962**, 315, 56–63.

[1612] J. W. Stout, H. Y. Lau, *J. Appl. Phys.* **1967**, 38, 1472–3.

[1613] W. H. Baur, S. Guggenheim, J.-C. Lin, *Acta Crystallogr. B* **1982**, 38, 351–5.

[1614] M. Vogel, W. Wenzel, *J. Chem. Phys.* **2005**, 123, 194110.

[1615] C. Cros, R. Feurer, M. Pouchard, *Mater. Res. Bull.* **1976**, 11, 117–24.

[1616] C. Cros, R. Feurer, M. Pouchard, *J. Fluorine Chem.* **1976**, 7, 605–18.

[1617] R. F. Williamson, W. O. J. Boo, *Inorg. Chem.* **1977**, 16, 646–8.

[1618] C. Cros, R. Feurer, M. Pouchard, P. Hagenmuller, *Mater. Res. Bull.* **1975**, 10, 383–91.

[1619] R. F. Williamson, W. O. J. Boo, *Inorg. Chem.* **1977**, 16, 649–51.

[1620] R. F. Williamson, W. O. J. Boo, *Inorg. Chem.* **1980**, 19, 31–4.

[1621] Q. Xu, F. Teng, D. Yu, L. Yang, Y. Teng, *J. Fluorine Chem.* **2016**, 188, 153–6.

[1622] O. Ruff, H. Lickfett, *Chem. Ber.* **1911**, 44, 2539–49.

[1623] R. G. Cavell, H. C. Clark, *J. Chem. Soc.* **1962**, 2692–8.

[1624] H. W. Roesky, O. Glemser, K.-H. Hellberg, *Chem. Ber.* **1966**, 99, 459–61.

[1625] O. Glemser, J. Wegener, R. Mews, *Chem. Ber.* **1967**, 100, 2474–83.

[1626] D. Yellin, J. Shamir, H. Selig, *Isr. J. Chem.* **1968**, 6, 505–6.

[1627] B. J. Sturm, C. W. Sheridan, P. H. Crayton, R. N. Vance jr., *Inorg. Synth.* **1963**, 87–92.

[1628] H. Funk, H. Böhland, *Z. Anorg. Allg. Chem.* **1964**, 334, 155–62.

[1629] K. H. Jack, V. Gutmann, *Acta Crystallogr.* **1951**, 4, 246–9.

[1630] P. Daniel, A. Bulou, M. Leblanc, M. Rousseau, J. Nouet, *Mater. Res. Bull.* **1990**, 25, 413–20.

[1631] V. N. Bukhmarina, A. Y. Gerasimov, Y. B. Predtechenskii, *Vib. Spectrosc.* **1992**, 4, 91–4.

[1632] V. G. Solomonik, J. E. Boggs, J. F. Stanton, *J. Phys. Chem. A* **1999**, 103, 838–40.

[1633] E. Petersen, *J. Prakt. Chem.* **1889**, 40, 44–62.

[1634] P. Woidy, F. Kraus, *Z. Naturforsch. B* **2015**, 70, 161.

[1635] E. Alter, R. Hoppe, *Z. Anorg. Allg. Chem.* **1975**, 412, 110–20.

[1636] R. S. Nyholm, A. G. Sharpe, *J. Chem. Soc.* **1952**, 3579–87.

[1637] J.-C. Cousseins, J.-C. Cretenet, *C. R. Acad. Sci., Ser. IIc: Chim.* **1967**, 265C, 1464–7.

[1638] B. N. Wani, U. R. K. Rao, *Synth. React. Inorg. Met.-Org. Chem.* **1991**, 21, 779–91.

[1639] U. Bentrup, *Thermochim. Acta* **1996**, 284, 397–406.

[1640] Y.-S. Hong, W. O. J. Boo, D. L. Mattern, *J. Solid State Chem.* **2010**, 183, 1805–10.

[1641] B. M. Wanklyn, *J. Inorg. Nucl. Chem.* **1965**, 27, 481–2.

[1642] R. Becker, W. Sawodny, *Z. Naturforsch. B* **1973**, 28, 360.

[1643] J.-C. Cretenet, *C. R. Acad. Sci., Ser. IIc: Chim.* **1969**, 268C, 945–7.

[1644] J. Ravez, D. Dumora, *C. R. Acad. Sci., Ser. IIc: Chim.* **1969**, 269C, 235–8.

[1645] J. Graulich, D. Babel, *Z. Anorg. Allg. Chem.* **2003**, 629, 1223–8.

[1646] U. Bentrup, *Z. Anorg. Allg. Chem.* **1993**, 619, 954–60.

[1647] A. Rahten, S. Milićev, *Thermochim. Acta* **1997**, 302, 137–41.

[1648] H. Li, G. Richter, J. Maier, *Adv. Mater.* **2003**, 15, 736–9.

[1649] M. Nishijima, I. D. Gocheva, S. Okada, T. Doi, J.-i. Yamaki, T. Nishida, *J. Power Sources* **2009**, 190, 558–62.

[1650] J. Krishna Murthy, U. Groß, S. Rüdiger, E. Ünveren, E. Kemnitz, *J. Fluorine Chem.* **2004**, 125, 937–49.

[1651] M. Bayard, M. Pouchard, P. Hagenmuller, A. Wold, *J. Solid State Chem.* **1975**, 12, 41–50.

[1652] W. E. Falconer, G. R. Jones, W. A. Sunder, I. Haigh, R. D. Peacock, *J. Inorg. Nucl. Chem.* **1973**, 35, 751–3.

[1653] Y.-P. Chang, L. Furness, W. Levason, G. Reid, W. Zhang, *J. Fluorine Chem.* **2016**, 191, 149–60.

[1654] S. Becker, B. G. Müller, *Angew. Chem.* **1990**, 102, 426–7.

[1655] H. Selig, B. Frlec, *J. Inorg. Nucl. Chem.* **1967**, 29, 1887–94.

[1656] W. Liebe, E. Weise, W. Klemm, *Z. Anorg. Allg. Chem.* **1961**, 311, 281–9.

[1657] S. Nakhal, D. Weber, M. Lerch, *Z. Naturforsch. B* **2013**, 68, 121.

[1658] L. Kolditz, V. Neumann, G. Kilch, *Z. Anorg. Allg. Chem.* **1963**, 325, 275–80.

[1659] R. Haiges, J. A. Boatz, K. O. Christe, *Angew. Chem., Int. Ed. Engl.* **2010**, 49, 8008–12.

[1660] T. Funaioli, F. Marchetti, G. Pampaloni, S. Zacchini, *Dalton Trans.* **2013**, 42, 14168–77.

[1661] M. Schabert, G. Pausewang, *Z. Naturforsch. B* **1985**, 40, 1437.

[1662] G. Pausewang, *Z. Anorg. Allg. Chem.* **1971**, 381, 189–97.

[1663] P. Bukove, S. Milićev, A. Demšar, L. Golič, *J. Chem. Soc., Dalton Trans.* **1981**, 1802–6.

[1664] A. Demšar, P. Bukovec, *Thermochim. Acta* **1987**, 115, 249–54.

[1665] A. Demšar, P. Bukovec, *Thermochim. Acta* **1988**, 133, 247–50.

[1666] A. Demšar, P. Bukovec, *Thermochim. Acta* **1988**, 131, 133–40.

[1667] A. Demšar, P. Bukovec, *Inorg. Chim. Acta* **1977**, 25, L121–L2.

[1668] H. J. Eméleus, V. Gutmann, *J. Chem. Soc.* **1949**, 2979–82.

[1669] L. E. Trevorrow, J. Fischer, R. K. Steunenberg, *J. Am. Chem. Soc.* **1957**, 79, 5167–8.

[1670] A. Šmalc, *Monatsh. Chem.* **1967**, 98, 163–4.

[1671] R. G. Cavell, H. C. Clark, *Inorg. Chem.* **1964**, 3, 1789–91.

[1672] H. H. Claassen, H. Selig, *J. Chem. Phys.* **1966**, 44, 4039–43.

[1673] M. J. Vasile, G. R. Jones, W. E. Falconer, *J. Chem. Soc., Chem. Commun.* **1971**, 1355–6.

[1674] K. Hagen, M. M. Gilbert, L. Hedberg, K. Hedberg, *Inorg. Chem.* **1982**, 21, 2690–3.

[1675] E. G. Hope, *J. Chem. Soc., Dalton Trans.* **1990**, 723–5.

[1676] A. J. Edwards, G. R. Jones, *J. Chem. Soc. A* **1969**, 1651–4.

[1677] I. R. Beattie, K. M. S. Livingston, G. A. Ozin, D. J. Reynolds, *J. Chem. Soc. A* **1969**, 958–65.

[1678] S. Brownstein, G. Latremouille, *Can. J. Chem.* **1974**, 52, 2236–41.

[1679] R. R. Holmes, R. M. Deiters, J. A. Golen, *Inorg. Chem.* **1969**, 8, 2612–20.

[1680] H. Selig, J. H. Holloway, J. Tyson, H. H. Claassen, *J. Chem. Phys.* **1970**, 53, 2559–64.

[1681] L. S. Bernstein, S. Abramowitz, I. W. Levin, *J. Chem. Phys.* **1976**, 64, 3228–36.

[1682] R. J. Gillespie, I. Bytheway, T.-H. Tang, R. F. W. Bader, *Inorg. Chem.* **1996**, 35, 3954–63.

[1683] S. Brownstein, *J. Fluorine Chem.* **1980**, 15, 539–40.

[1684] J. H. Canterford, T. A. O'Donnell, *Inorg. Chem.* **1967**, 6, 541–4.

[1685] J. H. Canterford, T. A. O'Donnell, *Inorg. Chem.* **1966**, 5, 1442–6.

[1686] F. Tamadon, K. Seppelt, *Angew. Chem., Int. Ed. Engl.* **2013**, 52, 767–9.

[1687] W. Kuhrt, R. Kreutz, H. Schindler, J. Massone, *Z. Naturforsch. B* **1972**, 27, 748.

[1688] B. R. Fowler, K. C. Moss, *J. Fluorine Chem.* **1980**, 15, 67–73.

[1689] W. Klemm, *Angew. Chem.* **1954**, 66, 468–74.

[1690] B. Cox, *J. Chem. Soc.* **1956**, 876–8.

[1691] R. D. W. Kemmitt, D. R. Russell, D. W. A. Sharp, *J. Chem. Soc.* **1963**, 4408–13.

[1692] H. C. Clark, H. J. Eméleus, *J. Chem. Soc.* **1958**, 190–5.

[1693] B. Žemva, J. Slivnik, S. Milićev, *J. Fluorine Chem.* **1977**, 9, 251–4.

[1694] H. P. Sampath Kumar, D. K. Padma, *J. Fluorine Chem.* **1991**, 51, 171–7.

[1695] V. V. Bardin, A. A. Avramenko, G. G. Furin, et al., *J. Fluorine Chem.* **1985**, 28, 37–45.

[1696] V. V. Bardin, S. G. Bardina, *J. Fluorine Chem.* **2004**, 125, 1411–4.

[1697] V. V. Bardin, A. A. Avramenko, G. G. Furin, et al., *J. Fluorine Chem.* **1990**, 49, 385–400.

[1698] W. W. Dukat, J. H. Holloway, E. G. Hope, M. R. Rieland, P. J. Townson, R. L. Powell, *J. Chem. Soc., Chem. Commun.* **1993**, 1429–30.

[1699] H. J. Eméleus, A. A. Woolf, *J. Chem. Soc.* **1950**, 164–8.

[1700] H. M. Haendler, S. F. Bartram, R. S. Becker, W. J. Bernard, S. W. Bukata, *J. Am. Chem. Soc.* **1954**, 76, 2177–8.

[1701] A. J. Edwards, D. R. Lloyd, *J. Chem. Soc., Chem. Commun.* **1972**, 719a–a.

[1702] H. Selig, H. H. Claassen, *J. Chem. Phys.* **1966**, 44, 1404–6.

[1703] R. J. H. Clark, P. D. Mitchell, *J. Chem. Soc., Dalton Trans.* **1972**, 2429–33.

[1704] R. J. H. Clark, D. M. Rippon, *Mol. Phys.* **1974**, 28, 305–19.

[1705] A. Almenningen, S. Samdal, D. Christen, *J. Mol. Struct.* **1978**, 48, 69–78.

[1706] A. J. Edwards, P. Taylor, *J. Chem. Soc., Chem. Commun.* **1970**, 1474b–5.

[1707] J. Supeł, U. Abram, A. Hagenbach, K. Seppelt, *Inorg. Chem.* **2007**, 46, 5591–5.

[1708] R. C. Hibbert, *J. Chem. Soc., Dalton Trans.* **1986**, 751–3.

[1709] M. F. Davis, W. Levason, J. Paterson, G. Reid, M. Webster, *Eur. J. Inorg. Chem.* **2008**, 2008, 802–11.

[1710] Ž. Zupanek, M. Tramšek, A. Kokalj, G. Tavčar, *Inorg. Chem.* **2018**, 57, 13866–79.

[1711] G. A. Kolta, D. W. A. Sharp, J. M. Winfield, *J. Fluorine Chem.* **1979**, 14, 153–62.

[1712] J. Weidlein, K. Dehnicke, *Z. Anorg. Allg. Chem.* **1966**, 348, 278–85.

[1713] J. C. Pérez-Flores, R. Villamor, D. Ávila-Brande, et al., *J. Mater. Chem. A* **2015**, 3, 20508–15.

[1714] K. Dehnicke, J. Weidlein, *Angew. Chem., Int. Ed. Engl.* **1966**, 5, 1041.

[1715] M. A. Cambaz, B. P. Vinayan, O. Clemens, et al., *Inorg. Chem.* **2016**, 55, 3789–96.

[1716] M. F. Davis, M. Jura, A. Leung, et al., *Dalton Trans.* **2008**, 6265–73.

[1717] G. W. Bushnell, K. C. Moss, *Can. J. Chem.* **1972**, 50, 3700–5.

[1718] H. Rieskamp, R. Mattes, *Z. Anorg. Allg. Chem.* **1973**, 401, 158–64.

[1719] M. Lozinšek, E. Goreshnik, B. Žemva, *Acta Chim. Slov.* **2014**, 61, 6.

[1720] M. Lozinšek, *Acta Chim. Slov.* **2015**, 62, 7.

[1721] J. A. S. Howell, K. C. Moss, *J. Chem. Soc. A* **1971**, 270–2.

[1722] P. Slota, G. Mitra, *J. Fluorine Chem.* **1975**, 5, 185–8.

[1723] G. Pausewang, K. Dehnicke, *Z. Anorg. Allg. Chem.* **1969**, 369, 265–77.

[1724] A. K. Sengupta, B. B. Bhaumik, *Z. Anorg. Allg. Chem.* **1972**, 390, 61–3.

[1725] A. K. Sengupta, B. B. Bhaumik, *Z. Anorg. Allg. Chem.* **1972**, 390, 311–5.

[1726] A. K. Sengupta, B. B. Bhaumik, *Z. Anorg. Allg. Chem.* **1971**, 384, 255–9.

[1727] M. Lozinšek, E. Goreshnik, B. Žemva, *Z. Anorg. Allg. Chem.* **2012**, 638, 2123–8.

[1728] T. Kanatani, K. Matsumoto, R. Hagiwara, *Eur. J. Inorg. Chem.* **2010**, 2010, 1049–55.

[1729] F. H. Aidoudi, P. J. Byrne, P. K. Allan, S. J. Teat, P. Lightfoot, R. E. Morris, *Dalton Trans.* **2011**, 40, 4324–31.

[1730] S. Rostamzadehmansor, G. Ebrahimzadehrajaei, S. Ghammamy, K. Mehrani, L. Saghatforoush, *J. Fluorine Chem.* **2008**, 129, 674–9.

[1731] R. Mattes, H. Rieskamp, *Z. Naturforsch. B* **1972**, 27, 1424.

[1732] E. Ahlborn, E. Diemann, A. Müller, *Z. Anorg. Allg. Chem.* **1972**, 394, 1–7.

[1733] R. R. Ryan, S. H. Mastin, M. J. Reisfeld, *Acta Crystallogr. B* **1971**, 27, 1270–4.

[1734] F. H. Aidoudi, C. Black, K. S. A. Arachchige, A. M. Z. Slawin, R. E. Morris, P. Lightfoot, *Dalton Trans.* **2014**, 43, 568–75.

[1735] R. Haiges, M. Vasiliu, D. A. Dixon, K. O. Christe, *Angew. Chem., Int. Ed. Engl.* **2015**, 54, 9101–5.

[1736] Z.-H. Shan, J. Liu, L.-M. Xu, Y.-F. Tang, J.-H. Chen, Z. Yang, *Org. Lett.* **2012**, 14, 3712–5.

[1737] Y.-H. Chen, R.-S. Tang, L.-Y. Chen, T.-H. Chuang, *J. Org. Chem.* **2019**, 84, 4501–6.
[1738] R. E. Damon, R. H. Schlessinger, J. F. Blount, *J. Org. Chem.* **1976**, 41, 3772–3.
[1739] A. J. Liepa, R. E. Summons, *J. Chem. Soc., Chem. Commun.* **1977**, 826–7.
[1740] J. Hartenstein, G. Satzinger, *Angew. Chem.* **1977**, 89, 739–40.
[1741] D. A. Burnett, D. J. Hart, *J. Org. Chem.* **1987**, 52, 5662–7.
[1742] A. J. Liepa, R. N. Nearn, D. M. J. Wright, *Aust. J. Chem.* **2004**, 57, 473–82.
[1743] M. J. Niphakis, G. I. Georg, *Org. Lett.* **2011**, 13, 196–9.
[1744] J.-B. Xia, K. W. Cormier, C. Chen, *Chem. Sci.* **2012**, 3, 2240–5.
[1745] K. H. Lau, D. L. Hildenbrand, *J. Chem. Phys.* **1979**, 71, 1572–7.
[1746] E. Capelli, R. J. M. Konings, *J. Fluorine Chem.* **2018**, 208, 55–64.
[1747] V. Kalamse, N. Wadnerkar, A. Chaudhari, *Int. J. Quant. Chem.* **2011**, 111, 2014–20.
[1748] K. F. Ng, W. Zou, W. Liu, A. S.-C. Cheung, *J. Chem. Phys.* **2017**, 146, 094308.
[1749] H. Schäfer, H. G. Schnering, K. J. Niehues, H. G. Nieder-Vahrenholz, *J. Less-Common Met.* **1965**, 9, 95–104.
[1750] G. Kliche, H. G. v. Schnering, *Z. Naturforsch. B* **1989**, 44, 74.
[1751] R. Knoll, J. Sokolovski, Y. BenHaim, et al., *Phys. B* **2006**, 381, 47–52.
[1752] J. Köhler, A. Simon, M.-H. Whangbo, *Z. Anorg. Allg. Chem.* **2009**, 635, 2396–8.
[1753] H. Schäfer, H. G. v. Schnering, A. Simon, et al., *J. Less-Common Met.* **1966**, 10, 154–5.
[1754] H. J. Eméleus, V. Gutmann, *J. Chem. Soc.* **1950**, 2115–8.
[1755] P. Ehrlich, F. Plöger, G. Pietzka, *Z. Anorg. Allg. Chem.* **1955**, 282, 19–23.
[1756] V. Gutman, K. H. Jack, *Acta Crystallogr.* **1951**, 4, 244–6.
[1757] M. Pouchard, M. R. Torki, G. Demazeau, P. Hagenmuller, *C. R. Acad. Sci., Ser. IIc: Chim.* **1971**, 273C, 1093–6.
[1758] B. Yang, J. Wang, X. Liu, M. Zhao, *PCCP, Phys. Chem. Chem. Phys.* **2018**, 20, 4781–6.
[1759] F. P. Gortsema, R. Didchenko, *Inorg. Chem.* **1965**, 4, 182–6.
[1760] J. Chassaing, D. Bizot, *J. Fluorine Chem.* **1980**, 16, 451–9.
[1761] F. P. Gortsema, J. B. Beal Jr, K. Schmidt, *Inorg. Synth.* **1973**, 105–9.
[1762] J. Bandemehr, M. Conrad, F. Kraus, *Acta Crystallogr., Sect. E* **2016**, 72, 1211–3.
[1763] F. E. Dickson, *J. Inorg. Nucl. Chem.* **1969**, 31, 2636–8.
[1764] M. B. de Bournonville, D. Bizot, J. Chassaing, M. Quarton, *J. Solid State Chem.* **1986**, 62, 212–9.
[1765] J. Chassaing, C. Monteil, D. Bizot, *J. Solid State Chem.* **1982**, 43, 327–33.
[1766] S. Llorente, F. Goubard, P. Gredin, D. Bizot, J. Chassaing, M. Quarton, *Z. Anorg. Allg. Chem.* **1998**, 624, 1538–42.
[1767] L. O. Gilpatrick, L. M. Toth, *Inorg. Chem.* **1974**, 13, 2242–5.
[1768] L. M. Toth, L. O. Gilpatrick, *Inorg. Chem.* **1976**, 15, 243–4.
[1769] A. J. Edwards, *J. Chem. Soc.* **1964**, 3714–8.
[1770] H. Selig, A. Reis, E. L. Gasner, *J. Inorg. Nucl. Chem.* **1968**, 30, 2087–90.
[1771] L. E. Alexander, I. R. Beattie, P. J. Jones, *J. Chem. Soc., Dalton Trans.* **1972**, 210–2.
[1772] J. Fawcett, A. J. Hewitt, J. H. Holloway, M. A. Stephen, *J. Chem. Soc., Dalton Trans.* **1976**, 2422–4.
[1773] J. Pauli, W. Storek, L. Riesel, *Z. Chem.* **1988**, 28, 226–7.
[1774] J. A. S. Howell, K. C. Moss, *J. Chem. Soc. A* **1971**, 2483–7.
[1775] Y. A. Buslayev, E. G. Ilyin, M. E. Ignatov, L. S. Butorina, T. A. Mastryukova, *J. Fluorine Chem.* **1978**, 12, 387–95.
[1776] Ž. Zupanek, M. Tramšek, A. Kokalj, G. Tavčar, *J. Fluorine Chem.* **2019**, 227, 109373.
[1777] A. J. Edwards, *J. Chem. Soc., Chem. Commun.* **1970**, 820.
[1778] A. J. Edwards, G. R. Jones, *J. Chem. Soc. A* **1970**, 1491–7.
[1779] M. C. Marignac, *Ann. Chim. Phys.* **1866**, 8, 5–75.

[1780]   J. A. S. Howell, K. C. Moss, *J. Chem. Soc. A* **1971**, 2481–3.

[1781]   R. E. Wilson, S. De Sio, V. Vallet, *Eur. J. Inorg. Chem.* **2016**, 2016, 5467–76.

[1782]   O. L. Keller, A. Chetham-Strode, *Inorg. Chem.* **1966**, 5, 367–72.

[1783]   K. B. Andersen, E. Christensen, R. W. Berg, N. J. Bjerrum, J. H. von Barner, *Inorg. Chem.* **2000**, 39, 3449–54.

[1784]   D. Bizot, H. Vestegaine, *J. Inorg. Nucl. Chem.* **1976**, 28, 67–8.

[1785]   J. S. Fordyce, R. L. Baum, *J. Chem. Phys.* **1966**, 44, 1159–65.

[1786]   A. F. Vik, V. Dracopoulos, G. N. Papatheodorou, T. Østvold, *J. Chem. Soc., Dalton Trans.* **2001**, 2164–72.

[1787]   M. Boča, J. Cibulková, B. Kubíková, M. Chrenková, V. Daněk, *J. Mol. Liq.* **2005**, 116, 29–36.

[1788]   K. J. Packer, E. L. Muetterties, *J. Am. Chem. Soc.* **1963**, 85, 3035–6.

[1789]   G. M. Brown, L. A. Walker, *Acta Crystallogr.* **1966**, 20, 220–9.

[1790]   M. F. Ghorab, J. M. Winfield, *J. Fluorine Chem.* **1990**, 49, 367–83.

[1791]   S. Brownstein, *J. Inorg. Nucl. Chem.* **1973**, 35, 3567–74.

[1792]   S. A. Baer, M. Lozinšek, F. Kraus, *Z. Anorg. Allg. Chem.* **2013**, 639, 2586–8.

[1793]   J. Sala-Pala, J. Y. Calves, J. E. Guerchais, S. Brownstein, J. C. Dewan, A. J. Edwards, *Can. J. Chem.* **1978**, 56, 1545–8.

[1794]   Y. A. Buslayev, Y. V. Kokunov, V. D. Kopanev, M. P. Gustyakova, *J. Inorg. Nucl. Chem.* **1974**, 36, 1569–74.

[1795]   E. Goreshnik, *J. Fluorine Chem.* **2019**, 226, 109352.

[1796]   H.-G. Nieder-Vahrenholz, H. Schäufer, *Z. Anorg. Allg. Chem.* **1987**, 544, 122–6.

[1797]   H. Schäfer, D. Bauer, W. Beckmann, et al., *Sci. Nat.* **1964**, 51, 241.

[1798]   J. Köhler, A. Simon, L. van Wüllen, et al., *Z. Anorg. Allg. Chem.* **2002**, 628, 2683–90.

[1799]   N. I. Giricheva, G. V. Girichev, *J. Mol. Struct.* **1999**, 484, 1–9.

[1800]   J. Sala-Pala, J. Y. Calves, J. E. Guerchais, *J. Inorg. Nucl. Chem.* **1975**, 37, 1294–6.

[1801]   W. Levason, G. Reid, J. Trayer, W. Zhang, *Dalton Trans.* **2014**, 43, 3649–59.

[1802]   R. Haiges, M. Vasiliu, D. A. Dixon, K. O. Christe, *Dalton Trans.* **2016**, 45, 10523–9.

[1803]   S. Andersson, A. Aström, *Acta Chem. Scand.* **1965**, 19, 2136–8.

[1804]   J. Dabachi, M. Body, C. Galven, F. Boucher, C. Legein, *Inorg. Chem.* **2017**, 56, 5219–32.

[1805]   L. K. Frevel, H. W. Rinn, *Acta Crystallogr.* **1956**, 9, 626–7.

[1806]   J. H. Von Barner, E. Christensen, N. J. Bjerrum, B. Gilbert, *Inorg. Chem.* **1991**, 30, 561–6.

[1807]   A. M. Srivastava, J. F. Ackerman, *Chem. Mater.* **1992**, 4, 1011–3.

[1808]   A. J. Norquist, C. L. Stern, K. R. Poeppelmeier, *Inorg. Chem.* **1999**, 38, 3448–9.

[1809]   V. Van, J. Madejová, A. Silný, V. Danek, *Chem. Pap.* **2000**, 54, 137–43.

[1810]   M. E. Welk, A. J. Norquist, C. L. Stern, K. R. Poeppelmeier, *Inorg. Chem.* **2000**, 39, 3946–7.

[1811]   G. Blasse, G. J. Dirksen, M. P. Crosnier-Lopez, J. L. Fourquet, *Solid State Commun.* **1994**, 90, 595–7.

[1812]   U. R. K. Rao, K. S. Venkateswarlu, B. N. Wani, *J. Fluorine Chem.* **1986**, 31, 29–35.

[1813]   V. I. Konstantinov, E. G. Polyakov, P. T. Stangrit, *Electrochim. Acta* **1978**, 23, 713–6.

[1814]   E. B. Merkulov, E. I. Voit, A. V. Gerasimenko, *J. Fluorine Chem.* **2019**, 226, 109357.

[1815]   R. Stomberg, *Acta Chem. Scand.* **1980**, 34 A, 193–8.

[1816]   R. Stomberg, *Acta Chem. Scand.* **1981**, 35 A, 389–94.

[1817]   R. Geetha, P. S. Rao, V. Babu, S. Subramanian, *Inorg. Chem.* **1991**, 30, 1630–5.

[1818]   T. Murase, H. Irie, K. Hashimoto, *J. Phys. Chem. B* **2005**, 109, 13420–3.

[1819]   C. Bohnke, O. Bohnke, J. L. Fourquet, *Mol. Cryst. Liq. Cryst. Sci. Technol., Sect. A* **1998**, 311, 23–9.

[1820]   C. Bohnke, J. L. Fourquet, N. Randrianantoandro, T. Brousse, O. Crosnier, *J. Solid State Electrochem.* **2001**, 5, 1–7.

[1821]   H. Ullah, K. Guerin, P. Bonnet, *Eur. J. Inorg. Chem.* **2019**, 2019, 230–6.

[1822] A. D. Pienaar, J. B. Wagener, P. L. Crouse, *Int. J. Miner. Process.* **2012**, 114–117, 7–10.
[1823] D. Watanabe, D. Akiyama, N. Sato, *J. Fluorine Chem.* **2018**, 216, 1–6.
[1824] R. A. Kent, J. L. Margrave, *J. Am. Chem. Soc.* **1965**, 87, 3582–5.
[1825] K. Katoh, T. Okabayashi, M. Tanimoto, Y. Sumiyoshi, Y. Endo, *J. Chem. Phys.* **2004**, 120, 7927–32.
[1826] M. Bencheikh, R. Koivisto, O. Launila, J. P. Flament, *J. Chem. Phys.* **1997**, 106, 6231–9.
[1827] B. J. Sturm, *Inorg. Chem.* **1962**, 1, 665–72.
[1828] H. Lux, G. Illmann, *Chem. Ber.* **1958**, 91, 2143–50.
[1829] P. E. Lim, J. W. Stout, *J. Chem. Phys.* **1975**, 63, 4886–902.
[1830] T. Chatterji, T. C. Hansen, *J. Phys.: Condens. Matter.* **2011**, 23, 276007.
[1831] B. Vest, P. Schwerdtfeger, M. Kolonits, M. Hargittai, *Chem. Phys. Lett.* **2009**, 468, 143–7.
[1832] H. Lux, L. Eberle, D. Sarre, *Chem. Ber.* **1964**, 97, 503–9.
[1833] A. D. Westland, *J. Inorg. Nucl. Chem.* **1973**, 35, 451–6.
[1834] R. Von der Mühll, D. Dumora, J. Ravez, P. Hagenmuller, *J. Solid State Chem.* **1970**, 2, 262–8.
[1835] H. G. von Schnering, B. Kolloch, A. Kolodziejczyk, *Angew. Chem.* **1971**, 83, 440.
[1836] A. Tressaud, J. M. Dance, J. Ravez, J. Portier, P. Hagenmuller, J. B. Goodenough, *Mater. Res. Bull.* **1973**, 8, 1467–77.
[1837] E. Banks, J. A. Deluca, O. Berkooz, *J. Solid State Chem.* **1973**, 6, 569–73.
[1838] J.-M. Dance, A. Tressaud, *C. R. Acad. Sci., Ser. IIc: Chim.* **1973**, 277C, 379–82.
[1839] H. V. Wartenberg, *Z. Anorg. Allg. Chem.* **1942**, 249, 100–12.
[1840] G. Tavčar, T. Skapin, *J. Fluorine Chem.* **2019**, 222–223, 81–9.
[1841] K. F. Zmbov, J. L. Margrave, *J. Inorg. Nucl. Chem.* **1967**, 29, 673–80.
[1842] M. Epple, W. Massa, *Z. Anorg. Allg. Chem.* **1978**, 444, 47–53.
[1843] G. Knoke, D. Babel, *Z. Naturforsch. B* **1975**, 30, 454.
[1844] D. Babel, G. Knoke, *Z. Anorg. Allg. Chem.* **1978**, 442, 151–62.
[1845] E. Baumgärtel, J. Teich, *Z. Anorg. Allg. Chem.* **1971**, 386, 279–84.
[1846] J. Teich, E. Baumgärtel, *Z. Anorg. Allg. Chem.* **1971**, 386, 285–7.
[1847] E. Baumgärtel, J. Teich, W. Müller, *Z. Anorg. Allg. Chem.* **1971**, 383, 113–9.
[1848] G. Brunton, *Mater. Res. Bull.* **1969**, 4, 621–6.
[1849] G. Siebert, R. Hoppe, *Z. Anorg. Allg. Chem.* **1972**, 391, 117–25.
[1850] D. Babel, R. Haegele, G. Pausewang, F. Wall, *Mater. Res. Bull.* **1973**, 8, 1371–82.
[1851] L. Helmholz, A. V. Guzzo, R. N. Sanders, *J. Chem. Phys.* **1961**, 35, 1349–52.
[1852] J. Ravez, R. g. Von Der Mühll, P. Hagenmuller, *J. Solid State Chem.* **1975**, 14, 20–4.
[1853] R. Wei, Q. Li, Y. Gong, et al., *J. Phys. Chem. A* **2017**, 121, 7603–12.
[1854] T. Xiao, G. Wang, Y. Liao, *Chem. Phys.* **2018**, 513, 182–7.
[1855] H. Wartenberg, *Z. Anorg. Allg. Chem.* **1941**, 247, 135–46.
[1856] E. G. Hope, P. J. Jones, W. Levason, J. S. Ogden, M. Tajik, J. W. Turff, *J. Chem. Soc., Dalton Trans.* **1985**, 1443–9.
[1857] O. Krämer, B. G. Müller, *Z. Anorg. Allg. Chem.* **1995**, 621, 1969–72.
[1858] P. Benkič, Z. Mazej, B. Žemva, *Angew. Chem., Int. Ed. Engl.* **2002**, 41, 1398–9.
[1859] H. C. Clark, Y. N. Sadana, *Can. J. Chem.* **1964**, 42, 50–6.
[1860] J. Jacobs, H. S. P. Mueller, H. Willner, E. Jacob, H. Buerger, *Inorg. Chem.* **1992**, 31, 5357–63.
[1861] E. Huss, W. Klemm, *Z. Anorg. Allg. Chem.* **1950**, 262, 25–32.
[1862] G. Siebert, R. Hoppe, *Sci. Nat.* **1971**, 58, 95–6.
[1863] G. Siebert, R. Hoppe, *Z. Anorg. Allg. Chem.* **1972**, 391, 126–36.
[1864] G. Siebert, R. Hoppe, *Z. Anorg. Allg. Chem.* **1972**, 391, 113–6.
[1865] Z. Mazej, E. Goreshnik, *Eur. J. Inorg. Chem.* **2008**, 2008, 1795–812.
[1866] B. Žemva, J. Zupan, J. Slivnik, *J. Inorg. Nucl. Chem.* **1973**, 35, 3941–2.

[1867] Z. Mazej, J. Darriet, J. Grannec, K. Lutar, A. Tressaud, B. Žemva, *J. Fluorine Chem.* **1999**, 99, 25–8.

[1868] W. V. Rochat, J. N. Gerlach, G. L. Gard, *Inorg. Chem.* **1970**, 9, 998–9.

[1869] A. J. Edwards, *Proc. Chem. Soc.* **1963**, 205.

[1870] O. Glemser, H. Roesky, K.-h. Hellberg, *Angew. Chem.* **1963**, 75, 346–7.

[1871] S. P. Mallela, J. M. Shreeve, D. D. DesMarteau, *Inorg. Synth.* **1992**, 124–7.

[1872] E. Jacob, H. Willner, *Chem. Ber.* **1990**, 123, 1319–21.

[1873] E. J. Jacob, L. Hedberg, K. Hedberg, H. Davis, G. L. Gard, *J. Phys. Chem.* **1984**, 88, 1935–6.

[1874] S. D. Brown, T. M. Loehr, G. L. Gard, *J. Chem. Phys.* **1976**, 64, 260–2.

[1875] H. Shorafa, K. Seppelt, *Z. Anorg. Allg. Chem.* **2009**, 635, 112–4.

[1876] T. A. O'Donnell, D. F. Stewart, *Inorg. Chem.* **1966**, 5, 1434–7.

[1877] J. Slivnik, B. žemva, *Z. Anorg. Allg. Chem.* **1971**, 385, 137–41.

[1878] S. D. Brown, G. L. Gard, *Inorg. Nucl. Chem. Lett.* **1975**, 11, 19–21.

[1879] R. Bougon, W. W. Wilson, K. O. Christe, *Inorg. Chem.* **1985**, 24, 2286–92.

[1880] S. D. Brown, T. M. Loehr, G. L. Gard, *J. Fluorine Chem.* **1976**, 7, 19–32.

[1881] T. Schlöder, F. Brosi, B. J. Freyh, T. Vent-Schmidt, S. Riedel, *Inorg. Chem.* **2014**, 53, 5820–9.

[1882] H. C. Clark, Y. N. Sadana, *Can. J. Chem.* **1964**, 42, 702–4.

[1883] P. J. Green, B. M. Johnson, T. M. Loehr, G. L. Gard, *Inorg. Chem.* **1982**, 21, 3562–5.

[1884] M. McHughes, R. D. Willett, H. B. Davis, G. L. Gard, *Inorg. Chem.* **1986**, 25, 426–7.

[1885] E. G. Hope, P. J. Jones, W. Levason, J. S. Ogden, M. Tajik, J. W. Turff, *J. Chem. Soc., Dalton Trans.* **1984**, 2445–7.

[1886] E. G. Hope, P. J. Jones, W. Levason, J. S. Ogden, M. Tajik, *J. Chem. Soc., Chem. Commun.* **1984**, 1355–6.

[1887] E. Miyoshi, Y. Sakai, A. Murakami, et al., *J. Chem. Phys.* **1988**, 89, 4193–8.

[1888] C. J. Marsden, P. P. Wolynec, *Inorg. Chem.* **1991**, 30, 1681–2.

[1889] A. Neuhaus, G. Frenking, C. Huber, J. Gauss, *Inorg. Chem.* **1992**, 31, 5355–6.

[1890] K. Pierloot, B. O. Roos, *Inorg. Chem.* **1992**, 31, 5353–4.

[1891] C. J. Marsden, D. Moncrieff, G. E. Quelch, *J. Phys. Chem.* **1994**, 98, 2038–43.

[1892] L. G. Vanquickenborne, A. E. Vinckier, K. Pierloot, *Inorg. Chem.* **1996**, 35, 1305–9.

[1893] A. Engelbrecht, A. V. Grosse, *J. Am. Chem. Soc.* **1952**, 74, 5262–4.

[1894] K. Wiechert, *Z. Anorg. Allg. Chem.* **1950**, 261, 310–23.

[1895] G. D. Flesch, H. J. Svec, *J. Am. Chem. Soc.* **1958**, 80, 3189–91.

[1896] H.-L. Krauss, F. Schwarzbach, *Chem. Ber.* **1961**, 94, 1205–7.

[1897] A. K. Brisdon, J. H. Holloway, E. G. Hope, *J. Fluorine Chem.* **1998**, 89, 35–7.

[1898] A. J. Edwards, W. E. Falconer, W. A. Sunder, *J. Chem. Soc., Dalton Trans.* **1974**, 541–2.

[1899] P. J. Green, G. L. Gard, *Inorg. Chem.* **1977**, 16, 1243–5.

[1900] S. D. Brown, P. J. Green, G. L. Gard, *J. Fluorine Chem.* **1975**, 5, 203–19.

[1901] K. O. Christe, W. W. Wilson, R. A. Bougon, *Inorg. Chem.* **1986**, 25, 2163–9.

[1902] E. G. Hope, P. J. Jones, W. Levason, J. S. Ogden, M. Tajik, J. W. Turff, *J. Chem. Soc., Dalton Trans.* **1985**, 529–33.

[1903] J. H. Holloway, E. G. Hope, P. J. Townson, R. L. Powell, *J. Fluorine Chem.* **1996**, 76, 105–7.

[1904] S. P. Mallela, J. n. M. Shreeve, *Organometallics* **1989**, 8, 2751–4.

[1905] G. L. Gard, S. M. Williamson, *Inorg. Synth.* **1986**, 67–9.

[1906] S. D. Brown, G. L. Gard, T. M. Loehr, *J. Chem. Phys.* **1976**, 64, 1219–22.

[1907] R. J. French, L. Hedberg, K. Hedberg, G. L. Gard, B. M. Johnson, *Inorg. Chem.* **1983**, 22, 892–5.

[1908] J. Besida, T. A. O'Donnell, P. G. Eller, *Can. J. Chem.* **1989**, 67, 2047–51.

[1909] P. J. Green, G. L. Gard, *Inorg. Nucl. Chem. Lett.* **1978**, 14, 179–82.

[1910] S. D. Brown, G. L. Gard, *Inorg. Chem.* **1973**, 12, 483–4.

[1911] A. G. M. Barrett, D. H. R. Barton, T. Tsushima, *J. Chem. Soc., Perkin Trans. 1* **1980**, 639–42.

[1912] D. L. Hildenbrand, *J. Chem. Phys.* **1976**, 65, 614–8.

[1913] K. G. Dyall, *J. Phys. Chem. A* **2000**, 104, 4077–83.

[1914] D. L. Hildenbrand, *J. Chem. Phys.* **1975**, 62, 3074–9.

[1915] A. Bensaoula, E. Grossman, A. Ignatiev, *J. Appl. Phys.* **1987**, 62, 4587–90.

[1916] G. M. Neumann, *J. Fluorine Chem.* **1972**, 1, 473–86.

[1917] H. Meinert, W. Schifferdecker, *Z. Chem.* **1979**, 19, 33–4.

[1918] D. E. LaValle, R. M. Steele, M. K. Wilkinson, H. L. Yakel, *J. Am. Chem. Soc.* **1960**, 82, 2433–4.

[1919] M. Fukutomi, J. D. Corbett, *J. Less-Common Met.* **1977**, 55, 125–30.

[1920] F. Averdunk, R. Hoppe, *J. Less-Common Met.* **1990**, 161, 135–40.

[1921] V. V. Sliznev, N. V. Belova, *J. Mol. Struct.* **2017**, 1132, 73–87.

[1922] R. Hoppe, K. Lehr, *Z. Anorg. Allg. Chem.* **1975**, 416, 240–50.

[1923] H. F. Priest, W. C. Schumb, *J. Am. Chem. Soc.* **1948**, 70, 3378–9.

[1924] R. D. Peacock, *Proc. Chem. Soc.* **1957**, 59.

[1925] R. T. Paine, L. B. Asprey, *Inorg. Chem.* **1974**, 13, 1529–31.

[1926] W. J. Casteel, D. H. Lohmann, N. Bartlett, *J. Fluorine Chem.* **2001**, 112, 165–71.

[1927] H. Meinert, A. Dimitrov, *ChemInform* **1976**, 7.

[1928] J. B. Bates, *Inorg. Nucl. Chem. Lett.* **1971**, 7, 957–60.

[1929] V. V. Sliznev, V. G. Solomonik, *J. Struct. Chem.* **2000**, 41, 11–8.

[1930] G. V. Girichev, N. I. Giricheva, O. G. Krasnova, *J. Mol. Struct.* **2001**, 567-568, 203–10.

[1931] H. Meinert, W. Schifferdecker, H. Sattler, *Z. Chem.* **1977**, 17, 310–1.

[1932] G. Brunton, *Mater. Res. Bull.* **1971**, 6, 555–60.

[1933] A. J. Edwards, B. R. Steventon, *J. Chem. Soc., Dalton Trans.* **1977**, 1860–2.

[1934] J. Angenault, J. C. Couturier, Y. Mary, M. Quarton, *J. Appl. Crystallogr.* **1987**, 20, 133.

[1935] J.-C. Couturier, J. Angenault, Y. Mary, M. Quarton, *J. Less-Common Met.* **1988**, 138, 71–7.

[1936] D. F. Stewart, T. A. O'Donnell, *Nature* **1966**, 210, 836.

[1937] T. O'Donnell, P. Wilson, *Aust. J. Chem.* **1968**, 21, 1415–9.

[1938] A. J. Edwards, R. D. Peacock, R. W. H. Small, *J. Chem. Soc.* **1962**, 4486–91.

[1939] M. Mercer, T. J. Ouellette, C. T. Ratcliffe, D. W. A. Sharp, *J. Chem. Soc. A* **1969**, 2532–4.

[1940] T. A. O'Donnell, D. F. Stewart, *J. Inorg. Nucl. Chem.* **1962**, 24, 309–14.

[1941] T. J. Ouellette, C. T. Ratcliffe, D. W. A. Sharp, A. M. Steven, F. Schreiner, *Inorg. Synth.* **1972**, 146–50.

[1942] J. Schröder, F. J. Grewe, *Angew. Chem.* **1968**, 80, 118–9.

[1943] J. Schröder, F. J. Grewe, *Chem. Ber.* **1970**, 103, 1536–46.

[1944] T. A. O'Donnell, T. E. Peel, *J. Inorg. Nucl. Chem.* **1976**, 28, 61–2.

[1945] A. J. Edwards, *J. Chem. Soc. A* **1969**, 909.

[1946] J. B. Bates, *Spectrochim. Acta, Part A* **1971**, 27, 1255–8.

[1947] R. D. Peacock, T. P. Sleight, *J. Fluorine Chem.* **1971**, 1, 243–5.

[1948] R. E. Stene, B. Scheibe, C. Pietzonka, A. J. Karttunen, W. Petry, F. Kraus, *J. Fluorine Chem.* **2018**, 211, 171–9.

[1949] T. J. Ouellette, C. T. Ratcliffe, D. W. A. Sharp, *J. Chem. Soc. A* **1969**, 2351–4.

[1950] E. I. Voit, A. V. Voit, V. K. Goncharuk, V. I. Sergienko, *J. Struct. Chem.* **1999**, 40, 380–6.

[1951] N. Acquista, S. Abramowitz, *J. Chem. Phys.* **1973**, 58, 5484–8.

[1952] N. I. Giricheva, O. G. Krasnova, G. V. Girichev, *J. Struct. Chem.* **1997**, 38, 54–61.

[1953] G. H. Cady, G. B. Hargreaves, *J. Chem. Soc.* **1961**, 1568–74.

[1954] R. T. Paine, L. A. Quarterman, *J. Inorg. Nucl. Chem.* **1976**, 28, 85–6.

[1955] S. Brownstein, *Can. J. Chem.* **1973**, 51, 2530–3.

[1956] G. B. Hargreaves, R. D. Peacock, *J. Chem. Soc.* **1957**, 4212–4.

[1957] G. B. Hargreaves, R. D. Peacock, *J. Chem. Soc.* **1958**, 2170–5.

[1958] G. B. Hargreaves, R. D. Peacock, *J. Chem. Soc.* **1958**, 3776–9.

[1959] G. B. Hargreaves, R. D. Peacock, *J. Chem. Soc.* **1958**, 4390–3.

[1960] J. Burgess, I. Haigh, R. D. Peacock, P. Taylor, *J. Chem. Soc., Dalton Trans.* **1974**, 1064–6.

[1961] J. Burgess, R. D. Peacock, *J. Fluorine Chem.* **1977**, 10, 479–86.

[1962] W. W. Wilson, K. O. Christe, *Inorg. Chem.* **1981**, 20, 4139–43.

[1963] L. Kolditz, U. Calov, *Z. Chem.* **1966**, 6, 431.

[1964] O. Ruff, F. Eisner, *Chem. Ber.* **1907**, 40, 2926–35.

[1965] S. Siegel, D. A. Northrop, *Inorg. Chem.* **1966**, 5, 2187–8.

[1966] G. S. Quiñones, G. Hägele, K. Seppelt, *Chem. Eur. J.* **2004**, 10, 4755–62.

[1967] J. C. Bailar, *J. Inorg. Nucl. Chem.* **1958**, 8, 165–75.

[1968] B. Frlec, H. H. Hyman, *Inorg. Chem.* **1967**, 6, 1596–8.

[1969] F. N. Tebbe, E. L. Muetterties, *Inorg. Chem.* **1968**, 7, 172–4.

[1970] A. M. Noble, J. M. Winfield, *Inorg. Nucl. Chem. Lett.* **1968**, 4, 339–42.

[1971] A. M. Bond, I. Irvine, T. A. O'Donnell, *Inorg. Chem.* **1977**, 16, 841–4.

[1972] A. Prescott, D. W. A. Sharp, J. M. Winfield, *J. Chem. Soc., Chem. Commun.* **1973**, 667–8.

[1973] A. Prescott, D. W. A. Sharp, J. M. Winfield, *J. Chem. Soc., Dalton Trans.* **1975**, 936–9.

[1974] G. M. Anderson, J. Iqbal, D. W. A. Sharp, J. M. Winfield, J. H. Cameron, A. G. McLeod, *J. Fluorine Chem.* **1984**, 24, 303–17.

[1975] K. H. Moock, M. H. Rock, *J. Chem. Soc., Dalton Trans.* **1993**, 2459–63.

[1976] H. Meinert, L. Friedrich, *Z. Chem.* **1975**, 15, 411–2.

[1977] A. Prescott, D. W. A. Sharp, J. M. Winfield, *J. Chem. Soc., Dalton Trans.* **1975**, 934–6.

[1978] A. Beuter, W. Kuhlmann, W. Sawodny, *J. Fluorine Chem.* **1975**, 6, 367–78.

[1979] J. R. Geichman, E. A. Smith, P. R. Ogle, *Inorg. Chem.* **1963**, 2, 1012–5.

[1980] K. Matsumoto, R. Hagiwara, R. Yoshida, et al., *Dalton Trans.* **2004**, 144–9.

[1981] M. F. Ghorab, J. M. Winfield, *J. Fluorine Chem.* **1993**, 62, 101–10.

[1982] S. El-Kurdi, A.-A. Al-Terkawi, B. M. Schmidt, A. Dimitrov, K. Seppelt, *Chem. Eur. J.* **2010**, 16, 595–9.

[1983] S. Brownstein, *Can. J. Chem.* **1978**, 56, 343–7.

[1984] D. Turnbull, S. D. Wetmore, M. Gerken, *Angew. Chem., Int. Ed. Engl.* **2019**, 58, 13035–8.

[1985] A. M. Noble, J. M. Winfield, *J. Chem. Soc. A* **1970**, 501–6.

[1986] A. M. Noble, J. M. Winfield, *J. Chem. Soc. A* **1970**, 2574–8.

[1987] A. Majid, R. R. McLean, T. J. Ouellette, D. W. A. Sharp, J. M. Winfield, *Inorg. Nucl. Chem. Lett.* **1971**, 7, 53–6.

[1988] A. Majid, D. W. A. Sharp, J. M. Winfield, I. Hanley, *J. Chem. Soc., Dalton Trans.* **1973**, 1876–8.

[1989] M. Harman, D. W. A. Sharp, J. M. Winfield, *Inorg. Nucl. Chem. Lett.* **1974**, 10, 183–5.

[1990] F. E. Brinckman, R. B. Johannesen, R. F. Hammerschmidt, L. B. Handy, *J. Fluorine Chem.* **1975**, 6, 427–36.

[1991] R. Craciun, D. Picone, R. T. Long, et al., *Inorg. Chem.* **2010**, 49, 1056–70.

[1992] G. W. Fraser, C. J. W. Gibbs, R. D. Peacock, *J. Chem. Soc. A* **1970**, 1708–11.

[1993] H. Meinert, L. Friedrich, W. Kohl, *Z. Chem.* **1975**, 15, 492.

[1994] B. G. Ward, F. E. Stafford, *Inorg. Chem.* **1968**, 7, 2569–73.

[1995] W. W. Wilson, K. O. Christe, R. Bougon, *Inorg. Synth.* **1986**, 37–8.

[1996] H. Selig, W. A. Sunder, F. C. Schilling, W. E. Falconer, *J. Fluorine Chem.* **1978**, 11, 629–35.

[1997] B. F. Hoskins, A. Linden, T. A. O'Donnell, *Inorg. Chem.* **1987**, 26, 2223–8.

[1998] A. J. Edwards, B. R. Steventon, *J. Chem. Soc. A* **1968**, 2503–10.

[1999] A. J. Edwards, G. R. Jones, *J. Chem. Soc. A* **1968**, 2074–8.

[2000] I. R. Beattie, D. J. Reynolds, *Chem. Commun.* **1968**, 1531–2.

[2001] M. J. Bennett, T. E. Haas, J. T. Purdham, *Inorg. Chem.* **1972**, 11, 207–8.

[2002] L. B. Asprey, R. R. Ryan, E. Fukushima, *Inorg. Chem.* **1972**, 11, 3122.

[2003] A. J. Edwards, G. R. Jones, B. R. Steventon, *Chem. Commun.* **1967**, 462–3.

[2004] R. T. Paine, R. S. McDowell, *Inorg. Chem.* **1974**, 13, 2366–70.

[2005] L. E. Alexander, I. R. Beattie, A. Bukovszky, P. J. Jones, C. J. Marsden, G. J. V. Schalkwyk, *J. Chem. Soc., Dalton Trans.* **1974**, 81–4.

[2006] A. G. Robiette, K. Hedberg, L. Hedberg, *J. Mol. Struct.* **1977**, 37, 105–12.

[2007] W. Levason, R. Narayanaswamy, J. S. Ogden, A. J. Rest, J. W. Turff, *J. Chem. Soc., Dalton Trans.* **1981**, 2501–7.

[2008] K. K. Banger, C. S. Blackman, A. K. Brisdon, *J. Chem. Soc., Dalton Trans.* **1996**, 2975–8.

[2009] P. A. Tucker, P. A. Taylor, J. H. Holloway, D. R. Russell, *Acta Crystallogr. B* **1975**, 31, 906–8.

[2010] J. H. Holloway, G. J. Schrobilgen, P. Taylor, *J. Chem. Soc., Chem. Commun.* **1975**, 40–1.

[2011] J. H. Holloway, G. J. Schrobilgen, *Inorg. Chem.* **1980**, 19, 2632–40.

[2012] L. Arnaudet, R. Bougon, B. Buu, et al., *Inorg. Chem.* **1989**, 28, 257–62.

[2013] L. Arnaudet, R. Bougon, B. Ban, et al., *Can. J. Chem.* **1990**, 68, 507–12.

[2014] L. Arnaudet, R. Bougon, B. Ban, M. Lance, W. C. Kaska, *J. Fluorine Chem.* **1991**, 53, 171–80.

[2015] R. Bougon, T. Bui Huy, P. Charpin, *Inorg. Chem.* **1975**, 14, 1822–30.

[2016] A. Beuter, W. Sawodny, *Z. Anorg. Allg. Chem.* **1976**, 427, 37–44.

[2017] K. O. Christe, W. W. Wilson, C. J. Schack, R. D. Wilson, R. Bougon, *Inorg. Synth.* **1986**, 39–48.

[2018] Y. Katayama, R. Hagiwara, Y. Ito, *J. Fluorine Chem.* **1995**, 74, 89–95.

[2019] R. E. Stene, B. Scheibe, A. J. Karttunen, W. Petry, F. Kraus, *Eur. J. Inorg. Chem.* **2019**, 2019, 3672–82.

[2020] K. Matsumoto, R. Hagiwara, *J. Fluorine Chem.* **2005**, 126, 1095–100.

[2021] R. Haiges, J. Skotnitzki, Z. Fang, D. A. Dixon, K. O. Christe, *Angew. Chem., Int. Ed. Engl.* **2015**, 54, 15550–5.

[2022] M. J. Atherton, J. H. Holloway, *J. Chem. Soc., Chem. Commun.* **1977**, 424–5.

[2023] M. J. Atherton, J. H. Holloway, *Inorg. Nucl. Chem. Lett.* **1978**, 14, 121–3.

[2024] H. Shorafa, H. Ficicioglu, F. Tamadon, F. Girgsdies, K. Seppelt, *Inorg. Chem.* **2010**, 49, 4263–7.

[2025] M. J. Atherton, J. H. Holloway, *J. Chem. Soc., Chem. Commun.* **1978**, 254–5.

[2026] V. M. Petrov, V. N. Petrova, N. I. Giricheva, G. V. Girichev, *J. Struct. Chem.* **1997**, 38, 318–22.

[2027] R. Haiges, J. Skotnitzki, Z. Fang, D. A. Dixon, K. O. Christe, *Angew. Chem., Int. Ed. Engl.* **2015**, 54, 9581–5.

[2028] R. Mattes, G. Müller, H. J. Becher, *Z. Anorg. Allg. Chem.* **1972**, 389, 177–87.

[2029] F. Brosi, T. Schlöder, A. Schmidt, H. Beckers, S. Riedel, *Dalton Trans.* **2016**, 45, 5038–44.

[2030] J. D. Tornero, J. Fayos, *Ferroelectr. Lett. Sect.* **1990**, 12, 85–9.

[2031] R. Hoppe, W. Liebe, W. Dähne, *Z. Anorg. Allg. Chem.* **1961**, 307, 276–89.

[2032] M. Conrad, C. Pietzonka, J. Bernzen, et al., *Z. Anorg. Allg. Chem.* **2018**, 644, 1557–61.

[2033] E. T. Keve, S. C. Abrahams, J. L. Bernstein, *J. Chem. Phys.* **1969**, 51, 4928–36.

[2034] T. Tsuboi, *Phase Transit.* **1989**, 18, 119–24.

[2035] M. Yoshimura, M. Hidaka, T. Mizushima, J. Sakurai, T. Tsuboi, W. Kleemann, *J. Magn. Magn. Mater.* **2006**, 299, 404–11.

[2036] M. Kristl, B. Dojer, N. Hojnik, A. Golobič, *J. Fluorine Chem.* **2014**, 166, 15–21.

[2037] M. Hidaka, H. Fujii, S. Maeda, *Phase Transit.* **1986**, 6, 101–14.

[2038] L. J. Lewis, Y. Lépine, *Phys. Rev. B, Condens. Matter* **1989**, 40, 3319–22.

[2039] Y. Ivanov, T. Nimura, K. Tanaka, *Acta Crystallogr. B* **2004**, 60, 359–68.

[2040] M. Guennou, P. Bouvier, G. Garbarino, J. Kreisel, E. K. H. Salje, *J. Phys.: Condens. Matter.* **2011**, 23, 485901.

[2041] A. Waśkowska, A. Ratuszna, *J. Solid State Chem.* **1998**, 137, 71–6.

[2042] S. Asbrink, A. Waskowska, H. G. Krane, L. Gerward, J. S. Olsen, *J. Appl. Crystallogr.* **1999**, 32, 174–7.

[2043] M. Welsch, S. Kummer-Dörner, B. Peschel, D. Babel, *Z. Anorg. Allg. Chem.* **1999**, 625, 1255–60.

[2044] Z. Mazej, *J. Fluorine Chem.* **2002**, 114, 75–80.

[2045] M. A. Hepworth, K. H. Jack, R. S. Nyholm, *Nature* **1957**, 179, 211–2.

[2046] E. J. P. Fear, J. Thrower, *J. Appl. Chem.* **1955**, 5, 353–8.

[2047] R. Hoppe, W. Dähne, W. Klemm, *Justus Liebigs Ann. Chem.* **1962**, 658, 1–5.

[2048] K.-H. Wandner, R. Hoppe, *Z. Anorg. Allg. Chem.* **1987**, 546, 113–21.

[2049] M. Molinier, W. Massa, S. Khairoun, A. Tressaud, J. L. Soubeyroux, *Z. Naturforsch. B* **1991**, 46, 1669.

[2050] W. Massa, *Z. Anorg. Allg. Chem.* **2007**, 633, 690–2.

[2051] P. Nuñez, A. Tressaud, J. Grannec, et al., *Z. Anorg. Allg. Chem.* **1992**, 609, 71–6.

[2052] M. Molinier, W. Massa, *Z. Naturforsch. B* **1992**, 47, 783.

[2053] U. Englich, C. Frommen, W. Massa, *J. Alloys Compd.* **1997**, 246, 155–65.

[2054] E. Dubler, L. Linowsky, J.-P. Matthieu, H.-R. Oswald, *Helv. Chim. Acta* **1977**, 60, 1589–600.

[2055] W. Massa, M. Steiner, *J. Solid State Chem.* **1980**, 32, 137–43.

[2056] F. Aguado, F. Rodriguez, P. Núñez, *Phys. Rev. B, Condens. Matter* **2007**, 76, 094417.

[2057] W. Massa, E. Schmidt Roland, *Z. Naturforsch. B* **1990**, 45, 593.

[2058] U. Bentrup, W. Massa, *Z. Anorg. Allg. Chem.* **1991**, 593, 207–16.

[2059] V. Kaučič, P. Bukovec, *J. Chem. Soc., Dalton Trans.* **1979**, 1512–5.

[2060] W. Massa, G. Baum, S. Drueeke, *Acta Crystallogr. Sect. C: Cryst. Struct. Commun.* **1988**, 44, 167–8.

[2061] P. Núñez, A. Tressaud, F. Hahn, et al., *Phys. Status Solidi A* **1991**, 127, 505–17.

[2062] P. Núñez, C. Elías, J. Fuentes, et al., *J. Chem. Soc., Dalton Trans.* **1997**, 4335–40.

[2063] U. Jacobs, L. Schröder, W. Massa, et al., *Z. Anorg. Allg. Chem.* **1998**, 624, 1471–6.

[2064] P. Núñez, J. C. Ruiz-Morales, A. D. Lozano-Gorrín, et al., *Dalton Trans.* **2004**, 273–8.

[2065] R. Stief, W. Massa, *Z. Anorg. Allg. Chem.* **2004**, 630, 2502–7.

[2066] R. Stief, W. Massa, J. Pebler, *Z. Anorg. Allg. Chem.* **2002**, 628, 2631–6.

[2067] T. Li, R. Clulow, A. J. Bradford, S. L. Lee, A. M. Z. Slawin, P. Lightfoot, *Dalton Trans.* **2019**, 48, 4784–7.

[2068] M. K. Chaudhuri, J. C. Das, H. S. Dasgupta, *J. Inorg. Nucl. Chem.* **1981**, 43, 85–7.

[2069] M. N. Bhattacharjee, M. K. Chaudhuri, H. S. Dasgupta, D. T. Khathing, *J. Chem. Soc., Dalton Trans.* **1981**, 2587–8.

[2070] M. N. Bhattacharjee, M. K. Chaudhuri, H. M. Marsden, *Inorg. Synth.* **1986**, 50–2.

[2071] J. R. Günter, J. P. Matthieu, H. R. Oswald, *Helv. Chim. Acta* **1978**, 61, 328–36.

[2072] A. J. Edwards, *J. Chem. Soc. A* **1971**, 2653–5.

[2073] D. R. Sears, J. L. Hoard, *J. Chem. Phys.* **1969**, 50, 1066–71.

[2074] P. Bukovec, V. Kaucic, *Acta Crystallogr. B* **1978**, 34, 3339–41.

[2075] J. Pebler, W. Massa, H. Lass, B. Ziegler, *J. Solid State Chem.* **1987**, 71, 87–94.

[2076] F. Hahn, W. Massa, *Z. Naturforsch. B* **1990**, 45, 1341.

[2077] W. Massa, V. Burk, *Z. Anorg. Allg. Chem.* **1984**, 516, 119–26.

[2078] U. Bentrup, L. Schröder, W. Massa, *Z. Naturforsch. B* **1992**, 47, 789.

[2079] A. Ahmadi, R. Stief, W. Massa, J. Pebler, *Z. Anorg. Allg. Chem.* **2001**, 627, 869–76.

[2080] U. Bentrup, K. Harms, W. Massa, J. Pebler, *Solid State Sci.* **2000**, 2, 373–7.

[2081] R. Stief, W. Massa, *Z. Anorg. Allg. Chem.* **2006**, 632, 797–800.

[2082] P. Nuñez, A. Tressaud, J. Darriet, et al., *J. Solid State Chem.* **1988**, 77, 240–9.

[2083] W. Massa, *Z. Anorg. Allg. Chem.* **1975**, 415, 254–62.

[2084] P. Bukovec, J. Šiftar, *Monatsh. Chem.* **1975**, 106, 1333–6.

[2085] U. Englich, W. Massa, A. Tressaud, *Acta Crystallogr. Sect. C: Cryst. Struct. Commun.* **1992**, 48, 6–8.

[2086] K. Wieghardt, H. Siebert, *Z. Anorg. Allg. Chem.* **1971**, 381, 12–20.

[2087] G. Rother, H. Worzala, U. Bentrup, *Z. Anorg. Allg. Chem.* **1998**, 624, 1706–11.

[2088] C. Elías, J. Fuentes, P. Núñez, V. D. Rodríguez, U. Jacobs, W. Massa, *Z. Anorg. Allg. Chem.* **1998**, 624, 2001–6.

[2089] R. Hoppe, W. Dähne, W. Klemm, *Sci. Nat.* **1961**, 48, 429.

[2090] H. Roesky, O. Glemser, *Angew. Chem.* **1963**, 75, 920–1.

[2091] T. L. Court, M. F. A. Dove, *J. Chem. Soc., Chem. Commun.* **1971**, 726.

[2092] K. Lutar, A. Jesih, B. Žemva, *Polyhedron* **1988**, 7, 1217–9.

[2093] B. G. Müller, M. Serafin, *Z. Naturforsch. B* **1987**, 42, 1102.

[2094] S. Nunziante Cesaro, J. V. Rau, N. S. Chilingarov, G. Balducci, L. N. Sidorov, *Inorg. Chem.* **2001**, 40, 179–81.

[2095] W. W. Wilson, K. O. Christe, R. Bougon, *Inorg. Synth.* **1986**, 48–50.

[2096] L. B. Asprey, M. J. Reisfeld, N. A. Matwiyoff, *J. Mol. Spectrosc.* **1970**, 34, 361–9.

[2097] L. Helmholz, M. E. Russo, *J. Chem. Phys.* **1973**, 59, 5455–70.

[2098] A. M. Black, C. D. Flint, *J. Chem. Soc., Dalton Trans.* **1974**, 977–81.

[2099] R. Hoppe, B. Hofmann, *Z. Anorg. Allg. Chem.* **1977**, 436, 65–74.

[2100] H. Bode, W. Wendt, *Z. Anorg. Allg. Chem.* **1952**, 269, 165–72.

[2101] R. Hoppe, K. H. Wandner, *J. Fluorine Chem.* **1983**, 23, 589–92.

[2102] R. Hoppe, K. Blinne, *Z. Anorg. Allg. Chem.* **1957**, 291, 269–75.

[2103] R. Hoppe, *J. Inorg. Nucl. Chem.* **1958**, 8, 437–40.

[2104] B. G. Müller, *J. Fluorine Chem.* **1981**, 17, 409–21.

[2105] S. N. Solov'yov, A. A. Firer, A. Y. Dupal, *Russ. J. Chem. Phys. A* **2009**, 83, 689–91.

[2106] A. A. Firer, S. N. Solov'ev, A. Y. Dupal, *Russ. J. Chem. Phys. A* **2009**, 83, 1236–8.

[2107] Z. Mazej, E. Goreshnik, Z. Jagličić, Y. Filinchuk, N. Tumanov, L. G. Akselrud, *Eur. J. Inorg. Chem.* **2017**, 2017, 2130–7.

[2108] J. Burgess, J. Fawcett, R. D. Peacock, K. Lutar, B. Žemva, *J. Fluorine Chem.* **1992**, 58, 53–8.

[2109] W. Sawodny, K. M. Rau, *J. Fluorine Chem.* **1993**, 61, 111–6.

[2110] T. C. Ehlert, M. Hsia, *J. Fluorine Chem.* **1972**, 2, 33–51.

[2111] A. Engelbrecht, A. V. Grosse, *J. Am. Chem. Soc.* **1954**, 76, 2042–5.

[2112] J. Aymonino Pedro, H. Schulze, A. Müller, *Z. Naturforsch. B* **1969**, 24, 1508.

[2113] W. L. Levason, J. S. Ogden, A. K. Saad, et al., *J. Fluorine Chem.* **1991**, 53, 43–51.

[2114] A. K. Brisdon, J. H. Holloway, E. G. Hope, P. J. Townson, W. Levason, J. S. Ogden, *J. Chem. Soc., Dalton Trans.* **1991**, 3127–32.

[2115] E. L. Varetti, A. Müller, *Z. Anorg. Allg. Chem.* **1978**, 442, 230–4.

[2116] A. K. Srivastava, N. Misra, *Mol. Phys.* **2014**, 112, 2820–6.

[2117] G. Balducci, M. Campodonico, G. Gigli, G. Meloni, S. N. Cesaro, *J. Chem. Phys.* **2002**, 117, 10613–20.

[2118] H. J. T. Preston, J. J. Kaufman, *Int. J. Quant. Chem.* **1977**, 12, 471–84.

[2119] O. Launila, A. M. James, B. Simard, *J. Mol. Spectrosc.* **1994**, 164, 559–69.

[2120] R. L. Johnson, B. Siegel, *J. Inorg. Nucl. Chem.* **1969**, 31, 2391–6.

[2121] G. Peters, W. Preetz, *Z. Naturforsch. B* **1979**, 34, 1767.

[2122] R. J. H. Clark, M. J. Stead, *Inorg. Chem.* **1983**, 22, 1214–20.

[2123] S. Mariappan Balasekaran, T. K. Todorova, C. T. Pham, et al., *Inorg. Chem.* **2016**, 55, 5417–21.

[2124] G. Henkel, G. Peters, W. Preetz, J. Skowronek, *Z. Naturforsch. B* **1990**, 45, 469.

[2125] S. M. Balasekaran, A. P. Sattelberger, A. Hagenbach, F. Poineau, *Inorg. Chem.* **2018**, 57, 319–25.

[2126] S. M. Balasekaran, A. P. Sattelberger, B. Noll, A. Hagenbach, F. Poineau, *Z. Anorg. Allg. Chem.* **2019**, 645, 27–30.

[2127] R. T. Paine, L. B. Asprey, *Inorg. Chem.* **1975**, 14, 1111–3.

[2128]   O. Ruff, W. Kwasnik, *Z. Anorg. Allg. Chem.* **1934**, 219, 65–81.

[2129]   G. B. Hargreaves, R. D. Peacock, *J. Chem. Soc.* **1960**, 1099–103.

[2130]   D. E. LaValle, R. M. Steele, W. T. Smith, *J. Inorg. Nucl. Chem.* **1966**, 28, 260–3.

[2131]   G. Brauer, H.-D. Allardt, *Z. Anorg. Allg. Chem.* **1962**, 316, 134–40.

[2132]   P. F. Weck, E. Kim, F. Poineau, E. E. Rodriguez, A. P. Sattelberger, K. R. Czerwinski, *Inorg. Chem.* **2009**, 48, 6555–8.

[2133]   K. Schwochau, W. Herr, *Angew. Chem.* **1963**, 75, 95.

[2134]   K. P. Dostal, M. Nagel, A. Freyer, *Isot. Environ. Health Stud.* **1984**, 20, 205–6.

[2135]   R. Alberto, G. Anderegg, *Polyhedron* **1985**, 4, 1067.

[2136]   S. M. Balasekaran, M. Molski, J. Spandl, A. Hagenbach, R. Alberto, U. Abram, *Inorg. Chem.* **2013**, 52, 7094–9.

[2137]   K. Schwochau, *Z. Naturforsch. A* **1964**, 19, 1237.

[2138]   H. Müller, M. Idilbi, *Fresenius' J. Anal. Chem.* **1998**, 360, 729–31.

[2139]   K. S. Pedersen, M. Sigrist, M. A. Sørensen, et al., *Angew. Chem., Int. Ed. Engl.* **2014**, 53, 1351–4.

[2140]   A. J. Edwards, D. Hugill, R. D. Peacock, *Nature* **1963**, 200, 672.

[2141]   J. K. Gibson, *J. Fluorine Chem.* **1991**, 55, 299–311.

[2142]   *J. Inorg. Nucl. Chem.* **1976**, 28, 231–2.

[2143]   D. Hugill, R. D. Peacock, *J. Chem. Soc. A* **1966**, 1339–41.

[2144]   W. A. Sunder, F. A. Stevie, *J. Fluorine Chem.* **1975**, 6, 449–63.

[2145]   J. H. Holloway, D. C. Puddick, G. M. Staunton, D. Brown, *Inorg. Chim. Acta* **1982**, 64, L209–L10.

[2146]   H. Selig, C. L. Chernick, J. G. Malm, *J. Inorg. Nucl. Chem.* **1961**, 19, 377.

[2147]   T. Drews, J. Supeł, A. Hagenbach, K. Seppelt, *Inorg. Chem.* **2006**, 45, 3782–8.

[2148]   J. G. Malm, H. Selig, *J. Inorg. Nucl. Chem.* **1961**, 20, 189–97.

[2149]   H. H. Claassen, H. Selig, J. G. Malm, *J. Chem. Phys.* **1962**, 36, 2888–90.

[2150]   M. J. Molski, K. Seppelt, *Dalton Trans.* **2009**, 3379–83.

[2151]   V. Boudon, M. Rotger, Y. He, H. Hollenstein, M. Quack, U. Schmitt, *J. Chem. Phys.* **2002**, 117, 3196–207.

[2152]   H. H. Claassen, G. L. Goodman, J. H. Holloway, H. Selig, *J. Chem. Phys.* **1970**, 53, 341–8.

[2153]   J. Shamir, J. G. Malm, *J. Inorg. Nucl. Chem.* **1976**, 28, 107–11.

[2154]   E. J. Jacob, L. S. Bartell, *J. Chem. Phys.* **1970**, 53, 2231–5.

[2155]   H. H. Claassen, J. G. Malm, H. Selig, *J. Chem. Phys.* **1962**, 36, 2890–2.

[2156]   G. R. Meredith, J. D. Webb, E. R. Bernstein, *Mol. Phys.* **1977**, 34, 995–1017.

[2157]   R. T. Paine, *Inorg. Chem.* **1973**, 12, 1457–8.

[2158]   J. H. Holloway, J. B. Raynor, *J. Chem. Soc., Dalton Trans.* **1975**, 737–41.

[2159]   E. Jacob, M. Fähnle, *Angew. Chem., Int. Ed. Engl.* **1976**, 15, 159–60.

[2160]   J. H. Holloway, H. Selig, *J. Inorg. Nucl. Chem.* **1968**, 30, 473–8.

[2161]   R. C. Burns, T. A. O'Donnell, *J. Inorg. Nucl. Chem.* **1980**, 42, 1285–91.

[2162]   M. Malischewski, M. Adelhardt, J. Sutter, K. Meyer, K. Seppelt, *Science* **2016**, 353, 678–82.

[2163]   B. v. Ahsen, C. Bach, H. Pernice, H. Willner, F. Aubke, *J. Fluorine Chem.* **2000**, 102, 243–52.

[2164]   D. M. Bruce, A. J. Hewitt, J. H. Holloway, R. D. Peacock, I. L. Wilson, *J. Chem. Soc., Dalton Trans.* **1976**, 2230–5.

[2165]   D. M. Bruce, J. H. Holloway, D. R. Russell, *J. Chem. Soc., Chem. Commun.* **1973**, 321–2.

[2166]   D. M. Bruce, J. H. Holloway, D. R. Russell, *J. Chem. Soc., Dalton Trans.* **1978**, 64–7.

[2167]   T. A. O'Donnell, K. A. Phillips, A. B. Waugh, *Inorg. Chem.* **1973**, 12, 1435–7.

[2168]   J. D. Webb, E. R. Bernstein, *J. Am. Chem. Soc.* **1978**, 100, 483–5.

[2169]   S. Adam, A. Ellern, K. Seppelt, *Chem. Eur. J.* **1996**, 2, 398–402.

[2170]   A. M. Affoune, J. Bouteillon, J. C. Poignet, *J. Appl. Electrochem.* **2002**, 32, 521–6.

[2171] J. H. Canterford, A. B. Waugh, *Inorg. Nucl. Chem. Lett.* **1971**, 7, 395–9.

[2172] J. Burgess, C. J. W. Fraser, I. Haigh, R. D. Peacock, *J. Chem. Soc., Dalton Trans.* **1973**, 501–4.

[2173] E. Jacob, *Angew. Chem., Int. Ed. Engl.* **1982**, 21, 142–3.

[2174] R. D. Peacock, D. F. Stewart, *Inorg. Nucl. Chem. Lett.* **1967**, 3, 255–6.

[2175] J. Fawcett, R. D. Peacock, D. R. Russell, *J. Chem. Soc., Chem. Commun.* **1982**, 958–9.

[2176] J. Burgess, J. Fawcett, R. D. Peacock, *J. Fluorine Chem.* **1984**, 24, 341–3.

[2177] J. Fawcett, R. D. Peacock, D. R. Russell, *J. Chem. Soc., Dalton Trans.* **1987**, 567–71.

[2178] A. J. Edwards, G. R. Jones, *J. Chem. Soc. A* **1968**, 2511–5.

[2179] A. J. Edwards, G. R. Jones, R. J. C. Sills, *Chem. Commun.* **1968**, 1177–8.

[2180] A. J. Edwards, G. R. Jones, R. J. C. Sills, *J. Chem. Soc. A* **1970**, 2521–3.

[2181] E. T. Paine, K. L. Treuil, F. E. Stafford, *Spectrochim. Acta, Part A* **1973**, 29, 1891–7.

[2182] E. J. Baran, *Monatsh. Chem.* **1978**, 109, 1337–41.

[2183] W. Kuhlmann, W. Sawodny, *J. Fluorine Chem.* **1977**, 9, 337–40.

[2184] J. Burgess, J. Fawcett, R. D. Peacock, D. Pickering, *J. Chem. Soc., Dalton Trans.* **1976**, 1363–4.

[2185] J. G. Malm, H. Selig, S. Fried, *J. Am. Chem. Soc.* **1960**, 82, 1510.

[2186] E. J. Jacob, L. S. Bartell, *J. Chem. Phys.* **1970**, 53, 2235–42.

[2187] H. H. Claassen, H. Selig, *J. Chem. Phys.* **1965**, 43, 103–5.

[2188] E. W. Kaiser, J. S. Muenter, W. Klemperer, W. E. Falconer, *J. Chem. Phys.* **1970**, 53, 53–5.

[2189] T. Vogt, A. N. Fitch, J. K. Cockcroft, *Science* **1994**, 263, 1265–7.

[2190] H. Selig, E. L. Gasner, *J. Inorg. Nucl. Chem.* **1968**, 30, 658–9.

[2191] J. Canterford, T. O'Donnell, A. Waugh, *Aust. J. Chem.* **1971**, 24, 243–7.

[2192] H. Selig, Z. Karpas, *Isr. J. Chem.* **1971**, 9, 53–6.

[2193] J. Baran Enrique, *Z. Naturforsch. A* **1976**, 31, 1733.

[2194] S. Riedel, M. Renz, M. Kaupp, *Inorg. Chem.* **2007**, 46, 5734–8.

[2195] N. LeBlond, H. P. A. Mercier, D. A. Dixon, G. J. Schrobilgen, *Inorg. Chem.* **2000**, 39, 4494–509.

[2196] R. Haiges. Preparation of Transition Metal Fluorides using ClF3. In: H. W. Roesky (ed.) Efficient Preparations of Fluorine Compounds: Wiley; 2013:100–7.

[2197] J. Burgess, J. Fawcett, N. Morton, R. D. Peacock, *J. Chem. Soc., Dalton Trans.* **1977**, 2149–50.

[2198] E. E. Aynsley, R. D. Peacock, P. L. Robinson, *J. Chem. Soc.* **1950**, 1622–4.

[2199] N. LeBlond, G. J. Schrobilgen, *Chem. Commun.* **1996**, 2479–80.

[2200] J. Supeł, R. Marx, K. Seppelt, *Z. Anorg. Allg. Chem.* **2005**, 631, 2979–86.

[2201] W. J. Casteel, D. A. Dixon, N. LeBlond, P. E. Lock, H. P. A. Mercier, G. J. Schrobilgen, *Inorg. Chem.* **1999**, 38, 2340–58.

[2202] I. R. Beattie, R. A. Crocombe, J. S. Ogden, *J. Chem. Soc., Dalton Trans.* **1977**, 1481–9.

[2203] N. LeBlond, G. J. Schrobilgen, *Inorg. Chem.* **2001**, 40, 1245–9.

[2204] W. J. Casteel, D. M. MacLeod, H. P. A. Mercier, G. J. Schrobilgen, *Inorg. Chem.* **1996**, 35, 7279–88.

[2205] N. LeBlond, D. A. Dixon, G. J. Schrobilgen, *Inorg. Chem.* **2000**, 39, 2473–87.

[2206] H. P. A. Mercier, G. J. Schrobilgen, *Inorg. Chem.* **1993**, 32, 145–51.

[2207] H. Selig, J. G. Malm, *J. Inorg. Nucl. Chem.* **1963**, 25, 349–51.

[2208] J. Binenboym, U. El-Gad, H. Selig, *Inorg. Chem.* **1974**, 13, 319–21.

[2209] E. J. Baran, *Spectrosc. Lett.* **1975**, 8, 599–603.

[2210] K. J. Franklin, C. J. L. Lock, B. G. Sayer, G. J. Schrobilgen, *J. Am. Chem. Soc.* **1982**, 104, 5303–6.

[2211] H. Selig, U. El-gad, *J. Inorg. Nucl. Chem.* **1973**, 35, 3517–22.

[2212] M. V. Ivanova, T. Köchner, H. P. A. Mercier, G. J. Schrobilgen, *Inorg. Chem.* **2013**, 52, 6806–19.

[2213] J. F. Lotspeich, A. Javan, A. Engelbrecht, *J. Chem. Phys.* **1959**, 31, 633–43.

[2214] F. E. Kühn, J. J. Haider, E. Herdtweck, et al., *Inorg. Chim. Acta* **1998**, 279, 44–50.

[2215]   M. V. Ivanova, H. P. A. Mercier, G. J. Schrobilgen, *J. Am. Chem. Soc.* **2015**, 137, 13398–413.

[2216]   M. C. Chakravorti, M. K. Chaudhuri, *Z. Anorg. Allg. Chem.* **1973**, 398, 221–4.

[2217]   B. Pouilly, J. Schamps, D. J. W. Lumley, R. F. Barrow, *J. Phys. B* **1978**, 11, 2281–7.

[2218]   B. Pouilly, J. Schamps, *J. Phys. B* **1978**, 11, 2289–99.

[2219]   M. Bencheikh, *J. Mol. Spectrosc.* **1997**, 183, 419–20.

[2220]   S. M. Kermode, J. M. Brown, *J. Mol. Spectrosc.* **2002**, 213, 158–69.

[2221]   C. W. Bauschlicher, *Chem. Phys.* **1996**, 211, 163–9.

[2222]   C. Koukounas, A. Mavridis, *J. Phys. Chem. A* **2008**, 112, 11235–50.

[2223]   R. S. Ram, P. F. Bernath, S. P. Davis, *J. Chem. Phys.* **1996**, 104, 6949–55.

[2224]   J. J. Harrison, J. M. Brown, M. A. Flory, P. M. Sheridan, S. K. McLamarrah, L. M. Ziurys, *J. Chem. Phys.* **2007**, 127, 194308.

[2225]   Z. Zhang, J. Guo, X. Yu, J. Zheng, Y. Chen, *J. Mol. Spectrosc.* **2007**, 244, 117–21.

[2226]   X. Zhang, J. Guo, T. Wang, L. Pei, Y. Chen, C. Chen, *J. Mol. Spectrosc.* **2003**, 220, 209–13.

[2227]   H. Wang, X. Zhuang, T. C. Steimle, *J. Chem. Phys.* **2009**, 131, 114315.

[2228]   B. Pinchemel, T. Hirao, P. F. Bernath, *J. Mol. Spectrosc.* **2002**, 215, 262–8.

[2229]   M. Benomier, A. van Groenendael, B. Pinchemel, T. Hirao, P. F. Bernath, *J. Mol. Spectrosc.* **2005**, 233, 244–55.

[2230]   W. Zou, W. Liu, *J. Chem. Phys.* **2006**, 124, 154312.

[2231]   D. L. Arsenault, D. W. Tokaryk, A. G. Adam, C. Linton, *J. Mol. Spectrosc.* **2016**, 324, 20–7.

[2232]   T. L. Court, M. F. A. Dove, *J. Chem. Soc., Dalton Trans.* **1973**, 1995–7.

[2233]   A. M. Bond, G. Hefter, *J. Inorg. Nucl. Chem.* **1972**, 34, 603–7.

[2234]   D. Babel, *Z. Anorg. Allg. Chem.* **1969**, 369, 117–30.

[2235]   A. Tressaud, R. De Pape, J. Portier, P. Hagenmuller, *C. R. Acad. Sci., Ser. IIc: Chim.* **1968**, 266C, 984–86.

[2236]   W. Rüdorff, G. Lincke, D. Babel, *Z. Anorg. Allg. Chem.* **1963**, 320, 150–70.

[2237]   W. Rüdorff, J. Kändler, G. Lincke, D. Babel, *Angew. Chem.* **1959**, 71, 672.

[2238]   W. Rüdorff, J. Kandler, D. Babel, *Z. Anorg. Allg. Chem.* **1962**, 317, 261–87.

[2239]   M. K. Chaudhuri, S. K. Ghosh, Z. Hiese, *J. Chem. Soc., Dalton Trans.* **1984**, 1763–4.

[2240]   C. Shen, L. C. Chacón, N. Rosov, S. H. Elder, J. C. Allman, N. Bartlett, *C. R. Acad. Sci., Ser. IIc: Chim.* **1999**, 2, 557–63.

[2241]   D. S. Crocket, R. A. Grossman, *Inorg. Chem.* **1964**, 3, 644–6.

[2242]   Y. Shirako, Y. G. Shi, A. Aimi, et al., *J. Solid State Chem.* **2012**, 191, 167–74.

[2243]   M. W. Shafer, T. R. McGuire, *J. Phys. Chem. Solids* **1969**, 30, 1989–97.

[2244]   E. T. Keve, S. C. Abrahams, J. L. Bernstein, *J. Chem. Phys.* **1970**, 53, 3279–87.

[2245]   J. E. Weidenborner, A. L. Bednowitz, *Acta Crystallogr. B* **1970**, 26, 1464–8.

[2246]   N. N. Greenwood, A. T. Howe, F. Ménil, *J. Chem. Soc. A* **1971**, 2218–24.

[2247]   T. W. Balcerek, L. Cathey, D. G. Karraker, *J. Inorg. Nucl. Chem.* **1978**, 40, 773–7.

[2248]   G. Heger, R. Viebahn-Hänsler, *Solid State Commun.* **1972**, 11, 1119–22.

[2249]   R. Haegele, W. Verscharen, D. Babel, J.-M. Dance, A. Tressaud, *J. Solid State Chem.* **1978**, 24, 77–84.

[2250]   J. Ferguson, E. R. Krausz, G. B. Robertson, H. J. Guggenheim, *Chem. Phys. Lett.* **1972**, 17, 551–3.

[2251]   A. D. Wesland, R. Hoppe, S. S. I. Kaseno, *Z. Anorg. Allg. Chem.* **1965**, 338, 319–31.

[2252]   D. Cao, D. Yin, D. Shi, Z. Fu, J. Zhang, C. Li, *Adv. Funct. Mater.* **2017**, 27, 1701130.

[2253]   T. Balić-Žunić, A. Garavelli, D. Mitolo, *Eur. J. Mineral.* **2018**, 30, 841–8.

[2254]   M. Burbano, M. Duttine, O. Borkiewicz, et al., *Inorg. Chem.* **2015**, 54, 9619–25.

[2255]   S. A. Baer, F. Kraus, *Z. Naturforsch. B* **2011**, 66, 865.

[2256]   E. G. Walton, P. J. Corvan, D. B. Brown, P. Day, *Inorg. Chem.* **1976**, 15, 1737–9.

[2257]   G. Brauer, M. Eichner, *Z. Anorg. Allg. Chem.* **1958**, 296, 13–9.

[2258] M. C. Chakravorti, G. V. B. Subrahamanyam, *J. Chem. Educ.* **1991**, 68, 961.

[2259] M. Vlasse, J. C. Massies, G. Demazeau, *J. Solid State Chem.* **1973**, 8, 109–13.

[2260] F. J. Brink, R. L. Withers, J. G. Thompson, *J. Solid State Chem.* **2000**, 155, 359–65.

[2261] G. Tobias, M. Armand, G. Rousse, et al., *Solid State Sci.* **2014**, 38, 55–61.

[2262] F. J. Brink, R. L. Withers, L. Norén, *J. Solid State Chem.* **2001**, 161, 31–7.

[2263] J. Zhu, D. Deng, *Angew. Chem., Int. Ed. Engl.* **2015**, 54, 3079–83.

[2264] L.-P. Wang, T.-S. Wang, X.-D. Zhang, et al., *J. Mater. Chem. A* **2017**, 5, 18464–8.

[2265] S. Y. Fu, Y. Z. Li, W. Chu, Y. M. Yang, D. G. Tong, Q. Le Zeng, *J. Mater. Chem. A* **2015**, 3, 16716–27.

[2266] M. Park, J.-H. Shim, H. Kim, H. Park, N. Kim, J. Kim, *J. Power Sources* **2018**, 396, 551–8.

[2267] J. Zhai, Z. Lei, K. Sun, *Chem. Eur. J.* **2019**, 25, 7733–9.

[2268] M. I. Nikitin, N. S. Chilingarov, A. S. Alikhanyan, *Russ. J. Inorg. Chem.* **2019**, 64, 377–82.

[2269] E. L. Osina, N. S. Chilingarov, S. B. Osin, E. V. Skokan, M. I. Nikitin, *Russ. J. Chem. Phys. A* **2019**, 93, 185–91.

[2270] H. Siebert, B. Breitenstein, *Z. Anorg. Allg. Chem.* **1970**, 379, 44–7.

[2271] B. Zemva, K. Lutar, L. Chacon, et al., *J. Am. Chem. Soc.* **1995**, 117, 10025–34.

[2272] A. L. Hector, E. G. Hope, W. Levason, M. T. Weller, *Z. Anorg. Allg. Chem.* **1998**, 624, 1982–8.

[2273] S. N. Solov'ev, A. A. Korunov, K. G. Zubkov, A. A. Firer, *Russ. J. Chem. Phys. A* **2012**, 86, 516–8.

[2274] N. Bartlett, R. D. Chambers, A. J. Roche, R. C. H. Spink, L. Chacón, J. M. Whalen, *Chem. Commun.* **1996**, 1049–50.

[2275] M. Tramsek, B. Zemva, *Acta Chim. Slov.* **2002**, 49, 209–20.

[2276] H. Remy, H. Busch, *Chem. Ber.* **1933**, 66, 961–9.

[2277] J. Lapasset, P. Sciau, J. Moret, N. Gros, *Acta Crystallogr. B* **1986**, 42, 258–62.

[2278] D. Dumora, R. Von der Mühll, J. Ravez, *Mater. Res. Bull.* **1971**, 6, 561–9.

[2279] J. L. Fourquet, H. Duroy, *J. Solid State Chem.* **1993**, 103, 353–8.

[2280] M. J. Portier, A. Tressaud, R. de Pape, P. Hagenmuller, *Mater. Res. Bull.* **1968**, 3, 433–6.

[2281] D. B. Shinn, D. S. Crocket, H. M. Haendler, *Inorg. Chem.* **1966**, 5, 1927–33.

[2282] M. Leblanc, G. Ferey, R. De Pape, J. Teillet, *Acta Crystallogr. Sect. C: Cryst. Struct. Commun.* **1985**, 41, 657–60.

[2283] A. Delunas, V. Maxia, F. Abbattista, D. Mazza, M. Vallino, *Nuovo Cimento B* **1987**, 9, 1275–83.

[2284] D. Babel, F. Wall, G. Heger, *Z. Naturforsch. B* **1974**, 29, 139.

[2285] M. Hidaka, I. G. Wood, B. M. Wanklyn, B. J. Garrard, *J. Phys. C, Solid State Phys.* **1979**, 12, 1799–807.

[2286] C. Pique, A. Bulou, M. C. Moron, R. Burriel, J. L. Fourquet, M. Rousseau, *J. Phys.: Condens. Matter.* **1990**, 2, 8277–92.

[2287] A. Desert, A. Bulou, M. Leblanc, J. Nouet, *J. Phys.: Condens. Matter.* **1998**, 10, 9067–79.

[2288] P. Lacorre, J. Pannetier, F. Averdunk, R. Hoppe, G. Ferey, *J. Solid State Chem.* **1989**, 79, 1–11.

[2289] J. Ravez, J. Grannec, R. von der Mühll, *C. R. Acad. Sci., Ser. IIc: Chim.* **1971**, 272C, 1042–4.

[2290] U. Bentrup, W. Massa, *Z. Naturforsch. B* **1991**, 46, 395.

[2291] S. Ghosh, A. K. Sengupta, B. B. Bhaumik, *J. Fluorine Chem.* **1996**, 76, 125–6.

[2292] J. Graulich, W. Massa, D. Babel, *Z. Anorg. Allg. Chem.* **2003**, 629, 365–7.

[2293] R. von der Mühll, F. Daut, J. Ravez, *J. Solid State Chem.* **1973**, 8, 206–12.

[2294] R. Von Der Muhll, S. Andersson, J. Galy, *Acta Crystallogr. B* **1971**, 27, 2345–53.

[2295] A. J. Edwards, *J. Chem. Soc., Dalton Trans.* **1972**, 816–8.

[2296] E. Herdtweck, W. Massa, *Z. Anorg. Allg. Chem.* **1989**, 579, 191–9.

[2297] J.-M. Le Meins, A. Hemon-Ribaud, G. Courbion, *Acta Crystallogr. Sect. C: Cryst. Struct. Commun.* **1997**, 53, 1165–6.

[2298] R. L. Carlin, R. Burriel, J. A. Rojo, F. Palacio, *Inorg. Chem.* **1984**, 23, 2213–5.

[2299] J. L. Fourquet, F. Plet, Y. Calage, *J. Solid State Chem.* **1988**, 74, 34–8.

[2300] R. Lück, U. Bentrup, R. Stösser, *Z. Anorg. Allg. Chem.* **1989**, 576, 215–24.

[2301] U. Bentrup, D.-H. Menz, *Z. Anorg. Allg. Chem.* **1990**, 591, 230–6.

[2302] A. Le Bail, *Powder Diffr.* **2014**, 29, 33–41.

[2303] M. Vlasse, G. Matejka, A. Tressaud, B. M. Wanklyn, *Acta Crystallogr. B* **1977**, 33, 3377–80.

[2304] L. Croguennec, P. Deniard, R. Brec, et al., *J. Solid State Chem.* **1997**, 131, 189–97.

[2305] D. Hanžel, A. Moljk, J. Slivnik, *Inorg. Nucl. Chem. Lett.* **1974**, 10, 1–5.

[2306] A. Hemon-Ribaud, J. M. Greneche, G. Courbion, *J. Solid State Chem.* **1994**, 112, 82–91.

[2307] H. Henkel, R. Hoppe, *Z. Anorg. Allg. Chem.* **1969**, 364, 253–62.

[2308] A. Tressaud, J. Portier, S. Shearer-Turrell, J.-L. Dupin, P. Hagenmuller, *J. Inorg. Nucl. Chem.* **1970**, 32, 2179–86.

[2309] A. Tressaud, J. Portier, R. de Pape, P. Hagenmuller, *J. Solid State Chem.* **1970**, 2, 269–77.

[2310] J. Grannec, J. Portier, M. Pouchard, P. Hagenmuller, *J. Inorg. Nucl. Chem.* **1976**, 28, 119–22.

[2311] L. Stein, J. M. Neil, G. R. Alms, *Inorg. Chem.* **1969**, 8, 2472–6.

[2312] R. Jesse, R. Hoppe, *Z. Anorg. Allg. Chem.* **1974**, 403, 143–8.

[2313] W. Viebahn, D. Babel, *Z. Anorg. Allg. Chem.* **1974**, 406, 38–44.

[2314] K. S. Raju, B. M. Wanklyn, *J. Cryst. Growth* **1993**, 128, 1078–80.

[2315] E. Alter, R. Hoppe, *Z. Anorg. Allg. Chem.* **1974**, 405, 167–75.

[2316] E. Alter, R. Hoppe, *Z. Anorg. Allg. Chem.* **1974**, 407, 305–12.

[2317] E. Alter, R. Hoppe, *Z. Anorg. Allg. Chem.* **1974**, 407, 313–8.

[2318] D. Reinen, C. Friebel, V. Propach, *Z. Anorg. Allg. Chem.* **1974**, 408, 187–204.

[2319] M. D. Meyers, F. A. Cotton, *J. Am. Chem. Soc.* **1960**, 82, 5027–30.

[2320] W. Klemm, W. Brandt, R. Hoppe, *Z. Anorg. Allg. Chem.* **1961**, 308, 179–89.

[2321] T. Schlöder, T. Vent-Schmidt, S. Riedel, *Angew. Chem., Int. Ed. Engl.* **2012**, 51, 12063–7.

[2322] J. V. Rau, S. Nunziante Cesaro, N. S. Chilingarov, G. Balducci, *Inorg. Chem.* **1999**, 38, 5695–7.

[2323] B. Zemva, K. Lutar, A. Jesih, W. J. Casteel, N. Bartlett, *J. Chem. Soc., Chem. Commun.* **1989**, 346–7.

[2324] R. Hoppe, *Recl. Trav. Chim. Pays-Bas.* **1956**, 75, 569–75.

[2325] J. W. Quail, G. A. Rivett, *Can. J. Chem.* **1972**, 50, 2447–50.

[2326] W. Klemm, E. Huss, *Z. Anorg. Allg. Chem.* **1949**, 258, 221–6.

[2327] R. Bougon, *C. R. Acad. Sci., Ser. IIc: Chim.* **1968**, 267C.

[2328] H. Bode, E. Voss, *Z. Anorg. Allg. Chem.* **1956**, 286, 136–41.

[2329] R. Hoppe, T. Fleischer, *J. Fluorine Chem.* **1978**, 11, 251–64.

[2330] S. N. Solov'ev, K. I. Shatalov, *Russ. J. Chem. Phys. A* **2009**, 83, 1053–4.

[2331] T. Fleischer, R. Hoppe, *Z. Anorg. Allg. Chem.* **1982**, 489, 7–10.

[2332] J. C. Taylor, P. W. Wilson, *J. Inorg. Nucl. Chem.* **1974**, 36, 1561–3.

[2333] H. Henkel, R. Hoppe, G. C. Allen, *J. Inorg. Nucl. Chem.* **1969**, 31, 3855–9.

[2334] N. Bartlett, *J. Fluorine Chem.* **2006**, 127, 1285–8.

[2335] A. Engelbrecht, E. Mayer, C. Pupp, *Monatsh. Chem.* **1964**, 95, 633–48.

[2336] W. W. Wilson, K. O. Christe, *Inorg. Chem.* **1984**, 23, 3261–2.

[2337] S. N. Solov'ev, K. I. Shatalov, A. Y. Dupal, *Russ. J. Chem. Phys. A* **2011**, 85, 1855.

[2338] A. Jesih, K. Lutar, I. Leban, B. Zemva, *Inorg. Chem.* **1989**, 28, 2911–4.

[2339] K. I. Shatalov, S. N. Solov'ev, *Russ. J. Chem. Phys. A* **2011**, 85, 331–3.

[2340] S. N. Solov'ev, K. I. Shatalov, A. Y. Dupal, *Russ. J. Chem. Phys. A* **2014**, 88, 893–5.

[2341] D. L. Hildenbrand, K. H. Lau, *J. Chem. Phys.* **1988**, 89, 5825–8.

[2342] R. Li, R. H. Jensen, W. J. Balfour, S. A. Shepard, A. G. Adam, *J. Chem. Phys.* **2004**, 121, 2591–7.

[2343] K. F. Ng, A. M. Southam, A. S. C. Cheung, *J. Mol. Spectrosc.* **2016**, 328, 32–6.

[2344] T. Okabayashi, T. Kurahara, E. Y. Okabayashi, M. Tanimoto, *J. Chem. Phys.* **2012**, 136, 174311.

[2345] N. Bartlett, R. Maitland, *Acta Crystallogr.* **1958**, 11, 747–8.

[2346] O. Ruff, E. Ascher, *Z. Anorg. Allg. Chem.* **1929**, 183, 193–213.

[2347] B. Müller, R. Hoppe, *Mater. Res. Bull.* **1972**, 7, 1297–306.

[2348] B. Müller, R. Hoppe, *Sci. Nat.* **1971**, 58, 268.

[2349] N. Bartlett, P. R. Rao, *Proc. Chem. Soc.* **1964**, 393–4.

[2350] B. G. Müller, *J. Fluorine Chem.* **1982**, 20, 291–9.

[2351] B. G. Müller, *Sci. Nat.* **1979**, 66, 519–20.

[2352] B. Bachmann, B. G. Müller, *Z. Anorg. Allg. Chem.* **1993**, 619, 387–91.

[2353] Z. Mazej, A. Tressaud, J. Darriet, *J. Fluorine Chem.* **2001**, 110, 139–43.

[2354] Z. Mazej, P. Benkič, A. Tressaud, B. Žemva, *Eur. J. Inorg. Chem.* **2004**, 2004, 1827–34.

[2355] R. P. Rao, R. C. Sherwood, N. Bartlett, *J. Chem. Phys.* **1968**, 49, 3728–30.

[2356] B. Bachmann, B. G. Müller, *Z. Anorg. Allg. Chem.* **1991**, 597, 9–18.

[2357] B. G. Müller, *Z. Anorg. Allg. Chem.* **1982**, 491, 245–52.

[2358] B. Bachmann, B. G. Müller, *Z. Anorg. Allg. Chem.* **1992**, 616, 7–13.

[2359] C. De Nadaï, A. Demourgues, P. Gravereau, J. Grannec, *J. Solid State Chem.* **1999**, 148, 242–9.

[2360] N. Bartlett, D. H. Lohmann, *J. Chem. Soc.* **1964**, 619–26.

[2361] R. Wesendrup, P. Schwerdtfeger, *Inorg. Chem.* **2001**, 40, 3351–4.

[2362] S. A. Brewer, K. S. Coleman, J. Fawcett, et al., *J. Chem. Soc., Dalton Trans.* **1995**, 1073–6.

[2363] S. A. Brewer, L. A. Buggey, J. H. Holloway, E. G. Hope, *J. Chem. Soc., Dalton Trans.* **1995**, 2941–4.

[2364] K. S. Coleman, J. H. Holloway, E. G. Hope, *J. Chem. Soc., Dalton Trans.* **1997**, 1713–8.

[2365] P. Barthazy, R. M. Stoop, M. Wörle, A. Togni, A. Mezzetti, *Organometallics* **2000**, 19, 2844–52.

[2366] D. Huang, P. R. Koren, K. Folting, E. R. Davidson, K. G. Caulton, *J. Am. Chem. Soc.* **2000**, 122, 8916–31.

[2367] D. Huang, K. B. Renkema, K. G. Caulton, *Polyhedron* **2006**, 25, 459–68.

[2368] J. Fawcett, D. A. J. Harding, E. G. Hope, K. Singh, G. A. Solan, *Dalton Trans.* **2009**, 6861–70.

[2369] K. S. Coleman, J. Fawcett, D. A. J. Harding, E. G. Hope, K. Singh, G. A. Solan, *Eur. J. Inorg. Chem.* **2010**, 2010, 4130–8.

[2370] E. E. Aynsley, R. D. Peacock, P. L. Robinson, *Chem. Ind.* **1952**, 1002.

[2371] M. A. Hepworth, K. H. Jack, R. D. Peacock, G. J. Westland, *Acta Crystallogr.* **1957**, 10, 63–9.

[2372] W. J. Casteel, A. P. Wilkinson, H. Borrmann, R. E. Serfass, N. Bartlett, *Inorg. Chem.* **1992**, 31, 3124–31.

[2373] A. L. Hector, W. Levason, M. T. Weiler, E. G. Hope, *J. Fluorine Chem.* **1997**, 84, 161–5.

[2374] P. L. Robinson, G. J. Westland, *J. Chem. Soc.* **1956**, 4481–7.

[2375] A. G. Sharpe, *J. Chem. Soc.* **1950**, 3444–50.

[2376] L. Grosse, R. Hoppe, *Z. Anorg. Allg. Chem.* **1987**, 552, 123–31.

[2377] R. D. Peacock, *J. Chem. Soc.* **1955**, 3291–2.

[2378] D. Naumann. Fluoride der Nebengruppenelemente. Fluor und Fluorverbindungen. Heidelberg: Steinkopff; 1980:103–57.

[2379] R. Domesle, R. Hoppe, *Z. Anorg. Allg. Chem.* **1983**, 501, 102–10.

[2380] V. Wilhelm, R. Hoppe, *Z. Anorg. Allg. Chem.* **1975**, 414, 91–6.

[2381] J. H. Holloway, R. D. Peacock, *J. Chem. Soc.* **1963**, 3892–3.

[2382] R. C. Burns, T. A. O'Donnell, *J. Inorg. Nucl. Chem.* **1980**, 42, 1613–9.

[2383] G. B. Hargreaves, R. D. Peacock, *J. Chem. Soc.* **1960**, 2618–20.

[2384] W. A. Sunder, W. E. Falconer, *Inorg. Nucl. Chem. Lett.* **1972**, 8, 537–40.

[2385] N. Bartlett, A. Tressaud, *C. R. Acad. Sci., Ser. IIc: Chim.* **1974**, 278C, 1501–4.

[2386] P. R. Rao, A. Tressaud, N. Bartlett, *J. Inorg. Nucl. Chem.* **1976**, 28, 23–8.

[2387] J. e. Slivnik, B. Žemva, B. Družina, *J. Fluorine Chem.* **1980**, 15, 351–2.

[2388] T. Drews, D. Rusch, S. Seidel, S. Willemsen, K. Seppelt, *Chem. Eur. J.* **2008**, 14, 4280–6.

[2389] A. F. Wright, B. E. F. Fender, N. Bartlett, K. Leary, *Inorg. Chem.* **1978**, 17, 748–9.

[2390] T. Sakurai, A. Takahashi, *J. Inorg. Nucl. Chem.* **1979**, 41, 681–5.

[2391] F. Aubke, *J. Fluorine Chem.* **1995**, 72, 195–201.

[2392] L. Graham, O. Graudejus, N. K. Jha, N. Bartlett, *Coord. Chem. Rev.* **2000**, 197, 321–34.

[2393] H. Henkel, R. Hoppe, *Z. Anorg. Allg. Chem.* **1968**, 359, 160–77.

[2394] W. Preetz, Y. Petros, *Angew. Chem., Int. Ed. Engl.* **1971**, 10, 936.

[2395] V. Wilhelm, R. Hoppe, *Z. Anorg. Allg. Chem.* **1974**, 405, 193–6.

[2396] V. Wilhelm, R. Hoppe, *Z. Anorg. Allg. Chem.* **1974**, 407, 13–22.

[2397] B. Cox, D. W. A. Sharp, A. G. Sharpe, *J. Chem. Soc.* **1956**, 1242–4.

[2398] V. Wilhelm, R. Hoppe, *Z. Anorg. Allg. Chem.* **1975**, 414, 130–6.

[2399] M. A. Hepworth, P. L. Robinson, G. J. Westland, *J. Chem. Soc.* **1954**, 4269–75.

[2400] M. A. Hepworth, R. D. Peacock, P. L. Robinson, *J. Chem. Soc.* **1954**, 1197–201.

[2401] S. I. Ivlev, A. J. Karttunen, M. R. Buchner, M. Conrad, R. V. Ostvald, F. Kraus, *Crystals* **2018**, 8, 11.

[2402] S. S. Rudel, T. G. Müller, F. Kraus, *Z. Anorg. Allg. Chem.* **2015**, 641, 298–303.

[2403] F. Kraus, *Acta Crystallogr., Sect. E* **2014**, 70, i43.

[2404] R. Hoppe, W. Klemm, *Z. Anorg. Allg. Chem.* **1952**, 268, 364–71.

[2405] A. Tressaud, N. Bartlett, *J. Solid State Chem.* **2001**, 162, 333–40.

[2406] D. H. Brown, K. R. Dixon, D. W. A. Sharp, *J. Chem. Soc. A* **1966**, 1244–6.

[2407] L. Kolditz, J. Gisbier, *Z. Anorg. Allg. Chem.* **1969**, 366, 265–73.

[2408] W. Preetz, Y. Petros, *Z. Anorg. Allg. Chem.* **1975**, 415, 15–24.

[2409] D. F. Evans, G. K. Turner, *J. Chem. Soc., Dalton Trans.* **1975**, 1238–43.

[2410] H.-L. Keller, H. Homborg, *Z. Anorg. Allg. Chem.* **1976**, 422, 261–5.

[2411] A. Tressaud, J.-M. Dance, P. Hagenmuller, *Isr. J. Chem.* **1978**, 17, 126–8.

[2412] J. H. Holloway, P. R. Rao, N. Bartlett, *Chem. Commun.* **1965**, 306–7.

[2413] N. Bartlett, D. H. Lohmann, *Proc. Chem. Soc.* **1960**, 14–5.

[2414] N. Bartlett, P. R. Rao, *Chem. Commun.* **1965**, 252–3.

[2415] J. H. Holloway, R. D. Peacock, R. W. H. Small, *J. Chem. Soc.* **1964**, 644–8.

[2416] S. J. Mitchell, J. H. Holloway, *J. Chem. Soc. A* **1971**, 2789–94.

[2417] B. K. Morrell, A. Zalkin, A. Tressaud, N. Bartlett, *Inorg. Chem.* **1973**, 12, 2640–4.

[2418] A. J. Hewitt, J. H. Holloway, R. D. Peacock, J. B. Raynor, I. L. Wilson, *J. Chem. Soc., Dalton Trans.* **1976**, 579–83.

[2419] A. Tressaud, P. Hagenmuller, *J. Fluorine Chem.* **2001**, 111, 221–5.

[2420] F. O. Sladky, P. A. Bulliner, N. Bartlett, *J. Chem. Soc. A* **1969**, 2179–88.

[2421] E. Weise, W. Klemm, *Z. Anorg. Allg. Chem.* **1955**, 279, 74–83.

[2422] J. E. Griffiths, W. A. Sunder, *J. Fluorine Chem.* **1975**, 6, 533–56.

[2423] A. K. Brisdon, J. H. Holloway, E. G. Hope, W. Levason, *Polyhedron* **1992**, 11, 7–11.

[2424] R. Bougon, W. V. Cicha, M. Lance, L. Meublat, M. Nierlich, J. Vigner, *Inorg. Chem.* **1991**, 30, 102–9.

[2425] P. Botkovitz, G. M. Lucier, R. P. Rao, N. Bartlett, *Acta Chim. Slov.* **1999**, 46, 141–54.

[2426] N. Bartlett, D. H. Lohmann, *Proc. Chem. Soc.* **1962**, 115–6.

[2427] N. Bartlett, D. H. Lohmann, *J. Chem. Soc.* **1962**, 5253–61.

[2428] S. S. Rudel, F. Kraus, *Z. Anorg. Allg. Chem.* **2015**, 641, 2404–7.

[2429] W. E. Falconer, F. J. DiSalvo, A. J. Edwards, J. E. Griffiths, W. A. Sunder, M. J. Vasile, *J. Inorg. Nucl. Chem.* **1976**, 28, 59–60.

[2430] V. B. Sokolov, Y. V. Drobyshevskii, V. N. Prusakov, A. V. Ryzhkov, S. S. Khoroshev, *Dokl. Akad. Nauk* **1976**, 229, 641–4.

[2431] O. Graudejus, S. H. Elder, G. M. Lucier, C. Shen, N. Bartlett, *Inorg. Chem.* **1999**, 38, 2503–9.

[2432] A. V. Wilson, T. Nguyen, F. Brosi, et al., *Inorg. Chem.* **2016**, 55, 1108–23.

[2433] S. Riedel, M. Kaupp, *Angew. Chem., Int. Ed. Engl.* **2006**, 45, 3708–11.

[2434] E. G. Rakov, A. V. Dzhalavyan, *Zh. Neorg. Khim.* **1987**, 32, 853–8.

[2435] H. H. Claassen, H. Selig, J. G. Malm, C. L. Chernick, B. Weinstock, *J. Am. Chem. Soc.* **1961**, 83, 2390–1.

[2436] C. L. Chernick, H. H. Claassen, B. Weinstock, *J. Am. Chem. Soc.* **1961**, 83, 3165–6.

[2437] A. K. Brisdon, P. J. Jones, W. Levason, et al., *J. Chem. Soc., Dalton Trans.* **1990**, 715–8.

[2438] B. Weinstock, J. G. Malm, E. E. Weaver, *J. Am. Chem. Soc.* **1961**, 83, 4310–7.

[2439] B. Weinstock, H. H. Claassen, C. L. Chernick, *J. Chem. Phys.* **1963**, 38, 1470–5.

[2440] N. Bartlett, *Angew. Chem.* **1968**, 80, 453–60.

[2441] D. K. Padma, R. D. Peacock, *J. Fluorine Chem.* **1981**, 17, 539–41.

[2442] H. Shorafa, D. Mollenhauer, B. Paulus, K. Seppelt, *Angew. Chem., Int. Ed. Engl.* **2009**, 48, 5845–7.

[2443] R. Craciun, R. T. Long, D. A. Dixon, K. O. Christe, *J. Phys. Chem. A* **2010**, 114, 7571–82.

[2444] H. Selig, W. A. Sunder, F. A. Disalvo, W. E. Falconer, *J. Fluorine Chem.* **1978**, 11, 39–50.

[2445] P. Huppmann, H. Labischinski, D. Lentz, H. Pritzkow, K. Seppelt, *Z. Anorg. Allg. Chem.* **1982**, 487, 7–25.

[2446] B. von Ahsen, M. Berkei, G. Henkel, H. Willner, F. Aubke, *J. Am. Chem. Soc.* **2002**, 124, 8371–9.

[2447] E. Bernhardt, C. Bach, B. Bley, et al., *Inorg. Chem.* **2005**, 44, 4189–205.

[2448] T. Sakurai, A. Takahashi, *J. Inorg. Nucl. Chem.* **1977**, 39, 427–9.

[2449] J. H. Holloway, R. D. Peacock, *J. Chem. Soc.* **1963**, 527–30.

[2450] L. Meublat, M. Lance, R. Bougon, *Can. J. Chem.* **1989**, 67, 1729–31.

[2451] H. Shorafa, K. Seppelt, *Z. Anorg. Allg. Chem.* **2007**, 633, 543–7.

[2452] E. G. Rakov, A. V. Dzhalavyan, *Zh. Neorg. Khim.* **1987**, 32, 853–8.

[2453] O. Ruff, J. Fischer, *Z. Anorg. Allg. Chem.* **1929**, 179, 161–85.

[2454] O. Glemser, H. W. Roesky, K.-H. Hellberg, H.-U. Werther, *Chem. Ber.* **1966**, 99, 2652–62.

[2455] H. Shorafa, K. Seppelt, *Inorg. Chem.* **2006**, 45, 7929–34.

[2456] R. Bougon, W. V. Cicha, J. Isabey, *J. Fluorine Chem.* **1994**, 67, 271–6.

[2457] S. Riedel, M. Kaupp, *Inorg. Chem.* **2006**, 45, 10497–502.

[2458] R. Burbank, *J. Appl. Crystallogr.* **1974**, 7, 41–4.

[2459] W. E. Falconer, F. J. Disalvo, J. E. Griffiths, F. A. Stevie, W. A. Sunder, M. J. Vasile, *J. Fluorine Chem.* **1975**, 6, 499–520.

[2460] N. Bartlett, N. K. Jha, *J. Chem. Soc. A* **1968**, 536–43.

[2461] N. Bartlett, J. Trotter, *J. Chem. Soc. A* **1968**, 543–7.

[2462] N. Bartlett, N. K. Jha, J. Trotter, *Proc. Chem. Soc.* **1962**, 277.

[2463] S. Riedel, *J. Fluorine Chem.* **2007**, 128, 938–42.

[2464] O. Ruff, F. W. Tschirch, *Chem. Ber.* **1913**, 46, 929–49.

[2465] B. Weinstock, J. G. Malm, *J. Am. Chem. Soc.* **1958**, 80, 4466–8.

[2466] M. A. Hepworth, P. L. Robinson, *J. Inorg. Nucl. Chem.* **1957**, 4, 24–9.

[2467] R. Bougon, B. Buu, K. Seppelt, *Chem. Ber.* **1993**, 126, 1331–6.

[2468] P. J. Jones, W. Levason, M. Tajik, *J. Fluorine Chem.* **1984**, 25, 195–201.

[2469] W. P. Griffith, *J. Chem. Soc. A* **1969**, 211–8.

[2470] M. Gerken, D. A. Dixon, G. J. Schrobilgen, *Inorg. Chem.* **2002**, 41, 259–77.

[2471] K. O. Christe, R. Bougon, *J. Chem. Soc., Chem. Commun.* **1992**, 1056a–a.

[2472] R. Bougon, *J. Fluorine Chem.* **1991**, 53, 419–27.

[2473] K. O. Christe, D. A. Dixon, H. G. Mack, et al., *J. Am. Chem. Soc.* **1993**, 115, 11279–84.

[2474] W. J. Casteel, D. A. Dixon, H. P. A. Mercier, G. J. Schrobilgen, *Inorg. Chem.* **1996**, 35, 4310–22.

[2475] J. Lin, Z. Zhao, C. Liu, et al., *J. Am. Chem. Soc.* **2019**, 141, 5409–14.

[2476] T. C. Ehlert, J. S. Wang, *J. Phys. Chem.* **1977**, 81, 2069–73.

[2477] M. Barber, J. W. Linnett, N. H. Taylor, *J. Chem. Soc.* **1961**, 3323–32.

[2478] P. Woidy, A. J. Karttunen, M. Widenmeyer, R. Niewa, F. Kraus, *Chem. Eur. J.* **2015**, 21, 3290–303.

[2479] J. J. Berzelius, *Ann. Phys.* **1824**, 77, 169–230.

[2480] F. Ebert, H. Woitinek, *Z. Anorg. Allg. Chem.* **1933**, 210, 269–72.

[2481] P. Schwerdtfeger, P. D. W. Boyd, G. A. Bowmaker, L. P. Aldridge, *Struct. Chem.* **1990**, 1, 405–15.

[2482] A. Walsh, C. R. A. Catlow, R. Galvelis, et al., *Chem. Sci.* **2012**, 3, 2565–9.

[2483] Z. Mazej, P. Benkič, *J. Fluorine Chem.* **2005**, 126, 803–8.

[2484] F. H. Jardine, L. Rule, A. G. Vohra, *J. Chem. Soc. A* **1970**, 238–40.

[2485] D. J. Gulliver, W. Levason, M. Webster, *Inorg. Chim. Acta* **1981**, 52, 153–9.

[2486] K. Köhler, H. Pritzkow, H. Lang, *J. Organomet. Chem.* **1998**, 553, 31–8.

[2487] D. P. Zaleski, S. L. Stephens, D. P. Tew, D. M. Bittner, N. R. Walker, A. C. Legon, *PCCP, Phys. Chem. Chem. Phys.* **2015**, 17, 19230–7.

[2488] X. Wang, L. Andrews, F. Brosi, S. Riedel, *Chem. Eur. J.* **2013**, 19, 1397–409.

[2489] J. M. Crabtree, C. S. Lees, K. Little, *J. Inorg. Nucl. Chem.* **1955**, 1, 213–7.

[2490] P. M. O'Donnell, A. E. Spakowski, *J. Electrochem. Soc.* **1964**, 111, 633–6.

[2491] R. L. Ritter, H. A. Smith, *J. Phys. Chem.* **1966**, 70, 805–14.

[2492] R. L. Ritter, H. A. Smith, *J. Phys. Chem.* **1967**, 71, 2036–43.

[2493] K. R. Muddukrishna, R. N. Singh, D. K. Padma, *J. Fluorine Chem.* **1992**, 57, 155–8.

[2494] D. W. Clack, W. T. Williams, *J. Inorg. Nucl. Chem.* **1973**, 35, 3535–9.

[2495] C. Billy, H. M. Haendler, *J. Am. Chem. Soc.* **1957**, 79, 1049–51.

[2496] R. A. Kent, J. D. McDonald, J. L. Margrave, *J. Phys. Chem.* **1966**, 70, 874–7.

[2497] S. Geller, W. L. Bond, *J. Chem. Phys.* **1958**, 29, 925–30.

[2498] S. C. Abrahams, E. Prince, *J. Chem. Phys.* **1962**, 36, 50–5.

[2499] S. Gifford, W. Cherry, J. Jecmen, M. Readnour, *Inorg. Chem.* **1974**, 13, 1434–7.

[2500] B. Frlec, D. Gantar, J. H. Holloway, *J. Fluorine Chem.* **1982**, 20, 385–96.

[2501] Z. Mazej, I. Arčon, P. Benkič, A. Kodre, A. Tressaud, *Chem. Eur. J.* **2004**, 10, 5052–8.

[2502] J. Portier, A. tressaud, J.-L. Dupin, *C. R. Acad. Sci., Ser. IIc: Chim.* **1970**, 270C, 216.

[2503] W. Rüdorff, D. Babel, *Sci. Nat.* **1962**, 49, 230.

[2504] J. Tong, C. Lee, M. H. Whangbo, R. K. Kremer, A. Simon, J. Köhler, *Solid State Sci.* **2010**, 12, 680–4.

[2505] J. C. T. Lee, S. Yuan, S. Lal, et al., *Nat. Phys.* **2011**, 8, 63.

[2506] D. Babel, *Z. Anorg. Allg. Chem.* **1965**, 336, 200–6.

[2507] D. Kissel, R. Hoppe, *Z. Anorg. Allg. Chem.* **1986**, 540, 135–41.

[2508] E. Herdtweck, D. Babel, *Z. Anorg. Allg. Chem.* **1981**, 474, 113–22.

[2509] R. Haegele, D. Babel, *Z. Anorg. Allg. Chem.* **1974**, 409, 11–22.

[2510] D. Dumora, C. Fouassier, R. von der Mühll, J. Ravez, P. Hagenmuller, *C. R. Acad. Sci., Ser. IIc: Chim.* **1971**, 273C, 247.

[2511] A. Chrétien, M. Samouël, *Monatsh. Chem.* **1972**, 103, 17–23.

[2512] H. G. von Schnering, *Z. Anorg. Allg. Chem.* **1973**, 400, 201–7.

[2513] C. Friebel, *Z. Naturforsch. B* **1974**, 29, 634.

[2514] D. Oelkrug, *Z. Phys. Chem.* **1967**, 56, 325.

[2515] C. De Nadaï, A. Demourgues, L. Lozano, P. Gravereau, J. Grannec, *J. Mater. Chem.* **1998**, 8, 2487–91.

[2516] V. Kaiser, D. Babel, *Z. Anorg. Allg. Chem.* **1991**, 595, 139–49.

[2517] V. Kaiser, P. Dahlke, D. Babel, *Z. Anorg. Allg. Chem.* **2002**, 628, 993–1000.

[2518] M. Samouël, A. de Kozak, J. Renaudin, G. Ferey, *Z. Anorg. Allg. Chem.* **1989**, 569, 169–76.

[2519] P. Gredin, A. De Kozak, M. Quarton, J. Renaudin, G. Ferey, *Z. Anorg. Allg. Chem.* **1993**, 619, 1088–94.
[2520] P. Gredin, G. Corbel, J. P. Wright, N. Dupont, A. de Kozak, *Z. Anorg. Allg. Chem.* **2003**, 629, 1960–4.
[2521] A. Le Lirzin, J. Darriet, A. Tressaud, D. Babel, *Z. Anorg. Allg. Chem.* **2008**, 634, 2737–9.
[2522] T. Ono, K. Morita, M. Yano, et al., *J. Phys. Conf. Ser.* **2009**, 145, 012005.
[2523] S. Kummer, W. Massa, D. Babel, *Z. Naturforsch. B* **1988**, 43, 694.
[2524] N. Dupont, P. Gredin, A. Caramanian, A. de Kozak, *J. Solid State Chem.* **1999**, 147, 657–63.
[2525] N. Dupont, P. Gredin, M. Samouël, A. d. Kozak, *Z. Anorg. Allg. Chem.* **2002**, 628, 191–7.
[2526] Z. Mazej, E. Goreshnik, *Eur. J. Inorg. Chem.* **2015**, 2015, 1453–6.
[2527] M. Ishizuka, M. Terai, M. Hidaka, S. Endo, I. Yamada, O. Shimomura, *Phys. Rev. B, Condens. Matter* **1998**, 57, 64–7.
[2528] C. R. Bhattacharjee, P. K. Choudhury, *Transit. Metal Chem.* **1998**, 23, 561–4.
[2529] K. M. Janmanchi, J. W. R. Dolbier, *Org. Process Res. Dev.* **2008**, 12, 349–54.
[2530] W. Biltz, E. Rahlfs, *Z. Anorg. Allg. Chem.* **1927**, 166, 351–76.
[2531] P. Woidy, W. Meng, F. Kraus, *Z. Naturforsch. B* **2014**, 69, 1.
[2532] Y.-H. Cui, M.-Z. Xue, Y.-N. Zhou, S.-M. Peng, X.-L. Wang, Z.-W. Fu, *Electrochim. Acta* **2011**, 56, 2328–35.
[2533] X. Hua, R. Robert, L.-S. Du, et al., *J. Phys. Chem. C* **2014**, 118, 15169–84.
[2534] T. Krahl, F. Marroquin Winkelmann, A. Martin, N. Pinna, E. Kemnitz, *Chem. Eur. J.* **2018**, 24, 7177–87.
[2535] N. Bartlett, G. Lucier, C. Shen, et al., *J. Fluorine Chem.* **1995**, 71, 163–4.
[2536] T. Fleischer, R. Hoppe, *Z. Anorg. Allg. Chem.* **1982**, 492, 76–82.
[2537] G. C. Allen, K. D. Warren, *Inorg. Chem.* **1969**, 8, 1895–901.
[2538] J. Grannec, P. Sorbe, J. Portier, P. Hagenmuller, *C. R. Acad. Sci., Ser. IIc: Chim.* **1975**, 280C, 45.
[2539] D. Kissel, R. Hoppe, *Z. Anorg. Allg. Chem.* **1986**, 532, 17–22.
[2540] A. Tressaud, J. Darriet, P. Lagassié, J. Grannec, P. Hagenmuller, *Mater. Res. Bull.* **1984**, 19, 983–8.
[2541] A. Hartung, D. Babel, *J. Fluorine Chem.* **1982**, 19, 369–78.
[2542] R. R. Jesse, R. Hoppe, *Z. Anorg. Allg. Chem.* **1977**, 428, 83–90.
[2543] R. Hoppe, G. Wingefeld, *Z. Anorg. Allg. Chem.* **1984**, 519, 189–94.
[2544] R. Hoppe, G. Wingefeld, *Z. Anorg. Allg. Chem.* **1984**, 519, 195–203.
[2545] W. Harnischmacher, R. Hoppe, *Angew. Chem.* **1973**, 85, 590.
[2546] K. O. Christe, W. W. Wilson, R. D. Wilson, *Inorg. Chem.* **1980**, 19, 3254–6.
[2547] D. Kissel, R. Hoppe, *Z. Anorg. Allg. Chem.* **1988**, 559, 40–8.
[2548] P. Sorbe, J. Grannec, J. Portier, P. Hagenmuller, *C. R. Acad. Sci., Ser. IIc: Chim.* **1976**, 282C, 663.
[2549] R. Hoppe, *Angew. Chem., Int. Ed. Engl.* **1981**, 20, 63–87.
[2550] P. M. O'Donnell, *J. Electrochem. Soc.* **1970**, 117, 1273–5.
[2551] A. Hettich, *Z. Anorg. Allg. Chem.* **1927**, 167, 67–74.
[2552] L. Poyer, M. Fielder, H. Harrison, et al., *Inorg. Synth.* **1957**, 5, 18–21.
[2553] W. Tong, G. G. Amatucci, *J. Power Sources* **2011**, 196, 1449–54.
[2554] A. Williams, *J. Phys.: Condens. Matter.* **1989**, 1, 2569–74.
[2555] G. Argay, I. Naray-Szabo, *Acta Chim. Acad. Sci. Hung.* **1966**, 49, 329.
[2556] H. Kawamura, I. Shirotani, H. Inokuchi, *Chem. Phys. Lett.* **1974**, 24, 549–50.
[2557] L. Pauling. The Nature of the Chemical Bond – An Introduction to Modern Structural Chemistry. 3 ed. Ithaca, United States: Cornell University Press; 1960.
[2558] K. Andres, N. A. Kuebler, M. B. Robin, *J. Phys. Chem. Solids* **1966**, 27, 1747–8.

[2559]   T. Flóra, I. Gaál, *Thermochim. Acta* **1973**, 7, 173–81.

[2560]   M. T. Beck, *Thermochim. Acta* **1974**, 9, 459–60.

[2561]   K. Seppelt. Preparation of Highly Active Silver Fluoride. In: H. W. Roesky (ed.) Efficient Preparations of Fluorine Compounds: Wiley; 2013:9–10.

[2562]   A. G. Sharpe, *J. Chem. Soc.* **1952**, 4538–9.

[2563]   W. T. Miller, R. J. Burnard, *J. Am. Chem. Soc.* **1968**, 90, 7367–8.

[2564]   C. Shen, B. Žemva, G. M. Lucier, O. Graudejus, J. A. Allman, N. Bartlett, *Inorg. Chem.* **1999**, 38, 4570–7.

[2565]   F. Kraus, S. A. Baer, M. B. Fichtl, *Eur. J. Inorg. Chem.* **2009**, 2009, 441–7.

[2566]   W. Grochala, Z. Mazej, *Philos. Trans. R. Soc. A* **2015**, 373.

[2567]   H. Jockusch, *Sci. Nat.* **1934**, 22, 561.

[2568]   P. Połczyński, R. Jurczakowski, A. Grzelak, E. Goreshnik, Z. Mazej, W. Grochala, *Chem. Eur. J.* **2019**, 25, 4927–30.

[2569]   A. Jesih, K. Lutar, B. Žemva, et al., *Z. Anorg. Allg. Chem.* **1990**, 588, 77–83.

[2570]   P. Fischer, D. Schwarzenbach, H. M. Rietveld, *J. Phys. Chem. Solids* **1971**, 32, 543–50.

[2571]   A. Grzelak, J. Gawraczyński, T. Jaroń, et al., *Inorg. Chem.* **2017**, 56, 14651–61.

[2572]   J. Romiszewski, W. Grochala, L. Z. Stolarczyk, *J. Phys.: Condens. Matter.* **2007**, 19, 116206.

[2573]   W. Grochala, R. G. Egdell, P. P. Edwards, Z. Mazej, B. Žemva, *ChemPhysChem* **2003**, 4, 997–1001.

[2574]   P. Fischer, G. Roult, D. Schwarzenbach, *J. Phys. Chem. Solids* **1971**, 32, 1641–7.

[2575]   H. V. Wartenberg, *Z. Anorg. Allg. Chem.* **1939**, 242, 406–12.

[2576]   B. Zemva, R. Hagiwara, W. J. Casteel, K. Lutar, A. Jesih, N. Bartlett, *J. Am. Chem. Soc.* **1990**, 112, 4846–9.

[2577]   G. Lucier, C. Shen, W. J. Casteel, L. Chacón, N. Barlett, *J. Fluorine Chem.* **1995**, 72, 157–63.

[2578]   P. Malinowski, Z. Mazej, W. Grochala, *Z. Anorg. Allg. Chem.* **2008**, 634, 2608–16.

[2579]   W. Grochala, *J. Fluorine Chem.* **2008**, 129, 82–90.

[2580]   J. Tong, F. Kraus, J. Köhler, A. Simon, J. Liu, M.-H. Whangbo, *Z. Anorg. Allg. Chem.* **2011**, 637, 1118–21.

[2581]   B. G. Müller, *Z. Anorg. Allg. Chem.* **1987**, 553, 196–204.

[2582]   R.-H. Odenthal, R. Hoppe, *Monatsh. Chem.* **1971**, 102, 1340–50.

[2583]   J. Tong, J. Köhler, A. Simon, C. Lee, M.-H. Whangbo, *Z. Anorg. Allg. Chem.* **2012**, 638, 1792–5.

[2584]   D. Kurzydłowski, Z. Mazej, W. Grochala, *Dalton Trans.* **2013**, 42, 2167–73.

[2585]   R.-H. Odenthal, R. Hoppe, *Z. Anorg. Allg. Chem.* **1971**, 385, 92–101.

[2586]   R.-H. Odenthal, R. Hoppe, *Sci. Nat.* **1970**, 57, 305–6.

[2587]   G. C. Allen, R. F. McMeeking, R. Hoppe, B. Müller, *J. Chem. Soc., Chem. Commun.* **1972**, 291–2.

[2588]   C. Friebel, D. Reinen, *Z. Anorg. Allg. Chem.* **1975**, 413, 51–60.

[2589]   G. C. Allen, R. F. McMeeking, *J. Chem. Soc., Dalton Trans.* **1976**, 1063–8.

[2590]   D. Kurzydłowski, T. Jaroń, A. Ozarowski, et al., *Inorg. Chem.* **2016**, 55, 11479–89.

[2591]   R.-H. Odenthal, D. Paus, R. Hoppe, *Z. Anorg. Allg. Chem.* **1974**, 407, 151–6.

[2592]   B. Zemva, K. Lutar, A. Jesih, et al., *J. Am. Chem. Soc.* **1991**, 113, 4192–8.

[2593]   R. Hoppe, B. Müller, *Sci. Nat.* **1969**, 56, 35.

[2594]   D. Gantar, I. Leban, B. Frlec, J. H. Holloway, *J. Chem. Soc., Dalton Trans.* **1987**, 2379–83.

[2595]   W. J. Casteel, G. Lucier, R. Hagiwara, H. Borrmann, N. Bartlett, *J. Solid State Chem.* **1992**, 96, 84–96.

[2596]   D. Gantar, B. Frlec, D. R. Russell, J. H. Holloway, *Acta Crystallogr. Sect. C: Cryst. Struct. Commun.* **1987**, 43, 618–20.

[2597]   O. Graudejus, A. P. Wilkinson, N. Bartlett, *Inorg. Chem.* **2000**, 39, 1545–8.

[2598]   R. Fischer, B. G. Müller, *Z. Anorg. Allg. Chem.* **2002**, 628, 2592–6.

[2599] W. Grochala, R. Hoffmann, *Angew. Chem., Int. Ed. Engl.* **2001**, 40, 2742–81.

[2600] Z. Mazej, T. Michałowski, E. A. Goreshnik, et al., *Dalton Trans.* **2015**, 44, 10957–68.

[2601] G. M. Lucier, J. M. Whalen, N. Bartlett, *J. Fluorine Chem.* **1998**, 89, 101–4.

[2602] A. L. Hector, W. Levason, M. T. Weller, E. G. Hope, *J. Fluorine Chem.* **1997**, 86, 105–8.

[2603] R. Hoppe, R. Homann, *Z. Anorg. Allg. Chem.* **1970**, 379, 193–8.

[2604] A. J. Edwards, R. G. Plevey, M. P. Steward, *J. Fluorine Chem.* **1971**, 1, 246–8.

[2605] R. Hoppe, R. Homann, *Sci. Nat.* **1966**, 53, 501.

[2606] P. Sorbe, J. Grannec, J. Portier, P. Hagenmuller, *J. Fluorine Chem.* **1978**, 11, 243–50.

[2607] K. L. Saenger, C. P. Sun, *Phys. Rev. A* **1992**, 46, 670–3.

[2608] D. Schröder, J. Hrušák, I. C. Tornieporth-Oetting, T. M. Klapötke, H. Schwarz, *Angew. Chem.* **1994**, 106, 223–5.

[2609] C. J. Evans, M. C. L. Gerry, *J. Am. Chem. Soc.* **2000**, 122, 1560–1.

[2610] D. Kurzydłowski, W. Grochala, *Z. Anorg. Allg. Chem.* **2008**, 634, 1082–6.

[2611] S. A. Cooke, M. C. L. Gerry, *J. Am. Chem. Soc.* **2004**, 126, 17000–8.

[2612] C. J. Evans, D. S. Rubinoff, M. C. L. Gerry, *PCCP, Phys. Chem. Chem. Phys.* **2000**, 2, 3943–8.

[2613] X. Wang, L. Andrews, K. Willmann, F. Brosi, S. Riedel, *Angew. Chem., Int. Ed. Engl.* **2012**, 51, 10628–32.

[2614] P. Schwerdtfeger, J. S. McFeaters, M. J. Liddell, J. Hrušák, H. Schwarz, *J. Chem. Phys.* **1995**, 103, 245–52.

[2615] E. K. Butler, B. J. Knurr, K. J. Manke, T. R. Vervoort, T. D. Varberg, *J. Phys. Chem. A* **2010**, 114, 4831–4.

[2616] D. Kurzydłowski, W. Grochala, *Chem. Commun.* **2008**, 1073–5.

[2617] R. Küster, K. Seppelt, *Z. Anorg. Allg. Chem.* **2000**, 626, 236–40.

[2618] F. G. Herring, G. Hwang, K. C. Lee, et al., *J. Am. Chem. Soc.* **1992**, 114, 1271–7.

[2619] S. H. Elder, G. M. Lucier, F. J. Hollander, N. Bartlett, *J. Am. Chem. Soc.* **1997**, 119, 1020–6.

[2620] L. B. Asprey, F. H. Kruse, K. H. Jack, R. Maitland, *Inorg. Chem.* **1964**, 3, 602–4.

[2621] A. G. Sharpe, *J. Chem. Soc.* **1949**, 2901–2.

[2622] M. F. A. Dove, P. Benkič, C. Platte, T. J. Richardson, N. Bartlett, *J. Fluorine Chem.* **2001**, 110, 83–6.

[2623] T. M. Klapötke. Laboratory-Scale Synthesis of Gold Trifluoride and Uranium Hexafluoride. In: H. W. Roesky (ed.) Efficient Preparations of Fluorine Compounds: Wiley; 2013:94–9.

[2624] I. C. Tornieporth-Oetting, T. M. Klapötke, *Chem. Ber.* **1995**, 128, 957–8.

[2625] F. W. B. Einstein, P. R. Rao, J. Trotter, N. Bartlett, *J. Chem. Soc. A* **1967**, 478–82.

[2626] B. Réffy, M. Kolonits, A. Schulz, T. M. Klapötke, M. Hargittai, *J. Am. Chem. Soc.* **2000**, 122, 3127–34.

[2627] U. Engelmann, B. G. Müller, *Z. Anorg. Allg. Chem.* **1991**, 598, 103–10.

[2628] A. J. Edwards, G. R. Jones, *J. Chem. Soc. A* **1969**, 1936–8.

[2629] M. O. Faltens, D. A. Shirley, *J. Chem. Phys.* **1970**, 53, 4249–64.

[2630] R. Schmidt, B. G. Müller, *Z. Anorg. Allg. Chem.* **2004**, 630, 2393–7.

[2631] J. Linnera, S. I. Ivlev, F. Kraus, A. J. Karttunen, *Z. Anorg. Allg. Chem.* **2019**, 645, 284–91.

[2632] B. G. Müller, *Z. Anorg. Allg. Chem.* **1987**, 555, 57–63.

[2633] H. Bialowons, B. G. Müller, *Z. Anorg. Allg. Chem.* **1997**, 623, 434–8.

[2634] R. Fischer, B. G. Müller, *Z. Anorg. Allg. Chem.* **1997**, 623, 1729–33.

[2635] H. Bialowons, B. G. Müller, *Z. Anorg. Allg. Chem.* **1997**, 623, 1719–22.

[2636] R. Schmidt, B. G. Müller, *Z. Anorg. Allg. Chem.* **1999**, 625, 605–8.

[2637] B. G. Müller, *Angew. Chem.* **1987**, 99, 685.

[2638] U. Engelmann, B. G. Müller, *Z. Anorg. Allg. Chem.* **1990**, 589, 51–61.

[2639] U. Engelmann, B. G. Müller, *Z. Anorg. Allg. Chem.* **1992**, 618, 43–52.

[2640] U. Engelmann, B. G. Müller, *Z. Anorg. Allg. Chem.* **1993**, 619, 1661–8.

[2641]   H. Fitz, B. G. Müller, *Z. Anorg. Allg. Chem.* **2002**, 628, 126–32.

[2642]   O. Graudejus, B. G. Müller, *Z. Anorg. Allg. Chem.* **1996**, 622, 187–90.

[2643]   R. Schmidt, B. G. Müller, *Z. Anorg. Allg. Chem.* **1999**, 625, 602–4.

[2644]   S. Seidel, K. Seppelt, *Science* **2000**, 290, 117–8.

[2645]   I.-C. Hwang, K. Seppelt, *Z. Anorg. Allg. Chem.* **2002**, 628, 765–9.

[2646]   M. A. Ellwanger, S. Steinhauer, P. Golz, et al., *Chem. Eur. J.* **2017**, 23, 13501–9.

[2647]   M. A. Ellwanger, S. Steinhauer, P. Golz, T. Braun, S. Riedel, *Angew. Chem., Int. Ed. Engl.* **2018**, 57, 7210–4.

[2648]   M. A. Ellwanger, C. von Randow, S. Steinhauer, et al., *Chem. Commun.* **2018**, 54, 9301–4.

[2649]   P. Huppmann, H. Hartl, K. Seppelt, *Z. Anorg. Allg. Chem.* **1985**, 524, 26–32.

[2650]   A. Pérez-Bitrián, M. Baya, J. M. Casas, A. Martín, B. Menjón, J. Orduna, *Angew. Chem., Int. Ed. Engl.* **2018**, 57, 6517–21.

[2651]   R. Kumar, A. Linden, C. Nevado, *J. Am. Chem. Soc.* **2016**, 138, 13790–3.

[2652]   K. Leary, N. Bartlett, *J. Chem. Soc., Chem. Commun.* **1972**, 903–4.

[2653]   J. Lin, S. Zhang, W. Guan, G. Yang, Y. Ma, *J. Am. Chem. Soc.* **2018**, 140, 9545–50.

[2654]   M. J. Vasile, T. J. Richardson, F. A. Stevie, W. E. Falconer, *J. Chem. Soc., Dalton Trans.* **1976**, 351–3.

[2655]   I.-C. Hwang, K. Seppelt, *Angew. Chem., Int. Ed. Engl.* **2001**, 40, 3690–3.

[2656]   J. Brunvoll, A. A. Ischenko, A. A. Ivanov, et al., *Acta Chem. Scand.* **1982**, 36a, 705–9.

[2657]   Z. Mazej, *J. Fluorine Chem.* **2004**, 125, 1723–33.

[2658]   J. E. Griffiths, W. A. Sunder, *Spectrochim. Acta, Part A* **1979**, 35, 1329–31.

[2659]   Z. Mazej, E. Goreshnik, *Solid State Sci.* **2006**, 8, 671–7.

[2660]   Z. Mazej, E. Goreshnik, G. Tavčar, *J. Fluorine Chem.* **2011**, 132, 686–9.

[2661]   A. J. Edwards, W. E. Falconer, J. E. Griffiths, W. A. Sunder, M. J. Vasile, *J. Chem. Soc., Dalton Trans.* **1974**, 1129–33.

[2662]   J. F. Lehmann, G. J. Schrobilgen, *J. Fluorine Chem.* **2003**, 119, 109–24.

[2663]   N. Bartlett, K. Leary, *Rev. Chim. Miner.* **1976**, 13, 82–97.

[2664]   A. A. Timakov, V. N. Prusakov, Y. V. Drobyshevskii, *Dokl. Akad. Nauk SSSR* **1986**, 291, 125–8.

[2665]   V. V. Ostropikov, E. G. Rakov, *Izv. Vyssh. Uchebn. Zaved., Khim. Khim. Tekhnol.* **1989**, 32, 3–17.

[2666]   S. Riedel, M. Kaupp, *Inorg. Chem.* **2006**, 45, 1228–34.

[2667]   D. Himmel, S. Riedel, *Inorg. Chem.* **2007**, 46, 5338–42.

[2668]   W. Baur, *Acta Crystallogr.* **1958**, 11, 488–90.

[2669]   T. C. Waddington. Lattice Energies and their Significance in Inorganic Chemistry. In: H. J. Emeléus, A. G. Sharpe (eds.) Advances in Inorganic Chemistry and Radiochemistry: Academic Press; 1959:157–221.

[2670]   R. Wagner, *Chem. Ber.* **1886**, 19, 896–8.

[2671]   D. J. Machin, R. L. Martin, R. S. Nyholm, *J. Chem. Soc.* **1963**, 1490–500.

[2672]   D. J. Machin, R. S. Nyholm, *J. Chem. Soc.* **1963**, 1500–5.

[2673]   A. P. Lane, D. W. A. Sharp, J. M. Barraclough, D. H. Brown, D. A. Paterson, *J. Chem. Soc. A* **1971**, 94–100.

[2674]   H. G. Schnering, P. Bleckmann, *Sci. Nat.* **1965**, 52, 538.

[2675]   H. G. v. Schnering, P. Bleckmann, *Sci. Nat.* **1968**, 55, 342–3.

[2676]   R. Almairac, H. N. Bordallo, A. Bulou, J. Nouet, *Phys. Rev. B, Condens. Matter* **1995**, 52, 9370–6.

[2677]   H. G. von Schnering, *Z. Anorg. Allg. Chem.* **1967**, 353, 13–25.

[2678]   C. Martineau, F. Fayon, C. Legein, et al., *Dalton Trans.* **2008**, 6150–8.

[2679]   H. Böhland, *Z. Chem.* **1963**, 3, 395–6.

[2680]   G. Meyer, N. Böhmer, *Z. Anorg. Allg. Chem.* **2000**, 626, 1332–4.

[2681]   B. Dojer, A. Golobič, Z. Jagličić, M. Kristl, M. Drofenik, *Monatsh. Chem.* **2012**, 143, 175–80.

[2682]   S. Yakovlev, M. Avdeev, M. Mezouar, *J. Solid State Chem.* **2009**, 182, 1545–9.

[2683]   O. J. Kleppa, M. Wakihara, *J. Inorg. Nucl. Chem.* **1976**, 38, 715–9.

[2684]   T. Fleischer, R. Hoppe, *Z. Anorg. Allg. Chem.* **1982**, 492, 83–94.

[2685]   A. Grzechnik, R. Kaindl, K. Friese, *J. Phys. Chem. Solids* **2007**, 68, 382–8.

[2686]   A. Grzechnik, J. M. Posse, W. Morgenroth, K. Friese, *J. Solid State Chem.* **2007**, 180, 1998–2003.

[2687]   H. M. Haendler, W. J. Bernard, *J. Am. Chem. Soc.* **1951**, 73, 5218–9.

[2688]   P. Nuka, *Z. Anorg. Allg. Chem.* **1929**, 180, 235–40.

[2689]   P. Rawat, R. Nagarajan, *J. Fluorine Chem.* **2016**, 182, 98–103.

[2690]   S. A. Polyshchuk, S. P. Kozerenko, Y. V. Gagarinsky, *J. Less-Common Met.* **1974**, 34, 261–6.

[2691]   F. Kraus, *Monatsh. Chem.* **2012**, 143, 1097–100.

[2692]   P. Glavič, J. Slivnik, A. Bole, *J. Inorg. Nucl. Chem.* **1975**, 37, 345–8.

[2693]   H. von Helmolt, *Z. Anorg. Allg. Chem.* **1893**, 3, 115–52.

[2694]   M. Rousseau, J. Y. Gesland, J. Julliard, J. Nouet, J. Zarembowitch, A. Zarembowitch, *Phys. Rev. B, Condens. Matter* **1975**, 12, 1579–90.

[2695]   A. Le Bail, J. L. Fourquet, J. Rubín, E. Palacios, J. Bartolomé, *Phys. B* **1990**, 162, 231–6.

[2696]   G. Jayaram, V. G. Krishnan, *Phys. Rev. B, Condens. Matter* **1995**, 51, 1294–6.

[2697]   R. Beaudoin, H. Ménard, *Can. J. Chem.* **1988**, 66, 236–41.

[2698]   M. Hidaka, S. Hosogi, M. Ono, K. Horai, *Solid State Commun.* **1977**, 23, 503–6.

[2699]   M. de Murcia, P. Bräunlich, M. Egee, G. Mary, *Solid State Commun.* **1980**, 34, 737–41.

[2700]   P. F. Weller, *Inorg. Chem.* **1965**, 4, 1545–51.

[2701]   R. E. Noftle, J. W. Green, S. K. Yarbro, *J. Fluorine Chem.* **1976**, 7, 221–7.

[2702]   E. Dorm, *J. Chem. Soc., Chem. Commun.* **1971**, 466–7.

[2703]   J. N. Blocher, E. H. Hall, *J. Phys. Chem.* **1959**, 63, 127–8.

[2704]   B. Brunetti, V. Piacente, A. Latini, P. Scardala, *J. Chem. Eng. Data* **2008**, 53, 2493–5.

[2705]   E. Schrötter, B. G. Müller, *Z. Anorg. Allg. Chem.* **1992**, 618, 54–9.

[2706]   M. Kaupp, H. G. von Schnering, *Inorg. Chem.* **1994**, 33, 4179–85.

[2707]   M.-s. Liao, Q.-e. Zhang, W. H. E. Schwarz, *Inorg. Chem.* **1995**, 34, 5597–605.

[2708]   M.-s. Liao, Q.-e. Zhang, *Comput. Theor. Chem.* **1995**, 358, 195–203.

[2709]   P. Schwerdtfeger, P. D. W. Boyd, S. Brienne, et al., *Inorg. Chim. Acta* **1993**, 213, 233–46.

[2710]   K. Brodersen, J. Hoffmann, *Z. Anorg. Allg. Chem.* **1980**, 469, 32–44.

[2711]   R. P. Rastogi, B. L. Dubey, N. D. Agrawal, *J. Inorg. Nucl. Chem.* **1975**, 37, 1167–72.

[2712]   K. Köhler, D. Breitinger, *Sci. Nat.* **1974**, 61, 684.

[2713]   I. I. Guerus, Y. L. Yagupolskii, *J. Organomet. Chem.* **1983**, 247, 81–7.

[2714]   M. H. Habibi, S. Farhadi, *Tetrahedron Lett.* **1999**, 40, 2821–4.

[2715]   F. Nerdel, *Sci. Nat.* **1952**, 39, 209–10.

[2716]   A. L. Henne, *J. Am. Chem. Soc.* **1938**, 60, 1569–71.

[2717]   E. L. Muetterties, Preparation of mercuric fluoride by reaction of mercuric oxide and hydrogen fluoride gas, Patent US19540418766, **1956**.

[2718]   M. Hostettler, D. Schwarzenbach, *C. R., Chim.* **2005**, 8, 147–56.

[2719]   S. Schyck, E. Evlyukhin, E. Kim, M. Pravica, *Chem. Phys. Lett.* **2019**, 724, 35–41.

[2720]   R. Hoppe, R. Homann, *Z. Anorg. Allg. Chem.* **1969**, 369, 212–6.

[2721]   J. R. DeBackere, H. P. A. Mercier, G. J. Schrobilgen, *Inorg. Chem.* **2015**, 54, 1606–26.

[2722]   Z. Mazej, E. A. Goreshnik, *J. Solid State Chem.* **2015**, 228, 53–9.

[2723]   I. D. Brown, R. J. Gillespie, K. R. Morgan, et al., *Inorg. Chem.* **1987**, 26, 689–93.

[2724]   I.-C. Hwang, S. Seidel, K. Seppelt, *Angew. Chem., Int. Ed. Engl.* **2003**, 42, 4392–5.

[2725]   E. L. Muetterties, D. D. Coffman, *J. Am. Chem. Soc.* **1958**, 80, 5914–8.

[2726]   S. Elsheimer, W. R. Dolbier, M. Murla, K. Seppelt, *J. Org. Chem.* **1984**, 49, 205–7.

[2727] A. Waterfeld, R. Mews, *J. Chem. Soc., Chem. Commun.* **1982**, 839–40.

[2728] A. Waterfeld, W. Isenberg, R. Mews, W. Clegg, G. M. Sheldrick, *Chem. Ber.* **1983**, 116, 724–31.

[2729] J. Barluenga, J. M. Martinez-Gallo, C. Nájera, M. Yus, *J. Chem. Soc., Chem. Commun.* **1985**, 1422–3.

[2730] M. Geisel, R. Mews, *Chem. Ber.* **1987**, 120, 1675–7.

[2731] R. Mews. Cesium, Mercury, and Silver Salts with Sulfur–Nitrogen–Fluorine Anions: Useful Transfer Reagents for NSF Building Blocks. In: H. W. Roesky (ed.) Efficient Preparations of Fluorine Compounds: Wiley; 2013:88–93.

[2732] H. Goldwhite, R. N. Haszeldine, R. N. Mukherjee, *J. Chem. Soc.* **1961**, 3825–7.

[2733] C. G. Krespan, *J. Org. Chem.* **1960**, 25, 105–7.

[2734] W. T. Miller, M. B. Freedman, J. H. Fried, H. F. Koch, *J. Am. Chem. Soc.* **1961**, 83, 4105–6.

[2735] W. T. Miller, M. B. Freedman, *J. Am. Chem. Soc.* **1963**, 85, 180–3.

[2736] B. L. Dyatkin, S. R. Sterlin, B. I. Martynov, E. I. Mysov, I. L. Knunyants, *Tetrahedron* **1971**, 27, 2843–9.

[2737] B. L. Dyatkin, S. R. Sterlin, L. G. Zhuravkova, B. I. Martynov, E. I. Mysov, I. L. Knunyants, *Tetrahedron* **1973**, 29, 2759–67.

[2738] H. J. Emeléus, G. L. Hurst, *J. Chem. Soc.* **1964**, 396–9.

[2739] J. S. Thrasher, J. B. Nielsen, S. G. Bott, D. J. McClure, S. A. Morris, J. L. Atwood, *Inorg. Chem.* **1988**, 27, 570–5.

[2740] E. L. Muetterties, Fluorine and sulfur compounds, Patent US19540451301, **1956**.

[2741] M. H. Habibi, T. E. Mallouk, *J. Fluorine Chem.* **1991**, 51, 291–4.

[2742] B. G. Müller, W. Klein, M. Jansen, *Z. Anorg. Allg. Chem.* **2004**, 630, 1123–5.

[2743] S. Riedel, M. Kaupp, P. Pyykkö, *Inorg. Chem.* **2008**, 47, 3379–83.

[2744] D. Samanta, P. Jena, *J. Am. Chem. Soc.* **2012**, 134, 8400–3.

[2745] J. Botana, X. Wang, C. Hou, et al., *Angew. Chem., Int. Ed. Engl.* **2015**, 54, 9280–3.

[2746] M. Kaupp, M. Dolg, H. Stoll, H. G. von Schnering, *Inorg. Chem.* **1994**, 33, 2122–31.

[2747] M. Kaupp, H. G. von Schnering, *Angew. Chem., Int. Ed. Engl.* **1993**, 32, 861–3.

[2748] X. Wang, L. Andrews, S. Riedel, M. Kaupp, *Angew. Chem., Int. Ed. Engl.* **2007**, 46, 8371–5.

[2749] A. Ghosh, J. Conradie, *Eur. J. Inorg. Chem.* **2016**, 2016, 2989–92.

[2750] R. E. Banks, B. E. Smart, J. C. Tatlow (eds.) Organofluorine Chemistry – Principles and Commercial Applications. 1 ed.: Springer US; 1994.

[2751] A. Maciejewski, *J. Photochem. Photobiol. A* **1990**, 51, 87–131.

[2752] R. Battino, H. L. Clever, *Chem. Rev.* **1966**, 66, 395–463.

[2753] E. Wilhelm, R. Battino, *Chem. Rev.* **1973**, 73, 1–9.

[2754] J. Sangster, *J. Phys. Chem. Ref. Data* **1989**, 18, 1111–229.

[2755] P. Kirsch. Applications of Organofluorine Compounds. Modern Fluoroorganic Chemistry: Wiley-VCH; 2004:203–77.

[2756] G. J. Puts, P. Crouse, B. M. Ameduri, *Chem. Rev.* **2019**, 119, 1763–805.

[2757] K. Müller, C. Faeh, F. Diederich, *Science* **2007**, 317, 1881–6.

[2758] S. Purser, P. R. Moore, S. Swallow, V. Gouverneur, *Chem. Soc. Rev.* **2008**, 37, 320–30.

[2759] W. K. Hagmann, *J. Med. Chem.* **2008**, 51, 4359–69.

[2760] P. Jeschke, *ChemBioChem* **2004**, 5, 570–89.

[2761] J. Wang, M. Sánchez-Roselló, J. L. Aceña, et al., *Chem. Rev.* **2014**, 114, 2432–506.

[2762] Y. Zhou, J. Wang, Z. Gu, et al., *Chem. Rev.* **2016**, 116, 422–518.

[2763] J. Han, A. M. Remete, L. S. Dobson, et al., *J. Fluorine Chem.* **2020**, 239, 109639.

[2764] J.-P. Bégué, D. Bonnet-Delpon, *J. Fluorine Chem.* **2006**, 127, 992–1012.

[2765] K. L. Kirk, *J. Fluorine Chem.* **2006**, 127, 1013–29.

[2766] P. T. Lowe, D. O'Hagan, *J. Fluorine Chem.* **2020**, 230, 109420.

[2767] R. Callejo, M. J. Corr, M. Yang, et al., *Chem. Eur. J.* **2016**, 22, 8137–51.

[2768] M. J. Corr, R. A. Cormanich, C. N. von Hahmann, et al., *Org. Biomol. Chem.* **2016**, 14, 211–9.

[2769] I. Tirotta, V. Dichiarante, C. Pigliacelli, et al., *Chem. Rev.* **2015**, 115, 1106–29.

[2770] J.-D. Yang, Y. Wang, X.-S. Xue, J.-P. Cheng, *J. Org. Chem.* **2017**, 82, 4129–35.

[2771] D. Meyer, H. Jangra, F. Walther, H. Zipse, P. Renaud, *Nat. Commun.* **2018**, 9, 4888.

[2772] J. H. Simons, W. J. Harland, *J. Electrochem. Soc.* **1949**, 95, 55.

[2773] N. V. Ignat'ev. 4 – Electrochemical Fluorination: A Powerful Tool for the Preparation of Organofluorine Compounds. In: H. Groult, F. R. Leroux, A. Tressaud (eds.) Modern Synthesis Processes and Reactivity of Fluorinated Compounds: Elsevier; 2017:71–123.

[2774] G. Alvernhe, A. Laurent, G. Haufe, *J. Fluorine Chem.* **1986**, 34, 147–56.

[2775] R. Franz, *J. Fluorine Chem.* **1980**, 15, 423–34.

[2776] M. B. Giudicelli, D. Picq, B. Veyron, *Tetrahedron Lett.* **1990**, 31, 6527–30.

[2777] O. E. Okoromoba, J. Han, G. B. Hammond, B. Xu, *J. Am. Chem. Soc.* **2014**, 136, 14381–4.

[2778] O. E. Okoromoba, G. B. Hammond, B. Xu, *Org. Lett.* **2015**, 17, 3975–7.

[2779] S. Liang, G. B. Hammond, B. Xu, *Chem. Eur. J.* **2017**, 23, 17850–61.

[2780] K. O. Christe, W. W. Wilson, R. D. Wilson, R. Bau, J. A. Feng, *J. Am. Chem. Soc.* **1990**, 112, 7619–25.

[2781] A. A. Kolomeitsev, F. U. Seifert, G.-V. Röschenthaler, *J. Fluorine Chem.* **1995**, 71, 47–9.

[2782] H. Sun, S. G. DiMagno, *J. Am. Chem. Soc.* **2005**, 127, 2050–1.

[2783] D. W. Kim, H.-J. Jeong, S. T. Lim, M.-H. Sohn, *Angew. Chem., Int. Ed. Engl.* **2008**, 47, 8404–6.

[2784] A. S. Pilcher, H. L. Ammon, P. DeShong, *J. Am. Chem. Soc.* **1995**, 117, 5166–7.

[2785] R. Z. Gnann, R. I. Wagner, K. O. Christe, R. Bau, G. A. Olah, W. W. Wilson, *J. Am. Chem. Soc.* **1997**, 119, 112–5.

[2786] A. Kornath, F. Neumann, H. Oberhammer, *Inorg. Chem.* **2003**, 42, 2894–901.

[2787] W. J. Middleton, *Org. Synth.* **1986**, 64, 221–3.

[2788] W. Heilemann, R. Mews, *Chem. Ber.* **1988**, 121, 461–3.

[2789] V. Bagchi, P. Paraskevopoulou, P. Das, et al., *J. Am. Chem. Soc.* **2014**, 136, 11362–81.

[2790] A. A. Kolomeitsev, G. Bissky, P. Kirsch, G. V. Röschenthaler, *J. Fluorine Chem.* **2000**, 103, 159–61.

[2791] A. A. Kolomeitsev, G. Bissky, J. Barten, N. Kalinovich, E. Lork, G.-V. Röschenthaler, *Inorg. Chem.* **2002**, 41, 6118–24.

[2792] C. Ni, M. Hu, J. Hu, *Chem. Rev.* **2015**, 115, 765–825.

[2793] W. J. Middleton, *J. Org. Chem.* **1975**, 40, 574–8.

[2794] L. N. Markovskij, V. E. Pashinnik, A. V. Kirsanov, *Synthesis* **1973**, 1973, 787–9.

[2795] G. S. Lal, G. P. Pez, R. J. Pesaresi, F. M. Prozonic, H. Cheng, *J. Org. Chem.* **1999**, 64, 7048–54.

[2796] F. Beaulieu, L.-P. Beauregard, G. Courchesne, M. Couturier, F. LaFlamme, A. L'Heureux, *Org. Lett.* **2009**, 11, 5050–3.

[2797] A. L'Heureux, F. Beaulieu, C. Bennett, et al., *J. Org. Chem.* **2010**, 75, 3401–11.

[2798] T. Umemoto, R. P. Singh, Y. Xu, N. Saito, *J. Am. Chem. Soc.* **2010**, 132, 18199–205.

[2799] T. Umemoto, R. P. Singh, *J. Fluorine Chem.* **2012**, 140, 17–27.

[2800] A. J. Cresswell, S. G. Davies, P. M. Roberts, J. E. Thomson, *Chem. Rev.* **2015**, 115, 566–611.

[2801] R. L. Pruett, J. T. Barr, K. E. Rapp, C. T. Bahner, J. D. Gibson, R. H. Lafferty, *J. Am. Chem. Soc.* **1950**, 72, 3646–50.

[2802] K. E. Rapp, J. T. Barr, R. L. Pruett, C. T. Bahner, J. D. Gibson, R. H. Lafferty, *J. Am. Chem. Soc.* **1952**, 74, 749–53.

[2803] T. Akio, I. Hiroshi, I. Nobuo, *Bull. Chem. Soc. Jpn.* **1979**, 52, 3377–80.

[2804] H. Koroniak, J. Walkowiak, K. Grys, A. Rajchel, A. Alty, R. Du Boisson, *J. Fluorine Chem.* **2006**, 127, 1245–51.

[2805] F. Munyemana, A.-M. Frisque-Hesbain, A. Devos, L. Ghosez, *Tetrahedron Lett.* **1989**, 30, 3077–80.

[2806] B. Haveaux, A. Dekoker, M. Rens, A. R. Sidani, J. Toye, L. Ghosez, *Org. Synth.* **1979**, 59, 26.

[2807] A. Devos, J. Remion, A.-M. Frisque-Hesbain, A. Colens, L. Ghosez, *J. Chem. Soc., Chem. Commun.* **1979**, 1180–1.

[2808] T. Furuya, T. Fukuhara, S. Hara, *J. Fluorine Chem.* **2005**, 126, 721–5.

[2809] H. Hayashi, H. Sonoda, K. Fukumura, T. Nagata, *Chem. Commun.* **2002**, 1618–9.

[2810] R. Pajkert, T. Böttcher, M. Ponomarenko, M. Bremer, G.-V. Röschenthaler, *Tetrahedron* **2013**, 69, 8943–51.

[2811] P. Tang, W. Wang, T. Ritter, *J. Am. Chem. Soc.* **2011**, 133, 11482–4.

[2812] T. Fujimoto, F. Becker, T. Ritter, *Org. Process Res. Dev.* **2014**, 18, 1041–4.

[2813] D. O'Hagan, H. Deng, *Chem. Rev.* **2015**, 115, 634–49.

[2814] O. Lerman, Y. Tor, D. Hebel, S. Rozen, *J. Org. Chem.* **1984**, 49, 806–13.

[2815] S. Rozen, Y. Menahem, *J. Fluorine Chem.* **1980**, 16, 19–31.

[2816] E. Differding, H. Ofner, *Synlett* **1991**, 1991, 187–9.

[2817] F. A. Davis, W. Han, *Tetrahedron Lett.* **1991**, 32, 1631–4.

[2818] S. Singh, D. D. DesMarteau, S. S. Zuberi, M. Witz, H. N. Huang, *J. Am. Chem. Soc.* **1987**, 109, 7194–6.

[2819] E. Differding, R. W. Lang, *Tetrahedron Lett.* **1988**, 29, 6087–90.

[2820] E. Differding, R. W. Lang, *Helv. Chim. Acta* **1989**, 72, 1248–52.

[2821] H. Meinert, *Z. Chem.* **1965**, 5, 64.

[2822] T. Umemoto, K. Tomita, *Tetrahedron Lett.* **1986**, 27, 3271–4.

[2823] R. E. Banks, R. A. Du Boisson, E. Tsiliopoulos, *J. Fluorine Chem.* **1986**, 32, 461–6.

[2824] R. E. Banks, M. K. Besheesh, S. N. Mohialdin-Khaffaf, I. Sharif, *J. Chem. Soc., Perkin Trans. 1* **1996**, 2069–76.

[2825] R. E. Banks, *J. Fluorine Chem.* **1998**, 87, 1–17.

[2826] S. Stavber, M. Zupan, A. J. Poss, G. A. Shia, *Tetrahedron Lett.* **1995**, 36, 6769–72.

[2827] S. Stavber, M. Zupan, *Synlett* **1996**, 1996, 693–4.

[2828] K. Ramig, *Synthesis* **2002**, 2002, 2627–31.

[2829] G. Paprott, S. Lehmann, K. Seppelt, *Chem. Ber.* **1988**, 121, 727–33.

[2830] D. M. Lemal, *Acc. Chem. Res.* **2001**, 34, 662–71.

[2831] T. I. Filyakova, A. Y. Zapevalov, M. I. Kodess, M. A. Kurykin, L. S. German, *Russ. Chem. Bull.* **1994**, 43, 1526–31.

[2832] E. D. Laganis, D. M. Lemal, *J. Am. Chem. Soc.* **1980**, 102, 6633–4.

[2833] G. Paprott, K. Seppelt, *J. Am. Chem. Soc.* **1984**, 106, 4060–1.

[2834] N. J. Maraschin, B. D. Catsikis, L. H. Davis, G. Jarvinen, R. J. Lagow, *J. Am. Chem. Soc.* **1975**, 97, 513–7.

[2835] J. L. Adcock, W. D. Evans, L. Heller-Grossman, *J. Org. Chem.* **1983**, 48, 4953–7.

[2836] J. L. Adcock, W. D. Evans, *J. Org. Chem.* **1984**, 49, 2719–23.

[2837] M. Rueda-Becerril, C. Chatalova Sazepin, J. C. T. Leung, et al., *J. Am. Chem. Soc.* **2012**, 134, 4026–9.

[2838] Y. Amaoka, M. Nagatomo, M. Inoue, *Org. Lett.* **2013**, 15, 2160–3.

[2839] X. Zhang, S. Guo, P. Tang, *Org. Chem. Front.* **2015**, 2, 806–10.

[2840] J.-j. Ma, W.-b. Yi, G.-p. Lu, C. Cai, *Org. Biomol. Chem.* **2015**, 13, 2890–4.

[2841] G. Yuan, F. Wang, N. A. Stephenson, et al., *Chem. Commun.* **2017**, 53, 126–9.

[2842] C. R. Pitts, B. Ling, R. Woltornist, R. Liu, T. Lectka, *J. Org. Chem.* **2014**, 79, 8895–9.

[2843] K. E. Danahy, J. C. Cooper, J. F. Van Humbeck, *Angew. Chem., Int. Ed. Engl.* **2018**, 57, 5134–8.

[2844] J.-B. Xia, C. Zhu, C. Chen, *J. Am. Chem. Soc.* **2013**, 135, 17494–500.

[2845] H. Egami, S. Masuda, Y. Kawato, Y. Hamashima, *Org. Lett.* **2018**, 20, 1367–70.

[2846] S. Bloom, J. L. Knippel, T. Lectka, *Chem. Sci.* **2014**, 5, 1175–8.

[2847] S. Bloom, M. McCann, T. Lectka, *Org. Lett.* **2014**, 16, 6338–41.

[2848] J.-B. Xia, C. Zhu, C. Chen, *Chem. Commun.* **2014**, 50, 11701–4.

[2849] C. W. Kee, K. F. Chin, M. W. Wong, C.-H. Tan, *Chem. Commun.* **2014**, 50, 8211–4.

[2850] C. R. Pitts, D. D. Bume, S. A. Harry, M. A. Siegler, T. Lectka, *J. Am. Chem. Soc.* **2017**, 139, 2208–11.

[2851] D. D. Bume, C. R. Pitts, F. Ghorbani, et al., *Chem. Sci.* **2017**, 8, 6918–23.

[2852] N. V. Ignat'ev, U. Welz-Biermann, U. Heider, et al., *J. Fluorine Chem.* **2003**, 124, 21–37.

[2853] F. G. Drakesmith, D. A. Hughes, *J. Appl. Electrochem.* **1979**, 9, 685–97.

[2854] T. Abe, H. Baba, E. Hayashi, S. Nagase, *J. Fluorine Chem.* **1983**, 23, 123–46.

[2855] T. Abe, E. Hayashi, H. Baba, S. Nagase, *J. Fluorine Chem.* **1984**, 25, 419–34.

[2856] P. Sartori, N. Ignat'ev, S. Datsenko, *J. Fluorine Chem.* **1995**, 75, 157–61.

[2857] P. Sartori, N. Ignat'ev, C. Jünger, R. Jüschke, P. Rieland, *J. Solid State Electrochem.* **1998**, 2, 110–6.

[2858] N. V. Ignat'ev, H. Willner, P. Sartori, *J. Fluorine Chem.* **2009**, 130, 1183–91.

[2859] S. Rozen, M. Brand, *J. Org. Chem.* **1981**, 46, 733–6.

[2860] P. A. Champagne, J. Desroches, J.-D. Hamel, M. Vandamme, J.-F. Paquin, *Chem. Rev.* **2015**, 115, 9073–174.

[2861] A. Piou, S. Celerier, S. Brunet, *J. Fluorine Chem.* **2012**, 134, 103–6.

[2862] A. Piou, S. Celerier, S. Brunet, *J. Fluorine Chem.* **2010**, 131, 1241–6.

[2863] L. Saint-Jalmes, *J. Fluorine Chem.* **2006**, 127, 85–90.

[2864] E. Fuglseth, T. H. K. Thvedt, M. F. Møll, B. H. Hoff, *Tetrahedron* **2008**, 64, 7318–23.

[2865] Y. He, X. Zhang, N. Shen, X. Fan, *J. Fluorine Chem.* **2013**, 156, 9–14.

[2866] Z. Chen, W. Zhu, Z. Zheng, X. Zou, *J. Fluorine Chem.* **2010**, 131, 340–4.

[2867] T. Kitamura, S. Kuriki, M. H. Morshed, Y. Hori, *Org. Lett.* **2011**, 13, 2392–4.

[2868] T. Kitamura, K. Muta, S. Kuriki, *Tetrahedron Lett.* **2013**, 54, 6118–20.

[2869] S. Suzuki, T. Kamo, K. Fukushi, et al., *Chem. Sci.* **2014**, 5, 2754–60.

[2870] K. Shibatomi, Y. Soga, A. Narayama, I. Fujisawa, S. Iwasa, *J. Am. Chem. Soc.* **2012**, 134, 9836–9.

[2871] N. N. Yarovenko, M. A. Raksha, V. N. Shemanina, A. S. Vasileva, *J. Gen. Chem. USSR (Engl. Transl.)* **1957**, 27, 2246.

[2872] S. Kobayashi, A. Yoneda, T. Fukuhara, S. Hara, *Tetrahedron* **2004**, 60, 6923–30.

[2873] A. Yoneda, T. Fukuhara, S. Hara, *Chem. Commun.* **2005**, 3589–90.

[2874] T. Furuya, T. Nomoto, T. Fukuhara, S. Hara, *J. Fluorine Chem.* **2009**, 130, 348–53.

[2875] F. Sladojevich, S. I. Arlow, P. Tang, T. Ritter, *J. Am. Chem. Soc.* **2013**, 135, 2470–3.

[2876] G. Bellavance, P. Dubé, B. Nguyen, *Synlett* **2012**, 2012, 569–74.

[2877] M. C. Pacheco, S. Purser, V. Gouverneur, *Chem. Rev.* **2008**, 108, 1943–81.

[2878] S. Das, S. Chandrasekhar, J. S. Yadav, R. Grée, *Tetrahedron Lett.* **2007**, 48, 5305–7.

[2879] M. M. Bio, M. Waters, G. Javadi, Z. J. Song, F. Zhang, D. Thomas, *Synthesis* **2008**, 2008, 891–6.

[2880] S. Bresciani, D. O'Hagan, *Tetrahedron Lett.* **2010**, 51, 5795–7.

[2881] R. D. P. Wadoux, X. Lin, N. S. Keddie, D. O'Hagan, *Tetrahedron: Asymmetry* **2013**, 24, 719–23.

[2882] R. Cheerlavancha, A. Lawer, M. Cagnes, M. Bhadbhade, L. Hunter, *Org. Lett.* **2013**, 15, 5562–5.

[2883] J. Venkiah, F. Cornille, C. Deshayes, A. Doutheau, *J. Fluorine Chem.* **1990**, 49, 183–95.

[2884] T. Kobayashi, M. Maeda, H. Komatsu, M. Kojima, *Chem. Pharm. Bull.* **1982**, 30, 3082–7.

[2885] S. Rozen, Y. Faust, H. Ben-Yakov, *Tetrahedron Lett.* **1979**, 20, 1823–6.

[2886] D. M. Walba, H. A. Razavi, N. A. Clark, D. S. Parmar, *J. Am. Chem. Soc.* **1988**, 110, 8686–91.

[2887] G. L. Hann, P. Sampson, *J. Chem. Soc., Chem. Commun.* **1989**, 1650–1.

[2888] M. K. Nielsen, C. R. Ugaz, W. Li, A. G. Doyle, *J. Am. Chem. Soc.* **2015**, 137, 9571–4.

[2889] K.-Y. Kim, B. C. Kim, H. B. Lee, H. Shin, *J. Org. Chem.* **2008**, 73, 8106–8.

[2890] S. S. Shinde, B. S. Lee, D. Y. Chi, *Org. Lett.* **2008**, 10, 733–5.

[2891] V. H. Jadhav, H.-J. Jeong, S. T. Lim, M.-H. Sohn, D. W. Kim, *Org. Lett.* **2011**, 13, 2502–5.

[2892] J. W. Lee, H. Yan, H. B. Jang, et al., *Angew. Chem., Int. Ed. Engl.* **2009**, 48, 7683–6.

[2893] V. H. Jadhav, S. H. Jang, H.-J. Jeong, et al., *Chem. Eur. J.* **2012**, 18, 3918–24.

[2894] D. E. Sood, S. Champion, D. M. Dawson, et al., *Angew. Chem., Int. Ed. Engl.* **2020**, 59, 8460–3.

[2895] B. P. Bandgar, V. T. Kamble, A. V. Biradar, *Monatsh. Chem.* **2005**, 136, 1579–82.

[2896] A. Sattler, G. Haufe, *J. Fluorine Chem.* **1994**, 69, 185–90.

[2897] J. Umezawa, O. Takahashi, K. Furuhashi, H. Nohira, *Tetrahedron: Asymmetry* **1994**, 5, 491–8.

[2898] R. Skupin, G. Haufe, *J. Fluorine Chem.* **1998**, 92, 157–65.

[2899] I. L. Knunyants, G. G. Yakobson. Syntheses of Fluoroorganic Compounds. Berlin, Heidelberg: Springer-Verlag; 1985.

[2900] H. Yoshino, K. Matsumoto, R. Hagiwara, Y. Ito, K. Oshima, S. Matsubara, *J. Fluorine Chem.* **2006**, 127, 29–35.

[2901] H. Amri, M. M. El Gateo, *J. Fluorine Chem.* **1990**, 46, 75–82.

[2902] M. Shimizu, H. Yoshioka, *Tetrahedron Lett.* **1988**, 29, 4101–4.

[2903] M. Shimizu, H. Yoshioka, *Tetrahedron Lett.* **1989**, 30, 967–70.

[2904] A. J. Cresswell, S. G. Davies, J. A. Lee, et al., *Org. Lett.* **2010**, 12, 2936–9.

[2905] T. Hamatani, S. Matsubara, H. Matsuda, M. Schlosser, *Tetrahedron* **1988**, 44, 2875–81.

[2906] S. Bruns, G. Haufe, *J. Fluorine Chem.* **2000**, 104, 247–54.

[2907] G. Haufe, S. Bruns, *Adv. Synth. Catal.* **2002**, 344, 165–71.

[2908] J. A. Kalow, A. G. Doyle, *J. Am. Chem. Soc.* **2010**, 132, 3268–9.

[2909] J. Zhu, G. C. Tsui, M. Lautens, *Angew. Chem., Int. Ed. Engl.* **2012**, 51, 12353–6.

[2910] W. Dmowski, *J. Fluorine Chem.* **1986**, 32, 255–82.

[2911] S. Rozen, E. Mishani, A. Bar-Haim, *J. Org. Chem.* **1994**, 59, 2918.

[2912] G. A. Olah, M. Nojima, I. Kerekes, *Synthesis* **1973**, 1973, 487–8.

[2913] O. Cohen, R. Sasson, S. Rozen, *J. Fluorine Chem.* **2006**, 127, 433–6.

[2914] J.-G. Kim, D. O. Jang, *Synlett* **2010**, 2010, 3049–52.

[2915] O. Roy, B. Marquet, J.-P. Alric, et al., *J. Fluorine Chem.* **2014**, 167, 74–8.

[2916] C. B. Murray, G. Sandford, S. R. Korn, D. S. Yufit, J. A. K. Howard, *J. Fluorine Chem.* **2005**, 126, 569–74.

[2917] P. Švec, A. Eisner, L. Kolářová, T. Weidlich, V. Pejchal, A. Růžička, *Tetrahedron Lett.* **2008**, 49, 6320–3.

[2918] M. Ochiai, A. Yoshimura, M. M. Hoque, T. Okubo, M. Saito, K. Miyamoto, *Org. Lett.* **2011**, 13, 5568–71.

[2919] S. C. Sondej, J. A. Katzenellenbogen, *J. Org. Chem.* **1986**, 51, 3508–13.

[2920] G. K. Surya Prakash, D. Hoole, V. P. Reddy, G. A. Olah, *Synlett* **1993**, 1993, 691–3.

[2921] C. York, G. K. Surya Prakash, G. A. Olah, *Tetrahedron* **1996**, 52, 9–14.

[2922] N. Yoneda, T. Fukuhara, *Chem. Lett.* **2001**, 30, 222–3.

[2923] A. Shinichi, Y. Norihiko, F. Tsuyoshi, H. Shoji, *Bull. Chem. Soc. Jpn.* **2002**, 75, 1597–603.

[2924] R. Sasson, A. Hagooly, S. Rozen, *Org. Lett.* **2003**, 5, 769–71.

[2925] V. P. Reddy, R. Alleti, M. K. Perambuduru, U. Welz-Biermann, H. Buchholz, G. K. S. Prakash, *Chem. Commun.* **2005**, 654–6.

[2926] E. Laurent, B. Marquet, C. Roze, F. Ventalon, *J. Fluorine Chem.* **1998**, 87, 215–20.

[2927] H. Ishii, N. Yamada, T. Fuchigami, *Tetrahedron* **2001**, 57, 9067–72.

[2928] S. Rozen. The Surprising Chemistry of Bromine Trifluoride. In: H. W. Roesky (ed.) Efficient Preparations of Fluorine Compounds: Wiley; 2013:154–72.

[2929] S. Rozen, E. Mishani, *J. Chem. Soc., Chem. Commun.* **1993**, 1761–2.

[2930] I. Ben-David, D. Rechavi, E. Mishani, S. Rozen, *J. Fluorine Chem.* **1999**, 97, 75–8.

[2931] O. Cohen, E. Mishani, S. Rozen, *Tetrahedron* **2010**, 66, 3579–82.

[2932] Y. Hagooly, R. Sasson, M. J. Welch, S. Rozen, *Eur. J. Org. Chem.* **2008**, 2008, 2875–80.

[2933] P. Kirsch, M. Bremer, A. Taugerbeck, T. Wallmichrath, *Angew. Chem., Int. Ed. Engl.* **2001**, 40, 1480–4.

[2934] D. V. Sevenard, P. Kirsch, E. Lork, G.-V. Röschenthaler, *Tetrahedron Lett.* **2003**, 44, 5995–8.

[2935] A. Hagooly, I. Ben-David, S. Rozen, *J. Org. Chem.* **2002**, 67, 8430–4.

[2936] P. Kirsch, A. Taugerbeck, *Eur. J. Org. Chem.* **2002**, 2002, 3923–6.

[2937] C. Schaffrath, S. L. Cobb, D. O'Hagan, *Angew. Chem., Int. Ed. Engl.* **2002**, 41, 3913–5.

[2938] X. Zhu, D. A. Robinson, A. R. McEwan, D. O'Hagan, J. H. Naismith, *J. Am. Chem. Soc.* **2007**, 129, 14597–604.

[2939] W. X. Zhang, L. Su, W. G. Hu, J. Zhou, *Synlett* **2012**, 23, 2413–5.

[2940] J. A. Kalow, D. E. Schmitt, A. G. Doyle, *J. Org. Chem.* **2012**, 77, 4177–83.

[2941] H. Park, D.-H. Yoon, H.-J. Ha, S. I. Son, W. K. Lee, *Bull. Korean Chem. Soc.* **2014**, 35, 699–700.

[2942] J. A. Kalow, A. G. Doyle, *Tetrahedron* **2013**, 69, 5702–9.

[2943] P. Kirsch, K. Tarumi, *Angew. Chem., Int. Ed. Engl.* **1998**, 37, 484–9.

[2944] P. Kirsch, *Liq. Cryst.* **1999**, 26, 449–52.

[2945] C. York, G. K. S. Prakash, G. A. Olah, *J. Org. Chem.* **1994**, 59, 6493–4.

[2946] A. E. Feiring. Addition of Halogen Fluorides to Unsaturated Systems. In: M. Hudlicky, A. E. Pavlath Chemistry of Organic Fluorine Compounds II – A Critical Review: Washington DC: American Chemical Society; 1995:61–9.

[2947] H. Yoshino, S. Matsubara, K. Oshima, K. Matsumoto, R. Hagiwara, Y. Ito, *J. Fluorine Chem.* **2005**, 126, 121–3.

[2948] P. Conte, B. Panunzi, M. Tingoli, *Tetrahedron Lett.* **2006**, 47, 273–6.

[2949] L. T. C. Crespo, R. d. S. Ribeiro, M. C. S. de Mattos, P. M. Esteves, *Synthesis* **2010**, 2010, 2379–82.

[2950] J. Walkowiak, B. Marciniak, H. Koroniak, *J. Fluorine Chem.* **2012**, 143, 287–91.

[2951] A. D. Dilman, P. A. Belyakov, M. I. Struchkova, D. E. Arkhipov, A. A. Korlyukov, V. A. Tartakovsky, *J. Org. Chem.* **2010**, 75, 5367–70.

[2952] S. Stavbar, M. Zupan, *J. Fluorine Chem.* **1978**, 12, 307–13.

[2953] C. Xue, X. Jiang, C. Fu, S. Ma, *Chem. Commun.* **2013**, 49, 5651–3.

[2954] D. L. S. Brahms, W. P. Dailey, *Chem. Rev.* **1996**, 96, 1585–632.

[2955] M. J. Tozer, T. F. Herpin, *Tetrahedron* **1996**, 52, 8619–83.

[2956] D. Seyferth, H. Dertouzos, R. Suzuki, J. Y. P. Mui, *J. Org. Chem.* **1967**, 32, 2980–4.

[2957] D. Seyferth, S. P. Hopper, K. V. Darragh, *J. Am. Chem. Soc.* **1969**, 91, 6536–7.

[2958] D. Seyferth, S. P. Hopper, *J. Org. Chem.* **1972**, 37, 4070–5.

[2959] J. M. Birchall, G. W. Cross, R. N. Haszeldine, *Proc. Chem. Soc.* **1960**, 81.

[2960] G. A. Wheaton, D. D. Burton, *J. Fluorine Chem.* **1976**, 8, 97–100.

[2961] G. A. Wheaton, D. J. Burton, *J. Org. Chem.* **1978**, 43, 2643–51.

[2962] Q. Chen, S. Wu, *J. Org. Chem.* **1989**, 54, 3023–7.

[2963] Q.-Y. Chen, S.-W. Wu, *J. Chem. Soc., Chem. Commun.* **1989**, 705–6.

[2964] W. R. Dolbier, C. R. Burkholder, *J. Org. Chem.* **1990**, 55, 589–94.

[2965] W. R. Dolbier, H. Wojtowicz, C. R. Burkholder, *J. Org. Chem.* **1990**, 55, 5420–2.

[2966] P. Erbes, W. Boland, *Helv. Chim. Acta* **1992**, 75, 766–72.

[2967] F. Tian, V. Kruger, O. Bautista, et al., *Org. Lett.* **2000**, 2, 563–4.

[2968] W. Xu, Q.-Y. Chen, *Org. Biomol. Chem.* **2003**, 1, 1151–6.

[2969] F. Wang, T. Luo, J. Hu, et al., *Angew. Chem., Int. Ed. Engl.* **2011**, 50, 7153–7.

[2970] U. Teruo, T. Ginjiro, *Bull. Chem. Soc. Jpn.* **1986**, 59, 3625–9.

[2971] G. S. Lal, *J. Org. Chem.* **1993**, 58, 2791–6.

[2972] Z.-Q. Xu, D. D. DesMarteau, Y. Gotoh, *J. Chem. Soc., Chem. Commun.* **1991**, 179–81.

[2973]   J.-C. Xiao, J. n. M. Shreeve, *J. Fluorine Chem.* **2005**, 126, 473–6.

[2974]   G. Stavber, M. Zupan, S. Stavber, *Tetrahedron Lett.* **2007**, 48, 2671–3.

[2975]   G. Stavber, S. Stavber, *Adv. Synth. Catal.* **2010**, 352, 2838–46.

[2976]   T. H. Krane Thvedt, E. Fuglseth, E. Sundby, B. H. Hoff, *Tetrahedron* **2009**, 65, 9550–6.

[2977]   G. Stavber, M. Zupan, S. Stavber, *Synlett* **2009**, 2009, 589–94.

[2978]   J. Liu, J. Chan, C. M. Bryant, et al., *Tetrahedron Lett.* **2012**, 53, 2971–5.

[2979]   T. Kitamura, K. Muta, K. Muta, *J. Org. Chem.* **2014**, 79, 5842–6.

[2980]   S. Sato, M. Yoshida, S. Hara, *Synthesis* **2005**, 2005, 2602–5.

[2981]   Q. Yang, L.-L. Mao, B. Yang, S.-D. Yang, *Org. Lett.* **2014**, 16, 3460–3.

[2982]   R. D. Chambers, A. M. Kenwright, M. Parsons, G. Sandford, J. S. Moilliet, *J. Chem. Soc., Perkin Trans. 1* **2002**, 2190–7.

[2983]   M. Meanwell, B. S. Adluri, Z. Yuan, et al., *Chem. Sci.* **2018**, 9, 5608–13.

[2984]   M. Meanwell, M. B. Nodwell, R. E. Martin, R. Britton, *Angew. Chem., Int. Ed. Engl.* **2016**, 55, 13244–8.

[2985]   H.-Q. Luo, T.-P. Loh, *Tetrahedron Lett.* **2009**, 50, 1554–6.

[2986]   M.-Y. Chang, N.-C. Lee, M.-F. Lee, Y.-P. Huang, C.-H. Lin, *Tetrahedron Lett.* **2010**, 51, 5900–3.

[2987]   J. Pavlinac, M. Zupan, S. Stavbar, *Molecules* **2009**, 14, 2394–409.

[2988]   L. Basset, A. Martin-Mingot, M.-P. Jouannetaud, J.-C. Jacquesy, *Tetrahedron Lett.* **2008**, 49, 1551–4.

[2989]   P. Kralj, M. Zupan, S. Stavber, *J. Org. Chem.* **2006**, 71, 3880–8.

[2990]   Y. Zhu, J. Han, J. Wang, et al., *Chem. Rev.* **2018**, 118, 3887–964.

[2991]   X. Yang, T. Wu, R. J. Phipps, F. D. Toste, *Chem. Rev.* **2015**, 115, 826–70.

[2992]   J.-A. Ma, D. Cahard, *Chem. Rev.* **2004**, 104, 6119–46.

[2993]   J.-A. Ma, D. Cahard, *Chem. Rev.* **2008**, 108, PR1–PR43.

[2994]   Y. Takeuchi, T. Suzuki, A. Satoh, T. Shiragami, N. Shibata, *J. Org. Chem.* **1999**, 64, 5708–11.

[2995]   N. Shibata, E. Suzuki, Y. Takeuchi, *J. Am. Chem. Soc.* **2000**, 122, 10728–9.

[2996]   D. Cahard, C. Audouard, J.-C. Plaquevent, L. Toupet, N. Roques, *Tetrahedron Lett.* **2001**, 42, 1867–9.

[2997]   W.-B. Yi, X. Huang, Z. Zhang, D.-R. Zhu, C. Cai, W. Zhang, *Beilstein J. Org. Chem.* **2012**, 8, 1233–40.

[2998]   J. Xu, Y. Hu, D. Huang, K.-H. Wang, C. Xu, T. Niu, *Adv. Synth. Catal.* **2012**, 354, 515–26.

[2999]   J. Novacek, M. Waser, *Eur. J. Org. Chem.* **2014**, 2014, 802–9.

[3000]   C.-L. Zhu, M. Maeno, F.-G. Zhang, et al., *Eur. J. Org. Chem.* **2013**, 2013, 6501–5.

[3001]   V. Rauniyar, A. D. Lackner, G. L. Hamilton, F. D. Toste, *Science* **2011**, 334, 1681–4.

[3002]   J. R. Wolstenhulme, J. Rosenqvist, O. Lozano, et al., *Angew. Chem., Int. Ed. Engl.* **2013**, 52, 9796–800.

[3003]   T. Ishimaru, N. Shibata, T. Horikawa, et al., *Angew. Chem., Int. Ed. Engl.* **2008**, 47, 4157–61.

[3004]   T. Fukuzumi, N. Shibata, M. Sugiura, S. Nakamura, T. Toru, *J. Fluorine Chem.* **2006**, 127, 548–51.

[3005]   R. J. Phipps, F. D. Toste, *J. Am. Chem. Soc.* **2013**, 135, 1268–71.

[3006]   S. Brandes, B. Niess, M. Bella, A. Prieto, J. Overgaard, K. A. Jørgensen, *Chem. Eur. J.* **2006**, 12, 6039–52.

[3007]   M. Marigo, D. Fielenbach, A. Braunton, A. Kjærsgaard, K. A. Jørgensen, *Angew. Chem., Int. Ed. Engl.* **2005**, 44, 3703–6.

[3008]   J. Franzén, M. Marigo, D. Fielenbach, T. C. Wabnitz, A. Kjærsgaard, K. A. Jørgensen, *J. Am. Chem. Soc.* **2005**, 127, 18296–304.

[3009]   D. D. Steiner, N. Mase, C. F. Barbas III, *Angew. Chem., Int. Ed. Engl.* **2005**, 44, 3706–10.

[3010]   T. D. Beeson, D. W. C. MacMillan, *J. Am. Chem. Soc.* **2005**, 127, 8826–8.

[3011]   D. Enders, M. R. M. Hüttl, *Synlett* **2005**, 2005, 0991-3.

[3012] P. Kwiatkowski, T. D. Beeson, J. C. Conrad, D. W. C. MacMillan, *J. Am. Chem. Soc.* **2011**, 133, 1738–41.
[3013] S. J. Shaw, D. A. Goff, L. A. Boralsky, M. Irving, R. Singh, *J. Org. Chem.* **2013**, 78, 8892–7.
[3014] Y.-h. Lam, K. N. Houk, *J. Am. Chem. Soc.* **2014**, 136, 9556–9.
[3015] X. Yang, R. J. Phipps, F. D. Toste, *J. Am. Chem. Soc.* **2014**, 136, 5225–8.
[3016] F. A. Davis, P. V. N. Kasu, *Tetrahedron Lett.* **1998**, 39, 6135–8.
[3017] F. A. Davis, W. Han, *Tetrahedron Lett.* **1992**, 33, 1153–6.
[3018] F. A. Davis, P. V. N. Kasu, G. Sundarababu, H. Qi, *J. Org. Chem.* **1997**, 62, 7546–7.
[3019] M. Ihara, T. Kai, N. Taniguchi, K. Fukumoto, *J. Chem. Soc., Perkin Trans. 1* **1990**, 2357–8.
[3020] M. Ihara, N. Taniguchi, T. Kai, K. Satoh, K. Fukumoto, *J. Chem. Soc., Perkin Trans. 1* **1992**, 221–7.
[3021] M. K. Edmonds, F. H. M. Graichen, J. Gardiner, A. D. Abell, *Org. Lett.* **2008**, 10, 885–7.
[3022] K. Nakayama, D. L. Browne, I. R. Baxendale, S. V. Ley, *Synlett* **2013**, 24, 1298–302.
[3023] H. Lubin, C. Dupuis, J. Pytkowicz, T. Brigaud, *J. Org. Chem.* **2013**, 78, 3487–92.
[3024] M. C. O'Reilly, C. W. Lindsley, *Tetrahedron Lett.* **2013**, 54, 3627–9.
[3025] K. Shibatomi, T. Okimi, Y. Abe, A. Narayama, N. Nakamura, S. Iwasa, *Beilstein J. Org. Chem.* **2014**, 10, 323–31.
[3026] O. O. Fadeyi, C. W. Lindsley, *Org. Lett.* **2009**, 11, 943–6.
[3027] H. Jiang, A. Falcicchio, K. L. Jensen, M. W. Paixão, S. Bertelsen, K. A. Jørgensen, *J. Am. Chem. Soc.* **2009**, 131, 7153–7.
[3028] W. Zi, Y.-M. Wang, F. D. Toste, *J. Am. Chem. Soc.* **2014**, 136, 12864–7.
[3029] M. Tredwell, K. Tenza, M. C. Pacheco, V. Gouverneur, *Org. Lett.* **2005**, 7, 4495–7.
[3030] G. Reginato, A. Mordini, A. Tenti, M. Valacchi, J. Broguiere, *Tetrahedron: Asymmetry* **2008**, 19, 2882–6.
[3031] M. Sawicki, A. Kwok, M. Tredwell, V. Gouverneur, *Beilstein J. Org. Chem.* **2007**, 3, 34.
[3032] G. T. Giuffredi, S. Purser, M. Sawicki, A. L. Thompson, V. Gouverneur, *Tetrahedron: Asymmetry* **2009**, 20, 910–20.
[3033] S. Purser, B. Odell, T. D. W. Claridge, P. R. Moore, V. Gouverneur, *Chem. Eur. J.* **2006**, 12, 9176–85.
[3034] S. Purser, C. Wilson, P. R. Moore, V. Gouverneur, *Synlett* **2007**, 2007, 1166–8.
[3035] Y.-h. Lam, C. Bobbio, I. R. Cooper, V. Gouverneur, *Angew. Chem., Int. Ed. Engl.* **2007**, 46, 5106–10.
[3036] Y.-h. Lam, M. N. Hopkinson, S. J. Stanway, V. Gouverneur, *Synlett* **2007**, 2007, 3022–6.
[3037] S. Boldon, J. E. Moore, V. Gouverneur, *Chem. Commun.* **2008**, 3622–4.
[3038] L. Carroll, M. C. Pacheco, L. Garcia, V. Gouverneur, *Chem. Commun.* **2006**, 4113–5.
[3039] L. Carroll, S. McCullough, T. Rees, T. D. W. Claridge, V. Gouverneur, *Org. Biomol. Chem.* **2008**, 6, 1731–3.
[3040] D. A. Petrone, J. Ye, M. Lautens, *Chem. Rev.* **2016**, 116, 8003–104.
[3041] R. Szpera, D. F. J. Moseley, L. B. Smith, A. J. Sterling, V. Gouverneur, *Angew. Chem., Int. Ed. Engl.* **2019**, 58, 14824–48.
[3042] J. Miró, C. del Pozo, *Chem. Rev.* **2016**, 116, 11924–66.
[3043] K. L. Hull, W. Q. Anani, M. S. Sanford, *J. Am. Chem. Soc.* **2006**, 128, 7134–5.
[3044] K. B. McMurtrey, J. M. Racowski, M. S. Sanford, *Org. Lett.* **2012**, 14, 4094–7.
[3045] R.-Y. Zhu, K. Tanaka, G.-C. Li, et al., *J. Am. Chem. Soc.* **2015**, 137, 7067–70.
[3046] Q. Zhang, X.-S. Yin, K. Chen, S.-Q. Zhang, B.-F. Shi, *J. Am. Chem. Soc.* **2015**, 137, 8219–26.
[3047] J. Miao, K. Yang, M. Kurek, H. Ge, *Org. Lett.* **2015**, 17, 3738–41.
[3048] Q. Zhu, D. Ji, T. Liang, X. Wang, Y. Xu, *Org. Lett.* **2015**, 17, 3798–801.
[3049] Y.-J. Mao, S.-J. Lou, H.-Y. Hao, D.-Q. Xu, *Angew. Chem., Int. Ed. Engl.* **2018**, 57, 14085–9.
[3050] H. Park, P. Verma, K. Hong, J.-Q. Yu, *Nat. Chem.* **2018**, 10, 755–62.

[3051] M.-G. Braun, A. G. Doyle, *J. Am. Chem. Soc.* **2013**, 135, 12990–3.

[3052] Y. Hamashima, M. Sodeoka, *J. Synth. Org. Chem. Jpn.* **2007**, 65, 1099–107.

[3053] T. Suzuki, T. Goto, Y. Hamashima, M. Sodeoka, *J. Org. Chem.* **2007**, 72, 246–50.

[3054] Y. Hamashima, M. Sodeoka, *Synlett* **2006**, 2006, 1467–78.

[3055] Y. Hamashima, T. Suzuki, H. Takano, et al., *Tetrahedron* **2006**, 62, 7168–79.

[3056] N.-R. Lee, S.-M. Kim, D. Y. Kim, *Bull. Korean Chem. Soc.* **2009**, 30, 829–36.

[3057] L. Hintermann, A. Togni, *Angew. Chem., Int. Ed. Engl.* **2000**, 39, 4359–62.

[3058] L. Hintermann, M. Perseghini, A. Togni, *Beilstein J. Org. Chem.* **2011**, 7, 1421–35.

[3059] A. Bertogg, L. Hintermann, D. P. Huber, M. Perseghini, M. Sanna, A. Togni, *Helv. Chim. Acta* **2012**, 95, 353–403.

[3060] K. Motoi, H. Shunsuke, K. Yuji, H. Shuichi, I. Toshiyuki, *Chem. Lett.* **2010**, 39, 466–7.

[3061] N. Shibata, J. Kohno, K. Takai, et al., *Angew. Chem., Int. Ed. Engl.* **2005**, 44, 4204–7.

[3062] K. Shibatomi, Y. Tsuzuki, S.-i. Nakata, Y. Sumikawa, S. Iwasa, *Synlett* **2007**, 2007, 0551-4.

[3063] S. Kazutaka, T. Yuya, I. Seiji, *Chem. Lett.* **2008**, 37, 1098–9.

[3064] M.-J. Cho, Y.-K. Kang, N.-R. Lee, D. Y. Kim, *Bull. Korean Chem. Soc.* **2007**, 28, 2191–2.

[3065] S. H. Kang, D. Y. Kim, *Adv. Synth. Catal.* **2010**, 352, 2783–6.

[3066] M. Frings, C. Bolm, *Eur. J. Org. Chem.* **2009**, 2009, 4085–90.

[3067] K. Balaraman, R. Vasanthan, V. Kesavan, *Tetrahedron: Asymmetry* **2013**, 24, 919–24.

[3068] J. Peng, D.-M. Du, *RSC Adv.* **2014**, 4, 2061–7.

[3069] K. Shibatomi, A. Narayama, Y. Soga, T. Muto, S. Iwasa, *Org. Lett.* **2011**, 13, 2944–7.

[3070] A. Narayama, K. Shibatomi, Y. Soga, T. Muto, S. Iwasa, *Synlett* **2013**, 24, 375–8.

[3071] D. S. Reddy, N. Shibata, J. Nagai, S. Nakamura, T. Toru, S. Kanemasa, *Angew. Chem., Int. Ed. Engl.* **2008**, 47, 164–8.

[3072] S. Suzuki, Y. Kitamura, S. Lectard, Y. Hamashima, M. Sodeoka, *Angew. Chem., Int. Ed. Engl.* **2012**, 51, 4581–5.

[3073] O. Jacquet, N. D. Clément, C. Blanco, M. M. Belmonte, J. Benet-Buchholz, P. W. N. M. van Leeuwen, *Eur. J. Org. Chem.* **2012**, 2012, 4844–52.

[3074] S.-M. Kim, Y.-K. Kang, M.-J. Cho, J.-Y. Mang, D. Y. Kim, *Bull. Korean Chem. Soc.* **2007**, 28, 2435–41.

[3075] J. Ramírez, D. P. Huber, A. Togni, *Synlett* **2007**, 2007, 1143–7.

[3076] J. Alvarado, A. T. Herrmann, A. Zakarian, *J. Org. Chem.* **2014**, 79, 6206–20.

[3077] T. Suzuki, Y. Hamashima, M. Sodeoka, *Angew. Chem., Int. Ed. Engl.* **2007**, 46, 5435–9.

[3078] T. Ishimaru, N. Shibata, D. S. Reddy, T. Horikawa, S. Nakamura, T. Toru, *Beilstein J. Org. Chem.* **2008**, 4, 16.

[3079] D. S. Reddy, N. Shibata, T. Horikawa, et al., *Chem. Asian J.* **2009**, 4, 1411–5.

[3080] S. Bloom, C. R. Pitts, D. C. Miller, et al., *Angew. Chem., Int. Ed. Engl.* **2012**, 51, 10580–3.

[3081] P. Xu, S. Guo, L. Wang, P. Tang, *Angew. Chem., Int. Ed. Engl.* **2014**, 53, 5955–8.

[3082] S. Bloom, C. R. Pitts, R. Woltornist, A. Griswold, M. G. Holl, T. Lectka, *Org. Lett.* **2013**, 15, 1722–4.

[3083] A. M. Hua, D. N. Mai, R. Martinez, R. D. Baxter, *Org. Lett.* **2017**, 19, 2949–52.

[3084] J.-B. Xia, Y. Ma, C. Chen, *Org. Chem. Front.* **2014**, 1, 468–72.

[3085] S. D. Halperin, H. Fan, S. Chang, R. E. Martin, R. Britton, *Angew. Chem., Int. Ed. Engl.* **2014**, 53, 4690–3.

[3086] M. B. Nodwell, A. Bagai, S. D. Halperin, R. E. Martin, H. Knust, R. Britton, *Chem. Commun.* **2015**, 51, 11783–6.

[3087] S. D. Halperin, D. Kwon, M. Holmes, et al., *Org. Lett.* **2015**, 17, 5200–3.

[3088] M. B. Nodwell, H. Yang, M. Čolović, et al., *J. Am. Chem. Soc.* **2017**, 139, 3595–8.

[3089] E. W. Webb, J. B. Park, E. L. Cole, et al., *J. Am. Chem. Soc.* **2020**, 142, 9493–500.

[3090] B. J. Groendyke, D. I. AbuSalim, S. P. Cook, *J. Am. Chem. Soc.* **2016**, 138, 12771–4.

[3091] H. Guan, S. Sun, Y. Mao, et al., *Angew. Chem., Int. Ed. Engl.* **2018**, 57, 11413–7.

[3092] W. Liu, J. T. Groves, *Angew. Chem., Int. Ed. Engl.* **2013**, 52, 6024–7.

[3093] X. Huang, W. Liu, H. Ren, R. Neelamegam, J. M. Hooker, J. T. Groves, *J. Am. Chem. Soc.* **2014**, 136, 6842–5.

[3094] W. Liu, X. Huang, J. T. Groves, *Nat. Protoc.* **2013**, 8, 2348–54.

[3095] W. Liu, X. Huang, M.-J. Cheng, R. J. Nielsen, W. A. Goddard, J. T. Groves, *Science* **2012**, 337, 1322–5.

[3096] Y. He, Z. Yang, R. T. Thornbury, F. D. Toste, *J. Am. Chem. Soc.* **2015**, 137, 12207–10.

[3097] T. Wu, G. Yin, G. Liu, *J. Am. Chem. Soc.* **2009**, 131, 16354–5.

[3098] C. Hou, P. Chen, G. Liu, *Angew. Chem., Int. Ed. Engl.* **2020**, 59, 2735–9.

[3099] Z. Yuan, H.-Y. Wang, X. Mu, P. Chen, Y.-L. Guo, G. Liu, *J. Am. Chem. Soc.* **2015**, 137, 2468–71.

[3100] H. Zhang, Y. Song, J. Zhao, J. Zhang, Q. Zhang, *Angew. Chem., Int. Ed. Engl.* **2014**, 53, 11079–83.

[3101] J. Saavedra-Olavarría, G. C. Arteaga, J. J. López, E. G. Pérez, *Chem. Commun.* **2015**, 51, 3379–82.

[3102] E. Emer, L. Pfeifer, J. M. Brown, V. Gouverneur, *Angew. Chem., Int. Ed. Engl.* **2014**, 53, 4181–5.

[3103] E. P. A. Talbot, T. d. A. Fernandes, J. M. McKenna, F. D. Toste, *J. Am. Chem. Soc.* **2014**, 136, 4101–4.

[3104] M.-G. Braun, M. H. Katcher, A. G. Doyle, *Chem. Sci.* **2013**, 4, 1216–20.

[3105] M. H. Katcher, A. G. Doyle, *J. Am. Chem. Soc.* **2010**, 132, 17402–4.

[3106] M. H. Katcher, A. Sha, A. G. Doyle, *J. Am. Chem. Soc.* **2011**, 133, 15902–5.

[3107] C. Hollingworth, A. Hazari, M. N. Hopkinson, et al., *Angew. Chem., Int. Ed. Engl.* **2011**, 50, 2613–7.

[3108] A. M. Lauer, J. Wu, *Org. Lett.* **2012**, 14, 5138–41.

[3109] J. J. Topczewski, T. J. Tewson, H. M. Nguyen, *J. Am. Chem. Soc.* **2011**, 133, 19318–21.

[3110] E. Benedetto, M. Tredwell, C. Hollingworth, T. Khotavivattana, J. M. Brown, V. Gouverneur, *Chem. Sci.* **2013**, 4, 89–96.

[3111] Z. Zhang, F. Wang, X. Mu, P. Chen, G. Liu, *Angew. Chem., Int. Ed. Engl.* **2013**, 52, 7549–53.

[3112] J. M. Larsson, S. R. Pathipati, K. J. Szabó, *J. Org. Chem.* **2013**, 78, 7330–6.

[3113] F. Yin, Z. Wang, Z. Li, C. Li, *J. Am. Chem. Soc.* **2012**, 134, 10401–4.

[3114] S. Mizuta, I. S. R. Stenhagen, M. O'Duill, et al., *Org. Lett.* **2013**, 15, 2648–51.

[3115] Z. Li, Z. Wang, L. Zhu, X. Tan, C. Li, *J. Am. Chem. Soc.* **2014**, 136, 16439–43.

[3116] S. Fan, C.-Y. He, X. Zhang, *Tetrahedron* **2010**, 66, 5218–28.

[3117] H. Amii, T. Kobayashi, Y. Hatamoto, K. Uneyama, *Chem. Commun.* **1999**, 1323–4.

[3118] M. Mae, H. Amii, K. Uneyama, *Tetrahedron Lett.* **2000**, 41, 7893–6.

[3119] H. Amii, K. Uneyama, *Chem. Rev.* **2009**, 109, 2119–83.

[3120] K. Manabu, I. Takashi, *Bull. Chem. Soc. Jpn.* **1990**, 63, 1185–90.

[3121] R. B. Greenwald, D. H. Evans, *J. Org. Chem.* **1976**, 41, 1470–2.

[3122] Y. Hanzawa, K.-i. Kawagoe, M. Ito, Y. Kobayashi, *Chem. Pharm. Bull.* **1987**, 35, 1633–6.

[3123] P. L. Coe, A. Sellars, J. C. Tatlow, G. Whittaker, H. C. Fielding, *J. Chem. Soc., Chem. Commun.* **1982**, 362–3.

[3124] R. N. Renaud, D. E. Sullivan, *Can. J. Chem.* **1972**, 50, 3084–5.

[3125] J.-C. Blazejewski, R. Dorme, C. Wakselman, *Synthesis* **1985**, 1985, 1120–1.

[3126] D. G. Naae, *J. Org. Chem.* **1980**, 45, 1394–401.

[3127] M. G. Campbell, T. Ritter, *Chem. Rev.* **2015**, 115, 612–33.

[3128] S. Preshlock, M. Tredwell, V. Gouverneur, *Chem. Rev.* **2016**, 116, 719–66.

[3129] T. C. Wilson, T. Cailly, V. Gouverneur, *Chem. Soc. Rev.* **2018**, 47, 6990–7005.

[3130] D. D. MacNicol, P. R. Mallison, A. Murphy, G. J. Sym, *Tetrahedron Lett.* **1982**, 23, 4131–4.

[3131] H. Sun, S. G. DiMagno, *Angew. Chem., Int. Ed. Engl.* **2006**, 45, 2720–5.

[3132] Y. F. Hu, J. Luo, C. X. Lü, *Chin. Chem. Lett.* **2010**, 21, 151–4.
[3133] L. J. Allen, J. M. Muhuhi, D. C. Bland, R. Merzel, M. S. Sanford, *J. Org. Chem.* **2014**, 79, 5827–33.
[3134] V. V. Grushin, W. J. Marshall, *Organometallics* **2008**, 27, 4825–8.
[3135] F. Effenberger, W. Streicher, *Chem. Ber.* **1991**, 124, 157–62.
[3136] D. J. Adams, J. H. Clark, *Chem. Soc. Rev.* **1999**, 28, 225–31.
[3137] S. Hiroshi, Y. Naoto, Y. Yasuo, F. Osamu, K. Yoshikazu, *Bull. Chem. Soc. Jpn.* **1990**, 63, 2010–7.
[3138] S. D. Kuduk, R. M. DiPardo, M. G. Bock, *Org. Lett.* **2005**, 7, 577–9.
[3139] P. LaBeaume, M. Placzek, M. Daniels, et al., *Tetrahedron Lett.* **2010**, 51, 1906–9.
[3140] H. Nemoto, T. Nishiyama, S. Akai, *Org. Lett.* **2011**, 13, 2714–7.
[3141] H. Nemoto, K. Takubo, K. Shimizu, S. Akai, *Synlett* **2012**, 23, 1978–84.
[3142] L. Garel, L. Saint-Jalmes, *Tetrahedron Lett.* **2006**, 47, 5705–8.
[3143] J. Heredia-Moya, K. L. Kirk, *J. Fluorine Chem.* **2007**, 128, 674–8.
[3144] Z.-q. Yu, Y.-w. Lv, C.-m. Yu, W.-k. Su, *Tetrahedron Lett.* **2013**, 54, 1261–3.
[3145] M. Döbele, S. Vanderheiden, N. Jung, S. Bräse, *Angew. Chem., Int. Ed. Engl.* **2010**, 49, 5986–8.
[3146] C.-K. Chu, J.-H. Kim, D.-W. Kim, K.-H. Chung, J. A. Katzenellenbogen, D.-Y. Chi, *Bull. Korean Chem. Soc.* **2005**, 26, 599–602.
[3147] Z. Zhu, N. L. Colbry, M. Lovdahl, et al., *Org. Process Res. Dev.* **2007**, 11, 907–9.
[3148] B. Wang, L. Qin, K. D. Neumann, S. Uppaluri, R. L. Cerny, S. G. DiMagno, *Org. Lett.* **2010**, 12, 3352–5.
[3149] B. Wang, R. L. Cerny, S. Uppaluri, J. J. Kempinger, S. G. DiMagno, *J. Fluorine Chem.* **2010**, 131, 1113–21.
[3150] P. Xu, D. Zhao, F. Berger, et al., *Angew. Chem., Int. Ed. Engl.* **2020**, 59, 1956–60.
[3151] T. Tian, W.-H. Zhong, S. Meng, X.-B. Meng, Z.-J. Li, *J. Org. Chem.* **2013**, 78, 728–32.
[3152] R. D. Chambers, M. A. Fox, G. Sandford, J. Trmcic, A. Goeta, *J. Fluorine Chem.* **2007**, 128, 29–33.
[3153] R. D. Chambers, J. Hutchinson, G. Sandford, *J. Fluorine Chem.* **1999**, 100, 63–73.
[3154] J. P. Alric, B. Marquet, T. Billard, B. R. Langlois, *J. Fluorine Chem.* **2005**, 126, 659–65.
[3155] R. D. Chambers, G. Sandford, J. Trmcic, T. Okazoe, *Org. Process Res. Dev.* **2008**, 12, 339–44.
[3156] D. Holling, G. Sandford, A. S. Batsanov, D. S. Yufit, J. A. K. Howard, *J. Fluorine Chem.* **2005**, 126, 1377–83.
[3157] J. Liu, J. R. Barrio, N. Satyamurthy, *J. Fluorine Chem.* **2006**, 127, 1175–87.
[3158] I. Vints, J. Gatenyo, S. Rozen, *J. Org. Chem.* **2013**, 78, 11794–7.
[3159] S. P. Anand, L. A. Quarterman, P. A. Christian, H. H. Hyman, R. Filler, *J. Org. Chem.* **1975**, 40, 3796–7.
[3160] I. Agranat, M. Rabinovitz, H. Selig, C.-H. Lin, *Synthesis* **1977**, 1977, 267–8.
[3161] A. P. Lothian, C. A. Ramsden, M. M. Shaw, R. G. Smith, *Tetrahedron* **2011**, 67, 2788–93.
[3162] M. R. P. Heravi, *J. Fluorine Chem.* **2008**, 129, 217–21.
[3163] B. Troegel, T. Lindel, *Org. Lett.* **2012**, 14, 468–71.
[3164] J. C. Sloop, J. L. Jackson, R. D. Schmidt, *Heteroat. Chem.* **2009**, 20, 341–5.
[3165] T. Umemoto, G. Tomizawa, *J. Org. Chem.* **1989**, 54, 1726–31.
[3166] A. B. Bueno, C. J. Flynn, J. Gilmore, et al., *Tetrahedron Lett.* **2005**, 46, 7769–71.
[3167] A. Nagaki, Y. Uesugi, H. Kim, J.-i. Yoshida, *Chem. Asian J.* **2013**, 8, 705–8.
[3168] K. Higashiguchi, K. Matsuda, Y. Asano, A. Murakami, S. Nakamura, M. Irie, *Eur. J. Org. Chem.* **2005**, 2005, 91–7.
[3169] P. H. Briner, M. C. T. Fyfe, P. Martin, P. J. Murray, F. Naud, M. J. Procter, *Org. Process Res. Dev.* **2006**, 10, 346–8.

[3170] H. J. Son, W. Wang, T. Xu, et al., *J. Am. Chem. Soc.* **2011**, 133, 1885–94.

[3171] F. Gohier, P. Frère, J. Roncali, *J. Org. Chem.* **2013**, 78, 1497–503.

[3172] G. Lai, T. Guo, *Synth. Commun.* **2007**, 38, 72–6.

[3173] M. Belley, Z. Douida, J. Mancuso, M. De Vleeschauwer, *Synlett* **2005**, 2005, 247–50.

[3174] A. K. Ghosh, P. Lagisetty, B. Zajc, *J. Org. Chem.* **2007**, 72, 8222–6.

[3175] S. Yamada, A. Gavryushin, P. Knochel, *Angew. Chem., Int. Ed. Engl.* **2010**, 49, 2215–8.

[3176] S. Yamada, P. Knochel, *Synthesis* **2010**, 2010, 2490–4.

[3177] P. Anbarasan, H. Neumann, M. Beller, *Angew. Chem., Int. Ed. Engl.* **2010**, 49, 2219–22.

[3178] P. Anbarasan, H. Neumann, M. Beller, *Chem. Asian J.* **2010**, 5, 1775–8.

[3179] C. Cazorla, E. Métay, B. Andrioletti, M. Lemaire, *Tetrahedron Lett.* **2009**, 50, 3936–8.

[3180] E. Bellamy, O. Bayh, C. Hoarau, F. Trécourt, G. Quéguiner, F. Marsais, *Chem. Commun.* **2010**, 46, 7043–5.

[3181] G. Sandford, *Tetrahedron* **2003**, 59, 437–54.

[3182] T. Furuya, H. M. Kaiser, T. Ritter, *Angew. Chem., Int. Ed. Engl.* **2008**, 47, 5993–6.

[3183] N. D. Ball, M. S. Sanford, *J. Am. Chem. Soc.* **2009**, 131, 3796–7.

[3184] S.-J. Lou, D.-Q. Xu, A.-B. Xia, et al., *Chem. Commun.* **2013**, 49, 6218–20.

[3185] X. Wang, T.-S. Mei, J.-Q. Yu, *J. Am. Chem. Soc.* **2009**, 131, 7520–1.

[3186] K. S. L. Chan, M. Wasa, X. Wang, J.-Q. Yu, *Angew. Chem., Int. Ed. Engl.* **2011**, 50, 9081–4.

[3187] C. Chen, C. Wang, J. Zhang, Y. Zhao, *J. Org. Chem.* **2015**, 80, 942–9.

[3188] S.-J. Lou, D.-Q. Xu, Z.-Y. Xu, *Angew. Chem., Int. Ed. Engl.* **2014**, 53, 10330–5.

[3189] S.-J. Lou, Q. Chen, Y.-F. Wang, et al., *ACS Catal.* **2015**, 5, 2846–9.

[3190] X.-Y. Chen, E. J. Sorensen, *J. Am. Chem. Soc.* **2018**, 140, 2789–92.

[3191] D. A. Watson, M. Su, G. Teverovskiy, et al., *Science* **2009**, 325, 1661–4.

[3192] T. Noël, T. J. Maimone, S. L. Buchwald, *Angew. Chem., Int. Ed. Engl.* **2011**, 50, 8900–3.

[3193] H. G. Lee, P. J. Milner, S. L. Buchwald, *Org. Lett.* **2013**, 15, 5602–5.

[3194] H. G. Lee, P. J. Milner, S. L. Buchwald, *J. Am. Chem. Soc.* **2014**, 136, 3792–5.

[3195] P. J. Milner, Y. Yang, S. L. Buchwald, *Organometallics* **2015**, 34, 4775–80.

[3196] A. C. Sather, H. G. Lee, V. Y. De La Rosa, Y. Yang, P. Müller, S. L. Buchwald, *J. Am. Chem. Soc.* **2015**, 137, 13433–8.

[3197] J. Wannberg, C. Wallinder, M. Ünlüsoy, C. Sköld, M. Larhed, *J. Org. Chem.* **2013**, 78, 4184–9.

[3198] A. R. Mazzotti, M. G. Campbell, P. Tang, J. M. Murphy, T. Ritter, *J. Am. Chem. Soc.* **2013**, 135, 14012–5.

[3199] K. Yamamoto, J. Li, J. A. O. Garber, et al., *Nature* **2018**, 554, 511–4.

[3200] M. A. Subramanian, L. E. Manzer, *Science* **2002**, 297, 1665.

[3201] T. Truong, K. Klimovica, O. Daugulis, *J. Am. Chem. Soc.* **2013**, 135, 9342–5.

[3202] S. J. Lee, K. J. Makaravage, A. F. Brooks, P. J. H. Scott, M. S. Sanford, *Angew. Chem., Int. Ed. Engl.* **2019**, 58, 3119–22.

[3203] N. Ichiishi, A. J. Canty, B. F. Yates, M. S. Sanford, *Org. Lett.* **2013**, 15, 5134–7.

[3204] P. S. Fier, J. Luo, J. F. Hartwig, *J. Am. Chem. Soc.* **2013**, 135, 2552–9.

[3205] Y. Ye, M. S. Sanford, *J. Am. Chem. Soc.* **2013**, 135, 4648–51.

[3206] Y. Ye, S. D. Schimler, P. S. Hanley, M. S. Sanford, *J. Am. Chem. Soc.* **2013**, 135, 16292–5.

[3207] A. Casitas, M. Canta, M. Solà, M. Costas, X. Ribas, *J. Am. Chem. Soc.* **2011**, 133, 19386–92.

[3208] B. Yao, Z.-L. Wang, H. Zhang, D.-X. Wang, L. Zhao, M.-X. Wang, *J. Org. Chem.* **2012**, 77, 3336–40.

[3209] P. S. Fier, J. F. Hartwig, *J. Am. Chem. Soc.* **2012**, 134, 10795–8.

[3210] X. Mu, H. Zhang, P. Chen, G. Liu, *Chem. Sci.* **2014**, 5, 275–80.

[3211] L. S. Sharninghausen, A. F. Brooks, W. P. Winton, K. J. Makaravage, P. J. H. Scott, M. S. Sanford, *J. Am. Chem. Soc.* **2020**, 142, 7362–7.

[3212] J. Zhang, H. Wang, S. Ren, W. Zhang, Y. Liu, *Org. Lett.* **2015**, 17, 2920–3.

[3213] T. Furuya, A. E. Strom, T. Ritter, *J. Am. Chem. Soc.* **2009**, 131, 1662–3.

[3214] P. Tang, T. Furuya, T. Ritter, *J. Am. Chem. Soc.* **2010**, 132, 12150–4.

[3215] T. Furuya, T. Ritter, *Org. Lett.* **2009**, 11, 2860–3.

[3216] S. R. Dubbaka, V. R. Narreddula, S. Gadde, T. Mathew, *Tetrahedron* **2014**, 70, 9676–81.

[3217] P. Tang, T. Ritter, *Tetrahedron* **2011**, 67, 4449–54.

[3218] P. S. Fier, J. F. Hartwig, *Science* **2013**, 342, 956–60.

[3219] J. Qian, Y. Liu, J. Zhu, B. Jiang, Z. Xu, *Org. Lett.* **2011**, 13, 4220–3.

[3220] T. Xu, G. Liu, *Org. Lett.* **2012**, 14, 5416–9.

[3221] O. Planas, F. Wang, M. Leutzsch, J. Cornella, **2019**.

[3222] H. Kobayashi, T. Sonoda, K. Takuma, N. Honda, T. Nakata, *J. Fluorine Chem.* **1985**, 27, 1–22.

[3223] R. D. Chambers, C. W. Hall, J. Hutchinson, R. W. Millar, *J. Chem. Soc., Perkin Trans. 1* **1998**, 1705–14.

[3224] L. Lochmann, J. Pospíšil, D. Lím, *Tetrahedron Lett.* **1966**, 7, 257–62.

[3225] M. Schlosser, *Pure Appl. Chem.* **1988**, 60, 1627–34.

[3226] N. J. R. van Eikema Hommes, P. von Ragué Schleyer, *Angew. Chem., Int. Ed. Engl.* **1992**, 31, 755–8.

[3227] V. Snieckus, *Chem. Rev.* **1990**, 90, 879–933.

[3228] P. L. Coe, A. J. Waring, T. D. Yarwood, *J. Chem. Soc., Perkin Trans. 1* **1995**, 2729–37.

[3229] G. Wittig, L. Pohmer, *Chem. Ber.* **1956**, 89, 1334–51.

[3230] A. K. Yudin, L. J. P. Martyn, S. Pandiaraju, J. Zheng, A. Lough, *Org. Lett.* **2000**, 2, 41–4.

[3231] M. Schlosser, *Eur. J. Org. Chem.* **2001**, 2001, 3975–84.

[3232] M. Schlosser, *Angew. Chem., Int. Ed. Engl.* **2005**, 44, 376–93.

[3233] H. W. Gschwend, H. R. Rodriguez, *Organic Reactions.* **1979**, 1–360.

[3234] T. Ahrens, J. Kohlmann, M. Ahrens, T. Braun, *Chem. Rev.* **2015**, 115, 931–72.

[3235] O. Eisenstein, J. Milani, R. N. Perutz, *Chem. Rev.* **2017**, 117, 8710–53.

[3236] M. F. Kuehnel, D. Lentz, T. Braun, *Angew. Chem., Int. Ed. Engl.* **2013**, 52, 3328–48.

[3237] M. K. Whittlesey, E. Peris, *ACS Catal.* **2014**, 4, 3152–9.

[3238] W. D. Jones, *Dalton Trans.* **2003**, 3991–5.

[3239] U. Jäger-Fiedler, M. Klahn, P. Arndt, et al., *J. Mol. Catal. A, Chem.* **2007**, 261, 184–9.

[3240] S. P. Reade, M. F. Mahon, M. K. Whittlesey, *J. Am. Chem. Soc.* **2009**, 131, 1847–61.

[3241] J.-H. Zhan, H. Lv, H. Yu, J.-L. Zhang, *Adv. Synth. Catal.* **2012**, 354, 1529–41.

[3242] H. Lv, Y.-B. Cai, J.-L. Zhang, *Angew. Chem., Int. Ed. Engl.* **2013**, 52, 3203–7.

[3243] J. Li, T. Zheng, H. Sun, X. Li, *Dalton Trans.* **2013**, 42, 13048–53.

[3244] H. Lv, J.-H. Zhan, Y.-B. Cai, Y. Yu, B. Wang, J.-L. Zhang, *J. Am. Chem. Soc.* **2012**, 134, 16216–27.

[3245] J. Vela, J. M. Smith, Y. Yu, et al., *J. Am. Chem. Soc.* **2005**, 127, 7857–70.

[3246] J. Krüger, J. Leppkes, C. Ehm, D. Lentz, *Chem. Asian J.* **2016**, 11, 3062–71.

[3247] G. Podolan, P. Jungk, D. Lentz, R. Zimmer, H.-U. Reissig, *Adv. Synth. Catal.* **2015**, 357, 3215–28.

[3248] G. Podolan, D. Lentz, H.-U. Reissig, *Angew. Chem., Int. Ed. Engl.* **2013**, 52, 9491–4.

[3249] M. Aizenberg, D. Milstein, *Science* **1994**, 265, 359–61.

[3250] M. Aizenberg, D. Milstein, *J. Am. Chem. Soc.* **1995**, 117, 8674–5.

[3251] B. L. Edelbach, W. D. Jones, *J. Am. Chem. Soc.* **1997**, 119, 7734–42.

[3252] A. D. Jaeger, C. Ehm, D. Lentz, *Chem. Eur. J.* **2018**, 24, 6769–77.

[3253] A. D. Jaeger, R. Walter, C. Ehm, D. Lentz, *Chem. Asian J.* **2018**, 13, 2908–15.

[3254] N. Y. Adonin, V. F. Starichenko, *J. Fluorine Chem.* **2000**, 101, 65–7.

[3255] S. S. Laev, V. U. Evtefeev, V. D. Shteingarts, *J. Fluorine Chem.* **2001**, 110, 43–6.

[3256] S. S. Laev, V. D. Shteingarts, *Tetrahedron Lett.* **1997**, 38, 3765–8.

[3257] S. S. Laev, V. D. Shteingarts, *J. Fluorine Chem.* **1999**, 96, 175–85.

[3258] T. Braun, S. Parsons, R. N. Perutz, M. Voith, *Organometallics* **1999**, 18, 1710–6.

[3259] D. Noveski, T. Braun, B. Neumann, A. Stammler, H.-G. Stammler, *Dalton Trans.* **2004**, 4106–19.
[3260] M. Arisawa, T. Suzuki, T. Ishikawa, M. Yamaguchi, *J. Am. Chem. Soc.* **2008**, 130, 12214–5.
[3261] M. Arisawa, T. Ichikawa, M. Yamaguchi, *Org. Lett.* **2012**, 14, 5318–21.
[3262] Q. Zhou, B. Zhang, T. Du, H. Gu, H. Jiang, R. Chen, *Tetrahedron* **2012**, 68, 4233–41.
[3263] C. Yu, C. Zhang, X. Shi, *Eur. J. Org. Chem.* **2012**, 2012, 1953–9.
[3264] L. Keyes, A. D. Sun, J. A. Love, *Eur. J. Org. Chem.* **2011**, 2011, 3985–94.
[3265] J. Zhang, J. Wu, Y. Xiong, S. Cao, *Chem. Commun.* **2012**, 48, 8553–5.
[3266] L. Sun, M. Rong, D. Kong, Z. Bai, Y. Yuan, Z. Weng, *J. Fluorine Chem.* **2013**, 150, 117–23.
[3267] P. A. Deck, M. M. Konaté, B. V. Kelly, C. Slebodnick, *Organometallics* **2004**, 23, 1089–97.
[3268] I. Yutaka, C. Naoto, Y. Shuhei, M. Shinji, *Chem. Lett.* **1998**, 27, 157–8.
[3269] M. I. Sladek, T. Braun, B. Neumann, H.-G. Stammler, *J. Chem. Soc., Dalton Trans.* **2002**, 297–9.
[3270] T. Braun, S. P. Foxon, R. N. Perutz, P. H. Walton, *Angew. Chem., Int. Ed. Engl.* **1999**, 38, 3326–9.
[3271] T. Braun, S. Rothfeld, V. Schorlemer, A. Stammler, H.-G. Stammler, *Inorg. Chem. Commun.* **2003**, 6, 752–5.
[3272] T. Braun, J. Izundu, A. Steffen, B. Neumann, H.-G. Stammler, *Dalton Trans.* **2006**, 5118–23.
[3273] R. J. Lindup, T. B. Marder, R. N. Perutz, A. C. Whitwood, *Chem. Commun.* **2007**, 3664–6.
[3274] M. Teltewskoi, J. A. Panetier, S. A. Macgregor, T. Braun, *Angew. Chem., Int. Ed. Engl.* **2010**, 49, 3947–51.
[3275] X. Zhang, S. Cao, *Tetrahedron Lett.* **2017**, 58, 375–92.
[3276] M. F. Kuehnel, D. Lentz. Preparation of Fluoroolefins. In: H. W. Roesky (ed.) Efficient Preparations of Fluorine Compounds: Wiley; 2013:315–33.
[3277] G. Chelucci, *Chem. Rev.* **2012**, 112, 1344–462.
[3278] H. Yanai, T. Taguchi, *Eur. J. Org. Chem.* **2011**, 2011, 5939–54.
[3279] S. Hara. Stereoselective Synthesis of Mono-fluoroalkenes. In: J. Wang (ed.) Stereoselective Alkene Synthesis: Berlin, Heidelberg: Springer Berlin Heidelberg; 2012:59–86.
[3280] G. Landelle, M. Bergeron, M.-O. Turcotte-Savard, J.-F. Paquin, *Chem. Soc. Rev.* **2011**, 40, 2867–908.
[3281] S. Fustero, A. Simón-Fuentes, P. Barrio, G. Haufe, *Chem. Rev.* **2015**, 115, 871–930.
[3282] H. Lu, H. B. Friedrich, D. J. Burton, *J. Fluorine Chem.* **1995**, 75, 83–6.
[3283] Y. Nakamura, K. Uneyama, *J. Org. Chem.* **2007**, 72, 5894–7.
[3284] W. J. Middleton, W. H. Sharkey, *J. Am. Chem. Soc.* **1959**, 81, 803–4.
[3285] R. Sauvêtre, J. F. Normant, *Tetrahedron Lett.* **1982**, 23, 4325–8.
[3286] S. Martin, R. Sauvêtre, J.-F. Normant, *Tetrahedron Lett.* **1982**, 23, 4329–32.
[3287] T. Okano, K. Ito, T. Ueda, H. Muramatsu, *J. Fluorine Chem.* **1986**, 32, 377–88.
[3288] A. Runge, W. W. Sander, *Tetrahedron Lett.* **1990**, 31, 5453–6.
[3289] T. Hanamoto, Y. Koga, T. Kawanami, H. Furuno, J. Inanaga, *Angew. Chem., Int. Ed. Engl.* **2004**, 43, 3582–4.
[3290] T. Hanamoto, Y. Koga, T. Kawanami, H. Furuno, J. Inanaga, *Tetrahedron Lett.* **2006**, 47, 493–5.
[3291] H. G. Viehe, R. Merényi, J. F. M. Oth, J. R. Senders, P. Valange, *Angew. Chem., Int. Ed. Engl.* **1964**, 3, 755–6.
[3292] G. B. Blackwell, R. N. Haszeldine, D. R. Taylor, *J. Chem. Soc., Perkin Trans. 1* **1982**, 2207–10.
[3293] M. Shiosaki, M. Unno, T. Hanamoto, *J. Org. Chem.* **2010**, 75, 8326–9.
[3294] J. Ichikawa, *J. Synth. Org. Chem. Jpn.* **2010**, 68, 1175–84.
[3295] K. Funabiki, K.-i. Sawa, K. Shibata, M. Matsui, *Synlett* **2002**, 2002, 1134–6.
[3296] H. M. Park, T. Uegaki, T. Konno, T. Ishihara, H. Yamanaka, *Tetrahedron Lett.* **1999**, 40, 2985–8.

[3297] K. Uneyama, F. Yan, S. Hirama, T. Katagiri, *Tetrahedron Lett.* **1996**, 37, 2045–8.

[3298] C. Portella, Y. Shermolovich, *Tetrahedron Lett.* **1997**, 38, 4063–4.

[3299] J. Ichikawa, Y. Wada, M. Fujiwara, K. Sakoda, *Synthesis* **2002**, 2002, 1917–36.

[3300] T. Yamaguchi, M. Irie, *J. Org. Chem.* **2005**, 70, 10323–8.

[3301] K. Yagi, C. F. Soong, M. Irie, *J. Org. Chem.* **2001**, 66, 5419–23.

[3302] A. E. Bayliff, M. R. Bryce, R. D. Chambers, *J. Chem. Soc., Perkin Trans. 1* **1987**, 763–7.

[3303] A. K. Brisdon, I. R. Crossley, K. R. Flower, R. G. Pritchard, J. E. Warren, *Angew. Chem., Int. Ed. Engl.* **2003**, 42, 2399–401.

[3304] Y. He, N. Shen, X. Fan, X. Zhang, *Tetrahedron* **2013**, 69, 8818–23.

[3305] T. Okuyama, M. Fujita, R. Gronheid, G. Lodder, *Tetrahedron Lett.* **2000**, 41, 5125–9.

[3306] M. Yoshida, A. Komata, S. Hara, *Tetrahedron* **2006**, 62, 8636–45.

[3307] T.-H. Nguyen, M. Abarbri, D. Guilloteau, S. Mavel, P. Emond, *Tetrahedron* **2011**, 67, 3434–9.

[3308] M. Yoshida, K. Osafune, S. Hara, *Synthesis* **2007**, 2007, 1542–6.

[3309] T. Kitamura, S. Mizuno, K. Muta, J. Oyamada, *J. Org. Chem.* **2018**, 83, 2773–8.

[3310] T. Dohi, D. Kato, R. Hyodo, D. Yamashita, M. Shiro, Y. Kita, *Angew. Chem., Int. Ed. Engl.* **2011**, 50, 3784–7.

[3311] K. Sano, T. Fukuhara, S. Hara, *J. Fluorine Chem.* **2009**, 130, 708–13.

[3312] S. H. Lee, J. Schwartz, *J. Am. Chem. Soc.* **1986**, 108, 2445–7.

[3313] G. Pelletier, L. Constantineau-Forget, A. B. Charette, *Chem. Commun.* **2014**, 50, 6883–5.

[3314] M. Jiang, M. Shi, *Tetrahedron* **2009**, 65, 5222–7.

[3315] W. Fu, G. Zou, M. Zhu, et al., *J. Fluorine Chem.* **2009**, 130, 996–1000.

[3316] C. Zhou, J. Li, B. Lü, C. Fu, S. Ma, *Org. Lett.* **2008**, 10, 581–3.

[3317] C. Zhou, Z. Ma, Z. Gu, C. Fu, S. Ma, *J. Org. Chem.* **2008**, 73, 772–4.

[3318] H. Cui, Z. Chai, G. Zhao, S. Zhu, *Chin. J. Chem.* **2009**, 27, 189–94.

[3319] M. C. Pacheco, V. Gouverneur, *Org. Lett.* **2005**, 7, 1267–70.

[3320] R. Ranjbar-Karimi, *Ultrason. Sonochem.* **2010**, 17, 768–9.

[3321] A. Macé, F. Tripoteau, Q. Zhao, et al., *Org. Lett.* **2013**, 15, 906–9.

[3322] M.-Y. Chang, M.-F. Lee, C.-H. Lin, N.-C. Lee, *Tetrahedron Lett.* **2011**, 52, 826–9.

[3323] Y. Shen, W. Qiu, *Tetrahedron Lett.* **1987**, 28, 4283–4.

[3324] Y. Shen, W. Qiu, *J. Chem. Soc., Chem. Commun.* **1987**, 703–4.

[3325] Y. Huang, Y. Shen, W. Ding, J. Zheng, *Tetrahedron Lett.* **1981**, 22, 5283–4.

[3326] Y. Shen, Y. Xin, W. Cen, Y. Huang, *Synthesis* **1984**, 1984, 35–7.

[3327] Y. Shen, W. Cen, Y. Huang, *Synthesis* **1985**, 1985, 159–60.

[3328] T. A. Hamlin, C. B. Kelly, R. M. Cywar, N. E. Leadbeater, *J. Org. Chem.* **2014**, 79, 1145–55.

[3329] J. P. Begue, D. Mesureur, *J. Fluorine Chem.* **1988**, 39, 271–82.

[3330] J.-P. Bégué, D. Mesureur, *Synthesis* **1989**, 1989, 309–12.

[3331] G. K. S. Prakash, Y. Wang, J. Hu, G. A. Olah, *J. Fluorine Chem.* **2005**, 126, 1361–7.

[3332] X.-P. Wang, J.-H. Lin, J.-C. Xiao, X. Zheng, *Eur. J. Org. Chem.* **2014**, 2014, 928–32.

[3333] M. Obayashi, E. Ito, K. Matsui, K. Kondo, *Tetrahedron Lett.* **1982**, 23, 2323–6.

[3334] M. L. Edwards, D. M. Stemerick, E. T. Jarvi, D. P. Matthews, J. R. McCarthy, *Tetrahedron Lett.* **1990**, 31, 5571–4.

[3335] H. Trabelsi, B. Bertaina, A. Cambon, *Can. J. Chem.* **1985**, 63, 426–31.

[3336] T. M. Trnka, M. W. Day, R. H. Grubbs, *Angew. Chem., Int. Ed. Engl.* **2001**, 40, 3441–4.

[3337] S. Imhof, S. Randl, S. Blechert, *Chem. Commun.* **2001**, 1692–3.

[3338] S. S. Salim, R. K. Bellingham, V. Satcharoen, R. C. D. Brown, *Org. Lett.* **2003**, 5, 3403–6.

[3339] V. De Matteis, F. L. van Delft, R. de Gelder, J. Tiebes, F. P. J. T. Rutjes, *Tetrahedron Lett.* **2004**, 45, 959–63.

[3340] V. De Matteis, F. L. van Delft, J. Tiebes, F. P. J. T. Rutjes, *Synlett* **2008**, 2008, 351–4.

[3341]   M. Marhold, A. Buer, H. Hiemstra, J. H. van Maarseveen, G. Haufe, *Tetrahedron Lett.* **2004**, 45, 57–60.

[3342]   J. Minville, J. Poulin, C. Dufresne, C. F. Sturino, *Tetrahedron Lett.* **2008**, 49, 3677–81.

[3343]   M. Marhold, C. Stillig, R. Fröhlich, G. Haufe, *Eur. J. Org. Chem.* **2014**, 2014, 5777–85.

[3344]   V. De Matteis, O. Dufay, D. C. J. Waalboer, F. L. van Delft, J. Tiebes, F. P. J. T. Rutjes, *Eur. J. Org. Chem.* **2007**, 2007, 2667–75.

[3345]   V. De Matteis, F. L. van Delft, H. Jakobi, S. Lindell, J. Tiebes, F. P. J. T. Rutjes, *J. Org. Chem.* **2006**, 71, 7527–32.

[3346]   J. A. Akana, K. X. Bhattacharyya, P. Müller, J. P. Sadighi, *J. Am. Chem. Soc.* **2007**, 129, 7736–7.

[3347]   B. C. Gorske, C. T. Mbofana, S. J. Miller, *Org. Lett.* **2009**, 11, 4318–21.

[3348]   F. Nahra, S. R. Patrick, D. Bello, et al., *ChemCatChem* **2015**, 7, 240–4.

[3349]   Y. Li, X. Liu, D. Ma, B. Liu, H. Jiang, *Adv. Synth. Catal.* **2012**, 354, 2683–8.

[3350]   M. A. Tius, J. K. Kawakami, *Synth. Commun.* **1992**, 22, 1461–71.

[3351]   M. A. Tius, J. K. Kawakami, *Synlett* **1993**, 1993, 207–8.

[3352]   M. A. Tius, J. K. Kawakami, *Tetrahedron* **1995**, 51, 3997–4010.

[3353]   T. de Haro, C. Nevado, *Chem. Commun.* **2011**, 47, 248–9.

[3354]   M. N. Hopkinson, G. T. Giuffredi, A. D. Gee, V. Gouverneur, *Synlett* **2010**, 2010, 2737–42.

[3355]   T. Xu, X. Mu, H. Peng, G. Liu, *Angew. Chem., Int. Ed. Engl.* **2011**, 50, 8176–9.

[3356]   G. Zhang, T. Xiong, Z. Wang, G. Xu, X. Wang, Q. Zhang, *Angew. Chem., Int. Ed. Engl.* **2015**, 54, 12649–53.

[3357]   C. Yao, S. Wang, J. Norton, M. Hammond, *J. Am. Chem. Soc.* **2020**, 142, 4793–9.

[3358]   S. B. Lang, R. J. Wiles, C. B. Kelly, G. A. Molander, *Angew. Chem., Int. Ed. Engl.* **2017**, 56, 15073–7.

[3359]   J. P. Phelan, S. B. Lang, J. Sim, et al., *J. Am. Chem. Soc.* **2019**, 141, 3723–32.

[3360]   Y. He, D. Anand, Z. Sun, L. Zhou, *Org. Lett.* **2019**, 21, 3769–73.

[3361]   D. Ding, Y. Lan, Z. Lin, C. Wang, *Org. Lett.* **2019**, 21, 2723–30.

[3362]   X. Lu, X.-X. Wang, T.-J. Gong, J.-J. Pi, S.-J. He, Y. Fu, *Chem. Sci.* **2019**, 10, 809–14.

[3363]   Z. Lin, Y. Lan, C. Wang, *ACS Catal.* **2019**, 9, 775–80.

[3364]   R. J. Wiles, J. P. Phelan, G. A. Molander, *Chem. Commun.* **2019**, 55, 7599–602.

[3365]   Y. Huang, T. Hayashi, *J. Am. Chem. Soc.* **2016**, 138, 12340–3.

[3366]   Z. Yang, M. Möller, R. M. Koenigs, *Angew. Chem., Int. Ed. Engl.* **2020**, 59, 5572–6.

[3367]   T. Ichitsuka, T. Fujita, J. Ichikawa, *ACS Catal.* **2015**, 5, 5947–50.

[3368]   Z. Zhang, Q. Zhou, W. Yu, et al., *Org. Lett.* **2015**, 17, 2474–7.

[3369]   G. Wu, Y. Deng, C. Wu, X. Wang, Y. Zhang, J. Wang, *Eur. J. Org. Chem.* **2014**, 2014, 4477–81.

[3370]   Q. Wang, C. Ni, M. Hu, et al., *Angew. Chem., Int. Ed. Engl.* **2020**, 59, 8507–11.

[3371]   B. M. Kraft, W. D. Jones, *J. Am. Chem. Soc.* **2002**, 124, 8681–9.

[3372]   J. Krüger, C. Ehm, D. Lentz, *Dalton Trans.* **2016**, 45, 16789–98.

[3373]   M. F. Kühnel, D. Lentz, *Angew. Chem., Int. Ed. Engl.* **2010**, 49, 2933–6.

[3374]   M. F. Kuehnel, P. Holstein, M. Kliche, et al., *Chem. Eur. J.* **2012**, 18, 10701–14.

Index